| | | |
|---|---|---|
| British thermal unit (60° F) | joule | +03 1.054 68 |
| bushel (U.S.) | meter³ | -02 3.523 907 |
| cable | meter | +02 2.194 56 |
| caliber | meter | -04 2.54 |
| calorie (International Steam Table) | joule | +00 4.1868 |
| calorie (mean) | joule | +00 4.190 02 |
| calorie (thermochemical) | joule | +00 4.184 |
| calorie (15° C) | joule | +00 4.185 80 |
| calorie (20° C) | joule | +00 4.181 90 |
| calorie (kilogram, International Steam Table) | joule | +03 4.186 8 |
| calorie (kilogram, mean) | joule | +03 4.190 02 |
| calorie (kilogram, thermochemical) | joule | +03 4.184 |
| carat (metric) | kilogram | -04 2.00 |
| Celsius (temperature) | kelvin | $t_K = t_C + 273.15$ |
| centimeter of mercury (0° C) | newton/meter² | +03 1.333 22 |
| centimeter of water (4° C) | newton/meter² | +01 9.806 38 |
| chain (engineer or ramden) | meter | +01 3.048 |
| chain (surveyor or gunter) | meter | +01 2.011 68 |
| circular mil | meter² | -10 5.067 074 |
| cord | meter³ | +00 3.624 556 |
| coulomb (international of 1948) | coulomb | -01 9.998 35 |
| cubit | meter | -01 4.572 |
| cup | meter³ | -04 2.365 882 |
| curie | disintegration/second | +10 3.70 |
| day (mean solar) | second (mean solar) | +04 8.64 |
| day (sidereal) | second (mean solar) | +04 8.616 409 |
| degree (angle) | radian | -02 1.745 329 |
| denier (international) | kilogram/meter | -07 1.00 |
| dram (avoirdupois) | kilogram | -03 1.771 845 |

| | | |
|---|---|---|
| gill (U.K.) | meter³ | -04 1.420 652 |
| gill (U.S.) | meter³ | -04 1.182 941 |
| grad | degree (angular) | -01 9.00 |
| grad | radian | -02 1.570 796 |
| grain | kilogram | -05 6.479 891 |
| gram | kilogram | -03 1.00 |
| hand | meter | -01 1.016 |
| hectare | meter² | +04 1.00 |
| henry (international of 1948) | henry | +00 1.000 495 |
| hogshead (U.S.) | meter³ | -01 2.384 809 |
| horsepower (550 foot lbf/second) | watt | +02 7.456 998 |
| horsepower (boiler) | watt | +03 9.809 50 |
| horsepower (electric) | watt | +02 7.46 |
| horsepower (metric) | watt | +02 7.354 99 |
| horsepower (U.K.) | watt | +02 7.457 |
| horsepower (water) | watt | +02 7.460 43 |
| hour (mean solar) | second (mean solar) | +03 3.60 |
| hour (sidereal) | second (mean solar) | +00 … 170 |
| hundredweight (long) | kilogram | …34 |
| hundredweight (short) | kilogram | …23 |
| inch | meter | |
| inch of mercury (32° F) | newton/met… | |
| inch of mercury (60° F) | newton/met… | …9 |
| inch of water (39.2° F) | newton/met… | |
| inch of water (60° F) | newton/met… | |
| joule (international of 1948) | joule | |
| kayser | 1/meter | |
| kilocalorie (International Steam Table) | joule | |
| kilocalorie (mean) | joule | |
| kilocalorie (thermochemical) | joule | …5 |
| kilogram mass | kilogram | |

# ENGINEERING MANUAL

# OTHER McGRAW-HILL HANDBOOKS OF INTEREST

*American Institute of Physics* • American Institute of Physics Handbook
*American Society of Mechanical Engineers* • ASME Handbooks
    Engineering Tables          Metals Engineering—Processes
    Metals Engineering—Design    Metals Properties
*Baumeister and Marks* • Standard Handbook for Mechanical Engineers
*Berry, Bollay, and Beers* • Handbook of Meteorology
*Blatz* • Radiation Hygiene Handbook
*Brady* • Materials Handbook
*Burington* • Handbook of Mathematical Tables and Formulas
*Burington and May* • Handbook of Probability and Statistics with Tables
*Callender* • Time-Saver Standards for Architectural Design Data
*Chow* • Handbook of Applied Hydrology
*Condon and Odishaw* • Handbook of Physics
*Considine* • Process Instruments and Controls Handbook
*Considine and Ross* • Handbook of Applied Instrumentation
*Dean* • Lange's Handbook of Chemistry
*Etherington* • Nuclear Engineering Handbook
*Fink* • Electronics Engineers' Handbook
*Fink and Carroll* • Standard Handbook for Electrical Engineers
*Flügge* • Handbook of Engineering Mechanics
*Grant* • Hackh's Chemical Dictionary
*Hamsher* • Communication System Engineering Handbook
*Harris and Crede* • Shock and Vibration Handbook
*Henney* • Radio Engineering Handbook
*Hicks* • Standard Handbook of Engineering Calculations
*Hunter* • Handbook of Semiconductor Electronics
*Huskey and Korn* • Computer Handbook
*Ireson* • Reliability Handbook
*Juran* • Quality Control Handbook
*Kaelble* • Handbook of X-rays
*Kallen* • Handbook of Instrumentation and Controls
*King and Brater* • Handbook of Hydraulics
*Klerer and Korn* • Digital Computer User's Handbook
*Koelle* • Handbook of Astronautical Engineering
*Korn and Korn* • Mathematical Handbook for Scientists and Engineers
*Landee, Davis, and Albrecht* • Electronic Designers' Handbook
*Lange* • Handbook of Chemistry
*Machol* • System Engineering Handbook
*Mantell* • Engineering Materials Handbook
*Markus* • Electronics and Nucleonics Dictionary
*Meites* • Handbook of Analytical Chemistry
*Merritt* • Standard Handbook for Civil Engineers
*Perry* • Chemical Engineers' Handbook
*Richey* • Agricultural Engineers' Handbook
*Rothbart* • Mechanical Design and Systems Handbook
*Streeter* • Handbook of Fluid Dynamics
*Truxal* • Control Engineers' Handbook
*Tuma* • Engineering Mathematics Handbook
*Tuma* • Handbook of Physical Calculations
*Tuma* • Technology Mathematics Handbook
*Urquhart* • Civil Engineering Handbook

# ENGINEERING MANUAL

*A Practical Reference of Design Methods
and Data in Building Systems, Chemical, Civil,
Electrical, Mechanical, and Environmental
Engineering and Energy Conversion*

EDITOR-IN-CHIEF

## ROBERT H. PERRY, Ph.D.

*Engineering and Economic Consultant*

THIRD EDITION

## McGRAW-HILL BOOK COMPANY

*New York    St. Louis    San Francisco    Auckland    Bogatá
Düsseldorf    Johannesburg    London    Madrid
Mexico    Montreal    New Delhi    Panama
Paris    São Paulo    Singapore
Sydney    Tokyo    Toronto*

Library of Congress Cataloging in Publication Data

Perry, Robert H
  Engineering manual.

  First ed., 1959, edited by J. H. Perry and
R. H. Perry.
  Includes index.
    1. Engineering—Handbooks, manuals, etc.
  I.  Perry, John Howard, 1895–1953, ed. Engineering
manual.  II.  Title.
  TA151.P645   1976      620'.002'02      76-12514
  ISBN  0-07-049476-2

    34567890   MUBP   78543210987

  *The editors for this book were Harold B. Crawford and Margaret Lamb,
and the production supervisor was Teresa F. Leaden.
It was set in 8A by Bi-Comp, Incorporated.*

  *It was printed by The Murray Printing Company and bound by The Book Press.*

# CONTENTS

# CONTRIBUTORS

**William F. Ames, M.S.** Professor, Department of Mechanics and Hydraulics, University of Iowa; Member, Society of Industrial and Applied Mathematics, American Mathematical Society

**Eugene A. Avallone, B.M.E., M.S., M.E., P.E.** Professor of Mechanical Engineering, The City College of the City University of New York; Member, American Society of Mechanical Engineers.

**John W. Bartlett, Ph.D.** Manager, Process Evaluation, Pacific Northwest Laboratories, Battelle Memorial Institute; Member, American Institute of Chemical Engineers, American Nuclear Society, American Association for the Advancement of Science

**Theodore Baumeister, B.S., M.E., P.E.** Stevens Professor Emeritus of Mechanical Engineering, Columbia University; Fellow, American Society of Mechanical Engineers

**Gary F. Bennett, Ph.D.** Professor of Biochemical Engineering, The University of Toledo; Member, American Institute of Chemical Engineers

**Austin E. Brant, Jr., B.C.E., M.S., P.E.** Partner, Tippetts-Abbett-Mc-Carthy-Stratton, Engineers and Architects; Fellow, American Society of Civil Engineers, Institute of Traffic Engineers; Member, Transportation Research Board, Operations Research Society of America.

**Gregory E. Brooks, B.C.E., M.C.E., P.E.** Chief Structural Engineer, Haines Lundberg & Wachler; Fellow, American Society of Civil Engineers; Member, American Concrete Institute, American Welding Society, Building Research Institute, National Society of Professional Engineers

**Burton B. Crocker, S.M., P.E.** Senior Engineering Fellow, Monsanto Co.; Fellow, American Institute of Chemical Engineers; Member, Air Pollution Control Association

**Leander Economides, B.S.M.E., P.E., R.A.** Economides & Goldberg, Consulting Engineers; Member, American Society of Heating, Air Conditioning and Refrigeration Engineers, Building Research Institute, American Society of Mechanical Engineers, American Institute of Architects

**Arthur E. Hoerl, M.A.** Professor, Department of Statistics and Computer Science, University of Delaware; Member, Society of Industrial and Applied Mathematics, American Statistical Association

**C. Michael Kelly, Ph.D.** Associate Professor, Air Products and Chemicals Professor of Chemical Engineering, Department of Chemical Engineering,

vi

Villanova University; Member, American Institute of Chemical Engineers, American Society for Engineering Education

**Peter E. Liley, Ph.D.** Senior Researcher, Center for Information and Numerical Data Analysis and Synthesis, and Professor, School of Mechanical Engineering, Purdue University; Member, Institute of Physics (London)

**Leonard S. Oberman, B.S.C.E., P.E.** Associate, Tippetts-Abbett-McCarthy-Stratton, Engineers and Architects; Fellow, American Society of Civil Engineers; Member, Permanent International Association of Navigation Congresses, American Road Builders Association

**Robert H. Perry, Ph.D.** Engineering and Economic Consultant

**Laurence J. White, B.S.** Marketing Services Manager, Chemical Engineering; Member, American Institute of Chemical Engineers, American Chemical Society, Air Pollution Control Association

**Robert E. White, D.Ch.E.** Professor and Chairman, Department of Chemical Engineering, Villanova University; Fellow, American Institute of Chemical Engineers; Member, American Society for Engineering Education

**D. L. Whitehead, M.S.** Manager, Engineering Laboratories, High Voltage Section, Westinghouse Electric Corporation; Fellow, Institute of Electrical and Electronics Engineers; Committee Member, National Electrical Manufacturers Association, American Standards Association

**Bronislaus F. Winckowski, B.E.E., P.E.** Chief Electrical Engineer, Haines Lundberg & Waehler; Senior Member, The Institute of Electrical and Electronics Engineers; Member, Illuminating Engineering Society, Building Research Institute, National Society of Professional Engineers

**Otto W. Witzell, Ph.D.** Dean of the Graduate School, Drexel Institute of Technology; Member, American Society for Mechanical Engineers, American Society for Engineering Education.

# PREFACE

The third edition of the *Engineering Manual* is consistent with the original intent of the book: To summarize practical, easily used design methods and requisite data across the spectrum of engineering. The work of revision emphasized incorporation of the latest design techniques, changes in applicable laws and codes, and orientation toward the changing nature of engineering resulting from the continuing evolution of the profession.

Each section has been written by a professional actively engaged in the *practice* of engineering. By and large, they are members of the industrial segment of the profession and even those authors with a formal university affiliation have extensive consulting practices. The emphasis throughout is on the *application* of technical knowledge and theoretical bases for the methods described are not presented. Reference should be made to specialized texts and handbooks for such information.

Two completely new sections have been added: Environmental Engineering and Energy Conversion. There is increasing awareness that pollution is of immediate concern, requiring the best efforts of engineers. The combined impacts of continued industrial development and population growth make environmental engineering a necessity. The new environmental section describes present and projected laws in the United States applicable to air and water pollution. Design techniques and equipment currently in use for wastewater treatment are described as are the technological aspects of air pollution control.

The Energy Conversion section replaces a former one titled Nuclear Engineering. The section now presents, as an integrated whole, the range of energy conversion techniques now applied and those anticipated in the immediate future. A discussion of energy sources, availability, and consumption patterns precedes subsections describing the fundamentals of energy conversion, fossil fuel power plants, and nuclear fuel power generation. The nuclear plant subsection is initiated by a review of the physical fundamentals, then design and control is discussed, followed by an analysis of materials of construction and of the nuclear fuel cycle. Other energy conversion subsections survey the engineering possibilities of fusion, solar power, magnetohydrodynamics, and fuel cells.

Building Systems Engineering was chosen as a more accurate description of the section which, in prior editions, was titled Architectural Engineering. The main subsections discuss the design of structural, mechanical, and electrical systems in the construction of buildings for a variety of uses. Changes in building codes and more complete research data required updating of much of the information. In addition, extensive new tables were added to allow precise calculation of heating and cooling loads for various construction modes and environmental conditions. Estimation of hot and cold water demands for different building uses is facilitated by extensive new data. New material regarding the electrical grounding of buildings was also added.

As there has been a rapid evolution of design methods and in the nature of equipment in use, it was necessary to completely rewrite the Chemical Engineering section. The section is divided into subsections entitled Diffusional Operations, Multiphase Contacting and Phase Distribution, Mechanical Separations and Phase Collection, Chemical Kinetics and Reactor Design, and Process Control. The last subsection is a new topic to the book and reflects the growing importance of sophisticated automation in modern industrial installations. Topics covered include control system analysis, feedback control systems, and the design and tuning of control systems.

The changes to the sections dealing with civil, electrical, and mechanical engineering were less extensive but involved important updating. New material was added to the civil engineering section regarding concrete structure design utilizing the now favored method of ultimate strengths, photogrammetry as a surveying technique, world standards for railroad track gages, and construction engineering. Included in the last subsection is a method for estimating benefit-cost ratios including environmental and social factors as well as economic ones.

High-voltage flashover has gained importance as a design consideration since the last edition and therefore new material on this topic has been added to the Electrical Engineering section.

The Mechanical Engineering section includes entirely new coverage of stress concentration and fatigue in machine design. New subsections on bearing selection and refrigeration have also been added.

Mathematical Tables and Mathematics, Section 1, has been modified to provide as convenient an array of conversion factors as possible. In addition to an alphabetical listing, fourteen new tables have been added in which the factors have been grouped by type (e.g., length, work, density, thermal conductivity, and viscosity). In mathematics, a completely new subsection has been added, Fitting Curves to Data. This information enables matching of functionally described curves to data of all types whether they are historical, reference based, or from other sources.

Section 2, the Engineering Core, presents principles and data fundamental to all the engineering disciplines, i.e., thermodynamics, fluid flow, and heat transfer. It is correctly expected that little change in the basic principles would occur. However, data have been updated and six new tables have been added which give typical in-service heat transfer coefficients for different types of installation.

The section devoted entirely to physical, chemical, and mechanical data, Section 3, has been revised in several respects. Updating has been extensive, and major additions involved thermodynamic data, transport properties, and the mechanical and corrosion properties of materials. For all data, several references are usually given to larger compilations.

The question of the extent to which the book should utilize metric units proved difficult since its publication occurs, as far as the United States is concerned, during the transition period between the use of different systems of units. Most practicing engineers still use the United States Customary System. However, the increasing use of metric units in universities and the internationalization of the products of engineering are rapidly changing current practice. As a result, this edition is a hybrid. In those professional areas where change to SI units is commonplace, those units have been used. In parts of the book, conversion factors are included in the appropriate text and in other sections only the customary units are employed. Extensive conversion tables between the two systems of units are located in Section 1, Mathematical Tables and Mathematics. For convenience, another table of conversion factors is displayed between the back covers.

The world is becoming acutely aware of a range of engineering problems that present threatening possibilities. There is no need to review them here. Some, usually people without appropriate experience, place the blame for this situation on technology. They arrive at the absurd conclusion that all technological development should be abandoned. However, it is obvious to the thoughtful that only by careful and sensible application of today's technology, and that to be developed in the future, can we find solutions to our difficulties. Not only solutions—but also opportunities. A book such as this can only describe tools for such thoughtful technological applications. I hope that we have been successful in that effort.

*Robert H. Perry*

SECTION 1

# MATHEMATICAL TABLES AND MATHEMATICS

**Arthur E. Hoerl, M.A.**; Professor, Department of Statistics and Computer Science, University of Delaware; Member, Society of Industrial and Applied Mathematics, American Statistical Association

**William F. Ames, M.S.**; Professor, Department of Mechanics and Hydraulics, University of Iowa; Member, Society of Industrial and Applied Mathematics, American Mathematical Society

**Peter E. Liley, Ph.D.**; Senior Researcher, Center for Information and Numerical Data Analysis and Synthesis, and Professor, School of Mechanical Engineering, Purdue University; Member, Institute of Physics (London)

**Robert H. Perry, Ph.D.**; Engineering and Economic Consultant

## CONTENTS

## Table 1-1. Alphabetical Listing of Common Conversions

| To convert from | To | Multiply by |
|---|---|---|
| Acres | Square feet | 43,560 |
| Acres | Square meters | 4074 |
| Acres | Square miles | 0.001563 |
| Acre-feet | Cubic meters | 1233 |
| Ampere-hours (absolute) | Coulombs (absolute) | 3600 |
| Angstrom units | Inches | $3.937 \times 10^{-9}$ |
| Angstrom units | Meters | $1 \times 10^{-10}$ |
| Angstrom units | Microns | $1 \times 10^{-4}$ |
| Atmospheres | Millimeters of mercury at 32°F. | 760 |
| Atmospheres | Dynes per square centimeter | $1.0133 \times 10^{6}$ |
| Atmospheres | Newtons per square meter | 101,325 |
| Atmospheres | Feet of water at 39.1°F. | 33.90 |
| Atmospheres | Grams per square centimeter | 1033.3 |
| Atmospheres | Inches of mercury at 32°F. | 29.921 |
| Atmospheres | Pounds per square foot | 2116.3 |
| Atmospheres | Pounds per square inch | 14.696 |
| Bags (cement) | Pounds (cement) | 94 |
| Barrels (cement) | Pounds (cement) | 376 |
| Barrels (oil) | Cubic meters | 0.15899 |
| Barrels (oil) | Gallons | 42 |
| Barrels (U.S. liquid) | Cubic meters | 0.11924 |
| Barrels (U.S. liquid) | Gallons | 31.5 |
| Barrels per day | Gallons per minute | 0.02917 |
| Bars | Atmospheres | 0.9869 |
| Bars | Newtons per square meter | $1 \times 10^{5}$ |
| Bars | Pounds per square inch | 14.504 |
| Board feet | Cubic feet | $\frac{1}{12}$ |
| Boiler horsepower | B.t.u. per hour | 33,480 |
| Boiler horsepower | Kilowatts | 9.803 |
| B.t.u. | Calories (gram) | 252 |
| B.t.u. | Centigrade heat units (c.h.u. or p.c.u.) | 0.55556 |
| B.t.u. | Foot-pounds | 777.9 |
| B.t.u. | Horsepower-hours | $3.929 \times 10^{-4}$ |
| B.t.u. | Joules | 1055.1 |
| B.t.u. | Liter-atmospheres | 10.41 |
| B.t.u. | Pounds carbon to $CO_2$ | $6.88 \times 10^{-5}$ |
| B.t.u. | Pounds water evaporated from and at 212°F. | 0.001036 |
| B.t.u. | Cubic foot-atmospheres | 0.3676 |
| B.t.u. | Kilowatt-hours | $2.930 \times 10^{-4}$ |
| B.t.u. per cubic foot | Joules per cubic meter | 37,260 |
| B.t.u. per hour | Watts | 0.29307 |
| B.t.u. per minute | Horsepower | 0.02357 |
| B.t.u. per pound | Joules per kilogram | 2326 |
| B.t.u. per pound per degree Fahrenheit | Calories per gram per degree centigrade | 1 |
| B.t.u. per pound per degree Fahrenheit | Joules per kilogram per degree Kelvin | 4186.8 |
| B.t.u. per second | Watts | 1054.4 |
| B.t.u. per square foot per hour | Joules per square meter per second | 3.1546 |
| B.t.u. per square foot per minute | Kilowatts per square foot | 0.1758 |
| B.t.u. per square foot per second for a temperature gradient of 1°F. per inch | Calories, gram (15°C.), per square centimeter per second for a temperature gradient of 1°C. per centimeter | 1.2405 |

## Table 1-1. Alphabetical Listing of Common Conversions (Continued)

| To convert from | To | Multiply by |
|---|---|---|
| B.t.u. (60°F.) per degree Fahrenheit | Calories per degree centigrade | 453.6 |
| Bushels (U.S. dry) | Cubic feet | 1.2444 |
| Bushels (U.S. dry) | Cubic meters | 0.03524 |
| Calories, gram | B.t.u. | $3.968 \times 10^{-3}$ |
| Calories, gram | Foot-pounds | 3.087 |
| Calories, gram | Joules | 4.1868 |
| Calories, gram | Liter-atmospheres | $4.130 \times 10^{-2}$ |
| Calories, gram | Horsepower-hours | $1.5591 \times 10^{-6}$ |
| Calories. gram, per gram per degree C. | Joules per kilogram per degree Kelvin | 4186.8 |
| Calories, kilogram | Kilowatt-hours | 0.0011626 |
| Calories, kilogram per second | Kilowatts | 4.185 |
| Candle power (spherical) | Lumens | 12.556 |
| Carats (metric) | Grams | 0.2 |
| Centigrade heat units | B.t.u. | 1.8 |
| Centimeters | Angstrom units | $1 \times 10^8$ |
| Centimeters | Feet | 0.03281 |
| Centimeters | Inches | 0.3937 |
| Centimeters | Meters | 0.01 |
| Centimeters | Microns | 10,000 |
| Centimeters of mercury at 0°C. | Atmospheres | 0.013158 |
| Centimeters of mercury at 0°C. | Feet of water at 39.1°F. | 0.4460 |
| Centimeters of mercury at 0°C. | Newtons per square meter | 1333.2 |
| Centimeters of mercury at 0°C. | Pounds per square foot | 27.845 |
| Centimeters of mercury at 0°C. | Pounds per square inch | 0.19337 |
| Centimeters per second | Feet per minute | 1.9685 |
| Centimeters of water at 4°C. | Newtons per square meter | 98.064 |
| Centistokes | Square meters per second | $1 \times 10^{-6}$ |
| Circular mils | Square centimeters | $5.067 \times 10^{-6}$ |
| Circular mils | Square inches | $7.854 \times 10^{-7}$ |
| Circular mils | Square mils | 0.7854 |
| Cords | Cubic feet | 128 |
| Cubic centimeters | Cubic feet | $3.532 \times 10^{-5}$ |
| Cubic centimeters | Gallons | $2.6417 \times 10^{-4}$ |
| Cubic centimeters | Ounces (U.S. fluid) | 0.03381 |
| Cubic centimeters | Quarts (U.S. fluid) | 0.0010567 |
| Cubic feet | Bushels (U.S.) | 0.8036 |
| Cubic feet | Cubic centimeters | 28,317 |
| Cubic feet | Cubic meters | 0.028317 |
| Cubic feet | Cubic yards | 0.03704 |
| Cubic feet | Gallons | 7.481 |
| Cubic feet | Liters | 28.316 |
| Cubic foot-atmospheres | Foot-pounds | 2116.3 |
| Cubic foot-atmospheres | Liter-atmospheres | 28.316 |
| Cubic feet of water (60°F.) | Pounds | 62.37 |
| Cubic feet per minute | Cubic centimeters per second | 472.0 |
| Cubic feet per minute | Gallons per second | 0.1247 |
| Cubic feet per second | Gallons per minute | 448.8 |
| Cubic feet per second | Million gallons per day | 0.64632 |
| Cubic inches | Cubic meters | $1.6387 \times 10^{-5}$ |
| Cubic yards | Cubic meters | 0.76456 |
| Curies | Disintegrations per minute | $2.2 \times 10^{12}$ |
| Curies | Coulombs per minute | $1.1 \times 10^{12}$ |
| Degrees | Radians | 0.017453 |
| Drams (apothecaries' or troy) | Grams | 3.888 |

## Table 1-1. Alphabetical Listing of Common Conversions (Continued)

| To convert from | To | Multiply by |
|---|---|---|
| Drams (avoirdupois) | Grams | 1.7719 |
| Dynes | Newtons | $1 \times 10^{-5}$ |
| Ergs | Joules | $1 \times 10^{-7}$ |
| Faradays | Coulombs (abs.) | 96,500 |
| Fathoms | Feet | 6 |
| Feet | Meters | 0.3048 |
| Feet per minute | Centimeters per second | 0.5080 |
| Feet per minute | Miles per hour | 0.011364 |
| Feet per (second)$^2$ | Meters per (second)$^2$ | 0.3048 |
| Feet of water at 39.2°F. | Newtons per square meter | 2989 |
| Foot-poundals | B.t.u. | $3.995 \times 10^{-5}$ |
| Foot-poundals | Joules | 0.04214 |
| Foot-poundals | Liter-atmospheres | $4.159 \times 10^{-4}$ |
| Foot-pounds | B.t.u. | 0.0012856 |
| Foot-pounds | Calories, gram | 0.3239 |
| Foot-pounds | Foot-poundals | 32.174 |
| Foot-pounds | Horsepower-hours | $5.051 \times 10^{-7}$ |
| Foot-pounds | Kilowatt-hours | $3.766 \times 10^{-7}$ |
| Foot-pounds | Liter-atmospheres | 0.013381 |
| Foot-pounds force | Joules | 1.3558 |
| Foot-pounds per second | Horsepower | 0.0018182 |
| Foot-pounds per second | Kilowatts | 0.0013558 |
| Furlongs | Miles | 0.125 |
| Gallons (U.S. liquid) | Barrels (U.S. liquid) | 0.03175 |
| Gallons | Cubic meters | 0.003785 |
| Gallons | Cubic feet | 0.13368 |
| Gallons | Gallons (Imperial) | 0.8327 |
| Gallons | Liters | 3.785 |
| Gallons | Ounces (U.S. fluid) | 128 |
| Gallons per minute | Cubic feet per hour | 8.021 |
| Gallons per minute | Cubic feet per second | 0.002228 |
| Grains | Grams | 0.06480 |
| Grains | Pounds | $^1/_{7000}$ |
| Grains per cubic foot | Grams per cubic meter | 2.2884 |
| Grains per gallon | Parts per million | 17.118 |
| Grams | Drams (avoirdupois) | 0.5644 |
| Grams | Drams (troy) | 0.2572 |
| Grams | Grains | 15.432 |
| Grams | Kilograms | 0.001 |
| Grams | Pounds (avoirdupois) | 0.0022046 |
| Grams | Pounds (troy) | 0.002679 |
| Grams per cubic centimeter | Pounds per cubic foot | 62.43 |
| Grams per cubic centimeter | Pounds per gallon | 8.345 |
| Grams per liter | Grains per gallon | 58.42 |
| Grams per liter | Pounds per cubic foot | 0.0624 |
| Grams per square centimeter | Pounds per square foot | 2.0482 |
| Grams per square centimeter | Pounds per square inch | 0.014223 |
| Hectares | Acres | 2.471 |
| Hectares | Square meters | 10,000 |
| Horsepower (British) | B.t.u. per minute | 42.42 |
| Horsepower (British) | B.t.u. per hour | 2545 |
| Horsepower (British) | Foot-pounds per minute | 33,000 |
| Horsepower (British) | Foot-pounds per second | 550 |
| Horsepower (British) | Watts | 745.7 |
| Horsepower (British) | Horsepower (metric) | 1.0139 |
| Horsepower (British) | Pounds carbon to $CO_2$ per hour | 0.175 |

## Table 1-1. Alphabetical Listing of Common Conversions (Continued)

| To convert from | To | Multiply by |
|---|---|---|
| Horsepower (British) | Pounds water evaporated per hour at 212°F | 2.64 |
| Horsepower (metric) | Foot-pounds per second | 542.47 |
| Horsepower (metric) | Kilogram-meters per second | 7.5 |
| Hours (mean solar) | Seconds | 3600 |
| Inches | Meters | 0.0254 |
| Inches of mercury at 60°F | Newtons per square meter | 3376.9 |
| Inches of water at 60°F | Newtons per square meter | 248.84 |
| Joules (absolute) | B.t.u. (mean) | $9.480 \times 10^{-4}$ |
| Joules (absolute) | Calories, gram (mean) | 0.2389 |
| Joules (absolute) | Cubic foot-atmospheres | 0.3485 |
| Joules (absolute) | Foot-pounds | 0.7376 |
| Joules (absolute) | Kilowatt-hours | $2.7778 \times 10^{-7}$ |
| Joules (absolute) | Liter-atmospheres | 0.009869 |
| Kilocalories | Joules | 4186.8 |
| Kilograms | Pounds (avoirdupois) | 2.2046 |
| Kilograms force | Newtons | 9.807 |
| Kilograms per square centimeter | Pounds per square inch | 14.223 |
| Kilometers | Miles | 0.6214 |
| Kilowatt-hours | B.t.u. | 3414 |
| Kilowatt-hours | Foot-pounds | $2.6552 \times 10^{6}$ |
| Kilowatts | Horsepower | 1.3410 |
| Knots (international) | Meters per second | 0.5144 |
| Knots (nautical miles per hour) | Miles per hour | 1.1516 |
| Lamberts | Candles per square inch | 2.054 |
| Liter-atmospheres | Cubic foot-atmospheres | 0.03532 |
| Liter-atmospheres | Foot-pounds | 74.74 |
| Liters | Cubic feet | 0.03532 |
| Liters | Cubic meters | 0.001 |
| Liters | Gallons | 0.26418 |
| Lumens | Watts | 0.001496 |
| Micromicrons | Microns | $1 \times 10^{-6}$ |
| Microns | Angstrom units | $1 \times 10^{4}$ |
| Microns | Meters | $1 \times 10^{-6}$ |
| Miles (nautical) | Feet | 6080 |
| Miles (nautical) | Miles (U.S. statute) | 1.1516 |
| Miles | Feet | 5280 |
| Miles | Meters | 1609.3 |
| Miles per hour | Feet per second | 1.4667 |
| Miles per hour | Meters per second | 0.4470 |
| Milliliters | Cubic centimeters | 1 |
| Millimeters | Meters | 0.001 |
| Millimeters of mercury at 0°C | Newtons per square meter | 133.32 |
| Millimicrons | Microns | 0.001 |
| Mils | Inches | 0.001 |
| Mils | Meters | $2.54 \times 10^{-5}$ |
| Minims (U.S.) | Cubic centimeters | 0.06161 |
| Minutes (angle) | Radians | $2.909 \times 10^{-4}$ |
| Minutes (mean solar) | Seconds | 60 |
| Newtons | Kilograms | 0.10197 |
| Ounces (avoirdupois) | Kilograms | 0.02835 |
| Ounces (avoirdupois) | Ounces (troy) | 0.9115 |
| Ounces (U.S. fluid) | Cubic meters | $2.957 \times 10^{-5}$ |
| Ounces (troy) | Ounces (apothecaries') | 1.000 |
| Pints (U.S. liquid) | Cubic meters | $4.732 \times 10^{-4}$ |
| Poundals | Newtons | 0.13826 |

## Table 1-1. Alphabetical Listing of Common Conversions (Continued)

| To convert from | To | Multiply by |
|---|---|---|
| Pounds (avoirdupois) | Grains | 7000 |
| Pounds (avoirdupois) | Kilograms | 0.45359 |
| Pounds (avoirdupois) | Pounds (troy) | 1.2153 |
| Pounds per cubic foot | Grams per cubic centimeter | 0.016018 |
| Pounds per cubic foot | Kilograms per cubic meter | 16.018 |
| Pounds per square foot | Atmospheres | $4.725 \times 10^{-4}$ |
| Pounds per square foot | Kilograms per square meter | 4.882 |
| Pounds per square inch | Atmospheres | 0.06805 |
| Pounds per square inch | Kilograms per square centimeter | 0.07031 |
| Pounds per square inch | Newtons per square meter | 6894.8 |
| Pounds force | Newtons | 4.4482 |
| Pounds force per square foot | Newtons per square meter | 47.88 |
| Pounds water evaporated from and at 212°F. | Horsepower-hours | 0.379 |
| Pound-centigrade units (p.c.u.) | B.t.u. | 1.8 |
| Quarts (U.S. liquid) | Cubic meters | $9.464 \times 10^{-4}$ |
| Radians | Degrees | 57.30 |
| Revolutions per minute | Radians per second | 0.10472 |
| Seconds (angle) | Radians | $4.848 \times 10^{-6}$ |
| Slugs | Gee pounds | 1 |
| Slugs | Kilograms | 14.594 |
| Slugs | Pounds | 32.17 |
| Square centimeters | Square feet | 0.0010764 |
| Square feet | Square meters | 0.0929 |
| Square feet per hour | Square meters per second | $2.581 \times 10^{-5}$ |
| Square inches | Square centimeters | 6.452 |
| Square inches | Square meters | $6.452 \times 10^{-4}$ |
| Square yards | Square meters | 0.8361 |
| Stokes | Square meters per second | $1 \times 10^{-4}$ |
| Tons (long) | Kilograms | 1016 |
| Tons (long) | Pounds | 2240 |
| Tons (metric) | Kilograms | 1000 |
| Tons (metric) | Pounds | 2204.6 |
| Tons (metric) | Tons (short) | 1.1023 |
| Tons (short) | Kilograms | 907.18 |
| Tons (short) | Pounds | 2000 |
| Tons (refrigeration) | B.t.u. per hour | 12,000 |
| Tons (British shipping) | Cubic feet | 42.00 |
| Tons (U.S. shipping) | Cubic feet | 40.00 |
| Torr (mm. mercury, 0°C.) | Newtons per square meter | 133.32 |
| Watts | B.t.u. per hour | 3.413 |
| Watts | Joules per second | 1 |
| Watts | Kilogram-meters per second | 0.10197 |
| Watt-hours | Joules | 3600 |
| Yards | Meters | 0.9144 |

## Table 1-2. Conversion Tables (By Type)

### Length Equivalents

| Centimeters | Inches | Feet | Yards | Meters | Chains | Kilometers | Miles |
|---|---|---|---|---|---|---|---|
| 1 | 0.3937 | 0.03281 | 0.01094 | 0.01 | $0.0_34971$ | $10^{-5}$ | $0.0_56214$ |
|  | $\bar{1}$.59517 | $\bar{2}$.51598 | $\bar{2}$.03886 | $\bar{2}$.00000 | $\bar{4}$.69644 | $\bar{5}$.00000 | $\bar{6}$.79335 |
| 2.540 | 1 | 0.08333 | 0.02778 | 0.0254 | 0.001263 | $0.0_1254$ | $0.0_11578$ |
| 0.40483 |  | $\bar{2}$.92082 | $\bar{2}$.44370 | $\bar{2}$.40483 | $\bar{3}$.10127 | $\bar{5}$.40483 | $\bar{5}$.19818 |
| 30.48 | 12 | 1 | 0.3333 | 0.3048 | 0.01515 | $0.0_33048$ | $0.0_31894$ |
| 1.48401 | 1.07918 |  | $\bar{1}$.52288 | $\bar{1}$.48401 | $\bar{2}$.18046 | $\bar{4}$.48401 | $\bar{4}$.27736 |
| 91.44 | 36 | 3 | 1 | 0.9144 | 0.04545 | $0.0_39144$ | $0.0_35682$ |
| 1.96114 | 1.55630 | 0.47712 |  | $\bar{1}$.96114 | $\bar{2}$.65758 | $\bar{4}$.96114 | $\bar{4}$.75449 |
| 100 | 39.37 | 3.281 | 1.0936 | 1 | 0.04971 | 0.001 | $0.0_36214$ |
| 2.00000 | 1.59517 | 0.51598 | 0.03886 |  | $\bar{2}$.69644 | $\bar{3}$.00000 | $\bar{4}$.79335 |
| 2012 | 792 | 66 | 22 | 20.12 | 1 | 0.02012 | 0.0125 |
| 3.30356 | 2.89873 | 1.81954 | 1.34242 | 1.30356 |  | $\bar{2}$.30356 | $\bar{2}$.09691 |
| 100000 | 39370 | 3281 | 1093.6 | 1000 | 49.71 | 1 | 0.6214 |
| 5.00000 | 4.59517 | 3.51598 | 3.03886 | 3.00000 | 1.69644 |  | $\bar{1}$.79335 |
| 160934 | 63360 | 5280 | 1760 | 1609 | 80 | 1.609 | 1 |
| 5.20665 | 4.80182 | 3.72263 | 3.24551 | 3.20665 | 1.90309 | 0.20665 |  |

Logarithms of the equivalents are given below the equivalent. In some cases in this table and in those that follow, the equivalents have been rounded off, while the logarithm corresponds to the equivalent carried to a greater number of decimal places.

Subscripts after any figure, $0_3$, $9_4$, etc., mean that that figure is to be repeated the indicated number of times.

### Area Equivalents
(1 hectare = 100 ares = 10,000 centiares or square meters)

| Square meters | Square inches | Square feet | Square yards | Square rods | Square chains | Roods | Acres | Square miles or sections |
|---|---|---|---|---|---|---|---|---|
| 1 | 1550 | 10.76 | 1.196 | 0.0395 | 0.002471 | $0.0_39884$ | $0.0_32471$ | $0.0_63861$ |
|  | 3.19033 | 1.03197 | 0.07773 | $\bar{2}$.59700 | $\bar{3}$.39288 | $\bar{3}$.99495 | $\bar{4}$.39288 | $\bar{7}$.58670 |
| $0.0_36452$ | 1 | 0.006944 | $0.0_37716$ | $0.0_42551$ | $0.0_51594$ | $0.0_66377$ | $0.0_61594$ | $0.0_92491$ |
| $\bar{4}$.80967 |  | $\bar{3}$.84164 | $\bar{4}$.88740 | $\bar{5}$.40667 | $\bar{6}$.20255 | $\bar{7}$.80461 | $\bar{7}$.20255 | $\overline{10}$.39637 |
| 0.09290 | 144 | 1 | 0.1111 | 0.003673 | $0.0_22296$ | $0.0_19183$ | $0.0_22296$ | $0.0_73587$ |
| $\bar{2}$.96803 | 2.15836 |  | $\bar{1}$.04576 | $\bar{3}$.56503 | $\bar{4}$.36091 | $\bar{5}$.96297 | $\bar{4}$.36091 | $\bar{8}$.55473 |
| 0.8361 | 1296 | 9 | 1 | 0.03306 | 0.002066 | $0.0_38264$ | 0.0002066 | $0.0_63228$ |
| $\bar{1}$.92227 | 3.11260 | 0.95424 |  | $\bar{2}$.51927 | $\bar{3}$.31515 | $\bar{4}$.91721 | $\bar{4}$.31515 | $\bar{7}$.50898 |
| 25.29 | 39204 | 272.25 | 30.25 | 1 | 0.0625 | 0.02500 | 0.00625 | $0.0_59766$ |
| 1.40300 | 4.59333 | 2.43497 | 1.48072 |  | $\bar{2}$.79588 | $\bar{2}$.39794 | $\bar{3}$.79588 | $\bar{6}$.98970 |
| 404.7 | 627264 | 4356 | 484 | 16 | 1 | 0.4 | 0.1 | 0.0001562 |
| 2.60712 | 5.79745 | 3.63909 | 2.68484 | 1.20412 |  | $\bar{1}$.60206 | $\bar{1}$.00000 | $\bar{4}$.19382 |
| 1012 | 1568160 | 10890 | 1210 | 40 | 2.5 | 1 | 0.25 | $0.0_33906$ |
| 3.00506 | 6.19539 | 4.03703 | 3.08278 | 1.60206 | 0.39794 |  | $\bar{1}$.39794 | $\bar{4}$.59176 |
| 4047 | 6272640 | 43560 | 4840 | 160 | 10 | 4 | 1 | 0.001562 |
| 3.60712 | 6.79745 | 4.63909 | 3.68484 | 2.20412 | 1.00000 | 0.60206 |  | $\bar{3}$.19382 |
| 2589988 |  | 27878400 | 3097600 | 102400 | 6400 | 2560 | 640 | 1 |
| 6.41330 |  | 7.44527 | 6.49102 | 5.01030 | 3.80618 | 3.40824 | 2.80618 |  |

## Table 1-2. Conversion Tables (By Type) (Continued)

### Volume and Capacity Equivalents*

| Cubic inches | Cubic feet | Cubic yards | U.S. Apothecary fluid ounces | U.S. quarts Liquid | U.S. quarts Dry | U.S. gallons | U.S. bushels | Cubic decimeters or liters |
|---|---|---|---|---|---|---|---|---|
| 1 | $0.0_35787$ $\bar{4}.76246$ | $0.0_42143$ $\bar{5}.33109$ | $0.5541$ $\bar{1}.74360$ | $0.01732$ $\bar{2}.23845$ | $0.01488$ $\bar{2}.17263$ | $0.0_24329$ $\bar{3}.63639$ | $0.0_34650$ $\bar{4}.66748$ | $0.01639$ $\bar{2}.21450$ |
| 1728 $3.23754$ | 1 | $0.03704$ $\bar{2}.56864$ | $957.5$ $2.98114$ | $29.92$ $1.47599$ | $25.71$ $1.41017$ | $7.481$ $0.87393$ | $0.8036$ $\bar{1}.90502$ | $28.32$ $1.45205$ |
| 46656 $4.66891$ | 27 $1.43136$ | 1 | $25853$ $4.41251$ | $807.9$ $2.90736$ | $694.3$ $2.84153$ | $202.2$ $2.30530$ | $21.70$ $1.33638$ | $764.6$ $2.88341$ |
| 1.805 $0.25640$ | $0.001044$ $\bar{3}.01886$ | $0.0_43868$ $\bar{5}.58749$ | 1 | $0.03125$ $\bar{2}.49485$ | $0.02686$ $\bar{2}.42903$ | $0.007812$ $\bar{3}.89279$ | $0.0_38392$ $\bar{4}.92388$ | $0.02957$ $\bar{2}.47091$ |
| 57.75 $1.76155$ | $0.03342$ $\bar{2}.52401$ | $0.001238$ $\bar{3}.09264$ | $32$ $1.50515$ | 1 | $0.8594$ $\bar{1}.93418$ | $0.25$ $\bar{1}.39794$ | $0.02686$ $\bar{2}.42903$ | $0.9464$ $\bar{1}.97606$ |
| 67.20 $1.82737$ | $0.03889$ $\bar{2}.58983$ | $0.001440$ $\bar{3}.15847$ | $37.24$ $1.57097$ | $1.164$ $0.06582$ | 1 | $0.2909$ $\bar{1}.46376$ | $0.03125$ $\bar{2}.49485$ | $1.101$ $0.04187$ |
| 231 $2.36361$ | $0.1337$ $\bar{1}.12607$ | $0.004951$ $\bar{3}.69470$ | $128$ $2.10721$ | $4$ $0.60206$ | $3.437$ $0.53624$ | 1 | $0.1074$ $\bar{1}.03109$ | $3.785$ $0.57812$ |
| 2150 $3.33252$ | $1.244$ $0.09498$ | $0.04609$ $\bar{2}.66362$ | $1192$ $3.07612$ | $37.24$ $1.57097$ | $32$ $1.50515$ | $9.309$ $0.96891$ | 1 | $35.24$ $1.54696$ |
| 61.02 $1.78550$ | $0.03531$ $\bar{2}.54795$ | $0.001308$ $\bar{3}.11659$ | $33.81$ $1.52909$ | $1.057$ $0.02394$ | $0.9081$ $\bar{1}.95812$ | $0.2642$ $\bar{1}.42188$ | $0.02838$ $\bar{2}.45297$ | 1 |

* The SI unit of volume is the cubic meter which may be calculated by multiplying cubic decimeters (obtained from this table) by 1,000 (i.e., $10^3$).

### Mass Equivalents

| Kilograms | Grains | Ounces Troy and apoth | Ounces Avoirdupois | Pounds Troy and apoth | Pounds Avoirdupois | Tons Short | Tons Long | Tons Metric |
|---|---|---|---|---|---|---|---|---|
| 1 | 15432 $4.18843$ | $32.15$ $1.50719$ | $35.27$ $1.54745$ | $2.6792$ $0.42801$ | $2.205$ $0.34333$ | $0.0_21102$ $\bar{3}.04230$ | $0.0_39842$ $\bar{4}.99309$ | $0.001$ $\bar{3}.00000$ |
| $0.0_46480$ $\bar{5}.81157$ | 1 | $0.0_22083$ $\bar{3}.31876$ | $0.0_22286$ $\bar{3}.35902$ | $0.0_31736$ $\bar{4}.23958$ | $0.0_31429$ $\bar{4}.15490$ | $0.0_71143$ $\bar{8}.85387$ | $0.0_76378$ $\bar{8}.80465$ | $0.0_76480$ $\bar{8}.81157$ |
| $0.03110$ $\bar{2}.49281$ | 480 $2.68124$ | 1 | $1.09714$ $0.04026$ | $0.08333$ $\bar{2}.92082$ | $0.06857$ $\bar{2}.83614$ | $0.0_43429$ $\bar{5}.53511$ | $0.0_43061$ $\bar{5}.48590$ | $0.0_43110$ $\bar{5}.49281$ |
| $0.02835$ $\bar{2}.45255$ | 437.5 $2.64098$ | $0.9115$ $\bar{1}.95974$ | 1 | $0.07595$ $\bar{2}.88056$ | $0.0625$ $\bar{2}.79588$ | $0.0_43125$ $\bar{5}.49485$ | $0.0_42790$ $\bar{5}.44563$ | $0.0_42835$ $\bar{5}.45255$ |
| $0.3732$ $\bar{1}.57199$ | 5760 $3.76042$ | $12$ $1.07918$ | $13.17$ $1.11944$ | 1 | $0.8229$ $\bar{1}.91532$ | $0.0_34114$ $\bar{4}.61429$ | $0.0_33673$ $\bar{4}.56508$ | $0.0_33732$ $\bar{4}.57199$ |
| $0.4536$ $\bar{1}.65667$ | 7000 $3.84510$ | $14.58$ $1.16386$ | $16$ $1.20412$ | $1.215$ $0.08468$ | 1 | $0.0005$ $\bar{4}.69897$ | $0.0_34464$ $\bar{4}.64975$ | $0.0_34536$ $\bar{4}.65667$ |
| $907.2$ $2.95770$ | $140_6$ $7.14613$ | $29167$ $4.46489$ | $320_3$ $4.50515$ | $2431$ $3.38571$ | $2000$ $3.30103$ | 1 | $0.8929$ $\bar{1}.95078$ | $0.9072$ $\bar{1}.95770$ |
| $1016$ $3.00691$ | $15680_4$ $7.19535$ | $32667$ $4.51411$ | $35840$ $4.55437$ | $2722$ $3.43492$ | $2240$ $3.35025$ | $1.12$ $0.04922$ | 1 | $1.016$ $0.00691$ |
| $1000$ $3.00000$ | $15432356$ $7.18843$ | $32151$ $4.50719$ | $35274$ $4.54745$ | $2679$ $3.42801$ | $2205$ $3.34333$ | $1.102$ $0.04230$ | $0.9842$ $\bar{1}.99309$ | 1 |

## Table 1-2. Conversion Tables (By Type) (Continued)

### Velocity Equivalents

| Centimeters per sec | Meters per sec | Meters per min | Kilometers per hr | Feet per sec | Feet per min | Miles per hr | Knots |
|---|---|---|---|---|---|---|---|
| 1 | 0.01<br>$\bar{2}$.00000 | 0.6<br>$\bar{1}$.77815 | 0.036<br>$\bar{2}$.55630 | 0.03281<br>$\bar{2}$.51598 | 1.9685<br>0.29414 | 0.02237<br>$\bar{2}$.34965 | 0.01944<br>$\bar{2}$.28866 |
| 100<br>2.00000 | 1 | 60<br>1.77815 | 3.6<br>0.55630 | 3.281<br>0.51598 | 196.85<br>2.29414 | 2.237<br>0.34965 | 1.944<br>0.28866 |
| 1.667<br>0.22185 | 0.01667<br>$\bar{2}$.22185 | 1 | 0.06<br>$\bar{2}$.77815 | 0.05468<br>$\bar{2}$.73783 | 3.281<br>0.51598 | 0.03728<br>$\bar{2}$.57150 | 0.03240<br>$\bar{2}$.51050 |
| 27.78<br>1.44370 | 0.2778<br>$\bar{1}$.44370 | 16.67<br>1.22185 | 1 | 0.9113<br>$\bar{1}$.95968 | 54.68<br>1.73783 | 0.6214<br>$\bar{1}$.79335 | 0.53996<br>$\bar{1}$.73236 |
| 30.48<br>1.48401 | 0.3048<br>$\bar{1}$.48401 | 18.29<br>1.26217 | 1.097<br>0.04032 | 1 | 60<br>1.77815 | 0.6818<br>$\bar{1}$.83367 | 0.59248<br>$\bar{1}$.77268 |
| 0.5080<br>1.70586 | 0.005080<br>$\bar{3}$.70586 | 0.3048<br>$\bar{1}$.48401 | 0.01829<br>$\bar{2}$.26217 | 0.01667<br>$\bar{2}$.22185 | 1 | 0.01136<br>$\bar{2}$.05553 | 0.00987<br>$\bar{3}$.99453 |
| 44.70<br>1.65035 | 0.4470<br>$\bar{1}$.65035 | 26.82<br>1.42850 | 1.609<br>0.20670 | 1.467<br>0.16633 | 88<br>1.94448 | 1 | 0.86898<br>$\bar{1}$.93901 |
| 51.44<br>1.71133 | 0.5144<br>$\bar{1}$.71133 | 30.87<br>1.48949 | 1.852<br>0.26764 | 1.688<br>0.22732 | 101.3<br>2.00547 | 1.151<br>1.06100 | 1 |

### Acceleration Equivalents

| Centimeters per sec per sec | Meters per sec per sec | Meters per hr per sec | Kilometers per hr per sec | Feet per hr per sec | Feet per sec per sec | Feet per min per min | Miles per hr per sec | Knots per sec |
|---|---|---|---|---|---|---|---|---|
| 1 | 0.01<br>$\bar{2}$.00000 | 36.00<br>1.55630 | 0.036<br>$\bar{2}$.55630 | 118.1<br>2.07225 | 0.03281<br>$\bar{2}$.51599 | 118.1<br>2.07225 | 0.02237<br>$\bar{2}$.34965 | 0.01944<br>$\bar{2}$.29865 |
| 100<br>2.00000 | 1 | 3600<br>3.55630 | 3.6<br>0.55630 | 11811<br>4.07225 | 3.281<br>0.51599 | 11811<br>4.07225 | 2.237<br>0.34965 | 1.944<br>0.29865 |
| 0.02778<br>$\bar{2}$.44370 | 0.0002778<br>$\bar{4}$.44370 | 1 | 0.001<br>$\bar{3}$.00000 | 3.281<br>0.51599 | 0.0009113<br>$\bar{4}$.95968 | 3.281<br>0.51599 | 0.0006214<br>$\bar{4}$.79325 | 0.0005400<br>$\bar{4}$.73235 |
| 27.78<br>1.44370 | 0.2778<br>$\bar{1}$.44370 | 1000<br>3.00000 | 1 | 3281<br>3.51599 | 0.9113<br>$\bar{1}$.95968 | 3281<br>3.51599 | 0.6214<br>$\bar{1}$.79335 | 0.5400<br>$\bar{1}$.73235 |
| 0.008467<br>$\bar{3}$.92771 | 0.00008467<br>$\bar{5}$.92771 | 0.3048<br>$\bar{1}$.48401 | 0.0003048<br>$\bar{4}$.48401 | 1 | 0.0002778<br>$\bar{4}$.44370 | 1 | 0.0001894<br>$\bar{4}$.27737 | 0.0001646<br>$\bar{4}$.21640 |
| 30.48<br>1.48401 | 0.3048<br>$\bar{1}$.48401 | 1097<br>3.04030 | 1.097<br>0.04030 | 3600<br>3.55630 | 1 | 3600<br>3.55630 | 0.6818<br>$\bar{1}$.83366 | 0.4572<br>$\bar{1}$.66008 |
| 0.008467<br>$\bar{3}$.92771 | 0.00008467<br>$\bar{5}$.92771 | 0.3048<br>$\bar{1}$.48401 | 0.0003048<br>$\bar{4}$.48401 | 1 | 0.0002778<br>$\bar{4}$.44370 | 1 | 0.001894<br>$\bar{4}$.27737 | 0.0001646<br>$\bar{4}$.21640 |
| 44.70<br>1.65035 | 0.4470<br>$\bar{1}$.65035 | 1609<br>3.20665 | 1.609<br>0.20665 | 5280<br>3.72263 | 1.467<br>0.13636 | 5280<br>3.72263 | 1 | 0.8690<br>$\bar{1}$.93901 |
| 51.44<br>1.71134 | 0.5144<br>$\bar{1}$.71134 | 1852<br>3.26764 | 1.852<br>0.26764 | 6076<br>3.78362 | 1.688<br>0.22732 | 6076<br>3.78362 | 1.151<br>0.06099 | 1 |

## Table 1-2. Conversion Tables (By Type) (Continued)

### Pressure Equivalents*

| Bars, megabaryes, or megadynes per cm² | Kilograms per cm² | Pounds per in.² | Short tons per ft² | Atmospheres | Columns of mercury at temperature 0°C and $g = 980.665$ cm per sec² | | Columns of water at temperature 15°C and $g = 980.665$ cm per sec² | | |
|---|---|---|---|---|---|---|---|---|---|
| | | | | | Meters | Inches | Meters | Inches | Feet |
| 1 | 1.0197<br>0.00848 | 14.50<br>$\bar{1}$.16148 | 1.044<br>0.01882 | 0.9869<br>$\bar{1}$.99427 | 0.7501<br>$\bar{1}$.87510 | 29.53<br>1.47025 | 10.21<br>1.00886 | 401.8<br>2.60402 | 33.49<br>1.52485 |
| 0.9807<br>$\bar{1}$.99152 | 1 | 14.22<br>1.15300 | 1.024<br>0.01034 | 0.9678<br>$\bar{1}$.98579 | 0.7356<br>$\bar{1}$.86662 | 28.96<br>1.46177 | 10.01<br>1.00038 | 394.1<br>2.59556 | 32.84<br>1.51636 |
| 0.06895<br>$\bar{2}$.83852 | 0.07031<br>$\bar{2}$.84700 | 1 | 0.072<br>$\bar{2}$.85733 | 0.06805<br>$\bar{2}$.83280 | 0.05171<br>$\bar{2}$.71360 | 2.036<br>0.30876 | 0.7037<br>$\bar{1}$.84738 | 27.70<br>1.44254 | 2.309<br>0.36336 |
| 0.9576<br>$\bar{1}$.98119 | 0.9765<br>$\bar{1}$.98966 | 13.89<br>1.14267 | 1 | 0.9451<br>$\bar{1}$.97547 | 0.7183<br>$\bar{1}$.85628 | 28.28<br>1.45143 | 9.774<br>0.99006 | 384.8<br>2.58521 | 32.07<br>1.50604 |
| 1.0133<br>0.00573 | 1.0332<br>0.01420 | 14.70<br>1.16722 | 1.058<br>0.02453 | 1 | 0.76<br>$\bar{1}$.88081 | 29.92<br>1.47598 | 10.34<br>1.01459 | 407.1<br>2.60975 | 33.93<br>1.53058 |
| 1.3332<br>0.12490 | 1.3595<br>0.13338 | 19.34<br>1.28640 | 1.392<br>0.14373 | 1.316<br>0.11919 | 1 | 39.37<br>1.59517 | 13.61<br>1.13378 | 535.7<br>2.72894 | 44.64<br>1.64976 |
| 0.03386<br>$\bar{2}$.52975 | 0.03453<br>$\bar{2}$.53823 | 0.4912<br>$\bar{1}$.69124 | 0.03536<br>$\bar{2}$.54857 | 0.03342<br>$\bar{2}$.52402 | 0.02540<br>$\bar{2}$.40484 | 1 | 0.3456<br>$\bar{1}$.53861 | 13.61<br>1.13378 | 1.134<br>0.05460 |
| 0.09798<br>$\bar{2}$.99114 | 0.09991<br>$\bar{2}$.99962 | 1.421<br>0.15262 | 0.1023<br>$\bar{1}$.00996 | 0.09670<br>$\bar{2}$.98541 | 0.07349<br>$\bar{2}$.86622 | 2.893<br>0.46139 | 1 | 39.37<br>1.59517 | 3.281<br>0.51598 |
| 0.002489<br>$\bar{3}$.39598 | 0.002438<br>$\bar{3}$.40446 | 0.03609<br>$\bar{2}$.55745 | 0.002599<br>$\bar{3}$.41479 | 0.002456<br>$\bar{3}$.39024 | 0.001867<br>$\bar{3}$.27106 | 0.07349<br>$\bar{2}$.86622 | 0.02540<br>$\bar{2}$.40484 | 1 | 0.08333<br>$\bar{2}$.92082 |
| 0.02986<br>$\bar{2}$.47516 | 0.03045<br>$\bar{2}$.48364 | 0.4331<br>$\bar{1}$.63663 | 0.03119<br>$\bar{2}$.49397 | 0.02947<br>$\bar{2}$.46942 | 0.02240<br>$\bar{2}$.35024 | 0.8819<br>$\bar{1}$.94540 | 0.3048<br>$\bar{1}$.48401 | 12<br>1.07918 | 1 |

* The SI unit of pressure is a pascal (newtons per square meter) which may be calculated by multiplying bars (obtained from this table) by 100,000 (i.e., $10^5$).

MATHEMATICAL TABLES AND MATHEMATICS

## Table 1-2. Conversion Tables (By Type) (Continued)

### Energy or Work Equivalents

| Joules | Kilogram-meters | Foot-pounds | Kilowatt-hours | Metric horse-power-hours | Horse-power-hours | Liter-atmospheres | Kilo-calories | British thermal units |
|---|---|---|---|---|---|---|---|---|
| 1 | 0.10197 | 0.7376 | $0.0_6 2778$ | $0.0_6 3777$ | $0.0_6 3725$ | 0.009869 | $0.0_3 2388$ | $0.0_3 9478$ |
|  | $\bar{1}.00848$ | $\bar{1}.86780$ | $\bar{7}.44370$ | $\bar{7}.57711$ | $\bar{7}.57113$ | $\bar{3}.99427$ | $\bar{4}.37809$ | $\bar{4}.97670$ |
| 9.80665 | 1 | 7.233 | $0.0_5 2724$ | $0.0_5 37037$ | $0.0_5 3653$ | 0.09678 | 0.002342 | 0.009295 |
| 0.9915207 |  | 0.85932 | $\bar{6}.43521$ | $\bar{6}.56863$ | $\bar{6}.56265$ | $\bar{2}.98579$ | $\bar{3}.36961$ | $\bar{3}.96825$ |
| 1.356 | 0.1383 | 1 | $0.0_6 3766$ | $0.0_6 51206$ | $0.0_6 50505$ | 0.01338 | $0.0_3 3238$ | 0.001285 |
| 0.13220 | $\bar{1}.14068$ |  | $\bar{7}.57590$ | $\bar{7}.70932$ | $\bar{7}.70333$ | $\bar{2}.12647$ | $\bar{4}.51029$ | $\bar{3}.10890$ |
| $3.600 \times 10^6$ | $3.671 \times 10^5$ | $2.655 \times 10^6$ | 1 | 1.3596 | 1.341 | 35528 | 859.9 | 3412 |
| 6.55630 | 5.56478 | 6.42410 |  | 0.13342 | 0.12743 | 4.55057 | 2.93443 | 3.53303 |
| $2.648 \times 10^6$ | 270000 | $1.9529 \times 10^6$ | 0.7355 | 1 | 0.9863 | 26131 | 632.4 | 2510 |
| 6.42288 | 5.43136 | 6.29068 | $\bar{1}.86658$ |  | $\bar{1}.99401$ | 4.41715 | 2.80098 | 3.39961 |
| $2.6845 \times 10^6$ | $2.7375 \times 10^5$ | $1.98 \times 10^6$ | 0.7457 | 1.0139 | 1 | 26493 | 641.2 | 2544 |
| 6.42887 | 5.43735 | 6.29667 | $\bar{1}.87356$ | 0.00598 |  | 4.42314 | 2.80699 | 3.40557 |
| 101.33 | 10.333 | 74.74 | $0.0_4 2815$ | $0.0_4 3827$ | $0.0_4 3775$ | 1 | 0.02420 | 0.09604 |
| 2.00573 | 1.01421 | 1.87353 | $\bar{5}.44952$ | $\bar{5}.58284$ | $\bar{5}.57686$ |  | $\bar{2}.38382$ | $\bar{2}.98246$ |
| 4187 | 426.9 | 3088 | 0.001163 | 0.001581 | 0.001560 | 41.32 | 1 | 3.968 |
| 3.62191 | 2.63036 | 3.48971 | $\bar{3}.06558$ | $\bar{3}.19902$ | $\bar{3}.19304$ | 1.61618 |  | 0.59861 |
| 1055 | 107.6 | 778.2 | $0.0_3 2931$ | $0.0_3 3985$ | $0.0_3 3930$ | 10.41 | 0.25200 | 1 |
| 3.02300 | 2.03178 | 2.89110 | $\bar{4}.46697$ | $\bar{4}.60042$ | $\bar{4}.59444$ | 1.01757 | $\bar{1}.40139$ |  |

### Power Equivalents*

| Horsepower | Kilowatts | Metric horsepower | Poncelets | Kg m per sec | Ft-lb per sec | Kilocalories per sec | Btu per sec |
|---|---|---|---|---|---|---|---|
| 1 | 0.7457 | 1.014 | 0.7604 | 76.04 | 550 | 0.1781 | 0.7068 |
|  | $\bar{1}.87256$ | 0.00599 | $\bar{1}.88105$ | 1.88105 | 2.74036 | $\bar{1}.25066$ | $\bar{1}.84936$ |
| 1.341 | 1 | 1.360 | 1.020 | 102.0 | 737.6 | 0.2388 | 0.9478 |
| 0.12743 |  | 0.13343 | 0.00848 | 2.00848 | 2.86780 | $\bar{1}.37813$ | $\bar{1}.97673$ |
| 0.9863 | 0.7355 | 1 | 0.75 | 75 | 542.5 | 0.1757 | 0.6971 |
| $\bar{1}.99402$ | $\bar{1}.86658$ |  | $\bar{1}.87506$ | 1.87506 | 2.73438 | $\bar{1}.24467$ | $\bar{1}.84328$ |
| 1.315 | 0.9807 | 1.333 | 1 | 100 | 723.3 | 0.2342 | 0.9295 |
| 0.11896 | $\bar{1}.99152$ | 0.12493 |  | 2.00000 | 2.85932 | $\bar{1}.36961$ | $\bar{1}.96825$ |
| 0.01315 | 0.009807 | 0.01333 | 0.01 | 1 | 7.233 | 0.002342 | 0.009295 |
| $\bar{2}.11896$ | $\bar{3}.99152$ | $\bar{2}.12493$ | $\bar{2}.00000$ |  | 0.85932 | $\bar{3}.36961$ | $\bar{3}.96825$ |
| 0.00182 | 0.001356 | 0.00184 | 0.00138 | 0.1383 | 1 | $0.0_3 3238$ | 0.001285 |
| $\bar{3}.25946$ | $\bar{3}.13220$ | $\bar{3}.26562$ | $\bar{3}.14067$ | $\bar{1}.14067$ |  | $\bar{4}.51029$ | $\bar{3}.10890$ |
| 5.615 | 4.187 | 5.692 | 4.269 | 426.9 | 3088 | 1 | 3.968 |
| 0.74934 | 0.62187 | 0.75530 | 0.63036 | 2.63036 | 3.48971 |  | 0.59861 |
| 1.415 | 1.055 | 1.434 | 1.076 | 107.6 | 778.2 | 0.2520 | 1 |
| 0.15074 | 0.02320 | 0.15668 | 0.03178 | 2.03178 | 2.89110 | $\bar{1}.40138$ |  |

* The SI unit of power is a watt which may be calculated by dividing kilowatts (obtained from this table) by 1,000 (i.e., $10^3$).

## Table 1-2. Conversion Tables (By Type) (Continued)

### Density Equivalents*

| Grams per cm³ | Lb per in.³ | Lb per ft³ | Short tons (2,000 lb) per yd³ | Lb per U.S. gal |
|---|---|---|---|---|
| 1 | 0.03613 $\overline{2}$.55787 | 62.43 1.79539 | 0.8428 $\overline{1}$.92572 | 8.345 0.92143 |
| 27.68 1.44217 | 1 | 1728 3.23754 | 23.33 1.36792 | 231 2.36361 |
| 0.01602 $\overline{2}$.20466 | 0.0₃5787 $\overline{4}$.76245 | 1 | 0.0135 $\overline{2}$.13033 | 0.1337 $\overline{1}$.12613 |
| 1.187 0.07428 | 0.04287 $\overline{2}$.63212 | 74.07 1.86964 | 1 | 9.902 0.99572 |
| 0.1198 $\overline{1}$.07855 | 0.004329 $\overline{3}$.63639 | 7.481 0.87396 | 0.1010 $\overline{1}$.00432 | 1 |

* The SI unit of density is kilograms per cubic meter which may be calculated from grams per cubic centimeter (obtained from this table) by multiplying by 1,000 (i.e., $10^3$).

### Thermal Conductivity Equivalents

| Cal/ (sec)(cm²)(cm/°C) | Watts/ (cm²)(cm/°C) | Cal/ (hr)(cm²)(cm/°C) | Btu/ (hr)(ft²)(ft/°F) | Btu/ (day)(ft²)(in./°F) |
|---|---|---|---|---|
| 1 | 4.187 | 3,600 | 241.9 | 69,670 |
| 0.2388 | 1 | 860 | 57.79 | 16,641 |
| 0.0002778 | 0.001163 | 1 | 0.0672 | 19.35 |
| 0.004134 | 0.01731 | 14.88 | 1 | 288 |
| 0.00001435 | 0.00006009 | 0.05167 | 0.00347 | 1 |

### Thermal Conductance Equivalents

| Cal/(sec)(cm²)(°C) | Watts/(cm²)(°C) | Cal/(hr)(cm²)(°C) | Btu/(hr)(ft²)(°F) | Btu/(day)(ft²)(°F) |
|---|---|---|---|---|
| 1 | 4.187 | 3,600 | 7,373 | 176,962 |
| 0.2388 | 1 | 860 | 1,761 | 42,267 |
| 0.0002778 | 0.001163 | 1 | 2.048 | 49.16 |
| 0.0001356 | 0.0005678 | 0.4882 | 1 | 24 |
| 0 000005651 | 0.00002366 | 0.02034 | 0.04167 | 1 |

### Heat Flow Equivalents

| Cal/(sec)(cm²) | Watts/cm² | Cal/(hr)(cm²) | Btu/(hr)(ft²) | Btu/(day)(ft²) |
|---|---|---|---|---|
| 1 | 4.187 | 3,600 | 13,272 | 318,531 |
| 0.2388 | 1 | 860 | 3,170 | 76,081 |
| 0.0002778 | 0.001163 | 1 | 3.687 | 88.48 |
| 0.00007535 | 0.0003154 | 0.2712 | 1 | 24 |
| 0.000003139 | 0.00001314 | 0.01130 | 0.04167 | 1 |

## Table 1-2. Conversion Tables (By Type) (Continued)

### Viscosity Equivalents

| Centipoise | Poise* | Gram force sec per cm² | Lb force sec per ft² | Lb mass per ft per hr | Slugs per ft per hr | Newton-sec per m² |
|---|---|---|---|---|---|---|
| 1 | 0.01 | 0.0000102 | 0.00002089 | 2.4191 | 0.07519 | 0.001 |
| 100 | 1 | 0.00102 | 0.002089 | 241.91 | 7.5188 | 0.1 |
| 98,070 | 980.7 | 1 | 2.0482 | 237,200 | 7.373 | 98.07 |
| 47,880 | 478.8 | 0.4882 | 1 | 115,800 | 3,600 | 47.88 |
| 0.4134 | 0.004134 | 0.000004215 | 0.000008634 | 1 | 0.03108 | 0.0004134 |
| 13.3 | 0.133 | 0.0001356 | 0.0002778 | 32.174 | 1 | 0.0133 |
| 1,000 | 10. | 0.0102 | 0.0209 | 2,419 | 75.19 | 1 |

\* Grams (mass) per centimeter per second.

## Table 1-3. Gas-constant Values

| Temp. Scale | Press. units | Vol. units | Wt. units | Energy units | $R$ |
|---|---|---|---|---|---|
| Kelvin.......... | ........ | .... | g moles | calories | 1.9872 |
| | ........ | .... | g moles | joules (abs) | 8.3144 |
| | ........ | .... | g moles | joules (int) | 8.3130 |
| | atm | cm³ | g moles | atm-cm³ | 82.057 |
| | atm | liters | g moles | atm-liters | 0.08205 |
| | mm Hg | liters | g moles | mm Hg–liters | 62.361 |
| | bar | liters | g moles | bar-liters | 0.08314 |
| | kg/cm² | liters | g moles | kg/(cm²)(liters) | 0.08478 |
| | atm | ft³ | lb moles | atm-ft³ | 1.314 |
| | mm Hg | ft³ | lb moles | mm Hg–ft³ | 998.9 |
| | ........ | .... | lb moles | chu or pcu | 1.9872 |
| Rankine......... | ........ | .... | lb moles | Btu | 1.9872 |
| | ........ | .... | lb moles | hp-hr | 0.0007805 |
| | ........ | .... | lb moles | kw-hr | 0.0005819 |
| | atm | ft³ | lb moles | atm-ft³ | 0.7302 |
| | in. Hg | ft³ | lb moles | in. Hg–ft³ | 21.85 |
| | mm Hg | ft³ | lb moles | mm Hg–ft³ | 555.0 |
| | lb/in.² abs | ft³ | lb moles | lb/(in.²)(ft³) | 10.73 |
| | lb/ft² abs | ft³ | lb moles | ft-lb | 1,545.0 |

## Table 1-4. Specific-gravity Conversions*

See Conversion Formulas at End of Table

| Sp gr | °Bé | °API | Lb/gal | Lb/ft³ | Sp gr | °Bé | °API | Lb/gal | Lb/ft³ |
|-------|-----|------|--------|--------|-------|-----|------|--------|--------|
| 0.60 | 103.33 | 104.33 | 4.993 | 37.35 | 0.80 | 45.00 | 45.38 | 6.661 | 49.83 |
| 0.61 | 99.51 | 100.47 | 5.076 | 37.97 | 0.81 | 42.84 | 43.19 | 6.744 | 50.45 |
| 0.62 | 95.81 | 96.73 | 5.160 | 38.60 | 0.82 | 40.73 | 41.06 | 6.827 | 51.07 |
| 0.63 | 92.22 | 93.10 | 5.243 | 39.22 | 0.83 | 38.67 | 38.98 | 6.911 | 51.70 |
| 0.64 | 88.75 | 89.59 | 5.321 | 39.85 | 0.84 | 36.67 | 36.95 | 6.994 | 52.32 |
| 0.65 | 85.38 | 86.19 | 5.410 | 40.47 | 0.85 | 34.71 | 34.97 | 7.978 | 52.94 |
| 0.66 | 82.12 | 82.89 | 5.493 | 41.09 | 0.86 | 32.79 | 33.03 | 7.161 | 53.57 |
| 0.67 | 78.96 | 79.69 | 5.577 | 41.72 | 0.87 | 30.92 | 31.14 | 7.244 | 54.19 |
| 0.68 | 75.88 | 76.59 | 5.660 | 42.34 | 0.88 | 29.09 | 29.30 | 7.328 | 54.82 |
| 0.69 | 72.90 | 73.57 | 5.743 | 42.96 | 0.89 | 27.30 | 27.49 | 7.411 | 55.44 |
| 0.70 | 70.00 | 70.64 | 5.827 | 43.59 | 0.90 | 25.56 | 25.72 | 7.491 | 56.06 |
| 0.71 | 67.18 | 67.80 | 5.910 | 44.21 | 0.91 | 23.85 | 23.99 | 7.578 | 56.69 |
| 0.72 | 64.44 | 65.03 | 5.994 | 44.83 | 0.92 | 22.17 | 22.30 | 7.661 | 57.31 |
| 0.73 | 61.78 | 62.34 | 6.077 | 45.46 | 0.93 | 20.54 | 20.65 | 7.745 | 57.93 |
| 0.74 | 59.19 | 59.72 | 6.160 | 46.08 | 0.94 | 18.94 | 19.03 | 7.828 | 58.56 |
| 0.75 | 56.67 | 57.17 | 6.234 | 46.71 | 0.95 | 17.37 | 17.45 | 7.911 | 59.18 |
| 0.76 | 54.21 | 54.68 | 6.327 | 47.33 | 0.96 | 15.83 | 15.90 | 7.995 | 59.81 |
| 0.77 | 51.82 | 52.27 | 6.410 | 47.95 | 0.97 | 14.33 | 14.38 | 8.078 | 60.42 |
| 0.78 | 49.49 | 49.91 | 6.494 | 48.58 | 0.98 | 12.86 | 12.89 | 8.162 | 61.05 |
| 0.79 | 47.22 | 47.61 | 6.577 | 49.20 | 0.99 | 11.41 | 11.43 | 8.250 | 61.68 |
|  |  |  |  |  | 1.00 | 10.00 | 10.00 | 8.328 | 62.30 |

| Sp gr | °Bé | °Tw | Lb/gal | Lb/ft³ | Sp gr | °Bé | °Tw | Lb/gal | Lb/ft³ |
|-------|-----|-----|--------|--------|-------|-----|-----|--------|--------|
| 1.01 | 1.44 | 2 | 8.412 | 62.92 | 1.31 | 34.31 | 62 | 10.913 | 81.63 |
| 1.02 | 2.84 | 4 | 8.495 | 63.55 | 1.32 | 35.15 | 64 | 10.997 | 82.26 |
| 1.03 | 4.22 | 6 | 8.578 | 64.17 | 1.33 | 35.98 | 66 | 11.080 | 82.88 |
| 1.04 | 5.58 | 8 | 8.662 | 64.80 | 1.34 | 36.79 | 68 | 11.163 | 83.50 |
| 1.05 | 6.91 | 10 | 8.745 | 65.41 | 1.35 | 37.59 | 70 | 11.247 | 84.13 |
| 1.06 | 8.21 | 12 | 8.829 | 66.04 | 1.36 | 38.38 | 72 | 11.330 | 84.75 |
| 1.07 | 9.49 | 14 | 8.912 | 66.67 | 1.37 | 39.16 | 74 | 11.414 | 85.37 |
| 1.08 | 10.74 | 16 | 8.995 | 67.29 | 1.38 | 39.93 | 76 | 11.497 | 86.00 |
| 1.09 | 11.97 | 18 | 9.079 | 67.91 | 1.39 | 40.68 | 78 | 11.560 | 86.62 |
| 1.10 | 13.18 | 20 | 9.162 | 68.54 | 1.40 | 41.43 | 80 | 11.664 | 87.25 |
| 1.11 | 14.37 | 22 | 9.246 | 69.16 | 1.41 | 42.16 | 82 | 11.747 | 87.88 |
| 1.12 | 15.54 | 24 | 9.329 | 69.79 | 1.42 | 42.89 | 84 | 11.830 | 88.49 |
| 1.13 | 16.68 | 26 | 9.412 | 70.41 | 1.43 | 43.60 | 86 | 11.914 | 89.12 |
| 1.14 | 17.81 | 28 | 9.496 | 71.03 | 1.44 | 44.31 | 88 | 11.997 | 89.74 |
| 1.15 | 18.91 | 30 | 9.579 | 71.66 | 1.45 | 45.00 | 90 | 12.081 | 90.36 |
| 1.16 | 20.00 | 32 | 9.662 | 72.28 | 1.46 | 45.68 | 92 | 12.164 | 90.99 |
| 1.17 | 21.07 | 34 | 9.746 | 72.90 | 1.47 | 46.36 | 94 | 12.247 | 91.61 |
| 1.18 | 22.12 | 36 | 9.829 | 73.53 | 1.48 | 47.03 | 96 | 12.331 | 92.24 |
| 1.19 | 23.15 | 38 | 9.913 | 74.15 | 1.49 | 47.68 | 98 | 12.414 | 92.86 |
| 1.20 | 24.17 | 40 | 9.996 | 74.78 | 1.50 | 48.33 | 100 | 12.498 | 93.49 |
| 1.21 | 25.17 | 42 | 10.079 | 75.40 | 1.51 | 48.97 | 102 | 12.581 | 94.11 |
| 1.22 | 26.15 | 44 | 10.163 | 76.02 | 1.52 | 49.61 | 104 | 12.664 | 94.79 |
| 1.23 | 27.11 | 46 | 10.246 | 76.65 | 1.53 | 50.23 | 106 | 12.748 | 95.36 |
| 1.24 | 28.06 | 48 | 10.330 | 77.27 | 1.54 | 50.84 | 108 | 12.831 | 95.98 |
| 1.25 | 29.00 | 50 | 10.413 | 77.89 | 1.55 | 51.45 | 110 | 12.914 | 96.61 |
| 1.26 | 29.92 | 52 | 10.496 | 78.51 | 1.56 | 52.05 | 112 | 12.998 | 97.23 |
| 1.27 | 30.83 | 54 | 10.580 | 79.14 | 1.57 | 52.64 | 114 | 13.081 | 97.85 |
| 1.28 | 31.72 | 56 | 10.663 | 79.76 | 1.58 | 53.23 | 116 | 13.165 | 98.48 |
| 1.29 | 32.60 | 58 | 10.746 | 80.38 | 1.59 | 53.81 | 118 | 13.248 | 99.10 |
| 1.30 | 33.46 | 60 | 10.830 | 81.01 | 1.60 | 54.38 | 120 | 13.331 | 99.73 |

## Table 1-4. Specific-gravity Conversions (Continued)

See Conversion Formulas at End of Table

| Sp gr | °Bé | °Tw | Lb/gal | Lb/ft³ | Sp gr | °Bé | °Tw | Lb/gal | Lb/ft³ |
|---|---|---|---|---|---|---|---|---|---|
| 1.61 | 54.94 | 122 | 13.415 | 100.35 | 1.81 | 64.89 | 162 | 15.082 | 112.82 |
| 1.62 | 55.49 | 124 | 13.498 | 100.97 | 1.82 | 65.33 | 164 | 15.166 | 113.45 |
| 1.63 | 56.04 | 126 | 13.582 | 101.60 | 1.83 | 65.77 | 166 | 15.249 | 114.07 |
| 1.64 | 56.59 | 128 | 13.665 | 102.22 | 1.84 | 66.20 | 168 | 15.333 | 114.70 |
| 1.65 | 57.12 | 130 | 13.748 | 102.84 | 1.85 | 66.62 | 170 | 15.416 | 115.31 |
| 1.66 | 57.65 | 132 | 13.832 | 103.47 | 1.86 | 67.04 | 172 | 15.499 | 115.94 |
| 1.67 | 58.17 | 134 | 13.915 | 104.09 | 1.87 | 67.46 | 174 | 15.583 | 116.56 |
| 1.68 | 58.69 | 136 | 13.998 | 104.72 | 1.88 | 67.87 | 176 | 15.666 | 117.19 |
| 1.69 | 59.20 | 138 | 14.082 | 105.34 | 1.89 | 68.28 | 178 | 15.750 | 117.81 |
| 1.70 | 59.71 | 140 | 14.165 | 105.96 | 1.90 | 68.68 | 180 | 15.832 | 118.43 |
| 1.71 | 60.20 | 142 | 14.249 | 106.59 | 1.91 | 69.08 | 182 | 15.916 | 119.06 |
| 1.72 | 60.70 | 144 | 14.332 | 107.21 | 1.92 | 69.48 | 184 | 16.000 | 110.68 |
| 1.73 | 61.18 | 146 | 14.415 | 107.83 | 1.93 | 69.87 | 186 | 16.083 | 120.31 |
| 1.74 | 61.67 | 148 | 14.499 | 108.46 | 1.94 | 70.26 | 188 | 16.166 | 120.93 |
| 1.75 | 62.14 | 150 | 14.582 | 109.08 | 1.95 | 70.64 | 190 | 16.250 | 121.56 |
| 1.76 | 62.61 | 152 | 14.665 | 109.71 | 1.96 | 71.02 | 192 | 16.333 | 122.18 |
| 1.77 | 63.08 | 154 | 14.749 | 110.32 | 1.97 | 71.40 | 194 | 16.417 | 122.80 |
| 1.78 | 63.54 | 156 | 14.832 | 110.95 | 1.98 | 71.77 | 196 | 16.500 | 123.43 |
| 1.79 | 63.99 | 158 | 14.916 | 111.58 | 1.99 | 72.14 | 198 | 16.583 | 124.05 |
| 1.80 | 64.44 | 160 | 14.999 | 112.20 | 2.00 | 72.50 | 200 | 16.667 | 124.68 |

NOTE: The conversion formulas are

$$°Bé = 145 - \frac{145}{sp\ gr} \text{ (heavier than } H_2O)$$

$$°Bé = \frac{140}{sp\ gr} - 130 \text{ (lighter than } H_2O)$$

$$°Tw = \frac{sp\ gr\ 60°/60°F - 1}{0.005}$$

$$°API = \frac{141.5}{sp\ gr} - 131.5$$

Pounds per gallon and pounds per cubic foot are at 60°F, weight in air.

* The SI units of kilograms per cubic meter may be calculated from pounds per cubic foot by multiplying by 16.018.

## Table 1-5. Temperature Conversion Table

The general formulas are

$$°F = (°C \times \tfrac{9}{5}) + 32 \quad \text{and} \quad °C = (°F - 32) \times \tfrac{5}{9}$$

The absolute temperature is defined by the statement that it is directly proportional to the pressure-volume product of an ideal gas. On the centigrade scale the absolute temperature is equal to $273.16 + t°C$. A temperature of $50°C$ is then equal to $273.16 + 50 = 323.16$ degrees centigrade absolute (°C abs) or Kelvin (°K). On the Fahrenheit scale the absolute temperature is equal to $460 + t°F$ and is termed degrees Rankine (°R). The numbers in bold-face type refer to the temperature (in either centigrade or Fahrenheit degrees) which it is desired to convert into the other scale. If converting from Fahrenheit degrees to centigrade degrees, the equivalent temperature is in the left column, while if converting from degrees centigrade to degrees Fahrenheit, the equivalent temperature is in the column on the right. Interpolation factors are printed in the last two columns.

GIVEN: 200°C.
REQUIRED: Equivalent temperatures in degrees Fahrenheit, Rankine, and Kelvin.
SOLUTION: 200°C = 392°F
200°C = 200 + 273.16 = 473.16°K
200°C = 392 + 460 = 852°R

| C | F | °R | C | | F | C | | F | C | | F |
|---|---|---|---|---|---|---|---|---|---|---|---|
| −262 | **−440** | | −15.0 | **5** | 41.0 | 12.8 | **55** | 131.0 | 66 | **150** | 302 |
| −257 | **−430** | | −14.4 | **6** | 42.8 | 13.3 | **56** | 132.8 | 71 | **160** | 320 |
| −251 | **−420** | | −13.9 | **7** | 44.6 | 13.9 | **57** | 134.6 | 77 | **170** | 338 |
| −246 | **−410** | | −13.3 | **8** | 46.4 | 14.4 | **58** | 136.4 | 82 | **180** | 356 |
| −240 | **−400** | | −12.8 | **9** | 48.2 | 15.0 | **59** | 138.2 | 88 | **190** | 374 |
| −234 | **−390** | | −12.2 | **10** | 50.0 | 15.6 | **60** | 140.0 | 93 | **200** | 392 |
| −229 | **−380** | | −11.7 | **11** | 51.8 | 16.1 | **61** | 141.8 | 99 | **210** | 410 |
| −223 | **−370** | | −11.1 | **12** | 53.6 | 16.7 | **62** | 143.6 | 104 | **220** | 428 |
| −218 | **−360** | | −10.6 | **13** | 55.4 | 17.2 | **63** | 145.4 | 110 | **230** | 446 |
| −212 | **−350** | | −10.0 | **14** | 57.2 | 17.8 | **64** | 147.2 | 116 | **240** | 464 |
| −207 | **−340** | | −9.44 | **15** | 59.0 | 18.3 | **65** | 149.0 | 121 | **250** | 482 |
| −201 | **−330** | | −8.89 | **16** | 60.8 | 18.9 | **66** | 150.8 | 127 | **260** | 500 |
| −196 | **−320** | | −8.33 | **17** | 62.6 | 19.4 | **67** | 152.6 | 132 | **270** | 518 |
| −190 | **−310** | | −7.78 | **18** | 64.4 | 20.0 | **68** | 154.4 | 138 | **280** | 536 |
| −184 | **−300** | | −7.22 | **19** | 66.2 | 20.6 | **69** | 156.2 | 143 | **290** | 554 |
| −179 | **−290** | | −6.67 | **20** | 68.0 | 21.1 | **70** | 158.0 | 149 | **300** | 572 |
| −173 | **−280** | | −6.11 | **21** | 69.8 | 21.7 | **71** | 159.8 | 154 | **310** | 590 |
| −169 | **−273** | −459.4 | −5.56 | **22** | 71.6 | 22.2 | **72** | 161.6 | 160 | **320** | 608 |
| −168 | **−270** | −454 | −5.00 | **23** | 73.4 | 22.8 | **73** | 163.4 | 166 | **330** | 626 |
| −162 | **−260** | −436 | −4.44 | **24** | 75.2 | 23.3 | **74** | 165.2 | 171 | **340** | 644 |
| −157 | **−250** | −418 | −3.89 | **25** | 77.0 | 23.9 | **75** | 167.0 | 177 | **350** | 662 |
| −151 | **−240** | −400 | −3.33 | **26** | 78.8 | 24.4 | **76** | 168.8 | 182 | **360** | 680 |
| −146 | **−230** | −382 | −2.78 | **27** | 80.6 | 25.0 | **77** | 170.6 | 188 | **370** | 698 |
| −140 | **−220** | −364 | −2.22 | **28** | 82.4 | 25.6 | **78** | 172.4 | 193 | **380** | 716 |
| −134 | **−210** | −346 | −1.67 | **29** | 84.2 | 26.1 | **79** | 174.2 | 199 | **390** | 734 |
| −129 | **−200** | −328 | −1.11 | **30** | 86.0 | 26.7 | **80** | 176.0 | 204 | **400** | 752 |
| −123 | **−190** | −310 | −0.56 | **31** | 87.8 | 27.2 | **81** | 177.8 | 210 | **410** | 770 |
| −118 | **−180** | −292 | 0 | **32** | 89.6 | 27.8 | **82** | 179.6 | 216 | **420** | 788 |
| −112 | **−170** | −274 | 0.56 | **33** | 91.4 | 28.3 | **83** | 181.4 | 221 | **430** | 806 |
| −107 | **−160** | −256 | 1.11 | **34** | 93.2 | 28.9 | **84** | 183.2 | 227 | **440** | 824 |
| −101 | **−150** | −238 | 1.67 | **35** | 95.0 | 29.4 | **85** | 185.0 | 232 | **450** | 842 |
| −95.6 | **−140** | −220 | 2.22 | **36** | 96.8 | 30.0 | **86** | 186.8 | 238 | **460** | 860 |
| −90.0 | **−130** | −202 | 2.78 | **37** | 98.6 | 30.6 | **87** | 188.6 | 243 | **470** | 878 |
| −84.4 | **−120** | −184 | 3.33 | **38** | 100.4 | 31.1 | **88** | 190.4 | 249 | **480** | 896 |
| −78.9 | **−110** | −166 | 3.89 | **39** | 102.2 | 31.7 | **89** | 192.2 | 254 | **490** | 914 |
| −73.3 | **−100** | −148 | 4.44 | **40** | 104.0 | 32.2 | **90** | 194.0 | 260 | **500** | 932 |
| −67.8 | **−90** | −130 | 5.00 | **41** | 105.8 | 32.8 | **91** | 195.8 | 266 | **510** | 950 |
| −62.2 | **−80** | −112 | 5.56 | **42** | 107.6 | 33.3 | **92** | 197.6 | 271 | **520** | 968 |
| −56.7 | **−70** | −94 | 6.11 | **43** | 109.4 | 33.9 | **93** | 199.4 | 277 | **530** | 986 |
| −51.1 | **−60** | −76 | 6.67 | **44** | 111.2 | 34.4 | **94** | 201.2 | 282 | **540** | 1004 |
| −45.6 | **−50** | −58 | 7.22 | **45** | 113.0 | 35.0 | **95** | 203.0 | 288 | **550** | 1022 |
| −40.0 | **−40** | −40 | 7.78 | **46** | 114.8 | 35.6 | **96** | 204.8 | 293 | **560** | 1040 |
| −34.4 | **−30** | −22 | 8.33 | **47** | 116.6 | 36.1 | **97** | 206.6 | 299 | **570** | 1058 |
| −28.9 | **−20** | −4 | 8.89 | **48** | 118.4 | 36.7 | **98** | 208.4 | 304 | **580** | 1076 |
| −23.3 | **−10** | 14 | 9.44 | **49** | 120.2 | 37.2 | **99** | 210.2 | 310 | **590** | 1094 |
| −17.8 | **0** | 32 | 10.0 | **50** | 122.0 | 38 | **100** | 212 | 316 | **600** | 1112 |
| −17.2 | **1** | 33.8 | 10.6 | **51** | 123.8 | 43 | **110** | 230 | 321 | **610** | 1130 |
| −16.7 | **2** | 35.6 | 11.1 | **52** | 125.6 | 49 | **120** | 248 | 327 | **620** | 1148 |
| −16.1 | **3** | 37.4 | 11.7 | **53** | 127.4 | 54 | **130** | 266 | 332 | **630** | 1166 |
| −15.6 | **4** | 39.2 | 12.2 | **54** | 129.2 | 60 | **140** | 284 | 338 | **640** | 1184 |

## Table 1-5. Temperature Conversion Table (Continued)

| C | F | C | F | C | F | C | F |
|---|---|---|---|---|---|---|---|
| 343 | 650 1202 | 704 | 1300 2372 | 1066 | 1950 3542 | 1371 | 2500 4532 |
| 349 | 660 1220 | 710 | 1310 2390 | 1071 | 1960 3560 | 1377 | 2510 4550 |
| 354 | 670 1238 | 716 | 1320 2408 | 1077 | 1970 3578 | 1382 | 2520 4568 |
| 360 | 680 1256 | 721 | 1330 2426 | 1082 | 1980 3596 | 1388 | 2530 4586 |
| 366 | 690 1274 | 727 | 1340 2444 | 1088 | 1990 3614 | 1393 | 2540 4604 |
| 371 | 700 1292 | 732 | 1350 2462 | 1093 | 2000 3632 | 1399 | 2550 4622 |
| 377 | 710 1310 | 738 | 1360 2480 | 1099 | 2010 3650 | 1404 | 2560 4640 |
| 382 | 720 1328 | 743 | 1370 2498 | 1104 | 2020 3668 | 1410 | 2570 4658 |
| 388 | 730 1346 | 749 | 1380 2516 | 1110 | 2030 3686 | 1416 | 2580 4676 |
| 393 | 740 1364 | 754 | 1390 2534 | 1116 | 2040 3704 | 1421 | 2590 4694 |
| 399 | 750 1382 | 760 | 1400 2552 | 1121 | 2050 3722 | 1427 | 2600 4712 |
| 404 | 760 1400 | 766 | 1410 2570 | 1127 | 2060 3740 | 1432 | 2610 4730 |
| 410 | 770 1418 | 771 | 1420 2588 | 1132 | 2070 3758 | 1438 | 2620 4748 |
| 416 | 780 1436 | 777 | 1430 2606 | 1138 | 2080 3776 | 1443 | 2630 4766 |
| 421 | 790 1454 | 782 | 1440 2624 | 1143 | 2090 3794 | 1449 | 2640 4784 |
| 427 | 800 1472 | 788 | 1450 2642 | 1149 | 2100 3812 | 1454 | 2650 4802 |
| 432 | 810 1490 | 793 | 1460 2660 | 1154 | 2110 3830 | 1460 | 2660 4820 |
| 438 | 820 1508 | 799 | 1470 2678 | 1160 | 2120 3848 | 1466 | 2670 4838 |
| 443 | 830 1526 | 804 | 1480 2696 | 1166 | 2130 3866 | 1471 | 2680 4856 |
| 449 | 840 1544 | 810 | 1490 2714 | 1171 | 2140 3884 | 1477 | 2690 4874 |
| 454 | 850 1562 | 816 | 1500 2732 | 1177 | 2150 3902 | 1482 | 2700 4892 |
| 460 | 860 1580 | 821 | 1510 2750 | 1182 | 2160 3920 | 1488 | 2710 4910 |
| 466 | 870 1598 | 827 | 1520 2768 | 1188 | 2170 3938 | 1493 | 2720 4928 |
| 471 | 880 1616 | 832 | 1530 2786 | 1193 | 2180 3956 | 1499 | 2730 4946 |
| 477 | 890 1634 | 838 | 1540 2804 | 1199 | 2190 3974 | 1504 | 2740 4964 |
| 482 | 900 1652 | 843 | 1550 2822 | 1204 | 2200 3992 | 1510 | 2750 4982 |
| 488 | 910 1670 | 849 | 1560 2840 | 1210 | 2210 4010 | 1516 | 2760 5000 |
| 493 | 920 1688 | 854 | 1570 2858 | 1216 | 2220 4028 | 1521 | 2770 5018 |
| 499 | 930 1706 | 860 | 1580 2876 | 1221 | 2230 4046 | 1527 | 2780 5036 |
| 504 | 940 1724 | 866 | 1590 2894 | 1227 | 2240 4064 | 1532 | 2790 5054 |
| 510 | 950 1742 | 871 | 1600 2912 | 1232 | 2250 4082 | 1538 | 2800 5072 |
| 516 | 960 1760 | 877 | 1610 2930 | 1238 | 2260 4100 | 1543 | 2810 5090 |
| 521 | 970 1778 | 882 | 1620 2948 | 1243 | 2270 4118 | 1549 | 2820 5108 |
| 527 | 980 1796 | 888 | 1630 2966 | 1249 | 2280 4136 | 1554 | 2830 5126 |
| 532 | 990 1814 | 893 | 1640 2984 | 1254 | 2290 4154 | 1560 | 2840 5144 |
| 538 | 1000 1832 | 899 | 1650 3002 | 1260 | 2300 4172 | 1566 | 2850 5162 |
| 543 | 1010 1850 | 904 | 1660 3020 | 1266 | 2310 4190 | 1571 | 2860 5180 |
| 549 | 1020 1868 | 910 | 1670 3038 | 1271 | 2320 4208 | 1577 | 2870 5198 |
| 554 | 1030 1886 | 916 | 1680 3056 | 1277 | 2330 4226 | 1582 | 2880 5216 |
| 560 | 1040 1904 | 921 | 1690 3074 | 1282 | 2340 4244 | 1588 | 2890 5234 |
| 566 | 1050 1922 | 927 | 1700 3092 | 1288 | 2350 4262 | 1593 | 2900 5252 |
| 571 | 1060 1940 | 932 | 1710 3110 | 1293 | 2360 4280 | 1599 | 2910 5270 |
| 577 | 1070 1958 | 938 | 1720 3128 | 1299 | 2370 4298 | 1604 | 2920 5288 |
| 582 | 1080 1976 | 943 | 1730 3146 | 1304 | 2380 4316 | 1610 | 2930 5306 |
| 588 | 1090 1994 | 949 | 1740 3164 | 1310 | 2390 4334 | 1616 | 2940 5324 |
| 593 | 1100 2012 | 954 | 1750 3182 | 1316 | 2400 4352 | 1621 | 2950 5342 |
| 599 | 1110 2030 | 960 | 1760 3200 | 1321 | 2410 4370 | 1627 | 2960 5360 |
| 604 | 1120 2048 | 966 | 1770 3218 | 1327 | 2420 4388 | 1632 | 2970 5378 |
| 610 | 1130 2066 | 971 | 1780 3236 | 1332 | 2430 4406 | 1638 | 2980 5396 |
| 616 | 1140 2084 | 977 | 1790 3254 | 1338 | 2440 4424 | 1643 | 2990 5414 |
| 621 | 1150 2102 | 982 | 1800 3272 | 1343 | 2450 4442 | 1649 | 3000 5432 |
| 627 | 1160 2120 | 988 | 1810 3290 | 1349 | 2460 4460 | | |
| 632 | 1170 2138 | 993 | 1820 3308 | 1354 | 2470 4478 | | |
| 638 | 1180 2156 | 999 | 1830 3326 | 1360 | 2480 4496 | | |
| 643 | 1190 2174 | 1004 | 1840 3344 | 1366 | 2490 4514 | | |
| 649 | 1200 2192 | 1010 | 1850 3362 | | | | |
| 654 | 1210 2210 | 1016 | 1860 3380 | | | | |
| 660 | 1220 2228 | 1021 | 1870 3398 | | | | |
| 666 | 1230 2246 | 1027 | 1880 3416 | | | | |
| 671 | 1240 2264 | 1032 | 1890 3434 | | | | |
| 677 | 1250 2282 | 1038 | 1900 3452 | | | | |
| 682 | 1260 2300 | 1043 | 1910 3470 | | | | |
| 688 | 1270 2318 | 1049 | 1920 3488 | | | | |
| 693 | 1280 2336 | 1054 | 1930 3506 | | | | |
| 699 | 1290 2354 | 1060 | 1940 3524 | | | | |

### Interpolation Factors

| | | | | | |
|---|---|---|---|---|---|
| 0.56 | 1 | 1.8 | 3.33 | 6 | 10.8 |
| 1.11 | 2 | 3.6 | 3.89 | 7 | 12.6 |
| 1.67 | 3 | 5.4 | 4.44 | 8 | 14.4 |
| 2.22 | 4 | 7.2 | 5.00 | 9 | 16.2 |
| 2.78 | 5 | 9.0 | 5.56 | 10 | 18.0 |

## Table 1-6. Wire and Sheet-metal Gauge Equivalents

Values in Approximate Decimals of an Inch.
As a number of gauges are in use for various shapes and metals, it is advisable to state the thickness in thousandths when specifying gauge number.
Metric wire gauge is ten times the diameter in millimeters.

| Gauge no. | (1)* | (2)* | (3)* | (4)* | (5)* | (6)* | Gauge no. |
|---|---|---|---|---|---|---|---|
| 0000000 | . . . . . . | 0.4900 | . . . . | . . . . . | 0.6666 | 0.500 | 0000000 |
| 000000 | . . . . . . | .4615 | . . . . | . . . . . | .6250 | .464 | 000000 |
| 00000 | . . . . . | .4305 | . . . . | . . . . . | .5883 | .432 | 00000 |
| 0000 | 0.460 | .3938 | 0.454 | . . . . . | .5416 | .400 | 0000 |
| 000 | .410 | .3625 | .425 | . . . . . | .5000 | .372 | 000 |
| 00 | .365 | .3310 | .380 | . . . . . | .4452 | .348 | 00 |
| 0 | .325 | .3065 | .340 | . . . . . | .3964 | .324 | 0 |
| 1 | .289 | .2830 | .300 | . . . . . | .3532 | .300 | 1 |
| 2 | .258 | .2625 | .284 | . . . . . | .3147 | .276 | 2 |
| 3 | .229 | .2437 | .259 | 0.239 | .2804 | .252 | 3 |
| 4 | .204 | .2253 | .238 | .224 | .2500 | .232 | 4 |
| 5 | .182 | .2070 | .220 | .209 | .2225 | .212 | 5 |
| 6 | .162 | .1920 | .203 | .194 | .1981 | .192 | 6 |
| 7 | .144 | .1770 | .180 | .179 | .1764 | .176 | 7 |
| 8 | .128 | .1620 | .165 | .164 | .1570 | .160 | 8 |
| 9 | .114 | .1483 | .148 | .150 | .1398 | .144 | 9 |
| 10 | .102 | .1350 | .134 | .135 | .1250 | .128 | 10 |
| 11 | .091 | .1205 | .120 | .120 | .1113 | .116 | 11 |
| 12 | .081 | .1055 | .109 | .105 | .0991 | .104 | 12 |
| 13 | .072 | .0915 | .095 | .090 | .0882 | .092 | 13 |
| 14 | .064 | .0800 | .083 | .075 | .0785 | .080 | 14 |
| 15 | .057 | .0720 | .072 | .067 | .0699 | .072 | 15 |
| 16 | .051 | .0625 | .065 | .060 | .0625 | .064 | 16 |
| 17 | .045 | .0540 | .058 | .054 | .0556 | .056 | 17 |
| 18 | .040 | .0475 | .049 | .0478 | .0495 | .048 | 18 |
| 19 | .036 | .0410 | .042 | .0418 | .0440 | .040 | 19 |
| 20 | .032 | .0348 | .035 | .0359 | .0392 | 036 | 20 |
| 21 | .0285 | .0317 | .032 | .0329 | .0349 | .032 | 21 |
| 22 | .0253 | .0286 | .028 | .0299 | .0313 | .028 | 22 |
| 23 | .0226 | .0258 | .025 | .0269 | .0278 | .024 | 23 |
| 24 | .0201 | .0230 | .022 | .0239 | .0248 | .022 | 24 |
| 25 | .0179 | .0204 | .020 | .0209 | .0220 | .020 | 25 |
| 26 | .0159 | .0181 | .018 | .0179 | .0196 | .018 | 26 |
| 27 | .0142 | .0173 | .016 | .0164 | .0175 | .0164 | 27 |
| 28 | .0126 | .0162 | .014 | .0149 | .0156 | .0148 | 28 |
| 29 | .0113 | .0150 | .013 | .0135 | .0139 | .0136 | 29 |
| 30 | .0100 | .0140 | .012 | .0120 | .0123 | .0124 | 30 |
| 31 | .0089 | .0132 | .010 | .0105 | .0110 | .0116 | 31 |
| 32 | .0080 | .0128 | .009 | .0097 | .0098 | .0108 | 32 |
| 33 | .0071 | .0118 | .008 | .0090 | .0087 | .0100 | 33 |
| 34 | .0063 | .0104 | .007 | .0082 | .0077 | .0092 | 34 |
| 35 | .0056 | .0095 | .005 | .0075 | .0069 | .0084 | 35 |
| 36 | .0050 | .0090 | .004 | .0067 | .0061 | .0076 | 36 |
| 37 | .0045 | .0085 | . . . . . | .0064 | .0054 | .0068 | 37 |
| 38 | .0040 | .0080 | . . . . . | .0060 | .0048 | .0060 | 38 |
| 39 | .0035 | .0075 | . . . . . | . . . . . | .0043 | .0052 | 39 |
| 40 | .0031 | .0070 | . . . . . | . . . . . | .0039 | .0048 | 40 |

* Gauges are arranged in columns as follows:
1. American (Awg) or Brown & Sharpe (B.&S.), for nonferrous wire and sheet; sometimes used for iron wire.
2. U.S. Steel Wire, Washburn & Moen, Roebling, or American Steel & Wire Co., for steel wire.
3. Birmingham, B.W.G., for steel wire and heat-exchanger tubing, or Stubs Iron Wire, for iron or brass wire; sometimes used for copper plate and for steel plate 12 gauge and heavier and for steel tubes.
4. U.S. Standard, for sheet and plate metal and wrought iron.
5. Standard Birmingham, B.G., for sheet and hoop metal.
6. Imperial Standard Wire Gauge, S.W.G., British legal standard for sheet metal.

## Table 1-7. Fundamental Physical Constants†

| Name | Symbol | Value | Units |
|------|--------|-------|-------|
| **Basic Constants** | | | |
| Velocity of light | $c$ | $2.99792458 \times 10^{10}$ | cm/sec |
| Gravitation constant | $g$ | $6.672 \times 10^{-8}$ | cm³/(g)(sec²) |
| Faraday | | 96.484.6 | coulombs |
| Absolute temperature of ice point, 0°C | $T_0°C$ | 273.16 | °K |
| Pressure-volume product for 1 mole gas at 0°C and zero pressure | $(pv)^{p-0}T_0°C$ | 2,271.160 | joules/mole |
| *or* | | | |
| Volume 1 mole gas at 0°C and 1 atm | | 22,411.5 | cm³ |
| Atomic weight oxygen | | 16 | |
| Planck's constant | $h$ | $6.626176 \times 10^{-27}$ | |
| Avogadro's number | $N$ | $6.02204 \times 10^{23}$ | molecules/g mole |
| Pi | $\pi$ | 3.14159265 | |
| Second | $''$ | 0.000004848136811095 | radian |
| Minute | $'$ | 0.0002908882 | radian |
| Degree | $°$ | 0.01745329 | radian |
| Radian | | 57.2957795 | deg |
| Napierian-logarithm base | $e$ | 2.71828183 | |
| **Derived and Experimental Constants** | | | |
| Liter | | 1,000.027 | cm³ |
| Calorie (gram): | | | |
| 20°C | | 4.181 | joule |
| 15°C | | 4.185 | joule |
| Mean | | 4.186 | joule |
| Btu | | | |
| 39°F | | 1,060.4 | joule |
| 60°F* | | 1,054.6 | joule |
| Mean | | 1,054.8 | joule |
| Standard gravity | $g_0$ | 980.665 | cm/sec² |
| Standard atmosphere | atm | 1,013.250 | dynes/cm² |
| Second (mean solar) | | 1.00273791 | sec (sidereal) |
| Joule | | $0.999835 \pm 0.000052$ | int joule |

\* Used in conversion tables.

† For conversion to other units, see Tables 1-1 and 1-2.

## Table 1-8. Five-place Common Logarithms of Numbers
### 100–155

| No. | L | 0 | 1 | 2 | 3 | 4 | 5 | 6 | 7 | 8 | 9 |
|---|---|---|---|---|---|---|---|---|---|---|---|
| 100 | 00 | 000 | 043 | 087 | 130 | 173 | 217 | 260 | 303 | 346 | 389 |
| 101 | | 432 | 475 | 518 | 561 | 604 | 647 | 689 | 732 | 775 | 817 |
| 102 | | 860 | 903 | 945 | 988 | *030 | *072 | *115 | *157 | *199 | *242 |
| 103 | 01 | 284 | 326 | 368 | 410 | 452 | 494 | 536 | 578 | 620 | 662 |
| 104 | | 703 | 745 | 787 | 828 | 870 | 912 | 953 | 995 | *036 | *078 |
| 105 | 02 | 119 | 160 | 202 | 243 | 284 | 325 | 366 | 408 | 449 | 490 |
| 106 | | 531 | 572 | 612 | 653 | 694 | 735 | 776 | 816 | 857 | 898 |
| 107 | | 938 | 979 | *019 | *060 | *100 | *141 | *181 | *222 | *262 | *302 |
| 108 | 03 | 342 | 383 | 423 | 463 | 503 | 543 | 583 | 623 | 663 | 703 |
| 109 | | 743 | 782 | 822 | 862 | 902 | 941 | 981 | *021 | *060 | *100 |
| 110 | 04 | 139 | 179 | 218 | 258 | 297 | 336 | 376 | 415 | 454 | 493 |
| 111 | | 532 | 571 | 610 | 650 | 689 | 727 | 766 | 805 | 844 | 883 |
| 112 | | 922 | 961 | 999 | *038 | *077 | *115 | *154 | *192 | *231 | *269 |
| 113 | 05 | 308 | 346 | 385 | 423 | 461 | 500 | 538 | 576 | 614 | 652 |
| 114 | | 690 | 729 | 767 | 805 | 843 | 881 | 918 | 956 | 994 | *032 |
| 115 | 06 | 070 | 108 | 145 | 183 | 221 | 258 | 296 | 333 | 371 | 408 |
| 116 | | 446 | 483 | 521 | 558 | 595 | 633 | 670 | 707 | 744 | 781 |
| 117 | | 819 | 856 | 893 | 930 | 967 | *004 | *041 | *078 | *115 | *151 |
| 118 | 07 | 188 | 225 | 262 | 298 | 335 | 372 | 408 | 445 | 482 | 518 |
| 119 | | 555 | 591 | 628 | 664 | 700 | 737 | 773 | 809 | 846 | 882 |
| 120 | | 918 | 954 | 990 | *027 | *063 | *099 | *135 | *171 | *207 | *243 |
| 121 | 08 | 279 | 314 | 350 | 386 | 422 | 458 | 493 | 529 | 565 | 600 |
| 122 | | 636 | 672 | 707 | 743 | 778 | 814 | 849 | 884 | 920 | 955 |
| 123 | | 991 | *026 | *061 | *096 | *132 | *167 | *202 | *237 | *272 | *307 |
| 124 | 09 | 342 | 377 | 412 | 447 | 482 | 517 | 552 | 587 | 621 | 656 |
| 125 | | 691 | 726 | 760 | 795 | 830 | 864 | 899 | 934 | 968 | *003 |
| 126 | 10 | 037 | 072 | 106 | 140 | 175 | 209 | 243 | 278 | 312 | 346 |
| 127 | | 380 | 415 | 449 | 483 | 517 | 551 | 585 | 619 | 653 | 687 |
| 128 | | 721 | 755 | 789 | 823 | 857 | 890 | 924 | 958 | 992 | *025 |
| 129 | 11 | 059 | 093 | 126 | 160 | 193 | 227 | 261 | 294 | 327 | 361 |
| 130 | | 394 | 428 | 461 | 494 | 528 | 561 | 594 | 628 | 661 | 694 |
| 131 | | 727 | 760 | 793 | 826 | 860 | 893 | 926 | 959 | 992 | *024 |
| 132 | 12 | 057 | 090 | 123 | 156 | 189 | 222 | 254 | 287 | 320 | 353 |
| 133 | | 385 | 418 | 450 | 483 | 516 | 548 | 581 | 613 | 646 | 678 |
| 134 | | 710 | 743 | 775 | 808 | 840 | 872 | 905 | 937 | 969 | *001 |
| 135 | 13 | 033 | 066 | 098 | 130 | 162 | 194 | 226 | 258 | 290 | 322 |
| 136 | | 354 | 386 | 418 | 450 | 481 | 513 | 545 | 577 | 609 | 640 |
| 137 | | 672 | 704 | 735 | 767 | 799 | 830 | 862 | 893 | 925 | 956 |
| 138 | | 988 | *019 | *051 | *082 | *114 | *145 | *176 | *208 | *239 | *270 |
| 139 | 14 | 301 | 333 | 364 | 395 | 426 | 457 | 489 | 520 | 551 | 582 |
| 140 | | 613 | 644 | 675 | 706 | 737 | 768 | 799 | 829 | 860 | 891 |
| 141 | | 922 | 953 | 983 | *014 | *045 | *076 | *106 | *137 | *168 | *198 |
| 142 | 15 | 229 | 259 | 290 | 320 | 351 | 381 | 412 | 442 | 473 | 503 |
| 143 | | 534 | 564 | 594 | 625 | 655 | 685 | 715 | 746 | 776 | 806 |
| 144 | | 836 | 866 | 897 | 927 | 957 | 987 | *017 | *047 | *077 | *107 |
| 145 | 16 | 137 | 167 | 197 | 227 | 256 | 286 | 316 | 346 | 376 | 406 |
| 146 | | 435 | 465 | 495 | 524 | 554 | 584 | 613 | 643 | 673 | 702 |
| 147 | | 732 | 761 | 791 | 820 | 850 | 879 | 909 | 938 | 967 | 997 |
| 148 | 17 | 026 | 056 | 085 | 114 | 143 | 173 | 202 | 231 | 260 | 289 |
| 149 | | 319 | 348 | 377 | 406 | 435 | 464 | 493 | 522 | 551 | 580 |
| 150 | | 609 | 638 | 667 | 696 | 725 | 754 | 782 | 811 | 840 | 869 |
| 151 | | 898 | 926 | 955 | 984 | *013 | *041 | *070 | *099 | *127 | *156 |
| 152 | 18 | 184 | 213 | 241 | 270 | 299 | 327 | 355 | 384 | 412 | 441 |
| 153 | | 469 | 498 | 526 | 554 | 583 | 611 | 639 | 667 | 696 | 724 |
| 154 | | 752 | 780 | 808 | 837 | 865 | 893 | 921 | 949 | 977 | *005 |
| 155 | 19 | 033 | 061 | 089 | 117 | 145 | 173 | 201 | 229 | 257 | 285 |
| No. | L | 0 | 1 | 2 | 3 | 4 | 5 | 6 | 7 | 8 | 9 |

**Proportional parts**

| | 44 | 43 | 42 |
|---|---|---|---|
| 1 | 4.4 | 4.3 | 4.2 |
| 2 | 8.8 | 8.6 | 8.4 |
| 3 | 13.2 | 12.9 | 12.6 |
| 4 | 17.6 | 17.2 | 16.8 |
| 5 | 22.0 | 21.5 | 21.0 |
| 6 | 26.4 | 25.8 | 25.2 |
| 7 | 30.8 | 30.1 | 29.4 |
| 8 | 35.2 | 34.4 | 33.6 |
| 9 | 39.6 | 38.7 | 37.8 |

| | 41 | 40 | 39 |
|---|---|---|---|
| 1 | 4.1 | 4.0 | 3.9 |
| 2 | 8.2 | 8.0 | 7.8 |
| 3 | 12.3 | 12.0 | 11.7 |
| 4 | 16.4 | 16.0 | 15.6 |
| 5 | 20.5 | 20.0 | 19.5 |
| 6 | 24.6 | 24.0 | 23.4 |
| 7 | 28.7 | 28.0 | 27.3 |
| 8 | 32.8 | 32.0 | 31.2 |
| 9 | 36.9 | 36.0 | 35.1 |

| | 38 | 37 | 36 |
|---|---|---|---|
| 1 | 3.8 | 3.7 | 3.6 |
| 2 | 7.6 | 7.4 | 7.2 |
| 3 | 11.4 | 11.1 | 10.8 |
| 4 | 15.2 | 14.8 | 14.4 |
| 5 | 19.0 | 18.5 | 18.0 |
| 6 | 22.8 | 22.2 | 21.6 |
| 7 | 26.6 | 25.9 | 25.2 |
| 8 | 30.4 | 29.6 | 28.8 |
| 9 | 34.2 | 33.3 | 32.4 |

| | 35 | 34 | 33 |
|---|---|---|---|
| 1 | 3.5 | 3.4 | 3.3 |
| 2 | 7.0 | 6.8 | 6.6 |
| 3 | 10.5 | 10.2 | 9.9 |
| 4 | 14.0 | 13.6 | 13.2 |
| 5 | 17.5 | 17.0 | 16.5 |
| 6 | 21.0 | 20.4 | 19.8 |
| 7 | 24.5 | 23.8 | 23.1 |
| 8 | 28.0 | 27.2 | 26.4 |
| 9 | 31.5 | 30.6 | 29.7 |

| | 32 | 31 | 30 |
|---|---|---|---|
| 1 | 3.2 | 3.1 | 3.0 |
| 2 | 6.4 | 6.2 | 6.0 |
| 3 | 9.6 | 9.3 | 9.0 |
| 4 | 12.8 | 12.4 | 12.0 |
| 5 | 16.0 | 15.5 | 15.0 |
| 6 | 19.2 | 18.6 | 18.0 |
| 7 | 22.4 | 21.7 | 21.0 |
| 8 | 25.6 | 24.8 | 24.0 |
| 9 | 28.8 | 27.9 | 27.0 |

**Proportional parts**

* Indicates change in the first two decimal places.

## Table 1-8. Five-place Common Logarithms of Numbers (Continued)
### 155–210

| No. | L | 0 | 1 | 2 | 3 | 4 | 5 | 6 | 7 | 8 | 9 |
|---|---|---|---|---|---|---|---|---|---|---|---|
| 155 | 19 | 033 | 061 | 089 | 117 | 145 | 173 | 201 | 229 | 257 | 285 |
| 156 | | 312 | 340 | 368 | 396 | 424 | 451 | 479 | 507 | 535 | 562 |
| 157 | | 590 | 618 | 645 | 673 | 700 | 728 | 756 | 783 | 811 | 838 |
| 158 | | 866 | 893 | 921 | 948 | 976 | *003 | *030 | *058 | *085 | *112 |
| 159 | 20 | 140 | 167 | 194 | 222 | 249 | 276 | 303 | 330 | 358 | 385 |
| 160 | | 412 | 439 | 466 | 493 | 520 | 548 | 575 | 602 | 629 | 656 |
| 161 | | 683 | 710 | 737 | 763 | 790 | 817 | 844 | 871 | 898 | 925 |
| 162 | | 952 | 978 | *005 | *032 | *059 | *085 | *112 | *139 | *165 | *192 |
| 163 | 21 | 219 | 245 | 272 | 299 | 325 | 352 | 378 | 405 | 431 | 458 |
| 164 | | 484 | 511 | 537 | 564 | 590 | 617 | 643 | 669 | 696 | 722 |
| 165 | | 748 | 775 | 801 | 827 | 854 | 880 | 906 | 932 | 958 | 985 |
| 166 | 22 | 011 | 037 | 063 | 089 | 115 | 141 | 168 | 194 | 220 | 246 |
| 167 | | 272 | 298 | 324 | 350 | 376 | 401 | 427 | 453 | 479 | 505 |
| 168 | | 531 | 557 | 583 | 608 | 634 | 660 | 686 | 712 | 737 | 763 |
| 169 | | 789 | 814 | 840 | 866 | 891 | 917 | 943 | 968 | 994 | *019 |
| 170 | 23 | 045 | 070 | 096 | 121 | 147 | 172 | 198 | 223 | 249 | 274 |
| 171 | | 300 | 325 | 350 | 376 | 401 | 426 | 452 | 477 | 502 | 528 |
| 172 | | 553 | 578 | 603 | 629 | 654 | 679 | 704 | 729 | 754 | 776 |
| 173 | | 805 | 830 | 855 | 880 | 905 | 930 | 955 | 980 | *005 | *030 |
| 174 | 24 | 055 | 080 | 105 | 130 | 155 | 180 | 204 | 229 | 254 | 279 |
| 175 | | 304 | 329 | 353 | 378 | 403 | 428 | 452 | 477 | 502 | 527 |
| 176 | | 551 | 576 | 601 | 625 | 650 | 674 | 699 | 724 | 748 | 773 |
| 177 | | 797 | 822 | 846 | 871 | 895 | 920 | 944 | 969 | 993 | *018 |
| 178 | 25 | 042 | 066 | 091 | 115 | 139 | 164 | 188 | 212 | 237 | 261 |
| 179 | | 285 | 310 | 334 | 358 | 382 | 406 | 431 | 455 | 479 | 503 |
| 180 | | 527 | 551 | 575 | 600 | 624 | 648 | 672 | 696 | 720 | 744 |
| 181 | | 768 | 792 | 816 | 840 | 864 | 888 | 912 | 935 | 959 | 983 |
| 182 | 26 | 007 | 031 | 055 | 079 | 102 | 126 | 150 | 174 | 198 | 221 |
| 183 | | 245 | 269 | 293 | 316 | 340 | 364 | 387 | 411 | 435 | 458 |
| 184 | | 482 | 505 | 529 | 553 | 576 | 600 | 623 | 647 | 670 | 694 |
| 185 | | 717 | 741 | 764 | 788 | 811 | 834 | 858 | 881 | 905 | 928 |
| 186 | | 951 | 975 | 998 | *021 | *045 | *068 | *091 | *114 | *138 | *161 |
| 187 | 27 | 184 | 207 | 231 | 254 | 277 | 300 | 323 | 346 | 370 | 393 |
| 188 | | 416 | 439 | 462 | 485 | 508 | 531 | 554 | 577 | 600 | 623 |
| 189 | | 646 | 669 | 692 | 715 | 738 | 761 | 784 | 807 | 830 | 853 |
| 190 | | 875 | 898 | 921 | 944 | 967 | 990 | *012 | *035 | *058 | *081 |
| 191 | 28 | 103 | 126 | 149 | 172 | 194 | 217 | 240 | 262 | 285 | 308 |
| 192 | | 330 | 353 | 375 | 398 | 421 | 443 | 466 | 488 | 511 | 533 |
| 193 | | 556 | 578 | 601 | 623 | 646 | 668 | 691 | 713 | 735 | 758 |
| 194 | | 780 | 803 | 825 | 847 | 870 | 892 | 914 | 937 | 959 | 981 |
| 195 | 29 | 003 | 026 | 048 | 070 | 092 | 115 | 137 | 159 | 181 | 203 |
| 196 | | 226 | 248 | 270 | 292 | 314 | 336 | 358 | 380 | 403 | 425 |
| 197 | | 447 | 469 | 491 | 513 | 535 | 557 | 579 | 601 | 623 | 645 |
| 198 | | 667 | 688 | 710 | 732 | 754 | 776 | 798 | 820 | 842 | 863 |
| 199 | | 885 | 907 | 929 | 951 | 973 | 994 | *016 | *038 | *060 | *081 |
| 200 | 30 | 103 | 125 | 146 | 168 | 190 | 211 | 233 | 255 | 276 | 298 |
| 201 | | 320 | 341 | 363 | 384 | 406 | 428 | 449 | 471 | 492 | 514 |
| 202 | | 535 | 557 | 578 | 600 | 621 | 643 | 664 | 685 | 707 | 728 |
| 203 | | 750 | 771 | 792 | 814 | 835 | 856 | 878 | 899 | 920 | 942 |
| 204 | | 963 | 984 | *006 | *027 | *048 | *069 | *091 | *112 | *133 | *154 |
| 205 | 31 | 175 | 197 | 218 | 239 | 260 | 281 | 302 | 323 | 345 | 366 |
| 206 | | 387 | 408 | 429 | 450 | 471 | 492 | 513 | 534 | 555 | 576 |
| 207 | | 597 | 618 | 639 | 660 | 681 | 702 | 723 | 744 | 765 | 785 |
| 208 | | 806 | 827 | 848 | 869 | 890 | 911 | 931 | 952 | 973 | 994 |
| 209 | 32 | 015 | 035 | 056 | 077 | 098 | 118 | 139 | 160 | 181 | 201 |
| 210 | | 222 | 243 | 263 | 284 | 305 | 325 | 346 | 366 | 387 | 408 |
| No. | L | 0 | 1 | 2 | 3 | 4 | 5 | 6 | 7 | 8 | 9 |

**Proportional parts**

| | 29 | 28 |
|---|---|---|
| 1 | 2.9 | 2.8 |
| 2 | 5.8 | 5.6 |
| 3 | 8.7 | 8.4 |
| 4 | 11.6 | 11.2 |
| 5 | 14.5 | 14.0 |
| 6 | 17.4 | 16.8 |
| 7 | 20.3 | 19.6 |
| 8 | 23.2 | 22.4 |
| 9 | 26.1 | 25.2 |

| | 27 | 26 |
|---|---|---|
| 1 | 2.7 | 2.6 |
| 2 | 5.4 | 5.2 |
| 3 | 8.1 | 7.8 |
| 4 | 10.8 | 10.4 |
| 5 | 13.5 | 13.0 |
| 6 | 16.2 | 15.6 |
| 7 | 18.9 | 18.2 |
| 8 | 21.6 | 20.8 |
| 9 | 24.3 | 23.4 |

| | 25 |
|---|---|
| 1 | 2.5 |
| 2 | 5.0 |
| 3 | 7.5 |
| 4 | 10.0 |
| 5 | 12.5 |
| 6 | 15.0 |
| 7 | 17.5 |
| 8 | 20.0 |
| 9 | 22.5 |

| | 24 |
|---|---|
| 1 | 2.4 |
| 2 | 4.8 |
| 3 | 7.2 |
| 4 | 9.6 |
| 5 | 12.0 |
| 6 | 14.4 |
| 7 | 16.8 |
| 8 | 19.2 |
| 9 | 21.6 |

| | 23 |
|---|---|
| 1 | 2.3 |
| 2 | 4.6 |
| 3 | 6.9 |
| 4 | 9.2 |
| 5 | 11.5 |
| 6 | 13.8 |
| 7 | 16.1 |
| 8 | 18.4 |
| 9 | 20.7 |

| | 22 |
|---|---|
| 1 | 2.2 |
| 2 | 4.4 |
| 3 | 6.6 |
| 4 | 8.8 |
| 5 | 11.0 |
| 6 | 13.2 |
| 7 | 15.4 |
| 8 | 17.6 |
| 9 | 19.8 |

\* Indicates change in the first two decimal places.

## Table 1-8. Five-place Common Logarithms of Numbers (Continued)
### 210–265

| No. | L | 0 | 1 | 2 | 3 | 4 | 5 | 6 | 7 | 8 | 9 |
|---|---|---|---|---|---|---|---|---|---|---|---|
| 210 | 32 | 222 | 243 | 263 | 284 | 305 | 325 | 346 | 366 | 387 | 408 |
| 211 |  | 428 | 449 | 469 | 490 | 511 | 531 | 552 | 572 | 593 | 613 |
| 212 |  | 634 | 654 | 675 | 695 | 715 | 736 | 756 | 777 | 797 | 818 |
| 213 |  | 838 | 858 | 879 | 899 | 919 | 940 | 960 | 980 | *001 | *021 |
| 214 | 33 | 041 | 062 | 082 | 102 | 122 | 143 | 163 | 183 | 203 | 224 |
| 215 |  | 244 | 264 | 284 | 304 | 325 | 345 | 365 | 385 | 405 | 425 |
| 216 |  | 445 | 465 | 486 | 506 | 526 | 546 | 566 | 586 | 606 | 626 |
| 217 |  | 646 | 666 | 686 | 706 | 726 | 746 | 766 | 786 | 806 | 826 |
| 218 |  | 846 | 866 | 885 | 905 | 925 | 945 | 965 | 985 | *005 | *025 |
| 219 | 34 | 044 | 064 | 084 | 104 | 124 | 143 | 163 | 183 | 203 | 223 |
| 220 |  | 242 | 262 | 282 | 301 | 321 | 341 | 361 | 380 | 400 | 420 |
| 221 |  | 439 | 459 | 479 | 498 | 518 | 537 | 557 | 577 | 596 | 616 |
| 222 |  | 635 | 655 | 674 | 694 | 713 | 733 | 753 | 772 | 792 | 811 |
| 223 |  | 830 | 850 | 869 | 889 | 908 | 928 | 947 | 967 | 986 | *005 |
| 224 | 35 | 025 | 044 | 064 | 083 | 102 | 122 | 141 | 160 | 180 | 199 |
| 225 |  | 218 | 238 | 257 | 276 | 295 | 315 | 334 | 353 | 372 | 392 |
| 226 |  | 411 | 430 | 449 | 468 | 488 | 507 | 526 | 545 | 564 | 583 |
| 227 |  | 603 | 622 | 641 | 660 | 679 | 698 | 717 | 736 | 755 | 774 |
| 228 |  | 793 | 813 | 832 | 851 | 870 | 889 | 908 | 927 | 946 | 965 |
| 229 |  | 984 | *003 | *021 | *040 | *059 | *078 | *097 | *116 | *135 | *154 |
| 230 | 36 | 173 | 192 | 211 | 229 | 248 | 267 | 286 | 305 | 324 | 342 |
| 231 |  | 361 | 380 | 399 | 418 | 436 | 455 | 474 | 493 | 511 | 530 |
| 232 |  | 549 | 568 | 586 | 605 | 624 | 642 | 661 | 680 | 698 | 717 |
| 233 |  | 736 | 754 | 773 | 791 | 810 | 829 | 847 | 866 | 884 | 903 |
| 234 |  | 922 | 940 | 959 | 977 | 996 | *014 | *033 | *051 | *070 | *088 |
| 235 | 37 | 107 | 125 | 144 | 162 | 181 | 199 | 218 | 236 | 254 | 273 |
| 236 |  | 291 | 310 | 328 | 346 | 365 | 383 | 401 | 420 | 438 | 457 |
| 237 |  | 475 | 493 | 511 | 530 | 548 | 566 | 585 | 603 | 621 | 639 |
| 238 |  | 658 | 676 | 694 | 712 | 731 | 749 | 767 | 785 | 803 | 822 |
| 239 |  | 840 | 858 | 876 | 894 | 912 | 931 | 949 | 967 | 985 | *003 |
| 240 | 38 | 021 | 039 | 057 | 075 | 093 | 112 | 130 | 148 | 166 | 184 |
| 241 |  | 202 | 220 | 238 | 256 | 274 | 292 | 310 | 328 | 346 | 364 |
| 242 |  | 382 | 399 | 417 | 435 | 453 | 471 | 489 | 507 | 525 | 543 |
| 243 |  | 561 | 579 | 596 | 614 | 632 | 650 | 668 | 686 | 703 | 721 |
| 244 |  | 739 | 757 | 775 | 792 | 810 | 828 | 846 | 863 | 881 | 899 |
| 245 |  | 917 | 934 | 952 | 970 | 987 | *005 | *023 | *041 | *058 | *076 |
| 246 | 39 | 094 | 111 | 129 | 146 | 164 | 182 | 199 | 217 | 235 | 252 |
| 247 |  | 270 | 287 | 305 | 322 | 340 | 358 | 375 | 393 | 410 | 428 |
| 248 |  | 445 | 463 | 480 | 498 | 515 | 533 | 550 | 568 | 585 | 602 |
| 249 |  | 620 | 637 | 655 | 672 | 690 | 707 | 724 | 742 | 759 | 777 |
| 250 |  | 794 | 811 | 829 | 846 | 863 | 881 | 898 | 915 | 933 | 950 |
| 251 |  | 967 | 985 | *002 | *019 | *037 | *054 | *071 | *088 | *106 | *123 |
| 252 | 40 | 140 | 157 | 175 | 192 | 209 | 226 | 243 | 261 | 278 | 295 |
| 253 |  | 312 | 329 | 346 | 364 | 381 | 398 | 415 | 432 | 449 | 466 |
| 254 |  | 483 | 500 | 518 | 535 | 552 | 569 | 586 | 603 | 620 | 637 |
| 255 |  | 654 | 671 | 688 | 705 | 722 | 739 | 756 | 773 | 790 | 807 |
| 256 |  | 824 | 841 | 858 | 875 | 892 | 909 | 926 | 943 | 960 | 976 |
| 257 |  | 993 | *010 | *027 | *044 | *061 | *078 | *095 | *111 | *128 | *145 |
| 258 | 41 | 162 | 179 | 196 | 212 | 229 | 246 | 263 | 280 | 296 | 313 |
| 259 |  | 330 | 347 | 364 | 380 | 397 | 414 | 430 | 447 | 464 | 481 |
| 260 |  | 497 | 514 | 531 | 547 | 564 | 581 | 597 | 614 | 631 | 647 |
| 261 |  | 664 | 681 | 697 | 714 | 731 | 747 | 764 | 780 | 797 | 814 |
| 262 |  | 830 | 847 | 863 | 880 | 896 | 913 | 929 | 946 | 963 | 979 |
| 263 |  | 996 | *012 | *029 | *045 | *062 | *078 | *095 | *111 | *127 | *144 |
| 264 | 42 | 160 | 177 | 193 | 210 | 226 | 243 | 259 | 275 | 292 | 308 |
| 265 |  | 325 | 341 | 357 | 374 | 390 | 406 | 423 | 439 | 456 | 472 |
| No. | L | 0 | 1 | 2 | 3 | 4 | 5 | 6 | 7 | 8 | 9 |

**Proportional parts**

| | 21 |
|---|---|
| 1 | 2.1 |
| 2 | 4.2 |
| 3 | 6.3 |
| 4 | 8.4 |
| 5 | 10.5 |
| 6 | 12.6 |
| 7 | 14.7 |
| 8 | 16.8 |
| 9 | 18.9 |

| | 20 |
|---|---|
| 1 | 2.0 |
| 2 | 4.0 |
| 3 | 6.0 |
| 4 | 8.0 |
| 5 | 10.0 |
| 6 | 12.0 |
| 7 | 14.0 |
| 8 | 16.0 |
| 9 | 18.0 |

| | 19 |
|---|---|
| 1 | 1.9 |
| 2 | 3.8 |
| 3 | 5.7 |
| 4 | 7.6 |
| 5 | 9.5 |
| 6 | 11.4 |
| 7 | 13.3 |
| 8 | 15.2 |
| 9 | 17.1 |

| | 18 |
|---|---|
| 1 | 1.8 |
| 2 | 3.6 |
| 3 | 5.4 |
| 4 | 7.2 |
| 5 | 9.0 |
| 6 | 10.8 |
| 7 | 12.6 |
| 8 | 14.4 |
| 9 | 16.2 |

* Indicates change in the first two decimal places.

## Table 1-8. Five-place Common Logarithms of Numbers (Continued)
### 265–320

| No. | L | 0 | 1 | 2 | 3 | 4 | 5 | 6 | 7 | 8 | 9 |
|---|---|---|---|---|---|---|---|---|---|---|---|
| 265 | 42 | 325 | 341 | 357 | 374 | 390 | 406 | 423 | 439 | 456 | 472 |
| 266 | | 488 | 504 | 521 | 537 | 553 | 570 | 586 | 602 | 619 | 635 |
| 267 | | 651 | 667 | 684 | 700 | 716 | 732 | 749 | 765 | 781 | 797 |
| 268 | | 813 | 830 | 846 | 862 | 878 | 894 | 911 | 927 | 943 | 959 |
| 269 | | 975 | 991 | *008 | *024 | *040 | *056 | *072 | *088 | *104 | *120 |
| 270 | 43 | 136 | 152 | 169 | 185 | 201 | 217 | 233 | 249 | 265 | 281 |
| 271 | | 297 | 313 | 329 | 345 | 361 | 377 | 393 | 409 | 425 | 441 |
| 272 | | 457 | 473 | 489 | 505 | 521 | 537 | 553 | 569 | 584 | 600 |
| 273 | | 616 | 632 | 648 | 664 | 680 | 696 | 712 | 727 | 743 | 759 |
| 274 | | 775 | 791 | 807 | 823 | 838 | 854 | 870 | 886 | 902 | 917 |
| 275 | | 933 | 949 | 965 | 981 | 996 | *012 | *028 | *044 | *059 | *075 |
| 276 | 44 | 091 | 107 | 122 | 138 | 154 | 170 | 185 | 201 | 217 | 232 |
| 277 | | 248 | 264 | 279 | 295 | 311 | 326 | 342 | 358 | 373 | 389 |
| 278 | | 404 | 420 | 436 | 451 | 467 | 483 | 498 | 514 | 529 | 545 |
| 279 | | 560 | 576 | 592 | 607 | 623 | 638 | 654 | 669 | 685 | 700 |
| 280 | | 716 | 731 | 747 | 762 | 778 | 793 | 809 | 824 | 840 | 855 |
| 281 | | 871 | 886 | 902 | 917 | 932 | 948 | 963 | 979 | 994 | *010 |
| 282 | 45 | 025 | 040 | 056 | 071 | 086 | 102 | 117 | 133 | 148 | 163 |
| 283 | | 179 | 194 | 209 | 225 | 240 | 255 | 271 | 286 | 301 | 317 |
| 284 | | 332 | 347 | 362 | 378 | 393 | 408 | 423 | 439 | 454 | 469 |
| 285 | | 484 | 500 | 515 | 530 | 545 | 561 | 576 | 591 | 606 | 621 |
| 286 | | 637 | 652 | 667 | 682 | 697 | 712 | 728 | 743 | 758 | 773 |
| 287 | | 788 | 803 | 818 | 834 | 849 | 864 | 879 | 894 | 909 | 924 |
| 288 | | 939 | 954 | 969 | 984 | *000 | *015 | *030 | *045 | *060 | *075 |
| 289 | 46 | 090 | 105 | 120 | 135 | 150 | 165 | 180 | 195 | 210 | 225 |
| 290 | | 240 | 255 | 270 | 285 | 300 | 315 | 330 | 345 | 359 | 374 |
| 291 | | 389 | 404 | 419 | 434 | 449 | 464 | 479 | 494 | 509 | 523 |
| 292 | | 538 | 553 | 568 | 583 | 598 | 613 | 627 | 642 | 657 | 672 |
| 293 | | 687 | 702 | 716 | 731 | 746 | 761 | 776 | 790 | 805 | 820 |
| 294 | | 835 | 850 | 864 | 879 | 894 | 909 | 923 | 938 | 953 | 967 |
| 295 | | 982 | 997 | *012 | *026 | *041 | *056 | *070 | *085 | *100 | *115 |
| 296 | 47 | 129 | 144 | 159 | 173 | 188 | 202 | 217 | 232 | 246 | 261 |
| 297 | | 276 | 290 | 305 | 319 | 334 | 349 | 363 | 378 | 392 | 407 |
| 298 | | 422 | 436 | 451 | 465 | 480 | 494 | 509 | 524 | 538 | 553 |
| 299 | | 567 | 582 | 596 | 611 | 625 | 640 | 654 | 669 | 683 | 698 |
| 300 | | 712 | 727 | 741 | 756 | 770 | 784 | 799 | 813 | 828 | 842 |
| 301 | | 857 | 871 | 886 | 900 | 914 | 929 | 943 | 958 | 972 | 986 |
| 302 | 48 | 001 | 015 | 029 | 044 | 058 | 073 | 087 | 101 | 116 | 130 |
| 303 | | 144 | 159 | 173 | 187 | 202 | 216 | 230 | 245 | 259 | 273 |
| 304 | | 287 | 302 | 316 | 330 | 344 | 359 | 373 | 387 | 402 | 416 |
| 305 | | 430 | 444 | 458 | 473 | 487 | 501 | 515 | 530 | 544 | 558 |
| 306 | | 572 | 586 | 601 | 615 | 629 | 643 | 657 | 671 | 686 | 700 |
| 307 | | 714 | 728 | 742 | 756 | 770 | 785 | 799 | 813 | 827 | 841 |
| 308 | | 855 | 869 | 883 | 897 | 911 | 926 | 940 | 954 | 968 | 982 |
| 309 | | 996 | *010 | *024 | *038 | *052 | *066 | *080 | *094 | *108 | *122 |
| 310 | 49 | 136 | 150 | 164 | 178 | 192 | 206 | 220 | 234 | 248 | 262 |
| 311 | | 276 | 290 | 304 | 318 | 332 | 346 | 360 | 374 | 388 | 402 |
| 312 | | 415 | 429 | 443 | 457 | 471 | 485 | 499 | 513 | 527 | 541 |
| 313 | | 554 | 568 | 582 | 596 | 610 | 624 | 638 | 651 | 665 | 679 |
| 314 | | 693 | 707 | 721 | 734 | 748 | 762 | 776 | 790 | 803 | 817 |
| 315 | | 831 | 845 | 859 | 872 | 886 | 900 | 914 | 927 | 941 | 955 |
| 316 | | 969 | 982 | 996 | *010 | *024 | *037 | *051 | *065 | *079 | *092 |
| 317 | 50 | 106 | 120 | 133 | 147 | 161 | 174 | 188 | 202 | 215 | 229 |
| 318 | | 243 | 256 | 270 | 284 | 297 | 311 | 325 | 338 | 352 | 365 |
| 319 | | 379 | 393 | 406 | 420 | 433 | 447 | 461 | 474 | 488 | 501 |
| 320 | | 515 | 529 | 542 | 556 | 569 | 583 | 596 | 610 | 623 | 637 |
| No. | L | 0 | 1 | 2 | 3 | 4 | 5 | 6 | 7 | 8 | 9 |

### Proportional parts

| | 17 | | 16 | | 15 | | 14 |
|---|---|---|---|---|---|---|---|
| 1 | 1.7 | 1 | 1.6 | 1 | 1.5 | 1 | 1.4 |
| 2 | 3.4 | 2 | 3.2 | 2 | 3.0 | 2 | 2.8 |
| 3 | 5.1 | 3 | 4.8 | 3 | 4.5 | 3 | 4.2 |
| 4 | 6.8 | 4 | 6.4 | 4 | 6.0 | 4 | 5.6 |
| 5 | 8.5 | 5 | 8.0 | 5 | 7.5 | 5 | 7.0 |
| 6 | 10.2 | 6 | 9.6 | 6 | 9.0 | 6 | 8.4 |
| 7 | 11.9 | 7 | 11.2 | 7 | 10.5 | 7 | 9.8 |
| 8 | 13.6 | 8 | 12.8 | 8 | 12.0 | 8 | 11.2 |
| 9 | 15.3 | 9 | 14.4 | 9 | 13.5 | 9 | 12.6 |

* Indicates change in the first two decimal places.

## Table 1-8. Five-place Common Logarithms of Numbers (Continued)
### 320–375

| No. | L | 0 | 1 | 2 | 3 | 4 | 5 | 6 | 7 | 8 | 9 |
|---|---|---|---|---|---|---|---|---|---|---|---|
| 320 | 50 | 515 | 529 | 542 | 556 | 569 | 583 | 596 | 610 | 623 | 637 |
| 321 | | 651 | 664 | 678 | 691 | 705 | 718 | 732 | 745 | 759 | 772 |
| 322 | | 786 | 799 | 813 | 826 | 840 | 853 | 866 | 880 | 893 | 907 |
| 323 | | 920 | 934 | 947 | 961 | 974 | 987 | *001 | *014 | *028 | *041 |
| 324 | 51 | 055 | 068 | 081 | 095 | 108 | 121 | 135 | 148 | 162 | 175 |
| 325 | | 188 | 202 | 215 | 228 | 242 | 255 | 268 | 282 | 295 | 308 |
| 326 | | 322 | 335 | 348 | 362 | 375 | 388 | 402 | 415 | 428 | 441 |
| 327 | | 455 | 468 | 481 | 495 | 508 | 521 | 534 | 548 | 561 | 574 |
| 328 | | 587 | 601 | 614 | 627 | 640 | 654 | 667 | 680 | 693 | 706 |
| 329 | | 720 | 733 | 746 | 759 | 772 | 786 | 799 | 812 | 825 | 838 |
| 330 | | 851 | 865 | 878 | 891 | 904 | 917 | 930 | 943 | 957 | 970 |
| 331 | | 983 | 996 | *009 | *022 | *035 | *048 | *061 | *075 | *088 | *101 |
| 332 | 52 | 114 | 127 | 140 | 153 | 166 | 179 | 192 | 205 | 218 | 231 |
| 333 | | 244 | 257 | 271 | 284 | 297 | 310 | 323 | 336 | 349 | 362 |
| 334 | | 375 | 388 | 401 | 414 | 427 | 440 | 453 | 466 | 479 | 492 |
| 335 | | 504 | 517 | 530 | 543 | 556 | 569 | 582 | 595 | 608 | 621 |
| 336 | | 634 | 647 | 660 | 673 | 686 | 699 | 711 | 724 | 737 | 750 |
| 337 | | 763 | 776 | 789 | 802 | 815 | 827 | 840 | 853 | 866 | 879 |
| 338 | | 892 | 905 | 917 | 930 | 943 | 956 | 969 | 982 | 994 | *007 |
| 339 | 53 | 020 | 033 | 046 | 058 | 071 | 084 | 097 | 110 | 122 | 135 |
| 340 | | 148 | 161 | 173 | 186 | 199 | 212 | 224 | 237 | 250 | 263 |
| 341 | | 275 | 288 | 301 | 314 | 326 | 339 | 352 | 365 | 377 | 390 |
| 342 | | 403 | 415 | 428 | 441 | 453 | 466 | 479 | 491 | 504 | 517 |
| 343 | | 529 | 542 | 555 | 567 | 580 | 593 | 605 | 618 | 631 | 643 |
| 344 | | 656 | 668 | 681 | 694 | 706 | 719 | 732 | 744 | 757 | 769 |
| 345 | | 782 | 795 | 807 | 820 | 832 | 845 | 857 | 870 | 883 | 895 |
| 346 | | 908 | 920 | 933 | 945 | 958 | 970 | 983 | 995 | *008 | *020 |
| 347 | 54 | 033 | 045 | 058 | 070 | 083 | 095 | 108 | 120 | 133 | 145 |
| 348 | | 158 | 170 | 183 | 195 | 208 | 220 | 233 | 245 | 258 | 270 |
| 349 | | 283 | 295 | 307 | 320 | 332 | 345 | 357 | 370 | 382 | 394 |
| 350 | | 407 | 419 | 432 | 444 | 456 | 469 | 481 | 494 | 506 | 518 |
| 351 | | 531 | 543 | 555 | 568 | 580 | 593 | 605 | 617 | 630 | 642 |
| 352 | | 654 | 667 | 679 | 691 | 704 | 716 | 728 | 741 | 753 | 765 |
| 353 | | 777 | 790 | 802 | 814 | 827 | 839 | 851 | 864 | 876 | 888 |
| 354 | | 900 | 913 | 925 | 937 | 949 | 962 | 974 | 986 | 998 | *011 |
| 355 | 55 | 023 | 035 | 047 | 060 | 072 | 084 | 096 | 108 | 121 | 133 |
| 356 | | 145 | 157 | 169 | 182 | 194 | 206 | 218 | 230 | 242 | 255 |
| 357 | | 267 | 279 | 291 | 303 | 315 | 328 | 340 | 352 | 364 | 376 |
| 358 | | 388 | 400 | 413 | 425 | 437 | 449 | 461 | 473 | 485 | 497 |
| 359 | | 509 | 522 | 534 | 546 | 558 | 570 | 582 | 594 | 606 | 618 |
| 360 | | 630 | 642 | 654 | 666 | 678 | 691 | 703 | 715 | 727 | 739 |
| 361 | | 751 | 763 | 775 | 787 | 799 | 811 | 823 | 835 | 847 | 859 |
| 362 | | 871 | 883 | 895 | 907 | 919 | 931 | 943 | 955 | 967 | 979 |
| 363 | | 991 | *003 | *015 | *027 | *038 | *050 | *062 | *074 | *086 | *098 |
| 364 | 56 | 110 | 122 | 134 | 146 | 158 | 170 | 182 | 194 | 205 | 217 |
| 365 | | 229 | 241 | 253 | 265 | 277 | 289 | 301 | 313 | 324 | 336 |
| 366 | | 348 | 360 | 372 | 384 | 396 | 407 | 419 | 431 | 443 | 455 |
| 367 | | 467 | 478 | 490 | 502 | 514 | 526 | 538 | 549 | 561 | 573 |
| 368 | | 585 | 597 | 608 | 620 | 632 | 644 | 656 | 667 | 679 | 691 |
| 369 | | 703 | 714 | 726 | 738 | 750 | 761 | 773 | 785 | 797 | 808 |
| 370 | | 820 | 832 | 844 | 855 | 867 | 879 | 891 | 902 | 914 | 926 |
| 371 | | 937 | 949 | 961 | 972 | 984 | 996 | *008 | *019 | *031 | *043 |
| 372 | 57 | 054 | 066 | 078 | 089 | 101 | 113 | 124 | 136 | 148 | 159 |
| 373 | | 171 | 183 | 194 | 206 | 217 | 229 | 241 | 252 | 264 | 276 |
| 374 | | 287 | 299 | 310 | 322 | 334 | 345 | 357 | 368 | 380 | 392 |
| 375 | | 403 | 415 | 426 | 438 | 449 | 461 | 473 | 484 | 496 | 507 |
| No. | L | 0 | 1 | 2 | 3 | 4 | 5 | 6 | 7 | 8 | 9 |

**Proportional parts**

| | 14 | | 13 | | 12 |
|---|---|---|---|---|---|
| 1 | 1.4 | 1 | 1.3 | 1 | 1.2 |
| 2 | 2.8 | 2 | 2.6 | 2 | 2.4 |
| 3 | 4.2 | 3 | 3.9 | 3 | 3.6 |
| 4 | 5.6 | 4 | 5.2 | 4 | 4.8 |
| 5 | 7.0 | 5 | 6.5 | 5 | 6.0 |
| 6 | 8.4 | 6 | 7.8 | 6 | 7.2 |
| 7 | 9.8 | 7 | 9.1 | 7 | 8.4 |
| 8 | 11.2 | 8 | 10.4 | 8 | 9.6 |
| 9 | 12.6 | 9 | 11.7 | 9 | 10.8 |

* Indicates change in the first two decimal places.

## Table 1-8. Five-place Common Logarithms of Numbers (Continued)
### 375–430

| No. | L | 0 | 1 | 2 | 3 | 4 | 5 | 6 | 7 | 8 | 9 | Proportional parts |
|---|---|---|---|---|---|---|---|---|---|---|---|---|
| 375 | 57 | 403 | 415 | 426 | 438 | 449 | 461 | 473 | 484 | 496 | 507 | |
| 376 | | 519 | 530 | 542 | 553 | 565 | 577 | 588 | 600 | 611 | 623 | |
| 377 | | 634 | 646 | 657 | 669 | 680 | 692 | 703 | 715 | 726 | 738 | |
| 378 | | 749 | 761 | 772 | 784 | 795 | 807 | 818 | 830 | 841 | 852 | |
| 379 | | 864 | 875 | 887 | 898 | 910 | 921 | 933 | 944 | 956 | 967 | |
| 380 | | 978 | 990 | *001 | *013 | *024 | *035 | *047 | *058 | *070 | *081 | |
| 381 | 58 | 093 | 104 | 115 | 127 | 138 | 149 | 161 | 172 | 184 | 195 | |
| 382 | | 206 | 218 | 229 | 240 | 252 | 263 | 275 | 286 | 297 | 309 | |
| 383 | | 320 | 331 | 343 | 354 | 365 | 377 | 388 | 399 | 411 | 422 | |
| 384 | | 433 | 444 | 456 | 467 | 478 | 490 | 501 | 512 | 524 | 535 | |
| 385 | | 546 | 557 | 569 | 580 | 591 | 602 | 614 | 625 | 636 | 647 | |
| 386 | | 659 | 670 | 681 | 692 | 704 | 715 | 726 | 737 | 749 | 760 | |
| 387 | | 771 | 782 | 794 | 805 | 816 | 827 | 838 | 850 | 861 | 872 | |
| 388 | | 883 | 894 | 906 | 917 | 928 | 939 | 950 | 961 | 973 | 984 | |
| 389 | | 995 | *006 | *017 | *028 | *040 | *051 | *062 | *073 | *084 | *095 | **11** |
| 390 | 59 | 106 | 118 | 129 | 140 | 151 | 162 | 173 | 184 | 195 | 207 | 1 \| 1.1 |
| 391 | | 218 | 229 | 240 | 251 | 262 | 273 | 284 | 295 | 306 | 318 | 2 \| 2.2 |
| 392 | | 329 | 340 | 351 | 362 | 373 | 384 | 395 | 406 | 417 | 428 | 3 \| 3.3 |
| 393 | | 439 | 450 | 461 | 472 | 483 | 494 | 506 | 517 | 528 | 539 | 4 \| 4.4 |
| 394 | | 550 | 561 | 572 | 583 | 594 | 605 | 616 | 627 | 638 | 649 | 5 \| 5.5 |
| 395 | | 660 | 671 | 682 | 693 | 704 | 715 | 726 | 737 | 748 | 759 | 6 \| 6.6 |
| 396 | | 770 | 780 | 791 | 802 | 813 | 824 | 835 | 846 | 857 | 868 | 7 \| 7.7 |
| 397 | | 879 | 890 | 901 | 912 | 923 | 934 | 945 | 956 | 966 | 977 | 8 \| 8.8 |
| 398 | | 988 | 999 | *010 | *021 | *032 | *043 | *054 | *065 | *076 | *086 | 9 \| 9.9 |
| 399 | 60 | 097 | 108 | 119 | 130 | 141 | 152 | 163 | 173 | 184 | 195 | |
| 400 | | 206 | 217 | 228 | 239 | 249 | 260 | 271 | 282 | 293 | 304 | |
| 401 | | 314 | 325 | 336 | 347 | 358 | 369 | 379 | 390 | 401 | 412 | |
| 402 | | 423 | 433 | 444 | 455 | 466 | 477 | 487 | 498 | 509 | 520 | |
| 403 | | 531 | 541 | 552 | 563 | 574 | 584 | 595 | 606 | 617 | 627 | |
| 404 | | 638 | 649 | 660 | 670 | 681 | 692 | 703 | 713 | 724 | 735 | |
| 405 | | 746 | 756 | 767 | 778 | 788 | 799 | 810 | 821 | 831 | 842 | |
| 406 | | 853 | 863 | 874 | 885 | 895 | 906 | 917 | 927 | 938 | 949 | |
| 407 | | 959 | 970 | 981 | 991 | *002 | *013 | *023 | *034 | *045 | *055 | |
| 408 | 61 | 066 | 077 | 087 | 098 | 109 | 119 | 130 | 140 | 151 | 162 | |
| 409 | | 172 | 183 | 194 | 204 | 215 | 225 | 236 | 247 | 257 | 268 | |
| 410 | | 278 | 289 | 300 | 310 | 321 | 331 | 342 | 352 | 363 | 374 | **10** |
| 411 | | 384 | 395 | 405 | 416 | 426 | 437 | 448 | 458 | 469 | 479 | 1 \| 1.0 |
| 412 | | 490 | 500 | 511 | 521 | 532 | 542 | 553 | 563 | 574 | 584 | 2 \| 2.0 |
| 413 | | 595 | 606 | 616 | 627 | 637 | 648 | 658 | 669 | 679 | 690 | 3 \| 3.0 |
| 414 | | 700 | 711 | 721 | 731 | 742 | 752 | 763 | 773 | 784 | 794 | 4 \| 4.0 |
| 415 | | 805 | 815 | 826 | 836 | 847 | 857 | 868 | 878 | 888 | 899 | 5 \| 5.0 |
| 416 | | 909 | 920 | 930 | 941 | 951 | 962 | 972 | 982 | 993 | *003 | 6 \| 6.0 |
| 417 | 62 | 014 | 024 | 034 | 045 | 055 | 066 | 076 | 086 | 097 | 107 | 7 \| 7.0 |
| 418 | | 118 | 128 | 138 | 149 | 159 | 170 | 180 | 190 | 201 | 211 | 8 \| 8.0 |
| 419 | | 221 | 232 | 242 | 252 | 263 | 273 | 284 | 294 | 304 | 315 | 9 \| 9.0 |
| 420 | | 325 | 335 | 346 | 356 | 366 | 377 | 387 | 397 | 408 | 418 | |
| 421 | | 428 | 439 | 449 | 459 | 469 | 480 | 490 | 500 | 511 | 521 | |
| 422 | | 531 | 542 | 552 | 562 | 572 | 583 | 593 | 603 | 614 | 624 | |
| 423 | | 634 | 644 | 655 | 665 | 675 | 685 | 696 | 706 | 716 | 726 | |
| 424 | | 737 | 747 | 757 | 767 | 778 | 788 | 798 | 808 | 818 | 829 | |
| 425 | | 839 | 849 | 859 | 870 | 880 | 890 | 900 | 910 | 921 | 931 | |
| 426 | | 941 | 951 | 961 | 972 | 982 | 992 | *002 | *012 | *022 | *033 | |
| 427 | 63 | 043 | 053 | 063 | 073 | 083 | 094 | 104 | 114 | 124 | 134 | |
| 428 | | 144 | 155 | 165 | 175 | 185 | 195 | 205 | 215 | 225 | 236 | |
| 429 | | 246 | 256 | 266 | 276 | 286 | 296 | 306 | 317 | 327 | 337 | |
| 430 | | 347 | 357 | 367 | 377 | 387 | 397 | 407 | 417 | 428 | 438 | |
| No. | L | 0 | 1 | 2 | 3 | 4 | 5 | 6 | 7 | 8 | 9 | Proportional parts |

* Indicates change in the first two decimal places.

## Table 1-8. Five-place Common Logarithms of Numbers (Continued)
### 430–485

| No. | L | 0 | 1 | 2 | 3 | 4 | 5 | 6 | 7 | 8 | 9 |
|---|---|---|---|---|---|---|---|---|---|---|---|
| 430 | 63 | 347 | 357 | 367 | 377 | 387 | 397 | 407 | 417 | 428 | 438 |
| 431 |  | 448 | 458 | 468 | 478 | 488 | 498 | 508 | 518 | 528 | 538 |
| 432 |  | 548 | 558 | 568 | 579 | 589 | 599 | 609 | 619 | 629 | 639 |
| 433 |  | 649 | 659 | 669 | 679 | 689 | 699 | 709 | 719 | 729 | 739 |
| 434 |  | 749 | 759 | 769 | 779 | 789 | 799 | 809 | 819 | 829 | 839 |
| 435 |  | 849 | 859 | 869 | 879 | 889 | 899 | 909 | 919 | 929 | 939 |
| 436 |  | 949 | 959 | 969 | 979 | 988 | 998 | *008 | *018 | *028 | *038 |
| 437 | 64 | 048 | 058 | 068 | 078 | 088 | 098 | 108 | 118 | 128 | 137 |
| 438 |  | 147 | 157 | 167 | 177 | 187 | 197 | 207 | 217 | 227 | 237 |
| 439 |  | 246 | 256 | 266 | 276 | 286 | 296 | 306 | 316 | 326 | 335 |
| 440 |  | 345 | 355 | 365 | 375 | 385 | 395 | 404 | 414 | 424 | 434 |
| 441 |  | 444 | 454 | 464 | 473 | 483 | 493 | 503 | 513 | 523 | 532 |
| 442 |  | 542 | 552 | 562 | 572 | 582 | 591 | 601 | 611 | 621 | 631 |
| 443 |  | 640 | 650 | 660 | 670 | 680 | 689 | 699 | 709 | 719 | 729 |
| 444 |  | 738 | 748 | 758 | 768 | 777 | 787 | 797 | 807 | 816 | 826 |
| 445 |  | 836 | 846 | 856 | 865 | 875 | 885 | 895 | 904 | 914 | 924 |
| 446 |  | 933 | 943 | 953 | 963 | 972 | 982 | 992 | *002 | *011 | *021 |
| 447 | 65 | 031 | 040 | 050 | 060 | 070 | 079 | 089 | 099 | 108 | 118 |
| 448 |  | 128 | 137 | 147 | 157 | 167 | 176 | 186 | 196 | 205 | 215 |
| 449 |  | 225 | 234 | 244 | 254 | 263 | 273 | 283 | 292 | 302 | 312 |
| 450 |  | 321 | 331 | 341 | 350 | 360 | 369 | 379 | 389 | 398 | 408 |
| 451 |  | 418 | 427 | 437 | 447 | 456 | 466 | 475 | 485 | 495 | 504 |
| 452 |  | 514 | 523 | 533 | 543 | 552 | 562 | 571 | 581 | 591 | 600 |
| 453 |  | 610 | 619 | 629 | 639 | 648 | 658 | 667 | 677 | 686 | 696 |
| 454 |  | 706 | 715 | 725 | 734 | 744 | 753 | 763 | 773 | 782 | 792 |
| 455 |  | 801 | 811 | 820 | 830 | 839 | 849 | 858 | 868 | 877 | 887 |
| 456 |  | 896 | 906 | 916 | 925 | 935 | 944 | 954 | 963 | 973 | 982 |
| 457 | 66 | 992 | *001 | *011 | *020 | *030 | *039 | *049 | *058 | *068 | *077 |
| 458 |  | 087 | 096 | 106 | 115 | 124 | 134 | 143 | 153 | 162 | 172 |
| 459 |  | 181 | 191 | 200 | 210 | 219 | 229 | 238 | 247 | 257 | 266 |
| 460 |  | 276 | 285 | 295 | 304 | 314 | 323 | 332 | 342 | 351 | 361 |
| 461 |  | 370 | 380 | 389 | 398 | 408 | 417 | 427 | 436 | 445 | 455 |
| 462 |  | 464 | 474 | 483 | 492 | 502 | 511 | 521 | 530 | 539 | 549 |
| 463 |  | 558 | 567 | 577 | 586 | 596 | 605 | 614 | 624 | 633 | 642 |
| 464 |  | 652 | 661 | 671 | 680 | 689 | 699 | 708 | 717 | 727 | 736 |
| 465 |  | 745 | 755 | 764 | 773 | 783 | 792 | 801 | 811 | 820 | 829 |
| 466 |  | 839 | 848 | 857 | 867 | 876 | 885 | 894 | 904 | 913 | 922 |
| 467 |  | 932 | 941 | 950 | 960 | 969 | 978 | 987 | 997 | *006 | *015 |
| 468 | 67 | 025 | 034 | 043 | 052 | 062 | 071 | 080 | 090 | 099 | 108 |
| 469 |  | 117 | 127 | 136 | 145 | 154 | 164 | 173 | 182 | 191 | 201 |
| 470 |  | 210 | 219 | 228 | 238 | 247 | 256 | 265 | 274 | 284 | 293 |
| 471 |  | 302 | 311 | 321 | 330 | 339 | 348 | 357 | 367 | 376 | 385 |
| 472 |  | 394 | 403 | 413 | 422 | 431 | 440 | 449 | 459 | 468 | 477 |
| 473 |  | 486 | 495 | 504 | 514 | 523 | 532 | 541 | 550 | 560 | 569 |
| 474 |  | 578 | 587 | 596 | 605 | 614 | 624 | 633 | 642 | 651 | 660 |
| 475 |  | 669 | 679 | 688 | 697 | 706 | 715 | 724 | 733 | 742 | 752 |
| 476 |  | 761 | 770 | 779 | 788 | 797 | 806 | 815 | 825 | 834 | 843 |
| 477 |  | 852 | 861 | 870 | 879 | 888 | 897 | 906 | 916 | 925 | 934 |
| 478 |  | 943 | 952 | 961 | 970 | 979 | 988 | 997 | *006 | *015 | *024 |
| 479 | 68 | 034 | 043 | 052 | 061 | 070 | 079 | 088 | 097 | 106 | 115 |
| 480 |  | 124 | 133 | 142 | 151 | 160 | 169 | 178 | 187 | 196 | 205 |
| 481 |  | 215 | 224 | 233 | 242 | 251 | 260 | 269 | 278 | 287 | 296 |
| 482 |  | 305 | 314 | 323 | 332 | 341 | 350 | 359 | 368 | 377 | 386 |
| 483 |  | 395 | 404 | 413 | 422 | 431 | 440 | 449 | 458 | 467 | 476 |
| 484 |  | 485 | 494 | 502 | 511 | 520 | 529 | 538 | 547 | 556 | 565 |
| 485 |  | 574 | 583 | 592 | 601 | 610 | 619 | 628 | 637 | 646 | 655 |
| No. | L | 0 | 1 | 2 | 3 | 4 | 5 | 6 | 7 | 8 | 9 |

**Proportional parts**

| | 10 |
|---|---|
| 1 | 1.0 |
| 2 | 2.0 |
| 3 | 3.0 |
| 4 | 4.0 |
| 5 | 5.0 |
| 6 | 6.0 |
| 7 | 7.0 |
| 8 | 8.0 |
| 9 | 9.0 |

| | 9 |
|---|---|
| 1 | 0.9 |
| 2 | 1.8 |
| 3 | 2.7 |
| 4 | 3.6 |
| 5 | 4.5 |
| 6 | 5.4 |
| 7 | 6.3 |
| 8 | 7.2 |
| 9 | 8.1 |

* Indicates change in the first two decimal places.

MATHEMATICAL TABLES AND MATHEMATICS

## Table 1-8. Five-place Common Logarithms of Numbers (Continued)
### 485–540

| No. | L | 0 | 1 | 2 | 3 | 4 | 5 | 6 | 7 | 8 | 9 |
|-----|-----|-----|-----|-----|-----|-----|-----|-----|-----|-----|-----|
| 485 | 68 | 574 | 583 | 592 | 601 | 610 | 619 | 628 | 637 | 646 | 655 |
| 486 | | 664 | 673 | 682 | 690 | 699 | 708 | 717 | 726 | 735 | 744 |
| 487 | | 753 | 762 | 771 | 780 | 789 | 797 | 806 | 815 | 824 | 833 |
| 488 | | 842 | 851 | 860 | 869 | 878 | 886 | 895 | 904 | 913 | 922 |
| 489 | | 931 | 940 | 949 | 958 | 966 | 975 | 984 | 993 | *002 | *011 |
| 490 | 69 | 020 | 028 | 037 | 046 | 055 | 064 | 073 | 082 | 090 | 099 |
| 491 | | 108 | 117 | 126 | 135 | 144 | 152 | 161 | 170 | 179 | 188 |
| 492 | | 197 | 205 | 214 | 223 | 232 | 241 | 249 | 258 | 267 | 276 |
| 493 | | 285 | 294 | 302 | 311 | 320 | 329 | 338 | 346 | 355 | 364 |
| 494 | | 373 | 381 | 390 | 399 | 408 | 417 | 425 | 434 | 443 | 452 |
| 495 | | 461 | 469 | 478 | 487 | 496 | 504 | 513 | 522 | 531 | 539 |
| 496 | | 548 | 557 | 566 | 574 | 583 | 592 | 601 | 609 | 618 | 627 |
| 497 | | 636 | 644 | 653 | 662 | 671 | 679 | 688 | 697 | 705 | 714 |
| 498 | | 723 | 732 | 740 | 749 | 758 | 767 | 775 | 784 | 793 | 801 |
| 499 | | 810 | 819 | 827 | 836 | 845 | 854 | 862 | 871 | 880 | 888 |
| 500 | | 897 | 906 | 914 | 923 | 932 | 940 | 949 | 958 | 966 | 975 |
| 501 | | 984 | 992 | *001 | *010 | *018 | *027 | *036 | *044 | *053 | *062 |
| 502 | 70 | 070 | 079 | 088 | 096 | 105 | 114 | 122 | 131 | 140 | 148 |
| 503 | | 157 | 165 | 174 | 183 | 191 | 200 | 209 | 217 | 226 | 234 |
| 504 | | 243 | 252 | 260 | 269 | 278 | 286 | 295 | 303 | 312 | 321 |
| 505 | | 329 | 338 | 346 | 355 | 364 | 372 | 381 | 389 | 398 | 406 |
| 506 | | 415 | 424 | 432 | 441 | 449 | 458 | 467 | 475 | 484 | 492 |
| 507 | | 501 | 509 | 518 | 526 | 535 | 544 | 552 | 561 | 569 | 578 |
| 508 | | 586 | 595 | 603 | 612 | 621 | 629 | 638 | 646 | 655 | 663 |
| 509 | | 672 | 680 | 689 | 697 | 706 | 714 | 723 | 731 | 740 | 749 |
| 510 | | 757 | 766 | 774 | 783 | 791 | 800 | 808 | 817 | 825 | 834 |
| 511 | | 842 | 851 | 859 | 868 | 876 | 885 | 893 | 902 | 910 | 919 |
| 512 | | 927 | 935 | 944 | 952 | 961 | 969 | 978 | 986 | 995 | *003 |
| 513 | 71 | 012 | 020 | 029 | 037 | 046 | 054 | 063 | 071 | 079 | 088 |
| 514 | | 096 | 105 | 113 | 122 | 130 | 139 | 147 | 155 | 164 | 172 |
| 515 | | 181 | 189 | 198 | 206 | 214 | 223 | 231 | 240 | 248 | 257 |
| 516 | | 265 | 273 | 282 | 290 | 299 | 307 | 315 | 324 | 332 | 341 |
| 517 | | 349 | 357 | 366 | 374 | 383 | 391 | 399 | 408 | 416 | 425 |
| 518 | | 433 | 441 | 450 | 458 | 467 | 475 | 483 | 492 | 500 | 508 |
| 519 | | 517 | 525 | 533 | 542 | 550 | 559 | 567 | 575 | 584 | 592 |
| 520 | | 600 | 609 | 617 | 625 | 634 | 642 | 650 | 659 | 667 | 675 |
| 521 | | 684 | 692 | 700 | 709 | 717 | 725 | 734 | 742 | 750 | 759 |
| 522 | | 767 | 775 | 784 | 792 | 800 | 809 | 817 | 825 | 834 | 842 |
| 523 | | 850 | 858 | 867 | 875 | 883 | 892 | 900 | 908 | 917 | 925 |
| 524 | | 933 | 941 | 950 | 958 | 966 | 975 | 983 | 991 | 999 | *008 |
| 525 | 72 | 016 | 024 | 032 | 041 | 049 | 057 | 066 | 074 | 082 | 090 |
| 526 | | 099 | 107 | 115 | 123 | 132 | 140 | 148 | 156 | 165 | 173 |
| 527 | | 181 | 189 | 198 | 206 | 214 | 222 | 230 | 239 | 247 | 255 |
| 528 | | 263 | 272 | 280 | 288 | 296 | 305 | 313 | 321 | 329 | 337 |
| 529 | | 346 | 354 | 362 | 370 | 378 | 387 | 395 | 403 | 411 | 419 |
| 530 | | 428 | 436 | 444 | 452 | 460 | 469 | 477 | 485 | 493 | 501 |
| 531 | | 509 | 518 | 526 | 534 | 542 | 550 | 559 | 567 | 575 | 583 |
| 532 | | 591 | 599 | 607 | 616 | 624 | 632 | 640 | 648 | 656 | 665 |
| 533 | | 673 | 681 | 689 | 697 | 705 | 713 | 722 | 730 | 738 | 746 |
| 534 | | 754 | 762 | 770 | 779 | 787 | 795 | 803 | 811 | 819 | 827 |
| 535 | | 835 | 844 | 852 | 860 | 868 | 876 | 884 | 892 | 900 | 908 |
| 536 | | 916 | 925 | 933 | 941 | 949 | 957 | 965 | 973 | 981 | 989 |
| 537 | | 997 | *006 | *014 | *022 | *030 | *038 | *046 | *054 | *062 | *070 |
| 538 | 73 | 078 | 086 | 094 | 102 | 111 | 119 | 127 | 135 | 143 | 151 |
| 539 | | 159 | 167 | 175 | 183 | 191 | 199 | 207 | 215 | 223 | 231 |
| 540 | | 239 | 247 | 255 | 264 | 272 | 280 | 288 | 296 | 304 | 312 |
| No. | L | 0 | 1 | 2 | 3 | 4 | 5 | 6 | 7 | 8 | 9 |

Proportional parts

| 9 | |
|---|---|
| 1 | 0.9 |
| 2 | 1.8 |
| 3 | 2.7 |
| 4 | 3.6 |
| 5 | 4.5 |
| 6 | 5.4 |
| 7 | 6.3 |
| 8 | 7.2 |
| 9 | 8.1 |

| 8 | |
|---|---|
| 1 | 0.8 |
| 2 | 1.6 |
| 3 | 2.4 |
| 4 | 3.2 |
| 5 | 4.0 |
| 6 | 4.8 |
| 7 | 5.6 |
| 8 | 6.4 |
| 9 | 7.2 |

* Indicates change in the first two decimal places.

## Table 1-8. Five-place Common Logarithms of Numbers (Continued)
### 540–595

| No. | L | 0 | 1 | 2 | 3 | 4 | 5 | 6 | 7 | 8 | 9 |
|---|---|---|---|---|---|---|---|---|---|---|---|
| 540 | 73 | 239 | 247 | 255 | 264 | 272 | 280 | 288 | 296 | 304 | 312 |
| 541 |   | 320 | 328 | 336 | 344 | 352 | 360 | 368 | 376 | 384 | 392 |
| 542 |   | 400 | 408 | 416 | 424 | 432 | 440 | 448 | 456 | 464 | 472 |
| 543 |   | 480 | 488 | 496 | 504 | 512 | 520 | 528 | 536 | 544 | 552 |
| 544 |   | 560 | 568 | 576 | 584 | 592 | 600 | 608 | 616 | 624 | 632 |
| 545 |   | 640 | 648 | 656 | 664 | 672 | 679 | 687 | 695 | 703 | 711 |
| 546 |   | 719 | 727 | 735 | 743 | 751 | 759 | 767 | 775 | 783 | 791 |
| 547 |   | 799 | 807 | 815 | 823 | 830 | 838 | 846 | 854 | 862 | 870 |
| 548 |   | 878 | 886 | 894 | 902 | 910 | 918 | 926 | 934 | 941 | 949 |
| 549 |   | 957 | 965 | 973 | 981 | 989 | 997 | *005 | *013 | *020 | *028 |
| 550 | 74 | 036 | 044 | 052 | 060 | 068 | 076 | 084 | 092 | 099 | 107 |
| 551 |   | 115 | 123 | 131 | 139 | 147 | 155 | 162 | 170 | 178 | 186 |
| 552 |   | 194 | 202 | 210 | 218 | 225 | 233 | 241 | 249 | 257 | 265 |
| 553 |   | 273 | 280 | 288 | 296 | 304 | 312 | 320 | 327 | 335 | 343 |
| 554 |   | 351 | 359 | 367 | 374 | 382 | 390 | 398 | 406 | 414 | 421 |
| 555 |   | 429 | 437 | 445 | 453 | 461 | 468 | 476 | 484 | 492 | 500 |
| 556 |   | 507 | 515 | 523 | 531 | 539 | 547 | 554 | 562 | 570 | 578 |
| 557 |   | 586 | 593 | 601 | 609 | 617 | 624 | 632 | 640 | 648 | 656 |
| 558 |   | 663 | 671 | 679 | 687 | 695 | 702 | 710 | 718 | 726 | 733 |
| 559 |   | 741 | 749 | 757 | 764 | 772 | 780 | 788 | 796 | 803 | 811 |
| 560 |   | 819 | 827 | 834 | 842 | 850 | 858 | 865 | 873 | 881 | 889 |
| 561 |   | 896 | 904 | 912 | 920 | 927 | 935 | 943 | 950 | 958 | 966 |
| 562 |   | 974 | 981 | 989 | 997 | *005 | *012 | *020 | *028 | *035 | *043 |
| 563 | 75 | 051 | 059 | 066 | 074 | 082 | 089 | 097 | 105 | 113 | 120 |
| 564 |   | 128 | 136 | 143 | 151 | 159 | 166 | 174 | 182 | 189 | 197 |
| 565 |   | 205 | 213 | 220 | 228 | 236 | 243 | 251 | 259 | 266 | 274 |
| 566 |   | 282 | 289 | 297 | 305 | 312 | 320 | 328 | 335 | 343 | 351 |
| 567 |   | 358 | 366 | 374 | 381 | 389 | 397 | 404 | 412 | 420 | 427 |
| 568 |   | 435 | 442 | 450 | 458 | 465 | 473 | 481 | 488 | 496 | 504 |
| 569 |   | 511 | 519 | 526 | 534 | 542 | 549 | 557 | 565 | 572 | 580 |
| 570 |   | 587 | 595 | 603 | 610 | 618 | 626 | 633 | 641 | 648 | 656 |
| 571 |   | 664 | 671 | 679 | 686 | 694 | 702 | 709 | 717 | 724 | 732 |
| 572 |   | 740 | 747 | 755 | 762 | 770 | 778 | 785 | 793 | 800 | 808 |
| 573 |   | 815 | 823 | 831 | 838 | 846 | 853 | 861 | 868 | 876 | 884 |
| 574 |   | 891 | 899 | 906 | 914 | 921 | 929 | 937 | 944 | 952 | 959 |
| 575 |   | 967 | 974 | 982 | 989 | 997 | *005 | *012 | *020 | *027 | *035 |
| 576 | 76 | 042 | 050 | 057 | 065 | 072 | 080 | 087 | 095 | 103 | 110 |
| 577 |   | 118 | 125 | 133 | 140 | 148 | 155 | 163 | 170 | 178 | 185 |
| 578 |   | 193 | 200 | 208 | 215 | 223 | 230 | 238 | 245 | 253 | 260 |
| 579 |   | 268 | 275 | 283 | 290 | 298 | 305 | 313 | 320 | 328 | 335 |
| 580 |   | 343 | 350 | 358 | 365 | 373 | 380 | 388 | 395 | 403 | 410 |
| 581 |   | 418 | 425 | 433 | 440 | 448 | 455 | 462 | 470 | 477 | 485 |
| 582 |   | 492 | 500 | 507 | 515 | 522 | 530 | 537 | 545 | 552 | 559 |
| 583 |   | 567 | 574 | 582 | 589 | 597 | 604 | 612 | 619 | 626 | 634 |
| 584 |   | 641 | 649 | 656 | 664 | 671 | 678 | 686 | 693 | 701 | 708 |
| 585 |   | 716 | 723 | 730 | 738 | 745 | 753 | 760 | 768 | 775 | 782 |
| 586 |   | 790 | 797 | 805 | 812 | 819 | 827 | 834 | 842 | 849 | 856 |
| 587 |   | 864 | 871 | 879 | 886 | 893 | 901 | 908 | 916 | 923 | 930 |
| 588 |   | 938 | 945 | 953 | 960 | 967 | 975 | 982 | 989 | 997 | *004 |
| 589 | 77 | 012 | 019 | 026 | 034 | 041 | 048 | 056 | 063 | 070 | 078 |
| 590 |   | 085 | 093 | 100 | 107 | 115 | 122 | 129 | 137 | 144 | 151 |
| 591 |   | 159 | 166 | 173 | 181 | 188 | 195 | 203 | 210 | 218 | 225 |
| 592 |   | 232 | 240 | 247 | 254 | 262 | 269 | 276 | 283 | 291 | 298 |
| 593 |   | 305 | 313 | 320 | 327 | 335 | 342 | 349 | 357 | 364 | 371 |
| 594 |   | 379 | 386 | 393 | 401 | 408 | 415 | 422 | 430 | 437 | 444 |
| 595 |   | 452 | 459 | 466 | 474 | 481 | 488 | 495 | 503 | 510 | 517 |
| No. | L | 0 | 1 | 2 | 3 | 4 | 5 | 6 | 7 | 8 | 9 |

**Proportional parts**

| | 9 |
|---|---|
| 1 | 0.9 |
| 2 | 1.8 |
| 3 | 2.7 |
| 4 | 3.6 |
| 5 | 4.5 |
| 6 | 5.4 |
| 7 | 6.3 |
| 8 | 7.2 |
| 9 | 8.1 |

| | 8 |
|---|---|
| 1 | 0.8 |
| 2 | 1.6 |
| 3 | 2.4 |
| 4 | 3.2 |
| 5 | 4.0 |
| 6 | 4.8 |
| 7 | 5.6 |
| 8 | 6.4 |
| 9 | 7.2 |

| | 7 |
|---|---|
| 1 | 0.7 |
| 2 | 1.4 |
| 3 | 2.1 |
| 4 | 2.8 |
| 5 | 3.5 |
| 6 | 4.2 |
| 7 | 4.9 |
| 8 | 5.6 |
| 9 | 6.3 |

* Indicates change in the first two decimal places.

## Table 1-8. Five-place Common Logarithms of Numbers (Continued)
### 595–650

| No. | L | 0 | 1 | 2 | 3 | 4 | 5 | 6 | 7 | 8 | 9 |
|-----|----|-----|-----|-----|-----|-----|-----|-----|-----|-----|-----|
| 595 | 77 | 452 | 459 | 466 | 474 | 481 | 488 | 495 | 503 | 510 | 517 |
| 596 |    | 525 | 532 | 539 | 546 | 554 | 561 | 568 | 576 | 583 | 590 |
| 597 |    | 597 | 605 | 612 | 619 | 627 | 634 | 641 | 648 | 656 | 663 |
| 598 |    | 670 | 677 | 685 | 692 | 699 | 706 | 714 | 721 | 728 | 735 |
| 599 |    | 743 | 750 | 757 | 764 | 772 | 779 | 786 | 793 | 801 | 808 |
| 600 |    | 815 | 822 | 830 | 837 | 844 | 851 | 859 | 866 | 873 | 880 |
| 601 |    | 887 | 895 | 902 | 909 | 916 | 924 | 931 | 938 | 945 | 952 |
| 602 |    | 960 | 967 | 974 | 981 | 989 | 996 | *003 | *010 | *017 | *025 |
| 603 | 78 | 032 | 039 | 046 | 053 | 061 | 068 | 075 | 082 | 089 | 097 |
| 604 |    | 104 | 111 | 118 | 125 | 132 | 140 | 147 | 154 | 161 | 168 |
| 605 |    | 176 | 183 | 190 | 197 | 204 | 211 | 219 | 226 | 233 | 240 |
| 606 |    | 247 | 254 | 262 | 269 | 276 | 283 | 290 | 297 | 305 | 312 |
| 607 |    | 319 | 326 | 333 | 340 | 347 | 355 | 362 | 369 | 376 | 383 |
| 608 |    | 390 | 398 | 405 | 412 | 419 | 426 | 433 | 440 | 447 | 455 |
| 609 |    | 462 | 469 | 476 | 483 | 490 | 497 | 505 | 512 | 519 | 526 |
| 610 |    | 533 | 540 | 547 | 554 | 561 | 569 | 576 | 583 | 590 | 597 |
| 611 |    | 604 | 611 | 618 | 625 | 633 | 640 | 647 | 654 | 661 | 668 |
| 612 |    | 675 | 682 | 689 | 696 | 704 | 711 | 718 | 725 | 732 | 739 |
| 613 |    | 746 | 753 | 760 | 767 | 774 | 781 | 789 | 796 | 803 | 810 |
| 614 |    | 817 | 824 | 831 | 838 | 845 | 852 | 859 | 866 | 873 | 880 |
| 615 |    | 888 | 895 | 902 | 909 | 916 | 923 | 930 | 937 | 944 | 951 |
| 616 |    | 958 | 965 | 972 | 979 | 986 | 993 | *000 | *007 | *014 | *021 |
| 617 | 79 | 029 | 036 | 043 | 050 | 057 | 064 | 071 | 078 | 085 | 092 |
| 618 |    | 099 | 106 | 113 | 120 | 127 | 134 | 141 | 148 | 155 | 162 |
| 619 |    | 169 | 176 | 183 | 190 | 197 | 204 | 211 | 218 | 225 | 232 |
| 620 |    | 239 | 246 | 253 | 260 | 267 | 274 | 281 | 288 | 295 | 302 |
| 621 |    | 309 | 316 | 323 | 330 | 337 | 344 | 351 | 358 | 365 | 372 |
| 622 |    | 379 | 386 | 393 | 400 | 407 | 414 | 421 | 428 | 435 | 442 |
| 623 |    | 449 | 456 | 463 | 470 | 477 | 484 | 491 | 498 | 505 | 512 |
| 624 |    | 518 | 525 | 532 | 539 | 546 | 553 | 560 | 567 | 574 | 581 |
| 625 |    | 588 | 595 | 602 | 609 | 616 | 623 | 630 | 637 | 644 | 651 |
| 626 |    | 657 | 664 | 671 | 678 | 685 | 692 | 699 | 706 | 713 | 720 |
| 627 |    | 727 | 734 | 741 | 748 | 754 | 761 | 768 | 775 | 782 | 789 |
| 628 |    | 796 | 803 | 810 | 817 | 824 | 831 | 837 | 844 | 851 | 858 |
| 629 |    | 865 | 872 | 879 | 886 | 893 | 900 | 906 | 913 | 920 | 927 |
| 630 |    | 934 | 941 | 948 | 955 | 962 | 969 | 975 | 982 | 989 | 996 |
| 631 | 80 | 003 | 010 | 017 | 024 | 030 | 037 | 044 | 051 | 058 | 065 |
| 632 |    | 072 | 079 | 085 | 092 | 099 | 106 | 113 | 120 | 127 | 134 |
| 633 |    | 140 | 147 | 154 | 161 | 168 | 175 | 182 | 188 | 195 | 202 |
| 634 |    | 209 | 216 | 223 | 229 | 236 | 243 | 250 | 257 | 264 | 271 |
| 635 |    | 277 | 284 | 291 | 298 | 305 | 312 | 318 | 325 | 332 | 339 |
| 636 |    | 346 | 353 | 359 | 366 | 373 | 380 | 387 | 393 | 400 | 407 |
| 637 |    | 414 | 421 | 428 | 434 | 441 | 448 | 455 | 462 | 468 | 475 |
| 638 |    | 482 | 489 | 496 | 502 | 509 | 516 | 523 | 530 | 536 | 543 |
| 639 |    | 550 | 557 | 564 | 570 | 577 | 584 | 591 | 598 | 604 | 611 |
| 640 |    | 618 | 625 | 632 | 638 | 645 | 652 | 659 | 665 | 672 | 679 |
| 641 |    | 686 | 693 | 699 | 706 | 713 | 720 | 726 | 733 | 740 | 747 |
| 642 |    | 754 | 760 | 767 | 774 | 781 | 787 | 794 | 801 | 808 | 814 |
| 643 |    | 821 | 828 | 835 | 841 | 848 | 855 | 862 | 868 | 875 | 882 |
| 644 |    | 889 | 895 | 902 | 909 | 916 | 922 | 929 | 936 | 943 | 949 |
| 645 |    | 956 | 963 | 969 | 976 | 983 | 990 | 996 | *003 | *010 | *017 |
| 646 | 81 | 023 | 030 | 037 | 043 | 050 | 057 | 064 | 070 | 077 | 084 |
| 647 |    | 090 | 097 | 104 | 111 | 117 | 124 | 131 | 137 | 144 | 151 |
| 648 |    | 158 | 164 | 171 | 178 | 184 | 191 | 198 | 204 | 211 | 218 |
| 649 |    | 224 | 231 | 238 | 245 | 251 | 258 | 265 | 271 | 278 | 285 |
| 650 |    | 291 | 298 | 305 | 311 | 318 | 325 | 331 | 338 | 345 | 351 |
| No. | L | 0 | 1 | 2 | 3 | 4 | 5 | 6 | 7 | 8 | 9 |

Proportional parts

| 8 | |
|---|-----|
| 1 | 0.8 |
| 2 | 1.6 |
| 3 | 2.4 |
| 4 | 3.2 |
| 5 | 4.0 |
| 6 | 4.8 |
| 7 | 5.6 |
| 8 | 6.4 |
| 9 | 7.2 |

| 7 | |
|---|-----|
| 1 | 0.7 |
| 2 | 1.4 |
| 3 | 2.1 |
| 4 | 2.8 |
| 5 | 3.5 |
| 6 | 4.2 |
| 7 | 4.9 |
| 8 | 5.6 |
| 9 | 6.3 |

* Indicates change in the first two decimal places.

## Table 1-8. Five-place Common Logarithms of Numbers (Continued)
### 650–705

| No. | L | 0 | 1 | 2 | 3 | 4 | 5 | 6 | 7 | 8 | 9 | Proportional parts |
|-----|---|---|---|---|---|---|---|---|---|---|---|--------------------|
| 650 | 81 | 291 | 298 | 305 | 311 | 318 | 325 | 331 | 338 | 345 | 351 | |
| 651 | | 358 | 365 | 371 | 378 | 385 | 391 | 398 | 405 | 411 | 418 | |
| 652 | | 425 | 431 | 438 | 445 | 451 | 458 | 465 | 471 | 478 | 485 | |
| 653 | | 491 | 498 | 505 | 511 | 518 | 525 | 531 | 538 | 544 | 551 | |
| 654 | | 558 | 564 | 571 | 578 | 584 | 591 | 598 | 604 | 611 | 618 | |
| 655 | | 624 | 631 | 637 | 644 | 651 | 657 | 664 | 671 | 677 | 684 | |
| 656 | | 690 | 697 | 704 | 710 | 717 | 723 | 730 | 737 | 743 | 750 | |
| 657 | | 757 | 763 | 770 | 776 | 783 | 790 | 796 | 803 | 809 | 816 | |
| 658 | | 823 | 829 | 836 | 842 | 849 | 856 | 862 | 869 | 875 | 882 | |
| 659 | | 889 | 895 | 902 | 908 | 915 | 921 | 928 | 935 | 941 | 948 | |
| 660 | | 954 | 961 | 968 | 974 | 981 | 987 | 994 | *000 | *007 | *014 | |
| 661 | 82 | 020 | 027 | 033 | 040 | 046 | 053 | 060 | 066 | 073 | 079 | |
| 662 | | 086 | 092 | 099 | 105 | 112 | 119 | 125 | 132 | 138 | 145 | |
| 663 | | 151 | 158 | 164 | 171 | 178 | 184 | 191 | 197 | 204 | 210 | |
| 664 | | 217 | 223 | 230 | 236 | 243 | 250 | 256 | 263 | 269 | 276 | |
| 665 | | 282 | 289 | 295 | 302 | 308 | 315 | 321 | 328 | 334 | 341 | |
| 666 | | 347 | 354 | 360 | 367 | 374 | 380 | 387 | 393 | 400 | 406 | |
| 667 | | 413 | 419 | 426 | 432 | 439 | 445 | 452 | 458 | 465 | 471 | |
| 668 | | 478 | 484 | 491 | 497 | 504 | 510 | 517 | 523 | 530 | 536 | |
| 669 | | 543 | 549 | 556 | 562 | 569 | 575 | 582 | 588 | 595 | 601 | |
| 670 | | 607 | 614 | 620 | 627 | 633 | 640 | 646 | 653 | 659 | 666 | |
| 671 | | 672 | 679 | 685 | 692 | 698 | 705 | 711 | 718 | 724 | 730 | |
| 672 | | 737 | 743 | 750 | 756 | 763 | 769 | 776 | 782 | 789 | 795 | |
| 673 | | 802 | 808 | 814 | 821 | 827 | 834 | 840 | 847 | 853 | 860 | |
| 674 | | 866 | 872 | 879 | 885 | 892 | 898 | 905 | 911 | 918 | 924 | |
| 675 | | 930 | 937 | 943 | 950 | 956 | 963 | 969 | 975 | 982 | 988 | |
| 676 | | 995 | *001 | *008 | *014 | *020 | *027 | *033 | *040 | *046 | *052 | |
| 677 | 83 | 059 | 065 | 072 | 078 | 085 | 091 | 097 | 104 | 110 | 117 | |
| 678 | | 123 | 129 | 136 | 142 | 149 | 155 | 161 | 168 | 174 | 181 | |
| 679 | | 187 | 193 | 200 | 206 | 213 | 219 | 225 | 232 | 238 | 245 | |
| 680 | | 251 | 257 | 264 | 270 | 276 | 283 | 289 | 296 | 302 | 308 | |
| 681 | | 315 | 321 | 327 | 334 | 340 | 347 | 353 | 359 | 366 | 372 | |
| 682 | | 378 | 385 | 391 | 398 | 404 | 410 | 417 | 423 | 429 | 436 | |
| 683 | | 442 | 448 | 455 | 461 | 468 | 474 | 480 | 487 | 493 | 499 | |
| 684 | | 506 | 512 | 518 | 525 | 531 | 537 | 544 | 550 | 556 | 563 | |
| 685 | | 569 | 575 | 582 | 588 | 594 | 601 | 607 | 613 | 620 | 626 | |
| 686 | | 632 | 639 | 645 | 651 | 658 | 664 | 670 | 677 | 683 | 689 | |
| 687 | | 696 | 702 | 708 | 715 | 721 | 727 | 734 | 740 | 746 | 753 | |
| 688 | | 759 | 765 | 771 | 778 | 784 | 790 | 797 | 803 | 809 | 816 | |
| 689 | | 822 | 828 | 835 | 841 | 847 | 853 | 860 | 866 | 872 | 879 | |
| 690 | | 885 | 891 | 898 | 904 | 910 | 916 | 923 | 929 | 935 | 942 | |
| 691 | | 948 | 954 | 960 | 967 | 973 | 979 | 986 | 992 | 998 | *004 | |
| 692 | 84 | 011 | 017 | 023 | 029 | 036 | 042 | 048 | 055 | 061 | 067 | |
| 693 | | 073 | 080 | 086 | 092 | 098 | 105 | 111 | 117 | 123 | 130 | |
| 694 | | 136 | 142 | 148 | 155 | 161 | 167 | 173 | 180 | 186 | 192 | |
| 695 | | 198 | 205 | 211 | 217 | 223 | 230 | 236 | 242 | 248 | 255 | |
| 696 | | 261 | 267 | 273 | 280 | 286 | 292 | 298 | 305 | 311 | 317 | |
| 697 | | 323 | 330 | 336 | 342 | 348 | 354 | 361 | 367 | 373 | 379 | |
| 698 | | 386 | 392 | 398 | 404 | 410 | 417 | 423 | 429 | 435 | 442 | |
| 699 | | 448 | 454 | 460 | 466 | 473 | 479 | 485 | 491 | 497 | 504 | |
| 700 | | 510 | 516 | 522 | 528 | 535 | 541 | 547 | 553 | 559 | 566 | |
| 701 | | 572 | 578 | 584 | 590 | 597 | 603 | 609 | 615 | 621 | 628 | |
| 702 | | 634 | 640 | 646 | 652 | 658 | 665 | 671 | 677 | 683 | 689 | |
| 703 | | 696 | 702 | 708 | 714 | 720 | 726 | 733 | 739 | 745 | 751 | |
| 704 | | 757 | 763 | 770 | 776 | 782 | 788 | 794 | 800 | 807 | 813 | |
| 705 | | 819 | 825 | 831 | 837 | 844 | 850 | 856 | 862 | 868 | 874 | |
| No. | L | 0 | 1 | 2 | 3 | 4 | 5 | 6 | 7 | 8 | 9 | Proportional parts |

7
| | |
|---|---|
| 1 | 0.7 |
| 2 | 1.4 |
| 3 | 2.1 |
| 4 | 2.8 |
| 5 | 3.5 |
| 6 | 4.2 |
| 7 | 4.9 |
| 8 | 5.6 |
| 9 | 6.3 |

6
| | |
|---|---|
| 1 | 0.6 |
| 2 | 1.2 |
| 3 | 1.8 |
| 4 | 2.4 |
| 5 | 3.0 |
| 6 | 3.6 |
| 7 | 4.2 |
| 8 | 4.8 |
| 9 | 5.4 |

* Indicates change in the first two decimal places.

## Table 1-8. Five-place Common Logarithms of Numbers (Continued)
### 705-760

| No. | L | 0 | 1 | 2 | 3 | 4 | 5 | 6 | 7 | 8 | 9 | Proportional parts |
|---|---|---|---|---|---|---|---|---|---|---|---|---|
| 705 | 84 | 819 | 825 | 831 | 837 | 844 | 850 | 856 | 862 | 868 | 874 | |
| 706 | | 880 | 887 | 893 | 899 | 905 | 911 | 917 | 924 | 930 | 936 | |
| 707 | | 942 | 948 | 954 | 960 | 967 | 973 | 979 | 985 | 991 | 997 | |
| 708 | 85 | 003 | 009 | 016 | 022 | 028 | 034 | 040 | 046 | 052 | 059 | |
| 709 | | 065 | 071 | 077 | 083 | 089 | 095 | 101 | 107 | 114 | 120 | |
| 710 | | 126 | 132 | 138 | 144 | 150 | 156 | 163 | 169 | 175 | 181 | |
| 711 | | 187 | 193 | 199 | 205 | 211 | 217 | 224 | 230 | 236 | 242 | |
| 712 | | 248 | 254 | 260 | 266 | 272 | 278 | 285 | 291 | 297 | 303 | |
| 713 | | 309 | 315 | 321 | 327 | 333 | 339 | 345 | 352 | 358 | 364 | |
| 714 | | 370 | 376 | 382 | 388 | 394 | 400 | 406 | 412 | 418 | 425 | |
| 715 | | 431 | 437 | 443 | 449 | 455 | 461 | 467 | 473 | 479 | 485 | |
| 716 | | 491 | 497 | 503 | 510 | 516 | 522 | 528 | 534 | 540 | 546 | |
| 717 | | 552 | 558 | 564 | 570 | 576 | 582 | 588 | 594 | 600 | 606 | |
| 718 | | 612 | 618 | 625 | 631 | 637 | 643 | 649 | 655 | 661 | 667 | |
| 719 | | 673 | 679 | 685 | 691 | 697 | 703 | 709 | 715 | 721 | 727 | |
| 720 | | 733 | 739 | 745 | 751 | 757 | 763 | 769 | 775 | 781 | 788 | |
| 721 | | 794 | 800 | 806 | 812 | 818 | 824 | 830 | 836 | 842 | 848 | |
| 722 | | 854 | 860 | 866 | 872 | 878 | 884 | 890 | 896 | 902 | 908 | |
| 723 | | 914 | 920 | 926 | 932 | 938 | 944 | 950 | 956 | 962 | 968 | |
| 724 | | 974 | 980 | 986 | 992 | 998 | *004 | *010 | *016 | *022 | *028 | |
| 725 | 86 | 034 | 040 | 046 | 052 | 058 | 064 | 070 | 076 | 082 | 088 | |
| 726 | | 094 | 100 | 106 | 112 | 118 | 124 | 130 | 136 | 141 | 147 | |
| 727 | | 153 | 159 | 165 | 171 | 177 | 183 | 189 | 195 | 201 | 207 | |
| 728 | | 213 | 219 | 225 | 231 | 237 | 243 | 249 | 255 | 261 | 267 | |
| 729 | | 273 | 279 | 285 | 291 | 297 | 303 | 308 | 314 | 320 | 326 | |
| 730 | | 332 | 338 | 344 | 350 | 356 | 362 | 368 | 374 | 380 | 386 | |
| 731 | | 392 | 398 | 404 | 410 | 416 | 421 | 427 | 433 | 439 | 445 | |
| 732 | | 451 | 457 | 463 | 469 | 475 | 481 | 487 | 493 | 499 | 504 | |
| 733 | | 510 | 516 | 522 | 528 | 534 | 540 | 546 | 552 | 558 | 564 | |
| 734 | | 570 | 576 | 581 | 587 | 593 | 599 | 605 | 611 | 617 | 623 | |
| 735 | | 629 | 635 | 641 | 646 | 652 | 658 | 664 | 670 | 676 | 682 | |
| 736 | | 688 | 694 | 700 | 705 | 711 | 717 | 723 | 729 | 735 | 741 | |
| 737 | | 747 | 753 | 759 | 764 | 770 | 776 | 782 | 788 | 794 | 800 | |
| 738 | | 806 | 812 | 817 | 823 | 829 | 835 | 841 | 847 | 853 | 859 | |
| 739 | | 864 | 870 | 876 | 882 | 888 | 894 | 900 | 906 | 911 | 917 | |
| 740 | | 923 | 929 | 935 | 941 | 947 | 953 | 958 | 964 | 970 | 976 | |
| 741 | | 982 | 988 | 994 | 999 | *005 | *011 | *017 | *023 | *029 | *035 | |
| 742 | 87 | 040 | 046 | 052 | 058 | 064 | 070 | 075 | 081 | 087 | 093 | |
| 743 | | 099 | 105 | 111 | 116 | 122 | 128 | 134 | 140 | 146 | 151 | |
| 744 | | 157 | 163 | 169 | 175 | 181 | 186 | 192 | 198 | 204 | 210 | |
| 745 | | 216 | 221 | 227 | 233 | 239 | 245 | 251 | 256 | 262 | 268 | |
| 746 | | 274 | 280 | 286 | 291 | 297 | 303 | 309 | 315 | 320 | 326 | |
| 747 | | 332 | 338 | 344 | 350 | 355 | 361 | 367 | 373 | 379 | 384 | |
| 748 | | 390 | 396 | 402 | 408 | 413 | 419 | 425 | 431 | 437 | 442 | |
| 749 | | 448 | 454 | 460 | 466 | 471 | 477 | 483 | 489 | 495 | 500 | |
| 750 | | 506 | 512 | 518 | 523 | 529 | 535 | 541 | 547 | 552 | 558 | |
| 751 | | 564 | 570 | 576 | 581 | 587 | 593 | 599 | 604 | 610 | 616 | |
| 752 | | 622 | 628 | 633 | 639 | 645 | 651 | 656 | 662 | 668 | 674 | |
| 753 | | 680 | 685 | 691 | 697 | 703 | 708 | 714 | 720 | 726 | 731 | |
| 754 | | 737 | 743 | 749 | 754 | 760 | 766 | 772 | 777 | 783 | 789 | |
| 755 | | 795 | 800 | 806 | 812 | 818 | 823 | 829 | 835 | 841 | 846 | |
| 756 | | 852 | 858 | 864 | 869 | 875 | 881 | 887 | 892 | 898 | 904 | |
| 757 | | 910 | 915 | 921 | 927 | 933 | 938 | 944 | 950 | 955 | 961 | |
| 758 | | 967 | 973 | 978 | 984 | 990 | 996 | *001 | *007 | *013 | *018 | |
| 759 | 88 | 024 | 030 | 036 | 041 | 047 | 053 | 059 | 064 | 070 | 076 | |
| 760 | | 081 | 087 | 093 | 099 | 104 | 110 | 116 | 121 | 127 | 133 | |
| No. | L | 0 | 1 | 2 | 3 | 4 | 5 | 6 | 7 | 8 | 9 | Proportional parts |

|  | 6 |
|---|---|
| 1 | 0.6 |
| 2 | 1.2 |
| 3 | 1.8 |
| 4 | 2.4 |
| 5 | 3.0 |
| 6 | 3.6 |
| 7 | 4.2 |
| 8 | 4.8 |
| 9 | 5.4 |

* Indicates change in the first two decimal places.

## Table 1-8. Five-place Common Logarithms of Numbers (Continued)
760-815

| No. | L | 0 | 1 | 2 | 3 | 4 | 5 | 6 | 7 | 8 | 9 | Proportional parts |
|-----|-----|-----|-----|-----|-----|-----|-----|-----|-----|-----|-----|-----|
| 760 | 88 | 081 | 087 | 093 | 099 | 104 | 110 | 116 | 121 | 127 | 133 | |
| 761 | | 138 | 144 | 150 | 156 | 161 | 167 | 173 | 178 | 184 | 190 | |
| 762 | | 196 | 201 | 207 | 213 | 218 | 224 | 230 | 235 | 241 | 247 | |
| 763 | | 252 | 258 | 264 | 270 | 275 | 281 | 287 | 292 | 298 | 304 | |
| 764 | | 309 | 315 | 321 | 326 | 332 | 338 | 343 | 349 | 355 | 360 | |
| | | | | | | | | | | | | 6 |
| 765 | | 366 | 372 | 378 | 383 | 389 | 395 | 400 | 406 | 412 | 417 | |
| 766 | | 423 | 429 | 434 | 440 | 446 | 451 | 457 | 463 | 468 | 474 | 1 \| 0.6 |
| 767 | | 480 | 485 | 491 | 497 | 502 | 508 | 514 | 519 | 525 | 530 | 2 \| 1.2 |
| 768 | | 536 | 542 | 547 | 553 | 559 | 564 | 570 | 576 | 581 | 587 | 3 \| 1.8 |
| 769 | | 593 | 598 | 604 | 610 | 615 | 621 | 627 | 632 | 638 | 643 | 4 \| 2.4 |
| | | | | | | | | | | | | 5 \| 3.0 |
| 770 | | 649 | 655 | 660 | 666 | 672 | 677 | 683 | 689 | 694 | 700 | 6 \| 3.6 |
| 771 | | 705 | 711 | 717 | 722 | 728 | 734 | 739 | 745 | 750 | 756 | 7 \| 4.2 |
| 772 | | 762 | 767 | 773 | 779 | 784 | 790 | 795 | 801 | 807 | 812 | 8 \| 4.8 |
| 773 | | 818 | 824 | 829 | 835 | 840 | 846 | 852 | 857 | 863 | 868 | 9 \| 5.4 |
| 774 | | 874 | 880 | 885 | 891 | 897 | 902 | 908 | 913 | 919 | 925 | |
| 775 | | 930 | 936 | 941 | 947 | 953 | 958 | 964 | 969 | 975 | 981 | |
| 776 | | 986 | 992 | 997 | *003 | *009 | *014 | *020 | *025 | *031 | *037 | |
| 777 | 89 | 042 | 048 | 053 | 059 | 064 | 070 | 076 | 081 | 087 | 092 | |
| 778 | | 098 | 104 | 109 | 115 | 120 | 126 | 131 | 137 | 143 | 148 | |
| 779 | | 154 | 159 | 165 | 170 | 176 | 182 | 187 | 193 | 198 | 204 | |
| 780 | | 209 | 215 | 221 | 226 | 232 | 237 | 243 | 248 | 254 | 260 | |
| 781 | | 265 | 271 | 276 | 282 | 287 | 293 | 298 | 304 | 310 | 315 | |
| 782 | | 321 | 326 | 332 | 337 | 343 | 348 | 354 | 360 | 365 | 371 | |
| 783 | | 376 | 382 | 387 | 393 | 398 | 404 | 409 | 415 | 421 | 426 | |
| 784 | | 432 | 437 | 443 | 448 | 454 | 459 | 465 | 470 | 476 | 481 | |
| | | | | | | | | | | | | 5 |
| 785 | | 487 | 493 | 498 | 504 | 509 | 515 | 520 | 526 | 531 | 537 | 1 \| 0.5 |
| 786 | | 542 | 548 | 553 | 559 | 564 | 570 | 575 | 581 | 586 | 592 | 2 \| 1.0 |
| 787 | | 597 | 603 | 609 | 614 | 620 | 625 | 631 | 636 | 642 | 647 | 3 \| 1.5 |
| 788 | | 653 | 658 | 664 | 669 | 675 | 680 | 686 | 691 | 697 | 702 | 4 \| 2.0 |
| 789 | | 708 | 713 | 719 | 724 | 730 | 735 | 741 | 746 | 752 | 757 | 5 \| 2.5 |
| | | | | | | | | | | | | 6 \| 3.0 |
| 790 | | 763 | 768 | 774 | 779 | 785 | 790 | 796 | 801 | 807 | 812 | 7 \| 3.5 |
| 791 | | 818 | 823 | 829 | 834 | 840 | 845 | 851 | 856 | 862 | 867 | 8 \| 4.0 |
| 792 | | 873 | 878 | 883 | 889 | 894 | 900 | 905 | 911 | 916 | 922 | 9 \| 4.5 |
| 793 | | 927 | 933 | 938 | 944 | 949 | 955 | 960 | 966 | 971 | 977 | |
| 794 | | 982 | 988 | 993 | 998 | *004 | *009 | *015 | *020 | *026 | *031 | |
| 795 | 90 | 037 | 042 | 048 | 053 | 059 | 064 | 069 | 075 | 080 | 086 | |
| 796 | | 091 | 097 | 102 | 108 | 113 | 119 | 124 | 129 | 135 | 140 | |
| 797 | | 146 | 151 | 157 | 162 | 168 | 173 | 179 | 184 | 189 | 195 | |
| 798 | | 200 | 206 | 211 | 217 | 222 | 227 | 233 | 238 | 244 | 249 | |
| 799 | | 255 | 260 | 266 | 271 | 276 | 282 | 287 | 293 | 298 | 304 | |
| 800 | | 309 | 314 | 320 | 325 | 331 | 336 | 342 | 347 | 352 | 358 | |
| 801 | | 363 | 369 | 374 | 380 | 385 | 390 | 396 | 401 | 407 | 412 | |
| 802 | | 417 | 423 | 428 | 434 | 439 | 445 | 450 | 455 | 461 | 466 | |
| 803 | | 472 | 477 | 482 | 488 | 493 | 499 | 504 | 509 | 515 | 520 | |
| 804 | | 526 | 531 | 536 | 542 | 547 | 553 | 558 | 563 | 569 | 574 | |
| 805 | | 580 | 585 | 590 | 596 | 601 | 607 | 612 | 617 | 623 | 628 | |
| 806 | | 634 | 639 | 644 | 650 | 655 | 660 | 666 | 671 | 677 | 682 | |
| 807 | | 687 | 693 | 698 | 704 | 709 | 714 | 720 | 725 | 730 | 736 | |
| 808 | | 741 | 747 | 752 | 757 | 763 | 768 | 773 | 779 | 784 | 789 | |
| 809 | | 795 | 800 | 806 | 811 | 816 | 822 | 827 | 832 | 838 | 843 | |
| 810 | | 849 | 854 | 859 | 865 | 870 | 875 | 881 | 886 | 891 | 897 | |
| 811 | | 902 | 907 | 913 | 918 | 924 | 929 | 934 | 940 | 945 | 950 | |
| 812 | | 956 | 961 | 966 | 972 | 977 | 982 | 988 | 993 | 998 | *004 | |
| 813 | 91 | 009 | 014 | 020 | 025 | 030 | 036 | 041 | 046 | 052 | 057 | |
| 814 | | 062 | 068 | 073 | 078 | 084 | 089 | 094 | 100 | 105 | 110 | |
| 815 | | 116 | 121 | 126 | 132 | 137 | 142 | 148 | 153 | 158 | 164 | |
| No. | L | 0 | 1 | 2 | 3 | 4 | 5 | 6 | 7 | 8 | 9 | Proportional parts |

* Indicates change in the first two decimal places.

## Table 1-8. Five-place Common Logarithms of Numbers (Continued)
### 815–870

| No. | L | 0 | 1 | 2 | 3 | 4 | 5 | 6 | 7 | 8 | 9 |
|---|---|---|---|---|---|---|---|---|---|---|---|
| 815 | 91 | 116 | 121 | 126 | 132 | 137 | 142 | 148 | 153 | 158 | 164 |
| 816 | | 169 | 174 | 180 | 185 | 190 | 196 | 201 | 206 | 212 | 217 |
| 817 | | 222 | 228 | 233 | 238 | 243 | 249 | 254 | 259 | 265 | 270 |
| 818 | | 275 | 281 | 286 | 291 | 297 | 302 | 307 | 312 | 318 | 323 |
| 819 | | 328 | 334 | 339 | 344 | 350 | 355 | 360 | 365 | 371 | 376 |
| 820 | | 381 | 387 | 392 | 397 | 403 | 408 | 413 | 418 | 424 | 429 |
| 821 | | 434 | 440 | 445 | 450 | 455 | 461 | 466 | 471 | 477 | 482 |
| 822 | | 487 | 492 | 498 | 503 | 508 | 514 | 519 | 524 | 529 | 535 |
| 823 | | 540 | 545 | 551 | 556 | 561 | 566 | 572 | 577 | 582 | 587 |
| 824 | | 593 | 598 | 603 | 609 | 614 | 619 | 624 | 630 | 635 | 640 |
| 825 | | 645 | 651 | 656 | 661 | 666 | 672 | 677 | 682 | 687 | 693 |
| 826 | | 698 | 703 | 709 | 714 | 719 | 724 | 730 | 735 | 740 | 745 |
| 827 | | 751 | 756 | 761 | 766 | 772 | 777 | 782 | 787 | 793 | 798 |
| 828 | | 803 | 808 | 814 | 819 | 824 | 829 | 834 | 840 | 845 | 850 |
| 829 | | 855 | 861 | 866 | 871 | 876 | 882 | 887 | 892 | 897 | 903 |
| 830 | | 908 | 913 | 918 | 924 | 929 | 934 | 939 | 944 | 950 | 955 |
| 831 | | 960 | 965 | 971 | 976 | 981 | 986 | 991 | 997 | *002 | *007 |
| 832 | 92 | 012 | 018 | 023 | 028 | 033 | 038 | 044 | 049 | 054 | 059 |
| 833 | | 065 | 070 | 075 | 080 | 085 | 091 | 096 | 101 | 106 | 111 |
| 834 | | 117 | 122 | 127 | 132 | 137 | 143 | 148 | 153 | 158 | 163 |
| 835 | | 169 | 174 | 179 | 184 | 189 | 195 | 200 | 205 | 210 | 215 |
| 836 | | 221 | 226 | 231 | 236 | 241 | 247 | 252 | 257 | 262 | 267 |
| 837 | | 273 | 278 | 283 | 288 | 293 | 298 | 304 | 309 | 314 | 319 |
| 838 | | 324 | 330 | 335 | 340 | 345 | 350 | 355 | 361 | 366 | 371 |
| 839 | | 376 | 381 | 387 | 392 | 397 | 402 | 407 | 412 | 418 | 423 |
| 840 | | 428 | 433 | 438 | 443 | 449 | 454 | 459 | 464 | 469 | 474 |
| 841 | | 480 | 485 | 490 | 495 | 500 | 505 | 511 | 516 | 521 | 526 |
| 842 | | 531 | 536 | 542 | 547 | 552 | 557 | 562 | 567 | 572 | 578 |
| 843 | | 583 | 588 | 593 | 598 | 603 | 609 | 614 | 619 | 624 | 629 |
| 844 | | 634 | 639 | 645 | 650 | 655 | 660 | 665 | 670 | 675 | 681 |
| 845 | | 686 | 691 | 696 | 701 | 706 | 711 | 717 | 722 | 727 | 732 |
| 846 | | 737 | 742 | 747 | 752 | 758 | 763 | 768 | 773 | 778 | 783 |
| 847 | | 788 | 793 | 799 | 804 | 809 | 814 | 819 | 824 | 829 | 834 |
| 848 | | 840 | 845 | 850 | 855 | 860 | 865 | 870 | 875 | 881 | 886 |
| 849 | | 891 | 896 | 901 | 906 | 911 | 916 | 921 | 927 | 932 | 937 |
| 850 | | 942 | 947 | 952 | 957 | 962 | 967 | 973 | 978 | 983 | 988 |
| 851 | | 993 | 998 | *003 | *008 | *013 | *018 | *024 | *029 | *034 | *039 |
| 852 | 93 | 044 | 049 | 054 | 059 | 064 | 069 | 075 | 080 | 085 | 090 |
| 853 | | 095 | 100 | 105 | 110 | 115 | 120 | 125 | 131 | 136 | 141 |
| 854 | | 146 | 151 | 156 | 161 | 166 | 171 | 176 | 181 | 186 | 192 |
| 855 | | 197 | 202 | 207 | 212 | 217 | 222 | 227 | 232 | 237 | 242 |
| 856 | | 247 | 252 | 258 | 263 | 268 | 273 | 278 | 283 | 288 | 293 |
| 857 | | 298 | 303 | 308 | 313 | 318 | 323 | 328 | 334 | 339 | 344 |
| 858 | | 349 | 354 | 359 | 364 | 369 | 374 | 379 | 384 | 389 | 394 |
| 859 | | 399 | 404 | 409 | 414 | 420 | 425 | 430 | 435 | 440 | 445 |
| 860 | | 450 | 455 | 460 | 465 | 470 | 475 | 480 | 485 | 490 | 495 |
| 861 | | 500 | 505 | 510 | 515 | 520 | 526 | 531 | 536 | 541 | 546 |
| 862 | | 551 | 556 | 561 | 566 | 571 | 576 | 581 | 586 | 591 | 596 |
| 863 | | 601 | 606 | 611 | 616 | 621 | 626 | 631 | 636 | 641 | 646 |
| 864 | | 651 | 656 | 661 | 666 | 671 | 677 | 682 | 687 | 692 | 697 |
| 865 | | 702 | 707 | 712 | 717 | 722 | 727 | 732 | 737 | 742 | 747 |
| 866 | | 752 | 757 | 762 | 767 | 772 | 777 | 782 | 787 | 792 | 797 |
| 867 | | 802 | 807 | 812 | 817 | 822 | 827 | 832 | 837 | 842 | 847 |
| 868 | | 852 | 857 | 862 | 867 | 872 | 877 | 882 | 887 | 892 | 897 |
| 869 | | 902 | 907 | 912 | 917 | 922 | 927 | 932 | 937 | 942 | 947 |
| 870 | | 952 | 957 | 962 | 967 | 972 | 977 | 982 | 987 | 992 | 997 |
| No. | L | 0 | 1 | 2 | 3 | 4 | 5 | 6 | 7 | 8 | 9 |

Proportional parts

| 6 | |
|---|---|
| 1 | 0.6 |
| 2 | 1.2 |
| 3 | 1.8 |
| 4 | 2.4 |
| 5 | 3.0 |
| 6 | 3.6 |
| 7 | 4.2 |
| 8 | 4.8 |
| 9 | 5.4 |

| 5 | |
|---|---|
| 1 | 0.5 |
| 2 | 1.0 |
| 3 | 1.5 |
| 4 | 2.0 |
| 5 | 2.5 |
| 6 | 3.0 |
| 7 | 3.5 |
| 8 | 4.0 |
| 9 | 4.5 |

*Indicates change in the first two decimal places.

## Table 1-8. Five-place Common Logarithms of Numbers (Continued)
### 870–925

| No. | L | 0 | 1 | 2 | 3 | 4 | 5 | 6 | 7 | 8 | 9 | Proportional parts |
|-----|----|-----|-----|-----|-----|-----|-----|-----|-----|-----|-----|---|
| 870 | 93 | 952 | 957 | 962 | 967 | 972 | 977 | 982 | 987 | 992 | 997 | |
| 871 | 94 | 002 | 007 | 012 | 017 | 022 | 027 | 032 | 037 | 042 | 047 | |
| 872 | | 052 | 057 | 062 | 067 | 072 | 077 | 082 | 087 | 091 | 096 | |
| 873 | | 101 | 106 | 111 | 116 | 121 | 126 | 131 | 136 | 141 | 146 | |
| 874 | | 151 | 156 | 161 | 166 | 171 | 176 | 181 | 186 | 191 | 196 | |
| 875 | | 201 | 206 | 211 | 216 | 221 | 226 | 231 | 236 | 240 | 245 | |
| 876 | | 250 | 255 | 260 | 265 | 270 | 275 | 280 | 285 | 290 | 295 | |
| 877 | | 300 | 305 | 310 | 315 | 320 | 325 | 330 | 335 | 340 | 345 | |
| 878 | | 349 | 354 | 359 | 364 | 369 | 374 | 379 | 384 | 389 | 394 | |
| 879 | | 399 | 404 | 409 | 414 | 419 | 424 | 429 | 433 | 438 | 443 | |
| 880 | | 448 | 453 | 458 | 463 | 468 | 473 | 478 | 483 | 488 | 493 | |
| 881 | | 498 | 503 | 507 | 512 | 517 | 522 | 527 | 532 | 537 | 542 | |
| 882 | | 547 | 552 | 557 | 562 | 567 | 571 | 576 | 581 | 586 | 591 | |
| 883 | | 596 | 601 | 606 | 611 | 616 | 621 | 626 | 630 | 635 | 640 | |
| 884 | | 645 | 650 | 655 | 660 | 665 | 670 | 675 | 680 | 685 | 689 | |
| 885 | | 694 | 699 | 704 | 709 | 714 | 719 | 724 | 729 | 734 | 738 | **5** |
| 886 | | 743 | 748 | 753 | 758 | 763 | 768 | 773 | 778 | 783 | 787 | 1 \| 0.5 |
| 887 | | 792 | 797 | 802 | 807 | 812 | 817 | 822 | 827 | 832 | 836 | 2 \| 1.0 |
| 888 | | 841 | 846 | 851 | 856 | 861 | 866 | 871 | 876 | 880 | 885 | 3 \| 1.5 |
| 889 | | 890 | 895 | 900 | 905 | 910 | 915 | 919 | 924 | 929 | 934 | 4 \| 2.0 |
| | | | | | | | | | | | | 5 \| 2.5 |
| | | | | | | | | | | | | 6 \| 3.0 |
| 890 | | 939 | 944 | 949 | 954 | 959 | 963 | 968 | 973 | 978 | 983 | 7 \| 3.5 |
| 891 | | 988 | 993 | 998 | *002 | *007 | *012 | *017 | *022 | *027 | *032 | 8 \| 4.0 |
| 892 | 95 | 036 | 041 | 046 | 051 | 056 | 061 | 066 | 071 | 075 | 080 | 9 \| 4.5 |
| 893 | | 085 | 090 | 095 | 100 | 105 | 109 | 114 | 119 | 124 | 129 | |
| 894 | | 134 | 139 | 143 | 148 | 153 | 158 | 163 | 168 | 173 | 177 | |
| 895 | | 182 | 187 | 192 | 197 | 202 | 207 | 211 | 216 | 221 | 226 | |
| 896 | | 231 | 236 | 240 | 245 | 250 | 255 | 260 | 265 | 270 | 274 | |
| 897 | | 279 | 284 | 289 | 294 | 299 | 303 | 308 | 313 | 318 | 323 | |
| 898 | | 328 | 332 | 337 | 342 | 347 | 352 | 357 | 361 | 366 | 371 | |
| 899 | | 376 | 381 | 386 | 390 | 395 | 400 | 405 | 410 | 415 | 419 | |
| 900 | | 424 | 429 | 434 | 439 | 444 | 448 | 453 | 458 | 463 | 468 | |
| 901 | | 472 | 477 | 482 | 487 | 492 | 497 | 501 | 506 | 511 | 516 | |
| 902 | | 521 | 525 | 530 | 535 | 540 | 545 | 550 | 554 | 559 | 564 | |
| 903 | | 569 | 574 | 578 | 583 | 588 | 593 | 598 | 602 | 607 | 612 | |
| 904 | | 617 | 622 | 626 | 631 | 636 | 641 | 646 | 650 | 655 | 660 | |
| 905 | | 665 | 670 | 674 | 679 | 684 | 689 | 694 | 698 | 703 | 708 | **4** |
| 906 | | 713 | 718 | 722 | 727 | 732 | 737 | 742 | 746 | 751 | 756 | 1 \| 0.4 |
| 907 | | 761 | 766 | 770 | 775 | 780 | 785 | 789 | 794 | 799 | 804 | 2 \| 0.8 |
| 908 | | 809 | 813 | 818 | 823 | 828 | 832 | 837 | 842 | 847 | 852 | 3 \| 1.2 |
| 909 | | 856 | 861 | 866 | 871 | 875 | 880 | 885 | 890 | 895 | 899 | 4 \| 1.6 |
| | | | | | | | | | | | | 5 \| 2.0 |
| 910 | | 904 | 909 | 914 | 918 | 923 | 928 | 933 | 938 | 942 | 947 | 6 \| 2.4 |
| 911 | | 952 | 957 | 961 | 966 | 971 | 976 | 980 | 985 | 990 | 995 | 7 \| 2.8 |
| 912 | | 999 | *004 | *009 | *014 | *019 | *023 | *028 | *033 | *038 | *042 | 8 \| 3.2 |
| 913 | 96 | 047 | 052 | 057 | 061 | 066 | 071 | 076 | 080 | 085 | 090 | 9 \| 3.6 |
| 914 | | 095 | 099 | 104 | 109 | 114 | 118 | 123 | 128 | 133 | 137 | |
| 915 | | 142 | 147 | 152 | 156 | 161 | 166 | 171 | 175 | 180 | 185 | |
| 916 | | 190 | 194 | 199 | 204 | 209 | 213 | 218 | 223 | 227 | 232 | |
| 917 | | 237 | 242 | 246 | 251 | 256 | 261 | 265 | 270 | 275 | 280 | |
| 918 | | 284 | 289 | 294 | 298 | 303 | 308 | 313 | 317 | 322 | 327 | |
| 919 | | 332 | 336 | 341 | 346 | 350 | 355 | 360 | 365 | 369 | 374 | |
| 920 | | 379 | 384 | 388 | 393 | 398 | 402 | 407 | 412 | 417 | 421 | |
| 921 | | 426 | 431 | 435 | 440 | 445 | 450 | 454 | 459 | 464 | 468 | |
| 922 | | 473 | 478 | 483 | 487 | 492 | 497 | 501 | 506 | 511 | 515 | |
| 923 | | 520 | 525 | 530 | 534 | 539 | 544 | 548 | 553 | 558 | 563 | |
| 924 | | 567 | 572 | 577 | 581 | 586 | 591 | 595 | 600 | 605 | 609 | |
| 925 | | 614 | 619 | 624 | 628 | 633 | 638 | 642 | 647 | 652 | 656 | |
| No. | L | 0 | 1 | 2 | 3 | 4 | 5 | 6 | 7 | 8 | 9 | Proportional parts |

* Indicates change in the first two decimal places.

## Table 1-8. Five-place Common Logarithms of Numbers (Continued)
### 925–980

| No. | L | 0 | 1 | 2 | 3 | 4 | 5 | 6 | 7 | 8 | 9 |
|---|---|---|---|---|---|---|---|---|---|---|---|
| 925 | 96 | 614 | 619 | 624 | 628 | 633 | 638 | 642 | 647 | 652 | 656 |
| 926 | | 661 | 666 | 670 | 675 | 680 | 685 | 689 | 694 | 699 | 703 |
| 927 | | 708 | 713 | 717 | 722 | 727 | 731 | 736 | 741 | 745 | 750 |
| 928 | | 755 | 759 | 764 | 769 | 774 | 778 | 783 | 788 | 792 | 797 |
| 929 | | 802 | 806 | 811 | 816 | 820 | 825 | 830 | 834 | 839 | 844 |
| 930 | | 848 | 853 | 858 | 862 | 867 | 872 | 876 | 881 | 886 | 890 |
| 931 | | 895 | 900 | 904 | 909 | 914 | 918 | 923 | 928 | 932 | 937 |
| 932 | | 942 | 946 | 951 | 956 | 960 | 965 | 970 | 974 | 979 | 984 |
| 933 | | 988 | 993 | 997 | *002 | *007 | *011 | *016 | *021 | *025 | *030 |
| 934 | 97 | 035 | 039 | 044 | 049 | 053 | 058 | 063 | 067 | 072 | 077 |
| 935 | | 081 | 086 | 090 | 095 | 100 | 104 | 109 | 114 | 118 | 123 |
| 936 | | 128 | 132 | 137 | 142 | 146 | 151 | 155 | 160 | 165 | 169 |
| 937 | | 174 | 179 | 183 | 188 | 192 | 197 | 202 | 206 | 211 | 216 |
| 938 | | 220 | 225 | 230 | 234 | 239 | 243 | 248 | 253 | 257 | 262 |
| 939 | | 267 | 271 | 276 | 280 | 285 | 290 | 294 | 299 | 304 | 308 |
| 940 | | 313 | 317 | 322 | 327 | 331 | 336 | 341 | 345 | 350 | 354 |
| 941 | | 359 | 364 | 368 | 373 | 377 | 382 | 387 | 391 | 396 | 400 |
| 942 | | 405 | 410 | 414 | 419 | 424 | 428 | 433 | 437 | 442 | 447 |
| 943 | | 451 | 456 | 460 | 465 | 470 | 474 | 479 | 483 | 488 | 493 |
| 944 | | 497 | 502 | 506 | 511 | 516 | 520 | 525 | 529 | 534 | 539 |
| 945 | | 543 | 548 | 552 | 557 | 562 | 566 | 571 | 575 | 580 | 585 |
| 946 | | 589 | 594 | 598 | 603 | 607 | 612 | 617 | 621 | 626 | 630 |
| 947 | | 635 | 640 | 644 | 649 | 653 | 658 | 663 | 667 | 672 | 676 |
| 948 | | 681 | 685 | 690 | 695 | 699 | 704 | 708 | 713 | 717 | 722 |
| 949 | | 727 | 731 | 736 | 740 | 745 | 750 | 754 | 759 | 763 | 768 |
| 950 | | 772 | 777 | 782 | 786 | 791 | 795 | 800 | 804 | 809 | 813 |
| 951 | | 818 | 823 | 827 | 832 | 836 | 841 | 845 | 850 | 855 | 859 |
| 952 | | 864 | 868 | 873 | 877 | 882 | 887 | 891 | 896 | 900 | 905 |
| 953 | | 909 | 914 | 918 | 923 | 928 | 932 | 937 | 941 | 946 | 950 |
| 954 | | 955 | 959 | 964 | 968 | 973 | 978 | 982 | 987 | 991 | 996 |
| 955 | 98 | 000 | 005 | 009 | 014 | 019 | 023 | 028 | 032 | 037 | 041 |
| 956 | | 046 | 050 | 055 | 059 | 064 | 069 | 073 | 078 | 082 | 087 |
| 957 | | 091 | 096 | 100 | 105 | 109 | 114 | 118 | 123 | 127 | 132 |
| 958 | | 137 | 141 | 146 | 150 | 155 | 159 | 164 | 168 | 173 | 177 |
| 959 | | 182 | 186 | 191 | 195 | 200 | 205 | 209 | 214 | 218 | 223 |
| 960 | | 227 | 232 | 236 | 241 | 245 | 250 | 254 | 259 | 263 | 268 |
| 961 | | 272 | 277 | 281 | 286 | 290 | 295 | 299 | 304 | 308 | 313 |
| 962 | | 318 | 322 | 327 | 331 | 336 | 340 | 345 | 349 | 354 | 358 |
| 963 | | 363 | 367 | 372 | 376 | 381 | 385 | 390 | 394 | 399 | 403 |
| 964 | | 408 | 412 | 417 | 421 | 426 | 430 | 435 | 439 | 444 | 448 |
| 965 | | 453 | 457 | 462 | 466 | 471 | 475 | 480 | 484 | 489 | 493 |
| 966 | | 498 | 502 | 507 | 511 | 516 | 520 | 525 | 529 | 534 | 538 |
| 967 | | 543 | 547 | 552 | 556 | 561 | 565 | 570 | 574 | 579 | 583 |
| 968 | | 588 | 592 | 597 | 601 | 605 | 610 | 614 | 619 | 623 | 628 |
| 969 | | 632 | 637 | 641 | 646 | 650 | 655 | 659 | 664 | 668 | 673 |
| 970 | | 677 | 682 | 686 | 691 | 695 | 700 | 704 | 709 | 713 | 717 |
| 971 | | 722 | 726 | 731 | 735 | 740 | 744 | 749 | 753 | 758 | 762 |
| 972 | | 767 | 771 | 776 | 780 | 785 | 789 | 793 | 798 | 802 | 807 |
| 973 | | 811 | 816 | 820 | 825 | 829 | 834 | 838 | 843 | 847 | 851 |
| 974 | | 856 | 860 | 865 | 869 | 874 | 878 | 883 | 887 | 892 | 896 |
| 975 | | 900 | 905 | 909 | 914 | 918 | 923 | 927 | 932 | 936 | 941 |
| 976 | | 945 | 949 | 954 | 958 | 963 | 967 | 972 | 976 | 981 | 985 |
| 977 | | 989 | 994 | 998 | *003 | *007 | *012 | *016 | *021 | *025 | *029 |
| 978 | 99 | 034 | 038 | 043 | 047 | 052 | 056 | 061 | 065 | 069 | 074 |
| 979 | | 078 | 083 | 087 | 092 | 096 | 100 | 105 | 109 | 114 | 118 |
| 980 | | 123 | 127 | 131 | 136 | 140 | 145 | 149 | 154 | 158 | 162 |
| No. | L | 0 | 1 | 2 | 3 | 4 | 5 | 6 | 7 | 8 | 9 |

Proportional parts

| | 5 |
|---|---|
| 1 | 0.5 |
| 2 | 1.0 |
| 3 | 1.5 |
| 4 | 2.0 |
| 5 | 2.5 |
| 6 | 3.0 |
| 7 | 3.5 |
| 8 | 4.0 |
| 9 | 4.5 |

| | 4 |
|---|---|
| 1 | 0.4 |
| 2 | 0.8 |
| 3 | 1.2 |
| 4 | 1.6 |
| 5 | 2.0 |
| 6 | 2.4 |
| 7 | 2.8 |
| 8 | 3.2 |
| 9 | 3.6 |

* Indicates change in the first two decimal places.

## Table 1-8. Five-place Common Logarithms of Numbers (Continued)
### 980-1000

| No. | L | 0 | 1 | 2 | 3 | 4 | 5 | 6 | 7 | 8 | 9 | Proportional parts | |
|-----|----|-----|-----|-----|-----|-----|-----|-----|-----|-----|-----|---|---|
| 980 | 99 | 123 | 127 | 131 | 136 | 140 | 145 | 149 | 154 | 158 | 162 | | |
| 981 | | 167 | 171 | 176 | 180 | 185 | 189 | 193 | 198 | 202 | 207 | | 5 |
| 982 | | 211 | 216 | 220 | 224 | 229 | 233 | 238 | 242 | 247 | 251 | | |
| 983 | | 255 | 260 | 264 | 269 | 273 | 277 | 282 | 286 | 291 | 295 | 1 | 0.5 |
| 984 | | 300 | 304 | 308 | 313 | 317 | 322 | 326 | 330 | 335 | 339 | 2 | 1.0 |
| | | | | | | | | | | | | 3 | 1.5 |
| 985 | | 344 | 348 | 352 | 357 | 361 | 366 | 370 | 374 | 379 | 383 | 4 | 2.0 |
| 986 | | 388 | 392 | 397 | 401 | 405 | 410 | 414 | 419 | 423 | 427 | 5 | 2.5 |
| 987 | | 432 | 436 | 441 | 445 | 449 | 454 | 458 | 463 | 467 | 471 | 6 | 3.0 |
| 988 | | 476 | 480 | 484 | 489 | 493 | 498 | 502 | 506 | 511 | 515 | 7 | 3.5 |
| 989 | | 520 | 524 | 528 | 533 | 537 | 542 | 546 | 550 | 555 | 559 | 8 | 4.0 |
| | | | | | | | | | | | | 9 | 4.5 |
| 990 | | 564 | 568 | 572 | 577 | 581 | 585 | 590 | 594 | 599 | 603 | | |
| 991 | | 607 | 612 | 616 | 621 | 625 | 629 | 634 | 638 | 642 | 647 | | 4 |
| 992 | | 651 | 656 | 660 | 664 | 669 | 673 | 677 | 682 | 686 | 691 | | |
| 993 | | 695 | 699 | 704 | 708 | 712 | 717 | 721 | 726 | 730 | 734 | 1 | 0.4 |
| 994 | | 739 | 743 | 747 | 752 | 756 | 760 | 765 | 769 | 774 | 778 | 2 | 0.8 |
| | | | | | | | | | | | | 3 | 1.2 |
| 995 | | 782 | 787 | 791 | 795 | 800 | 804 | 808 | 813 | 817 | 822 | 4 | 1.6 |
| 996 | | 826 | 830 | 835 | 839 | 843 | 848 | 852 | 856 | 861 | 865 | 5 | 2.0 |
| 997 | | 870 | 874 | 878 | 883 | 887 | 891 | 896 | 900 | 904 | 909 | 6 | 2.4 |
| 998 | | 913 | 917 | 922 | 926 | 930 | 935 | 939 | 944 | 948 | 952 | 7 | 2.8 |
| 999 | | 957 | 961 | 965 | 970 | 974 | 978 | 983 | 987 | 991 | 996 | 8 | 3.2 |
| | | | | | | | | | | | | 9 | 3.6 |
| 1000 | 00 | 000 | 004 | 009 | 013 | 017 | 022 | 026 | 030 | 035 | 039 | | |
| No. | L | 0 | 1 | 2 | 3 | 4 | 5 | 6 | 7 | 8 | 9 | Proportional parts | |

* Indicates change in the first two decimal places.

## Table 1-9. Natural Trigonometric Functions and Their Logarithms

| Deg | Radians | Nat sin | Log sin | Nat cos | Log cos | Nat tan | Log tan | Nat cot | Log cot | Radians | Deg |
|---|---|---|---|---|---|---|---|---|---|---|---|
| 0° 00′ | 0.0000 | 0.0000 |  | 1.0000 | 0.0000 | 0.0000 |  |  |  | 1.5708 | 90° 00′ |
| 10 | .0029 | .0029 | 7.4637 | 1.0000 | 0.0000 | .0029 | 7.4637 | 343.77 | 2.5363 | 1.5679 | 50 |
| 20 | .0058 | .0058 | 7.7648 | 1.0000 | 0.0000 | .0058 | 7.7648 | 171.89 | 2.2352 | 1.5650 | 40 |
| 30 | .0087 | .0087 | 7.9408 | 1.0000 | 0.0000 | .0087 | 7.9409 | 114.59 | 2.0591 | 1.5621 | 30 |
| 40 | .0116 | .0116 | 8.0658 | 0.9999 | 0.0000 | .0116 | 8.0658 | 85.940 | 1.9342 | 1.5592 | 20 |
| 50 | .0145 | .0145 | 8.1627 | .9999 | 0.0000 | .0146 | 8.1627 | 68.750 | 1.8373 | 1.5563 | 10 |
| 1° 00′ | .0175 | .0175 | 8.2419 | .9998 | 9.9999 | .0175 | 8.2419 | 57.290 | 1.7581 | 1.5533 | 89° 00′ |
| 10 | .0204 | .0204 | 8.3088 | .9998 | 9.9999 | .0204 | 8.3089 | 49.104 | 1.6911 | 1.5504 | 50 |
| 20 | .0233 | .0233 | 8.3668 | .9997 | 9.9999 | .0233 | 8.3669 | 42.964 | 1.6331 | 1.5475 | 40 |
| 30 | .0262 | .0262 | 8.4179 | .9997 | 9.9999 | .0262 | 8.4181 | 38.188 | 1.5819 | 1.5446 | 30 |
| 40 | .0291 | .0291 | 8.4637 | .9996 | 9.9998 | .0291 | 8.4639 | 34.368 | 1.5362 | 1.5417 | 20 |
| 50 | .0320 | .0320 | 8.5050 | .9995 | 9.9998 | .0320 | 8.5053 | 31.242 | 1.4947 | 1.5388 | 10 |
| 2° 00′ | .0349 | .0349 | 8.5428 | .9994 | 9.9997 | .0349 | 8.5431 | 28.636 | 1.4569 | 1.5359 | 88° 00′ |
| 10 | .0378 | .0378 | 8.5776 | .9993 | 9.9997 | .0378 | 8.5779 | 26.432 | 1.4221 | 1.5330 | 50 |
| 20 | .0407 | .0407 | 8.6097 | .9992 | 9.9996 | .0408 | 8.6101 | 24.542 | 1.3899 | 1.5301 | 40 |
| 30 | .0436 | .0436 | 8.6397 | .9991 | 9.9996 | .0437 | 8.6401 | 22.904 | 1.3599 | 1.5272 | 30 |
| 40 | .0465 | .0465 | 8.6677 | .9989 | 9.9995 | .0466 | 8.6682 | 21.470 | 1.3318 | 1.5243 | 20 |
| 50 | .0495 | .0494 | 8.6940 | .9988 | 9.9995 | .0495 | 8.6945 | 20.206 | 1.3055 | 1.5213 | 10 |
| 3° 00′ | .0524 | .0523 | 8.7188 | .9986 | 9.9994 | .0524 | 8.7194 | 19.081 | 1.2806 | 1.5184 | 87° 00′ |
| 10 | .0553 | .0552 | 8.7423 | .9985 | 9.9993 | .0553 | 8.7429 | 18.075 | 1.2571 | 1.5155 | 50 |
| 20 | .0582 | .0581 | 8.7645 | .9983 | 9.9993 | .0582 | 8.7653 | 17.169 | 1.2348 | 1.5126 | 40 |
| 30 | .0611 | .0611 | 8.7857 | .9981 | 9.9992 | .0612 | 8.7865 | 16.350 | 1.2135 | 1.5097 | 30 |
| 40 | .0640 | .0640 | 8.8059 | .9980 | 9.9991 | .0641 | 8.8067 | 15.605 | 1.1933 | 1.5068 | 20 |
| 50 | .0669 | .0669 | 8.8251 | .9978 | 9.9990 | .0670 | 8.8261 | 14.924 | 1.1739 | 1.5039 | 10 |
| 4° 00′ | .0698 | .0698 | 8.8436 | .9976 | 9.9989 | .0699 | 8.8446 | 14.301 | 1.1554 | 1.5010 | 86° 00′ |
| 10 | .0727 | .0727 | 8.8613 | .9974 | 9.9989 | .0729 | 8.8624 | 13.727 | 1.1376 | 1.4981 | 50 |
| 20 | .0756 | .0756 | 8.8783 | .9971 | 9.9988 | .0758 | 8.8795 | 13.197 | 1.1205 | 1.4952 | 40 |
| 30 | .0785 | .0785 | 8.8946 | .9969 | 9.9987 | .0787 | 8.8960 | 12.706 | 1.1040 | 1.4923 | 30 |
| 40 | .0814 | .0814 | 8.9104 | .9967 | 9.9986 | .0816 | 8.9119 | 12.251 | 1.0882 | 1.4893 | 20 |
| 50 | .0844 | .0843 | 8.9256 | .9964 | 9.9985 | .0846 | 8.9272 | 11.826 | 1.0728 | 1.4864 | 10 |
| 5° 00′ | .0873 | .0872 | 8.9403 | .9962 | 9.9983 | .0875 | 8.9420 | 11.430 | 1.0581 | 1.4835 | 85° 00′ |
| 10 | .0902 | .0901 | 8.9545 | .9959 | 9.9982 | .0904 | 8.9563 | 11.059 | 1.0437 | 1.4806 | 50 |
| 20 | .0931 | .0930 | 8.9683 | .9957 | 9.9981 | .0934 | 8.9701 | 10.712 | 1.0299 | 1.4777 | 40 |
| 30 | .0960 | .0959 | 8.9816 | .9954 | 9.9980 | .0963 | 8.9836 | 10.385 | 1.0164 | 1.4748 | 30 |
| 40 | .0989 | .0987 | 8.9945 | .9951 | 9.9979 | .0992 | 8.9966 | 10.078 | 1.0034 | 1.4719 | 20 |
| 50 | .1018 | .1016 | 9.0070 | .9948 | 9.9978 | .1022 | 9.0093 | 9.7882 | 0.9907 | 1.4690 | 10 |

| Deg | Radians | Log tan | Nat tan | Log cot | Nat cot | Log sin | Nat sin | Log cos | Nat cos | Radians | Deg |
|---|---|---|---|---|---|---|---|---|---|---|---|
| 84° 00' | 1.4661 | 9784 | 9.5144 | 9.0216 | .1051 | 9.9976 | .9945 | 9.0192 | .1045 | .1047 | 6° 00' |
| 50 | 1.4632 | 9664 | 9.2553 | 9.0336 | .1081 | 9.9975 | .9942 | 9.0311 | .1074 | .1076 | 10 |
| 40 | 1.4603 | 9547 | 9.0098 | 9.0453 | .1110 | 9.9973 | .9939 | 9.0426 | .1103 | .1105 | 20 |
| 30 | 1.4573 | 9433 | 8.7769 | 9.0567 | .1139 | 9.9972 | .9936 | 9.0539 | .1132 | .1134 | 30 |
| 20 | 1.4544 | 9323 | 8.5556 | 9.0678 | .1169 | 9.9971 | .9932 | 9.0648 | .1161 | .1164 | 40 |
| 10 | 1.4515 | 9214 | 8.3450 | 9.0786 | .1198 | 9.9969 | .9929 | 9.0755 | .1190 | .1193 | 50 |
| 83° 00' | 1.4486 | .9109 | 8.1443 | 9.0891 | .1228 | 9.9968 | .9926 | 9.0859 | .1219 | .1222 | 7° 00' |
| 50 | 1.4457 | .9005 | 7.9530 | 9.0995 | .1257 | 9.9966 | .9922 | 9.0961 | .1248 | .1251 | 10 |
| 40 | 1.4428 | .8904 | 7.7704 | 9.1096 | .1287 | 9.9964 | .9918 | 9.1060 | .1276 | .1280 | 20 |
| 30 | 1.4399 | .8806 | 7.5958 | 9.1194 | .1317 | 9.9963 | .9914 | 9.1157 | .1305 | .1309 | 30 |
| 20 | 1.4370 | .8709 | 7.4287 | 9.1291 | .1346 | 9.9961 | .9911 | 9.1252 | .1334 | .1338 | 40 |
| 10 | 1.4341 | .8615 | 7.2687 | 9.1385 | .1376 | 9.9959 | .9907 | 9.1345 | .1363 | .1367 | 50 |
| 82° 00' | 1.4312 | .8522 | 7.1154 | 9.1478 | .1405 | 9.9958 | .9903 | 9.1436 | .1392 | .1396 | 8° 00' |
| 50 | 1.4283 | .8431 | 6.9682 | 9.1569 | .1435 | 9.9956 | .9899 | 9.1525 | .1421 | .1425 | 10 |
| 40 | 1.4254 | .8342 | 6.8269 | 9.1658 | .1465 | 9.9954 | .9894 | 9.1612 | .1449 | .1454 | 20 |
| 30 | 1.4224 | .8255 | 6.6912 | 9.1745 | .1495 | 9.9952 | .9890 | 9.1697 | .1478 | .1484 | 30 |
| 20 | 1.4195 | .8169 | 6.5606 | 9.1831 | .1524 | 9.9950 | .9886 | 9.1781 | .1507 | .1513 | 40 |
| 10 | 1.4166 | .8085 | 6.4348 | 9.1915 | .1554 | 9.9948 | .9881 | 9.1863 | .1536 | .1542 | 50 |
| 81° 00' | 1.4137 | .8003 | 6.3138 | 9.1997 | .1584 | 9.9946 | .9877 | 9.1943 | .1564 | .1571 | 9° 00' |
| 50 | 1.4108 | .7922 | 6.1970 | 9.2078 | .1614 | 9.9944 | .9872 | 9.2022 | .1593 | .1600 | 10 |
| 40 | 1.4079 | .7842 | 6.0844 | 9.2158 | .1644 | 9.9942 | .9868 | 9.2100 | .1622 | .1629 | 20 |
| 30 | 1.4050 | .7764 | 5.9758 | 9.2236 | .1673 | 9.9940 | .9863 | 9.2176 | .1651 | .1658 | 30 |
| 20 | 1.4021 | .7687 | 5.8708 | 9.2313 | .1703 | 9.9938 | .9858 | 9.2251 | .1679 | .1687 | 40 |
| 10 | 1.3992 | .7611 | 5.7694 | 9.2389 | .1733 | 9.9936 | .9853 | 9.2324 | .1708 | .1716 | 50 |
| 80° 00' | 1.3963 | .7537 | 5.6713 | 9.2463 | .1763 | 9.9934 | .9848 | 9.2397 | .1737 | .1745 | 10° 00' |
| 50 | 1.3934 | .7464 | 5.5764 | 9.2536 | .1793 | 9.9931 | .9843 | 9.2468 | .1765 | .1774 | 10 |
| 40 | 1.3904 | .7391 | 5.4845 | 9.2609 | .1823 | 9.9929 | .9838 | 9.2538 | .1794 | .1804 | 20 |
| 30 | 1.3875 | .7320 | 5.3955 | 9.2680 | .1853 | 9.9927 | .9833 | 9.2606 | .1822 | .1833 | 30 |
| 20 | 1.3846 | .7250 | 5.3093 | 9.2750 | .1884 | 9.9924 | .9827 | 9.2674 | .1851 | .1862 | 40 |
| 10 | 1.3817 | .7181 | 5.2257 | 9.2819 | .1914 | 9.9922 | .9822 | 9.2741 | .1880 | .1891 | 50 |
| 79° 00' | 1.3788 | .7114 | 5.1446 | 9.2887 | .1944 | 9.9920 | .9816 | 9.2806 | .1908 | .1920 | 11° 00' |
| 50 | 1.3759 | .7047 | 5.0658 | 9.2954 | .1974 | 9.9917 | .9811 | 9.2871 | .1937 | .1949 | 10 |
| 40 | 1.3730 | .6981 | 4.9894 | 9.3020 | .2004 | 9.9915 | .9805 | 9.2934 | .1965 | .1978 | 20 |
| 30 | 1.3701 | .6915 | 4.9152 | 9.3085 | .2035 | 9.9912 | .9799 | 9.2997 | .1994 | .2007 | 30 |
| 20 | 1.3672 | .6851 | 4.8430 | 9.3149 | .2065 | 9.9909 | .9793 | 9.3058 | .2022 | .2036 | 40 |
| 10 | 1.3643 | .6788 | 4.7729 | 9.3212 | .2095 | 9.9907 | .9788 | 9.3119 | .2051 | .2065 | 50 |
| 78° 00' | 1.3614 | .6725 | 4.7046 | 9.3275 | .2126 | 9.9904 | .9782 | 9.3179 | .2079 | .2094 | 12° 00' |
| Deg | Radians | Log tan | Nat tan | Log cot | Nat cot | Log sin | Nat sin | Log cos | Nat cos | Radians | Deg |

## Table 1-9. Natural Trigonometric Functions and Their Logarithms (Continued)

| Deg | Radians | Nat sin | Log sin | Nat cos | Log cos | Nat tan | Log tan | Nat cot | Log cot | Radians | Deg |
|---|---|---|---|---|---|---|---|---|---|---|---|
| 12° 00' | .2094 | .2079 | 9.3179 | .9782 | 9.9904 | .2126 | 9.3225 | 4.7046 | .6725 | 1.3614 | 78° 00' |
| 10 | .2123 | .2108 | 9.3238 | .9775 | 9.9901 | .2156 | 9.3337 | 4.6383 | .6664 | 1.3584 | 50 |
| 20 | .2153 | .2136 | 9.3296 | .9769 | 9.9899 | .2186 | 9.3397 | 4.5736 | .6603 | 1.3555 | 40 |
| 30 | .2182 | .2164 | 9.3353 | .9763 | 9.9896 | .2217 | 9.3458 | 4.5107 | .6542 | 1.3526 | 30 |
| 40 | .2211 | .2193 | 9.3410 | .9757 | 9.9893 | .2248 | 9.3517 | 4.4494 | .6483 | 1.3497 | 20 |
| 50 | .2240 | .2221 | 9.3466 | .9750 | 9.9890 | .2278 | 9.3576 | 4.3897 | .6424 | 1.3468 | 10 |
| 13° 00' | .2269 | .2250 | 9.3521 | .9744 | 9.9887 | .2309 | 9.3634 | 4.3315 | .6366 | 1.3439 | 77° 00' |
| 10 | .2298 | .2278 | 9.3575 | .9737 | 9.9884 | .2339 | 9.3691 | 4.2747 | .6309 | 1.3410 | 50 |
| 20 | .2327 | .2306 | 9.3629 | .9730 | 9.9881 | .2370 | 9.3748 | 4.2193 | .6252 | 1.3381 | 40 |
| 30 | .2356 | .2335 | 9.3682 | .9724 | 9.9878 | .2401 | 9.3804 | 4.1653 | .6197 | 1.3352 | 30 |
| 40 | .2385 | .2363 | 9.3734 | .9717 | 9.9875 | .2432 | 9.3859 | 4.1126 | .6141 | 1.3323 | 20 |
| 50 | .2414 | .2391 | 9.3786 | .9710 | 9.9872 | .2462 | 9.3914 | 4.0611 | .6086 | 1.3294 | 10 |
| 14° 00' | .2443 | .2419 | 9.3837 | .9703 | 9.9869 | .2493 | 9.3968 | 4.0108 | .6032 | 1.3265 | 76° 00' |
| 10 | .2473 | .2447 | 9.3887 | .9696 | 9.9866 | .2524 | 9.4021 | 3.9617 | .5979 | 1.3235 | 50 |
| 20 | .2502 | .2476 | 9.3937 | .9689 | 9.9863 | .2555 | 9.4074 | 3.9136 | .5926 | 1.3206 | 40 |
| 30 | .2531 | .2504 | 9.3986 | .9682 | 9.9859 | .2586 | 9.4127 | 3.8667 | .5873 | 1.3177 | 30 |
| 40 | .2560 | .2532 | 9.4035 | .9674 | 9.9856 | .2617 | 9.4178 | 3.8208 | .5822 | 1.3148 | 20 |
| 50 | .2589 | .2560 | 9.4083 | .9667 | 9.9853 | .2648 | 9.4230 | 3.7760 | .5770 | 1.3119 | 10 |
| 15° 00' | .2618 | .2588 | 9.4130 | .9659 | 9.9849 | .2680 | 9.4281 | 3.7321 | .5720 | 1.3090 | 75° 00' |
| 10 | .2647 | .2616 | 9.4177 | .9652 | 9.9846 | .2711 | 9.4331 | 3.6891 | .5669 | 1.3061 | 50 |
| 20 | .2676 | .2644 | 9.4223 | .9644 | 9.9843 | .2742 | 9.4381 | 3.6471 | .5619 | 1.3032 | 40 |
| 30 | .2705 | .2672 | 9.4269 | .9636 | 9.9839 | .2773 | 9.4430 | 3.6059 | .5570 | 1.3003 | 30 |
| 40 | .2734 | .2700 | 9.4314 | .9629 | 9.9836 | .2805 | 9.4479 | 3.5656 | .5521 | 1.2974 | 20 |
| 50 | .2763 | .2728 | 9.4359 | .9621 | 9.9832 | .2836 | 9.4527 | 3.5261 | .5473 | 1.2945 | 10 |
| 16° 00' | .2793 | .2756 | 9.4403 | .9613 | 9.9828 | .2868 | 9.4575 | 3.4874 | .5425 | 1.2915 | 74° 00' |
| 10 | .2822 | .2784 | 9.4447 | .9605 | 9.9825 | .2899 | 9.4622 | 3.4495 | .5378 | 1.2886 | 50 |
| 20 | .2851 | .2812 | 9.4491 | .9596 | 9.9821 | .2931 | 9.4669 | 3.4124 | .5331 | 1.2857 | 40 |
| 30 | .2880 | .2840 | 9.4533 | .9588 | 9.9817 | .2962 | 9.4716 | 3.3759 | .5284 | 1.2828 | 30 |
| 40 | .2909 | .2868 | 9.4576 | .9580 | 9.9814 | .2994 | 9.4762 | 3.3402 | .5238 | 1.2799 | 20 |
| 50 | .2938 | .2896 | 9.4618 | .9572 | 9.9810 | .3026 | 9.4808 | 3.3052 | .5192 | 1.2770 | 10 |
| 17° 00' | .2967 | .2924 | 9.4659 | .9563 | 9.9806 | .3057 | 9.4853 | 3.2709 | .5147 | 1.2741 | 73° 00' |
| 10 | .2996 | .2952 | 9.4701 | .9555 | 9.9802 | .3089 | 9.4898 | 3.2371 | .5102 | 1.2712 | 50 |
| 20 | .3025 | .2979 | 9.4741 | .9546 | 9.9798 | .3121 | 9.4943 | 3.2041 | .5057 | 1.2683 | 40 |
| 30 | .3054 | .3007 | 9.4781 | .9537 | 9.9794 | .3153 | 9.4987 | 3.1716 | .5013 | 1.2654 | 30 |
| 40 | .3083 | .3035 | 9.4821 | .9528 | 9.9790 | .3185 | 9.5031 | 3.1397 | .4969 | 1.2625 | 20 |
| 50 | .3113 | .3063 | 9.4861 | .9520 | 9.9786 | .3217 | 9.5075 | 3.1084 | .4925 | 1.2595 | 10 |

| Deg | Radians | Log tan | Nat tan | Log cot | Nat cot | Log sin | Nat sin | Log cos | Nat cos | Radians | Deg |
|---|---|---|---|---|---|---|---|---|---|---|---|
| 72° 00' | 1.2566 | .4882 | 3.0777 | 9.5118 | .3249 | 9.9782 | .9511 | 9.4900 | .3090 | .3142 | 18° 00' |
| 50 | 1.2537 | .4839 | 3.0475 | 9.5161 | .3281 | 9.9778 | .9502 | 9.4939 | .3118 | .3171 | 10 |
| 40 | 1.2508 | .4797 | 3.0178 | 9.5203 | .3314 | 9.9774 | .9492 | 9.4977 | .3145 | .3200 | 20 |
| 30 | 1.2479 | .4755 | 2.9887 | 9.5245 | .3346 | 9.9770 | .9483 | 9.5015 | .3173 | .3229 | 30 |
| 20 | 1.2450 | .4713 | 2.9600 | 9.5287 | .3378 | 9.9765 | .9474 | 9.5052 | .3201 | .3258 | 40 |
| 10 | 1.2421 | .4672 | 2.9319 | 9.5329 | .3411 | 9.9761 | .9465 | 9.5090 | .3228 | .3287 | 50 |
| 71° 00' | 1.2392 | .4630 | 2.9042 | 9.5370 | .3443 | 9.9757 | .9455 | 9.5126 | .3256 | .3316 | 19° 00' |
| 50 | 1.2363 | .4589 | 2.8770 | 9.5411 | .3476 | 9.9752 | .9446 | 9.5163 | .3283 | .3345 | 10 |
| 40 | 1.2334 | .4549 | 2.8502 | 9.5451 | .3509 | 9.9748 | .9436 | 9.5199 | .3311 | .3374 | 20 |
| 30 | 1.2305 | .4509 | 2.8239 | 9.5492 | .3541 | 9.9744 | .9426 | 9.5235 | .3338 | .3403 | 30 |
| 20 | 1.2275 | .4469 | 2.7980 | 9.5532 | .3574 | 9.9739 | .9417 | 9.5271 | .3366 | .3432 | 40 |
| 10 | 1.2246 | .4429 | 2.7725 | 9.5571 | .3607 | 9.9734 | .9407 | 9.5306 | .3393 | .3462 | 50 |
| 70° 00' | 1.2217 | .4389 | 2.7475 | 9.5611 | .3640 | 9.9730 | .9397 | 9.5341 | .3420 | .3491 | 20° 00' |
| 50 | 1.2188 | .4350 | 2.7228 | 9.5650 | .3673 | 9.9725 | .9387 | 9.5375 | .3448 | .3520 | 10 |
| 40 | 1.2159 | .4311 | 2.6985 | 9.5689 | .3706 | 9.9721 | .9377 | 9.5409 | .3475 | .3549 | 20 |
| 30 | 1.2130 | .4273 | 2.6746 | 9.5727 | .3739 | 9.9716 | .9367 | 9.5443 | .3502 | .3578 | 30 |
| 20 | 1.2101 | .4234 | 2.6511 | 9.5766 | .3772 | 9.9711 | .9357 | 9.5477 | .3529 | .3607 | 40 |
| 10 | 1.2072 | .4196 | 2.6279 | 9.5804 | .3805 | 9.9706 | .9346 | 9.5510 | .3557 | .3636 | 50 |
| 69° 00' | 1.2043 | .4158 | 2.6051 | 9.5842 | .3839 | 9.9702 | .9336 | 9.5543 | .3584 | .3665 | 21° 00' |
| 50 | 1.2014 | .4121 | 2.5826 | 9.5879 | .3872 | 9.9697 | .9325 | 9.5576 | .3611 | .3694 | 10 |
| 40 | 1.1985 | .4083 | 2.5605 | 9.5917 | .3906 | 9.9692 | .9315 | 9.5609 | .3638 | .3723 | 20 |
| 30 | 1.1956 | .4046 | 2.5387 | 9.5954 | .3939 | 9.9687 | .9304 | 9.5641 | .3665 | .3752 | 30 |
| 20 | 1.1926 | .4009 | 2.5172 | 9.5991 | .3973 | 9.9682 | .9294 | 9.5673 | .3692 | .3782 | 40 |
| 10 | 1.1897 | .3972 | 2.4960 | 9.6028 | .4007 | 9.9677 | .9283 | 9.5704 | .3719 | .3811 | 50 |
| 68° 00' | 1.1868 | .3936 | 2.4751 | 9.6064 | .4040 | 9.9672 | .9272 | 9.5736 | .3746 | .3840 | 22° 00' |
| 50 | 1.1839 | .3900 | 2.4545 | 9.6100 | .4074 | 9.9667 | .9261 | 9.5767 | .3773 | .3869 | 10 |
| 40 | 1.1810 | .3864 | 2.4342 | 9.6136 | .4108 | 9.9661 | .9250 | 9.5798 | .3800 | .3898 | 20 |
| 30 | 1.1781 | .3828 | 2.4142 | 9.6172 | .4142 | 9.9656 | .9239 | 9.5828 | .3827 | .3927 | 30 |
| 20 | 1.1752 | .3792 | 2.3945 | 9.6208 | .4176 | 9.9651 | .9228 | 9.5859 | .3854 | .3956 | 40 |
| 10 | 1.1723 | .3757 | 2.3750 | 9.6243 | .4211 | 9.9646 | .9216 | 9.5889 | .3881 | .3985 | 50 |
| 67° 00' | 1.1694 | .3722 | 2.3559 | 9.6279 | .4245 | 9.9640 | .9205 | 9.5919 | .3907 | .4014 | 23° 00' |
| 50 | 1.1665 | .3687 | 2.3369 | 9.6314 | .4279 | 9.9635 | .9194 | 9.5948 | .3934 | .4043 | 10 |
| 40 | 1.1636 | .3652 | 2.3183 | 9.6348 | .4314 | 9.9629 | .9182 | 9.5978 | .3961 | .4072 | 20 |
| 30 | 1.1606 | .3617 | 2.2998 | 9.6383 | .4348 | 9.9624 | .9171 | 9.6007 | .3988 | .4102 | 30 |
| 20 | 1.1577 | .3583 | 2.2817 | 9.6418 | .4383 | 9.9619 | .9159 | 9.6036 | .4014 | .4131 | 40 |
| 10 | 1.1548 | .3548 | 2.2637 | 9.6452 | .4418 | 9.9613 | .9147 | 9.6065 | .4041 | .4160 | 50 |
| 66° 00' | 1.1519 | .3514 | 2.2460 | 9.6486 | .4452 | 9.9607 | .9136 | 9.6093 | .4067 | .4189 | 24° 00' |
| Deg | Radians | Log tan | Nat tan | Log cot | Nat cot | Log sin | Nat sin | Log cos | Nat cos | Radians | Deg |

Table 1-9. Natural Trigonometric Functions and Their Logarithms (Continued)

| Deg | Radians | Nat sin | Log sin | Nat cos | Log cos | Nat tan | Log tan | Nat cot | Log cot | Radians | Deg |
|---|---|---|---|---|---|---|---|---|---|---|---|
| 24° 00' | 0.4189 | 0.4067 | 9.6093 | 0.9136 | 9.9607 | 0.4452 | 9.6486 | 2.2460 | 0.3514 | 1.1519 | 66° 00' |
| 10 | .4218 | .4094 | 9.6121 | .9124 | 9.9602 | .4487 | 9.6520 | 2.2286 | .3480 | 1.1490 | 50 |
| 20 | .4247 | .4120 | 9.6149 | .9112 | 9.9596 | .4522 | 9.6554 | 2.2113 | .3447 | 1.1461 | 40 |
| 30 | .4276 | .4147 | 9.6177 | .9100 | 9.9590 | .4557 | 9.6587 | 2.1943 | .3413 | 1.1432 | 30 |
| 40 | .4305 | .4173 | 9.6205 | .9088 | 9.9584 | .4592 | 9.6620 | 2.1775 | .3380 | 1.1403 | 20 |
| 50 | .4334 | .4200 | 9.6232 | .9075 | 9.9579 | .4628 | 9.6654 | 2.1609 | .3346 | 1.1374 | 10 |
| 25° 00' | .4363 | .4226 | 9.6260 | .9063 | 9.9573 | .4663 | 9.6687 | 2.1445 | .3313 | 1.1345 | 65° 00' |
| 10 | .4392 | .4253 | 9.6287 | .9051 | 9.9567 | .4699 | 9.6720 | 2.1283 | .3280 | 1.1316 | 50 |
| 20 | .4422 | .4279 | 9.6313 | .9038 | 9.9561 | .4734 | 9.6752 | 2.1123 | .3248 | 1.1286 | 40 |
| 30 | .4451 | .4305 | 9.6340 | .9026 | 9.9555 | .4770 | 9.6785 | 2.0965 | .3215 | 1.1257 | 30 |
| 40 | .4480 | .4331 | 9.6366 | .9013 | 9.9549 | .4806 | 9.6817 | 2.0809 | .3183 | 1.1228 | 20 |
| 50 | .4509 | .4358 | 9.6392 | .9001 | 9.9543 | .4841 | 9.6850 | 2.0655 | .3150 | 1.1199 | 10 |
| 26° 00' | .4538 | .4384 | 9.6418 | .8988 | 9.9537 | .4877 | 9.6882 | 2.0503 | .3118 | 1.1170 | 64° 00' |
| 10 | .4567 | .4410 | 9.6444 | .8975 | 9.9530 | .4913 | 9.6914 | 2.0353 | .3086 | 1.1141 | 50 |
| 20 | .4596 | .4436 | 9.6470 | .8962 | 9.9524 | .4950 | 9.6946 | 2.0204 | .3054 | 1.1112 | 40 |
| 30 | .4625 | .4462 | 9.6495 | .8949 | 9.9518 | .4986 | 9.6977 | 2.0057 | .3023 | 1.1083 | 30 |
| 40 | .4654 | .4488 | 9.6521 | .8936 | 9.9512 | .5022 | 9.7009 | 1.9912 | .2991 | 1.1054 | 20 |
| 50 | .4683 | .4514 | 9.6546 | .8923 | 9.9505 | .5059 | 9.7040 | 1.9768 | .2960 | 1.1025 | 10 |
| 27° 00' | .4712 | .4540 | 9.6571 | .8910 | 9.9499 | .5095 | 9.7072 | 1.9626 | .2928 | 1.0996 | 63° 00' |
| 10 | .4741 | .4566 | 9.6595 | .8897 | 9.9492 | .5132 | 9.7103 | 1.9486 | .2897 | 1.0966 | 50 |
| 20 | .4771 | .4592 | 9.6620 | .8884 | 9.9486 | .5169 | 9.7134 | 1.9347 | .2866 | 1.0937 | 40 |
| 30 | .4800 | .4618 | 9.6644 | .8870 | 9.9479 | .5206 | 9.7165 | 1.9210 | .2835 | 1.0908 | 30 |
| 40 | .4829 | .4643 | 9.6668 | .8857 | 9.9473 | .5243 | 9.7196 | 1.9074 | .2805 | 1.0879 | 20 |
| 50 | .4858 | .4669 | 9.6692 | .8843 | 9.9466 | .5280 | 9.7226 | 1.8940 | .2774 | 1.0850 | 10 |
| 28° 00' | .4887 | .4695 | 9.6716 | .8830 | 9.9459 | .5317 | 9.7257 | 1.8807 | .2743 | 1.0821 | 62° 00' |
| 10 | .4916 | .4720 | 9.6740 | .8816 | 9.9453 | .5355 | 9.7287 | 1.8676 | .2713 | 1.0792 | 50 |
| 20 | .4945 | .4746 | 9.6763 | .8802 | 9.9446 | .5392 | 9.7318 | 1.8546 | .2683 | 1.0763 | 40 |
| 30 | .4974 | .4772 | 9.6787 | .8788 | 9.9439 | .5430 | 9.7348 | 1.8418 | .2652 | 1.0734 | 30 |
| 40 | .5003 | .4797 | 9.6810 | .8774 | 9.9432 | .5467 | 9.7378 | 1.8291 | .2622 | 1.0705 | 20 |
| 50 | .5032 | .4823 | 9.6833 | .8760 | 9.9425 | .5505 | 9.7408 | 1.8165 | .2592 | 1.0676 | 10 |
| 29° 00' | .5061 | .4848 | 9.6856 | .8746 | 9.9418 | .5543 | 9.7438 | 1.8041 | .2563 | 1.0647 | 61° 00' |
| 10 | .5091 | .4874 | 9.6878 | .8732 | 9.9411 | .5581 | 9.7467 | 1.7917 | .2533 | 1.0617 | 50 |
| 20 | .5120 | .4899 | 9.6901 | .8718 | 9.9404 | .5619 | 9.7497 | 1.7796 | .2503 | 1.0588 | 40 |
| 30 | .5149 | .4924 | 9.6923 | .8704 | 9.9397 | .5658 | 9.7526 | 1.7675 | .2474 | 1.0559 | 30 |
| 40 | .5178 | .4950 | 9.6946 | .8689 | 9.9390 | .5696 | 9.7556 | 1.7556 | .2444 | 1.0530 | 20 |
| 50 | .5207 | .4975 | 9.6968 | .8675 | 9.9383 | .5735 | 9.7585 | 1.7438 | .2415 | 1.0501 | 10 |

| Deg | Radians | Log tan | Nat tan | Log cot | Nat cot | Log sin | Nat sin | Log cos | Nat cos | Radians | Deg |
|---|---|---|---|---|---|---|---|---|---|---|---|
| 60° 00' | 1.0472 | .2386 | 1.7321 | 9.7614 | .5774 | 9.9375 | .8660 | 9.6990 | .5000 | .5236 | 30° 00' |
| 50 | 1.0443 | .2357 | 1.7205 | 9.7644 | .5812 | 9.9368 | .8646 | 9.7012 | .5025 | .5265 | 10 |
| 40 | 1.0414 | .2328 | 1.7090 | 9.7673 | .5851 | 9.9361 | .8631 | 9.7033 | .5050 | .5294 | 20 |
| 30 | 1.0385 | .2299 | 1.6977 | 9.7702 | .5891 | 9.9353 | .8616 | 9.7055 | .5075 | .5323 | 30 |
| 20 | 1.0356 | .2270 | 1.6864 | 9.7730 | .5930 | 9.9346 | .8602 | 9.7076 | .5100 | .5352 | 40 |
| 10 | 1.0327 | .2241 | 1.6753 | 9.7759 | .5969 | 9.9338 | .8587 | 9.7097 | .5125 | .5381 | 50 |
| 59° 00' | 1.0297 | .2212 | 1.6643 | 9.7788 | .6009 | 9.9331 | .8572 | 9.7118 | .5150 | .5411 | 31° 00' |
| 50 | 1.0268 | .2184 | 1.6534 | 9.7816 | .6048 | 9.9323 | .8557 | 9.7139 | .5175 | .5440 | 10 |
| 40 | 1.0239 | .2155 | 1.6426 | 9.7845 | .6088 | 9.9315 | .8542 | 9.7160 | .5200 | .5469 | 20 |
| 30 | 1.0210 | .2127 | 1.6319 | 9.7873 | .6128 | 9.9308 | .8526 | 9.7181 | .5225 | .5498 | 30 |
| 20 | 1.0181 | .2099 | 1.6213 | 9.7902 | .6168 | 9.9300 | .8511 | 9.7201 | .5250 | .5527 | 40 |
| 10 | 1.0152 | .2070 | 1.6107 | 9.7930 | .6208 | 9.9292 | .8496 | 9.7222 | .5275 | .5556 | 50 |
| 58° 00' | 1.0123 | .2042 | 1.6003 | 9.7958 | .6249 | 9.9284 | .8481 | 9.7242 | .5299 | .5585 | 32° 00' |
| 50 | 1.0094 | .2014 | 1.5900 | 9.7986 | .6289 | 9.9276 | .8465 | 9.7262 | .5324 | .5614 | 10 |
| 40 | 1.0065 | .1986 | 1.5798 | 9.8014 | .6330 | 9.9268 | .8450 | 9.7282 | .5348 | .5643 | 20 |
| 30 | 1.0036 | .1958 | 1.5697 | 9.8042 | .6371 | 9.9260 | .8434 | 9.7302 | .5373 | .5672 | 30 |
| 20 | 1.0007 | .1930 | 1.5597 | 9.8070 | .6412 | 9.9252 | .8418 | 9.7322 | .5398 | .5701 | 40 |
| 10 | 0.9977 | .1903 | 1.5497 | 9.8098 | .6453 | 9.9244 | .8403 | 9.7342 | .5422 | .5730 | 50 |
| 57° 00' | .9948 | .1875 | 1.5399 | 9.8125 | .6494 | 9.9236 | .8387 | 9.7361 | .5446 | .5760 | 33° 00' |
| 50 | .9919 | .1847 | 1.5301 | 9.8153 | .6536 | 9.9228 | .8371 | 9.7381 | .5471 | .5789 | 10 |
| 40 | .9890 | .1820 | 1.5204 | 9.8180 | .6577 | 9.9219 | .8355 | 9.7400 | .5495 | .5818 | 20 |
| 30 | .9861 | .1792 | 1.5108 | 9.8208 | .6619 | 9.9211 | .8339 | 9.7419 | .5519 | .5847 | 30 |
| 20 | .9832 | .1765 | 1.5013 | 9.8235 | .6661 | 9.9203 | .8323 | 9.7438 | .5544 | .5876 | 40 |
| 10 | .9803 | .1737 | 1.4919 | 9.8263 | .6703 | 9.9194 | .8307 | 9.7457 | .5568 | .5905 | 50 |
| 56° 00' | .9774 | .1710 | 1.4826 | 9.8290 | .6745 | 9.9186 | .8290 | 9.7476 | .5592 | .5934 | 34° 00' |
| 50 | .9745 | .1683 | 1.4733 | 9.8317 | .6788 | 9.9177 | .8274 | 9.7494 | .5616 | .5963 | 10 |
| 40 | .9716 | .1656 | 1.4641 | 9.8344 | .6830 | 9.9169 | .8258 | 9.7513 | .5640 | .5992 | 20 |
| 30 | .9687 | .1629 | 1.4550 | 9.8371 | .6873 | 9.9160 | .8241 | 9.7531 | .5664 | .6021 | 30 |
| 20 | .9657 | .1602 | 1.4460 | 9.8398 | .6916 | 9.9151 | .8225 | 9.7550 | .5688 | .6050 | 40 |
| 10 | .9628 | .1575 | 1.4370 | 9.8425 | .6959 | 9.9143 | .8208 | 9.7568 | .5712 | .6080 | 50 |
| 55° 00' | .9599 | .1548 | 1.4282 | 9.8452 | .7002 | 9.9134 | .8192 | 9.7586 | .5736 | .6109 | 35° 00' |
| 50 | .9570 | .1521 | 1.4193 | 9.8479 | .7046 | 9.9125 | .8175 | 9.7604 | .5760 | .6138 | 10 |
| 40 | .9541 | .1494 | 1.4106 | 9.8506 | .7089 | 9.9116 | .8158 | 9.7622 | .5783 | .6167 | 20 |
| 30 | .9512 | .1467 | 1.4020 | 9.8533 | .7133 | 9.9107 | .8141 | 9.7640 | .5807 | .6196 | 30 |
| 20 | .9483 | .1441 | 1.3934 | 9.8559 | .7177 | 9.9098 | .8124 | 9.7657 | .5831 | .6225 | 40 |
| 10 | .9454 | .1414 | 1.3848 | 9.8586 | .7221 | 9.9089 | .8107 | 9.7675 | .5854 | .6254 | 50 |
| 54° 00' | .9425 | .1387 | 1.3764 | 9.8613 | .7265 | 9.9080 | .8090 | 9.7692 | .5878 | .6283 | 36° 00' |
| Deg | Radians | Log tan | Nat tan | Log cot | Nat cot | Log sin | Nat sin | Log cos | Nat cos | Radians | Deg |

Table 1-9. Natural Trigonometric Functions and Their Logarithms (Continued)

| Deg | Radians | Nat sin | Log sin | Nat cos | Log cos | Nat tan | Log tan | Nat cot | Log cot | Radians | Deg |
|---|---|---|---|---|---|---|---|---|---|---|---|
| 36° 00' | 0.6283 | 0.5878 | 9.7692 | 0.8090 | 9.9080 | 0.7265 | 9.8613 | 1.3764 | 0.1387 | 0.9425 | 54° 0' |
| 10 | .6312 | .5901 | 9.7710 | .8073 | 9.9070 | .7310 | 9.8639 | 1.3680 | .1361 | .9396 | 50 |
| 20 | .6341 | .5925 | .7727 | .8056 | .9061 | .7355 | .8666 | 1.3597 | .1334 | .9367 | 40 |
| 30 | .6370 | .5948 | .7744 | .8039 | .9052 | .7400 | .8692 | 1.3514 | .1308 | .9338 | 30 |
| 40 | .6400 | .5972 | .7761 | .8021 | .9042 | .7445 | .8719 | 1.3432 | .1282 | .9308 | 20 |
| 50 | .6429 | .5995 | .7778 | .8004 | .9033 | .7490 | .8745 | 1.3351 | .1255 | .9279 | 10 |
| 37° 00' | .6458 | .6018 | 9.7795 | .7986 | 9.9024 | .7536 | 9.8771 | 1.3270 | .1229 | .9250 | 53° 00' |
| 10 | .6487 | .6041 | .7811 | .7969 | .9014 | .7581 | .8797 | 1.3190 | .1203 | .9221 | 50 |
| 20 | .6516 | .6065 | .7828 | .7951 | .9004 | .7627 | .8824 | 1.3111 | .1176 | .9192 | 40 |
| 30 | .6545 | .6088 | .7845 | .7934 | .8995 | .7673 | .8850 | 1.3032 | .1150 | .9163 | 30 |
| 40 | .6574 | .6111 | .7861 | .7916 | .8985 | .7720 | .8876 | 1.2954 | .1124 | .9134 | 20 |
| 50 | .6603 | .6134 | .7877 | .7898 | .8975 | .7766 | .8902 | 1.2876 | .1098 | .9105 | 10 |
| 38° 00' | .6632 | .6157 | 9.7893 | .7880 | 9.8965 | .7813 | 9.8928 | 1.2799 | .1072 | .9076 | 52° 00' |
| 10 | .6661 | .6180 | .7910 | .7862 | .8955 | .7860 | .8954 | 1.2723 | .1046 | .9047 | 50 |
| 20 | .6690 | .6202 | .7926 | .7844 | .8946 | .7907 | .8980 | 1.2647 | .1020 | .9018 | 40 |
| 30 | .6720 | .6225 | .7942 | .7826 | .8935 | .7954 | .9006 | 1.2572 | .0994 | .8988 | 30 |
| 40 | .6749 | .6248 | .7957 | .7808 | .8925 | .8002 | .9032 | 1.2497 | .0968 | .8959 | 20 |
| 50 | .6778 | .6271 | .7973 | .7790 | .8915 | .8050 | .9058 | 1.2423 | .0942 | .8930 | 10 |
| 39° 00' | .6807 | .6293 | 9.7989 | .7772 | 9.8905 | .8098 | 9.9084 | 1.2349 | .0916 | .8901 | 51° 00' |
| 10 | .6836 | .6316 | .8004 | .7753 | .8895 | .8146 | .9110 | 1.2276 | .0891 | .8872 | 50 |
| 20 | .6865 | .6338 | .8020 | .7735 | .8884 | .8195 | .9135 | 1.2203 | .0865 | .8843 | 40 |
| 30 | .6894 | .6361 | .8035 | .7716 | .8874 | .8243 | .9161 | 1.2131 | .0839 | .8814 | 30 |
| 40 | .6923 | .6383 | .8050 | .7698 | .8864 | .8292 | .9187 | 1.2059 | .0813 | .8785 | 20 |
| 50 | .6952 | .6406 | .8066 | .7679 | .8853 | .8342 | .9213 | 1.1988 | .0788 | .8756 | 10 |
| 40° 00' | .6981 | .6428 | 9.8081 | .7660 | 9.8843 | .8391 | 9.9238 | 1.1918 | .0762 | .8727 | 50° 00' |
| 10 | .7010 | .6450 | .8096 | .7642 | .8832 | .8441 | .9264 | 1.1847 | .0736 | .8698 | 50 |
| 20 | .7039 | .6472 | .8111 | .7623 | .8821 | .8491 | .9289 | 1.1778 | .0711 | .8668 | 40 |
| 30 | .7069 | .6495 | .8125 | .7604 | .8811 | .8541 | .9315 | 1.1709 | .0685 | .8639 | 30 |
| 40 | .7098 | .6517 | .8140 | .7585 | .8800 | .8591 | .9341 | 1.1640 | .0659 | .8610 | 20 |
| 50 | .7127 | .6539 | .8155 | .7566 | .8789 | .8642 | .9366 | 1.1572 | .0634 | .8581 | 10 |
| 41° 00' | .7156 | .6561 | 9.8169 | .7547 | 9.8778 | .8693 | 9.9392 | 1.1504 | .0608 | .8552 | 49° 00' |
| 10 | .7185 | .6583 | .8184 | .7528 | .8767 | .8744 | .9417 | 1.1436 | .0583 | .8523 | 50 |
| 20 | .7214 | .6604 | .8198 | .7509 | .8756 | .8796 | .9443 | 1.1369 | .0557 | .8494 | 40 |
| 30 | .7243 | .6626 | .8213 | .7490 | .8745 | .8847 | .9468 | 1.1303 | .0532 | .8465 | 30 |
| 40 | .7272 | .6648 | .8227 | .7470 | .8733 | .8899 | .9494 | 1.1237 | .0507 | .8436 | 20 |
| 50 | .7301 | .6670 | .8241 | .7451 | .8722 | .8952 | .9519 | 1.1171 | .0481 | .8407 | 10 |

| Deg | Radians | Log tan | Nat tan | Log cot | Nat cot | Log sin | Nat sin | Log cos | Nat cos | Radians | Deg |
|---|---|---|---|---|---|---|---|---|---|---|---|
| 42° 00' | .7330 | .0456 | 1.1106 | 9.9544 | .9004 | 9.8711 | .7431 | 9.8255 | .6691 | .8378 | 48° 00' |
| 10 | .7359 | .0430 | 1.1041 | 9.9570 | .9057 | 9.8699 | .7412 | 9.8269 | .6713 | .8348 | 50 |
| 20 | .7389 | .0405 | 1.0977 | 9.9595 | .9110 | 9.8688 | .7392 | 9.8283 | .6734 | .8319 | 40 |
| 30 | .7418 | .0380 | 1.0913 | 9.9621 | .9163 | 9.8676 | .7373 | 9.8297 | .6756 | .8290 | 30 |
| 40 | .7447 | .0354 | 1.0850 | 9.9646 | .9217 | 9.8665 | .7353 | 9.8311 | .6777 | .8261 | 20 |
| 50 | .7476 | .0329 | 1.0786 | 9.9671 | .9271 | 9.8653 | .7333 | 9.8324 | .6799 | .8232 | 10 |
| 43° 00' | .7505 | .0303 | 1.0724 | 9.9697 | .9325 | 9.8641 | .7314 | 9.8338 | .6820 | .8203 | 47° 00' |
| 10 | .7534 | .0278 | 1.0661 | 9.9722 | .9380 | 9.8630 | .7294 | 9.8351 | .6841 | .8174 | 50 |
| 20 | .7563 | .0253 | 1.0599 | 9.9747 | .9435 | 9.8618 | .7274 | 9.8365 | .6862 | .8145 | 40 |
| 30 | .7592 | .0228 | 1.0538 | 9.9773 | .9490 | 9.8606 | .7254 | 9.8378 | .6884 | .8116 | 30 |
| 40 | .7621 | .0202 | 1.0477 | 9.9798 | .9545 | 9.8594 | .7234 | 9.8391 | .6905 | .8087 | 20 |
| 50 | .7650 | .0177 | 1.0416 | 9.9823 | .9601 | 9.8582 | .7214 | 9.8405 | .6926 | .8058 | 10 |
| 44° 00' | .7679 | .0152 | 1.0355 | 9.9848 | .9657 | 9.8569 | .7193 | 9.8418 | .6947 | .8029 | 46° 00' |
| 10 | .7709 | .0126 | 1.0295 | 9.9874 | .9713 | 9.8557 | .7173 | 9.8431 | .6968 | .7999 | 50 |
| 20 | .7738 | .0101 | 1.0236 | 9.9899 | .9770 | 9.8545 | .7153 | 9.8444 | .6988 | .7970 | 40 |
| 30 | .7767 | .0076 | 1.0176 | 9.9924 | .9827 | 9.8532 | .7133 | 9.8457 | .7009 | .7941 | 30 |
| 40 | .7796 | .0051 | 1.0117 | 9.9950 | .9884 | 9.8520 | .7112 | 9.8469 | .7030 | .7912 | 20 |
| 50 | .7825 | .0025 | 1.0058 | 9.9975 | .9942 | 9.8507 | .7092 | 9.8482 | .7051 | .7883 | 10 |
| 45° 00' | .7854 | .0000 | 1.0000 | 0.0000 | 1.0000 | 9.8495 | .7071 | 9.8495 | .7071 | .7854 | 45° 00' |
| Deg | Radians | Log tan | Nat tan | Log cot | Nat cot | Log sin | Nat sin | Log cos | Nat cos | Radians | Deg |

## Table 1-10. Values and Logarithms of Exponential and Hyperbolic Functions

| $x$ | $e^x$ Value | $e^x$ $\log_{10}$ | $e^{-x}$ (value) | $\sinh x$ Value | $\sinh x$ $\log_{10}$ | $\cosh x$ Value | $\cosh x$ $\log_{10}$ | $\tanh x$ (value) |
|---|---|---|---|---|---|---|---|---|
| 0.00 | 1.0000 | 0.00000 | 1.00000 | 0.0000 | $-\infty$ | 1.0000 | 0.00000 | 0.00000 |
| 0.01 | 1.0101 | .00434 | 0.99005 | .0100 | $\bar{2}.00001$ | 1.0001 | .00002 | .01000 |
| 0.02 | 1.0202 | .00869 | .98020 | .0200 | $\bar{2}.30106$ | 1.0002 | .00009 | .02000 |
| 0.03 | 1.0305 | .01303 | .97045 | .0300 | $\bar{2}.47719$ | 1.0005 | .00020 | .02999 |
| 0.04 | 1.0408 | .01737 | .96079 | .0400 | $\bar{2}.60218$ | 1.0008 | .00035 | .03998 |
| 0.05 | 1.0513 | .02171 | .95123 | .0500 | $\bar{2}.69915$ | 1.0013 | .00054 | .04996 |
| 0.06 | 1.0618 | .02606 | .94176 | .0600 | $\bar{2}.77841$ | 1.0018 | .00078 | .05993 |
| 0.07 | 1.0725 | .03040 | .93239 | .0701 | $\bar{2}.84545$ | 1.0025 | .00106 | .06989 |
| 0.08 | 1.0833 | .03474 | .92312 | .0801 | $\bar{2}.90355$ | 1.0032 | .00139 | .07983 |
| 0.09 | 1.0942 | .03909 | .91393 | .0901 | $\bar{2}.95483$ | 1.0041 | .00176 | .08976 |
| 0.10 | 1.1052 | .04343 | .90484 | .1002 | $\bar{1}.00072$ | 1.0050 | .00217 | .09967 |
| 0.11 | 1.1163 | .04777 | .89583 | .1102 | $\bar{1}.04227$ | 1.0061 | .00262 | .10956 |
| 0.12 | 1.1275 | .05212 | .88692 | .1203 | $\bar{1}.08022$ | 1.0072 | .00312 | .11943 |
| 0.13 | 1.1388 | .05646 | .87809 | .1304 | $\bar{1}.11517$ | 1.0085 | .00366 | .12927 |
| 0.14 | 1.1503 | .06080 | .86936 | .1405 | $\bar{1}.14755$ | 1.0098 | .00424 | .13909 |
| 0.15 | 1.1618 | .06514 | .86071 | .1506 | $\bar{1}.17772$ | 1.0113 | .00487 | .14889 |
| 0.16 | 1.1735 | .06949 | .85214 | .1607 | $\bar{1}.20597$ | 1.0128 | 00554 | .15865 |
| 0.17 | 1.1853 | .07383 | .84366 | .1708 | $\bar{1}.23254$ | 1.0145 | .00625 | .16838 |
| 0.18 | 1.1972 | .07817 | .83527 | .1810 | $\bar{1}.25762$ | 1.0162 | .00700 | .17808 |
| 0.19 | 1.2092 | .08252 | .82696 | .1911 | $\bar{1}.28136$ | 1.0181 | .00779 | .18775 |
| 0.20 | 1.2214 | .08686 | .81873 | .2013 | $\bar{1}.30392$ | 1.0201 | .00863 | .19738 |
| 0.21 | 1.2337 | .09120 | .81058 | .2115 | $\bar{1}.32541$ | 1.0221 | 00951 | .20697 |
| 0.22 | 1.2461 | .09554 | .80252 | .2218 | $\bar{1}.34592$ | 1.0243 | .01043 | .21652 |
| 0.23 | 1.2586 | .09989 | .79453 | .2320 | $\bar{1}.36555$ | 1.0266 | .01139 | .22603 |
| 0.24 | 1.2712 | .10423 | .78663 | .2423 | $\bar{1}.38437$ | 1.0289 | .01239 | .23550 |
| 0.25 | 1.2840 | .10857 | .77880 | .2526 | $\bar{1}.40245$ | 1.0314 | .01343 | .24492 |
| 0.26 | 1.2969 | .11292 | .77105 | .2629 | $\bar{1}.41986$ | 1.0340 | .01452 | .25430 |
| 0.27 | 1.3100 | .11726 | .76338 | .2733 | $\bar{1}.43663$ | 1.0367 | .01564 | .26362 |
| 0.28 | 1.3231 | .12160 | .75578 | .2837 | $\bar{1}.45282$ | 1.0395 | .01681 | .27291 |
| 0.29 | 1.3364 | .12595 | .74826 | .2941 | $\bar{1}.46847$ | 1.0423 | .01801 | .28213 |
| 0.30 | 1.3499 | .13029 | .74082 | .3045 | $\bar{1}.48362$ | 1.0453 | .01926 | .29131 |
| 0.31 | 1.3634 | .13463 | .73345 | .3150 | $\bar{1}.49830$ | 1.0484 | .02054 | .30044 |
| 0.32 | 1.3771 | .13897 | .72615 | .3255 | $\bar{1}.51254$ | 1.0516 | .02187 | .30951 |
| 0.33 | 1.3910 | .14332 | .71892 | .3360 | $\bar{1}.52637$ | 1.0549 | .02323 | .31852 |
| 0.34 | 1.4049 | .14766 | .71177 | .3466 | $\bar{1}.53981$ | 1.0584 | .02463 | .32748 |
| 0.35 | 1.4191 | .15200 | .70469 | .3572 | $\bar{1}.55290$ | 1.0619 | .02607 | .33638 |
| 0.36 | 1.4333 | .15635 | .69768 | .3678 | $\bar{1}.56564$ | 1.0655 | .02755 | .34521 |
| 0.37 | 1.4477 | .16069 | .69073 | .3785 | $\bar{1}.57807$ | 1.0692 | .02907 | .35399 |
| 0.38 | 1.4623 | .16503 | .68386 | .3892 | $\bar{1}.59019$ | 1.0731 | .03063 | .36271 |
| 0.39 | 1.4770 | .16937 | .67706 | .4000 | $\bar{1}.60202$ | 1.0770 | .03222 | .37136 |
| 0.40 | 1.4918 | .17372 | .67032 | .4108 | $\bar{1}$ 61358 | 1.0811 | .03385 | .37995 |
| 0.41 | 1.5068 | .17806 | .66365 | .4216 | $\bar{1}.62488$ | 1.0852 | .03552 | .38847 |
| 0.42 | 1.5220 | .18240 | .65705 | .4325 | $\bar{1}.63594$ | 1.0895 | .03723 | .39693 |
| 0.43 | 1 5373 | .18675 | .65051 | .4434 | $\bar{1}.64677$ | 1.0939 | .03897 | .40532 |
| 0.44 | 1.5527 | .19109 | .64404 | .4543 | $\bar{1}.65738$ | 1.0984 | .04075 | .41364 |
| 0.45 | 1.5683 | .19543 | .63763 | .4653 | $\bar{1}.66777$ | 1.1030 | .04256 | .42190 |
| 0.46 | 1.5841 | .19978 | .63128 | .4764 | $\bar{1}.67797$ | 1.1077 | .04441 | .43008 |
| 0.47 | 1.6000 | .20412 | .62500 | .4875 | $\bar{1}.68797$ | 1.1125 | .04630 | .43820 |
| 0.48 | 1.6161 | .20846 | .61878 | .4986 | $\bar{1}.69779$ | 1.1174 | .04822 | .44624 |
| 0.49 | 1.6323 | .21280 | .61263 | .5098 | $\bar{1}.70744$ | 1.1225 | .05018 | .45422 |
| 0.50 | 1.6487 | .21715 | .60653 | .5211 | $\bar{1}.71692$ | 1.1276 | .05217 | .46212 |

## Table 1-10. Values and Logarithms of Exponential and Hyperbolic Functions (Continued)

| $x$ | $e^x$ Value | $e^x$ $\log_{10}$ | $e^{-x}$ (value) | $\sinh x$ Value | $\sinh x$ $\log_{10}$ | $\cosh x$ Value | $\cosh x$ $\log_{10}$ | $\tanh x$ (value) |
|---|---|---|---|---|---|---|---|---|
| 0.50 | 1.6487 | 0.21715 | 0.60653 | 0.5211 | 1̄.71692 | 1.1276 | 0.05217 | 0.46212 |
| 0.51 | 1.6653 | .22149 | .60050 | 0.5324 | 1̄.72624 | 1.1329 | .05419 | .46995 |
| 0.52 | 1.6820 | .22583 | .59452 | 0.5438 | 1̄.73540 | 1.1383 | .05625 | .47770 |
| 0.53 | 1.6989 | .23018 | .58860 | 0.5552 | 1̄.74442 | 1.1438 | .05834 | .48538 |
| 0.54 | 1.7160 | .23452 | .58275 | 0.5666 | 1̄.75330 | 1.1494 | .06046 | .49299 |
| 0.55 | 1.7333 | .23886 | .57695 | 0.5782 | 1̄.76204 | 1.1551 | .06262 | .50052 |
| 0.56 | 1.7507 | .24320 | .57121 | 0.5897 | 1̄.77065 | 1.1609 | .06481 | .50798 |
| 0.57 | 1.7683 | .24755 | .56553 | 0.6014 | 1̄.77914 | 1.1669 | .06703 | .51536 |
| 0.58 | 1.7860 | .25189 | .55990 | 0.6131 | 1̄.78751 | 1.1730 | .06929 | .52267 |
| 0.59 | 1.8040 | .25623 | .55433 | 0.6248 | 1̄.79576 | 1.1792 | .07157 | .52990 |
| 0.60 | 1.8221 | .26058 | .54881 | 0.6367 | 1̄.80390 | 1.1855 | .07389 | .53705 |
| 0.61 | 1.8404 | .26492 | .54335 | 0.6485 | 1̄.81194 | 1.1919 | .07624 | .54413 |
| 0.62 | 1.8589 | .26926 | .53794 | 0.6605 | 1̄.81987 | 1.1984 | .07861 | .55113 |
| 0.63 | 1.8776 | .27361 | .53259 | 0.6725 | 1̄.82770 | 1.2051 | .08102 | .55805 |
| 0.64 | 1.8965 | .27795 | .52729 | 0.6846 | 1̄.83543 | 1.2119 | .08346 | .56490 |
| 0.65 | 1.9155 | .28229 | .52205 | 0.6967 | 1̄.84308 | 1.2188 | .08593 | .57167 |
| 0.66 | 1.9348 | .28664 | .51685 | 0.7090 | 1̄.85063 | 1.2258 | .08843 | .57836 |
| 0.67 | 1.9542 | .29098 | .51171 | 0.7213 | 1̄.85809 | 1.2330 | .09095 | .58498 |
| 0.68 | 1.9739 | .29532 | .50662 | 0.7336 | 1̄.86548 | 1.2402 | .09351 | .59152 |
| 0.69 | 1.9937 | .29966 | .50158 | 0.7461 | 1̄.87278 | 1.2476 | .09609 | .59798 |
| 0.70 | 2.0138 | .30401 | .49659 | 0.7586 | 1̄.88000 | 1.2552 | .09870 | .60437 |
| 0.71 | 2.0340 | .30835 | .49164 | 0.7712 | 1̄.88715 | 1.2628 | .10134 | .61068 |
| 0.72 | 2.0544 | .31269 | .48675 | 0.7838 | 1̄.89423 | 1.2706 | .10401 | .61691 |
| 0.73 | 2.0751 | .31703 | .48191 | 0.7966 | 1̄.90123 | 1.2785 | .10670 | .62307 |
| 0.74 | 2.0959 | .32138 | .47711 | 0.8094 | 1̄.90817 | 1.2865 | .10942 | .62915 |
| 0.75 | 2.1170 | .32572 | .47237 | 0.8223 | 1̄.91504 | 1.2947 | .11216 | .63515 |
| 0.76 | 2.1383 | .33006 | .46767 | 0.8353 | 1̄.92185 | 1.3030 | .11493 | .64108 |
| 0.77 | 2.1598 | .33441 | .46301 | 0.8484 | 1̄.92859 | 1.3114 | .11773 | .64693 |
| 0.78 | 2.1815 | .33875 | .45841 | 0.8615 | 1̄.93527 | 1.3199 | .12055 | .65271 |
| 0.79 | 2.2034 | .34309 | .45384 | 0.8748 | 1̄.94190 | 1.3286 | .12340 | .65841 |
| 0.80 | 2.2255 | .34744 | .44933 | 0.8881 | 1̄.94846 | 1.3374 | .12627 | .66404 |
| 0.81 | 2.2479 | .35178 | .44486 | 0.9015 | 1̄.95498 | 1.3464 | .12917 | .66959 |
| 0.82 | 2.2705 | .35612 | .44043 | 0.9150 | 1̄.96144 | 1.3555 | .13209 | .67507 |
| 0.83 | 2.2933 | .36046 | .43605 | 0.9286 | 1̄.96784 | 1.3647 | .13503 | .68048 |
| 0.84 | 2.3164 | .36481 | .43171 | 0.9423 | 1̄.97420 | 1.3740 | .13800 | .68581 |
| 0.85 | 2.3396 | .36915 | .42741 | 0.9561 | 1̄.98051 | 1.3835 | .14099 | .69107 |
| 0.86 | 2.3632 | .37349 | .42316 | 0.9700 | 1̄.98677 | 1.3932 | .14400 | .69626 |
| 0.87 | 2.3869 | .37784 | .41895 | 0.9840 | 1̄.99299 | 1.4029 | .14704 | .70137 |
| 0.88 | 2.4109 | .38218 | .41478 | 0.9981 | 1̄.99916 | 1.4128 | .15009 | .70642 |
| 0.89 | 2.4351 | .38652 | .41066 | 1.0122 | 0.00528 | 1.4229 | .15317 | .71139 |
| 0.90 | 2.4596 | .39087 | .40657 | 1.0265 | .01137 | 1.4331 | .15627 | .71630 |
| 0.91 | 2.4843 | .39521 | .40252 | 1.0409 | .01741 | 1.4434 | .15939 | .72113 |
| 0.92 | 2.5093 | .39955 | .39852 | 1.0554 | .02341 | 1.4539 | .16254 | .72590 |
| 0.93 | 2.5345 | .40389 | .39455 | 1.0700 | .02937 | 1.4645 | .16570 | .73059 |
| 0.94 | 2.5600 | .40824 | .39063 | 1.0847 | .03530 | 1.4753 | .16888 | .73522 |
| 0.95 | 2.5857 | .41258 | .38674 | 1.0995 | .04119 | 1.4862 | .17208 | .73978 |
| 0.96 | 2.6117 | .41692 | .38289 | 1.1144 | .04704 | 1.4973 | .17531 | .74428 |
| 0.97 | 2.6379 | .42127 | .37908 | 1.1294 | .05286 | 1.5085 | .17855 | .74870 |
| 0.98 | 2.6645 | .42561 | .37531 | 1.1446 | .05864 | 1.5199 | .18181 | .75307 |
| 0.99 | 2.6912 | .42995 | .37158 | 1.1598 | .06439 | 1.5314 | .18509 | .75736 |
| 1.00 | 2.7183 | .43429 | .36788 | 1.1752 | .07011 | 1.5431 | .18839 | .76159 |

## Table 1-10. Values and Logarithms of Exponential and Hyperbolic Functions (Continued)

| $x$ | $e^x$ Value | $e^x$ log₁₀ | $e^{-x}$ (value) | sinh $x$ Value | sinh $x$ log₁₀ | cosh $x$ Value | cosh $x$ log₁₀ | tanh $x$ (value) |
|---|---|---|---|---|---|---|---|---|
| 1.00 | 2.7183 | 0.43429 | 0.36788 | 1.1752 | 0.07011 | 1.5431 | 0.18839 | 0.76159 |
| 1.01 | 2.7456 | .43864 | .36422 | 1.1907 | .07580 | 1.5549 | .19171 | .76576 |
| 1.02 | 2.7732 | .44298 | .36060 | 1.2063 | .08146 | 1.5669 | .19504 | .76987 |
| 1.03 | 2.8011 | .44732 | .35701 | 1.2220 | .08708 | 1.5790 | .19839 | .77391 |
| 1.04 | 2.8292 | .45167 | .35345 | 1.2379 | .09268 | 1.5913 | .20176 | .77789 |
| 1.05 | 2.8577 | .45601 | .34994 | 1.2539 | .09825 | 1.6038 | .20515 | .78181 |
| 1.06 | 2.8864 | .46035 | .34646 | 1.2700 | .10379 | 1.6164 | .20855 | .78566 |
| 1.07 | 2.9154 | .46470 | .34301 | 1.2862 | .10930 | 1.6292 | .21197 | .78946 |
| 1.08 | 2.9447 | .46904 | .33960 | 1.3025 | .11479 | 1.6421 | .21541 | .79320 |
| 1.09 | 2.9743 | .47338 | .33622 | 1.3190 | .12025 | 1.6552 | .21886 | .79688 |
| 1.10 | 3.0042 | .47772 | .33287 | 1.3356 | .12569 | 1.6685 | .22233 | .80050 |
| 1.11 | 3.0344 | .48207 | .32956 | 1.3524 | .13111 | 1.6820 | .22582 | .80406 |
| 1.12 | 3.0649 | .48641 | .32628 | 1.3693 | .13649 | 1.6956 | .22931 | .80757 |
| 1.13 | 3.0957 | .49075 | .32303 | 1.3863 | .14186 | 1.7093 | .23283 | .81102 |
| 1.14 | 3.1268 | .49510 | .31982 | 1.4035 | .14720 | 1.7233 | .23636 | .81441 |
| 1.15 | 3.1582 | .49944 | .31664 | 1.4208 | .15253 | 1.7374 | .23990 | .81775 |
| 1.16 | 3.1899 | .50378 | .31349 | 1.4382 | .15783 | 1.7517 | .24346 | .82104 |
| 1.17 | 3.2220 | .50812 | .31037 | 1.4558 | .16311 | 1.7662 | .24703 | .82427 |
| 1.18 | 3.2544 | .51247 | .30728 | 1.4735 | .16836 | 1.7808 | .25062 | .82745 |
| 1.19 | 3.2871 | .51681 | .30422 | 1.4914 | .17360 | 1.7957 | .25422 | .83058 |
| 1.20 | 3.3201 | .52115 | .30119 | 1.5095 | .17882 | 1.8107 | .25784 | .83365 |
| 1.21 | 3.3535 | .52550 | .29820 | 1.5276 | .18402 | 1.8258 | .26146 | .83668 |
| 1.22 | 3.3872 | .52984 | .29523 | 1.5460 | .18920 | 1.8412 | .26510 | .83965 |
| 1.23 | 3.4212 | .53418 | .29229 | 1.5645 | .19437 | 1.8568 | .26876 | .84258 |
| 1.24 | 3.4556 | .53853 | .28938 | 1.5831 | .19951 | 1.8725 | .27242 | .84546 |
| 1.25 | 3.4903 | .54287 | .28650 | 1.6019 | .20464 | 1.8884 | .27610 | .84828 |
| 1.26 | 3.5254 | .54721 | .28365 | 1.6209 | .20975 | 1.9045 | .27979 | .85106 |
| 1.27 | 3.5609 | .55155 | .28083 | 1.6400 | .21485 | 1.9208 | .28349 | .85380 |
| 1.28 | 3.5966 | .55590 | .27804 | 1.6593 | .21993 | 1.9373 | .28721 | .85648 |
| 1.29 | 3.6328 | .56024 | .27527 | 1.6788 | .22499 | 1.9540 | .29093 | .85913 |
| 1.30 | 3.6693 | .56458 | .27253 | 1.6984 | .23004 | 1.9709 | .29467 | .86172 |
| 1.31 | 3.7062 | .56893 | .26982 | 1.7182 | .23507 | 1.9880 | .29842 | .86428 |
| 1.32 | 3.7434 | .57327 | .26714 | 1.7381 | .24009 | 2.0053 | .30217 | .86678 |
| 1.33 | 3.7810 | .57761 | .26448 | 1.7583 | .24509 | 2.0228 | .30594 | .86925 |
| 1.34 | 3.8190 | .58195 | .26185 | 1.7786 | .25008 | 2.0404 | .30972 | .87167 |
| 1.35 | 3.8574 | .58630 | .25924 | 1.7991 | .25505 | 2.0583 | .31352 | .87405 |
| 1.36 | 3.8962 | .59064 | .25666 | 1.8198 | .26002 | 2.0764 | .31732 | .87639 |
| 1.37 | 3.9354 | .59498 | .25411 | 1.8406 | .26496 | 2.0947 | .32113 | .87869 |
| 1.38 | 3.9749 | .59933 | .25158 | 1.8617 | .26990 | 2.1132 | .32495 | .88095 |
| 1.39 | 4.0149 | .60367 | .24908 | 1.8829 | .27482 | 2.1320 | .32878 | .88317 |
| 1.40 | 4.0552 | .60801 | .24660 | 1.9043 | .27974 | 2.1509 | .33262 | .88535 |
| 1.41 | 4.0960 | .61236 | .24414 | 1.9259 | .28464 | 2.1700 | .33647 | .88749 |
| 1.42 | 4.1371 | .61670 | .24171 | 1.9477 | .28952 | 2.1894 | .34033 | .88960 |
| 1.43 | 4.1787 | .62104 | .23931 | 1.9697 | .29440 | 2.2090 | .34420 | .89167 |
| 1.44 | 4.2207 | .62538 | .23693 | 1.9919 | .29926 | 2.2288 | .34807 | .89370 |
| 1.45 | 4.2631 | .62973 | .23457 | 2.0143 | .30412 | 2.2488 | .35196 | .89569 |
| 1.46 | 4.3060 | .63407 | .23224 | 2.0369 | .30896 | 2.2691 | .35585 | .89765 |
| 1.47 | 4.3492 | .63841 | .22993 | 2.0597 | .31379 | 2.2896 | .35976 | .89958 |
| 1.48 | 4.3929 | .64276 | .22764 | 2.0827 | .31862 | 2.3103 | .36367 | .90147 |
| 1.49 | 4.4371 | .64710 | .22537 | 2.1059 | .32343 | 2.3312 | .36759 | .90332 |
| 1.50 | 4.4817 | .65144 | .22313 | 2.1293 | .32823 | 2.3524 | .37151 | .90515 |

## Table 1-10. Values and Logarithms of Exponential and Hyperbolic Functions (Continued)

| $x$ | $e^x$ | | $e^{-x}$ (value) | sinh $x$ | | cosh $x$ | | tanh $x$ (value) |
|---|---|---|---|---|---|---|---|---|
| | Value | log₁₀ | | Value | log₁₀ | Value | log₁₀ | |
| 1.50 | 4.4817 | 0.65144 | 0.22313 | 2.1293 | 0.32823 | 2.3524 | 0.37151 | 0.90515 |
| 1.51 | 4.5267 | .65578 | .22091 | 2.1529 | .33303 | 2.3738 | .37545 | .90694 |
| 1.52 | 4.5722 | .66013 | .21871 | 2.1768 | .33781 | 2.3955 | .37939 | .90870 |
| 1.53 | 4.6182 | .66447 | .21654 | 2.2008 | .34258 | 2.4174 | .38334 | .91042 |
| 1.54 | 4.6646 | .66881 | .21438 | 2.2251 | .34735 | 2.4395 | .38730 | .91212 |
| 1.55 | 4.7115 | .67316 | .21225 | 2.2496 | .35211 | 2.4619 | .39126 | .91379 |
| 1.56 | 4.7588 | .67750 | .21014 | 2.2743 | .35686 | 2.4845 | .39524 | .91542 |
| 1.57 | 4.8066 | .68184 | .20805 | 2.2993 | .36160 | 2.5073 | .39921 | .91703 |
| 1.58 | 4.8550 | .68619 | .20598 | 2.3245 | .36633 | 2.5305 | .40320 | .91860 |
| 1.59 | 4.9037 | .69053 | .20393 | 2.3499 | .37105 | 2.5538 | .40719 | .92015 |
| 1.60 | 4.9530 | .69487 | .20190 | 2.3756 | .37577 | 2.5775 | .41119 | .92167 |
| 1.61 | 5.0028 | .69921 | .19989 | 2.4015 | .38048 | 2.6013 | .41520 | .92316 |
| 1.62 | 5.0531 | .70356 | .19790 | 2.4276 | .38518 | 2.6255 | .41921 | .92462 |
| 1.63 | 5.1039 | .70790 | .19593 | 2.4540 | .38987 | 2.6499 | .42323 | .92606 |
| 1.64 | 5.1552 | .71224 | .19398 | 2.4806 | .39456 | 2.6746 | .42725 | .92747 |
| 1.65 | 5.2070 | .71659 | .19205 | 2.5075 | .39923 | 2.6995 | .43129 | .92886 |
| 1.66 | 5.2593 | .72093 | .19014 | 2.5346 | .40391 | 2.7247 | .43532 | .93022 |
| 1.67 | 5.3122 | .72527 | .18825 | 2.5620 | .40857 | 2.7502 | .43937 | .93155 |
| 1.68 | 5.3656 | .72961 | .18637 | 2.5896 | .41323 | 2.7760 | .44341 | .93286 |
| 1.69 | 5.4195 | .73396 | .18452 | 2.6175 | .41788 | 2.8020 | .44747 | .93415 |
| 1.70 | 5.4739 | .73830 | .18268 | 2.6456 | .42253 | 2.8283 | .45153 | .93541 |
| 1.71 | 5.5290 | .74264 | .18087 | 2.6740 | .42717 | 2.8549 | .45559 | .93665 |
| 1.72 | 5.5845 | .74699 | .17907 | 2.7027 | .43180 | 2.8818 | .45966 | .93736 |
| 1.73 | 5.6407 | .75133 | .17728 | 2.7317 | .43643 | 2.9090 | .46374 | .93906 |
| 1.74 | 5.6973 | .75567 | .17552 | 2.7609 | .44105 | 2.9364 | .46782 | .94023 |
| 1.75 | 5.7546 | .76002 | .17377 | 2.7904 | .44567 | 2.9642 | .47191 | .94138 |
| 1.76 | 5.8124 | .76436 | .17204 | 2.8202 | .45028 | 2.9922 | .47600 | .94250 |
| 1.77 | 5.8709 | .76870 | .17033 | 2.8503 | .45488 | 3.0206 | .48009 | .94361 |
| 1.78 | 5.9299 | .77304 | .16864 | 2.8806 | .45948 | 3.0492 | .48419 | .94470 |
| 1.79 | 5.9895 | .77739 | .16696 | 2.9112 | .46408 | 3.0782 | .48830 | .94576 |
| 1.80 | 6.0496 | .78173 | .16530 | 2.9422 | .46867 | 3.1075 | .49241 | .94681 |
| 1.81 | 6.1104 | .78607 | .16365 | 2.9734 | .47325 | 3.1371 | .49652 | .94783 |
| 1.82 | 6.1719 | .79042 | .16203 | 3.0049 | .47783 | 3.1669 | .50064 | .94884 |
| 1.83 | 6.2339 | .79476 | .16041 | 3.0367 | .48241 | 3.1972 | .50476 | .94983 |
| 1.84 | 6.2965 | .79910 | .15882 | 3.0689 | .48698 | 3.2277 | .50889 | .95080 |
| 1.85 | 6.3598 | .80344 | .15724 | 3.1013 | .49154 | 3.2585 | .51302 | .95175 |
| 1.86 | 6.4237 | .80779 | .15567 | 3.1340 | .49610 | 3.2897 | .51716 | .95268 |
| 1.87 | 6.4883 | .81213 | .15412 | 3.1671 | .50066 | 3.3212 | .52130 | .95359 |
| 1.88 | 6.5535 | .81647 | .15259 | 3.2005 | .50521 | 3.3530 | .52544 | .95449 |
| 1.89 | 6.6194 | .82082 | .15107 | 3.2341 | .50976 | 3.3852 | .52959 | .95537 |
| 1.90 | 6.6859 | .82516 | .14957 | 3.2682 | .51430 | 3.4177 | .53374 | .95624 |
| 1.91 | 6.7531 | .82950 | .14808 | 3.3025 | .51884 | 3.4506 | .53789 | .95709 |
| 1.92 | 6.8210 | .83385 | .14661 | 3.3372 | .52338 | 3.4838 | .54205 | .95792 |
| 1.93 | 6.8895 | .83819 | .14515 | 3.3722 | .52791 | 3.5173 | .54621 | .95873 |
| 1.94 | 6.9588 | .84253 | .14370 | 3.4075 | .53244 | 3.5512 | .55038 | .95953 |
| 1.95 | 7.0287 | .84687 | .14227 | 3.4432 | .53696 | 3.5855 | .55455 | .96032 |
| 1.96 | 7.0993 | .85122 | .14086 | 3.4792 | .54148 | 3.6201 | .55872 | .96109 |
| 1.97 | 7.1707 | .85556 | .13946 | 3.5156 | .54600 | 3.6551 | .56290 | .96185 |
| 1.98 | 7.2427 | .85990 | .13807 | 3.5523 | .55051 | 3.6904 | .56707 | .96259 |
| 1.99 | 7.3155 | .86425 | .13670 | 3.5894 | .55502 | 3.7261 | .57126 | .96331 |
| 2.00 | 7.3891 | .86859 | .13534 | 3.6269 | .55953 | 3.7622 | .57544 | .96403 |

## Table 1-10. Values and Logarithms of Exponential and Hyperbolic Functions (Continued)

| $x$ | $e^x$ Value | $e^x$ log$_{10}$ | $e^{-x}$ (value) | sinh $x$ Value | sinh $x$ log$_{10}$ | cosh $x$ Value | cosh $x$ log$_{10}$ | tanh $x$ (value) |
|---|---|---|---|---|---|---|---|---|
| 2.00 | 7.3891 | 0.86859 | 0.13534 | 3.6269 | 0.55953 | 3.7622 | 0.57544 | 0.96403 |
| 2.01 | 7.4633 | 0.87293 | .13399 | 3.6647 | .56403 | 3.7987 | .57963 | .96473 |
| 2.02 | 7.5383 | 0.87727 | .13266 | 3.7028 | .56853 | 3.8355 | .58382 | .96541 |
| 2.03 | 7.6141 | 0.88162 | .13134 | 3.7414 | .57303 | 3.8727 | .58802 | .96609 |
| 2.04 | 7.6906 | 0.88596 | .13003 | 3.7803 | .57753 | 3.9103 | .59221 | .96675 |
| 2.05 | 7.7679 | 0.89030 | .12873 | 3.8196 | .58202 | 3.9483 | .59641 | .96740 |
| 2.06 | 7.8460 | 0.89465 | .12745 | 3.8593 | .58650 | 3.9867 | .60061 | .96803 |
| 2.07 | 7.9248 | 0.89899 | .12619 | 3.8993 | .59099 | 4.0255 | .60482 | .96865 |
| 2.08 | 8.0045 | 0.90333 | .12493 | 3.9398 | .59547 | 4.0647 | .60903 | .96926 |
| 2.09 | 8.0849 | 0.90768 | .12369 | 3.9806 | .59995 | 4.1043 | .61324 | .96986 |
| 2.10 | 8.1662 | 0.91202 | .12246 | 4.0219 | .60443 | 4.1443 | .61745 | .97045 |
| 2.11 | 8.2482 | 0.91636 | .12124 | 4.0635 | .60890 | 4.1847 | .62167 | .97103 |
| 2.12 | 8.3311 | 0.92070 | .12003 | 4.1056 | .61337 | 4.2256 | .62589 | .97159 |
| 2.13 | 8.4149 | 0.92505 | .11884 | 4.1480 | .61784 | 4.2669 | .63011 | .97215 |
| 2.14 | 8.4994 | 0.92939 | .11765 | 4.1909 | .62231 | 4.3085 | .63433 | .97269 |
| 2.15 | 8.5849 | 0.93373 | .11648 | 4.2342 | .62677 | 4.3507 | .63856 | .97323 |
| 2.16 | 8.6711 | 0.93808 | .11533 | 4.2779 | .63123 | 4.3932 | .64278 | .97375 |
| 2.17 | 8.7583 | 0.94242 | .11418 | 4.3221 | .63569 | 4.4362 | .64701 | .97426 |
| 2.18 | 8.8463 | 0.94676 | .11304 | 4.3666 | .64015 | 4.4797 | .65125 | .97477 |
| 2.19 | 8.9352 | 0.95110 | .11192 | 4.4116 | .64460 | 4.5236 | .65548 | .97526 |
| 2.20 | 9.0250 | 0.95545 | .11080 | 4.4571 | .64905 | 4.5679 | .65972 | .97574 |
| 2.21 | 9.1157 | 0.95979 | .10970 | 4.5030 | .65350 | 4.6127 | .66396 | .97622 |
| 2.22 | 9.2073 | 0.96413 | .10861 | 4.5494 | .65795 | 4.6580 | .66820 | .97668 |
| 2.23 | 9.2999 | 0.96848 | .10753 | 4.5962 | .66240 | 4.7037 | .67244 | .97714 |
| 2.24 | 9.3933 | 0.97282 | .10646 | 4.6434 | .66684 | 4.7499 | .67668 | .97759 |
| 2.25 | 9.4877 | 0.97716 | .10540 | 4.6912 | .67128 | 4.7966 | .68093 | .97803 |
| 2.26 | 9.5831 | 0.98151 | .10435 | 4.7394 | .67572 | 4.8437 | .68518 | .97846 |
| 2.27 | 9.6794 | 0.98585 | .10331 | 4.7880 | .68016 | 4.8914 | .68943 | .97888 |
| 2.28 | 9.7767 | 0.99019 | .10228 | 4.8372 | .68459 | 4.9395 | .69368 | .97929 |
| 2.29 | 9.8749 | 0.99453 | .10127 | 4.8868 | .68903 | 4.9881 | .69794 | .97970 |
| 2.30 | 9.9742 | 0.99888 | .10026 | 4.9370 | .69346 | 5.0372 | .70219 | .98010 |
| 2.31 | 10.074 | 1.00322 | .09926 | 4.9876 | .69789 | 5.0868 | .70645 | .98049 |
| 2.32 | 10.176 | 1.00756 | .09827 | 5.0387 | .70232 | 5.1370 | .71071 | .98087 |
| 2.33 | 10.278 | 1.01191 | .09730 | 5.0903 | .70675 | 5.1876 | .71497 | .98124 |
| 2.34 | 10.381 | 1.01625 | .09633 | 5.1425 | .71117 | 5.2388 | .71923 | .98161 |
| 2.35 | 10.486 | 1.02059 | .09537 | 5.1951 | .71559 | 5.2905 | .72349 | .98197 |
| 2.36 | 10.591 | 1.02493 | .09442 | 5.2483 | .72002 | 5.3427 | .72776 | .98233 |
| 2.37 | 10.697 | 1.02928 | .09348 | 5.3020 | .72444 | 5.3954 | .73203 | .98267 |
| 2.38 | 10.805 | 1.03362 | .09255 | 5.3562 | .72885 | 5.4487 | .73630 | .98301 |
| 2.39 | 10.913 | 1.03796 | .09163 | 5.4109 | .73327 | 5.5026 | .74056 | .98335 |
| 2.40 | 11.023 | 1.04231 | .09072 | 5.4662 | .73769 | 5.5569 | .74484 | .98367 |
| 2.41 | 11.134 | 1.04665 | .08982 | 5.5221 | .74210 | 5.6119 | .74911 | .98400 |
| 2.42 | 11.246 | 1.05099 | .08892 | 5.5785 | .74652 | 5.6674 | .75338 | .98431 |
| 2.43 | 11.359 | 1.05534 | .08804 | 5.6354 | .75093 | 5.7235 | .75766 | .98462 |
| 2.44 | 11.473 | 1.05968 | .08716 | 5.6929 | .75534 | 5.7801 | .76194 | .98492 |
| 2.45 | 11.588 | 1.06402 | .08629 | 5.7510 | .75975 | 5.8373 | .76621 | .98522 |
| 2.46 | 11.705 | 1.06836 | .08543 | 5.8097 | .76415 | 5.8951 | .77049 | .98551 |
| 2.47 | 11.822 | 1.07271 | .08458 | 5.8689 | .76856 | 5.9535 | .77477 | .98579 |
| 2.48 | 11.941 | 1.07705 | .08374 | 5.9288 | .77296 | 6.0125 | .77906 | .98607 |
| 2.49 | 12.061 | 1.08139 | .08291 | 5.9892 | .77737 | 6.0721 | .78334 | .98635 |
| 2.50 | 12.182 | 1.08574 | .08208 | 6.0502 | .78177 | 6.1323 | .78762 | .98661 |

## Table 1-10. Values and Logarithms of Exponential and Hyperbolic Functions (Continued)

| $x$ | $e^x$ | | $e^{-x}$ (value) | $\sinh x$ | | $\cosh x$ | | $\tanh x$ (value) |
|---|---|---|---|---|---|---|---|---|
| | Value | $\log_{10}$ | | Value | $\log_{10}$ | Value | $\log_{10}$ | |
| 2.50 | 12.182 | 1.08574 | 0.08208 | 6.0502 | 0.78177 | 6.1323 | 0.78762 | 0.98661 |
| 2.51 | 12.305 | 1.09008 | .08127 | 6.1118 | .78617 | 6.1931 | .79191 | .98688 |
| 2.52 | 12.429 | 1.09442 | .08046 | 6.1741 | .79057 | 6.2545 | .79619 | .98714 |
| 2.53 | 12.554 | 1.09877 | .07966 | 6.2369 | .79497 | 6.3166 | .80048 | .98739 |
| 2.54 | 12.680 | 1.10311 | .07887 | 6.3004 | .79937 | 6.3793 | .80477 | .98764 |
| 2.55 | 12.807 | 1.10745 | .07808 | 6.3645 | .80377 | 6.4426 | .80906 | .98788 |
| 2.56 | 12.936 | 1.11179 | .07730 | 6.4293 | .80816 | 6.5066 | .81335 | .98812 |
| 2.57 | 13.066 | 1.11614 | .07654 | 6.4946 | .81256 | 6.5712 | .81764 | .98835 |
| 2.58 | 13.197 | 1.12048 | .07577 | 6.5607 | .81695 | 6.6365 | .82194 | .98858 |
| 2.59 | 13.330 | 1.12482 | .07502 | 6.6274 | .82134 | 6.7024 | .82623 | .98881 |
| 2.60 | 13.464 | 1.12917 | .07427 | 6.6947 | .82573 | 6.7690 | .83052 | .98903 |
| 2.61 | 13.599 | 1.13351 | .07353 | 6.7628 | .83012 | 6.8363 | .83482 | .98924 |
| 2.62 | 13.736 | 1.13785 | .07280 | 6.8315 | .83451 | 6.9043 | .83912 | .98946 |
| 2.63 | 13.874 | 1.14219 | .07208 | 6.9008 | .83890 | 6.9729 | .84341 | .98966 |
| 2.64 | 14.013 | 1.14654 | .07136 | 6.9709 | .84329 | 7.0423 | .84771 | .98987 |
| 2.65 | 14.154 | 1.15088 | .07065 | 7.0417 | .84768 | 7.1123 | .85201 | .99007 |
| 2.66 | 14.296 | 1.15522 | .06995 | 7.1132 | .85206 | 7.1831 | .85631 | .99026 |
| 2.67 | 14.440 | 1.15957 | .06925 | 7.1854 | .85645 | 7.2546 | .86061 | .99045 |
| 2.68 | 14.585 | 1.16391 | .06856 | 7.2583 | .86083 | 7.3268 | .86492 | .99064 |
| 2.69 | 14.732 | 1.16825 | .06788 | 7.3319 | .86522 | 7.3998 | .86922 | .99083 |
| 2.70 | 14.880 | 1.17260 | .06721 | 7.4063 | .86960 | 7.4735 | .87352 | .99101 |
| 2.71 | 15.029 | 1.17694 | .06654 | 7.4814 | .87398 | 7.5479 | .87783 | .99118 |
| 2.72 | 15.180 | 1.18128 | .06587 | 7.5572 | .87836 | 7.6231 | .88213 | .99136 |
| 2.73 | 15.333 | 1.18562 | .06522 | 7.6338 | .88274 | 7.6991 | .88644 | .99153 |
| 2.74 | 15.487 | 1.18997 | .06457 | 7.7112 | .88712 | 7.7758 | .89074 | .99170 |
| 2.75 | 15.643 | 1.19431 | .06393 | 7.7894 | .89150 | 7.8533 | .89505 | .99186 |
| 2.76 | 15.800 | 1.19865 | .06329 | 7.8683 | .89588 | 7.9316 | .89936 | .99202 |
| 2.77 | 15.959 | 1.20300 | .06266 | 7.9480 | .90026 | 8.0106 | .90367 | .99218 |
| 2.78 | 16.119 | 1.20734 | .06204 | 8.0285 | .90463 | 8.0905 | .90798 | .99233 |
| 2.79 | 16.281 | 1.21168 | .06142 | 8.1098 | .90901 | 8.1712 | .91229 | .99248 |
| 2.80 | 16.445 | 1.21602 | .06081 | 8.1919 | .91339 | 8.2527 | .91660 | .99263 |
| 2.81 | 16.610 | 1.22037 | .06020 | 8.2749 | .91776 | 8.3351 | .92091 | .99278 |
| 2.82 | 16.777 | 1.22471 | .05961 | 8.3586 | .92213 | 8.4182 | .92522 | .99292 |
| 2.83 | 16.945 | 1.22905 | .05901 | 8.4432 | .92651 | 8.5022 | .92953 | .99306 |
| 2.84 | 17.116 | 1.23340 | .05843 | 8.5287 | .93088 | 8.5871 | .93385 | .99320 |
| 2.85 | 17.288 | 1.23774 | .05784 | 8.6150 | .93525 | 8.6728 | .93816 | .99333 |
| 2.86 | 17.462 | 1.24208 | .05727 | 8.7021 | .93963 | 8.7594 | .94247 | .99346 |
| 2.87 | 17.637 | 1.24643 | .05670 | 8.7902 | .94400 | 8.8469 | .94679 | .99359 |
| 2.88 | 17.814 | 1.25077 | .05613 | 8.8791 | .94837 | 8.9352 | .95110 | .99372 |
| 2.89 | 17.993 | 1.25511 | .05558 | 8.9689 | .95274 | 9.0244 | .95542 | .99384 |
| 2.90 | 18.174 | 1.25945 | .05502 | 9.0596 | .95711 | 9.1146 | .95974 | .99396 |
| 2.91 | 18.357 | 1.26380 | .05448 | 9.1512 | .96148 | 9.2056 | .96405 | .99408 |
| 2.92 | 18.541 | 1.26814 | .05393 | 9.2437 | .96584 | 9.2976 | .96837 | .99420 |
| 2.93 | 18.728 | 1.27248 | .05340 | 9.3371 | .97021 | 9.3905 | .97269 | .99431 |
| 2.94 | 18.916 | 1.27683 | .05287 | 9.4315 | .97458 | 9.4844 | .97701 | .99443 |
| 2.95 | 19.106 | 1.28117 | .05234 | 9.5268 | .97895 | 9.5791 | .98133 | .99454 |
| 2.96 | 19.298 | 1.28551 | .05182 | 9.6231 | .98331 | 9.6749 | .98565 | .99464 |
| 2.97 | 19.492 | 1.28985 | .05130 | 9.7203 | .98763 | 9.7716 | .98997 | .99475 |
| 2.98 | 19.688 | 1.29420 | .05079 | 9.8185 | .99205 | 9.8693 | .99429 | .99485 |
| 2.99 | 19.886 | 1.29854 | .05029 | 9.9177 | .99641 | 9.9680 | .99861 | .99496 |
| 3.00 | 20.086 | 1.30288 | .04979 | 10.018 | 1.00078 | 10.068 | 1.00293 | .99505 |

### Table 1-10. Values and Logarithms of Exponential and Hyperbolic Functions (Continued)

| $x$ | $e^x$ | | $e^{-x}$ (value) | $\sinh x$ | | $\cosh x$ | | $\tanh x$ (value) |
|---|---|---|---|---|---|---|---|---|
| | Value | $\log_{10}$ | | Value | $\log_{10}$ | Value | $\log_{10}$ | |
| 3.00 | 20.086 | 1.30288 | 0.04979 | 10.018 | 1.00078 | 10.068 | 1.00293 | 0.99505 |
| 3.05 | 21.115 | 1.32460 | .04736 | 10.534 | 1.02259 | 10.581 | 1.02454 | 0.99552 |
| 3.10 | 22.198 | 1.34631 | .04505 | 11.076 | 1.04440 | 11.122 | 1.04616 | 0.99595 |
| 3.15 | 23.336 | 1.36803 | .04285 | 11.647 | 1.06620 | 11.690 | 1.06779 | 0.99633 |
| 3.20 | 24.533 | 1.38974 | .04076 | 12.246 | 1.08799 | 12.287 | 1.08943 | 0.99668 |
| 3.25 | 25.790 | 1.41146 | .03877 | 12.876 | 1.10977 | 12.915 | 1.11108 | 0.99700 |
| 3.30 | 27.113 | 1.43317 | .03688 | 13.538 | 1.13155 | 13.575 | 1.13273 | 0.99728 |
| 3.35 | 28.503 | 1.45489 | .03508 | 14.234 | 1.15332 | 14.269 | 1.15439 | 0 99754 |
| 3.40 | 29.964 | 1.47660 | .03337 | 14.965 | 1.17509 | 14.999 | 1.17605 | 0.99777 |
| 3.45 | 31.500 | 1.49832 | .03175 | 15.734 | 1.19685 | 15.766 | 1.19772 | 0.99799 |
| 3.50 | 33.115 | 1.52003 | .03020 | 16.543 | 1.21860 | 16.573 | 1.21940 | 0.99818 |
| 3.55 | 34.813 | 1.54175 | .02872 | 17.392 | 1.24036 | 17.421 | 1.24107 | 0.99835 |
| 3.60 | 36.598 | 1.56346 | .02732 | 18.286 | 1.26211 | 18.313 | 1.26275 | 0.99851 |
| 3.65 | 38.475 | 1.58517 | .02599 | 19.224 | 1.28385 | 19.250 | 1.28444 | 0.99865 |
| 3.70 | 40.447 | 1.60689 | .02472 | 20.211 | 1.30559 | 20.236 | 1.30612 | 0.99878 |
| 3.75 | 42.521 | 1.62860 | .02352 | 21.249 | 1.32733 | 21.272 | 1.32781 | 0.99889 |
| 3.80 | 44.701 | 1.65032 | .02237 | 22.339 | 1.34907 | 22.362 | 1.34951 | 0.99900 |
| 3.85 | 46.993 | 1.67203 | .02128 | 23.486 | 1.37081 | 23.507 | 1.37120 | 0.99909 |
| 3.90 | 49.402 | 1.69375 | .02024 | 24.691 | 1.39254 | 24.711 | 1.39290 | 0.99918 |
| 3.95 | 51.935 | 1.71546 | .01925 | 25.958 | 1.41427 | 25.977 | 1.41459 | 0.99926 |
| 4.00 | 54.598 | 1.73718 | .01832 | 27.290 | 1.43600 | 27.308 | 1.43629 | 0.99933 |
| 4.10 | 60.340 | 1.78061 | .01657 | 30.162 | 1.47946 | 30.178 | 1.47970 | 0.99945 |
| 4.20 | 66.686 | 1.82404 | .01500 | 33.336 | 1.52291 | 33.351 | 1.52310 | 0.99955 |
| 4.30 | 73.700 | 1.86747 | .01357 | 36.843 | 1.56636 | 36.857 | 1.56652 | 0.99963 |
| 4.40 | 81.451 | 1.91090 | .01227 | 40.719 | 1.60980 | 40.732 | 1.60993 | 0.99970 |
| 4.50 | 90.017 | 1.95433 | .01111 | 45.003 | 1.65324 | 45.014 | 1.65335 | 0.99975 |
| 4.60 | 99.484 | 1.99775 | .01005 | 49.737 | 1.69668 | 49.747 | 1.69677 | 0.99980 |
| 4.70 | 109.95 | 2.04118 | .00910 | 54.969 | 1.74012 | 54.978 | 1.74019 | 0.99983 |
| 4.80 | 121.51 | 2.08461 | .00823 | 60.751 | 1.78355 | 60.759 | 1.78361 | 0.99986 |
| 4.90 | 134.29 | 2.12804 | .00745 | 67.141 | 1.82699 | 67.149 | 1.82704 | 0.99989 |
| 5.00 | 148.41 | 2.17147 | .00674 | 74.203 | 1.87042 | 74.210 | 1.87046 | 0.99991 |
| 5.10 | 164.02 | 2.21490 | .00610 | 82.008 | 1.91389 | 82.014 | 1.91389 | 0.99993 |
| 5.20 | 181.27 | 2.25833 | .00552 | 90.633 | 1.95729 | 90.639 | 1.95731 | 0.99994 |
| 5.30 | 200.34 | 2.30176 | .00499 | 100.17 | 2.00074 | 100.17 | 2.00074 | 0.99995 |
| 5.40 | 221.41 | 2.34519 | .00452 | 110.70 | 2.04415 | 110.71 | 2.04417 | 0.99996 |
| 5.50 | 244.69 | 2.38862 | .00409 | 122.34 | 2.08758 | 122.35 | 2.08760 | 0.99997 |
| 5.60 | 270.43 | 2.43205 | .00370 | 135.21 | 2.13101 | 135.22 | 2.13103 | 0.99997 |
| 5.70 | 298.87 | 2.47548 | .00335 | 149.43 | 2.17444 | 149.44 | 2.17445 | 0.99998 |
| 5.80 | 330.30 | 2.51891 | .00303 | 165.15 | 2.21787 | 165.15 | 2.21788 | 0.99998 |
| 5.90 | 365.04 | 2.56234 | .00274 | 182.52 | 2.26130 | 182.52 | 2.26131 | 0.99998 |
| 6.00 | 403.43 | 2.60577 | .00248 | 201.71 | 2.30473 | 201.72 | 2.30474 | 0.99999 |
| 6.25 | 518.01 | 2.71434 | .00193 | 259.01 | 2.41331 | 259.01 | 2.41331 | 0.99999 |
| 6.50 | 665.14 | 2.82291 | .00150 | 332.57 | 2.52188 | 332.57 | 2.52189 | 1.00000 |
| 6.75 | 854.06 | 2.93149 | .00117 | 427.03 | 2.63046 | 427.03 | 2.63046 | 1.00000 |
| 7.00 | 1096.6 | 3.04006 | .00091 | 548.32 | 2.73903 | 548.32 | 2.73903 | 1.00000 |
| 7.50 | 1808.0 | 3.25721 | .00055 | 904.02 | 2.95618 | 904.02 | 2.95618 | 1.00000 |
| 8.00 | 2981.0 | 3.47436 | .00034 | 1490.5 | 3.17333 | 1490.5 | 3.17333 | 1.00000 |
| 8.50 | 4914.8 | 3.69150 | .00020 | 2457.4 | 3.39047 | 2457.4 | 3.39047 | 1.00000 |
| 9.00 | 8103.1 | 3.90865 | .00012 | 4051.5 | 3.60762 | 4051.5 | 3.60762 | 1.00000 |
| 9.50 | 13360. | 4.12580 | .00007 | 6679.9 | 3.82477 | 6679.9 | 3.82477 | 1.00000 |
| 10.00 | 22026. | 4.34294 | .00005 | 11013. | 4.04191 | 11013. | 4.04191 | 1.00000 |

## Table 1-11. Compound Interest Factors*

For examples demonstrating use see end of table.

| | Single payment | | Uniform annual series | | | | |
|---|---|---|---|---|---|---|---|
| | Compound-amount factor | Present-worth factor | Sinking-fund factor | Capital-recovery factor | Compound-amount factor | Present-worth factor | |
| $n$ | Given $P$, to find $S$ $(1+i)^n$ | Given $S$, to find $P$ $\dfrac{1}{(1+i)^n}$ | Given $S$, to find $R$ $\dfrac{i}{(1+i)^n-1}$ | Given $P$, to find $R$ $\dfrac{i(1+i)^n}{(1+i)^n-1}$ | Given $R$, to find $S$ $\dfrac{(1+i)^n-1}{i}$ | Given $R$, to find $P$ $\dfrac{(1+i)^n-1}{i(1+i)^n}$ | $n$ |

3 per cent Compound Interest Factors

| $n$ | | | | | | | $n$ |
|---|---|---|---|---|---|---|---|
| 1 | 1.030 | 0.9709 | 1.00000 | 1.03000 | 1.000 | 0.971 | 1 |
| 2 | 1.061 | .9426 | 0.49261 | 0.52261 | 2.030 | 1.913 | 2 |
| 3 | 1.093 | .9151 | .32353 | .35353 | 3.091 | 2.829 | 3 |
| 4 | 1.126 | .8885 | .23903 | .26903 | 4.184 | 3.717 | 4 |
| 5 | 1.159 | .8626 | .18835 | .21835 | 5.309 | 4.580 | 5 |
| 6 | 1.194 | .8375 | .15460 | .18460 | 6.468 | 5.417 | 6 |
| 7 | 1.230 | .8131 | .13051 | .16051 | 7.662 | 6.230 | 7 |
| 8 | 1.267 | .7894 | .11246 | .14246 | 8.892 | 7.020 | 8 |
| 9 | 1.305 | .7664 | .09843 | .12843 | 10.159 | 7.786 | 9 |
| 10 | 1.344 | .7441 | .08723 | .11723 | 11.464 | 8.530 | 10 |
| 11 | 1.384 | .7224 | .07808 | .10808 | 12.808 | 9.253 | 11 |
| 12 | 1.426 | .7014 | .07046 | .10046 | 14.192 | 9.954 | 12 |
| 13 | 1.469 | .6810 | .06403 | .09403 | 15.618 | 10.635 | 13 |
| 14 | 1.513 | .6611 | .05853 | .08853 | 17.086 | 11.296 | 14 |
| 15 | 1.558 | .6419 | .05377 | .08377 | 18.599 | 11.938 | 15 |
| 16 | 1.605 | .6232 | .04961 | .07961 | 20.157 | 12.561 | 16 |
| 17 | 1.653 | .6050 | .04595 | .07595 | 21.762 | 13.166 | 17 |
| 18 | 1.702 | .5874 | .04271 | .07271 | 23.414 | 13.754 | 18 |
| 19 | 1.754 | .5703 | .03981 | .06981 | 25.117 | 14.324 | 19 |
| 20 | 1.806 | .5537 | .03722 | .06722 | 26.870 | 14.877 | 20 |
| 21 | 1.860 | .5375 | .03487 | .06487 | 28.676 | 15.415 | 21 |
| 22 | 1.916 | .5219 | .03275 | .06275 | 30.537 | 15.937 | 22 |
| 23 | 1.974 | .5067 | .03081 | .06081 | 32.453 | 16.444 | 23 |
| 24 | 2.033 | .4919 | .02905 | .05905 | 34.426 | 16.936 | 24 |
| 25 | 2.094 | .4776 | .02743 | .05743 | 36.459 | 17.413 | 25 |
| 26 | 2.157 | .4637 | .02594 | .05594 | 38.553 | 17.877 | 26 |
| 27 | 2.221 | .4502 | .02456 | .05456 | 40.710 | 18.327 | 27 |
| 28 | 2.288 | .4371 | .02329 | .05329 | 42.931 | 18.764 | 28 |
| 29 | 2.357 | .4243 | .02211 | .05211 | 45.219 | 19.188 | 29 |
| 30 | 2.427 | .4120 | .02102 | .05102 | 47.575 | 19.600 | 30 |
| 31 | 2.500 | .4000 | .02000 | .05000 | 50.003 | 20.000 | 31 |
| 32 | 2.575 | .3883 | .01905 | .04905 | 52.503 | 20.389 | 32 |
| 33 | 2.652 | .3770 | .01816 | .04816 | 55.078 | 20.766 | 33 |
| 34 | 2.732 | .3660 | .01732 | .04732 | 57.730 | 21.132 | 34 |
| 35 | 2.814 | .3554 | .01654 | .04654 | 60.462 | 21.487 | 35 |
| 40 | 3.262 | .3066 | .01326 | .04326 | 75.401 | 23.115 | 40 |
| 45 | 3.782 | .2644 | .01079 | .04079 | 92.720 | 24.519 | 45 |
| 50 | 4.384 | .2281 | .00887 | .03887 | 112.797 | 25.730 | 50 |
| 55 | 5.082 | .1968 | .00735 | .03735 | 136.072 | 26.774 | 55 |
| 60 | 5.892 | .1697 | .00613 | .03613 | 163.053 | 27.676 | 60 |
| 65 | 6.830 | .1464 | .00515 | .03515 | 194.333 | 28.453 | 65 |
| 70 | 7.918 | .1263 | .00434 | .03434 | 230.594 | 29.123 | 70 |
| 75 | 9.179 | .1089 | .00367 | .03367 | 272.631 | 29.702 | 75 |
| 80 | 10.641 | .0940 | .00311 | .03311 | 321.363 | 30.201 | 80 |
| 85 | 12.336 | .0811 | .00265 | .03265 | 377.857 | 30.631 | 85 |
| 90 | 14.300 | .0699 | .00226 | .03226 | 443.349 | 31.002 | 90 |
| 95 | 16.578 | .0603 | .00193 | .03193 | 519.272 | 31.323 | 95 |
| 100 | 19.219 | .0520 | .00165 | .03165 | 607.288 | 31.599 | 100 |

* For high interest rates, see Tables 1-12 and 1-13.

## Table 1-11. Compound Interest Factors (Continued)

For examples demonstrating use see end of table.

| | Single payment | | Uniform annual series | | | | |
|---|---|---|---|---|---|---|---|
| | Compound-amount factor | Present-worth factor | Sinking-fund factor | Capital-recovery factor | Compound-amount factor | Present-worth factor | |
| $n$ | Given $P$, to find $S$ $(1+i)^n$ | Given $S$, to find $P$ $\dfrac{1}{(1+i)^n}$ | Given $S$, to find $R$ $\dfrac{i}{(1+i)^n-1}$ | Given $P$, to find $R$ $\dfrac{i(1+i)^n}{(1+i)^n-1}$ | Given $R$, to find $S$ $\dfrac{(1+i)^n-1}{i}$ | Given $R$, to find $P$ $\dfrac{(1+i)^n-1}{i(1+i)^n}$ | $n$ |

4 per cent Compound Interest Factors

| $n$ | | | | | | | $n$ |
|---|---|---|---|---|---|---|---|
| 1 | 1.040 | 0.9615 | 1.00000 | 1.04000 | 1.000 | 0.962 | 1 |
| 2 | 1.082 | .9246 | 0.49020 | 0.53020 | 2.040 | 1.886 | 2 |
| 3 | 1.125 | .8890 | .32035 | .36035 | 3.122 | 2.775 | 3 |
| 4 | 1.170 | .8548 | .23549 | .27549 | 4.246 | 3.630 | 4 |
| 5 | 1.217 | .8219 | .18463 | .22463 | 5.416 | 4.452 | 5 |
| 6 | 1.265 | .7903 | .15076 | .19076 | 6.633 | 5.242 | 6 |
| 7 | 1.316 | .7599 | .12661 | .16661 | 7.898 | 6.002 | 7 |
| 8 | 1.369 | .7307 | .10853 | .14853 | 9.214 | 6.733 | 8 |
| 9 | 1.423 | .7026 | .09449 | .13449 | 10.583 | 7.435 | 9 |
| 10 | 1.480 | .6756 | .08329 | .12329 | 12.006 | 8.111 | 10 |
| 11 | 1.539 | .6496 | .07415 | .11415 | 13.486 | 8.760 | 11 |
| 12 | 1.601 | .6246 | .06655 | .10655 | 15.026 | 9.385 | 12 |
| 13 | 1.665 | .6006 | .06014 | .10014 | 16.627 | 9.986 | 13 |
| 14 | 1.732 | .5775 | .05467 | .09467 | 18.292 | 10.563 | 14 |
| 15 | 1.801 | .5553 | .04994 | .08994 | 20.024 | 11.118 | 15 |
| 16 | 1.873 | .5339 | .04582 | .08582 | 21.825 | 11.652 | 16 |
| 17 | 1.948 | .5134 | .04220 | .08220 | 23.698 | 12.166 | 17 |
| 18 | 2.026 | .4936 | .03899 | .07899 | 25.645 | 12.659 | 18 |
| 19 | 2.107 | .4746 | .03614 | .07614 | 27.671 | 13.134 | 19 |
| 20 | 2.191 | .4564 | .03358 | .07358 | 29.778 | 13.590 | 20 |
| 21 | 2.279 | .4388 | .03128 | .07128 | 31.969 | 14.029 | 21 |
| 22 | 2.370 | .4220 | .02920 | .06920 | 34.248 | 14.451 | 22 |
| 23 | 2.465 | .4057 | .02731 | .06731 | 36.618 | 14.857 | 23 |
| 24 | 2.563 | .3901 | .02559 | .06559 | 39.083 | 15.247 | 24 |
| 25 | 2.666 | .3751 | .02401 | .06401 | 41.646 | 15.622 | 25 |
| 26 | 2.772 | .3607 | .02257 | .06257 | 44.312 | 15.983 | 26 |
| 27 | 2.883 | .3468 | .02124 | .06124 | 47.084 | 16.330 | 27 |
| 28 | 2.999 | .3335 | .02001 | .06001 | 49.968 | 16.663 | 28 |
| 29 | 3.119 | .3207 | .01888 | .05888 | 52.966 | 16.984 | 29 |
| 30 | 3.243 | .3083 | .01783 | .05783 | 56.085 | 17.292 | 30 |
| 31 | 3.373 | .2965 | .01686 | .05686 | 59.328 | 17.588 | 31 |
| 32 | 3.508 | .2851 | .01595 | .05595 | 62.701 | 17.874 | 32 |
| 33 | 3.648 | .2741 | .01510 | .05510 | 66.210 | 18.148 | 33 |
| 34 | 3.794 | .2636 | .01431 | .05431 | 69.858 | 18.411 | 34 |
| 35 | 3.946 | .2534 | .01358 | .05358 | 73.652 | 18.665 | 35 |
| 40 | 4.801 | .2083 | .01052 | .05052 | 95.026 | 19.793 | 40 |
| 45 | 5.841 | .1712 | .00826 | .04826 | 121.029 | 20.720 | 45 |
| 50 | 7.107 | .1407 | .00655 | .04655 | 152.667 | 21.482 | 50 |
| 55 | 8.646 | .1157 | .00523 | .04523 | 191.159 | 22.109 | 55 |
| 60 | 10.520 | .0951 | .00420 | .04420 | 237.991 | 22.623 | 60 |
| 65 | 12.799 | .0781 | .00339 | .04339 | 294.968 | 23.047 | 65 |
| 70 | 15.572 | .0642 | .00275 | .04275 | 364.290 | 23.395 | 70 |
| 75 | 18.945 | .0528 | .00223 | .04223 | 448.631 | 23.680 | 75 |
| 80 | 23.050 | .0434 | .00181 | .04181 | 551.245 | 23.915 | 80 |
| 85 | 28.044 | .0357 | .00148 | .04148 | 676.090 | 24.109 | 85 |
| 90 | 34.119 | .0293 | .00121 | .04121 | 827.983 | 24.267 | 90 |
| 95 | 41.511 | .0241 | .00099 | .04099 | 1,012.785 | 24.398 | 95 |
| 100 | 50.505 | .0198 | .00081 | .04081 | 1,237.624 | 24.505 | 100 |

## Table 1-11. Compound Interest Factors (Continued)

For examples demonstrating use see end of table.

| n | Single payment | | Uniform annual series | | | | n |
|---|---|---|---|---|---|---|---|
| | Compound-amount factor | Present-worth factor | Sinking-fund factor | Capital-recovery factor | Compound-amount factor | Present-worth factor | |
| | Given $P$, to find $S$ $(1+i)^n$ | Given $S$, to find $P$ $\dfrac{1}{(1+i)^n}$ | Given $S$, to find $R$ $\dfrac{i}{(1+i)^n-1}$ | Given $P$, to find $R$ $\dfrac{i(1+i)^n}{(1+i)^n-1}$ | Given $R$, to find $S$ $\dfrac{(1+i)^n-1}{i}$ | Given $R$, to find $P$ $\dfrac{(1+i)^n-1}{i(1+i)^n}$ | |

5 per cent Compound Interest Factors

| n | | | | | | | n |
|---|---|---|---|---|---|---|---|
| 1 | 1.050 | 0.9524 | 1.00000 | 1.05000 | 1.000 | 0.952 | 1 |
| 2 | 1.103 | .9070 | 0.48780 | 0.53780 | 2.050 | 1.859 | 2 |
| 3 | 1.158 | .8638 | .31721 | .36721 | 3.153 | 2.723 | 3 |
| 4 | 1.216 | .8227 | .23201 | .28201 | 4.310 | 3.546 | 4 |
| 5 | 1.276 | .7835 | .18097 | .23097 | 5.526 | 4.329 | 5 |
| 6 | 1.340 | .7462 | .14702 | .19702 | 6.802 | 5.076 | 6 |
| 7 | 1.407 | .7107 | .12282 | .17282 | 8.142 | 5.786 | 7 |
| 8 | 1.477 | .6768 | .10472 | .15472 | 9.549 | 6.463 | 8 |
| 9 | 1.551 | .6446 | .09069 | .14069 | 11.027 | 7.108 | 9 |
| 10 | 1.629 | .6139 | .07950 | .12950 | 12.578 | 7.722 | 10 |
| 11 | 1.710 | .5847 | .07039 | .12039 | 14.207 | 8.306 | 11 |
| 12 | 1.796 | .5568 | .06283 | .11283 | 15.917 | 8.863 | 12 |
| 13 | 1.886 | .5303 | .05646 | .10646 | 17.713 | 9.394 | 13 |
| 14 | 1.980 | .5051 | .05102 | .10102 | 19.599 | 9.899 | 14 |
| 15 | 2.079 | .4810 | .04634 | .09634 | 21.579 | 10.380 | 15 |
| 16 | 2.183 | .4581 | .04227 | .09227 | 23.657 | 10.838 | 16 |
| 17 | 2.292 | .4363 | .03870 | .08870 | 25.840 | 11.274 | 17 |
| 18 | 2.407 | .4155 | .03555 | .08555 | 28.132 | 11.690 | 18 |
| 19 | 2.527 | .3957 | .03275 | .08275 | 30.539 | 12.085 | 19 |
| 20 | 2.653 | .3769 | .03024 | .08024 | 33.066 | 12.462 | 20 |
| 21 | 2.786 | .3589 | .02800 | .07800 | 35.719 | 12.821 | 21 |
| 22 | 2.925 | .3418 | .02597 | .07597 | 38.505 | 13.163 | 22 |
| 23 | 3.072 | .3256 | .02414 | .07414 | 41.430 | 13.489 | 23 |
| 24 | 3.225 | .3101 | .02247 | .07247 | 44.502 | 13.799 | 24 |
| 25 | 3.386 | .2953 | .02095 | .07095 | 47.727 | 14.094 | 25 |
| 26 | 3.556 | .2812 | .01956 | .06956 | 51.113 | 14.375 | 26 |
| 27 | 3.733 | .2678 | .01829 | .06829 | 54.669 | 14.643 | 27 |
| 28 | 3.920 | .2551 | .01712 | .06712 | 58.403 | 14.898 | 28 |
| 29 | 4.116 | .2429 | .01605 | .06605 | 62.323 | 15.141 | 29 |
| 30 | 4.322 | .2314 | .01505 | .06505 | 66.439 | 15.372 | 30 |
| 31 | 4.538 | .2204 | .01413 | .06413 | 70.761 | 15.593 | 31 |
| 32 | 4.765 | .2099 | .01328 | .06328 | 75.299 | 15.803 | 32 |
| 33 | 5.003 | .1999 | .01249 | .06249 | 80.064 | 16.003 | 33 |
| 34 | 5.253 | .1904 | .01176 | .06176 | 85.067 | 16.193 | 34 |
| 35 | 5.516 | .1813 | .01107 | .06107 | 90.320 | 16.374 | 35 |
| 40 | 7.040 | .1420 | .00828 | .05828 | 120.800 | 17.159 | 40 |
| 45 | 8.985 | .1113 | .00626 | .05626 | 159.700 | 17.774 | 45 |
| 50 | 11.467 | .0872 | .00478 | .05478 | 209.348 | 18.256 | 50 |
| 55 | 14.636 | .0683 | .00367 | .05367 | 272.713 | 18.633 | 55 |
| 60 | 18.679 | .0535 | .00283 | .05283 | 353.584 | 18.929 | 60 |
| 65 | 23.840 | .0419 | .00219 | .05219 | 456.798 | 19.161 | 65 |
| 70 | 30.426 | .0329 | .00170 | .05170 | 588.529 | 19.343 | 70 |
| 75 | 38.833 | .0258 | .00132 | .05132 | 756.654 | 19.485 | 75 |
| 80 | 49.561 | .0202 | .00103 | .05103 | 971.229 | 19.596 | 80 |
| 85 | 63.254 | .0158 | .00080 | .05080 | 1,245.087 | 19.684 | 85 |
| 90 | 80.730 | .0124 | .00063 | .05063 | 1,594.607 | 19.752 | 90 |
| 95 | 103.035 | .0097 | .00049 | .05049 | 2,040.694 | 19.806 | 95 |
| 100 | 131.501 | .0076 | .00038 | .05038 | 2,610.025 | 19.848 | 100 |

## Table 1-11. Compound Interest Factors (Continued)

For examples demonstrating use see end of table.

| n | Single payment | | Uniform annual series | | | | n |
|---|---|---|---|---|---|---|---|
| | Compound-amount factor | Present-worth factor | Sinking-fund factor | Capital-recovery factor | Compound-amount factor | Present-worth factor | |
| | Given $P$, to find $S$ $(1 + i)^n$ | Given $S$, to find $P$ $\dfrac{1}{(1 + i)^n}$ | Given $S$, to find $R$ $\dfrac{i}{(1 + i)^n - 1}$ | Given $P$, to find $R$ $\dfrac{i(1 + i)^n}{(1 + i)^n - 1}$ | Given $R$, to find $S$ $\dfrac{(1 + i)^n - 1}{i}$ | Given $R$, to find $P$ $\dfrac{(1 + i)^n - 1}{i(1 + i)^n}$ | |

6 per cent Compound Interest Factors

| n | | | | | | | n |
|---|---|---|---|---|---|---|---|
| 1 | 1.060 | 0.9434 | 1.00000 | 1.06000 | 1.000 | 0.943 | 1 |
| 2 | 1.124 | .8900 | 0.48544 | 0.54544 | 2.060 | 1.833 | 2 |
| 3 | 1.191 | .8396 | .31411 | .37411 | 3.184 | 2.673 | 3 |
| 4 | 1.262 | .7921 | .22859 | .28859 | 4.375 | 3.465 | 4 |
| 5 | 1.338 | .7473 | .17740 | .23740 | 5.637 | 4.212 | 5 |
| 6 | 1.419 | .7050 | .14336 | .20336 | 6.975 | 4.917 | 6 |
| 7 | 1.504 | .6651 | .11914 | .17914 | 8.394 | 5.582 | 7 |
| 8 | 1.594 | .6274 | .10104 | .16104 | 9.897 | 6.210 | 8 |
| 9 | 1.689 | .5919 | .08702 | .14702 | 11.491 | 6.802 | 9 |
| 10 | 1.791 | .5584 | .07587 | .13587 | 13.181 | 7.360 | 10 |
| 11 | 1.898 | .5268 | .06679 | .12679 | 14.972 | 7.887 | 11 |
| 12 | 2.012 | .4970 | .05928 | .11928 | 16.870 | 8.384 | 12 |
| 13 | 2.133 | .4688 | .05296 | .11296 | 18.882 | 8.853 | 13 |
| 14 | 2.261 | .4423 | .04758 | .10758 | 21.015 | 9.295 | 14 |
| 15 | 2.397 | .4173 | .04296 | .10296 | 23.276 | 9.712 | 15 |
| 16 | 2.540 | .3936 | .03895 | .09895 | 25.673 | 10.106 | 16 |
| 17 | 2.693 | .3714 | .03544 | .09544 | 28.213 | 10.477 | 17 |
| 18 | 2.854 | .3503 | .03236 | .09236 | 30.906 | 10.828 | 18 |
| 19 | 3.026 | .3305 | .02962 | .08962 | 33.760 | 11.158 | 19 |
| 20 | 3.207 | .3118 | .02718 | .08718 | 36.786 | 11.470 | 20 |
| 21 | 3.400 | .2942 | .02500 | .08500 | 39.993 | 11.764 | 21 |
| 22 | 3.604 | .2775 | .02305 | .08305 | 43.392 | 12.042 | 22 |
| 23 | 3.820 | .2618 | .02128 | .08128 | 46.996 | 12.303 | 23 |
| 24 | 4.049 | .2470 | .01968 | .07968 | 50.816 | 12.550 | 24 |
| 25 | 4.292 | .2330 | .01823 | .07823 | 54.865 | 12.783 | 25 |
| 26 | 4.549 | .2198 | .01690 | .07690 | 59.156 | 13.003 | 26 |
| 27 | 4.822 | .2074 | .01570 | .07570 | 63.706 | 13.211 | 27 |
| 28 | 5.112 | .1956 | .01459 | .07459 | 68.528 | 13.406 | 28 |
| 29 | 5.418 | .1846 | .01358 | .07358 | 73.640 | 13.591 | 29 |
| 30 | 5.743 | .1741 | .01265 | .07265 | 79.058 | 13.765 | 30 |
| 31 | 6.088 | .1643 | .01179 | .07179 | 84.802 | 13.929 | 31 |
| 32 | 6.453 | .1550 | .01100 | .07100 | 90.899 | 14.084 | 32 |
| 33 | 6.841 | .1462 | .01027 | .07027 | 97.343 | 14.230 | 33 |
| 34 | 7.251 | .1379 | .00960 | .06960 | 104.184 | 14.368 | 34 |
| 35 | 7.686 | .1301 | .00897 | .06897 | 111.435 | 14.498 | 35 |
| 40 | 10.286 | .0972 | .00646 | .06646 | 154.762 | 15.046 | 40 |
| 45 | 13.765 | .0727 | .00470 | .06470 | 212.744 | 15.456 | 45 |
| 50 | 18.420 | .0543 | .00344 | .06344 | 290.336 | 15.762 | 50 |
| 55 | 24.650 | .0406 | .00254 | .06254 | 394.172 | 15.991 | 55 |
| 60 | 32.988 | .0303 | .00188 | .06188 | 533.128 | 16.161 | 60 |
| 65 | 44.145 | .0227 | .00139 | .06139 | 719.083 | 16.289 | 65 |
| 70 | 59.076 | .0169 | .00103 | .06103 | 967.932 | 16.385 | 70 |
| 75 | 79.057 | .0126 | .00077 | .06077 | 1,300.949 | 16.456 | 75 |
| 80 | 105.796 | .0095 | .00057 | .06057 | 1,746.600 | 16.509 | 80 |
| 85 | 141.579 | .0071 | .00043 | .06043 | 2,342.982 | 16.549 | 85 |
| 90 | 189.465 | .0053 | .00032 | .06032 | 3,141.075 | 16.579 | 90 |
| 95 | 253.546 | .0039 | .00024 | .06024 | 4,209.104 | 16.601 | 95 |
| 100 | 339.302 | .0029 | .00018 | .06018 | 5,638.368 | 16.618 | 100 |

## Table 1-11. Compound Interest Factors (Continued)

For examples demonstrating use see end of table.

| n | Single payment | | Uniform annual series | | | | n |
|---|---|---|---|---|---|---|---|
| | Compound-amount factor | Present-worth factor | Sinking-fund factor | Capital-recovery factor | Compound-amount factor | Present-worth factor | |
| | Given $P$, to find $S$ $(1+i)^n$ | Given $S$, to find $P$ $\dfrac{1}{(1+i)^n}$ | Given $S$, to find $R$ $\dfrac{i}{(1+i)^n-1}$ | Given $P$, to find $R$ $\dfrac{i(1+i)^n}{(1+i)^n-1}$ | Given $R$, to find $S$ $\dfrac{(1+i)^n-1}{i}$ | Given $R$, to find $P$ $\dfrac{(1+i)^n-1}{i(1+i)^n}$ | |

### 8 per cent Compound Interest Factors

| n | | | | | | | n |
|---|---|---|---|---|---|---|---|
| 1 | 1.080 | 0.9259 | 1.00000 | 1.08000 | 1.000 | 0.926 | 1 |
| 2 | 1.166 | .8573 | 0.48077 | 0.56077 | 2.080 | 1.783 | 2 |
| 3 | 1.260 | .7938 | .30803 | .38803 | 3.246 | 2.577 | 3 |
| 4 | 1.360 | .7350 | .22192 | .30192 | 4.506 | 3.312 | 4 |
| 5 | 1.469 | .6806 | .17046 | .25046 | 5.867 | 3.993 | 5 |
| 6 | 1.587 | .6302 | .13632 | .21632 | 7.336 | 4.623 | 6 |
| 7 | 1.714 | .5835 | .11207 | .19207 | 8.923 | 5.206 | 7 |
| 8 | 1.851 | .5403 | .09401 | .17401 | 10.637 | 5.747 | 8 |
| 9 | 1.999 | .5002 | .08008 | .16008 | 12.488 | 6.247 | 9 |
| 10 | 2.159 | .4632 | .06903 | .14903 | 14.487 | 6.710 | 10 |
| 11 | 2.332 | .4289 | .06008 | .14008 | 16.645 | 7.139 | 11 |
| 12 | 2.518 | .3971 | .05270 | .13270 | 18.977 | 7.536 | 12 |
| 13 | 2.720 | .3677 | .04652 | .12652 | 21.495 | 7.904 | 13 |
| 14 | 2.937 | .3405 | .04130 | .12130 | 24.215 | 8.244 | 14 |
| 15 | 3.172 | .3152 | .03683 | .11683 | 27.152 | 8.559 | 15 |
| 16 | 3.426 | .2919 | .03298 | .11298 | 30.324 | 8.851 | 16 |
| 17 | 3.700 | .2703 | .02963 | .10963 | 33.750 | 9.122 | 17 |
| 18 | 3.996 | .2502 | .02670 | .10670 | 37.450 | 9.372 | 18 |
| 19 | 4.316 | .2317 | .02413 | .10413 | 41.446 | 9.604 | 19 |
| 20 | 4.661 | .2145 | .02185 | .10185 | 45.762 | 9.818 | 20 |
| 21 | 5.034 | .1987 | .01983 | .09983 | 50.423 | 10.017 | 21 |
| 22 | 5.437 | .1839 | .01803 | .09803 | 55.457 | 10.201 | 22 |
| 23 | 5.871 | .1703 | .01642 | .09642 | 60.893 | 10.371 | 23 |
| 24 | 6.341 | .1577 | .01498 | .09498 | 66.765 | 10.529 | 24 |
| 25 | 6.848 | .1460 | .01368 | .09368 | 73.106 | 10.675 | 25 |
| 26 | 7.396 | .1352 | .01251 | .09251 | 79.954 | 10.810 | 26 |
| 27 | 7.988 | .1252 | .01145 | .09145 | 87.351 | 10.935 | 27 |
| 28 | 8.627 | .1159 | .01049 | .09049 | 95.339 | 11.051 | 28 |
| 29 | 9.317 | .1073 | .00962 | .08962 | 103.966 | 11.158 | 29 |
| 30 | 10.063 | .0994 | .00883 | .08883 | 113.283 | 11.258 | 30 |
| 31 | 10.868 | .0920 | .00811 | .08811 | 123.346 | 11.350 | 31 |
| 32 | 11.737 | .0852 | .00745 | .08745 | 134.214 | 11.435 | 32 |
| 33 | 12.676 | .0789 | .00685 | .08685 | 145.951 | 11.514 | 33 |
| 34 | 13.690 | .0730 | .00630 | .08630 | 158.627 | 11.587 | 34 |
| 35 | 14.785 | .0676 | .00580 | .08580 | 172.317 | 11.655 | 35 |
| 40 | 21.725 | .0460 | .00386 | .08386 | 259.057 | 11.925 | 40 |
| 45 | 31.920 | .0313 | .00259 | .08259 | 386.506 | 12.108 | 45 |
| 50 | 46.902 | .0213 | .00174 | .08174 | 573.770 | 12.233 | 50 |
| 55 | 68.914 | .0145 | .00118 | .08118 | 848.923 | 12.319 | 55 |
| 60 | 101.257 | .0099 | .00080 | .08080 | 1,253.213 | 12.377 | 60 |
| 65 | 148.780 | .0067 | .00054 | .08054 | 1,847.248 | 12.416 | 65 |
| 70 | 218.606 | .0046 | .00037 | .08037 | 2,720.080 | 12.443 | 70 |
| 75 | 321.205 | .0031 | .00025 | .08025 | 4,002.557 | 12.461 | 75 |
| 80 | 471.955 | .0021 | .00017 | .08017 | 5,886.935 | 12.474 | 80 |
| 85 | 693.456 | .0014 | .00012 | .08012 | 8,655.706 | 12.482 | 85 |
| 90 | 1,018.915 | .0010 | .00008 | .08008 | 12,723.939 | 12.488 | 90 |
| 95 | 1,497.121 | .0007 | .00005 | .08005 | 18,701.507 | 12.492 | 95 |
| 100 | 2,199.761 | .0005 | .00004 | .08004 | 27,484.516 | 12.494 | 100 |

## Table 1-11. Compound Interest Factors (Continued)

For examples demonstrating use see end of table.

| | Single payment | | Uniform annual series | | | | |
|---|---|---|---|---|---|---|---|
| | Compound-amount factor | Present-worth factor | Sinking-fund factor | Capital-recovery factor | Compound-amount factor | Present-worth factor | |
| $n$ | Given $P$, to find $S$ $(1+i)^n$ | Given $S$, to find $P$ $\dfrac{1}{(1+i)^n}$ | Given $S$, to find $R$ $\dfrac{i}{(1+i)^n-1}$ | Given $P$, to find $R$ $\dfrac{i(1+i)^n}{(1+i)^n-1}$ | Given $R$, to find $S$ $\dfrac{(1+i)^n-1}{i}$ | Given $R$, to find $P$ $\dfrac{(1+i)^n-1}{i(1+i)^n}$ | $n$ |
| 10 per cent Compound Interest Factors | | | | | | | |
| 1 | 1.100 | 0.9091 | 1.00000 | 1.10000 | 1.000 | 0.909 | 1 |
| 2 | 1.210 | .8264 | 0.47619 | 0.57619 | 2.100 | 1.736 | 2 |
| 3 | 1.331 | .7513 | .30211 | .40211 | 3.310 | 2.487 | 3 |
| 4 | 1.464 | .6830 | .21547 | .31547 | 4.641 | 3.170 | 4 |
| 5 | 1.611 | .6209 | .16380 | .26380 | 6.105 | 3.791 | 5 |
| 6 | 1.772 | .5645 | .12961 | .22961 | 7.716 | 4.355 | 6 |
| 7 | 1.949 | .5132 | .10541 | .20541 | 9.487 | 4.868 | 7 |
| 8 | 2.144 | .4665 | .08744 | .18744 | 11.436 | 5.335 | 8 |
| 9 | 2.358 | .4241 | .07364 | .17364 | 13.579 | 5.759 | 9 |
| 10 | 2.594 | .3855 | .06275 | .16275 | 15.937 | 6.144 | 10 |
| 11 | 2.853 | .3505 | .05396 | .15396 | 18.531 | 6.495 | 11 |
| 12 | 3.138 | .3186 | .04676 | .14676 | 21.384 | 6.814 | 12 |
| 13 | 3.452 | .2897 | .04078 | .14078 | 24.523 | 7.103 | 13 |
| 14 | 3.797 | .2633 | .03575 | .13575 | 27.975 | 7.367 | 14 |
| 15 | 4.177 | .2394 | .03147 | .13147 | 31.772 | 7.606 | 15 |
| 16 | 4.595 | .2176 | .02782 | .12782 | 35.950 | 7.824 | 16 |
| 17 | 5.054 | .1978 | .02466 | .12466 | 40.545 | 8.022 | 17 |
| 18 | 5.560 | .1799 | .02193 | .12193 | 45.599 | 8.201 | 18 |
| 19 | 6.116 | .1635 | .01955 | .11955 | 51.159 | 8.365 | 19 |
| 20 | 6.727 | .1486 | .01746 | .11746 | 57.275 | 8.514 | 20 |
| 21 | 7.400 | .1351 | .01562 | .11562 | 64.002 | 8.649 | 21 |
| 22 | 8.140 | .1228 | .01401 | .11401 | 71.403 | 8.772 | 22 |
| 23 | 8.954 | .1117 | .01257 | .11257 | 79.543 | 8.883 | 23 |
| 24 | 9.850 | .1015 | .01130 | .11130 | 88.497 | 8.985 | 24 |
| 25 | 10.835 | .0923 | .01017 | .11017 | 98.347 | 9.077 | 25 |
| 26 | 11.918 | .0839 | .00916 | .10916 | 109.182 | 9.161 | 26 |
| 27 | 13.110 | .0763 | .00826 | .10826 | 121.100 | 9.237 | 27 |
| 28 | 14.421 | .0693 | .00745 | .10745 | 134.210 | 9.307 | 28 |
| 29 | 15.863 | .0630 | .00673 | .10673 | 148.631 | 9.370 | 29 |
| 30 | 17.449 | .0573 | .00608 | .10608 | 164.494 | 9.427 | 30 |
| 31 | 19.194 | .0521 | .00550 | .10550 | 181.943 | 9.479 | 31 |
| 32 | 21.114 | .0474 | .00497 | .10497 | 201.138 | 9.526 | 32 |
| 33 | 23.225 | .0431 | .00450 | .10450 | 222.252 | 9.569 | 33 |
| 34 | 25.548 | .0391 | .00407 | .10407 | 245.477 | 9.609 | 34 |
| 35 | 28.102 | .0356 | .00369 | .10369 | 271.024 | 9.644 | 35 |
| 40 | 45.259 | .0221 | .00226 | .10226 | 442.593 | 9.779 | 40 |
| 45 | 72.890 | .0137 | .00139 | .10139 | 718.905 | 9.863 | 45 |
| 50 | 117.391 | .0085 | .00086 | .10086 | 1,163.909 | 9.915 | 50 |
| 55 | 189.059 | .0053 | .00053 | .10053 | 1,880.591 | 9.947 | 55 |
| 60 | 304.482 | .0033 | .00033 | .10033 | 3,034.816 | 9.967 | 60 |
| 65 | 490.371 | .0020 | .00020 | .10020 | 4,893.707 | 9.980 | 65 |
| 70 | 789.747 | .0013 | .00013 | .10013 | 7,887.470 | 9.987 | 70 |
| 75 | 1,271.895 | .0008 | .00008 | .10008 | 12,708.954 | 9.992 | 75 |
| 80 | 2,048.400 | .0005 | .00005 | .10005 | 20,474.002 | 9.995 | 80 |
| 85 | 3,298.969 | .0003 | .00003 | .10003 | 32,979.690 | 9.997 | 85 |
| 90 | 5,313.023 | .0002 | .00002 | .10002 | 53,120.226 | 9.998 | 90 |
| 95 | 8,556.676 | .0001 | .00001 | .10001 | 85,556.760 | 9.999 | 95 |
| 100 | 13,780.612 | .0001 | .00001 | .10001 | 137,796.123 | 9.999 | 100 |

## Table 1-11. Compound Interest Factors (Continued)

### Examples in Use of Tables

GIVEN: $2,500 is invested now at 5 per cent.
REQUIRED: Accumulated value in 10 years (i.e., the amount of a given principal).

SOLUTION:
$$S = P(1 + i)^n = \$2,500 \times 1.05^{10}$$
$$\text{Compound-amount factor} = (1 + i)^n = 1.05^{10} = 1.629$$
$$S = \$2,500 \times 1.629 = \$4,062.50$$

GIVEN: $19,500 will be required in 5 years to replace equipment now in use.
REQUIRED: With interest available at 3 per cent, what sum must be deposited in the bank at present to provide the required capital (i.e., the principal which will amount to a given sum).

SOLUTION:
$$P = S \frac{1}{(1 + i)^n} = \$19,500 \frac{1}{1.03^5}$$
$$\text{Present-worth factor} = 1/(1 + i)^n = 1/1.03^5 = 0.8626$$
$$P = \$19,500 \times 0.8626 = \$16,821$$

GIVEN: $50,000 will be required in 10 years to purchase equipment.
REQUIRED: With interest available at 4 per cent, what sum must be deposited each year to provide the required capital (i.e., the annuity which will amount to a given fund).

SOLUTION:
$$R = S \frac{i}{(1 + i)^n - 1} = \$50,000 \frac{0.04}{1.04^{10} - 1}$$
$$\text{Sinking-fund factor} = \frac{i}{(1 + i)^n - 1} = \frac{0.04}{1.04^{10} - 1} = 0.08329$$
$$R = \$50,000 \times 0.08329 = \$4,164$$

GIVEN: $20,000 is invested at 10 per cent interest.
REQUIRED: Annual sum that can be withdrawn over a 20-year period (i.e., the annuity provided by a given capital).

SOLUTION:
$$R = P \frac{i(1 + i)^n}{(1 + i)^n - 1} = \$20,000 \frac{0.10 \times 1.10^{20}}{1.10^{20} - 1}$$
$$\text{Capital-recovery factor} = \frac{i(1 + i)^n}{(1 + i)^n - 1} = \frac{0.10 \times 1.10^{20}}{1.10^{20} - 1} = 0.11746$$
$$R = \$20,000 \times 0.11746 = \$2,349.20$$

GIVEN: $500 is invested each year at 8 per cent interest.
REQUIRED: Accumulated value in 15 years (i.e., amount of an annuity).

SOLUTION:
$$S = R \frac{(1 + i)^n - 1}{i} = \$500 \frac{1.08^{15} - 1}{0.08}$$
$$\text{Compound-amount factor} = \frac{(1 + i)^n - 1}{i} = \frac{1.08^{15} - 1}{0.08} = 27.152$$
$$S = \$500 \times 27.152 = \$13,576$$

GIVEN: $8,000 is required annually for 25 years.
REQUIRED: Sum that must be deposited now at 6 per cent interest.

SOLUTION:
$$P = R \frac{(1 + i)^n - 1}{i(1 + i)^n} = \$8,000 \frac{1.06^{25} - 1}{0.06 \times 1.06^{25}}$$
$$\text{Present-worth factor} = \frac{(1 + i)^n - 1}{i(1 + i)^n} = \frac{1.06^{25} - 1}{0.06 \times 1.06^{25}} = 12.783$$
$$P = \$8,000 \times 12.78 = \$102,264$$

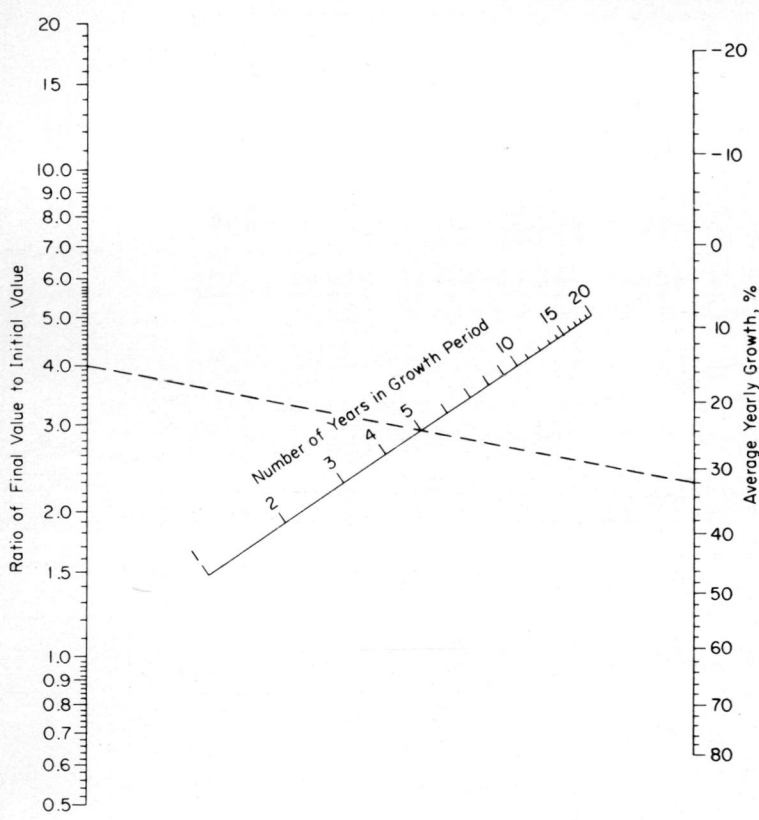

FIG. 1-1.   Nomograph for solving growth-rate problems.   Basis: Compound interest formula, $P_2 = P_1(1 + i)^n$.   Example: Sales in 1967 are \$40,000; sales in 1972 are \$160,000.   Ratio is 4.0; average growth rate is 32 per cent per year.

Table 1-12.  Single-payment Present Worth Factors for High Interest Rates

| n | 10% | 12% | 15% | 17% | 20% | 25% | 30% | 35% | 40% | 45% | 50% |
|---|---|---|---|---|---|---|---|---|---|---|---|
| 1 | 0.9091 | 0.8929 | 0.8696 | 0.8547 | 0.8333 | 0.8000 | 0.7692 | 0.7407 | 0.7143 | 0.6897 | 0.6667 |
| 2 | 0.8264 | 0.7972 | 0.7561 | 0.7305 | 0.6944 | 0.6400 | 0.5917 | 0.5487 | 0.5102 | 0.4756 | 0.4444 |
| 3 | 0.7513 | 0.7118 | 0.6575 | 0.6244 | 0.5787 | 0.5120 | 0.4552 | 0.4064 | 0.3644 | 0.3280 | 0.2963 |
| 4 | 0.6830 | 0.6355 | 0.5718 | 0.5337 | 0.4823 | 0.4096 | 0.3501 | 0.3011 | 0.2603 | 0.2262 | 0.1975 |
| 5 | 0.6209 | 0.5674 | 0.4972 | 0.4561 | 0.4019 | 0.3277 | 0.2693 | 0.2230 | 0.1859 | 0.1560 | 0.1317 |
| 6 | 0.5645 | 0.5066 | 0.4323 | 0.3898 | 0.3349 | 0.2621 | 0.2072 | 0.1652 | 0.1328 | 0.1076 | 0.0878 |
| 7 | 0.5132 | 0.4523 | 0.3759 | 0.3332 | 0.2791 | 0.2097 | 0.1594 | 0.1224 | 0.0949 | 0.0742 | 0.0585 |
| 8 | 0.4665 | 0.4039 | 0.3269 | 0.2848 | 0.2326 | 0.1678 | 0.1226 | 0.0906 | 0.0678 | 0.0512 | 0.0390 |
| 9 | 0.4241 | 0.3606 | 0.2843 | 0.2434 | 0.1938 | 0.1342 | 0.0943 | 0.0671 | 0.0484 | 0.0353 | 0.0260 |
| 10 | 0.3855 | 0.3220 | 0.2472 | 0.2080 | 0.1615 | 0.1074 | 0.0725 | 0.0497 | 0.0346 | 0.0243 | 0.0173 |
| 11 | 0.3505 | 0.2875 | 0.2149 | 0.1778 | 0.1346 | 0.0859 | 0.0558 | 0.0368 | 0.0247 | 0.0168 | 0.0116 |
| 12 | 0.3186 | 0.2567 | 0.1869 | 0.1520 | 0.1122 | 0.0687 | 0.0429 | 0.0273 | 0.0176 | 0.0116 | 0.0077 |
| 13 | 0.2897 | 0.2292 | 0.1625 | 0.1299 | 0.0935 | 0.0550 | 0.0330 | 0.0202 | 0.0126 | 0.0080 | 0.0051 |
| 14 | 0.2633 | 0.2046 | 0.1413 | 0.1110 | 0.0779 | 0.0440 | 0.0254 | 0.0150 | 0.0090 | 0.0055 | 0.0034 |
| 15 | 0.2394 | 0.1827 | 0.1229 | 0.0949 | 0.0649 | 0.0352 | 0.0195 | 0.0111 | 0.0064 | 0.0038 | 0.0023 |
| 16 | 0.2176 | 0.1631 | 0.1069 | 0.0811 | 0.0541 | 0.0281 | 0.0150 | 0.0082 | 0.0046 | 0.0026 | 0.0015 |
| 17 | 0.1978 | 0.1456 | 0.0929 | 0.0693 | 0.0451 | 0.0225 | 0.0116 | 0.0061 | 0.0033 | 0.0018 | 0.0010 |
| 18 | 0.1799 | 0.1300 | 0.0808 | 0.0593 | 0.0376 | 0.0180 | 0.0089 | 0.0045 | 0.0023 | 0.0012 | 0.0007 |
| 19 | 0.1635 | 0.1161 | 0.0703 | 0.0506 | 0.0313 | 0.0144 | 0.0068 | 0.0033 | 0.0017 | 0.0009 | 0.0005 |
| 20 | 0.1486 | 0.1037 | 0.0611 | 0.0433 | 0.0261 | 0.0115 | 0.0053 | 0.0025 | 0.0012 | 0.0006 | 0.0003 |
| 21 | 0.1351 | 0.0926 | 0.0531 | 0.0370 | 0.0217 | 0.0092 | 0.0040 | 0.0018 | 0.0009 | 0.0004 | 0.0002 |
| 22 | 0.1228 | 0.0826 | 0.0462 | 0.0316 | 0.0181 | 0.0074 | 0.0031 | 0.0014 | 0.0006 | 0.0003 | 0.0001 |
| 23 | 0.1117 | 0.0738 | 0.0402 | 0.0270 | 0.0151 | 0.0059 | 0.0024 | 0.0010 | 0.0004 | 0.0002 | 0.0001 |
| 24 | 0.1015 | 0.0659 | 0.0349 | 0.0231 | 0.0126 | 0.0047 | 0.0018 | 0.0007 | 0.0003 | 0.0001 | |
| 25 | 0.0923 | 0.0588 | 0.0304 | 0.0197 | 0.0105 | 0.0038 | 0.0014 | 0.0006 | 0.0002 | 0.0001 | |
| 30 | 0.0573 | 0.0334 | 0.0151 | 0.0090 | 0.0042 | 0.0012 | 0.0004 | 0.0001 | | | |
| 35 | 0.0356 | 0.0189 | 0.0075 | 0.0041 | 0.0017 | 0.0004 | 0.0001 | | | | |
| 40 | 0.0221 | 0.0108 | 0.0037 | 0.0019 | 0.0007 | 0.0001 | | | | | |
| 45 | 0.0137 | 0.0061 | 0.0019 | 0.0009 | 0.0003 | | | | | | |
| 50 | 0.0085 | 0.0035 | 0.0009 | 0.0004 | 0.0001 | | | | | | |
| 60 | 0.0033 | 0.0011 | 0.0002 | | | | | | | | |
| 70 | 0.0013 | 0.0004 | 0.0001 | | | | | | | | |
| 80 | 0.0005 | 0.0001 | | | | | | | | | |
| 90 | 0.0002 | | | | | | | | | | |
| 100 | 0.0001 | | | | | | | | | | |

## Table 1-13. Capital Recovery Factors for High Interest Rates

| n | 10% | 12% | 15% | 17% | 20% | 25% | 30% | 35% | 40% | 45% | 50% |
|---|---|---|---|---|---|---|---|---|---|---|---|
| 1 | 1.10000 | 1.12000 | 1.15000 | 1.17000 | 1.20000 | 1.25000 | 1.30000 | 1.35000 | 1.40000 | 1.45000 | 1.50000 |
| 2 | 0.57619 | 0.59170 | 0.61512 | 0.63083 | 0.65455 | 0.69444 | 0.73478 | 0.77553 | 0.81667 | 0.85816 | 0.90000 |
| 3 | 0.40211 | 0.41635 | 0.43798 | 0.45257 | 0.47473 | 0.51230 | 0.55063 | 0.58966 | 0.62936 | 0.66966 | 0.71053 |
| 4 | 0.31547 | 0.32923 | 0.35027 | 0.36453 | 0.38629 | 0.42344 | 0.46163 | 0.50076 | 0.54077 | 0.58156 | 0.62308 |
| 5 | 0.26380 | 0.27741 | 0.29832 | 0.31256 | 0.33438 | 0.37185 | 0.41058 | 0.45046 | 0.49136 | 0.53318 | 0.57583 |
| 6 | 0.22961 | 0.24323 | 0.26424 | 0.27861 | 0.30071 | 0.33882 | 0.37839 | 0.41926 | 0.46126 | 0.50426 | 0.54812 |
| 7 | 0.20541 | 0.21912 | 0.24036 | 0.25495 | 0.27742 | 0.31634 | 0.35687 | 0.39880 | 0.44192 | 0.48607 | 0.53108 |
| 8 | 0.18744 | 0.20130 | 0.22285 | 0.23769 | 0.26061 | 0.30040 | 0.34192 | 0.38489 | 0.42907 | 0.47427 | 0.52030 |
| 9 | 0.17364 | 0.18768 | 0.20957 | 0.22469 | 0.24808 | 0.28876 | 0.33124 | 0.37519 | 0.42034 | 0.46646 | 0.51335 |
| 10 | 0.16275 | 0.17698 | 0.19925 | 0.21466 | 0.23852 | 0.28007 | 0.32346 | 0.36832 | 0.41432 | 0.46123 | 0.50882 |
| 11 | 0.15396 | 0.16842 | 0.19107 | 0.20676 | 0.23110 | 0.27349 | 0.31773 | 0.36339 | 0.41013 | 0.45768 | 0.50585 |
| 12 | 0.14676 | 0.16144 | 0.18448 | 0.20047 | 0.22526 | 0.26845 | 0.31345 | 0.35982 | 0.40718 | 0.45527 | 0.50388 |
| 13 | 0.14078 | 0.15568 | 0.17911 | 0.19538 | 0.22062 | 0.26454 | 0.31024 | 0.35722 | 0.40510 | 0.45362 | 0.50258 |
| 14 | 0.13575 | 0.15087 | 0.17469 | 0.19123 | 0.21689 | 0.26150 | 0.30782 | 0.35532 | 0.40363 | 0.45249 | 0.50172 |
| 15 | 0.13147 | 0.14682 | 0.17102 | 0.18782 | 0.21388 | 0.25912 | 0.30598 | 0.35393 | 0.40259 | 0.45172 | 0.50114 |
| 16 | 0.12782 | 0.14339 | 0.16795 | 0.18500 | 0.21144 | 0.25724 | 0.30458 | 0.35290 | 0.40185 | 0.45118 | 0.50076 |
| 17 | 0.12466 | 0.14046 | 0.16537 | 0.18266 | 0.20944 | 0.25576 | 0.30351 | 0.35214 | 0.40132 | 0.45081 | 0.50051 |
| 18 | 0.12193 | 0.13794 | 0.16319 | 0.18071 | 0.20781 | 0.25459 | 0.30269 | 0.35158 | 0.40094 | 0.45056 | 0.50034 |
| 19 | 0.11955 | 0.13576 | 0.16134 | 0.17907 | 0.20646 | 0.25366 | 0.30207 | 0.35117 | 0.40067 | 0.45039 | 0.50023 |
| 20 | 0.11746 | 0.13388 | 0.15976 | 0.17769 | 0.20536 | 0.25292 | 0.30159 | 0.35087 | 0.40048 | 0.45027 | 0.50015 |
| 21 | 0.11562 | 0.13224 | 0.15842 | 0.17653 | 0.20444 | 0.25233 | 0.30122 | 0.35064 | 0.40034 | 0.45018 | 0.50010 |
| 22 | 0.11401 | 0.13081 | 0.15727 | 0.17555 | 0.20369 | 0.25186 | 0.30094 | 0.35048 | 0.40024 | 0.45013 | 0.50007 |
| 23 | 0.11257 | 0.12956 | 0.15628 | 0.17472 | 0.20307 | 0.25148 | 0.30072 | 0.35035 | 0.40017 | 0.45009 | 0.50004 |
| 24 | 0.11130 | 0.12846 | 0.15543 | 0.17402 | 0.20255 | 0.25119 | 0.30055 | 0.35026 | 0.40012 | 0.45006 | 0.50003 |
| 25 | 0.11017 | 0.12750 | 0.15470 | 0.17342 | 0.20212 | 0.25095 | 0.30043 | 0.35019 | 0.40009 | 0.45004 | 0.50002 |
| 30 | 0.10608 | 0.12414 | 0.15230 | 0.17154 | 0.20085 | 0.25031 | 0.30011 | 0.35004 | 0.40002 | 0.45001 | 0.50000 |
| 35 | 0.10369 | 0.12232 | 0.15113 | 0.17070 | 0.20034 | 0.25010 | 0.30003 | 0.35001 | 0.40000 | 0.45000 | 0.50000 |
| 40 | 0.10226 | 0.12130 | 0.15056 | 0.17032 | 0.20014 | 0.25003 | 0.30001 | 0.35000 | 0.40000 | 0.45000 | 0.50000 |
| 45 | 0.10139 | 0.12074 | 0.15028 | 0.17015 | 0.20005 | 0.25001 | 0.30000 | 0.35000 | 0.40000 | 0.45000 | 0.50000 |
| 50 | 0.10086 | 0.12042 | 0.15014 | 0.17007 | 0.20002 | 0.25000 | 0.30000 | 0.35000 | 0.40000 | 0.45000 | 0.50000 |
| 60 | 0.10033 | 0.12013 | 0.15003 | 0.17001 | 0.20000 | 0.25000 | 0.30000 | 0.35000 | 0.40000 | 0.45000 | 0.50000 |
| 70 | 0.10013 | 0.12004 | 0.15001 | 0.17000 | 0.20000 | 0.25000 | 0.30000 | 0.35000 | 0.40000 | 0.45000 | 0.50000 |
| 80 | 0.10005 | 0.12001 | 0.15000 | 0.17000 | 0.20000 | 0.25000 | 0.30000 | 0.35000 | 0.40000 | 0.45000 | 0.50000 |
| 90 | 0.10002 | 0.12000 | 9.15000 | 0.17000 | 0.20000 | 0.25000 | 0.30000 | 0.35000 | 0.40000 | 0.45000 | 0.50000 |
| 100 | 0.10001 | 0.12000 | 0.15000 | 0.17000 | 0.20000 | 0.25000 | 0.30000 | 0.35000 | 0.40000 | 0.45000 | 0.50000 |

Table 1-14A. $F$ Distribution, Upper 5% Points ($F_{.95}$)

| Degrees of freedom for denominator | Degrees of freedom for numerator | | | | | | | | | | | | | | | | | | |
|---|---|---|---|---|---|---|---|---|---|---|---|---|---|---|---|---|---|---|---|
| | 1 | 2 | 3 | 4 | 5 | 6 | 7 | 8 | 9 | 10 | 12 | 15 | 20 | 24 | 30 | 40 | 60 | 120 | ∞ |
| 1 | 161 | 200 | 216 | 225 | 230 | 234 | 237 | 239 | 241 | 242 | 244 | 246 | 248 | 249 | 250 | 251 | 252 | 253 | 254 |
| 2 | 18.5 | 19.0 | 19.2 | 19.2 | 19.3 | 19.3 | 19.4 | 19.4 | 19.4 | 19.4 | 19.4 | 19.4 | 19.4 | 19.5 | 19.5 | 19.5 | 19.5 | 19.5 | 19.5 |
| 3 | 10.1 | 9.55 | 9.28 | 9.12 | 9.01 | 8.94 | 8.89 | 8.85 | 8.81 | 8.79 | 8.74 | 8.70 | 8.66 | 8.64 | 8.62 | 8.59 | 8.57 | 8.55 | 8.53 |
| 4 | 7.71 | 6.94 | 6.59 | 6.39 | 6.26 | 6.16 | 6.09 | 6.04 | 6.00 | 5.96 | 5.91 | 5.86 | 5.80 | 5.77 | 5.75 | 5.72 | 5.69 | 5.66 | 5.63 |
| 5 | 6.61 | 5.79 | 5.41 | 5.19 | 5.05 | 4.95 | 4.88 | 4.82 | 4.77 | 4.74 | 4.68 | 4.62 | 4.56 | 4.53 | 4.50 | 4.46 | 4.43 | 4.40 | 4.37 |
| 6 | 5.99 | 5.14 | 4.76 | 4.53 | 4.39 | 4.28 | 4.21 | 4.15 | 4.10 | 4.06 | 4.00 | 3.94 | 3.87 | 3.84 | 3.81 | 3.77 | 3.74 | 3.70 | 3.67 |
| 7 | 5.59 | 4.74 | 4.35 | 4.12 | 3.97 | 3.87 | 3.79 | 3.73 | 3.68 | 3.64 | 3.57 | 3.51 | 3.44 | 3.41 | 3.38 | 3.34 | 3.30 | 3.27 | 3.23 |
| 8 | 5.32 | 4.46 | 4.07 | 3.84 | 3.69 | 3.58 | 3.50 | 3.44 | 3.39 | 3.35 | 3.28 | 3.22 | 3.15 | 3.12 | 3.08 | 3.04 | 3.01 | 2.97 | 2.93 |
| 9 | 5.12 | 4.26 | 3.86 | 3.63 | 3.48 | 3.37 | 3.29 | 3.23 | 3.18 | 3.14 | 3.07 | 3.01 | 2.94 | 2.90 | 2.86 | 2.83 | 2.79 | 2.75 | 2.71 |
| 10 | 4.96 | 4.10 | 3.71 | 3.48 | 3.33 | 3.22 | 3.14 | 3.07 | 3.02 | 2.98 | 2.91 | 2.85 | 2.77 | 2.74 | 2.70 | 2.66 | 2.62 | 2.58 | 2.54 |
| 11 | 4.84 | 3.98 | 3.59 | 3.36 | 3.20 | 3.09 | 3.01 | 2.95 | 2.90 | 2.85 | 2.79 | 2.72 | 2.65 | 2.61 | 2.57 | 2.53 | 2.49 | 2.45 | 2.40 |
| 12 | 4.75 | 3.89 | 3.49 | 3.26 | 3.11 | 3.00 | 2.91 | 2.85 | 2.80 | 2.75 | 2.69 | 2.62 | 2.54 | 2.51 | 2.47 | 2.43 | 2.38 | 2.34 | 2.30 |
| 13 | 4.67 | 3.81 | 3.41 | 3.18 | 3.03 | 2.92 | 2.83 | 2.77 | 2.71 | 2.67 | 2.60 | 2.53 | 2.46 | 2.42 | 2.38 | 2.34 | 2.30 | 2.25 | 2.21 |
| 14 | 4.60 | 3.74 | 3.34 | 3.11 | 2.96 | 2.85 | 2.76 | 2.70 | 2.65 | 2.60 | 2.53 | 2.46 | 2.39 | 2.35 | 2.31 | 2.27 | 2.22 | 2.18 | 2.13 |
| 15 | 4.54 | 3.68 | 3.29 | 3.06 | 2.90 | 2.79 | 2.71 | 2.64 | 2.59 | 2.54 | 2.48 | 2.40 | 2.33 | 2.29 | 2.25 | 2.20 | 2.16 | 2.11 | 2.07 |
| 16 | 4.49 | 3.63 | 3.24 | 3.01 | 2.85 | 2.74 | 2.66 | 2.59 | 2.54 | 2.49 | 2.42 | 2.35 | 2.28 | 2.24 | 2.19 | 2.15 | 2.11 | 2.06 | 2.01 |
| 17 | 4.45 | 3.59 | 3.20 | 2.96 | 2.81 | 2.70 | 2.61 | 2.55 | 2.49 | 2.45 | 2.38 | 2.31 | 2.23 | 2.19 | 2.15 | 2.10 | 2.06 | 2.01 | 1.96 |
| 18 | 4.41 | 3.55 | 3.16 | 2.93 | 2.77 | 2.66 | 2.58 | 2.51 | 2.46 | 2.41 | 2.34 | 2.27 | 2.19 | 2.15 | 2.11 | 2.06 | 2.02 | 1.97 | 1.92 |
| 19 | 4.38 | 3.52 | 3.13 | 2.90 | 2.74 | 2.63 | 2.54 | 2.48 | 2.42 | 2.38 | 2.31 | 2.23 | 2.16 | 2.11 | 2.07 | 2.03 | 1.98 | 1.93 | 1.88 |
| 20 | 4.35 | 3.49 | 3.10 | 2.87 | 2.71 | 2.60 | 2.51 | 2.45 | 2.39 | 2.35 | 2.28 | 2.20 | 2.12 | 2.08 | 2.04 | 1.99 | 1.95 | 1.90 | 1.84 |
| 21 | 4.32 | 3.47 | 3.07 | 2.84 | 2.68 | 2.57 | 2.49 | 2.42 | 2.37 | 2.32 | 2.25 | 2.18 | 2.10 | 2.05 | 2.01 | 1.96 | 1.92 | 1.87 | 1.81 |
| 22 | 4.30 | 3.44 | 3.05 | 2.82 | 2.66 | 2.55 | 2.46 | 2.40 | 2.34 | 2.30 | 2.23 | 2.15 | 2.07 | 2.03 | 1.98 | 1.94 | 1.89 | 1.84 | 1.78 |
| 23 | 4.28 | 3.42 | 3.03 | 2.80 | 2.64 | 2.53 | 2.44 | 2.37 | 2.32 | 2.27 | 2.20 | 2.13 | 2.05 | 2.01 | 1.96 | 1.91 | 1.86 | 1.81 | 1.76 |
| 24 | 4.26 | 3.40 | 3.01 | 2.78 | 2.62 | 2.51 | 2.42 | 2.36 | 2.30 | 2.25 | 2.18 | 2.11 | 2.03 | 1.98 | 1.94 | 1.89 | 1.84 | 1.79 | 1.73 |
| 25 | 4.24 | 3.39 | 2.99 | 2.76 | 2.60 | 2.49 | 2.40 | 2.34 | 2.28 | 2.24 | 2.16 | 2.09 | 2.01 | 1.96 | 1.92 | 1.87 | 1.82 | 1.77 | 1.71 |
| 30 | 4.17 | 3.32 | 2.92 | 2.69 | 2.53 | 2.42 | 2.33 | 2.27 | 2.21 | 2.16 | 2.09 | 2.01 | 1.93 | 1.89 | 1.84 | 1.79 | 1.74 | 1.68 | 1.62 |
| 40 | 4.08 | 3.23 | 2.84 | 2.61 | 2.45 | 2.34 | 2.25 | 2.18 | 2.12 | 2.08 | 2.00 | 1.92 | 1.84 | 1.79 | 1.74 | 1.69 | 1.64 | 1.58 | 1.51 |
| 60 | 4.00 | 3.15 | 2.76 | 2.53 | 2.37 | 2.25 | 2.17 | 2.10 | 2.04 | 1.99 | 1.92 | 1.84 | 1.75 | 1.70 | 1.65 | 1.59 | 1.53 | 1.47 | 1.39 |
| 120 | 3.92 | 3.07 | 2.68 | 2.45 | 2.29 | 2.18 | 2.09 | 2.02 | 1.96 | 1.91 | 1.83 | 1.75 | 1.66 | 1.61 | 1.55 | 1.50 | 1.43 | 1.35 | 1.25 |
| ∞ | 3.84 | 3.00 | 2.60 | 2.37 | 2.21 | 2.10 | 2.01 | 1.94 | 1.88 | 1.83 | 1.75 | 1.67 | 1.57 | 1.52 | 1.46 | 1.39 | 1.32 | 1.22 | 1.00 |

* Interpolation should be performed using reciprocals of the degrees of freedom.

## Table 1-14B. F Distribution, Upper 1% Points ($F_{.99}$)

| Degrees of freedom for denominator | \ Degrees of freedom for numerator | | | | | | | | | | | | | | | | | |
|---|---|---|---|---|---|---|---|---|---|---|---|---|---|---|---|---|---|---|
| | **1** | **2** | **3** | **4** | **5** | **6** | **7** | **8** | **9** | **10** | **15** | **20** | **24** | **30** | **40** | **60** | **120** | **∞** |
| 1 | 4052 | 5000 | 5403 | 5625 | 5764 | 5859 | 5928 | 5982 | 6023 | 6056 | 6157 | 6209 | 6235 | 6261 | 6287 | 6313 | 6339 | 6366 |
| 2 | 98.5 | 99.0 | 99.2 | 99.2 | 99.3 | 99.3 | 99.4 | 99.4 | 99.4 | 99.4 | 99.4 | 99.4 | 99.5 | 99.5 | 99.5 | 99.5 | 99.5 | 99.5 |
| 3 | 34.1 | 30.8 | 29.5 | 28.7 | 28.2 | 27.9 | 27.7 | 27.5 | 27.3 | 27.2 | 26.9 | 26.7 | 26.6 | 26.5 | 26.4 | 26.3 | 26.2 | 26.1 |
| 4 | 21.2 | 18.0 | 16.7 | 16.0 | 15.5 | 15.2 | 15.0 | 14.8 | 14.7 | 14.5 | 14.2 | 14.0 | 13.9 | 13.8 | 13.7 | 13.7 | 13.6 | 13.5 |
| 5 | 16.3 | 13.3 | 12.1 | 11.4 | 11.0 | 10.7 | 10.5 | 10.3 | 10.2 | 10.1 | 9.72 | 9.55 | 9.47 | 9.38 | 9.29 | 9.20 | 9.11 | 9.02 |
| 6 | 13.7 | 10.9 | 9.78 | 9.15 | 8.75 | 8.47 | 8.26 | 8.10 | 7.98 | 7.87 | 7.56 | 7.40 | 7.31 | 7.23 | 7.14 | 7.06 | 6.97 | 6.88 |
| 7 | 12.2 | 9.55 | 8.45 | 7.85 | 7.46 | 7.19 | 6.99 | 6.84 | 6.72 | 6.62 | 6.31 | 6.16 | 6.07 | 5.99 | 5.91 | 5.82 | 5.74 | 5.65 |
| 8 | 11.3 | 8.65 | 7.59 | 7.01 | 6.63 | 6.37 | 6.18 | 6.03 | 5.91 | 5.81 | 5.52 | 5.36 | 5.28 | 5.20 | 5.12 | 5.03 | 4.95 | 4.86 |
| 9 | 10.6 | 8.02 | 6.99 | 6.42 | 6.06 | 5.80 | 5.61 | 5.47 | 5.35 | 5.26 | 4.96 | 4.81 | 4.73 | 4.65 | 4.57 | 4.48 | 4.40 | 4.31 |
| 10 | 10.0 | 7.56 | 6.55 | 5.99 | 5.64 | 5.39 | 5.20 | 5.06 | 4.94 | 4.85 | 4.56 | 4.41 | 4.33 | 4.25 | 4.17 | 4.08 | 4.00 | 3.91 |
| 11 | 9.65 | 7.21 | 6.22 | 5.67 | 5.32 | 5.07 | 4.89 | 4.74 | 4.63 | 4.54 | 4.25 | 4.10 | 4.02 | 3.94 | 3.86 | 3.78 | 3.69 | 3.60 |
| 12 | 9.33 | 6.93 | 5.95 | 5.41 | 5.06 | 4.82 | 4.64 | 4.50 | 4.39 | 4.30 | 4.01 | 3.86 | 3.78 | 3.70 | 3.62 | 3.54 | 3.45 | 3.36 |
| 13 | 9.07 | 6.70 | 5.74 | 5.21 | 4.86 | 4.62 | 4.44 | 4.30 | 4.19 | 4.10 | 3.82 | 3.66 | 3.59 | 3.51 | 3.43 | 3.34 | 3.25 | 3.17 |
| 14 | 8.86 | 6.51 | 5.56 | 5.04 | 4.69 | 4.46 | 4.28 | 4.14 | 4.03 | 3.94 | 3.66 | 3.51 | 3.43 | 3.35 | 3.27 | 3.18 | 3.09 | 3.00 |
| 15 | 8.68 | 6.36 | 5.42 | 4.89 | 4.56 | 4.32 | 4.14 | 4.00 | 3.89 | 3.80 | 3.52 | 3.37 | 3.29 | 3.21 | 3.13 | 3.05 | 2.96 | 2.87 |
| 16 | 8.53 | 6.23 | 5.29 | 4.77 | 4.44 | 4.20 | 4.03 | 3.89 | 3.78 | 3.69 | 3.41 | 3.26 | 3.18 | 3.10 | 3.02 | 2.93 | 2.84 | 2.75 |
| 17 | 8.40 | 6.11 | 5.19 | 4.67 | 4.34 | 4.10 | 3.93 | 3.79 | 3.68 | 3.59 | 3.31 | 3.16 | 3.08 | 3.00 | 2.92 | 2.83 | 2.75 | 2.65 |
| 18 | 8.29 | 6.01 | 5.09 | 4.58 | 4.25 | 4.01 | 3.84 | 3.71 | 3.60 | 3.51 | 3.23 | 3.08 | 3.00 | 2.92 | 2.84 | 2.75 | 2.66 | 2.57 |
| 19 | 8.18 | 5.93 | 5.01 | 4.50 | 4.17 | 3.94 | 3.77 | 3.63 | 3.52 | 3.43 | 3.15 | 3.00 | 2.92 | 2.84 | 2.76 | 2.67 | 2.58 | 2.49 |
| 20 | 8.10 | 5.85 | 4.94 | 4.43 | 4.10 | 3.87 | 3.70 | 3.56 | 3.46 | 3.37 | 3.09 | 2.94 | 2.86 | 2.78 | 2.69 | 2.61 | 2.52 | 2.42 |
| 21 | 8.02 | 5.78 | 4.87 | 4.37 | 4.04 | 3.81 | 3.64 | 3.51 | 3.40 | 3.31 | 3.03 | 2.88 | 2.80 | 2.72 | 2.64 | 2.55 | 2.46 | 2.36 |
| 22 | 7.95 | 5.72 | 4.82 | 4.31 | 3.99 | 3.76 | 3.59 | 3.45 | 3.35 | 3.26 | 2.98 | 2.83 | 2.75 | 2.67 | 2.58 | 2.50 | 2.40 | 2.31 |
| 23 | 7.88 | 5.66 | 4.76 | 4.26 | 3.94 | 3.71 | 3.54 | 3.41 | 3.30 | 3.21 | 2.93 | 2.78 | 2.70 | 2.62 | 2.54 | 2.45 | 2.35 | 2.26 |
| 24 | 7.82 | 5.61 | 4.72 | 4.22 | 3.90 | 3.67 | 3.50 | 3.36 | 3.26 | 3.17 | 2.89 | 2.74 | 2.66 | 2.58 | 2.49 | 2.40 | 2.31 | 2.21 |
| 25 | 7.77 | 5.57 | 4.68 | 4.18 | 3.86 | 3.63 | 3.46 | 3.32 | 3.22 | 3.13 | 2.85 | 2.70 | 2.62 | 2.53 | 2.45 | 2.36 | 2.27 | 2.17 |
| 30 | 7.56 | 5.39 | 4.51 | 4.02 | 3.70 | 3.47 | 3.30 | 3.17 | 3.07 | 2.98 | 2.70 | 2.55 | 2.47 | 2.39 | 2.30 | 2.21 | 2.11 | 2.01 |
| 40 | 7.31 | 5.18 | 4.31 | 3.83 | 3.51 | 3.29 | 3.12 | 2.99 | 2.89 | 2.80 | 2.52 | 2.37 | 2.29 | 2.20 | 2.11 | 2.02 | 1.92 | 1.80 |
| 60 | 7.08 | 4.98 | 4.13 | 3.65 | 3.34 | 3.12 | 2.95 | 2.82 | 2.72 | 2.63 | 2.35 | 2.20 | 2.12 | 2.03 | 1.94 | 1.84 | 1.73 | 1.60 |
| 120 | 6.85 | 4.79 | 3.95 | 3.48 | 3.17 | 2.96 | 2.79 | 2.66 | 2.56 | 2.47 | 2.19 | 2.03 | 1.95 | 1.86 | 1.76 | 1.66 | 1.53 | 1.38 |
| ∞ | 6.63 | 4.61 | 3.78 | 3.32 | 3.02 | 2.80 | 2.64 | 2.51 | 2.41 | 2.32 | 2.04 | 1.88 | 1.79 | 1.70 | 1.59 | 1.47 | 1.32 | 1.00 |

\* Interpolation should be performed using reciprocals of the degrees of freedom.

## Table 1-15. Values of $t$

| df | $t_{.40}$ | $t_{.30}$ | $t_{.20}$ | $t_{.10}$ | $t_{.05}$ | $t_{.025}$ | $t_{.01}$ | $t_{.005}$ |
|----|-----------|-----------|-----------|-----------|-----------|------------|-----------|------------|
| 1  | 0.325 | 0.727 | 1.376 | 3.078 | 6.314 | 12.706 | 31.821 | 63.657 |
| 2  | .289  | .617  | 1.061 | 1.886 | 2.920 | 4.303  | 6.965  | 9.925  |
| 3  | .277  | .584  | 0.978 | 1.638 | 2.353 | 3.182  | 4.541  | 5.841  |
| 4  | .271  | .569  | .941  | 1.533 | 2.132 | 2.776  | 3.747  | 4.604  |
| 5  | .267  | .559  | .920  | 1.476 | 2.015 | 2.571  | 3.365  | 4.032  |
| 6  | .265  | .553  | .906  | 1.440 | 1.943 | 2.447  | 3.143  | 3.707  |
| 7  | .263  | .549  | .896  | 1.415 | 1.895 | 2.365  | 2.998  | 3.499  |
| 8  | .262  | .546  | .889  | 1.397 | 1.860 | 2.306  | 2.896  | 3.355  |
| 9  | .261  | .543  | .883  | 1.383 | 1.833 | 2.262  | 2.821  | 3.250  |
| 10 | .260  | .542  | .879  | 1.372 | 1.812 | 2.228  | 2.764  | 3.169  |
| 11 | .260  | .540  | .876  | 1.363 | 1.796 | 2.201  | 2.718  | 3.106  |
| 12 | .259  | .539  | .873  | 1.356 | 1.782 | 2.179  | 2.681  | 3.055  |
| 13 | .259  | .538  | .870  | 1.350 | 1.771 | 2.160  | 2.650  | 3.012  |
| 14 | .258  | .537  | .868  | 1.345 | 1.761 | 2.145  | 2.624  | 2.977  |
| 15 | .258  | .536  | .866  | 1.341 | 1.753 | 2.131  | 2.602  | 2.947  |
| 16 | .258  | .535  | .865  | 1.337 | 1.746 | 2.120  | 2.583  | 2.921  |
| 17 | .257  | .534  | .863  | 1.333 | 1.740 | 2.110  | 2.567  | 2.898  |
| 18 | .257  | .534  | .862  | 1.330 | 1.734 | 2.101  | 2.552  | 2.878  |
| 19 | .257  | .533  | .861  | 1.328 | 1.729 | 2.093  | 2.539  | 2.861  |
| 20 | .257  | .533  | .860  | 1.325 | 1.725 | 2.086  | 2.528  | 2.845  |
| 21 | .257  | .532  | .859  | 1.323 | 1.721 | 2.080  | 2.518  | 2.831  |
| 22 | .256  | .532  | .858  | 1.321 | 1.717 | 2.074  | 2.508  | 2.819  |
| 23 | .256  | .532  | .858  | 1.319 | 1.714 | 2.069  | 2.500  | 2.807  |
| 24 | .256  | .531  | .857  | 1.318 | 1.711 | 2.064  | 2.492  | 2.797  |
| 25 | .256  | .531  | .856  | 1.316 | 1.708 | 2.060  | 2.485  | 2.787  |
| 26 | .256  | .531  | .856  | 1.315 | 1.706 | 2.056  | 2.479  | 2.779  |
| 27 | .256  | .531  | .855  | 1.314 | 1.703 | 2.052  | 2.473  | 2.771  |
| 28 | .256  | .530  | .855  | 1.313 | 1.701 | 2.048  | 2.467  | 2.763  |
| 29 | .256  | .530  | .854  | 1.311 | 1.699 | 2.045  | 2.462  | 2.756  |
| 30 | .256  | .530  | .854  | 1.310 | 1.697 | 2.042  | 2.457  | 2.750  |
| 40 | .255  | .529  | .851  | 1.303 | 1.684 | 2.021  | 2.423  | 2.704  |
| 60 | .254  | .527  | .848  | 1.296 | 1.671 | 2.000  | 2.390  | 2.660  |
| 120| .254  | .526  | .845  | 1.289 | 1.658 | 1.980  | 2.358  | 2.617  |
| ∞  | .253  | .524  | .842  | 1.282 | 1.645 | 1.960  | 2.326  | 2.576  |

Above values refer to a single tail outside the indicated limit of $t$. For example, for 95 percent of the area to be between $-t$ and $+t$ in a two-tailed $t$ distribution, use the values for $t_{0.025}$ or 2.5 percent for each tail.

## Table 1-16. Mathematical Signs and Symbols

| | |
|---|---|
| $\pm(\mp)$ | plus or minus (minus or plus) |
| : | divided by, ratio sign |
| : : | proportional sign |
| $<$ | less than |
| $\not<$ | not less than |
| $>$ | greater than |
| $\not>$ | not greater than |
| $\cong$ | approximately equals, congruent |
| $\sim$ | similar to |
| $\rightleftharpoons$ | equivalent to |
| $\neq$ | not equal to |
| $\doteq$ | approaches, is approximately equal to |
| $\propto$ | varies as |
| $\infty$ | infinity |
| $\therefore$ | therefore |
| $\sqrt{\phantom{x}}$ | square root |
| $\sqrt[3]{\phantom{x}}$ | cube root |
| $\sqrt[n]{\phantom{x}}$ | $n$th root |
| $\angle$ | angle |
| $\perp$ | perpendicular to |
| $\parallel$ | parallel to |
| $|x|$ | numerical value of $x$ |
| log or log₁₀ | common logarithm or Briggsian logarithm |
| logₑ or ln | natural logarithm or hyperbolic logarithm or Napierian logarithm |
| e | base (2.718) of natural system of logarithms |
| $a°$ | an angle $a$ degrees |
| $a'$ | $a$ prime, an angle $a$ minutes |
| $a''$ | $a$ double prime, an angle $a$ seconds, $a$ second |
| sin | sine |
| cos | cosine |
| tan | tangent |
| ctn or cot | cotangent |
| sec | secant |
| csc | cosecant |
| vers | versed sine |
| covers | coversed sine |
| exsec | exsecant |
| $\sin^{-1}$ | anti sine or angle whose sine is |
| sinh | hyperbolic sine |
| cosh | hyperbolic cosine |
| tanh | hyperbolic tangent |
| $\sinh^{-1}$ | anti hyperbolic sine or angle whose hyperbolic sine is |
| $f(x)$ or $\phi(x)$ | function of $x$ |
| $\Delta x$ | increment of $x$ |
| $\Sigma$ | summation of |
| $dx$ | differential of $x$ |
| $dy/dx$ or $y'$ | derivative of $y$ with respect to $x$ |
| $d^2y/dx^2$ or $y''$ | second derivative of $y$ with respect to $x$ |
| $d^ny/dx^n$ | $n$th derivative of $y$ with respect to $x$ |
| $\partial y/\partial x$ | partial derivative of $y$ with respect to $x$ |
| $\partial^n y/\partial x^n$ | $n$th partial derivative of $y$ with respect to $x$ |
| $\dfrac{\partial^n y}{\partial x\,\partial y}$ | $n$th partial derivative with respect to $x$ and $y$ |
| $\int$ | integral of |

## Table 1-16. Mathematical Signs and Symbols (Continued)

$\displaystyle\int_a^b$    integral between the limits $a$ and $b$

$\dot{y}$    first derivative of $y$ with respect to time

$\ddot{y}$    second derivative of $y$ with respect to time

$\Delta$ or $\nabla^2$    the "Laplacian"

$$\left(\frac{\partial^2}{\partial x^2} + \frac{\partial^2}{\partial y^2} + \frac{\partial^2}{\partial z^2}\right)$$

$\delta$    sign of a variation

$\oint$    sign of integration around a closed path

## Table 1-17. Common Abbreviations and Symbols for Units*

| | | | | |
|---|---|---|---|---|
| Absolute | abs | Gallon | gal |
| Alternating current (as adjective) | a-c | Gallons per minute | gal/min |
| Ampere | A | Gallons per second | gal/s |
| Ampere-hour | Ah | Grain | gr |
| Ampere per meter | A/m | Gram | g |
| Angstrom | Å | Gram-calorie | g-cal |
| Antilogarithm | antilog | Henry | H |
| Atomic mass unit | u | Hertz | Hz |
| Atomic weight | at. wt. | High pressure (as adjective) | h-p |
| Atmosphere (standard) | atm | Horsepower | hp |
| Average | avg | Horsepower hour | hp-h |
| | | Hour | h |
| Barometer (or bar) | bar | | |
| Barrel | bbl | Inch | in |
| Baumé | Bé | | |
| Boiler pressure | bp | Joule | J |
| Brake horsepower | bhp | Joule per kelvin | J/K |
| British thermal unit | Btu | | |
| | | Kelvin | K† |
| Calorie (thermochemical) | cal | Kilocycle per second | kHz |
| Candela | cd | Kilogram | kg |
| Candlepower | cp | Kilograms per second | kg/s |
| Centimeter | cm | Kilohertz | kHz |
| Centipoise | cP | Kilometer | km |
| Centistokes | cSt | Kilovolt | kV |
| Cologarithm | colog | Kilowatt | kW |
| Coulomb | C | | |
| Cubic foot | ft³ | Latitude | lat |
| Cubic feet per minute | ft³/min | Linear foot | lin ft |
| Cubic feet per second | ft³/s | Liter | l |
| Cubic inch | in³ | Logarithm | log |
| Cubic meter | m³ | Low pressure (as adjective) | l-p |
| | | Lumen | lm |
| Decibel | dB | Lux | lx |
| Degree Celsius | °C | | |
| Degree Fahrenheit | °F | Maximum | max |
| Degree Kelvin (see Kelvin) | | Maxwell | Mx |
| | | Mean effective pressure | mep |
| Degree Rankine | °R | Meter | m |
| Direct current (as adjective) | d-c | Micro | μ |
| | | Mho (siemens) | S |
| Dyne | dyn | Micromicron | μμm |
| | | Micron | μm |
| Efficiency | eff | Miles per hour | mi/h |
| Electron volt | eV | Milliampere | mA |
| Elevation | el | Millimeter | mm |
| | | Millimicron (nanometer) | nm |
| Farad | F | Minimum | min. |
| Feet per minute | ft/min | Minute | min |
| Feet per second | ft/s | Mole | mol |
| Foot | ft | | |

* Extracted (in part) from American National Standard. Letter Symbols for Units Used in Science and Technology, ANSI Y10.19-1969. American Society of Mechanical Engineers.
† NOTE: Not °K, degree sign not shown for degrees Kelvin.

## Table 1-17. Common Abbreviations and Symbols for Units (Continued)

| | | | |
|---|---|---|---|
| Nanometer............. | nm | Specific gravity.......... | sp gr |
| Newton................ | N | Specific heat............. | sp ht |
| Newtons per square meter | N/m² | Square foot............. | ft² |
| | | Square inch............. | in² |
| Ohm.................. | Ω | Square meter............ | m² |
| | | Stokes.................. | St |
| Parts per million........ | ppm | | |
| Pascal................. | Pa | | |
| Pound................. | lb | Ton (2,000 lb)........... | ton |
| Pounds per square foot... | lbf/ft² | Tonne (1,000 kg)........ | t |
| Pounds per square inch... | lbf/in² | | |
| Pounds per square inch, | | Volt..................... | V |
|   absolute............. | psia | Volt-ampere............. | VA |
| Pounds per square inch, | | Volt/meter............. | V/m |
|   gage.................. | psig | | |
| | | Watt................... | W |
| Revolutions per minute... | r/min | Watt per meter kelvin.... | W/(m·K) |
| Revolutions per second... | r/s | Watthour............... | Wh |
| | | Weber.................. | Wb |
| Quart.................. | qt | Weight................. | wt |
| | | | |
| Second................. | s | | |
| Siemens................ | S | Yard.................. | yd |

## Table 1-18. Greek Alphabet

| | | | | | | |
|---|---|---|---|---|---|---|
| Alpha | = A, $\alpha$ | = A, a | Nu | = N, $\nu$ | = N, n |
| Beta | = B, $\beta$ | = B, b | Xi | = $\Xi$, $\xi$ | = X, x |
| Gamma | = $\Gamma$, $\gamma$ | = G, g | Omicron | = O, $o$ | = O, o |
| Delta | = $\Delta$, $\delta$ | = D, d | Pi | = $\Pi$, $\pi$ | = P, p |
| Epsilon | = E, $\epsilon$ | = E, e | Rho | = P, $\rho$ | = R, r |
| Zeta | = Z, $\zeta$ | = Z, z | Sigma | = $\Sigma$, $\sigma$ | = S, s |
| Eta | = H, $\eta$ | = E, e | Tau | = T, $\tau$ | = T, t |
| Theta | = $\Theta$, $\theta$ | = Th, th | Upsilon | = $\Upsilon$, $\upsilon$ | = U, u |
| Iota | = I, $\iota$ | = I, i | Phi | = $\Phi$, $\phi$ | = Ph, ph |
| Kappa | = K, $\kappa$ | = K, k | Chi | = X, $\chi$ | = Ch, ch |
| Lambda | = $\Lambda$, $\lambda$ | = L, l | Psi | = $\Psi$, $\psi$ | = Ps, ps |
| Mu | = M, $\mu$ | = M, m | Omega | = $\Omega$, $\omega$ | = O, o |

## Table 1-19. Russian Alphabet

| Block | Italic | Transliteration |
|---|---|---|
| А а | *А а* | A, a |
| Б б | *Б б* | B, b |
| В в | *В в* | V, v |
| Г г | *Г г* | G, g |
| Д д | *Д д* | D, d |
| Е е | *Е е* | Ye, ye; E, e* |
| Ё ё | *Ё ё* | Zh, zh |
| Ж ж | *Ж ж* | Z, z |
| З з | *З з* | I, i |
| И и | *И и* | Y, y |
| Й й | *Й й* | K, k |
| К к | *К к* | L, l |
| Л л | *Л л* | M, m |
| М м | *М м* | N, n |
| Н н | *Н н* | O, o |
| О о | *О о* | P, p |
| П п | *П п* | R, r |
| Р р | *Р р* | S, s |
| С с | *С с* | T, t |
| Т т | *Т т* | U, u |
| У у | *У у* | F, f |
| Ф ф | *Ф ф* | Kh, kh |
| Х х | *Х х* | Ts, ts |
| Ц ц | *Ц ц* | Ch, ch |
| Ч ч | *Ч ч* | Sh, sh |
| Ш ш | *Ш ш* | Shch, shch |
| Щ щ | *Щ щ* | ʺ |
| Ъ ъ | *Ъ ъ* | Y, y |
| Ы ы | *Ы ы* | ʹ |
| Ь ь | *Ь ь* | E, e |
| Э э | *Э э* | Yu, yu |
| Ю ю | *Ю ю* | Ya, ya |
| Я я | *Я я* | |

* *ye* initially, after vowels, and after  ,  ; *e* elsewhere.  When written as ë in Russian, transliterate as yë or ë.  The use of diacritical marks is preferred, but such marks may be omitted when expediency dictates.

## MATHEMATICS

### 1-20. Miscellaneous Constants and Identities

$\pi$ = 3.1415926536 (pi)  
$e$ = 2.7182818285 (Napierian)  
$\gamma$ = 0.5772156649 (Euler)  
log $x$ = 0.4342944819 ln $x$

ln $\pi$ = 1.1447298858  
log $\pi$ = 0.4971498727  
log (log $e$) = 9.637784311–10  
ln $x$ = 2.302585093 log $x$

$\sqrt{2}$ = 1.4142135624  
$\sqrt{3}$ = 1.7320508076  
$\sqrt{5}$ = 2.2360679775  
$\sqrt{e}$ = 1.6487212707  
$\sqrt{\pi}$ = 1.7724538509  
$\sqrt{2\pi}$ = 2.5066282746

$\sqrt{6}$ = 2.4494897428  
$\sqrt{7}$ = 2.6457513110  
$\sqrt{8}$ = 2.8284271247  
$\sqrt{\pi/2}$ = 1.2533141373  
$\sqrt{\pi/3}$ = 1.0233267079  
$\sqrt{0.1}$ = 0.3162277660

$\sqrt{10}$ = 3.1622776602  
$\sqrt{\gamma}$ = 0.7597471059  
$\sqrt{1/\pi}$   0.5641895835  
ln 2 = 0.6931471806  
ln 10 = 2.302585093  
log $e$ = 0.4342944819

1 radian = 57.2957795131 degrees  
1 minute = 0.0002908882 radian

1 degree = 0.0174532925 radian  
1 second = 0.0000048481 radian

*Indeterminants*     *Example*

$a^0 = 1$     $(a > 0)$

$0^a = 0$     $(a > 0)$

$a^\infty = 0$     $(a < 1)$

$a^\infty = \infty$     $(a > 1)$  
$\infty^a = 0$     $(a < 0)$

$\infty^a = \infty$     $(a > 0)$

$1^0 = 1$

$(\infty)(0)$     $\lim_{x \to \infty} xe^{-x} = 0$

$0^0$     $\lim_{x \to 0} x^x = 1$

$\infty^0$     $\lim_{x \to 1} \left( \dfrac{1}{1 - x} \right)^{1-x} = 1$

$1^\infty$     $\lim_{x \to 1} (1 - x)^{1/x} = e^{-1}$

$\infty - \infty$     $\lim_{x \to \pi/2} (\sec x - \tan x) = 0$

$0/0$     $\lim_{x \to 0} \dfrac{\sin x}{x} = 1$

$\infty/\infty$     $\lim_{x \to \pi/2} \dfrac{\ln (\pi/2 - x)}{\tan x} = \infty$

$a^{-n} = 1/a^n$  
$(ab)^n = a^n b^n$

$(a^n)^m = a^{nm}$  
$\sqrt[n]{a} = a^{1/n}$

$\sqrt[m]{\sqrt[n]{a}} = a^{1/mn}$  
$\sqrt[n]{a^m} = a^{m/n}$

For $\dfrac{a}{b} = \dfrac{c}{d}$  then  $\dfrac{a + b}{b} = \dfrac{c + d}{d}$  $\dfrac{a - b}{b} = \dfrac{c - d}{d}$

$\dfrac{a - b}{a + b} = \dfrac{c - d}{c + d}$  $\dfrac{a + b}{a - b} = \dfrac{c + d}{c - d}$

$\log ab = \log a + \log b$  
$\log \dfrac{a}{b} = \log a - \log b$  
For $S = x^n$  
$\quad = e^{n \ln x}$

$\log a^n = n \log a$  
$\log \sqrt[n]{a} = \dfrac{1}{n} \log a$  
$S = \sqrt{x}$  
$S_{i+1} = \dfrac{1}{2} \left( \dfrac{x}{S_i} + S_i \right)$  
By iteration

### Approximations ($\cong$) When $x$ Is Small

$$\frac{1}{1 \pm x} \cong 1 \mp x \qquad\qquad \sqrt{1 \pm x} \cong 1 \pm \frac{x}{2}$$

$$\frac{1 + y}{1 \pm x} \cong 1 + y \mp x(1 + y) \qquad (1 \pm x)^{-n} \cong 1 \mp nx$$

$$(1 \pm x)^n \cong 1 \pm nx \qquad\qquad (1 \pm x)^{-1/2} \cong 1 \mp x/2$$

$$(a \pm x)^2 \cong a^2 \pm 2ax \qquad\qquad e^x \cong 1 + x$$

$$\sin x \cong x \quad (x \text{ in radians}) \qquad \tan x \cong x$$

$$\sqrt{y(y + x)} \cong \frac{2y + x}{2} \qquad \sqrt{y^2 + x^2} \cong y + \frac{x^2}{2y} \quad \left(\frac{x}{y} \text{ small}\right)$$

### General

$$n! \cong e^{-n} n^n \sqrt{2\pi n}$$

$$n! \cong \sqrt{2\pi} \left(\frac{\sqrt{n^2 + n + \frac{1}{6}}}{e}\right)^{n + \frac{1}{2}}$$

## 1-21. Mensuration

Let $A$ denote areas and $V$ volumes in the following definitions.

**Plane Geometric Figures with Straight Boundaries.** *Triangles* (see also Trigonometry). $A = \frac{1}{2}bh$, where $b = $ base, $h = $ altitude.

*Rectangle.* $A = ab$, where $a$ and $b = $ lengths of sides.

*Parallelogram* (opposite sides parallel). $A = ah = ab \sin \alpha$, where $a$, $b = $ lengths of sides, $h = $ height, and $\alpha = $ angle between sides.

*Trapezoid* (four sides, two parallel). $A = \frac{1}{2}(a + b)h$, where $a$ and $b = $ lengths of parallel sides, and $h = $ height.

*Quadrilateral* (four sides). $A = \frac{1}{2}ab \sin \theta$, where $a$, $b = $ lengths of diagonals, and acute angle between them is $\theta$.

*Regular Polygon of $n$ sides.* $A = \frac{1}{4}nl^2 \cot (180°/n)$, where $l = $ length of each side.

**Inscribed and Circumscribed Circles with Regular Polygons of $n$ Sides.** Let $l = $ length of one side, and $n = $ number of sides.

| Figure | $n$ | Area | Radius of circumscribed circle | Radius of inscribed circle |
|---|---|---|---|---|
| Equilateral triangle.... | 3 | $0.4330l^2$ | $0.5774l$ | $0.2887l$ |
| Square.............. | 4 | $1.0000l^2$ | $0.7071l$ | $0.5000l$ |
| Pentagon........... | 5 | $1.7205l^2$ | $0.8507l$ | $0.6882l$ |
| Hexagon............ | 6 | $2.5981l^2$ | $1.0000l$ | $0.8660l$ |
| Heptagon........... | 7 | $3.6339l^2$ | $1.1523l$ | $1.0383l$ |
| Octagon............ | 8 | $4.8284l^2$ | $1.3065l$ | $1.2071l$ |

**Plane Geometric Figures with Curved Boundaries.** *Circle.* Let $C = $ circumference, $r = $ radius, $D = $ diameter, $A = $ area, $S = $ arc length subtended by angle $\theta$ in radians, $l$ = cord length subtended by $\theta$, $H$ = rise, $d = r - H$.

$$C = 2\pi r = \pi D \qquad (\pi = 3.14159 \ldots)$$
$$S = r\theta = \tfrac{1}{2}D\theta$$
$$l = 2\sqrt{r^2 - d^2} = 2r\sin\frac{\theta}{2} = 2d\tan\frac{\theta}{2}$$
$$H = r - d$$
$$d = \frac{1}{2}\sqrt{4r^2 - l^2} = \frac{1}{2}l\cot\frac{\theta}{2}$$
$$\theta = \frac{S}{r} = 2\cos^{-1}\frac{d}{r} = 2\sin^{-1}\frac{l}{D}$$
$$A\ (\text{circle}) = \pi r^2 = \tfrac{1}{4}\pi D^2$$
$$A\ (\text{sector}) = \tfrac{1}{2}rS = \tfrac{1}{2}r^2\theta$$
$$A\ (\text{segment}) = A\ (\text{sector}) - A\ (\text{triangle}) = \tfrac{1}{2}r^2(\theta - \sin\theta)$$
$$= r^2\cos^{-1}\frac{r - H}{r} - (r - H)\sqrt{2rH - H^2}$$

**Ring** (area between two circles of radius $r_1$ and $r_2$). (The circles need not be concentric, but one of the circles must enclose the other.)

$$A = \pi(r_1 + r_2)(r_1 - r_2), \qquad r_1 > r_2$$

**Ellipse.** Let the semiaxes of the ellipse be $a$ and $b$. $A = \pi ab$, $C = 4aE(k)$, $k = 1 - b^2/a^2$, and $E(k)$ is the (tabulated) complete elliptic integral of the first kind.

**Solid Geometric Figures with Plane Boundaries.** **Cube.** Volume = $a^3$; total surface area = $6a^2$; diagonal = $a\sqrt{3}$, where $a$ = length of one side of cube.

**Rectangular Parallelepiped.** Volume = $abc$; surface area = $2(ab + ac + bc)$; diagonal = $\sqrt{a^2 + b^2 + c^2}$, where $a$, $b$, $c$ = lengths of sides.

**Prism.** Volume = (area of base) × (altitude); lateral surface area = (perimeter of right section) × (lateral edge).

**Pyramid.** Volume = $\tfrac{1}{3}$(area of base) × (altitude); lateral area of regular pyramid = $\tfrac{1}{2}$ (perimeter of base) × (slant height) = $\tfrac{1}{2}$ (number of sides) × (length of one side) × (slant height).

**Volume and Surface Area of Regular Polyhedra with Edge $l$**

| Type of surface | Name | Volume | Surface area |
|---|---|---|---|
| 4 equilateral triangles | Tetrahedron | $0.1179l^3$ | $1.7321l^2$ |
| 6 squares | Hexahedron | $1.0000l^3$ | $6.0000l^2$ |
| 8 equilateral triangles | Octahedron | $0.4714l^3$ | $3.4641l^2$ |
| 12 pentagons | Dodecahedron | $7.6631l^3$ | $20.6458l^2$ |
| 20 equilateral triangles | Icosahedron | $2.1817l^3$ | $8.6603l^2$ |

**Solids Bounded by Curved Surfaces.** **Cylinder.** $V$ = (area of base) × (altitude); lateral surface area = (perimeter of right section) × (lateral edge).

*Right Circular Cylinder.* $V = \pi$ (radius)$^2$ × (altitude); lateral surface area = $2\pi$ (radius) × (altitude).

*Truncated Right Circular Cylinder.* $V = \pi r^2 h$; lateral area = $2\pi rh$; $h = \frac{1}{2}(h_1 + h_2)$.

*Hollow Cylinder.* Volume = $\pi h(R^2 - r^2)$, where $r$ and $R$ = internal and external radii, and $h$ = height of cylinder.

**Sphere**

$V$ (sphere) = $\frac{4}{3}\pi R^3 = \frac{1}{6}\pi D^3$

$V$ (spherical sector) = $\frac{2}{3}\pi R^2 h = \frac{1}{6}\pi D^2 h$

$V$ (spherical segment of one base) = $\frac{1}{6}\pi h_1(3r_1^2 + h_1^2)$

$A$ (sphere) = $4\pi R^2 = \pi D^2$

$A$ (zone) = $2\pi Rh = \pi Dh$

**Cone.**  $V = \frac{1}{3}$(area of base) × (altitude).

*Right Circular Cone.*  $V = (\pi/3)r^2 h$, where $h$ = altitude, and $r$ = radius of base; curved surface area = $\pi r \sqrt{r^2 + h^2}$, curved surface of *frustum of a right cone* = $\pi(r_1 + r_2)\sqrt{h^2 + (r_1 - r_2)^2}$, where $r_1$, $r_2$ = radii of base and top, respectively, and $h$ = altitude; volume of frustum of a right cone = $\pi(h/3)(r_1^2 + r_1 r_2 + r_2^2) = \frac{1}{3}(A_1 + A_2 + \sqrt{A_1 A_2})$, where $A_1$ = area of base and $A_2$ = area of top.

**Ellipsoid.**  $V = \frac{4}{3}\pi abc$, where $a$, $b$, $c$ = lengths of semiaxes.

**Torus** (obtained by rotating a circle of radius $r$ about a line whose distance is $R > r$ from the center of the circle). $V = 2\pi^2 R r^2$; surface area = $4\pi^2 Rr$.

**Miscellaneous Formulas** (see also section on calculus).  *Volume of a Solid of Revolution* (the solid generated by rotating a plane area about the $x$ axis).

$$V = \pi \int_a^b [f(x)]^2 \, dx,$$ where $y = f(x)$ = equation of place curve, and $a \le x \le b$.

*Area of a Surface of Revolution.*  $S = 2\pi \int_a^b y \, ds$, where

$$ds = \sqrt{1 + \left(\frac{dy}{dx}\right)^2} \, dx$$

and $y = f(x)$ is the equation of the plane curve rotated about the $x$ axis to generate the surface.

*Area Bounded by $f(x)$, the $x$ Axis, and the Lines $x = a$, $x = b$*

$$A = \int_a^b f(x) \, dx$$

*Length of Arc of a Plane Curve*

If $y = f(x)$:    Length of arc $s = \int_a^b \sqrt{1 + \left(\frac{dy}{dx}\right)^2} \, dx$

If $x = g(y)$:    Length of arc $s = \int_c^d \sqrt{1 + \left(\frac{dx}{dy}\right)^2} \, dy$

If $x = f(t)$:    $y = g(t)$, $s = \int_{t_0}^{t_1} \sqrt{\left(\frac{dx}{dt}\right)^2 + \left(\frac{dy}{dt}\right)^2} \, dt$

**The Theorem of Pappus** (for volumes and areas of surfaces of revolution). (1) If a plane area is revolved about a line which lies in its plane but does not intersect the area, then the volume generated is equal to the product of the area and the distance traveled by the area's center of gravity. (2) If an arc of a plane curve is revolved about a line which lies in its plane but does not intersect the arc, then the surface area generated by the arc is equal to the product of the length of the arc and the distance traveled by its center of gravity.

**Irregular Areas.** Let $y_0$, $y_1$, . . . , $y_n$ be the lengths of a series of equally spaced parallel chords, and $h$ be their distance apart. The area of the figure is given approximately by any of the following equations:

$$A_T = \frac{h}{2}\left[(y_0 + y_n) + 2(y_1 + y_2 + \cdots + y_{n-1})\right] \qquad \text{(trapezoidal rule)}$$

$$A_S = \frac{h}{3}\left[(y_0 + y_n) + 4(y_1 + y_3 + y_5 + \cdots + y_{n-1})\right.$$
$$\left. + 2(y_2 + y_4 + \cdots + y_{n-2})\right] \qquad \text{($n$ even, Simpson's rule)}$$

$$A_W = \frac{3h}{8}\left[(y_0 + y_n) + 3(y_1 + y_2 + y_4 + y_5 + \cdots) + 2(y_3 + y_6 + \cdots)\right]$$
$$\left(\text{Weddle's } \frac{3}{8} \text{ rule}\right)$$

The greater the value of $n$, the greater the accuracy of approximation.

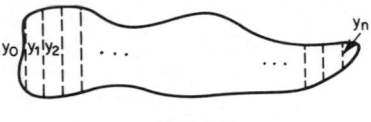

Fig. 1-2

## 1-22. Dimensions, Units, and Dimensionless Groups

Properties of the physical world are expressed in dimensioned form, and dimensions are expressed in units. Currently, three systems of units are in common use.

**Engineering System.** Mass $M$, time $t$, length $L$, and force $F$ are considered primary and independent dimensions. None of the four need be defined in terms of the others. However, force $F$ and mass $M$ are related by Newton's second law of motion $F = (1/g_c)Ma$, where $a$ = acceleration resulting from application of force $F$ on mass $M$, with dimension $L/t^2$. For this relation to be dimensionally consistent, a constant $(1/g_c)$ must be introduced *for this system*. The constant $g_c$ has the dimension $ML/Ft^2$, and is numerically equal to the acceleration of gravity at sea level.

**Absolute System.** Mass $M$, time $t$, and length $L$ are considered primary dimensions having independent units. Force is defined in terms of these three, and has the dimension $ML/t^2$. Since force and mass are not considered to be independent dimensionally, no relating dimensional constant is required.

The International System of Units (SI) is like the absolute metric system in that length, time, and mass are independent dimensions and force is defined in terms of those three units. The unit of length is the meter, mass is the kilogram, force is the newton (N), time is the second, and energy (including work and heat) is the joule (J). A newton is the force which when applied to a mass of one kilogram gives it an acceleration of one meter per second per second. A joule is that work done when the point of application of a force of one newton is displaced through a distance of one meter in the direction of the force.

**Gravitational System.** Force $F$, time $t$, and length $L$ are considered primary dimensions having independent units. Mass is defined in terms of these three, and has the dimension $Ft^2/L$. As force and mass are not considered to be independent dimensionally, no relating constant is required.

The three systems are compared and dimensions and units for both the English and metric systems given in Tables 1-20 and 1-21.

**Dimensionless Groups.** The use of dimensionless numbers or groups of variables is fundamental to many engineering studies. This results in connection with, for example, model and scale-up studies. If all variables affecting a particular operation are known, dimensional analysis can be employed to indicate the groups of variables that are important. Consequently, the number of experiments is reduced by dimensional analysis, since only the effect of each grouping need be explored, rather than the effect of each individual variable. There are a large number of useful dimensionless groups; so only the more commonly occurring ones are given in Table 1-22. The reader is referred to Boucher and Alves, "A Tabulation of Dimensionless Groups," *Chem. Eng. Progr.*, **55**, 55–64 (September, 1959) for additional listings.

## Table 1-20. Metric System of Units

| Quantity | Absolute system* | | Gravitational system | | Engineering system | |
|---|---|---|---|---|---|---|
| | Dimension | Unit | Dimension | Unit | Dimension | Unit |
| Time....... | $t$ | Second | $t$ | Second | $t$ | Second |
| Length...... | $L$ | Centimeter | $L$ | Centimeter | $L$ | Centimeter |
| Mass....... | $M$ | Gram | $Ft^2/L$ | Gram | $M$ | Gram |
| Force....... | $ML/t^2$ | Dyne | $F$ | Gram force | $F$ | Gram force |
| Work....... | $ML^2/t^2$ | Dyne-centimeter (erg) | $FL$ | Centimeter-gram force | $FL$ | Centimeter-gram force |
| Temperature | $T$ | Degree centigrade absolute or degree Kelvin | $T$ | Degree centigrade absolute or degree Kelvin | $T$ | Degree centigrade absolute or degree Kelvin |
| Heat........ | $ML^2/t^2$ | Dyne-centimeter (erg) | $FL$ | Centimeter-gram | $Q$ | Calorie |
| $g_c$.......... | None | None | None | None | $ML/Ft^2$ | 980.665 (g mass) (cm)/ (g force) (sec²) |
| $J$.......... | None | None | None | None | $FL/Q$ | 42,699 cm-g force/cal |

\* See text for an explanation of the similarities and differences of this system and the International System (SI) of units.

## Table 1-21. English System of Units

| Quantity | Absolute system | | Gravitational system | | Engineering system | |
|---|---|---|---|---|---|---|
| | Dimension | Unit | Dimension | Unit | Dimension | Unit |
| Time....... | $t$ | Second | $t$ | Second | $t$ | Second |
| Length...... | $L$ | Foot | $L$ | Foot | $L$ | Foot |
| Mass....... | $M$ | Pound mass | $Ft^2/L$ | Slug | $M$ | Pound mass |
| Force....... | $ML/t^2$ | Poundal | $F$ | Pound force | $F$ | Pound force |
| Work....... | $ML^2/t^2$ | Foot-poundal | $FL$ | Foot-pound force | $FL$ | Foot-pound force |
| Temperature | $T$ | Degree Fahrenheit absolute or degree Rankine | $T$ | Degree Fahrenheit absolute or degree Rankine | $T$ | Degree Fahrenheit absolute |
| Heat........ | $ML^2/t^2$ | Foot-poundal | $FL$ | Foot-pound force | $Q$ | Btu |
| $g_c$.......... | None | None | None | None | $ML/Ft^2$ | 32.174 (lb mass)(ft)/ (lb force) (sec²) |
| $J$........... | None | None | None | None | $FL/Q$ | 778.26 ft-lb force/Btu |

# Table 1-22. Dimensionless Groups

*Dimensions Used*

$[F]$ force, $[H]$ heat, $[L]$ length, $[M]$ mass, $[Q]$ electric charge, $[T]$ temperature, $[t]$ time.

*General Nomenclature*

$c_p$ = specific heat, $[H/MT]$; $g_L$ = gravitational constant, $[L/t^2]$; $g_c$ = conversion factor, $[LM/Ft^2]$; $h$ = heat transfer coefficient, $[H/L^2Tt]$; $k$ = thermal conductivity, $[H/LTt]$; $V$ = fluid velocity, $[L/t]$; $\beta$ = expansion coefficient, $[1/T]$; $\Delta T$ = temperature difference, $[T]$; $\mu$ = viscosity, $[M/Lt]$; $\rho$ = density, $[M/L^3]$; $\sigma$ = surface tension, $[F/L]$; $\tau_w$ = shear stress at wall, $[F/L^2]$.

| Name | Symbol | Formula | Special nomenclature |
|---|---|---|---|
| *Fluid mechanics* | | | |
| Bingham number............ | $N_{\mathrm{Bm}}$ | $\dfrac{\tau_y g_c L}{\mu_p V}$ | $L$ = width of channel, $[L]$<br>$\mu_p$ = coeff. of rigidity, $[M/tL]$<br>$\tau_y$ = yield stress, $[F/L^2]$<br>$E_b$ = bulk modulus of fluid, $[F/L^2]$. |
| Cauchy number............ | $N_c$ | $\dfrac{\rho V^2}{g_c E_b}$ | |
| Drag coefficient........... | $C_d$ | $\dfrac{(\rho - \rho')L g_L L}{\rho V^2}$ | $L$ = characteristic dimension of object, $[L]$<br>$\rho$ = object density, $[M/L^3]$<br>$\rho'$ = density of surrounding fluid, $[M/L^3]$ |
| Elasticity number.......... | $N_{\mathrm{El}}$ | $\dfrac{t_r \mu}{\rho L^2}$ | $L$ = radius of pipe, $[L]$<br>$t_r$ = relaxation time, $[t]$ |
| Euler number............. | $N_{\mathrm{Eu}}$ | $\dfrac{g_c(\Delta p_F/\rho)}{V^2}$ | $\dfrac{\Delta p_F}{\rho}$ = friction head, $[LF/M]$ |
| Fanning friction factor...... | $f$ | $\dfrac{g_c D(\Delta p_F/\rho)}{2V^2 L}$ | $L$ = length of pipe, $[L]$<br>$D$ = diameter of cross section, $[L]$<br>$\Delta p_F/\rho$ = friction head, $[LF/M]$ |
| Froude number............ | $N_{\mathrm{Fr}}$ | $\dfrac{V^2}{g_L L}$ | $L$ = characteristic system dimension, $[L]$ |
| Kármán number........... | $N_{\mathrm{K}}$ | $g_c \rho D^3(-dp/dL)$ | $D$ = pipe diameter, $[L]$<br>$dp/dL$ = pressure gradient, $[F/L^3]$ |
| Knudsen number.......... | $N_{\mathrm{Kn}}$ | $\dfrac{\lambda}{L}$ | $L$ = characteristic system dimension, $[L]$<br>$\lambda$ = length of mean free path, $[L]$ |
| Mach number............. | $N_{\mathrm{Ma}}$ | $\dfrac{V}{V_c}$ | $V_c$ = velocity of sound in fluid, $[L/t]$ |

| | | | |
|---|---|---|---|
| Pipeline parameter.......... | $\rho_n$ | $\dfrac{aV_0}{2g_cH}$ | $a$ = water-hammer wave velocity, $[L/t]$<br>$H$ = static head, $[LF/M]$<br>$V_0$ = initial velocity, $[L/t]$ |
| Poiseuille number.......... | $N_{Po}$ | $\dfrac{g_cD^2(-dp/dL)}{\mu V}$ | $D$ = diameter of round pipe, $[L]$<br>$dp/dL$ = pressure gradient, $[F/L^3]$ |
| Ratio of specific heats.......... | $\gamma$ | $\dfrac{c_p}{c_v}$ | $c_p$ = specific heat at constant pressure, $[H/MT]$<br>$c_v$ = specific heat at constant volume, $[H/MT]$ |
| Reynolds number.......... | $N_{Re}$ | $\dfrac{LV\rho}{\mu}$ | $L$ = characteristic system dimension, $[L]$ |
| Weber number.......... | $N_{We}$ | $\dfrac{V^2\rho L}{g_c\sigma}$ | $L$ = characteristic system dimension, $[L]$<br>$\sigma$ = surface tension, $[F/L]$ |
| *Heat transfer* | | | |
| Biot number.......... | $N_{Bi}$ | $\dfrac{hr_m}{k}$ | $r_m$ = distance from midpoint to surface, $[L]$ |
| Fourier number.......... | $N_{Fo}$ | $\dfrac{kt}{\rho c_p r_m^2}$ | $r_m$ = distance from midpoint to surface<br>$t$ = elapsed time, $[t]$ |
| Graetz number.......... | $N_{Gz}$ | $\dfrac{wc_p}{kL}$ | $L$ = length of heat-transfer channel, $[L]$<br>$w$ = mass flow rate, $[M/t]$ |
| Grashof number.......... | $N_{Gr}$ | $\dfrac{L^3\rho^2g_cL\beta\,\Delta T}{\mu^2}$ | $L$ = height of surface, $[L]$<br>$\Delta T$ = temperature difference across film, $[T]$<br>$\beta$ = expansion coefficient, $[1/T]$ |
| Nusselt number.......... | $N_{Nu}$ | $\dfrac{hL}{k}$ | $L$ = heat-transfer-path characteristic length, $[L]$ |
| Peclet number.......... | $N_{Pe}$ | $\dfrac{LV\rho c_p}{k}$ | $L$ = characteristic system dimension, $[L]$ |
| Prandtl number.......... | $N_{Pr}$ | $\dfrac{c_p\mu}{k}$ | |
| *Mass transfer* | | | |
| Lewis number.......... | $N_{Le}$ | $\dfrac{k}{\rho c_p D_v}$ | $D_v$ = molecular diffusivity, $[L^2/t]$ |
| Peclet number.......... | $N_{Pe'}$ | $\dfrac{LV}{D'}$ | $L$ = characteristic length, $[L]$<br>$D'$ = characteristic diffusion coeff., $[L^2/t]$ |

## Table 1-22. Dimensionless Groups (Continued)

| Name | Symbol | Formula | Special nomenclature |
|---|---|---|---|
| Schmidt number............. | $N_{Sc}$ | $\dfrac{\mu}{\rho D_V}$ | $D_V$ = molecular diffusivity, $[L^2/t]$ |
| Sherwood number............. | $N_{Sh}$ | $\dfrac{k_c L}{D_V}$ | $L$ = characteristic dimension, $[L]$<br>$k_c$ = mass transfer coeff., $[L/t]$ |
| Rayleigh number............. | $R'$ | $\dfrac{L^3 \rho^2 g_L \beta c_p \, \Delta T}{\mu k}$ | $L$ = surface height, $[L]$<br>$\Delta T$ = temperature difference across film, $[T]$<br>$\beta$ = expansion coefficient, $[1/T]$ |
| *Chemical reaction* | | | |
| Arrhenius group............. | | $\dfrac{E}{RT}$ | $E$ = activation energy, $[LF/M]$<br>$R$ = gas constant, $[LF/MT]$<br>$T$ = absolute temperature, $[T]$ |
| Damkohler group I............. | | $\dfrac{UL}{VC_A}$ | $C_A$ = concentration, $[M/L^3]$<br>$k$ = reaction rate constant, $[(L^3/M)^{n-1}1/t]$ |
| Damkohler group II............. | | $\dfrac{UL^2}{DvC_A}$ | $n$ = reaction order<br>$U$ = reaction rate, $[M/L^3 t]$<br>$D_V$ = molecular diffusivity, $[L^2/t]$ |
| Damkohler group III............. | | $\dfrac{DvC_A}{QUL}$ | $L$ = characteristic dimension, $[L]$ |
| Damkohler group IV............. | | $\dfrac{\rho c_p VT}{QUL^2}$ $\dfrac{QUL^2}{kT}$ | $Q$ = heat generated, $[H/M]$<br>$T$ = temperature, $[T]$ |

## 1-23. Logarithmic and Exponential Relationships

### Definitions

Log base $e = \log_e = \ln$      $\ln X = Z \to X = e^Z$
Log base $10 = \log_{10} = \log$      $\log X = Z \to X = 10^Z$
Log base $p = \log_p$      $\log_p X = Z \to X = p^Z$

$p$ any positive base except 0 or 1

Antilog$_p$, or $\log_p^{-1}$, is that number whose log value is stated for $\log_p X = Z$, then given $Z$, find $X \to X = \log^{-1} Z$.

### Conversion

For $\log_p X = Z \to X = p^Z$
$$= e^{Z \ln p} \quad \text{since } p = e^{\ln p}$$
$$= 10^{Z \log p} \quad \text{since } p = 10^{\log p}$$

### Identities

$$\log_p XY = \log_p X + \log_p Y \qquad \log_p X^{-n} = -n \log_p X$$
$$\log_p \frac{X}{Y} = \log_p X - \log_p Y \qquad \log_p \sqrt[n]{X} = \frac{1}{n} \log_p X$$
$$\log_p X^n = n \log_p X \qquad \log_p \sqrt[n]{X^m} = \frac{m}{n} \log_p X$$

$$\log X = (\log e) \ln X \quad \text{since} \quad X = e^{\ln X}$$
$$\log X = (\ln X)(\log e)$$
$$= 0.4342944819 \ln X$$
$$\ln X = (\ln 10) \log X$$
$$= 2.302585093 \log X$$

### Series

$$\ln x = \begin{cases} \dfrac{x-1}{x} + \dfrac{1}{2}\left(\dfrac{x-1}{x}\right)^2 + \dfrac{1}{3}\left(\dfrac{x-1}{x}\right)^3 + \cdots + \dfrac{1}{n}\left(\dfrac{x-1}{x}\right)^n + \cdots \\ \hfill x > 0.5 \\[2mm] (x-1) - \tfrac{1}{2}(x-1)^2 + \tfrac{1}{3}(x-1)^3 + \cdots + \dfrac{(-1)^{n+1}}{n}(x-1)^n + \cdots \\ \hfill 0 < x \leq 2.0 \\[2mm] 2\left[\dfrac{x-1}{x+1} + \dfrac{1}{3}\left(\dfrac{x-1}{x+1}\right)^3 + \dfrac{1}{5}\left(\dfrac{x-1}{x+1}\right)^5 + \cdots + \dfrac{1}{2n+1}\left(\dfrac{x-1}{x+1}\right)^{2n+1} + \cdots\right] \\ \hfill x > 0 \end{cases}$$

**Calculation Example.** Find $Z = A^B = (0.000273)^{0.074}$.

*By Logarithmic Transformation*

Define $\log Z = 0.074 \log 0.000273$
$$= 0.074 \log [(2.73)(0.0001)]$$
$$= 0.074 (\log 2.73 + \log 0.0001)$$
$$= 0.074 (0.43616 - 4)$$
$$= 0.03225 - 0.296$$
$$= -0.26375$$
$$= 0.73625 - 1$$
$$Z = \log^{-1} 0.73625 / \log^{-1} 1$$

where $\log^{-1} 0.73625 = 5.448 \qquad \log^{-1} 1 = 10$
$\therefore \qquad\qquad Z = 5.448/10 = 0.5448$

**By Exponential Table**

For $Z = (0.000273)^{0.074}$

$$= e^{0.074 \ln 0.000273}$$

$$= e^{0.074 \,(\ln 0.001 + \ln 0.273)}$$     (most ln tables have minimum arguments to

$$= e^{0.074(-8.20604)}$$     0.001)

$$= e^{-0.60725}$$

Most exponential tables are in increments of 0.01.

$\therefore$          $e^{-0.60} = 0.548812$          $e^{-0.61} = 0.543351$

and          $Z = 0.548812 - 0.725(0.548812 - 0.543351)$

$$= 0.5448$$

## 1-24. Analytic Geometry

The basic concept of analytic geometry is the establishment of a one-to-one correspondence between the points of the plane and number pair $(x,y)$. This correspondence may be done in a number of ways. The *rectangular*, or *cartesian*, coordinate system consists of two straight lines intersecting at right angles. A point is designated by $(x,y)$, where $x$ (the abscissa) is the distance of the point from the $y$ axis measured parallel to the $x$ axis. The ordinate $y$ is the distance of the point from the $x$ axis. The *quadrants* are labeled I, II, III, and IV.

Another common coordinate system is the **polar** system. In this system the position of a point is designated by the pair $(r,\theta)$, where

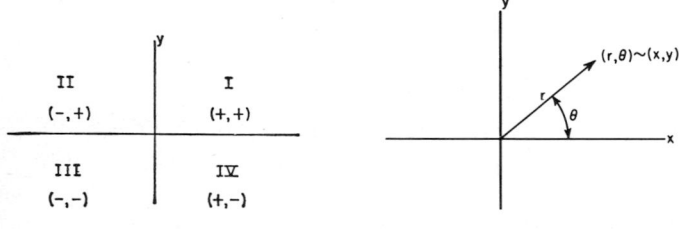

FIG. 1-3                    FIG. 1-4

$$r^2 = x^2 + y^2$$

and          $x = r \cos \theta$          $y = r \sin \theta$

**Distance.**   The distance between two points $(x_1,y_1)$, $(x_2,y_2)$ is given by

$$d = [(x_1 - x_2)^2 + (y_1 - y_2)^2]^{1/2}$$

or          $d = [r_1^2 + r_2^2 - 2r_1r_2 \cos (\theta_1 - \theta_2)]^{1/2}$

**Straight Line.**   The general equation of a straight line is given by

$$Ax + By + C = 0$$

If a straight line passes through two known points $(x_1,y_1)$ and $(x_2,y_2)$, then

$$y - y_1 = \frac{y_2 - y_1}{x_2 - x_1} (x - x_1)$$

where          $\dfrac{y_2 - y_1}{x_2 - x_1} = m$     ($m$ = slope)

Two straight lines are parallel if, and only if, they have the same slope. Two lines are perpendicular if, and only if, the product of their slopes is $-1$ (excluding lines parallel to the coordinate axes).

**Conic Sections.** The curves included in this group are obtained from plane sections of the cone. They include the circle, ellipse, parabola, hyperbola, and degeneratively, the point and straight line. A *conic* is the locus of a point whose distance from a fixed point called the *focus* is in a constant ratio to its distance from a fixed line called a *directrix*. The ratio is the *eccentricity e*. If $e = 0$, the conic is a circle; if $0 < e < 1$, the conic is an ellipse; if $e = 1$, the conic is a parabola; if $e > 1$, the conic is a hyperbola. Every conic section is representable by an equation of second degree. Conversely, every equation of second degree in two variables represents a conic. The general equation of the second degree is

$$Ax^2 + Bxy + Cy^2 + Dx + Ey + F = 0$$

Let $\Delta$ represent the determinant

$$\Delta = \begin{vmatrix} 2A & B & D \\ B & 2C & E \\ D & E & 2F \end{vmatrix}$$

and
$$Q = D^2 + E^2 - 4(A + C)F$$

| $B^2 - 4AC < 0$ | $B^2 - 4AC = 0$ | $B^2 - 4AC > 0$ |
|---|---|---|
| $A\Delta < 0$  $A \ne C$ ellipse | | |
| $\Delta \ne 0$   $A\Delta < 0$  $A = C$ circle | Parabola | Hyperbola |
| $A\Delta > 0$  no locus | | |

| $\Delta = 0$ | Point | $Q > 0$  2 parallel lines | 2 intersecting lines |
|---|---|---|---|
| | | $Q = 0$  1 straight line | |
| | | $Q < 0$  no locus | |

## 1-25. Algebra

**Addition and Subtraction.** Only like terms are added or subtracted:

$$(3 - 5x) + (3x + 4x^2) + (5 + 2x^2) = (3 + 5) + (-5 + 3)x + (4 + 2)x^2$$
$$= 8 - 2x + 6x^2$$

**Multiplication.** Carried out term by term and corresponding terms are combined:

$$(3xy + 6x)(3y + 2y^2 + 2xy) = 3xy(3y + 2y^2 + 2xy) + 6x(3y + 2y^2 + 2xy)$$
$$= (9xy^2 + 6xy^3 + 6x^2y^2)$$
$$+ (18xy + 12xy^2 + 12x^2y)$$
$$= 18xy + 21xy^2 + 6xy^3 + 6x^2y^2 + 12x^2y$$

**Division.** Division is carried out with like terms in the same fashion as arithmetic division.

| Divisor | Dividend | Quotient |
|---|---|---|
| $e^x + 1 )$ | $\overline{\phantom{xx}3e^{2x} + e^x + 1}$ | $3e^x - 2$ |
| | $-3e^{2x} - 3e^x$ | |
| | $\overline{\phantom{xxx}- 2e^x + 1}$ | |
| | $+ 2e^x + 2$ | |
| | $\overline{\phantom{xxxx}+3}$  remainder | |

**Operations with Zero.** All operations except division can be carried out with zero and a finite number $a \neq 0$:

$$a + 0 = 0; \quad a(0) = 0; \quad 0/a = 0; \quad a^0 = 1; \quad 0^a = 0 \quad (a > 0)$$

**Factoring**

$$x^2 - y^2 = (x - y)(x + y) \qquad x^2 + y^2 = (x + yi)(x - yi) \qquad i = \sqrt{-1}$$
$$x^3 - y^3 = (x - y)(x^2 + xy + y^2) \qquad x^3 + y^3 = (x + y)(x^2 - xy + y^2)$$
$$x^4 - y^4 = (x - y)(x + y)(x^2 + y^2)$$

**Quadratic Equation**

$$ax^2 + bx + c = 0 \qquad x = \frac{-b \pm \sqrt{b^2 - 4ac}}{2a}$$

If $b^2 - 4ac > 0$, the roots are real and unequal.
If $b^2 - 4ac = 0$, the roots are real and equal.
If $b^2 - 4ac < 0$, the roots are imaginary.

**Roots for Third- or Higher-order Equations.** See numerical methods, page 1-99.

**Permutation.** A specific sequencing of elements of a set. The number of permutations of $n$ things taken $r$ at a time is written

$$P(n,r) = \frac{n!}{(n - r)!} = n(n - 1)(n - 2) \cdots (n - r + 1)$$

The total permutations of $(1,2,3)$ is

$$P(3,3) = \frac{3!}{0!} = 6, \quad \text{that is,} \quad (123), (132), (213), (231), (321), (312)$$

**Combination.** A specific subset of a set without regard to sequence. The number of combinations of $n$ things taken $r$ at a time is written

$$_nC_r \quad \text{or} \quad C(n,r) = \frac{n!}{(n - r)!(r)!}$$

The number of combinations of three elements from a set of five is given by $C(5,3) = \frac{5!}{2!3!} = 10$. From the set $(1,2,3,4,5)$ the combinations are $(1,2,3)$, $(1,2,4)$, $(1,2,5)$, $(1,3,4)$, $(1,3,5)$, $(1,4,5)$, $(2,3,4)$, $(2,3,5)$, $(2,4,5)$, $(3,4,5)$.

**Binomial Theorem.** If $n$ is a positive integer,

$$(x + y)^n = x^n + nx^{n-1}y + \frac{n(n - 1)}{2!} x^{n-2}y^2 + \cdots + {_nC_r}x^{n-r}y^r + \cdots + y^n$$

where

$$_nC_r = \frac{n!}{(n - r)!(r)!} \qquad n! = 1 \times 2 \times 3 \times \cdots \times n, \, 0! = 1$$

## 1-26. Trigonometry
### Right Triangle

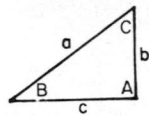

$\sin B = b/a \qquad \cos B = c/a \qquad \tan B = b/c$
$\csc B = 1/\sin B \qquad \sec B = 1/\cos B \qquad \cot B = 1/\tan B$
$A + B + C = 180°$
Inverse function $\theta = \sin^{-1} x$, or the angle $\theta$ whose sine is $x$

Fig. 1-5. Right triangle.

### Oblique Triangle

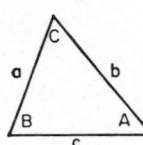

$$\frac{\sin A}{\sin B} = \frac{a}{b} \qquad a^2 = b^2 + c^2 - 2bc \cos A$$

$$\frac{a + b}{a - b} = \frac{\tan \dfrac{A + B}{2}}{\tan \dfrac{A - B}{2}}$$

Fig. 1-6. Oblique triangle.

$$\text{Area} = \frac{bc}{2} \sin A = \sqrt{s(s - a)(s - b)(s - c)}$$

where $2s = a + b + c$

### Relations between Functions of a Single Angle
$\sin^2 \theta + \cos^2 \theta = 1 \qquad (1 + \tan^2 \theta) \cos^2 \theta = 1$
$\sin 2\theta = 2 \sin \theta \cos \theta \qquad \cos 2\theta = 2 \cos^2 \theta - 1$
$\sin 3\theta = 3 \sin \theta - 4 \sin^3 \theta \qquad \cos 3\theta = 4 \cos^3 \theta - 3 \cos \theta$
$\sin n\theta = 2 \sin (n - 1)\theta \cos \theta - \sin (n - 2)\theta$
$\cos n\theta = 2 \cos (n - 1)\theta \cos \theta - \cos (n - 2)\theta$

$$\sin \frac{x}{2} = \pm \sqrt{0.5(1 - \cos x)} \qquad \cos x/2 = \pm \sqrt{0.5(1 + \cos x)}$$

$$\tan x/2 = \pm \sqrt{\frac{1 - \cos x}{1 + \cos x}} \qquad \begin{array}{l} \text{(signs according to quadrant of } x/2 \text{ for asso-} \\ \text{ciated function)} \end{array}$$

### Functions of Two Angles
$\sin (x \pm y) = \sin x \cos y \pm \cos x \sin y$
$\cos (x \pm y) = \cos x \cos y \mp \sin x \sin y$

$$\tan (x \pm y) = \frac{\tan x \pm \tan y}{1 \mp \tan x \tan y} \qquad \sin x \pm \sin y = 2 \sin \frac{x \pm y}{2} \cos \frac{x \mp y}{2}$$

$$\cos x + \cos y = 2 \cos \frac{x + y}{2} \cos \frac{x - y}{2}$$

$$\cos x - \cos y = -2 \sin \frac{x + y}{2} \sin \frac{x - y}{2}$$

$$\tan x \pm \tan y = \frac{\sin (x \pm y)}{\cos x \cos y} \qquad 2 \sin x \sin y = \cos (x - y) - \cos (x + y)$$

$2 \cos x \cos y = \cos (x - y) + \cos (x + y)$
$2 \sin x \cos y = \sin (x + y) + \sin (x - y)$

### Magnitude of Trigonometric Functions

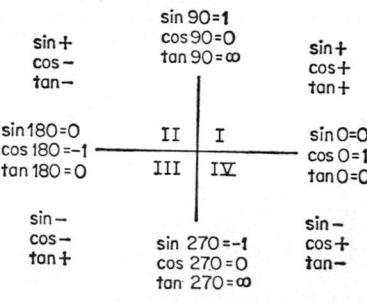

Fig. 1-7

**Hyperbolic Trigonometry.** The hyperbolic functions are certain combinations of exponentials $e^x$ and $e^{-x}$.

$$\cosh x = 0.5(e^x + e^{-x}) \qquad \sinh x = 0.5(e^x - e^{-x})$$

$$\sinh x + \cosh x = e^x \qquad \tanh x = \frac{e^x - e^{-x}}{e^x + e^{-x}}$$

## 1-27. Differential Calculus

**Definition of a Function.** A function is a quantity which takes on a definite value, or values, when other quantities are specified. If $y = e^x$, then $y = f(x)$; the value of $f(x)$ when $x$ has the value $a$ is represented by $f(a)$.

**Definition of Limit.** The statement that the *limit* of the function $f(x)$, as $x$ approaches $a$, is the number $N$ is expressed by writing

$$\lim_{x \to a} f(x) = N$$

and $f(x)$ can be calculated as close to $N$ as desired by making $x$ sufficiently close to $a$. No restriction is placed on $f(x)$ when $x = a$. For example,

$$\lim_{x \to 0} \frac{\sin x}{x} = 1$$

even though $f(x) = (\sin x)/x$ is undefined at $x = 0$.

**Operations with Limits.** The following operations are valid for $x \to a$:

(1) $\lim bf(x) = b \lim f(x)$

(2) $\lim [f(x) + g(x)] = \lim f(x) + \lim g(x)$

(3) $\lim [f(x)g(x)] = \lim f(x) \cdot \lim g(x)$

(4) $\lim \dfrac{f(x)}{g(x)} = \dfrac{\lim f(x)}{\lim g(x)} \qquad$ if $\lim_{x \to a} g(x) \neq 0$

**Definition of Continuity.** A function $f(x)$ is *continuous* at the point $x = a$ if

$$\lim_{h \to 0} [f(a + h) - f(a)] = 0$$

Discontinuities are classified into three types:

Finite: $y = \dfrac{\sin x}{x}$ at $x = 0$

Infinite: $y = \dfrac{1}{x}$ at $x = 0$

Jump: $y = \dfrac{10}{1 + e^{1/x}}$ $\begin{cases} \text{at} & x = 0^+, & y = 0^+ \\ \text{at} & x = 0, & y = 0 \\ \text{at} & x = 0^-, & y = 10 \end{cases}$

**Definition of Derivative.** The function $f(x)$ has a derivative at $x = a$, which can be denoted as $f'(a)$, if

$$\lim_{h \to 0} \frac{f(a + h) - f(a)}{h}$$

exists. This requires continuity at $x = a$. Conversely, a function may be continuous but not have a derivative. The derivative function

$$f'(x) = \frac{df}{dx} = \lim_{h \to 0} \frac{f(x + h) - f(x)}{h}$$

**Differential Operations.** The following differential operations are valid; where $f, g, \ldots$, are functions of $x$, $c$ is a constant; $e$ is the base of the natural logarithms.

$$\frac{dc'}{dx} = 0 \qquad \frac{dx}{dx} = 1 \qquad\qquad \frac{dy}{dx} = \frac{1}{dx/dy}$$

$$\frac{d}{dx}(f + g) = \frac{df}{dx} + \frac{dg}{dx} \qquad\qquad \frac{d}{dx}(f \cdot g) = f\frac{dg}{dx} + g\frac{df}{dx}$$

$$\frac{d}{dx}f^n = nf^{n-1}\frac{df}{dx} \qquad\qquad \frac{df}{dx} = \frac{df}{dv} \cdot \frac{dv}{dx}$$

$$\frac{d}{dx}\frac{f}{g} = \frac{g\,df/dx - f\,dg/dx}{g^2} \qquad\qquad \frac{da^x}{dx} = a^x \ln a$$

$$\frac{df^g}{dx} = gf^{g-1}\frac{df}{dx} + f^g \ln f \frac{dg}{dx}$$

**Differentials**

| | | |
|---|---|---|
| $de^x = e^x\,dx$ | $d \ln x = \dfrac{1}{x}\,dx$ | $d \log x = \dfrac{\log e}{x}\,dx$ |
| $d \sin x = \cos x\,dx$ | $d \cos x = -\sin x\,dx$ | $d \tan x = \sec^2 x\,dx$ |
| $d \cot x = -\csc^2 x\,dx$ | $d \sec x = \tan x \sec x\,dx$ | $d \csc x = -\cot x \csc x\,dx$ |
| $d \sin^{-1} x = (1 - x^2)^{-1/2}\,dx$ | $d \cos^{-1} x = -(1 - x^2)^{-1/2}\,dx$ | $d \tan^{-1} x = (1 + x^2)^{-1}\,dx$ |
| $d \cot^{-1} x = -(1 + x^2)^{-1}\,dx$ | $d \sec^{-1} x = x^{-1}(x^2 - 1)^{-1/2}\,dx$ | $d \csc^{-1} x = -x^{-1}(x^2 - 1)^{-1/2}\,dx$ |
| $d \sinh x = \cosh x\,dx$ | $d \cosh x = \sinh x\,dx$ | $d \tanh x = \text{sech}^2 x\,dx$ |
| $d \coth x = -\text{csch}^2 x\,dx$ | $d \,\text{sech}\, x = -\text{sech}\, x \tanh x\,dx$ | $d \,\text{csch}\, x = -\text{csch}\, x \coth x\,dx$ |
| $d \sinh^{-1} x = (x^2 + 1)^{-1/2}\,dx$ | $d \cosh^{-1} x = (x^2 - 1)^{-1/2}\,dx$ | $d \tanh^{-1} x = (1 - x^2)^{-1}\,dx$ |
| $d \coth^{-1} x = -(x^2 - 1)^{-1}\,dx$ | $d \,\text{sech}^{-1} x = x(1 - x^2)^{-1/2}\,dx$ | $d\,\text{csch}^{-1} x = -x^{-1}(x^2 + 1)^{-1/2}\,dx$ |

**Higher Differentials.** The first derivative of $f(x)$ with respect to $x$ is denoted as $f'(x)$ or $df/dx$. The derivative of the first derivative is denoted as $f''$, $f^{(2)}$, or $d^2f/dx^2$; similarly for the higher-order derivatives. The differential formulas can be successively applied for the higher-order derivatives.

**Indeterminate Forms—L'Hôpital's Theorem.** Forms of type $0/0$, $\infty/\infty$, $0 \cdot \infty$, etc., are called *indeterminates*. To find the limiting values of the corresponding functions, L'Hôpital's theorem can be used: If the ratio of two functions $f(x)$ and $g(x)$ is undefined at $x = a$, then the limit of the quotient is equal to the limit of the quotient of their separate derivatives if the limit exists.

For $y = (1 - x)^{1/x}$, then $\ln y = \dfrac{1}{x} \ln (1 - x)$:

$$\lim_{x \to 0} (\ln y) = \lim_{x \to 0} \frac{\ln (1 - x)}{x} = \frac{\lim (d/dx) \ln (1 - x)}{\lim (d/dx)x} = \frac{\lim -1/(1 - x)}{1} = -1$$

Therefore $\lim\limits_{x \to 0} y = e^{-1}$.

**Partial Derivative.** The abbreviation $Z = f(x,y)$ implies $Z$ is a function of the two independent variables $x$ and $y$. The derivative of $Z$ with respect to $x$, treating $y$ as a constant, is called the partial derivative with respect to $x$ and is usually denoted as $\partial Z/\partial x$, or $\partial f/\partial x$, or simply $f_x$. Partial differentiation is applied like full differentiation for single functions.

**Partial Differentiation for Composite Functions.** A function of any number of variables $x$, $y$, $z$, which are in turn functions of other independent variables $r$, $s$, $t$, are called composite functions.

1. For $f(x,y)$ with $x = g(t)$ and $y = h(t)$, then the total differential

$$\frac{df}{dt} = \frac{\partial f}{\partial x}\frac{dx}{dt} + \frac{\partial f}{\partial y}\frac{dy}{dt}$$

2. For $x = g(t,s)$ and $y = h(t,s)$,

$$\frac{\partial f}{\partial t} = \frac{\partial f}{\partial x}\frac{\partial x}{\partial t} + \frac{\partial f}{\partial y}\frac{\partial y}{\partial t}$$

The cross-partial derivative

$$\frac{\partial^2 f}{\partial t\, \partial s} = \frac{\partial^2 f}{\partial x^2}\frac{\partial x}{\partial s}\frac{\partial x}{\partial t} + \frac{\partial^2 f}{\partial x\, \partial y}\left(\frac{\partial x}{\partial s}\frac{\partial y}{\partial t} + \frac{\partial x}{\partial t}\frac{\partial y}{\partial s}\right) + \frac{\partial^2 f}{\partial y^2}\frac{\partial y}{\partial s}\frac{\partial y}{\partial t} + \frac{\partial f}{\partial x}\frac{\partial^2 x}{\partial s\, \partial t} + \frac{\partial f}{\partial y}\frac{\partial^2 y}{\partial s\, \partial t}$$

The second partial

$$\frac{\partial^2 f}{\partial t^2} = \frac{\partial^2 f}{\partial x^2}\left(\frac{\partial x}{\partial t}\right)^2 + 2\frac{\partial^2 f}{\partial x\, \partial y}\frac{\partial x}{\partial t}\frac{\partial y}{\partial t} + \frac{\partial^2 f}{\partial y^2}\left(\frac{\partial y}{\partial t}\right)^2 + \frac{\partial f}{\partial x}\frac{\partial^2 x}{\partial t^2} + \frac{\partial f}{\partial y}\frac{\partial^2 y}{\partial t^2}$$

## 1-28. Integral Calculus

**Indefinite Integral.** If $f'(x)\, dx$ is the differential of $f(x)$, the integral of $f'(x)\, dx$ is $f(x)$. Symbolically,

$$\int f'(x)\, dx = f(x) + C$$

where $C$ is a constant of integration. The following relationships hold (the constant of integration $C$ is implied):

$$\int u\, dv = uv - \int v\, du \qquad\qquad \int a\, dv = a \int dv$$

$$\int (du + dv + dw) = \int du + \int dv + \int dw$$

$$\int v^n\, dv = \frac{v^{n+1}}{n+1} \qquad n \neq -1 \qquad\qquad \int \frac{dv}{v} = \ln v$$

$$\int a^v\, dv = \frac{a^v}{\ln a} \qquad\qquad \int e^v\, dv = e^v$$

$$\int \sin v\, dv = -\cos v \qquad\qquad \int \cos v\, dv = \sin v$$

$$\int \sec^2 v\, dv = \tan v \qquad\qquad \int \csc^2 v\, dv = -\cot v$$

$$\int \sec v \tan v\, dv = \sec v \qquad\qquad \int \csc v \cot v\, dv = -\csc v$$

$$\int \frac{dv}{v^2 + a^2} = \frac{1}{a} \tan^{-1} \frac{v}{a} \qquad\qquad \int \frac{dv}{(a^2 - v^2)^{1/2}} = \sin^{-1} \frac{v}{a}$$

$$\int \frac{dv}{v^2 - a^2} = \frac{1}{2a} \ln \frac{v - a}{v + a} \qquad\qquad \int \frac{dv}{(v^2 \pm a^2)^{1/2}} = \ln(v + \sqrt{v^2 \pm a^2})$$

$$\int \sec v\, dv = \ln (\sec v + \tan v) \qquad\qquad \int \csc v\, dv = \ln (\csc v - \cot v)$$

**Methods of Integration.** Only a relatively small proportion of integrands can be directly integrated. For these, trial and error is required to redefine the integrand to a recognizable form for direct integration. Several general procedures are useful:

*Direct Formula.* This technique is applicable for transformation of the integrand to a tabled form.

For $\int x \sqrt{ax^2 + b}\, dx$, let $v = ax^2 + b \rightarrow dv = 2ax\, dx$.

$$\therefore \quad \int x \sqrt{ax^2 + b}\, dx = \frac{1}{2a} \int (ax^2 + b)^{1/2} (2ax\, dx) = \frac{1}{2a} \int v^{1/2}\, dv$$

$$= \frac{1}{2a} \frac{v^{3/2}}{3/2} + C = \frac{2}{6a} (ax^2 + b)^{3/2} + C$$

*Trigonometric Substitution.* This technique is particularly effective for integrands which are in the form of radicals. For these,

$$(x^2 - a^2)^{1/n} \qquad \text{let } x = a \sec \theta$$
$$(x^2 + a^2)^{1/n} \qquad \text{let } x = a \tan \theta$$
$$(a^2 - x^2)^{1/n} \qquad \text{let } x = a \sin \theta$$

*Algebraic Substitution.* Functions containing elements of the type $(a + bx)^{1/n}$ are best handled by the algebraic transformation $y^n = a + bx$.

*Partial Fractions.* Rational functions are of the type $f(x)/g(x)$, where $f(x)$ and $g(x)$ are polynomial expressions of degree $m$ and $n$, respectively. If $m \geq n$, algebraic division can be carried out to define a numerator which is at least one degree less than the denominator. The denominator function is then factored into linear terms and, if necessary, quadratic terms for complex roots. These are expanded to linear and, if necessary, quadratic terms:

$$\frac{8x^2 + 3}{x^3 - x^2 - x - 2} = \frac{8x^2 + 3}{(x^2 + x + 1)(x - 2)} = \frac{Ax + B}{x^2 + x + 1} + \frac{C}{x - 2}$$

Then

$$(Ax + B)(x - 2) + C(x^2 + x + 1) = 8x^2 + 3$$

or

$$A + C = 8 \qquad B + C - 2A = 0 \qquad C - 2B = 3$$

∴

$$A = 3 \qquad B = 1 \qquad C = 5$$

and

$$\frac{8x^2 + 3}{x^3 - x^2 - x - 2} = \frac{3x + 1}{x^2 + x + 1} + \frac{5}{x - 2}$$

Algebraic integration formulas can then be used.

*Parts.* For trigonometric and exponential function, integration by parts is very useful:

$$\int u \, dv = uv - \int v \, du$$

*Series Expansion.* When an explicit function cannot be found, the integration sometimes can be carried out by expanding the integrand into a power series.

**Definite Integral.** For the indefinite integral,

$$\int f(x) \, dx = F(x)$$

then for the definite integral,

$$\int_a^b f(x) \, dx = F(b) - F(a)$$

The definite integral is a function of the limits of integration and any variable coefficients of the integrand. It is not a function of the dummy variable of integration $x$:

$$\int_a^b f(x) \, dx \neq F(x)$$

*Properties*

$$\int_a^b cf(x) \, dx = c \int_a^b f(x) \, dx \qquad \int_a^b f(x) \, dx = - \int_b^a f(x) \, dx$$

$$\int_a^b [f(x) + g(x)] \, dx = \int_a^b f(x) \, dx + \int_a^b g(x) \, dx$$

$$\int_a^b f(x) \, dx = \int_a^c f(x) \, dx + \int_c^b f(x) \, dx$$

$$\int_a^b f(x) \, dx = (b - a)f(\xi) \qquad a \le \xi \le b$$

$$\frac{\partial}{\partial b} \int_a^b f(x) \, dx = f(b) \qquad \frac{\partial}{\partial a} \int_a^b f(x) \, dx = -f(a)$$

$$\frac{\partial}{\partial \alpha} \int_a^b f(x,\alpha) \, dx = \int_a^b \frac{\partial f(x,\alpha)}{\partial \alpha} \, dx$$

$$\int_a^b dx \int_c^d f(x,y) \, dy = \int_c^d dy \int_a^b f(x,y) \, dx$$

**Methods of Integration.** All the methods of integration for the indefinite integral can be used for the definite integral. In addition, several others are available.

**Change of Variable.** This substitution is basically the same as indefinite integrals, with the proper change in the limits of integration. For $x = \phi(t)$,

$$\int_a^b f(x) \, dx = \int_{t_0}^{t_1} f[\phi(t)]\phi' \, dt$$

where $\phi(t_1) = b$ and $\phi(t_0) = a$.

**Differentiation.** Definite integrals can be partially differentiated in respect to the limits or variable parameters contained in the integrand function when the corresponding defined functions satisfy certain properties. See, for example, Courant, "Differential and Integral Calculus," vol. 2, Interscience Publishers, Inc., 1936.

**Integration.** Comparable theory to that for differentiation of a definite integral also applies for integration of a definite integral.

**Complex Variable.** Certain definite integrals can be evaluated more readily by the technique of complex variable integration. See, for example, Copson, "Theory of Functions of a Complex Variable," Oxford University Press, 1935.

**Numerical Integration.** A numerical evaluation of a definite integral can be carried out by using Simpson's or Weddle's rule as shown on page 1–106.

### 1-29. Infinite Series

**Definitions.** A succession of numbers or terms which are formed according to some definite rule is called a *sequence*. The indicated sum of a sequence is called a *series*. If the terms of the sequence are variable, the series is called a *power series*. For the geometric progression,

$$S_n = a + ar + ar^2 + \cdots + ar^n$$

Then
$$S_n = a \frac{1 - r^n}{1 - r}$$

As $n \to \infty$, then

$$S = \lim_{n \to \infty} S_n = a \lim_{n \to \infty} \frac{1 - r^n}{1 - r}$$

The series is said to *converge* if the limit of $S_n$ approaches a fixed finite value. Otherwise the series is *divergent*.

For $\quad -1 < r < 1 \qquad \lim\limits_{n \to \infty} S_n \qquad$ converges

For $\quad r \geq 1, r < -1 \qquad \lim\limits_{n \to \infty} S_n \qquad$ diverges (unbounded sum)

For $\quad r = -1 \qquad\quad \lim\limits_{n \to \infty} S_n \qquad$ diverges (sum does not exist)

The latter series is called an *oscillating divergent series*. For the new series

$$S = 1 + \frac{a}{2} + \frac{a^2}{3} + \frac{a^3}{4} + \frac{a^4}{5} + \cdots + \frac{a^n}{n+1} + \cdots$$

$S = \log 2$ for $a = -1$ and $S$ diverges for $a = +1$. This series is called a *conditionally convergent* series. Conversely,

$$e^z = 1 + \frac{x}{1!} + \frac{x^2}{2!} + \frac{x^3}{3!} + \cdots + \frac{x^n}{n!} + \cdots$$

converges for any negative or positive value of $|x| < \infty$, and therefore is said to *converge absolutely*.

## Operations with Infinite Series

1. The convergence or divergence of an infinite series is unaffected by the removal of a finite number of finite terms.

2. If a series is conditionally convergent, its sum can be made to take on any arbitrary value by a suitable rearrangement of the series.

3. A series of positive terms, if convergent, has a sum independent of the order of the terms.

4. An oscillatory series always can be made to converge by grouping the terms in brackets.

5. A power series can be inverted, provided the first-degree term is not zero.

6. Two series may be added or subtracted term by term provided each is a convergent series.

7. A power series may be integrated termwise to represent the integral of the function within an interval of convergence.

8. A power series may be differentiated termwise to represent the differential function within the same region of convergence of the function.

**Tests for Convergence and Divergence.** In general, the problem of determining whether or not a given series will converge can require ingenuity. There is no all-inclusive test which can be applied to all series. There are many special tests.

*Comparison Test.* A series will converge if the absolute value of each term is less than the corresponding term of a known convergent series.

*nth-term Test.* A series is divergent unless the $n$th term of the series approaches zero as $n$ approaches infinity.

*Ratio Test.* If the absolute ratio of the $(n + 1)$ term divided by the $n$th term, as $n$ approaches infinity, approaches:

1. A number less than 1, the series is convergent.
2. A number greater than 1, the series is divergent.
3. A number equal to 1, the test is inconclusive.

**Summation Test.** If the partial summation $S_n$ of a series converges as $n$ becomes unbounded, the series converges, and conversely, it diverges if $S_n$ diverges.

**Alternating-series Test.** If the terms of a series are alternately positive and negative and never increase in numerical value, the series will converge, provided that the terms tend to zero as a limit.

**Cauchy's Root Test.** If the $n$th root of the $n$th-absolute-value term, as $n$ becomes infinite, approaches

1. A number less than 1, the series is convergent.
2. A number greater than 1, the series is divergent.
3. A number equal to 1, the test is inconclusive.

**Maclaurin's Integral Test.** A series $\Sigma a_n$ converges or diverges with the integral $\int_1^\infty f(x)\, dx$, where $f(n) = a_n$ and $f(x)$ is defined and continuous for $1 \leq x < \infty$ and $\lim\limits_{x \to \infty} f(x) = 0$.

### Series Summation and Identities

$$1 + 2 + 3 + \cdots + n = \frac{n(n + 1)}{2}$$

$$1^2 + 2^2 + 3^2 + \cdots + n^2 = \frac{n(n + 1)(2n + 1)}{6}$$

$$1^3 + 2^3 + 3^3 + \cdots + n^3 = \frac{n^2(n + 1)^2}{4}$$

$$1^4 + 2^4 + 3^4 + \cdots + n^4 = \frac{n(n + 1)(2n + 1)(3n^2 + 3n - 1)}{30}$$

Arithmetic progression:

$$\sum_{m=1}^n [a + (m - 1)d] = na + \tfrac{1}{2}n(n - 1)d$$

Geometric progression:

$$\sum_{m=1}^n ar^m = a\frac{1 - r^n}{1 - r}$$

Harmonic progression:

$$\sum_{m=1}^n \frac{1}{a + md} = S_n \qquad \text{(no general summation formula is known)}$$

Binomial series:

$$(x + y)^n = x^n + nx^{n-1}y + \frac{n(n - 1)}{2!}\, x^{n-2}y^2 + \cdots + y^n$$

$$(1 \pm x)^n = 1 \pm nx + \frac{n(n - 1)}{2!}\, x^2 \pm \frac{n(n - 1)(n - 2)}{3!}\, x^3 + \cdots \qquad x^2 < 1$$

Taylor's series:

$$f(x + h) = f(h) + xf'(h) + \frac{x^2}{2!}f''(h) + \frac{x^3}{3!}f'''(h) + \cdots$$

Maclaurin's series:

$$f(x) = f(0) + xf'(0) + \frac{x^2}{2!}f''(0) + \frac{x^3}{3!}f'''(h) + \cdots$$

Exponential series:

$$e^x = 1 + x + \frac{x^2}{2!} + \frac{x^3}{3!} + \cdots \qquad -\infty < x < \infty$$

Trigonometric series:

$$\sin x = x - \frac{x^3}{3!} + \frac{x^5}{5!} + \cdots + (-1)^n \frac{x^{2n+1}}{(2n+1)!} + \cdots \qquad -\infty < x < \infty$$

$$\cos x = 1 - \frac{x^2}{2!} + \frac{x^4}{4!} + \cdots + (-1)^n \frac{x^{2n}}{(2n)!} + \cdots \qquad -\infty < x < \infty$$

## 1-30. Vectors

A vector quantity has magnitude and direction; a scalar quantity has magnitude only. Common vector quantities are acceleration, alternating currents, and voltages, force, and velocity. A vector can be represented graphically by a straight line with an arrowhead, as in Fig. 1-8. (Length

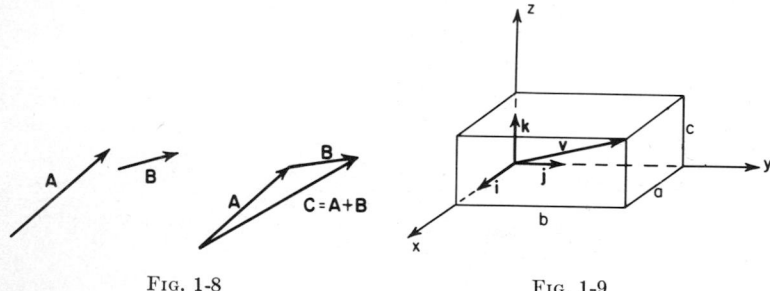

Fig. 1-8             Fig. 1-9

represents magnitude; direction is determined from the position of the line and the arrowhead.)

Vectors are usually indicated by boldface type (**A**), or by an arrow over the symbol $(\overrightarrow{A})$, or by a bar $(\overline{A})$.

**Representation.** A vector **V** in three dimensions can be represented by its projections along three mutually perpendicular lines, the $x$, $y$ and $z$ axes. Vectors of unit magnitude, directed in the positive sense along these three axes, are denoted by **i**, **j** and **k**, respectively. If $a$, $b$, and $c$ represent the lengths of the projections of **V** along these axes, we may represent **V** as **V** $= a\mathbf{i} + b\mathbf{j} + c\mathbf{k}$ (Fig. 1-8). The length (magnitude) of **V** is

$$\mathbf{V} = (a^2 + b^2 + c^2)^{1/2}.$$

**Algebra.** *Equality.* **A** = **B** if and only if both have the same magnitude and the same direction.

*Addition and Subtraction.* **A** + **B** = **B** + **A** (commutative law); **A** + **B** + **C** = (**A** + **B**) + **C** = **A** + (**B** + **C**) (associative law). If **A** = $a_1$**i** + $a_2$**j** + $a_3$**k**, **B** = $b_1$**i** + $b_2$**j** + $b_3$**k**, **A** ± **B** = $(a_1 \pm b_1)$**i** + $(a_2 \pm b_2)$**j** + $(a_3 \pm b_3)$**k**.

*Product of Vector V and Scalar s.* $s$**V** = **V**$s$ = $(sa)$**i** + $(sb)$**j** + $(sc)$**k**.

*Scalar Product of Two Vectors $V_1$, $V_2$.* The scalar (dot or inner) product, indicated by **V**$_1$ · **V**$_2$, is a scalar defined by **V**$_1$ · **V**$_2$ = $|$**V**$_1||$**V**$_2|$ cos $\theta$, where $\theta$ = angle between the vectors. **V**$_1$ · **V**$_2$ = $a_1a_2 + b_1b_2 + c_1c_2$; (**V**$_1$ + **V**$_2$) · **V**$_3$ = **V**$_1$ · **V**$_3$ + **V**$_2$ · **V**$_3$; **V**$_1$ · (**V**$_2$ + **V**$_3$) = **V**$_1$ · **V**$_2$ + **V**$_1$ · **V**$_3$ = (**V**$_2$ + **V**$_3$) · **V**$_1$ (commutative); **i** · **i** = **j** · **j** = **k** · **k** = 1; **i** · **j** = **i** · **k** = **j** · **k** = 0.

*Vector Product.* With reference to Fig. 1-9, the vector (outer) product of **V**$_1$ and **V**$_2$ is defined as the vector **V** = **V**$_1$ × **V**$_2$, **V**$_1$ × **V**$_2$ = $(b_1c_2 - b_2c_1)$**i** +

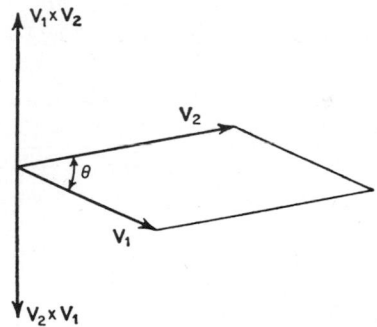

Fɪɢ. 1-10

$(c_1a_2 - c_2a_1)$**j** + $(a_1b_2 - a_2b_1)$**k** illustrated in the figure. $|$**V**$|$ = $|$**V**$_1||$**V**$_2|$ sin $\theta$; **V**$_1$ × **V**$_2$ = −**V**$_2$ × **V**$_1$; **V**$_1$ × (**V**$_2$ + **V**$_3$) = **V**$_1$ × **V**$_2$ + **V**$_1$ × **V**$_3$; (**V**$_1$ + **V**$_2$) × **V**$_3$ = **V**$_1$ × **V**$_3$ + **V**$_2$ × **V**$_3$; **i** × **i** = **j** × **j** = **k** × **k** = 0; **i** = **j** × **k** = −**k** × **j**; **j** = **k** × **i** = −**i** × **k**; **k** = **i** × **j** = −**j** × **i**; **V** × **V** = 0.

*Multiple Products.* (1) **A**(**B** · **C**); here **B** · **C** is a scalar, so that **A**(**B** · **C**) is a vector parallel to $A$. Clearly, **A**(**B** · **C**) ≠ (**A** · **B**)**C**. (2) **A** · (**B** × **C**) = **B** · (**C** × **A**) = **C** · (**A** × **B**). (3) **A** × (**B** × **C**) = **B**(**A** · **C**) − **C**(**A** · **B**).

## 1-31. Complex Numbers

The complex-number system $z$ is defined to be made up of an ordered pair of real numbers $x$ and $y$ denoted as $z = x + yi$ and satisfying certain laws of operation. The symbol $i$ is called the imaginary unit and represents $\sqrt{-1}$, where $i^2 = -1$ by definition. If $y \neq 0$, then $x + yi$ is called an imaginary number. Also, the number $x - yi$ is called its complex conjugate. The laws of operation are defined as follows:

**Addition.** $(x_1 + y_1i) + (x_2 + y_2i) = (x_1 + x_2) + (y_1 + y_2)i$

**Multiplication.** $(x_1 + y_1i)(x_2 + y_2i) = (x_1x_2 - y_1y_2) + (x_1y_2 + x_2y_1)i$

**Division.**  $\dfrac{x_1 + y_1 i}{x_2 + y_2 i} = \dfrac{x_1 x_2 + y_1 y_2}{x_2^2 + y_2^2} + \dfrac{x_2 y_1 - x_1 y_2}{x_2^2 + y_2^2} i$   $y_2 \neq 0$

It should be noted that numbers of the type $\sqrt{-a}$ should be reduced to the form $x + yi$, or in this instance, to $0 + \sqrt{a}\, i$.  Otherwise

$$\sqrt{-8}\ \sqrt{-2} = \sqrt{16} = 4,$$

which is not true.  For this case with $\sqrt{-8} = \sqrt{8}\, i$ and $\sqrt{-2} = \sqrt{2}\, i$, it follows that $\sqrt{-8}\ \sqrt{-2} = \sqrt{8}\, i\ \sqrt{2}\, i = -4$, since $i^2 = -1$.

Utilizing the rectangular coordinates $OX$ and $OY$, we represent the complex numbers $z = x + yi$ in the complex or $Z$ plane by plotting the ordered pair $(x,y)$.  Thus the absolute value or modulus of the number $z$ is given by $r = \sqrt{x^2 + y^2}$, the distance of the point $P$ from the origin.  Also the trigonometric form of $x + yi$ is given by $x + yi = r(\cos A + i \sin A)$, where the angle $A$ is measured counterclockwise from $OX$ to $OP$.

The roots of complex numbers are readily found by means of De Moivre's theorem which states that, if $n$ is any positive whole number, then

$$\cos nA + i \sin nA = (\cos A + i \sin A)^n$$

or   $\cos A + i \sin A = (\cos (1/n)\, A + i \sin (1/n)\, A)^n$

The cubic roots of $\tfrac{1}{2}\sqrt{3} + \tfrac{1}{2}i$ can, therefore, be readily found by writing the three equations corresponding to $\tfrac{1}{2}\sqrt{3} + \tfrac{1}{2}i = \cos 30° + i \sin 30°$ and the formula

$$\cos A + i \sin A = [\cos \tfrac{1}{3}(a + 360j) + i \sin \tfrac{1}{3}(A + 360j)]^3 \qquad j = 0, 1, 2$$

Thus:

$$\cos 30 + i \sin 30 = (\cos 10° + i \sin 10°)^3$$
$$\cos 30 + i \sin 30 = [\cos \tfrac{1}{3}(30 + 360) + i \sin \tfrac{1}{3}(30 + 360)]^3$$
$$= (\cos 130 + i \sin 130)^3$$
$$\cos 30 + i \sin 30 = [\cos \tfrac{1}{3}(30 + 720) + i \sin \tfrac{1}{3}(30 + 720)]^3$$
$$= (\cos 250 + i \sin 250)^3$$

and the cube roots

$$\sqrt[3]{\tfrac{1}{2}\sqrt{3} + \tfrac{1}{2}i} = 0.98481 + 0.17365i\,(\cos 10° = 0.98481, \sin 10° = 0.17365)$$
$$= -0.64279 + 0.76604i$$
$$= -0.34202 - 0.93969i$$

which satisfy

$$(x^3 - \tfrac{1}{2}\sqrt{3} - \tfrac{1}{2}i) = (x - 0.98481 - 0.17365i)(x + 0.64279 - 0.76604i)$$
$$(x + 0.34202 + 0.93969i)$$

Similarly, the roots to the polynomial equation $x^3 - 1 = 0$ are found to be $1$, $-\tfrac{1}{2} + \tfrac{1}{2}\sqrt{3}\, i$, $-\tfrac{1}{2} - \tfrac{1}{2}\sqrt{3}\, i$.

Certain types of elementary functions of a complex variable find important uses.  These are

$$e^z = e^{x+iy} = e^x e^{yi} = e^x(\cos y + i \sin y)$$
$$e^{-yi} = \cos y - i \sin y$$
$$z = x + yi = r(\cos \theta + i \sin \theta) = re^{i\theta}$$
$$\cos z = \tfrac{1}{2}e^{iz} + \tfrac{1}{2}e^{-iz}$$
$$\log z = \log r + i(\theta + 2n\pi) \qquad \text{(multiply valued)}$$
$$z^b = e^{b \log z}$$

Thus to find $(-i)^i$ it follows:

$$(-i)^i = e^{i \log (-i)} = e^{i \log 1 - i\pi/2}$$
$$= e^{i(-i\pi/2)}$$
$$= e^{\pi/2}$$

which is the principal value of $(-i)^i$ by virtue of the principal value $\sin (\pi/2) = 1$.

The definite integral of the function $f(x)$ is a number given by $\int_b^a f(x)\, dx$, which measures the area under the curve $f(x)$ between $b$ and $a$; the complex integral $\int_b^a f(z)\, dz$ does not. Rather this integral is a line integral over a specified path between $a$ and $b$. This follows from the fact that, in the complex plane corresponding to each point $(x,y)$, the function $f(x + iy)$ has a value (real or imaginary), and therefore $\int_b^a f(z)\, dz$ does not have meaning except in the sense of the specified path from $f(b)$ to $f(a)$ in terms of the values $f(z)$.

The selection of different paths will usually give different values to $\int_b^a f(z)\, dz$. Thus, caution should be exercised in carrying out complex integration, since the standard integration formulas do not hold. Nevertheless, this very fact is one of the reasons why complex-variable theory has been found to be extremely useful in engineering and physics in addition to the field of mathematics itself.

## 1-32. Determinants and Matrices

**Determinants.** Consider the system of two linear equations

$$a_{11}x_1 + a_{12}x_2 = b_1$$
$$a_{21}x_1 + a_{22}x_2 = b_2$$

If the first equation is multiplied by $a_{22}$ and the second by $-a_{12}$ and the results are added, we obtain

$$(a_{11}a_{22} - a_{21}a_{12})x_1 = b_1a_{22} - b_2a_{12}$$

The expression $a_{11}a_{22} - a_{21}a_{12}$ may be represented by the symbol

$$\begin{vmatrix} a_{11} & a_{12} \\ a_{21} & a_{22} \end{vmatrix} = a_{11}a_{22} - a_{21}a_{12}$$

This symbol is called a *determinant* of second order.  The *square* array of $n^2$ quantities $a_{ij}$, where $i = 1, \ldots, n$ is the row index, and $j = 1, \ldots, n$ is the column index, written in the form

$$A = |a_{ij}| = \begin{vmatrix} a_{11} & a_{12} & a_{13} & \cdots & a_{1n} \\ a_{21} & a_{22} & a_{23} & \cdots & a_{2n} \\ \cdots\cdots\cdots\cdots\cdots\cdots\cdots \\ a_{n1} & a_{n2} & a_{n3} & \cdots & a_{nn} \end{vmatrix}$$

is called a *determinant*.  The $n^2$ quantities $a_{ij}$ are called the *elements* of the determinant.  In the determinant $A$, let the $i$th row and $j$th column be deleted and a new determinant be formed having $n - 1$ rows and columns.  This new determinant is called the *minor* of $a_{ij}$, denoted $M_{ij}$.  The *cofactor* $A_{ij}$ of the element $a_{ij}$ is the signed minor of $a_{ij}$, determined by the rule $A_{ij} = (-1)^{i+j} M_{ij}$.

### Fundamental Properties of Determinants

1. The value of a determinant $A$ is not changed if the rows and columns are interchanged.

2. If the elements of one row (or one column) of a determinant are all zero, the value of $A$ is zero.

3. If the elements of one row (or column) of a determinant are multiplied by the same constant factor, the value of the determinant is multiplied by this factor.

4. If one determinant is obtained from another by interchanging any two rows (or columns), the value of either is the negative of the value of the other.

5. If two rows (or columns) of a determinant are identical, the value of the determinant is zero.

6. If two determinants are identical except for one row (or column), the sum of their values is given by a single determinant, obtained by adding corresponding elements of dissimilar rows (or columns) and leaving unchanged the remaining elements.

7. The value of a determinant is not changed if, to the elements of any row (or column), is added a constant multiple of the corresponding elements of any other row (or column).

8. If all elements but one in a row (or column) are zero, the value of the determinant is the product of that element times its cofactor.

The evaluation of determinants, using the definition, is quite laborious.  The labor can be reduced by applying the fundamental properties just outlined.

*Example:* Evaluate $\begin{vmatrix} 2 & 1 & 4 & 3 \\ -1 & 4 & 2 & 1 \\ 5 & 6 & 7 & 2 \\ 1 & 3 & 4 & 5 \end{vmatrix}$

The aim is to transform the determinant so that all elements but one in a given row (or column) are zero, *without changing the determinant* value.  This may be done by utilizing property 7.  Selecting the element 1 in the fourth column, add $-2$ times the fourth column to the third column; then $-4$ times the fourth column to the second column; then add the fourth column to the first column; the result is

$$A = \begin{vmatrix} 5 & -11 & -2 & 3 \\ 0 & 0 & 0 & 1 \\ 7 & -2 & 3 & 2 \\ 6 & -17 & -6 & 5 \end{vmatrix} = 1 \begin{vmatrix} 5 & -11 & -2 \\ 7 & -2 & 3 \\ 6 & -17 & -6 \end{vmatrix}$$

by property 8. Property 7 is now used on this $3 \times 3$ determinant. Subtract the elements of the first row from the third row. The result is

$$A = \begin{vmatrix} 5 & -11 & -2 \\ 7 & -2 & 3 \\ 1 & -6 & -4 \end{vmatrix}$$

Now add $-7$ times the third row to the second row, then $-5$ times the third row to the first row resulting in

$$A = \begin{vmatrix} 0 & 19 & 18 \\ 0 & 40 & 31 \\ 1 & -6 & -4 \end{vmatrix} = \begin{vmatrix} 19 & 18 \\ 40 & 31 \end{vmatrix} = -131$$

**Matrices.** A rectangular array of $mn$ quantities, arranged in $m$ rows and $n$ columns,

$$A = (a_{ij}) = \begin{bmatrix} a_{11} & \cdots & a_{1n} \\ a_{21} & \cdots & a_{2n} \\ \cdots\cdots\cdots\cdots \\ a_{m1} & \cdots & a_{mn} \end{bmatrix}$$

is called a *matrix*. The *elements* $a_{ij}$ may be numbers, functions, etc. The notation $a_{ij}$ means the element in the $i$th row and $j$th column; $i$ is called the *row index*, $j$ the *column index*. If $m = n$, the matrix is said to be *square* and of *order n*. A matrix, even if it is square, *does not have a numerical value*, as a determinant does. However, if the matrix $A$ is square, a determinant can be formed which has the same elements as the matrix $A$. This is called the *determinant of the matrix*, and written det $(A)$ or $|A|$. If $A$ is square and det $(A) \neq 0$, $A$ is said to be *nonsingular;* if det $(A) = 0$, $A$ is said to be *singular*. A matrix $A$ has *rank r* if, and only if, it has a nonvanishing determinant of order $r$ and no nonvanishing determinant of order $> r$.

*Algebra of Matrices.* Let $A = (a_{ij})$, $B = (b_{ij})$.

*Equality.* Two matrices $A$ and $B$ are *equal* $(=)$ if and only if they are identical; that is, they have the same number of rows and the same number of columns and equal corresponding elements $(a_{ij} = b_{ij}$ for all $i$ and $j)$.

*Addition and Subtraction.* The operations of *addition* $(+)$ and subtraction $(-)$ of two or more matrices is possible if and only if they have the same number of rows and columns. Thus $A \pm B = (a_{ij} \pm b_{ij})$, that is, addition and subtraction are of corresponding elements.

*Multiplication.* Let $A = (a_{ij})$, $i = 1, \ldots, m_1$; $j = 1, \ldots, m_2$. Let $B = (b_{ij})$, $i = 1, \ldots, n_1$; $j = 1, \ldots, n_2$. The *product AB* is defined if, and only if, the *number* of *columns* of $A$ $(m_2)$ equals the number of rows of $B$ $(n_1)$, that is, $n_1 = m_2$. For two such matrices the product $P = AB$ is *defined* by summing the element-by-element products of a row of $A$ by a column of $B$.

This is the *row-by-column rule.* Thus $p_{ij} = \sum\limits_{k=1}^{n_1} a_{ij}b_{kj}$. The resulting matrix has $m_1$ rows and $n_2$ columns. It is helpful to remember that the element $p_{ij}$ is formed from the $i$th row of the first matrix and the $j$th column of second matrix. In general, a matrix product is *not commutative;* that is, $AB \neq BA$.

**Inverse of a Matrix.** A *square* matrix $A$ is said to have an *inverse* if there exists a matrix $B$ such that $AB = BA = I$, where $I$ is the *identity* matrix of order $n$:

$$\begin{bmatrix} 1 & 0 & \cdots & \cdots & 0 \\ 0 & 1 & \cdot & \cdot & \\ \cdot & & & & \\ \cdot & & & & \\ \cdot & & & 1 & 0 \\ 0 & \cdots & \cdots & 0 & 1 \end{bmatrix}$$

The inverse $B$ is a square matrix of the order of $A$, designated by $A^{-1}$. Thus $AA^{-1} = A^{-1}A = I$. A square matrix $A$ has an inverse if, and only if, $A$ is *nonsingular.*

**Transposition.** The matrix obtained from $A$ by interchanging the rows and columns of $A$ is called the transpose of $A$, written $A'$ or $A^T$.

**Scalar Multiplication.** For $c$ any real or complex number and $A$ any matrix, $cA = (ca_{ij})$.

*Special Square Matrices*

1. A triangular matrix is a matrix all of whose elements above or below the *main diagonal* (set of elements $a_{11}, \ldots, a_{nn}$) are zero. If $A$ is triangular, $\det(A) = a_{11} \cdot a_{22} \cdot \cdots \cdot a_{nn}$.

2. A *diagonal matrix* is one such that all elements both above and below the main diagonal are zero (that is, $a_{ij} = 0$ for all $i \neq j$). If all diagonal elements are equal, the matrix is called *scalar.* If $A$ is diagonal, $A = (a_{ii})$, $A^{-1} = (1/a_{ii})$.

3. If $a_{ij} = a_{ji}$ for all $i$ and $j$ (that is, $A = A^T$), the matrix is *symmetric.*

4. If $a_{ij} = -a_{ji}$ for $i \neq j$, but the $a_{ii}$ are not all zero, the matrix is *skew.*

5. If $a_{ij} = -a_{ji}$ for all $i$ and $j$ (that is, $a_{ii} = 0$), the matrix is *skew symmetric.*

6. If $A^T = A^{-1}$, the matrix $A$ is *orthogonal.*

7. If the matrix $A^* = (\bar{a}_{ij})^T$, $\bar{a}_{ij} = $ complex conjugate of $a_{ij}$, $A^*$ is the *associate* of $A$.

8. If $A = A^{-1}$, $A$ is *involutory.*

9. If $A = A^*$, $A$ is *hermitian.*

10. If $A = -A^*$, $A$ is *skew hermitian.*

11. If $A^{-1} = (A^*)$, $A$ is *unitary.*

If $A$ is any matrix, then $AA^T$ and $A^TA$ are *square symmetric* matrices, usually of different order.

*Important Relations*

(1) $(A^{-1})^{-1} = A$.   (2) $(A^T)^T = A$.   (3) $(AB)^{-1} = B^{-1}A^{-1}$.   (4) $(AB)^T = B^TA^T$.   (5) $(A^{-1})^T = (A^T)^{-1}$.   (6) $(ABC)^{-1} = C^{-1}B^{-1}A^{-1}$.   (7) $ABC = A[BC]$

$= (AB)C$ provided all products have meaning. (8) $A^n = A \cdot \cdots \cdot A$, $n$ times. (9) $(A + B)(A - B) = A^2 + BA - AB + B^2 \neq A^2 + B^2$ since $AB \neq BA$.

## 1-33. Interpolation

The practicing engineer finds many opportunities to refer to tables as sources of information. Consequently, interpolation, or that process of "reading between the lines of a table," is useful.

**Linear Interpolation.** If a function $f(x)$ is approximately linear in a certain range, the ratio $\dfrac{f(x_1) - f(x_0)}{x_1 - x_0} = f[x_0,x_1]$ is approximately independent of $x_0$, $x_1$ in the range. The linear approximation to the function $f(x)$, $x_0 < x < x_1$, then leads to the interpolation formula

$$f(x) \approx f(x_0) + (x - x_0)f[x_0,x_1] \approx f(x_0) + \frac{x - x_0}{x_1 - x_0}[f(x_1) - f(x_0)]$$

$$\approx \frac{1}{x_1 - x_0}[(x_1 - x)f(x_0) - (x_0 - x)f(x_1)]$$

## 1-34. Numerical Methods

**Numerical Solution of Linear Equations.** The methods described here are concerned with a set of $n$ linear equations in $n$ unknowns $x_1, x_2, \ldots, x_n$, expressed in the form

$$
\begin{aligned}
a_{11}x_1 + a_{12}x_2 + a_{13}x_3 + \cdots + a_{1n}x_n &= b_1 \\
a_{21}x_1 + a_{22}x_2 + a_{23}x_3 + \cdots + a_{2n}x_n &= b_2 \\
\cdots\cdots\cdots\cdots\cdots\cdots\cdots\cdots\cdots\cdots\cdots & \\
a_{n1}x_1 + a_{n2}x_2 + a_{n3}x_3 + \cdots + a_{nn}x_n &= b_n
\end{aligned}
\tag{1-1}
$$

where the $n^2$ coefficients $a_{ij}$ and the $n$ right-hand members are given. Equation (1-1) may be written in matrix form as

$$AX = B \tag{1-2}$$

where

$$
A = \begin{bmatrix} a_{11}a_{12} \cdot \cdot \cdot a_{1n} \\ a_{21}a_{22} \cdot \cdot \cdot a_{2n} \\ \cdots\cdots\cdots \\ a_{n1}a_{n2} \cdot \cdot \cdot a_{nn} \end{bmatrix}
\qquad
X = \begin{bmatrix} x_1 \\ x_2 \\ \cdot \\ \cdot \\ \cdot \\ x_n \end{bmatrix}
\qquad
B = \begin{bmatrix} b_1 \\ b_2 \\ \cdot \\ \cdot \\ \cdot \\ b_n \end{bmatrix}
$$

and in the terminology $a_{ij}$, $i$ = row index, $j$ = column index. The problem, find the values of $x_1, x_2, \ldots, x_n$ satisfying equation (1-1), may be accomplished numerically from the form (1-1) or from (1-2) by matrix-inversion techniques. In either case the methods are *direct* (meaning "once through") or *iterative* (repeated) procedures.

*Direct Methods for Solving* **(1-1).** Suppose that the $b_j$ are not all zero and that the determinant of $A \neq 0$. Then (1-1) has a unique solution.

*Gauss Reduction.* This method is the simplest practical method for solving (1-1). It consists of dividing the first equation by $a_{11}$ (if $a_{11} = 0$, reorder the equations) and using the result to eliminate $x_1$ from all succeeding equations. Next, the modified second equation is divided by $a'_{22}$ (if $a'_{22} = 0$, a renumbering of equations and/or variables may again be necessary) and the resulting equation is used to eliminate $x_2$ from the succeeding equations. This elimination is done $n$ times. The result is of the *triangular* form.

$$x_1 + a'_{12}x_2 + a'_{13}x_3 + \cdots + a'_{1n}x_n = b'_1$$
$$x_2 + a'_{23}x_3 + \cdots + a'_{2n}x_n = b'_n$$
$$\cdots\cdots\cdots\cdots\cdots\cdots\cdots\cdots$$
$$x_{n-1} + a'_{n-1,n}x_n = b'_{n-1}$$
$$x_n = b'_n$$

where the $a'_{ij}$ and $b'_j$ represent the specific numerical values obtained by the above process. The solution is then obtained by working backward from the last equation.

*The Crout Reduction.* A modification of the Gauss procedure which is well adapted for use on desk calculators and digital computers is a method devised by Crout. Recording of intermediate steps is minimized in this procedure. The Crout algorithm is summarized by the equations

$$a'_{ij} = a_{ij} - \sum_{k=1}^{j-1} a'_{ik}a'_{kj} \qquad\qquad i \geq j$$

$$a'_{ij} = \frac{1}{a'_{ii}}\left[ a_{ij} - \sum_{k=1}^{i-1} a'_{ik}a'_{kj} \right] \qquad i < j \qquad\qquad (1\text{-}3)$$

$$b'_i = \frac{1}{a'_{ii}}\left[ b_i - \sum_{k=1}^{i-1} a'_{ik}b'_k \right]$$

and finally the solution

$$x_i = b'_i - \sum_{k=i+1}^{n} a'_{ik}x_k \qquad\qquad (1\text{-}4)$$

and $i$ and $j$ run from 1 to $n$ unless other restrictions are present.

*Iterative Methods for Solving* (1-1). In certain systems, for example, in the least-squares problems of statistics, it often happens that the diagonal elements (the elements $a_{ii}$) of (1-1) dominate strongly over the other elements. In these cases iterative methods may be used to solve the linear system (1-1). The more the diagonal terms dominate, the more rapidly the process converges, and is in many cases superior to the direct processes.

*Iteration in Total Steps.* Referring to the linear system (1-1), the first set of approximate values is obtained by taking into account only the dominant diagonal terms in each equation. The approximate values are then inserted into the full system to obtain the second approximation. And so on. If the system has been rewritten so that the diagonal terms dominate, then the procedure is to rewrite it as

$$x_1 = \frac{1}{a_{11}} (b_1 - a_{12}x_2 - a_{13}x_3 - \cdots - a_{1n}x_n)$$

$$x_2 = \frac{1}{a_{22}} (b_2 - a_{21}x_1 - a_{23}x_3 - \cdots - a_{2n}x_n) \qquad (1\text{-}5)$$

$$\cdots\cdots\cdots\cdots\cdots\cdots\cdots\cdots\cdots$$

$$x_n = \frac{1}{a_{nn}} (b_1 - a_{n1}x_1 - a_{n2}x_2 - \cdots - a_{nn-1}x_{n-1})$$

The initial approximation is

$$x_1{}^{(0)} = \frac{b_1}{a_{11}},\, x_2{}^{(0)} = \frac{b_2}{a_{22}},\, \ldots,\, x_n{}^{(0)} = \frac{b_n}{a_{nn}} \qquad (1\text{-}6)$$

The next approximation is obtained by inserting the initial approximations in (1-5) and repeating until the successive approximations agree to within a specified tolerance.

*Iteration in Single Steps.* In this method a diagonal unknown, say $x_4$, is computed approximately, neglecting all others. This value is inserted into all other equations, and from one of them, an approximation for a second diagonal element is obtained. And so forth. Thus at every step all unknowns are computed by means of all components already known.

***Matrix Inversion.*** In some problems, such as those encountered in statistical regression analysis, it is essential that the system (1-1) be solved by matrix inversion of (1-2). Thus $X = A^{-1}B$, where $A^{-1}$ is the inverse of $A$, defined in the section on algebra.

The number of methods for inverting matrices are many and varied. The methods previously described may be continued to obtain the inverse of the matrix. None of these processes are given here, but the reader is referred to Bodewig, "Matrix Calculus," North Holland Publishing Company, Amsterdam, 1956.

**Numerical Solution of Nonlinear Equations in One Variable.** *Special Methods for Polynomials.* Consider a polynomial equation of degree $n$,

$$P(x) = a_0x^n + a_1x^{n-1} + a_2x^{n-2} + \cdots + a_{n-1}x + a_n = 0 \qquad (1\text{-}7)$$

with real coefficients. $P(x)$ has exactly $n$ roots, which may be real or complex. If all the coefficients of $P(x)$ are integers, then any rational root, say $r/s$ ($r$, $s$ integers, having no common divisors), of $P(x)$ must be such that $r$ is an integral divisor of $a_n$ and $s$ is an integral divisor of $a_0$. Further, any polynomial with rational coefficients may be converted into one with integral coefficients by multiplying by the lowest common multiple of the denominators of the coefficients. In addition to these results one can obtain an upper and lower bound for the real roots by the following device: If $a_0 > 0$ in Eq. (1-7), and if in (1-7) the first negative coefficient is preceded by $k$ coefficients which are positive or zero, and if $G$ is the greatest of the absolute values of the negative coefficients, then each real root is less than $1 + \sqrt[k]{G/a_0}$.

A lower bound to the real roots may be found by applying the criterion to the equation $P(-x)$.

*Descartes Rule.* The number of positive real roots of a polynomial with real coefficients is either equal to the number of changes in sign $v$ or is less than $v$ by a positive even integer. The number of negative roots of $f(x)$ is either equal to the number of variations of sign of $f(-x)$ or is less than this by a positive even integer.

**General Methods for Nonlinear Equations.** *Successive Substitutions.* Let $f(x) = 0$ be the nonlinear equation to be solved. If this is rewritten as $x = F(x)$, then an iterative scheme can be set up in the form $x_{k+1} = F(x_k)$. To start the iteration, an initial guess must be obtained graphically or otherwise. The convergence or divergence of the procedure depends upon the method of writing $x = F(x)$, of which there will usually be several forms. If $a$ is a root of $f(x) = 0$, then a sufficient condition for convergence is that $|F'(x)| < 1$ in that interval about $a$ in which the iteration proceeds. The process is called *first-order* since the error at the $(k + 1)$st step is proportional to the first power of the error at the $k$th step.

*Methods of Perturbation.* Let $f(x) = 0$ be the equation. In general, the iterative relation is

$$x_{k+1} = x_k - \frac{f(x_k)}{\alpha_k} \tag{1-8}$$

where the iteration begins with $x_0$ as an initial approximation and $\alpha_k$ as some functional.

1. THE NEWTON-RAPHSON PROCEDURE. This variant chooses $\alpha_k = f'(x_k)$, where $f' = df/dx$ and geometrically consists of replacing the graph of $f(x)$ by the tangent line at $x = x_k$ in each successive step. If $f'(x)$ and $f''(x)$ have the same sign throughout an interval $a \leq x \leq b$ containing the solution, with $f(a)$, $f(b)$ of opposite signs, then the process converges, starting from any $x_0$ in the interval $a \leq x \leq b$. The process is second-order.

2. THE METHOD OF FALSE POSITION. This variant is commenced by finding $x_0$ and $x_1$ such that $f(x_0)$, $f(x_1)$ are of opposite signs. Then $\alpha_1 = $ slope of secant line joining $(x_0, f(x_0))$ and $(x_1, f(x_1))$ so that

$$x_2 = x_1 - \frac{x_1 - x_0}{f(x_1) - f(x_0)} f(x_1) \tag{1-9}$$

In each following step $\alpha_k$ is the slope of the line joining $(x_k, f(x_k))$ to the most recently determined point, where $f(x_j)$ has the opposite sign from that of $f(x_k)$. This method is of first order. Both of these processes can be immediately generalized to two or more simultaneous equations. For the Newton-Raphson process let the two equations be $f(x,y) = 0$, $g(x,y) = 0$. Begin with an initial approximation $(x_0, y_0)$ and then solve successively the linear equations

$$\Delta x_k \frac{\partial f}{\partial x}(x_k, y_k) + \Delta y_k \frac{\partial f}{\partial y}(x_k, y_k) = -f(x_k, y_k)$$

$$\Delta x_k \frac{\partial g}{\partial x}(x_k, y_k) + \Delta y_k \frac{\partial g}{\partial y}(x_k, y_k) = -g(x_k, y_k) \tag{1-10}$$

for $\Delta x_k$ and $\Delta y_k$. Then the $k + 1$ approximation is given from $x_{k+1} = x_k + \Delta x_k$, $y_{k+1} = y_k + \Delta y_k$. A modification consists in solving Eqs. (1-10), with $(x_k, y_k)$ replaced by $(x_0, y_0)$ (or other suitable pair later on in the iteration) in the derivatives. This means the derivatives (and therefore the coefficients of $\Delta x_k$, $\Delta y_k$) are independent of $k$. Hence the results become

$$\Delta x_k = \frac{-f(x_k,y_k)\, \dfrac{\partial g}{\partial y}\,(x_0,y_0) + g(x_k,y_k)\, \dfrac{\partial f}{\partial y}\,(x_0,y_0)}{\dfrac{\partial f}{\partial x}\,(x_0,y_0)\,\dfrac{\partial g}{\partial y}\,(x_0,y_0) - \dfrac{\partial f}{\partial y}\,(x_0,y_0)\,\dfrac{\partial g}{\partial x}\,(x_0,y_0)}$$

$$(1\text{-}11)$$

$$\Delta y_k = \frac{-g(x_k,y_k)\, \dfrac{\partial f}{\partial x}\,(x_0,y_0) + f(x_k,y_k)\, \dfrac{\partial g}{\partial x}\,(x_0,y_0)}{\dfrac{\partial f}{\partial x}\,(x_0,y_0)\,\dfrac{\partial g}{\partial y}\,(x_0,y_0) - \dfrac{\partial f}{\partial y}\,(x_0,y_0)\,\dfrac{\partial g}{\partial x}\,(x_0,y_0)}$$

and $x_{k+1} = \Delta x_k + x_k$, $y_{k+1} = \Delta y_k + y_k$. Such an alteration of the basic technique reduces the rapidity of convergence.

**Numerical Differentiation.** Numerical differentiation should be avoided wherever possible, particularly when data are empirical and subject to appreciable observation errors. Errors in data can affect numerical derivatives quite strongly; i.e., differentiation is a roughening process. When such a calculation must be made, it is usually desirable first to *smooth* the data to a certain extent.

*The Use of the Interpolation Formula.* If the data are given over equidistant values of the independent variable $x$, an interpolation formula such as the Newton formula (see Hildebrand) may be used and the resulting formula differentiated analytically. If the independent variable is not at equidistant values, then Lagrange's formulas must be used. By differentiating *three-* and *five-point* Lagrange interpolation formulas, the following differentiation formulas result for equally spaced tabular points:

*Three-point Formula.* Let $x_0$, $x_1$, $x_2$ be the three points.

$$f'(x_0) = \frac{1}{2h}\left[-3f(x_0) + 4f(x_1) - f(x_2)\right] + \frac{h^2}{3}f'''(\epsilon)$$

$$f'(x_1) = \frac{1}{2h}\left[-f(x_0) + f(x_2)\right] - \frac{h^2}{6}f'''(\epsilon)$$

$$f'(x_2) = \frac{1}{2h}\left[f(x_0) - 4f(x_1) + 3f(x_2)\right] + \frac{h^2}{3}f'''(\epsilon)$$

where the last term is an error term, $\min\limits_{j} x_j < \epsilon < \max\limits_{j} x_j$.

*Five-point Formula.* Let $x_0$, $x_1$, $x_2$, $x_3$, $x_4$ be the five values of the equally spaced independent variable and $f_j = f(x_j)$.

$$f'(x_0) = \frac{1}{12h} \left[ -25f_0 + 48f_1 - 36f_2 + 16f_3 - 3f_4 \right] + \frac{h^4}{5} f^{(v)}(\epsilon)$$

$$f'(x_1) = \frac{1}{12h} \left[ -3f_2 - 10f_1 + 18f_2 - 6f_3 + f_4 \right] - \frac{h^4}{20} f^{(v)}(\epsilon)$$

$$f'(x_2) = \frac{1}{12h} \left[ f_0 - 8f_1 + 8f_3 - f_4 \right] + \frac{h^4}{30} f^{(v)}(\epsilon)$$

$$f'(x_3) = \frac{1}{12h} \left[ -f_0 + 6f_1 - 18f_2 + 10f_3 + 3f_4 \right] - \frac{h^4}{20} f^{(v)}(\epsilon)$$

$$f'(x_4) = \frac{1}{12h} \left[ 3f_0 - 16f_1 + 36f_2 - 48f_3 + 25f_4 \right] + \frac{h^4}{5} f^{(v)}(\epsilon)$$

and the last term is again an error term.

**Numerical Integration.** A multitude of formulas have been developed to accomplish numerical integration, which consists of computing the value of a definite integral from a set of numerical values of the integrand.

*Newton-Cotes Integration Formulas (Equally Spaced Ordinates) for Functions of One Variable.* The definite integral $\int_a^b f(x)\, dx$ is to be evaluated.

*Trapezoidal Rule.* This formula consists of subdividing the interval $a \leq x \leq b$ into $n$ subintervals $a$ to $a + h$, $a + h$ to $a + 2h$, . . . , and replacing the graph of $f(x)$ by the result of joining the ends of adjacent ordinates by line segments. If $f_j = f(x_j) = f(a + jh)$, $f_0 = f(a)$, $f_n = f(b)$, the integration formula is

$$\int_a^b f(x)\, dx = \frac{h}{2} \left[ f_0 + 2f_1 + 2f_2 + \cdots + 2f_{n-1} + f_n \right] + E_n$$

where

$$|E_n| = \frac{nh^3}{12} |f''(\epsilon)| = \frac{(b-a)^3}{12n^2} |f''(\epsilon)| \qquad a < \epsilon < b$$

This procedure is not of high accuracy. However, if $f''(x)$ is continuous in $a < x < b$, the error goes to zero as $1/n^2$, $n \to \infty$.

*Parabolic Rule (Simpson's Rule).* This procedure consists of subdividing the interval $a < x < b$ into $n/2$ subintervals, each of length $2h$, where $n$ is an *even* integer. Using the notation as above, the integration formula is

$$\int_a^b f(x)\, dx$$

$$= \frac{h}{3} \left[ f_0 + 4f_1 + 2f_2 + 4f_3 + \cdots + 4f_{n-3} + 2f_{n-2} + 4f_{n-1} + f_n \right] + E_n$$

where

$$|E_n| = \frac{nh^5}{180} |f^{(1v)}(\epsilon)| = \frac{(b-a)^5}{180n^4} |f^{(1v)}(\epsilon)| \qquad a < \epsilon < b$$

This method approximates $f(x)$ by a parabola on each subinterval. This rule is more accurate than the trapezoidal rule unless $f(x)$ has wild behavior on $a \leq x \leq b$. It is the most widely used integration formula.

**Numerical Solution of Ordinary Differential Equations.** By a numerical solution of a differential equation is meant a table of values of the function $y$ and its derivatives over only a limited part of the range of the independent variable. Every differential equation of order $n$ can be rewritten as $n$ first-order differential equations. Therefore the methods given below will be for first-order equations, and the generalization to simultaneous systems will be developed later.

*Modified Adam's Method.* Let the first-order differential equation be $dy/dx = f(x,y)$ with the initial condition $(x_0,y_0)$; that is, $y = y_0$ when $x = x_0$. The procedure is as follows:

*Step* 1: From the given initial conditions $(x_0,y_0)$, compute

$$y'_0 = f(x_0,y_0) \qquad \text{and} \qquad y''_0 = \frac{\partial f(x_0,y_0)}{\partial x} + \frac{\partial f(x_0,y_0)}{\partial y}\, y'_0$$

Then determine

$$y_1 = y_0 + hy'_0 + \frac{h^2}{2}\, y''_0$$

where $h$ = subdivision of independent variable.

*Step* 2: Determine $y'_1 = f(x_1,y_1)$, $(x_1 = x_0 + h)$. These prepare us for the *Predictor Steps*

*Step* 3: $(y_{n+1})_1 = y_n + \dfrac{h}{24}\,[55y'_n - 59y'_{n-1} + 37y'_{n-2} - 9y'_{n-3}]$ where $y'_n$,

$y'_{n-1}$, etc., are calculated in Step 1.

*Step* 4: $(y'_{n+1})_1 = f[x_{n+1}, (y_{n+1})_1]$

*Corrector Steps*

*Step* 5: $(y_{n+1})_2 = y_n + \dfrac{h}{24}\,[9(y'_{n+1})_1 + 19y'_n - 5y'_{n-1} + y'_{n-2}]$

*Step* 6: $(y'_{n+1})_2 = f[x_{n+1}, (y_{n+1})_2]$

*Step* 7: Iterate Steps 5 and 6 if necessary.

*Runge-Kutta Methods.* These methods are self-starting and are inherently stable. Third- and fourth-order procedures are given below for $dy/dx = f(x,y)$, $h$ = interval size.

*Third-order* (error $\approx h^4$)

$$k_0 = hf(x_n,y_n)$$
$$k_1 = hf(x_n + \tfrac{1}{2}h,\, y_n + \tfrac{1}{2}k_0)$$
$$k_2 = hf(x_n + h,\, y_n + 2k_1 - k_0)$$

and
$$y_{n+1} = y_n + \tfrac{1}{6}(k_0 + 4k_1 + k_2)$$

for all $n \geq 0$, with initial condition $(x_0,y_0)$.

*Fourth-order* (error $\approx h^5$)

$$k_0 = hf(x_n,y_n)$$
$$k_1 = hf(x_n + \tfrac{1}{2}h,\, y_n + \tfrac{1}{2}k_0)$$
$$k_2 = hf(x_n + \tfrac{1}{2}h,\, y_n + \tfrac{1}{2}k_1)$$
$$k_3 = hf(x_n + h,\, y_n + k_2)$$
$$y_{n+1} = y_n + \tfrac{1}{6}(k_0 + 2k_1 + 2k_2 + k_3)$$

*Equations of Higher Order and Simultaneous Differential Equations.* Any differential equation of second or higher order can be reduced to a simultaneous system of first-order equations by the introduction of auxiliary variables. A fourth-order Runge-Kutta procedure for

$$\frac{dx}{dt} = f(t,x,y) \qquad \frac{dy}{dt} = g(t,x,y)$$

is given below.

Starting at the initial conditions $x_0$, $y_0$, $t_0$, the next values $x_1$, $y_1$ are computed via the equations below (where $\Delta t = h$, $t_j = h + t_{j-1}$).

$$k_0 = hf(t_0,x_0,y_0) \qquad\qquad l_0 = hg(t_0,x_0,y_0)$$

$$k_1 = hf\left(t_0 + \frac{h}{2},\, x_0 + \frac{k_0}{2},\, y_0 + \frac{l_0}{2}\right) \quad l_1 = hg\left(t_0 + \frac{h}{2},\, x_0 + \frac{k_0}{2},\, y_0 + \frac{l_0}{2}\right)$$

$$k_2 = hf\left(t_0 + \frac{h}{2},\, x_0 + \frac{k_1}{2},\, y_0 + \frac{l_1}{2}\right) \quad l_2 = hg\left(t_0 + \frac{h}{2},\, x_0 + \frac{k_1}{2},\, y_0 + \frac{l_1}{2}\right)$$

$$k_3 = hf(t_0 + h,\, x_0 + k_2,\, y_0 + l_2) \qquad l_3 = hg(t_0 + h,\, x_0 + k_2,\, y_0 + l_2)$$

and

$$x_1 = x_0 + \tfrac{1}{6}[k_0 + 2k_1 + 2k_2 + k_3] \qquad y_1 = y_0 + \tfrac{1}{6}[l_0 + 2l_1 + 2l_2 + l_3]$$

To continue the computation, replace $t_0$, $x_0$, $y_0$, in the above formulas, by $t_1 = t_0 + h$, $x_1$, $y_1$, just calculated. Extension of this method to more than two equations follows precisely this same pattern.

## 1-35. Fitting Curves to Data

It is on occasion necessary to functionally represent empirical data. When this is required, the empirical data are generally fitted to the model by the method of least squares: The determination of that set of coefficients ($a$, $b$, $c$, etc.) which minimizes the sum of squared differences between the observed data $y$ and model response $\hat{Y}$ over the same arguments $x$.

If the functional form of the model is known, then only the computational fit is required. If not, then an appropriate model must be found and subsequently fitted. To this end Figs. 1-11 through 1-34 can be used. A rectangular plot of the empirical data, with a smooth curve drawn through it, can be compared to various families of curves. That family which has similar trends can be fitted to the data. In general, only those models which have consistent asymptotic trends of the physical system should be selected.

It should be noted that all families depicted are invariant to scaling and therefore only the trends are relevant. For example, $y = 3x^2e^{-0.2x}$ plotted on $0 \le x \le 1$ is the same as $y = 0.03z^2e^{-0.02z}$ plotted on $0 \le z \le 10$; the same is true for the $y$ scale.

Most large-scale computational centers have nonlinear fitting programs. The following material is intended for those analysts who choose to do their

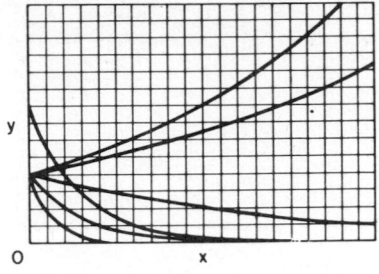

Fig. 1-11. The family of curves $y = ae^{bx}$.

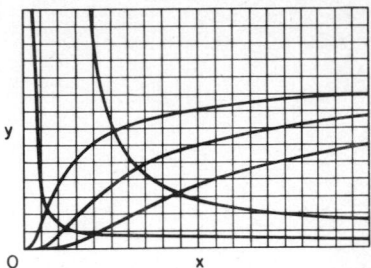

Fig. 1-12. The family of curves $y = ae^{b/x}$.

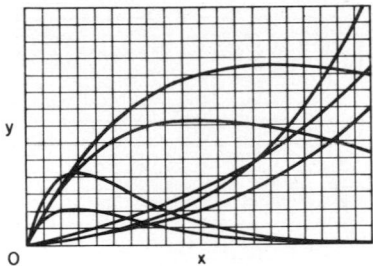

Fig. 1-13. The family of curves $y = axe^{bx}$.

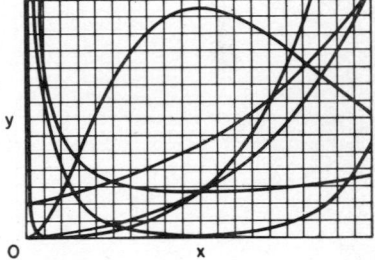

Fig. 1-14. The family of curves $y = ax^be^{cx}$.

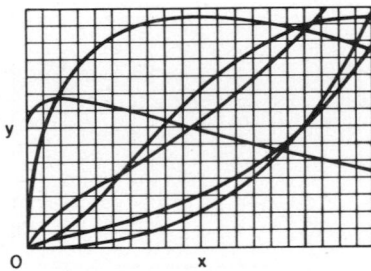

Fig. 1-15. The family of curves $y = ax^be^{cx}$.

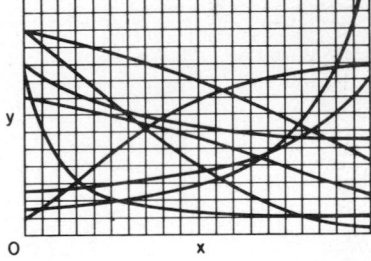

Fig. 1-16. The family of curves $y = ae^{bcx}$.

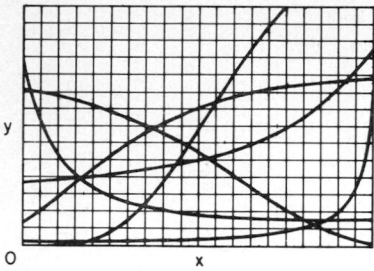

Fig. 1-17. The family of curves $y = ae^{bcx}$.

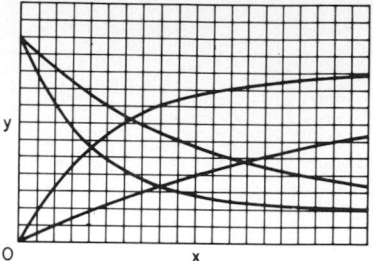

Fig. 1-18. The family of curves $y = a + be^{cx}$.

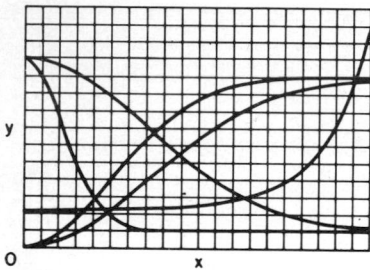

Fig. 1-19. The family of curves $y = a + be^{cx^2}$.

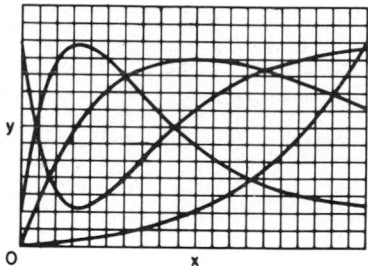

Fig. 1-20. The family of curves $y = a + bxe^{cx}$.

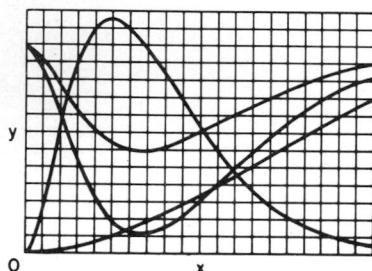

Fig. 1-21. The family of curves $y = a + bx^2e^{cx}$.

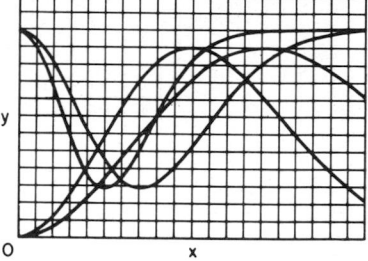

Fig. 1-22. The family of curves $y = a + bxe^{cx^2}$.

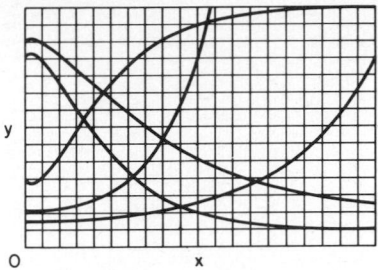

FIG. 1-23. The family of curves $y = a + bx^{cx}$.

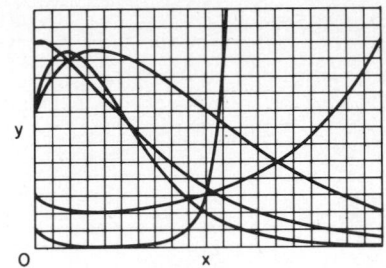

FIG. 1-24. The family of curves $y = a(bx)^{cx}$.

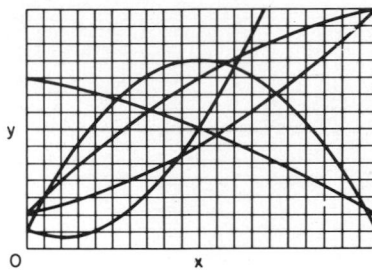

FIG. 1-25. The family of curves $y = a + bx + cx^2$.

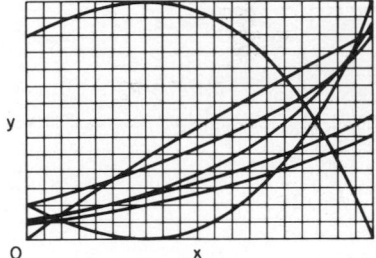

FIG. 1-26. The family of curves $y = a + bx + ce^{dx}$.

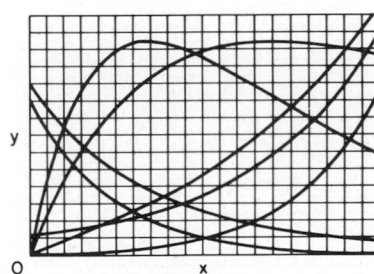

FIG. 1-27. The family of curves $y = ae^{bx} + ce^{dx}$.

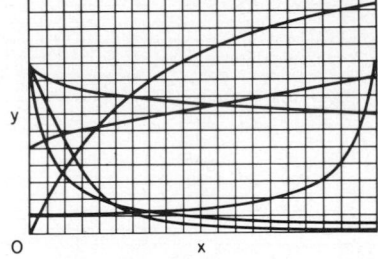

FIG. 1-28. The family of curves $y = a/(1 + bx^c)$.

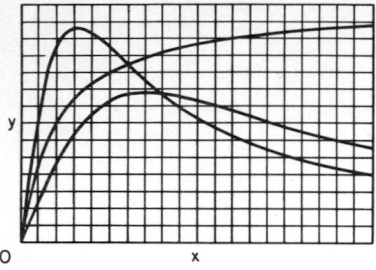

FIG. 1-29. The family of curves $y = ax/(1 + bx)$ and $y = ax/(a + bx^2)$.

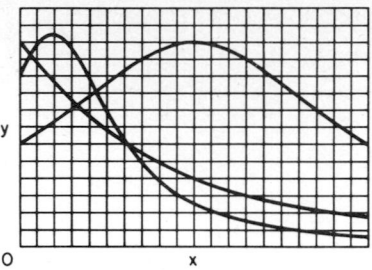

FIG. 1-30. The family of curves $y = a/(1 + bx + cx^2)$.

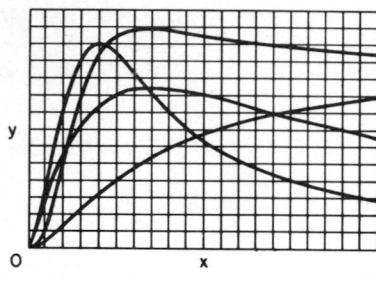

FIG. 1-31. The family of curves $y = ax/(1 + bx + cx^2)$ and $y = ax^2/(1 + bx + cx^2)$.

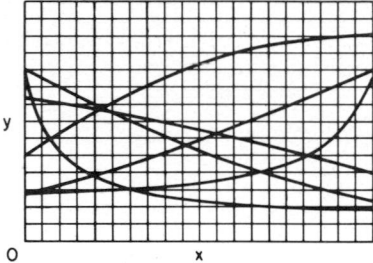

FIG. 1-32. The family of curves $y = a/(1 + be^{cx})$.

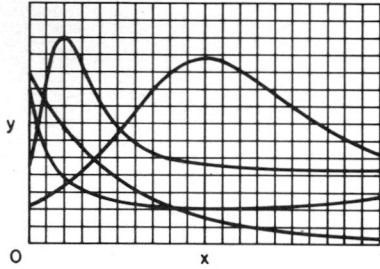

FIG. 1-33. The family of curves $y = a/(1 + bxe^{cx})$.

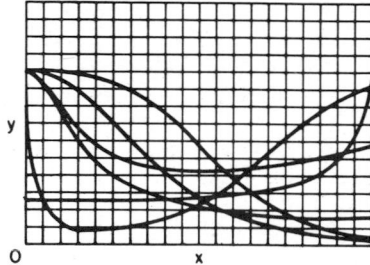

FIG. 1-34. The family of curves $y = a/(1 + bx^2 e^{cx})$.

own computations or who cannot avail themselves of the service but do have access to modest-sized office computers.

Three types of applications will be considered: linear, one-parameter nonlinear, and two-parameter nonlinear least squares.

**Linear Least Squares.** Given the data of Table 1-23, fit the linear model

### Table 1-23. Linear Data

| $x$ | $y$ | $x$ | $y$ | $x$ | $y$ |
|-----|-----|-----|-----|-----|-----|
| 0.8 | 0.59 | 2.0 | 1.63 | 3.2 | 1.81 |
| 1.1 | 0.80 | 2.3 | 1.60 | 3.5 | 2.17 |
| 1.4 | 1.20 | 2.6 | 1.55 | 3.8 | 2.40 |
| 1.7 | 1.11 | 2.9 | 2.09 | 4.1 | 2.48 |
|     |     |     |     | 4.4 | 2.57 |

$Y = a + bx$. The least-squares coefficients are those values of $a$ and $b$ which minimize the residual sum of squares

$$\text{RES.S.S.} = \text{S} = \sum_{i=1}^{n} (y_i - a - bx_i)^2$$

These can be analytically determined and are given by the solution to the normal equations:

$$na + (\Sigma x)b = \Sigma y$$
$$(\Sigma x)a + (\Sigma x^2)b = \Sigma xy$$

For the above data this results in

$$\hat{Y} = 0.3061 + 0.5332x$$

with a RES.S.S. $= 0.2583$ or residual variance of $(0.2583)/(13 - 2) = 0.023$.

**One-parameter Least Squares.** To fit the cooling rate data of Table 1-24 to the model $Y = a + be^{cx}$ requires an iterative computation. For this

### Table 1-24. Cooling Rate Data

| $x$ | $y$ | $x$ | $y$ | $x$ | $y$ |
|-----|-----|-----|-----|-----|-----|
| 0 | 92.0 | 3 | 74.5 | 10 | 53.5 |
| 1 | 85.5 | 5 | 67.0 | 15 | 45.0 |
| 2 | 79.5 | 7 | 60.5 | 20 | 39.5 |

purpose a first guess or approximation to $c$ is defined and an interval of $c \pm \Delta c$ selected; for the above, say $-0.10 \pm 0.03$. Given $c$, define $z = e^{cx}$, then the model $Y = a + bz$ can be fitted by linear least squares and the RES.S.S. for that value $c$ determined. For the prior data these result in the values

| $i$ | $c_i$ | $S_i = \text{RES.S.S}$ |
|---|---|---|
| 1 | $-0.13$ | 7.1802 |
| 2 | $-0.10$ | 3.1552 |
| 3 | $-0.07$ | 25.6330 |

The new center $c_5$ is given by

$$c_5 = c_2 + \left(\frac{\Delta c}{2}\right) \frac{S_1 - S_3}{S_1 - 2S_2 + S_3}$$
$$= -0.10 - 0.0104 = -0.1104$$

If $c_5$ had exceeded the bound $c_2 \pm 2\Delta c$, then $c_5 = c_2 \pm 2\Delta c$ in the appropriate direction should be used as the new center point. Since the optimum $c$ is bracketed, the new interval can be halved. For this:

| $i$ | $c_i$ | $S_i$ |
|---|---|---|
| 1 | $-0.1254$ | 5.0733 |
| 2 | $-0.1104$ | 1.8376 |
| 3 | $-0.0954$ | 4.7433 |

A final summary for

$$c_5 = -0.1104 + 0.000403$$
$$= -0.109997 \qquad S = 1.832046$$

or simply $c_5 = -0.110 \qquad S = 1.8320$

would be sufficient.

$$\hat{Y} = 33.60 + 57.845e^{-0.110x} \qquad 0 \le x \le 20$$

**Two-parameter Least Squares.** A fairly efficient procedure for the solution of two nonlinear parameter problems is a relaxation procedure based on successive one-parameter estimates. In this regard, two such applications will be considered. The first involves a heart-rate response, in time, under stimulus. For this the model

$$Y = b_0 + b_1 e^{c_1 x} + b_2 e^{c_2 x}$$

is to be fitted to the data in Table 1-25.

### Table 1-25. Heart-rate Response

| $x$ | $y$ | $x$ | $y$ | $x$ | $y$ | $x$ | $y$ |
|---|---|---|---|---|---|---|---|
| 0 | 91 | 1.4 | 137 | 2.8 | 141 | 4.2 | 153 |
| 0.2 | 113 | 1.6 | 142 | 3.0 | 144 | 4.4 | 151 |
| 0.4 | 125 | 1.8 | 140 | 3.2 | 148 | 4.6 | 154 |
| 0.6 | 128 | 2.0 | 143 | 3.4 | 148 | 4.8 | 155 |
| 0.8 | 126 | 2.2 | 141 | 3.6 | 150 | 5.0 | 157 |
| 1.0 | 130 | 2.4 | 144 | 3.8 | 150 | | |
| 1.2 | 131 | 2.6 | 142 | 4.0 | 152 | | |

Given values $c_1$ and $c_2$, define $z_1 = e^{c_1 x}$ and $z_2 = e^{c_2 x}$. The corresponding three linear simultaneous equations for the least-squares solution are:

$$nb_0 + (\Sigma z_1)b_1 + (\Sigma z_2)b_2 = \Sigma y$$
$$(\Sigma z_1)b_0 + (\Sigma z_1^2)b_1 + (\Sigma z_1 z_2)b_2 = \Sigma z_1 y$$
$$(\Sigma z_2)b_0 + (\Sigma z_1 z_2)b_1 + (\Sigma z_2^2)b_2 = \Sigma z_2 y$$

The residual sum of squares is computed by

$$\text{RES.S.S.} = \Sigma(y - b_0 - b_1 z_1 - b_2 z_2)^2$$

or if a large number of significant digits are used

$$\text{RES.S.S.} = \Sigma y^2 - b_0 \Sigma y - b_1 \Sigma z_1 y - b_2 \Sigma z_2 y$$

As a first guess, the values $c_{21} = -0.4$ and $c_{22} = -3$ were assumed. Table 1-26 contains a summary of the successive approximations.

### Table 1-26. Iteration Summary

| $c_{i1}$ | $c_{i2}$ | $S$ | $c_{i1}$ | $c_{i2}$ | $S$ | $c_{i1}$ | $c_{i2}$ | $S$ |
|---|---|---|---|---|---|---|---|---|
| −0.5 | −3.0 | 223.27 | −0.154 | −5.46 | 110.31 | −0.218 | −5.33 | 107.94 |
| −0.4 | ...... | 193.46 | ...... | −4.46 | 111.29 | −0.193 | ...... | 108.12 |
| −0.3 | ...... | 168.44 | ...... | −3.46 | 127.25 | −0.168 | ...... | 109.06 |
| −0.2 | −5.0 | 108.68 | −0.204 | −5.02 | 108.65 | −0.211 | −5.58 | 107.84 |
| | −3.0 | 149.43 | −0.154 | ...... | 109.46 | ...... | −5.33 | 107.92 |
| | −1.0 | 368.66 | −0.104 | ...... | 113.24 | ...... | −5.08 | 108.53 |
| −0.3 | −4.46 | 122.55 | −0.193 | −5.52 | 108.26 | −0.224 | −5.49 | 107.78 |
| −0.2 | ...... | 112.44 | ...... | −5.02 | 108.57 | −0.211 | ...... | 107.81 |
| −0.1 | ...... | 112.87 | ...... | −4.52 | 111.49 | −0.199 | ...... | 108.04 |

Finally, with one additional trial, the rounded solution

$$Y = 172.11 - 50.04e^{-0.22x} - 31.185e^{-5.57x}$$

with RES.S.S. = 107.75 and a corresponding residual standard deviation of $[107.75/(26 - 21)]^{1/2} = 2.27$ was obtained.

**Two-parameter Growth Curve.** On occasion it is of interest to fit growth rate data. For this purpose the Gompertz function is most effective:

$$Y = ab^{c^x}$$

or in log form:

$$\ln Y = \ln a + c^x \ln b$$
$$W = A + Bc^x$$

As an example of fitting this model consider the United States production of rayon* given in Table 1-27.

#### Table 1-27. Rayon Filament and Staple

| Year | Prod.* | Year | Prod.* | Year | Prod.* | Year | Prod.* |
|------|--------|------|--------|------|--------|------|--------|
| 1920 | 10.0   | 1933 | 174.5  | 1946 | 623.9  | 1959 | 867.2  |
| 1921 | 14.9   | 1934 | 172.5  | 1947 | 693.4  | 1960 | 740.2  |
| 1922 | 24.0   | 1935 | 206.3  | 1948 | 746.8  | 1961 | 793.2  |
| 1923 | 34.8   | 1936 | 224.7  | 1949 | 674.1  | 1962 | 920.4  |
| 1924 | 36.2   | 1937 | 254.8  | 1950 | 815.8  | 1963 | 979.3  |
| 1925 | 49.4   | 1938 | 207.9  | 1951 | 865.4  | 1964 | 1005.9 |
| 1926 | 60.1   | 1939 | 276.6  | 1952 | 806.3  | 1965 | 1081.8 |
| 1927 | 70.4   | 1940 | 327.7  | 1953 | 876.7  | 1966 | 1064.7 |
| 1928 | 91.4   | 1941 | 392.8  | 1954 | 820.4  | 1967 | 912.5  |
| 1929 | 113.5  | 1942 | 438.1  | 1955 | 972.8  | 1968 | 1104.4 |
| 1930 | 117.9  | 1943 | 468.1  | 1956 | 897.8  | 1969 | 1078.0 |
| 1931 | 136.1  | 1944 | 511.9  | 1957 | 877.6  | 1970 | 875.0  |
| 1932 | 117.5  | 1945 | 577.9  | 1958 | 737.3  | 1971 | 915.1  |

\* In millions of pounds.

If in the late thirties (see Mills) one were to fit these data say through 1937, by the methods previously outlined, the two-parameter nonlinear solution would converge to

$$\hat{Y} = 906.5(0.017)^{0.935^x} \qquad x = \text{year} - 1920$$

with $S = 1,613$.

In fitting the two nonlinear coefficients the $S$-space (the residual sum of squares as a function of assumed coefficient values $b$ and $c$ with $a$ the linear least-squares solution) contours will tend to have unusual shapes as indicated in Fig. 1-35. This is one reason why it is advantageous to have a high degree of interaction between machine and analyst.

To obtain reasonable first approximations, the log form can be used to estimate $b$ by iterating on $c$ alone. However, in general, these values should not be used for the model unless the log response is the way in which the response is to be used. Otherwise the exponentiated log estimates of the response can tend to be much farther removed than the standard form.

---

* Department of Commerce Statistical Bulletin No. 363, Table 356. Miles, F. C., Statistical Methods (1938). H. Holt Co. N.Y.

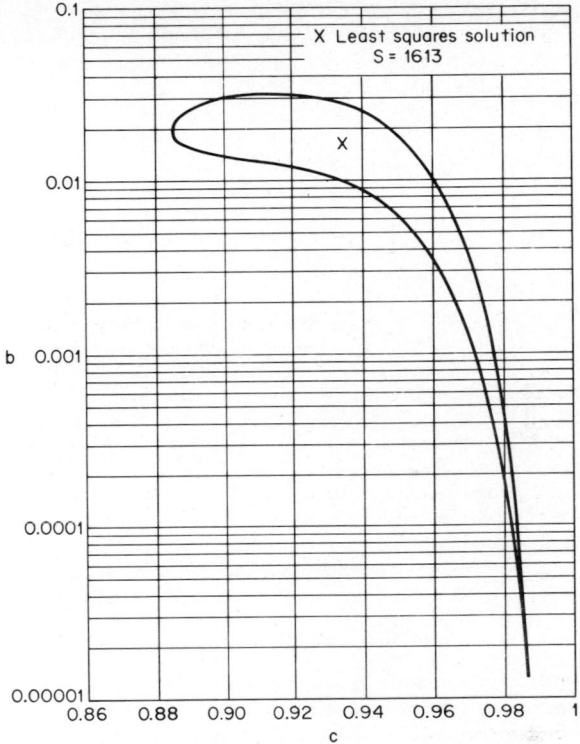

Fig. 1-35. Contour $S = 2400$, 1920–1937 data.

For example, the back-calculated residual sum of squares in the alternative form shows:

| Model | $S$ for $\hat{Y}$ | $S$ for $\ln \hat{Y}$ | |
|---|---|---|---|
| Original form | 1.613 | 2.231 | $(a = 906.5, b = .017, c = .935)$ |
| Log form | 0.3896 | 0.1452 | $(a = 414.3, b = 0.026, c = 0.898)$. |

If in the late forties data through 1948 were fitted to the Gompertz function, the curious result

$$\hat{Y} = 71,650(0.00033)^{0.98^x} \qquad \text{with } S = 9,831$$

would have been computed. This would indicate an asymptote of 71,650 million pounds. At that point, one would go back and analyze the data for anomalies.

For growth rate data a very effective technique is to plot the cumulative response average by year (see Fig. 1-36). It would be noted that 1938 was essentially a lost year. That is, the cumulative average drops severely and

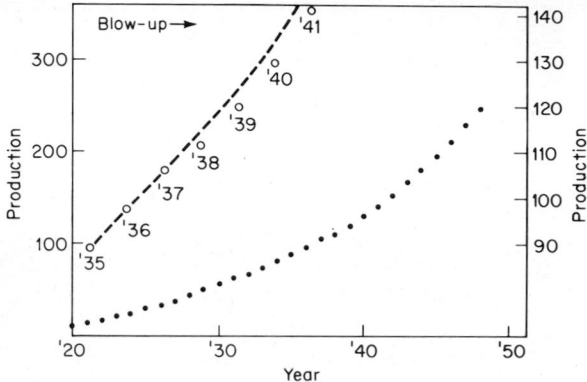

FIG. 1-36. Cumulative average production.

the whole subsequent pattern is offset by 1 year. This would suggest the exclusion of the 1938 data and the reduction, by 1 for $x$, for the subsequent years. Thus

$$Y = 17,901(0.0009)^{0.972^x} \qquad \text{with } S^* = 3,815$$

**Orthogonal Polynomials.** When highly precise measurements, precise interpolation near the edge of a table, or some complex mathematical function needs to be approximated for a computer program, the use of orthogonal polynomials might be justified. For this, successively higher terms in a general polynomial expansion are fitted to the data until sufficient reliability is achieved. The successively higher polynomial terms being orthogonal to prior terms negates having to recompute them.

For ready computation, it is convenient to rely on either a 10- or 11-point fit. For this, the three Tables 1-28, 1-29, and 1-30, summarize the necessary computational requirements. As an example, consider the gamma function

$$\Gamma(n) = \int_0^\infty x^{n-1} e^{-x}\, dx \qquad n > -2$$

over the argument $1 \le n \le 2$. To eight significant digits Table 1-31 includes values of the gamma function obtained by numerical integration.

Using a data layout sheet, similar to Tables 1-28 and 1-29, with the convention $Z_0 \equiv 1$, the quantity $\Sigma y$ is tabled in the first block. Since $\Sigma Z_0^2 = n$, the $b_0$ regression coefficient is $\bar{y}$. The successive regression coefficients, for say column $k$, are given by $\Sigma Z_k y / \Sigma Z_k^2$ with associated regression sum-of-

* A reduction ostensibly due to the elimination of 1938.

## Table 1-28. Orthogonal Polynomials for 10 Points

| Point | $y$ | $Z_1$ | $Z_2$ | $Z_3$ | $Z_4$ | $Z_5$ | $Z_6$ | $Z_7$ | $Z_8$ | $Z_9$ |
|---|---|---|---|---|---|---|---|---|---|---|
| 1 | | −9 | 6 | −42 | 18 | −6 | 3 | −9 | 1 | −1 |
| 2 | | −7 | 2 | 14 | −22 | 14 | −11 | 47 | −7 | 9 |
| 3 | | −5 | −1 | 35 | −17 | −1 | 10 | −86 | 20 | −36 |
| 4 | | −3 | −3 | 31 | 3 | −11 | 6 | 42 | −28 | 84 |
| 5 | | −1 | −4 | 12 | 18 | −6 | −8 | 56 | 14 | −126 |
| 6 | | 1 | −4 | −12 | 18 | 6 | −8 | −56 | 14 | 126 |
| 7 | | 3 | −3 | −31 | 3 | 11 | 6 | −42 | −28 | −84 |
| 8 | | 5 | −1 | −35 | −17 | 1 | 10 | 86 | 20 | 36 |
| 9 | | 7 | 2 | −14 | −22 | −14 | −11 | −47 | −7 | −9 |
| 10 | | 9 | 6 | 42 | 18 | 6 | 3 | 9 | 1 | 1 |
| $\Sigma Zy$ | | | | | | | | | | |
| $\Sigma Z^2$ | | 330 | 132 | 8580 | 2860 | 780 | 660 | 29,172 | 2,860 | 48,620 |
| $b$ | | | | | | | | | | |
| REG | | | | | | | | | | |
| RES | | | | | | | | | | |

## Table 1-29. Orthogonal Polynomials for 11 Points

| Point | $y$ | $Z_1$ | $Z_2$ | $Z_3$ | $Z_4$ | $Z_5$ | $Z_6$ | $Z_7$ | $Z_8$ | $Z_9$ |
|---|---|---|---|---|---|---|---|---|---|---|
| 1 | | −5 | 15 | −30 | 6 | −3 | 15 | −5 | 5 | −1 |
| 2 | | −4 | 6 | 6 | −6 | 6 | −48 | 23 | −31 | 8 |
| 3 | | −3 | −1 | 22 | −6 | 1 | 29 | −33 | 73 | −27 |
| 4 | | −2 | −6 | 23 | −1 | −4 | 36 | 2 | −68 | 48 |
| 5 | | −1 | −9 | 14 | 4 | −4 | −12 | 28 | −14 | −42 |
| 6 | | 0 | −10 | 0 | 6 | 0 | −40 | 0 | 70 | 0 |
| 7 | | 1 | −9 | −14 | 4 | 4 | −12 | −28 | −14 | 42 |
| 8 | | 2 | −6 | −23 | −1 | 4 | 36 | −2 | −68 | −48 |
| 9 | | 3 | −1 | −22 | −6 | −1 | 29 | 33 | 73 | 27 |
| 10 | | 4 | 6 | −6 | −6 | −6 | −48 | −23 | −31 | −8 |
| 11 | | 5 | 15 | 30 | 6 | 3 | 15 | 5 | 5 | 1 |
| $\Sigma Zy$ | | | | | | | | | | |
| $\Sigma Z^2$ | | 110 | 858 | 4 290 | 286 | 156 | 11,220 | 4,862 | 27,170 | 4,862 |
| $b$ | | | | | | | | | | |
| REG | | | | | | | | | | |
| RES | | | | | | | | | | |

## Table 1-30. Definition of Terms

| Term | 10 points | 11 points |
|---|---|---|
| $Z_1$ | $\dfrac{9}{X_M - X_m}[2X - X_M - X_m]$ | $\dfrac{5}{X_M - X_m}[2X - X_M - X_m]$ |
| $Z_2$ | $\frac{1}{6}(Z_1^2 - 33)$ | $Z_1^2 - 10$ |
| $Z_3$ | $\frac{1}{5}(5Z_2Z_1 - 16Z_1)$ | $\frac{1}{6}(5Z_2Z_1 - 39Z_1)$ |
| $Z_4$ | $\frac{1}{8}(Z_3Z_1 - 39Z_2)$ | $\frac{1}{10}(Z_3Z_1 - 6Z_2)$ |
| $Z_5$ | $\frac{1}{25}(3Z_4Z_1 - 8Z_3)$ | $\frac{1}{10}(3Z_4Z_1 - 2Z_3)$ |
| $Z_6$ | $\frac{1}{48}(11Z_5Z_1 - 24Z_4)$ | $\frac{1}{3}(11Z_5Z_1 - 20Z_4)$ |
| $Z_7$ | $\frac{1}{7}(13Z_6Z_1 - 48Z_5)$ | $\frac{1}{42}(13Z_6Z_1 - 255Z_5)$ |
| $Z_8$ | $\frac{1}{48}(5Z_7Z_1 - 119Z_6)$ | $\frac{1}{4}(5Z_7Z_1 - 7Z_6)$ |
| $Z_9$ | $\frac{1}{6}(17Z_8Z_1 - 16Z_7)$ | $\frac{1}{45}(17Z_8Z_1 - 76Z_7)$ |

Where $X_M \sim$ upper limit of $X$
$X_m \sim$ lower limit of $X$

### Table 1-31. Gamma Function

| $n$ | $\Gamma(n)$ | $n$ | $\Gamma(n)$ | $n$ | $\Gamma(n)$ |
|---|---|---|---|---|---|
| 1.0 | 1.0 | 1.35 | 0.89115144 | 1.70 | 0.90863873 |
| 1.05 | 0.97350426 | 1.40 | 0.88726382 | 1.75 | 0.91906253 |
| 1.10 | 0.95135077 | 1.45 | 0.88566138 | 1.80 | 0.93138377 |
| 1.15 | 0.93304093 | 1.50 | 0.88622692 | 1.85 | 0.94561118 |
| 1.20 | 0.91816874 | 1.55 | 0.88886835 | 1.90 | 0.96176583 |
| 1.25 | 0.90640248 | 1.60 | 0.89351535 | 1.95 | 0.97988065 |
| 1.30 | 0.89747070 | 1.65 | 0.90011682 | 2.00 | 1.0 |

squares $(\Sigma Z_k y)^2/\Sigma Z_k{}^2$. The new residual S.S. is the prior RES.S.S. minus the new REG.S.S. Successive terms are added until the desired accuracy has been achieved.

For the prior gamma function an abbreviation of these computations, using an 11-point fit, is given in Table 1-32, for the $11 - n$ values 1.0, 1.1, . . . , 2.0.

### Table 1-32. Gamma Function Fit

| $i$ | $\Sigma Z_i y$ | $b_i$ | RES.S.S.* |
|---|---|---|---|
| | | | 9.5427 E   0 |
| 0 | 1.0236 E + 1 | 9.3053 E − 1 | 1.7992 E − 1 |
| 1 | 1.0989 E − 1 | 9.9903 E − 4 | 1.7882 E − 1 |
| 2 | 3.9032 E   0 | 4.5492 E − 3 | 1.2559 E − 4 |
| 3 | −6.9761 E − 1 | −1.6261 E − 4 | 1.2146 E − 5 |
| 4 | 5.8354 E − 2 | 2.0404 E − 4 | 2.4012 E − 7 |
| 5 | −6.0271 E − 3 | −3.6835 E − 5 | 7.2644 E − 9 |
| 6 | 8.9383 E − 3 | 7.9664 E − 7 | 1.4386 E − 10 |
| 7 | −8.2931 E − 4 | −1.7057 E − 7 | 2.4040 E − 12 |
| 8 | 2.5409 E − 4 | 9.3519 E − 9 | 2.7807 E − 14 |

* After the fifth or sixth polynomial, it is necessary to compute $\Sigma(y - b_0 - b_1 Z_1 \cdots - b_k Z_k)^2$ directly rather than by formula.

As a computational check and to detect oscillation if it occurs, it is always advisable to use the derived polynomial to interpolate for intermediate arguments. Using the polynomial term generator in Table 1-30 and correcting for degrees of freedom (for the model 11-1-$k$ and 10 for the intermediate data), the corresponding error variances are consistent as indicated in Table 1-33.

### Table 1-33. Interpolation Check Calculation

| $i$ | Model RES variance | Interpolated variance |
|---|---|---|
| 1 | 1.99 E − 3 | 1.13 E − 3 |
| 2 | 1.57 E − 5 | 7.69 E − 6 |
| 3 | 1.74 E − 6 | 8.49 E − 7 |
| 4 | 4.00 E − 8 | 2.17 E − 8 |
| 5 | 1.45 E − 9 | 9.51 E − 10 |
| 6 | 3.60 E − 11 | 3.25 E − 11 |
| 7 | 8.01 E − 13 | 1.28 E − 12 |
| 8 | 9.27 E − 14 | 5.91 E − 14 |

## 1.36 Time-series Smoothing and Forecasting

A time series is a chronologically ordered set of measurements recording quantitatively the changes of an observed phenomenon with time. The set of measurements might be a record of total annual rainfall from 1880 to 1974 in New York City or the daily production of a chemical process during the past year. Functionally, a time series can be defined as

$$y(t) = Y(t) + \epsilon(t)$$

where $y(t)$ is the observed response, $Y(t)$ the true signal, and $\epsilon(t)$ a random component, all over time $t$.

Sources of random variation can be the introduction of external causes. For example, the rayon production of Table 1-27 is a time series (whose true signal is basically Gompertz in nature through say 1951) in which the significant deviation in 1938 due to a recessional influence can be interpreted as a random fluctuation from the normal growth trend. The true signal $Y(t)$ itself is generally nonanalytic and may only be at best a piecewise function. Here the commercialization of nylon caused a discontinuation of the growth rate trend in the beginning of the fifties or thereabouts.

In some ways smoothing and forecasting are intermixed. In many situations smoothing is carried out to detect the fundamental nature of $Y(t)$ and therefore serve as a guide for the projection. On the other hand, smoothing has virtue in its own right in the sense of a potential base for understanding the nature of the response. This would be the situation in studying annual rainfall trends in which it would be recognized that past performance alone would be totally inadequate for predicting the ensuing year's annual rainfall. Here, the random component totally dominates the signal during any one year.

In general, the more widely used techniques for smoothing and forecasting are:

Smoothing
   1. Functional
   2. Simple moving average
   3. Triangular-weighted moving average

Forecasting
   1. Functional
   2. Box-Jenkins procedure
   3. Exponential smoothing (Holt-Winters method)
   4. Autoregressive
   5. Diagnostic analysis

For smoothing, some specialized applications are amenable to empirical approximation. For example, the rayon production data can be smoothed effectively by fitting the Gompertz function through 1951. However, most applications are nonanalytic in nature and these can be approached through moving averages. For these, an odd window size (number of points in the average) is usually selected and the inclusive points adjacent to the midvalue are weighted and averaged. This transformation of measurements to a smoothed value is accomplished through the following relationship:

$$\hat{Y}_i = \sum_j w_j y_{i+j}$$

$$j = -\frac{n-1}{2}, -\frac{n-3}{2}, \cdots 0, \cdots, \frac{n-3}{2}, \frac{n-1}{2}$$

$$i = \frac{n+1}{2}, \frac{n+3}{2}, \cdots, \left(N - \frac{m-1}{2}\right)$$

$n$ = window size 3, 5, $\cdots$

$N$ = sample number observation

For a simple moving average the $w_j$ weights are simply $1/n$ and for triangular weights they are $\left[\dfrac{2(n+1-2j)}{(n+1)2}\right]$. Given $n = 5$, then with $N = 7$, the three respective averages would be determined for simple and triangular weighting as follows:

| $i$ | $Y_i$ | Simple | | | Triangular | | |
|---|---|---|---|---|---|---|---|
| | | First | Second | Third | First | Second | Third |
| 1 | $y_1$ | 0.2 | ... | ... | $\frac{1}{9}$ | | |
| 2 | $y_2$ | 0.2 | 0.2 | ... | $\frac{2}{9}$ | $\frac{1}{9}$ | |
| 3 | $y_3$ | 0.2 | 0.2 | 0.2 | $\frac{3}{9}$ | $\frac{2}{9}$ | $\frac{1}{9}$ |
| 4 | $y_4$ | 0.2 | 0.2 | 0.2 | $\frac{2}{9}$ | $\frac{3}{9}$ | $\frac{2}{9}$ |
| 5 | $y_5$ | 0.2 | 0.2 | 0.2 | $\frac{1}{9}$ | $\frac{2}{9}$ | $\frac{3}{9}$ |
| 6 | $y_6$ | ... | 0.2 | 0.2 | | $\frac{1}{9}$ | $\frac{2}{9}$ |
| 7 | $y_7$ | | | 0.2 | | | $\frac{1}{9}$ |
| $\Sigma wy$ | ... | $\hat{Y}_3$ | $\hat{Y}_4$ | $\hat{Y}_5$ | $\hat{Y}_3$ | $\hat{Y}_4$ | $\hat{Y}_5$ |

A difficult problem is in selecting the appropriate window size for a particular application. In general, this involves a balance between the reduction of variability associated with $\epsilon$ (averaging dampens the magnitude of the random variation) and the increased biasing of the weighted signal. In estimating $Y(t)$, $y(t)$ is the best unbiased estimate. However, to reduce the variance, a weighted average of $y(t-1)$, $y(t)$, and $y(t+1)$ could be used and unless $Y(t)$ is constant or linear, within plus and minus one time increment, the resultant estimate of $Y(t)$ will be biased. As the window size increases this bias tends to increase at the same time the variance tends to decrease. The appropriate window size is the best balance between these.

As a guide for selecting window size a moving average of adjacent points only can be used and the sample sum-of-squares between midpoint estimate and observed midvalue computed over the same data set. The window with minimum sum-of-squares would be selected.   For example, using Table 1-27 data, with $n = 3$, assuming a maximum $n$ of say 13 to define the first common point $y_7$

$$Y_7 = \frac{49.4 + 70.4}{2} = 59.9$$

$$y_7 = 60.1 \qquad \text{with SE} = (59.9 - 60.1)^2$$

(SE $\sim$ square error) through

$$Y_{46} = \frac{1005.9 + 1064.7}{2} = 1035.3$$

$$y_{46} = 1081.8 \qquad \text{with SE} = (1035.3 - 1081.8)^2$$

It should be recognized that the sum-of-squares quantity does not carry the same degree of weight for the early data as the latter data.   The magnitude of differences for the 1920 data is appreciably less than for the 1960s.   To circumvent this problem, logarithmic values can be used, if desired.   For the sample data, the respective average square (sum-of-squares divided by sample size) are:

| $n$ | Average square arithmetic | Logarithmic |
|---|---|---|
| 3 | 2,853.3 | 0.010356 |
| 5 | 2,319.3 | 0.007882 |
| 7 | 2,580.2 | 0.009252 |
| 9 | 2,927.0 | 0.010042 |
| 11 | 3,647.9 | 0.011774 |
| 13 | 4,127.5 | 0.013341 |

In short-term forecasing of rayon production, or similar applications, a successive updated growth curve can be computed and used as the basis for the forecast.   However, even here there would be no substitute for a market expert's probable assessment of a changing market in the early fifties.   Prior to the nylon crunch the "Gompertz" projection would probably be superior—after it would be a disaster.   However, these are very specialized forecasting applications.   In general, most forecasting applications do not have as such a sound structure-growth phenomenon.

For the general forecasting application the Box-Jenkins procedure is very effective.   However, it requires a fairly elaborate packaged computer program as well as a high degree of analyst interaction for its effective use.   The respective references should be consulted.

An extension of the so-called exponential smoothing method developed by Holt-Winters is useful.   The basic exponential smoothing procedure states that the best estimate of $Y(t)$ is a weighted average of the observation at time $t$, $y(t)$, and the last averaged estimate $\bar{y}(t - 1)$, that is,

$$\bar{y}(t) = ay(t) + (1 - a)\bar{y}(t - 1)$$

Expanding this,

$$\bar{y}(t) = ay(t) + a(1-a)y(t-1) + a(1-a)^2 y(t-2) + \cdots$$

Again, depending on the relative magnitude of variance (variance of $\epsilon$) and squared bias, $a$ will vary from almost zero to almost 1. A problem is what value to select for $a$.

To start the process the first estimate of $Y$ at time 1 can be $\bar{y}(1) = y(1)$. Thus

$$\bar{y}(2) = ay(2) + (1-a)\bar{y}(1)$$

and so forth. For forecasting, $\bar{y}(2)$ is used as the estimate of $Y(3)$ and so on.

With basic exponential smoothing, an estimate of $a$ can be derived following the general procedure of smoothing when sample data are available. Over the sample data, and several selected values of $a$ (say 0.1, 0.3, 0.5, 0.7, and 0.9), the average square between estimate and actual can be computed. That value of $a$ corresponding to the minimum average square would be selected. In general, these numbers tend to be relatively insensitive to values of $a$.

A Holt-Winter extension of exponential smoothing introduces a trend correction

$$T(t) = b[\bar{y}(t) - \bar{y}(t-1)] + (1-b)T(t-1)$$

In addition, if necessary, this can be extended to include a seasonal correction. With trend only, the smoothed series is given by

$$\bar{y}(t) = ay(t) + (1-a)[\bar{y}(t-1) + T(t-1)]$$

The resultant forecast for $t+1$ is

$$\hat{Y}(t+1) = \bar{y}(t) + T(t)$$

Here, sample data are used to simultaneously determine reasonable values for both $a$ and $b$.

Under an autoregressive technique a new response $Z(t)$ based on differences

$$Z(t) = y(t) - y(t-1)$$

is defined and the model

$$\hat{Z}(t) = b_0 + \sum_{i=1}^{P} b_i Z(t-i)$$

successively fitted for $P = 1, 2, 3, \ldots$. Using prior history, a plot of residual mean square (residual sum-of-squares corrected for the number of terms in the model) against $P$ for values of 1 through say 6 or 7 can then be used as a guide for selecting an appropriate number of terms. The forecast for the future would then be determined through the identity

$$\hat{y}(t + 1) = y(t) + \hat{Z}(t + 1)$$

based on the least-squares coefficients $b_i$ computed from the historic data.

The above forecasting techniques have been generalized through various computer packages and verge on full automation even though there remains a high degree of manual interaction, specification, and selection, for example with the Box-Jenkins package. On the other hand, another option is a detailed study or diagnostic analysis of particular applications. An example of this is a study of the monthly prescriptions filled by a pharmacy in a small college town. These are summarized in Table 1-34.

## Table 1-34. Monthly Prescriptions

| Month | 1970 | 1971 | 1972 | 1973 | 1974 |
|-------|------|------|------|------|------|
| Jan. | ..... | 3,644 | 3,946 | 3,991 | 3,893 |
| Feb. | ..... | 3,568 | 4,140 | 4,274 | 4,949 |
| Mar. | ..... | 3,781 | 4,260 | 4,328 | 4,623 |
| Apr. | ..... | 4,146 | 3,637 | 4,387 | 4,410 |
| May | ..... | 3,420 | 3,864 | 3,945 | 4,142 |
| Jun. | ..... | 3,310 | 3,470 | 3,297 | 3,294 |
| Jul. | ..... | 3,341 | 3,316 | 3,279 | 3,409 |
| Aug. | ..... | 3,339 | 3,335 | 3,322 | 3,419 |
| Sept. | ..... | 3,839 | 4,193 | 4,338 | 4,424 |
| Oct. | ..... | 4,374 | 4,349 | 4,689 | 4,604 |
| Nov. | ..... | 4,183 | 4,144 | 4,324 | 4,441 |
| Dec. | 4,056 | 4,030 | 4,184 | 3,931 | 4,404 |

As a forecasting problem and due to the obvious seasonal variation, a 12-month simple moving average was computed. Based on a plot of these data through November 1973, a linear trend for the observed period is indicated and therefore the linear model $(a + bt)$ was fitted to the moving average data. For the least-squares fit the model

$$Y = 3763.2 + 10.49t \qquad t = \text{months from May 1971}$$

with a residual standard deviation of 20.5 was computed.

A comparison between the projected average and the subsequent smoothed sample data through December 1974 verified the linear projection at least to that date:

| Through month | Projected | Observed | Through month | Projected | Observed |
|---------------|-----------|----------|---------------|-----------|----------|
| Dec. 1973 | 4,036 | 4,009 | Jul. | 4,109 | 4,110 |
| Jan. 1974 | 4,046 | 4,001 | Aug. | 4,120 | 4,118 |
| Feb. | 4,057 | 4,057 | Sept. | 4,130 | 4,126 |
| Mar. | 4,068 | 4,081 | Oct. | 4,141 | 4,118 |
| Apr. | 4,078 | 4,083 | Nov. | 4,151 | 4,128 |
| May | 4,088 | 4,100 | Dec. | 4,162 | 4,168 |
| Jun. | 4,099 | 4,100 | | | |

The root-mean-square difference between the smoothed projected and observed prescription orders is 18.0 (compared to the prior sample value 20.5).

For this particular application, the cyclical pattern is basically consistent. However, it does illustrate that cyclical patterns are not necessarily trigonometric or analytic in nature. An examination of the monthly trend averaged over the four sample years (excluding December 1974) shows the following cycle (which includes a linear component of 12 $\times$ 10.49 spread over the twelve months).

| Month | 4-year average |
|-------|------------|
| Dec | 4,050 |
| Jan | 3,869 |
| Feb | 4,233 |
| Mar | 4,248 |
| Apr | 4,145 |
| May | 3,843 |
| Jun | 3,343 |
| Jul | 3,336 |
| Aug | 3,354 |
| Sept | 4,199 |
| Oct | 4,504 |
| Nov | 4,270 |
| Total | 3,949 |

Here, the data show the effects of the college terms in addition to the seasonal effects:

| | |
|---|---|
| Winter semester | Sep.–Dec. |
| Vacation and mini semester* | Jan. |
| Spring semester | Feb.–May |
| Vacation and summer school* | Jun.–Aug. |

\* Appreciably lower enrollment.

For a month-to-month forecast, the model would be evaluated for say the last month of data, December 1974, with $t = 43$, and the corresponding monthly deviation (month average minus total average of 3,949) successively added (and then subtracted) through November. The following year cycle starting in December would include the addition of 125 added to each month—plus an additional 125 for each successive year.

In a similar way, if the general growth had been "Gompertz" (which would be the case over a sufficiently long period of growth) this would have been borne out by the 12-month simple moving average. If applicable, the Gompertz function would be fitted to the 12-month moving average to develop general trend. A monthly forecast then can be derived by averaging monthly residuals (differences between model back-calculation and observation) and combining these with the future months back-calculation in much the same way as the previous monthly forecast.

When the trend is real but random, the monthly cycle estimate and forecast become increasingly more difficult and virtually impossible when coupled with consistent monthly cycles and especially so coupled with varying monthly cycles.

# THE ENGINEERING CORE: THERMODYNAMICS, FLUID FLOW, HEAT TRANSFER

**Otto W. Witzell, Ph.D.**; Dean of the Graduate School, Drexel Institute of Technology; Member, American Society for Mechanical Engineers, American Society for Engineering Education. (Thermodynamics)

**Robert H. Perry, Ph.D.**; Engineering and Economic Consultant. (Fluid Flow and Heat Transfer)

## CONTENTS

# THERMODYNAMICS

## 2-1. Energy Transformations

Transfers of energy are expressed by formulations of the first law of thermodynamics. Several formulations are available, and depend on the manner in which mass enters or leaves the system. Any of the terms in the equations may be obtained, provided all the others are known.

**Closed-system Energy Formulation.** In a thermodynamic system in which no mass crosses the boundaries, the energy conservation principle takes the form

$$Q_{in} - Q_{out} = E_{final} - E_{initial} + W_{out} - W_{in} \tag{2-1}$$

where $Q$ is the heat transfer, $W$ is the work transfer, and $E$ is the stored energy, defined as $U + KE + PE$, with $U$ as the internal energy, KE as the kinetic energy, and PE as the potential energy. Examples of closed systems include batch kettles, calorimeters, and single-expansion or -compression processes.

**Steady-flow-system Energy Formulation.** For steady-flow systems the equation representing energy interchanges in the system must be written as follows:

$$Q_{in} - Q_{out} + \frac{V_1{}^2}{2} + y_1 + H_1 = \frac{V_2{}^2}{2} + y_2 + H_2 + W_{out} - W_{in} \tag{2-2}$$

Here $V$ is the velocity, $y$ is elevation above an arbitrary datum, and $H$ is the enthalpy, defined as $U + pv$, with $p$ as absolute pressure, $v$ as specific volume. Examples of steady-flow machines are boilers, turbines, gas engines, pumps, compressors, refrigerator systems, heat exchangers, throttle valves, and gas burners.

**Work.** In a closed system the work is often evidenced by pressure-volume changes. Specifically, if there are no changes in either kinetic or potential energies, and no frictional effects are present, the work can be evaluated by the expression

$$W = \int_{initial}^{final} p \, dv \tag{2-3}$$

To complete the evaluation of work from this expression, knowledge of the relation between $p$ and $v$ must be available.

*Example* 1: One pound of an ideal gas expands at constant temperature so that $pv =$ constant from an initial condition of 125 psia and 1.5 ft³ to a final condition of 25 psia. How much work is done?

$$W = \int_{initial}^{final} p \, dv = \int_{initial}^{final} \frac{C}{v} \, dv = C \int_{initial}^{final} \frac{dv}{v} = p_i v_i \int_{initial}^{final} \frac{dv}{v}$$

$$= p_i v_i \ln \frac{v_f}{v_i}$$

The final volume can be found from

$$p_i v_i = p_f v_f \quad \text{or} \quad v_f = \frac{p_i v_i}{p_f}$$

Then

$$W = p_i v_i \ln \frac{p_i}{p_f} = 125 \times 144 \times 1.5 \ln {}^{125}\!/_{25}$$

$$= 43{,}500 \text{ ft-lb} = 55.9 \text{ Btu}$$

In the case of a steady-flow system, the work must be evaluated using an expression

$$W = \int_{\text{initial}}^{\text{final}} - v \, dp \tag{2-4}$$

The same restrictions apply as in the case of the closed system; i.e., kinetic and potential changes, as well as frictional effects, must be absent.

**Heat Transfer.** In systems which undergo reversible processes, the heat transfer can be expressed as

$$Q = \int_{\text{initial}}^{\text{final}} T \, dS \tag{2-5}$$

where $T$ is the absolute temperature, and $S$ is the entropy.

**Combined First and Second Laws.** If Eq. (2-1) is written

$$Q_{\text{net}} - W_{\text{net}} = E_{\text{final}} - E_{\text{initial}}$$

substitution of Eqs. (2-3) and (2-5) results in an equation which expresses the joint statements of both laws:

$$T \, dS = dU + p \, dv \tag{2-6}$$

## 2-2. Properties

Several convenient relations between properties allow the evaluation of some of the terms in the conservation equations. As an example, the internal energy and enthalpy may be related to the specific heats.

**Internal Energy.** The specific heat at constant volume is defined by the equation

$$C_V = \frac{\partial U}{\partial T}\bigg|_v \tag{2-7}$$

This expression can be simplified and rearranged to allow determination of the internal energy. Where temperature changes are small, and for the case where either the system volume is constant or the system material is an ideal gas,

$$U_f - U_i = C_V[T_f - T_i] \tag{2-8}$$

**Enthalpy.** The enthalpy is related to the specific heat at constant pressure by

$$C_p = \frac{\partial H}{\partial T}\bigg|_p \tag{2-9}$$

Again, for small temperature changes and where either the pressure of the system is constant or the material is an ideal gas,

$$H_f - H_i = C_p[T_f - T_i] \tag{2-10}$$

In many cases changes of enthalpy or internal energy are desired over ranges of temperature such that the specific heat is not constant. In such cases, it becomes necessary to obtain information as to the variation of the specific heat with the temperature, so that mathematical evaluation of these functions can be obtained. Tabulations and formulations for the variation of specific heats of various substances with temperature are available in such sources as the JANAF Thermochemical Tables, published at the Thermo Laboratory, Dow Chemical Co.

## 2-3. Equations of State

**Ideal Gases.** The ideal-gas law represents the property relation for many permanent gases over large ranges of pressure and temperature. It is best expressed in equation form as

$$pv = nRT \tag{2-11}$$

In this equation $n$ represents the number of moles of gas, and $R$ is the universal gas constant. The value of this constant is 1.986 Btu/(mole)(°R), or 1,545 ft-lb/(mole)(°R). The pressure must be expressed as absolute pressure, and similarly, the temperature must also be absolute temperature.

The number of moles of gas is related to the mass of the gas by the expression

$$n = m/M \tag{2-12}$$

where $M$ represents the molecular weight. The molecular weights of various substances, as well as some of the other physical characteristics, are given in Table 2-1. Additional data may be found in Sec. 3.

**Real Gases.** In many cases, particularly at high pressure, the ideal-gas law will not adequately represent the relationship between the properties. A simple and useful modification of the gas law to include a factor called the compressibility factor results in

$$pv = CnRT \tag{2-13}$$

where $C$ is the compressibility factor, depending on the material, temperature, and pressure.

An approximation called the law of corresponding states makes it possible to determine the compressibility factor in terms of the temperature, pressure, and critical temperature and pressure of the material. We define the reduced temperature and pressure as

$$T_R = T/T_c \quad \text{and} \quad p_R = p/p_c \tag{2-14}$$

where $T_c$ and $p_c$ are the critical values. Figure 3-11 shows how the compressibility factor depends on the reduced temperature and pressure. Tabular values of the critical properties are shown in Table 2-1.

## Table 2-1. Properties of Gases*

| Gas | Symbol | Approx. mol. wt. | Critical pressure, psia | Critical temp., °F | Enthalpy of formation at 25°C, kcal/mole | Total enthalpy at 25°C, kcal/mole | Entropy at 25°C, cal/mole | Log of equil. const. formation at 25°C |
|---|---|---|---|---|---|---|---|---|
| Acetylene............ | $C_2H_2$ | 26 | 911 | 96.3 | 54.19 | 308.0 | 48.00 | 36.04 |
| Air................ | ..... | 29 | 546 | −220.3 | ..... | 3.8 | 45.97 | 2.907 |
| Ammonia............ | $NH_3$ | 17 | 1640 | 270.3 | −11.04 | 95.0 | 36.98 | 0 |
| Argon............... | $Ar$ | 40 | 706 | −187.7 | 0 | 1.5 | 64.34 | −22.714 |
| Benzene............. | $C_6H_6$ | 78 | 702 | 551.4 | 19.82 | 781.2 | 74.10 | 2.752 |
| Butane.............. | $C_4H_{10}$ | 58 | 530 | 307.4 | −29.81 | 686.1 | 51.07 | 69.095 |
| Carbon dioxide....... | $CO_2$ | 44 | 1073 | 88.0 | −94.05 | 2.24 | 47.21 | 24.029 |
| Carbon monoxide..... | $CO$ | 28 | 515 | −220.3 | −26.42 | 67.82 | 71.92 | 75.280 |
| Dichlorodifluoromethane (R12)...... | $CCl_2F_2$ | 121 | 597 | 233.6 | −112.00 | 118.83 | 54.85 | 5.761 |
| Ethane.............. | $C_2H_6$ | 30 | 718 | 90.0 | −20.24 | 372.4 | 52.45 | 11.934 |
| Ethylene............ | $C_2H_4$ | 28 | 748 | 49.3 | 12.50 | 335.7 | 50.13 | 0 |
| Helium.............. | $He$ | 4 | 33 | −450.2 | 0 | 1.48 | 30.13 | −1.532 |
| Heptane............. | $C_7H_{16}$ | 100 | 394 | 517.1 | −44.89 | 1147 | 101.64 | 0 |
| Hydrogen............ | $H_2$ | 2 | 188 | −399.8 | 0 | 69.4 | 31.21 | 8.902 |
| Methane............. | $CH_4$ | 16 | 674 | −116.5 | −17.89 | 213.2 | 44.50 | 0 |
| Nitrogen............ | $N_2$ | 28 | 493 | −232.8 | 0 | 3.8 | 45.77 | 0 |
| Oxygen.............. | $O_2$ | 32 | 731 | −181.8 | 0 | 4.11 | 49.00 | 0 |
| Propane............. | $C_3H_8$ | 44 | 632 | 206.3 | −24.82 | 529.5 | 64.51 | 4.037 |
| Water............... | $H_2O$ | 18 | 3106 | 705.5 | −57.80 | 13.7 | 45.11 | 40.048 |

* Additional physical, chemical, and thermodynamic data are given in Sec. 3.

*Example* 2: Calculate the volume of 1 mole of refrigerant 12 (dichloro-difluoromethane) at 1,500 psia and 350°F.

The critical properties are $t_c = 233.6°F$, $p_c = 597$ psia. The reduced properties are

$$T_R = \frac{350 + 460}{234 + 460} = 1.17 \qquad p_R = \frac{1,500}{597} = 2.51$$

From Fig. 3-11 $C = 0.5$, and the volume can be obtained from

$$v = \frac{CnRT}{p} = \frac{0.5 \times 1.0 \times 1,545 \times 810}{1,500 \times 144} = 2.9 \text{ ft}^3$$

**Tabulations.** Where accurate information is desired and where complicated equations of state must be used, it is more convenient, in many cases, to tabulate the relationships between the properties. Such tabulations are available for water vapor, various refrigerants, and for some other gases near their saturation conditions. Tables 3-28 and 3-29 show the properties of water vapor.

*Example* 3: How much heat must be added to 1 lb of water to change it from a saturated liquid to a saturated vapor at a constant temperature of 212°F?

The heat transfer can be obtained from $Q = \int T \, dS$ and since $T = $ constant

$$Q = T(S_{\text{final}} - S_{\text{initial}})$$

From the tables,

$$S_{\text{final}} = 1.7566 \qquad \text{and} \qquad S_{\text{initial}} = 0.3120$$

so that

$$Q = (212 + 460)(1.7566 - 0.3120) = 1070 \text{ Btu}$$

Since the pressure is constant during the boiling process, we can also calculate the heat transfer from the energy statement, which, when simplified to include only the terms of significance, is

$$Q - W = H_{\text{final}} - H_{\text{initial}}$$

The work can be obtained from $W = \int -v \, dp = 0$, so that $Q = H_{\text{final}} - H_{\text{initial}} = 1,150.4 - 180 = 1,070.4$ Btu.

## 2-4. Gas Mixtures

**Ideal-gas Mixtures.** In a mixture of ideal-gas components, the total pressure can be regarded as being equal to the sum of the individual pressures of each component. These individual pressures are called partial pressures, and can be determined from the expression

$$p_i = x_i p_T \qquad\qquad (2\text{-}15)$$

where $p_i$ = partial pressure of component,
  $x_i$ = mole fraction = $n_i/n_T$
  $p_T$ = mixture pressure

The components then will each obey the ideal-gas law so that

$$p_i v = n_i R T \tag{2-16}$$

**Real-gas Mixtures.** Approximations to the $pVT$ behavior of real-gas mixtures may be made using a generalized compressibility factor obtained from the reduced pressure and temperature of the mixture. These properties of the system are expressed in terms of the critical properties of the components of the mixture as follows:

$$p_R = \frac{p_{\text{mix}}}{p_{C,\text{mix}}} \qquad T_R = \frac{T_{\text{mix}}}{T_{C,\text{mix}}} \tag{2-17}$$

where $\qquad\qquad P_{C,\text{mix}} = \Sigma x_i P_{C,i}$

and $\qquad\qquad T_{C,\text{mix}} = \Sigma x_i T_{C,i}$

With these reduced coordinates, the value of the compressibility factor can again be obtained from Fig. 3-11

## 2-5. Systems Containing More Than One Phase

In cases where several phases exist in equilibrium together, the thermodynamic condition of equilibrium is described in terms of a system property called the Gibbs function, and defined by

$$G = H - TS \tag{2-18}$$

In the case where a single component exists in several phases, such as ice and steam, the equilibrium condition results in

$$g_I = g_{II} = g_N \tag{2-19}$$

where $g_I$ represents the specific Gibbs function for phase I.

The different conditions of existence are best represented graphically by a phase diagram and pressure-volume diagram. Typical examples of these diagrams are given in Fig. 2-1.

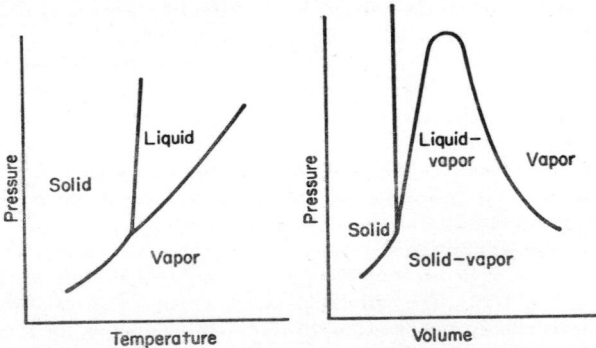

Fig. 2-1. Phase and $PV$ diagram.

**Vapor Pressure.** The vapor pressure or saturation pressure at any given temperature can often be obtained from property tables, or may be approximated by an expression of the form

$$\ln p = A/T + B \tag{2-20}$$

where $A$ and $B$ are constants that depend on the material. If pressure-temperature information is available for two points on any specific vapor curve, then the vapor pressure at any other temperature can be computed.

*Example* 4: Determine the saturation pressure of refrigerant 12 at 100°F. The boiling point is $-20°F$. Data available are

At $p = 14.7$ psia      $t = -20°F$
At $p = 597$ psia      $t = 234°F$      (critical values, Table 2-1)

Substitution of these values into the equation gives

$$\ln 14.7 = \frac{A}{440} + B$$

$$\ln 597 = \frac{A}{694} + B$$

from which $A = 4,440$, $B = 12.79$.

We can now calculate the saturation pressure from

$$\ln p = -4,440/560 + 12.79$$
$$p = 130 \text{ psia}$$

**Escaping Tendency and Fugacity.** In the case where the phases are not in equilibrium, the rate at which the system tends to equilibrium is partially determined by a generalized function called the escaping tendency. To describe this condition in a system containing several components and several phases, a necessary condition for equilibrium is that the escaping tendency must be the same throughout the system. One convenient measure of the escaping tendency is called the fugacity. For the gas phase the escaping tendency, and therefore the fugacity, may often be represented satisfactorily by the pressure of the component under consideration.

Where the gas phase deviates considerably from ideal behavior, fugacity must be used as a true measure of the escaping tendency, rather than the pressure. The fugacity of a substance is a property of that substance, and tabulations of it may be made in the same fashion as other properties of the materials. For example, the rate at which water vapor will penetrate a porous membrane is dependent on the fugacity of the water vapor on each side of the membrane.

**Single-component Systems.** A convenient generalized chart of the fugacity coefficient $f/p$ as it depends on the pressure and temperature of the system is shown in Fig. 3-12   The correlation is again made in terms of the reduced pressure and reduced temperature, in the same manner as for the compressibility factor.

*Example* 5: Compare the fugacity to the pressure for saturated water vapor at 3,000 psia.  The critical properties are 3,106 psia and 705°F.  The reduced properties are

$$p_R = \frac{3,000}{3,106} = 0.966 \qquad T_R = \frac{695 + 460}{705 + 460} = 0.991$$

From Fig. 3-12 we find $f/p = 0.65$.
Thus the fugacity is $0.65 \times 3,000 = 1,950$ psi.

**Systems of Several Components.**  The fugacity of a component of a mixture of gases can often be satisfactorily approximated by an expression

$$f_i = x_i f_{pure} \tag{2-21}$$

where $f_i$ = fugacity of component in mixture
$\quad x_i$ = mole fraction = $n_i/n_T$
$\quad f_{pure}$ = fugacity of pure component at pressure and temperature of mixture

*Chemical Equilibrium.*  Equilibrium in a chemically reacting system is describable in terms of the equilibrium constant.  Consider the reaction

$$CO + \frac{1}{2}O_2 \rightarrow CO_2 \tag{2-22}$$

at equilibrium.  Two expressions involving the equilibrium constant for this typical reaction are useful.

$$-\ln K = \frac{G_{final} - G_{initial}}{RT} \tag{2-23}$$

$$K = \frac{f_{CO_2}}{f_{CO}(f_{O_2})^{\frac{1}{2}}} \tag{2-24}$$

where $f_{CO_2}$, $f_{CO}$, and $f_{O_2}$ are the fugacities of each of these respective components as they exist in the equilibrium mixture.  The use of the approximation from Eq. (2-21) results in a more manageable relation for Eq. (2-24).

$$K = \frac{\dfrac{f_{CO_2, pure}}{p}}{\dfrac{f_{CO, pure}}{p}\left(\dfrac{f_{O_2, pure}}{p}\right)^{\frac{1}{2}}} \left[\frac{x_{CO_2}}{x_{CO}(x_{O_2})^{\frac{1}{2}}}\right] p^{-\frac{1}{2}} \tag{2-25}$$

The coefficients $f/p$ can now be obtained from Fig. 3-12.  A further simplification is possible if all components of the equilibrium mixture can be considered as ideal gases.  Under this condition $f/p = 1$, and

$$K = \frac{x_{CO_2}}{x_{CO}(x_{O_2})^{\frac{1}{2}}} p^{-\frac{1}{2}} \tag{2-26}$$

The equilibrium constant is a temperature function, and tabulations of this function for various reactions and temperatures may be found in the literature.  Knowledge of the equilibrium constant makes possible the complete determination of the composition of the equilibrium mixture.  The major portion of the data available for the equilibrium constants is so arranged that the pressure in Eqs. (2-23) and (2-24) must be introduced in atmospheres.  A typical tabulation of this sort is the JANAF Thermochemical Tables, published at the Thermo Laboratory, Dow Chemical Company.

*Example* 6: A gas stream containing equimolar proportions of CO and $H_2O$ is allowed to come to equilibrium at 1,000°K and 5 atm pressure. What is the composition if only CO, $CO_2$, $H_2O$, and $H_2$ are the important components?

The reaction at equilibrium will be

$$CO + H_2O \leftrightarrows CO_2 + H_2$$

The JANAF tables give the equilibrium constant of formation, i.e., that specific reaction in which the tabulated component is formed from its elements, with these elements occurring in their natural stable state at normal atmospheric conditions.

We can obtain values of the equilibrium constants as follows:

Reaction 1:                     $C_{gr} + O_2 \rightarrow CO_2$
Reaction 2:                     $C_{gr} + \frac{1}{2}O_2 \rightarrow CO$
Reaction 3:                     $H_2 + \frac{1}{2}O_2 \rightarrow H_2O$

The desired reaction

Reaction 4:               $CO + H_2O \rightarrow CO_2 + H_2$

can be obtained by subtracting reactions 2 and 3 from 1. The equilibrium constant for reaction 4 is

$$K_4 = \frac{K_1}{K_2 K_3}$$

since the equilibrium constant is exponentially dependent on the Gibbs function. At 1,000°K

$$\log K_1 = 20.680 \qquad \log K_2 = 10.459 \qquad \log K_3 = 10.062$$

and therefore

$$\log K_4 = 0.159 \qquad \text{and} \qquad K_4 = 1.442$$

If we now assume that all components at equilibrium are ideal gases, the fugacity coefficients will all be identical and

$$1.442 = \frac{(x_{CO_2})(x_{H_2})}{(x_{CO})(x_{H_2O})}$$

or

(*a*)                     $$1.442 = \frac{(n_{CO_2})(n_{H_2})}{(n_{CO})(n_{H_2O})}$$

We can now write an expression for the reaction, starting with CO and $H_2O$ and ending at equilibrium.

$$CO + H_2O \rightarrow n_{CO}CO + n_{H_2O}H_2O + n_{CO_2}CO_2 + n_{H_2}H_2$$

From this equation we can write three others, one for each elemental type.

(*b*)          For carbon:     $1 = n_{CO} + n_{CO_2}$
(*c*)          For oxygen:     $2 = n_{CO} + n_{H_2O} + n_{CO_2}$
(*d*)          For hydrogen:   $1 = n_{H_2O} + n_{H_2}$

We now have four equations, (*a*) to (*d*), which can be used to determine exactly $n_{CO}$, $n_{H_2O}$, $n_{CO_2}$, and $n_{H_2}$. A trial-and-error solution gives

$$n_{CO} = 0.454 \qquad n_{CO_2} = 0.526$$
$$n_{H_2O} = 0.454 \qquad n_{H_2} = 0.546$$

**Heterogeneous Systems.** Equilibrium determinations involving more than one phase may also be approximated very satisfactorily by using the mole fraction of the particular component in the phase under consideration for the fugacity in Eq. (2-24). For example, consider the equilibrium reaction

$$C_{gr} + O_2 \rightarrow CO_2$$

Since the condensed phase contains only one component, the mole fraction of the graphitic carbon is unity, and the equilibrium constant for the reaction is

$$K = \frac{x_{CO_2}}{x_{O_2}}$$

if both gas components can be considered ideal.

## 2-6. Thermochemistry

**Property Evaluations.** It is desirable to be able to obtain numerical values for various chemical properties of materials involved in processes where chemical reactions occur. For convenience, tabulations of such properties as enthalpy, internal energy, Gibbs function, and the equilibrium constant are made in terms of the reaction of formation.

The formation reaction is the reaction having the component under consideration as the only product, with each reactant in its elemental form and with these elements in their natural stable state at normal atmospheric conditions. For example, the enthalpy of formation of carbon dioxide at 25°C is expressed as follows:

$$C_{gr} + O_2 \rightarrow CO_2 \qquad H_f = -94 \text{ kcal/mole } CO_2 \qquad (2\text{-}27)$$

The negative sign indicates that the reaction is exothermic; i.e., it releases energy. The enthalpies of formation of the elements in their naturally occurring form must then be zero at all temperatures. Table 2-1 lists enthalpies and equilibrium constants of formation for some simple common gases at 25°C. Extensive tabulations of the enthalpy, Gibbs function, and equilibrium constant of formation, as they depend on temperature, can be found in the JANAF Thermochemical Tables. The enthalpy change of reaction for any reaction may be found from the enthalpies of formation of all the components that enter into the reaction.

*Example* 7: It is desired to find the enthalpy change for the following reaction at 25°C:

$$CO + \tfrac{1}{2}O_2 \rightarrow CO_2 \qquad (2\text{-}28)$$

The enthalpy change is found from

$$H_f - H_i = \Sigma H_{form,products} - \Sigma H_{form,reactants} \qquad (2\text{-}29)$$

or

$$H_f - H_i = H_{form,CO_2} - H_{form,CO} - \tfrac{1}{2}H_{form,O_2} \qquad (2\text{-}30)$$

From Table 2-1,

$$H_f - H_i = -94.05 - [-26.42 - \tfrac{1}{2}(O)] = -67.63 \text{ kcal/mole}$$

The same general rule also applies for the internal-energy change of reaction and change in Gibbs function for the reaction. If we represent all these properties by the generalized symbol $R$, then

$$\Delta R_{\text{reaction}} = \Sigma R_{\text{form,products}} - \Sigma R_{\text{form,reactants}} \tag{2-31}$$

As previously indicated, the equilibrium constant of reaction is related to those of formation by

$$K_{\text{reaction}} = \frac{\pi(K_{\text{form,products}})}{\pi(K_{\text{form,reactants}})} \tag{2-32}$$

An approximation to the variation of the equilibrium constant with temperature can be made from the equation

$$\ln K = \frac{A}{T} + B \tag{2-33}$$

where $A$ and $B$ are constants. Figure 2-2 shows the equilibrium constants for several components as they depend on the inverse temperature.

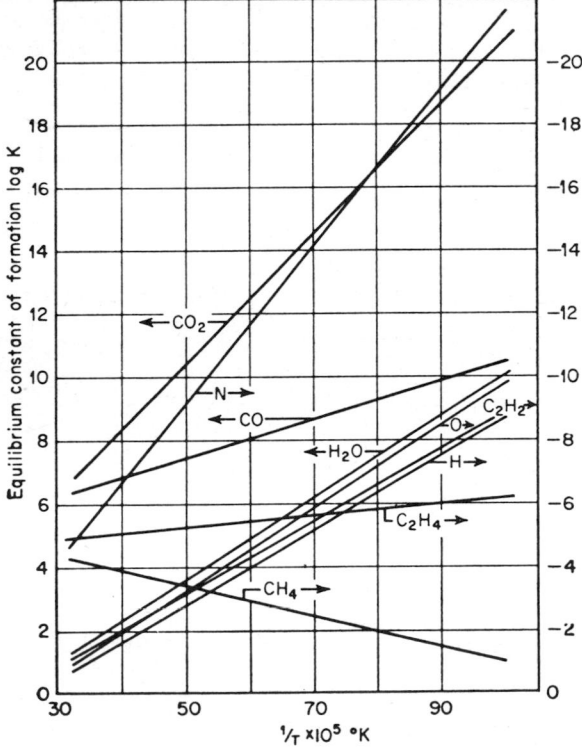

Fig. 2-2. Equilibrium constant of formation.

The entropy change for a reaction may be found from the absolute entropies of the components involved in the reaction. These absolute entropies can be obtained from specific-heat data and the use of the third law of thermodynamics, which, in effect, specifies the state of the component when its entropy is zero. Table 2-1 indicates some of these absolute entropies at 25°C, and a more complete tabulation is given in the JANAF tables.

Since the entropy is an additive property, the change in entropy for any reaction may be found from the absolute entropies of the components of the reaction in the same manner as from the enthalpies of formation. For example, for Eq. (2-28) at 25°C,

$$\Delta S_{reaction} = S_{CO_2} - S_{CO} - \tfrac{1}{2}S_{O_2}$$
$$\Delta S_{reaction} = 51.1 - 47.2 - \tfrac{1}{2}(49.0) = -20.6 \; cal/(mole)(°K) \quad (2\text{-}34)$$

All components, including the elemental forms, have nonzero values.

It is also convenient to tabulate the total enthalpy of various substances measured above some arbitrary datum. Tabulations of this sort are available, and Table 2-1 shows typical values of this property. Care must be exercised in the use of tabulated total enthalpy values since no universally accepted standard datum is in use. The data in various sources are thus often mutually exclusive.

The total enthalpies may also be used to find the enthalpy change of reaction. As for the entropy for Eq. (2-28) at 25°C,

$$\Delta H_{reaction} = H_{total,CO_2} - H_{total,CO} - \tfrac{1}{2}H_{total,O_2}$$
$$\Delta H_{reaction} = 2.24 - 67.82 - \tfrac{1}{2}(4.11) = -67.63 \; kcal/mole \quad (2\text{-}35)$$

Here also it is important to recognize that the total enthalpies of all components, including the elemental forms, have nonzero values. The variation of the total enthalpy and absolute entropy with temperature for a representative group of gases is shown in Table 2-2.

It is possible to relate most of the thermochemical properties to each other so that limited data can often be used to construct more extensive information. Some of these formulas have been given elsewhere, but are repeated here for convenience. For any reaction at a given temperature

$$\ln K = -\frac{\Delta G}{RT} \quad (2\text{-}23)$$

$$\Delta G = \Delta H - T\,\Delta S \quad (2\text{-}36)$$
$$\Delta H = \Sigma H_{form,products} - \Sigma H_{form,reactants} \quad (2\text{-}29)$$

or

$$\Delta H = \Sigma H_{T\,products} - \Sigma H_{T\,reactants} \quad (2\text{-}37)$$
$$\Delta S = \Sigma S_{products} - \Sigma S_{reactants} \quad (2\text{-}38)$$

For any component

$$C_p = \frac{\Delta H}{\Delta T} \quad \text{approximately} \quad (2\text{-}39)$$

**Flame Temperature at Constant Pressure.** The temperature of the flame in a combustion reaction can be satisfactorily approximated by assuming

## Table 2-2. Enthalpies and Entropies of Gases*

| Gas | Symbol | Enthalpy, kcal/mole | | | | | Entropy, cal/mole | | | | |
|---|---|---|---|---|---|---|---|---|---|---|---|
| | | 1000°K | 1500°K | 2000°K | 2500°K | 3000°K | 1000°K | 1500°K | 2000°K | 2500°K | 3000°K |
| Acetylene | $C_2H_2$ | 317.9 | 326.5 | 336.0 | 346.0 | 356.3 | 64.32 | 71.34 | 76.78 | 81.22 | 84.97 |
| Ammonia | $NH_3$ | 102.5 | 109.7 | 117.8 | 126.5 | 135.6 | 58.39 | 64.17 | 68.83 | 72.71 | 76.01 |
| Argon | Ar | 5.0 | 7.5 | 9.9 | 12.4 | 14.9 | 42.99 | 45.01 | 46.44 | 47.55 | 48.45 |
| Benzene | $C_6H_6$ | 808.0 | 835.1 | | | | 106.73 | 128.68 | | | |
| Butane | $C_4H_{10}$ | 704.1 | 712.7 | | | | 120.31 | 144.22 | | | |
| Carbon dioxide | $CO_2$ | 10.2 | 17.0 | 24.1 | 31.4 | 38.8 | 64.34 | 69.82 | 73.90 | 77.15 | 79.85 |
| Carbon monoxide | CO | 73.0 | 77.1 | 81.4 | 85.8 | 90.2 | 56.03 | 59.35 | 61.81 | 63.76 | 65.37 |
| Dichlorodifluoromethane | $CCl_2F_2$ | 134.4 | 146.8 | 159.5 | 172.2 | 185.1 | 97.92 | 107.98 | 115.27 | 120.96 | 125.63 |
| Ethane | $C_2H_6$ | 381.2 | 397.8 | 416.4 | 436.2 | 456.6 | 79.39 | 92.46 | | | |
| Ethylene | $C_2H_4$ | 347.9 | 360.3 | 374.0 | 388.5 | 403.4 | 72.06 | 82.02 | 89.90 | 96.36 | 101.79 |
| Helium | He | 4.97 | 7.45 | 9.94 | 12.4 | 14.9 | 36.15 | 38.16 | 39.59 | 40.70 | 41.61 |
| Heptane | $C_7H_{16}$ | 1178.0 | 1192.3 | | | | 180.32 | 220.45 | | | |
| Hydrogen | $H_2$ | 74.3 | 78.1 | 82.1 | 86.3 | 90.6 | 39.70 | 42.72 | 45.00 | 46.88 | 48.47 |
| Methane | $CH_4$ | 222.3 | 231.9 | 242.7 | 254.3 | 266.3 | 59.14 | 66.84 | 73.08 | 78.23 | 82.60 |
| Nitrogen | $N_2$ | 8.9 | 13.0 | 17.2 | 21.6 | 26.0 | 54.51 | 57.78 | 60.22 | 62.16 | 63.77 |
| Oxygen | $O_2$ | 9.5 | 13.8 | 18.3 | 22.8 | 27.6 | 58.19 | 61.66 | 64.21 | 66.25 | 67.97 |
| Propane | $C_3H_8$ | 543.4 | 550.2 | | | | 99.77 | 118.29 | | | |
| Water | $H_2O$ | 19.9 | 25.2 | 31.1 | 37.4 | 43.9 | 55.59 | 59.86 | 63.23 | 66.03 | 68.42 |

* Additional thermodynamic data may be found in Sec. 3.

that the enthalpy or internal-energy change for the reaction from initial to final temperature is zero.

*Example* 8: What temperature can be expected from the reaction of carbon monoxide with air at 1 atm pressure if the reactants are at 25°C?

Since the composition of air is 1 molar part oxygen and 3.78 molar parts nitrogen, the reaction is

$$CO + \tfrac{1}{2}(O_2 + 3.78N_2) = CO_2 + 1.89N_2$$

The enthalpy of the reactants is

$$H_{reactant} = H_{CO} + \tfrac{1}{2}H_{O_2} + 1.89H_{N_2}$$

At 25°C,

$$H_{reactant} = 67.82 + \tfrac{1}{2}(4.11) + 1.89 \times 3.77 = 77.0 \text{ kcal}$$

By trial and error we find that, at $T = 2650°K$,

$$H_{products} = H_{CO_2} + 1.89H_{N_2} = 76.9 \text{ kcal}$$

Thus the flame temperature will be 2650°K.

*Example* 9: In some cases the flame temperature becomes excessive to the point where dissociation reactions play an important part. What temperature can be expected from the reaction

$$CO + H_2 + O_2 \rightarrow \text{products}$$

at 1 atm and initially at 25°C?

We assume that the equilibrium products contain CO, $CO_2$, $H_2$, $H_2O$, and $O_2$. Thus the reaction is

$$CO + H_2 + O_2 \rightarrow n_{CO}CO + n_{CO_2}CO_2 + n_{H_2}H_2 + n_{H_2O}H_2O + n_{O_2}O_2$$

The equilibrium equations are obtained as follows:

(a) $\qquad\qquad C_{gr} + \tfrac{1}{2}O_2 \rightarrow CO \qquad K_A$
(b) $\qquad\qquad C_{gr} + O_2 \rightarrow CO_2 \qquad K_B$
(c) $\qquad\qquad H_2 + \tfrac{1}{2}O_2 \rightarrow H_2O \qquad K_C$

We can combine Eqs. (a) and (b) to give

(d) $\qquad\qquad CO + \tfrac{1}{2}O_2 \rightarrow CO_2 \qquad K_D = \dfrac{K_B}{K_A}$

The final statement is the flame-temperature statement:

$$H_{reactants} = H_{products}$$

The set of six simultaneous equations for solution are

(1) $\qquad\qquad K_D = \dfrac{n_{CO_2}}{n_{CO}(n_{O_2})^{\frac{1}{2}}} \, (n_T)^{\frac{1}{2}}$

(2) $\qquad\qquad K_C = \dfrac{n_{H_2O}}{n_{H_2}(n_{O_2})^{\frac{1}{2}}} \, (n_T)^{\frac{1}{2}}$

(3) $\qquad\qquad 1 = n_{CO} + n_{CO_2}$
(4) $\qquad\qquad 3 = n_{CO} + 2n_{CO_2} + n_{H_2O} + 2n_O$
(5) $\qquad\qquad 1 = n_{H_2} + n_{H_2O}$

(6)
$$[n_{CO}H_{CO} + n_{CO_2}H_{CO_2} + n_{H_2}H_{H_2} + n_{H_2O}H_{H_2O} + n_{O_2}H_{O_2}]_{\text{flame temperature}} =$$
$$[H_{CO} + H_{H_2} + H_{O_2}]_{25°C}$$

where $n_T = n_{CO} + n_{CO_2} + n_{H_2} + n_{H_2O} + n_{O_2}$.

If a trial flame temperature is assumed, the equilibrium constants and enthalpies are known, and allow solution for the five composition variables in any five of the equations. The assumed temperature can then be checked from the sixth equation. The results are

$$n_{CO} = 0.633$$
$$n_{CO_2} = 0.368$$
$$n_{H_2} = 0.183$$
$$n_{H_2O} = 0.817$$
$$n_{O_2} = 0.407$$

Substitution of these values into the enthalpy equation shows that

$$H_{\text{reactant}} = H_{\text{products}} = 141.4 \text{ kcal}$$
$$T = 3225°K$$

**Heat Released.** The energy released as heat in a combustion reaction may be calculated in a manner similar to that for the flame temperature. In this case, the final temperature must be specified, and is not a variable. All equations shown in Example 9 are applicable, with the exception of the enthalpy equation. If the reacting gas flow is steady, the energy equation results in

$$Q = H_{\text{final}} - H_{\text{initial}} \tag{2-40}$$

Thus the composition must first be determined, so that the heat transfer may then be calculated.

**Constant-volume Reactions.** In instances of this sort where combustion reactions take place in fixed volumes, such as bombs or cylinders of slow-moving internal-combustion engines, a simple modification of the above equations involving only the substitution of the internal energy for the enthalpy is required.

## 2-7. Mixtures of Ideal Gases and Vapors

**Humidity.** The relative humidity of an ideal-gas vapor mixture is defined by the equation

$$\text{RH} = \frac{p_v}{p_{\text{sat}}} \tag{2-41}$$

where $p_v$ is the actual pressure of the vapor phase, and $p_{\text{sat}}$ is the pressure exerted by the same vapor when saturated at the mixture temperature. For low vapor pressures the vapor may often be considered an ideal gas, so that

$$\text{RH} = \frac{v_{\text{sat}}}{v_V} \tag{2-42}$$

The absolute humidity, or specific humidity, is the ratio of the mass of vapor in the mixture to the mass of the dry gas.

$$\text{SH} = \frac{m_v}{m_g} \tag{2-43}$$

Observations of the concentrations of the two components are usually made in terms of three temperatures: (1) the dew-point temperature, (2) the wet-bulb temperature, and (3) the dry-bulb temperature.

The dew point is the temperature at which condensation of the vapor takes place if the mixture is cooled at constant pressure.

The wet-bulb temperature is the temperature achieved by the mixture if it is saturated by evaporating liquid into it so that the latent heat of vaporization comes from the mixture, thereby depressing its temperature.

The dry-bulb temperature is the normal temperature of the mixture.

*Example* 10: Calculate the relative humidity of an air–water vapor mixture having a dry-bulb temperature of 80°F and a wet-bulb temperature of 72°F.

The specific humidity may be found from

$$\text{SH} = \frac{c_{p_\text{gas}}(t_w - t_d) + \text{SH}_w L_w}{H_{v.d} - H_{L.w}} \tag{2-44}$$

where subscripts $w$ and $d$ refer to the wet- and dry-bulb temperatures, $L$ is the latent heat of the vapor, and $H_v$ and $H_L$ represent, respectively, the enthalpies of the saturated vapor and its corresponding saturated liquid. The specific humidity at the wet-bulb temperature is found from the definition and the assumption of ideal-gas behavior for both components.

$$\text{SH} = \frac{m_v}{m_g} = \frac{M_v}{M_g}\frac{p_v}{p_g} = \frac{18}{29}\left(\frac{0.39}{14.7 - 0.39}\right) = 0.0168 \text{ lb water vapor/lb air}$$

Then

$$\text{SH}_D = \frac{0.24(72 - 80) + 0.0168 \times 1{,}053}{1{,}097 - 40} = 0.150 \text{ lb water vapor/lb air}$$

The vapor pressure at the dry-bulb condition can be found by the equation above.

$$\text{SH} = \frac{M_v p_v}{M_g p_g} = \frac{18}{29}\left(\frac{p_v}{p - p_v}\right) \qquad \text{or} \qquad p_v = \frac{p}{(0.622/\text{SH}) + 1}$$

from which

$$p_v = \frac{14.7}{(0.622/0.0150) + 1} = 0.346 \text{ psia}$$

and

$$\text{RH} = \frac{p_v}{p_\text{sat}} = \frac{0.346}{0.507} = 69.2\%$$

*Example* 11: Calculate the relative humidity of an air–water vapor mixture having a dry-bulb temperature of 80°F and a dew point of 74°F.

From the dew point we find the pressure of the vapor to be 0.416 psia. Thus

$$\text{RH} = \frac{0.416}{0.507} = 82\%$$

**The Psychrometric Chart.** Solutions to the problems of air–water vapor mixtures are readily available in chart form. Figure 4-14 shows a psychrometric chart for this system. The solutions to Examples 10 and 11 can be taken directly off the chart.

For problems involving the change of conditions, heat exchanges may also be obtained by computing the change in enthalpy read from the psychrometric chart.

*Example* 12: It is desired to reduce the temperature of air at 80°F dry-bulb and 72°F wet-bulb to a dry-bulb temperature of 65°F. How much heat must be removed for each pound of air cooled?

$$Q = H_f - H_i = 30 - 35.8 = -5.8 \text{ Btu/lb air}$$

## 2-8. Processes

It is possible to tabulate the equations applicable for the determination of heat, work, and some selected property changes for the simplest processes, where an ideal gas is the medium and in which the conditions throughout the process are idealized. These relations are shown in Table 2-3, and the following example shows the manner in which they have been obtained.

### Table 2-3. Relations for Ideal-gas Processes
Reversible Processes with Constant Specific Heats

| Process | Isothermal $T = $ const | Constant pressure $p = $ const | Constant volume $v = $ const | Isentropic $S = $ const |
|---|---|---|---|---|
| $pvT$ relations | $pv = $ const | $\dfrac{v}{T} = $ const | $\dfrac{p}{T} = $ const | $pv^k = $ const $\quad Tv^{k-1} = $ const $\quad \dfrac{p^{(k-1)/k}}{T} = $ const |
| Nonflow work $-\displaystyle\int p\,dv$ | $pv \ln \dfrac{v_f}{v_i}$ | $p\,\Delta v$ | $0$ | $nC_V\,\Delta T$ |
| Steady-flow work $-\displaystyle\int v\,dp$ | $pv \ln \dfrac{p_i}{p_f}$ | $0$ | $v\,\Delta p$ | $nC_p\,\Delta T$ |
| Heat $\displaystyle\int T\,dS$ | $pv \ln \dfrac{p_i}{p_f}$ | $nC_p\,\Delta T$ | $nC_V\,\Delta T$ | $0$ |
| $\Delta U$ | $0$ | $nC_V\,\Delta T$ | $nC_V\,\Delta T$ | $nC_V\,\Delta T$ |
| $\Delta H$ | $0$ | $nC_p\,\Delta T$ | $nC_p\,\Delta T$ | $nC_p\,\Delta T$ |
| $\Delta S$ | $nR \ln \dfrac{p_i}{p_f}$ | $nC\, p \ln \dfrac{T_f}{T_i}$ | $nC_V \ln \dfrac{T_f}{T_i}$ | $0$ |

*Example* 13: It is desired to obtain expressions for the heat transfer, work, internal-energy change, enthalpy change, and entropy change for a process which takes place in a steady-flow frictionless machine at constant pressure.

The material may be considered to be an ideal gas. It is further to be assumed that kinetic-energy and potential-energy changes between the inlet and outlet gas streams can be neglected and only one inlet and one outlet stream are present.

The energy statement that applies is Eq. (2-2), which, in consideration of the limitations of the problem, may be written

$$Q_{net} - W_{net} = H_{final} - H_{initial}$$

The work may be obtained from Eq. (2-4).

$$W = \int - v \, dp = 0 \qquad \text{for constant pressure condition}$$

We have, then,

$$Q_{net} = H_{final} - H_{initial}$$

For an ideal gas we find from Eq. (2-10) that

$$H_f - H_i = C_p(T_f - T_i)$$

so that

$$Q = C_p(T_f - T_i)$$

The internal-energy change can be obtained from Eq. (2-8).

$$U_f - U_i = C_v(T_f - T_i)$$

The entropy change can be obtained by using Eq. (2-6).

$$T \, dS = dU + p \, dv$$

In this case, since $p$ is constant, we can write

$$S_f - S_i = \int \frac{dU}{T} + p \int dv \qquad (2\text{-}45)$$

or

$$S_f - S_i = C_V \ln \frac{T_f}{T_i} + R \ln \frac{v_f}{v_i} \qquad (2\text{-}46)$$

We can write the second term as follows:

$$R \ln \frac{v_f}{v_i} = R \ln \frac{T_f}{T_i}$$

so that

$$S_f - S_i = (C_V + R) \ln \frac{T_f}{T_i}$$

Recognizing that, for an ideal gas, $C_V + R = C_p$,                          (2-47)

$$S_f - S_i = C_p \ln \frac{T_f}{T_i}$$

## 2-9. Ideal-gas Power Cycles

The analysis of real power cycles can often be approximated by idealized cycles, using ideal gases in a piston-cylinder arrangement with no transfer of mass. Several such approximations are of interest.

**Carnot Cycle.** The cycle consists of four processes, two isothermals and two isentropics, as shown in Fig. 2-3. A detailed analysis of the cycle shows that

$$Q_{net} = W_{net} = (T_3 - T_4)R \ln \frac{p_2}{p_3}$$

$$Q_{12} = Q_{34} = 0$$

$$Q_{23} = W_{23} = p_2 v_2 \ln \frac{p_2}{p_3}$$

$$W_{12} = C_V(T_2 - T_1)$$
$$W_{34} = C_V(T_4 - T_3)$$
$$U_3 - U_2 = 0 \qquad U_1 - U_4 = 0$$

$$Q_{41} = W_{41} = p_4 v_4 \ln \frac{p_4}{p_1}$$

$$\text{eff} = \frac{T_2 - T_1}{T_2} = \frac{T_3 - T_4}{T_3} \tag{2-48}$$

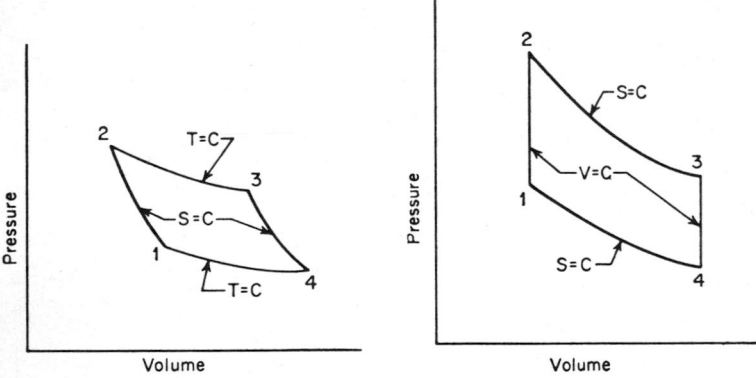

FIG. 2-3. Carnot cycle.　　　　FIG. 2-4. Otto cycle.

**Otto Cycle.** This cycle consists of two isentropics and two constant-volume processes as shown in Fig. 2-4, and is often used as a representation of a spark-ignition engine.

The cycle analysis indicates

$$Q_{net} = W_{net} = C_V[(T_2 + T_4) - (T_1 + T_3)]$$
$$Q_{12} = C_V(T_2 - T_1)$$
$$Q_{34} = C_V(T_4 - T_3)$$
$$Q_{23} = Q_{41} = 0$$
$$W_{12} = W_{34} = 0$$
$$W_{41} = C_V(T_4 - T_1)$$
$$W_{23} = C_V(T_3 - T_2)$$

$$\text{eff} = 1 - \frac{T_4}{T_1} = 1 - \left(\frac{V_4}{V_1}\right)^{k-1} \tag{2-49}$$

**Diesel Cycle.** The cycle consists of two isentropic, one constant-volume, and one constant-pressure process, as shown in Fig. 2-5. It is representative of a diesel engine.

Results of the cycle analysis are

$$Q_{\text{net}} = W_{\text{net}} = C_p(T_3 - T_2) - C_V(T_4 - T_1)$$
$$Q_{23} = C_p(T_3 - T_2)$$
$$Q_{41} = C_V(T_4 - T_1)$$
$$Q_{12} = Q_{34} = 0$$
$$W_{23} = p_2(v_3 - v_2)$$
$$W_{41} = 0$$
$$W_{12} = C_V(T_2 - T_1)$$
$$W_{34} = C_V(T_4 - T_3)$$

$$\text{eff} = 1 - \frac{1}{R} \frac{T_4 - T_1}{T_3 T_2} \tag{2-50}$$

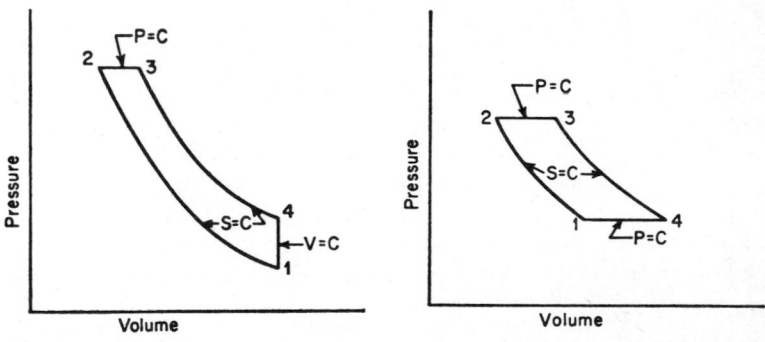

FIG. 2-5. Diesel cycle. $\qquad$ FIG. 2-6. Brayton cycle.

**Brayton Cycle.** This cycle consists of two isentropic and two constant-pressure processes. It is representative of the gas turbine, and is shown in Fig. 2-6.

Results of the cycle analysis are

$$Q_{\text{net}} = W_{\text{net}} = C_p[(T_4 + T_2) - (T_3 + T_1)]$$
$$Q_{23} = Q_{41} = 0$$
$$Q_{34} = C_p(T_4 - T_3)$$
$$Q_{12} = C_p(T_2 - T_1)$$
$$W_{12} = p_1(v_2 - v_1)$$
$$W_{34} = p_3(v_4 - v_3)$$
$$W_{23} = C_V(T_3 - T_2)$$
$$W_{41} = C_V(T_1 - T_4)$$

$$\text{eff} = 1 - \frac{T_1 - T_2}{T_4 - T_3} \tag{2-51}$$

**Stirling Cycle.** The Stirling cycle consists of two isothermal and two constant-volume processes. Practical examples of this cycle have been developed recent'y, and depend on regeneration to achieve practical efficiencies. Figure 2-7 shows this cycle. Practical cycles use the heat rejected in process 41 to partly regenerate the gas during process 23. Cycle analysis must depend on the degree of regeneration.

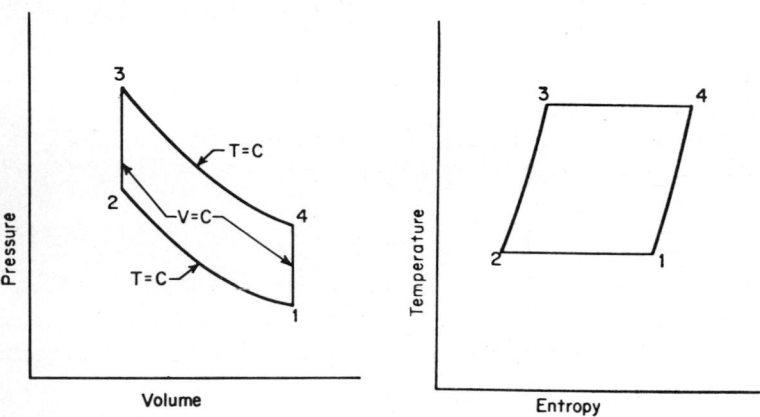

FIG. 2-7. *PV* diagram for Stirling cycle.     FIG. 2-8. Stirling cycle.

*Example* 14: Calculate the ideal cycle efficiency for a Stirling cycle, using air in which the maximum and minimum temperatures are 2000°R and 500°R and 30 per cent of the heat rejected is regained by regeneration.

The compression ratio is 6.

The temperature-entropy diagram is useful for representation of the problem as shown in Fig. 2-8.

The heat rejected in process 41 is

$$Q_{41} = C_V(T_4 - T_1) = 0.17 \times 1500 = 255 \text{ Btu}$$

The heat supplied in process 23 is

$$Q_{23} = 0.70C_v(T_3 - T_2) = 0.7 \times 255 = 179 \text{ Btu}$$

The heat supplied in process 34 is

$$Q_{34} = RT_3 \ln \frac{p_3}{p_4} = RT_3 \ln \frac{v_1}{v_2} = \frac{1545}{29} \times \frac{2000}{778} \ln 6 = 245 \text{ Btu}$$

The total heat supply is therefore

$$Q_{23} + Q_{34} = 179 + 245 = 424 \text{ Btu}$$

The net work done as a result of processes 12 and 34 is

$$W_{12} = RT_1 \ln \frac{p_1}{p_2} = -RT_1 \ln \frac{p_2}{p_1}$$

$$W_{34} = RT_3 \ln \frac{p_3}{p_4}$$

Reference to Fig. 2-7 shows that

$$\frac{p_3}{p_4} = \frac{p_2}{p_1} = \frac{v_1}{v_2} = \frac{v_4}{v_3}$$

The total net work is therefore

$$W_{\text{net}} = R(T_3 - T_1) \ln \frac{p_3}{p_4} = \frac{1545}{29} \times \frac{1500}{778} \ln 6 = 184 \text{ Btu}$$

The efficiency is $^{184}/_{424} = 43.4$ per cent.

## 2-10. Vapor Cycles

**Rankine Cycle.** The Rankine cycle is the ideal representation for the vapor power cycle. It consists of five processes, two isothermals, two isentropics,

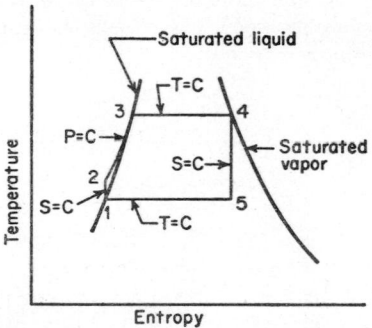

Fig. 2-9. Rankine cycle.

and one constant pressure. The cycle is shown in Fig. 2-9, and the corresponding apparatus diagram in Fig. 2-10.

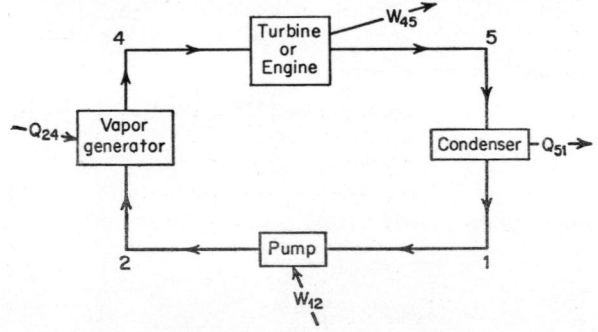

Fig. 2-10. Vapor power system.

The results of a cycle analysis show

$$Q_{24} = H_4 - H_2$$
$$Q_{51} = H_1 - H_5$$
$$W_{45} = H_5 - H_4$$
$$W_{12} = H_1 - H_2 = v'p_1 - p_2) \quad \text{approx.}$$

$$\text{eff} = \frac{(H_5 - H_4) + W_{12}}{H_4 - H_2} \tag{2-52}$$

In most cases $W_{12}$ is small, so that

$$\text{eff} = \frac{H_5 - H_4}{H_4 - H_2} \tag{2-53}$$

**Compression Refrigeration Cycle.**   The idealized compression refrigeration cycle consists of two constant-pressure processes, an isentropic process, and an irreversible throttling process.   The cycle is shown in Fig. 2-11, and the corresponding apparatus in Fig. 2-12.

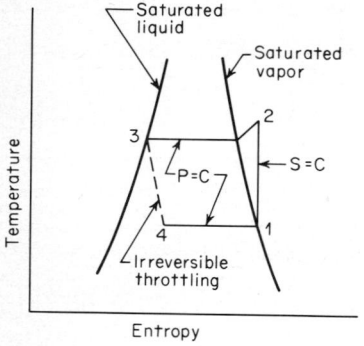

FIG. 2-11.   Compression refrigeration cycle.

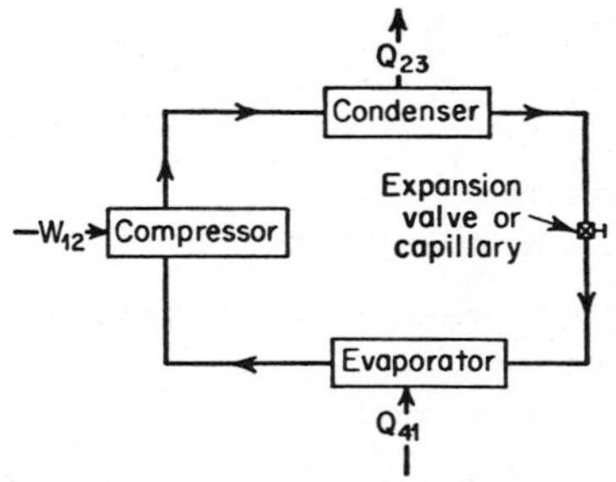

FIG. 2-12.   Compression refrigeration system.

The equations for the cycle are

$$Q_{12} = Q_{34} = 0$$
$$Q_{23} = H_3 - H_2$$
$$Q_{14} = H_4 - H_1$$
$$W_{23} = W_{41} = W_{34} = 0$$
$$H_3 = H_4$$

The coefficient of performance is

$$\text{C.O.P.} = \frac{Q_{41}}{W_{12}} = \frac{H_1 - H_4}{H_2 - H_1} \tag{2-54}$$

## FLUID FLOW

### 2-11. Nomenclature

$A$ = flow-path area normal to flow, ft$^2$

$A_f$ = wall-effect factor, dimensionless (see Table 2-9)

$a$ = area of jet, ft$^2$

$\bar{a}$ = transverse tube pitch, ft

$b$ = longitudinal tube pitch, ft

$C$ = pitot-tube coefficient, dimensionless

$C_c$ = jet area/orifice area, dimensionless

$C_D$ = drag coefficient, dimensionless

$C_d$ = coefficient of discharge for orifices

$CR$ = critical-pressure ratio, dimensionless

$c_p$ = specific heat at constant pressure, Btu/(lb)(°F)

$c_v$ = specific heat at constant volume, Btu/(lb)(°F)

$D$ = diameter of pipe, ft (except in Fig. 2-26, in.)

$D_c$ = tube clearance, ft

$D_e$ = hydraulic diameter, ft

$D_p$ = particle diameter, ft

$D_t$ = tube diameter, ft

$F$ = conversion factor, feet of fluid to inches of water = $\frac{1}{12} \times \bar{w}_W/\bar{w}_{HG}$

$\bar{F}$ = force or resistance to force, lb force

$f$ = friction factor, dimensionless

$f_c$ = friction factor for tower packings (Fig. 2-29)

$G$ = quantity of flow, lb/sec

$g$ = gravitational acceleration, ft/sec$^2$ (32.2 at sea level)

$g_c$ = conversion factor, 32.2 (lb mass)(ft)/(lb force)(sec$^2$)

$H$ = static head, ft fluid; height above sea level, ft; stack height, ft; with weirs, height above crest, ft

$H_m$ = head in manometric fluid, ft

$\Delta H$ = orifice differential, ft fluid flowing

$\Delta H_f$ = head loss due to friction, ft

$h$ = enthalpy, Btu/lb

$\bar{h}$ = weir approach velocity head, ft

$h'_w$ = effective draft, in. H$_2$O

$K$ = equivalent number velocity heads (Table 2-7); or elevation, ft

$k$ = ratio of specific heats, $c_p/c_v$

$L$ = length or linear dimension (with weirs, width), ft

$\bar{L} = N^b$

$m$ = hydraulic radius, ft

$N$ = number of rows of tubes in direction of flow

$N_{\text{Re}}$ = Reynolds number ($= D v \bar{w}/\mu$), dimensionless

$p$ = static pressure, psf

$p_i$ = impact pressure, psf

$\Delta p_f$ = pressure drop due to friction, psf

$Q$ = quantity of flow, cfs

$R$ = manometer reading, ft

$S$ = cross-sectional area, ft$^2$

$\overline{S}$ = specific surface, ft$^2$ particle surface/ft$^3$ bed

$T$ = absolute temperature, °R

$t$ = time, sec

$v$ = fluid velocity, fps

$v_c$ = acoustical velocity in fluid, fps

$v_0$ = superficial velocity in packed bed or local velocity as measured by pitot tube, fps

$v_{max}$ = velocity through minimum area of flow, fps

$\overline{v}$ = specific volume (= $1/\overline{w}$), ft$^3$/lb

$W_a$ = weight in air, lb

$W_f$ = weight in nonair fluid, lb

$\overline{W}_m$ = density of manometric fluid, lb/ft$^3$

$\overline{W}_0$ = density of working fluid, lb/ft$^3$ (also $\overline{W}_A$, $\overline{W}_B$)

$\overline{w}$ = density (= $1/\overline{v}$), lb/ft$^3$

$\overline{w}_{CA}$ = density of column of cold air outside stack, lb/ft$^3$

$\overline{w}_{HG}$ = density of column of hot gas inside stack, lb/ft$^3$

$\overline{w}_W$ = density of water, 62.4 lb/ft$^3$

$Z$ = elevational head, ft

$\phi$ = a mathematical function of —; in Eq. (2-121) and Fig. 2-29, $\phi$ = particle-shape factor

$\mu$ = viscosity of fluid, lb/(ft)(sec)

$\epsilon$ = fraction voids, dimensionless, or in Fig. 2-25, roughness factor

## 2-12. Hydrostatics

*Absolute pressure* equals gauge pressure plus atmospheric pressure.

*Pascal's principle.* An increase in the pressure at any point in a fluid is attended by an equal pressure at every other point in the fluid.

*Static pressure* at the base of a vertical column of a fluid of uniform density exceeds that at the top by $H\overline{w}g/g_c$ lb force/ft$^2$. Thus, the pressure equivalent to a head of 18 in. Hg, sp gr 13.6, assuming $g/g_c = 1$, is[1]

$$(18 \times 13.6 \times 62.3)/(12 \times 144) = 8.83 \text{ psi}$$

The absolute pressure and density of the atmosphere at a height $H$ ft above sea level is calculated by the formulas

$$p = p_1(1 - 0.00000687H)^{5.256} \qquad (2\text{-}55)$$
$$\overline{w} = \overline{w}_1(1 - 0.00000687H)^{4.256} \qquad (2\text{-}56)$$

where the subscript 1 refers to sea-level conditions.

**Buoyancy.** *Archimedes' Principle.* The resultant pressure of a fluid on a body immersed in it acts vertically upward through the center of gravity of the displaced fluid and is equal to the weight of the fluid displaced. The upward force is called the *buoyancy.*

---

[1] For nomenclature see p. 2-25.

Specific gravity and volume of solids by immersion

$$\text{Specific gravity} = W_a/(W_a - W_f) \qquad (2\text{-}57)$$

**Pressure Gauges.** *U Tube.* (Fig. 2-13). When the interface between the mercury and the fluid for which the pressure is wanted is $K$ ft below the point of attachment $A$ and gives a reading of $H_m$ ft

$$p_A = (H_m \overline{W}_m - K \overline{W}_A) \qquad (2\text{-}58)$$

At $A$ $\qquad\qquad H_A = H_m \overline{W}_m / \overline{W}_A - K \qquad (2\text{-}59)$

where $\overline{W}_A$ = weight density of fluid at $A$, lb/ft³

$\overline{W}_m$ = density manometric fluid, lb/ft³

$p_A$ = gauge pressure at $A$, psf

**The Differential U Tube.**[1] Figure 2-14 shows the difference between the taps $A$ and $B$ to be

$$p_A - p_B = H_m(\overline{W}_m - \overline{W}_A) + K_A \overline{W}_A - K_B \overline{W}_B \qquad (2\text{-}60)$$

where $K_A$, $K_B$ = vertical distances of upper mercury surface above $A$ and $B$, ft

$\overline{W}_A$, $\overline{W}_B$ = weight densities of fluid at $A$ and $B$, lb/ft³

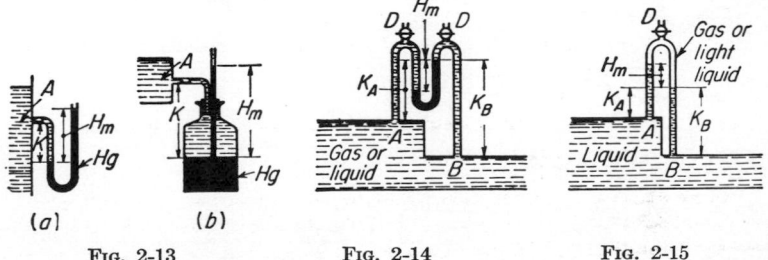

FIG. 2-13          FIG. 2-14          FIG. 2-15

If the differential is that caused by an orifice or other device measuring the flow of a liquid, the orifice differential

$$\Delta H = (p_1 v_1 - p_2 v_2) + (Z_1 - Z_2) = H_m(\overline{W}_m / \overline{W}_A - 1) \qquad (2\text{-}61)$$

For gases, except at very high pressures, $\overline{W}_A$ and $\overline{W}_B$ are so small compared with $\overline{W}_m$ that Eq. (8-135) reduces to

$$p_A - p_B = H_m \overline{W}_m \qquad (2\text{-}62)$$

**The Inverted Differential U Tube** (Fig. 2-15)

$$p_A - p_B = H_m(\overline{W}_A - \overline{W}_m) + K_A \overline{W}_A - K_B \overline{W}_B \qquad (2\text{-}63)$$

If the gauge is indicating the orifice differential of a head meter operating with a liquid,

$$\Delta H = H_m(1 - \overline{W}_m / \overline{W}_A) \qquad (2\text{-}64)$$

[1] For nomenclature see p. 2–25.

**Closed U Tubes.**   These measure directly the absolute pressure $p$ of a fluid (Fig. 2-16).

$$p = H_m \overline{W}_m \qquad (2\text{-}65)$$

where $\overline{W}_m = $ lb mass/ft³ = weight density of manometric fluid
$H_m = $ ft manometric fluid

For liquids and gases under very high pressures, the quantity $K\overline{W}_0$ should be subtracted from Eq. (2-65).

### Multiplying Gauges[1]

**Inclined U Tube** (Fig. 2-18).   If the reading is $R$ (feet), the formula $H_m = (R - R_0) \sin \theta$ is substituted in Eq. (2-60).   $R_0 = $ the zero reading.

**The Draft Gauge.**   Formulas are applied as with an inclined U tube, above (Fig. 2-17).

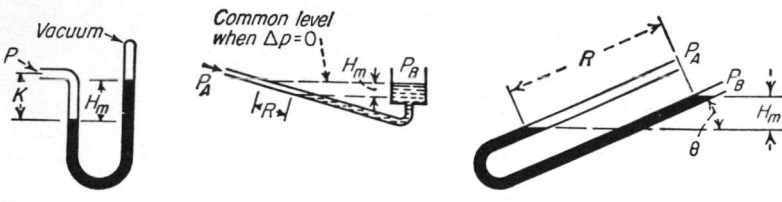

FIG. 2-16            FIG. 2-17                   FIG. 2-18

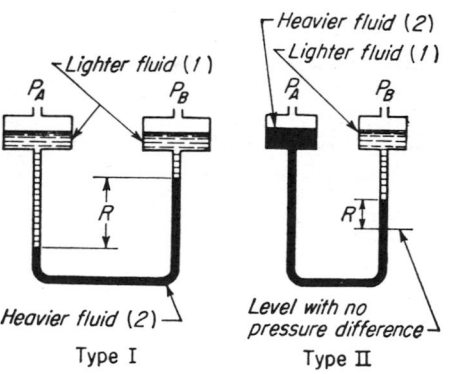

Type I                 Type II

FIG. 2-19

**Two-fluid U Tubes** (Fig. 2-19)

For Type I:

$$p_A - p_B = R - R_0[\overline{W}_2 - \overline{W}_1 + (a/A)\overline{W}_1] \qquad (2\text{-}66)$$

For Type II:

$$p_A - p_B = R[\overline{W}_2 - \overline{W}_1 + (a/A)(\overline{W}_2 + \overline{W}_1)] \qquad (2\text{-}67)$$

[1] For nomenclature see p. 2–25.

where $A$ = cross-sectional area of each reservoir, ft²

$a$ = cross-sectional area of the tube forming the U, ft²

## 2-13. Flow Measurement

**Nozzle and Orifice Flow.**[1]  *Free Spouting Jet Velocities.*  *Expansive Fluid, General.*  The ideal velocity is derived from

$$v_2{}^2/2g_c - v_1{}^2/2g_c = (h_1 - h_2)778 \tag{2-68}$$

If velocity of approach ($v_1$) is zero or negligibly small, then

$$v_2{}^2/2g_c = (h_1 - h_2)778 \tag{2-69}$$

or jet velocity

$$v = 223.7 \sqrt{h_1 - h_2} \tag{2-70}$$

For ideal flow without friction and losses the difference in enthalpy is found by use of a Mollier chart with isentropic expansion between initial and final conditions.  (This is equal to the Rankine cycle work.)  If the path of the expansion is other than isentropic, Eq. (2-70) is equally applicable when the appropriate values of $h_1$ and $h_2$ are used.

*Expansive Fluids, Fixed and Perfect Gases.*  If the gas laws of Boyle and Charles can be applied with sufficient accuracy, then the ideal velocity Eq. (2-68) becomes

$$v = 8.02 \sqrt{[k/(k-1)]p_1\bar{v}_1[1 - (p_2/p_1)^{(k-1)/k}]} \tag{2-71}$$

*Nonexpansive Fluids.*  For a nonexpansive fluid like water,

$$v_2{}^2/2g_c - v_1{}^2/2g_c = (p_1 - p_2)\bar{v} \tag{2-72}$$

If velocity of approach $v_1$ is zero or negligibly small, then Eq. (2-72) becomes

$$v_2{}^2/2g_c = (p_1 - p_2)\bar{v} \tag{2-73}$$

and the jet velocity,

$$v = 8.02 \sqrt{(p_1 - p_2)\bar{v}}$$
$$= 8.02 \sqrt{\Delta H} \tag{2-74}$$

$\Delta H$ is the static head on the orifice measured in feet of fluid flowing.  If manometric fluid is different from the fluid moving through the orifice, then

$$H = H_m x \overline{W}_m / \overline{W}_0 \tag{2-75}$$

Equations (2-72) to (2-75) are applicable to cases of expansive fluids where the density change between initial and final conditions is negligibly small (less than 2 per cent for many engineering calculations).

*Continuity Equation*

$$Q = Av \tag{2-76}$$

or

$$G = Av\bar{w} \tag{2-77}$$

**Quantity Flow through Orifices and Nozzles, Ideal Conditions.**  There are no limitations on the applicability of flow equations to nonexpansive fluids, but with expansive fluids a limitation on quantity is imposed by the sonic

[1] For nomenclature see p. 2–25.

or acoustic barrier. Thus, if the velocity from Eq. (2-70) or (2-71) is used in the continuity equation (2-76) or (2-77), for an expansive fluid, this critically limits the flow, so that the critical ratio $CR$ between initial and final pressures is defined by

$$CR = p_2/p_1 = [2/(k + 1)]^{(k/k-1)} \qquad (2\text{-}78)$$

For critical ratios of some common fluids see Table 2-4. This critical ratio is used in Eqs. (2-70) and (2-71) to determine the sonic velocity. If the over-all expansion ratio is such that the final pressure is lower than that given by Eq. (2-78), then $p_2$ in Eq. (2-71) is made equal to the critical value from Eq. (2-78), and the result is used in the continuity equation (2-76) or

### Table 2-4

| Fluid | CR |
|---|---|
| Air ($k = 1.4$)........ | 0.53 |
| Wet steam.......... | .58 |
| Superheated steam... | .55 |

(2-77), together with the throat area, to determine the volume or weight flow. For a streamlined nozzle with expansion beyond the critical, the weight flow is determined by these values for sonic conditions and constitutes the maximum obtainable. Final pressures, less than the critical, cannot increase the weight flow. The consequences of using these phenomena are the following several equations for flow:

When expansion is beyond the critical ($CR \approx 0.5$) through a 1-in.$^2$ orifice,

Saturated steam:

$$G = p_1/70 \qquad\qquad \text{Napier} \qquad (2\text{-}79)$$
$$G = p_1^{0.97}/60 \qquad\qquad \text{Grashof} \qquad (2\text{-}80)$$
$$G = p_1(16.3 - 0.96 \log p_1)/1{,}000 \quad \text{Rateau} \qquad (2\text{-}81)$$

Superheated steam:

$$G = p_1^{0.97}/60(1 + 0.00065 \times \text{deg superheat}) \qquad (2\text{-}82)$$

Wet steam:

$$G = p_1^{0.97}/60 \sqrt{\text{dryness fraction}} \qquad (2\text{-}83)$$

Air:

$$G = p_1/1.9 \sqrt{T_1} \qquad \text{Fliegner} \qquad (2\text{-}84)$$

### Orifice and Nozzle Coefficients

A nozzle or orifice coefficient, $C_d$, must be used as a multiplier in Eqs. (2-70), (2-71), and (2-74) for real-flow situations. For bellmouth or well-rounded nozzles, $C_d$ is essentially unity. For square or sharp-edged orifices, $C_d$ is given by Fig. 2-20 as a function of the location of the pipe taps and the ratio of orifice to pipe diameter $\beta$. The continuity equations (2-76) and (2-77) give actual volume or weight flow.

**Pitot Tube.** If one side of an ordinary differential manometer is connected to the lead from the impact opening and the other leg to the lead from

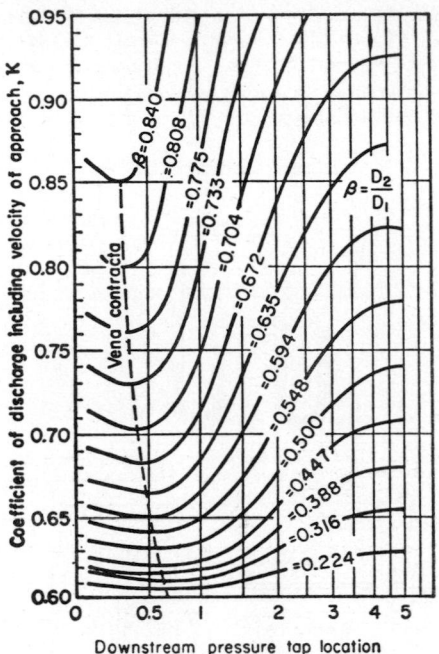

FIG. 2-20. Coefficient of discharge for square-edged circular orifices for $N_{Re} >$ 30,000, with the upstream tap located between one and two pipe diameters from the orifice plate.

the static taps, the manometer will automatically record the velocity pressure $(p_i - p)$ (Figs. 2-21 and 2-22). The local velocity is computed as follows:[1]

1. For liquids or for gases (up to 200 fps)

$$v_0 = C \sqrt{2g_c(p_i - p)/\overline{W}_0} \qquad (2\text{-}85)$$

2. For gases (200 fps to sonic velocity)

$$v_0 = C \sqrt{\frac{2g_c k}{k - 1} \frac{p}{\overline{W}_0} \left[ \left(\frac{p_i}{p}\right)^{\frac{k-1}{k}} - 1 \right]} \qquad (2\text{-}86)$$

where $C$ is a coefficient, dimensionless, which ranges from 0.98 to 1.00 for a well-made pitot tube.

The pitot tube measures only the local velocity, which means that a traverse has to be made if the flow rate or average velocity in a duct is to be determined by this means. For circular pipe, the most common traverse is the

[1] For nomenclature see p. 2–25.

10-point. Readings are taken at the following distances from the wall: $0.026D$, $0.082D$, $0.146D$, $0.226D$, $0.342D$, $0.658D$, $0.774D$, $0.854D$, $0.918D$, and $0.974D$, where $D$ = pipe diameter. An average of the velocities at these points will give a mean velocity which is theoretically only 0.3 per cent high for a normal velocity distribution.

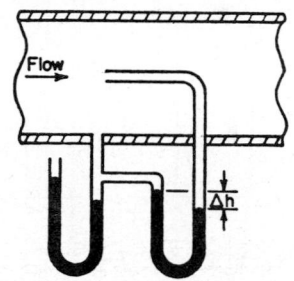

FIG. 2-21. Pitot tube with sidewall static tap.

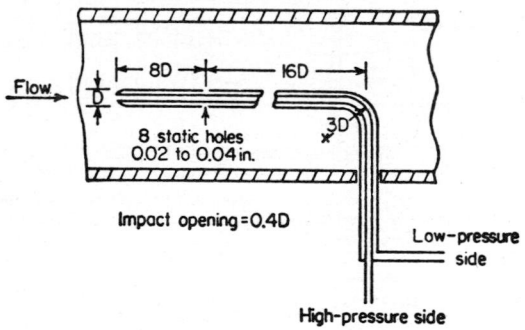

FIG. 2-22. Pitot-static tube.

**Weirs.** *Rectangular Weirs.* The Francis formula for discharge through rectangular weirs takes the following forms:[1]

1. For a suppressed weir, i.e., a weir so designed by rounding the edges that no contraction of the discharging sheet of liquid occurs,

$$Q = 3.33LH^{3/2} \qquad (2\text{-}87)$$

2. For a suppressed weir considering velocity of approach

$$Q = 3.33L[(H + \bar{h})^{3/2} - \bar{h}^{3/2}] \qquad (2\text{-}88)$$

3. For a contracted, i.e., sharp-edged, weir [see (1)],

$$Q = 3.33(L - 0.2H)H^{3/2} \qquad (2\text{-}89)$$

[1] For nomenclature see p. 2-25.

4. For a contracted weir considering velocity of approach

$$Q = 3.33(L - 0.2H)[(H + \bar{h})^{3⁄2} - \bar{h}^{3⁄2}]  \tag{2-90}$$

The Francis formula agrees with experiment within 3 per cent or less, if (1) $L$ is greater than $2H$, (2) height of crest above bottom of channel is at least $3H$, and (3) $H$ is not less than 0.3 ft. Narrow rectangular notches ($H > L$) have been found to give about 93 per cent of the discharge given by the Francis formula. Thus

$$q = 3.10LH^{3⁄2}  \tag{2-91}$$

### Triangular or Notch Weirs

For square-edged notches:  $q = 2.48H^{5⁄2}/\tan \alpha$     (2-92)
For $\theta = 90°$, $\alpha = 45°$:  $q = 2.48H^{5⁄2}$     (2-93)
For $\theta = 60°$, $\alpha = 60°$:  $q = 1.43H^{5⁄2}$     (2-94)

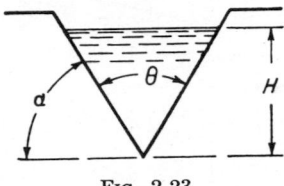

FIG. 2-23

**Flow Resistance.** The theoretical force of a jet is[1]

$$\bar{F} = (\bar{w}/g_c)av^2  \tag{2-95}$$

The actual force, resistance, or drag, in pounds is

$$\bar{F} = C_D av^2 \bar{w}/g_c$$

$$= \bar{W} \frac{av^2}{g_c} \phi_1 \left( \frac{Dv\bar{w}}{\mu} \right) \phi_2 \frac{v}{\sqrt{Dg}} \phi_3 \frac{v}{v_c}  \tag{2-96}$$

where $Dv\bar{w}/\mu$ = Reynolds number     (2-97)
    $v/\sqrt{Dg}$ = Froude number     (2-98)
    $v/v_c$ = Mach number     (2-99)

Drag coefficient is a function of each of the flow criteria given in Eqs. (2-97), (2-98), and (2-99). These criteria are usually dimensionless for maximum convenience and utility. The Reynolds number is significant in cases of full immersion or completely enclosed flow, as with aircraft wings, pipes, nozzles, pumps, fans. The Froude number is significant in cases of simultaneous motion through two fluids where there are a surface of discontinuity, gravity forces, and wave-making effects, as with ship's hulls. The Mach number is significant in supersonics, as with projectiles and jet propulsion.

**Rotameters.** The rotameter, an example of which is shown in Fig. 2-24, has become one of the most popular flowmeters in industry. It consists

[1] For nomenclature see p. 2–25.

essentially of a plummet, or "float," which is free to move up or down in a vertical, slightly tapered tube having its small end down. The fluid enters the lower end of the tube and causes the float to rise until the annular area between the float and the wall of the tube is such that the pressure drop across this constriction is just sufficient to support the float. Typically, the tapered tube is of glass and carries etched upon it a nearly linear scale on which the position of the float may be visually noted as an indication of the flow.

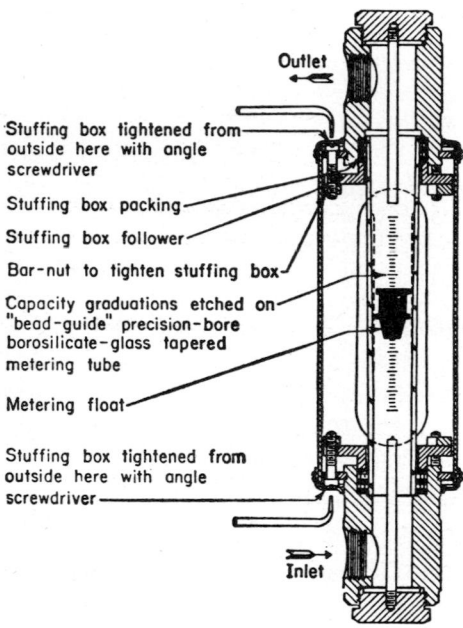

Outlet

Stuffing box tightened from outside here with angle screwdriver

Stuffing box packing

Stuffing box follower

Bar-nut to tighten stuffing box

Capacity graduations etched on "bead-guide" precision-bore borosilicate-glass tapered metering tube

Metering float

Stuffing box tightened from outside here with angle screwdriver

Inlet

Fig. 2-24. Rotameter.

Interchangeable precision-bore glass tubes and metal metering tubes are available. Rotameters have proved satisfactory both for gases and for liquids at high and at low pressures. A single instrument can readily cover a tenfold range of flow, and by providing "floats" of different densities, a 200-fold range is practicable.

Rotameters require no straight runs of pipe before or after the point of installation. Pressure losses are substantially constant over the whole flow range. However, most modern rotameters are precision-made, so that their performance closely corresponds to a master calibration plot for the type in question. Such a plot is supplied with the meter upon purchase.

Flow rate through a rotameter can be obtained from

$$G = Q\overline{W}_0 = KD_f \sqrt{\frac{W_f(\rho_f - \rho)\rho}{\rho_f}} \qquad (2\text{-}100)$$

where $K$ = flow parameter, $\text{ft}^{1/2}/\text{sec}$

$D_f$ = float diameter at constriction, ft

$W_f$ = float weight, lb

$\rho_f$ = float density, $\text{lb/ft}^3$

and other terms are as defined on page 2-25. The appropriate value of $K$ is obtained from a composite correlation of $K$ versus the parameters corresponding to the float shape being used.

The ratio of flow rates for two different fluids $A$ and $B$ at the same rotameter reading is given by

$$\frac{w_A}{w_B} = \frac{K_A}{K_B} \sqrt{\frac{(\rho_f - \rho_A)\rho_A}{(\rho_f - \rho_B)\rho_B}} \qquad (2\text{-}101)$$

A measure of self-compensation, with respect to weight rate of flow, for fluid-density changes can be introduced through the use of a float with a density twice that of the fluid being metered, in which case an increase of 10 per cent in $\rho$ will produce a decrease of only 0.5 per cent in $w$ for the same reading. The extent of immunity to changes in fluid viscosity depends upon the shape of the float.

## 2-14. Fluid Dynamics

**Bernoulli's Equation.**[1] Applied to flow of a noncompressible fluid, such as water, where $\bar{v}_1 = \bar{v}_2$, through a system, Bernoulli's equation is as follows:

$$\frac{Z_1 g}{g_c} + \frac{v_1{}^2}{2g_c} + p_1\bar{v}_1 = \frac{Z_2 g}{g_c} + \frac{v_2{}^2}{2g_c} + \Delta H_f + p_2\bar{v}_2 \qquad (2\text{-}102)$$

For compressible fluids the quantity $(p_2 - p_1)\bar{v}$ must be evaluated as an integral, $\int_1^2 \bar{v} \, dp$, which is dependent on the nature of the path followed as the fluid flows through the system. For isothermal flow

$$\int_1^2 \bar{v} \, dp = -\frac{RT}{M} \ln \frac{\bar{v}_2}{\bar{v}_1} = -\frac{RT}{M} \ln \frac{p_1}{p_2} \qquad (2\text{-}103)$$

**Flow of Fluids in Pipes.** Pressure drop due to friction $\Delta H_f$ in circular pipes of diameter $D$ and length $L$ is given by

$$\Delta H_f = fLv^2/2g_c D \qquad (2\text{-}104)$$

Values of $f$ may be obtained from Fig. 2-25. The value of $\epsilon$ to be used in calculating the relative roughness is obtained from Table 2-5.

[1] For nomenclature see page 2–25.

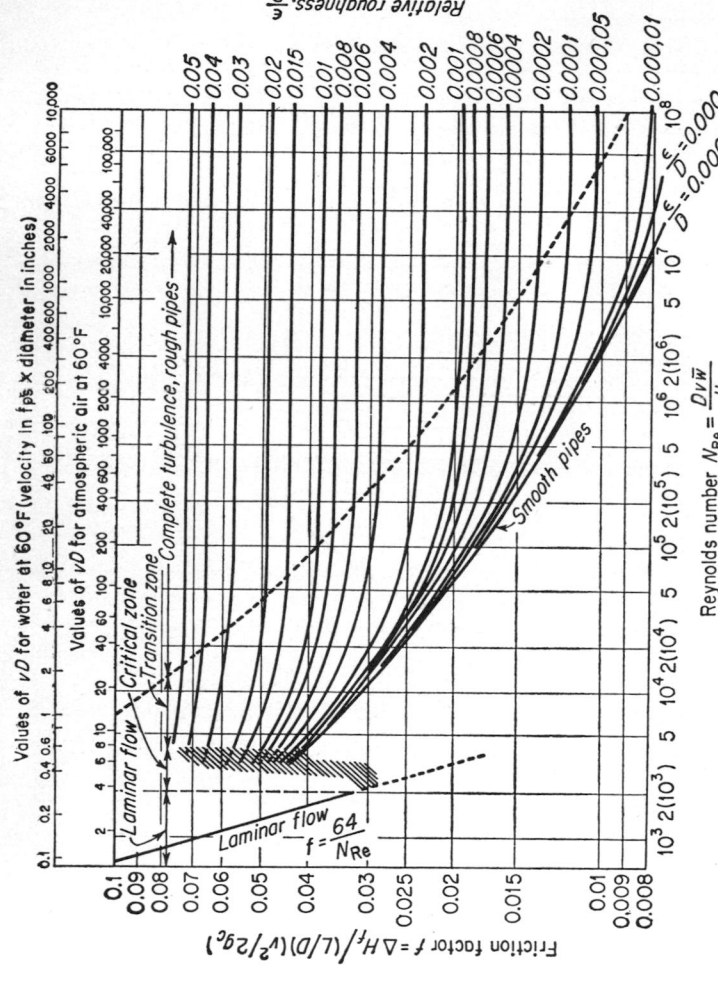

Fig. 2-25. Source: T. Baumeister (ed.), "Standard Handbook for Mechanical Engineers," 7th ed., p. 3-60, McGraw-Hill Book Company, New York, 1967.

2–36

## Table 2-5. Absolute Roughness Classification of Pipe Surfaces for Selection of Friction Factor $f$ in Fig. 2-25

| Commercial pipe surf. (new) | Abs roughness $\epsilon$, ft | Commercial pipe surf. (new) | Abs roughness $\epsilon$, ft |
|---|---|---|---|
| Glass, drawn brass, copper, lead... | Smooth | Cast iron.................. | 0.00085 |
| Wrought iron, steel.............. | 0.00015 | Wood stave................ | 0.0006–0.003 |
| Asphalted cast iron.............. | .0004 | Concrete.................. | 0.001–0.01 |
| Galvanized iron................. | .0005 | Riveted steel.............. | 0.003–0.03 |

SOURCE: T. Baumeister (ed.), "Standard Handbook for Mechanical Engineers," 7th ed., p. 3–59, McGraw-Hill Book Company, New York, 1967.

A convenient nomograph, Fig. 2-26, may be used to solve pipe flow-friction-loss problems. Table 2-6 is used to obtain the necessary coordinates for use of this nomograph.

## Table 2-6. Coordinates for Liquids and Aqueous Solutions

| | X | Y | | X | Y |
|---|---|---|---|---|---|
| Acetaldehyde................... | −0.3 | 3.7 | Formic acid.............. | 1.5 | 4.5 |
| Acetic acid, 100%.............. | 1.0 | 4.0 | Glycerol, 100%........... | 6.9 | 1.8 |
| Acetic acid, 77%............... | 2.6 | 3.8 | Glycerol, 50%........... | 3.0 | 3.7 |
| Acetic anhydride............... | 0.7 | 4.3 | Hydrochloric acid, 31.5%....... | 1.1 | 4.2 |
| Acetone, 100%................. | 0.9 | 3.4 | Linseed oil, raw........... | 3.4 | 1.8 |
| Acetone, 35%.................. | 2.7 | 3.7 | Mercury.................. | See chart | |
| Ammonia, anhydrous........... | 0.9 | 3.6 | Methanol, 100%........... | 0.8 | 3.3 |
| Ammonia, 26%................. | 1.9 | 3.6 | Methanol, 40%........... | 2.8 | 3.6 |
| Aniline....................... | 2.5 | 3.4 | Methyl acetate........... | 0.0 | 4.2 |
| Benzene...................... | 0.6 | 3.6 | Methyl chloride........... | −0.8 | 4.3 |
| Butanol...................... | 2.6 | 2.6 | Nitric acid, 95%........... | 0.8 | 5.8 |
| Calcium chloride brine, 25%...... | 2.6 | 4.2 | Nitric acid, 60%........... | 1.5 | 4.8 |
| Carbon disulfide............... | 0.0 | 5.6 | Nitrobenzene............. | 1.7 | 4.4 |
| Carbon tetrachloride........... | 0.7 | 6.0 | Octane................... | 0.4 | 2.7 |
| Chloroform................... | 0.0 | 6.0 | Phenol................... | 2.4 | 3.4 |
| Chlorosulfonic acid............. | 1.5 | 5.8 | Propionic acid............ | 0.6 | 3.8 |
| Cyclohexanol................. | 5.3 | 2.2 | Sodium chloride brine, 25%..... | 2.1 | 4.4 |
| Diphenyl..................... | 0.0 | 3.5 | Sodium hydroxide, 50%........ | 5.3 | 3.7 |
| Ethyl acetate................. | 0.2 | 3.9 | Sulfur dioxide............. | −0.2 | 6.1 |
| Ethyl alcohol, 95%............. | 1.9 | 3.0 | Sulfuric acid, 110%........... | 3.7 | 4.7 |
| Ethyl alcohol, 45%............. | 3.6 | 3.4 | Sulfuric acid, 98%........... | 3.5 | 4.8 |
| Ethyl chloride................. | 0.2 | 4.3 | Sulfuric acid, 78%........... | 3.2 | 4.8 |
| Ethyl ether................... | −0.3 | 3.2 | Tetrachloroethylene.......... | 0.3 | 6.2 |
| Ethylene glycol............... | 3.5 | 2.9 | Toluene.................. | 0.4 | 3.6 |
| Fluorocarbon F-11............. | 0.0 | 6.2 | Trichloroethylene......... | 0.1 | 5.9 |
| Fluorocarbon F-12............. | −1.2 | 5.9 | Turpentine............... | 1.1 | 3.1 |
| Fluorocarbon F-21............. | −0.4 | 5.9 | Vinyl acetate............. | 0.4 | 4.2 |
| Fluorocarbon F-22............. | −1.7 | 5.5 | Water................... | 2.0 | 4.2 |
| Fluorocarbon F-113............ | 0.9 | 6.2 | | | |

The economic pipe diameter for Schedule 40 steel pipe may be found using Fig. 2-27. For pressure drop with noncircular pipe or duct use Eq. (2-104), incorporating the mean hydraulic radius $m$ defined as

$$m = \frac{\text{cross-sectional area of pipe, ft}^2}{\text{wetted perimeter of pipe, ft}} \qquad (2\text{-}105)$$

in place of diameter $D$.

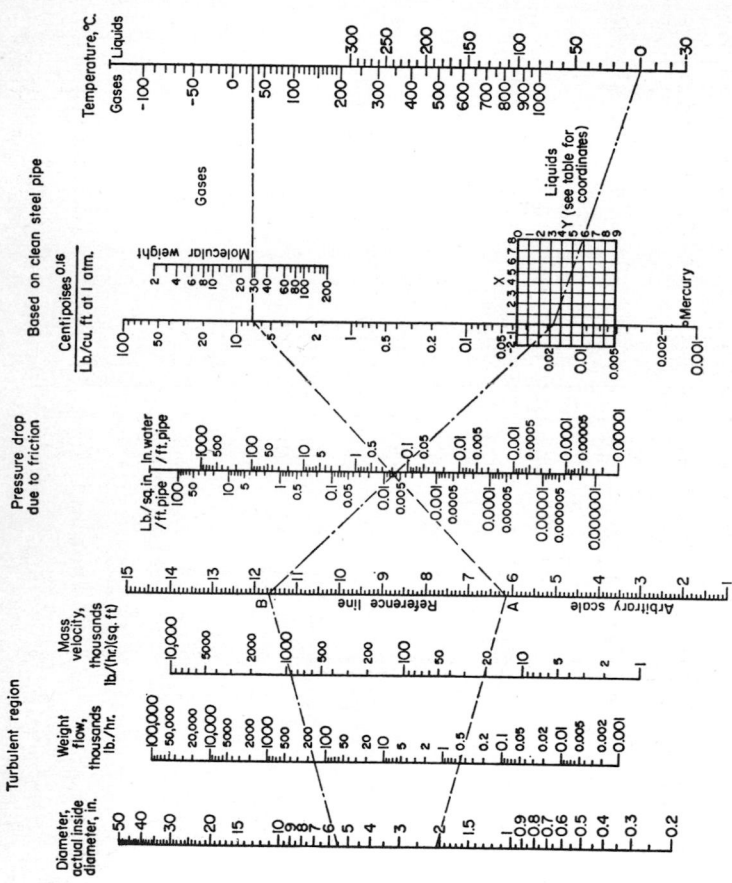

Fig. 2-26. Pipe-flow chart. For gases, multiply calculated pressure drop by the absolute pressure of the gas in atmospheres. See Table 2-6 for coordinates. [*Perry and Chilton (eds.), "Chemical Engineers' Handbook," 5th ed., McGraw-Hill Book Company, New York, 1973.*]

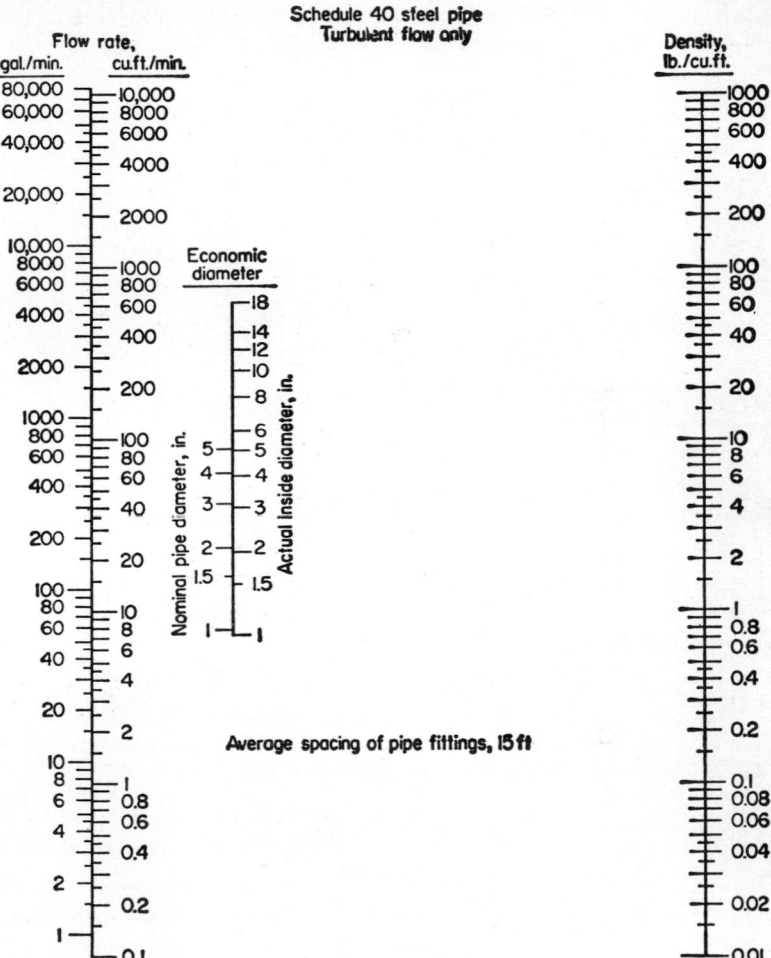

FIG. 2-27. Economic pipe diameter. Connect values of flow rate and density to obtain economic diameter. [*Perry and Chilton* (eds.), *"Chemical Engineers' Handbook,"* 5th ed., *McGraw-Hill Book Company, New York,* 1973.]

For circular pipes $$m = D/4 \qquad (2\text{-}106)$$

whence, substituting in Eq. (2-104)

$$\Delta H_f = fLv^2/8mg_c \qquad (2\text{-}107)$$

*Friction Loss through Fittings.* The friction loss due to screwed fittings and valves is given in Table 2-7 in terms of equivalent length in number of

## Table 2-7. Friction Loss of Screwed Fittings, Valves, Etc.

| Fitting | Equiv length in pipe diam, $L$ | No. "velocity heads," $K$ |
|---|---|---|
| 45° elbows........................... | 15 | 0.3 |
| 90° elbows, standard radius............ | 32 | 0.74 |
| 90° elbows, medium radius............. | 26 | 0.60 |
| 90° elbows, long sweep................ | 20 | 0.46 |
| 90° square elbow..................... | 60 | 1.3 |
| 180° close return bends............... | 75 | 1.7 |
| 180° medium-radius return bends....... | 50 | 1.2 |
| Tee (used as elbow, entering run)....... | 60 | 1.3 |
| Tee (used as elbow, entering branch).... | 90 | 1.9 |
| Couplings........................... | Negligible | |
| Unions.............................. | Negligible | |
| Gate valves, open..................... | 7 | 0.13 |
| Globe valves, open.................... | 300 | 6.0 |
| Angle valves, open ................... | 170 | 3.0 |
| Water meters, disk.................... | 400 | 8.0 |
| Water meters, piston.................. | 600 | 12.0 |
| Water meters, impulse wheel........... | 300 | 6.0 |

pipe diameters and in terms of number of velocity heads. These values are used in conjunction with Eq. (2-104) or Eq. (2-108).

$$\Delta H_f = Kv^2/2g_c \tag{2-108}$$

**Miscellaneous Pressure Losses.** *Expansion Loss.* For the sudden expansion from an area $S_1$ to an area $S_2$ with an average linear velocity from $v_1$ to $v_2$, the frictional loss of mechanical energy is[1]

$$\Delta H_f = (v_1 - v_2)^2/2g_c = (v_1{}^2/2g_c)(1 - S_1/S_2)^2 \tag{2-109}$$

This formula is exact for liquids in turbulent motion and is generally used for gases at moderate velocities. For viscous flow in the smaller pipe this formula result should be doubled. For discharge into a large tank,

$$\Delta H_f = v_1{}^2/2g_c \tag{2-110}$$

For a uniformly divergent duct, for an angle of divergence of $\alpha = 7.5$ to $35°$,

$$\Delta H_f = 3.50[\tan (\alpha/2)]^{1.22}[(v_1 - v_2)^2/2g_c] \tag{2-111}$$

*Contraction Loss.* For turbulent flow in the smaller pipe, at a sharp-edged entrance to a pipeline, there is a sudden cross-sectional reduction

$$\Delta H_f = K'v_2{}^2/2g_c$$

where $v_2$ = linear velocity in smaller pipe, fps

$$K' = 0.4(1.25 - S_2/S_1) \qquad \text{for } S_2/S_1 < 0.715$$
$$= 0.75(1 - S_2/S_1) \qquad \text{for } S_2/S_1 > 0.715$$

**Flow of Water in Closed Conduits.** For determining the head loss due to friction for the flow of *water under pressure in conduits*, the Hazen-Williams formula may be used

[1] For nomenclature see p. 2–25.

$$v = 1.318CR^{0.63}S^{0.54} \qquad (2\text{-}112)$$
$$h_f = Sl \qquad (2\text{-}113)$$

where $h_f$ = head loss due to friction, ft-lb force/lb mass

$S$ = hydraulic gradient, ft head loss/ft length

$l$ = length of conduit, ft

$v$ = average velocity of flow, fps

$R$ = hydraulic radius (equal to cross-sectional area of conduit divided by perimeter), ft

$C$ = Hazen-Williams coefficient

Values of $C$ commonly used for design purposes for various types of conduit material are given in Table 2-8.

### Table 2-8. Values of $C$ in Hazen-Williams Formula

| Type of pipe | $C^*$ | Type of pipe | $C^*$ |
|---|---|---|---|
| Cement-asbestos............. | 140 | Cast iron or wrought iron..... | 100 |
| Asphalt-lined iron or steel...... | 140 | Welded or seamless steel...... | 100 |
| Copper or brass.............. | 130 | Concrete.................... | 100 |
| Lead, tin, or glass............ | 130 | Corrugated steel............. | 60 |
| Wood stave.................. | 110 | | |

* Values of $C$ commonly used for design. The value of $C$ for pipes made of corrosive materials decreases as the age of the pipe increases; the values given are those that apply at an age of 15 to 20 years. For example, the value of $C$ for cast-iron pipes 30 in. in diameter or greater at various ages is approximately as follows: new, 130; 5 years old, 120; 10 years old, 115; 20 years old, 100; 30 years old, 90; 40 years old, 80; and 50 years old, 75. The value of $C$ for smaller-size pipes decreases at a more rapid rate.

For circular conduits, the Hazen-Williams formula may be written as

$$Q = 0.4322CD^{2.63}S^{0.54} \qquad (2\text{-}114)$$

where $Q$ = quantity of flow, cfs

$D$ = diameter of circular conduit, ft

Figure 2-28 is a nomograph for the solution of Eq. (2-112).

**Flow through Beds of Solids.** *Granular Solids.* For the flow of a single fluid through a bed of uniform solid granular particles, the pressure drop is given by[1]

$$\Delta p_f = fL\bar{w}v_0^2 A_f/2g_cD_p \qquad (2\text{-}115)$$

and $\qquad N_{Re} = D_pv_0\bar{w}/\mu \qquad (2\text{-}116)$

where $\qquad f = 3,400/N_{Re} \qquad$ for $N_{Re} < 40$ (viscous) $\qquad (2\text{-}117)$

$\qquad f = 152/N_{Re}^{0.15} \qquad$ for $N_{Re} > 40$ (turbulent) $\qquad (2\text{-}118)$

For values of $A_f$ see Table 2-9.

*Tower Packings.* For solid packings the equations given under granular solids can be used as an approximation if the void content falls between 35 to 45 per cent. If estimates can be made of the percentage of voids and

[1] For nomenclature see p. 2-25.

Equation: $v = 1.318\ CR^{0.63}\ S^{0.54}$

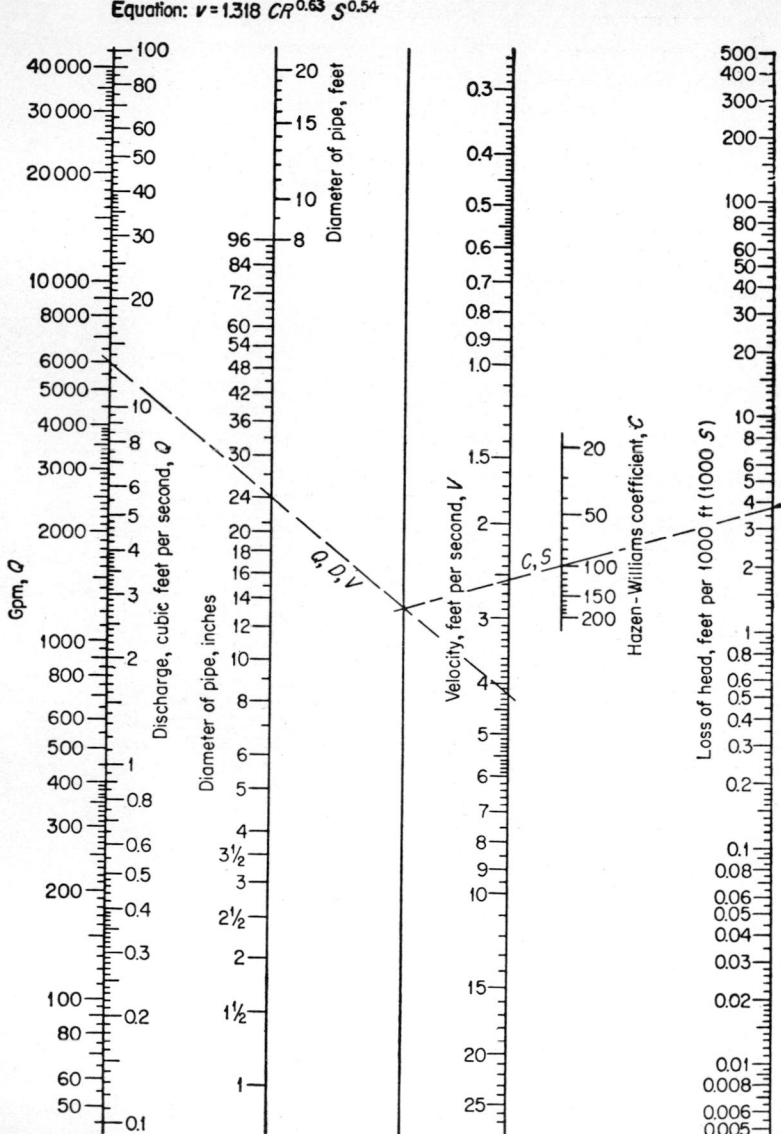

FIG. 2-28.  Nomograph for solution of Hazen-Williams equation for circular conduits flowing full of water.

### Table 2-9

| $D_p/D_t$ | $A_f$ | | $D_p/D_t$ | $A_f$ | |
|---|---|---|---|---|---|
| | Viscous | Turbulent | | Viscous | Turbulent |
| 0.0 | 1.0 | 1.0 | 0.15 | 0.77 | 0.65 |
| .05 | 0.90 | 0.84 | .25 | .74 | .57 |
| .10 | .83 | .72 | | | |

specific surface, the following equations can be used for both solid and hollow packings.

$$\Delta p_f = f_c L \overline{S} \overline{w} v_0^2 / g_c \tag{2-119}$$

and
$$N_{Re} = \overline{w} v_0 / \mu \overline{S} \tag{2-120}$$

The value of $f_c$ may be obtained from Fig. 2-29.

$$\overline{S} = 6(1 - \epsilon)/\phi D_p \tag{2-121}$$

where $\phi$, the particle-shape factor, equals 1 for spheres and 0.3 for berl saddles.

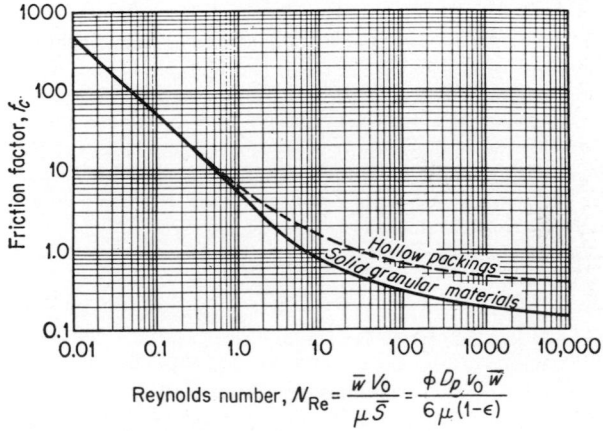

FIG. 2-29. Friction factors for flow through tower packings as a function of Reynolds number.

**Flow across Tube Banks.** *Viscous Flow.* The pressure drop across a bank of tubes under viscous-flow conditions[1] ($D_e v_{max} \overline{w}/\mu < 150$) is given by

$$\Delta p_f = 53 \mu \overline{L} v_{max} / g_c D_e^2 \tag{2-122}$$

where
$$D_e = (4\bar{a}b - \pi D_t^2)/\pi D_t \tag{2-123}$$

These equations apply primarily to banks of tubes in equilateral staggered

[1] For nomenclature see p. 2–25.

arrangement.   For staggered-square arrangement multiply the above pressure drop by 2; for an in-line square arrangement, by 1.5.

**Turbulent Flow.**   The pressure drop across a bank of tubes under turbulent-flow conditions $(D_c v_{max} \bar{w} / \mu > 40)$ can be estimated from[1]

$$\Delta p_f = f N \bar{w} v^2_{max} / 2g_c \tag{2-124}$$

where
$$f = 3.0 (D_c v_{max} \bar{w} / \mu)^{-0.2} \tag{2-125}$$

These equations apply primarily to banks of tubes in staggered equilateral arrangement.   They will give conservative answers for in-line arrangements.

**Draft in Stacks and Chimneys.**[2]   The theoretical draft caused by a stack of height $H$, in feet, is at the base of the vertical column[1]

$$h_w = (H/5.2)(\bar{w}_{CA} - \bar{w}_{HG}) \tag{2-126}$$

The actual, net, or effective draft which is available for overcoming the resistance of any connected load such as a furnace, boiler, still, or heater, is

$$h'_w = \frac{H}{5.2} (\bar{w}_{CA} - \bar{w}_{HG}) - \frac{1}{F} \frac{v^2}{2g_c} - \frac{\text{friction losses}}{F} \tag{2-127}$$

where $F$ = conversion factor from feet of fluid to inches of water =

$$(1/12)(\bar{w}_w / \bar{w}_{HG}) \tag{2-128}$$

Friction losses are given in feet of fluid flowing and must include allowance for pressure drop in straight run of stack and flue, plus losses due to valves, fittings, and changes of section (Table 2-7).

**Flow in Open Channels and Nonpressure Flow in Closed Conduits.**   For determining the head loss due to friction in open channels and for the *nonpressure flow* of water in noncircular conduits or for partial flow in all types of conduits, Manning's formula is generally used.   For use *with circular conduits flowing full*, the formula may be written as

$$h_f = C_f(l/D)(v^2/2g_c) \tag{2-129}$$
$$C_f = 185 n^2 / D^{1/3} \tag{2-130}$$

where $h_f$ = head loss due to friction, ft-lb force/lb mass
$D$ = diameter of circular conduit, ft (*Note:* Hydraulic radius $R$ for a circular conduit running full = $D/4$)
$n$ = coefficient of roughness for closed conduits (Table 2-10)
$l$ = length of conduit, ft
$v$ = average velocity of flow, fps
$g_c$ = 32.2 (lb mass)(ft)/(lb force)(sec²)

For long conduits where head losses other than the friction head loss are negligible, the Manning formula can be written as

$$Q = (0.4632/n) D^{8/3} S^{1/2} = \text{conveyance factor} \times S^{1/2} \tag{2-131}$$

[1] For nomenclature see p. 2–25.
[2] See also Sect. 4.

where $Q$ = quantity of flow, cfs

$S = h_f/l$ = hydraulic gradient, ft head loss/ft length

## Table 2-10. Roughness Coefficients (Manning's $n$) for Closed Conduits

| Type of conduit | | | Manning's $n$ | |
|---|---|---|---|---|
| | | | Good construction[*] | Fair construction[*] |
| Concrete pipe.................................. | | | 0.013 | 0.015 |
| Corrugated metal pipe or pipe arch, 2⅔- by ½-in. corrugation, riveted: | | | | |
| Plain...................................... | | | .024 | |
| Paved invert: | | | | |
| Per cent of circumference paved........ | 25 | 50 | | |
| Depth of flow: | | | | |
| Full.............................. | 0.021 | 0.018 | | |
| 0.8$D$............................. | .021 | .016 | | |
| 0.6$D$............................. | .019 | .013 | | |
| Vitrified clay pipe.......................... | | | .012 | .014 |
| Cast-iron pipe, uncoated.................... | | | .013 | |
| Steel pipe.................................. | | | .011 | |
| Brick...................................... | | | .014 | .017 |
| Monolithic concrete: | | | | |
| Wood forms, rough........................ | | | .015 | .017 |
| Wood forms, smooth....................... | | | .012 | .014 |
| Steel forms.............................. | | | .012 | .013 |
| Cemented-rubble masonry walls: | | | | |
| Concrete floor and top.................... | | | .017 | .022 |
| Natural floor............................ | | | .019 | .025 |
| Laminated treated wood.................... | | | .015 | .017 |
| Vitrified-clay liner plates................. | | | .015 | |

[*] For poor-quality construction use larger values of $n$.

The head loss due to friction for the flow of water *in an open channel* (*including a closed conduit not flowing full*) is

$$h_f = C_f(l/R)(v^2/2a_c) \qquad (2\text{-}132)$$

The coefficient $C_f$ is given by the Manning formula as

$$C_f = 29.15n^2/R^{1/3} \qquad (2\text{-}133)$$

where $R = A/P$

$n$ = coefficient of roughness (Table 2-12), dimensionless

$R$ = hydraulic radius, ft

$A$ = cross-sectional area of flow, ft²

$P$ = wetted perimeter, ft

$h_f$ = head loss due to friction, ft-lb force/lb mass

$l$ = length of channel or conduit, ft

$v$ = average velocity of flow, fps

$g_c$ = 32.2 (lb mass)(ft)/(lb force)(sec²)

Table 2-12 is taken from "Hydraulic Charts," prepared by the U.S. Bureau of Public Roads.

Equation: $v = \dfrac{1.486}{n} R^{2/3} S^{1/2}$

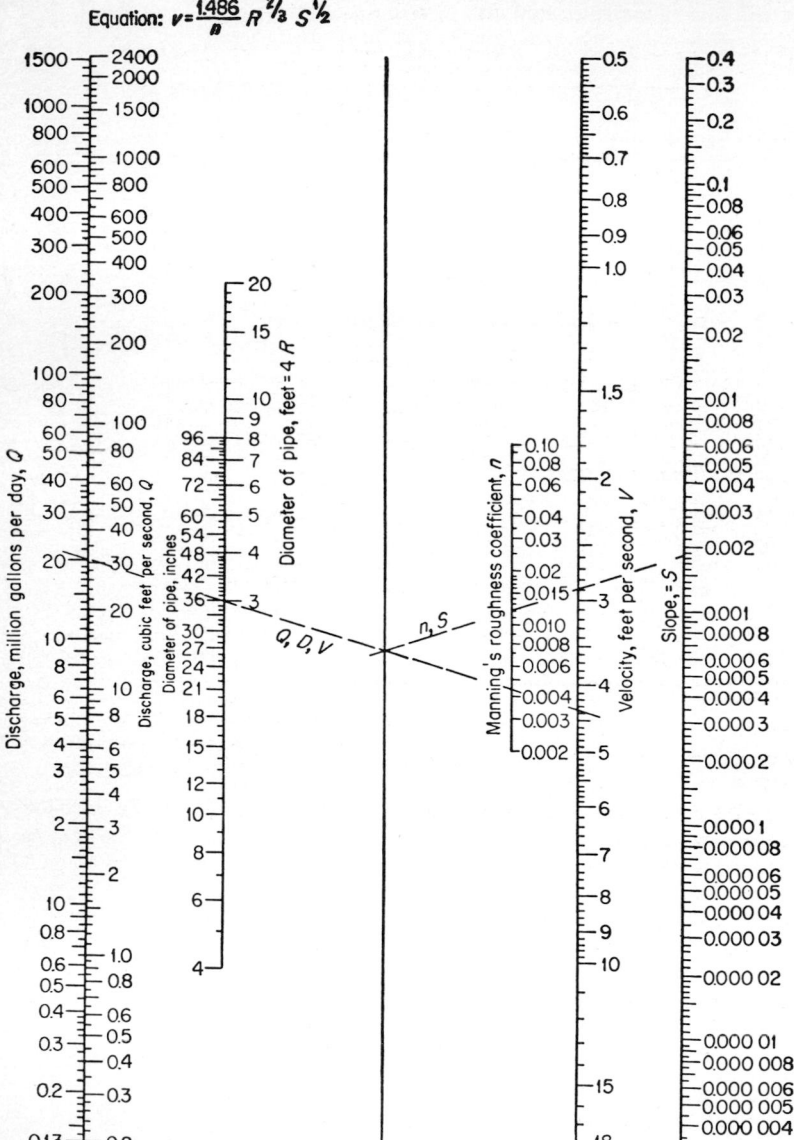

FIG. 2-30. Nomograph for solution of Manning equation for circular conduits for nonpressure flow of water.

## Table 2-11. Conveyance Factors

To obtain $Q$ for any diameter pipe, multiply the figure shown under the proper value of $n$ by the square root of the hydraulic gradient expressed in feet of head loss per foot of length.

| Diam, in. | Area, ft | Conveyance factors | | | | | |
|---|---|---|---|---|---|---|---|
| | | $n = 0.011$ | $n = 0.013$ | $n = 0.015$ | $n = 0.017$ | $n = 0.019$ | $n = 0.021$ |
| 6 | 0.196 | 6.62 | 5.60 | 4.85 | 4.28 | 3.83 | 3.47 |
| 8 | 0.349 | 14.32 | 12.12 | 10.50 | 9.27 | 8.29 | 7.50 |
| 10 | 0.545 | 25.80 | 21.83 | 18.92 | 16.70 | 14.94 | 13.52 |
| 12 | 0.785 | 42.15 | 35.66 | 30.91 | 27.27 | 24.40 | 22.08 |
| 15 | 1.227 | 76.46 | 64.70 | 56.07 | 49.48 | 44.27 | 40.05 |
| 18 | 1.767 | 124.2 | 105.1 | 91.04 | 80.33 | 71.88 | 65.03 |
| 21 | 2.405 | 187.1 | 158.3 | 137.2 | 121.1 | 108.3 | 98.01 |
| 24 | 3.142 | 267.4 | 226.2 | 196.1 | 173.0 | 154.8 | 140.0 |
| 27 | 3.976 | 365.8 | 309.6 | 268.3 | 236.7 | 211.8 | 191.6 |
| 30 | 4.909 | 484.7 | 410.1 | 355.5 | 313.6 | 280.6 | 253.9 |
| 36 | 7.069 | 788 | 667 | 578 | 510 | 456.1 | 412.7 |
| 42 | 9.621 | 1,189 | 1,006 | 872 | 770 | 688 | 623 |
| 48 | 12.566 | 1,698 | 1,436 | 1,245 | 1,098 | 983 | 889 |
| 54 | 15.904 | 2,325 | 1,967 | 1,705 | 1,504 | 1,346 | 1,218 |
| 60 | 19.635 | 3,077 | 2,604 | 2,256 | 1,991 | 1,781 | 1,612 |
| 66 | 23.758 | 3,967 | 3,357 | 2,909 | 2,567 | 2,297 | 2,078 |
| 72 | 28.274 | 5,004 | 4,234 | 3,669 | 3,238 | 2,897 | 2,621 |
| 84 | 38.495 | 7,550 | 6,390 | 5,540 | 4,885 | 4,370 | 3,954 |
| 96 | 50.27 | 10,780 | 9,120 | 7,900 | 6,970 | 6,240 | 5,640 |
| 108 | 63.62 | 14,760 | 12,490 | 10,820 | 9,550 | 8,540 | 7,730 |
| 120 | 78.54 | 19,540 | 16,540 | 14,330 | 12,650 | 11,320 | 10,240 |

For long channels where head losses other than the friction head loss are negligible, the Manning formula can be written as:

$$v = (1.486/n)R^{\frac{2}{3}}S^{\frac{1}{2}} \qquad (2\text{-}134)$$

where $S = h_f/l =$ hydraulic gradient, ft head loss/ft length
Figure 2-31 is a nomograph for the solution of Eq. (2-134).

**Head Loss at Bend in Channel Alignment.** The head loss due to turbulence at a bend in channel alignment is generally taken into account by an increase in the value of the coefficient of roughness $n$, depending in amount on the degree of curvature (Table 2-12).

**Critical Depth.** The critical depth for a given flow of water in a channel is the depth for which the total head (total head equals depth plus velocity head) is a minimum; for a given total head, the discharge is therefore a maximum at the critical depth. For a trapezoidal channel cross section, the critical depth is

$$D_c = 4BH_t/(5B + b) \qquad (2\text{-}135)$$

where $D_c =$ critical depth, ft
$H_t =$ total head $= d + v^2/2g$, where $d$ is the measured depth, ft
$B =$ width of channel at water surface, ft
$b =$ width of channel at bottom, ft
For a rectangular channel, the maximum discharge is

Equation: $V = \dfrac{1.486}{n} R^{2/3} S^{1/2}$

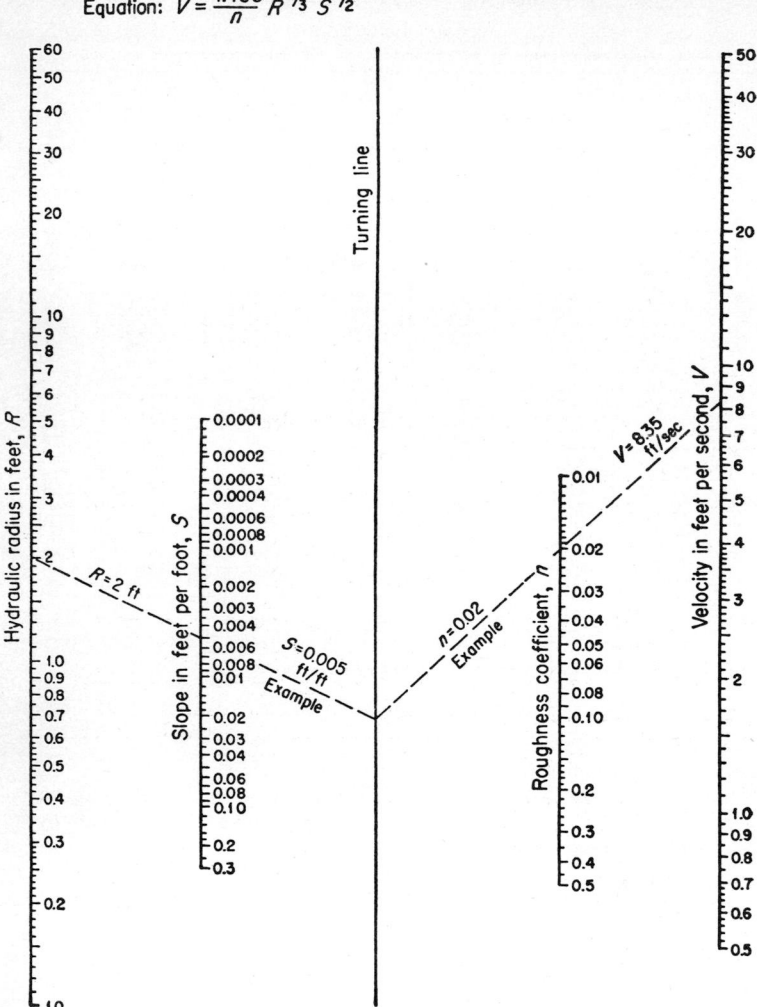

FIG. 2-31. Nomograph for solution of Manning equation [Eq. (2-134)] for flow of water in open channels.

## Table 2-12. Roughness Coefficients (Manning's $n$) for Open Channels

| Channel description | Manning's $n$ | |
| --- | --- | --- |
| | Good construction[a] | Fair construction[a] |
| **Open Channels, Nonvegetated Lining, Straight Alignment[b]** | | |
| Concrete, with all surfaces: | | |
|   Formed, no finish | 0.013 | 0.017 |
|   Trowel finish | .012 | .014 |
|   Float finish | .013 | .015 |
|     Some gravel on bottom | .015 | .017 |
|   Gunite, good section | .016 | .019 |
|   Gunite, wavy section | .018 | .022 |
| Concrete bottom, float-finished, sides of: | | |
|   Dressed stone in mortar | .015 | .017 |
|   Random stone in mortar | .017 | .020 |
|   Cement rubble masonry | .020 | .025 |
|   Plastered | .016 | .020 |
|   Dry rubble (riprap) | .020 | .030 |
| Gravel bottom, sides of: | | |
|   Formed concrete | .017 | .020 |
|   Random stone in mortar | .020 | .023 |
|   Dry rubble (riprap) | .023 | .033 |
| Brick | .014 | .017 |
| Asphalt: | | |
|   Smooth | .013 | |
|   Rough | .016 | |
| Planed wood, clean | .011 | .013 |
| Concrete-lined excavated rock: | | |
|   Good section | .017 | .020 |
|   Irregular section | .022 | .027 |
| Flumes (steep slope)[b] | | |
| **Open Channels, Excavated, Straight Alignment[c], Natural Lining** | | |
| Earth, uniform section (best): | | |
|   Clean, recently completed | 0.016 | 0.018 |
|   Clean, after weathering | .018 | .020 |
|   With short grass, few weeds | .022 | .027 |
|   Gravel, uniform section, clean | .022 | .025 |
| Earth, fairly uniform section: | | |
|   No vegetation | .022 | .025 |
|   Grass, some weeds | .025 | .030 |
|   Dense weeds or aquatic plants in deep channels | .030 | .035 |
|   Sides clean, gravel bottom | .025 | .030 |
|   Sides clean, cobble bottom | .030 | .040 |
| Dragline-excavated or dredged: | | |
|   No vegetation | .028 | .033 |
|   Light brush on banks | .035 | .050 |
| Rock: | | |
|   Based on design section | .035 | |
|   Based on actual mean section | | |
|     a. Smooth and uniform | .035 | .040 |
|     b. Jagged and irregular | .040 | .045 |
| Channels not maintained, weeds and brush uncut: | Fair condition | Poor condition |
| Dense weeds, high as flow depth | 0.08 | 0.12 |
| Clean bottom, brush on sides | .05 | .08 |
|   highest stage of flow | .07 | .11 |
| Dense brush, high stage | .10 | .14 |

## Table 2-12. Roughness Coefficients (Manning's $n$) for Open Channels (Continued)

Highway Ditches and Swales with Maintained Vegetation

| | Manning's $n$ | | | |
|---|---|---|---|---|
| | Depth 0.7 ft | | Depth 0.7–1.5 ft | |
| Type of vegetation | Velocity of flow | | | |
| | 2 fps | 6 fps | 2 fps | 6 fps |
| Bermuda; Kentucky bluegrass; buffalo: | | | | |
| Mowed to 2 in. | 0.07 | 0.045 | 0.05 | 0.035 |
| Length 4–6 in. | .09 | .05 | .06 | .04 |
| Good stand, any grass: | | | | |
| Length 12 in. | .18 | .09 | .12 | .07 |
| Length 24 in. | .30 | .15 | .20 | .10 |
| Fair stand, any grass: | | | | |
| Length 12 in. | .14 | .08 | .10 | .06 |
| Length 24 in. | .25 | .13 | .17 | .09 |

Street and Expressway Gutters[d]

| Construction | Manning's $n$ |
|---|---|
| Concrete gutter, troweled finish. | 0.012 |
| Asphalt pavement: | |
| Smooth texture. | .013 |
| Rough texture. | .016 |
| Concrete gutter with asphalt pavement: | |
| Smooth. | .013 |
| Rough. | .015 |
| Concrete pavement: | |
| Float finish. | .014 |
| Broom finish. | .016 |
| Rough. | .020 |

| Channel description | Range in $n$ |
|---|---|

Natural Stream Channels, Minor Streams[e]
(Surface width at flood stage less than 100 ft)

| Channel description | Range in $n$ |
|---|---|
| Fairly regular section: | |
| Some grass and weeds, little or no brush. | 0.030–0.035 |
| Dense growth of weeds, depth of flow materially greater than weed height. | 0.035–0.05 |
| Some weeds, light brush on banks. | 0.035–0.05 |
| Some weeds, heavy brush on banks. | 0.05 –0.07 |
| Some weeds, dense willows on banks. | 0.06 –0.08 |
| For trees within channel with branches submerged at high stage, increase all above values by. | 0.01 –0.02 |
| Irregular section, with pools, slight channel meander: | |
| For channels listed above, increase all values by about. | 0.01 –0.02 |
| Mountain streams; no vegetation in channel, banks usually steep, trees and brush along banks submerged at high stage: | |
| Bottom of gravel, cobbles, and few boulders. | 0.04 –0.05 |
| Bottom of cobbles with large boulders. | 0.05 –0.07 |

Natural Stream Channels, Major Streams
(Surface width at flood stage greater than 100 ft)

Roughness coefficient is usually less than for minor streams of similar description on account of less effective resistance offered by irregular banks or vegetation on banks. The value of $n$ for larger streams of most regular section, with no boulders or brush, may be in the range of from 0.028 to 0.033.

## Table 2-12. Roughness Coefficients (Manning's $n$) for Open Channels (Continued)

| Channel description | Range in $n$ |
|---|---|
| Natural Stream Channels, Flood Plains Adjacent to Natural Streams | |
| Pasture, no brush: | |
| Short grass | 0.030–0.035 |
| High grass | 0.035–0.05 |
| Cultivated areas: | |
| No crop | 0.03 –0.04 |
| Mature row crops | 0.035–0.045 |
| Mature field crops | 0.04 –0.05 |
| Heavy weeds, scattered brush | 0.05 –0.07 |
| Light brush and trees:[f] | |
| Winter | 0.05 –0.06 |
| Summer | 0.06 –0.08 |
| Medium to dense brush:[f] | |
| Winter | 0.07 –0.11 |
| Summer | 0.10 –0.16 |
| Dense willows, summer, not bent over by current | 0.15 –0.20 |
| Cleared land with tree stumps, 100–150 per acre: | |
| No sprouts | 0.04 –0.05 |
| With heavy growth of sprouts | 0.06 –0.08 |
| Heavy stand of timber, a few down trees, little undergrowth: | |
| Flood depth below branches | 0.10 –0.12 |
| Flood depth reaches branches ($n$ increases with depth) | 0.12 –0.16 |

[a] For poor-quality construction, use larger values of $n$.

[b] With steep slopes, depth of flow will generally be greater than computed by the usual methods for open channels, because of air entrainment and additional resistance offered by air in contact with high-velocity flow. An approximate depth may be calculated by increasing $n$ for the flume material involved by 20 to 30 %.

[c] With channel of alignment other than straight, loss of head by resistance forces will be increased. A small increase in value of $n$ may be made to allow for additional loss of energy.

[d] For gutters with small slopes where sediment may accumulate, increase values of $n$ by 0.002 to 0.005.

[e] The values of $n$ shown are principally derived from measurements made on fairly short but straight reaches of natural streams. Where slopes calculated from flood elevations along a considerable length of channel involving meanders and bends, are to be used in velocity calculations, the value of $n$ must be increased by about 3 to 15 per cent to provide for the additional loss of energy caused by bends.

[f] The presence of foliage on trees and brush under flood stage will materially increase the value of $n$. For trees in channel or on banks, and for brush on banks where submergence of branches increases with depth of flow, $n$ will increase with rising stage.

$$Q_{\max \text{ at } D_c} = 3.087 b H_t^{3/2} \quad \text{cfs} \qquad (2\text{-}136)$$

For a triangular channel, the maximum discharge is

$$Q_{\max \text{ at } D_c} = 1.435 B H_t^{3/2} \quad \text{cfs} \qquad (2\text{-}137)$$

## HEAT TRANSFER

There are three fundamental types of heat transfer: conduction, convection, and radiation.

*Conduction* is the transfer of heat from one part of a body to another part of the same body, or from one body to another in physical contact with it, without appreciable displacement of the particles of the body.

*Convection* is the transfer of heat from one point to another within a fluid, gas, or liquid by the mixing of one portion of the fluid with another. In natural convection, the motion of the fluid is entirely the result of differences in density resulting from temperature differences; in forced convection, the motion is produced by mechanical means. When the forced velocity is relatively low, it should be realized that "free convection" factors, such as density and temperature difference, may have an important influence.

*Radiation* is the transfer of heat from one body to another, not in contact with it, by means of wave motion through space.

All three types of heat transfer may occur at the same time, and it is advisable to consider the possibility of heat transfer by each type in any particular case.

### 2-15. Conduction

Fourier's law is the fundamental differential equation for heat transfer by conduction

$$dQ/d\theta = -kA \, dt/dx \qquad (2\text{-}138)$$

where $dQ/d\theta$ = rate of heat flow, Btu/hr
      $A$ = area at right angles to direction of heat flow, ft²
   $-dt/dx$ = rate of change of temperature with distance in direction of heat flow, i.e., temperature gradient, °F/ft

The factor $k$ is called the thermal conductivity, is dependent upon the material through which the heat is flowing and upon the temperature, and has the units Btu per hour per square foot per degree Fahrenheit per foot. (See page 3-23 for data on thermal conductivities.)

**Steady Flow of Heat.** For the steady flow of heat, the term $dQ/d\theta$ in Eq. (2-138) is constant and may be replaced by $Q/\theta$ or $q$. If $k$ and $A$ are independent of $t$ and $x$, Eq. (2-138) may be expressed as

$$q = kA(t_1 - t_2)/(x_2 - x_1) = kA \, \Delta t/x \qquad (2\text{-}139)$$

where $\Delta t$ = difference in temperatures
    $x$ = distance between points 1 and 2

Usually the thermal conductivity $k$ is not constant, but is a function of the temperature. In most cases, over the ranges of values used, the relation is linear. Integration of Eq. (2-138), with $k$ linear in $t$, gives

$$q = k_{avg}A \, \Delta t/x \qquad (2\text{-}140)$$

where $k_{avg}$ = arithmetic average thermal conductivity between temperatures $t_1$ and $t_2$

In case the cross-sectional area $A$ varies with the distance $x$, $A$ may be expressed in terms of $x$ in order to integrate Eq. (2-138). This may, however, lead to complicated expressions, and it is customary to use Eq. (2-140), substituting the proper average value of $A$:

1. A flat wall of constant area, $A_{avg} = A_1 = A_2$     (2-141)
2. Area is proportional to first power of distance, as for insulated pipes

$$A_{avg} = (A_2 - A_1)/2.3 \log (A_2/A_1) \qquad (2\text{-}142)$$

3. Area is proportional to square of distance, as in a hollow sphere

$$A_{avg} = \sqrt{A_1 A_2} \qquad (2\text{-}143)$$

*Conduction through Several Bodies in Series.* Since the heat flow through each of several walls must be the same,

$$q = k_1 A_1 \,\Delta t_1/x_1 = k_2 A_2 \,\Delta t_2/x_2 = k_3 A_3 \,\Delta t_3/x_3 \qquad (2\text{-}144)$$

if we let $R_1 = x_1/k_1 A_1$, $R_2 = x_2/k_2 A_2$, etc., then

$$q(R_1 + R_2 + R_3) = \Delta t_1 + \Delta t_2 + \Delta t_3 = \Sigma \,\Delta t \qquad (2\text{-}145)$$
$$q = \Sigma \,\Delta t/R_T = (t_1 - t_4)/R_T \qquad (2\text{-}146)$$

where $R_T$ is the over-all resistance and is the sum of the individual resistances in series. Then

$$R_T = R_1 + R_2 + \cdots + R_n \qquad (2\text{-}147)$$

*Conduction through Several Bodies in Parallel.* For $n$ resistances in parallel, the rates of heat flow are additive.

$$\begin{aligned}
q &= \Delta t/R_1 + \Delta t/R_2 + \cdots + \Delta t/R_n \\
&= (1/R_1 + 1/R_2 + \cdots + 1/R_n)\,\Delta t \\
&= (C_1 + C_2 + \cdots + C_n)\,\Delta t = \Sigma C\,\Delta t \qquad (2\text{-}148)
\end{aligned}$$

where $R_1$ to $R_n$ = individual resistances

$C_1$ to $C_n$ = individual conductances, $C = kA/x$

**Heat Transfer in the Unsteady State** (Heating and Cooling of Solids). In problems involving conduction of heat in the transient state, the temperature of the body varies with both time and the position of points in the body, and the mathematical relations are complicated. However, the basic differential equations for conduction have been integrated for various shapes and boundary conditions (Figs. 2-32 to 2-35) and the results may be plotted as curves involving four ratios defined as follows:

$$Y = (t' - t)/(t' - t_b) \qquad (2\text{-}149)$$
$$X = k\theta/\rho c_p r_m^2 \qquad (2\text{-}150)$$
$$m = k/h_T r_m \qquad (2\text{-}151)$$
$$n = r/r_m \qquad (2\text{-}152)$$

where $t'$ = temperature of surroundings, °F

$t_b$ = initial uniform temperature of body, °F

$t$ = temperature at given point in body at time $\theta$ (hours) measured from start of heating or cooling operations, °F

$k$ = uniform thermal conductivity of body, Btu/(hr)(ft²)(°F/ft)

$\rho$ = uniform density of body, lb/ft³

$c_p$ = specific heat of body, Btu/(lb)(°F)

$h_T$ = coefficient of total heat transfer between surroundings and surface of body, Btu/(hr)(°F)(ft²)

$r$ = distance, in direction of heat conduction, from mid-point or mid-plane of body to point under consideration, ft

$r_m$ = radius of sphere or cylinder, one-half the thickness of a slab heated from both faces, the total thickness of a slab heated from one face and insulated perfectly at the other, ft

$x$ = distance, in direction of heat conduction, from surface of semi-infinite body to point under consideration, ft

With the infinite slab, in the early stages of the operation where Fig. 2-35 gives insufficient precision, Fig. 2-34 may be used for points near the surface.

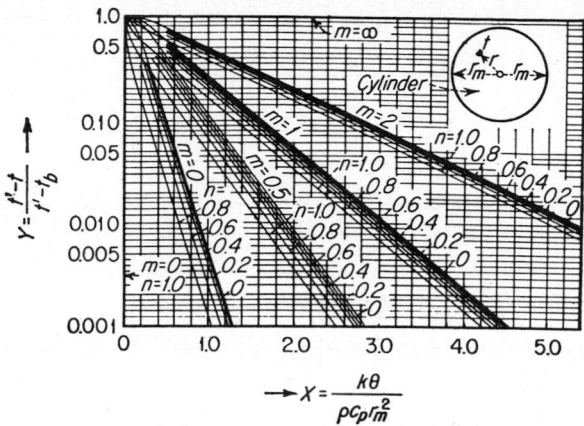

FIG. 2-32. Heating and cooling of a solid cylinder having infinite ratio of length to diameter.

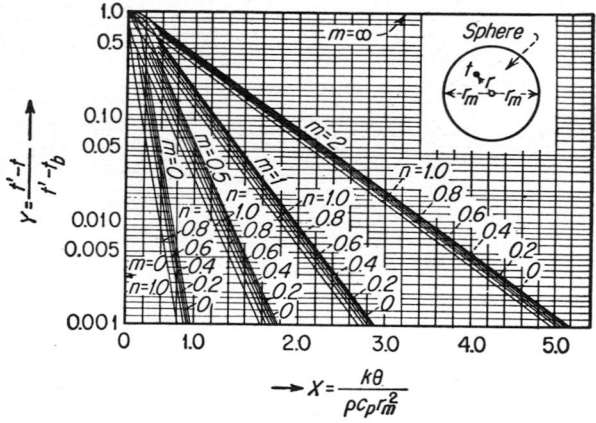

FIG. 2-33. Heating and cooling of a solid sphere.

For a brick-shaped solid having the dimensions $2r_{m1}$, $2r_{m2}$, and $2r_{m3}$ the value of $Y$ at a given time and position may be evaluated as follows: $Y$ equals the product $Y_1 Y_2 Y_3$, where $Y_1$ is evaluated from Fig. 2-34 at $X_1 = k\theta/\rho c_p r_{m1}^2$, $n_1 = r_1/r_{m1}$, and $m_1 = k/h_T r_{m1}$. Similarly, $Y_2$ and $Y_3$ are read for the same $\theta$ at $X_2$, $n_2$, and $m_2$ and at $X_3$, $n_3$, and $m_3$, corresponding to $r_{m2}$ and $r_{m3}$.

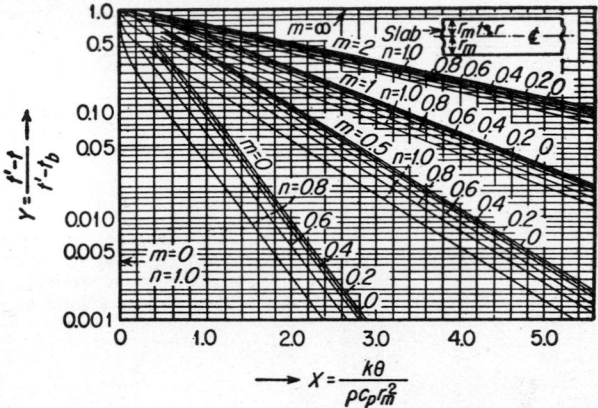

Fig. 2-34. Heating and cooling of a solid slab having a large face area relative to that of the edges.

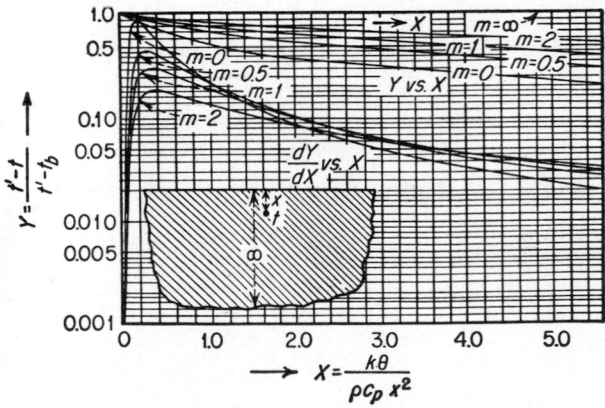

Fig. 2-35. Heating and cooling of a solid of infinite thickness, neglecting edge effects. (This may be used as an approximation in the zone near the surface of a body of finite thickness.)

## 2-16. Convection

**Coefficients of Heat Transfer.** In commercial heat-transfer equipment it is not convenient to measure tube-wall temperatures such as $t_3$ and $t_4$ in Fig. 2-36, and hence the rate of heat transfer is not easily calculated on the basis of conduction as expressed in Eq. (5-14) and as discussed in the previous section.

$$q = (k/x)A_{\text{avg}}(t_3 - t_4) = (k/x)A_{\text{avg}}\,\Delta t, \qquad (2\text{-}153)$$

where $k$ = thermal conductivity
       $x$ = thickness of tube wall

The over-all performance is thus expressed as an over-all coefficient of heat

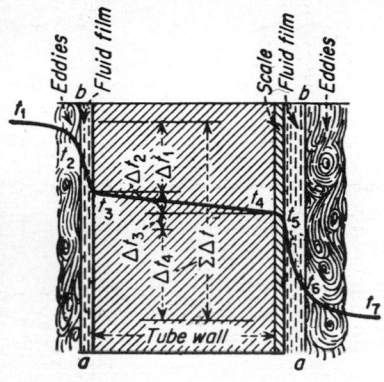

FIG. 2-36. Temperature gradients for steady flow of heat by conduction and convection from a warmer to a colder fluid separated by a solid wall.

transfer $U$ (Btu per hour per square foot per degree Fahrenheit) based on a convenient area such as the inside area $A_i$ (square feet), the outside area $A_0$ (square feet), or an average of these $A_{avg}$. Whence by definition

$$q = UA_{avg}(t_1 - t_7) = UA_{avg}\Sigma\,\Delta t \qquad (2\text{-}154)$$

The rate of heat transfer through each of the fluid resistances on either side of the tube wall is equal to that shown in Eqs. (2-153) and (2-154) and may be expressed as

$$q = h_iA_i(t_1 - t_3) = h_0A_0(t_5 - t_7) \qquad (2\text{-}155)$$

where $h_i$ and $h_0$ = film coefficient of heat transfer inside and outside tube wall, Btu/(hr)(ft²)(°F)

Similarly, the rate of heat transfer through a layer of scale on the tube wall may be expressed as

$$q = h_dA_0(t_4 - t_5) \qquad (2\text{-}156)$$

As was previously noted, it is most convenient to use the over-all coefficient of heat transfer and an over-all $\Delta t$ such as $t_1 - t_7$. To do this the individual coefficients are combined by basing the over-all coefficient on one area arbitrarily, let us say in this case $A_0$.

$$1/U = A_0/A_ih_i + xA_0/kA_{avg} + 1/h_d + 1/h_0 \qquad (2\text{-}157)$$

The second term in the Eq. (2-157), $xA_0/kA_{avg}$, is usually negligible, because of the relatively minor contribution of the tube wall to the over-all resistance to heat transfer. So the general procedure to evaluate the over-all coefficient $U$ is to evaluate the individual coefficients of heat transfer by the methods of Tables 2-13 and 2-14, use an appropriate scale coefficient from Table 2-15, and combine these with the appropriate physical data (thermal conductivity and dimensions) on the heat-exchanger tubing by means of Eq. (2-157). Also, typical in-service over-all heat transfer coefficients are given in Tables 2-16 through 2-21.

**Mean Temperature Difference.** For parallel or counterflow of fluids

$$q = UA\,\Delta t_{mean} = UA(\Delta t_1 - \Delta t_2)/\ln(\Delta t_1/\Delta t_2) \qquad (2\text{-}158)$$

where the right-hand term, excluding $UA$, is the logarithmic mean of the terminal temperature differences, in degrees Fahrenheit, of the exchanger.

*Multipass and Cross-flow Exchangers.* Here the flow is neither parallel nor countercurrent; the logarithmic-mean temperature difference does not apply. For these exchangers,

$$q = U_m A\,\Delta t'_{mean} = U_m A Y\,\Delta t_{mean} \qquad (2\text{-}159)$$

where $Y$ is obtained from Figs. 2-37 or 2-38 and $\Delta t_{mean}$ is as defined in Eq. (2-158). If one of the temperatures remains constant, as in a condenser or in an evaporative cooler, Eq. (2-158) applies for parallel flow, counterflow, multipass, and cross flow.

If $U$ varies considerably with temperature, the apparatus should be visualized as divided into stages, in each of which variation of $U$ with temperature or temperature difference is linear. Then for parallel or counterflow operation

$$q = A(\Delta t_1 U_2 - \Delta t_2 U_1)/\ln(\Delta t_1 U_2/\Delta t_2 U_1) \qquad (2\text{-}160)$$

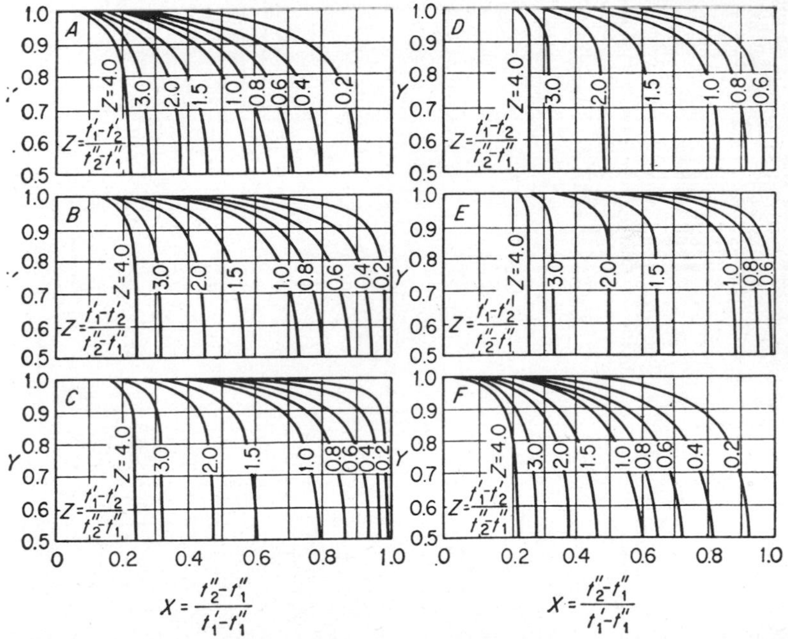

FIG. 2-37. Mean temperature difference in reversed-current exchangers. (Shell side well mixed at a given cross section.) (*A*) One shell pass and 2, 4, 6, tube passes. (*B*) Two shell passes and 4, 8, tube passes. (*C*) Three shell passes and 6, 12, tube passes. (*D*) Four shell passes and 8, 16, tube passes. (*E*) Six shell passes and 12, 24, 36, tube passes. (*F*) One shell pass and 3 tube passes. [*Perry and Chilton (eds.), "Chemical Engineers' Handbook," 5th ed., McGraw-Hill Book Company, New York, 1973.*]

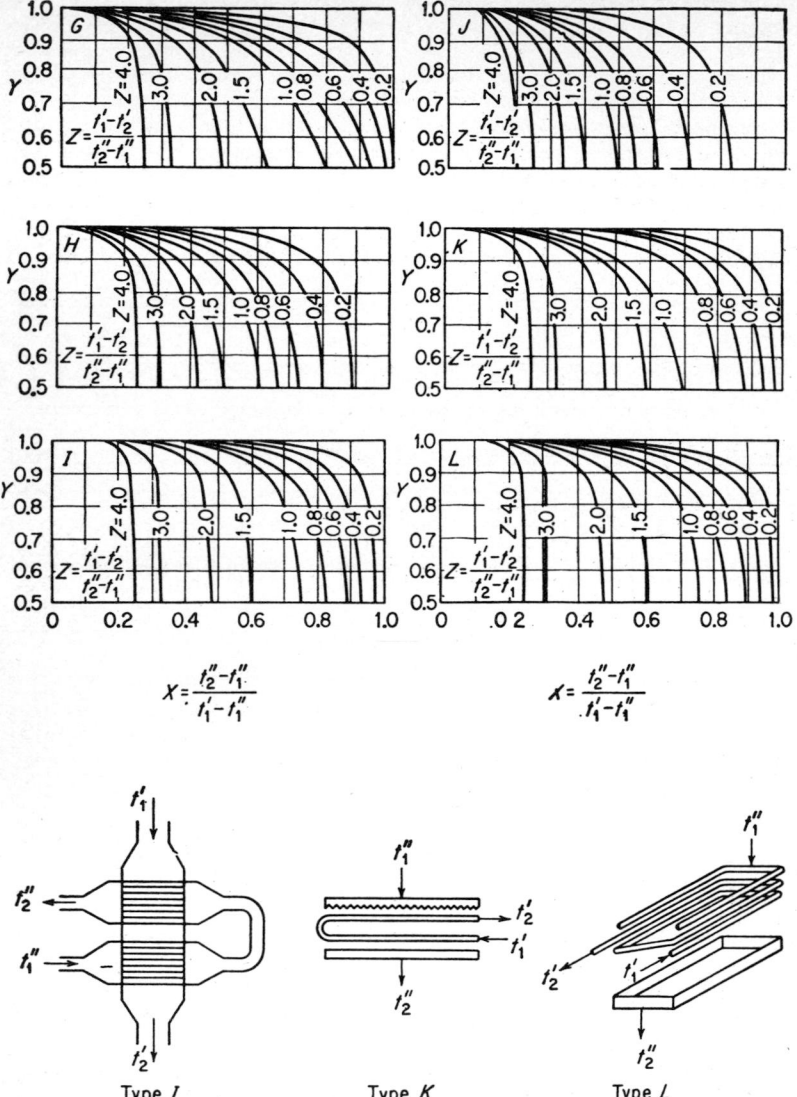

$$X = \frac{t_2'' - t_1''}{t_1' - t_1''} \qquad\qquad X = \frac{t_2'' - t_1''}{t_1' - t_1''}$$

Type $I$       Type $K$       Type $L$

Fig. 2-38. Mean temperature difference in cross-flow exchangers. ($G$) Cross flow, both fluids unmixed, 1 tube pass. ($H$) Cross flow, shell fluid mixed, 1 tube pass. ($I$) Cross flow, shell fluid mixed, 2 tube passes, shell fluid flows across second and first passes in series. ($J$) Cross flow, shell fluid mixed, 2 tube passes, shell fluid flows over first and second passes in series. ($K$) Cross flow (drip type), 2 horizontal passes with U-bend connections (trombone type). ($L$) Cross flow (drip type), helical coils with 2 turns. [*Perry and Chilton* (eds.), "*Chemical Engineers' Handbook*," 5th ed., McGraw-Hill Book Company, New York, 1973.]

**Conversion Factors for Coefficients of Heat Transfer.** Throughout this chapter, values of $h$ and $U$ are expressed in Btu per hour per square foot per degree Fahrenheit. Conversion factors to other units are listed in Table 1-2.

## 2-17. Radiant-heat Transmission

**General.** If two small bodies of areas $A_1$ and $A_2$ (in square feet) are placed in a large evacuated enclosure perfectly insulated externally, then, when the system has come to thermal equilibrium, the bodies will emit radiation at the rates $A_1W_1$ and $A_2W_2$, respectively, where $W$ is the total emissive power: energy per unit time per unit area of the surface (Btu per hour per square foot) emitted throughout the hemisphere above each element of surface. Let the energy impinging on unit area of any small body in the enclosure, due to radiation from the walls of the latter, be $I$ Btu/(hr)(ft²). If the bodies have absorptivities (fraction of incident radiation which is absorbed) of $\alpha_1$ and $\alpha_2$, then energy balances on the bodies will have the form $IA_1\alpha_1 = A_1W_1$ and $IA_2\alpha_2 = A_2W_2$, from which

$$W_1/\alpha_1 = W_2/\alpha_2 = W_x/\alpha_x$$

where $x$ is *any* body. This generalization, that at thermal equilibrium the ratio of the emissive power of a surface to its absorptivity is the same for all bodies, is *Kirchhoff's law.* The relation

$$W_B = \sigma T^4 \tag{2-161}$$

is the *Stefan-Boltzmann law,* and the proportionality constant $\sigma$ is the Stefan-Boltzmann constant:

$$0.173 \times 10^{-8} \text{ Btu/(ft}^2\text{)(hr)(}^\circ\text{R}^4\text{)}$$
$$5.67 \times 10^{-5} \text{ ergs/(cm}^2\text{)(sec)(}^\circ\text{K}^4\text{)}$$
$$4.88 \times 10^{-8} \text{ kg-cal/(m}^2\text{)(hr)(}^\circ\text{K}^4\text{)}$$

If $W_{B\lambda}$ is the *monochromatic emissive power* at wavelength $\lambda$ (centimeters) such that $W_{B\lambda}\, d\lambda$ is the energy emitted from a surface per unit area per unit time in the wavelength interval $\lambda$ to $d\lambda$, the relation among $W_{B\lambda}$ (Btu per hour per square foot per centimeter), $\lambda$, and $T$ is given by *Planck's law*

$$W_{B\lambda} = c_1\lambda^{-5}/(e^{c_2/\lambda t} - 1) \tag{2-162}$$

where $c_1 = 1.176 \times 10^{-8}$ Btu/(ft²)(hr)(cm⁴) or $0.885 \times 10^{-12}$ cal-cm²/sec
$c_2 = 2.58$ cm-°R or 1.433 cm-°K

If $\alpha_\lambda$ is a constant independent of $\lambda$, the surface is called *gray* and its total absorptivity $\alpha$ will be independent of the spectral-energy distribution of the incident radiation; then $\alpha_{1.2} = \alpha_{1.1} = \epsilon_1$; that is, emissivity $\epsilon$ may be used in substitution for $\alpha$ even though the temperatures of the incident radiation and the receiver are not the same.

**Radiation between Surfaces of Solids Separated by a Nonabsorbing Medium.** The net loss of energy by radiation from a body at temperature $T_1$ in *black* surroundings at $T_2$ is given by

$$q_{1,\text{net}} = 0.173A_1[\epsilon_1(T_1/100)^4 - \alpha_{1.2}(T_2/100)^4] \tag{2-163}$$

where $A_1$ is in square feet and $T$ is in degrees Rankine.

When $\alpha_{1.2} = \epsilon_1$, that is, when the body is gray, this simplifies to

$$q_{1,\text{net}} = 0.173A_1\epsilon_1[(T_1/100)^4 - (T_2/100)^4] \tag{2-164}$$

Values for the emissivity of various surfaces are given in Table 2-22 page 2-79.

## Table 2-13. Film Coefficients for Liquids

Units: Btu/(hr)(ft2)(°F)

To obtain the desired film coefficient, a base factor corresponding to the liquid and temperature under consideration is taken from the proper table and multiplied by a correction factor read from the nomograph accompanying that table. The following assumptions apply in each case: (1) the system is in equilibrium, that is, there is no change in temperature gradient with time, (2) radiation is negligible or has been taken into account by other calculations, (3) film temperature is defined as the arithmetic average of the temperatures of the retaining wall and the main body of the liquid. Wall temperature, generally not known, can be estimated or calculated by trial and error. Values of base factors in italics are extrapolated from physical properties.

### Case 1. Base Factors for Liquids Heated Inside Horizontal or Vertical Tubes, Turbulent Flow

| Average liquid temp, °F | 0 | 50 | 100 | 150 | 200 | 250 |
|---|---|---|---|---|---|---|
| Acetic acid, 100% | .... | 117 | 97.2 | 101 | 105 | 109 |
| Acetic acid, 50% | .... | .... | 156 | 180 | 203 | 228 |
| Acetone | 104 | 122 | 134 | 137 | 139 | 142 |
| Ammonia | 350 | 425 | 507 | 699 | 690 | 790 |
| Amyl acetate | 65.0 | 66.2 | 67.7 | 71.8 | 78.5 | 86.0 |
| Amyl alcohol, iso | 21.6 | 35.8 | 52.7 | 73.3 | 96.0 | 118 |
| Aniline | .... | 43.8 | 58.4 | 76.5 | 99.2 | 123 |
| Benzene | 139 | 75.6 | 94.5 | 108 | 121 | 154 |
| Brine, Ca Cl2, 25% | 31.2 | 190 | 257 | 332 | 420 | 517 |
| n-Butyl alcohol | 114 | 45.5 | 62.4 | 83.0 | 107 | 133 |
| Carbon disulfide | 57.4 | 119 | 125 | 129 | 132 | 135 |
| Carbon tetrachloride | 64.6 | 69.2 | 78.6 | 82.6 | 85.8 | 88.2 |
| Chlorobenzene | 126 | 73.3 | 78.8 | 80.5 | 82.0 | 82.8 |
| Ethyl acetate | 58.0 | 126 | 125 | 123 | 122 | 121 |
| Ethyl alcohol, 100% | 61.6 | 73.6 | 92.3 | 112 | 132 | 151 |
| Ethyl alcohol, 40% | 97.8 | 104 | 162 | 228 | 292 | 389 |
| Ethyl bromide | 71.4 | 104 | 110 | 114 | 119 | 122 |
| Ethylene glycol | 100 | 105 | 158 | 222 | 299 | 380 |
| Ethyl ether | 59.0 | 115 | 123 | 130 | 137 | 144 |
| Glycerol, 50% | 81.4 | 90.5 | 131 | 182 | 248 | 302 |
| Heptane | 85.8 | 87.0 | 94.7 | 102 | 112 | 122 |
| Hexane | 83.0 | 93.8 | 102 | 109 | 114 | 117 |
| Methyl alcohol, 100% | 86.0 | 110 | 126 | 138 | 149 | 160 |
| Methyl alcohol, 90% | 64.0 | 114 | 136 | 154 | 172 | 188 |
| Methyl alcohol, 40% | 72.0 | 110 | 164 | 213 | 264 | 312 |
| n-Octane | 103 | 79.0 | 85.9 | 92.0 | 97.0 | 102 |
| n-Pentane | 25.7 | 105 | 110 | 115 | 118 | 121 |
| Propyl alcohol, iso | .... | 49.3 | 71.5 | 94.5 | 117 | 139 |
| Sulfur dioxide | 167 | 171 | 175 | 180 | 182 | 194 |
| Sulfuric acid, 60% | 77.3 | 65.9 | 79.4 | 94.5 | 110 | 129 |
| Toluene | .... | 86.9 | 96.6 | 104 | 112 | 119 |
| Water | .... | 225 | 322 | 408 | 492 | 608 |

### Case 2. Base Factors for Liquids Cooled Inside Horizontal or Vertical Tubes, Turbulent Flow

| Average liquid temp, °F | 0 | 50 | 100 | 150 | 200 | 250 |
|---|---|---|---|---|---|---|
| Acetic acid, 100% | .... | 85.2 | 72.2 | 75.0 | 78.7 | 81.6 |
| Acetic acid, 50% | .... | 105 | 122 | 153 | 179 | 201 |
| Acetone | 88.4 | .... | 118 | 121 | 124 | 126 |
| Ammonia | 314 | 380 | 507 | 650 | 797 | 938 |
| Amyl acetate | 48.1 | 50.7 | 52.3 | 53.3 | 54.4 | 55.5 |
| Amyl alcohol, iso | .... | 22.7 | 36.0 | 53.0 | 72.3 | 91.8 |
| Aniline | .... | 31.6 | 45.2 | 63.3 | 86.9 | 110 |
| Benzene | 106 | 61.7 | 79.1 | 93.5 | 107 | 180 |
| Brine, Ca Cl2, 25% | 19.2 | 152 | 217 | 312 | 397 | 510 |
| n-Butyl alcohol | 103 | 31.4 | 45.6 | 64.0 | 84.9 | 107 |
| Carbon disulfide | .... | 116 | 116 | 121 | 121 | 128 |
| Carbon tetrachloride | 40.4 | 55.8 | 67.3 | 72.0 | 76.0 | 78.7 |
| Chlorobenzene | 54.6 | 57.5 | 60.7 | 63.8 | 65.0 | 65.6 |
| Ethyl acetate | 83.3 | 84.1 | 84.1 | 84.1 | 83.3 | 82.4 |
| Ethyl alcohol, 100% | 41.2 | 54.8 | 71.1 | 89.0 | 108 | 128 |
| Ethyl alcohol, 40% | .... | 70.4 | 121 | 176 | 230 | 315 |
| Ethyl bromide | 84.9 | 92.9 | 99.9 | 106 | 112 | 116 |
| Ethylene glycol, 50% | 44.2 | 71.7 | 120 | 183 | 261 | 315 |
| Ethyl ether | 86.0 | 99.0 | 109 | 118 | 126 | 135 |
| Glycerol, 50% | 54.9 | 59.5 | 94.5 | 134 | 179 | 222 |
| Heptane | 66.8 | 74.2 | 81.5 | 88.8 | 97.0 | 104 |
| Hexane | 70.7 | 78.5 | 87.2 | 95.2 | 102 | 108 |
| Methyl alcohol, 100% | 65.4 | 88.5 | 112 | 118 | 132 | 146 |
| Methyl alcohol, 90% | 68.5 | 92.0 | 112 | 128 | 142 | 156 |
| Methyl alcohol, 40% | 58.2 | 80.0 | 127 | 177 | 236 | 292 |
| n-Octane | 56.0 | 63.8 | 70.9 | 77.2 | 83.2 | 88.0 |
| n-Pentane | 82.8 | 89.6 | 96.4 | 101 | 106 | 110 |
| Propyl alcohol, iso | .... | 32.4 | 51.5 | 71.7 | 92.7 | 116 |
| Sulfur dioxide | 150 | 155 | 161 | 166 | 174 | 178 |
| Sulfuric acid, 60% | .... | 35.3 | 54.4 | 66.9 | 74.0 | 77.7 |
| Toluene | 61.8 | 71.5 | 81.4 | 90.3 | 97.6 | 103 |
| Water | .... | 153 | 273 | 355 | 427 | 483 |

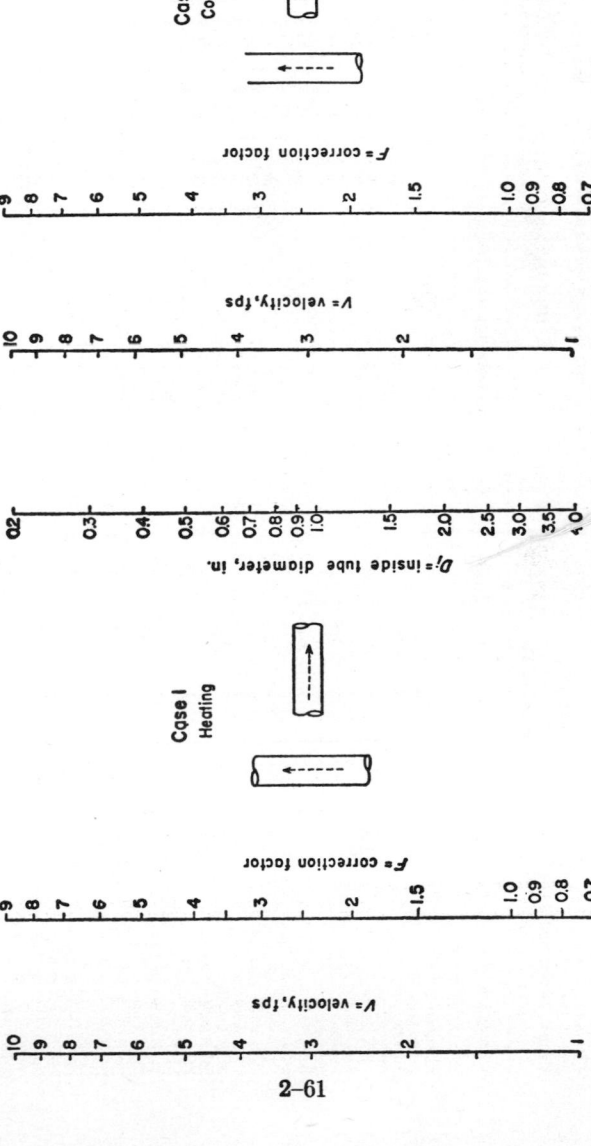

$D_i$ = inside tube diameter, in.

Case 2
Cooling

$F$ = correction factor

$V$ = velocity, fps

$D_i$ = inside tube diameter, in.

Case I
Heating

$F$ = correction factor

$V$ = velocity, fps

2-61

# Table 2-13. Film Coefficients for Liquids (Continued)

## Case 3. Liquids Heated or Cooled Outside Tube Bundles, Direction of Flow Parallel to Tubes

Film coefficients for liquids flowing outside tube bundles and in a direction parallel to the tubes can be determined from Cases 1 or 2 for liquids inside tubes, if an equivalent inside diameter is used in determining the correction factor. An equivalent diameter can be calculated by $d_e = 4A/P$, where $d_e$ is the equivalent ID in inches, $A$ is the cross-sectional area between tubes in square inches, and $P$ is the sum of the tube-perimeter segments forming the cross-section boundary, in inches.

## Case 4. Base Factors for Liquids Heated or Cooled Outside Single Tubes, Direction of Flow Normal to Tube

| Average film temp, °F | 0 | 50 | 100 | 150 | 200 | 250 |
|---|---|---|---|---|---|---|
| Acetic acid, 100% | | | 142 | 136 | 131 | 125 |
| Acetic acid, 50% | | 214 | 260 | 292 | 310 | 321 |
| Acetone | 165 | 174 | 184 | 186 | 187 | 189 |
| Ammonia | 486 | 548 | 616 | 685 | 758 | 887 |
| Amyl acetate | 114 | 106 | 97.9 | 91.0 | 84.3 | 76.5 |
| Amyl alcohol, iso | 54.0 | 73.0 | 94.9 | 118 | 140 | 163 |
| Aniline | | 97.0 | 116 | 139 | 164 | 194 |
| Benzene | | 124 | 140 | 152 | 163 | 174 |
| Brine, Ca Cl2, 25% | 264 | 335 | 419 | 508 | 617 | 734 |
| n-Butyl alcohol | 98.5 | 100 | 112 | 136 | 167 | 206 |
| Carbon disulfide | 164 | 166 | 169 | 171 | 173 | 173 |
| Carbon tetrachloride | | 105 | 114 | 116 | 117 | 118 |
| Chlorobenzene | 115 | 112 | 109 | 106 | 103 | 102 |
| Ethyl acetate | 154 | 145 | 137 | 129 | 119 | 111 |
| Ethyl alcohol, 100% | 108 | 127 | 146 | 165 | 183 | 199 |
| Ethyl alcohol, 40% | 130 | 199 | 277 | 355 | 430 | 508 |
| Ethyl bromide | 131 | 137 | 142 | 144 | 146 | 147 |
| Ethylene glycol, 50% | 147 | 209 | 283 | 362 | 447 | 545 |
| Ethyl ether | 146 | 154 | 161 | 169 | 175 | 182 |
| Glycerol, 50% | 147 | 192 | 249 | 331 | 451 | 565 |
| Heptane | 126 | 133 | 139 | 143 | 147 | 151 |
| Hexane | 147 | 134 | 141 | 147 | 151 | 155 |
| Methyl alcohol, 100% | 159 | 170 | 187 | 198 | 206 | 212 |
| Methyl alcohol, 90% | 132 | 186 | 209 | 226 | 238 | 261 |
| Methyl alcohol, 40% | 117 | 201 | 264 | 317 | 359 | 397 |
| n-Octane | 139 | 124 | 129 | 135 | 140 | 146 |
| n-Pentane | 62.5 | 91.0 | 118 | 143 | 152 | 154 |
| Propyl alcohol, iso | 230 | 225 | 223 | 221 | 163 | 180 |
| Sulfur dioxide | | | 223 | 221 | 221 | 218 |
| Sulfuric acid, 60% | | 110 | 137 | 150 | 164 | 176 |
| Toluene | 128 | 135 | 142 | 148 | 152 | 165 |
| Water | | 382 | 497 | 525 | 645 | 700 |

## Case 5. Base Factors for Liquids Heated Outside Single Horizontal Tubes, Natural Convection

| Average film temp, °F | 0 | 50 | 100 | 150 | 200 | 250 |
|---|---|---|---|---|---|---|
| Acetic acid, 100% | | | 19.8 | 18.8 | 17.8 | 17.8 |
| Acetone | | 27.1 | 28.0 | 28.2 | 28.6 | 28.8 |
| Ammonia | 85.5 | 96.3 | 108 | 120 | 132 | 144 |
| Benzene | | 24.1 | 20.3 | 21.9 | 23.5 | 25.0 |
| Carbon disulfide | 23.9 | 24.7 | 24.6 | 24.8 | 25.1 | 25.3 |
| Carbon tetrachloride | | 15.5 | 16.6 | 16.5 | 16.3 | 15.9 |
| Chlorobenzene | 16.4 | 15.7 | 14.8 | 14.0 | 13.4 | 13.0 |
| Ethyl acetate | 23.5 | 21.7 | 20.1 | 18.5 | 16.8 | 15.6 |
| Ethyl alcohol, 100% | 15.6 | 18.1 | 20.6 | 23.1 | 25.3 | 27.4 |
| Ethyl alcohol, 40% | 17.9 | 26.6 | 36.5 | 47.8 | 57.8 | 23.6 |
| Ethyl bromide | 20.1 | 20.7 | 21.1 | 21.4 | 21.6 | 21.9 |
| Ethyl ether | 22.3 | 23.9 | 24.8 | 25.6 | 26.4 | 22.9 |
| Ethyl iodide | 19.4 | 20.0 | 20.4 | 21.0 | 21.6 | 30.9 |
| Heptane | 18.7 | 19.9 | 21.0 | 21.8 | 24.1 | 38.4 |
| Hexane | 22.5 | 25.2 | 27.1 | 28.7 | 29.8 | 20.5 |
| Methyl alcohol, 100% | 21.5 | 25.9 | 29.1 | 31.8 | 34.1 | 24.4 |
| Methyl alcohol, 90% | 17.0 | 18.1 | 18.5 | 19.2 | 20.0 | 35.2 |
| n-Octane | 21.9 | 22.3 | 22.8 | 23.3 | 24.0 | 21.7 |
| n-Pentane | 21.9 | 18.5 | 36.5 | 36.0 | 35.5 | 35.2 |
| Sulfur dioxide | 37.9 | 37.2 | 15.5 | 19.2 | 22.9 | 24.4 |
| Sulfuric acid, 98% | | 11.8 | 15.6 | 18.2 | 20.3 | 21.7 |
| Sulfuric acid, 60% | 18.1 | 12.5 | 15.6 | 18.2 | 21.5 | 21.7 |
| Toluene | | 19.1 | 20.1 | 20.9 | 21.5 | 21.7 |
| Water | | 36.8 | 47.9 | 55.0 | 60.2 | 65.0 |

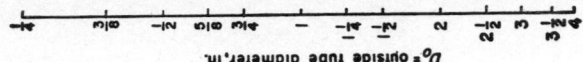

$D_o$ = outside tube diameter, in.

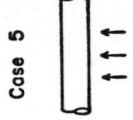

Case 5

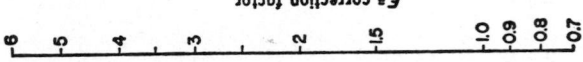

F = correction factor

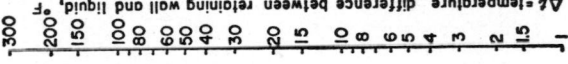

$\Delta t$ = temperature difference between retaining wall and liquid, °F

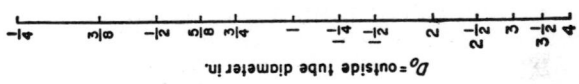

$D_o$ = outside tube diameter, in.

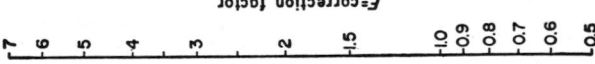

F = correction factor

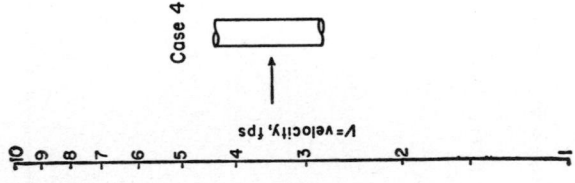

Case 4

V = velocity, fps

2-63

# Table 2-13. Film Coefficients for Liquids (Continued)

## Case 6. Base Factors for Liquids Heated Inside or Outside Vertical Tubes or on Vertical Plates, Low Velocities or Natural Convection Only

| Average film temp, °F | 0 | 50 | 100 | 150 | 200 | 250 |
|---|---|---|---|---|---|---|
| Acetic acid, 100% | 21.0 | | 15.6 | 15.5 | 15.6 | 16.5 |
| Acetone | 75.0 | 22.6 | 24.1 | 24.8 | 25.3 | 25.8 |
| Ammonia | | 88.5 | 103 | 118 | 138 | 165 |
| Benzene | 19.7 | 14.4 | 16.4 | 18.2 | 20.0 | 21.7 |
| Carbon disulfide | 11.4 | 20.3 | 21.0 | 21.6 | 22.2 | 22.8 |
| Carbon tetrachloride | 12.0 | 12.7 | 13.8 | 14.2 | 14.4 | 14.8 |
| Chlorobenzene | | 12.0 | 12.0 | 11.8 | 11.8 | 11.7 |
| Ethyl acetate | 18.6 | 18.1 | 17.4 | 16.6 | 16.9 | 16.0 |
| Ethyl alcohol, 100% | 10.8 | 13.2 | 15.9 | 18.6 | 21.2 | 23.9 |
| Ethyl alcohol, 40% | 10.2 | 17.4 | 26.2 | 25.1 | 45.4 | |
| Ethyl bromide | 17.1 | 18.0 | 18.8 | 19.4 | 19.8 | 20.2 |
| Ethyl ether | 19.4 | 21.0 | 17.7 | 19.8 | 21.8 | 24.6 |
| Ethyl iodide | 12.6 | 15.4 | 16.6 | 17.2 | 18.5 | 24.0 |
| Heptane | 14.8 | 15.6 | 17.8 | 18.9 | 19.4 | 19.9 |
| Hexane | 16.6 | 16.7 | 21.9 | 23.9 | 25.9 | 27.4 |
| Methyl alcohol, 100% | 16.7 | 19.5 | 22.6 | 25.4 | 27.8 | 30.3 |
| Methyl alcohol, 90% | 15.1 | 19.4 | 21.9 | 25.1 | 27.4 | 30.3 |
| n-Octane | 12.7 | 13.9 | 14.8 | 15.8 | 16.7 | 17.7 |
| n-Pentane | | 19.3 | 20.7 | 20.7 | 21.1 | 21.5 |
| Sulfur dioxide | 33.6 | 33.6 | 33.6 | 33.6 | 33.6 | 33.6 |
| Sulfuric acid, 98% | | 7.2 | 9.4 | 11.6 | 14.7 | 16.9 |
| Sulfuric acid, 60% | 13.7 | 8.5 | 11.2 | 13.3 | 18.5 | 15.3 |
| Toluene | | 14.9 | 16.1 | 17.3 | 18.5 | 19.6 |
| Water | | 23.3 | 31.9 | 38.0 | 42.4 | 47.9 |

ΔL = temperature difference between retaining wall and liquid, °F

Scale: 1.5  2  3  4  5  6  8  10  15  20  30  40  60  80  100  150  200  300

F = correction factor (1  2  3  4  5  6  7)

## Case 7. Liquids Heated or Cooled Outside Tube Bundles, Direction of Flow Normal to Tubes

Use data for Case 4 and multiply answer by 1.2 for tubes in line or by 1.3 for staggered tubes. For determining correction factor use velocity at narrowest section between tubes. For baffled heat exchangers, where a small part of the flow is parallel to tubes, use Case 4 data directly, but evaluate correction factor at velocity between tubes at widest part of shell, calculated as $144Q \div (d_s - nd)l$, where $Q$ is rate of flow in cubic feet per second, $d_s$ is inside shell diameter in inches, $n$ is number of tubes across wide part of shell, and $l$ is distance between baffles.

## Case 8. Liquids Heated or Cooled in Annular Spaces, Turbulent Flow

Use data for Case 1 (heating) or Case 2 (cooling), but substitute an equivalent diameter equal to $(d_1^2 - d_2^2) \div d_1$ in determining correction factor on nomograph. $d_1$ is inside diameter of outer pipe and $d_2$ is outside diameter of inner pipe.

## Case 9. Liquids Heated or Cooled Inside Coils, Turbulent Flow

Use data for Case 1 (heating) or Case 2 (cooling) and multiply answer by 1.2

## Case 10. Liquids Heated or Cooled Outside Coils, Natural or Forced Convection

Use data for Case 4 (forced convection) or Case 5 (natural convection).
SOURCE: H. J. Stoever, Chem. & Met. Eng., 51(5): (1944). Reproduced by permission of H. J. Stoever.

## Table 2-14. Film Coefficients for Gases, Condensing Vapors, and Boiling Liquids

Units: Btu/(hr)(ft²)(°F)

Film coefficients for gases, condensing vapors, and boiling liquids are obtained in the same way as was explained for liquids, Table 5-1.

### Case II. Base Factors for Gases Heated or Cooled Inside Horizontal or Vertical Tubes, Turbulent Flow

| Average gas temp, °F | -100 | 0 | 100 | 200 | 300 | 400 | 500 |
|---|---|---|---|---|---|---|---|
| Acetone | .... | .... | 3.98 | 4.58 | 5.26 | 6.33 | *7.36* |
| Acetylene | 4.47 | 4.90 | 5.32 | 5.75 | 6.13 | 6.55 | *4.59* |
| Air | 3.52 | 3.76 | 3.92 | 4.08 | 4.19 | 4.27 | 7.64 |
| Ammonia | 4.87 | 5.68 | 6.25 | 6.69 | 7.06 | *7.56* | 7.00 |
| Benzene | .... | .... | 3.61 | 4.33 | 5.20 | 6.10 | 7.03 |
| Butane | 2.57 | 4.61 | 5.48 | 5.98 | 6.42 | 6.74 | 3.83 |
| Carbon dioxide | 3.53 | 2.89 | 3.14 | 3.33 | 3.49 | 3.64 | 3.64 |
| Carbon monoxide | 3.77 | 3.77 | 3.97 | 4.17 | 4.33 | 4.49 | 1.92 |
| Chlorine | 1.69 | 1.66 | 1.73 | 1.78 | 1.83 | 1.88 | *1.98* |
| Chloroform | .... | .... | 1.90 | 2.07 | 2.26 | 2.45 | 2.64 |
| Ethane | 4.06 | 4.83 | 5.55 | 6.28 | 7.00 | 5.75 | 8.46 |
| Ethyl acetate | .... | .... | 3.81 | 4.49 | 5.14 | 5.85 | 6.52 |
| Ethyl alcohol | .... | .... | 3.50 | 3.75 | 3.89 | 4.03 | 6.06 |
| Ethyl chloride | .... | .... | 5.22 | 5.75 | 6.53 | 6.90 | *4.15* |
| Ethylene | 3.89 | 5.08 | 5.19 | 5.75 | 6.35 | 7.47 | 7.46 |
| Ethyl ether | .... | 4.56 | 5.30 | 5.93 | 6.68 | 8.78 | 8.78 |
| Helium | *20.6* | *21.2* | *21.9* | *22.6* | *23.9* | *25.9* | *24.6* |
| Hydrogen | *45.1* | *47.7* | *49.6* | *51.6* | *53.6* | *66.6* | *57.5* |
| Hydrogen sulfide | 2.68 | 2.90 | 3.15 | 3.37 | 3.60 | 8.79 | 9.98 |
| Methane | 6.68 | 7.49 | 8.06 | 8.47 | 8.79 | 9.11 | 9.55 |
| Methyl chloride | 1.88 | 2.36 | 2.85 | 3.31 | 3.74 | 4.16 | *4.56* |
| Nitric oxide | 3.43 | 3.62 | 3.77 | 3.89 | 3.99 | *4.07* | 4.15 |
| Nitrogen | 3.83 | 3.95 | 4.11 | 4.24 | 4.35 | 4.41 | 4.49 |
| Nitrous oxide | 2.87 | 2.94 | 3.00 | 3.06 | *3.12* | *3.18* | *3.24* |
| Oxygen | 3.38 | 3.57 | 3.71 | 3.82 | *3.93* | 4.01 | 4.09 |
| Pentane, iso | .... | 4.73 | 5.37 | 6.13 | 7.00 | 8.02 | 6.64 |
| Steam | .... | 4.73 | 5.37 | 5.82 | 6.18 | 6.41 | 6.64 |
| Sulfur dioxide | .... | 1.88 | 2.00 | 2.10 | 2.29 | 2.28 | 2.36 |

$D_i$ = inside tube diameter, in. (scale: 0.2 – 4.0)

F = correction factor

$$G = \frac{\text{rate of flow, lb/sec}}{\text{cross-sect. area, sq ft}} = \frac{1}{v_g} = \text{velocity} \times \text{density, (fps} \times \text{lb/ft}^3)$$

Case II

2–65

## Table 2-14. Film Coefficients for Gases, Condensing Vapors, and Boiling Liquids (Continued)

**Case 12. Base Factors for Gases Heated or Cooled Outside Single Tubes, Direction of Flow Normal to Tube, Turbulent Flow**

| Average film temp, °F | −100 | 0 | 100 | 200 | 300 | 400 | 500 |
|---|---|---|---|---|---|---|---|
| Acetone | *7.16* | 8.54 | 6.69 | 8.09 | 9.63 | 11.4 | *13.1* |
| Acetylene | 6.97 | 7.71 | 9.92 | 11.2 | *12.4* | *13.6* | *15.6* |
| Air | 8.34 | 10.4 | 8.38 | 8.96 | 9.46 | 9.88 | *10.2* |
| Ammonia | | | 12.1 | 13.4 | 14.6 | 15.7 | *16.8* |
| Benzene | | | 6.16 | 7.70 | 12.4 | 11.5 | *15.4* |
| Butane | 4.62 | 8.05 | 9.47 | 10.0 | 12.4 | 13.9 | 15.4 |
| Carbon dioxide | 6.91 | 5.57 | 6.32 | 6.95 | 7.59 | 8.15 | 8.73 |
| Carbon monoxide | *2.86* | 7.68 | 8.42 | 9.10 | 9.79 | 10.4 | 11.0 |
| Chlorine | | 3.12 | 3.39 | 3.63 | 3.86 | 4.06 | *4.27* |
| Chloroform | | | 3.46 | 3.94 | 4.46 | 5.02 | *5.53* |
| Ethane | *6.82* | 8.46 | 10.2 | 12.3 | 14.4 | | |
| Ethyl acetate | | | 6.42 | 7.85 | 9.38 | 10.0 | *12.6* |
| Ethyl alcohol | | | 9.10 | 9.75 | 10.5 | 11.1 | *11.7* |
| Ethyl chloride | | *4.80* | 5.59 | 6.30 | 6.98 | 7.16 | *8.57* |
| Ethylene | *6.40* | 8.02 | 9.54 | 11.0 | 12.5 | 13.9 | *16.4* |
| Ethyl ether | | 7.78 | 8.95 | 10.3 | 12.0 | 14.1 | *16.1* |
| Helium | 41.8 | 45.2 | 48.1 | *50.9* | 53.3 | 66.3 | 58.6 |
| Hydrogen | 75.5 | 82.4 | 88.8 | *94.2* | 97.7 | 101 | 105 |
| Hydrogen sulfide | 4.94 | 5.55 | 6.10 | 6.59 | *7.02* | 7.38 | *7.75* |
| Methane | 7.61 | 8.32 | 8.95 | 9.57 | 10.1 | 10.7 | 11.3 |
| Methyl chloride | *2.98* | 4.09 | 5.25 | 6.30 | 7.40 | 8.57 | |
| Nitric oxide | 6.64 | 7.36 | 8.00 | 8.56 | 9.04 | 9.44 | *9.84* |
| Nitrogen | 7.38 | 8.06 | 8.67 | 9.18 | 9.63 | 9.87 | *10.2* |
| Nitrous oxide | 5.49 | 5.72 | 5.90 | 6.01 | 6.32 | 6.49 | *6.66* |
| Oxygen | 6.83 | 7.54 | 8.12 | 8.60 | 9.09 | 9.50 | *9.89* |
| Pentane, iso | | 7.40 | 8.91 | 10.6 | 12.5 | 14.4 | |
| Steam | | | 9.97 | 11.1 | 12.1 | 13.1 | 14.0 |
| Sulfur dioxide | | 3.42 | 3.83 | 4.21 | 4.52 | 4.79 | 5.10 |

**Case 13. Base Factors for Gases Heated Outside Single Horizontal Tubes, Natural Convection**

| Average film temp, °F | −100 | 0 | 100 | 200 | 300 | 400 | 500 |
|---|---|---|---|---|---|---|---|
| Acetone | 0.83 | | 0.94 | 1.06 | 1.19 | 1.32 | *1.45* |
| Acetylene | 0.91 | 0.90 | 0.96 | 1.02 | 1.07 | 1.13 | *1.18* |
| Air | 0.82 | 0.88 | 0.88 | 0.87 | 0.85 | 0.84 | *0.83* |
| Ammonia | | 0.88 | 0.94 | 0.99 | 1.04 | 1.08 | *1.12* |
| Benzene | | | 1.00 | 1.17 | 1.36 | 1.58 | *1.80* |
| Butane | 0.71 | 1.25 | 1.36 | 1.47 | 1.58 | 1.68 | *1.78* |
| Carbon dioxide | 0.86 | 0.76 | 0.80 | 0.82 | 0.84 | 0.86 | *0.87* |
| Carbon monoxide | | 0.86 | 0.86 | 0.85 | 0.84 | 0.83 | *0.82* |
| Chlorine | | 0.56 | 0.55 | 0.55 | 0.65 | 0.54 | *0.63* |
| Chloroform | | | 0.72 | 0.77 | 0.82 | 0.86 | *0.91* |
| Ethane | 0.85 | 0.95 | 1.06 | 1.16 | 1.26 | 1.36 | *1.46* |
| Ethyl acetate | | | 1.12 | 1.29 | 1.45 | 1.60 | *1.76* |
| Ethyl alcohol | | | 1.17 | 1.16 | 1.13 | 1.12 | *1.10* |
| Ethyl chloride | *0.79* | 0.86 | 0.84 | 0.89 | 0.94 | 0.98 | *1.08* |
| Ethylene | | 0.86 | 0.95 | 1.04 | 1.11 | 1.19 | *1.26* |
| Ethyl ether | | | 1.42 | 1.57 | 1.72 | 1.88 | *1.86* |
| Helium | 2.02 | 1.94 | 1.88 | *1.84* | 1.89 | | *2.02* |
| Hydrogen | 2.44 | 2.39 | 2.34 | 2.30 | 2.26 | *2.21* | *2.16* |
| Hydrogen sulfide | | 0.66 | 0.67 | 0.68 | 0.68 | | |
| Methane | *1.04* | 1.15 | 1.21 | 1.27 | 1.33 | 1.39 | *1.44* |
| Methyl chloride | | | 0.71 | 0.79 | 0.87 | 0.95 | *1.03* |
| Nitric oxide | 0.86 | 0.86 | 0.85 | 0.84 | 0.83 | 0.81 | *0.79* |
| Nitrogen | 0.92 | 0.91 | 0.89 | 0.87 | 0.85 | 0.84 | *0.81* |
| Nitrous oxide | 0.86 | 0.79 | 0.75 | 0.71 | 0.68 | 0.65 | |
| Oxygen | 0.92 | 0.91 | 0.90 | 0.89 | 0.87 | 0.85 | *0.84* |
| Pentane, iso | | 1.26 | 1.40 | 1.56 | 1.74 | 1.42 | *2.14* |
| Steam | | | 0.80 | 0.81 | 0.83 | 0.85 | *0.86* |
| Sulfur dioxide | | | 0.59 | 0.60 | 0.61 | 0.61 | *0.61* |

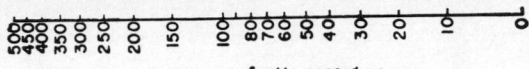

$P$= pressure, psig

500 450 400 350 300 250 200 150 100 80 70 60 50 40 30 20 10 0

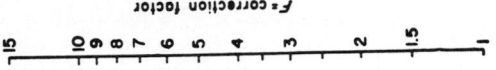

$F$= correction factor

15 10 9 8 7 6 5 4 3 2 1.5 1

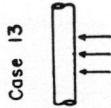

Case 13

$\dfrac{\Delta t}{D_o}=\dfrac{\text{temperature difference between retaining wall and gas, °F}}{\text{outside tube diameter, in.}}$

1,000 800 600 500 400 300 200 150 100 80 60 50 40 30 20 15 10

¼ ⅜ ½ ⅝ ¾ 1 1¼ 1½ 2 2½ 3 3½ 4

$D_o$= outside tube diameter, in.

Case 12

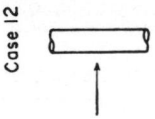

$F$= correction factor

15 10 8 6 5 4 3 2 1.5 .10 .08 .06 .05 .04 .03 .02 0.15

$G=\dfrac{\text{rate of flow, lb/sec}}{\text{cross-sect. area ft}^2}=V/\rho=\text{velocity}\times\text{density},(\text{fps}\times\text{lb/ft}^3)$

50 40 30 20 15 10 9 8 6 5 4 3 2 1.5 1.0 .8 .6 .5 .4 .3 .2 0.1

## Table 2-14. Film Coefficients for Gases, Condensing Vapors, and Boiling Liquids (Continued)

**Case 14. Gases Heated or Cooled Outside Tube Bundles, Direction of Flow Parallel to Tubes**

Use Case 11 data, but substitute an equivalent diameter in determining correction factor as in Case 3.

**Case 15. Gases Heated or Cooled Outside Tube Bundles, Direction of Flow Normal to Tubes**

See Case 7. Instead of using data for Case 4 as directed in Case 7, use data from Case 12 and correct in the same way.

**Case 16. Gases Heated or Cooled in Annular Spaces, Turbulent Flow**

Use data for Case 11 but substitute an equivalent diameter as described in Case 8.

**Case 17. Gases Heated or Cooled Outside Coils, Natural or Forced Convection**

Coefficients for these cases are approximately the same as for Cases 12 or 13, and these data should be used.

### Case 18. Base Factors for Gases Heated Inside or Outside Vertical Tubes or on Vertical Plates, Natural Convection

| Average film temp, °F | −100 | 0 | 100 | 200 | 300 | 400 | 500 |
|---|---|---|---|---|---|---|---|
| Acetone | 0.56 | 0.55 | 0.65 | 0.69 | 0.73 | 0.78 | 0.83 |
| Acetylene | 0.53 | 0.49 | 0.54 | 0.53 | 0.63 | 0.63 | 0.62 |
| Air | 0.50 | 0.50 | 0.46 | 0.43 | 0.41 | 0.39 | 0.38 |
| Ammonia | .... | .... | 0.50 | 0.49 | 0.49 | 0.48 | 0.48 |
| Benzene | .... | .... | 0.72 | 0.79 | 0.88 | 0.97 | 1.06 |
| Butane | .... | 0.83 | 0.93 | 0.94 | 0.96 | 0.97 | 0.98 |
| Carbon dioxide | 0.49 | 0.48 | 0.47 | 0.46 | 0.45 | 0.44 | 0.43 |
| Carbon monoxide | 0.51 | 0.48 | 0.45 | 0.42 | 0.40 | 0.38 | 0.36 |
| Chlorine | .... | 0.38 | 0.36 | 0.34 | 0.32 | 0.31 | 0.30 |
| Chloroform | .... | .... | 0.53 | 0.53 | 0.54 | 0.55 | 0.56 |
| Ethane | 0.57 | 0.59 | 0.61 | 0.63 | 0.65 | 0.67 | 0.69 |
| Ethyl acetate | .... | .... | 0.83 | 0.89 | 0.95 | 1.02 | 1.06 |
| Ethyl alcohol | .... | .... | 0.76 | 0.71 | 0.67 | 0.62 | 0.58 |
| Ethyl chloride | .... | .... | 0.56 | 0.55 | 0.54 | 0.54 | 0.53 |
| Ethylene | 0.52 | 0.54 | 0.55 | 0.56 | 0.58 | 0.59 | 0.60 |
| Ethyl ether | 0.52 | 0.54 | 1.03 | 1.05 | 1.11 | 1.17 | 1.24 |
| Helium | 0.83 | 0.74 | 0.69 | 0.65 | 0.62 | 0.72 | 0.68 |
| Hydrogen | 1.04 | 0.95 | 0.88 | 0.82 | 0.77 | 0.72 | .... |
| Hydrogen sulfide | .... | 0.41 | 0.39 | 0.37 | 0.35 | .... | .... |
| Methane | 0.46 | 0.45 | 0.44 | 0.43 | 0.42 | 0.41 | 0.40 |
| Methyl chloride | .... | 0.41 | 0.45 | 0.48 | 0.51 | 0.54 | 0.55 |
| Nitric oxide | 0.52 | 0.48 | 0.45 | 0.42 | 0.40 | 0.37 | 0.40 |
| Nitrogen | 0.54 | 0.49 | 0.46 | 0.43 | 0.40 | 0.38 | 0.35 |
| Nitrous oxide | 0.55 | 0.49 | 0.44 | 0.40 | 0.37 | 0.34 | 0.36 |
| Oxygen | 0.54 | 0.50 | 0.47 | 0.44 | 0.41 | 0.39 | 0.37 |
| Pentane, iso. | .... | 0.99 | 1.01 | 1.05 | 1.12 | 1.20 | 1.30 |
| Steam | .... | .... | 0.44 | 0.42 | 0.41 | 0.40 | 0.39 |
| Sulfur dioxide | .... | .... | 0.38 | 0.36 | 0.35 | 0.34 | 0.33 |

### Case 19. Base Factors for Condensation of Pure Saturated Vapors on Horizontal Tubes

| Temperature of condensate film (assume equal to tube wall), °F | 50 | 100 | 150 | 200 | 250 | 300 |
|---|---|---|---|---|---|---|
| Acetic acid | .... | 511 | 495 | 470 | 424 | 373 |
| Acetone | 772 | 789 | 805 | 805 | 795 | 780 |
| Ammonia | 2,768 | 3,145 | 3,459 | 3,711 | 3,875 | 3,965 |
| Aniline | 275 | 405 | 544 | 685 | 830 | 977 |
| Benzene | 554 | 609 | 658 | 706 | 755 | 798 |
| Carbon disulfide | 924 | 933 | 933 | 924 | 905 | 868 |
| Carbon tetrachloride | 551 | 580 | 569 | 488 | .... | .... |
| Chloroform | 735 | 791 | 847 | 895 | 950 | 897 |
| Ethyl acetate | 702 | 772 | 835 | 889 | 936 | 990 |
| Ethyl alcohol | 496 | 556 | 618 | 678 | 745 | 807 |
| Ethyl ether | 620 | 646 | 665 | 678 | 691 | 705 |
| Heptane | 488 | 537 | 580 | 607 | 628 | 645 |
| Hexane | 525 | 552 | 576 | 592 | 608 | 614 |
| Methyl alcohol | 695 | 772 | 850 | 920 | 972 | 1,030 |
| Octane | 489 | 513 | 538 | 554 | 575 | 585 |
| Propyl alcohol, iso | 284 | 400 | 488 | 548 | 596 | 632 |
| Steam | 1,830 | 2,440 | 3,020 | 3,590 | 4,120 | 4,660 |
| Sulfur dioxide | 1,260 | 1,200 | 1,115 | 1,010 | 900 | 780 |

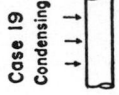

$W$ = rate of condensation per sq ft of tube surface, lb/(ft²)(hr)

2  3  4  5  6  7  8  10  15  20  30  40  50  60  70  80  100  150  200

Case 19
Condensing

$F$ = correction factor

2.5  2.0  1.5  1.0  0.8  0.7  0.6  0.5  0.4  0.3  0.20  0.15  0.10

$ND_o$ = number of tubes arranged directly over each other X outside tube diameter, in.

1/4  3/8  5/16  1/2  3/4  1  1¼  1½  2  3  4  5  6  8  10  15  20  30  40  50  60  80

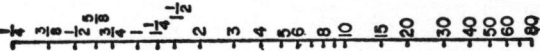

$P$ = pressure, psig

500  450  400  350  300  250  200  150  100  80  70  60  50  40  30  20  10  0

$F$ = correction factor

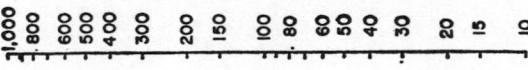

50  40  30  20  15  10  8  7  6  5  4  3  2  1.5  1

Case 18

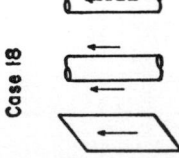

$\Delta t$ = temperature difference between retaining wall and gas, °F

1,000  800  600  500  400  300  200  150  100  80  60  50  40  30  20  15  10

# Table 2-14. Film Coefficients for Gases, Condensing Vapors, and Boiling Liquids (Continued)

**Case 20. Liquids Boiling on Horizontal or Vertical Plates**

Use nomograph for Case 20. For water boiling at pressures other than atmospheric multiply by the following correction factor:

| Abs press., atm.... | 0.2 | 0.4 | 0.6 | 0.8 | 1.0 | 2.0 | 4.0 | 6.0 | 8.0 | 10.0 | 15.0 |
|---|---|---|---|---|---|---|---|---|---|---|---|
| Correction factor... | 0.62 | 0.78 | 0.88 | 0.94 | 1.00 | 1.16 | 1.32 | 1.40 | 1.46 | 1.51 | 1.60 |

**Case 21. Gases Heated or Cooled Inside Coils, Turbulent Flow**

Use the following equation: $h_{coils} = [1 + 3.54 \,(d/d_c)] \times h_{tubes}$, where $d$ is the inside diameter of the pipe or tube in inches, $d_c$ is the diameter of the coil in inches, and $h_{tubes}$ is the straight-tube coefficient, from Case 11.

**Case 22. Air Heated on Horizontal Plates, Natural Convection**

For large plates (3 ft² or more) get coefficient from Case 18 and multiply by 1.27 for horizontal plates facing downward or by 0.67 for plates facing upward. If radiation is an important factor, a combined coefficient can be obtained by adding the product of the emissivity of the surface and the radiation coefficient, which is defined as $h_r = 0.173 \times 10^{-8} \,(T_1{}^4 - T_2{}^4) \div (T_1 - T_2)$, where $T_1$ is the temperature of the emitting surface and $T_2$ is the temperature of the absorbing surface, both in degrees Fahrenheit absolute.

**Case 23. Liquids Boiling Inside Tubes**

When liquid moves by natural convection only, use coefficient from Case 20 and multiply by 1.25. If liquid moves at high velocity, use coefficient from Case 1, since conditions then are as if there were no evaporation.

SOURCE: H. J. Stoever, *Chem. & Met. Eng.*, 51(5): (1944). Reproduced by permission of H. J. Stoever.

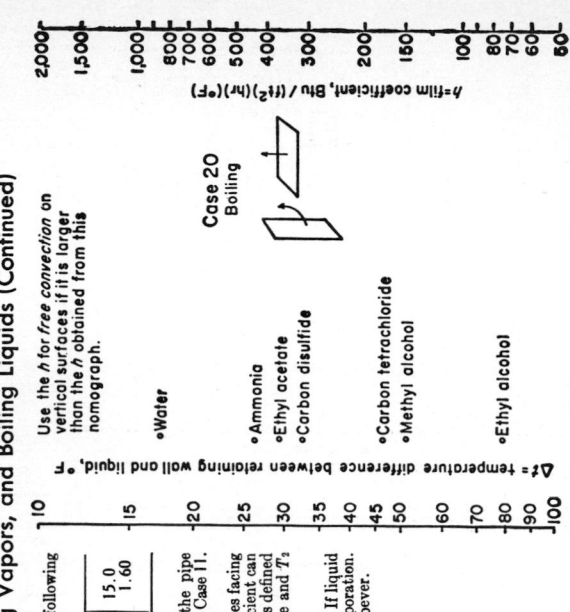

Use the $h$ for *free convection* on vertical surfaces if it is larger than the $h$ obtained from this nomograph.

Case 20
Boiling

$h$ = film coefficient, Btu / (ft²)(hr)(°F)

$\Delta t$ = temperature difference between retaining wall and liquid, °F

- Water
- Ammonia
- Ethyl acetate
- Carbon disulfide
- Carbon tetrachloride
- Methyl alcohol
- Ethyl alcohol

## Table 2-15. Heat Transfer Coefficients for Deposits*

B.t.u./(hr.)(sq. ft.)(°F.)

### Water

| Temperature of heating medium . . . | Up to 240°F. | | 240°–400°F. | |
|---|---|---|---|---|
| Temperature of water . . . . . . . . | 125°F. or less | | Above 125°F. | |
| Water velocity, ft./sec. . . . . . . . . | ≤3 | >3 | ≤3 | >3 |
| Distilled . . . . . . . . . . . . . . . | 2000 | 2000 | 2000 | 2000 |
| Sea water . . . . . . . . . . . . . . | 2000 | 2000 | 1000 | 1000 |
| Treated boiler feed water . . . . . . | 1000 | 2000 | 1000 | 1000 |
| Treated makeup for cooling tower . . | 1000 | 1000 | 500 | 500 |
| City, well, Great Lakes. . . . . . . . | 1000 | 1000 | 500 | 500 |
| Brackish, clean river water . . . . . . | 500 | 1000 | 330 | 500 |
| River water, muddy, silty . . . . . . . | 330 | 500 | 250 | 330 |
| Hard (over 15 g./gal.) . . . . . . . . | 330 | 330 | 200 | 200 |
| Chicago Sanitary Canal . . . . . . . | 125 | 170 | 100 | 125 |

### Chemicals

Inorganic:
  Gases (oil-bearing or dirty) . . . . . . . . . . . . . . . . . . . . 500
  Liquids (heating or vaporization) . . . . . . . . . . . . . . . . 500
  Refrigerant brines . . . . . . . . . . . . . . . . . . . . . . . . .1000
Organic:
  Gases
    Process . . . . . . . . . . . . . . . . . . . . . . . . . . . . . .1000
    Utility (oil-bearing, refrigerant, etc.) . . . . . . . . . . . . . . 500
    Condensing vapors (condensers) . . . . . . . . . . . . . . . .1000
  Liquids
    Process . . . . . . . . . . . . . . . . . . . . . . . . . . . . . .1000
    Vaporizing liquids (reboilers) . . . . . . . . . . . . . . . . . . 500
    Heat-transfer media . . . . . . . . . . . . . . . . . . . . . . .1000
    Refrigerant liquids . . . . . . . . . . . . . . . . . . . . . . . .1000
    Polymer-forming liquids . . . . . . . . . . . . . . . . . . . . 200
    Oils (vegetable and heavy gas oil) . . . . . . . . . . . . . . 330
    Asphalt and residuum . . . . . . . . . . . . . . . . . . . . . 100

* Rearranged and reproduced, by permission, from Standards of Tubular Manufacturers Association, New York, 1968 and reprinted from Perry and Chilton (eds.), "Chemical Engineers' Handbook," 5th ed., p. 10–38, McGraw-Hill Book Company, New York, 1973.

## Table 2-16. Typical Over-all Heat Transfer Coefficients in Tubular Heat Exchangers

$U = B.t.u./(°F.)(sq. ft.)(hr.)$

| Shell side | Tube side | Design $U$ | Includes total dirt |
|---|---|---|---|
| Liquid-liquid media | | | |
| Aroclor 1248 | Jet fuels | 100–150 | 0.0015 |
| Cutback asphalt | Water | 10–20 | .01 |
| Demineralized water | Water | 300–500 | .001 |
| Ethanol amine (MEA or DEA) 10–25% solutions | Water or DEA, or MEA solutions | 140–200 | .003 |
| Fuel oil | Water | 15–25 | .007 |
| Fuel oil | Oil | 10–15 | .008 |
| Gasoline | Water | 60–100 | .003 |
| Heavy oils | Heavy oils | 10–40 | .004 |
| Heavy oils | Water | 15–50 | .005 |
| Hydrogen-rich reformer stream | Hydrogen-rich reformer stream | 90–120 | .002 |
| Kerosene or gas oil | Water | 25–50 | .005 |
| Kerosene or gas oil | Oil | 20–35 | .005 |
| Kerosene or jet fuels | Trichlorethylene | 40–50 | .0015 |
| Jacket water | Water | 230–300 | .002 |
| Lube oil (low viscosity) | Water | 25–50 | .002 |
| Lube oil (high viscosity) | Water | 40–80 | .003 |
| Lube oil | Oil | 11–20 | .006 |
| Naphtha | Water | 50–70 | .005 |
| Naphtha | Oil | 25–35 | .005 |
| Organic solvents | Water | 50–150 | .003 |
| Organic solvents | Brine | 35–90 | .003 |
| Organic solvents | Organic solvents | 20–60 | .002 |
| Tall oil derivatives, vegetable oil, etc. | Water | 20–50 | .004 |
| Water | Caustic soda solutions (10–30%) | 100–250 | .003 |
| Water | Water | 200–250 | .003 |
| Wax distillate | Water | 15–25 | .005 |
| Wax distillate | Oil | 13–23 | .005 |
| Condensing vapor-liquid media | | | |
| Alcohol vapor | Water | 100–200 | .002 |
| Asphalt (450°F.) | Dowtherm vapor | 40–60 | .006 |
| Dowtherm vapor | Tall oil and derivatives | 60–80 | .004 |

NC = non-condensable gas present.
V = vacuum.
A = atmospheric pressure.
Dirt (or fouling factor) units are (hr.)(sq. ft.)(°F.)/B.t.u.
SOURCE: Perry and Chilton (eds.), "Chemical Engineers' Handbook," 5th ed., p. 10–39, McGraw-Hill Book Company, New York, 1973.

## Table 2-16. Typical Over-all Heat Transfer Coefficients in Tubular Heat Exchangers (Continued)

$U = \text{B.t.u.}/(°F.)(\text{sq. ft.})(\text{hr.})$

| Shell side | Tube side | Design $U$ | Includes total dirt |
|---|---|---|---|
| Dowtherm vapor | Dowtherm liquid | 80–120 | .0015 |
| Gas-plant tar | Steam | 40–50 | .0055 |
| High-boiling hydrocarbons V | Water | 20–50 | .003 |
| Low-boiling hydrocarbons A | Water | 80–200 | .003 |
| Hydrocarbon vapors (partial condenser) | Oil | 25–40 | .004 |
| Organic solvents A | Water | 100–200 | .003 |
| Organic solvents high NC, A | Water or brine | 20–60 | .003 |
| Organic solvents low NC, V | Water or brine | 50–120 | .003 |
| Kerosene | Water | 30–65 | .004 |
| Kerosene | Oil | 20–30 | .005 |
| Naphtha | Water | 50–75 | .005 |
| Naphtha | Oil | 20–30 | .005 |
| Stabilizer reflux vapors | Water | 80–120 | .003 |
| Steam | Feed water | 400–1000 | .0005 |
| Steam | No. 6 fuel oil | 15–25 | .0055 |
| Steam | No. 2 fuel oil | 60–90 | .0025 |
| Sulfur dioxide | Water | 150–200 | .003 |
| Tall-oil derivatives, vegetable oils (vapor) | Water | 20–50 | .004 |
| Water | Aromatic vapor-stream azeotrope | 40–80 | .005 |
| Gas-liquid media | | | |
| Air, $N_2$, etc. (compressed) | Water or brine | 40–80 | .005 |
| Air, $N_2$, etc., A | Water or brine | 10–50 | .005 |
| Water or brine | Air, $N_2$ (compressed) | 20–40 | .005 |
| Water or brine | Air, $N_2$, etc., A | 5–20 | .005 |
| Water | Hydrogen containing natural-gas mixtures | 80–125 | .003 |
| Vaporizers | | | |
| Anhydrous ammonia | Steam condensing | 150–300 | .0015 |
| Chlorine | Steam condensing | 150–300 | .0015 |
| Chlorine | Light heat-transfer oil | 40–60 | .0015 |
| Propane, butane, etc. | Steam condensing | 200–300 | .0015 |
| Water | Steam condensing | 250–400 | .0015 |

NC = non-condensable gas present.
V = vacuum.
A = atmospheric pressure.
Dirt (or fouling factor) units are (hr.)(sq. ft.)(°F.)/B.t.u.
SOURCE: Perry and Chilton (eds.), "Chemical Engineers' Handbook," 5th ed., p. 10–39, McGraw-Hill Book Company, New York, 1973.

## Table 2-17. Typical Over-all Heat Transfer Coefficients in Refinery Service

B.t.u./(°F.)(sq. ft.)(hr.)

| Fluid | API gravity | Fouling factor (one stream) | Reboiler, steam-heated | Condenser, water-cooled* | Exchangers, liquid to liquid (tube-side fluid designation appears below) | | | Reboiler (heating liquid designated below) | | | Condenser (cooling liquid designated below) | | | |
|---|---|---|---|---|---|---|---|---|---|---|---|---|---|---|
| | | | | | C | G | H | C | G† | K | D | F | G | J |
| A  Propane | ... | 0.001 | 160 | 95 | 85 | 85 | 80 | 110 | 95 | 35 | | | | |
| B  Butane | ... | .001 | 155 | 90 | 80 | 75 | 75 | 105 | 90 | 35 | | | | |
| C  400°F. end-point gasoline | 50 | .001 | 120 | 80 | 70 | 65 | 60 | 65 | 50 | 30 | 80 | 55 | 40 | 30 |
| D  Virgin light naphtha | 70 | .001 | 140 | 85 | 70 | 55 | 55 | 75 | 60 | 35 | 75 | 50 | 35 | 30 |
| E  Virgin heavy naphtha | 45 | .001 | 95 | 75 | 65 | 55 | 50 | 55 | 45 | 30 | 70 | 50 | 35 | 30 |
| F  Kerosene | 40 | .001 | 85 | 60 | 60 | 55 | 50 | ... | 45 | 25 | ... | 45 | 30 | 30 |
| G  Light gas oil | 30 | .002 | 70 | 50 | 60 | 50 | 50 | 50 | 40 | 25 | 70 | 40 | 30 | 20 |
| H  Heavy gas oil | 22 | .003 | 60 | 45 | 55 | 50 | 45 | | 40 | 20 | 70 | | | |
| J  Reduced crude | 17 | .005 | ... | ... | 55 | 45 | 40 | | | | | | | |
| K  Heavy fuel oil (tar) | 10 | .005 | ... | ... | 50 | 40 | 35 | | | | | | | |

Fouling factor, water side 0.002.

Heating or cooling streams are shown at top of columns as C, D, F, G, etc.

*Cooler, water-cooled, rates are about 5 per cent lower.

†With heavy gas oil (H) as heating medium, rates are about 5 per cent lower.

SOURCE: Perry and Chilton (eds.), "Chemical Engineers' Handbook," 5th ed., p. 10–40, McGraw-Hill Book Company, New York, 1973.

## Table 2-18. Over-all Coefficients for Air-cooled Exchangers on Bare-tube Basis

B.t.u./(°F.)(sq. ft.)(hr.)

| Condensing | Coefficient | Liquid cooling | Coefficient |
|---|---|---|---|
| Ammonia . . . . . . | 110 | Engine-jacket water | 125 |
| Freon-12 . . . . . . | 70 | Fuel oil . . . . . . | 25 |
| Gasoline . . . . . . | 80 | Light gas oil . . . . | 65 |
| Light hydrocarbons . | 90 | Light hydrocarbons . | 85 |
| Light naphtha . . . | 75 | Light naphtha . . . | 70 |
| Heavy naphtha . . . | 65 | Reformer liquid | |
| Reformer reactor | | streams . . . . . . | 70 |
| effluent . . . . . . | 70 | Residuum . . . . . . | 15 |
| Low-pressure steam | 135 | Tar . . . . . . . . . | 7 |
| Overhead vapors . . | 65 | | |

| Gas cooling | Operating pressure, lb./sq. in. gage | Pressure drop, lb./sq. in. | Coefficient |
|---|---|---|---|
| Air or flue gas . . . . . | 50 | 0.1 to 0.5 | 10 |
| | 100 | 2 | 20 |
| | 100 | 5 | 30 |
| Hydrocarbon gas . . . . | 35 | 1 | 35 |
| | 125 | 3 | 55 |
| | 1000 | 5 | 80 |
| Ammonia reactor stream | . . . . | . . . | 85 |

Bare-tube external surface is 0.262 sq. ft./lineal ft.
Fin-tube surface/bare-tube surface ratio is 16.9.

SOURCE: Perry and Chilton (eds.), "Chemical Engineers' Handbook," 5th ed., p. 10-40, McGraw-Hill Book Company, New York, 1973.

## Table 2-19. Panel Coils Immersed in Liquid. Over-all Average Heat Transfer Coefficients*

$U$ expressed in B.t.u./(hr.)(sq. ft.)(°F.)

| Hot side | Cold side | Clean-surface coefficients | | Design coefficients, considering usual fouling in this service | |
|---|---|---|---|---|---|
| | | Natural convection | Forced convection | Natural convection | Forced convection |
| Heating applications: | | | | | |
| Steam | Watery solution | 250–500 | 300–550 | 100–200 | 150–275 |
| Steam | Light oils | 50–70 | 110–140 | 40–45 | 60–110 |
| Steam | Medium lube oil | 40–60 | 100–130 | 35–40 | 50–100 |
| Steam | Bunker C or No. 6 fuel oil | 20–40 | 70–90 | 15–30 | 60–80 |
| Steam | Tar or asphalt | 15–35 | 50–70 | 15–25 | 40–60 |
| Steam | Molten sulfur | 35–45 | 45–55 | 20–35 | 35–45 |
| Steam | Molten paraffin | 35–45 | 45–55 | 25–35 | 40–50 |
| Steam | Air or gases | 2–4 | 5–10 | 1–3 | 4–8 |
| Steam | Molasses or corn sirup | 20–40 | 70–90 | 15–30 | 60–80 |
| High temperature hot water | Watery solutions | 115–140 | 200–250 | 70–100 | 110–160 |
| High temperature heat-transfer oil | Tar or asphalt | 12–30 | 45–65 | 10–20 | 30–50 |
| Dowtherm or Aroclor | Tar or asphalt | 15–30 | 50–60 | 12–20 | 30–50 |
| Cooling applications: | | | | | |
| Water | Watery solution | 110–135 | 195–245 | 65–95 | 105–155 |
| Water | Quench oil | 10–15 | 25–45 | 7–10 | 15–25 |
| Water | Medium lube oil | 8–12 | 20–30 | 5–8 | 10–20 |
| Water | Molasses or corn sirup | 7–10 | 18–26 | 4–7 | 8–15 |
| Water | Air or gases | 2–4 | 5–10 | 1–3 | 4–8 |
| Freon or ammonia | Watery solution | 35–45 | 60–90 | 20–35 | 40–60 |
| Calcium or sodium brine | Watery solution | 100–120 | 175–200 | 50–75 | 80–125 |

*Tranter Manufacturing, Inc.

SOURCE: Perry and Chilton (eds.), "Chemical Engineers' Handbook," 5th ed., p. 10–40, McGraw-Hill Book Company, New York, 1973.

## Table 2-20. Jacketed Vessels.  Over-all Coefficients

$U$ expressed in B.t.u./(hr.)(sq. ft.)(°F.)

| Fluid in-side jacket | Fluid in vessel | Wall material | Agitation | $U$ |
|---|---|---|---|---|
| Steam | Water | Enameled C. I.* | 0–400 r.p.m. | 96–120 |
| Steam | Milk | Enameled C. I. | None | 200 |
| Steam | Milk | Enameled C. I. | Stirring | 300 |
| Steam | Milk boiling | Enameled C. I. | None | 500 |
| Steam | Milk | Enameled C. I. | 200 r.p.m. | 86 |
| Steam | Fruit slurry | Enameled C. I. | None | 33–90 |
| Steam | Fruit slurry | Enameled C. I. | Stirring | 154 |
| Steam | Water | C. I. and loose lead lining | Agitated | 4–9 |
| Steam | Water | C. I. and loose lead lining | None | 3 |
| Steam | Boiling $SO_2$ | Steel | None | 60 |
| Steam | Boiling water | Steel | None | 187 |
| Hot water | Warm water | Enameled C. I. | None | 70 |
| Cold water | Cold water | Enameled C. I. | None | 43 |
| Ice water | Cold water | Stoneware | Agitated | 7 |
| Ice water | Cold water | Stoneware | None | 5 |
| Brine, low velocity | Nitration slurry | .................. | 35–58 r.p.m. | 32–60 |
| Water | Sodium al-coholate solution | "Frederking" (cast-in-coil) | Agitated, baffled | 80 |
| Steam | Evaporating water | Copper | ............ | 381 |
| Steam | Evaporating water | Enamelware | ............ | 36.7 |
| Steam | Water | Copper | None | 148 |
| Steam | Water | Copper | Simple stirring | 244 |
| Steam | Boiling water | Copper | None | 250 |
| Steam | Paraffin wax | Copper | None | 27.4 |
| Steam | Paraffin wax | Cast iron | Scraper | 107 |
| Water | Paraffin wax | Copper | None | 24.4 |
| Water | Paraffin wax | Cast iron | Scraper | 72.3 |
| Steam | Solution | Cast iron | Double scrapers | 175–210 |
| Steam | Slurry | Cast iron | Double scrapers | 160–175 |
| Steam | Paste | Cast iron | Double scrapers | 125–150 |
| Steam | Lumpy mass | Cast iron | Double scrapers | 75–96 |
| Steam | Powder (5% moisture) | Cast iron | Double scrapers | 41–51 |

*C. I. = cast iron.

SOURCE: Perry and Chilton (eds.), "Chemical Engineers' Handbook," 5th ed., p. 10-42, McGraw-Hill Book Company, New York, 1973.

## Table 2-21. External Coils. Typical Over-all Coefficients*

$U$ expressed in B.t.u./(hr.)(sq. ft.)(°F.)

| Type of coil | Coil spacing, in. | Fluid in coil | Fluid in vessel | Temp. range, °F. | $U$‡ without cement | $U$ with heat-transfer cement |
|---|---|---|---|---|---|---|
| ⅜ in. o.d. copper tubing attached with bands at 24-in. spacing | 2 | 5 to 50 lb./sq. in. gage steam | Water under light agitation | 158–210 | 1–5 | 42–46 |
| | 3⅛ | | | 158–210 | 1–5 | 50–53 |
| | 6¼ | | | 158–210 | 1–5 | 60–64 |
| | 12½ or greater | | | 158–210 | 1–5 | 69–72 |
| ⅜ in. o.d. copper tubing attached with bands at 24-in. spacing | 2 | 50 lb./sq. in. gage steam | No. 6 fuel oil under light agitation | 158–258 | 1–5 | 20–30 |
| | 3⅛ | | | 158–258 | 1–5 | 25–38 |
| | 6¼ | | | 158–240 | 1–5 | 30–40 |
| | 12½ or greater | | | 158–238 | 1–5 | 35–46 |
| Panel coils | | 50 lb./sq. in. gage steam | Boiling water | 212 | 29 | 48–54 |
| | | Water | Water | 158–212 | 8–30 | 19–48 |
| | | Water | No. 6 fuel oil | 228–278 | 6–15 | 24–56 |
| | | | Water | 130–150 | 7 | 15 |
| | | | No. 6 fuel oil | 130–150 | 4 | 9–19 |

\*Data courtesy of Thermon Manufacturing Co.
†External surface of tubing or side of panel coil facing tank.
‡For tubing, the coefficients are more dependent upon tightness of the coil against the tank than upon either fluid. The low end of the range is recommended.

source: Perry and Chilton (eds.), "Chemical Engineers' Handbook," 5th ed., p. 10–42, McGraw-Hill Book Company, New York, 1973.

## Table 2-22. The Normal Total Emissivity of Various Surfaces

### Metals and Their Oxides

| Surface | t, °F* | Emissivity* |
|---|---|---|
| **Aluminum:** | | |
| Highly polished plate, 98.3 % pure | 440–1,070 | 0.039–0.057 |
| Polished plate | 73 | 0.040 |
| Rough plate | 78 | 0.055 |
| Oxidized at 1110°F | 390–1,110 | 0.11–0.19 |
| Al-surfaced roofing | 100 | 0.216 |
| Calorized surfaces, heated at 1110°F: | | |
| Copper | 390–1,110 | 0.18–0.19 |
| Steel | 390–1,110 | 0.52–0.57 |
| **Brass:** | | |
| Highly polished: | | |
| 73.2 Cu, 26.7 Zn | 476–674 | 0.028–0.031 |
| 62.4 Cu, 36.8 Zn, 0.4 Pb, 0.3 Al | 494–710 | 0.033–0.037 |
| 82.9 Cu, 17.0 Zn | 530 | 0.030 |
| Polished | 100–600 | 0.096 |
| Rolled plate, natural surface | 72 | 0.06 |
| Rolled plate, rubbed with coarse emery | 72 | 0.20 |
| Dull plate | 120–660 | 0.22 |
| Oxidized by heating at 1110°F | 390–1,110 | 0.61–0.59 |
| Chromium (see nickel alloys for Ni-Cr steels) | 100–1,000 | 0.08–0.26 |
| **Copper:** | | |
| Carefully polished electrolytic | 176 | 0.018 |
| Plate, heated long time, covered with thick oxide layer | 77 | 0.78 |
| Plate heated at 1110°F | 390–1,110 | 0.57 |
| Cuprous oxide | 1,470–2,010 | 0.66–0.54 |
| Molten copper | 1,970–2,330 | 0.16–0.13 |
| **Gold:** | | |
| Pure, highly polished | 440–1,160 | 0.018–0.035 |
| **Iron and steel:** | | |
| Metallic surfaces (or very thin oxide layer): | | |
| Electrolytic iron, highly polished | 350–440 | 0.052–0.064 |
| Polished iron | 800–1,880 | 0.144–0.377 |
| Oxidized surfaces: | | |
| Iron plate, pickled, then rusted red | 68 | 0.612 |
| Iron plate, pickled, then completely rusted | 67 | 0.685 |
| Rolled sheet steel | 70 | 0.657 |
| Oxidized iron | 212 | 0.736 |
| Cast iron, oxidized at 1100°F | 390–1,110 | 0.64–0.78 |
| Steel, oxidized at 1100°F | 390–1,110 | 0.79 |
| Smooth oxidized electrolytic iron | 260–980 | 0.78–0.82 |
| Iron oxide | 930–2,190 | 0.85–0.89 |
| Rough ingot iron | 1,700–2,040 | 0.87–0.95 |
| Wrought iron, dull oxidized | 70–680 | 0.94 |
| Steel plate, rough | 100–700 | 0.94–0.97 |
| High-temp alloy steels (see nickel alloys) | | |
| Molten metal: | | |
| Cast iron | 2,370–2,550 | 0.29 |
| Mild steel | 2,910–3,270 | 0.28 |
| **Lead:** | | |
| Pure (99.96 %), unoxidized | 260–440 | 0.057–0.075 |
| Gray oxidized | 75 | 0.281 |
| Oxidized at 390°F | 390 | 0.63 |
| Mercury | 32–212 | 0.09–0.12 |
| Molybdenum filament | 1,340–4,700 | 0.096–0.292 |
| Monel metal, oxidized at 1110°F | 390–1,110 | 0.41–0.46 |
| **Nickel:** | | |
| Electroplated on polished iron, then polished | 74 | 0.045 |
| Technically pure (98.9 Ni + Mn), polished | 440–710 | 0.07–0.087 |
| Electroplated on pickled iron, not polished | 68 | 0.11 |
| Wire | 368–1,844 | 0.096–0.186 |
| Plate, oxidized by heating at 1110°F | 390–1,110 | 0.37–0.48 |
| Nickel oxide | 1,200–2,290 | 0.59–0.86 |
| **Nickel alloys:** | | |
| Chromnickel | 125–1,894 | 0.64–0.76 |
| Nickelin (18–32 Ni, 55–68 Cu, 20 Zn), gray oxidized | 70 | 0.262 |
| **Platinum:** | | |
| Pure, polished plate | 440–1,160 | 0.054–0.104 |
| Strip | 1,700–2,960 | 0.12 –0.17 |
| Filament | 80–2,240 | 0.036–0.192 |
| Wire | 440–2,510 | 0.073–0.182 |

### Metals and Their Oxides

| Surface | $t$, °F* | Emissivity* |
|---|---|---|
| Silver: | | |
| Polished, pure | 440–1,160 | 0.0198–0.0324 |
| Polished | 100–700 | 0.0221–0.0312 |
| Steel, see iron. | | |
| Tantalum filament | 2,420–5,430 | 0.194–0.31 |
| Tin, bright tinned iron sheet | 76 | 0.043 and 0.064 |
| Tungsten: | | |
| Filament, aged | 80–6,000 | 0.032–0.35 |
| Filament | 6,000 | 0.39 |
| Zinc: | | |
| Commercial, 99.1%, polished | 440–620 | 0.045–0.053 |
| Oxidized by heating at 750°F | 750 | 0.11 |
| Galvanized sheet iron, fairly bright | 82 | 0.228 |
| Galvanized sheet iron, gray oxidized | 75 | 0.276 |

### Refractories, Building Materials, Paints, and Miscellaneous

| Surface | $t$, °F | Emissivity |
|---|---|---|
| Asbestos: | | |
| Board | 74 | 0.96 |
| Paper | 100–700 | 0.93–0.945 |
| Brick: | | |
| Red, rough, but no gross irregularities | 70 | 0.93 |
| Silica, unglazed, rough | 1,832 | 0.80 |
| Silica, glazed, rough | 2,012 | 0.85 |
| Grog brick, glazed | 2,012 | 0.75 |
| See also refractory materials | | |
| Carbon: | | |
| T carbon (Gebr. Siemens) 0.9% ash. This started with emissivity at 260°F of 0.72, but on heating changed to values given | 260–1,160 | 0.81–0.79 |
| Carbon filament | 1,900–2,560 | 0.526 |
| Candle soot | 206–520 | 0.952 |
| Lampblack, waterglass coating | 209–362 | 0.959–0.947 |
| | 260–440 | 0.957–0.952 |
| Thin layer on iron plate | 69 | 0.927 |
| Thick coat | 68 | 0.967 |
| Lampblack, 0.003 in. or thicker | 100–700 | 0.945 |
| Enamel, white fused, on iron | 66 | 0.897 |
| Glass, smooth | 72 | 0.937 |
| Gypsum, 0.02 in. thick on smooth or blackened plate | 70 | 0.903 |
| Marble, light gray, polished | 72 | 0.931 |
| Oak, planed | 70 | 0.895 |
| Oil layers on polished nickel (lub. oil) | 68 | |
| Polished surface, alone | .......... | 0.045 |
| +0.001-in. oil | .......... | 0.27 |
| +0.002-in. oil | .......... | 0.46 |
| +0.005-in. oil | .......... | 0.72 |
| ∞ thick oil layer | .......... | 0.82 |
| Oil layers on aluminum foil (linseed oil): | | |
| Aluminum foil | 212 | 0.087 |
| +1 coat oil | 212 | 0.561 |
| +2 coats oil | 212 | 0.574 |
| Paints, lacquers, varnishes | | |
| Snow-white enamel varnish on rough iron plate | 73 | 0.906 |
| Black shiny lacquer, sprayed on iron | 76 | 0.875 |
| Oil paints, sixteen different, all colors | 212 | 0.92–0.96 |
| Aluminum paints and lacquers | | |
| 10% Al, 22% lacquer body, on rough or smooth surface | 212 | 0.52 |
| Paper, thin: | | |
| Pasted on tinned iron plate | 66 | 0.924 |
| Pasted on rough iron plate | 66 | 0.929 |
| Pasted on black lacquered plate | 66 | 0.944 |

### Table 2-22. The Normal Total Emissivity of
### Various Surfaces (Continued)
Refractories, Building Materials, Paints and Miscellaneous

| Surface | $t$, °F* | Emissivity* |
|---------|---------|-------------|
| Plaster, rough lime......................................... | 50–190 | 0.91 |
| Porcelain, glazed.......................................... | 72 | 0.924 |
| Quartz, rough, fused....................................... | 70 | 0.932 |
| Refractory materials, 40 different........................ | 1,110–1,830 | |
|   Poor radiators........................................ | .......... | 0.65–0.70 to 0.75 |
|   Good radiators........................................ | .......... | 0.80–0.85 to 0.85–0.90 |
| Roofing paper............................................. | 69 | 0.91 |
| Rubber: | | |
|   Hard, glossy plate..................................... | 74 | 0.945 |
|   Soft, gray, rough (reclaimed).......................... | 76 | 0.859 |
| Serpentine, polished...................................... | 74 | 0.900 |
| Water..................................................... | 32–212 | 0.95–0.963 |

* When two temperatures and two emissivities are given, they correspond, first to first and second to second, and linear interpolation is permissible.

# PHYSICAL, CHEMICAL, AND MECHANICAL PROPERTIES

**Peter E. Liley, Ph.D., D.I.C.;** Senior Researcher, Center for Numerical Data Analysis and Synthesis and Professor, School of Mechanical Engineering, Purdue University; Member, Institute of Physics, London

## CONTENTS

# PHYSICAL PROPERTIES OF ELEMENTS AND COMPOUNDS
## Abbreviations used in Tables 3–1 and 3–2

| | | | |
|---|---|---|---|
| alk. | alkali (i.e., aqueous NaOH) | g. | gas |
| aq. | aqueous | gly. | glycerol (glycerin) |
| bz. | benzene | h. | hot |
| c. | crystalline | i. | insoluble |
| cc | cubic centimeter | l. | liquid |
| chl. | chloroform | s. | soluble |
| conc. | concentrated | sl. | slightly |
| d. | decomposes | subl. | sublimes |
| dil. | dilute | v. | very |
| et. | ethyl ether | ∞ | soluble in all proportions |

*Formula weights* are based upon the International Atomic Weights of 1941 and are computed to the nearest hundredth. This basis was retained, as later adjustments produce, in general, only minor changes in formula weights. The 1971 values of atomic weights can be found in "Pure and Applied Chemistry," vol. 30, pp. 639–649, 1972.

*Melting point* is recorded in certain cases as d. 82 to indicate that decomposition occurs at 82°F. Where a value such as $-2H_2O$, 82 is given, it indicates loss of 2 moles of water per formula weight of the compound at a temperature of 82°F.

*Boiling point* is given at atmospheric pressure unless otherwise indicated, thus $82^{15\,mm}$ indicates the boiling point is 82°F when the pressure is 15 mm Hg.

*Solubility* is given in parts by weight (of the formula shown at the extreme left) per 100 parts by weight of the solvent; the small superscript indicates the temperature. In the case of gases the solubility is often expressed in the manner of $5cc^{70}$, which indicates that, at 70°F, 5 cc of the gas is soluble in 100 g of the solvent. The symbols of the common mineral acids represent dilute aqueous solutions of the acids.

## GENERAL REFERENCES

Considerations of reader interest, space availability, the system(s) of units employed, copyright considerations, etc., have all influenced the revision of material in previous editions for the present edition. Reference is made at numerous places to various specialized works and also, when appropriate, to more general works. A listing of general works may be useful to readers in need of further information.

Selected Values of Chemical Thermodynamic Properties, *N.B.S. Circ.* 500 plus additional *Tech. Notes* TN 270-1, 1965 *ff*. El-Sabban and Scott, *Bur. Mines Bull.* 654 (1970); reviews chemical thermodynamic properties of 25 organic compounds and lists 27 other general references, plus specific references. Stull, Prophet, et al., "JANAF Thermochemical Tables," 2d ed., NSRDS-NBS-37, 1971. Hultgren, Orr, et al., "Selected Values of Thermodynamic Properties of Metals and Alloys," John Wiley & Sons, Inc., New

York, 1963. Hilsenrath et al., Tables of Thermal Properties of Gases, *N.B.S. Circ.* 564 (1955). "International Critical Tables," McGraw-Hill Book Company, New York. D'Ans and Lax, "Handbook for Chemists and Physicists" (in German), Springer-Verlag, Berlin (often referred to as D'Ans-Lax). "Selected Values of Properties of Hydrocarbons and Related Compounds," American Petroleum Institute Research Project 44, Carnegie Institute of Technology, Pittsburgh, Pa.; continued at Thermodynamics Research Center, Texas A and M University, College Station, Texas, 1972. Timmermans, "Physico-chemical Constants of Pure Organic Compounds," Elsevier Publishing Company, Amsterdam, 1950. Kubaschewski and Evans, "Metallurgical Thermochemistry," John Wiley & Sons, Inc., New York, 1956. "Handbook of Fundamentals," ASHRAE, New York, 1972. "Thermodynamic Properties of Refrigerants," New York, 1968. "Thermophysical Properties of Refrigerants," American Society of Heating, Refrigeration and Air Conditioning Engineers, New York, 1973. Nesmeyanov, "Vapor Pressure of the Elements," Academic Press, New York, 1963; and Moscow original, 1961. Jordan, "Vapor Pressure of Organic Compounds," Interscience, New York, 1954. Hala, Pick, et al., "Vapor Liquid Equilibria," Pergamon Press, New York, 1958. Reisman, "Phase Equilibria," Academic Press, New York, 1970. King, "Phase Equilibrium in Mixtures," Pergamon Press, New York, 1969. Techo, "Bibliography of Thermodynamic Networks of Pure Substances," M.S. Thesis, Georgia Institute of Technology, 1958. Thermodynamic charts for 13 materials, Institute of International Refrigeration, 177 Boulevard Malesherbes, Paris. Gallant, *Series in Hydrocarbon Processing*, **44**(7) (1965-1969), listing physical properties of hydrocarbons. Landolt-Börnstein, "Eigenschaften der Materie in Ihren Aggregatzuständen," two volumes on transport phenomena, 1968 and 1969; other volumes should also be consulted for thermodynamic properties, magnetic properties, etc. Linke and Seidell, "Solubilities of Inorganic and Metalorganic Compounds," various volumes, D. Van Nostrand Company, Princeton, N.J. Stephen and Stephen, "Solubilities of Inorganic and Organic Compounds," The Macmillan Company, New York, 1963. Touloukian et al., "Thermophysical Properties of Matter," TPRC Data Series, Plenum Publishing Corp., New York (thermal conductivity, specific heat, radiative properties of solids, liquids and gases; future volumes include viscosity). Reid and Sherwood, "Properties of Gases and Liquids," McGraw-Hill Book Company, New York, 1966. Janz, "Estimation of Thermodynamic Properties of Organic Compounds," Academic Press, New York, 1958. Canjar and Manning, "Thermodynamic Properties and Reduced Correlations for Gases," Gulf, Houston, 1967. Kaye and Laby, "Tables of Physical and Chemical Constants," 12th ed., John Wiley & Sons, Inc., New York, 1960 and later editions. Mathews, *Chem. Revs.*, **72**, 71 (1972) extensively surveys values of the critical parameters for inorganic substances and compounds. Kudchaker, Alani, and Zwolinski, *Chem. Revs.*, **68**, 659 (1968) extensively review and recommend values of the critical parameters for organic compounds.

## Table 3-1. Physical Properties of Inorganic Elements and Compounds

| Compound | Formula | Mol wt | Sp gr, 60/60° | Mp, °F | Heat of fusion, Btu/lb | Bp, °F | Heat of vaporization at bp, Btu/lb |
|---|---|---|---|---|---|---|---|
| Aluminum | Al | 26.97 | 2.702 | 1220 | 170.2 | 3272 | 3591 |
| Ammonia | $NH_3$ | 17.03 | 0.771 | −107.9 | 142.9 | −28.03 | 589.9 |
| Ammonium bicarbonate | $NH_4HCO_3$ | 79.06 | 1.58 | 225.5 | | subl. | |
| Ammonium carbonate | $(NH_4)_2CO_3 \cdot H_2O$ | 114.11 | | d. 136 | | | |
| Ammonium chloride | $NH_4Cl$ | 53.05 | 1.527 | subl. 635 | | | |
| Ammonium formate | $NH_4CHO_2$ | 63.06 | 1.266 | 241 | | d. 356 | |
| Ammonium hydroxide | $NH_4OH$ | 35.05 | | −107 | | | |
| Ammonium nitrate | $NH_4NO_3$ | 80.05 | 1.725 | 337 | 32.8 | d. 410 | |
| Ammonium phosphate: | | | | | | | |
|  Monobasic | $(NH_4)_2HPO_4$ | 132.11 | 1.619 | d. | | | |
|  Dibasic | $NH_4H_2PO_4$ | 115.08 | 1.803 | | | | |
|  Meta | $(NH_4)_4P_4O_{12}$ | 388.08 | 2.21 | | | | |
| Ammonium sulfamate | $NH_4SO_3NH_2$ | 114.12 | | 270 | | | |
| Ammonium sulfate | $(NH_4)_2SO_4$ | 132.14 | 1.769 | d. 212 | | d. 320 | |
| Antimony | Sb | 121.76 | 6.684 | 1167 | 70.5 | 2516 | 671 |
| Argon | A | 39.94 | | −308.5 | 13.07 | −302.2 | 71.66 |
| Arsenic | As | 299.64 | 5.7 | subl. 1497 (30 atm) | 36.4 | subl. 1139 | 199 |
| Arsenious oxide | $As_4O_6$ | 395.64 | 3.865 | subl. | | 208.4 | 65.06 |
| Barium | Ba | 137.36 | 3.5 | 1562 | 18.34 | d. 2642 | 467.4 |
| Barium carbonate | $BaCO_3$ | 197.37 | 4.29 | | | d. | |
| Barium nitrate | $Ba(NO_3)_2$ | 261.38 | 3.244 | 1098 | 40.6 | | |
| Barium sulfate | $BaSO_4$ | 233.42 | 4.499 | d. 2876 | 74.8 | | |
| Beryllium | Be | 9.02 | 1.816 | 2343 | 499. | 5013 | |
| Bismuth | Bi | 209.00 | 9.80 | 520 | 21.57 | 2642 | 397 |
| Boric acid | $H_3BO_3$ | 61.84 | 1.435 | d. 365 | | d. | |
| Boron | B | 10.82 | 2.32 | 4172 | | 4622 | |
| Bromine | $Br_2$ | 159.83 | 3.119 | 19.04 | 29.05 | 137.8 | 83.6 |
| Cadmium | Cd | 112.41 | 8.65 | 610 | 23.4 | 1413 | 382.2 |
| Calcium | Ca | 40.08 | 1.55 | 1490 | 100.1 | 2192 | 1,643 |
| Calcium acetate | $Ca(C_2H_3O_2)_2 \cdot H_2O$ | 176.18 | | d. | | | |
| Calcium carbonate | $CaCO_3$ | 100.09 | 2.711 | 2442 | 228.3 | | |
| Calcium chloride | $CaCl_2$ | 110.99 | 2.152 | 1422 | 98.9 | >2900 | |
| Calcium fluoride | $CaF_2$ | 78.08 | 3.180 | 2426 | 94.5 | | |
| Calcium hydroxide | $Ca(OH)_2$ | 74.10 | 2.2 | $-H_2O$ 1075 | 39.0 | | |
| Calcium nitrate | $Ca(NO_3)_2$ | 236.16 | 1.82 | 108.9 | | d. 392 | |
| Calcium phosphate: | | | | | | | |
|  Monobasic | $CaH_4(PO_4)_2 \cdot H_2O$ | 252.09 | 2.22 | $-H_2O$ 212 | | | |
|  Dibasic | $CaHPO_4 \cdot 2 H_2O$ | 172.10 | 2.306 | | | | |
|  Tribasic | $Ca_3(PO_4)_2$ | 310.20 | 3.14 | d. 3038 | | | |

## Table 3-1. Physical Properties of Inorganic Elements and Compounds (Continued)

| Compound | Heat of formation Btu/lb mole at 77°F | Free energy of formation Btu/lb mole at 77°F | Solubility, g/100 ml solvent | | | Critical temp, °F | Critical press., atm | Critical density lb/ft³ |
|---|---|---|---|---|---|---|---|---|
| | | | Water | Alcohol | Ether, etc. | | | |
| Aluminum | 0 (c.) | 0 (c.) | i. | i. | s. HCl, $H_2SO_4$, alk. | | | |
| Ammonia | −19,728 (g.) | −6.665 (g.) | 89.9[32] | 13.2[86] | s. et. | 270.1 | 111.3 | 14.7 |
| Ammonium bicarbonate | −402,120 (aq.) | −295,380 (aq.) | 11.9[32] | i. | i. acetone | | | |
| Ammonium carbonate | −135,400 (c.) | −87,460 (c.) | 100[59] | i. | i. $CS_2$, $NH_3$ | | | |
| Ammonium chloride | | | 29.7[52] | 0.06[46] | s. $NH_3$ | | | |
| Ammonium formate | | | 102[22] | s. | s. $NH_3$ | | | |
| Ammonium hydroxide | −157,660 (aq.) | | s. | | 17.1[88] methanol; i. et. | | | |
| Ammonium nitrate | −157,320 (c.) | | 118.3[32] | 3.8[86] | | | | |
| Ammonium phosphate: | | | | | | | | |
|   Monobasic | | | 42.9[32] | i. | i. acetone | | | |
|   Dibasic | | | 22.7[52] | | i. acetone | | | |
|   Meta. | | | s. | | | | | |
| Ammonium sulfamate | | | 134[22] | i. | i. $NH_3$, acetone | | | |
| Ammonium sulfate | −507,130 (c.) | −387,108 (c.) | 76.6[52] | i. | s. hot conc. $H_2SO_4$ | | | |
| Antimony | 0 (c.) | 0 (c.) | i. | | | | | |
| Argon | 0 (g.) | 0 (g.) | 5.06 cc[32] | i. | | −188 | 48.0 | 33.2 |
| Arsenic | 0 (c.) | 0 (g.) | i. | i. | s. $HNO_3$, hot alk. | | | |
| Arsenious oxide | −554,760 (c.) | −485,280 (c.) | sl.s. | d. | i. et. | | | |
| Barium | 0 (c.) | 0 (c.) | d. | i. | | | | |
| Barium carbonate | −511,560 (c.) | −488,520 (c.) | 0.0022[54] | i. | s. dil. acid | | | |
| Barium nitrate | −426,582 (c.) | −341,890 (c.) | 5.0[32] | i. | s. $HNO_3$ | | | |
| Barium sulfate | −612,360 (c.) | −564,120 (c.) | 0.000115[22] | i. | 0.24[7] et, 22.2[96] gly. | | | |
| Beryllium | 0 (c.) | 0 | i. | | | | | |
| Bismuth | 0 (c.) | 0 (c.) | i. | | s. $HNO_3$ | | | |
| Boric acid | 0 (c.) | 0 (c.) | 266[32] | s. | s. et, alk. | | | |
| Boron | 0 (l.) | 0 (l.) | i. | i. | s. acid | | | |
| Bromine | 0 (c.) | 0 (c.) | 4.22[32] | s. | s. acid | 592 | 102 | 73.7 |
| Cadmium | 0 (c.) | 0 (c.) | d. | sl.s. | s. acid, $NH_4Cl$ | | | |
| Calcium | 0 (c.) | 0 (c.) | 5.2[25] | sl.s. | | | | |
| Calcium acetate | −655,380 (aq.) | −560,340 (aq.) | | s. | | | | |
| Calcium carbonate | −521,100 (c.) | −487,440 (c.) | 0.0014[77] | | | | | |
| Calcium chloride | −343,100 (c.) | −323,600 (c.) | 59.5[32] | s. | s. $NH_4Cl$ | | | |
| Calcium fluoride | −522,400 (c.) | | 0.0016[64] | | | | | |
| Calcium hydroxide | −424,040 (c.) | −385,000 (c.) | 0.185[52] | | | | | |
| Calcium nitrate | −403,290 (c.) | −319,280 (c.) | 266[32] | | | | | |
| Calcium phosphate: | | | | | | | | |
|   Monobasic | | | 0.02[78] | | | | | |
|   Dibasic | | | 0.0025 | | | | | |
|   Tribasic | | | | i. | i. acetic acid | | | |

# Table 3-1. Physical Properties of Inorganic Elements and Compounds (Continued)

| Compound | Formula | Mol wt | Sp gr 60/60° | Mp, °F | Heat of fusion, Btu/lb | Bp, °F | Heat of vaporization at bp, Btu/lb |
|---|---|---|---|---|---|---|---|
| Calcium sulfate (anhydrite) | $CaSO_4$ | 136.14 | 2.96 | 2642 | 88.6 | −2 H₂O 325 | |
| Calcium sulfate (gypsum) | $CaSO_4 \cdot 2\,H_2O$ | 172.17 | 2.32 | −1½ H₂O 262 | | | |
| Carbon, amorphous | $C$ | 12.01 | 1.8–2.1 | >6300 | | 7600 | |
| Carbon, graphite | $C$ | 12.01 | 2.26 | >6300 | 1649 | 7600 | |
| Carbon dioxide | $CO_2$ | 44.01 | 1.261 | −69.9 | 77.7 | subl. −109.3 | 246.6 |
| Carbon disulfide | $CS_2$ | 76.13 | 1.261 | −163.5 | 24.8 | 115.3 | 92.8 |
| Carbon monoxide | $CO$ | 28.01 | $0.814^{-219}$ | −341 | 12.85 | −314 | 123.8 |
| Cerium | $Ce$ | 140.13 | 6.9 | 1193 | 27.2 | 2552 | |
| Chlorine | $Cl_2$ | 70.91 | $1.56^{-28.5}$ | −150.9 | 38.9 | −30.3 | |
| Chromium | $Cr$ | 52.01 | 7.1 | 2939 | 136.0 | 3992 | |
| Chromium trioxide | $CrO_3$ | 100.01 | 2.70 | d. 386.6 | | | |
| Cobalt | $Co$ | 58.94 | 8.9 | 2696 | 111.8 | 5250 | 2,061 |
| Copper | $Cu$ | 63.59 | 8.92 | 1981 | 88.03 | 4172 | |
| Cupric oxide | $CuO$ | 79.57 | 6.40 | d. 1880 | 63.8 | | |
| Cupric sulfate | $CuSO_4 \cdot 5\,H_2O$ | 249.71 | 2.286 | −4 H₂O 230 | | −5 H₂O 480 | |
| Ferric chloride | $FeCl_3 \cdot 6\,H_2O$ | 270.32 | | 99 | | 536 | |
| Ferric sulfate | $Fe_2(SO_4)_3$ | 399.88 | 3.097 | d. 900 | | | |
| Gold | $Au$ | 197.02 | 19.3 | 1945 | 27.6 | 4700 | 747 |
| Hydrazine | $N_2H_4$ | 32.05 | 1.011 | 34.5 | | 236.3 | |
| Hydrazine hydrate | $N_2H_4 \cdot H_2O$ | 50.06 | 1.03 | −40 | | 245.3 | |
| Hydrobromic acid | $HBr$ | 80.92 | | −123 | 12.8 | −89 | 93.6 |
| Hydrochloric acid | $HCl$ | 36.47 | | −168 | 23.49 | −121 | 190.5 |
| Hydrocyanic acid | $HCN$ | 27.03 | 0.697 | 6.8 | 133.8 | 78.8 | 401.4 |
| Hydrofluoric acid | $HF$ | 20.01 | 0.988 | −117 | 98.4 | 66.9 | 671 |
| Hydrogen | $H_2$ | 2.016 | $0.0709^{-423}$ | −434.4 | 25.00 | −422.9 | 192.8 |
| Hydrogen peroxide | $H_2O_2$ | 34.02 | 1.438 | 30.4 | 133 | 304.5 | 543 |
| Hydrogen sulfide | $H_2S$ | 34.08 | | −117.2 | 30 | −75.3 | 235.7 |
| Hydroxylamine | $NH_2OH$ | 33.03 | 1.35 | 93.2 | | | |
| Iodine | $I_2$ | 253.84 | 4.93 | 236.3 | 25.9 | 363.8 | 73.7 |
| Iron | $Fe$ | 55.85 | 7.86 | 2795 | 86.4 | 5432 | 2926 |
| Lead | $Pb$ | 207.21 | 11.337 | 621.5 | 10.6 | 2948 | 365 |
| Lead chromate | $PbCrO_4$ | 323.22 | 6.12 | 1551 | | d. | |
| Lead formate | $Pb(HCO_2)_2$ | 297.25 | 4.56 | d. 375 | | | |
| Lead nitrate | $Pb(NO_3)_2$ | 331.23 | 4.53 | d. 880 | | | |
| Lithium | $Li$ | 6.94 | 0.53 | 367 | 285 | 2435 | 8,364 |
| Magnesium | $Mg$ | 24.32 | 1.74 | 1204 | 160 | 2030 | 2,407 |
| Magnesium sulfate | $MgSO_4 \cdot 7\,H_2O$ | 246.49 | 1.68 | d. 158 | | | |
| Manganese | $Mn$ | 54.93 | 7.20 | 2300 | 113 | 3450 | 1,807 |
| Mercury | $Hg$ | 200.61 | 13.546 | −38 | 5.00 | 674.4 | 125.4 |
| Molybdenum | $Mo$ | 95.95 | 10.2 | 4750 | 124 | 6690 | 2,400 |
| Nickel | $Ni$ | 58.69 | 8.90 | 2645 | 129 | 5250 | 2,677 |
| Nickel sulfate | $NiSO_4 \cdot 6\,H_2O$ | 262.85 | 2.07 | | | −6 H₂O 536 | |
| Nitric acid | $HNO_3$ | 63.02 | 1.52 | −44 | 17.1 | 187 | |
| Nitrogen | $N_2$ | 28.02 | $0.808^{-220}$ | −345.8 | 11.05 | −320 | 85.8 |
| Nitric oxide | $NO$ | 30.01 | $1.269^{-228.4}$ | −258 | 33.0 | −240 | 198.3 |

## Table 3-1. Physical Properties of Inorganic Elements and Compounds (Continued)

| Compound | Heat of formation Btu/lb mole at 77°F | Free energy of formation at 77°F | Solubility, g/100 ml solvent | | | Critical temp, °F | Critical press, atm | Critical density lb/ft³ |
|---|---|---|---|---|---|---|---|---|
| | | | Water | Alcohol | Ether, etc. | | | |
| Calcium sulfate (anhydrite) | −605,840 (c.) | −557,640 (c.) | 0.298$^{68}$ | .... | s. acid | | | |
| Calcium sulfate (gypsum) | −862,790 (c.) | −765,850 (c.) | 0.223$^{22}$ | .... | s. acid | | | |
| Carbon, amorphous | 0 | 0 | i. | i. | i. | | | |
| Carbon, graphite | 0 | 0 | i. | i. | i. | | | |
| Carbon, graphite (dioxide) | −169,294 (g.) | −169,668 (g.) | 179.7 cc$^{32}$ | .... | s. acid, alk. | 87.8 | 72.9 | 29.2 |
| Carbon disulfide | +50,600 (g.) | +29,030 (g.) | 0.2$^{33}$ | s. | s. et. | 221 | 61 | 18.8 |
| Carbon monoxide | −47,549 (g.) | −59,054 (g.) | 0.0044$^{22}$ | i. | s. $Cu_2Cl_2$ | −220 | 34.5 | 35.8 |
| Cerium | 0 (c.) | 0 (c.) | 1.46$^{22}$ | .... | s. dil. acid | | | |
| Chlorine | 0 (g.) | 0 (g.) | i. | s. | s. alk. | 291 | 76.1 | |
| Chromium | 0 (c.) | 0 (c.) | i. | .... | s. HCl | | | |
| Chromium trioxide | −250,740 (c.) | | 164.9$^{22}$ | s. | s. $H_2SO_4$, et. | | | |
| Cobalt | 0 (c.) | 0 (c.) | i. | .... | s. acid | | | |
| Copper | 0 (c.) | 0 (c.) | i. | .... | s. $HNO_3$ | | | |
| Cupric oxide | −69,300 (c.) | −57,400 (aq.) | i. | .... | s. acid | | | |
| Cupric sulfate | −361,400 (aq.) | −288,340 (aq.) | 24.3$^{22}$ | 1.$^{37.4}$ | s. acetone | | | |
| Ferric chloride | −231,300 (aq.) | −173,700 (aq.) | 246$^{22}$ | s. | i. $H_2SO_4$, $NH_3$ | | | |
| Ferric sulfate | −1,176,000 (aq.) | −960,100 (aq.) | sl.s. | i. | s. aqua regia | | | |
| Gold | 0 (c.) | 0 (c.) | i. | s. | | | | |
| Hydrazine | +21,710 (l.) | | ∞ | ∞ | i. et. | 716 | 145 | |
| Hydrazine hydrate | −104,330 (l.) | | ∞ | ∞ | i. et. | | | |
| Hydrobromic acid | −15,590 (g.) | −22,900 (g.) | 221$^{22}$ | s. | s. et. | 194.0 | 84.0 | 26 |
| Hydrochloric acid | −39,710 (g.) | −41,000 (g.) | 82.3$^{22}$ | s. | ∞ et. | 124.5 | 81.5 | 12.2 |
| Hydrocyanic acid | +55,980 (g.) | +50,290 (g.) | ∞ | ∞ | ∞ et. | 362.3 | 53.2 | |
| Hydrofluoric acid | −115,600 (g.) | −116,500 (g.) | ∞ | | sl.s. Fe, Pd, Pt | 446.4 | 12.8 | 1.94 |
| Hydrogen | 0 (g.) | 0 (g.) | 2.1 cc$^{22}$ | | i. et. | −399.8 | | |
| Hydrogen peroxide | −81,290 (l.) | −50,810 | ∞ | 9.54 cc$^{59}$ | s. $CS_2$ | | | |
| Hydrogen sulfide | −8,590 (g.) | −14,130 (g.) | 437 cc$^{59}$ | s. | s. acid | 212.7 | 88.9 | 21.8 |
| Hydroxylamine | | | s. | s. | s. et. | | | |
| Iodine | 0 (c.) | 0 (c.) | 0.0162$^{22}$ | i. | s. acid; i. alk. | 954 | | |
| Iron | 0 (c.) | 0 (c.) | i. | | s. $HNO_3$ | | | |
| Lead | 0 (c.) | 0 (c.) | i. | | s. acid, alk. | | | |
| Lead chromate | | | 0.00007$^{68}$ | | | | | |
| Lead formate | | | 1.6$^{1}$ | i. | | | | |
| Lead nitrate | −192,380 (c.) | | 38.8$^{22}$ | 8.8$^{77}$ | s. acid, $NH_3$ | | | |
| Lithium | 0 (c.) | 0 (c.) | d. | | s. acid | | | |
| Magnesium | 0 (c.) | 0 (c.) | d. | | s. dil. acid | | | |
| Magnesium sulfate | −585,700 (aq.) | −510,980 (aq.) | 72.4$^{22}$ | s. | s. $HNO_3$ | | | |
| Manganese | 0 (l.) | 0 (l.) | d. | | s. hot conc. $H_2SO_4$ | | | |
| Mercury | 0 (l.) | 0 (l.) | i. | | s. dil. $HNO_3$ | | | |
| Molybdenum | 0 (c.) | 0 (c.) | i. | | v.s. $NH_4OH$ | | | |
| Nickel | 0 (c.) | 0 (c.) | | | | | | |
| Nickel sulfate | −416,340 (aq.) | −337,680 (aq.) | 131$^{122}$ | v.s. | | | | |
| Nitric acid | −74,450 (l.) | −34,290 (l.) | ∞ | explodes | | | | |
| Nitrogen | 0 (g.) | 0 (g.) | 2.35 cc$^{32}$ | sl.s. | | −232.6 | 33.5 | 14.4 |
| Nitric oxide | +38,880 (g.) | +37,294 (g.) | 7.34 cc$^{22}$ | 26.6 cc | | | | |

## Table 3-1. Physical Properties of Inorganic Elements and Compounds (Continued)

| Compound | Formula | Mol wt | Sp gr 60°/60° | Mp, °F | Heat of fusion, Btu/lb | Bp, °F | Heat of vaporization at bp, Btu/lb |
|---|---|---|---|---|---|---|---|
| Nitrogen dioxide | $NO_2$ | 46.01 | $1.448^{68}$ | 15.3 | 108.4 | 70.3 | 137.7 |
| Nitrous oxide | $N_2O$ | 44.02 | $1.226^{-128}$ | −152.1 | 63.91 | −131 | 161.5 |
| Oxygen | $O_2$ | 32.00 | $1.14^{-297}$ | −361.1 | 5.96 | −297 | 91.63 |
| Ozone | $O_3$ | 48.00 | $1.71^{-297}$ | −420 | | −170 | 108 |
| Phosphoric acid, meta | $HPO_3$ | 79.99 | 2.2–2.5 | subl. | | | |
| Phosphoric acid, ortho | $H_3PO_4$ | 98.00 | 1.834 | 108.23 | 46.3 | −½ $H_2O$ 415 | 182 |
| Phosphorous acid, yellow | $P_4$ | 123.92 | 1.82 | 111.4 | 8.93 | 536 | 131 |
| Phosphoric oxide | $P_2O_5$ | 141.96 | 2.387 | subl. 480 | 11.3 | | 986 |
| Platinum | $Pt$ | 195.23 | 21.45 | 3190 | 43.3 | 7770 | 921 |
| Potassium | $K$ | 39.10 | 0.86 | 144 | 26.4 | 1400 | |
| Potassium sulfate | $K_2SO_4$ | 174.25 | 2.662 | | 83.7 | | |
| Silicic acid, ortho | $H_4SiO_4$ | 99.09 | 1.576 | | | | |
| Silicon | $Si$ | 28.06 | 2.4* | 2590 | 607.5 | 4700 | |
| Silicon chloride | $SiCl_4$ | 169.89 | 1.50 | −94 | 19.5 | 135.7 | 72.7 |
| Silver | $Ag$ | 107.88 | 10.5 | 1761 | 45.0 | 3542 | 1,013 |
| Silver nitrate | $AgNO_3$ | 169.89 | 4.352 | 414 | 29.2 | d. 820 | |
| Sodium | $Na$ | 22.997 | 0.97 | 207.5 | 49.3 | 1616 | 1,810 |
| Sodium bicarbonate | $NaHCO_3$ | 84.01 | 2.20 | −$CO_2$ 520 | | | |
| Sodium carbonate (sal soda) | $Na_2CO_3 \cdot 10\ H_2O$ | 286.16 | 1.46 | | | | |
| Sodium carbonate (soda ash) | $Na_2CO_3$ | 106.00 | 2.533 | 1564 | 119 | d. | 1,257 |
| Sodium chloride | $NaCl$ | 58.45 | 2.163 | 1473 | 222 | 2575 | |
| Sodium hydroxide | $NaOH$ | 40.00 | 2.13 | 605 | 90 | 2534 | |
| Sodium nitrate | $NaNO_3$ | 85.01 | 2.257 | 586 | 79.6 | d. 716 | |
| Sodium silicate, meta | $Na_2SiO_3$ | 122.05 | | 1990 | 152 | | |
| Sodium silicate, ortho | $Na_4SiO_4$ | 184.05 | | 1864 | | | |
| Sodium sulfate | $Na_2SO_4 \cdot 10\ H_2O$ | 322.21 | 1.464 | 90.3 | | −10 $H_2O$ 212 | |
| Strontium nitrate | $Sr(NO_3)_2$ | 211.65 | 2.986 | 1058 | | | |
| Strontium sulfate | $SrSO_4$ | 183.69 | 3.96 | 28.76 | | | |
| Sulfamic acid | $NH_2SO_3H$ | 97.09 | 2.03 | d. 400 | | | |
| Sulfur, amorphous | $S$ | 32.06 | 2.046 | 248 | | 832.3 | 141.8 |
| Sulfur, monoclinic | $S_8$ | 256.48 | 1.96 | 246 | 23.76 | 832.3 | |
| Sulfur, rhombic | $S_8$ | 256.48 | 2.07 | 235 | 49.7 | 832.3 | 167.5 |
| Sulfur dioxide | $SO_2$ | 64.06 | 1.434 | −103.9 | 43.3 | 14.0 | |
| Sulfuric acid | $H_2SO_4$ | 98.08 | 1.834 | 50.88 | | d. 644 | |
| Sulfurous acid | $H_2SO_3$ | 66.08 | 0.834 | | | | |
| Tantalum | $Ta$ | 180.88 | 16.6 | 5160 | 45.5 | >7400 | |
| Tellurium | $Te$ | 127.64 | α 6.24, β 6.00 | 845 | | 2534 | |

## Table 3-1. Physical Properties of Inorganic Elements and Compounds (Continued)

| Compound | Heat of formation Btu/lb mole at 77°F | Free energy of formation Btu/lb mole at 77°F | Solubility, g/100 ml solvent — Water | Alcohol | Ether, etc. | Critical temp, °F | Critical press., atm | Critical density lb/ft³ |
|---|---|---|---|---|---|---|---|---|
| Nitrogen dioxide | +14,330 (g.) | +22,068 (g.) | d. | s. | s. $HNO_2$ | | | |
| Nitrous oxide | +35,190 (g.) | +44,680 (g.) | $130.52\ cc^{32}$ | sl.s. | s. $H_2SO_4$ | | | |
| Oxygen | 0 (g.) | 0 (g.) | $4.89\ cc^{32}$ | | | −181.1 | 50.1 | 26 |
| Ozone | +60,980 (g.) | +69,950 (g.) | $0.494\ cc^{32}$ | | | 23 | 67 | 33.5 |
| Phosphoric acid, meta | ........ | ........ | s. | | | | | |
| Phosphoric acid, ortho | −536,780 (aq.) | +486,000 (aq.) | $234^{078}$ | 0.04 | | | | |
| Phosphorous, yellow | 0 (c.) | 0 (c.) | 0.0003 | | $1,000^{60}\ CS_2$ | | | |
| Phosphoric oxide | −648,000 (c.) | ........ | forms $H_3PO_4$ | | i. $NH_3$ | | | |
| Platinum | 0 (c.) | 0 (c.) | i. | s. | s. aqua regia | | | |
| Potassium | 0 (c.) | 0 (c.) | d. | i. | s. acid | | | |
| Potassium sulfate | −605,660 (aq.) | −559,730 (aq.) | $7.35^{32}$ | i. | i. acetone | | | |
| Silicic acid, ortho | −613,100 (c.) | ........ | sl.s. | | s. alk. | | | |
| Silicon | ........ | 0 (c.) | i. | d. | | | | |
| Silicon chloride | −270,000 (l.) | −241,000 (l.) | i. | | d. conc. $H_2SO_4$ | 451 | | |
| Silver | 0 (c.) | 0 (c.) | i. | v.sl.s. | s. $HNO$ | | | |
| Silver nitrate | −52,900 (c.) | −13,790 (c.) | $122^{32}$ | d. | s. gly. | | | |
| Sodium | ........ | ........ | forms NaOH | i. | i. ba. | | | |
| Sodium bicarbonate | −399,800 (aq.) | −365,166 (aq.) | $6.9^{32}$ | i. | i. et. | | | |
| Sodium carbonate (sal soda) | −495,230 (aq.) | −452,450 (aq.) | $21.5^{32}$ | i. | i. conc. HCl | | | |
| Sodium carbonate (soda ash) | −485,030 (c.) | −449,190 (c.) | $7.1^{32}$ | sl.s. | v.s. et. | | | |
| Sodium chloride | −176,980 (c.) | −165,409 (c.) | $35.7^{32}$ | v.s. | s. $NH_3$ | | | |
| Sodium hydroxide | −183,550 (c.) | −163,080 (c.) | $42^{33}$ | sl.s. | | | | |
| Sodium nitrate | −210,780 (c.) | −157,720 (c.) | $73^{32}$ | i. | | | | |
| Sodium silicate, meta | ........ | ........ | s. | | | | | |
| Sodium silicate, ortho | ........ | ........ | s. | i. | | | | |
| Sodium sulfate | −594,900 (c.) | −544,280 (c.) | $36^{59}$ | i. | s. $NH_3$ | | | |
| Strontium nitrate | −411,710 (aq.) | −334,260 (aq.) | $40^{32}$ | 0.012 | sl.s. acids | | | |
| Strontium sulfate | −621,000 (aq.) | −556,700 (aq.) | $0.011^{33}$ | i. | sl.s. acetone; i. et. | | | |
| Sulfamic acid | ........ | ........ | $20^{32}$ | sl.s. | sl.s. $CS_2$ | | | |
| Sulfur, amorphous | ........ | ........ | i. | | s. $CS_2$ | | | |
| Sulfur, monoclinic | −128 (c.) | −41 (c.) | i. | | $24^{22}\ CS_2$ | | | |
| Sulfur, rhombic | ........ | ........ | s. | s. | s. acetone | 1,904 | 116 | |
| Sulfur dioxide | −127,690 (g.) | −129,025 (g.) | $28.8^{32}$ | d. | | 315.5 | 77.8 | 32.7 |
| Sulfuric acid | −381,650 (aq.) | ........ | ∞ | s. | s. et. | | | |
| Sulfurous acid | −264,380 (aq.) | −231,370 (aq.) | s. | | i. $HNO_3$ | | | |
| Tantalum | 0 (c.) | 0 (c.) | i. | | i. $HNO_3$ | | | |
| Tellurium | 0 (c.) | 0 (c.) | i. | | s. $HNO_3$ | | | |

## Table 3-1. Physical Properties of Inorganic Elements and Compounds (Continued)

| Compound | Formula | Mol wt | Sp gr 60°/60° | Mp, °F | Heat of fusion, Btu/lb | Bp, °F | Heat of vaporization at bp, Btu/lb |
|---|---|---|---|---|---|---|---|
| Tin............ | Sn | 118.70 | 7.31 | 449.3 | 6.52 | 4100 | 258 |
| Titanium....... | Ti | 47.90 | 4.50 | 3270 | ........ | >5400 | |
| Titanium dioxide.... | TiO$_2$ | 79.90 | 4.26 | d.2980 | 257 | | |
| Tungsten........ | W | 183.92 | 19.3 | 6100 | 82.2 | 10,650 | 1,722 |
| Uranium........ | U | 238.07 | 18.7 | <3362 | 143.5 | | |
| Water.......... | H$_2$O | 18.016 | 1.00 | 32 | | 212 | 972 |
| Water, heavy..... | D$_2$O | 20.029 | 1.107 | 38.88 | | 214.56 | |
| Zinc........... | Zn | 65.38 | 7.140 | 787 | 43.9 | 1665 | 755 |
| Zinc oxide....... | ZnO | 81.38 | 5.47 | >3270 | 98.9 | | |
| Zirconium....... | Zr | 91.22 | 6.4 | 3100 | ........ | >5250 | |

Table 3-1. Physical Properties of Inorganic Elements and Compounds (Continued)

| Compound | Heat of formation Btu/lb mole at 77°F | Free energy of formation Btu/lb mole at 77°F | Solubility, g/100 ml solvent | | | Critical temp, °F | Critical press., atm | Critical density lb/ft³ |
|---|---|---|---|---|---|---|---|---|
| | | | Water | Alcohol | Ether, etc. | | | |
| Tin................. | 0 (c.) | 0 (c.) | i. | ..... | s. HCl | | | |
| Titanium............ | 0 (c.) | 0 (c.) | i. | ..... | s. acids | | | |
| Titanium dioxide..... | −405,000 (c.) | −381,400 (c.) | i. | ..... | s. $H_2SO_4$ | | | |
| Tungsten............ | 0 (c.) | 0 (c.) | i. | ..... | s. hot conc. KOH | | | |
| Uranium............. | −122,971 (l.) | −102,042 (l.) | i. | ..... | s. acid; i. alk. | | | |
| Water............... | | | ∞ | ∞ | sl.s. et. | 705.6 | 218.3 | 20. |
| Water, heavy........ | | | ∞ | ∞ | sl.s. et. | | | |
| Zinc................ | 0 (c.) | 0 (c.) | i. 0.00042[64] | ..... | s. acids, alk. | | | |
| Zinc oxide.......... | −150,050 (c.) | −137,140 (c.) | i. | ..... | s. acids, alk. | | | |
| Zirconium........... | 0 (c.) | 0 (c.) | i. | ..... | s. HF | | | |

## Table 3-2. Physical Properties of Organic Compounds

| Compound | Formula | Mol. wt | Sp gr, 60°/60° | Mp, °F | Heat of fusion, Btu/lb | Bp, °F | Heat of vaporization at bp, Btu/lb |
|---|---|---|---|---|---|---|---|
| Acetic acid | $CH_3COOH$ | 60.05 | 1.049 | 61.9 | 84.0 | 244.6 | 174.2 |
| Acetic anhydride | $(CH_3CO)_2O$ | 102.09 | 1.087 | −99.6 | | 284.0 | 119.1 |
| Acetone | $CH_3COCH_3$ | 58.08 | 0.792 | −139.0 | 42.2 | 141.8 | 223 |
| Acetylene | $CH{:}CH$ | 26.04 | Gas | −114 | 41.5 | −118.5 subl. | |
| Acrylonitrile | $CH_2{:}CHCN$ | 53.06 | 0.797 | −115.6 | | 173.0 | |
| Adipic acid | $COOH(CH_2)_4COOH$ | 146.14 | 1.366 | 306 | | 509.0 | |
| Aniline | $C_6H_5NH_2$ | 93.12 | 1.022 | −11.2 | 48.8 | 363.9 | 186.6 |
| Anthracene | $C_6H_4{:}(CH)_2{:}C_6H_4$ | 178.22 | 1.25 | 423 | 69.7 | 669 | |
| Benzene | $C_6H_6$ | 78.11 | 0.879 | 41.9 | 54.2 | 176.2 | 169.5 |
| Butadiene-1,3 | $(CH{:}CH_2)_2$ | 54.09 | 0.65 | −164.0 | | 24.1 | |
| Butane | $CH_3(CH_2)_2CH_3$ | 58.12 | 0.60 | −217 | 34.5 | 31.1 | 165.8 |
| Butyl cellosolve | $C_4H_9OCH_2CH_2OH$ | 118.17 | 0.903 | | | 339 | |
| Carbon dioxide | $CO_2$ | 44.01 | Gas | −69.9 | 77.6 | −109.3 subl. | 246.6 |
| Carbon monoxide | $CO$ | 28.01 | Gas | −341 | 12.8 | −310 | 93.8 |
| Carbon tetrachloride | $CCl_4$ | 153.84 | 1.595 | −9.1 | 74.8 | 168.8 | 83.6 |
| Cellosolve | $C_2H_5OCH_2CH_2OH$ | 90.12 | 0.931 | | | 275.2 | |
| Cellulose | $(C_6H_{10}O_5)_x$ | $(162.14)_x$ | 1.27–1.61 | | | | |
| o-Cresol | $CH_3C_6H_4OH$ | 108.13 | 1.047 | 86 | | 376.7 | 181 |
| m-Cresol | $CH_3C_6H_4OH$ | 108.13 | 1.034 | 52.7 | | 397 | |
| p-Cresol | $CH_3C_6H_4OH$ | 108.13 | 1.035 | 96.8 | 47.3 | 396.5 | |
| Cyclohexane | $C_6H_{12}$ | 84.16 | 0.779 | 43.8 | 13.6 | 177.3 | 153.7 |
| Cyclohexanol | $C_6H_{11}OH$ | 100.16 | 0.962 | 75.2 | 7.5 | 322.7 | 194.8 |
| Cyclohexene | $C_6H_{10}$ | 82.14 | 0.810 | −154.7 | | 181.4 | |
| Dimethyl amine | $(CH_3)_2NH$ | 45.08 | 0.680 | −140.8 | | 45.3 | 145.4 |
| Dimethyl aniline | $(CH_3)_2NC_6H_5$ | 121.18 | 0.956 | 36.5 | | 379 | |
| Dimethyl phthalate | $C_6H_4(CO_2CH_3)_2$ | 194.18 | 1.189 | | | 536 | |
| Diphenyl | $(C_6H_5)_2$ | 154.20 | 1.18 | 158 | 51.8 | 490.1 | |
| Diphenyl amine | $(C_6H_5)_2NH$ | 169.22 | 1.159 | 127 | 45.4 | 576 | |
| Ethyl acetate | $CH_3CO_2C_2H_5$ | 88.10 | 0.901 | −116.4 | 51.2 | 170.8 | 183.6 |
| Ethyl benzene | $C_2H_5C_6H_5$ | 106.16 | 0.867 | −138.9 | 37.1 | 277.2 | 145.8 |
| Ethyl chloride | $C_2H_5Cl$ | 64.52 | 0.916 | −218 | | 55.4 | 166.5 |
| Ethylene | $C_2H_4$ | 28.05 | Gas | −272.4 | 51.4 | −153.7 | 207.6 |
| Ethylene dibromide | $CH_2BrCH_2Br$ | 187.88 | 2.180 | 50 | 24.3 | 268.7 | 83.2 |
| Ethylene oxide | $(CH_2)_2{>}O$ | 44.05 | 0.887 | −168.4 | | 56.3 | 249.4 |
| Formaldehyde | $HCHO$ | 30.3 | Gas | −133.6 | | 5.8 | |
| Formamide | $HCONH_2$ | 45.04 | 1.139 | 35.6 | | 379.4 | |

## Table 3-2. Physical Properties of Organic Compounds (Continued)

| Compound | Heat of formation, Btu/lb mole at 77°F | Free energy of formation, Btu/lb mole at 77°F | Solubility, g/100 ml solvent | | | Critical temp, °F | Critical press., atm | Critical density, lb/ft³ |
|---|---|---|---|---|---|---|---|---|
| | | | Water | Alcohol | Ether, etc. | | | |
| Acetic acid | -209,160 | -168,408 | ∞ | ∞ | ∞ et., i. CS$_2$ | 610.9 | 57.1 | 21.9 |
| Acetic anhydride | | | 13.6 d. | ∞ | ∞ et.; s. chl.; s. bz. | 565 | 46.2 | |
| Acetone | -106,780 | -66,890 | ∞ | ∞ | ∞ et.; s. chl. | 455.9 | 46.6 | 17.0 |
| Acetylene | | | 100 cc[64] | 600 cc[64] | 2,500 cc acetone[64]; s. bz.; s. chl. | 97 | 61.6 | 14.4 |
| Acrylonitrile | | | | | ∞ et. | | | |
| Adipic acid | -423,920 | -318,910 | 1.5[59] | v.s. | v.sl. et. | | | |
| Aniline | | | 3.4[8] | ∞ | ∞ et., bz. | 798.1 | 52.3 | 21.2 |
| Anthracene | +21,090 | +53,560 | i. | 0.076[61] | 1.189 et.; 1.767 chl. | | | |
| Benzene | | | 0.082[72] | ∞ | ∞ et.; s. chl. | 552 | 48.6 | 18.7 |
| Butadiene-1,3 | | | i. | v.s. | ∞ et. | 306 | 42.7 | 15.3 |
| Butane | -53,660 (g.) | -6,760 (g.) | 15 cc[83] | 1,813 cc[83] | 2,980 cc et.[83] | 305.5 | 37.5 | 14.2 |
| Butyl cellosolve | | | ∞ | ∞ | | | | |
| Carbon dioxide | -169,294 (g.) | -169,668 (g.) | 90.1 cc[86] | 31 cc[59] | s. bz. | 87.8 | 72.9 | 29.2 |
| Carbon monoxide | -47,550 (g.) | -59,050 (g.) | 3.5 cc[32] | 20 cc[68] | ∞ et., bz. | -220 | 34.5 | 18.8 |
| Carbon tetrachloride | | | 0.08[88] | ∞ | | 541.8 | 45.0 | 34.8 |
| Cellosolve | | | ∞ | ∞ | i. et. | | | |
| Cellulose | | | i. | i. | | | | |
| o-Cresol | | | 3.1[104] | ∞ >86 | ∞ >88 et. | 792 | 49.4 | 22 |
| m-Cresol | | | 2.35[68] | ∞ >97 | ∞ et.>97 | 810 | 45.0 | |
| p-Cresol | | | 2.4[104] | s. | ∞ et. | 799 | 50.8 | |
| Cyclohexane | -67,210 | +11,500 | i. | s. | s. et. | 536 | 40.0 | 17.0 |
| Cyclohexanol | | | 5.67[59] | s. | v.s. et. | | | |
| Cyclohexene | | | i. | s. | s. et. | | | |
| Dimethyl amine | | | v.s. | | s. et. | 328.1 | 52.4 | |
| Dimethyl aniline | | | i. | | | 778.5 | 35.8 | |
| Dimethyl phthalate | | | 0.43 | | s. et. | | | |
| Diphenyl | | | i. | 10 | v.s. et. | | | |
| Diphenyl amine | | | 0.03[77] | 44 | ∞ et. | | | |
| Ethyl acetate | -199,300 | -137,000 | 8.5[49] | ∞ | ∞ et. | 482.2 | 37.8 | 19.2 |
| Ethyl benzene | -5,360 | +51,500 | 0.015[9] | ∞ | s. et. | 655.5 | 38 | |
| Ethyl chloride | | | 0.45[32] | ∞ | ∞ et. | 369.0 | 52 | |
| Ethylene | +22,490 (g.) | +29,310 (g.) | 26 cc[32] | 360 cc | v.s. et. | 48.6 | 50.0 | 14.2 |
| Ethylene dibromide | | | 0.43[36] | ∞ | ∞ et. | | | |
| Ethylene oxide | -29,000 | -12,500 | ∞ | ∞ | v.s. et. | 383 | 71.0 | 20 |
| Formaldehyde | -50,920 (g.) | -48,380 (g.) | v.s. | v.s. | v.s. et. | | | |
| Formamide | -80,350 | -65,880 | ∞ | ∞ | v.sl.s. et. | | | |

# Table 3-2. Physical Properties of Organic Compounds (Continued)

| Compound | Formula | Mol. wt | Sp gr 60°/60° | Mp, °F | Heat of fusion, Btu/lb | Bp, °F | Heat of vaporisation at bp, Btu/lb |
|---|---|---|---|---|---|---|---|
| Furfural | $C_4H_3OCHO$ | 96.08 | 1.159 | −37.7 | | 323.1 | 193.5 |
| Glycerol | $CH_2OHCHOHCH_2OH$ | 92.09 | 1.260 | 64.2 | 85.5 | 554 | |
| Glycerol nitrate | $CH_2NO_2CHNO_2CH_2NO_2$ | 227.09 | 1.601 | 55.9 | | 320$^{16\ mm}$ | |
| Glycerol tristearate | $(CH_3(CH_2)_{16}CO_2)_3C_3H_5$ | 891.45 | 0.862 | 159.4 | | | |
| Glycol | $(CH_2OH)_2$ | 62.07 | 1.113 | 3.9 | 77.8 | 387.3 | 143.9 |
| Hexamethylene-tetramine | $(CH_2)_6N_4$ | 140.19 | | subl. | | | |
| n-Hexane | $C_6H_{14}$ | 86.17 | 0.659 | −139.6 | 65.0 | 155.7 | |
| Isoprene | $CH_2{:}CH(CH_3){:}CH_2$ | 68.11 | 0.681 | −230.8 | | 93.3 | |
| Ketene | $H_2C{:}CO$ | 42.04 | Gas | −240 | | −69 | |
| Lauryl alcohol | $CH_3(CH_2)_{10}CH_2OH$ | 186.33 | 0.831 | 75 | | 495 | |
| Linoleic acid | $C_{17}H_{31}COOH$ | 280.44 | 0.903 | 14.9 | | 445 | |
| Maleic anhydride | $(CHCO)_2 > O$ | 98.06 | 1.5 | 137 | | 396 | |
| Methane | $CH_4$ | 16.04 | Gas | −296.5 | 25.2 | −258.6 | 219.2 |
| Methyl alcohol | $CH_3OH$ | 32.04 | 0.792 | −144 | 42.7 | 148.5 | 473.0 |
| Methyl amine | $CH_3NH_2$ | 31.06 | 0.699 | −134.5 | | 19.9 | |
| Methyl cellosolve | $CH_3OCH_2CH_2OH$ | 76.09 | 0.965 | | | 256 | |
| Naphthalene | $C_{10}H_8$ | 128.16 | 1.145 | 176.4 | 64.8 | 424.2 | 135.9 |
| α-Naphthol | $C_{10}H_7OH$ | 114.16 | 1.224 | 205 | 70.1 | 534 | |
| β-Naphthol | $C_{10}H_7OH$ | 114.16 | 1.217 | 252 | 56.3 | 546 | |
| Nitrobenzene | $C_6H_5NO_2$ | 123.11 | 1.205 | 42.3 | 40.5 | 411.6 | 142.4 |
| n-Octane | $C_8H_{18}$ | 114.22 | 0.703 | −70.2 | 77.7 | 258.2 | 131.6 |
| n-Octyl alcohol | $C_8H_{17}OH$ | 130.22 | 0.827 | 3.2 | | 382 | 175.5 |
| Oxalic acid | $(COOH)_2{\cdot}2H_2O$ | 126.07 | 1.653 | 214.7 | | subl. | |
| n-Pentane | $C_5H_{12}$ | 72.15 | 0.630 | −201.5 | | 96.9 | |
| Phenol | $C_6H_5OH$ | 94.11 | 1.071 | 108.5 | 50.2 | 358.5 | 153.6 |
| Phthalic anhydride | $C_6H_4 < (CO)_2 > O$ | 148.11 | 1.527 | 267.4 | 52.2 | 544.1 | |
| Propane | $C_3H_8$ | 44.09 | Gas | −305.7 | 34.4 | −43.7 | 183.0 |
| Propylene | $C_3H_6$ | 42.08 | Gas | −301.4 | 30.7 | −53.9 | 118.2 |
| Styrene | $C_6H_5CH{:}CH_2$ | 104.14 | 0.911 | −23.1 | | 293.4 | 156.2 |
| Toluene | $C_6H_5CH_3$ | 92.13 | 0.866 | −139 | 30.9 | 231.1 | |
| Urea | $H_2NCONH_2$ | 60.06 | 1.355 | 270.9 | | d. | |
| Vinyl (poly-) | $(CH_3CO_2CH{:}CH)_x$ | $(86.09)_x$ | 1.19 | 210–260 | | | |
| Vinyl chloride | $CH_2{:}CHCl$ | 62.50 | 0.908 | −256 | | 10.4 | |
| o-Xylene | $C_6H_4(CH_3)_2$ | 106.16 | 0.881 | −133 | 55.1 | 291.9 | 149.2 |
| m-Xylene | $C_6H_4(CH_3)_2$ | 106.16 | 0.867 | −54.2 | 46.9 | 282.4 | 147.6 |
| p-Xylene | $C_6H_4(CH_3)_2$ | 106.16 | 0.861 | 55.9 | 69.3 | 281.0 | 146.2 |

## Table 3-2. Physical Properties of Organic Compounds (Continued)

| Compound | Heat of formation, Btu/lb mole at 77°F | Free energy of formation, Btu/lb mole at 77°F | Solubility, g/100 ml solvent | | | Critical temp, °F | Critical press, atm | Critical density, lb/ft³ |
|---|---|---|---|---|---|---|---|---|
| | | | Water | Alcohol | Ether, etc. | | | |
| Furfural | | | 9.1[15] | ∞ | ∞ et. | | | |
| Glycerol | −286,520 | −204,600 | ∞ | ∞ | i. et. | | | |
| Glycerol nitrate | | | 0.18[88] | 50[88] | ∞ et. | | | |
| Glycerol tristearate | | | ∞ | s.h. | s.h. et. | | | |
| Glycol | −194,270 | −137,620 | ∞ | | 1.0 et. | 454.5 | 29.9 | 14.6 |
| Hexamethylenetetramine | −55,030 (g.) | −53,040 (g.) | 81[64] | 3 | v.sl.s. et. | | | |
| n-Hexane | −85,540 | −1,640 | 0.014[49] | 50[91] | ∞ et. | | | |
| Isoprene | | | d. | d. | ∞ et. | | | |
| Ketene | −26,600 (g.) | −25,740 (g.) | d. | s. | s. et. | | | |
| Lauryl alcohol | | | i. | s. | s. et. | | | |
| Linoleic acid | | | i. | s. | s. et. | | | |
| Maleic anhydride | −32,200 (g.) | −21,850 (g.) | 16.3[38] | 47 cc[88] | 104 cc[50] et. | | | |
| Methane | −102,670 | −71,640 | 0.4 cc[88] | ∞ | ∞ et. | −115.8 | 45.8 | 10.1 |
| Methyl alcohol | −12,060 (g.) | +11,880 (g.) | ∞ | v.s. | ∞ et. | 464.0 | 78.5 | 17.0 |
| Methyl amine | | | v.s. | ∞ | v.s. et. | 314.4 | 73.6 | |
| Methyl cellosolve | | | ∞ | 9.5[68] | v.s. et. | | | |
| Naphthalene | | | 0.003[77] | v.s. | v.s. et. | | | |
| α-Naphthol | | | sl.s.h. | v.s. | ∞ et. | | | |
| β-Naphthol | | | 0.07[47] | sl.s. | s. et. | | | |
| Nitrobenzene | −107,530 | +3,190 | 0.196[3] | ∞ | ∞ et. | 565.2 | 24.6 | 14.5 |
| n-Octane | | | 0.002[81] | s. | s. et. | | | |
| n-Octyl alcohol | | | 0.054[77] | ∞ | ∞ et. | | | |
| Oxalic acid | | | s. | ∞ | 1.3 ct. | | | |
| n-Pentane | −74,450 | −3,980 | 0.036[61] | s. | ∞ et. | 385.9 | 33.3 | 14.5 |
| Phenol | −68,040 | −19,840 | 8.2[99] | s. | ∞ ct. | 786.6 | 60.5 | |
| Phthalic anhydride | | | v.l.s. | s. | sl.s. et. | | | |
| Propane | −44,680 (g.) | −10,110 (g.) | 6.5 cc[64] | 1,200 cc | v.s. et. | 206.2 | 42.0 | 13.7 |
| Propylene | +8,780 (g.) | +26,940 (g.) | 44.6 cc | ∞ | | 197.2 | 45.6 | 14.5 |
| Styrene | +5,160 | +31,110 | v.l.s. | 20[68] | ∞ et. | 609.4 | 41.6 | 18 |
| Toluene | | | 0.05[91] | s. | ∞ et. | | | |
| Urea | −143,340 | −84,810 | 100[63] | s. | sl.s. et. | | | |
| Vinyl (poly-) | | | i. | | | | | |
| Vinyl chloride | | | sl.s. | s. | v.s. et. | | | |
| o-Xylene | −10,510 | +47,470 | i. | s. | ∞ et. | 677.1 | 36.9 | |
| m-Xylene | −10,940 | +46,310 | i. | s. | ∞ et. | 655 | 36 | |
| p-Xylene | −10,510 | +47,360 | i. | s. | v.s. et. | 653 | 35 | |

## Table 3-3. Densities of Miscellaneous Materials†

Approximate Values at Ordinary Temperature

| Name | Sp gr | Lb/ft³ | Name | Sp gr | Lb/ft³ |
|---|---|---|---|---|---|
| Aluminum bronze | 7.7 | 481 | Marble* | 2.6–2.86 | 170 |
| Anthracite | 1.4–1.7 | 97 | Mica | 2.65–3.2 | 182 |
| Asbestos* | 2.1–2.8 | 153 | Oats, bulk | 0.51 | 32 |
| Asphalt | 1.1–1.5 | 81 | Oil: | | |
| Ashes (cinders) | | 45 | Vegetable | 0.91–0.94 | 58 |
| Barytes* | 4.5 | 281 | Fuel | 1.0 | 63 |
| Bituminous | 1.2–1.5 | 84 | Lubricant | 0.9 | 56 |
| Bluestone* | 2.5–2.6 | 159 | Paper | 0.70–1.15 | 58 |
| Borax* | 1.7–1.8 | 109 | Paraffin | 0.87–0.91 | 56 |
| Brass (70 Cu, 30 Zn) | 8.53 | 532 | Phosphate rock (apatite)* | 3.2 | 200 |
| Brick, common | 1.8–2.0 | 120 | Pitch | 1.07–1.15 | 69 |
| Bronze (90 Cu, 10 Zn) | 8.80 | 550 | Plaster: | | |
| Cast iron | 7.2 | 450 | On lath | | 100 |
| Cement, loose | | 90 | On masonry | | 60 |
| Cement, set | 2.7–3.2 | 183 | Potatoes, piled | 0.67 | 44 |
| Charcoal | 0.4 | 25 | Pumice, natural* | 0.37–0.9 | 40 |
| Clay: | | | Rubber: | | |
| Dry | 1.0 | 63 | Goods | 1.0–2.0 | 93 |
| Damp, plastic | 1.8 | 110 | Raw | 0.92–0.96 | 59 |
| Coke | 1.0–1.4 | 75 | Salt, granulated, piled | 0.77 | 48 |
| Concrete: | | | Sand: | | |
| Plain | | 144 | Dry | | 100 |
| Reinforced | | 150 | Wet | | 120 |
| Cinder | | 100 | Sandstone* | 2.0–2.6 | 143 |
| Cork | 0.24 | 15 | Shale and slate* | 2.6–2.9 | 172 |
| Corn, bulk | 0.73 | 45 | Slag, blast furnace | 2.5–3.0 | 172 |
| Cotton, flax, hemp | 1.47–1.50 | 93 | Snow, loose | 0.125 | 8 |
| Earth: | | | Stainless steel (18:8) | 7.93 | 493 |
| Dry, loose | 1.2 | 76 | Steel | 7.87 | 490 |
| Dry, packed | 1.5 | 95 | Sugar | 1.61 | 100 |
| Moist, loose | 1.3 | 78 | Tile, hollow | | 55 |
| Moist, packed | 1.6 | 96 | Water: | | |
| Flour, loose | 0.4–0.5 | 28 | Fresh | 1.00 | 62.3 |
| Gasoline | 0.75 | 46.8 | Salt | 1.02 | 64 |
| Glass, common | 2.4–2.8 | 162 | Wheat, bulk | 0.77 | 48 |
| Granite* | 2.6–2.7 | 165 | Wood, seasoned: | | |
| Gravel: | | | Birch | 0.71 | 44 |
| Dry | | 100 | Cedar | 0.35 | 22 |
| Wet | | 120 | Cypress | 0.48] | 30 |
| Greenstone (trap)* | 2.8–3.2 | 187 | Elm | 0.56 | 35 |
| Gypsum* | 2.3–2.8 | 159 | Mahogany | 0.56–0.85 | 44 |
| Hay and straw (bales) | 0.32 | 20 | Maple, white | 0.53 | 33 |
| Hematite (iron ore) | 5.2 | 325 | Oak, red or black | 0.64–0.71 | 42 |
| Leather | 0.86–1.02 | 59 | Oak, white | 0.77 | 48 |
| Lignite | 1.1–1.4 | 78 | Pine, white | 0.43 | 27 |
| Limestone* | 2.1–2.86 | 155 | Pine, yellow | 0.71 | 44 |
| Limonite (iron ore) | 3.6–4.0 | 237 | Redwood | 0.42 | 26 |
| Magnesite* | 3.0 | 187 | Spruce | 0.45 | 28 |
| Magnetite (iron ore) | 4.9–5.2 | 315 | Walnut | 0.59 | 37 |

* Density for the mineral is specified. Most minerals, when quarried and piled are about 35 to 45 per cent less dense. Masonry is generally 5 to 10 per cent less dense than the mineral.

† A more extensive tabulation appears on pp. 571–572 of "Handbook of Fundamentals," ASHRAE, New York, 1973.

# SPECIFIC HEATS

Specific heat = P.c.u. / (lb)(deg.C) = Btu /(lb)(deg F)
= calories /(gm)(deg C)

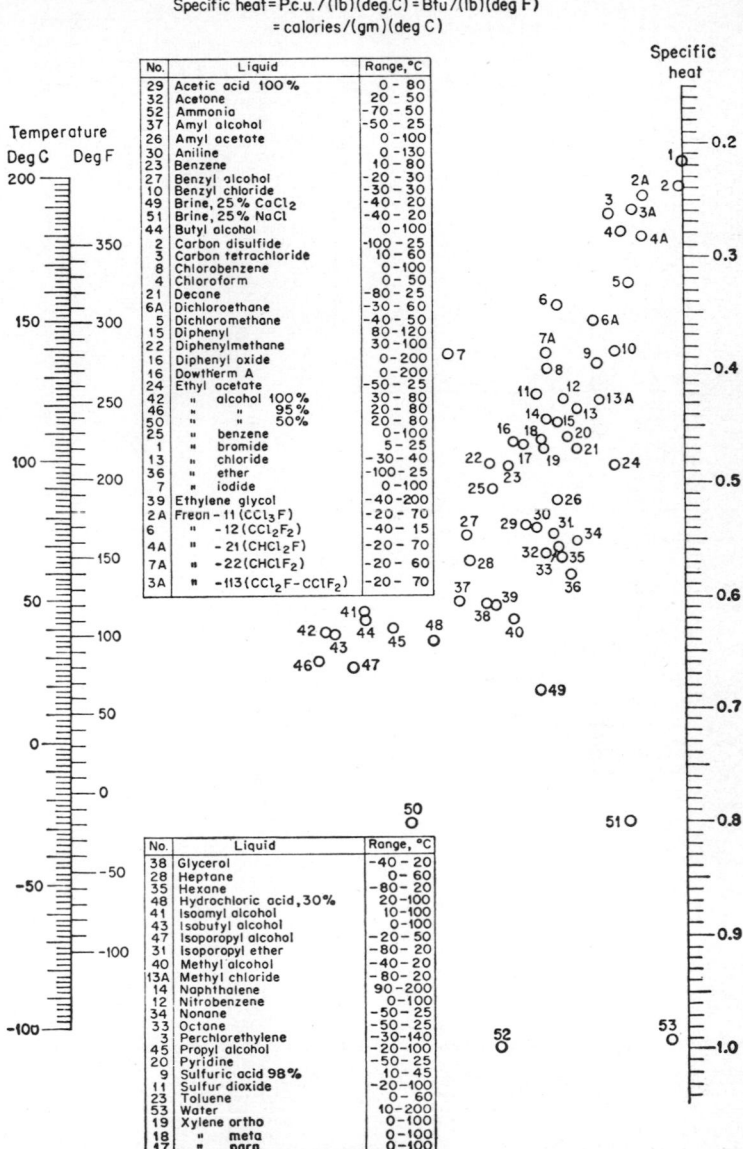

| No. | Liquid | Range,°C |
|---|---|---|
| 29 | Acetic acid 100% | 0 - 80 |
| 32 | Acetone | 20 - 50 |
| 52 | Ammonia | -70 - 50 |
| 37 | Amyl alcohol | -50 - 25 |
| 26 | Amyl acetate | 0 -100 |
| 30 | Aniline | 0 -130 |
| 23 | Benzene | 10 - 80 |
| 27 | Benzyl alcohol | -20 - 30 |
| 10 | Benzyl chloride | -30 - 30 |
| 49 | Brine, 25% CaCl₂ | -40 - 20 |
| 51 | Brine, 25% NaCl | -40 - 20 |
| 44 | Butyl alcohol | 0 -100 |
| 2 | Carbon disulfide | -100 - 25 |
| 3 | Carbon tetrachloride | 10 - 60 |
| 8 | Chlorobenzene | 0 -100 |
| 4 | Chloroform | 0 - 50 |
| 21 | Decane | -80 - 25 |
| 6A | Dichloroethane | -30 - 60 |
| 5 | Dichloromethane | -40 - 50 |
| 15 | Diphenyl | 80 -120 |
| 22 | Diphenylmethane | 30 -100 |
| 16 | Diphenyl oxide | 0 -200 |
| 16 | Dowtherm A | 0 -200 |
| 24 | Ethyl acetate | -50 - 25 |
| 42 | "    alcohol 100% | 30 - 80 |
| 46 | "        "    95% | 20 - 80 |
| 50 | "        "    50% | 20 - 80 |
| 25 | "    benzene | 0 -100 |
| 1 | "    bromide | 5 - 25 |
| 13 | "    chloride | -30 - 40 |
| 36 | "    ether | -100 - 25 |
| 7 | "    iodide | 0 -100 |
| 39 | Ethylene glycol | -40 -200 |
| 2A | Freon - 11 (CCl₃F) | -20 - 70 |
| 6 | "    - 12 (CCl₂F₂) | -40 - 15 |
| 4A | "    - 21 (CHCl₂F) | -20 - 70 |
| 7A | "    -22(CHClF₂) | -20 - 60 |
| 3A | "    -113(CCl₂F-CClF₂) | -20 - 70 |

| No. | Liquid | Range, °C |
|---|---|---|
| 38 | Glycerol | -40 - 20 |
| 28 | Heptane | 0 - 60 |
| 35 | Hexane | -80 - 20 |
| 48 | Hydrochloric acid,30% | 20 -100 |
| 41 | Isoamyl alcohol | 10 -100 |
| 43 | Isobutyl alcohol | 0 -100 |
| 47 | Isopropyl alcohol | -20 - 50 |
| 31 | Isoporopyl ether | -80 - 20 |
| 40 | Methyl alcohol | -40 - 20 |
| 13A | Methyl chloride | -80 - 20 |
| 14 | Naphthalene | 90 -200 |
| 12 | Nitrobenzene | 0 -100 |
| 34 | Nonane | -50 - 25 |
| 33 | Octane | -50 - 25 |
| 3 | Perchlorethylene | -30 -140 |
| 45 | Propyl alcohol | -20 -100 |
| 20 | Pyridine | -50 - 25 |
| 9 | Sulfuric acid 98% | -10 - 45 |
| 11 | Sulfur dioxide | -20 -100 |
| 23 | Toluene | 0 - 60 |
| 53 | Water | 10 -200 |
| 19 | Xylene ortho | 0 -100 |
| 18 | "    meta | 0 -100 |
| 17 | "    para | 0 -100 |

FIG. 3-1. Specific heats of liquids.   (See note below Table 3-4A.)

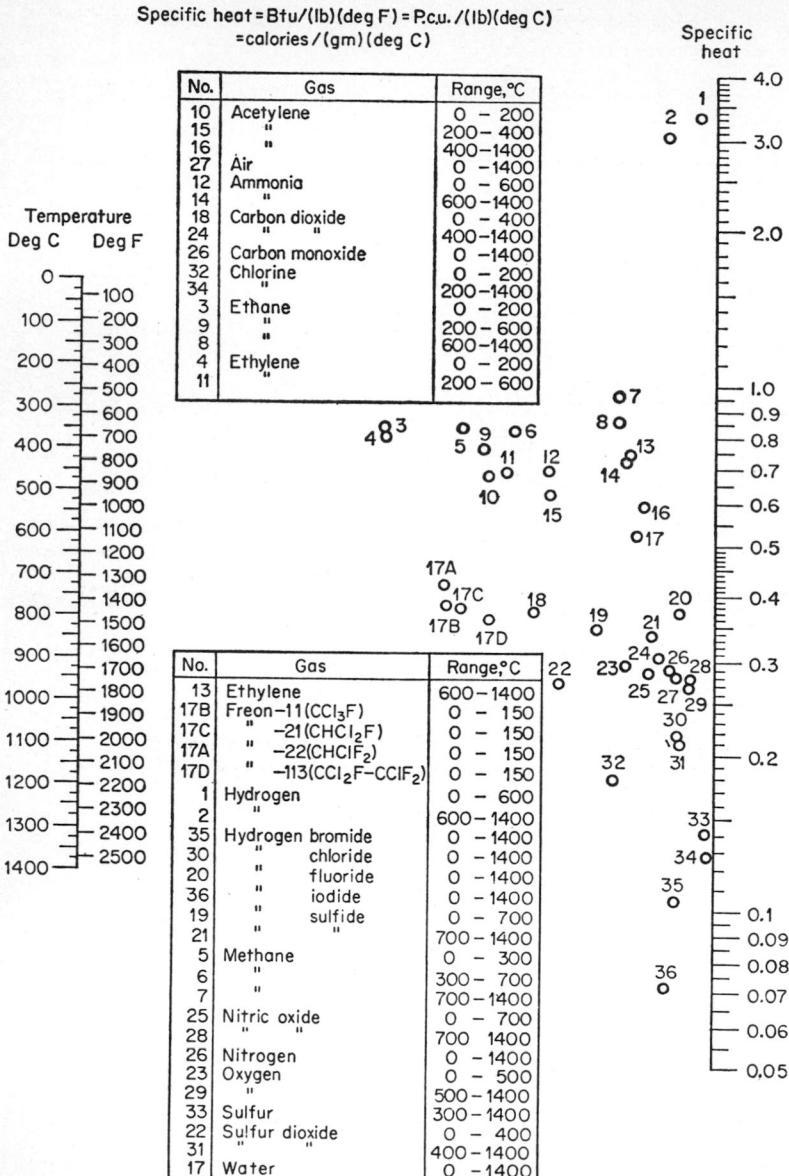

FIG. 3-2. Specific heats $(C_p)$ of gases at 1 atm pressure. (See note below Table 3-4A.)

## Table 3-4A. Specific Heats of Miscellaneous Materials*

| Material | Sp ht, cal/(g)(°C) or Btu/(lb)(°F) | Material | Sp ht, cal/(g)(°C) or Btu/(lb)(°F) |
|---|---|---|---|
| Alumina.................... | 0.2 (212°F) | Gypsum................... | 0.26 |
| | .274 (2730°F) | Iron...................... | .117 (212°F) |
| Aluminum.................. | .223 (212°F), | | .151 (2372°F) |
| | .259 (1832°F) | Kerosene................. | .47 |
| Asbestos.................. | .25 | Lead..................... | .032 (212°F) |
| Asphalt................... | .22 | | .033 (1652°F) |
| Bakelite.................. | .3–0.4 | Limestone................ | .22 |
| Brickwork................ | .2 | Litharge................. | .055 |
| Carbon................... | .168 (80–170°F) | Magnesia................. | .234 (212°F) |
| | .314 (100–1630°F) | | .188 (2730°F) |
| | .387 (130–2640°F) | Marble................... | .21 |
| Cellulose................. | .32 | Nickel.................... | .113 (212°F) |
| Charcoal, wool............. | .24 | | .143 (2372°F) |
| Clay...................... | .22 | Plastics.................. | .3–0.45 |
| Coal..................... | .26–0.37 | Polyethylene plastics......... | .55 |
| Coal tar.................. | .35 (100°F), | Porcelain................. | .18 (68–212°F) |
| | .45 (400°F) | | .22 (68–930°F) |
| Cobalt................... | .108 (212°F), | Pyrites, copper............. | .13 |
| | .176 (2372°F) | Pyrites, iron.............. | .14 |
| Coke..................... | .265 (70–750°F) | Pyroxylin plastics........... | .34–0.38 |
| | .403 (70–2400°F) | Quartz................... | .17 (32°F) |
| Concrete.................. | .156 | | .28 (660°F) |
| Copper................... | .094 (212°F), | Rubber, vulcanized......... | .42 |
| | .118 (2372°F) | Sand..................... | .19 |
| Fireclay brick.............. | .2 (212°F), | Silica.................... | .32 |
| | .3 (2730°F) | Silk..................... | .33 |
| Fluorspar................. | .21 | Steel.................... | .12 |
| Gasoline.................. | .53 | Stone.................... | .2, approx. |
| Glass: | | Turpentine................ | .42 |
| Crown.................. | .16–0.20 | Wood: | |
| Flint................... | .12 | Oak................... | .57 |
| Pyrex.................. | .20 | Most others............... | .45–0.65 |
| Fiber insulating wool....... | .20 | Wool.................... | .33 |
| Granite................... | .20 | Zinc..................... | .096 (212°F) |
| Graphite.................. | .165 (80–170°F) | | .126 (1652°F) |
| | .39 (130–2650°F) | | |

* A more extensive tabulation is given on pp. 571–572 of "Handbook of Fundamentals," ASHRAE, New York, 1973.

**Note to Figs. 3-1 and 3-2.** Touloukian and Makita, "Thermophysical Properties of Matter," vol. 6, Plenum Publishing Corp., New York, extensively review specific heat data for some 70 fluids in liquid and gaseous states. The 1976 ASHRAE bulletin, "Thermophysical Properties of Refrigerants," tabulates values for 38 common refrigerants. Thinh, Duran, et al., *Hydrocarbon Process.*, **50**, 98(1971) give equation fits for 408 hydrocarbons and related compounds. Extensive tables are also to be found in Perry and Chilton (eds.), "Chemical Engineers Handbook," 5th ed., McGraw-Hill Book Company, New York, 1973.

## Table 3-4B. Specific Heats of Foodstuffs*

| Material | % $H_2O$ | $c < T_f$ | $L_f$ | $c > T_f$ | Material | % $H_2O$ | $c < T_f$ | $L_f$ | $c > T_f$ |
|---|---|---|---|---|---|---|---|---|---|
| Vegetables: | | | | | Misc. Groceries: | | | | |
| Beans, dried... | 12.5 | 1.00 | 42 | 1.046 | Bread......... | 35 | 1.42 | 116 | 2.93 |
| Broccoli....... | 89.9 | 1.97 | 302 | 3.85 | Butter........ | 15 | 1.42 | 35 | 2.68 |
| Brussel sprouts | 84.9 | 1.92 | 284 | 3.68 | Cheese, | | | | |
| Cabbage...... | 92.4 | 1.97 | 307 | 3.93 | American.... | 60 | 1.67 | 200 | 2.93 |
| Carrots....... | 88.2 | 1.92 | 293 | 3.77 | Eggs.......... | | 1.67 | 232 | 3.18 |
| Corn, dried.... | 10.5 | 0.96 | 35 | 1.17 | Flour......... | 14 | 1.17 | | 1.59 |
| Lettuce....... | 94.8 | 2.01 | 316 | 4.02 | Ice cream..... | 62 | 1.88 | 223 | 3.26 |
| Onions........ | 87.5 | 1.92 | 288 | 3.77 | Margarine..... | 16 | 1.05 | 51 | 1.34 |
| Potatoes, | | | | | Milk.......... | 88 | 2.05 | 288 | 3.89 |
| white....... | 77.8 | 1.80 | 258 | 3.43 | Fruits: | | | | |
| Tomatoes, | | | | | Apples........ | 84 | 1.88 | 281 | 3.59 |
| ripening..... | 94.1 | 2.01 | 312 | 3.97 | Bananas...... | 75 | 1.76 | 251 | 3.48 |
| Fish and Meats: | | | | | Grapefruit.... | 89 | 1.92 | 293 | 3.81 |
| Bacon........ | 20 | 1.26 | 67 | 2.09 | Lemons....... | 89 | 1.92 | 295 | 3.85 |
| Beef, fresh lean | 68 | 1.67 | 233 | 3.22 | Oranges....... | 87 | 1.92 | 288 | 3.77 |
| Fish, cod..... | | 2.05 | 277 | 3.77 | Peaches ...... | 87 | 1.92 | 296 | 3.77 |
| Fish, frozen... | 70 | 1.72 | 235 | 3.18 | Pears......... | 83 | 1.88 | 274 | 3.60 |
| Hams......... | 60 | 1.59 | 202 | 2.85 | Strawberries... | 90 | 1.97 | 300 | 3.85 |
| Pork.......... | 60 | 1.59 | 202 | 2.85 | Watermelon... | 92 | 2.01 | 307 | 3.85 |
| Poultry, frozen | 74 | 1.55 | 247 | 3.31 | | | | | |

*$c < T_f$: Specific heat below freezing point, kJ(kg · K); $L_f$: enthalpy of fusion, kJ/kg; $c > T_f$: specific heat above freezing point, kJ/(kg · K). The fusion temperature varies from about 268 to 272 K.

## Table 3-5. Specific Heats of Aqueous Solutions at 68°F

| Substance and formula | Wt, % | Sp ht, cal/(g)(°C) or Btu/(lb)(°F) | Substance and formula | Wt, % | Sp ht, cal/(g)(°C) or Btu/(lb)(°F) |
|---|---|---|---|---|---|
| Acetic acid, $CH_3COOH$....... | 20 | 0.91* | Phosphoric acid, $H_3PO_4$ | | |
| | 60 | 0.73* | (Continued)............... | 40 | 0.70 |
| | 80 | 0.63* | | 56 | 0.60 |
| | 100 | 0.54* | | 72 | 0.50 |
| Ammonia, $NH_3$.............. | 35 | 1.00 | | 88 | 0.43 |
| | 72 | 0.99 | Potassium chloride, KCl....... | 4 | 0.95 |
| Aniline, $C_6H_5NH_2$............ | 94 | 0.58 | | 14 | 0.83 |
| | 98 | 0.53 | | 20 | 0.78 |
| | 100 | 0.50 | | 25 | 0.73 |
| Copper sulfate, $CuSO_4$........ | 2 | 0.98 | Potassium hydroxide, KOH.... | 2 | 0.97 |
| | 4 | 0.95 | | 5 | 0.93 |
| | 15 | 0.85 | | 14 | 0.81 |
| Ethyl alcohol, $C_2H_5OH$........ | 10 | 1.02 | | 24 | 0.75 |
| | 25 | 1.03 | n-Propyl alcohol, $C_3H_7OH$..... | 5 | 1.02 |
| | 60 | 0.86 | | 15 | 1.06 |
| | 80 | 0.73 | | 30 | 1.03 |
| | 100 | 0.58 | | 50 | 0.90 |
| Glycerol, $C_3H_5(OH)_3$.......... | 10 | 0.96† | | 70 | 0.78 |
| | 20 | 0.93† | | 90 | 0.65 |
| | 40 | 0.85† | | 100 | 0.57 |
| | 60 | 0.77† | Sodium carbonate, $Na_2CO_3$.... | 3 | 0.96‡ |
| | 80 | 0.67† | | 8 | 0.92‡ |
| | 100 | 0.56† | | 12 | 0.90‡ |
| Hydrochloric acid, HCl........ | 17 | 0.74 | Sodium chloride, NaCl........ | 1 | 0.99 |
| | 29 | 0.63 | | 8 | 0.91 |
| | 34 | 0.59 | | 25 | 0.81 |
| Methyl alcohol, $CH_3OH$....... | 10 | 1.0 | Sodium hydroxide, NaOH..... | 2 | 0.97 |
| | 20 | 0.98 | | 18 | 0.84 |
| | 40 | 0.92 | | 29 | 0.80 |
| | 60 | 0.81 | | 58 | 0.78 |
| | 80 | 0.71 | Sulfuric acid, $H_2SO_4$......... | 2 | 0.98 |
| | 100 | 0.60 | | 15 | 0.88 |
| Nitric acid, $HNO_3$............ | 10 | 0.90 | | 28 | 0.78 |
| | 30 | 0.73 | | 40 | 0.68 |
| | 50 | 0.65 | | 65 | 0.50 |
| | 70 | 0.62 | | 85 | 0.44 |
| | 90 | 0.52 | | 100 | 0.34 |
| Phosphoric acid, $H_3PO_4$....... | 5 | 0.97 | Zinc sulfate, $ZnSO_4$.......... | 4 | 0.95 |
| | 14 | 0.90 | | 15 | 0.84 |
| | 26 | 0.80 | | | |

* 100°F.  † 60°F.  ‡ 63.5°F.

## Table 3-6. Specific Heats and Their Ratio for Air*

| Pressure, bar† | | Temperature, K | | | | | | | | | | |
|---|---|---|---|---|---|---|---|---|---|---|---|---|
| | | 100 | 200 | 300 | 400 | 500 | 600 | 800 | 1000 | 1500 | 2000 | 2500 |
| 0.1 | $C_p$ | 1.005 | 1.003 | 1.005 | 1.014 | 1.030 | 1.051 | 1.099 | 1.142 | 1.231 | | |
| | $C_v$ | ..... | ..... | 0.718 | 0.727 | 0.742 | 0.764 | 0.812 | 0.855 | 0.944 | | |
| | $\gamma$ | ..... | ..... | 1.400 | 1.395 | 1.387 | 1.376 | 1.354 | 1.337 | 1.304 | 1.267 | 1.191 |
| 1 | $C_p$ | 1.032 | 1.007 | 1.007 | 1.014 | 1.030 | 1.051 | 1.099 | 1.141 | 1.231 | 1.34 | 1.67 |
| | $C_v$ | 0.736 | 0.717 | 0.718 | 0.727 | 0.743 | 0.764 | 0.812 | 0.855 | 0.944 | 1.051 | 1.361 |
| | $\gamma$ | 1.402 | 1.402 | 1.401 | 1.396 | 1.387 | 1.376 | 1.354 | 1.336 | 1.304 | 1.275 | 1.227 |
| 10 | $C_p$ | 2.041 | 1.049 | 1.021 | 1.021 | 1.034 | 1.055 | 1.100 | 1.142 | 1.231 | 1.32 | 1.49 |
| | $C_v$ | ..... | 0.728 | 0.721 | 0.727 | 0.744 | 0.764 | 0.811 | 0.854 | 0.944 | 1.034 | 1.195 |
| | $\gamma$ | ..... | 1.441 | 1.418 | 1.404 | 1.392 | 1.379 | 1.355 | 1.336 | 1.305 | 1.277 | 1.247 |
| 20 | $C_p$ | 2.01 | 1.10 | 1.04 | 1.03 | 1.04 | 1.06 | 1.10 | 1.14 | 1.23 | 1.32 | 1.46 |
| | $C_v$ | ..... | 0.739 | 0.723 | 0.728 | 0.745 | 0.763 | 0.811 | 0.853 | 0.944 | 1.034 | 1.169 |
| | $\gamma$ | ..... | 1.488 | 1.438 | 1.413 | 1.398 | 1.383 | 1.358 | 1.337 | 1.305 | 1.277 | 1.249 |
| 40 | $C_p$ | 1.96 | 1.22 | 1.07 | 1.04 | 1.05 | 1.06 | 1.11 | 1.15 | 1.23 | 1.32 | 1.45 |
| | $C_v$ | ..... | 0.755 | 0.726 | 0.729 | 0.747 | 0.763 | 0.809 | 0.853 | 0.945 | 1.034 | 1.159 |
| | $\gamma$ | ..... | 1.616 | 1.476 | 1.431 | 1.412 | 1.389 | 1.362 | 1.339 | 1.305 | 1.277 | 1.251 |
| 60 | $C_p$ | 1.92 | 1.36 | 1.10 | 1.06 | 1.06 | 1.07 | 1.12 | 1.16 | 1.23 | 1.32 | 1.44 |
| | $C_v$ | ..... | 0.771 | 0.728 | 0.730 | 0.749 | 0.763 | 0.809 | 0.853 | 0.945 | 1.034 | 1.150 |
| | $\gamma$ | ..... | 1.764 | 1.516 | 1.449 | 1.426 | 1.395 | 1.365 | 1.341 | 1.305 | 1.277 | 1.252 |
| 80 | $C_p$ | 1.88 | 1.51 | 1.13 | 1.07 | 1.07 | 1.08 | 1.11 | 1.15 | 1.23 | 1.32 | 1.43 |
| | $C_v$ | ..... | 0.786 | 0.731 | 0.731 | 0.751 | 0.763 | 0.808 | 0.853 | 0.945 | 1.033 | 1.141 |
| | $\gamma$ | ..... | 1.921 | 1.556 | 1.466 | 1.440 | 1.401 | 1.369 | 1.343 | 1.305 | 1.278 | 1.253 |
| 100 | $C_p$ | 1.85 | 1.65 | 1.16 | 1.09 | 1.07 | 1.08 | 1.11 | 1.15 | 1.23 | 1.32 | 1.43 |
| | $C_v$ | ..... | 0.799 | 0.733 | 0.731 | 0.753 | 0.763 | 0.807 | 0.852 | 0.946 | 1.033 | 1.141 |
| | $\gamma$ | ..... | 2.065 | 1.590 | 1.484 | 1.453 | 1.407 | 1.372 | 1.345 | 1.306 | 1.278 | 1.253 |

* Specific heats in kJ/(kg · K). $C_p$ = specific heat at constant pressure, $C_v$ = specific heat at constant volume, $\gamma = C_p/C_v$.
† To convert to pascals (newtons per square meter) multiply by $10^5$.

### Table 3-7. $C_p/C_v$: Ratios of Specific Heats of Gases at 1 Atm Pressure

| Compound | Formula | Temp, °F | Ratio of sp ht, $\gamma = C_p/C_v$ | Compound | Formula | Temp, °F | Ratio of sp ht, $\gamma = C_p/C_v$ |
|---|---|---|---|---|---|---|---|
| Acetaldehyde | $C_2H_4O$ | 86 | 1.14 | Hydrogen cyanide | HCN | 149 | 1.31 |
| Acetic acid | $C_2H_4O_2$ | 275 | 1.15 | | | 284 | 1.28 |
| Acetylene | $C_2H_2$ | 59 | 1.26 | | | 410 | 1.24 |
| | | -96 | 1.31 | Hydrogen iodide | HI | 68-212 | 1.40 |
| Air | ...... | 1697 | 1.36 | Hydrogen sulfide | $H_2S$ | 59 | 1.32 |
| | | 63 | 1.403 | | | -49 | 1.30 |
| | | -108 | 1.408 | | | -71 | 1.29 |
| | | -180 | 1.415 | Iodine | $I_2$ | 365 | 1.30 |
| Ammonia | $NH_3$ | 59 | 1.310 | Isobutane | $C_4H_{10}$ | 59 | 1.11 |
| Argon | A | 59 | 1.668 | Krypton | Kr | 66 | 1.68 |
| | | -292 | 1.76 | Mercury | Hg | 680 | 1.67 |
| | | 32-212 | 1.67 | Methane | $CH_4$ | 1112 | 1.113 |
| Benzene | $C_6H_6$ | 194 | 1.10 | | | 572 | 1.15 |
| Bromine | $Br_2$ | 68-662 | 1.32 | | | 59 | 1.31 |
| Carbon dioxide | $CO_2$ | 59 | 1.304 | | | -112 | 1.34 |
| | | -103 | 1.37 | | | -175 | 1.41 |
| Carbon disulfide | $CS_2$ | 212 | 1.21 | Methyl acetate | $C_3H_6O_2$ | 59 | 1.14 |
| Carbon monoxide | CO | 59 | 1.404 | Methyl alcohol | $CH_4O$ | 171 | 1.203 |
| | | -292 | 1.41 | Methyl ether | $C_2H_6O$ | 43-86 | 1.11 |
| Chlorine | $Cl_2$ | 59 | 1.355 | Methylal | $C_3H_8O_2$ | 55 | 1.06 |
| Chloroform | $CHCl_3$ | 212 | 1.15 | | | 104 | 1.09 |
| Cyanogen | $(CN)_2$ | 59 | 1.256 | Neon | Ne | 66 | 1.64 |
| Cyclohexane | $C_6H_{12}$ | 176 | 1.08 | Nitric oxide | NO | 59 | 1.400 |
| Dichlorodifluor-methane | $CCl_2F_2$ | 77 | 1.139 | | | -49 | 1.39 |
| Ethane | $C_2H_6$ | 212 | 1.19 | | | -112 | 1.38 |
| | | 59 | 1.22 | Nitrogen | $N_2$ | 59 | 1.404 |
| | | -116 | 1.28 | | | -294 | 1.47 |
| Ethyl alcohol | $C_2H_6O$ | 194 | 1.13 | Nitrous oxide | $N_2O$ | 212 | 1.28 |
| Ethyl ether | $C_4H_{10}O$ | 95 | 1.08 | | | 59 | 1.303 |
| | | 176 | 1.086 | | | -22 | 1.31 |
| Ethylene | $C_2H_4$ | 212 | 1.18 | | | -94 | 1.34 |
| | | 59 | 1.255 | Oxygen | $O_2$ | 59 | 1.401 |
| | | -132 | 1.35 | | | -105 | 1.415 |
| Helium | He | -292 | 1.660 | | | -294 | 1.45 |
| n-Hexane | $C_6H_{14}$ | 176 | 1.08 | n-Pentane | $C_5H_{12}$ | 187 | 1.086 |
| Hydrogen | $H_2$ | 59 | 1.410 | Phosphorus | P | 572 | 1.17 |
| | | -105 | 1.453 | Potassium | K | 1562 | 1.77 |
| | | -294 | 1.597 | Sodium | Na | 1382-1688 | 1.68 |
| Hydrogen bromide | HBr | 68 | 1.42 | Sulfur dioxide | $SO_2$ | 59 | 1.29 |
| Hydrogen chloride | HCl | 59 | 1.41 | Xenon | Xe | 66 | 1.66 |
| | | 212 | 1.40 | | | | |

### Table 3-8. Ratios of Specific Heats $(C_p/C_v)$ of Argon and Nitrogen at High Pressure

| Gas | Pressure, atm | Temperature, K | | | | | | |
|---|---|---|---|---|---|---|---|---|
| | | 250 | 300 | 400 | 500 | 600 | 800 | 1000 |
| Argon | 1 | 1.671 | 1.670 | 1.668 | 1.668 | 1.667 | 1.667 | 1.667 |
| | 10 | 1.713 | 1.697 | 1.682 | 1.676 | 1.673 | 1.669 | 1.668 |
| | 100 | 2.25 | 1.96 | 1.80 | 1.753 | 1.721 | 1.693 | 1.680 |
| Nitrogen | 1 | 1.402 | 1.401 | 1.398 | 1.391 | 1.382 | 1.360 | 1.341 |
| | 10 | 1.427 | 1.417 | 1.406 | 1.396 | 1.385 | 1.360 | 1.341 |
| | 100 | 1.68 | 1.566 | 1.480 | 1.437 | 1.410 | 1.372 | 1.347 |

## Table 3-9. Thermal Conductivity—Temperature Table for Solids*

Temperature, K

| Substance | 10 | 20 | 40 | 60 | 80 | 100 | 200 | 300 | 400 | 500 | 600 | 800 | 1000 | 1200 | 1400 |
|---|---|---|---|---|---|---|---|---|---|---|---|---|---|---|---|
| Alumina | 7 | 32 | 121 | 174 | 160 | 125 | 55 | 36 | 26 | 20 | 16 | 10 | 8 | 7 | 6 |
| Aluminum | 38,000 | 13,500 | 2,300 | 850 | 380 | 300 | 237 | 237 | 240 | 237 | 232 | 220 | 93 | 99 | 105 |
| Antimony | 470 | 230 | 110 | 80 | 60 | 48 | 32 | 26 | 22 | 20 | | | | | 25 |
| Beryllium oxide | 47 | 196 | 810 | 1,400 | 1,650 | 1,490 | 480 | 272 | 196 | 146 | 111 | 70 | 47 | 33 | |
| Bismuth | 240 | 100 | 45 | 31 | 24 | 22 | 18 | 16 | 14 | 12 | | | | | |
| Boron | 165 | 305 | 400 | 327 | 230 | 170 | 45 | 25 | 15 | 12 | | | | | |
| Cadmium | 900 | 250 | 150 | 120 | 110 | 110 | 105 | 104 | 101 | 99 | | | | | |
| Chromium | 400 | 570 | 450 | 250 | 180 | 158 | 111 | 90 | 87 | 85 | 81 | 71 | 65 | 62 | 61 |
| Cobalt | 250 | 450 | 380 | 250 | 190 | 160 | 120 | 100 | 85 | 70 | | | | | |
| Constantan | 4 | 9 | 16 | 18 | 19 | 20 | 23 | 25 | 27 | 30 | | | | | |
| Copper | 19,000 | 10,700 | 2,100 | 850 | 570 | 483 | 413 | 398 | 392 | 388 | 383 | 371 | 357 | 342 | |
| Gallium | 2,200 | 640 | 250 | 200 | 170 | 140 | 100 | 85 | | | | | | | |
| Gold | 2,800 | 1,500 | 520 | 380 | 350 | 345 | 327 | 315 | 312 | 309 | 304 | 292 | 278 | 262 | |
| Graphite[a] | 27 | 108 | 135 | 81 | 54 | 39 | 15 | 10 | 7 | 5 | 4 | 3 | 3 | 2 | 2 |
| Graphite[b] | 81 | 420 | 1,630 | 2,980 | 4,290 | 4,980 | 3,250 | 2,000 | 1,460 | 1,140 | 930 | 680 | 530 | 440 | 370 |
| Hastelloy | 1 | 3 | 4 | 5 | 6 | 7 | 9 | 10 | 11 | 13 | | | | | |
| Inconel | 2 | 4 | 8 | 10 | 11 | 11 | 14 | 15 | | | | | | | |
| Iridium | 1,300 | 1,900 | 750 | 360 | 230 | 172 | 147 | 145 | 143 | 140 | | | | | |
| Iron | 710 | 1,000 | 560 | 270 | 170 | 132 | 94 | 80 | 69 | 61 | 55 | 43 | 33 | 28 | 31 |
| Lead | 175 | 57 | 43 | 42 | 41 | 40 | 37 | 35 | 34 | 33 | 31 | 19 | 22 | 24 | 26 |
| Magnesium | 1,200 | 1,300 | 620 | 290 | 190 | 169 | 159 | 156 | 153 | 151 | 149 | 146 | 84 | 98 | 112 |
| Magnesium oxide | 1,100 | 3,100 | 2,200 | 950 | 460 | 260 | 75 | 48 | 36 | 27 | 21 | 13 | 10 | 8 | 7 |
| Manganese | 2 | 2 | 4 | 5 | 5 | 6 | 7 | 8 | 9 | 9 | | | | | |
| Manganin | 2 | 4 | 9 | 11 | 13 | 13 | 17 | 22 | 28 | 34 | 40 | | | | |
| Mercury | 54 | 40 | 35 | 33 | 33 | 32 | 32 | 8 | 10 | 11 | 12 | | | | |
| Molybdenum | 150 | 280 | 350 | 250 | 210 | 179 | 143 | 138 | 134 | 130 | 126 | 118 | 112 | 105 | 100 |
| Nickel | 2,600 | 1,700 | 570 | 290 | 200 | 158 | 106 | 91 | 80 | 72 | 66 | 67 | 72 | 76 | 80 |
| Nylon | 0.04 | 0.10 | 0.17 | 0.20 | 0.23 | 0.25 | 0.28 | 0.30 | | | | | | | |
| Palladium | 1,200 | 610 | 160 | 100 | 88 | 80 | 78 | 78 | 79 | 80 | | | | | |
| Platinum | 1,200 | 490 | 130 | 92 | 82 | 79 | 75 | 73 | 72 | 72 | 72 | 73 | 78 | 78 | 81 |
| PTFE[c] | 0.94 | 1.43 | 1.94 | 2.1 | 2.15 | 2.16 | 2.20 | 2.25 | 2.3 | 2.5 | | | | | |
| Pyrex | 0.12 | 0.20 | 0.33 | 0.42 | 0.51 | 0.57 | 0.88 | 1.1 | 1.6 | 2.1 | | | | | |
| Quartz | 1,200 | 480 | 82 | 40 | 30 | | | 150 | 145 | 140 | | | | | |
| Rhodium | 2,900 | 3,900 | 1,000 | 370 | 250 | 190 | 160 | | | | | | | | |
| Rubber | | | 0.13 | 0.15 | 0.16 | 0.17 | 0.20 | 0.22 | 0.24 | 0.25 | | | | | |
| Selenium (axis) | 140 | 57 | 25 | 15 | 10 | 8 | 6 | 4 | 3 | 2 | 2 | | | | |
| Silica | | | | | | | | 1.34 | 1.52 | 1.70 | 1.87 | 2.22 | 2.60 | | |

| | | | | | | | | | | | | | | | |
|---|---|---|---|---|---|---|---|---|---|---|---|---|---|---|---|
| Silver | 16,500 | 5,200 | 1,100 | 630 | 500 | 430 | 425 | 424 | 420 | 413 | 405 | 389 | 374 | 358 | |
| Tantalum | 108 | 146 | 88 | 68 | 62 | 59 | 58 | 57 | 58 | 58 | 59 | 59 | 60 | 61 | 62 |
| Tellurium | 300 | 93 | 29 | 17 | 13 | 11 | 6 | 4 | 3 | 3 | | | | | |
| Tin | | 320 | 130 | 101 | 90 | 84 | 72 | 67 | 62 | 60 | 19 | | | | |
| Titanium | 14 | 28 | 39 | 37 | 33 | 31 | 26 | 21 | 20 | 20 | | | | | |
| Tungsten | | | 880 | 330 | 310 | 280 | 190 | 180 | 170 | 150 | 140 | | | | |
| Uranium | | | | 20 | 22 | 23 | 26 | 28 | 30 | 32 | | | | | |
| Zinc | | | | 150 | 135 | 130 | 123 | 120 | 116 | 110 | 110 | | | | |
| Zirconium | 100 | 110 | 59 | 42 | 38 | 34 | 25 | 23 | 22 | 21 | 21 | | | | |

[a] Parallel to basal plane.
[b] Perpendicular to basal plane.
[c] Also known as "Teflon," etc.
* Especially at low temperatures, the thermal conductivity can often be markedly reduced by even small traces of impurities. These tables, for the highest purity specimens available, should thus be used with caution in applications with commercial materials. Thermal conductivities tabulated in W/(m · K).

### Table 3-10. Thermal Conductivities of Alloy Steels

| Type of steel | $k$, Btu/(hr)(ft²)(°F/ft) | |
|---|---|---|
| | 400°F | 1000°F |
| Carbon | 26.8 | 23.2 |
| 2 Cr, 0.5 Mo | 16.9 | 16.9 |
| 5 Cr, 0.5 Mo | 15.8 | 16.3 |
| 5 Cr, 0.5 Mo, 1.5 Si | 13.0 | 14.8 |
| 18 Cr, 8 Ni (stainless 304, 316, 321, 347) | 10.9 | 13.7 |
| 25 Cr, 20 Ni (stainless 310) | 9.5 | 12.9 |

### Table 3-11. Thermal Conductivities of Building and Insulating Materials*

| Material | Apparent density $\rho$, lb/ft³ at room temp | $t$, °F | $k$, Btu/(hr)(ft²)(°F/ft) |
|---|---|---|---|
| Aerogel, silica, opacified | 8.5 | 248 | 0.013 |
| | | 554 | 0.026 |
| Aluminum foil (7 air spaces per 2.5 in.) | 0.2 | 100 | 0.025 |
| | | 350 | 0.038 |
| Asbestos | 29 | −300 | 0.055 |
| | 29 | 32 | 0.090 |
| | 36 | 32 | 0.087 |
| | 36 | 400 | 0.121 |
| | 36 | 800 | 0.130 |
| | 44 | −300 | 0.100 |
| | 44 | 32 | 0.135 |
| Cement boards | 120 | 68 | 0.43 |
| Felt: | | | |
| 40 laminations/in | .... | 100 | 0.033 |
| 40 laminations/in | .... | 500 | 0.048 |
| 20 laminations/in | .... | 100 | 0.045 |
| 20 laminations/in | .... | 500 | 0.065 |
| Sheets | 55.5 | 124 | 0.096 |
| Slate | 112 | 32 | 0.087 |
| | 112 | 140 | 0.114 |
| Asphalt | 132 | 68 | 0.43 |
| Bricks: | | | |
| Alumina (92–99% Al₂O₃ by wt) fused | .... | 800 | 1.8 |
| Alumina (64–65% Al₂O₃ by wt) | .... | 2400 | 2.7 |
| Building brickwork | .... | 68 | 0.4 |
| Carbon | 96.7 | .... | 3.0 |
| Chrome (32% Cr₂O₃ by wt) | 200 | 392 | 0.67 |
| | 200 | 1200 | 0.85 |
| | 200 | 2400 | 1.0 |
| Diatomaceous earth: | | | |
| 1600°F service | .... | 300 | 0.058 |
| 1600°F service | .... | 1000 | 0.073 |
| 2500°F service | .... | 300 | 0.135 |
| 2500°F service | .... | 1000 | 0.163 |
| 2500°F service | .... | 2000 | 0.203 |
| Fire clay (Missouri) | .... | 392 | 0.58 |
| | .... | 1112 | 0.85 |
| | .... | 1832 | 0.95 |
| | .... | 2550 | 1.02 |
| Kaolin insulating brick | 27 | 932 | 0.15 |
| | 27 | 2100 | 0.26 |
| Kaolin insulating firebrick | 19 | 392 | 0.050 |
| | 19 | 1400 | 0.113 |
| Magnesite (86.8% MgO, 6.3% Fe₂O₃, 3% CaO, 2.6% SiO₂ by wt) | 158 | 400 | 2.2 |
| | 158 | 1200 | 1.6 |
| | 158 | 2200 | 1.1 |

* A more extensive table appears on pp. 571–572 of "Handbook of Fundamentals," ASHRAE, New York, 1973.

## Table 3-11. Thermal Conductivities of Building and Insulating Materials (Continued)

| Material | Apparent density $\rho$, lb/ft³ at room temp | $t$, °F | $k$, Btu/(hr)(ft²)(°F/ft) |
|---|---|---|---|
| Silicon carbide, recrystallized................... | 129 | 1112 | 10.7 |
| | 129 | 1832 | 8.0 |
| | 129 | 2550 | 6.3 |
| Calcium carbonate: | | | |
| Natural...... | 162 | 86 | 1.3 |
| White marble..... | ...... | ...... | 1.7 |
| Chalk.... | 96 | ...... | 0.4 |
| Calcium sulfate (4H₂O): | | | |
| Artificial............... | 84.6 | 104 | 0.22 |
| Plaster: | | | |
| Artificial...... | 132 | 167 | 0.43 |
| Building....... | 77.9 | 77 | 0.25 |
| Cambric, varnished........ | ...... | 100 | 0.091 |
| Cardboard, corrugated........ | ...... | ...... | 0.037 |
| Celluloid....... | 87.3 | 86 | 0.12 |
| Charcoal flakes..... | 11.9 | 176 | 0.043 |
| Clinker, granular........ | ...... | 0–1200 | 0.27 |
| Coke, petroleum ...... | ...... | 212 | 3.4 |
| | | 932 | 2.9 |
| 20–100 mesh...... | 62 | 750 | 0.55 |
| Coke, powdered...... | ...... | 32–212 | 0.11 |
| Concrete: | | | |
| Cinder...... | ...... | ...... | 0.20 |
| Stone...... | ...... | ...... | 0.54 |
| 1:4 dry...... | ...... | ...... | 0.44 |
| Cotton...... | 5 | 60 | 0.033 |
| Cotton wool...... | 5 | 86 | 0.024 |
| Cork: | | | |
| Board...... | 10 | 86 | 0.025 |
| Regranulated...... | 8.1 | 86 | 0.026 |
| Diatomaceous earth, powder...... | 18 | 100 | 0.039 |
| | 18 | 500 | 0.051 |
| | 18 | 1000 | 0.068 |
| Enamel, silicate...... | 38 | ...... | 0.5–0.75 |
| Felt, wool...... | 20.6 | 86 | 0.03 |
| Fiber insulating board...... | 14.8 | 70 | 0.028 |
| Glass: | | | |
| Borosilicate type...... | 139 | ...... | 0.63 |
| Window...... | ...... | ...... | 0.3–0.61 |
| Soda...... | ...... | ...... | 0.3–0.44 |
| Glass, insulating: | | | |
| Cellular, slab...... | 9 | 75 | 0.035 |
| Fiberglass wool...... | 3 | 100 | 0.0225 |
| | 3 | 300 | 0.0342 |
| | 9 | 100 | 0.0188 |
| | 9 | 500 | 0.0375 |
| Granite...... | ...... | ...... | 1.0–2.3 |
| Graphite: | | | |
| Longitudinal...... | ...... | 68 | 95. |
| Powdered (through 100 mesh)...... | 30 | 104 | 0.104 |
| Gypsum: | | | |
| Plaster...... | ...... | ...... | 0.27 |
| Cellular...... | 8 | ...... | 0.029 |
| | 30 | ...... | 0.083 |
| Powdered...... | 26–34 | ...... | 0.043–0.05 |
| Hair felt between layers of paper...... | 17 | 86 | 0.021 |
| Ice...... | 57.5 | 32 | 1.3 |
| Kapok...... | 0.88 | 68 | 0.020 |
| Lampblack...... | 10 | 104 | 0.038 |
| Lava...... | ...... | ...... | 0.49 |
| Leather, sole...... | 62.4 | ...... | 0.092 |
| Limestone (15.3% H₂O by vol)...... | 103 | 73 | 0.54 |
| Linen...... | ...... | 86 | 0.05 |

## Table 3-11. Thermal Conductivities of Building and Insulating Materials (Continued)

| Material | Apparent density $\rho$, lb/ft³ at room temp | $t$, °F | $k$, Btu/(hr)(ft²) (°F/ft) |
|---|---|---|---|
| Magnesia: | | | |
|   Powdered............................ | 49.7 | 117 | 0.35 |
|   85%............................... | 13 | 100 | 0.034 |
| | 13 | 400 | 0.040 |
| Magnesium oxide, compressed.................. | 49.9 | 68 | 0.32 |
| Marble........................................ | ..... | ..... | 1.2–1.7 |
| Mica (perpendicular to planes)................. | ..... | 122 | 0.25 |
| Mill shavings................................. | ..... | ..... | 0.033–0.05 |
| Mineral wool: | | | |
|   Fibrous............................ | 9.4 | 86 | 0.0225 |
| | 19.7 | 86 | 0.024 |
|   Block, with binder.................. | 16.7 | ..... | 0.031 |
|   Blanket............................ | 4.5 | ..... | 0.022 |
| Paper......................................... | ..... | ..... | 0.075 |
| Paper and asbestos fiber with emulsified asphalt binder. | 4.2 | 94 | 0.023 |
| Paraffin wax.................................. | ..... | 32 | 0.14 |
| Porcelain..................................... | ..... | 392 | 0.88 |
| Portland cement (see also concrete)............. | ..... | 194 | 0.17 |
| Pumice stone.................................. | ..... | 70–150 | 0.14 |
| Pyroxylin plastics............................. | ..... | ..... | 0.075 |
| Rock wool.................................... | ..... | 100 | 0.030 |
| | ..... | 600 | 0.057 |
| Rubber: | | | |
|   Hard.............................. | 74.8 | 32 | 0.087 |
|   Para............................... | ..... | 70 | 0.109 |
|   Soft............................... | ..... | 70 | 0.075–0.092 |
| Sand, dry..................................... | 94.6 | 68 | 0.19 |
| Sandstone.................................... | 140 | 104 | 1.06 |
| Sawdust...................................... | 12 | 70 | 0.034 |
| Silk.......................................... | 6.3 | 60 | 0.026 |
|   Varnished.......................... | ..... | 100 | .096 |
| Slag wool..................................... | 12 | 86 | .022 |
| Slate......................................... | ..... | 200 | .86 |
| Snow......................................... | 34.7 | 32 | .27 |
| Sugar-cane-fiber insulation blocks encased in asphalt membrane. | 13.8 | 70 | .025 |
| Wallboard: | | | |
|   Insulating type..................... | 14.8 | 70 | .028 |
|   Stiff pasteboard.................... | 43 | 86 | .04 |
| Wood shavings................................ | 8.8 | 86 | .034 |
| Wood: | | | |
|   Douglas fir, 0% moisture............ | 30 | 75 | .056 |
|   Longleaf yellow pine, 0% moisture.... | 40 | 75 | .072 |
| Wood lath and plaster......................... | ..... | 70 | .27 |
| Wood pulp in sheet form....................... | 16.5 | ..... | .028 |
| Wool, animal................................. | 6.9 | 86 | .021 |

## Table 3-12. Thermal Conductivities of Liquids*

| Liquid | $k$, Btu/(hr)(ft²)(°F/ft) | | | |
|---|---|---|---|---|
| | 50°F | 100°F | 200°F | 300°F |
| Acetic acid: | | | | |
| 100%............................... | 0.099 | | | |
| 50%................................ | 0.20 | | | |
| Ammonia........................... | 0.29 | | | |
| Ammonia, aqueous, 26%.............. | 0.253 | 0.274 | 0.315 | |
| Amyl acetate....................... | 0.083 | | | |
| n-Amyl alcohol..................... | 0.096 | 0.094 | 0.0895 | |
| Isoamyl alcohol.................... | 0.089 | 0.088 | 0.087 | |
| Aniline............................ | 0.100 | | | |
| Benzene............................ | 0.096 | 0.091 | 0.081 | |
| Bromobenzene....................... | 0.075 | 0.074 | 0.071 | |
| Butyl acetate...................... | 0.078 | 0.074 | 0.068 | |
| n-Butyl alcohol.................... | 0.098 | 0.097 | 0.094 | |
| Isobutyl alcohol................... | 0.091 | | | |
| Calcium chloride brine: | | | | |
| 30%............................... | ...... | 0.32 | | |
| 15%............................... | | 0.34 | | |
| Carbon disulfide................... | 0.095 | 0.092 | 0.086 | |
| Carbon tetrachloride............... | 0.067 | 0.065 | 0.062 | |
| Chloroform......................... | 0.073 | 0.071 | 0.067 | |
| Decane............................. | 0.0865 | 0.0843 | 0.0805 | 0.0775 |
| Dibutyl phthalate.................. | 0.079 | 0.078 | 0.073 | |
| Dichlorodifluoromethane............ | 0.054 | 0.048 | 0.035 | |
| Dowtherm A......................... | ...... | 0.082 | 0.079 | 0.077 |
| Dowtherm E......................... | | 0.073 | 0.071 | 0.064 |
| Ethane............................. | 0.056 | | | |
| Ethyl acetate...................... | 0.085 | 0.080 | 0.068 | |
| Ethyl alcohol...................... | 0.106 | 0.092 | 0.068 | |
| Ethyl benzene...................... | 0.088 | 0.085 | 0.077 | |
| Ethyl ether........................ | ...... | 0.080 | 0.077 | |
| Ethylene glycol.................... | 0.145 | 0.144 | 0.142 | 0.140 |
| Glycerine (USP).................... | 0.156 | 0.159 | 0.164 | |
| Heptane............................ | 0.083 | 0.0815 | 0.0765 | 0.071 |
| Hexane............................. | 0.0820 | 0.0795 | 0.0745 | 0.068 |
| Mercury............................ | 4.7 | ...... | ...... | 6.7 |
| Methyl alcohol..................... | 0.128 | 0.117 | 0.095 | |
| Nitrobenzene....................... | 0.097 | 0.093 | 0.089 | |
| Nitromethane....................... | 0.128 | 0.124 | 0.115 | |
| Octane............................. | 0.0845 | 0.0825 | 0.078 | 0.073 |
| Olive oil (USP).................... | 0.097 | 0.096 | 0.094 | |
| Paraldehyde........................ | 0.086 | 0.083 | 0.0785 | |
| Pentane............................ | 0.080 | 0.0775 | 0.071 | 0.0615 |
| Propane............................ | 0.072 | 0.0675 | 0.056 | |
| n-Propyl alcohol................... | 0.101 | 0.099 | 0.094 | |
| Propylene glycol................... | 0.116 | 0.115 | 0.113 | 0.110 |
| Sodium............................. | ...... | ...... | 14.8 | 15.0 |
| Sodium chloride brine: | | | | |
| 25%............................... | 0.34 | | | |
| 12.5%............................. | 0.35 | | | |
| Sulfuric acid: | | | | |
| 90%............................... | 0.22 | | | |
| 60%............................... | 0.26 | | | |
| 30%............................... | 0.31 | | | |
| Sulfur dioxide..................... | 0.116 | 0.108 | | |
| Trichlorethylene................... | 0.072 | 0.068 | 0.060 | |
| Toluene............................ | 0.087 | 0.086 | 0.084 | |
| Vinyl acetate...................... | 0.095 | 0.083 | 0.075 | |
| Water.............................. | 0.331 | 0.363 | 0.393 | 0.395 |
| Dow Corning silicone DC-200 (500,000 centistokes)........ | ...... | 0.090 | 0.085 | |
| GE silcones SF-96: | | | | |
| 40 centistokes.................. | ...... | 0.085 | 0.080 | |
| 100 centistokes................. | ...... | 0.086 | 0.081 | |
| 300 centistokes................. | | 0.090 | 0.084 | |
| 1,000 centistokes............... | ...... | 0.091 | 0.086 | |

NOTE: Viscosities, where specified, are at 85°F.

* For thermal conductivities of the elements in solid, liquid, or gaseous form see Ho, Powell, and Liley, *J. Phys. Chem. Ref. Data*, **1**, 279 (1972). Values for 38 refrigerants appear in abbreviated form in "Handbook of Fundamentals," ASHRAE, New York, 1973 and in more detail in the 1973 "Thermophysical Properties of Refrigerants Bulletin." Missenard, *Cahiers. Thermique*, **C2**, Paris, 1971 extensively surveys organic materials.

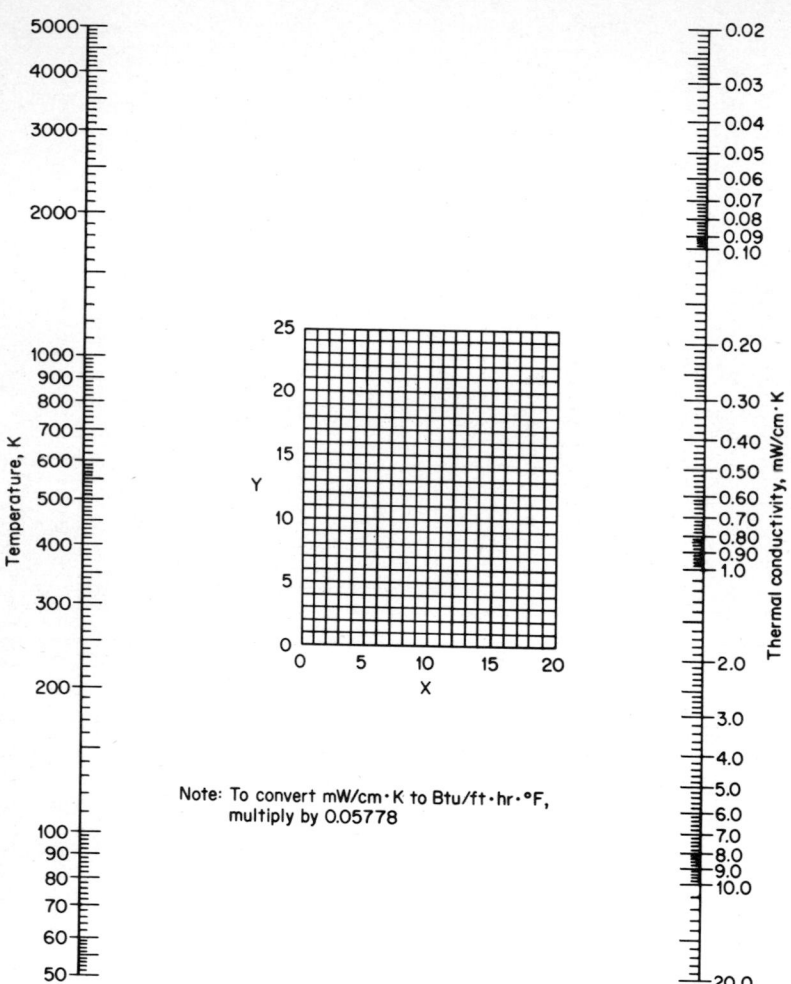

FIG. 3-3. Thermal conductivity of gases at 1 atm. (See coordinates on facing page.)

| No. | Gas or vapor | Temperature range | X | Y | No. | Gas or vapor | Temperature range | X | Y |
|-----|--------------|-------------------|---|---|-----|--------------|-------------------|---|---|
| 1 | Acetone | 250– 500°K | 3.7 | 14.8 | 25 | Freon 22 | 250– 500 | 6.5 | 16.6 |
| 2 | Acetylene | 200– 600 | 7.5 | 13.5 | 26 | Freon 113 | 250– 400 | 4.7 | 17.0 |
| 3a | Air | 50– 250 | 12.4 | 13.9 | 27a | Helium | 50– 500 | 17.0 | 2.5 |
| 3b | Air | 250–1,000 | 14.7 | 15.0 | 27b | Helium | 500–5,000 | 15.0 | 3.0 |
| 3c | Air | 1,000–1,500 | 17.1 | 14.5 | 28a | n-Heptane | 250– 600 | 4.0 | 14.8 |
| 4 | Ammonia | 200– 900 | 8.5 | 12.6 | 28b | n-Heptane | 600–1,000 | 6.9 | 14.9 |
| 5a | Argon | 50– 250 | 12.5 | 16.5 | 29 | n-Hexane | 250–1,000 | 3.7 | 14.0 |
| 5b | Argon | 250–5,000 | 15.4 | 18.1 | 30a | Hydrogen | 50– 250 | 13.2 | 1.2 |
| 6 | Benzene | 250– 600 | 2.8 | 14.2 | 30b | Hydrogen | 250–1,000 | 15.7 | 1.3 |
| 7 | Boron Trifluoride | 250– 400 | 12.4 | 16.4 | 30c | Hydrogen | 1,000–2,000 | 13.7 | 2.7 |
| 8 | Bromine | 250– 350 | 10.1 | 23.6 | 31 | Hydrogen | | | |
| 9 | n-Butane | 250– 500 | 5.6 | 14.1 | | Chloride | 200– 700 | 12.2 | 18.5 |
| 10 | i-Butane | 250– 500 | 5.7 | 14.0 | 32 | Krypton | 100– 700 | 13.7 | 21.8 |
| 11a | Carbon Dioxide | 200– 700 | 8.7 | 15.5 | 33a | Methane | 100– 300 | 11.2 | 11.7 |
| 11b | Carbon Dioxide | 700–1,200 | 13.3 | 15.4 | 33b | Methane | 300–1,000 | 8.5 | 11.0 |
| 12a | Carbon Monoxide | 80– 300 | 12.3 | 14.2 | 34 | Methyl Alcohol | 300– 500 | 5.0 | 14.3 |
| 12b | Carbon Monoxide | 300–1,200 | 15.2 | 15.2 | 35 | Methyl Chloride | 250– 700 | 4.7 | 15.7 |
| 13 | Carbon | | | | 36a | Neon | 50– 250 | 15.2 | 10.2 |
| | Tetrachloride | 250– 500 | 9.4 | 21.0 | 36b | Neon | 250–5,000 | 17.2 | 11.0 |
| 14 | Chlorine | 200– 700 | 10.8 | 20.1 | 37 | Nitric Oxide | 100–1,000 | 13.2 | 14.8 |
| 15a | Deuterium | 50– 100 | 12.7 | 17.3 | 38a | Nitrogen | 50– 250 | 12.5 | 14.0 |
| 15b | Deuterium | 100– 400 | 14.5 | 19.3 | 38b | Nitrogen | 250–1,500 | 15.8 | 15.3 |
| 16 | Ethane | 200–1,000 | 5.4 | 12.6 | 38c | Nitrogen | 1,500–3,000 | 12.5 | 16.5 |
| 17a | Ethyl Alcohol | 250– 350 | 2.0 | 13.0 | 39a | Nitrous Oxide | 200– 500 | 8.4 | 15.0 |
| 17b | Ethyl Alcohol | 350– 500 | 7.7 | 15.2 | 39b | Nitrous Oxide | 500–1,000 | 11.5 | 15.5 |
| 18 | Ethyl Ether | 250– 500 | 5.3 | 14.1 | 40a | Oxygen | 50– 300 | 12.2 | 13.8 |
| 19 | Ethylene | 200– 450 | 3.9 | 12.3 | 40b | Oxygen | 300–1,500 | 14.5 | 14.8 |
| 20a | Fluorine | 80– 600 | 12.3 | 13.8 | 41 | Pentane | 250– 500 | 5.0 | 14.1 |
| 20b | Fluorine | 600– 800 | 18.7 | 13.8 | 42a | Propane | 200– 300 | 2.7 | 12.0 |
| 21 | Freon 11 | 250– 500 | 7.5 | 19.0 | 42b | Propane | 300– 500 | 6.3 | 13.7 |
| 22 | Freon 12 | 250– 500 | 6.8 | 17.5 | 43 | Sulfur Dioxide | 250– 900 | 9.2 | 18.5 |
| 23 | Freon 13 | 250– 500 | 7.5 | 16.5 | 44 | Toluene | 250– 600 | 6.4 | 14.6 |
| 24 | Freon 21 | 250– 450 | 6.2 | 17.5 | 45 | Xenon | 150– 700 | 13.3 | 25.0 |

Fig. 3-3. Coordinates.

## Table 3-13. Thermal Conductivity of Air as a Function of Pressure and Temperature*

| Pressure, bars† | Temperature, K | | | | | | | |
|---|---|---|---|---|---|---|---|---|
| | 300 | 350 | 400 | 450 | 500 | 600 | 700 | 800 |
| 1 | 0.0262 | 0.0300 | 0.0338 | 0.0373 | 0.0407 | 0.0469 | 0.0524 | 0.0573 |
| 10 | 0.0266 | 0.0304 | 0.0341 | 0.0376 | 0.0410 | 0.0471 | 0.0526 | 0.0574 |
| 50 | 0.0284 | 0.0318 | 0.0354 | 0.0387 | 0.0420 | 0.0479 | 0.0532 | 0.0580 |
| 100 | 0.0314 | 0.0341 | 0.0373 | 0.0403 | 0.0433 | 0.0490 | 0.0541 | 0.0588 |
| 150 | 0.0349 | 0.0367 | 0.0394 | 0.0420 | 0.0449 | 0.0502 | 0.0552 | 0.0596 |
| 200 | 0.0385 | 0.0395 | 0.0418 | 0.0440 | 0.0465 | 0.0515 | 0.0562 | 0.0605 |
| 250 | 0.0420 | 0.0423 | 0.0441 | 0.0460 | 0.0483 | 0.0529 | 0.0573 | 0.0614 |
| 300 | 0.0455 | 0.0451 | 0.0464 | 0.0480 | 0.0501 | 0.0543 | 0.0584 | 0.0624 |
| 400 | 0.0522 | 0.0505 | 0.0509 | 0.0519 | 0.0535 | 0.0571 | 0.0608 | 0.0644 |
| 500 | 0.0584 | 0.0555 | 0.0552 | 0.0558 | 0.0568 | 0.0598 | 0.0642 | 0.0674 |
| 600 | 0.0644 | 0.0603 | 0.0594 | 0.0594 | 0.0600 | 0.0624 | 0.0653 | 0.0684 |
| 700 | 0.0703 | 0.0651 | 0.0636 | 0.0629 | 0.0632 | 0.0650 | 0.0675 | 0.0704 |
| 800 | 0.0760 | 0.0700 | 0.0676 | 0.0662 | 0.0662 | 0.0676 | 0.0697 | 0.0722 |
| 900 | 0.0811 | 0.0749 | 0.0716 | 0.0694 | 0.0694 | 0.0695 | 0.0719 | 0.0740 |
| 1000 | 0.0862 | 0.0798 | 0.0756 | 0.0726 | 0.0718 | 0.0725 | 0.0741 | 0.0758 |

* In W/(m · K).   Condensed from N. B. Vargaftik, "Heat Conductivity of Gases and Liquids," FTD-MT-24-133-71 (AD 736963), 1971.
† To convert to pascals (newtons per square meter) mutiply by $10^5$.

## Table 3-14. Thermal Conductivity of Water and Steam*

| Pressure, bars† | Temperature, K | | | | | | | |
|---|---|---|---|---|---|---|---|---|
| | 300 | 400 | 500 | 600 | 700 | 800 | 900 | 1000 |
| 1 | 0.614 | 0.0268 | 0.0358 | 0.0464 | 0.0581 | 0.0710 | 0.0843 | 0.0981 |
| 10 | 0.615 | 0.689 | 0.0380 | 0.0474 | 0.0590 | 0.0717 | 0.0851 | 0.0988 |
| 20 | 0.616 | 0.689 | 0.0402 | 0.0485 | 0.0599 | 0.0726 | 0.0859 | 0.0996 |
| 40 | 0.617 | 0.690 | 0.644 | 0.0516 | 0.0620 | 0.0744 | 0.0877 | 0.101 |
| 60 | 0.619 | 0.692 | 0.646 | 0.0561 | 0.0645 | 0.0764 | 0.0895 | 0.103 |
| 80 | 0.620 | 0.693 | 0.648 | 0.0628 | 0.0672 | 0.0785 | 0.0914 | 0.105 |
| 100 | 0.622 | 0.694 | 0.651 | 0.0730 | 0.0704 | 0.0807 | 0.0934 | 0.107 |
| 200 | 0.630 | 0.701 | 0.661 | 0.516 | 0.0966 | 0.0974 | 0.107 | 0.119 |
| 300 | 0.637 | 0.707 | 0.672 | 0.546 | 0.168 | 0.123 | 0.122 | 0.134 |
| 400 | 0.644 | 0.713 | 0.681 | 0.571 | 0.289 | 0.154 | 0.137 | 0.144 |
| 500 | 0.650 | 0.719 | 0.689 | 0.588 | 0.365 | 0.178 | 0.147 | 0.150 |

* In W/(m · K).
† To convert to pascals (newtons per square meter) multiply by $10^5$.

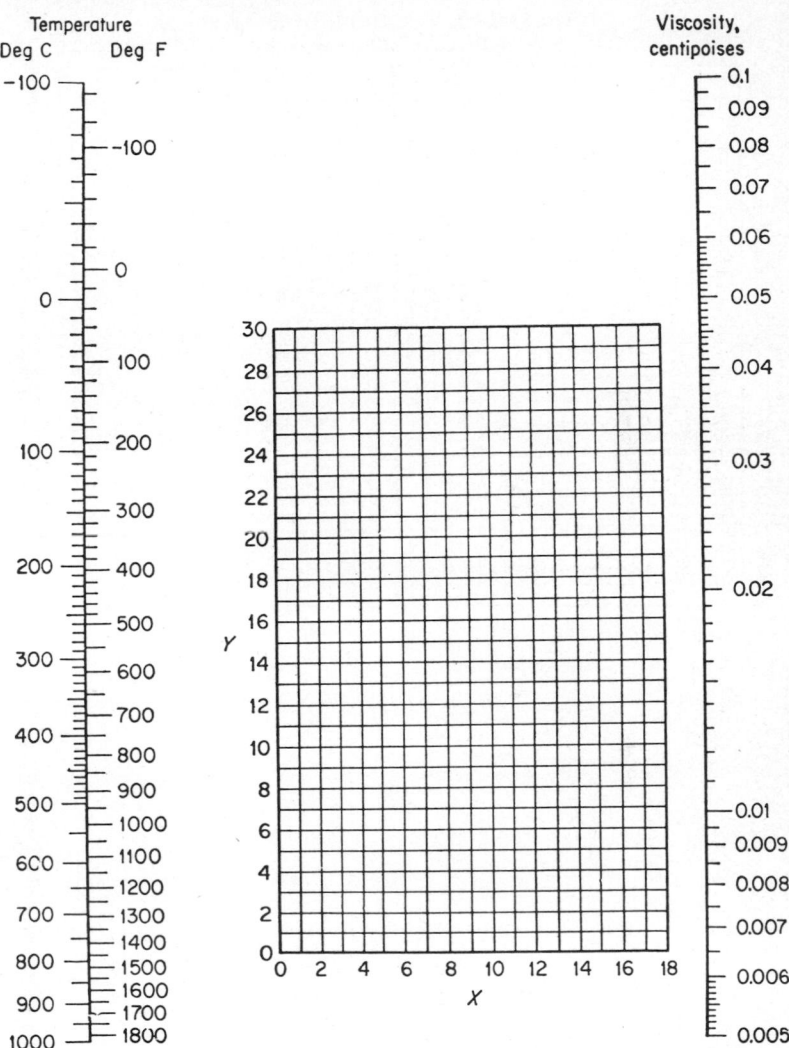

F$_{\text{IG}}$. 3-4. Viscosities of gases at 1 atm.    For coordinates, see Table 3-15.

## Table 3-15. Viscosities of Gases

Coordinates for Use with Fig. 3-4

| No. | Gas | X | Y | No. | Gas | X | Y |
|---|---|---|---|---|---|---|---|
| 1 | Acetic acid................ | 7.7 | 14.3 | 29 | Freon-113................. | 11.3 | 14.0 |
| 2 | Acetone................... | 8.9 | 13.0 | 30 | Helium................... | 10.9 | 20.5 |
| 3 | Acetylene................. | 9.8 | 14.9 | 31 | Hexane................... | 8.6 | 11.8 |
| 4 | Air...................... | 11.0 | 20.0 | 32 | Hydrogen................. | 11.2 | 12.4 |
| 5 | Ammonia................. | 8.4 | 16.0 | 33 | $3H_2 + 1N_2$............... | 11.2 | 17.2 |
| 6 | Argon.................... | 10.5 | 22.4 | 34 | Hydrogen bromide.......... | 8.8 | 20.9 |
| 7 | Benzene.................. | 8.5 | 13.2 | 35 | Hydrogen chloride......... | 8.8 | 18.7 |
| 8 | Bromine.................. | 8.9 | 19.2 | 36 | Hydrogen cyanide.......... | 9.8 | 14.9 |
| 9 | Butene................... | 9.2 | 13.7 | 37 | Hydrogen iodide........... | 9.0 | 21.3 |
| 10 | Butylene................. | 8.9 | 13.0 | 38 | Hydrogen sulfide........... | 8.6 | 18.0 |
| 11 | Carbon dioxide............ | 9.5 | 18.7 | 39 | Iodine................... | 9.0 | 18.4 |
| 12 | Carbon disulfide.......... | 8.0 | 16.0 | 40 | Mercury.................. | 5.3 | 22.9 |
| 13 | Carbon monoxide.......... | 11.0 | 20.0 | 41 | Methane.................. | 9.9 | 15.5 |
| 14 | Chlorine................. | 9.0 | 18.4 | 42 | Methyl alcohol............ | 8.5 | 15.6 |
| 15 | Chloroform............... | 8.9 | 15.7 | 43 | Nitric oxide.............. | 10.9 | 20.5 |
| 16 | Cyanogen................. | 9.2 | 15.2 | 44 | Nitrogen................. | 10.6 | 20.0 |
| 17 | Cyclohexane.............. | 9.2 | 12.0 | 45 | Nitrosyl chloride.......... | 8.0 | 17.6 |
| 18 | Ethane................... | 9.1 | 14.5 | 46 | Nitrous oxide............. | 8.8 | 19.0 |
| 19 | Ethyl acetate............. | 8.5 | 13.2 | 47 | Oxygen.................. | 11.0 | 21.3 |
| 20 | Ethyl alcohol............. | 9.2 | 14.2 | 48 | Pentane.................. | 7.0 | 12.8 |
| 21 | Ethyl chloride............ | 8.5 | 15.6 | 49 | Propane.................. | 9.7 | 12.9 |
| 22 | Ethyl ether.............. | 8.9 | 13.0 | 50 | Propyl alcohol............ | 8.4 | 13.4 |
| 23 | Ethylene................. | 9.5 | 15.1 | 51 | Propylene................ | 9.0 | 13.8 |
| 24 | Fluorine................. | 7.3 | 23.8 | 52 | Sulfur dioxide............ | 9.6 | 17.0 |
| 25 | Freon-11................. | 10.6 | 15.1 | 53 | Toluene.................. | 8.6 | 12.4 |
| 26 | Freon-12................. | 11.1 | 16.0 | 54 | 2,3,3-Trimethylbutane....... | 9.5 | 10.5 |
| 27 | Freon-21................. | 10.8 | 15.3 | 55 | Water................... | 8.0 | 16.0 |
| 28 | Freon-22................. | 10.1 | 17.0 | 56 | Xenon................... | 9.3 | 23.0 |

## Table 3-16. Viscosity of Steam (Micropoises)*

Viscosity, lb./(ft.)(hr.) $\times 10^{-2}$

| Pressure, lb./sq. in. abs. | Temp., °F. | | | | | | | |
|---|---|---|---|---|---|---|---|---|
| | −100 | −50 | 0 | 50 | 100 | 150 | 200 | 250 |
| 200 | 3.27 | 3.64 | 3.98 | 4.29 | 4.57 | 4.78 | 5.12 | 5.45 |
| 400 | 3.39 | 3.73 | 4.06 | 4.36 | 4.63 | 4.86 | 5.19 | 5.51 |
| 600 | 3.54 | 3.83 | 4.14 | 4.43 | 4.69 | 4.94 | 5.25 | 5.56 |
| 800 | 3.72 | 3.95 | 4.22 | 4.50 | 4.76 | 5.02 | 5.31 | 5.61 |
| 1,000 | 3.90 | 4.07 | 4.31 | 4.58 | 4.84 | 5.10 | 5.38 | 5.67 |
| 1,200 | 4.08 | 4.20 | 4.42 | 4.66 | 4.92 | 5.16 | 5.44 | 5.72 |
| 1,400 | 4.26 | 4.35 | 4.54 | 4.77 | 5.00 | 5.24 | 5.50 | 5.77 |
| 1,600 | 4.47 | 4.55 | 4.68 | 4.87 | 5.08 | 5.31 | 5.57 | 5.84 |
| 1,800 | 4.70 | 4.75 | 4.83 | 5.00 | 5.17 | 5.39 | 5.63 | 5.90 |
| 2,000 | 5.10 | 4.95 | 4.97 | 5.10 | 5.27 | 5.47 | 5.70 | 5.97 |
| 2,500 | 6.05 | 5.52 | 5.36 | 5.38 | 5.51 | 5.68 | 5.87 | 6.07 |
| 3,000 | 6.82 | 6.14 | 5.77 | 5.70 | 5.76 | 5.91 | 6.06 | 6.19 |
| 3,500 | 7.62 | 6.76 | 6.23 | 6.06 | 6.06 | 6.12 | 6.25 | 6.43 |
| 4,000 | 8.35 | 7.34 | 6.65 | 6.42 | 6.38 | 6.42 | 6.43 | 6.63 |
| 4,500 | 9.10 | 7.91 | 7.09 | 6.76 | 6.68 | 6.69 | 6.71 | 6.82 |
| 5,000 | 9.88 | 8.49 | 7.55 | 7.16 | 6.99 | 6.99 | 6.97 | 7.02 |
| 6,000 | 11.35 | 9.66 | 8.39 | 7.90 | 7.66 | 7.52 | 7.43 | 7.60 |
| 7,000 | 12.83 | 10.78 | 9.17 | 8.61 | 8.26 | 8.03 | 7.92 | 8.10 |
| 8,000 | 14.56 | 11.94 | 10.16 | 9.42 | 8.89 | 8.56 | 8.39 | 8.52 |
| 9,000 | 16.09 | 12.94 | 11.08 | 10.11 | 9.46 | 9.07 | 8.83 | 8.90 |
| 10,000 | 17.70 | 14.03 | 11.85 | 10.79 | 10.10 | 9.65 | 9.37 | 9.17 |

*Compiled by P. E. Liley. For tables in SI units from 90° to 1300°K., 1 to 1000 bars, see Vasserman, Kazavchinskii, and Rabinovich, "Thermophysical Properties of Air and Air Components," Moscow, 1966, and NBS-NSF trans. TT 70-50095, 1971. This source contains a discussion of present accuracy.

## Table 3-17. Viscosity of Air*

| Pressure, bars | Temp., °C. | | | | | | | | | | |
|---|---|---|---|---|---|---|---|---|---|---|---|
| | 0 | 50 | 100 | 150 | 200 | 250 | 300 | 400 | 500 | 600 | 700 |
| 1 | 17,500 | 5440 | 121 | 142 | 162 | 182 | 203 | 243 | 284 | 325 | 365 |
| 50 | 17,500 | 5450 | 2800 | 1820 | 1350 | 1070 | 206 | 250 | 289 | 329 | 369 |
| 100 | 17,500 | 5450 | 2810 | 1830 | 1360 | 1080 | 905 | 258 | 295 | 334 | 374 |
| 150 | 17,400 | 5460 | 2820 | 1840 | 1370 | 1100 | 917 | 269 | 302 | 340 | 379 |
| 200 | 17,400 | 5460 | 3830 | 1860 | 1380 | 1110 | 930 | 286 | 311 | 346 | 384 |
| 250 | 17,400 | 5470 | 2840 | 1870 | 1390 | 1120 | 943 | 321 | 321 | 353 | 389 |
| 300 | 17,400 | 5470 | 2850 | 1880 | 1400 | 1130 | 955 | 458 | 334 | 361 | 395 |
| 350 | 17,300 | 5480 | 2860 | 1890 | 1420 | 1150 | 968 | 573 | 349 | 369 | 401 |
| 400 | 17,300 | 5480 | 2870 | 1900 | 1430 | 1160 | 981 | 628 | 369 | 379 | 408 |
| 450 | 17,300 | 5490 | 2880 | 1910 | 1440 | 1170 | 993 | 664 | 393 | 389 | 415 |
| 500 | 17,200 | 5490 | 2890 | 1920 | 1450 | 1180 | 1010 | 693 | 421 | 401 | 423 |

*Copyright 1968 by the American Society of Mechanical Engineers. Reproduced by permission.

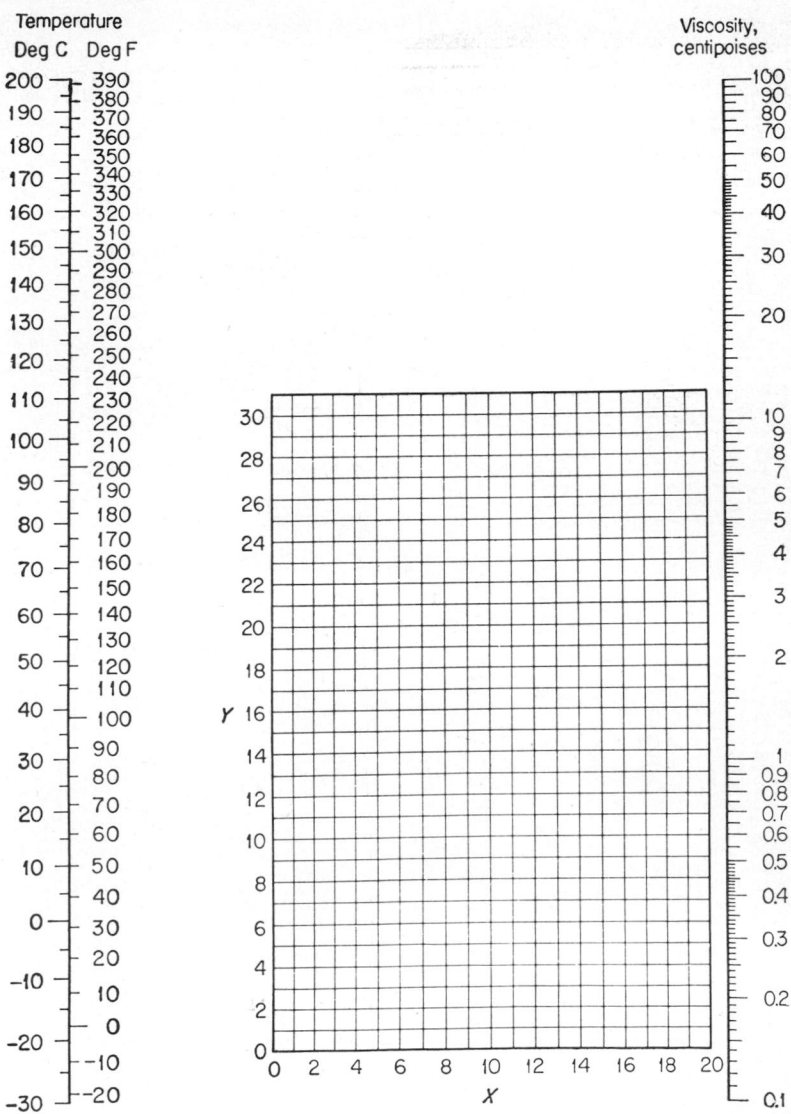

Fig. 3-5. Viscosities of liquids at 1 atm. For coordinates, see Table 3-18.

## Table 3-18. Viscosities of Liquids

Coordinates for Use with Fig. 3-5

| No. | Liquid | X | Y | No. | Liquid | X | Y |
|-----|--------|---|---|-----|--------|---|---|
| 1 | Acetaldehyde | 15.2 | 4.8 | 56 | Freon-22 | 17.2 | 4.7 |
| 2 | Acetic acid, 100% | 12.1 | 14.2 | 57 | Freon-13 | 12.5 | 11.4 |
| 3 | 70% | 9.5 | 17.0 | 58 | Glycerol, 100% | 2.0 | 30.0 |
| 4 | Acetic anhydride | 12.7 | 12.8 | 59 | 50% | 6.9 | 19.6 |
| 5 | Acetone, 100% | 14.5 | 7.2 | 60 | Heptene | 14.1 | 8.4 |
| 6 | 35% | 7.9 | 15.0 | 61 | Hexane | 14.7 | 7.0 |
| 7 | Allyl alcohol | 10.2 | 14.3 | 62 | Hydrochloric acid, 31.5% | 13.0 | 16.6 |
| 8 | Ammonia, 100% | 12.6 | 2.0 | 63 | Isobutyl alcohol | 7.1 | 18.0 |
| 9 | 26% | 10.1 | 13.9 | 64 | Isobutyric acid | 12.2 | 14.4 |
| 10 | Amyl acetate | 11.8 | 12.5 | 65 | Isopropyl alcohol | 8.2 | 16.0 |
| 11 | Amyl alcohol | 7.5 | 18.4 | 66 | Kerosene | 10.2 | 16.9 |
| 12 | Aniline | 8.1 | 18.7 | 67 | Linseed oil, raw | 7.5 | 27.2 |
| 13 | Anisole | 12.3 | 13.5 | 68 | Mercury | 18.4 | 16.4 |
| 14 | Arsenic trichloride | 13.9 | 14.5 | 69 | Methanol, 100% | 12.4 | 10.5 |
| 15 | Benzene | 12.5 | 10.9 | 70 | 90% | 12.3 | 11.8 |
| 16 | Brine, CaCl₂, 25% | 6.6 | 15.9 | 71 | 40% | 7.8 | 15.5 |
| 17 | NaCl, 25% | 10.2 | 16.6 | 72 | Methyl acetate | 14.2 | 8.2 |
| 18 | Bromine | 14.2 | 13.2 | 73 | Methyl chloride | 15.0 | 3.8 |
| 19 | Bromotoluene | 20.0 | 15.9 | 74 | Methyl ethyl ketone | 13.9 | 8.6 |
| 20 | Butyl acetate | 12.3 | 11.0 | 75 | Naphthalene | 7.9 | 18.1 |
| 21 | Butyl alcohol | 8.6 | 17.2 | 76 | Nitric acid, 95% | 12.8 | 13.8 |
| 22 | Butyric acid | 12.1 | 15.3 | 77 | 60% | 10.8 | 17.0 |
| 23 | Carbon dioxide | 11.6 | 0.3 | 78 | Nitrobenzene | 10.6 | 16.2 |
| 24 | Carbon disulfide | 16.1 | 7.5 | 79 | Nitrotoluene | 11.0 | 17.0 |
| 25 | Carbon tetrachloride | 12.7 | 13.1 | 80 | Octane | 13.7 | 10.0 |
| 26 | Chlorobenzene | 12.3 | 12.4 | 81 | Octyl alcohol | 6.6 | 21.1 |
| 27 | Chloroform | 14.4 | 10.2 | 82 | Pentachloroethane | 10.9 | 17.3 |
| 28 | Chlorosulfonic acid | 11.2 | 18.1 | 83 | Pentane | 14.9 | 5.2 |
| 29 | Chlorotoluene, ortho | 13.0 | 13.3 | 84 | Phenol | 6.9 | 20.8 |
| 30 | meta | 13.3 | 12.5 | 85 | Phosphorus tribromide | 13.8 | 16.7 |
| 31 | para | 13.3 | 12.5 | 86 | Phosphorus trichloride | 16.2 | 10.9 |
| 32 | Cresol, meta | 2.5 | 20.8 | 87 | Propionic acid | 12.8 | 13.8 |
| 33 | Cyclohexanol | 2.9 | 24.3 | 88 | Propyl alcohol | 9.1 | 16.5 |
| 34 | Dibromoethane | 12.7 | 15.8 | 89 | Propyl bromide | 14.5 | 9.6 |
| 35 | Dichloroethane | 13.2 | 12.2 | 90 | Propyl chloride | 14.4 | 7.5 |
| 36 | Dichloromethane | 14.6 | 8.9 | 91 | Propyl iodide | 14.1 | 11.6 |
| 37 | Diethyl oxalate | 11.0 | 16.4 | 92 | Sodium | 16.4 | 13.9 |
| 38 | Dimethyl oxalate | 12.3 | 15.8 | 93 | Sodium hydroxide, 50% | 3.2 | 25.8 |
| 39 | Diphenyl | 12.0 | 18.3 | 94 | Stannic chloride | 13.5 | 12.8 |
| 40 | Dipropyl oxalate | 10.3 | 17.7 | 95 | Sulfur dioxide | 15.2 | 7.1 |
| 41 | Ethyl acetate | 13.7 | 9.1 | 96 | Sulfuric acid, 110% | 7.2 | 27.4 |
| 42 | Ethyl alcohol, 100% | 10.5 | 13.8 | 97 | 98% | 7.0 | 24.8 |
| 43 | 95% | 9.8 | 14.3 | 98 | 60% | 10.2 | 21.3 |
| 44 | 40% | 6.5 | 16.6 | 99 | Sulfuryl chloride | 15.2 | 12.4 |
| 45 | Ethyl benzene | 13.2 | 11.5 | 100 | Tetrachloroethane | 11.9 | 15.7 |
| 46 | Ethyl bromide | 14.5 | 8.1 | 101 | Tetrachloroethylene | 14.2 | 12.7 |
| 47 | Ethyl chloride | 14.8 | 6.0 | 102 | Titanium tetrachloride | 14.4 | 12.3 |
| 48 | Ethyl ether | 14.5 | 5.3 | 103 | Toluene | 13.7 | 10.4 |
| 49 | Ethyl formate | 14.2 | 8.4 | 104 | Trichloroethylene | 14.8 | 10.5 |
| 50 | Ethyl iodide | 14.7 | 10.3 | 105 | Turpentine | 11.5 | 14.9 |
| 51 | Ethylene glycol | 6.0 | 23.6 | 106 | Vinyl acetate | 14.0 | 8.8 |
| 52 | Formic acid | 10.7 | 15.8 | 107 | Water | 10.2 | 13.0 |
| 53 | Freon-11 | 14.4 | 9.0 | 108 | Xylene, ortho | 13.5 | 12.1 |
| 54 | Freon-12 | 16.8 | 5.6 | 109 | meta | 13.9 | 10.6 |
| 55 | Freon-21 | 15.7 | 7.5 | 110 | para | 13.9 | 10.9 |

## Table 3-19. Diffusivities of Pairs of Gases and Vapors (1 Atm)*

$D_v$ in $Cm^2$/Sec (To convert to square feet per hour, multiply by 0.388.)

| Substance | Temp., °C | Air | A | $H_2$ | $O_2$ | $N_2$ | $CO_2$ |
|---|---|---|---|---|---|---|---|
| Acetic acid | 0 | 0.1064 | ...... | 0.416 | ...... | ...... | 0.0716 |
| Acetone | 0 | .109 | ...... | .361 | | | |
| n-Amyl alcohol | 0 | .0589 | ...... | .235 | ...... | ...... | .0422 |
| sec-Amyl alcohol | 30 | .072 | | | | | |
| Amyl butyrate | 0 | .040 | | | | | |
| Amyl formate | 0 | .0543 | | | | | |
| i-Amyl formate | 0 | .058 | | | | | |
| Amyl isobutyrate | 0 | .0419 | ...... | .171 | | | |
| Amyl propionate | 0 | .046 | ...... | .1914 | ...... | ...... | .0347 |
| Aniline | 0 | .0610 | | | | | |
| | 30 | .075 | | | | | |
| Anthracene | 0 | .0421 | | | | | |
| Argon | 20 | ...... | ...... | ...... | | 0.194 | |
| Benzene | 0 | .077 | ...... | .306 | 0.0797 | ...... | .0528 |
| Benzidine | 0 | .0298 | | | | | |
| Benzyl chloride | 0 | .066 | | | | | |
| n-Butyl acetate | 0 | .058 | | | | | |
| i-Butyl acetate | 0 | .0612 | ...... | .2364 | ...... | ...... | .0425 |
| n-Butyl alcohol | 0 | .0703 | ...... | .2716 | ...... | ...... | .0476 |
| | 30 | .088 | | | | | |
| i-Butyl alcohol | 0 | .0727 | ...... | .2771 | ...... | ...... | .0483 |
| Butyl amine | 0 | .0821 | | | | | |
| i-Butyl amine | 0 | .0853 | | | | | |
| i-Butyl butyrate | 0 | .0468 | ...... | .185 | ...... | ...... | .0327 |
| i-Butyl formate | 0 | .0705 | | | | | |
| i-Butyl isobutyrate | 0 | .0457 | ...... | .191 | ...... | ...... | .0364 |
| i-Butyl propionate | 0 | .0529 | ...... | .203 | ...... | ...... | .0366 |
| i-Butyl valerate | 0 | .0424 | ...... | .173 | ...... | ...... | .0308 |
| Butyric acid | 0 | .067 | ...... | .264 | ...... | ...... | .0476 |
| i-Butyric acid | 0 | .0679 | ...... | .271 | ...... | ...... | .0471 |
| Cadmium | 0 | ...... | ...... | ...... | ...... | .17 | |
| Caproic acid | 0 | .050 | | | | | |
| i-Caproic acid | 0 | .0513 | | | | | |
| Carbon dioxide | 0 | .138 | ...... | .550 | .139 | | |
| | 20 | ...... | ...... | ...... | ...... | .163 | |
| | 500 | ...... | ...... | ...... | .9 | | |
| Carbon disulfide | 0 | .0892 | ...... | .369 | ...... | ...... | .063 |
| Carbon monoxide | 0 | ...... | ...... | .651 | .185 | ...... | .137 |
| | 450 | ...... | ...... | ...... | 1.0 | | |
| Carbon tetrachloride | 0 | ...... | ...... | .293 | 0.0636 | | |
| Chlorobenzene | 30 | .075 | | | | | |
| Chloroform | 0 | .091 | | | | | |
| Chloropicrin | 25 | .088 | | | | | |
| m-Chlorotoluene | 0 | .054 | | | | | |
| o-Chlorotoluene | 0 | .059 | | | | | |
| p-Chlorotoluene | 0 | .051 | | | | | |
| Cyanogen chloride | 0 | .111 | | | | | |
| Cyclohexane | 15 | | 0.0719 | .319 | .0744 | .0760 | |
| | 45 | .086 | | | | | |
| n-Decane | 90 | ...... | ...... | .306 | ...... | .0841 | |
| Diethylamine | 0 | .0884 | | | | | |
| 2,3-Dimethyl butane | 15 | | .0657 | .301 | .0753 | .0751 | |
| Diphenyl | 0 | .0610 | | | | | |
| n-Dodecane | 126 | ...... | ...... | .308 | ...... | .0813 | |
| Ethane | 0 | ...... | ...... | .459 | | | |
| Ethanol | 0 | ...... | ...... | .377 | ...... | ...... | .0686 |
| Ether (diethyl) | 0 | .0778 | ...... | .298 | ...... | ...... | .0546 |
| Ethyl acetate | 0 | .0715 | ...... | .273 | ...... | ...... | .0487 |
| | 30 | .089 | | | | | |
| Ethyl alcohol | 0 | .102 | ...... | .375 | ...... | ...... | .0685 |
| Ethyl benzene | 0 | .0658 | | | | | |
| Ethyl n-butyrate | 0 | .0579 | ...... | .224 | ...... | ...... | .0407 |
| Ethyl i-butyrate | 0 | .0591 | ...... | .229 | ...... | ...... | .0413 |
| Ethylene | 0 | ...... | ...... | .486 | | | |
| Ethyl formate | 0 | .0840 | ...... | .337 | ...... | ...... | .0573 |
| Ethyl propionate | 0 | .068 | ...... | .236 | ...... | ...... | .0450 |

* In this table are a representative selection of diffusion coefficients. As general references, Hirschfelder, Curtiss, and Bird, "Molecular Theory of Gases and Liquids," Wiley, New York, 1964; Chapman and Cowling, "The Mathematical Theory of Non-uniform Gases," Cambridge University Press, New York, 1970; Reid and Sherwood, "The Properties of Gases and Liquids," McGraw-Hill, New York, 1964: Bretsznajder, "Prediction of Transport and Other Physical Properties of Fluids," Pergamon, New York, 1971, may be found useful. The most exhaustive recent compilation for gases is by Mason and Marrero, *J. Phys. Chem. Ref. Data*, 1 (1972).

## Table 3-19. Diffusivities of Pairs of Gases and Vapors (1 Atm) (Continued)

$D_v$ in Cm²/Sec (To convert to square feet per hour, multiply by 0.388.)

| Substance | Temp., °C | Air | A | H₂ | O₂ | N₂ | CO₂ |
|---|---|---|---|---|---|---|---|
| Ethyl valerate | 0 | 0.0512 | ...... | 0.205 | ...... | ...... | 0.0367 |
| Eugenol | 0 | .0377 | | | | | |
| Formic acid | 0 | .1308 | | .510 | | | .0874 |
| Helium | 0 | ...... | 0.641 | | | | |
| | 20 | ...... | | | | 0.705 | |
| n-Hexane | 15 | ...... | .0663 | .290 | 0.0753 | .0757 | |
| Hexyl alcohol | 0 | .0499 | | .200 | | | .0351 |
| Hydrogen | 0 | .611 | | | .697 | .674 | .550 |
| | 25 | ...... | | | | | .646 |
| | 500 | ...... | | | 4.2 | | |
| Hydrogen cyanide | 0 | .173 | | | | | |
| Hydrogen peroxide | 60 | .188 | | | | | |
| Iodine | 0 | .07 | | | | .070 | |
| Mercury | 0 | .112 | | .53 | | .13 | |
| Mesitylene | 0 | .056 | | | | | |
| Methane | 500 | ...... | | | 1.1 | | |
| Methyl acetate | 0 | .084 | | .333 | | | .0567 |
| Methyl alcohol | 0 | .132 | | .506 | | | .0879 |
| Methyl butyrate | 0 | .0633 | | .242 | | | .0446 |
| Methyl i-butyrate | 0 | .0639 | | .257 | | | .0451 |
| Methyl cyclopentane | 15 | ...... | .0731 | .318 | 0.0742 | .0758 | |
| Methyl formate | 0 | .0872 | | | | | |
| Methyl propionate | 0 | .0735 | | .295 | | | .0528 |
| Methyl valerate | 0 | .0569 | | | | | |
| Naphthalene | 0 | .0513 | | | | | |
| Nitrogen | 0 | ...... | | | .181 | | |
| | 25 | ...... | | | | | .165 |
| Nitrous oxide | 0 | ...... | | .535 | | | .096 |
| n-Octane | 0 | .0505 | | | | | |
| | 30 | ...... | .0642 | .271 | .0705 | .0710 | |
| Oxygen | 0 | .178 | | .697 | | .181 | .139 |
| Phosgene | 0 | .095 | | | | | |
| Propionic acid | 0 | .0829 | | .330 | | | .0588 |
| Propyl acetate | 0 | .067 | | | | | |
| n-Propyl alcohol | 0 | .085 | | .315 | | | .0577 |
| i-Propyl alcohol | 0 | .0818 | | | | | |
| | 30 | .101 | | | | | |
| n-Propyl benzene | 0 | .0481 | | | | | |
| i-Propyl benzene | 0 | .0489 | | | | | |
| n-Propyl bromide | 0 | .085 | | | | | |
| i-Propyl bromide | 0 | .0902 | | | | | |
| Propyl butyrate | 0 | .0530 | | .206 | | | .0364 |
| Propyl formate | 0 | .0712 | | .281 | | | .0490 |
| n-Propyl iodide | 0 | .079 | | | | | |
| i-Propyl iodide | 0 | .0802 | | | | | |
| n-Propyl isobutyrate | 0 | .0549 | | .212 | | | .0388 |
| i-Propyl isobutyrate | 0 | .059 | | | | | |
| Propyl propionate | 0 | .057 | | .212 | | | .0395 |
| Propyl valerate | 0 | .0466 | | .189 | | | .0341 |
| Safrol | 0 | .0434 | | | | | |
| i-Safrol | 0 | .0455 | | | | | |
| Sulfur hexafluoride | 25 | ...... | | .418 | | | |
| Toluene | 0 | .076 | .071 | | | | |
| | 30 | .088 | | | | | |
| Trimethyl carbinol | 0 | .087 | | | | | |
| 2,2,4-Trimethyl pentane | 30 | ...... | .0618 | .288 | .0688 | .0705 | |
| 2,2,3-Trimethyl heptane | 90 | ...... | | .270 | | .0684 | |
| n-Valeric acid | 0 | .050 | | | | | |
| i-Valeric acid | 0 | .0544 | | .212 | | | .0376 |
| Water | 0 | .220 | | .75 | | | .138 |
| | 450 | ...... | | | 1.3 | | |

## Table 3-20. Diffusivities in Liquids (25°C)

Dilute solutions and 1 atm unless otherwise noted; use $D_L\mu/T$ = constant to estimate effect of temperature; * indicates that reference gives effect of concentration.

| Solute | Solvent | $D_L \times 10^5$, cm²/sec † | Estimated possible error, ± % |
|---|---|---|---|
| Acetal*.................................... | Ethanol | 1.25 | 5 |
| Acetamide*................................ | Ethanol | 0.68 | 5 |
| Acetamide*................................ | Water | 1.19 | 3 |
| Acetic acid............................... | Acetone | 3.31 | |
| Acetic acid............................... | Benzene | 2.11 | |
| Acetic acid............................... | Carbon tetrachloride | 1.49 | |
| Acetic acid............................... | Ethylene glycol | 0.13 | |
| Acetic acid............................... | Toluene | 2.26 | |
| Acetic acid*.............................. | Water | 1.24 | 3 |
| Acetonitrile.............................. | Water | 1.66 | 5 |
| Acetylene................................. | Water | 1.78, 2.11 | |
| Allyl alcohol*............................ | Ethanol | 1.06 | 5 |
| Allyl alcohol............................. | Water | 1.19 | 6 |
| Ammonia*................................. | Water | 1.7, 2.0, 2.3 | |
| i-Amyl alcohol*........................... | Ethanol | 0.87 | 5 |
| i-Amyl alcohol............................ | Water | 1.0 | 8 |
| Benzene................................... | Carbon tetrachloride | 1.53 | |
| Benzene (50 mole %)....................... | n-Decane | 1.72 | |
| Benzene (50 mole %)....................... | 2,4-Dimethyl pentane | 2.49 | |
| Benzene (50 mole %)....................... | n-Dodecane | 1.40 | |
| Benzene (50 mole %)....................... | n-Heptane | 2.47 | |
| Benzene (50 mole %)....................... | n-Hexadecane | 0.96 | |
| Benzene (50 mole %)....................... | n-Octadecane | 0.86 | |
| Benzoic acid.............................. | Acetone | 2.62 | |
| Benzoic acid.............................. | Benzene | 1.38 | |
| Benzoic acid.............................. | Carbon tetrachloride | 0.91 | |
| Benzoic acid.............................. | Ethylene glycol | 0.043 | |
| Benzoic acid.............................. | Toluene | 1.49 | |
| Bromine................................... | Benzene | 2.7 | |
| Bromine................................... | Carbon disulfide | 4.1 | |
| Bromine................................... | Water | 1.3 | |
| Bromobenzene............................. | Benzene | 2.30 | |
| Bromoform*............................... | Acetone | 2.90 | |
| Bromoform................................ | i-Amyl alcohol | 0.53 | |
| Bromoform................................ | Ethanol | 1.08 | 5 |
| Bromoform*............................... | Ethyl ether | 3.62 | |
| Bromoform................................ | Methanol | 2.20 | |
| Bromoform................................ | n-Propanol | 0.94 | |
| n-Butanol................................ | Water | 0.96 | 5 |
| Caffeine.................................. | Water | 0.63 | 6 |
| Carbon dioxide............................ | Ethanol | 4.0 | 6 |
| Carbon dioxide............................ | Water | 1.96 | 1 |
| Carbon disulfide (50 mole %, 200 atm)........ | n-Butanol | 3.57 | |
| Carbon disulfide (50 mole %, 200 atm)........ | i-Butanol | 2.42 | |
| Carbon disulfide (50 mole %, 218 atm)........ | Chlorobenzene | 3.00 | |
| Carbon disulfide (50 mole %, 200 atm)........ | 2,4-Dimethyl pentane | 3.63 | |
| Carbon disulfide (50 mole %, 100 atm)........ | n-Heptane | 3.0 | |
| Carbon disulfide (50 mole %, 50 atm)......... | Methyl cyclohexane | 3.5 | |
| Carbon disulfide (50 mole %, 200 atm)........ | n-Octane | 3.10 | |
| Carbon disulfide (50 mole %)................. | Toluene | 2.06 | |
| Carbon tetrachloride...................... | Benzene | 2.04 | 3 |
| Carbon tetrachloride*..................... | Cyclohexane | 1.49 | 2 |
| Carbon tetrachloride...................... | Decalin | 0.776 | 2 |
| Carbon tetrachloride...................... | Dioxane | 1.02 | 2 |
| Carbon tetrachloride*..................... | Ethanol | 1.50 | 2 |
| Carbon tetrachloride...................... | n-Heptane | 3.17 | 2 |
| Carbon tetrachloride...................... | Kerosene | 0.961 | 2 |
| Carbon tetrachloride...................... | Methanol | 2.30 | 2 |
| Carbon tetrachloride...................... | i-Octane | 2.57 | 2 |
| Carbon tetrachloride...................... | Tetralin | 0.735 | 2 |
| Chloral*.................................. | Ethanol | 0.68 | 5 |
| Chloral hydrate........................... | Water | 0.77 | 7 |
| Chlorine.................................. | Water | 1.44 | 4 |
| Chlorobenzene............................ | Benzene | 2.66 | |

† To convert to square feet per hour, multiply by 0.388.

## Table 3-20. Diffusivities in Liquids (25°C) (Continued)

| Solute | Solvent | $D_L \times 10^5$, cm²/sec † | Estimated possible error, ± % |
|---|---|---|---|
| Chloroform | Benzene | 2.50 | 6 |
| Chloroform | Ethanol | 1.38 | 3 |
| Cinnamic acid | Acetone | 2.41 | |
| Cinnamic acid | Benzene | 1.12 | |
| Cinnamic acid | Carbon tetrachloride | 0.76 | |
| Cinnamic acid | Toluene | 2.41 | |
| 1,1'-Dichloropropanol | Water | 1.0 | 6 |
| Dicyanodiamide* | Water | 1.18 | 4 |
| Diethyl ether | Benzene | 2.73 | |
| Diethyl ether | Water | 0.85 | |
| 2,4-Dimethyl pentane (50 mole %) | n-Dodecane | 1.44 | |
| 2,4-Dimethyl pentane (50 mole %) | n-Hexadecane | 0.88 | |
| Ethanol* | Water | 1.28 | 4 |
| Ethyl acetate | Ethyl benzoate | 0.94 | |
| Ethylene dichloride | Benzene | 2.8 | |
| Formic acid | Acetone | 3.77 | |
| Formic acid | Benzene | 2.28 | |
| Formic acid | Carbon tetrachloride | 1.89 | |
| Formic acid | Ethylene glycol | 0.094 | |
| Formic acid | Toluene | 2.65 | |
| Formic acid | Water | 1.37 | 10 |
| Glucose | Water | 0.69 | 6 |
| Glycerol | i-Amyl alcohol | 0.12 | |
| Glycerol | Ethanol | 0.56 | |
| Glycerol* | Water | 0.94 | 6 |
| n-Heptane (50 mole %) | n-Dodecane | 1.58 | |
| n-Heptane (50 mole %) | n-Hexadecane | 1.00 | |
| n-Heptane (50 mole %) | n-Octadecane | 0.92 | |
| n-Heptane (50 mole %) | n-Tetradecane | 1.29 | |
| Hexamethylene tetramine | Water | 0.67 | |
| Hydrogen chloride* | Water | 3.10 | 3 |
| Hydrogen | Water | 5.85 (4.4?) | |
| Hydrogen sulfide | Water | 1.61 | |
| Hydroquinone* | Ethanol | 0.53 | 5 |
| Hydroquinone* | Water | 0.88, 1.12 | |
| Iodine | Acetic acid | 1.13 | |
| Iodine | Anisole | 1.25 | |
| Iodine | Benzene | 1.98 | |
| Iodine | Bromobenzene | 1.25 | 10 |
| Iodine | Carbon disulfide | 3.2 | |
| Iodine | Carbon tetrachloride | 1.45 | 8 |
| Iodine | Chloroform | 2.30 | 3 |
| Iodine | Cyclohexane | 1.80 | |
| Iodine | Dioxane | 1.07 | |
| Iodine* | Ethanol | 1.30 | |
| Iodine | Ethyl acetate | 2.2 | |
| Iodine | Ethyl ether | 3.61 | |
| Iodine | Ethylene bromide | 0.93 | |
| Iodine | n-Heptane | 3.4, 2.5 | |
| Iodine | n-Hexane | 4.15 | |
| Iodine | Mesitylene | 1.49 | |
| Iodine | Methanol | 1.74 | |
| Iodine | Methyl cyclohexane | 2.1 | |
| Iodine | n-Octane | 2.76 | |
| Iodine | Tetrabromoethane | 2.0 | |
| Iodine | n-Tetradecane | 0.96 | |
| Iodine | Toluene | 2.1 | |
| Iodine | m-Xylene | 1.82 | |
| Iodobenzene | Ethanol | 1.09 | 3 |
| Lactose* | Water | 0.49 | 5 |
| Maltose* | Water | 0.48 | 5 |
| Mannitol* | Water | 0.65 | 5 |
| Methanol | Water | 1.6 | |
| Nicotine* | Water | 0.60 | 8 |
| Nitric acid* | Water | 2.98 | 2 |
| Nitrobenzene | Carbon tetrachloride | 1.00 | |
| Nitrogen | Water | 1.9 | |
| Nitrous oxide | Water | 1.8 | |
| Oxalic acid* | Water | 1.61 | 2 |
| Oxygen | Glycerol*-water (106 poise) | 0.24 | |

† To convert to square feet per hour, multiply by 0.388.

## Table 3-20. Diffusivities in Liquids (25°C) (Continued)

| Solute | Solvent | $D_L \times 10^6$, cm²/sec † | Estimated possible error, ± % |
|---|---|---|---|
| Oxygen | Sucrose*-water (125 poise) | 0.25 | |
| Oxygen | Water | 2.5 | 20 |
| Pentaerythritol* | Water | 0.77 | 4 |
| Phenol | i-Amyl alcohol | 0.2 | |
| Phenol | Benzene | 1.68 | |
| Phenol | Carbon disulfide | 3.7 | |
| Phenol | Chloroform | 2.0 | |
| Phenol | Ethanol | 0.89 | |
| Phenol | Ethyl ether | 3.9 | |
| n-Propanol | Water | 1.1 | |
| Pyridine* | Ethanol | 1.24 | 3 |
| Pyridine | Water | 0.76 | 7 |
| Pyrogallol | Water | 0.74 | 7 |
| Raffinose* | Water | 0.41 | 4 |
| Resorcinol* | Ethanol | 0.46 | 5 |
| Resorcinol* | Water | 0.87 | 4 |
| Saccharose* | Water | 0.49 | 4 |
| Stearic acid* | Ethanol | 0.65 | 5 |
| Succinic acid* | Water | 0.94 | |
| Sucrose | Water | 0.56 | 6 |
| Sulfur dioxide | Water | 1.7 | |
| Sulfuric acid* | Water | 1.97 | 3 |
| Tartaric acid* | Water | 0.80 | 10 |
| 1,1,2,2-Tetrabromoethane | 1,1,2,2-Tetrachloroethane | 0.61 | 4 |
| Toluene | n-Decane | 2.09 | |
| Toluene | n-Dodecane | 1.38 | |
| Toluene | n-Heptane | 3.72 | |
| Toluene | n-Hexane | 4.21 | |
| Toluene | n-Tetradecane | 1.02 | |
| Urea | Ethanol | 0.73 | |
| Urea | Water | 1.37 | 2 |
| Urethane | Water | 1.06 | |
| Water | Glycerol | 0.021 | |

† To convert to square feet per hour, multiply by 0.388.

# VAPOR PRESSURES

## Table 3-21. Vapor Pressures of Inorganic Compounds, above 1 Atm*

| Name | Formula | Pressure, atm Temperature, °C | | | | | | | | | Critical point | |
|---|---|---|---|---|---|---|---|---|---|---|---|---|
| | | 1 | 2 | 5 | 10 | 20 | 30 | 40 | 50 | 60 | $t$, °C | $P$, atm |
| Ammonia | $NH_3$ | −33.6 | −18.7 | +4.7 | 25.7 | 50.1 | 66.1 | 78.9 | 89.3 | 98.3 | 132.4 | 111.5 |
| Carbon monoxide | CO | −191.3 | −183.5 | −170.7 | −161.0 | −149.7 | −141.9 | | | | −138.7 | 34.6 |
| dioxide | $CO_2$ | −78.2 | −69.1 | −56.7 | −39.5 | −18.9 | −5.3 | +5.9 | 14.9 | 22.4 | 31.1 | 73.0 |
| disulfide | $CS_2$ | 46.5 | 69.1 | 104.8 | 136.3 | 175.5 | 201.5 | 222.8 | 240.0 | 256.0 | 273.0 | 72.9 |
| Chlorine | $Cl_2$ | −33.8 | −16.9 | +10.3 | 35.6 | 65.0 | 84.8 | 101.6 | 115.2 | 127.1 | 144.0 | 76.1 |
| para-Hydrogen | $H_2$ | −252.5 | −250.2 | −246.0 | −241.8 | | | | | | −240.0 | 12.8 |
| Hydrogen bromide | HBr | −66.5 | −51.5 | −29.1 | −8.4 | +16.8 | 33.9 | 48.1 | 60.0 | 70.6 | 90.0 | 84.4 |
| chloride | HCl | −84.8 | −71.4 | −50.5 | −31.7 | −8.8 | +5.9 | 17.8 | 27.9 | 36.2 | 51.4 | 81.6 |
| cyanide | HCN | 25.9 | 45.8 | 75.8 | 102.7 | 135.0 | 153.8 | 169.9 | 183.5 | | 183.5 | 50.0 |
| Water | $H_2O$ | 100.0 | 120.1 | 152.4 | 180.5 | 213.1 | 234.6 | 251.1 | 264.7 | 276.5 | 374.2 | 218.0 |
| Hydrogen sulfide | $H_2S$ | −60.4 | −45.9 | −22.3 | 0.4 | 25.5 | 41.9 | 55.8 | 66.7 | 76.3 | 100.3 | 88.9 |
| Krypton | Kr | −152.0 | −143.5 | −130.0 | −118.0 | −101.7 | −88.8 | −78.4 | −66.5 | | −63 | 54 |
| Nitrogen | $N_2$ | −195.8 | −189.2 | −179.1 | −169.8 | −157.6 | −148.3 | | | | −147.2 | 33.5 |
| Oxygen | $O_2$ | −183.1 | −176.0 | −164.5 | −153.2 | −140.0 | −130.7 | −124.1 | | | −118.9 | 49.7 |
| Sulfur dioxide | $SO_2$ | −10.0 | +6.3 | 32.1 | 55.5 | 83.8 | 102.6 | 118.0 | 130.2 | 141.7 | 157.2 | 77.7 |
| trioxide | $SO_3$ | 44.8 | 60.0 | 82.5 | 104.0 | 138.0 | 157.8 | 175.0 | 187.8 | 198.0 | 218.3 | 83.6 |

SOURCE: Compiled from the extended tables by D. R. Stull, *Ind. Eng. Chem.*, **39**, 517 (1947).

* T. Boublik, V. Fried, and E. Hala, "The Vapor Pressure of Pure Substances," Elsevier Publishing Company, Amsterdam, 1973 is recommended for more recent information.

## Table 3-22. Vapor Pressures of Inorganic Compounds, up to 1 Atm*

| Compound | | Pressure, mm Hg — Temperature, °C | | | | | | | | | | Melting point, °C |
|---|---|---|---|---|---|---|---|---|---|---|---|---|
| Name | Formula | 1 | 5 | 10 | 20 | 40 | 60 | 100 | 200 | 400 | 760 | |
| Ammonia | NH₃ | −109.1 | −97.5 | −91.9 | −85.8 | −79.2 | −74.3 | −68.4 | −57.0 | −45.4 | −33.6 | −77.7 |
| heavy | ND₃ | — | — | — | — | — | −74.0 | −67.4 | −57.0 | −45.4 | −33.4 | −74.0 |
| Carbon (graphite) | C | 3586 | 3828 | 3946 | 4069 | 4196 | 4273 | 4373 | 4516 | 4660 | 4827 | — |
| dioxide | CO₂ | −134.3 | −124.4 | −119.5 | −114.4 | −108.6 | −104.8 | −100.2 | −93.0 | −85.7 | −78.2 | −57.5 |
| disulfide | CS₂ | −73.8 | −54.3 | −44.7 | −34.3 | −22.5 | −15.3 | −5.1 | 10.4 | 28.0 | 46.5 | −110.8 |
| monoxide | CO | −222.0 | −217.2 | −215.0 | −212.8 | −210.0 | −208.1 | −205.7 | −201.3 | −196.3 | −191.3 | −205. |
| Chlorine | Cl₂ | −118.0 | −106.7 | −101.6 | −93.3 | −84.5 | −79.0 | −71.7 | −60.2 | −47.3 | −33.8 | −100.7 |
| fluoride | ClF | — | — | — | — | — | — | — | — | — | −100.5 | −145 |
| Hydrogen | H₂ | −263.3 | −261.9 | −261.3 | −260.4 | −259.6 | −258.9 | −257.9 | −256.3 | −254.5 | −252.5 | −259.1 |
| Hydrogen bromide | HBr | −138.8 | −127.4 | −121.8 | −115.4 | −108.3 | −103.8 | −97.7 | −88.1 | −78.0 | −66.5 | −87.0 |
| chloride | HCl | −150.8 | −140.7 | −135.6 | −130.0 | −123.8 | −119.6 | −114.0 | −105.2 | −95.3 | −84.8 | −114.3 |
| cyanide | HCN | −71.0 | −55.3 | −47.7 | −39.7 | −30.9 | −25.1 | −17.8 | −5.3 | 10.2 | 25.9 | −13.2 |
| sulfide | H₂S | −134.3 | −122.4 | −116.3 | −109.7 | −102.3 | −97.9 | −91.6 | −82.3 | −71.8 | −60.4 | −85.5 |
| Krypton | Kr | −199.3 | −191.3 | −187.2 | −182.9 | −178.4 | −175.7 | −171.8 | −165.9 | −159.0 | −152.0 | −156.7 |
| Nitrogen | N₂ | −226.1 | −221.3 | −219.1 | −216.8 | −214.0 | −212.3 | −209.7 | −205.6 | −200.9 | −195.8 | −210.0 |
| Nitric oxide | NO | −184.5 | −180.6 | −178.2 | −175.3 | −171.7 | −168.9 | −166.0 | −162.3 | −156.8 | −151.7 | −161 |
| Nitrogen dioxide | NO₂ | −55.6 | −42.7 | −36.7 | −30.4 | −23.9 | −19.9 | −14.7 | −5.0 | 8.0 | 21.0 | −9.3 |
| Oxygen | O₂ | −219.1 | −213.4 | −210.6 | −207.5 | −204.1 | −201.9 | −198.8 | −194.0 | −188.8 | −183.1 | −218.7 |
| Sulfur | S | 183.8 | 223.0 | 243.8 | 264.7 | 288.3 | 305.5 | 327.2 | 359.7 | 399.6 | 444.6 | 112.8 |
| dioxide | SO₂ | −95.5 | −83.0 | −76.8 | −69.7 | −60.5 | −54.6 | −46.9 | −35.1 | −23.0 | −10.0 | −73.2 |
| trioxide (α) | SO₃ | −39.0 | −23.7 | −16.5 | −9.1 | −1.0 | 4.0 | 10.5 | 20.5 | 32.6 | 44.8 | 16.8 |
| trioxide (β) | SO₃ | −34.0 | −19.2 | −12.3 | −4.9 | 3.2 | 8.0 | 14.3 | 23.7 | 32.6 | 44.8 | 32.6 |
| trioxide (γ) | SO₃ | −15.3 | −2.0 | 4.3 | 11.1 | 17.9 | 21.4 | 28.0 | 35.8 | 44.0 | 51.6 | 62.1 |
| Water | H₂O | −17.3 | 1.2 | 11.2 | 22.1 | 34.0 | 41.5 | 51.6 | 66.5 | 83.0 | 100.0 | 0.0 |

* See footnote to Table 3-21.

## Table 3-23. Vapor Pressures of Organic Compounds, above 1 Atm*

| Compound | | Pressure, atm (Temperature, °C) | | | | | | | | | Critical point | |
| Name | Formula | 1 | 2 | 5 | 10 | 20 | 30 | 40 | 50 | 60 | $t_c$, °C | $P_c$, atm |
|---|---|---|---|---|---|---|---|---|---|---|---|---|
| Acetic acid | $C_2H_4O_2$ | 118.1 | 143.5 | 180.3 | 214.0 | 252.0 | 276.5 | 297.0 | 312.5 | | 321.6 | 57.2 |
| anhydride | $C_4H_6O_3$ | 139.6 | 162.0 | 194.0 | 221.5 | 253.0 | 272.8 | 288.5 | | | 296 | 46 |
| Acetone | $C_3H_6O$ | 56.5 | 78.6 | 113.0 | 144.5 | 181.0 | 205.0 | 214.5 | | | 235.0 | 47.0 |
| Acetylene | $C_2H_2$ | −84.0 | −71.6 | −50.2 | −32.7 | −10.0 | +4.8 | 16.8 | 26.8 | 34.8 | 36.0 | 62.0 |
| Allene (propadiene) | $C_3H_4$ | −35.0 | −18.4 | +8.0 | 33.3 | 64.5 | 85.5 | 103.5 | 118.0 | | 120.7 | 51.8 |
| Aniline | $C_6H_7N$ | 184.4 | 212.8 | 254.8 | 292.7 | 342.0 | 375.5 | 400.0 | 422.4 | | 426 | 52.4 |
| Benzene | $C_6H_6$ | 80.1 | 103.8 | 142.5 | 178.8 | 221.5 | 249.5 | 272.3 | 290.3 | | 290.5 | 50.1 |
| Bromobenzene | $C_6H_5Br$ | 156.2 | 186.2 | 232.5 | 274.5 | 327.0 | 359.8 | 387.5 | | | 397 | 44.6 |
| 1,3-Butadiene | $C_4H_6$ | −4.5 | +15.3 | 47.0 | 76.0 | 114.0 | 139.8 | 158.0 | | | 161.8 | 42.6 |
| iso-Butane (2-methylpropane) | $C_4H_{10}$ | −11.7 | +7.5 | 39.0 | 66.8 | 99.5 | 120.5 | | | | 134.0 | 37.0 |
| n-Butane | $C_4H_{10}$ | −0.5 | +18.8 | 50.0 | 79.5 | 116.0 | 140.6 | | | | 152.8 | 36.0 |
| iso-Butyl alcohol (2-methylpropanol-1) | $C_4H_{10}O$ | 108.0 | 127.3 | 156.2 | 182.0 | 212.5 | 232.0 | 251.0 | | | 265 | 48 |
| n-Butyl alcohol (1-butanol) | $C_4H_{10}O$ | 117.5 | 139.8 | 172.5 | 203.0 | 237.0 | 259.0 | 277.0 | | | 287 | 48.4 |
| sec-Butyl alcohol (2-butanol) | $C_4H_{10}O$ | 99.5 | 118.2 | 147.5 | 172.0 | 204.0 | 230.0 | 251.0 | | | 265 | 48 |
| tert-Butyl alcohol (trimethyl carbinol) | $C_4H_{10}O$ | 82.9 | 102.0 | 130.0 | 154.2 | 184.5 | 207.0 | 222.5 | | | 235 | 49 |
| iso-Butyl formate | $C_5H_{10}O_2$ | 98.2 | 121.8 | 157.8 | 192.4 | 234.0 | 261.0 | | | | 278.0 | 38.0 |
| Butyric acid | $C_4H_8O_2$ | 163.5 | 188.3 | 225.0 | 257.0 | 295.0 | 319.0 | 338.0 | 352.0 | | 355 | 52.0 |
| iso-Butyric acid | $C_4H_8O_2$ | 154.5 | 179.8 | 217.0 | 250.0 | 289.0 | 315.0 | 336.0 | | | 336 | 40.0 |
| Carbon dioxide | $CO_2$ | −78.2 | −69.1 | −56.7 | −39.5 | −18.9 | −5.3 | +5.9 | 14.9 | 22.4 | 31.1 | 73.0 |
| disulfide | $CS_2$ | 46.5 | 69.1 | 104.8 | 136.3 | 175.5 | 201.5 | 222.8 | 240.0 | 256.2 | 273.0 | 72.9 |
| monoxide | $CO$ | −191.3 | −183.5 | −170.7 | −161.0 | −149.7 | −141.9 | | | | −138.7 | 34.6 |
| tetrachloride | $CCl_4$ | 76.7 | 102.0 | 141.7 | 178.0 | 222.0 | 251.2 | 276.0 | | | 283.1 | 45.0 |
| Chlorobenzene | $C_6H_5Cl$ | 132.2 | 160.2 | 205.0 | 245.3 | 292.8 | 324.4 | 349.8 | | | 359.2 | 44.6 |
| Chlorodifluoromethane | $CHClF_2$ | −40.8 | −24.7 | +0.3 | 24.0 | 52.0 | 70.3 | 85.3 | | | 96 | 48.7 |
| Chloroform (trichloromethane) | $CHCl_3$ | 61.3 | 83.9 | 120.0 | 152.3 | 191.8 | 216.5 | 237.5 | 254.0 | | 260 | 54.9 |
| 1-Chloro-1,2,2-trifluoroethane | $C_2ClF_3$ | −27.9 | −11.1 | +15.5 | 40.0 | 71.1 | 91.9 | | | | 107.0 | 39.0 |
| Chlorotrifluoromethane | $CClF_3$ | −81.2 | −66.7 | −42.7 | −18.5 | +12.0 | 34.8 | 52.8 | | | 53 | 40.3 |
| Cyanogen | $C_2N_2$ | −21.0 | −4.4 | +21.4 | 44.6 | 72.6 | 91.6 | 106.5 | | | 126.6 | 58.2 |
| Cyclohexane | $C_6H_{12}$ | 80.7 | 106.0 | 146.4 | 184.0 | 228.4 | 257.5 | | | | 279.9 | 40.3 |
| 1,2-Dibromoethane | $C_2H_4Br_2$ | 131.5 | 157.7 | 200.0 | 237.0 | 269.0 | 286.0 | 295.0 | 300.0 | 304.5 | 309.8 | 70.6 |
| Dichlorodifluoromethane | $CCl_2F_2$ | −29.8 | −12.2 | +16.1 | 42.4 | 74.0 | 95.6 | | | | 111.5 | 39.6 |
| 1,1-Dichloroethane | $C_2H_4Cl_2$ | 57.3 | 80.2 | 117.3 | 150.3 | 192.7 | 220.0 | 243.0 | 261.5 | | 261.5 | 53.0 |
| 1,2-Dichloroethane | $C_2H_4Cl_2$ | 83.7 | 108.1 | 147.8 | 183.5 | 226.5 | 254.0 | 272.0 | 285.0 | | 288.4 | 53.0 |
| cis-1,2-Dichloroethylene | $C_2H_2Cl_2$ | 59.0 | 82.1 | 119.3 | 152.3 | 194.0 | 221.5 | 244.5 | 260.0 | | 271.0 | 57.9 |

| Name | Formula | | | | | | | | | | $t_c$ | $P_c$ |
|---|---|---|---|---|---|---|---|---|---|---|---|---|
| trans-1,2-Dichloroethylene | $C_2H_2Cl_2$ | 47.8 | 69.8 | 104.0 | 135.7 | 174.0 | 199.8 | 220.0 | 236.5 | | 243.3 | 54.5 |
| Dichlorofluoromethane | $CHCl_2F$ | 8.9 | 28.4 | 59.0 | 87.0 | 121.2 | 144.0 | 162.6 | 177.5 | | 178.5 | 51.0 |
| 1,2-Dichloro-1,1,2,2-tetrafluoroethane | $C_2Cl_2F_4$ | 3.5 | 27.8 | 54.0 | 82.3 | 117.5 | 140.9 | | | | 145.7 | 32.3 |
| Diethylamine | $C_4H_{11}N$ | 55.5 | | 113.0 | 145.3 | 184.5 | 210.0 | | | | 223.3 | 36.6 |
| Diethyl ether | $C_4H_{10}O$ | 34.6 | 56.0 | 90.0 | 122.0 | 159.0 | 183.3 | | | | 193.8 | 35.5 |
| Diethyl sulfide | $C_4H_{10}S$ | 88.0 | | 153.9 | 190.2 | 234.0 | 263.0 | | | | 283.8 | 39.1 |
| Dimethylamine | $C_2H_7N$ | 7.4 | 25.0 | 53.9 | 80.0 | 111.7 | 132.2 | 149.8 | | | 164.5 | 52.4 |
| 2,3-Dimethylbutane | $C_6H_{14}$ | 58.0 | 82.0 | 120.3 | 155.7 | 198.7 | 225.5 | | | | 227.4 | 30.7 |
| Dimethyl ether | $C_2H_6O$ | -23.7 | | 20.8 | 45.5 | 75.7 | 96.0 | 112.1 | 125.2 | | 126.9 | 52.0 |
| Dimethyl oxalate | $C_4H_6O_4$ | 163.3 | 189.6 | 228.7 | | | | | | | 260 | 9.5 |
| Dimethyl sulfide | $C_2H_6S$ | 36.0 | 57.8 | 92.3 | 124.5 | 163.8 | 188.5 | 206.5 | 220.0 | | 229.9 | 54.6 |
| n-Dodecane | $C_{12}H_{26}$ | 216.2 | 249.2 | 300.0 | 345.8 | | | | | | 385 | 17.5 |
| Ethane | $C_2H_6$ | -88.6 | | -52.8 | -32.0 | -14.2 | +10.0 | 23.6 | | | 32.3 | 48.2 |
| Ethyl acetate | $C_4H_8O_2$ | 77.1 | 97.5 | 136.6 | 169.7 | 209.5 | 235.0 | 218.0 | 230.0 | 242.0 | 250.1 | 37.9 |
| Ethyl alcohol (ethanol) | $C_2H_6O$ | 78.4 | 100.6 | 126.0 | 151.8 | 183.0 | 203.0 | 218.0 | 230.0 | | 243.5 | 63.1 |
| Ethylamine | $C_2H_7N$ | 16.6 | 35.7 | 65.3 | 91.8 | 124.0 | 146.0 | 163.0 | 176.0 | | 183.2 | 55.5 |
| Ethyl benzene | $C_8H_{10}$ | 136.2 | 163.5 | 207.5 | 246.3 | 294.5 | 326.5 | | | | 346.4 | 38.1 |
| Ethyl bromide | $C_2H_5Br$ | 38.4 | 60.2 | 95.0 | 126.8 | 164.3 | 188.0 | 206.5 | 220.0 | 229.5 | 230.8 | 61.5 |
| Ethyl chloride | $C_2H_5Cl$ | 12.3 | 32.5 | 64.0 | 92.6 | 127.3 | 149.5 | 167.0 | 180.5 | | 187.2 | 52.0 |
| Ethyl fluoride | $C_2H_5F$ | -32.0 | 16.7 | 7.7 | 30.2 | 57.5 | 75.7 | 90.0 | | | 102.2 | 49.6 |
| Ethyl formate | $C_3H_6O_2$ | 54.3 | 76.0 | 110.5 | 142.2 | 180.0 | 205.0 | 225.0 | | | 235.3 | 46.8 |
| Ethyl isobutyrate | $C_6H_{12}O_2$ | 110.1 | 135.5 | 174.2 | 210.0 | 253.0 | 280.0 | | 220.0 | | 280.0 | 30.0 |
| Ethyl mercaptan (ethanethiol) | $C_2H_6S$ | 35.0 | 56.6 | 90.7 | 121.9 | 159.5 | 184.3 | 204.7 | | | 225.5 | 54.2 |
| Ethyl methyl ether | $C_3H_8O$ | 7.5 | 26.5 | 56.4 | 84.0 | 108.0 | 141.4 | 160.0 | | | 164.7 | 43.4 |
| Ethyl propionate | $C_5H_{10}O_2$ | 99.1 | 123.8 | 162.7 | 197.8 | 240.0 | 264.5 | | | | 272.8 | 33.2 |
| Ethyl propyl ether | $C_5H_{12}O$ | 61.7 | 83.3 | 123.1 | 156.2 | 197.2 | 223.0 | | | | 227.4 | 32.1 |
| Ethylene | $C_2H_4$ | -103.7 | -90.8 | -71.1 | -52.8 | -29.1 | -14.2 | -1.5 | +8.9 | 9.6 | 9.6 | 50.7 |
| Fluorobenzene | $C_6H_5F$ | 84.7 | 109.9 | 148.5 | 184.4 | 227.6 | 257.0 | 279.3 | | | 286.5 | 44.7 |
| n-Heptane | $C_7H_{16}$ | 98.4 | 124.8 | 165.7 | 202.8 | 247.5 | | | | | 266.8 | 26.9 |
| n-Hexane | $C_6H_{14}$ | 68.7 | 93.0 | 131.7 | 166.6 | 209.4 | 234.8 | | | | 234.8 | 29.6 |
| Hydrogen cyanide (hydrocyanic acid) | $CHN$ | 25.9 | 45.8 | 75.8 | 102.7 | 135.0 | 153.8 | 169.9 | 183.5 | | 183.5 | 50.0 |
| Iodobenzene | $C_6H_5I$ | 188.6 | 220.0 | 270.0 | 315.7 | 371.5 | 406.0 | 437.2 | | | 448 | 44.7 |
| Methane | $CH_4$ | -161.5 | -152.3 | -138.3 | -124.8 | -108.5 | -96.3 | -86.3 | | | -82.1 | 45.8 |
| Methyl acetate | $C_3H_6O_2$ | 57.8 | 79.5 | 113.1 | 144.2 | 181.0 | 205.0 | 225.0 | 214.0 | 224.0 | 233.7 | 46.3 |
| Methyl acetylene (propyne) | $C_3H_4$ | -23.3 | 7.1 | 19.5 | 43.8 | 74.0 | 94.0 | 111.5 | 133.7 | 144.6 | 128 | 52.8 |
| Methyl alcohol | $CH_4O$ | 64.7 | 112.5 | 112.5 | 138.0 | 167.8 | 186.5 | 203.5 | 214.0 | | 240.0 | 78.7 |
| Methylamine | $CH_5N$ | -6.3 | 10.1 | 36.0 | 59.5 | 87.8 | 106.3 | 121.8 | 133.7 | | 156.9 | 73.6 |
| Methyl bromide | $CH_3Br$ | 3.6 | 23.3 | 54.8 | 84.0 | 121.7 | 147.5 | 170.2 | 190.0 | | 194 | 51.6 |
| Methyl butyrate | $C_5H_{10}O_2$ | 102.3 | 127.5 | 166.7 | 203.0 | 244.5 | 272.0 | | | | 281.2 | 34.2 |
| Methyl chloride | $CH_3Cl$ | -24.0 | 6.4 | 22.0 | 47.3 | 77.3 | 97.5 | 113.8 | 126.0 | 137.5 | 143.8 | 65.8 |
| Methyl fluoride | $CH_3F$ | -78.2 | 64.5 | 42.0 | 21.0 | 2.6 | 15.5 | 26.5 | 36.0 | 43.5 | 44.9 | 62.0 |
| Methyl formate | $C_2H_4O_2$ | 32.0 | 51.9 | 83.5 | 112.0 | 147.2 | 169.7 | 188.5 | 213.0 | | 214.0 | 59.1 |
| Methyl iodide | $CH_3I$ | 42.4 | 65.5 | 101.8 | 138.0 | 176.5 | 206.0 | 228.5 | 248.0 | | 255 | 54.6 |
| Methyl isobutyrate | $C_5H_{10}O_2$ | 92.6 | 116.7 | 155.2 | 190.2 | 232.0 | 259.5 | | | 185.0 | 267.5 | 33.9 |
| Methyl mercaptan (methanethiol) | $CH_4S$ | 6.8 | 26.1 | 55.9 | 83.4 | 117.5 | 140.0 | 157.7 | 172.0 | | 196.8 | 71.4 |
| Methyl propionate | $C_4H_8O_2$ | 79.8 | 103.0 | 139.8 | 172.6 | 212.5 | 239.0 | | | | 257.4 | 39.3 |

* See footnote to Table 3-21.

## Table 3-23. Vapor Pressures of Organic Compounds, above 1 Atm (Continued)

| Compound | | Pressure, atm — Temperature, °C | | | | | | | | | Critical point | |
|---|---|---|---|---|---|---|---|---|---|---|---|---|
| Name | Formula | 1 | 2 | 5 | 10 | 20 | 30 | 40 | 50 | 60 | $t$, °C | $P_c$, atm |
| n-Octane | $C_8H_{18}$ | 125.6 | 152.7 | 196.2 | 235.8 | 281.4 | ..... | ..... | ..... | ..... | 296.2 | 24.7 |
| iso-Pentane (2-methylbutane) | $C_5H_{12}$ | 27.8 | 48.8 | 82.8 | 114.5 | 154.0 | 180.3 | ..... | ..... | ..... | 187.8 | 32.8 |
| n-Pentane | $C_5H_{12}$ | 36.1 | 58.0 | 92.4 | 124.7 | 164.3 | 191.3 | ..... | ..... | ..... | 197.2 | 33.0 |
| neo-Pentane (2,2-dimethylpropane) | $C_5H_{12}$ | +9.5 | 29.5 | 61.1 | 90.7 | 127.6 | 152.5 | ..... | ..... | ..... | 159.0 | 33.0 |
| Phenol | $C_6H_6O$ | 181.9 | 208.0 | 248.2 | 283.8 | 328.7 | 358.0 | 382.1 | 400.0 | 418.7 | 419 | 60.5 |
| Phosgene (carbonyl chloride) | $CCl_2O$ | 8.3 | 27.3 | 57.2 | 85.0 | 119.0 | 141.8 | 159.8 | 174.0 | ..... | 181.7 | 56.0 |
| Propane | $C_3H_8$ | −42.1 | −25.6 | +1.4 | 26.9 | 58.1 | 78.7 | 94.8 | ..... | ..... | 96.8 | 42.0 |
| Propionic acid | $C_3H_6O_2$ | 141.1 | 160.0 | 186.0 | 203.5 | 220.0 | 228.0 | 233.0 | 238.0 | ..... | 239.5 | 53.0 |
| Propyl acetate | $C_5H_{10}O_2$ | 101.8 | 126.8 | 165.7 | 200.5 | 242.8 | 269.0 | ..... | ..... | ..... | 276.2 | 33.2 |
| iso-Propyl alcohol (2-propanol) | $C_3H_8O$ | 82.5 | 101.3 | 130.2 | 155.7 | 186.0 | 205.0 | 220.2 | 232.0 | ..... | 235 | 53 |
| n-Propyl alcohol (1-propanol) | $C_3H_8O$ | 97.8 | 117.0 | 149.0 | 177.0 | 210.8 | 232.3 | 250.0 | ..... | ..... | 263.7 | 49.9 |
| Propylamine | $C_3H_9N$ | 48.5 | 69.8 | 102.8 | 133.4 | 170.0 | 194.3 | 214.5 | ..... | ..... | 223.8 | 46.8 |
| Propyl formate | $C_4H_8O_2$ | 81.3 | 104.3 | 142.0 | 176.4 | 217.5 | 245.0 | ..... | ..... | ..... | 264.8 | 39.5 |
| Propylene | $C_3H_6$ | −47.7 | −31.4 | −4.8 | +19.8 | 49.5 | 70.0 | 85.0 | ..... | ..... | 91.4 | 45.4 |
| Tetramethylsilane | $C_4H_{12}Si$ | 27.0 | 48.0 | 82.0 | 113.0 | 152.0 | 178.0 | ..... | ..... | ..... | 185 | 33 |
| Toluene | $C_7H_8$ | 110.6 | 136.5 | 178.0 | 215.8 | 262.5 | 292.8 | 319.0 | ..... | ..... | 320.6 | 41.6 |
| Trichlorofluoromethane | $CCl_3F$ | 23.7 | 44.1 | 77.3 | 108.2 | 146.7 | 172.0 | 194.0 | ..... | ..... | 198.0 | 43.2 |

SOURCE: Compiled from the extended tables by D. R. Stull, *Ind. Eng. Chem.*, 39, 517 (1947). For data on gasoline and aircraft fuels see Hibbard, *NACA Research Mem.* E56I21, 1956 (declassified 1958). Extensive data for aqueous solutions of ethylene glycol, diethylene glycol, triethylene glycol, and propylene glycol from −20 to 300°F are contained in "Glycols," Union Carbide Corp. publ. F4763F, 1958. For vapor-pressure curves of the Freon compounds to 300°F, 1,000 psia, see E. I. du Pont De Nemours & Co., Inc., *Tech. Bull.* B-2, 1957; for methane data see Johnson (ed.), WADD-TR-56-60, 1960.

## Table 3-24. Vapor Pressure of Organic Compounds, up to 1 Atm*

| Compound | | Pressure, mm Hg | | | | | | | | | | Melting point, °C |
|---|---|---|---|---|---|---|---|---|---|---|---|---|
| Name | Formula | Temperature, °C | | | | | | | | | | |
| | | 1 | 5 | 10 | 20 | 40 | 60 | 100 | 200 | 400 | 760 | |
| Acetic acid | $C_2H_4O_2$ | −17.2 | +6.3 | 17.5 | 29.9 | 43.0 | 51.7 | 63.0 | 80.0 | 99.0 | 118.1 | 16.7 |
| anhydride | $C_4H_6O_3$ | −1.7 | 24.8 | 36.0 | 48.3 | 62.1 | 70.8 | 82.2 | 100.0 | 119.8 | 139.6 | −73 |
| Acetylene | $C_2H_2$ | −142.9 | −133.0 | −128.2 | −122.8 | −116.7 | −112.8 | −107.9 | −100.3 | −92.0 | −84.0 | −81.5 |
| Allene (propadiene) | $C_3H_4$ | −120.6 | −108.0 | −101.0 | −93.4 | −85.2 | −78.8 | −72.5 | −61.3 | −48.5 | −35.0 | −136 |
| Aniline | $C_6H_7N$ | 34.8 | 57.9 | 69.4 | 82.0 | 96.7 | 106.0 | 119.9 | 140.1 | 161.9 | 184.4 | −6.2 |
| Benzene | $C_6H_6$ | −36.7 | −19.6 | −11.5 | −2.6 | 7.6 | 15.4 | 26.1 | 42.2 | 60.6 | 80.1 | +5.5 |
| Bromobenzene | $C_6H_5Br$ | +2.9 | 27.8 | 40.0 | 53.8 | 68.6 | 78.1 | 90.8 | 110.1 | 132.3 | 156.2 | −30.7 |
| 1-3-Butadiene | $C_4H_6$ | −102.8 | −87.6 | −79.2 | −71.0 | −61.3 | −55.1 | −46.8 | −33.9 | −19.3 | −4.5 | −108.9 |
| n-Butane | $C_4H_{10}$ | −101.5 | −85.7 | −77.8 | −68.9 | −59.1 | −52.8 | −44.2 | −31.2 | −16.3 | −0.5 | −135 |
| iso-Butane (2-methylpropane) | $C_4H_{10}$ | −109.2 | −94.1 | −86.4 | −77.9 | −68.4 | −62.4 | −54.1 | −41.5 | −27.1 | −11.7 | −145 |
| n-Butyl alcohol | $C_4H_{10}O$ | −1.2 | 20.0 | 30.2 | 41.5 | 53.4 | 60.3 | 70.1 | 84.3 | 100.8 | 117.5 | −79.9 |
| iso-Butyl alcohol | $C_4H_{10}O$ | −9.0 | 11.6 | 21.7 | 32.4 | 44.1 | 51.7 | 61.5 | 75.9 | 91.4 | 108.0 | −108 |
| sec-Butyl alcohol | $C_4H_{10}O$ | −12.2 | 7.2 | 16.9 | 27.3 | 38.1 | 45.2 | 54.1 | 67.9 | 83.9 | 99.5 | −114.7 |
| tert-Butyl alcohol | $C_4H_{10}O$ | −20.4 | +3.0 | 10.8 | 20.4 | 24.1 | 31.0 | 43.4 | 52.7 | 68.0 | 82.9 | +25.3 |
| iso-Butyl formate | $C_5H_{10}O_2$ | −32.7 | −11.4 | 0.8 | 14.3 | 24.1 | 32.4 | 43.4 | 60.0 | 79.0 | 98.2 | −95.3 |
| Butyric acid | $C_4H_8O_2$ | 25.5 | 49.8 | 61.5 | 74.0 | 88.0 | 96.5 | 108.0 | 125.5 | 144.5 | 163.5 | −7.9 |
| iso-Butyric acid | $C_4H_8O_2$ | 14.7 | 39.3 | 51.2 | 64.0 | 77.8 | 86.3 | 98.0 | 115.8 | 134.5 | 154.5 | −47 |
| Carbon dioxide | $CO_2$ | −134.3 | −124.4 | −119.5 | −114.4 | −108.6 | −104.8 | −100.2 | −93.0 | −85.7 | −78.2 | −57.5 |
| disulfide | $CS_2$ | −73.8 | −54.3 | −44.7 | −34.3 | −22.5 | −15.3 | −5.1 | 10.4 | 28.0 | 46.5 | −110.8 |
| monoxide | $CO$ | −222.0 | −217.2 | −215.0 | −212.8 | −210.0 | −208.1 | −205.7 | −201.3 | −196.3 | −191.3 | −205.0 |
| Carbon tetrachloride | $CCl_4$ | −50.0 | −30.0 | −19.6 | −8.2 | 4.3 | 12.3 | 23.0 | 38.3 | 57.8 | 76.7 | −22.6 |
| Chlorobenzene | $C_6H_5Cl$ | −13.0 | +10.6 | 22.2 | 35.3 | 49.7 | 58.3 | 70.7 | 89.4 | 110.0 | 132.2 | −45.2 |
| Chlorodifluoromethane | $CHClF_2$ | −122.8 | −110.2 | −103.7 | −96.5 | −88.6 | −83.4 | −76.4 | −65.8 | −53.6 | −40.8 | −160 |
| Chloroform (trichloromethane) | $CHCl_3$ | −58.0 | −39.1 | −29.7 | −19.0 | −7.1 | 0.5 | 10.4 | 25.9 | 42.7 | 61.3 | −63.5 |
| 1-Chloro-1,2,2-trifluoroethylene | $C_2ClF_3$ | −116.0 | −102.5 | −95.9 | −88.2 | −79.7 | −74.1 | −66.7 | −55.0 | −41.7 | −27.9 | −157.5 |
| Chlorotrifluoromethane | $CClF_3$ | −149.5 | −139.2 | −134.1 | −128.5 | −121.9 | −117.3 | −111.7 | −102.5 | −92.7 | −81.2 | |
| Cyanogen | $C_2N_2$ | −95.8 | −83.2 | −76.8 | −70.1 | −62.7 | −57.9 | −51.8 | −42.0 | −33.0 | −21.0 | −34.4 |
| Cyclohexane | $C_6H_{12}$ | −45.3 | −25.4 | −15.9 | −5.0 | 6.7 | 14.7 | 25.5 | 42.0 | 60.8 | 80.7 | +6.6 |
| Dichlorofluoromethane | $CHCl_2F$ | −91.3 | −75.5 | −67.5 | −58.6 | −48.8 | −42.6 | −33.9 | −20.9 | −6.2 | 8.9 | −135 |
| 1,2-Dichloro-1,1,2,2-tetrafluoroethane | $C_2Cl_2F_4$ | −95.4 | −80.0 | −72.3 | −63.5 | −53.7 | −47.5 | −39.1 | −26.3 | −12.0 | 3.5 | 94 |
| Diethylamine | $C_4H_{11}N$ | | | −23.5 | −14.0 | −3.0 | 4.0 | 14.8 | 29.5 | 46.5 | 55.5 | −38.9 |
| Diethyl ether | $C_4H_{10}O$ | −74.3 | −56.9 | −48.1 | −38.5 | −27.7 | −21.8 | −11.5 | 2.2 | 17.9 | 34.6 | −116.3 |
| sulfide | $C_4H_{10}S$ | −39.6 | −18.6 | −8.0 | 3.5 | 16.1 | 24.2 | 35.0 | 51.3 | 69.7 | 88.0 | −99.5 |
| Dimethylamine | $C_2H_7N$ | −87.7 | −72.2 | −64.6 | −56.0 | −46.7 | −40.7 | −32.6 | −20.4 | −7.1 | 7.4 | 96 |
| 2,3-Dimethylbutane | $C_6H_{14}$ | −63.6 | −44.5 | −34.9 | −24.1 | −12.4 | −4.9 | 5.4 | 21.1 | 39.0 | 58.0 | −128.4 |
| Dimethylether | $C_2H_6O$ | −115.7 | −103.9 | −98.2 | −92.1 | −85.2 | −80.6 | −74.7 | −63.9 | −50.8 | −23.7 | −138.5 |
| oxalate | $C_4H_6O_4$ | 20.0 | 49.2 | 62.8 | 76.2 | 92.8 | 104.8 | 123.3 | 143.3 | 163.3 | | |
| sulfide | $C_2H_6S$ | −75.6 | −58.0 | −49.2 | −39.4 | −28.4 | −21.4 | −12.0 | 2.6 | 18.7 | 36.0 | −83.2 |
| n-Dodecane | $C_{12}H_{26}$ | 47.8 | 75.8 | 90.0 | 104.6 | 121.7 | 132.1 | 146.2 | 167.2 | 191.0 | 216.2 | −9.6 |
| Ethane | $C_2H_6$ | −159.5 | −148.5 | −142.9 | −136.7 | −129.8 | −125.4 | −119.3 | −110.2 | −99.7 | −88.6 | −183.2 |
| Ethyl acetate | $C_4H_8O_2$ | −43.4 | −13.5 | 13.5 | 8.0 | 9.1 | 16.6 | 27.0 | 42.0 | 59.3 | 77.1 | −82.4 |
| Ethyl alcohol (ethanol) | $C_2H_6O$ | −31.3 | −12.0 | 2.3 | 8.0 | 19.0 | 26.0 | 34.9 | 48.4 | 63.5 | 78.4 | −112 |
| Ethylamine | $C_2H_7N$ | −82.3 | −66.4 | −58.3 | −48.6 | −39.8 | −33.4 | −25.1 | −12.3 | +2.0 | 16.6 | −80.6 |

* See footnote to Table 3-21.

3–51

## Table 3-24. Vapor Pressure of Organic Compounds, up to 1 Atm (Continued)

| Name | Formula | Pressure, mm Hg — Temperature, °C | | | | | | | | | | Melting point, °C |
|---|---|---|---|---|---|---|---|---|---|---|---|---|
| | | 1 | 5 | 10 | 20 | 40 | 60 | 100 | 200 | 400 | 760 | |
| Ethylbenzene | $C_8H_{10}$ | 9.8 | +13.9 | 25.9 | 38.6 | 52.8 | 61.8 | 74.1 | 92.7 | 113.8 | 136.2 | −94.9 |
| Ethyl bromide | $C_2H_5Br$ | −74.3 | −56.4 | −47.5 | −37.8 | −26.7 | −19.5 | −10.0 | +4.5 | 21.0 | 38.4 | −117.8 |
| chloride | $C_2H_5Cl$ | −89.8 | −73.9 | −65.8 | −56.8 | −47.0 | −40.6 | −32.0 | −18.6 | −3.9 | +12.3 | −139 |
| Ethylene | $C_2H_4$ | −168.3 | −158.3 | −153.2 | −147.6 | −141.3 | −137.3 | −131.8 | −123.4 | −113.9 | −103.7 | −169 |
| Ethyl fluoride | $C_2H_5F$ | −117.0 | −103.8 | −97.7 | −90.0 | −81.8 | −76.4 | −69.3 | −58.0 | −45.5 | −32.0 | |
| formate | $C_3H_6O_2$ | −60.5 | −42.2 | −33.0 | −22.7 | −11.5 | −4.3 | +5.4 | 20.0 | 37.1 | 54.3 | −79 |
| mercaptan (ethanethiol) | $C_2H_6S$ | −76.7 | −59.1 | −50.2 | −40.7 | −29.8 | −22.4 | −13.0 | −1.5 | +17.7 | 35.0 | |
| methyl ether | $C_2H_6O$ | −91.0 | −75.6 | −67.8 | −59.1 | −49.4 | −43.3 | −34.8 | −22.0 | −7.8 | +7.5 | −121 |
| propionate | $C_5H_{10}O_2$ | −28.0 | −7.2 | +3.4 | 14.3 | 27.2 | 35.1 | 45.2 | 61.7 | 79.8 | 99.1 | −72.6 |
| propyl ether | $C_6H_{14}O$ | −64.3 | −45.0 | −35.0 | −24.0 | −12.0 | −4.0 | +6.8 | 23.3 | 41.6 | 61.7 | |
| Fluorobenzene | $C_6H_5F$ | −43.4 | −22.8 | −12.4 | −1.2 | +9.5 | 19.6 | 30.4 | 47.2 | 65.7 | 84.7 | −42.1 |
| n-Heptane | $C_7H_{16}$ | −34.0 | −12.7 | −2.1 | +9.5 | 22.3 | 30.6 | 41.8 | 58.7 | 78.0 | 98.4 | −90.6 |
| n-Hexane | $C_6H_{14}$ | −53.9 | −34.5 | −25.0 | −14.1 | −2.3 | +5.4 | 15.8 | 31.6 | 49.6 | 68.7 | −95.3 |
| Hydrogen cyanide (hydrocyanic acid) | $CHN$ | −71.0 | −55.3 | −47.7 | −39.7 | −30.9 | −25.1 | −17.8 | −5.3 | +10.2 | 25.9 | −13.2 |
| Iodobenzene | $C_6H_5I$ | 24.1 | 55.0 | 64.0 | 78.3 | 94.4 | 105.0 | 118.3 | 139.8 | 163.9 | 188.6 | −28.5 |
| Methane | $CH_4$ | −205.9 | −199.0 | −195.5 | −191.8 | −187.7 | −185.1 | −181.4 | −175.5 | −168.8 | −161.5 | −182.5 |
| Methyl acetate | $C_3H_6O_2$ | −57.2 | −38.6 | −29.3 | −19.1 | −7.9 | −0.5 | +9.4 | 24.0 | 40.0 | 57.8 | −98.7 |
| acetylene (propyne) | $C_3H_4$ | −111.0 | −97.5 | −90.5 | −82.9 | −74.3 | −68.8 | −61.3 | −49.8 | −37.2 | −23.3 | −102.7 |
| alcohol (methanol) | $CH_4O$ | −44.0 | −25.3 | −16.2 | −6.0 | +3.0 | 12.1 | 21.2 | 34.8 | 49.9 | 64.7 | −97.8 |
| Methylamine | $CH_5N$ | −95.8 | −81.3 | −73.8 | −65.9 | −56.9 | −51.3 | −43.7 | −32.4 | −19.7 | −6.3 | −93.5 |
| Methyl bromide | $CH_3Br$ | −96.3 | −80.6 | −72.8 | −64.0 | −54.2 | −48.0 | −39.4 | −26.5 | −11.9 | +3.6 | −93 |
| n-butyrate | $C_5H_{10}O_2$ | −26.8 | −5.0 | +2.9 | 8.4 | 16.7 | 21.0 | 29.6 | 37.4 | 48.0 | 63.1 | |
| isobutyrate | $C_5H_{10}O_2$ | −34.1 | −13.0 | −2.9 | +8.4 | 21.0 | 28.9 | 39.6 | 55.7 | 73.6 | 92.6 | −84.7 |
| chloride | $CH_3Cl$ | −74.2 | −57.0 | −48.6 | −38.0 | −26.0 | −18.7 | −8.0 | +8.0 | 25.3 | 24.0 | −97.7 |
| fluoride | $CH_3F$ | −147.3 | −137.0 | −131.6 | −125.9 | −119.1 | −115.0 | −109.0 | −99.9 | −89.5 | −78.2 | |
| formate | $C_2H_4O_2$ | −57.0 | −44.2 | −36.6 | −28.7 | −19.1 | −12.9 | −5.4 | +8.0 | 16.0 | 32.0 | −99.8 |
| iodide | $CH_3I$ | −28.5 | −12.6 | −5.4 | +2.4 | 14.7 | 21.9 | 29.0 | 44.2 | 61.8 | 42.4 | −64.4 |
| propionate | $C_4H_8O_2$ | −15.0 | +5.9 | 14.7 | 25.3 | 34.2 | 45.1 | 53.8 | 66.8 | 82.0 | 79.8 | −87.5 |
| n-Octane | $C_8H_{18}$ | −14.0 | +8.3 | 19.2 | 31.5 | 45.1 | 53.8 | 65.7 | 83.6 | 104.0 | 125.6 | −56.8 |
| n-Pentane | $C_5H_{12}$ | −76.6 | −62.5 | −50.1 | −40.2 | −29.2 | −22.2 | −12.6 | +1.9 | 18.5 | 36.1 | −129.7 |
| iso-Pentane (2-methylbutane) | $C_5H_{12}$ | −82.9 | −65.8 | −57.0 | −47.3 | −35.5 | −29.6 | −20.2 | −5.9 | +10.5 | 27.8 | −159.7 |
| neo-Pentane (2,2-dimethylpropane) | $C_5H_{12}$ | −102.0 | −85.4 | −76.7 | −67.2 | −56.1 | −49.0 | −39.1 | −23.7 | −7.1 | +9.5 | −16.6 |
| Phenol | $C_6H_6O$ | 40.1 | 62.5 | 73.8 | 86.0 | 100.1 | 108.4 | 121.4 | 139.0 | 160.0 | 181.9 | 40.6 |
| Phosgene (carbonyl chloride) | $CCl_2O$ | −92.9 | −77.0 | −69.3 | −60.3 | −50.3 | −44.0 | −35.6 | −22.3 | −7.6 | +8.3 | −104 |
| Propane | $C_3H_8$ | −128.9 | −115.4 | −108.5 | −100.9 | −92.4 | −87.0 | −79.6 | −68.4 | −55.6 | −42.1 | −187.1 |
| Propionic acid | $C_3H_6O_2$ | 4.6 | 28.0 | 39.7 | 52.0 | 65.8 | 74.1 | 85.8 | 102.5 | 122.0 | 141.1 | −22 |
| n-Propyl acetate | $C_5H_{10}O_2$ | −26.7 | −5.4 | +5.0 | 16.0 | 28.8 | 37.0 | 47.8 | 64.0 | 82.0 | 101.8 | −92.5 |
| alcohol (1-propanol) | $C_3H_8O$ | 15.0 | 14.7 | 5.0 | 25.3 | 36.4 | 43.5 | 52.8 | 66.8 | 82.0 | 97.8 | −127 |
| iso-Propyl alcohol (2-propanol) | $C_3H_8O$ | −26.1 | +7.0 | 14.7 | 23.8 | 16.0 | 30.5 | 39.5 | 53.0 | 67.8 | 82.5 | −85.8 |
| n-Propylamine | $C_3H_9N$ | −64.4 | −46.3 | −37.2 | −27.1 | −16.0 | −9.0 | +0.5 | 15.0 | 31.5 | 48.5 | −83 |
| n-Propyl formate | $C_4H_8O_2$ | −31.9 | −12.7 | −2.9 | +1.7 | 10.8 | 18.8 | 29.5 | 45.3 | 62.6 | 81.3 | −92.9 |
| Propylene | $C_3H_6$ | −112.1 | −97.7 | −91.3 | −83.4? | −81.8? | −18.8 | −29.5 | −45.3 | −62.6 | 47.7 | −185 |
| Toluene | $C_7H_8$ | −26.7 | −4.4 | +6.4 | 18.4 | 31.8 | 40.3 | 51.9 | 69.5 | 89.5 | 81.3 | −95.0 |
| Trichlorofluoromethane | $CCl_3F$ | −84.3 | −67.6 | −59.0 | −49.7 | −39.0 | −32.3 | −23.0 | −9.1 | +6.8 | 23.7 | |

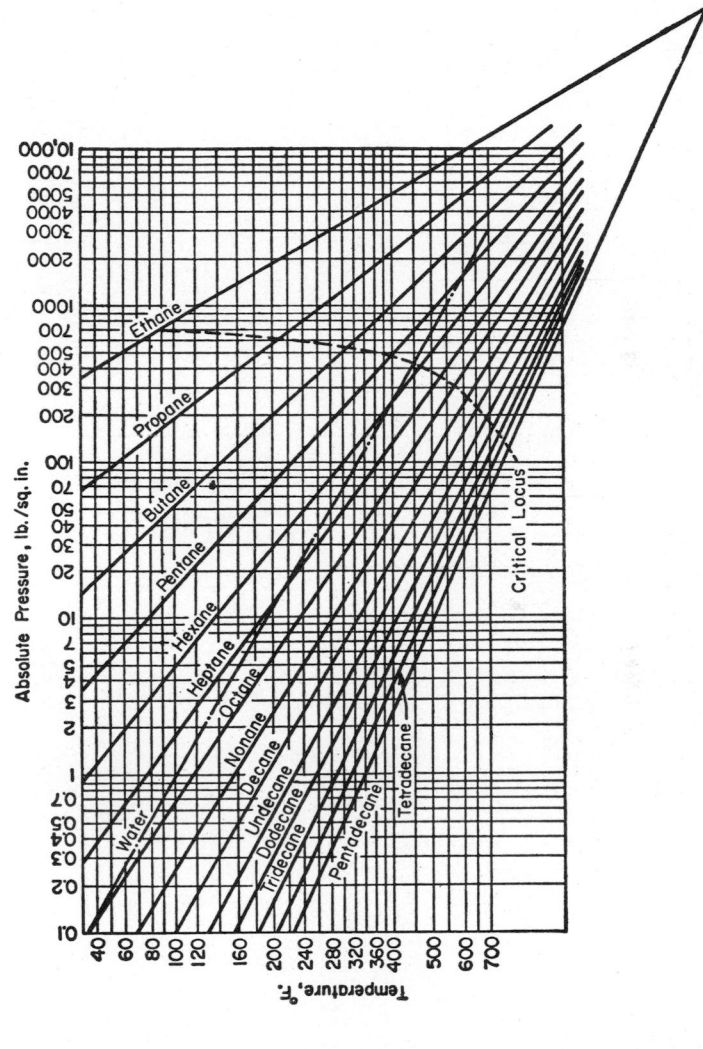

FIG. 3-6. Cox chart of vapor pressures of normal paraffin hydrocarbons. (*Sage and Lacey, "Volumetric and Phase Behavior of Hydrocarbons," Stanford University Press, Stanford, Calif., 1939.*)

## LATENT HEATS

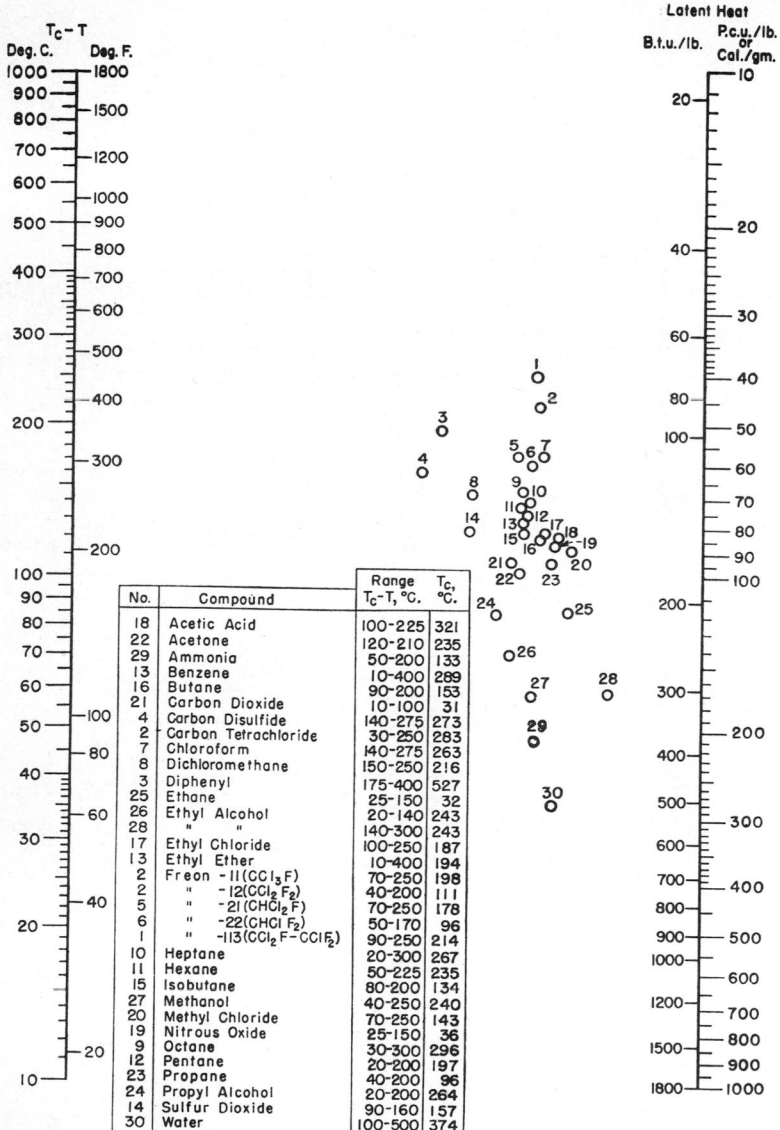

| No. | Compound | Range $T_c-T$, °C. | $T_c$, °C. |
|---|---|---|---|
| 18 | Acetic Acid | 100-225 | 321 |
| 22 | Acetone | 120-210 | 235 |
| 29 | Ammonia | 50-200 | 133 |
| 13 | Benzene | 10-400 | 289 |
| 16 | Butane | 90-200 | 153 |
| 21 | Carbon Dioxide | 10-100 | 31 |
| 4 | Carbon Disulfide | 140-275 | 273 |
| 2 | Carbon Tetrachloride | 30-250 | 283 |
| 7 | Chloroform | 140-275 | 263 |
| 8 | Dichloromethane | 150-250 | 216 |
| 3 | Diphenyl | 175-400 | 527 |
| 25 | Ethane | 25-150 | 32 |
| 26 | Ethyl Alcohol | 20-140 | 243 |
| 28 | "       " | 140-300 | 243 |
| 17 | Ethyl Chloride | 100-250 | 187 |
| 13 | Ethyl Ether | 10-400 | 194 |
| 2 | Freon -11(CCl$_3$F) | 70-250 | 198 |
| 2 | "    -12(CCl$_2$F$_2$) | 40-200 | 111 |
| 5 | "    -21(CHCl$_2$F) | 70-250 | 178 |
| 6 | "    -22(CHClF$_2$) | 50-170 | 96 |
| 1 | "    -113(CCl$_2$F-CClF$_2$) | 90-250 | 214 |
| 10 | Heptane | 20-300 | 267 |
| 11 | Hexane | 50-225 | 235 |
| 15 | Isobutane | 80-200 | 134 |
| 27 | Methanol | 40-250 | 240 |
| 20 | Methyl Chloride | 70-250 | 143 |
| 19 | Nitrous Oxide | 25-150 | 36 |
| 9 | Octane | 30-300 | 296 |
| 12 | Pentane | 20-200 | 197 |
| 23 | Propane | 40-200 | 96 |
| 24 | Propyl Alcohol | 20-200 | 264 |
| 14 | Sulfur Dioxide | 90-160 | 157 |
| 30 | Water | 100-500 | 374 |

Fig. 3-7. Latent heat of vaporization. In using this nomograph it should be noted that the ordinate represents the difference between the actual and the critical temperature and not the actual temperature of the fluid.

## THERMODYNAMIC PROPERTIES

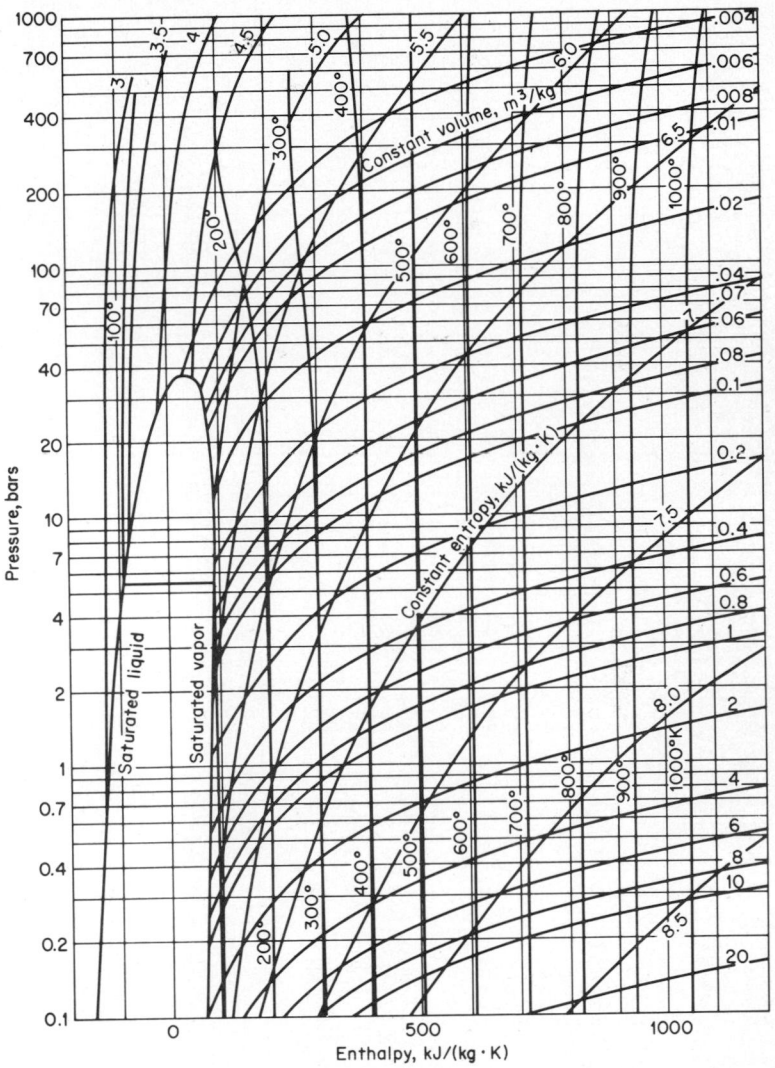

FIG. 3-8. Enthalpy–log pressure diagram for air.

## Table 3-25. Ideal Gas Thermophysical Properties of Air

| Property* | Temperature, K | | | | | | | | | |
|---|---|---|---|---|---|---|---|---|---|---|
| | 100 | 200 | 300 | 400 | 500 | 600 | 700 | 800 | 900 | 1000 |
| $v$(m³/kg)† | 0.281 | 0.573 | 0.861 | 1.148 | 1.436 | 1.723 | 2.010 | 2.297 | 2.585 | 2.872 |
| $h$(kJ/kg) | 100.0 | 200.1 | 300.5 | 401.2 | 503.4 | 607.5 | 713.8 | 822.5 | 933.5 | 1047 |
| $s$(kJ/kg·K)† | 5.759 | 6.463 | 6.871 | 7.161 | 7.389 | 7.579 | 7.743 | 7.888 | 8.019 | 8.138 |
| $C_p$(kJ/kg·K) | 1.000 | 1.002 | 1.007 | 1.014 | 1.030 | 1.051 | 1.075 | 1.099 | 1.121 | 1.141 |
| $C_v$(kJ/kg·K) | 0.714 | 0.715 | 0.718 | 0.727 | 0.743 | 0.764 | 0.788 | 0.813 | 0.834 | 0.854 |
| $\gamma$ | 1.400 | 1.401 | 1.400 | 1.395 | 1.387 | 1.376 | 1.364 | 1.353 | 1.344 | 1.336 |
| $\mu$(Ns/m²)10⁵‡ | 0.711 | 1.325 | 1.846 | 2.301 | 2.701 | 3.058 | 3.388 | 3.698 | 3.981 | 4.244 |
| $k$(W/m·K)10⁴‡ | 0.940 | 1.814 | 2.631 | 3.362 | 4.036 | 4.660 | 5.234 | 5.768 | 6.289 | 6.808 |
| Pr | 0.756 | 0.730 | 0.707 | 0.694 | 0.689 | 0.690 | 0.696 | 0.705 | 0.710 | 0.711 |
| $\bar{v}_s$(m/s) | 200.5 | 283.7 | 347.3 | 400.3 | 446.3 | 487.0 | 523.7 | 557.6 | 589.4 | 619.5 |

* $v$ = specific volume, $h$ = enthalpy, $s$ = entropy, $C_p$ = specific heat at constant pressure, $C_v$ = specific heat at constant volume, $\gamma$ = specific heat ratio, $\mu$ = viscosity, $k$ = thermal conductivity, Pr = Prandtl number, $v_s$ = velocity of sound.
† As volume and entropy both become infinite at zero pressure, the tabulated values are for a pressure of 1 bar.
‡ To obtain values $k$ or $\mu$ divide the tabulated $k$ values by 100 and the tabulated $\mu$ values by 10⁵; e.g., at 100 K, $k$ = 0.00940 W/(m·K) and $\mu$ = 0.00000711 Nsm⁻²

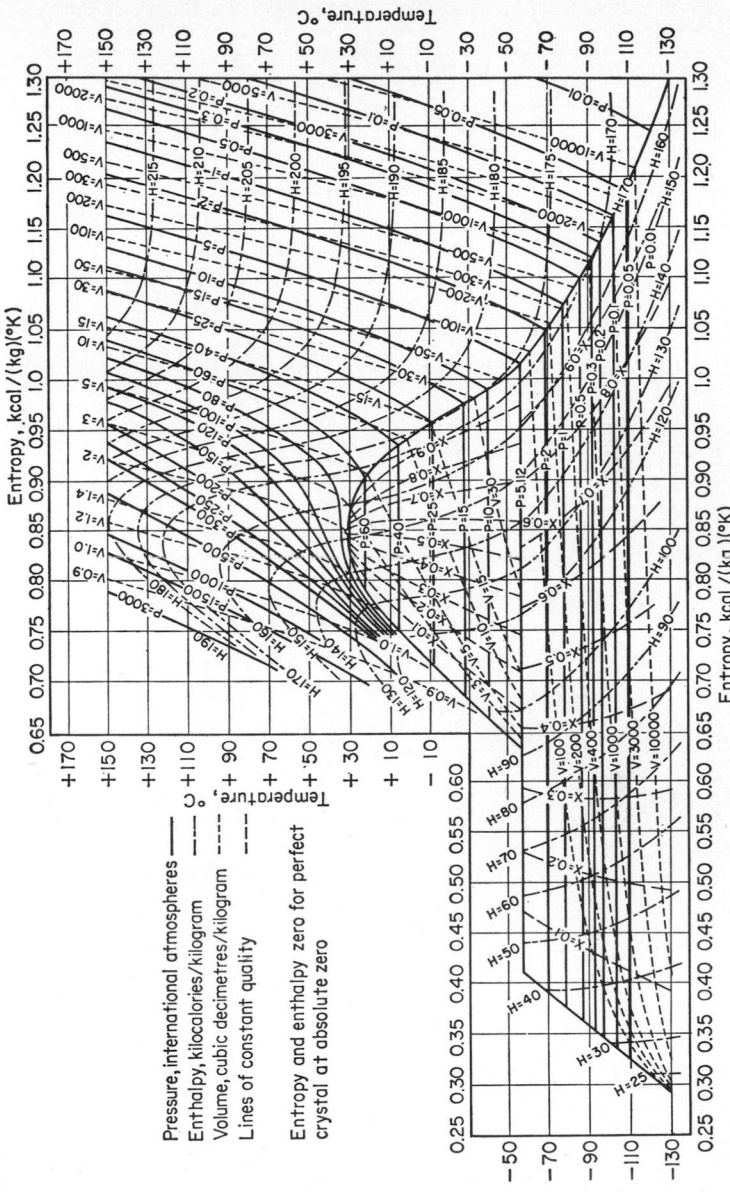

Fig. 3-9. Temperature-entropy diagram for carbon dioxide.

3–57

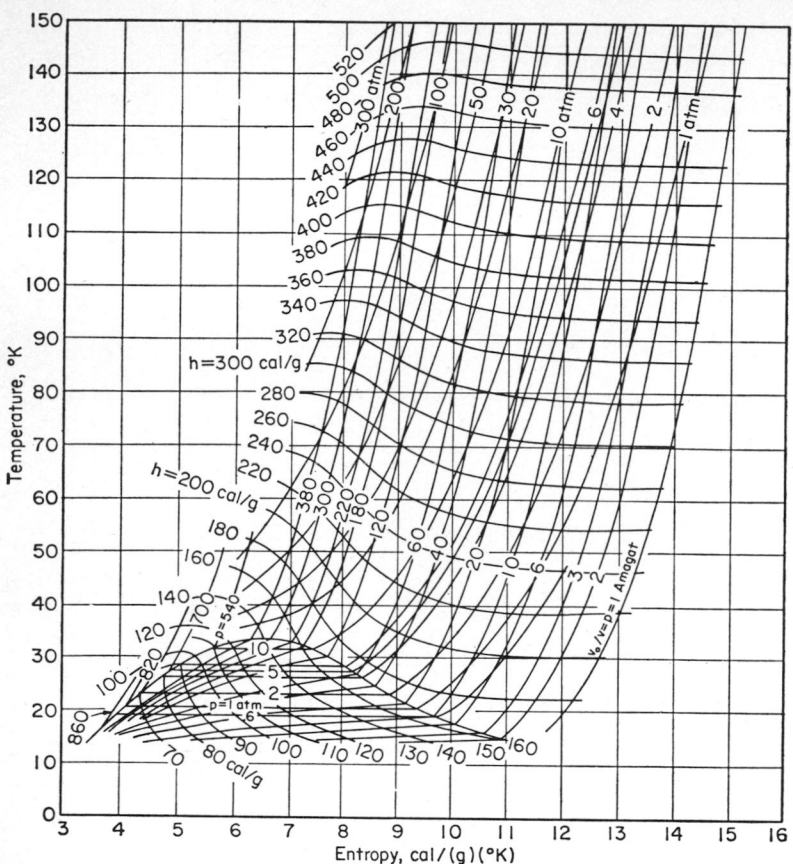

FIG. 3-10a. Temperature-entropy diagram for hydrogen (0 to 150 K).

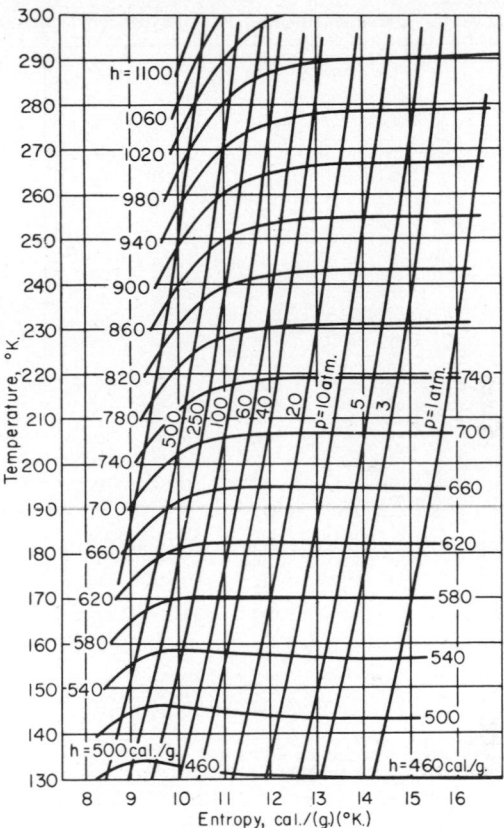

FIG. 3-10b. Temperature-entropy diagram for hydrogen (130 to 300 K). [*Woolley, Scott, and Brickwedde, J. Res. N.B.S.*, **41**, 379 (1948).]

## Table 3-26. Thermodynamic Properties of Mercury*

| Temp., K | Pressure, bars† | Specific volume, m³/kg | | Enthalpy, kJ/kg | | Entropy, kJ/kg · K | |
|---|---|---|---|---|---|---|---|
| | | Liquid | Vapor | Liquid | Vapor | Liquid | Vapor |
| 250 | 2.212.−8 | 7.325.−5 | 5.56.+6 | 39.81 | 347.49 | 0.3544 | 1.5943 |
| 260 | 6.926.−8 | 7.339.−5 | 1.80.+6 | 41.22 | 348.53 | 0.3597 | 1.5453 |
| 270 | 1.990.−7 | 7.352.−5 | 6.37.+5 | 42.63 | 349.57 | 0.3649 | 1.5028 |
| 280 | 4.956.−7 | 7.365.−5 | 2.44.+5 | 44.03 | 350.60 | 0.3699 | 1.4651 |
| 290 | 1.237.−6 | 7.378.−5 | 1.00.+5 | 45.43 | 351.64 | 0.3748 | 1.4309 |
| 300 | 2.903.−6 | 7.392.−5 | 4.38.+4 | 46.83 | 352.67 | 0.3795 | 1.3993 |
| 310 | 6.438.−6 | 7.405.−5 | 2.03.+4 | 48.22 | 353.71 | 0.3841 | 1.3698 |
| 320 | 1.357.−5 | 7.418.−5 | 9880 | 49.61 | 354.75 | 0.3885 | 1.3423 |
| 330 | 2.732.−5 | 7.432.−5 | 5030 | 50.99 | 355.78 | 0.3928 | 1.3165 |
| 340 | 5.516.−5 | 7.445.−5 | 2679 | 52.38 | 356.82 | 0.3969 | 1.2925 |
| 350 | 1.022.−4 | 7.459.−5 | 1480 | 53.75 | 357.86 | 0.4009 | 1.2699 |
| 360 | 1.829.−4 | 7.472.−5 | 847 | 55.13 | 358.89 | 0.4048 | 1.2487 |
| 370 | 3.169.−4 | 7.486.−5 | 500 | 56.50 | 359.93 | 0.4085 | 1.2287 |
| 380 | 5.334.−4 | 7.499.−5 | 304 | 57.87 | 360.96 | 0.4122 | 1.2099 |
| 390 | 8.733.−4 | 7.513.−5 | 185 | 59.24 | 362.00 | 0.4157 | 1.1921 |
| 400 | 1.394.−3 | 7.526.−5 | 120 | 60.61 | 363.04 | 0.4192 | 1.1754 |
| 450 | 1.053.−2 | 7.595.−5 | 18.0 | 67.41 | 368.21 | 0.4354 | 1.1037 |
| 500 | 5.261.−2 | 7.664.−5 | 3.98 | 74.19 | 373.38 | 0.4495 | 1.0479 |
| 559 | 0.1917 | 7.735.−5 | 1.18 | 80.95 | 378.53 | 0.4624 | 1.0035 |
| 600 | 0.5695 | 7.807.−5 | 0.432 | 87.72 | 383.64 | 0.4741 | 0.9698 |
| 650 | 1.6836 | 7.881.−5 | 0.187 | 94.51 | 388.68 | 0.4850 | 0.9376 |
| 700 | 3.1527 | 7.957.−5 | 0.092 | 101.34 | 392.62 | 0.4949 | 0.9127 |
| 750 | 6.197 | 8.036.−5 | 0.049 | 108.24 | 398.41 | 0.5046 | 0.8915 |
| 800 | 11.18 | 8.118.−5 | 0.029 | 115.23 | 403.04 | 0.5136 | 0.8733 |
| 850 | 18.82 | 8.203.−5 | 0.018 | 122.31 | 407.44 | 0.5221 | 0.8576 |
| 900 | 29.88 | 8.292.−5 | 0.012 | 129.53 | 411.61 | 0.5302 | 0.8437 |
| 950 | 45.23 | 8.385.−5 | 0.008 | 136.89 | 415.49 | 0.5381 | 0.8313 |
| 1000 | 65.74 | 8.482.−5 | 0.006 | 144.42 | 419.08 | 0.5456 | 0.8203 |

\* The notation 5.293.−7; 2.44. +5; etc. signifies $5.293 \times 10^{-7}$; $2.44. \times 10^5$; etc.
† To convert to pascals (newtons per square meter) multiply by $10^5$.

## Table 3-27. Thermodynamic Properties of Sodium*

| Temp., K. | Pressure, barst | Specific volume, m³/kg | | Enthalpy, kJ/kg | | Entropy, kJ/kg · K | |
|---|---|---|---|---|---|---|---|
| | | Liquid | Vapor | Liquid | Vapor | Liquid | Vapor |
| 380 | 2.631.−10 | 1.081.−3 | 5.222.+9 | 500 | 5003 | 2.853 | 14.703 |
| 400 | 1.385.−9 | 1.086.−3 | 1.044.+9 | 527 | 5020 | 2.924 | 14.156 |
| 450 | 4.594.−8 | 1.100.−3 | 3.537.+7 | 595 | 5062 | 3.084 | 13.010 |
| 500 | 7.523.−7 | 1.114.−3 | 2.395.+6 | 662 | 5101 | 3.225 | 12.102 |
| 550 | 7.377.−6 | 1.129.−3 | 2.679.+5 | 728 | 5137 | 3.351 | 11.366 |
| 600 | 4.926.−5 | 1.145.−3 | 4.359.+4 | 793 | 5168 | 3.464 | 10.756 |
| 650 | 2.448.−4 | 1.160.−3 | 9452 | 858 | 5196 | 3.567 | 10.242 |
| 700 | 9.649.−4 | 1.177.−3 | 2566 | 922 | 5220 | 3.662 | 9.803 |
| 750 | 3.160.−3 | 1.194.−3 | 833.2 | 985 | 5241 | 3.749 | 9.424 |
| 800 | 8.904.−3 | 1.211.−3 | 312.8 | 1048 | 5260 | 3.831 | 9.095 |
| 850 | 2.217.−2 | 1.229.−3 | 132.3 | 1111 | 5277 | 3.907 | 8.808 |
| 900 | 4.980.−2 | 1.247.−3 | 61.78 | 1174 | 5292 | 3.979 | 8.555 |
| 950 | 0.1025 | 1.271.−3 | 31.36 | 1237 | 5307 | 4.046 | 8.331 |
| 1000 | 0.1963 | 1.286.−3 | 17.08 | 1299 | 5322 | 4.111 | 8.134 |
| 1100 | 0.6002 | 1.327.−3 | 6.023 | 1426 | 5352 | 4.231 | 7.801 |
| 1200 | 1.5037 | 1.372.−3 | 2.572 | 1554 | 5386 | 4.343 | 7.536 |
| 1300 | 3.2454 | 1.419.−3 | 1.264 | 1685 | 5420 | 4.448 | 7.321 |
| 1400 | 6.2538 | 1.469.−3 | 0.691 | 1820 | 5453 | 4.548 | 7.142 |
| 1500 | 11.014 | 1.523.−3 | 0.409 | 1959 | 5481 | 4.644 | 6.991 |
| 1600 | 18.02 | 1.581.−3 | 0.259 | 2104 | 5504 | 4.737 | 6.862 |
| 1700 | 27.78 | 1.643.−3 | 0.171 | 2255 | 5524 | 4.828 | 6.751 |
| 1800 | 40.91 | 1.713.−3 | 0.118 | 2410 | 5540 | 4.916 | 6.654 |
| 1900 | 57.97 | 1.792.−3 | 8.35.−2 | 2570 | 5552 | 5.000 | 6.570 |
| 2000 | 79.49 | 1.884.−3 | 6.09.−2 | 2734 | 5558 | 5.083 | 6.495 |
| 2100 | 106.0 | 1.993.−3 | 4.54.−2 | 2905 | 5558 | 5.164 | 6.427 |
| 2200 | 137.9 | 2.123.−3 | 3.44.−2 | 3085 | 5550 | 5.244 | 6.365 |
| 2300 | 175.6 | 2.285.−3 | 2.64.−2 | 3275 | 5531 | 5.325 | 6.306 |
| 2400 | 219.5 | 2.493.−3 | 2.04.−2 | 3480 | 5497 | 5.407 | 6.248 |
| 2500 | 269.8 | 2.781.−3 | 1.57.−2 | 3708 | 5441 | 5.495 | 6.188 |
| 2600 | 327.0 | 3.230.−3 | 1.19.−2 | 3977 | 5342 | 5.594 | 6.119 |
| 2700 | 390.9 | 4.201.−3 | 8.33.−3 | 4369 | 5122 | 5.733 | 6.012 |
| 2733‡ | 413.6 | 5.501.−3 | 5.50.−3 | 4773 | 4773 | | |

Reproduced and converted from A. Padilla, Argonne National Laboratory Report ANL8095, 1974 and private communication, August 1974.

* The notation 2.631.−10, 5.222.+9, etc. signifies $2.631 \times 10^{-10}$, $5,222 \times 10^9$ etc.
† To convert to pascals (newtons per square meter) multiply by $10^5$.
‡ Critical temperature.

## Table 3-28. Properties of Saturated Steam, Pressure Table

| Abs press., lb/in.² $p$ | Temp, °F $t$ | Volume, ft³/lb | | Enthalpy, Btu/lb | | Entropy, Btu/(lb)(°R) | | Internal energy, Btu/lb | |
|---|---|---|---|---|---|---|---|---|---|
| | | Liquid $v_f$ | Vapor $v_g$ | Liquid $h_f$ | Vapor $h_g$ | Liquid $s_f$ | Vapor $s_g$ | Liquid $u_f$ | Vapor $u_g$ |
| 1 | 101.74 | 0.01614 | 333.6 | 69.70 | 1,106.0 | 0.1326 | 1.9782 | 69.70 | 1,044.3 |
| 2 | 126.08 | .01623 | 173.73 | 93.99 | 1,116.3 | 0.1749 | 1.9200 | 93.98 | 1,051.9 |
| 3 | 141.48 | .01630 | 118.71 | 109.37 | 1,122.6 | 0.2008 | 1.8863 | 109.36 | 1,056.7 |
| 4 | 152.97 | .01636 | 90.63 | 120.86 | 1,127.3 | 0.2198 | 1.8625 | 120.85 | 1,060.2 |
| 5 | 162.24 | .01640 | 73.52 | 130.13 | 1,131.1 | 0.2347 | 1.8441 | 130.12 | 1,063.1 |
| 6 | 170.06 | .01645 | 61.98 | 137.96 | 1,134.2 | 0.2472 | 1.8292 | 137.94 | 1,065.4 |
| 7 | 176.85 | .01649 | 53.64 | 144.76 | 1,136.9 | 0.2581 | 1.8167 | 144.74 | 1,067.4 |
| 8 | 182.86 | .01653 | 47.34 | 150.79 | 1,139.3 | 0.2674 | 1.8057 | 150.77 | 1,069.2 |
| 9 | 188.28 | .01656 | 42.40 | 156.22 | 1,141.4 | 0.2759 | 1.7962 | 156.19 | 1,070.8 |
| 10 | 193.21 | .01659 | 38.42 | 161.17 | 1,143.3 | 0.2835 | 1.7876 | 161.14 | 1,072.2 |
| 14,696 | 212.00 | .01672 | 26.80 | 180.07 | 1,150.4 | 0.3120 | 1.7566 | 180.02 | 1,077.5 |
| 15 | 213.03 | .01672 | 26.29 | 181.11 | 1,150.8 | 0.3135 | 1.7549 | 181.06 | 1,077.8 |
| 20 | 227.96 | .01683 | 20.089 | 196.16 | 1,156.3 | 0.3356 | 1.7319 | 196.10 | 1,081.9 |
| 25 | 240.07 | .01692 | 16.303 | 208.42 | 1,160.6 | 0.3533 | 1.7139 | 208.34 | 1,085.1 |
| 30 | 250.33 | .01701 | 13.746 | 218.82 | 1,164.1 | 0.3680 | 1.6993 | 218.73 | 1,087.8 |
| 35 | 259.28 | .01708 | 11.898 | 227.91 | 1,167.1 | 0.3807 | 1.6870 | 227.80 | 1,090.1 |
| 40 | 267.25 | .01715 | 10.498 | 236.03 | 1,169.7 | 0.3919 | 1.6763 | 235.90 | 1,092.0 |
| 45 | 274.44 | .01721 | 9.401 | 243.36 | 1,172.0 | 0.4019 | 1.6669 | 243.22 | 1,093.7 |
| 50 | 281.01 | .01727 | 8.515 | 250.09 | 1,174.1 | 0.4110 | 1.6585 | 249.93 | 1,095.3 |
| 55 | 287.07 | .01732 | 7.787 | 256.30 | 1,175.9 | 0.4193 | 1.6509 | 256.12 | 1,096.7 |
| 60 | 292.71 | .01738 | 7.175 | 262.09 | 1,177.6 | 0.4270 | 1.6438 | 261.90 | 1,097.9 |
| 65 | 297.97 | .01743 | 6.655 | 267.50 | 1,179.1 | 0.4342 | 1.6374 | 267.29 | 1,099.1 |
| 70 | 302.92 | .01748 | 6.206 | 272.61 | 1,180.6 | 0.4409 | 1.6315 | 272.38 | 1,100.2 |
| 75 | 307.60 | .01753 | 5.816 | 277.43 | 1,181.9 | 0.4472 | 1.6259 | 277.19 | 1,101.2 |
| 80 | 312.03 | .01757 | 5.472 | 282.02 | 1,183.1 | 0.4531 | 1.6207 | 281.76 | 1,102.1 |
| 85 | 316.25 | .01761 | 5.168 | 286.39 | 1,184.2 | 0.4587 | 1.6158 | 286.11 | 1,102.9 |
| 90 | 320.27 | .01766 | 4.896 | 290.56 | 1,185.3 | 0.4641 | 1.6112 | 290.27 | 1,103.7 |
| 95 | 324.12 | .01770 | 4.652 | 294.56 | 1,186.2 | 0.4692 | 1.6068 | 294.25 | 1,104.5 |
| 100 | 327.81 | .01774 | 4.432 | 298.40 | 1,187.2 | 0.4740 | 1.6026 | 298.08 | 1,105.2 |
| 110 | 334.77 | .01782 | 4.049 | 305.66 | 1,188.9 | 0.4832 | 1.5948 | 305.30 | 1,106.5 |
| 120 | 341.25 | .01789 | 3.728 | 312.44 | 1,190.4 | 0.4916 | 1.5878 | 312.05 | 1,107.6 |
| 130 | 347.32 | .01796 | 3.455 | 318.81 | 1,191.7 | 0.4995 | 1.5812 | 318.38 | 1,108.6 |
| 140 | 353.02 | .01802 | 3.220 | 324.82 | 1,193.0 | 0.5069 | 1.5751 | 324.35 | 1,109.6 |
| 150 | 358.42 | .01809 | 3.015 | 330.51 | 1,194.1 | 0.5138 | 1.5694 | 330.01 | 1,110.5 |
| 160 | 363.53 | .01815 | 2.834 | 335.93 | 1,195.1 | 0.5204 | 1.5640 | 335.39 | 1,111.2 |
| 170 | 368.41 | .01822 | 2.675 | 341.09 | 1,196.0 | 0.5266 | 1.5590 | 340.52 | 1,111.9 |
| 180 | 373.06 | .01827 | 2.532 | 346.03 | 1,196.9 | 0.5325 | 1.5542 | 345.42 | 1,112.5 |
| 190 | 377.51 | .01833 | 2.404 | 350.79 | 1,197.6 | 0.5381 | 1.5497 | 350.15 | 1,113.1 |
| 200 | 381.79 | .01839 | 2.288 | 355.36 | 1,198.4 | 0.5435 | 1.5453 | 354.68 | 1,113.7 |
| 250 | 400.95 | .01865 | 1.8438 | 376.00 | 1,201.1 | 0.5675 | 1.5263 | 375.14 | 1,115.8 |
| 300 | 417.33 | .01890 | 1.5433 | 393.84 | 1,202.8 | 0.5879 | 1.5104 | 392.79 | 1,117.1 |
| 350 | 431.72 | .01913 | 1.3260 | 409.69 | 1,203.9 | 0.6056 | 1.4966 | 408.45 | 1,118.0 |
| 400 | 444.59 | .0193 | 1.1613 | 424.0 | 1,204.5 | 0.6214 | 1.4844 | 422.6 | 1,118.5 |
| 450 | 456.28 | .0195 | 1.0320 | 437.2 | 1,204.6 | 0.6356 | 1.4734 | 435.5 | 1,118.7 |
| 500 | 467.01 | .0197 | 0.9278 | 499.4 | 1,204.4 | 0.6487 | 1.4634 | 447.6 | 1,118.6 |
| 550 | 476.94 | .0199 | .8424 | 460.8 | 1,203.9 | 0.6608 | 1.4542 | 458.8 | 1,118.2 |
| 600 | 486.21 | .0201 | .7698 | 471.6 | 1,203.2 | 0.6720 | 1.4454 | 469.4 | 1,117.7 |
| 650 | 494.90 | .0203 | .7083 | 481.8 | 1,202.3 | 0.6826 | 1.4374 | 479.4 | 1,117.1 |
| 700 | 503.10 | .0205 | .6554 | 491.5 | 1,201.2 | 0.6925 | 1.4296 | 488.8 | 1,116.3 |
| 750 | 510.86 | .0207 | .6092 | 500.8 | 1,200.0 | 0.7019 | 1.4223 | 598.0 | 1,115.4 |

## Table 3-28. Properties of Saturated Steam, Pressure Table (Continued)

| Abs press., lb/in.² $p$ | Temp, °F $t$ | Volume, ft³/lb | | Enthalpy, Btu/lb | | Entropy, Btu/ (lb)(°R) | | Internal energy, Btu/lb | |
|---|---|---|---|---|---|---|---|---|---|
| | | Liquid $v_f$ | Vapor $v_g$ | Liquid $h_f$ | Vapor $h_g$ | Liquid $s_f$ | Vapor $s_g$ | Liquid $u_f$ | Vapor $u_g$ |
| 800 | 518.23 | .0209 | .5687 | 509.7 | 1,198.6 | 0.7108 | 1.4153 | 506.6 | 1,114.4 |
| 850 | 525.26 | .0210 | .5327 | 518.3 | 1,197.1 | 0.7194 | 1.4085 | 515.0 | 1,113.3 |
| 900 | 531.98 | .0212 | .5006 | 526.6 | 1,195.4 | 0.7275 | 1.4020 | 523.1 | 1,112.1 |
| 950 | 538.43 | .0214 | .4717 | 534.6 | 1,193.7 | 0.7355 | 1.3957 | 530.9 | 1,110.8 |
| 1,000 | 544.61 | .0216 | .4456 | 542.4 | 1,191.8 | 0.7430 | 1.3897 | 538.4 | 1,109.4 |
| 1,100 | 556.31 | .0220 | .4001 | 557.4 | 1,187.8 | 0.7575 | 1.3780 | 552.9 | 1,106.4 |
| 1,200 | 567.22 | .0223 | .3619 | 571.7 | 1,183.4 | 0.7711 | 1.3667 | 566.7 | 1,103.0 |
| 1,300 | 577.46 | .0227 | .3293 | 585.4 | 1,178.6 | 0.7840 | 1.3559 | 580.0 | 1,099.4 |
| 1,400 | 587.10 | .0231 | .3012 | 598.7 | 1,173.4 | 0.7963 | 1.3454 | 592.7 | 1,095.4 |
| 1,500 | 596.23 | .0235 | .2765 | 611.6 | 1,167.9 | 0.8082 | 1.3351 | 605.1 | 1,091.2 |
| 2,000 | 635.82 | .0257 | .1878 | 671.7 | 1,135.1 | 0.8619 | 1.2849 | 662.2 | 1,065.6 |
| 2,500 | 668.13 | .0287 | .1307 | 730.6 | 1,091.1 | 0.9126 | 1.2322 | 717.3 | 1,030.6 |
| 3,000 | 695.36 | .0346 | .0858 | 802.5 | 1,020.3 | 0.9731 | 1.1615 | 783.4 | 972.7 |
| 3,206.2 | 705.40 | .0503 | .0503 | 902.7 | 902.7 | 1.0580 | 1.0580 | 872.9 | 872.9 |

# Table 3-29. Properties of Superheated Steam

| Abs press, psi (sat. temp) — v, h, s | 200°F | 300°F | 400°F | 500°F | 600°F | 700°F | 800°F | 900°F | 1000°F | 1100°F | 1200°F | 1400°F | 1600°F |
|---|---|---|---|---|---|---|---|---|---|---|---|---|---|
| 1 (101.74) v* / h* / s* | 392.6 / 1150.4 / 2.0512 | 452.3 / 1195.8 / 2.1153 | 512.0 / 1241.7 / 2.1720 | 571.6 / 1288.3 / 2.2233 | 631.2 / 1335.7 / 2.2702 | 690.8 / 1383.8 / 2.3137 | 750.4 / 1432.8 / 2.3542 | 809.9 / 1482.7 / 2.3923 | 869.5 / 1533.5 / 2.4283 | 929.1 / 1585.2 / 2.4625 | 988.7 / 1637.7 / 2.4952 | 1107.8 / 1745.7 / 2.5566 | 1227.0 / 1857.5 / 2.6137 |
| 5 (162.24) v / h / s | 78.16 / 1148.8 / 1.8718 | 90.25 / 1195.0 / 1.9370 | 102.26 / 1241.2 / 1.9942 | 114.22 / 1288.0 / 2.0456 | 126.16 / 1335.4 / 2.0927 | 138.10 / 1383.6 / 2.1361 | 150.03 / 1432.7 / 2.1767 | 161.95 / 1482.6 / 2.2148 | 173.87 / 1533.4 / 2.2509 | 185.79 / 1585.1 / 2.2851 | 197.71 / 1637.7 / 2.3178 | 221.6 / 1745.7 / 2.3792 | 245.4 / 1857.4 / 2.4363 |
| 10 (193.21) v / h / s | 38.85 / 1146.6 / 1.7927 | 45.00 / 1193.9 / 1.8595 | 51.04 / 1240.6 / 1.9172 | 57.05 / 1287.5 / 1.9689 | 63.03 / 1335.1 / 2.0160 | 69.01 / 1383.4 / 2.0596 | 74.98 / 1432.5 / 2.1002 | 80.95 / 1482.4 / 2.1383 | 86.92 / 1533.2 / 2.1744 | 92.88 / 1585.0 / 2.2086 | 98.84 / 1637.6 / 2.2413 | 110.77 / 1745.6 / 2.3028 | 122.69 / 1857.3 / 2.3598 |
| 14.696 (212.00) v / h / s | ..... | 30.53 / 1192.8 / 1.8160 | 34.68 / 1239.9 / 1.8743 | 38.78 / 1287.1 / 1.9261 | 42.86 / 1334.8 / 1.9734 | 46.94 / 1383.2 / 2.0170 | 51.00 / 1432.3 / 2.0576 | 55.07 / 1482.3 / 2.0958 | 59.13 / 1533.1 / 2.1319 | 63.19 / 1584.8 / 2.1662 | 67.25 / 1637.5 / 2.1989 | 75.37 / 1745.5 / 2.2603 | 83.48 / 1857.3 / 2.3174 |
| 20 (227.96) v / h / s | ..... | 22.36 / 1191.6 / 1.7808 | 25.43 / 1239.2 / 1.8396 | 28.46 / 1286.6 / 1.8918 | 31.47 / 1334.4 / 1.9392 | 34.47 / 1382.9 / 1.9829 | 37.46 / 1432.1 / 2.0235 | 40.45 / 1482.1 / 2.0618 | 43.44 / 1533.0 / 2.0978 | 46.42 / 1584.7 / 2.1321 | 49.41 / 1637.4 / 2.1648 | 55.37 / 1745.4 / 2.2263 | 61.34 / 1857.2 / 2.2834 |
| 40 (267.25) v / h / s | ..... | 11.040 / 1186.8 / 1.6994 | 12.628 / 1236.5 / 1.7608 | 14.168 / 1284.8 / 1.8140 | 15.688 / 1333.1 / 1.8619 | 17.198 / 1381.9 / 1.9058 | 18.702 / 1431.3 / 1.9467 | 20.20 / 1481.4 / 1.9850 | 21.70 / 1532.4 / 2.0212 | 23.20 / 1584.3 / 2.0555 | 24.69 / 1637.0 / 2.0883 | 27.68 / 1745.1 / 2.1498 | 30.66 / 1857.0 / 2.2069 |
| 60 (292.71) v / h / s | ..... | 7.259 / 1181.6 / 1.6492 | 8.357 / 1233.6 / 1.7135 | 9.403 / 1283.0 / 1.7678 | 10.427 / 1331.8 / 1.8162 | 11.441 / 1380.9 / 1.8605 | 12.449 / 1430.5 / 1.9015 | 13.452 / 1480.8 / 1.9400 | 14.454 / 1531.9 / 1.9762 | 15.453 / 1583.8 / 2.0106 | 16.451 / 1636.6 / 2.0434 | 18.446 / 1744.8 / 2.1049 | 20.44 / 1856.7 / 2.1621 |
| 80 (312.03) v / h / s | ..... | ..... | 6.220 / 1230.7 / 1.6791 | 7.020 / 1281.1 / 1.7346 | 7.797 / 1330.5 / 1.7836 | 8.562 / 1379.9 / 1.8281 | 9.322 / 1429.7 / 1.8694 | 10.077 / 1480.1 / 1.9079 | 10.830 / 1531.3 / 1.9442 | 11.582 / 1583.4 / 1.9787 | 12.332 / 1636.2 / 2.0115 | 13.830 / 1744.5 / 2.0721 | 15.325 / 1856.5 / 2.1303 |
| 100 (327.81) v / h / s | ..... | ..... | 4.937 / 1227.6 / 1.6518 | 5.589 / 1279.1 / 1.7085 | 6.218 / 1329.1 / 1.7581 | 6.835 / 1378.9 / 1.8029 | 7.446 / 1428.9 / 1.8443 | 8.052 / 1479.5 / 1.8829 | 8.656 / 1530.8 / 1.9193 | 9.259 / 1582.9 / 1.9538 | 9.860 / 1635.7 / 1.9867 | 11.060 / 1744.2 / 2.0484 | 12.258 / 1856.2 / 2.1056 |
| 120 (341.25) v / h / s | ..... | ..... | 4.081 / 1224.4 / 1.6287 | 4.636 / 1277.2 / 1.6869 | 5.165 / 1327.7 / 1.7370 | 5.683 / 1377.8 / 1.7822 | 6.195 / 1428.1 / 1.8237 | 6.702 / 1478.8 / 1.8625 | 7.207 / 1530.2 / 1.8990 | 7.710 / 1582.4 / 1.9335 | 8.212 / 1635.3 / 1.9664 | 9.214 / 1743.9 / 2.0281 | 10.213 / 1856.0 / 2.0854 |

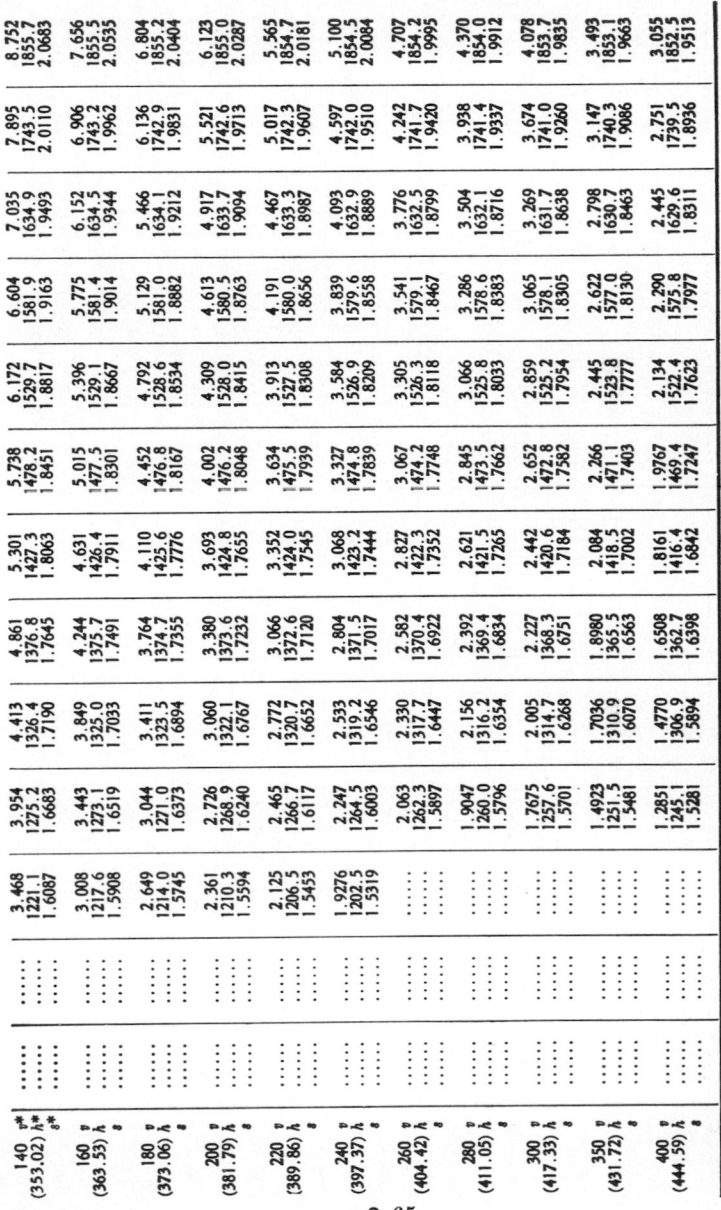

| p (t) | | | | | | | | | | | | | |
|---|---|---|---|---|---|---|---|---|---|---|---|---|---|
| **140**<br>(353.02)<br>v*<br>h*<br>s* | 8.752<br>1855.7<br>2.0683 | 7.895<br>1743.5<br>2.0110 | 7.035<br>1634.9<br>1.9493 | 6.604<br>1581.9<br>1.9163 | 6.172<br>1529.7<br>1.8817 | 5.738<br>1478.2<br>1.8451 | 5.301<br>1427.3<br>1.8063 | 4.861<br>1376.8<br>1.7645 | 4.413<br>1326.4<br>1.7190 | 3.954<br>1275.2<br>1.6683 | 3.468<br>1221.1<br>1.6087 | ..... | ..... |
| **160**<br>(363.55)<br>v<br>h<br>s | 7.656<br>1855.5<br>2.0535 | 6.906<br>1743.2<br>1.9962 | 6.152<br>1634.5<br>1.9344 | 5.775<br>1581.4<br>1.9014 | 5.396<br>1529.1<br>1.8667 | 5.015<br>1477.5<br>1.8301 | 4.631<br>1426.4<br>1.7911 | 4.244<br>1375.7<br>1.7491 | 3.849<br>1325.0<br>1.7033 | 3.443<br>1273.1<br>1.6519 | 3.008<br>1217.6<br>1.5908 | ..... | ..... |
| **180**<br>(373.06)<br>v<br>h<br>s | 6.804<br>1855.2<br>2.0404 | 6.136<br>1742.9<br>1.9831 | 5.466<br>1634.1<br>1.9212 | 5.129<br>1581.0<br>1.8882 | 4.792<br>1528.6<br>1.8534 | 4.452<br>1476.8<br>1.8167 | 4.110<br>1425.6<br>1.7776 | 3.764<br>1374.7<br>1.7355 | 3.411<br>1323.5<br>1.6894 | 3.044<br>1271.0<br>1.6373 | 2.649<br>1214.0<br>1.5745 | ..... | ..... |
| **200**<br>(381.79)<br>v<br>h<br>s | 6.123<br>1855.0<br>2.0287 | 5.521<br>1742.6<br>1.9713 | 4.917<br>1633.7<br>1.9094 | 4.613<br>1580.5<br>1.8763 | 4.309<br>1528.0<br>1.8415 | 4.002<br>1476.2<br>1.8048 | 3.693<br>1424.8<br>1.7655 | 3.380<br>1373.6<br>1.7232 | 3.060<br>1322.1<br>1.6767 | 2.726<br>1268.9<br>1.6240 | 2.361<br>1210.3<br>1.5594 | ..... | ..... |
| **220**<br>(389.86)<br>v<br>h<br>s | 5.565<br>1854.7<br>2.0181 | 5.017<br>1742.3<br>1.9607 | 4.467<br>1633.3<br>1.8987 | 4.191<br>1580.0<br>1.8656 | 3.913<br>1527.5<br>1.8308 | 3.634<br>1475.5<br>1.7939 | 3.352<br>1424.0<br>1.7545 | 3.066<br>1372.6<br>1.7120 | 2.772<br>1320.7<br>1.6652 | 2.465<br>1266.7<br>1.6117 | 2.125<br>1206.5<br>1.5453 | ..... | ..... |
| **240**<br>(397.37)<br>v<br>h<br>s | 5.100<br>1854.5<br>2.0084 | 4.597<br>1742.0<br>1.9510 | 4.093<br>1632.9<br>1.8889 | 3.839<br>1579.6<br>1.8558 | 3.584<br>1526.9<br>1.8209 | 3.327<br>1474.8<br>1.7839 | 3.068<br>1423.2<br>1.7444 | 2.804<br>1371.5<br>1.7017 | 2.533<br>1319.2<br>1.6546 | 2.247<br>1264.5<br>1.6003 | 1.9276<br>1202.5<br>1.5319 | ..... | ..... |
| **260**<br>(404.42)<br>v<br>h<br>s | 4.707<br>1854.2<br>1.9995 | 4.242<br>1741.7<br>1.9420 | 3.776<br>1632.5<br>1.8799 | 3.541<br>1579.1<br>1.8467 | 3.305<br>1526.3<br>1.8118 | 3.067<br>1474.2<br>1.7748 | 2.827<br>1422.3<br>1.7352 | 2.582<br>1370.4<br>1.6922 | 2.330<br>1317.7<br>1.6447 | 2.063<br>1262.3<br>1.5897 | ..... | ..... | ..... |
| **280**<br>(411.05)<br>v<br>h<br>s | 4.370<br>1854.0<br>1.9912 | 3.938<br>1741.4<br>1.9337 | 3.504<br>1632.1<br>1.8716 | 3.286<br>1578.6<br>1.8383 | 3.066<br>1525.8<br>1.8033 | 2.845<br>1473.5<br>1.7662 | 2.621<br>1421.5<br>1.7265 | 2.392<br>1369.4<br>1.6834 | 2.156<br>1316.2<br>1.6354 | 1.9047<br>1260.0<br>1.5796 | ..... | ..... | ..... |
| **300**<br>(417.33)<br>v<br>h<br>s | 4.078<br>1853.7<br>1.9835 | 3.674<br>1741.0<br>1.9260 | 3.269<br>1631.7<br>1.8638 | 3.065<br>1578.1<br>1.8305 | 2.859<br>1525.2<br>1.7954 | 2.652<br>1472.8<br>1.7582 | 2.442<br>1420.6<br>1.7184 | 2.227<br>1368.3<br>1.6751 | 2.005<br>1314.7<br>1.6268 | 1.7675<br>1257.6<br>1.5701 | ..... | ..... | ..... |
| **350**<br>(431.72)<br>v<br>h<br>s | 3.493<br>1853.1<br>1.9663 | 3.147<br>1740.3<br>1.9086 | 2.798<br>1630.7<br>1.8463 | 2.622<br>1577.0<br>1.8130 | 2.445<br>1523.8<br>1.7777 | 2.266<br>1471.1<br>1.7403 | 2.084<br>1418.5<br>1.7002 | 1.8980<br>1365.5<br>1.6563 | 1.7036<br>1310.9<br>1.6070 | 1.4923<br>1251.5<br>1.5481 | ..... | ..... | ..... |
| **400**<br>(444.59)<br>v<br>h<br>s | 3.055<br>1852.5<br>1.9513 | 2.751<br>1739.5<br>1.8936 | 2.445<br>1629.6<br>1.8311 | 2.290<br>1575.8<br>1.7977 | 2.134<br>1522.4<br>1.7623 | 1.9767<br>1469.4<br>1.7247 | 1.8161<br>1416.4<br>1.6842 | 1.6508<br>1362.7<br>1.6398 | 1.4770<br>1306.9<br>1.5894 | 1.2851<br>1245.1<br>1.5281 | ..... | ..... | ..... |

Table 3-29. Properties of Superheated Steam (Continued)

| Abs press., psi (sat. temp) | | 500°F | 550°F | 600°F | 620°F | 640°F | 660°F | 680°F | 700°F | 800°F | 900°F | 1000°F | 1200°F | 1400°F | 1600°F |
|---|---|---|---|---|---|---|---|---|---|---|---|---|---|---|---|
| 450 (456.28) | v* | 1.1231 | 1.2155 | 1.3005 | 1.3332 | 1.3652 | 1.3967 | 1.4278 | 1.4584 | 1.6074 | 1.7516 | 1.8928 | 2.170 | 2.443 | 2.714 |
| | h* | 1238.4 | 1272.0 | 1302.8 | 1314.6 | 1325.2 | 1337.5 | 1348.8 | 1359.9 | 1414.3 | 1467.7 | 1521.0 | 1628.6 | 1738.7 | 1851.9 |
| | s* | 1.5095 | 1.5437 | 1.5735 | 1.5845 | 1.5951 | 1.6054 | 1.6153 | 1.6250 | 1.6699 | 1.7108 | 1.7485 | 1.8177 | 1.8803 | 1.9381 |
| 500 (467.01) | v | 0.9927 | 1.0800 | 1.1591 | 1.1893 | 1.2188 | 1.2478 | 1.2763 | 1.3044 | 1.4405 | 1.5715 | 1.6996 | 1.9504 | 2.197 | 2.442 |
| | h | 1231.3 | 1266.8 | 1298.6 | 1310.7 | 1322.6 | 1334.2 | 1345.7 | 1357.0 | 1412.1 | 1466.0 | 1519.6 | 1627.6 | 1737.9 | 1851.3 |
| | s | 1.4919 | 1.5280 | 1.5588 | 1.5701 | 1.5810 | 1.5915 | 1.6016 | 1.6115 | 1.6571 | 1.6982 | 1.7363 | 1.8056 | 1.8683 | 1.9262 |
| 550 (476.94) | v | 0.8852 | 0.9686 | 1.0431 | 1.0714 | 1.0989 | 1.1259 | 1.1523 | 1.1783 | 1.3038 | 1.4241 | 1.5414 | 1.7706 | 1.9957 | 2.219 |
| | h | 1223.7 | 1261.2 | 1294.3 | 1306.8 | 1318.9 | 1330.8 | 1342.5 | 1354.0 | 1409.9 | 1464.3 | 1518.2 | 1626.6 | 1737.1 | 1850.6 |
| | s | 1.4751 | 1.5131 | 1.5451 | 1.5568 | 1.5680 | 1.5787 | 1.5890 | 1.5991 | 1.6452 | 1.6868 | 1.7250 | 1.7946 | 1.8575 | 1.9155 |
| 600 (486.21) | v | 0.7947 | 0.8753 | 0.9463 | 0.9729 | 0.9988 | 1.0241 | 1.0489 | 1.0732 | 1.1899 | 1.3013 | 1.4096 | 1.6208 | 1.8279 | 2.033 |
| | h | 1215.7 | 1255.5 | 1289.9 | 1302.7 | 1315.2 | 1327.4 | 1339.3 | 1351.1 | 1407.7 | 1462.5 | 1516.7 | 1625.5 | 1736.3 | 1850.0 |
| | s | 1.4586 | 1.4990 | 1.5323 | 1.5443 | 1.5558 | 1.5667 | 1.5773 | 1.5875 | 1.6343 | 1.6762 | 1.7147 | 1.7846 | 1.8476 | 1.9056 |
| 700 (503.10) | v | ..... | 0.7277 | 0.7934 | 0.8177 | 0.8411 | 0.8639 | 0.8860 | 0.9077 | 1.0108 | 1.1082 | 1.2024 | 1.3853 | 1.5641 | 1.7405 |
| | h | ..... | 1243.2 | 1280.6 | 1294.3 | 1307.5 | 1320.3 | 1332.8 | 1345.0 | 1403.2 | 1459.0 | 1513.9 | 1623.5 | 1734.8 | 1848.8 |
| | s | ..... | 1.4722 | 1.5084 | 1.5212 | 1.5333 | 1.5449 | 1.5559 | 1.5665 | 1.6147 | 1.6573 | 1.6963 | 1.7666 | 1.8299 | 1.8881 |
| 800 (518.23) | v | ..... | 0.6154 | 0.6779 | 0.7006 | 0.7223 | 0.7433 | 0.7635 | 0.7833 | 0.8763 | 0.9633 | 1.0470 | 1.2088 | 1.3662 | 1.5214 |
| | h | ..... | 1229.8 | 1270.7 | 1285.4 | 1299.4 | 1312.9 | 1325.9 | 1338.6 | 1398.6 | 1455.4 | 1511.0 | 1621.4 | 1733.2 | 1847.5 |
| | s | ..... | 1.4467 | 1.4863 | 1.5000 | 1.5129 | 1.5250 | 1.5366 | 1.5476 | 1.5972 | 1.6407 | 1.6801 | 1.7510 | 1.8146 | 1.8729 |
| 900 (531.98) | v | ..... | 0.5264 | 0.5873 | 0.6089 | 0.6294 | 0.6491 | 0.6680 | 0.6863 | 0.7716 | 0.8506 | 0.9262 | 1.0714 | 1.2124 | 1.3509 |
| | h | ..... | 1215.0 | 1260.1 | 1275.3 | 1290.9 | 1305.1 | 1318.8 | 1332.1 | 1393.9 | 1451.8 | 1508.1 | 1619.3 | 1731.6 | 1846.3 |
| | s | ..... | 1.4216 | 1.4653 | 1.4800 | 1.4938 | 1.5066 | 1.5187 | 1.5303 | 1.5814 | 1.6257 | 1.6656 | 1.7371 | 1.8009 | 1.8595 |
| 1000 (544.61) | v | ..... | 0.4533 | 0.5140 | 0.5350 | 0.5546 | 0.5733 | 0.5912 | 0.6084 | 0.6878 | 0.7604 | 0.8294 | 0.9615 | 1.0893 | 1.2146 |
| | h | ..... | 1198.3 | 1248.8 | 1265.9 | 1281.9 | 1297.0 | 1311.4 | 1325.3 | 1389.2 | 1448.2 | 1505.1 | 1617.3 | 1730.0 | 1845.0 |
| | s | ..... | 1.3961 | 1.4450 | 1.4610 | 1.4757 | 1.4893 | 1.5021 | 1.5141 | 1.5670 | 1.6121 | 1.6525 | 1.7245 | 1.7886 | 1.8474 |
| 1100 (556.31) | v | ..... | ..... | 0.4532 | 0.4738 | 0.4929 | 0.5110 | 0.5281 | 0.5445 | 0.6191 | 0.6866 | 0.7503 | 0.8716 | 0.9885 | 1.031 |
| | h | ..... | ..... | 1236.7 | 1255.3 | 1272.4 | 1288.5 | 1303.7 | 1318.3 | 1384.3 | 1444.5 | 1502.2 | 1615.2 | 1728.4 | 1843.8 |
| | s | ..... | ..... | 1.4251 | 1.4425 | 1.4583 | 1.4728 | 1.4862 | 1.4989 | 1.5535 | 1.5995 | 1.6405 | 1.7130 | 1.7775 | 1.8363 |
| 1200 (567.22) | v | ..... | ..... | 0.4016 | 0.4222 | 0.4410 | 0.4586 | 0.4752 | 0.4909 | 0.5617 | 0.6250 | 0.6843 | 0.7967 | 0.9046 | 1.0101 |
| | h | ..... | ..... | 1223.5 | 1243.9 | 1262.4 | 1279.6 | 1295.7 | 1311.0 | 1379.3 | 1440.7 | 1499.2 | 1613.1 | 1726.9 | 1842.5 |
| | s | ..... | ..... | 1.4062 | 1.4243 | 1.4413 | 1.4568 | 1.4710 | 1.4843 | 1.5409 | 1.5879 | 1.6293 | 1.7025 | 1.7672 | 1.8263 |
| 1400 (587.10) | v | ..... | ..... | 0.3174 | 0.3390 | 0.3580 | 0.3753 | 0.3912 | 0.4062 | 0.4714 | 0.5281 | 0.5805 | 0.6789 | 0.7727 | 0.8640 |
| | h | ..... | ..... | 1193.0 | 1218.4 | 1240.4 | 1260.3 | 1278.5 | 1295.5 | 1369.1 | 1433.1 | 1493.2 | 1608.9 | 1723.7 | 1840.0 |
| | s | ..... | ..... | 1.3639 | 1.3837 | 1.4079 | 1.4258 | 1.4419 | 1.4567 | 1.5177 | 1.5666 | 1.6093 | 1.6836 | 1.7489 | 1.8083 |

| Press., psia (sat. temp.) | Property | | | | | | | | | | | |
|---|---|---|---|---|---|---|---|---|---|---|---|---|
| 1600 (604.90) | v* | 0.2733 | 0.2936 | 0.3112 | 0.3271 | 0.3417 | 0.4034 | 0.4553 | 0.5027 | 0.5906 | 0.6738 | 0.7545 |
| | h* | 1187.8 | 1215.2 | 1238.7 | 1259.6 | 1278.7 | 1358.4 | 1425.3 | 1487.0 | 1604.6 | 1720.5 | 1837.5 |
| | s* | 1.3489 | 1.3741 | 1.3952 | 1.4137 | 1.4303 | 1.4964 | 1.5476 | 1.5914 | 1.6669 | 1.7328 | 1.7926 |
| 1800 (621.03) | v | | 0.2407 | 0.2597 | 0.2760 | 0.2907 | 0.3502 | 0.3986 | 0.4421 | 0.5218 | 0.5968 | 0.6693 |
| | h | | 1185.1 | 1214.0 | 1238.5 | 1260.3 | 1347.2 | 1417.4 | 1480.8 | 1600.4 | 1717.3 | 1835.0 |
| | s | | 1.3377 | 1.3638 | 1.3855 | 1.4044 | 1.4765 | 1.5301 | 1.5752 | 1.6520 | 1.7185 | 1.7786 |
| 2000 (635.82) | v | | 0.1936 | 0.2161 | 0.2337 | 0.2489 | 0.3074 | 0.3532 | 0.3935 | 0.4668 | 0.5352 | 0.6011 |
| | h | | 1145.6 | 1184.9 | 1214.8 | 1240.0 | 1335.5 | 1409.2 | 1474.5 | 1596.1 | 1714.1 | 1832.5 |
| | s | | 1.2945 | 1.3300 | 1.3564 | 1.3783 | 1.4576 | 1.5139 | 1.5603 | 1.6384 | 1.7055 | 1.7660 |
| 2500 (668.13) | v | | | | 0.1484 | 0.1686 | 0.2294 | 0.2710 | 0.3061 | 0.3678 | 0.4244 | 0.4784 |
| | h | | | | 1132.3 | 1176.8 | 1303.6 | 1387.8 | 1458.4 | 1585.3 | 1706.1 | 1826.2 |
| | s | | | | 1.2687 | 1.3073 | 1.4127 | 1.4772 | 1.5273 | 1.6088 | 1.6775 | 1.7389 |
| 3000 (695.36) | v | | | | | 0.0984 | 0.1760 | 0.2159 | 0.2476 | 0.3018 | 0.3505 | 0.3966 |
| | h | | | | | 1060.7 | 1267.2 | 1365.0 | 1441.8 | 1574.3 | 1698.0 | 1819.9 |
| | s | | | | | 1.1966 | 1.3690 | 1.4439 | 1.4984 | 1.5837 | 1.6540 | 1.7163 |
| 3206.2 (705.40) | v | | | | | | 0.1583 | 0.1981 | 0.2288 | 0.2806 | 0.3267 | 0.3703 |
| | h | | | | | | 1250.5 | 1355.2 | 1434.7 | 1569.8 | 1694.6 | 1817.2 |
| | s | | | | | | 1.3508 | 1.4309 | 1.4874 | 1.5742 | 1.6452 | 1.7080 |
| 3500 | v | | | | | 0.0306 | 0.1364 | 0.1762 | 0.2058 | 0.2546 | 0.2977 | 0.3381 |
| | h | | | | | 780.5 | 1224.9 | 1340.7 | 1424.5 | 1563.3 | 1689.8 | 1813.6 |
| | s | | | | | 0.9519 | 1.3241 | 1.4127 | 1.4723 | 1.5615 | 1.6336 | 1.6968 |
| 4000 | v | | | | | 0.0287 | 0.1052 | 0.1462 | 0.1743 | 0.2192 | 0.2581 | 0.2943 |
| | h | | | | | 763.8 | 1174.8 | 1314.4 | 1406.8 | 1552.2 | 1681.7 | 1807.2 |
| | s | | | | | 0.9347 | 1.2757 | 1.3827 | 1.4482 | 1.5417 | 1.6154 | 1.6795 |
| 4500 | v | | | | | 0.0276 | 0.0798 | 0.1226 | 0.1500 | 0.1917 | 0.2273 | 0.2602 |
| | h | | | | | 753.5 | 1113.9 | 1286.5 | 1388.4 | 1540.8 | 1673.5 | 1800.9 |
| | s | | | | | 0.9235 | 1.2204 | 1.3529 | 1.4253 | 1.5235 | 1.5990 | 1.6640 |
| 5000 | v | | | | | 0.0268 | 0.0593 | 0.1036 | 0.1303 | 0.1696 | 0.2027 | 0.2329 |
| | h | | | | | 746.4 | 1047.1 | 1256.5 | 1369.5 | 1529.5 | 1665.3 | 1794.5 |
| | s | | | | | 0.9152 | 1.1622 | 1.3231 | 1.4034 | 1.5066 | 1.5839 | 1.6499 |
| 5500 | v | | | | | 0.0262 | 0.0462 | 0.0880 | 0.1143 | 0.1516 | 0.1825 | 0.2106 |
| | h | | | | | 741.3 | 985.0 | 1224.1 | 1349.3 | 1518.2 | 1637.0 | 1788.1 |
| | s | | | | | 0.9090 | 1.1093 | 1.2930 | 1.3821 | 1.4908 | 1.5699 | 1.6369 |

\* $v$ = volume, ft³/lb
$h$ = enthalpy, Btu/lb
$s$ = entropy, Btu/(lb)(°R)

SOURCE: Abridged from Keenan and Keyes, "Thermodynamic Properties of Steam," John Wiley & Sons, Inc., New York, 1936. Copyright, 1937, by Joseph H. Keenan and Frederick G. Keyes. Newer tables [by Keenan, Keyes, Hill, and Moore, 1969] show some differences.

## Table 3-30. Properties of Saturated Solid/Vapor Water*

| Temp., °F. | Pressure, lb./sq. in. abs. | Volume, cu. ft./lb. | | Enthalpy, B.t.u./lb. | | Entropy, B.t.u./(lb.)(°F.) | |
|---|---|---|---|---|---|---|---|
| | | Solid | Vapor | Solid | Vapor | Solid | Vapor |
| −160 | 4.949. − 8 | 0.01722 | 3.607. + 9 | −222.05 | 990.38 | −0.4907 | 3.5549 |
| −150 | 1.620. − 7 | 0.01723 | 1.139. + 9 | −218.82 | 994.80 | −0.4801 | 3.4387 |
| −140 | 4.928. − 7 | 0.01724 | 3.864. + 8 | −215.49 | 999.21 | −0.4695 | 3.3301 |
| −130 | 1.403. − 6 | 0.01725 | 1.400. + 8 | −212.08 | 1003.63 | −0.4590 | 3.2284 |
| −120 | 3.757. − 6 | 0.01726 | 5.386. + 7 | −208.58 | 1008.05 | −0.4485 | 3.1330 |
| −110 | 9.517. − 6 | 0.01728 | 2.189. + 7 | −204.98 | 1012.47 | −0.4381 | 3.0434 |
| −100 | 2.291. − 5 | 0.01729 | 9.352. + 6 | −201.28 | 1016.89 | −0.4277 | 2.9591 |
| −90 | 5.260. − 5 | 0.01730 | 4.186. + 6 | −197.49 | 1021.31 | −0.4173 | 2.8796 |
| −80 | 1.157. − 4 | 0.01731 | 1.955. + 6 | −193.60 | 1025.73 | −0.4069 | 2.8045 |
| −70 | 2.443. − 4 | 0.01732 | 9.501. + 5 | −189.61 | 1030.15 | −0.3965 | 2.7336 |
| −60 | 4.972. − 4 | 0.01734 | 4.788. + 5 | −185.52 | 1034.58 | −0.3862 | 2.6664 |
| −50 | 9.776. − 4 | 0.01735 | 2.496. + 5 | −181.34 | 1039.00 | −0.3758 | 2.6028 |
| −45 | 1.354. − 3 | 0.01736 | 1.824. + 5 | −179.21 | 1041.21 | −0.3707 | 2.5723 |
| −40 | 1.861. − 3 | 0.01737 | 1.343. + 5 | −177.06 | 1043.42 | −0.3655 | 2.5425 |
| −35 | 2.540. − 3 | 0.01737 | 9.961. + 4 | −174.88 | 1045.63 | −0.3604 | 2.5135 |
| −30 | 3.440. − 3 | 0.01738 | 7.441. + 4 | −172.68 | 1047.84 | −0.3552 | 2.4853 |
| −25 | 4.627. − 3 | 0.01739 | 5.596. + 4 | −170.46 | 1050.05 | −0.3501 | 2.4577 |
| −20 | 6.181. − 3 | 0.01739 | 4.237. + 4 | −168.21 | 1052.26 | −0.3449 | 2.4308 |
| −15 | 8.204. − 3 | 0.01740 | 3.228. + 4 | −165.94 | 1054.47 | −0.3398 | 2.4046 |
| −10 | 1.082. − 2 | 0.01741 | 2.475. + 4 | −163.65 | 1056.67 | −0.3347 | 2.3791 |
| −5 | 1.419. − 2 | 0.01741 | 1.909. + 4 | −161.33 | 1058.88 | −0.3295 | 2.3541 |
| 0 | 1.849. − 2 | 0.01742 | 1.481. + 4 | −158.98 | 1061.09 | −0.3244 | 2.3297 |
| 5 | 2.396. − 2 | 0.01743 | 1.155. + 4 | −156.61 | 1063.29 | −0.3193 | 2.3039 |
| 10 | 3.087. − 2 | 0.01744 | 9.060. + 3 | −154.22 | 1065.50 | −0.3142 | 2.2827 |
| 15 | 3.957. − 2 | 0.01744 | 7.144. + 3 | −151.80 | 1067.70 | −0.3090 | 2.2600 |
| 16 | 4.156. − 2 | 0.01745 | 6.817. + 3 | −151.32 | 1068.14 | −0.3080 | 2.2555 |
| 18 | 4.581. − 2 | 0.01745 | 6.210. + 3 | −150.34 | 1069.02 | −0.3060 | 2.2466 |
| 20 | 5.045. − 2 | 0.01745 | 5.662. + 3 | −149.36 | 1069.90 | −0.3039 | 2.2378 |
| 22 | 5.552. − 2 | 0.01746 | 5.166. + 3 | −148.38 | 1070.38 | −0.3019 | 2.2291 |
| 24 | 6.105. − 2 | 0.01746 | 4.717. + 3 | −147.39 | 1071.66 | −0.2998 | 2.2205 |
| 26 | 6.708. − 2 | 0.01746 | 4.311. + 3 | −146.40 | 1072.53 | −0.2978 | 2.2119 |
| 28 | 7.365. − 2 | 0.01746 | 3.943. + 3 | −145.40 | 1073.41 | −0.2957 | 2.2034 |
| 30 | 8.080. − 2 | 0.01747 | 3.608. + 3 | −144.40 | 1074.29 | −0.2937 | 2.1950 |
| 31 | 8.461. − 2 | 0.01747 | 3.453. + 3 | −143.90 | 1074.73 | −0.2927 | 2.1908 |
| 32 | 8.858. − 2 | 0.01747 | 3.305. + 3 | −143.40 | 1075.16 | −0.2916 | 2.1867 |

The notation 4.949. − 8, 3.607. + 9, etc., means $4.949 \times 10^{-8}$, $3.607 \times 10^9$, etc.
*Condensed from A.S.H.R.A.E. "Fundamentals," 1967 and 1972. Reproduced by permission. For table in metric units see *Kholod. Tekh.*, **121**, 79 (1956). The validity of many standard reference tables has been critically reviewed by Jancso, Pupezin, and van Hook, *J. Phys. Chem.*, **74**, 2984 (1970). This source is recommended for further study.

## Table 3-31. Properties of Water and Steam at High Pressure*

| Pressure, kB. | | Temp., °C. | | | | | | | | |
|---|---|---|---|---|---|---|---|---|---|---|
| | | 0 | 50 | 100 | 150 | 200 | 250 | 300 | 350 | 400 |
| 1 | v | 0.9567 | 0.9733 | 1.0002 | 1.0362 | 1.0821 | 1.1400 | 1.2140 | 1.3115 | 1.4444 |
| | h | 95 | 294 | 495 | 698 | 903 | 1113 | 1329 | 1555 | 1794 |
| | s | -0.010 | 0.658 | 1.238 | 1.748 | 2.206 | 2.628 | 3.022 | 3.400 | 3.768 |
| 2 | v | 0.9244 | 0.9434 | 0.9678 | 0.9978 | 1.0338 | 1.0767 | 1.1276 | 1.1879 | 1.2595 |
| | h | 181 | 375 | 572 | 769 | 967 | 1167 | 1369 | 1575 | 1783 |
| | s | -0.039 | 0.615 | 1.179 | 1.675 | 2.117 | 2.518 | 2.888 | 3.232 | 3.554 |
| 4 | v | 0.8785 | 0.8978 | 0.9190 | 0.9429 | 0.9695 | 0.9989 | 1.0316 | 1.0679 | 1.1082 |
| | h | 344 | 532 | 723 | 914 | 1105 | 1295 | 1485 | 1676 | 1867 |
| | s | -0.111 | 0.530 | 1.081 | 1.561 | 1.986 | 2.368 | 2.716 | 3.035 | 3.330 |
| 6 | v | 0.8455 | 0.8643 | 0.8833 | 0.9031 | 0.9247 | 0.9482 | 0.9732 | 0.9999 | 1.0285 |
| | h | 507 | 682 | 873 | 1060 | 1246 | 1432 | 1617 | 1802 | 1985 |
| | s | -0.127 | 0.450 | 0.998 | 1.469 | 1.886 | 2.259 | 2.597 | 2.906 | 3.189 |
| 8 | v | | 0.8380 | 0.8558 | 0.8728 | 0.8908 | 0.9102 | 0.9308 | 0.9526 | 0.9754 |
| | h | | 826 | 1020 | 1205 | 1389 | 1572 | 1754 | 1935 | 2115 |
| | s | | 0.370 | 0.927 | 1.393 | 1.804 | 2.171 | 2.504 | 2.807 | 3.084 |
| 10 | v | | 0.8165 | 0.8335 | 0.8486 | 0.8642 | 0.8806 | 0.8980 | 0.9163 | 0.9356 |
| | h | | 966 | 1165 | 1350 | 1532 | 1712 | 1892 | 2071 | 2248 |
| | s | | 0.290 | 0.863 | 1.328 | 1.735 | 2.098 | 2.426 | 2.725 | 2.999 |
| 20 | v | | | 0.7601 | 0.7705 | 0.7807 | 0.7909 | 0.8013 | 0.8120 | 0.8229 |
| | h | | | 1862 | 2052 | 2231 | 2408 | 2583 | 2755 | 2926 |
| | s | | | 0.605 | 1.084 | 1.485 | 1.840 | 2.159 | 2.448 | 2.711 |
| 50 | v | | | | | 0.6732 | 0.6791 | 0.6850 | 0.6911 | |
| | h | | | | | 4341 | 4512 | 4681 | 4848 | |
| | s | | | | | 1.388 | 1.699 | 1.981 | 2.239 | |

*Extracted from Table 13 of Juza, "An Equation of State for Water and Steam. Steam Tables in the Critical Region and in the Range from 1000 to 100,000 bars," Academia, Prague, 1966. (Pressure in kilo-bars, volume in cu. cm/g, enthalpy in J/g, entropy in J/g. °K., temperature in °C.) The original tables give values for 0(25)500(50)1000°C., 1 to 100 kB. See also Grindley and Lind, *J. Chem. Phys.*, **54**, 3983 (1971) (to 8 kB., 150°C.); Walker, NOLTR-66-217 (AD 651105), (1967) (12 functions calculated from Kell and Whalley *Phil. Trans. Roy. Soc.*, **A258**, 565 (1965), to 150°C., 1 kB.), Snay and Butler, NAVORD *Rept.* 4181, 1965 (AD 118927) and Gleyzal and Snay, NAVORD *Rept.* 6749, 1959 extend to 10$^6$kB, 10$^6$°K. See also Sharp, UCRL 7118, 1962; Howard, UCRL 6455T, 1961, and Chapin, *J. Chem. Eng. Data*, **15**, 323 (1970).

## Table 3-32. Properties of Refrigerants—Saturated State

| Refrig. No.† | Temp., °F | Pressure, psia | Specific volume, ft³/lb | | Enthalpy, Btu/lb | | Entropy, Btu/(lb)(°F) | |
|---|---|---|---|---|---|---|---|---|
| | | | Liquid | Vapor | Liquid | Vapor | Liquid | Vapor |
| 11 | −168 | 0.0008 | 0.0091 | 26983 | −26.06 | 73.12 | −0.0741 | 0.2659 |
| | −160 | 0.0016 | 0.0091 | 14835 | −24.41 | 73.92 | −0.0686 | 0.2596 |
| | −120 | 0.0217 | 0.0093 | 1224 | −16.20 | 78.10 | −0.0428 | 0.2348 |
| | −80 | 0.1579 | 0.0096 | 187.6 | −8.08 | 82.54 | −0.0202 | 0.2185 |
| | −40 | 0.7432 | 0.0059 | 43.92 | 0.00 | 87.19 | 0.0000 | 0.2078 |
| | 0 | 2.561 | 0.0102 | 13.87 | 8.10 | 91.97 | 0.0184 | 0.2009 |
| | 40 | 7.022 | 0.0105 | 5.430 | 16.26 | 96.79 | 0.0354 | 0.1966 |
| | 80 | 16.23 | 0.0109 | 2.489 | 24.53 | 101.58 | 0.0513 | 0.1941 |
| | 120 | 32.94 | 0.0113 | 1.282 | 32.91 | 106.26 | 0.0662 | 0.1928 |
| | 160 | 60.45 | 0.0118 | 0.7204 | 41.43 | 110.76 | 0.0804 | 0.1922 |
| | 200 | 102.5 | 0.0124 | 0.4311 | 50.18 | 114.99 | 0.0939 | 0.1921 |
| | 240 | 163.4 | 0.0131 | 0.2695 | 59.29 | 118.83 | 0.1071 | 0.1922 |
| | 280 | 247.6 | 0.0140 | 0.1727 | 68.00 | 122.09 | 0.1203 | 0.1921 |
| | 320 | 360.7 | 0.0153 | 0.1107 | 79.77 | 124.34 | 0.1340 | 0.1912 |
| | 360 | 508.8 | 0.0177 | 0.0670 | 92.81 | 124.25 | 0.1498 | 0.1881 |
| | 388* | 639.5 | 0.0289 | 0.0289 | 112.08 | 112.08 | 0.1722 | 0.1722 |
| 12 | −152 | 0.1380 | 0.0096 | 197.6 | −23.11 | 60.63 | −0.0639 | 0.2082 |
| | −120 | 0.6419 | 0.0098 | 46.74 | −16.57 | 64.05 | −0.0437 | 0.1936 |
| | −80 | 2.881 | 0.0102 | 11.53 | −8.35 | 68.47 | −0.0209 | 0.1814 |
| | −40 | 9.308 | 0.0106 | 3.875 | 0.00 | 72.91 | 0.0000 | 0.1737 |
| | 0 | 23.85 | 0.0110 | 1.609 | 8.52 | 77.27 | 0.0193 | 0.1689 |
| | 40 | 51.67 | 0.0116 | 0.7736 | 17.27 | 81.44 | 0.0375 | 0.1659 |
| | 80 | 98.87 | 0.0123 | 0.4114 | 26.37 | 85.28 | 0.0548 | 0.1639 |
| | 120 | 172.4 | 0.0132 | 0.2333 | 36.01 | 88.61 | 0.0717 | 0.1624 |
| | 160 | 279.8 | 0.0145 | 0.1361 | 46.63 | 91.01 | 0.0889 | 0.1605 |
| | 200 | 430.1 | 0.0167 | 0.0767 | 59.20 | 91.28 | 0.1079 | 0.1565 |
| | 234* | 596.9 | 0.0287 | 0.0287 | 78.86 | 78.86 | 0.1359 | 0.1359 |
| 13 | −200 | 0.433 | 0.0095 | 61.33 | −34.55 | 38.55 | −0.1008 | 0.1807 |
| | −160 | 3.104 | 0.0099 | 9.750 | −26.08 | 42.73 | −0.0721 | 0.1575 |
| | −120 | 12.48 | 0.0105 | 2.681 | −17.67 | 46.80 | −0.0459 | 0.1439 |
| | −80 | 36.98 | 0.0111 | 0.9689 | −9.05 | 50.62 | −0.0223 | 0.1349 |
| | −40 | 87.43 | 0.0119 | 0.4234 | 0.00 | 54.02 | 0.0000 | 0.1287 |
| | 0 | 176.8 | 0.0130 | 0.2066 | 10.05 | 56.69 | 0.0223 | 0.1238 |
| | 40 | 319.6 | 0.0148 | 0.1046 | 21.37 | 57.82 | 0.0452 | 0.1181 |
| | 80 | 535.5 | 0.0205 | 0.0413 | 38.53 | 52.09 | 0.0767 | 0.1019 |
| | 84* | 561.3 | 0.0277 | 0.0277 | 45.27 | 45.27 | 0.0890 | 0.0890 |
| 21 | −40 | 1.358 | 0.0106 | 32.09 | 0.00 | 114.56 | 0.0000 | 0.2730 |
| | 0 | 4.582 | 0.0109 | 10.35 | 9.44 | 119.37 | 0.0214 | 0.2606 |
| | 40 | 12.32 | 0.0113 | 4.130 | 19.04 | 124.19 | 0.0414 | 0.2519 |
| | 80 | 27.96 | 0.0118 | 1.923 | 29.03 | 128.98 | 0.0606 | 0.2458 |
| | 120 | 55.75 | 0.0123 | 1.001 | 39.46 | 133.53 | 0.0791 | 0.2414 |
| | 160 | 100.6 | 0.0128 | 0.565 | 50.43 | 137.69 | 0.0972 | 0.2381 |
| 22 | −155 | 0.2088 | 0.0101 | 180.8 | −27.10 | 86.95 | −0.0751 | 0.2992 |
| | −120 | 1.0954 | 0.0105 | 38.28 | −19.19 | 91.02 | −0.0506 | 0.2739 |
| | −80 | 4.7822 | 0.0109 | 9.695 | −9.84 | 95.71 | −0.0246 | 0.2534 |
| | −40 | 15.222 | 0.0114 | 3.296 | 0.00 | 100.26 | 0.0000 | 0.2389 |
| | 0 | 38.657 | 0.0119 | 1.372 | 10.41 | 104.47 | 0.0236 | 0.2282 |

## Table 3-32. Properties of Refrigerants—
## Saturated State (Continued)

| Refrig. No. | Temp., °F | Pressure, psia | Specific volume, ft³/lb | | Enthalpy, Btu/lb | | Entropy, Btu/(lb)(°F) | |
|---|---|---|---|---|---|---|---|---|
| | | | Liquid | Vapor | Liquid | Vapor | Liquid | Vapor |
| | 40 | 83.21 | 0.0126 | 0.6575 | 21.42 | 108.14 | 0.0463 | 0.2199 |
| | 80 | 158.3 | 0.0135 | 0.3462 | 33.11 | 111.05 | 0.0685 | 0.2129 |
| | 120 | 247.6 | 0.0147 | 0.1924 | 45.71 | 112.78 | 0.0904 | 0.2061 |
| | 160 | 444.5 | 0.0166 | 0.1070 | 59.95 | 112.26 | 0.1133 | 0.1978 |
| | 200 | 686.4 | 0.0224 | 0.0474 | 80.86 | 102.85 | 0.1446 | 0.1779 |
| | 205* | 721.9 | 0.0305 | 0.0305 | 91.33 | 91.33 | 0.1602 | 0.1602 |
| 113 | 0 | 0.838 | 0.0097 | 31.31 | 7.98 | 78.89 | 0.0182 | 0.1725 |
| | 40 | 2.655 | 0.0099 | 10.68 | 16.16 | 84.65 | 0.0352 | 0.1723 |
| | 80 | 6.902 | 0.0103 | 4.392 | 24.63 | 90.51 | 0.0515 | 0.1736 |
| | 120 | 15.40 | 0.0106 | 2.078 | 33.48 | 96.41 | 0.0673 | 0.1758 |
| | 160 | 30.44 | 0.0111 | 1.094 | 42.74 | 102.29 | 0.0287 | 0.1788 |
| | 200 | 54.66 | 0.0115 | 0.624 | 52.45 | 108.07 | 0.0978 | 0.1821 |
| | 250 | 101.8 | 0.0122 | 0.339 | 62.87 | 113.88 | 0.1123 | 0.1842 |
| | 300 | 175.6 | 0.0132 | 0.190 | 75.82 | 120.15 | 0.1297 | 0.1881 |
| | 350 | 283.3 | 0.0147 | 0.105 | 90.12 | 124.80 | 0.1477 | 0.1905 |
| | 400 | 434.3 | 0.0182 | 0.049 | 105.78 | 122.62 | 0.1659 | 0.1855 |
| | 417* | 498.9 | 0.0280 | 0.028 | 109.49 | 109.49 | 0.1699 | 0.1699 |
| 114 | −120 | 0.076 | 0.0092 | 281.1 | −15.45 | 53.18 | −0.0408 | 0.1613 |
| | −80 | 0.462 | 0.0095 | 51.50 | −7.93 | 58.47 | −0.0199 | 0.1550 |
| | −40 | 1.906 | 0.0098 | 13.71 | 0.00 | 64.06 | 0.0000 | 0.1527 |
| | 0 | 5.949 | 0.0102 | 4.754 | 8.42 | 69.88 | 0.0191 | 0.1528 |
| | 40 | 15.08 | 0.0106 | 1.996 | 17.38 | 75.81 | 0.0378 | 0.1547 |
| | 80 | 32.66 | 0.0110 | 0.9628 | 26.87 | 81.76 | 0.0560 | 0.1577 |
| | 120 | 62.70 | 0.0116 | 0.5134 | 36.86 | 87.61 | 0.0737 | 0.1613 |
| | 160 | 109.6 | 0.0123 | 0.2934 | 47.30 | 93.21 | 0.0910 | 0.1650 |
| | 200 | 178.4 | 0.0133 | 0.1745 | 58.19 | 98.34 | 0.1077 | 0.1686 |
| | 240 | 275.3 | 0.0148 | 0.1039 | 69.72 | 102.51 | 0.1243 | 0.1712 |
| | 280 | 410.7 | 0.0181 | 0.0548 | 83.30 | 103.63 | 0.1426 | 0.1701 |

Table 3-32 copyright by E. I. du Pont de Nemours & Co.   Reprinted with permission of the copyright owner.
* Critical temperature.
† Refrigerant No. 11, trichlorofluoromethane; 12, dichlorodifluoromethane; 13, chlorotrifluoromethane; 21, dichlorofluoromethane; 22, chlorodifluoromethane; 113, trichlorotrifluoroethane; 114, dichlorotetrafluoroethane.

## Table 3-33. Properties of Refrigerants—Superheated State
### (at atmospheric pressure)

| Refrig. No.‡ | Temp., °F | Specific volume, ft³/lb | Enthalpy, Btu/lb | Entropy, Btu/(lb)(°F) |
|---|---|---|---|---|
| 11 | 75ˢ | 2.731 | 100.97 | 0.1943 |
|  | 100 | 2.877 | 104.53 | 0.2008 |
|  | 150 | 3.163 | 111.73 | 0.2132 |
|  | 200 | 3.443 | 119.11 | 0.2248 |
|  | 250 | 3.719 | 126.66 | 0.2358 |
|  | 300 | 3.993 | 134.36 | 0.2463 |
|  | 350 | 4.265 | 142.20 | 0.2563 |
| 12 | −22ˢ | 2.532 | 74.93 | 0.1712 |
|  | 0 | 2.676 | 77.92 | 0.1779 |
|  | 50 | 3.003 | 85.02 | 0.1925 |
|  | 100 | 3.321 | 92.37 | 0.2063 |
|  | 150 | 3.634 | 99.98 | 0.2193 |
|  | 200 | 3.944 | 107.84 | 0.2317 |
|  | 250 | 4.252 | 115.93 | 0.2435 |
| 13* | −100 | 2.368 | 49.29 | 0.1475 |
|  | 0 | 3.098 | 63.05 | 0.1810 |
|  | 100 | 3.801 | 78.10 | 0.2106 |
|  | 200 | 4.495 | 94.58 | 0.2376 |
|  | 300 | 5.185 | 112.32 | 0.2627 |
|  | 400 | 5.877 | 131.11 | 0.2859 |
|  | 500 | 6.546 | 150.71 | 0.3075 |
| 21† | 46ˢ | 3.667 | 124.88 | 0.2509 |
|  | 50 | 3.700 | 125.45 | 0.2520 |
|  | 100 | 4.085 | 132.38 | 0.2649 |
|  | 150 | 4.468 | 139.60 | 0.2773 |
|  | 200 | 4.850 | 147.20 | 0.2893 |
|  | 250 | 5.230 | 155.14 | 0.3009 |
|  | 300 | 5.609 | 163.41 | 0.3122 |
| 22 | −41ˢ | 3.406 | 100.11 | 0.2393 |
|  | 0 | 3.782 | 106.11 | 0.2529 |
|  | 50 | 4.228 | 113.64 | 0.2685 |
|  | 150 | 5.100 | 129.68 | 0.2972 |
|  | 200 | 5.530 | 138.19 | 0.3106 |
|  | 250 | 5.959 | 147.03 | 0.3236 |
| 113† | 115ˢ | 2.271 | 95.66 | 0.1755 |
|  | 150 | 2.421 | 101.27 | 0.1850 |
|  | 200 | 2.635 | 109.51 | 0.1980 |
|  | 250 | 2.848 | 118.02 | 0.2105 |
|  | 300 | 3.060 | 126.84 | 0.2225 |
|  | 350 | 3.271 | 135.94 | 0.2340 |
|  | 400 | 3.481 | 145.31 | 0.2453 |
| 114* | 40ˢ | 2.006 | 75.77 | 0.1547 |
|  | 50 | 2.052 | 77.47 | 0.1581 |
|  | 100 | 2.276 | 85.98 | 0.1740 |
|  | 150 | 2.496 | 94.87 | 0.1892 |
|  | 200 | 2.714 | 104.10 | 0.2037 |
|  | 250 | 2.931 | 113.64 | 0.2177 |
|  | 300 | 3.146 | 123.45 | 0.2310 |

Table 3-33 copyright E. I. du Pont de Nemours & Co.   Reprinted with permission of the copyright owner.
s = saturation condition.
* At 15 psia.
† At 14 psia
‡ See footnote to Table 3-32 for refrigerant names.

## REDUCED PROPERTIES

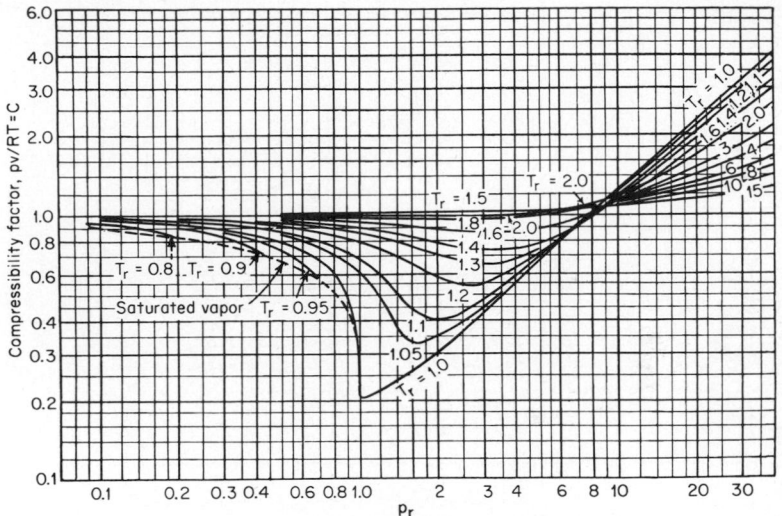

Fig. 3-11. Compressibility factors of gases.

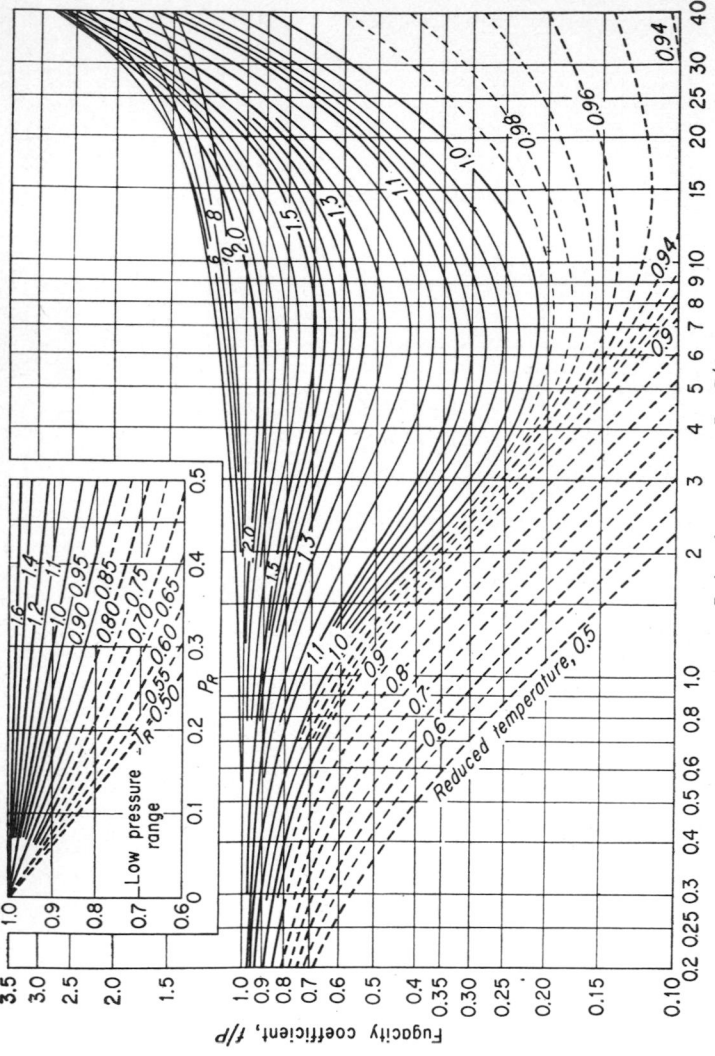

Fig. 3-12. Fugacity coefficients of gases. Chao, Greenkorn, et al., *AIChE J.*, **17**, 353 (1971) tabulated $f/p$ as a function of reduced pressure and temperature. These tables were extended by Chao and Greenkorn in "Thermodynamics of Fluids," Marcel Dekker, Inc., New York, 1975.

## LINEAR AND CUBICAL EXPANSION

GAS DENSITY — Based on the assumption that the gas laws hold over the ranges of temperature and pressure below (Example shown for air at 100°F and 50 lb/sq in. gage)

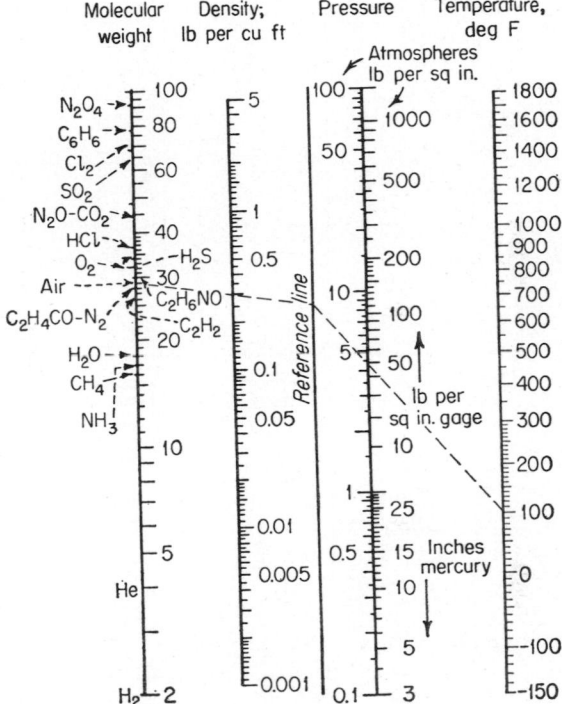

FIG. 3-13. Gas-density nomograph.

## Table 3-34. Expansion Coefficients of Solids*

| Substance | Temperature or temp. range (°F) | CLE × 10⁶/°F | CCE × 10⁶/°F |
|---|---|---|---|
| Aluminum | 68 | 12.4 | |
| Aluminum | 572 | 15.8 | |
| Antimony | 32–212 | ..... | 17.6 |
| Amber | 32–194 | 34 | |
| Bakelite, bleached | 68–140 | 12 | |
| Beryl | 32–212 | ..... | 0.58 |
| Brass, cast | 32–212 | 10.4 | |
| Bronze(75Cu,25Sn) | 62–212 | 10.2 | |
| Carbon(graphite) | 122 | 3.3 | |
| Chromium | 140 | 3.8 | |
| Copper | 32–212 | 9.3 | 27.8 |
| Copper | 392 | 9.4 | |
| Fluorspar, CaF₂ | 32–212 | 10.8 | |
| Glass: | | | |
|   Flint | 122–140 | 4.4 | |
|   Hard | 32–212 | ..... | 11.9 |
|   Plate | 32–212 | 4.95 | |
|   Quartz | 60–930 | 0.32 | |
|   Silica | 32–176 | ..... | 0.72 |
|   Tube | 32–212 | 4.95 | 15.3 |
| Ice | −15 | 28 | 62.5 |
| Iron | 32–212 | 6.6 | 19.7 |
| Lead | 32–212 | ..... | 46.5 |
| Lead | 536 | 19.0 | |
| Limestone | 77–212 | 5 | |
| Marble | 59–212 | 6.5 | |
| Monel metal | 77–212 | 7.8 | |
| Nickel | 68 | 7.0 | |
| Paraffin | 32–60 | 59 | |
| Paraffin | 60–100 | 72 | |
| Paraffin | 100–120 | 265 | |
| Platinum | 68 | 4.9 | |
| Quartz, parallel to axis | 32–176 | 4.4 | |
| Quartz, perpendicular to axis | 32–176 | 7.4 | 21.4 |
| Rubber, hard | 32 | 38 | |
| Silver | 68 | 10.5 | |
| Silver | 32–212 | 12.8 | 38.2 |
| Solder(2Pb,1Sn) | 32–212 | 14 | |
| Steel | 0–400 | 6.8 | |
| Tin | 32–212 | 12.8 | 38.2 |
| Wood, parallel to fiber: | | | |
|   Mahogany | 36 | 2.0 | |
|   Maple | 36 | 3.5 | |
|   Oak | 36 | 2.7 | |
|   Pine | 36 | 3.0 | |
| Wood, perpendicular to fiber: | | | |
|   Mahogany | 36 | 2.2 | |
|   Maple | 36 | 2.7 | |
|   Oak | 36 | 3.0 | |
|   Pine | 36 | 1.9 | |
| Zinc | 32–212 | ..... | 49.5 |
| Zinc | 68–482 | 22.1 | |

\* CLE = coefficient of linear expansion, CCE = coefficient of cubical expansion. While simple theory predicts that the CCE should be three times the CLE, discrepancies may occur due to experimental error in determining one or both coefficients, isotropic effects, etc.

## Table 3-35. Cubical Expansion of Liquids

| Liquid | Coef/°F × 10³ | Liquid | Coef/°F × 10³ |
|---|---|---|---|
| Acetic acid | 0.595 | Hydrochloric acid, 33.2% | 0.253 |
| Acetone | .825 | Mercury | .109 |
| Alcohol, ethyl | .622 | Olive oil | .400 |
| Alcohol, methyl | .666 | Pentane (93°API) | .74 |
| Benzene | .686 | Petroleum: | |
| Bromine | .630 | 15°API | .35 |
| Calcium chloride: | | 35°API | .44 |
| 5.8% solution | .139 | Phenol | .605 |
| 40.9% solution | .260 | Sodium chloride, 20.6% solution | .230 |
| Carbon disulfide | .676 | Sulfuric acid: | |
| Carbon tetrachloride | .686 | 10.9% | .215 |
| Chloroform | .707 | 100.0% | .31 |
| Ether | .920 | Turpentine | .541 |
| Glycerin | .280 | Water | .115 |

* For an example of use, see Table 3-31.

## Table 3-36. General Corrosion Properties of Some Metals and Alloys*

Ratings: 0 unsuitable. Not available in form required or not suitable for fabrication requirements or not suitable for corrosion conditions.
1 poor to fair.
2 fair. For mild conditions or where periodic replacement is possible. Restricted use.
3 fair to good.
4 good. Suitable when superior alternatives are uneconomic.
5 good to excellent.
6 normally excellent.
Small variations in service conditions may appreciably affect corrosion resistance. Choice of materials is therefore guided wherever possible by a combination of experience and laboratory and site tests.

| | Liquids | | | | | | | | | | | Gases | | | | |
| | Non-oxidizing or reducing media | | | | Oxidizing media | | | Natural waters | | | | Common industrial media | | | | |
| | | | Alkaline solutions, e.g. | | | | | Fresh-water supplies | | Sea water | | Steam | | Furnace gases with incidental sulfur content | | |
| Material | Acid solutions, excluding hydrochloric, e.g., phosphoric, sulfuric, most conditions, many organics | Neutral solutions e.g., many non-oxidizing salt solutions, chlorides, sulfates | Caustic and mild alkalies, excluding ammonium hydroxide | Ammonium hydroxide and amines | Acid solutions, e.g., nitric | Neutral or alkaline solutions, e.g., persulfates, peroxides, chromates | Pitting media,† acid ferric chloride solutions | Static or slow-moving | Turbulent | Static or slow-moving | Turbulent | Moist, condensate | Dry at high temp., promoting slight dissociation | Reducing, e.g., heat-treatment furnace gases | Oxidizing, e.g., flue gases | Ambient air, city or industrial |
|---|---|---|---|---|---|---|---|---|---|---|---|---|---|---|---|---|
| Cast iron, flake graphite, plain or low alloy | 1 | 3 | 4 | 5 | 0 | 4 | 0 | 4 | 3 | 4 | 2 | 4 | 4 | 1 | 1 | 3 |
| Ductile iron (higher strength and hardness may be attained by composition and heat-treatment or both) | 1 | 3 | 4 | 5 | 0 | 4 | 0 | 4 | 4 | 4 | 3 | 4 | 4 | 1 | 1 | 3 |
| Ni-Resist corrosion-resistant cast iron, type 1 (14 Ni; 7 Cu; 2 Cr; bal. Fe) | 4 | 5 | 5 | 5 | 0 | 5 | 0 | 5 | 5 | 5 | 5 | 5 | 5 | 3 | 2 | 4 |

| Material | | | | | | | | | | | | | | | |
|---|---|---|---|---|---|---|---|---|---|---|---|---|---|---|---|
| Ni-Resist corrosion-resistant cast iron, type 2 Cu free (20–30 Ni; 2–3 Cr; bal. Fe) | 4 | 5 | 5 | 6 | 0 | 5 | 0 | 5 | 5 | 5 | 5 | 5 | 3 | 2 | 4 |
| Ni-Resist corrosion-resistant cast iron, ductile (24 Ni; bal. Fe) | 4 / 6 | 5 / 6 | 5 / 2 | 6 / 5 | 0 / 6 | 5 / 6 | 0 / 3 | 5 / 5 | 5 / 5 | 5 / 5 | 5 / 6 | 5 / 6 | 3 / 4 | 2 / 3 | 4 / 6 |
| 14% silicon iron | 1 | 3 | 4 | 5 | 0 | 4 | 0 | 4 | 4 | 2 | 4 | 4 | 1 | 1 | 3 |
| Mild steel, also low-alloy irons and steels | 2 | 4 | 4 | 6 | 5 | 6 | 0 | 4 | 1 | 4 | 5 | 6 | 3 | 2 | 4 |
| Stainless steel, ferritic 17% Cr type | 3 | 4 | 5 | 6 | 6 | 6 | 0 | 5 | 2 | 5 | 6 | 6 | 2 | 3 | 5 |
| Stainless steel, austenitic 18 Cr; 8 Ni type | 4 | 5 | 5 | 6 | 5 | 6 | 1 | 5 | 3 | 5 | 6 | 6 | 2 | 4 | 6 |
| Stainless steel, austenitic 18 Cr; 12 Ni; 2.5 Mo type | 5 | 6 | 5 | 6 | 5 | 6 | 2 | 6 | 4 | 6 | 6 | 6 | 2 | 4 | 6 |
| Stainless steel, austenitic 20 Cr; 29 Ni; 2.5 Mo; 3.5 Cu type | 5 | 6 | 5 | 6 | 5 | 6 | 2 | 6 | 4 | 6 | 6 | 6 | 2 | 4 | 6 |
| Ni-o-nel nickel-iron-chromium alloy (40 Ni; 21 Cr; 3 Mo; 1.5 Cu; bal. Fe) | 6 | 6 | 5 | 6 | 5 | 6 | 2 | 6 | 4 | 6 | 6 | 6 | 2 | 5 | 6 |
| Hastelloy alloy C[a] (55 Ni; 17 Mo; 16 Cr; 6 Fe; 4 W) | 5 | 6 | 5 | 6 | 4 | 6 | 5 | 6 | 6 | 6 | 6 | 6 | 3 | 4 | 6 |
| Hastelloy alloy B[b] (61 Ni; 28 Mo; 6 Fe) | 6 | 5 | 4 | 4 | 0 | 3 | 0 | 6 | 0 | 4 | 6 | 5 | 3 | 2 | 5 |
| Hastelloy alloy D (82 Ni; 9 Si; 3 Cu) | 6 | 6 | 3 | 4 | 2 | 5 | 1 | 6 | 2 | 6 | 6 | 6 | 4 | 2 | 6 |
| Inconel nickel-chromium alloy (78 Ni; 15 Cr; 7 Fe) | 3 | 6 | 6 | 6 | 3 | 6 | 1 | 6 | 4 | 6 | 6 | 6 | 2 | 4 | 6 |

[a] Also Chlorimet 3.    [b] Also Chlorimet 2.

## Table 3-36. General Corrosion Properties of Some Metals and Alloys* (Continued)

Ratings: 0 unsuitable.  Not available in form required or not suitable for fabrication requirements or not suitable for corrosion conditions.
1 poor to fair.
2 fair.  For mild conditions or where periodic replacement is possible.  Restricted use.
3 fair to good.
4 good.  Suitable when superior alternatives are uneconomic.
5 good to excellent.
6 normally excellent.
Small variations in service conditions may appreciably affect corrosion resistance.  Choice of materials is therefore guided wherever possible by a combination of experience and laboratory and site tests.

| Material | Non-oxidizing or reducing media | | | | Oxidizing media | | | Natural waters | | | | Steam | | Gases — Common industrial media | | |
| --- | --- | --- | --- | --- | --- | --- | --- | --- | --- | --- | --- | --- | --- | --- | --- | --- |
| | | | Alkaline solutions, e.g. | | | | | Fresh-water supplies | | Sea water | | | | Furnace gases with incidental sulfur content | | |
| | Acid solutions, excluding hydrochloric, e.g., phosphoric, sulfuric, most conditions, many organics | Neutral solutions, e.g., many non-oxidizing salt solutions, chlorides, sulfates | Caustic and mild alkalies, excluding ammonium hydroxide | Ammonium hydroxide and amines | Acid solutions, e.g., nitric | Neutral or alkaline solutions, e.g., persulfates, peroxides, chromates | Pitting media,† acid ferric chloride solutions | Static or slow-moving | Turbulent | Static or slow-moving | Turbulent | Moist, condensate | Dry at high temp., promoting slight dissociation | Reducing, e.g., heat-treatment furnace gases | Oxidizing, e.g., flue gases | Ambient air, city or industrial |
| Copper-nickel alloys up to 30% nickel | 4 | 5 | 5 | 0 | 0 | 4 | 1 | 6 | 6 | 6 | 6 | 6 | 5 | 2 | 2 | 5 |
| Monel 400 nickel-copper alloy (66 Ni; 30 Cu; 2 Fe) | 5 | 6 | 6 | 1 | 0 | 5 | 1 | 6 | 6 | 4 | ·· | 6 | 6 | 2 | 3 | 5 |
| Alloy 505 nickel-copper cast alloy (66 Ni; 30 Cu; 4 Si) | 5 | 6 | 6 | 1 | 0 | 5 | — | 6 | 6 | 4 | 6 | 6 | 6 | 2 | 3 | 5 |
| Monel K-500 age hardenable Ni-Cu alloy (67 Ni; 30 Cu; 3 Al) | 5 | 6 | 6 | 1 | 0 | 5 | 1 | 6 | 6 | 4 | 6 | 6 | 6 | 2 | 3 | 5 |

| | 1 | 2 | 3 | 4 | 5 | 6 | 7 | 8 | 9 | 10 | 11 | 12 | 13 | 14 | 15 | 16 |
|---|---|---|---|---|---|---|---|---|---|---|---|---|---|---|---|---|
| A nickel—commercial (99.4 Ni) | 4 | 5 | 6 | 1 | 0 | 5 | 0 | 6 | 6 | 3 | 5 | 6 | 6 | 2 | 2 | 4 |
| Copper and silicon bronze | 4 | 4 | 4 | 0 | 0 | 4 | 0 | 6 | 5 | 4 | 1 | 5 | 5 | 2 | 2 | 5 |
| Aluminum brass (76 Cu; 22 Zn; 2 Al) | 3 | 4 | 2 | 0 | 0 | 3 | 0 | 6 | 6 | 4 | 5 | 6 | 5 | 2 | 2 | 5 |
| Nickel-aluminum-bronze (80 Cu; 10 Al; 5 Ni; 5 Fe) | 4 | 4 | 2 | 0 | 0 | 3 | 0 | 6 | 6 | 4 | 5 | 6 | 5 | 2 | 3 | 5 |
| Bronze, type A (88 Cu; 5 Sn; 5 Ni; 2 Zn) | 4 | 5 | 4 | 0 | 0 | 4 | 0 | 6 | 6 | 5 | 5 | 6 | 5 | 2 | 2 | 5 |
| Aluminum and its alloys | 1 | 3 | 0 | 6 | 0–5 | 0–4 | 0 | 4 | 5 | 0–5 | 4 | 5 | 2 | 5 | 4 | 5 |
| Lead, chemical or antimonial | 5 | 5 | 2 | 2 | 0 | 2 | 0 | 6 | 5 | 5 | 3 | 2 | 0 | 4 | 3 | 5 |
| Silver | 4 | 6 | 6 | 0 | 0 | 2 | 0 | 6 | 6 | 5 | 5 | 6 | 5 | 4 | 4 | 4 |
| Titanium | 3 | 6 | 2 | 6 | 6 | 6 | 6 | 6 | 6 | 6 | 6 | 6 | 5 | 3 | 5 | 6 |
| Zirconium | 3 | 6 | 2 | 6 | 6 | 6 | 2 | 6 | 6 | 6 | 6 | 6 | 6 | 3 | 5 | 6 |

3–81

## Table 3-36. General Corrosion Properties of Some Metals and Alloys* (Continued)

Ratings: 0 unsuitable. Not available in form required or not suitable for fabrication requirements or not suitable for corrosion conditions.
1 poor to fair.
2 fair. For mild conditions or where periodic replacement is possible. Restricted use.
3 fair to good.
4 good. Suitable when superior alternatives are uneconomic.
5 good to excellent.
6 normally excellent.
Small variations in service conditions may appreciably affect corrosion resistance. Choice of materials is therefore guided wherever possible by a combination of experience and laboratory and site tests.

| Material | Gases (Cont'd.) | | | | Available forms | Cold formability in wrought and clad form | Weldability | Max. strength annealed condition ×1000 lb./sq. in. | Coeff. of thermal expansion millionths per °F. 70°-212°F. | Remarks¶ |
| | Halogens and derivatives | | | | | | | | | |
| | Halogens | | Halide acids, moist, e.g., hydrochloric hydrolysis products of organic halides | Hydrogen halides, dry,‡ e.g., dry hydrogen chloride, °F. | | | | | | |
| | Moist, e.g., chlorine below dew point | Dry, e.g., fluorine above dew point | | | | | | | | |
| Cast iron, flake graphite, plain or low alloy | 0 | 2 | 0 | 2 < 400<br>1 < 750 | Cast | No | Fair§ | 45 | 6.7 | |
| Ductile iron (higher strength and hardness may be attained by composition and heat-treatment or both) | 0 | 2 | 0 | 2 < 400<br>1 < 750 | Cast | No | Good§ | 67 | 7.5 | |
| Ni-Resist corrosion-resistant cast iron, type 1 (14 Ni; 7 Cu; 2 Cr; bal. Fe) | 0 | 2 | 3 | 3 < 400<br>2 < 750 | Cast | No | Good§ | 22-31 | 10.3 | |
| Ni-Resist corrosion-resistant cast iron, type 2 Cu free (20-30 Ni; 2-3 Cr; bal. Fe) | 0 | 2 | 3 | 3 < 400<br>2 < 750 | Cast | No | Good§ | 22-31 | 9.6 | Type 3 Ni-Resist has same corrosion resistance |

| Material | | | | Form | | | | | Remarks |
|---|---|---|---|---|---|---|---|---|---|
| Ni-Resist corrosion-resistant cast iron, ductile (24 Ni; bal. Fe)........ | 0 | 2 | 3 | 3 < 400<br>2 < 750 | Cast | No | Good§ | 56 | 10.4 | Very brittle, susceptible to cracking by mechanical and thermal shock |
| 14% silicon iron......... | 0 | 0 | 4 | 1 < 400 | Cast | No | No | 22 | 7.4 | High strengths obtainable by alloying; also improved atmospheric corrosion resistance. See A.S.T.M. specifications for particular grade |
| Mild steel, also low-alloy irons and steels | 0 | 3 | 0 | 3 < 400<br>1 < 750 | Wrought, cast | Good | Good | 67 | 6.7 | A.I.S.I. type 430 |
| Stainless steel, ferritic 17% Cr type | 0 | 2 | 0 | 2 < 400 | Wrought, cast, clad | Good | Good§ | 78 | 6.0 | A.S.T.M. corrosion- and heat-resisting steels A.I.S.I. type 304 |
| Stainlesssteel, austenitic 18 Cr; 8 Ni types | 0 | 2 | 0 | 3 < 400 | Wrought, cast, clad | Good | Good | 90 | 9.6 | A.S.T.M. corrosion- and heat-resisting steels. Stabilized or ELC types used for welding |
| Stainless steel, austenitic 18 Cr; 12 Ni; 2.5 Mo | 0 | 3 | 2 | 4 < 400<br>3 < 750 | Wrought, cast, clad | Good | Good | 90 | 8.9 | A.I.S.I. type 316 A.S.T.M. corrosion- and heat-resisting steel. ELC type used for welding |
| Stainless steel, austenitic 20 Cr; 29 Ni; 2.5 Mo; 3.5 Cu type | 1 | 3 | 3 | 4 < 400<br>2 < 750 | Wrought, cast | Good | Good | 90 | 9.4 | A.C.I. CH-7M. Good resistance to sulfuric, phosphoric, and fatty acids at elevated temperatures |
| Ni-o-nel nickel-iron-chromium alloy (40 Ni; 21 Cr; 3 Mo; 1.5 Cu; bal. Fe) | 2 | 3 | 3 | 4 < 400<br>3 < 750 | Wrought, cast, clad | Good | Good | 100 | 7.3 | Special alloy with good resistance to sulfuric phosphoric, and fatty acids. Resistant to chlorides in some environments |
| Hastelloy alloy C[a] (55 Ni; 17 Mo;16 Cr; 6 Fe; 4 W) | 5 | 4 | 4 | 4 < 750<br>3 < 900 | Wrought, cast, clad | Fair | Good | 145 | 6.3 | Excellent resistance to wet chlorine gas and sodium hypochlorite solutions |
| Hastelloy alloy B[b] (61 Ni; 28 Mo; 6 Fe) | 1 | 3 | 5 | 4 < 750<br>3 < 900 | Wrought, cast, clad | Fair | Good | 135 | 5.6 | Resistant to solutions of hydrochloric and sulfuric acids |
| Hastelloy alloy D (82 Ni; 9 Si; 3 Cu) | 1 | 1 | 2 | 3 < 400<br>2 < 750<br>1 < 900 | Cast | No | § | 90–110 | 6.1 | Greatest application in hot concentrated solutions of sulfuric acid |
| Inconel nickel-chromium alloy (78 Ni; 15 Cr; 7 Fe) | 2 | 5 | 3 | 5 < 400<br>4 < 900 | Wrought, cast, clad | Good | Good | 90 | 8.9 | Wide application in food and pharmaceutical industries |

* Data courtesy of International Nickel Co.
† On unsuitable materials these media may promote potentially dangerous pitting.
‡‡ Temperatures are approximate.
§ Special precautions required.
¶ Many of these materials are suitable for resisting dry corrosion at elevated temperatures.

SOURCE: Perry, and Chilton (eds.). Chemical Engineers' Handbook, 5th ed., pp. 23-34 to 23-37, McGraw-Hill Book Company, New York, 1973.

a Also Chlorimet 3.
b Also Chlorimet 2.

## Table 3-36. General Corrosion Properties of Some Metals and Alloys* (Continued)

Ratings: 0 unsuitable. Not available in form required or not suitable for fabrication requirements or not suitable for corrosion conditions.
1 poor to fair.
2 fair. For mild conditions or where periodic replacement is possible. Restricted use.
3 fair to good.
4 good. Suitable when superior alternatives are uneconomic.
5 good to excellent.
6 normally excellent.
Small variations in service conditions may appreciably affect corrosion resistance. Choice of materials is therefore guided wherever possible by a combination of experience and laboratory and site tests.

| Material | Gases (Cont'd.) Halogens and derivatives | | | | Available forms | Cold form-ability in wrought and clad form | Weld-ability | Max. strength annealed condition ×1000 lb./sq. in. | Coeff. of thermal expansion, millionths per °F. 70°-212°F. | Remarks¶ |
|---|---|---|---|---|---|---|---|---|---|---|
| | Halogens | | Halide acids, moist, e.g., hydrochloric hydrolysis products of organic halides | Hydrogen halides, dry,‡ e.g., dry hydrogen chloride, °F. | | | | | | |
| | Moist, e.g., chlorine below dew point | Dry, e.g., fluorine above dew point | | | | | | | | |
| Copper-nickel alloys up to 30% nickel | 1 | 5 | 2 | 4 < 400<br>3 < 750 | Wrought, cast, clad | Good | Good | 38-62 | 9.3-8.5 | High-iron types excellent for resisting high-velocity effects in condenser tubes |
| Monel 400 nickel-copper alloy (66 Ni; 30 Cu; 2 Fe) | 2 | 6 | 3 | 4 < 400<br>3 < 750<br>2 < 900 | Wrought, cast, clad | Good | Good | 77 | 7.5 | Widely used for sulfuric acid pickling equipment. Also for propeller shafts in motor boats. Take precautions to avoid sulfur attack during fabrication |
| Alloy 505 nickel-copper cast alloy (66 Ni; 30 Cu; 4 Si) | 2 | 4 | 3 | 6 < 400<br>3 < 750<br>2 < 900 | Cast | No | No | 100 | 8.8 | Non-galling characteristics. Excellent for bearings or bushings. High strength developed by heat-treatment |
| Monel K-500 age hardenable Ni-Cu alloy (67 Ni; 30 Cu; 3 Al) | 2 | 6 | 3 | 6 < 400<br>3 < 750<br>2 < 900 | Wrought, cast | Fair | Good | 99-155 | 7.4 | High strength obtainable by heat-treatment. Take precautions to avoid sulfur attack during fabrication |
| A nickel—commercial (99.4 Ni) | 2 | 6 | 2 | 6 < 400<br>5 < 750<br>4 < 900 | Wrought, cast, clad | Good | Good | 54 | 6.6 | Widely used for hot concentrated caustic solutions. Take precautions to avoid sulfur attack during fabrication |

| Material | | | | Form | | | | Density | Remarks |
|---|---|---|---|---|---|---|---|---|---|
| Copper and silicon bronze | 0 | 5 | 2 | 3 < 400 2 < 750 | Wrought, cast, clad | Excellent | Fair | 29 | 9.3–9.5 | Unsuitable for hot concentrated mineral acids or for high-velocity HF |
| Aluminum brass (76 Cu; 22 Zn; 2 Al) | 0 | 4 | 2 | 2 < 400 | Wrought, cast | Good | Fair | 60 | 10.3 | May develop localized corrosion in sea water |
| Nickel-aluminum-bronze (80 Cu; 10 Al; 5 Ni; 5 Fe) | 0 | 4 | 3 | 3 < 400 2 < 750 | Wrought, cast | Good | Fair | 60–80 | 9.4 | Ship propellers an excellent application |
| Bronze, type A (88 Cu; 5 Sn; 5 Ni; 2 Zn) | 0 | 4 | 3 | 3 < 400 2 < 750 | Cast | No | § | 45 | 11.0 | High strengths obtainable by heat-treatment. Not susceptible to dezincification |
| Aluminum and its alloys | 0 | 6 | 0 | 3 < 400 1 < 750 | Wrought, cast, clad | Good | Good | 9–90 | 11.5–13.7 | Extent of corrosion dependent upon type and concentration of acidic ions. Wide range of mechanical properties obtainable by alloying and heat-treatment |
| Lead, chemical or antimonial | 0 | 1 | 3 | 0 | Wrought, cast, clad | Excellent | Good | 2 | 16.4–15.1 | High purity "chemical lead" preferred for most applications |
| Silver | 5 | 5 | 3 | 4 < 400 2 < 750 | Wrought, cast, clad | Excellent | Good | 21 | 10.6 | Used as a lining |
| Titanium | 6 | 0 | 1 | 0 | Wrought, cast | Fair | Good § | 6–90 | 5.0 | Red fuming HNO₃ may indicate explosions. |
| Zirconium | 6 | 1 | 6 | 0 | Wrought, cast | Fair | Good § | | | Good resistance to solutions containing chlorides |

* Data courtesy of International Nickel Co.
† On unsuitable materials these media may promote potentially dangerous pitting.
‡ Temperatures are approximate.
§ Special precautions required.
¶ Many of these materials are suitable for resisting dry corrosion at elevated temperatures.

SOURCE: Perry and Chilton (eds.), Chemical Engineers' Handbook, 5th ed, pp. 23-34 to 23-37, McGraw-Hill Book Company, New York, 1973.

a Also Chlorimet 3.
b Also Chlorimet 2.

## Table 3-37. Chemical Resistance of Important Plastics

Ratings are for Long-term Exposures at Ambient Temperatures (Less Than 100°F)

| | Poly-propylene poly-ethylene | CAB* | ABS† | PVC‡ | Saran§ | Polyester glass¶ | Epoxy glass | Phenolic asbestos | Fluoro-carbons | Chlorinated polyether (Penton) | Poly-carbonate |
|---|---|---|---|---|---|---|---|---|---|---|---|
| 10% $H_2SO_4$ | Excel. | Good | Excel. | Excel. | Excel. | Excel. | Excel. | Excel. | Excel. | Excel. | Excel. |
| 50% $H_2SO_4$ | Excel. | Poor | Excel. | Excel. | Excel. | Good | Excel. | Excel. | Excel. | Excel. | Excel. |
| 10% HCl | Excel. | Excel. | Excel. | Excel. | Excel. | Excel. | Excel. | Excel. | Excel. | Excel. | Excel. |
| 10% $HNO_3$ | Excel. | Poor | Good | Excel. | Excel. | Good | Good | Fair | Excel. | Excel. | Excel. |
| 10% Acetic | Excel. | Good | Excel. | Excel. | Excel. | Excel. | Excel. | Excel. | Excel. | Excel. | Excel. |
| 10% NaOH | Excel. | Fair | Excel. | Good | Fair | Fair | Excel. | Poor | Excel. | Excel. | Excel. |
| 50% NaOH | Excel. | Poor | Excel. | Excel. | Fair | Poor | Good | Poor | Excel. | Excel. | Excel. |
| $NH_4OH$ | Excel. | Poor | Excel. | Excel. | Poor | Fair | Excel. | Poor | Excel. | Excel. | Excel. |
| NaCl | Excel. | Excel. | Excel. | Excel. | Excel. | Excel. | Excel. | Excel. | Excel. | Excel. | Excel. |
| $FeCl_3$ | Excel. | Excel. | Excel. | Excel. | Excel. | Excel. | Excel. | Excel. | Excel. | Excel. | Excel. |
| $CuSO_4$ | Excel. | Excel. | Excel. | Excel. | Excel. | Excel. | Excel. | Excel. | Excel. | Excel. | Excel. |
| $NH_4NO_3$ | Excel. | Excel. | Excel. | Excel. | Excel. | Excel. | Excel. | Good | Excel. | Excel. | Excel. |
| Wet $H_2S$ | Excel. | Excel. | Excel. | Excel. | Excel. | Excel. | Excel. | Excel. | Excel. | Excel. | |
| Wet $Cl_2$ | Poor | Poor | Excel. | Good | Poor | Poor | Poor | Excel. | Excel. | Excel. | |
| Wet $SO_2$ | Excel. | Poor | Excel. | Excel. | Good | Excel. | Excel. | Excel. | Excel. | Excel. | |
| Gasoline | Poor | Excel. | Excel. | Excel. | Excel. | Good | Excel. | Excel. | Excel. | Excel. | Excel. |
| Benzene | Poor | Poor | Poor | Poor | Fair | Good | Excel. | Excel. | Excel. | Fair | Fair |
| $CCl_4$ | Poor | Poor | Poor | Fair | Fair | Excel. | Good | Excel. | Excel. | Fair | Poor |
| Acetone | Poor | Poor | Poor | Poor | Fair | Poor | Good | Poor | Excel. | Good | Good |
| Alcohol | Poor | Poor | Excel. | Excel. | Excel. | Excel. | Excel. | Excel. | Excel. | Excel. | Excel. |

* Cellulose acetate butyrate.
† Acrylonitrile butadiene styrene polymer.
‡ Polyvinyl chloride, type I.
§ Chemical resistance of Saran-lined pipe is superior to extruded Saran in some environments.
¶ Refers to general-purpose polyesters. Special polyesters have superior resistance, particularly in alkalies.
SOURCE: Perry and Chilton, "Chemical Engineer's Handbook," Sec. 23. Copyright © 1973, McGraw-Hill, Inc.

## MECHANICAL PROPERTIES
### Table 3-38. Chemical Ranges and Limits of AISI Carbon Steels

| AISI No. | Chemical composition limits, % | | | |
|---|---|---|---|---|
| | C | Mn | P, max. | S, max. |
| C 1008 | 0.10 max. | 0.25/0.50 | 0.040 | 0.050 |
| C 1010 | .08/0.13 | 0.30/0.60 | .040 | .050 |
| C 1011 | .08/0.13 | 0.60/0.90 | .040 | .050 |
| C 1012 | .10/0.15 | 0.30/0.60 | .040 | .050 |
| C 1015 | .13/0.18 | 0.30/0.60 | .040 | .050 |
| C 1016 | .13/0.18 | 0.60/0.90 | .040 | .050 |
| C 1017 | .15/0.20 | 0.30/0.60 | .040 | .050 |
| C 1018 | .15/0.20 | 0.60/0.90 | .040 | .050 |
| C 1019 | .15/0.20 | 0.70/1.00 | .040 | .050 |
| C 1020 | .18/0.23 | 0.30/0.60 | .040 | .050 |
| C 1021 | .18/0.23 | 0.60/0.90 | .040 | .050 |
| C 1022 | .18/0.23 | 0.70/1.00 | .040 | .050 |
| C 1023 | .20/0.25 | 0.30/0.60 | .040 | .050 |
| C 1024 | .19/0.25 | 1.35/1.65 | .040 | .050 |
| C 1025 | .22/0.28 | 0.30/0.60 | .040 | .050 |
| C 1026 | .22/0.28 | 0.60/0.90 | .040 | .050 |
| C 1027 | .22/0.29 | 1.20/1.50 | .040 | .050 |
| C 1029 | .25/0.31 | 0.60/0.90 | .040 | .050 |
| C 1030 | .28/0.34 | 0.60/0.90 | .040 | .050 |
| C 1031 | .28/0.34 | 0.30/0.60 | .040 | .050 |
| C 1033 | .30/0.36 | 0.70/1.00 | .040 | .050 |
| C 1035 | .32/0.38 | 0.60/0.90 | .040 | .050 |
| C 1036 | .30/0.37 | 1.20/1.50 | .040 | .050 |
| C 1037 | .32/0.38 | 0.70/1.00 | .040 | .050 |
| C 1038 | .35/0.42 | 0.60/0.90 | .040 | .050 |
| C 1039 | .37/0.44 | 0.70/1.00 | .040 | .050 |
| C 1040 | .37/0.44 | 0.60/0.90 | .040 | .050 |
| C 1041 | .36/0.44 | 1.35/1.65 | .040 | .050 |
| C 1042 | .40/0.47 | 0.60/0.90 | .040 | .050 |
| C 1043 | .40/0.47 | 0.70/1.00 | .040 | .050 |
| C 1045 | .43/0.50 | 0.60/0.90 | .040 | .050 |
| C 1046 | .43/0.50 | 0.70/1.00 | .040 | .050 |
| C 1049 | .46/0.53 | 0.60/0.90 | .040 | .050 |
| C 1050 | .48/0.55 | 0.60/0.90 | .040 | .050 |
| C 1051 | .45/0.56 | 0.85/1.15 | .040 | .050 |
| C 1052 | .47/0.55 | 1.20/1.50 | .040 | .050 |
| C 1053 | .48/0.55 | 0.70/1.00 | .040 | .050 |
| C 1055 | .50/0.60 | 0.60/0.90 | .040 | .050 |
| C 1060 | .55/0.65 | 0.60/0.90 | .040 | .050 |
| C 1070 | .65/0.75 | 0.60/0.90 | .040 | .050 |
| C 1078 | .72/0.85 | 0.30/0.60 | .040 | .050 |
| C 1080 | .75/0.88 | 0.60/0.90 | .040 | .050 |
| C 1084 | .80/0.93 | 0.60/0.90 | .040 | .050 |
| C 1085 | .80/0.93 | 0.70/1.00 | .040 | .050 |
| C 1086 | .80/0.93 | 0.30/0.50 | .040 | .050 |
| C 1090 | .85/0.98 | 0.60/0.90 | .040 | .050 |
| C 1095 | .90/1.03 | 0.30/0.50 | .040 | .050 |

## Table 3-38. References

N. E. Woldman and R. C. Gibbons (Eds.), "Engineering Alloys," Van Nostrand Reinhold, 5th ed., 1427 pp.,1973 list some 36,000 alloys with various mechanical properties, including tensile strength, yield point, percentage elongation, percentage reduction of area, Brinell hardness number, etc.    W. Gardner (Ed.), "Chemical Synonyms and Trade Names," 6th ed., 635 pp., CRC Press, 1968 and Zimmerman and Levine, "Handbook of Material Trade Names," basic volume plus four supplemental volumes, Industrial Research Service Inc., Dover, N.H., 1953–1965 may also be found useful.    For a listing of Soviet alloys, the Defense Intelligence Agency Report ST-HB-01-1-65-INT(AD822518), 239 pp., 1966 is recommended.    The following compendia, etc., may also be found useful: S. L. Hoyt (Ed.), Metal Properties, "ASME Handbook," 440 pp:, McGraw-Hill Book Company, New York, 1954; T. Lyman (Ed.), "Metals Handbook, vol. I, Properties and Selection of Metals," 8th ed., American Society for Metals, Metals Park, Ohio, 1961; N. I. Sax, "Dangerous Properties of Industrial Materials," 3d ed., 1251 pp., Reinhold Publishing Corp., New York, 1968 (includes some general property information); C. J. Smithells, "Metals Reference Book," 2d ed., 2 vols., 984 pp., Butterworth Scientific Publications, London, 1955 (see especially vol. 2); R. Hultgren, P. D. Desai, et al., "Selected Values of the Thermodynamic Properties of the Elements," 636 pp., American Society for Metals, Metals Park, Ohio, 1973; R. Hultgren, P. D. Desai, et al., "Selected Values of the Thermodynamic Properties of Binary Alloys," 1435 pp., American Society for Metals, Metals Park, Ohio, 1973; W. B. Pearson, "Handbook of Lattice Spacings and Structures of Metals and Alloys," 1446 pp., Pergamon Press, New York, 1967.

## Table 3-39. Chemical Ranges and Limits of AISI Standard Alloys

Open-hearth and electric-furnace alloy steels, bars, billets, blooms, and slabs
Ranges and limits apply to steel not exceeding 200 in.[2] in cross-sectional area.

| AISI No. | Chemical composition ranges and limits, % | | | | | | | |
| --- | --- | --- | --- | --- | --- | --- | --- | --- |
| | C | Mn | P, max. | S, max. | Si | Ni | Cr | Mo |
| 1330 | 0.28/0.33 | 1.60/1.90 | 0.040 | 0.040 | 0.20/0.35 | | | |
| 1335 | .33/0.38 | 1.60/1.90 | .040 | .040 | .20/0.35 | | | |
| 1340 | .38/0.43 | 1.60/1.90 | .040 | .040 | .20/0.35 | | | |
| 1345 | .43/0.48 | 1.60/1.90 | .040 | .040 | .20/0.35 | | | |
| 3140 | .38/0.43 | 0.70/0.90 | .040 | .040 | .20/0.35 | 1.10/1.40 | 0.55/0.75 | |
| E3310 | .08/0.13 | .45/0.60 | .025 | .025 | .20/0.35 | 3.25/3.75 | 1.40/1.75 | |
| 4012 | .09/0.14 | .75/1.00 | .040 | .040 | .20/0.35 | ........ | ........ | 0.15/0.25 |
| 4023 | .20/0.25 | .70/0.90 | .040 | .040 | .20/0.35 | ........ | ........ | .20/0.30 |
| 4024 | .20/0.25 | .70/0.90 | .040 | .035/.050 | .20/0.35 | ........ | ........ | .20/0.30 |
| 4027 | .25/0.30 | .70/0.90 | .040 | .040 | .20/0.35 | ........ | ........ | .20/0.30 |
| 4028 | .25/0.30 | .70/0.90 | .040 | .035/.050 | .20/0.35 | ........ | ........ | .20/0.30 |
| 4037 | .35/0.40 | .70/0.90 | .040 | .040 | .20/0.35 | ........ | ........ | .20/0.30 |
| 4042 | .40/0.45 | .70/0.90 | .040 | .040 | .20/0.35 | ........ | ........ | .20/0.30 |
| 4047 | .45/0.50 | .70/0.90 | .040 | .040 | .20/0.35 | ........ | ........ | .20/0.30 |
| 4063 | .60/0.67 | .75/1.00 | .040 | .040 | .20/0.35 | ........ | ........ | .20/0.30 |
| 4118 | .18/0.23 | .70/0.90 | .040 | .040 | .20/0.35 | ........ | 0.40/0.60 | .08/0.15 |
| 4130 | .28/0.33 | .40/0.60 | .040 | .040 | .20/0.35 | ........ | .80/1.10 | .15/0.25 |
| 4135 | .33/0.38 | .70/0.90 | .040 | .040 | .20/0.35 | ........ | .80/1.10 | .15/0.25 |
| 4137 | .35/0.40 | .70/0.90 | .040 | .040 | .20/0.35 | ........ | .80/1.10 | .15/0.25 |
| 4140 | .38/0.43 | .75/1.00 | .040 | .040 | .20/0.35 | ........ | .80/1.10 | .15/0.25 |
| 4142 | .40/0.45 | .75/1.00 | .040 | .040 | .20/0.35 | ........ | .80/1.10 | .15/0.25 |
| 4145 | .43/0.48 | .75/1.00 | .040 | .040 | .20/0.35 | ........ | .80/1.10 | .15/0.25 |
| 4147 | .45/0.50 | .75/1.00 | .040 | .040 | .20/0.35 | ........ | .80/1.10 | .15/0.25 |
| 4150 | .48/0.53 | .75/1.00 | .040 | .040 | .20/0.35 | ........ | .80/1.10 | .15/0.25 |
| 4320 | .17/0.22 | .45/0.65 | .040 | .040 | .20/0.35 | 1.65/2.00 | .40/0.60 | .20/0.30 |
| 4337 | .35/0.40 | .60/0.80 | .040 | .040 | .20/0.35 | 1.65/2.00 | .70/0.90 | .20/0.30 |
| E4337 | .35/0.40 | .65/0.85 | .025 | .025 | .20/0.35 | 1 65/2.00 | .70/0.90 | .20/0.30 |
| 4340 | .38/0.43 | .60/0.80 | .040 | .040 | .20/0.35 | 1.65/2.00 | .70/0.90 | .20/0.30 |
| E4340 | .38/0.43 | .65/0.85 | .025 | .025 | .20/0.35 | 1.65/2.00 | .70/0.90 | .20/0.30 |
| 4422 | .20/0.25 | .70/0.90 | .040 | .040 | .20/0.35 | ........ | ........ | .35/0.45 |
| 4427 | .24/0.29 | .70/0.90 | .040 | .040 | .20/0.35 | ........ | ........ | .35/0.45 |
| 4520 | .18/0.23 | .45/0.65 | .040 | .040 | .20/0.35 | ........ | ........ | .45/0.60 |
| 4615 | .13/0.18 | .45/0.65 | .040 | .040 | .20/0.35 | 1.65/2.00 | ........ | .20/0.30 |
| 4617 | .15/0.20 | .45/0.65 | .040 | .040 | .20/0.35 | 1.65/2.00 | ........ | .20/0.30 |
| 4620 | .17/0.22 | .45/0.65 | .040 | .040 | .20/0.35 | 1.65/2.00 | ........ | .20/0.30 |
| 4621 | .18/0.23 | .70/0.90 | .040 | .040 | .20/0.35 | 1.65/2.00 | ........ | .20/0.30 |
| 4718 | .16/0.21 | .70/0.90 | .040 | .040 | .20/0.35 | 0.90/1.20 | .35/0.55 | .30/0.40 |
| 4720 | .17/0.22 | .50/0.70 | .040 | .040 | .20/0.35 | 0.90/1.20 | .35/0.55 | .15/0.25 |
| 4815 | .13/0.18 | .40/0.60 | .040 | .040 | .20/0.35 | 3.25/3.75 | ........ | .20/0.30 |
| 4817 | .15/0.20 | .40/0.60 | .040 | .040 | .20/0.35 | 3.25/3.75 | ........ | .20/0.30 |
| 4820 | .18/0.23 | .50/0.70 | .040 | .040 | .20/0.35 | 3.25/3.75 | ........ | .20/0.30 |
| 5015 | .12/0.17 | .30/0.50 | .040 | .040 | .20/0.35 | ........ | .30/0.50 | |
| 5046 | .43/0.50 | .75/1.00 | .040 | .040 | .20/0.35 | ........ | .20/0.35 | |
| 5115 | .13/0.18 | .70/0.90 | .040 | .040 | .20/0.35 | ........ | .70/0.90 | |
| 5120 | .17/0.22 | .70/0.90 | .040 | .040 | .20/0.35 | ........ | .70/0.90 | |
| 5130 | .28/0.33 | .70/0.90 | .040 | .040 | .20/0.35 | ........ | .80/1.10 | |
| 5132 | .30/0.35 | .60/0.80 | .040 | .040 | .20/0.35 | ........ | .75/1.00 | |
| 5135 | .33/0.38 | .60/0.80 | .040 | .040 | .20/0.35 | ........ | .80/1.05 | |
| 5140 | .38/0.43 | .70/0.90 | .040 | .040 | .20/0.35 | ........ | .70/0.90 | |
| 5145 | .43/0.48 | .70/0.90 | .040 | .040 | .20/0.35 | ........ | .70/0.90 | |

## Table 3-39. Chemical Ranges and Limits of AISI Standard Alloys (Continued)

| AISI No. | Chemical composition ranges and limits, % | | | | | | | |
|---|---|---|---|---|---|---|---|---|
| | C | Mn | P, max. | S, max. | Si | Ni | Cr | Mo |
| 5147 | 0.45/0.52 | 0.70/0.95 | 0.040 | 0.040 | 0.20/0.35 | ........ | 0.85/1.15 | |
| 5150 | .48/0.53 | .70/0.90 | .040 | .040 | 0.20/0.35 | ........ | 0.70/0.90 | |
| 5155 | .50/0.60 | .70/0.90 | .040 | .040 | 0.20/0.35 | ........ | 0.70/0.90 | |
| 5160 | .55/0.65 | .75/1.00 | .040 | .040 | 0.20/0.35 | ........ | 0.70/0.90 | |
| E50100 | .95/1.10 | .25/0.45 | .025 | .025 | 0.20/0.35 | ........ | 0.40/0.60 | |
| E51100 | .95/1.10 | .25/0.45 | .025 | .025 | 0.20/0.35 | ........ | 0.90/1.15 | |
| E52100 | .95/1.10 | .25/0.45 | .025 | .025 | 0.20/0.35 | ........ | 1.30/1.60 | |
| | | | | | | | | V |
| 6118 | .16/0.21 | .50/0.70 | .040 | .040 | 0.20/0.35 | ........ | 0.50/0.70 | .10/0.15 |
| 6120 | .17/0.22 | .70/0.90 | .040 | .040 | 0.20/0.35 | ........ | 0.70/0.90 | .10 min. |
| 6150 | .48/0.53 | .70/0.90 | .040 | .040 | 0.20/0.35 | ........ | 0.80/1.10 | .15 min. |
| | | | | | | | | Mo |
| 8115 | .13/0.18 | .70/0.90 | .040 | .040 | 0.20/0.35 | 0.20/0.40 | 0.30/0.50 | .08/0.15 |
| 8615 | .13/0.18 | .70/0.90 | .040 | .040 | 0.20/0.35 | 0.40/0.70 | 0.40/0.60 | .15/0.25 |
| 8617 | .15/0.20 | .70/0.90 | .040 | .040 | 0.20/0.35 | 0.40/0.70 | 0.40/0.60 | .15/0.25 |
| 8620 | .18/0.23 | .70/0.90 | .040 | .040 | 0.20/0.35 | 0.40/0.70 | 0.40/0.60 | .15/0.25 |
| 8622 | .20/0.25 | .70/0.90 | .040 | .040 | 0.20/0.35 | 0.40/0.70 | 0.40/0.60 | .15/0.25 |
| 8625 | .23/0.28 | .70/0.90 | .040 | .040 | 0.20/0.35 | 0.40/0.70 | 0.40/0.60 | .15/0.25 |
| 8627 | .25/0.30 | .70/0.90 | .040 | .040 | 0.20/0.35 | 0.40/0.70 | 0.40/0.60 | .15/0.25 |
| 8630 | .28/0.33 | .70/0.90 | .040 | .040 | 0.20/0.35 | 0.40/0.70 | 0.40/0.60 | .15/0.25 |
| 8637 | .35/0.40 | .75/1.00 | .040 | .040 | 0.20/0.35 | 0.40/0.70 | 0.40/0.60 | .15/0.25 |
| 8640 | .38/0.43 | .75/1.00 | .040 | .040 | 0.20/0.35 | 0.40/0.70 | 0.40/0.60 | .15/0.25 |
| 8642 | .40/0.45 | .75/1.00 | .040 | .040 | 0.20/0.35 | 0.40/0.70 | 0.40/0.60 | .15/0.25 |
| 8645 | .43/0.48 | .75/1.00 | .040 | .040 | 0.20/0.35 | 0.40/0.70 | 0.40/0.60 | .15/0.25 |
| 8650 | .48/0.53 | .75/1.00 | .040 | .040 | 0.20/0.35 | 0.40/0.70 | 0.40/0.60 | .15/0.25 |
| 8655 | .50/0.60 | .75/1.00 | .040 | .040 | 0.20/0.35 | 0.40/0.70 | 0.40/0.60 | .15/0.25 |
| 8660 | .55/0.65 | .75/1.00 | .040 | .040 | 0.20/0.35 | 0.40/0.70 | 0.40/0.60 | .15/0.25 |
| 8720 | .18/0.23 | .70/0.90 | .040 | .040 | 0.20/0.35 | 0.40/0.70 | 0.40/0.60 | .20/0.30 |
| 8735 | .33/0.38 | .75/1.00 | .040 | .040 | 0.20/0.35 | 0.40/0.70 | 0.40/0.60 | .20/0.30 |
| 8740 | .38/0.43 | .75/1.00 | .040 | .040 | 0.20/0.35 | 0.40/0.70 | 0.40/0.60 | .20/0.30 |
| 8742 | .40/0.45 | .75/1.00 | .040 | .040 | 0.20/0.35 | 0.40/0.70 | 0.40/0.60 | .20/0.30 |
| 8822 | .20/0.25 | .75/1.00 | .040 | .040 | 0.20/0.35 | 0.40/0.70 | 0.40/0.60 | .30/0.40 |
| 9255 | .05/0.60 | .70/0.95 | .040 | .040 | 1.80/2.20 | | | |
| 9260 | .55/0.65 | .70/1.00 | .040 | .040 | 1.80/2.20 | | | |
| 9262 | .55/0.65 | .75/1.00 | .040 | .040 | 1.80/2.20 | ........ | 0.25/0.40 | |
| E9310 | .08/0.13 | .45/0.65 | .025 | .025 | 0.20/0.35 | 3.00/3.50 | 1.00/1.40 | .08/0.15 |
| 9840 | .38/0.43 | .70/0.90 | .040 | .040 | 0.20/0.35 | 0.85/1.15 | 0.70/0.90 | .20/0.30 |
| 9850 | .48/0.53 | .70/0.90 | .040 | .040 | 0.20/0.35 | 0.85/1.15 | 0.70/0.90 | .20/0.30 |

NOTES:

1. Grades shown with prefix letter E generally are manufactured by the basic electric-furnace process. All others are normally manufactured by the basic open-hearth process, but may be manufactured by the basic electric-furnace process with adjustments in phosphorus and sulfur.

2. The phosphorus and sulfur limitations for each process are as follows:

| | |
|---|---|
| Basic electric furnace | 0.025 max. % |
| Basic open hearth | .040 max. % |
| Acid electric furnace | .050 max. % |
| Acid open hearth | .050 max. % |

3. Minimum silicon limit for acid open-hearth or acid electric-furnace alloy steel is 0.15 per cent.

4. Small quantities of certain elements are present in alloy steels which are not specified or required. These elements are considered as incidental, and may be present to the following maximum amounts: copper, 0.35 per cent; nickel, 0.25 per cent; chromium, 0.20 per cent; and molybdenum, 0.06 per cent.

5. Where minimum and maximum sulfur content is shown, it is indicative of resulfurized steels.

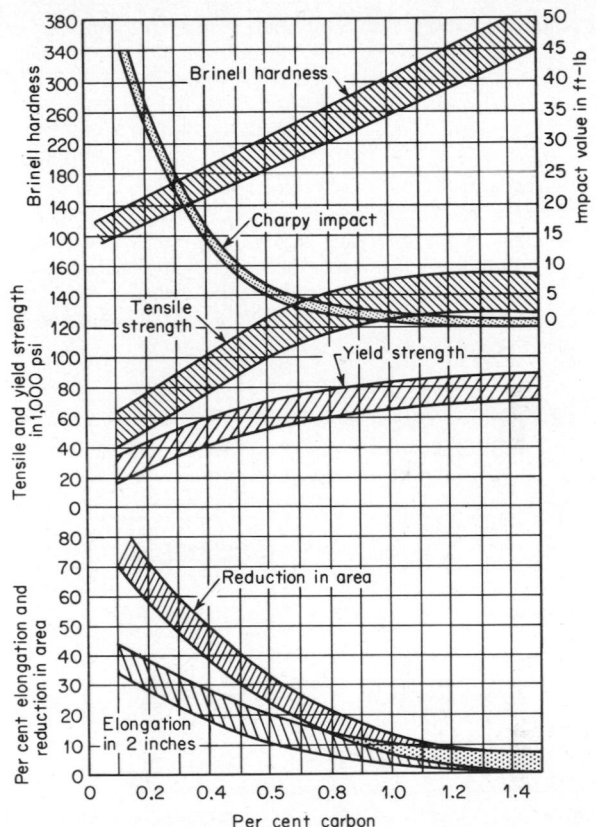

FIG. 3-14. Effect of carbon content on mechanical properties of hot-worked steels. (After F. T. Sisco, *Alloys of Iron and Carbon*, vol. 2, McGraw-Hill Book Company, New York.) Properties of annealed and normalized steels would be approximately the same as hot-worked steels.

Caution should be exercised in design under any extreme conditions. While the effect of pressure is easily visualized, the embrittling effect of low temperature is not so well recognized. Some metals even exhibit the effect at temperatures near the ice point. Low-temperature toughness can be measured conveniently by a notched-bar impact test. The data so obtained are sensitive to a number of structural and compositional variables, giving wide spreads in the measured parameters. Low-temperature design should therefore be based on as complete a set of input information as possible and should also allow generous safety factors. An alternative course would be to employ materials that do not become brittle. The concise discussion on pages 322–323 of R. B. Scott, "Cryogenic Engineering," D. Van Nostrand, Co., Inc., Princeton, 1959 is recommended as one source for further information.

### Table 3-40. Typical Mechanical Properties of Some AISI Steels with Various Heat-treatments

Sections up to 1½ in. Diam or Thickness

| Draw temp, °F | Tensile strength, kpsi | Yield strength, kpsi | Reduction of area, % | Elong. in 2 in., % | Brinell hardness | Tensile strength, kpsi | Yield strength, kpsi | Reduction of area, % | Elong. in 2 in., % | Brinell hardness |
|---|---|---|---|---|---|---|---|---|---|---|
| | AISI C 1040 quenched in water at 1500°F | | | | | AISI 1340 normalized at 1585°F, quenched in oil at 1550°F | | | | |
| 600 | 125 | 104 | 46 | 11 | 260 | 227 | 206 | 43 | 11 | 448 |
| 800 | 119 | 91 | 53 | 13 | 250 | 181 | 166 | 51 | 13 | 372 |
| 1000 | 110 | 78 | 58 | 15 | 220 | 140 | 121 | 58 | 17.5 | 297 |
| 1100 | 108 | 71 | 60 | 17 | 216 | 125 | 103 | 62 | 20 | 270 |
| 1200 | 104 | 66 | 62 | 20 | 210 | 115 | 88 | 65 | 23 | 250 |
| 1300 | 98 | 60 | 64 | 22 | 205 | 110 | 78 | 68 | 25.5 | 234 |
| | AISI 2340 normalized at 1600°F, quenched in oil at 1425°F | | | | | AISI 3140 normalized at 1600°F, quenched in oil at 1500°F | | | | |
| 600 | 222 | 205 | 43 | 11 | 437 | 228 | 209 | 42 | 11 | 448 |
| 800 | 180 | 165 | 50 | 14 | 372 | 187 | 168 | 51 | 13 | 372 |
| 1000 | 139 | 122 | 58 | 19 | 297 | 140 | 128 | 59 | 18 | 352 |
| 1100 | 121 | 108 | 62 | 22 | 270 | 125 | 112 | 63 | 21.5 | 332 |
| 1200 | 110 | 95 | 65 | 25 | 250 | 112 | 100 | 66 | 24 | 297 |
| 1300 | 99 | 85 | 67 | 27 | 240 | 105 | 90 | 68 | 27.5 | 283 |

## Table 3-40. Typical Mechanical Properties of Some AISI Steels with Various Heat-treatments (Continued)

| Draw temp, °F | Tensile strength, kpsi | Yield strength, kpsi | Reduction of area, % | Elong. in 2 in., % | Brinell hardness | Tensile strength, kpsi | Yield strength, kpsi | Reduction of area, % | Elong. in 2 in., % | Brinell hardness |
|---|---|---|---|---|---|---|---|---|---|---|
| AISI 4042 normalized at 1600°F, quenched in oil at 1500°F | | | | | | AISI 4140 normalized at 1600°F, quenched in oil at 1500°F | | | | |
| 600 | 231 | 210 | 41 | 12 | 448 | 225 | 208 | 42.5 | 10 | 426 |
| 800 | 175 | 158 | 50 | 14 | 372 | 180 | 163 | 49 | 13 | 372 |
| 1000 | 140 | 125 | 58 | 19 | 297 | 135 | 120 | 57 | 18 | 283 |
| 1100 | 125 | 110 | 62 | 23 | 260 | 120 | 105 | 61 | 20 | 250 |
| 1200 | 113 | 99 | 65 | 26 | 234 | 108 | 195 | 62 | 22.5 | 228 |
| 1300 | 105 | 92 | 68 | 30 | 210 | 100 | 88 | 63 | 25 | 216 |
| AISI 4340 normalized at 1600°F, quenched in oil at 1525°F | | | | | | AISI 4640 normalized at 1600°F, quenched in oil at 1500°F | | | | |
| 600 | 250 | 230 | 40 | 9 | 484 | 225 | 208 | 43 | 11 | 448 |
| 800 | 211 | 200 | 44 | 10 | 426 | 182 | 168 | 50 | 13 | 372 |
| 1000 | 173 | 160 | 52 | 12.5 | 352 | 141 | 125 | 58 | 19 | 283 |
| 1100 | 158 | 140 | 56 | 15 | 313 | 125 | 109 | 62 | 22.5 | 260 |
| 1200 | 140 | 123 | 60 | 18 | 283 | 110 | 93 | 64 | 26 | 234 |
| 1300 | 123 | 108 | 63 | 22.5 | 250 | 100 | 81 | 66 | 27.5 | 222 |
| AISI 5140 normalized at 1575°F, quenched in oil at 1500°F | | | | | | AISI 8640 normalized at 1600°F, quenched in oil at 1525°F | | | | |
| 600 | 232 | 211 | 43 | 11 | 448 | 240 | 220 | 42 | 10 | 472 |
| 800 | 190 | 162 | 49 | 12.5 | 372 | 202 | 188 | 45 | 12.5 | 415 |
| 1000 | 140 | 124 | 58 | 17.5 | 283 | 165 | 148 | 53 | 16 | 332 |
| 1100 | 123 | 108 | 62 | 21 | 250 | 145 | 130 | 57 | 18 | 297 |
| 1200 | 110 | 95 | 65 | 24 | 228 | 130 | 113 | 61 | 21 | 283 |
| 1300 | 100 | 88 | 68 | 29 | 210 | 116 | 100 | 63 | 22.5 | 250 |

SOURCE: T. Baumeister (ed.), "Standard Handbook for Mechanical Engineers," 7th ed., p. 6-36, McGraw-Hill Book Company, 1967. Used by permission of McGraw-Hill Book Company.

# Table 3-41. Properties of Metals and Alloys

| Material | Nominal composition (essential elements), % | Form and condition | Typical mechanical properties | | | | Typical physical constants | | | | | | | |
|---|---|---|---|---|---|---|---|---|---|---|---|---|---|---|
| | | | Yield strength (0.2% offset), 1000 lb./sq. in. | Tensile strength, 1000 lb./sq. in. | Elongation in 2 in., % | Hardness, Brinell | Density, lb./cu. in. | Specific gravity | Melting point, °F. | Specific heat (32°-212°F.), B.t.u./(lb.)(°F.) | Thermal expansion coefficient (32°-212°F.), $\times 10^{-6}$ in./(in.)(°F.) | Thermal conductivity (32°-212°F.), B.t.u./(sq. ft.)(hr.)(°F./in.) | Electrical resistivity (68°F.), ohms/cir. mil ft. | Tensile modulus of elasticity $\times 10^6$ lb./sq. in. |
| **Low-alloy Irons and Steels** | | | | | | | | | | | | | | |
| Carbon steel[a] A.I.S.I.-S.A.E. 1020 | Fe bal., Mn 0.45, Si 0.25, C 0.20 | Annealed<br>Hot-rolled<br>Hardened (water quench 1000°F. temper) | 38<br>42<br>62 | 65<br>68<br>90 | 30<br>32<br>25 | 130<br>135<br>179 | 0.284 | 7.86 | 2760 | 0.107 | 6.7 | 360 | 60 | 30 |
| 300-M[b] | C 0.43, Mn 0.80, Si 1.60, Ni 1.85, Cr 0.85, Mo 0.38, V 0.08 | Hardened (oil quench, 600°F. temper) | 240 | 290 | 10. | 535 | 0.283 | 7.84 | 2740 | 0.107 | 6.5 | 400 | 70 | 30 |
| Wrought iron | Fe bal., Slag 2.5 | Hot-rolled | 30 | 48 | 30 (in 8 in.) | 100 | 0.278 | 7.70 | 2750 | 0.11 | 6.35 | 418 | 70 | 29 |
| Ingot iron | Fe 99.9 plus | Hot-rolled<br>Annealed | 29<br>19 | 45<br>38 | 26<br>45 | 90<br>67 | 0.284 | 7.86 | 2795 | 0.108 | 6.8 | 490 | 57 | 30.1 |
| Cast gray iron | C 3.4, Si 1.8, Mn 0.5, Fe bal. | Cast (as cast) | ...... | 25 min. | 0.5 max. | 180 | 0.260 | 7.20 | 2150 | ...... | 6.7 | 310 | 400 | 13 ±1.5 |
| Malleable iron | C 2.5, Si 1, Mn 0.55 max | Cast (annealed) | 33 | 52 | 12 | 130 | 0.264 | 7.32 | 2250 | 0.122 | 6.6 | ...... | 180 | 25 |
| Ni-Tensyliron | C 2.7, Si 1.8, Mn 0.8, Ni 2.3, Cr 0.3, Fe bal., Mo 0.4 | Cast (as cast)<br>Cast (heat-treat) | 30<br>40 | 60<br>80 | ......<br>...... | 260<br>390 | 0.260 | 7.20 | 2150 | ...... | 6.5 | 320 | ...... | 20 ±1.5 |

| Material | Condition | | | | | | | | | | | | |
|---|---|---|---|---|---|---|---|---|---|---|---|---|---|
| Ductile iron (Mg-containing)<br>C 3.4, Si 2.5, Mn 0.40, P 0.1 max., Ni 0–1, Mg 0.06, Fe bal. | Cast (as cast)<br>Cast (quench, temper) | 53<br>68<br>108 | 70<br>90<br>135 | 18<br>7<br>5 | 170<br>235<br>310 | 0.26 | 7.2 | 2100 | ..... | 7.5 | 228 | 360 | 25 |
| Ductile iron (Mg-containing) (heat-resistant)<br>C 3.3, Si 4.3, Mn 0.4, P 0.1 max., Ni 0–1, Mg 0.06, Fe bal. | Cast (annealed, as cast) | 60 | 80 | 10 | 220 | 0.26 | 7.2 | 2100 | ..... | ..... | 200 | 437 | |
| Ni-Resist ductile iron (Mg-containing)<br>C 2.8, Si 2.5, Mn 1, P 0.2 max., Ni 20, Cr 2, Mg 0.1, Fe bal. | Cast (as cast) | 35 | 60 | 10 | 175 | 0.268 | 7.4 | 2250 | ..... | 10.4 | ..... | 610 | 18.5 |
| Ni-Hard type 2<br>C 2.7, Si 0.6, Mn 0.5, Ni 4.5, Cr 2.0, Fe bal. | Sand-cast<br>Chill-cast (temper) | .....<br>..... | 55<br>75 | .....<br>..... | 550<br>625 | 0.275 | 7.70 | 2150 | ..... | 4.8 | 99 | ..... | 25 |
| Ni-Hard type 1<br>C 3.5, Si 0.6, Mn 0.5, Ni 4.5, Cr 2.0, Fe bal. | Sand-cast<br>Chill-cast (temper) | .....<br>..... | 40<br>50 | .....<br>..... | 600<br>700 | 0.275 | 7.70 | 2150 | ..... | 4.8 | 99 | ..... | 25 |
| Ni-Resist type 1<br>C 2.8, Si 2.0, Mn 1.2, Ni 15.5, Cr 2.5, Cu 6.5, Fe bal. | Cast (as cast) | ..... | 27 | 2 | 150 | 0.264 | 7.3 | 2250 | 0.110 | 10.3 | 276 | 842 | 13.0 ±1.5 |
| Ni-Resist type 2<br>C 2.8, Si 2.0, Mn 1.0, Ni 20.0, Cr 2.5, Fe bal. | Cast (as cast) | ..... | 27 | 2 | 140 | 0.264 | 7.3 | 2250 | 0.116 | 9.6 | 276 | 1023 | 15.6 ±1.5 |
| Ni-Resist type 3<br>C 2.6 max, Si 1.5, Mn 0.6, Ni 30.0, Cr 3.5, Fe bal. | Cast | ..... | 30 | 2 | 140 | 0.268 | 7.4 | 2250 | 0.111 | 5.3 | 273 | ..... | 15.2 ±1.5 |
| Ni-Resist type 4<br>C 2.6 max, Si 5.5, Mn 0.6, Ni 30.0, Cr 5.5, Fe bal. | Cast | ..... | 30 | 2 | 180 | 0.268 | 7.4 | 2200 | 0.120 | 7.3 | 261 | 962 | 15.0 ±1.5 |

[a] Carbon-steel mechanical properties are strongly influenced by the carbon level; the physical properties will not be appreciably changed.
[b] Type of ultra-high-strength steel, attaining tensile strengths of 220,000 to 300,000 lb./sq. in.

# Table 3-41. Properties of Metals and Alloys (Continued)

| Material | Nominal composition (essential elements), % | Form and condition | Typical mechanical properties | | | | Typical physical constants | | | | | | | |
|---|---|---|---|---|---|---|---|---|---|---|---|---|---|---|---|
| | | | Yield strength (0.2% offset), 1000 lb./sq. in. | Tensile strength, 1000 lb./sq. in. | Elongation in 2 in., % | Hardness, Brinell | Density, lb./cu. in. | Specific gravity | Melting point, °F. | Specific heat (32°-212°F.), B.t.u./(lb.)(°F.) | Thermal expansion coefficient (32°-212°F.) $\times 10^{-6}$ in./(in.)(°F.) | Thermal conductivity (32°-212°F.) B.t.u./(sq. ft.)(hr.)(°F./in.) | Electrical resistivity (68°F.) ohms/cir. mil ft. | Tensile modulus of elasticity $\times 10^6$ lb./sq. in. |
| Ni-Resist type D-2 | C 2.90 max, Si 2.5, Mn 1.0, P 0.2 max, Ni 20.0, Cr 2.5 | Cast | 34 | 62 | 14 | 160 | 0.268 | 7.41 | 2250 | ..... | 10.4 | 93 | 614 | 17.5 ±1.5 |
| Ni-Resist type D-3 | C 2.60 max, Si 2.5, Mn 0.6, P 0.2 max, Ni 30.0, Cr 3.5 | Cast | 35 | 61 | 12 | 150 | 0.27 | 7.45 | 2250 | ..... | 7.0 | ..... | ..... | 14.0 ±1.5 |
| Ni-Resist type D-4 | C 2.60 max, Si 5.5, Mn 0.6, P 0.2 max, Ni 30.0, Cr 5.5 | Cast | 41 | 66 | 2.5 | 190 | 0.27 | 7.45 | 2200 | ..... | 8.0 | ..... | ..... | 13.0 ±1.5 |
| **Wrought Stainless Steels** | | | | | | | | | | | | | | |
| Stainless steel type 201 | C 0.15 max, Mn 5.5-7.5, Cr 16.0-18.0, Ni 3.5-5.5, N 0.25 max. | Mill-annealed strip | 50 | 115 | 60 | 194 | 0.28 | 7.7 | 2550-2650 | 0.12 | ..... | 113 | 414 | 28.6 |
| Stainless steel type 202 | C 0.15 max, Mn 7.5-10.0, Cr 17.0-19.0, Ni 4.0-6.0, N 0.25 max. | Mill-annealed strip | 50 | 100 | 60 | 184 | 0.28 | 7.7 | 2550-2650 | 0.12 | ..... | 113 | 414 | 28.6 |

| Material | Composition | Condition | | | | | | | | | | | |
|---|---|---|---|---|---|---|---|---|---|---|---|---|---|
| Stainless steel type 301 | Fe bal., Cr 17, Ni 7, C 0.08–0.20 | Annealed | 30 | 100 | 72 | 160 | 0.29 | 8.02 | 2550–2590 | 0.12 | 9.4 | 112.8 | 435 | 28 |
| | | Cold-rolled^c | up to 165 | up to 200 | 15^d | 385 | | | | | | | | |
| Stainless steel type 302 | Fe bal., Cr 18, Ni 8, C 0.08–0.20 | Annealed | 30 | 90 | 60 | 160 | 0.29 | 8.02 | 2550–2590 | 0.12 | 9.6 | 112.8 | 435 | 28 |
| | | Cold-rolled^c | up to 165 | up to 190 | 8^d | up to 400 | | | | | | | | |
| Stainless steel type 304 | Fe bal., Cr 19, Ni 9.0, C 0.08 max. | Annealed | 30 | 85 | 62 | 160 | 0.29 | 8.02 | 2550–2650 | 0.12 | 9.6 | 113 | 435 | 28 |
| | | Cold-rolled^d | up to 160 | up to 185 | 8^d | up to 400 | | | | | | | | |
| Stainless steel type 304L | Fe bal., Cr 19, Ni 10, C 0.03 max. | Annealed | 30 | 80 | 60 | 150 | 0.29 | 8.02 | 2550–2650 | 0.12 | 9.6 | 113 | 435 | 28 |
| | | Cold-drawn | 95 | 125 | 25 | 277 | | | | | | | | |
| Stainless steel type 309 | Fe bal., Cr 23, Ni 13, C 0.20 max. | Annealed | 30 | 82 | 50 | 165 | 0.29 | 8.02 | 2550–2650 | 0.12 | 8.3 | 96 | 470 | 29 |
| | | Cold-rolled^c | up to 120 | up to 150 | 4^d | 275 | | | | | | | | |
| Stainless steel type 310 | Fe bal., Cr 25, Ni 20, C 0.25 max. | Annealed | 40 | 100 | 50 | 165 | 0.29 | 8.02 | 2550–2650 | 0.12 | 8.0 | 96 | 470 | 29 |
| Stainless steel type 316 | Fe bal., Cr 18 Ni 11, Mo 2.5, C 0.10 max. | Annealed | 30 | 90 | 50 | 165 | 0.29 | 8.02 | 2500–2550 | 0.12 | 8.9 | 113 | 445 | 28 |
| | | Cold-rolled^e | up to 120 | up to 150 | 8^d | 275 | | | | | | | | |
| Stainless steel type 316L | Fe bal., Cr 17, Ni 12, C 0.03 max., Mo 2 | Annealed | 30 | 80 | 60 | 150 | 0.29 | 8.02 | 2500–2550 | 0.12 | 8.9 | 113 | 445 | 28 |
| | | Cold-drawn | 60 | 90 | 45 | 190 | | | | | | | | |
| Stainless steel types 321 and 347 (321 has Ti) (347 has Cb) | Fe bal., Cr 18, Ni 10, C 0.10 max., Ti 4 × carbon min. or Cb 8 × carbon min. | Annealed | 30 | 85 | 50 | 160 | 0.286 | 7.92 | 2550–2600 | 0.12 | 9.3 | 110 | 435 | 28 |
| | | Cold-rolled | up to 120 | up to 150 | 5^d | 300 | | | | | | | | |

^c The cold-rolled properties depend upon composition; types 302 and 304 are not rolled often in excess of 175,000 lb./sq. in. tensile strength.
^d The values for elongation (% in 2 in.) are obtainable in the steel cold-rolled to *the maximum stated* yield strength and tensile strength. For lower values of tensile strength, elongation will be correspondingly higher.
^e Types 316, 321, and 374 are used chiefly in the annealed condition.

Table 3-41. Properties of Metals and Alloys (Continued)

| Material | Nominal composition (essential elements), % | Form and condition | Typical mechanical properties | | | | Typical physical constants | | | | | | | |
|---|---|---|---|---|---|---|---|---|---|---|---|---|---|---|
| | | | Yield strength (0.2% offset), 1000 lb./sq. in. | Tensile strength, 1000 lb./sq. in. | Elongation in 2 in., % | Hardness, Brinell | Density, lb./cu. in. | Specific gravity | Melting point, °F. | Specific heat (32°–212°F.), B.t.u./(lb.)(°F.) | Thermal expansion coefficient (32°–212°F.), $\times 10^{-6}$ in./(in.)(°F.) | Thermal conductivity (32°–212°F.), B.t.u./(sq. ft.)(hr.)(°F./in.) | Electrical resistivity (68°F.), ohms/cir. mil ft. | Tensile modulus of elasticity $\times 10^6$ lb./sq. in. |
| Stainless steel type 330 | Fe bal., Ni 36, Cr 16 | Hot-rolled | 55 | 100 | 35 | 200 | 0.284 | 7.86 | 2515 | 0.11 | 6.3 / 8.8 (68%–932°F.) | 90 | 600 | |
| | | Cold-drawn (anneal) | | 80 | | | | | | | | | | |
| | | Cold-drawn (heat-treated) | | 150 | | | | | | | | | | |
| Stainless steel AM 350 | Fe bal., Cr 17, Ni 4, Mo 3, C 0.08 | Annealed | 45 | 156 | 21 | 205 | 0.286 | | | | 9.0 | | | 30 |
| | | Hardened | 153 | 195 | 12 | 382 | | | | | | | | |
| Stainless steel type 410 | Fe bal., Cr 12.5, C 0.15 max. | Annealed | 40 | 75 | 30 | 150 | 0.28 | 7.75 | 2700–2790 | 0.11 | 5.5 | 173 | 340 | 29 |
| | | Heat-treated | 115 | 150 | 15 | 300 | | | | | | | | |
| Stainless steel type 414 | Fe bal., Cr 12.5, Ni 2.5 | Annealed | 80 | 100 | 22 | 217 | 0.28 | 7.75 | 2600–2700 | 0.11 | 6.1 | 173 | 420 | 29 |
| | | Heat-treated | 150 | 200 | 17 | 387 | | | | | | | | |
| Stainless steel type 420 | Fe bal., Cr 13, C 0.35 | Annealed | 60 | 98 | 28 | 180 | 0.28 | 7.75 | 2650–2750 | 0.11 | 5.7 | 173 | 330 | 29 |
| | | Heat-treated | 200 | 250 | 8 | 480 | | | | | | | | |
| Stainless steel type 430 | Fe bal., Cr 16, C 0.12 max. | Annealed | 40 | 70 | 35 | 165 | 0.28 | 7.75 | 2600–2750 | 0.11 | 6.0 | 180 | 360 | 29 |
| | | Cold-rolled | 95 | 110 | 10 | 225 | | | | | | | | |
| Stainless steel type 431 | Fe bal., Cr 16, Ni 2 | Annealed | 85 | 120 | 25 | 250 | 0.280 | 7.75 | 2600–2700 | 0.11 | 6.5 | 140 | 430 | 29 |
| | | Heat-treated | 150 | 195 | 20 | 400 | | | | | | | | |
| Stainless steel type 446 | Fe bal., Cr 25, C 0.35 max. | Annealed | 50 | 80 | 30 | 165 | 0.27 | 7.45 | 2600–2750 | 0.12 | 5.8 | 145 | 405 | 29 |

| Material | Composition | Condition | | | | | | | | | | | | |
|---|---|---|---|---|---|---|---|---|---|---|---|---|---|---|
| Stainless steel 17-4 PH | Fe bal., Cr 17, Ni 4, Cu 4, Co 0.35, C 0.07 | Annealed / Hardened | 110 / 180 | 150 / 195 | 12 / 13 | 363 / 404 | 0.28 | ..... | ..... | ..... | 6 | 124 | ..... | 28.5 |
| Stainless steel 17-7 PH | Fe bal., Cr 17, Ni 7, Al 1, C 0.09 | Annealed / Hardened | 40 / 185 | 130 / 200 | 30 / 9 | 165 / 404 | 0.282 | ..... | ..... | ..... | 8.5 | 128 | ..... | 29 |
| Stainless steel HNM | Fe bal., Cr 18, Mn 3.5, Ni 0.5, C 0.30 | Annealed / Hardened | 56 / 124 | 116 / 168 | 57 / 19 | 192 / 352 | 0.284 | ..... | ..... | ..... | 8.4 | ..... | ..... | 29 |
| Stainless steel, stainless W | Fe bal., Cr 17, Ni 7, Ti 0.7, Al 0.2, C 0.07 | Annealed / Hardened | 75–115 / 150–185 | 120–150 / 170–210 | 8–15 / 8–16 | 255 / 365 | 0.28 | ..... | ..... | ..... | 6 | ..... | ..... | 28 |
| Carpenter stainless No. 20[f] | C 0.07 max, Mn 0.75, Si 1.00, Cr 20.00, Ni 29.00, Mo 2.00 min, Cu 3.00 min | Annealed | 35 | 85 | 50 | 160 | 0.289 | 8.02 | ..... | 0.12 | 9.4 | 145.2 (212°F.) | 451 | 28.0 |
| **Cast Stainless Steels** | | | | | | | | | | | | | | |
| Cast 12 Cr Alloy (CA-15) (C means corrosion-resistant casting, Alloy Casting Institute designations) | C 0.15 max, Mn 1.00 max, Si 1.50 max, Cr 11.5–14, Ni 1.00 max, Fe bal. | Air-cooled from 1800°F. Tempered at 600°F. / Air-cooled from 1800°F. Tempered at 1400°F. | 150 / 75 | 200 / 100 | 7 / 30 | 390 / 185 | 0.275 | 7.61 | 2750 | 0.11 | 6.4 (70°–1000°F.) | 174 (212°F.) | 468 | 29 |
| Cast 12 Cr Alloy (CA-40) | C 0.20–0.40, Mn 1.00 max, Si 1.50 max, Cr 11.5–14, Ni 1.00 max, Fe bal. | Air-cooled from 1800°F. Tempered at 600°F. / Air-cooled from 1800°F. Tempered at 1400°F. | 165 / 67 | 220 / 110 | 1 / 18 | 470 / 212 | 0.275 | 7.61 | 2725 | 0.11 | 6.4 (70°–1000°F.) | 174 (212°F.) | 456 | 29 |

[f] Carpenter stainless No. 20-Cb same composition except for columbium + tantalum, 8 times the carbon minimum.

Table 3-41. Properties of Metals and Alloys (Continued)

| Material | Nominal composition (essential elements), % | Form and condition | Typical mechanical properties | | | | Typical physical constants | | | | | | | | |
|---|---|---|---|---|---|---|---|---|---|---|---|---|---|---|---|
| | | | Yield strength (0.2% offset), 1000 lb./sq. in. | Tensile strength, 1000 lb./sq. in. | Elongation in 2 in., % | Hardness, Brinell | Density, lb./cu. in. | Specific gravity | Melting point, °F. | Specific heat (32°–212°F.), B.t.u./(lb.)(°F.) | Thermal expansion coefficient × 10⁻⁶ in./(in.)(°F.) (32°–212°F.) | Thermal conductivity (32°–212°F.) B.t.u./(sq. ft.)(hr.)(°F./in.) | Electrical resistivity (68°F.) ohms/cir. mil ft. | Tensile modulus of elasticity × 10⁶ lb./sq. in. |
| Cast 20 Cr Alloy (CB-30) | C 0.30 max., Mn 1.00 max., Si 1.00 max., Cr 18–22, Ni 2 max. | Annealed | 60 | 95 | 15 | 195 | 0.272 | 7.53 | 2725 | 0.11 | 6.5 (70°–1000°F.) | 153.6 (212°F.) | 456 | 29 |
| Cast 28-4 Alloy (CC-50) | C 0.50 max., Mn 1.00 max., Si 1.00 max., Cr 26–30, Ni 4.00 max., Fe bal. | As-cast / Air-cooled from 1900°F. | 65 / 65 | 70 / 97 | 2 / 18 | 212 / 210 | 0.272 | 7.53 | 2725 | 0.12 | 6.4 (70°–1000°F.) | 151.2 (212°F.) | 462 | 29 |
| Cast 29-9 Alloy (CE-30) | C 0.30 max., Mn 1.50 max., Si 2.00 max., Cr 26–30, Ni 8–11, Fe bal. | As-cast | 60 | 95 | 15 | 170 | 0.277 | 7.67 | 2650 | 0.14 | 9.6 (70°–1000°F.) | ..... | 510 | 25 |
| Cast 20-10 Alloy (CF-8) | C 0.08 max., Mn 1.50 max., Si 2.00 max., Cr 18–21, Ni 8–11, Fe bal. | Water-quenched (1950°–2050°F.) | 37 | 77 | 55 | 140 | 0.280 | 7.75 | 2600 | 0.12 | 10.0 (70°–1000°F.) | 110.4 (212°F.) | 457.2 | 28 |
| Cast 20-10 Alloy (CF-20) | C 0.20 max., Mn 1.50 max., Si 2.00 max., Cr 18–21, Ni 8–11, Fe bal. | Water-quenched (above 2000°F.) | 36 | 77 | 50 | 163 | 0.280 | 7.75 | 2575 | 0.12 | 10.4 (70°–1000°F.) | 110.4 (212°F.) | 467.4 | 28 |

| Alloy | Composition | Condition | | | | | | | | | | | | |
|---|---|---|---|---|---|---|---|---|---|---|---|---|---|---|
| Cast 20-10-2.5 Alloy (CF-8M) | C 0.08 max., Mn 1.50 max., Si 1.50 max., Cr 18-21, Ni 9-12, Mo 2.5, Fe bal. | Water-quenched (1950°-2050°F.) | 42 | 80 | 50 | 156-170 | 0.280 | 7.75 | 2550 | 0.12 | 9.7 (70°-1000°F.) | 112.8 (212°F.) | 492 | 28 |
| Cast 18-38 Alloy (HU) (H means heat-resistant) | C 0.35-0.75, Mn 2.00 max., Si 2.50 max., Cr 17-21, Ni 37-41, Fe bal. | As-cast | 40 | 70 | 9 | 170 | | | | | | | | |
| | | Aged | 43 | 73 | 5 | 190 | 0.290 | 8.03 | 2450 | 0.11 | 9.7 (70°-2000°F.) | 106.8 (70°-1000°F.) | 630 | 27 |
| Cast 12-60 Alloy (HW) | C 0.35-0.75, Mn 2.00 max., Si 2.50 max., Cr 10-14, Ni 58-62, Fe bal. | As-cast | 36 | 68 | 4 | 185 | | | | | | | | |
| | | Aged | 52 | 84 | 4 | 205 | 0.294 | 8.14 | 2350 | 0.11 | 9.2 (70°-2000°F.) | 92.4 (70°-212°F.) | 672 | 25 |
| Cast 15-65 Alloy (HX) | C 0.35-0.75, Mn 2.00 max., Si 2.50 max., Cr 15-19, Ni 64-68, Fe bal. | As-cast | 36 | 65 | 9 | 176 | | | | | | | | |
| | | Aged | 44 | 73 | 9 | 185 | 0.294 | 8.14 | 2350 | 0.11 | 9.5 (70°-2000°F.) | ..... | ..... | 25 |
| Cast 9 Cr Alloy (HA) | C 0.20 max., Mn 0.35-0.65 max., Si 1.00 max., Cr 8-10, Fe bal. | Annealed | 65 | 95 | 23 | 180 | | | | | | | | |
| | | Normalized | 81 | 107 | 21 | 220 | 0.279 | 7.72 | 2750 | 0.11 | 7.5 (70°-1200°F.) | 204 (70°-1000°F.) | 420 | 29 |
| Cast 28-4 Alloy (HC) | C 0.50 max., Mn 1.00 max., Si 2.00 max., Cr 26-30, Ni 4 max., Fe bal. | As-cast | 65 | 70 | 2 | 190 | | | | | | | | |
| | | Aged | 80 | 115 | 18 | | 0.272 | 7.53 | 2725 | 0.12 | 7.7 (70°-2000°F.) | 214.8 (70°-1000°F.) | 462 | 29 |
| Cast 28-7 Alloy (HD) | C 0.50 max., Mn 1.50 max., Si 2.00 max., Cr 26-30, Ni 4-7, Fe bal. | As-cast | 48 | 85 | 16 | 190 | 0.274 | 7.58 | 2700 | 0.12 | 9.2 (70°-2000°F.) | 214.8 (70°-1000°F.) | 486 | 27 |

# Table 3-41. Properties of Metals and Alloys (Continued)

| Material | Nominal composition (essential elements), % | Form and condition | Typical mechanical properties | | | | Typical physical constants | | | | | | | |
|---|---|---|---|---|---|---|---|---|---|---|---|---|---|---|
| | | | Yield strength (0.2% offset), 1000 lb./sq. in. | Tensile strength, 1000 lb./sq. in. | Elongation in 2 in., % | Hardness, Brinell | Density, lb./cu. in. | Specific gravity | Melting point, °F. | Specific heat (32°-212°F.), B.t.u./(lb.)(°F.) | Thermal expansion coefficient (32°-212°F.) $\times 10^{-6}$ in./(in.)(°F.) | Thermal conductivity (32°-212°F.) B.t.u./(sq. ft.)(hr.)(°F.)(°F./in.) | Electrical resistivity (68°F.) ohms/cir. mil ft. | Tensile modulus of elasticity $\times 10^6$ lb./sq. in. |
| Cast 29-9 Alloy (HE) | C 0.20-0.50, Mn 2.00 max., Si 2.00 max., Cr 26-30, Ni 8-11, Fe bal. | As-cast | 45 | 95 | 20 | 200 | 0.277 | 7.68 | 2650 | 0.14 | 11.1 (70°-2000°F.) | 120 (70°-1500°F.) | 510 | 25 |
| | | Aged | 55 | 90 | 10 | 270 | | | | | | | | |
| Cast 21-10 Alloy (HF) | C 0.20-0.40, Mn 2.00 max., Si 2.00 max., Cr 19-23, Ni 9-12, Fe bal. | As-cast | 45 | 85 | 35 | 165 | 0.280 | 7.75 | 2550 | 0.12 | 10.9 (70°-2000°F.) | 160.8 (70°-1000°F.) | 480 | 28 |
| | | Aged | 50 | 100 | 25 | 190 | | | | | | | | |
| Cast 25-12 Alloy (HH) | C 0.20-0.50, Mn 2.00 max., Si 2.00 max., Cr 24-28, Ni 11-14, Fe bal. | As-cast—Type 1<br>Type 2 | 50<br>40 | 80<br>85 | 25<br>15 | 185<br>180 | 0.279 | 7.72 | 2500 | 0.12 | 10.8 (70°-2000°F.) | 130.8 (70°-1000°F.) | 450–510 | 27 |
| | | Aged—Type 1<br>Type 2 | 55<br>45 | 86<br>92 | 11<br>8 | 200<br>200 | | | | | | | | |
| Cast 28-15 Alloy (HI) | C 0.20-0.50, Mn 2.00 max., Si 2.00 max., Cr 26-30, Ni 14-18, Fe bal. | As-cast | 45 | 80 | 12 | 180 | 0.279 | 7.72 | 2550 | 0.12 | 10.8 (70°-2000°F.) | 98.4 (70°-1000°F.) | ..... | 27 |
| | | Aged | 65 | 90 | 6 | 200 | | | | | | | | |
| Cast 25-20 Alloy (HK) | C 0.20-0.60, Mn 2.00 max., Si 2.00 max., Cr 24-28, Ni 18-22, Fe bal. | As-Cast | 50 | 75 | 17 | 170 | 0.280 | 7.75 | 2550 | 0.12 | 10.1 (70°-2000°F.) | 130.8 (70°-1000°F.) | 540 | 29 |
| | | Aged | 50 | 85 | 10 | 190 | | | | | | | | |

| Alloy | Composition | Condition | | | | | | | | | | | |
|---|---|---|---|---|---|---|---|---|---|---|---|---|---|
| Cast 30-20 Alloy (HL) | C 0.20–0.60, Mn 2.00 max, Si 2.00 max, Cr 28–32, Ni 18–22, Fe bal. | As-cast | 52 | 82 | 19 | 192 | 0.279 | 7.72 | 2600 | 0.12 | 10.1 (70°–2000°F.) | 130.8 (70°–1000°F.) | 574 | 29 |
| Cast 20-25 Alloy (HN) | C 0.20–0.50, Mn 2.00 max, Si 2.00 max, Cr 19–23, Ni 23–27, Fe bal. | As-cast | 38 | 68 | 17 | 160 | 0.283 | 7.83 | 2500 | 0.11 | ..... | ..... | ..... | 27 |
| Cast 15-35 Alloy (HT) | C 0.35–0.75, Mn 2.00 max, Si 2.50 max, Cr 13–17, Ni 33–37, Fe bal. | As-cast | 40 | 70 | 10 | 180 | 0.286 | 7.96 | 2425 | 0.11 | 9.8 (70°–2000°F.) | 136.8 (70°–1000°F.) | 600 | 27 |
| | | Aged | 45 | 75 | 5 | 200 | | | | | | | | |
| Cast 20-10-2.5 Alloy (CF-12M) | C 0.12 max, Mn 1.50 max, Si 1.50 max, Cr 18–21, Ni 9–12, Mo 2,5, Fe bal. | Water-quenched (from above 2000°F.) | 42 | 80 | 50 | 156–170 | 0.280 | 7.75 | 2550 | 0.12 | 9.7 (70°–1000°F.) | 112.8 (212°F.) | 492 | 28 |
| Cast 20-10 Cb Alloy (CF-8C) | C 0.08 max, Mn 1.50 max, Si 1.50 max, Cr 18–21, Ni 9–12, Cb or Cb-Ta, Fe bal. | Water-quenched (1950°–2050°F.) | 38 | 77 | 39 | 149 | 0.280 | 7.75 | 2600 | 0.12 | 10.3 (70°–1000°F.) | 111.6 (212°F.) | 426 | 28 |
| Cast 20-10 Alloy (CF-16F) | C 0.16 max, Mn 1.50 max, Si 2.00 max, Cr 18–21, Ni 9–12, Mo 1.5 max, Fe bal. | Water-quenched (from above 2000°F.) | 40 | 77 | 52 | 150 | 0.280 | 7.75 | 2550 | 0.12 | 9.9 (70°–1000°F.) | 112.8 (212°F.) | 432 | 28 |
| Cast 25-12 Alloy (CH-20) | C 0.20 max, Mn 1.50 max, Si 2.00 max, Cr 22–26, Ni 12–15, Fe bal. | Water-quenched (from above 2000°F.) | 50 | 88 | 38 | 190 | 0.279 | 7.72 | 2600 | 0.12 | 9.6 (70°–1000°F.) | 98.4 (212°F.) | 504 | 28 |

## Table 3-41. Properties of Metals and Alloys (Continued)

| Material | Nominal composition (essential elements), % | Form and condition | Typical mechanical properties | | | | Typical physical constants | | | | | | | |
|---|---|---|---|---|---|---|---|---|---|---|---|---|---|---|
| | | | Yield strength (0.2% offset), 1000 lb./sq. in. | Tensile strength, 1000 lb./sq. in. | Elongation in 2 in., % | Hardness, Brinell | Density, lb./cu. in. | Specific gravity | Melting point, °F. | Specific heat (32°-212°F.), B.t.u./(lb.)(°F.) | Thermal expansion coefficient (32°-212°F.) $\times 10^{-6}$ in./(in.)(°F.) | Thermal conductivity (32°-212°F.) B.t.u./(sq. ft.)(hr.)(°F./in.) | Electrical resistivity (68°F.) ohms/cir. mil ft. | Tensile modulus of elasticity $\times 10^6$ lb./sq. in. |
| Cast 25-20 Alloy (CK-20) | C 0.20 max., Mn 1.50 max., Si 2.00 max., Cr 23-27, Ni 19-22, Fe bal. | Water-quenched (from 2100°F.) | 38 | 76 | 37 | 144 | 0.280 | 7.75 | 2600 | 0.12 | 9.2 (70°-1000°F.) | 98.4 (212°F.) | 540 | 29 |
| Cast 25-20 Alloy (CN-7M) | C 0.07 max., Mn 1.50 max., Si 100, Cr 18-22, Ni 21-31, Fe bal. | As-cast | 30 | 65 | 30-45 | 130-150 | 0.287 | 8.02 | 2650 | 0.12 | 9.7 (70°-1000°F.) | 145.2 (212°F.) | 537.6 | 30 |
| **Other Cast Alloys** | | | | | | | | | | | | | | |
| Durimet 20 | C 0.07, Mn 1.5, Si 1.5, Cr 20.0, Ni 29.0, Mo 2.0, Cu 3.0 | Cast, annealed | 30 | 65 | 48 | 130 | 0.286 | ..... | ..... | ..... | 8.6 | 145 (212°F.) | ..... | ..... |
| Iron-silicon Alloy (Duriron) | Si 14.50, C 0.85, Mn 0.65, Mo nil, Fe bal. | Cast only | ..... | 16 | Nil | 520 | 0.255 | 7.0 | 2300 | 0.13 | 7.4 (68°-392°F.) | ..... | ..... | 23 |
| Fe-Si-Mo Alloy (Durichlor 51) | Si 14.50, C 0.85, Mn 0.65, Cr 4.5, Fe bal. | Cast only | ..... | 16 | Nil | 520 | 0.255 | 7.0 | 2300 | 0.13 | 7.2 | ..... | ..... | 23 |

| Alloy | Composition | Condition | | | | | | | | | | | | |
|---|---|---|---|---|---|---|---|---|---|---|---|---|---|---|
| Ni-Mo Alloy (Chlorimet 2) | Ni 62.00, Mo 32.00 Fe 6.00 max, Si 1.00, C 0.10 | Cast | 55 | 80 | 5 | 230 | 0.333 | 9.24 | 2460 | ..... | 4.7 | ..... | ..... | ..... | 27 |
| Ni-Cr-Mo Alloy (Chlorimet 3) | Ni 60.00, Cr 18.00, Mo 18.00, Fe 7.5 max., Si 1.00, C 0.07 | Cast only | 50 | 75 | 10 | 220 | 0.235 | 8.94 | 2380 | 0.092 | 7.0 (68°–392°F.) | ..... | ..... | 24.5 |
| Illium G | Ni bal., Cr 22.5, Fe 6.5, Mo 6.4, Cu 6.5, Mn, Si | As-cast | 50 | 68 | 32 | 200 | 0.310 | 8.31 | 2375 | 0.105 | 7.5 | 84 | 735 | 24 |
| Illium 98 | Ni 55.00, Cr 28.00, Mo 8.5, Cu 5.5, Mn 1.25, Fe 1.00 | Cast | ..... | 54 | 18 | 155 | | | | | | | | |

**Nickel Alloys**

| Alloy | Composition | Condition | | | | | | | | | | | | |
|---|---|---|---|---|---|---|---|---|---|---|---|---|---|---|
| Nickel (pure) | Ni 99.99 | Annealed | 8.5 | 46 | 30 | ..... | 0.322 | 8.91 | 2650 | 0.11 | 7.4 | 543 | 41 | 30 |
| Nickel (cast) | Ni 95.6, Cu 0.5, Fe 0.5, Mn 0.8, Si 1.5, C 0.8 | As-cast | 25 | 57 | 22 | 110 | 0.301 | 8.34 | 2450–2600 | 0.13 | 8.85 (70°–1400°F.) | 410 | 125 | 21.5 |
| Nickel 200 | Ni(+Co) 99.40, C 0.06, Mn 0.25, Fe 0.15, S 0.005, Si 0.05, Cu 0.05 | Annealed<br>Hot-rolled<br>Cold-drawn<br>Cold-rolled | 20<br>25<br>70<br>95 | 70<br>75<br>95<br>105 | 40<br>40<br>25<br>5 | 100<br>110<br>170<br>210 | 0.321 | 8.89 | 2615–2635 | 0.13 | 6.6 | 420 | 53<br>57 | 30 |
| Low-carbon nickel 201 | Ni(+Co) 99.50, C 0.02, Mn 0.20, Fe 0.15, S 0.005, Si 0.05, Cu 0.05 | Annealed<br>Hot-rolled<br>Cold-rolled | 15<br>25<br>65 | 60<br>60<br>95 | 50<br>45<br>15 | 90<br>105<br>150 | 0.321 | 8.89 | 2615–2635 | 0.11 | 7.2 | 420 | 50 | 30 |
| Nickel 212 | Ni(+Co) 97.85, C 0.05, Mn 1.95, Fe 0.05, S 0.005, Si 0.04, Cu 0.03 | Annealed<br>Hot-rolled<br>Cold-drawn | 35<br>50<br>80 | 75<br>90<br>100 | 40<br>35<br>25 | 140<br>150<br>190 | 0.319 | 8.86 | 2600 | 0.11 | 7.4 | 335 | 85 | 30 |
| Nickel 211 | Ni(+Co) 95.00, C 0.10, Mn 4.75, Fe 0.05, S 0.005, Si 0.05, Cu 0.02 | Annealed<br>Hot-rolled<br>Cold-drawn | 35<br>50<br>80 | 75<br>90<br>100 | 40<br>35<br>25 | 140<br>150<br>190 | 0.315 | 8.78 | 2600 | 0.11 | 7.4 | 335 | 110 | 30 |

## Table 3-41. Properties of Metals and Alloys (Continued)

| Material | Nominal composition (essential elements), % | Form and condition | Yield strength (0.2% offset), 1000 lb./sq. in. | Tensile strength, 1000 lb./sq. in. | Elongation in 2 in., % | Hardness, Brinell | Density, lb./cu. in. | Specific gravity | Melting point, °F. | Specific heat (32°-212°F.), B.t.u./(lb.)(°F.) | Thermal expansion coefficient (32°-212°F.), × 10⁻⁶ in./(in.)(°F.) | Thermal conductivity (32°-212°F.), B.t.u./(sq. ft.)(hr.)(°F./in.) | Electrical resistivity (68°F.), ohms/cir. mil ft. | Tensile modulus of elasticity × 10⁶ lb./sq. in. |
|---|---|---|---|---|---|---|---|---|---|---|---|---|---|---|
| Duranickel | Ni(+Co) 93.90, C 0.15, Mn 0.25, Fe 0.15, S 0.005, Si 0.55, Cu 0.05, Al 4.50, Ti 0.45 | Annealed | 45 | 100 | 40 | 160 | 0.298 | 8.26 | 2550–2620 | 0.104 | 7.2 | 128 | 280 | 30 |
|  |  | Annealed age-hardened | 125 | 170 | 25 | 330 |  |  |  | ..... | ..... | 137 | 260 |  |
|  |  | Spring |  | 175 | 5 | 320 |  |  |  |  |  |  |  |  |
|  |  | Spring, age-hardened |  | 205 | 10 | 370 |  |  |  |  |  |  |  |  |
| Permanickel | Ni(+Co) 98.65, C 0.25, Mn 0.10, Fe 0.10, S 0.005, Si 0.06, Cu 0.02, Ti 0.45, Mg 0.35 | Annealed | 45 | 105 | 45 | 160 | 0.316 | 8.75 | 2550–2620 | 0.106 (70°–750° F.) | 7.2 | 400 | 94.5 | 30 |
|  |  | Annealed, age-hardened | 125 | 175 | 25 | 325 |  |  |  |  |  |  |  |  |
|  |  | Spring |  | 180 | 5 |  |  |  |  |  |  |  |  |  |
|  |  | Spring, age-hardened | 195 | 210 | 10 |  |  |  |  |  |  |  |  |  |
| Nickel 205 Electronic grade | Ni(+Co) 99.55, C 0.09, Mn 0.20, Fe 0.05, S 0.005, Si 0.05, Cu 0.02 | Annealed | 20 | 70 | 40 | 100 | 0.321 | 8.89 | 2615–2635 | 0.11 | 7.2 |  | 57 | 30 |
| Nickel 220 A.S.T.M. B239, grade 11 | Ni(+Co) 99.65, C 0.06, Mn 0.10, Fe 0.05, S 0.005, Si 0.05, Cu 0.02, Mg 0.04 | Annealed | 20 | 70 | 40 | 100 | 0.0321 | 8.89 | 2615–2635 | 0.11 | 7.2 | 420 | 57 | 30 |
| Nickel 225 | Ni(+Co) 99.50, C 0.07, Mn 0.10, S 0.005, Si 0.20, Cu 0.02, Fe 0.05 | Annealed | 20 | 70 | 40 | 100 | 0.321 | 8.89 | 2615–2635 | 0.11 | 7.2 | 420 | 57 | 30 |

| Material | Composition | Condition | | | | | | | | | | | | |
|---|---|---|---|---|---|---|---|---|---|---|---|---|---|---|
| Nickel 213 | N(+Co) 94.2, Cu 0.5, Fe 0.5, Mn 0.8, Si 1.5, C 1.5 | As-cast | 30 | 55 | 20 | 105 | 0.301 | 8.34 | 2400–2600 | ..... | 8.75 (70°–1400°F.) | ..... | 125 | 22 |
| Nickel 305 | Ni(+Co) 91.5, Cu 0.5, Fe 0.5, Mn 0.8, Si 6.0, C 0.8 | As-cast | 62 | 85 | 2 | 220 | 0.288 | 8.01 | 2400–2600 | ..... | 8.76 (70°–1400°F.) | ..... | 180 | 24 |
| | | Annealed, aged | 65 | 90 | 2 | 240 | | | | | | | | |
| Hastelloy Alloy B | Ni bal., Mo 28, Fe 5, Mn, Si ..... | Sand-cast (anneal) | 50 | 80 | 8 | 199 | ..... | ..... | ..... | ..... | ..... | ..... | ..... | 26.5 |
| | | Rolled (anneal) | 56 | 120 | 50 | 215 | 0.334 | 9.24 | 2410–2460 | 0.091 | 5.6 | 72 | 812 | 30.8 |
| | | Investment cast | 54 | 85 | 14 | 209 | | | | | | | | 28.5 |
| Hastelloy Alloy C | Ni bal, Mo 16, Cr 16, Fe 5, W 4, Mn, Si ..... | Sand-cast (anneal) | 50 | 78 | 5 | 199 | ..... | ..... | ..... | ..... | ..... | ..... | ..... | 26 |
| | | Rolled (anneal) | 71 | 130 | 45 | 204 | 0.323 | 8.94 | 2320–2380 | 0.092 | 6.3 | 61 | 834 | 30.9 |
| | | Investment cast | 50 | 80 | 10 | 215 | | | | | | | | 24.5 |
| Hastelloy Alloy D | Ni bal. Si 10, Cu 3, Mn | Sand-cast (anneal) | 118 | 118 | 0–2 | 321 | 0.282 | 7.80 | 2030–2050 | 0.108 | 6.1 | 145 (at 72°F.) | 680 | 28.9 |
| Hastelloy Alloy G | Ni 44, Cr 22, Fe 20, Mo 6.5, Cb + Ta 2.1, Cu 2.0, C 0.05 max., W 1 max. | Sheet | 46.2 | 102 | 61 | B-84 Rockwell | 0.30 | ..... | 2300–2450 | 0.93–0.109 | 7.5 | 75 | ..... | 27.8 |
| | | Plate | 45.0 | 99.6 | 62 | | | | | | | | | |
| | | Sand-cast | 38.8 | 87.5 | 30 | | | | | | | | | |
| Hastelloy Alloy X | Co 1.5 max., Fe 18.5, Cr 22.0, Mo 9.0, W 0.6, C 0.15 max. (wrought), C 0.20 max. (cast), Ni bal. | Wrought sheet Mill-annealed | 52 | 113.2 | 41.0 | 194 | 0.297 | 8.23 | 2350 | 0.1046 | 7.90 (70°–600°F.) | 62.8 | 712 | 28.6 |
| | | | ..... | 67.0 | 17.0 | 172 | | | | | | | | |
| | | As investment cast | 46.5 | | | | | | | | | | | |

# Table 3-41. Properties of Metals and Alloys (Continued)

| Material | Nominal composition (essential elements), % | Form and condition | Typical mechanical properties | | | | Typical physical constants | | | | | | | |
|---|---|---|---|---|---|---|---|---|---|---|---|---|---|---|
| | | | Yield strength (0.2% offset), 1000 lb./sq. in. | Tensile strength, 1000 lb./sq. in. | Elongation in 2 in., % | Hardness, Brinell | Density, lb./cu. in. | Specific gravity | Melting point, °F. | Specific heat (32°–212°F.), B.t.u./(lb.)(°F.) | Thermal expansion coefficient (32°–212°F.) × $10^{-6}$ in./(in.)(°F.) | Thermal conductivity (32°–212°F.) B.t.u./ (sq. ft.)(hr.)(°F./in.) | Electrical resistivity (68°F.) ohms/cir. mil ft. | Tensile modulus of elasticity × $10^6$ lb./sq. in. |
| Incoloy Alloy 800 | Ni(+Co) 31.90, C 0.04, Mn 0.75, Fe 46.05, S 0.007, Si 0.35, Cu 0.27, Cr 20.60 | Annealed | 40 | 90 | 40 | 150 | 0.290 | 8.01 | 2540–2600 | 0.120 | 8.2 | 85 | 600 (76°F.) | 28.5 |
| Incoloy Alloy 801 | Ni(+Co) 32.20, C 0.04, Mn 0.85, Fe 44.60, S 0.007, Si 0.35, Cu 0.15, Cr 20.80, Ti 1.00 | Annealed | 50 | 90 | 40 | 160 | 0.287 | 7.99 | ..... | ..... | ..... | ..... | ..... | 28.5 |
| Incoloy Alloy 804 | Ni(+Co) 42.10, C 0.06, Mn 0.85, Fe 25.60, S 0.007, Si 0.60, Cu 0.40, Cr 29.70, Al 0.25, Ti 0.40 | Annealed | 45 | 100 | 40 | 175 | 0.286 | 7.92 | ..... | ..... | 8.0 | ..... | 635 | 32.0 |
| Incoloy Alloy 825 | Ni 40, Cr 21, Fe 31, Mo 3.0, Cu 1.75, Mn 0.60, Si 0.40, C 0.05 | Annealed Cold-drawn | 45 ..... | 95 155 | 40 10 | 185 255 | 0.294 | 8.14 | ..... | ..... | ..... | ..... | ..... | 28.5 |

| Alloy | Composition | Condition | | | | | | | | | | | | |
|---|---|---|---|---|---|---|---|---|---|---|---|---|---|---|
| Incoloy Alloy 901 | Ni(+Co) 42.65, C 0.05, Mn 0.45, Fe 33.90, S 0.010, Si 0.40, Cu 0.10, Cr 13.45, Al 0.25, Ti 2.50, Mo 6.20 | Annealed / Annealed, aged | 45 / 105 | 110 / 165 | 45 / 26 | 160 / 310 | 0.297 | 8.23 | ..... | ..... | 7.75 | 93 | ..... / 662 (76°F.) | 30.0 |
| Nimonic 75 | Ni(+Co) 77.40, C 0.10, Mn 0.45, Fe 0.50, S 0.007, Si 0.45, Cu 0.05, Cr 20.50, Al 0.15, Ti 0.35 | Annealed | 55 | 115 | 40 | 168 | 0.302 | 8.35 | 2530–2590 | 0.11 | 6.8 | 86.5 | 655 | 27 |
| Nimonic 80A | N(+Co) 74.45, C 0.05, Mn 0.55, Fe 0.55, S 0.007, Si 0.20, Cu 0.05, Cr 20.45, Al 1.25, Ti 2.40 | Annealed | 60 | 115 | 60 | 185 | 0.296 | 8.25 | 2530–2590 | 0.10 | 6.5 | 84.5 | 745 | 31 |
| Nimonic 90 | Ni(+Co) 57.00, C 0.05, Mn 0.50, Fe 0.45, S 0.007, Si 0.20, Cu 0.05, Cr 20.55, Al 1.65, Ti 2.60, Co 16.90 | Annealed | 90 | 155 | ..... | 260 | 0.298 | 8.25 | 2470–2530 | ..... | 6.5 | 86.5 | 690 | 31 |
| Inconel (wrought) Alloy 600 | Ni(+Co) 76.40, C 0.04, Mn 0.20, Fe 7.20, S 0.007, Si 0.20, Cu 0.10, Cr 15.85 | Annealed / Cold-drawn | 35 / 100 | 90 / 130 | 45 / 20 | 150 / 200 | 0.304 | 8.43 | 2540–2600 | 0.109 | 7.0 | 104 / 113 (212°F.) | 623 (76°F.) | 31.0 |
| Inconel (cast) Alloy 610 | Ni(+Co) 71.5, Cu 0.5, Fe 8.0, Mn 1.0, Si 2.0, Cr 16.0, C 0.20 | As-cast | 38 | 80 | 15 | 175 | 0.300 | 8.30 | 2540–2600 | 0.11 | 8.92 (70°–1400°F.) | 104 | 700 | 23 |

## Table 3-41. Properties of Metals and Alloys (Continued)

| Material | Nominal composition (essential elements), % | Form and condition | Typical mechanical properties | | | | Typical physical constants | | | | | | | |
|---|---|---|---|---|---|---|---|---|---|---|---|---|---|---|
| | | | Yield strength (0.2% offset), 1000 lb./sq. in. | Tensile strength, 1000 lb./sq. in. | Elongation in 2 in., % | Hardness, Brinell | Density, lb./cu. in. | Specific gravity | Melting point, °F. | Specific heat (32°-212°F.), B.t.u./(lb.)(°F.) | Thermal expansion coefficient (32°-212°F.), $\times 10^{-6}$ in./(in.)(°F.) | Thermal conductivity (32°-212°F.), B.t.u./(sq. ft.)(hr.)(°F./in.) | Electrical resistivity (68°F.), ohms/cir. mil ft. | Tensile modulus of elasticity, $\times 10^6$ lb./sq. in. |
| Inconel Alloy 625 | Ni(+Co) 62.59, C 0.05, Mn 0.55, Fe 6.85, S 0.007, Si 0.35, Cu 0.05, Cr 20, Al 0.15, Ti 0.30, Cb(+Ta) 3.95 | Annealed | 75 | 125 | 50 | 176 | 0.298 | 8.44 | ..... | ..... | 7.0 | 75 | 776 (76°F.) | 30.0 |
| Inconel Alloy 700 | Ni 45.0, C 0.16, Mn 0.10, Fe 7.0, S 0.008, Si 0.25, Cr 15.0, Al 3.0, Ti 2.20, Mo 3.0, Co 28.0 | Hot-rolled | 106 | 168 | ..... | 321 | 0.295 | 8.17 | 2450-2600 | 0.11 | 6.8 | 86 | 777 | 32 |
| | | Heat-treated | ..... | | | | | | | | 9.3 (70°-1600°F.) | | | |
| Inconel Alloy 702 | Ni(+Co) 78.00, C 0.02, Mn 0.05, Fe 0.30, S 0.007, Si 0.15, Cu 0.05, Cr 15.85, Al 3.00, Ti 0.60 | Annealed | 45 | 105 | 55 | 150 | 0.302 | 8.37 | ..... | ..... | 6.95 | ..... | ..... | 31.5 |
| Inconel Alloy 705 weldable | Ni(+Co) 68.5, Cu 0.5, Fe 9.0, Mn 1.0, Si 1.6, Cr 15.5, C 0.20, Cb added | As-cast | 38 | 80 | 15 | 175 | 0.300 | 8.30 | 2540-2600 | 0.11 | 8.92 (70°-1400°F.) | 104 | 700 | 23 |

| Alloy | Composition | Condition | | | | | | | | | | | | |
|---|---|---|---|---|---|---|---|---|---|---|---|---|---|---|
| Inconel Alloy 705 | Ni(+Co) 68.0, Cu 0.5, Fe 8.0, Mn 1.0, Si 5.5, Cr 15.5, C 0.20 | As-cast | 90 | 95 | 2 | 340 | 0.292 | 8.06 | 2540–2600 | .... | 9.20 (70°–1100°F.) | .... | 700 | 25 |
| | | Annealed and aged | 95 | 100 | 2 | 340 | | | | | | | | |
| Inconel Alloy 713 | Ni(+Co) bal., Cr 13.0, C 0.13, Mo 4.5, Cb 2.0, Al 6.0, Ti 06 | Investment-cast | 102 | 120 | 6 | .... | 0.286 | 7.91 | 2300–2350 | .... | 6.1 | | | |
| Inconel Alloy 721 | Ni(+Co) 71.40, C 0.03, Mn 2.20, Fe 6.70, S 0.007, Si 0.10, Cu 0.04, Cr 16.40, Al 0.05, Ti 3.05 | Annealed | 45 | 110 | 50 | 160 | 0.298 | 8.25 | .. | .... | .... | .... | .... | 31.0 |
| | | Annealed, age-hardened | 100 | 145 | 7 | 300 | | | | | | | | |
| Inconel Alloy 722 | Ni(+Co) 74.35, C 0.05, Mn 0.60, Fe 6.50, S 0.007, Si 0.20, Cu 0.05, Cr 15.20, Al 0.60, Ti 2.40 | Annealed | 45 | 110 | 50 | 160 | 0.298 | 8.31 | 2540–2600 | 0.105 | 7.6 | 102 | 750 | 31.0 |
| | | Annealed, age-hardened | 95 | 155 | 30 | 275 | | | | | | | 728 (76°F.) | |
| Inconel Alloy X-750 | Ni(+Co) 72.85, C 0.04, Mn 0.65, Fe 6.80, S 0.007, Si 0.30, Cu 0.05, Cr 15.15, Al 0.75, Ti 2.50, Cb(+Ta) 0.85 | Annealed | 50 | 115 | 50 | 150 | 0.298 | 8.25 | 2540–2600 | 0.105 | 6.7 | 102 | 735 (76°F.) | 31.0 |
| | | Annealed, age-hardened | 115 | 175 | 25 | 300 | | | | | | 212°F. | | |
| Monel Alloy 400 | Ni(+Co) 66.15, C 0.12, Mn 0.90, Fe 1.35, S 0.005, Si 0.15, Cu 31.30 | Annealed | 35 | 75 | 40 | 125 | 0.319 | 8.83 | 2370–2460 | 0.127 | 7.5 | 174 (212°F.) | 290 | 26 |
| | | Hot-rolled | 50 | 90 | 35 | 150 | | | | | | | | |
| | | Cold-drawn | 80 | 110 | 25 | 190 | | | | | | | | |
| | | Cold-rolled | 100 | 110 | 5 | 240 | | | | | | | | |
| Monel Alloy 401 | Ni(+Co) 44.0, Cu bal., Fe 1.0, Mn 0.8, C 0.10 | Annealed | 20 | 75 | 51 | 140 | 0.321 | .... | 2400–2450 | 0.13 | 7.6 | 133 (70°F.) | 300 | |

## Table 3-41. Properties of Metals and Alloys (Continued)

| Material | Nominal composition (essential elements), % | Form and condition | Yield strength (0.2% offset), 1000 lb./sq. in. | Tensile strength, 1000 lb./sq. in. | Elongation in 2 in., % | Hardness, Brinell | Density, lb./cu. in. | Specific gravity | Melting point, °F. | Specific heat (at 32°-212°F.), B.t.u./(lb.)(°F.) | Thermal expansion coefficient (32°-212°F.), × 10⁻⁶ in./(in.)(°F.) | Thermal conductivity (32°-212°F.) B.t.u./(sq. ft.)(hr.)(°F./in.) | Electrical resistivity (68°F.), ohms/cir. mil ft. | Tensile modulus of elasticity × 10⁶ lb./sq. in. |
|---|---|---|---|---|---|---|---|---|---|---|---|---|---|---|
| Monel Alloy R-405 | Ni(+Co) 66.35, C 0.18, Mn 0.90, Fe 1.35, S 0.050, Si 0.15, Cu 31.00 | Hot-rolled | 45 | 85 | 35 | 145 | 0.319 | 8.84 | 2370–2460 | 0.127 | 7.8 | 180 | 290 | 26 |
|  |  | Cold-drawn | 75 | 100 | 25 | 200 |  |  |  |  |  |  |  |  |
| Monel Alloy 402 (obsolete) | Ni(+Co) 58.10, C 0.12, Mn 0.90, Fe 1.20, S 0.005, Si 0.10, Cu 39.55 | Hot-rolled | 65 | 90 | 35 | 175 | 0.320 | 8.85 | ..... | ..... | ..... | ..... | 305 | 26 |
|  |  | Cold-drawn | 85 | 95 | 25 | 215 |  |  |  |  |  |  |  |  |
| Monel Alloy 404 | Ni(+Co) 55.85, C 0.12, Mn 0.10, Fe 0.15, S 0.006, Si 0.10, Cu 44 | Cold-drawn | 24 | 65 | 50 | 200 | 0.320 | ..... | ..... | ..... | 7.7 (80°-200°F.) | 172 | 319 | 25.2 |
| Monel Alloy 411 weldable (Alloy 19) Centrifugal castings only | Ni(+Co) 62.0, Cu 31.5, Fe 2.0, Mn 0.8, Si 1.5, C 0.20, Cb added | As-cast | 35 | 75 | 35 | 140 | 0.312 | 8.63 | 2400–2450 | 0.13 | 9.17 (70°-1100°F.) | 186 | ..... | 23 |
| Monel Alloy K-500 | Ni(+Co) 65.25, C 0.15, Mn 0.60, Fe 1.00, S 0.005, Si 0.15, Cu 29.60, Al 2.75, Ti 0.45 | Annealed | 45 | 100 | 40 | 155 | 0.305 | 8.47 | 2400–2460 | 0.127 | 7.4 | 130 | 350 | 26 |
|  |  | Annealed, age-hardened | 100 | 155 | 25 | 270 |  |  |  |  |  |  |  |  |
|  |  | Spring | 140 | 150 | 5 | 300 |  |  |  |  |  |  |  |  |
|  |  | Spring, age-hardened | 160 | 185 | 10 | 335 |  |  |  |  |  |  |  |  |

| Material | Composition | Condition | | | | | | | | | | | | |
|---|---|---|---|---|---|---|---|---|---|---|---|---|---|---|
| Monel Alloy 506 (Alloy 16), centrifugal castings only | Ni(+Co) 63.0, Cu 30.5, Fe 1.5, Mn 0.8, Si 3.2, C 0.10 | As-cast | 70 | 115 | 10 | 265 | 0.305 | 8.48 | 2350–2400 | 0.13 | 8.93 (70°–1100°F.) | 145 | 370 | 24 |
| Monel Alloy 505 (Alloy 17), centrifugal castings only | Ni+(Co) 63.0, Cu 29.5, Fe 2.0, Mn 0.8, Si 4.0, C 0.08 | Annealed | 75 | 110 | 8 | 220 | 0.302 | 8.36 | 2250–2350 | 0.13 | 8.87 (70°–1000°F.) | 136 | 380 | 24 |
| | | As-cast, or annealed and aged | 110 | 135 | 2 | 340 | 0.302 | 8.36 | 2250–2350 | 0.13 | 8.87 | 136 | 380 | 24 |
| Monel Alloy 502 | Ni(+Co) 64.75, C 0.25, Mn 0.60, Fe 1.00, S 0.005, Si 0.15, Cu 29.85, Al 2.85, Ti 0.45 | Hot-finished, annealed | 35 | 95 | 48 | 160 | 0.305 | 8.44 | 2400–2460 | 0.127 | 7.0 | 130 | 350 | 26.0 |
| | | Hot-finished, annealed, age-hardened | 90 | 145 | 25 | 275 | | | | | | | | |

**Aluminum and Alloys**

| Aluminum Alloy No. | Composition | Condition | | | | | | | | | | | | |
|---|---|---|---|---|---|---|---|---|---|---|---|---|---|---|
| 1100 | Al 99 plus | Annealed-0 | 5 | 13 | 45 | 23 | .... | .... | .... | .... | .... | 1540 | 18 | 10 |
| | | Cold-rolled-H14 | 17 | 18 | 20 | 32 | 0.098 | 2.71 | 1190–1215 | 0.23 | 13.1 (68°F.) | .... | 19 | |
| | | Cold-rolled-H18 | 22 | 24 | 15 | 44 | .... | .... | .... | .... | .... | 1510 (68°F.) | | |
| 3003 | Al bal., Mn 1.2 | Annealed-0 | 6 | 16 | 40 | 28 | .... | .... | .... | .... | .... | 1340 | 21 | 10 |
| | | Cold-rolled-H14 | 21 | 22 | 16 | 40 | 0.099 | 2.73 | 1190–1210 | 0.23 | 12.9 (68°F.) | 1100 | 25 | |
| | | Cold-rolled-H18 | 27 | 29 | 10 | 55 | .... | .... | .... | .... | .... | 1075 (68°F.) | 26 | |
| 5052 | Al bal., Mg 2.5, Cr 0.25 | Annealed-0 | 13 | 28 | 30 | 47 | .... | .... | .... | .... | .... | 960 (68°F.) | 30 | 10.2 |
| | | Cold-rolled and stabilized-H34 | 31 | 38 | 14 | 68 | 0.097 | 2.68 | 1100–1200 | 0.23 | 13.2 (68°F.) | .... | .... | |
| | | Cold-rolled and stabilized-H38 | 37 | 42 | 8 | 77 | .... | .... | .... | .... | .... | 960 (68°F.) | 30 | |
| 5086 | Al bal., Mn 0.5, Mg 4.0, C 0.15 | Annealed-0 | 17 | 38 | 22 | .... | 0.96 | 2.66 | 1085–1185 | .... | .... | 940 (77°F.) | .... | 10.2 |
| | | H34 | 37 | 47 | 10 | | | | | | | | | |
| 6063 | Al bal., Si 0.4, Mg 0.7 | Annealed-0 | 7 | 13 | 30 | 25 | | | | | | | | 10.0 |
| | | Artificially aged-T5 | 21 | 27 | 12 | 60 | 0.098 | 2.70 | 1140–1205 | | 13.0 (68°F.) | 1390 (68°F.) | 20 | |
| | | Heat-treated and artificially aged-T6 | 31 | 35 | 12 | 73 | | | | | | | | |

# Table 3-41. Properties of Metals and Alloys (Continued)

| Material | Nominal composition (essential elements), % | Form and condition | Typical mechanical properties | | | | Typical physical constants | | | | | | | |
|---|---|---|---|---|---|---|---|---|---|---|---|---|---|---|
| | | | Yield strength (0.2% offset), 1000 lb./sq. in. | Tensile strength, 1000 lb./sq. in. | Elongation in 2 in., % | Hardness, Brinell | Density, lb./cu. in. | Specific gravity | Melting point, °F. | Specific heat (32°–212°F.), B.t.u./(lb.)(°F.) | Thermal expansion coefficient (32°–212°F.), $\times 10^{-6}$ in./(in.)(°F.) | Thermal conductivity (32°–212°F.), B.t.u./(sq. ft.)(hr.)(°F./in.) | Electrical resistivity (68°F.), ohms/cir. mil ft. | Tensile modulus of elasticity $\times 10^6$ lb./sq. in. |
| 7075 | Al bal., Zn 5.6, Cu 1.6, Mg 2.5, Cr 0.3 | Annealed-0 | 15 | 33 | 17 | 60 | 0.101 | 2.80 | 890–1180 | 0.23 | 12.9 (68°F.) | 840 (68°F.) | 34 | 10.4 |
| | | Heat-treated and artificially aged-T6 | 73 | 83 | 11 | 150 | | | | | | | | |
| 380 | Al bal., Cu 3.5, Si 9.0 | Die-cast-F | 26 | 43 | 2.0 | ..... | 0.098 | 2.72 | 1100–1100 | ..... | 11.6 (68°F.) | 670 (68°F.) | 45 | 10.3 |
| 43 | Al bal., Si 5.0 | Sand-cast-F | 8 | 19 | 8 | 40 | 0.097 | 2.69 | 1065–1170 | 0.23 | 12.3 (68°F.) | 990 (68°F.) | 27 | 10.3 |
| | | Permanent mold-cast-F | 9 | 23 | 10 | 45 | | | | | | | | |
| | | Die-cast-F | 16 | 30 | 7 | | | | | | | | | |
| 195 | Al bal., Cu 4.5, Si 0.8 | Sand-cast; heat-treated-4 | 16 | 32 | 8.5 | 60 | 0.102 | 2.81 | 970–1190 | 0.23 | 12.7 (68°F.) | 960 (68°F.) 960 (68°F.) | 30 | 10.3 |
| | | Sand-cast; heat-treated and artificially aged-T6 | 24 | 36 | 5 | 75 | ..... | ..... | ..... | ..... | ..... | | | |

## Copper and Alloys

| Material | Nominal composition (essential elements), % | Form and condition | Yield strength (0.2% offset), 1000 lb./sq. in. | Tensile strength, 1000 lb./sq. in. | Elongation in 2 in., % | Hardness, Brinell | Density, lb./cu. in. | Specific gravity | Melting point, °F. | Specific heat (32°–212°F.), B.t.u./(lb.)(°F.) | Thermal expansion coefficient (32°–212°F.), $\times 10^{-6}$ in./(in.)(°F.) | Thermal conductivity (32°–212°F.), B.t.u./(sq. ft.)(hr.)(°F./in.) | Electrical resistivity (68°F.), ohms/cir. mil ft. | Tensile modulus of elasticity $\times 10^6$ lb./sq. in. |
|---|---|---|---|---|---|---|---|---|---|---|---|---|---|---|
| Nickel silver 18% (wrought) 752 65–18 | Cu 65, Zn 17, Ni 18 | Annealed | 25 | 58 | 40 | 70 | 0.316 | 8.75 | 2030 | 0.09 | 9.0 | 230 | 173 | 18 |
| | | Cold-rolled (HT)[a] | 70 | 85 | 4 | 170 | | | | | | | | |
| | | Cold-drawn wire (HT) | ..... | 105 | | | | | | | | | | |

| Material | Composition | Condition | | | | | | | | | | | | |
|---|---|---|---|---|---|---|---|---|---|---|---|---|---|---|
| Nickel silver 10% (wrought) 740 65-10 | Cu 65, Zn 25, Ni 10 | Annealed | 20 | 55 | 45 | 60 | 0.313 | 8.67 | 1870 | 0.09 | 9.0 | 320 | 125 | 17.5 |
| | | Cold-rolled (HT) | 70 | 88 | 5 | 180 | | | | | | | | |
| | | Cold-drawn wire (HT) | ...... | 110 | | | | | | | | | | |
| Nickel silver 20% (cast) (11A) | Cu bal., Ni 20, Zn 6, Pb 5, Sn 4 | Cast (as cast) | 25 | 40 | 15 | 85 | 0.318 | 8.8 | 1980 | 0.12 | 10.0 | 120 | 210 | 15 |
| Cupronickel 10% 706 | Cu 88.35, Ni 10, Fe 1.25, Mn 0.4 | Annealed | 22 | 44 | 45 | ..... | 0.323 | 8.94 | 2090 | 0.09 | 9.3 | 310 | 113 | 18 |
| | | Cold-drawn tube | 57 | 60 | 15 | | | | | | | | | |
| Cupronickel 30% 715 | Cu 68.90, Ni 30, Mn 0.60, Fe 0.50 | Annealed | 22 | 55 | 45 | 70 | 0.323 | 8.94 | 2240 | 0.09 | 8.5 | 200 | 220 | 22 |
| | | Cold-drawn | 60 | 75 | 20 | 150 | | | | | | | | |
| | | Cold-rolled | 70 | 77 | 5 | 155 | | | | | | | | |
| Cupronickel 55-45 (Constantan) | Cu 55, Ni 45 | Annealed | 30 | 60 | 45 | ..... | 0.321 | 8.89 | 2300 | ..... | 8.1 | 155 | 290 | 24 |
| | | Cold-drawn | 50 | 65 | 30 | | | | | | | | | |
| | | Cold-rolled | 65 | 85 | 20 | | | | | | | | | |
| Copper 102 | Cu 99.9 plus | Annealed | 10 | 32 | 45 | 42 | 0.322 | 8.91 | 1980 | 0.092 | 9.3 | 2700 | 10.3 | 17 |
| | | Cold-drawn | 40 | 45 | 15 | 90 | | | | | | | | |
| | | Cold-rolled (HT) | 40 | 46 | 5 | 100 | | | | | | | | |
| Red brass (wrought) 230 | Cu 85, Zn 15 | Annealed | 15 | 40 | 50 | 50 | 0.316 | 8.75 | 1875 | 0.09 | 9.8 | 1100 | 28 | 17 |
| | | Cold-drawn | 55 | 70 | 15 | 120 | | | | | | | | |
| | | Cold-rolled | 60 | 75 | 7 | 135 | | | | | | | | |
| Red brass (cast) | Cu 85, Zn 5, Pb 5, Sn 5 | Cast (as cast) | 17 | 35 | 25 | 60 | 0.317 | 8.75 | 1810–1840 | ..... | 10.2 | 500 | 63 | 13 |
| Gilding metal 210 | Cu 95.0, Zn 5.0 | Cold-rolled | 50 | 56 | 5 | 114 | 0.320 | ..... | 1950 | ..... | 10.0 | 1600 | ..... | 17 |
| Commercial bronze 220 | Cu 90.0, Zn 10.0 | Cold-rolled | 54 | 61 | 5 | 125 | 0.318 | ..... | 1910 | ..... | 10.2 | 1300 | ..... | 17 |
| Cartridge 70-30 brass 260 | Cu 70.0, Zn 30.0 | Cold-rolled | 63 | 76 | 8 | 155 | 0.308 | ..... | 1750 | ..... | 11.1 | 840 | ..... | 16 |
| Architectural bronze 385 | Cu 57.0, Zn 40.0, Pb 3.0 | Annealed | 20 | 60 | 30 | 95 | 0.306 | ..... | 1630 | ..... | 11.6 | 850 | ..... | 14 |
| Phosphor bronze 10% 524 | Cu 90, Sn 10, P 0.25 | Spring temper | ...... | 122 | 4 | 241 | 0.317 | ..... | 1830 | ..... | 10.2 | 350 | ..... | 16 |
| Phosphor bronze 5% 510 | Cu 94.75, Sn 5, P 0.25 | Annealed | 20 | 50 | 50 | 60 | 0.320 | 8.86 | 1920 | 0.09 | 9.4 | 480 | 69 | 16 |
| | | Cold-drawn wire (HT) | ...... | 130 | 2 | ..... | | | | | | | | |
| | | Cold-rolled (HT) | 65 | 80 | 8 | 160 | | | | | | | | |

a Hard temper (HT). Hard temper and heat-treated (strip) (HT, HT).

## Table 3-41. Properties of Metals and Alloys (Continued)

| Material | Nominal composition (essential elements), % | Form and condition | Yield strength (0.2% offset), 1000 lb./sq. in. | Tensile strength, 1000 lb./sq. in. | Elongation in 2 in., % | Hardness, Brinell | Density, lb./cu. in. | Specific gravity | Melting point, °F. | Specific heat (32°-212°F.), B.t.u./(lb.)(°F.) | Thermal expansion coefficient (32°-212°F.), $\times 10^{-6}$ in./(in.)(°F.) | Thermal conductivity (32°-212°F.), B.t.u./(sq. ft.)(hr.)(°F./in.) | Electrical resistivity (68°F.), ohms/cir. mil ft. | Tensile modulus of elasticity, $\times 10^6$ lb./sq. in. |
|---|---|---|---|---|---|---|---|---|---|---|---|---|---|---|
| Aluminum brass | Cu 76.0, Zn 22.0, Al 2.0, As trace | Annealed | 27 | 60 | 55 | 82 | 0.301 | ..... | 1780 | ..... | 10.3 | 700 | ..... | 16 |
| Yellow brass (high brass 268) | Cu 65, Zn 35 | Annealed | 18 | 48 | 60 | 55 | 0.306 | 8.47 | 1710 | 0.09 | 10.5 | 830 | 40 | 15 |
| | | Cold-drawn | 55 | 70 | 15 | 115 | | | | | | | | |
| | | Cold-rolled (HT) | 60 | 74 | 10 | 180 | | | | | | | | |
| Naval brass 464 | Cu 60, Zn 39.25, Sn 0.75 | Annealed | 22 | 56 | 40 | 90 | 0.304 | 8.41 | 1625 | 0.09 | 11.0 | 810 | 40 | 15 |
| | | Cold-drawn | 40 | 65 | 35 | 150 | | | | | | | | |
| Admiralty brass 443 (inhibited) | Cu 71, Zn 28, Sn 1, As, Sb, or P present | Annealed | 20 | 53 | 65 | 60 | 0.308 | 8.53 | 1720 | 0.09 | 10.2 | 770 | 42 | 16 |
| Muntz metal 280 | Cu 60, Zn 40 | Annealed | 20 | 54 | 45 | 80 | 0.303 | 8.39 | 1660 | 0.09 | 10.8 | 870 | 37 | 15 |
| Manganese bronze 675 | Cu 58.5, Zn 39.2, Fe 1, Sn 1, Mn 0.3 | Annealed | 30 | 60 | 30 | 95 | 0.302 | 8.36 | 1645 | 0.09 | 11.2 | 770 | 45 | 15 |
| | | Cold-drawn | 50 | 80 | 20 | 180 | | | | | | | | |
| High-silicon bronze A 655 | Cu 96, Si 3, Mn, Zn, or Fe | Annealed | 22 | 58 | 60 | 70 | 0.308 | 8.53 | 1865 | 0.09 | 9.5 | 225 | 160 | 15 |
| | | Cold-drawn | 60 | 90 | 20 | 180 | | | | | | | | |
| | | Cold-rolled | 60 | 95 | 7 | 190 | | | | | | | | |
| Low-silicon bronze B 651 | Cu 96, Si 0.8-2.0, Mn 0.7 max, Fe 0.8 max. | Annealed | 15 | 40 | 50 | F55 B80 (Rockwell) | 0.316 | ..... | ..... | 0.09 | 9.9 | 360 | | |
| | | Hardened | 55 | 70 | 15 | | | | | | | | | |
| Aluminum bronze 612 | Cu 92, Al 8 | Annealed | 25 | 70 | 60 | 80 | 0.281 | 7.78 | 1900 | 0.09 | 9.2 | 490 | 70 | 17 |
| | | Hard | 65 | 105 | 7 | 210 | | | | | | | | |

| Material | Condition | | | | | | | | | | | | Ref. |
|---|---|---|---|---|---|---|---|---|---|---|---|---|---|
| Ni-Vee bronze type A<br>Cu 88, Ni 5, Sn 5, Zn 2 | As-cast | 22 | 50 | 40 | 85 | 0.32 | 8.8 | 1950 | ... | 11 | ... | ... | 15 |
| | Tempered | 40 | 65 | 10 | 130 | 0.32 | 8.8 | 1950 | ... | 11 | ... | ... | 15 |
| | Heat-treated | 55 | 85 | 10 | 180 | 0.32 | 8.8 | 1950 | ... | 11 | ... | ... | 15 |
| Ni-Vee bronze type B<br>Cu 87, Ni 5, Sn 5, Pb 1, Zn 2 | As-cast | 20 | 45 | 30 | 80 | 0.32 | 8.8 | 1925 | ... | 10 | ... | ... | 13 |
| | Tempered | 30 | 60 | 8 | 120 | ... | ... | ... | ... | ... | ... | ... | 14 |
| Copper beryllium 172<br>Be 1.9, Co 0.25, Cu bal. | Annealed (SA)[i] | ... | 70 | 45 | B60 (Rockwell) | ... | ... | ... | ... | ... | ... | ... | 17 |
| | Annealed (SA, HT) | ... | 175 | 6 | C38 | 0.298 | 8.26 | 1600–1800 | 0.10[A] | 9.3 | 750–900 | ... | 19 |
| | Cold-rolled (HT) | ... | 110 | 5 | B99 | ... | ... | ... | ... | ... | ... | ... | 17 |
| | Cold-rolled (HT, HT) | ... | 200 | 2 | C42 | ... | ... | ... | ... | ... | ... | ... | 19 |
| Copper beryllium 170<br>Be 1.7, Co 0.25, Cu bal. | Annealed (SA) | ... | ... | ... | ... | ... | ... | ... | ... | ... | ... | ... | 17 |
| | Annealed (SA, HT) | ... | 70 | 45 | B60 | 0.298 | 8.26 | 1600–1800 | ... | 9.3 | 750–900 | ... | 19 |
| | Cold-rolled (HT) | ... | 165 | 6 | C35 | ... | ... | ... | ... | ... | ... | ... | 17 |
| | Cold-rolled (HT, HT) | ... | ... | ... | ... | ... | ... | ... | ... | ... | ... | ... | 19 |
| Copper beryllium 177<br>Be 0.55, Co 2.5, Cu bal. | Annealed (SA) | ... | 45 | 27 | B30 | ... | ... | ... | ... | ... | ... | ... | 16 |
| | Annealed (SA, HT) | ... | 110 | 9 | B96 | 0.316 | 8.75 | 1885–1955 | ... | 9.8 (68°–392°F.) | 1450–1800 | ... | 18 |
| | Cold-rolled (HT) | ... | 78 | 6 | B75 | ... | ... | ... | ... | ... | ... | ... | 16 |
| | Cold-rolled (HT, HT) | ... | 120 | 8 | B98 | ... | ... | ... | ... | ... | ... | ... | 18 |
| Copper beryllium 176<br>Be 0.38, Co 1.55, Ag 1.0, Cu bal. | Annealed (SA) | ... | 45 | 27 | B30 | ... | ... | ... | ... | ... | ... | ... | 16 |
| | Annealed (SA, HT) | ... | 110 | 15 | B96 | 0.316 | 8.75 | 1850–1930 | ... | 9.8 (68°–392°F.) | 1500–1700 | ... | 18 |
| | Cold-rolled (HT) | ... | 70 | 13 | B73 | ... | ... | ... | ... | ... | ... | ... | 16 |
| | Cold-rolled (HT, HT) | ... | 120 | 14 | B99 | ... | ... | ... | ... | ... | ... | ... | 18 |

[A] cal./g./°C. (30°–100°C.).
[i] Solution annealed (SA). Solution annealed, heat-treated (SA, HT).

Table 3-41. Properties of Metals and Alloys (Continued)

| Material | Nominal composition (essential elements), % | Form and condition | Typical mechanical properties | | | | Typical physical constants | | | | | | | |
| --- | --- | --- | --- | --- | --- | --- | --- | --- | --- | --- | --- | --- | --- | --- |
| | | | Yield strength (0.2% offset), 1000 lb./sq. in. | Tensile strength, 1000 lb./sq. in. | Elongation in 2 in., % | Hardness, Brinell | Density, lb./cu. in. | Specific gravity | Melting point, °F. | Specific heat (32°–212°F.), B.t.u./(lb.)(°F.) | Thermal expansion coefficient (32°–212°F.), $\times 10^{-6}$ in./(in.)(°F.) | Thermal conductivity (32°–212°F.), B.t.u./(sq. ft.)(hr.)(°F./in.) | Electrical resistivity (68°F.), ohms/cir. mil ft. | Tensile modulus of elasticity $\times 10^6$ lb./sq. in. |
| **Lead and Alloys** | | | | | | | | | | | | | | |
| Chemical lead | Pb 99.9, Cu 0.06, Bi 0.005 max. | Rolled | 1.9 | 2.5 | 50 | 5 | 0.410 | 11.35 | 621 | 0.030 | 16.4 | 240 | 124 | 2 |
| Antimonial lead | Pb 94, Sb 6 | Cast | | 6.8 | 22 | 12 | 0.393 | 10.90 | 554 | 0.032 | 15.1 | 200 | 140 | 3 |
| | | Rolled | | 4.1 | 47 | 9 | | | | | | | | |
| Tellurium lead | Pb 99.85, Te 0.04, Cu 0.06 | Rolled | 2.2 | 3 | 45 | 6 | 0.410 | 11.35 | 621 | 0.030 | 16.4 | 240 | 124 | 2 |
| Soft solder 50-50 | Sn 50, Pb 50 | Cast | | 6.8 | 50 | 14 | 0.321 | 8.89 | 421 | 0.051 | 13.1 | 310 | 93 | |
| Soft solder 60-40 | Sn 60, Pb 40 | Cast | | 7.1 | 45 | 15 | 0.306 | 8.50 | 374 | 0.055 | 12.2 | 330 | 90 | |
| **Magnesium Alloys** | | | | | | | | | | | | | | |
| Magnesium alloy AZ92A | Mg bal., Al 9.0, Zn 2.0, Mn 0.10 min. | Sand-cast (as cast) | 14 | 24 | 6 | 50 | 0.066 | 1.83 | 1100 | 0.245 | 14.5 | 360 | 84 | 6.5 |
| | | Sand-cast (solution heat-treated) | 14 | 40 | 12 | 55 | | | | | | 310 | 101 | |
| | | Sand-cast (solution heat-treated and aged) | 19 | 40 | 5 | 83 | | | | | | 410 | 74 | |
| | | Sand-cast (age-hardened) | 16 | 30 | 4 | 66 | | | | | | 410 | 74 | |

| Alloy | Composition | Condition | | | | | | | | | | | | |
|---|---|---|---|---|---|---|---|---|---|---|---|---|---|---|
| Magnesium alloy AZ31B | Mg bal., Al 3.0, An 1.0, Mn 0.20 min. | Rolled-plate (strain hardened then partially annealed) | 24 | 37 | 18 |  | 0.064 | 1.77 | 1160 | 0.245 | 14.5 | 540 | 55 | 6.5 |
|  |  | Rolled-sheet (strain hardened then partially annealed) | 32 | 42 | 15 | 73 |  |  |  |  |  | 540 | 55 |  |
|  |  | Annealed | 22 | 37 | 21 | 56 |  |  |  |  |  | 540 | 55 |  |
|  |  | Extruded | 28 | 38 | 14 |  |  |  |  |  |  | 540 | 55 |  |
| Magnesium alloy AZ80A | Mg bal., Al 8.5, Zn 0.5, Mn 0.15 min. | Extruded | 36 | 49 | 11 | 60 | 0.065 | 1.80 | 1130 | 0.245 | 14.5 | 350 | 87 | 6.5 |
|  |  | Extruded (age-hardened) | 39 | 53 | 6 | 82 |  |  |  |  |  |  |  |  |
|  |  | Forged (age-hardened) | 34 | 50 | 6 | 72 |  |  |  |  |  |  |  |  |
| Magnesium alloy AZ91A and AZ91B | Mg bal., Al 9.0, Zn 0.6, Mn 0.13 min. | Die-cast (as cast) | 22 | 33 | 3 | 67 | 0.065 | 1.81 | 1105 | 0.245 | 14.5 | 370 | 83 | 6.5 |
| Magnesium alloy AZ91C | Mg bal., Al 8.7, Zn 0.7, Mn 0.13 min. | Sand-cast (as cast) | 14 | 24 | 2 | 52 |  |  |  |  |  | 370 | 82 |  |
|  |  | Sand-cast (solution heat-treated) | 14 | 40 | 11 | 55 | 0.065 | 1.81 | 1105 | 0.245 | 14.5 | 320 | 97 | 6.5 |
|  |  | Sand-cast (solution heat-treated and aged) | 19 | 40 | 5 | 73 |  |  |  |  |  | 390 | 78 |  |
| Magnesium alloy EZ33A | Mg bal., Zn 2.6, Zr 0.7, other elements 3.0 | Sand-cast (age-hardened) | 15 | 23 | 3 | 50 | 0.066 | 1.83 | 1189 | 0.245 | 14.5 | 690 | 42 | 6.5 |
| Magnesium alloy HK31A | Mg bal., Th 3.0, Zr 0.7 | Sand-cast (solution heat-treated and aged) | 15 | 30 | 8 | 55 | 0.065 | 1.79 | 1200 | 0.245 | 14.5 | 640 | 46 | 6.5 |
|  |  | Rolled-sheet (strain-hardened then partially annealed) | 29 | 37 | 8 | 57 |  |  |  |  |  | 780 | 37 |  |

# Table 3-41. Properties of Metals and Alloys (Continued)

| Material | Nominal composition (essential elements), % | Form and condition | Typical mechanical properties | | | | Typical physical constants | | | | | | | |
|---|---|---|---|---|---|---|---|---|---|---|---|---|---|---|
| | | | Yield strength (0.2% offset), 1000 lb./sq. in. | Tensile strength, 1000 lb./sq. in. | Elongation in 2 in., % | Hardness, Brinell | Density, lb./cu. in. | Specific gravity | Melting point, °F. | Specific heat (32°-212°F.), B.t.u./(lb.)(°F.) | Thermal expansion coefficient (32°-212°F.), × 10⁻⁶ in./(in.)(°F.) | Thermal conductivity (32°-212°F.), B.t.u./(sq. ft.)(hr.)(°F./in.) | Electrical resistivity (68°F.), ohms/cir. mil ft. | Tensile modulus of elasticity × 10⁶ lb./sq. in. |
| Magnesium alloy HZ32A | Mg bal., Th 3.0, Zn 2.1, Zr 0.7 | Sand-cast (age-hardened) | 14 | 29 | 7 | 57 | 0.066 | 1.83 | 1198 | 0.245 | 14.5 | 740 | 39 | 6.5 |
| Magnesium alloy ZK60A | Mg bal., Zn 5.7, Zr 0.55 | Extruded / Extruded (age-hardened) / Forged | 37 / 43 / 38 | 49 / 52 / 49 | 14 / 12 / 13 | 75 / 82 | ..... / 0.066 | ..... / 1.83 | 1175 | ..... / 0.245 | 14.5 | 800 / 850 | 36 / 34 | 6.5 |
| **Titanium and Alloys** | | | | | | | | | | | | | | |
| Titanium (comercially pure) | Ti bal., Fe 0.2 max., N₂ 0.05 max., C 0.08 max., H₂ 0.015 max., O₂ 0.4 | Annealed | 75 | 85 | 23 | 200 | 0.163 | 4.54 | 3100 | 0.129 / 0.125 | 5.0 (70°-1000°F.) | 114 | 370 | 15 |
| Titanium-chromium-iron-molybdenum alloy (Ti 5Al-25.Sn) | Ti bal., Al 5, Sn 2.5, Fe 0.5 max., C 0.05 max., H₂ 0.015 max., N₂ 0.05 max. | Annealed | 120 | 127 | 12 | 34 (Rockwell C) | 0.162 | 4.5 | 3000 | 0.125 | 5.3 (70°-1000°F.) | 54 | 1015 | 16 |
| Titanium-aluminum-vanadium alloy (Ti-6 Al-4V) | Ti bal., Al 6.0, V 4.0, Fe 0.25 max., C 0.08 max., H₂ 0.0125 max., N₂ 0.05 max. | Annealed / Heat-treated | 130 / 145 | 140 / 155 | 14 / 12 | 32 / 37 (Rockwell C) | 0.161 | 4.46 | 3000 | 0.135 | 5.3 (70°-1000°F.) | 50 | 1020 | 16.5 |

## Other Non-ferrous Alloys

| Alloy | Composition | As-cast | | | | | | | | | | | | |
|---|---|---|---|---|---|---|---|---|---|---|---|---|---|---|
| Antimony | Sb 100 | ..... | 1.56 | 42 | | | 0.249 | 6.62 | 1166 | 0.092 | 5.5 | 131 | 234 | 11.3 |
| Bi-Pb-In-Cd-Sn alloy (Cerrolow-117) | Bi 44.70, Pb 22.60, Sn 8.30, Cd 5.30, In 19.10 | Cast | 5.4 | 1.5 | | 12 | 0.32 | 8.85 | 117 | 0.035 | 13.8 | ..... | 0.347 | |
| Bi-Pb-Sn-Cd alloy (Cerrobend) | Bi 50.00, Pb 26.70, Sn 13.30, Cd 10.00 | Cast | 5.99 | 200 | | 9.2 | 0.399 | 9.38 | 158 | 0.040 | 12.2 | | 0.434 | |
| Bi-Pb alloy (Cerrobase) | Bi 55.50, Pb 44.50 | Cast | 6.4 | 60–70 | | 10.2 | 0.380 | 10.5 | 255 | 0.03+ | 11.6 | | 0.182 | |
| Bi-Sn alloy (Cerrotru) | Bi 58.00, Sn 42.00 | Cast | 8.0 | 200 | | 22 | 0.315 | 8.72 | 281 | 0.045 | 8.3 | | 0.520 | |
| Columbium | ..... | ..... | 35 | 40 | ..... | ..... | 0.31 | | 4474 | 0.06 | 3.8 | 375 | | |
| Columbium 1Zr | Zr 0.8–1.2, Cb bal. | ..... | 135 | 48 | ..... | ..... | 0.31 | | 4350 | | | | | |
| Columbium F48 | W 13.5–16.5, Mn 4.5–5.5, Zr 0.85–1.15, Cb bal. | ..... | 110 | 120 | ..... | ..... | 0.34 | | 4500 | | | | | |
| Columbium FS82 | Ta 33, Zr 0.8, Cb bal. | ..... | 50 | 68 | ..... | ..... | 0.37 | | 4550 | | | | | |
| Columbium D31 | Ti 10, Mo 10, Cb bal. | ..... | 90 | 100 | ..... | ..... | 0.29 | | 4100 | | | | | |
| Gold | Au 100 | Hard / Annealed | 30 / 17.5 | | | 2 / 40 | 0.692 | 19.3 | 1945 | 0.056 | 7.9 | 2060 | 14.7 | 10.8 |
| 18K white gold | Au 75, Ni 18.5, Zn 5.25, Cu 1.25 | Hard / Annealed | ..... / 93 | 178 / 115 | 323 / 211 | | 0.53 | 14.61 | 1886 | ...... | ...... | | 210.3 | 16.0 |
| Haynes Stellite Alloy 21 | C 0.25, Cr 28, Ni 2.5, Mo 5.5, Co bal. | As investment cast | 82.0 | 103 | 313 max | 8.0 | 0.299 | 8.30 | 2465 | 0.1006 | 7.83 (70–600°F.) | 100.6 (at 392°F.) | 527 | 36.0 |

Table 3-41. Properties of Metals and Alloys (Continued)

| Material | Nominal composition (essential elements), % | Form and condition | Typical mechanical properties | | | | Typical physical constants | | | | | | | |
| --- | --- | --- | --- | --- | --- | --- | --- | --- | --- | --- | --- | --- | --- | --- |
| | | | Yield strength (0.2% offset), 1000 lb./sq. in. | Tensile strength, 1000 lb./sq. in. | Elongation in 2 in., % | Hardness, Brinell | Density, lb./cu. in. | Specific gravity | Melting point, °F. | Specific heat (32°–212°F.), B.t.u./(lb.)(°F.) | Thermal expansion coefficient (32°–212°F.), $\times 10^{-6}$ in./(in.)(°F.) | Thermal conductivity (32°–212°F.), B.t.u./(sq. ft.)(hr.)(°F./in.) | Electrical resistivity (68°F.), ohms/cir. mil ft. | Tensile modulus of elasticity $\times 10^6$ lb./sq. in. |
| Hayes Stellite Alloy 31 (X-40 Cast) | C 0.50, Cr 25.5, Ni 11, W 7.5, Co bal. | As investment cast | 80.0 | 113 | 8.0 | 313 max. | 0.311 | 8.61 | 2500 | 0.0981 | 7.84 (70°–600°F.) | 102.7 (at 392°F.) | ..... | 36.0 |
| Haynes Stellite Alloy 25 | C 0.15 max, Cr 20.0, Ni 10.0, W 15.0, Mn 1.5, Co bal. | Wrought sheet Mill annealed | 63 | 140 | 60.0 | 244 | 0.330 | 9.15 | 2425–2570 | 0.0924 | 7.61 (70°–600°F.) | 64.9 | 532 | 34.2 |
| Haynes Stellite Alloy 36 | C 0.40, Cr 19, Ni 10, W 15.0, Mn 1.5, Co bal. | As investment cast | 90 | 103 | 5.0 | 298 max. | 0.326 | 9.04 | 2535 | 0.092 | 7.65 (70°–600°F.) | 65 | 532 | 33.8 |
| Iridium | Ir 100 | Annealed | ..... | 36 | ..... | 175 | 0.808 | 22.4 | 4449 | ..... | 3.7 | 407 | 29.4 | 75 |
| Molybdenum | Mo 99.9 plus | As-rolled Stress-relieved Recrystallized | 75 75 50 | 100 100 70 | 30 30 45 | 250 240 190 | 0.369 | 10.22 | 4730 | 0.061 | 2.67 | 900 | 31.3 | 46.0 |
| 0.5% Titanium molybdenum alloy | Mo bal., Ti 0.5 | As-rolled Stress-relieved Recrystallized | 90 90 60 | 120 120 80 | 30 30 40 | 290 280 200 | 0.369 | 10.22 | 4730 | 0.061 | 3.06 | 816 | 31.3 | 46.0 |

| Alloy | Composition | Condition | 58 | 118 | 49 | 194 | 0.296 | 8.20 | 2340–2400 | 0.104 | ..... | 101 (at 392°F.) | 560 | 29 |
|---|---|---|---|---|---|---|---|---|---|---|---|---|---|---|
| Multimet N-155 Alloy | Ni 19.0–21.0, Co 18.5–21.0, Cr 20.0–22.5, Mo 2.5–3.5, W 2.0–3.0, Fe bal., C 0.08–0.16, N 0.10–0.20, Cb + Ta 0.75–1.25 | Mill annealed sheet | 54 | 111 | 55 | 189 | | | | | | | | |
| | | Mill annealed bar | 54 | 98 | 23 | 180 | | | | | | | | |
| | | Sand-cast | 58 | 101 | 31 | 180 max. | | | | | | | | |
| | | An investment cast | | | | | | | | | | | | |
| Platinum | Pt (commercial) | Hard | ..... | 65 | 2 | 101 | 0.773 | 21.4 | 3215 | 0.057 | 5.0 | 465 | 65 | 22.0 |
| | | Annealed | ..... | 27 | 28 | 65 | | | | | | | | |
| Platinum-iridium | Pt 90, Ir 10 | Hard | ..... | 80 | 2 | 169 | 0.776 | 21.5 | 3299 | ..... | 5.0 | ..... | 146 | 25.0 |
| | | Annealed | 34 | 53 | 23 | 104 | | | | | | | | |
| Platinum-rhodium | Pt 90, Rh 10 | Hard | ..... | 93 | 3 | 169 | 0.720 | 19.93 | 3353 | ..... | ..... | ..... | 117 | 21.2 |
| | | Annealed | 18.3 | 50 | 36 | 79 | | | | | | | | |
| Platinum-ruthenium | Pt 90, Ru 10 | Hard | ..... | 145 | 2 | 210 | 0.713 | 19.8 | 3344 | ..... | ..... | ..... | 255 | 31.5 |
| | | Annealed | 47.6 | 91 | 28 | 156 | | | | | | | | |
| Palladium | Pd (commercial) | Hard | ..... | 55 | ... | 91 | 0.433 | 12.0 | 2829 | ..... | 6.5 | 488 | 63.5 | 16.3 |
| | | Annealed | 7.6 | 30 | 30 | 47 | | | | | | | | |
| Palladium-ruthenium | Pd 95.5, Ru 4.5 | Hard | ..... | 132 | 3 | 184 | 0.433 | 12.0 | ..... | ..... | ..... | ..... | 145.5 | 20.4 |
| | | Annealed | 51 | 85 | 26 | 120 | | | | | | | | |
| Palladium-silver | Pd 60, Ag 40 | Hard | 94 | 100 | ... | 176 | 0.415 | 11.5 | 2530° F. liquidus | ..... | ..... | ..... | 264.2 | 22.2 |
| | | Annealed | 15 | 47 | 40 | 87 | | | 2425° F. solidus | | | | | |
| Palladium alloy 934 | Pd 35, Pt 10, Au 10, Ag 30, Cu 15 | Annealed | 61 | 96 | 24 | 180 | 0.430 | 11.9 | 1985 | ..... | ..... | ..... | 214.8 | |
| | | Heat-treated | 125 | 146 | 10 | Aged 280 | | | | | | | | |
| Rhodium | Rh 100 | Annealed | ..... | 80 | ..... | 119 | 0.448 | 12.44 | 3571 | ..... | 4.6 | 611 | 27 | 50 |
| Silver (pure) | Ag 99.9 plus | Annealed | 12 | 23 | 45 | 30 | 0.379 | 10.50 | 1760 | 0.056 | 10.6 | 2900 | 9.8 | 10.5 |
| | | Cold-rolled | 38 | 43 | 6 | 90 | | | | | | | | |
| Sterling silver | Ag 92.5, Cu bal. | Hard | 50 | 64 | 4 | 125 | 0.376 | ..... | 1635 | ..... | 10.5 | 2510 | 12.08 | 10.5 |
| | | Annealed | 20 | 41 | 26 | 65 | | | | | | | | |
| Silver, coin | Ag 90, Cu bal. | Hard | 53 | 65 | 4 | 125 | 0.374 | ..... | 1615 | ..... | 10.5 | 2490 | 12.13 | 11 |
| | | Annealed | 23 | 42 | 26 | 70 | | | | | | | | |
| Tantalum | Ta 99.9 plus | Annealed sheet | 45 | 60 | 37 | 55 | 0.60 | 16.6 | 5425 | 0.036 | 3.6 | 377 | 74.6 | |
| | | Unannealed sheet | 100 | 110 | 3 | 123 | | | | | | | | |
| Tantalum 10W | W 10, Ta bal. | Annealed | 158 | 160 | ..... | ..... | 0.61 | ..... | 5516 | ..... | ..... | ..... | | |
| Tin | Sn 100 | As-cast | ..... | 2.1 | 70 | 3.9 | 0.263 | 7.29 | 449 | 0.0954 | 12.8 | 428 | 66 | 6 |

## Table 3-41. Properties of Metals and Alloys (Continued)

| Material | Nominal composition (essential elements), % | Form and condition | Typical mechanical properties | | | | Typical physical constants | | | | | | | |
|---|---|---|---|---|---|---|---|---|---|---|---|---|---|---|
| | | | Yield strength (0.2% offset), 1000 lb./sq. in. | Tensile strength, 1000 lb./sq. in. | Elongation in 2 in., % | Hardness, Brinell | Density, lb./cu. in. | Specific gravity | Melting point, °F. | Specific heat (32°–212°F.), B.t.u./(lb.)(°F.) | Thermal expansion coefficient (32°–212°F.), $\times 10^{-6}$ in./(in.)(°F.) | Thermal conductivity (32°–212°F.), B.t.u./(sq. ft.)(hr.)(°F./in.) | Electrical resistivity (68°F.), ohms/cir. mil ft. | Tensile modulus of elasticity $\times 10^6$ lb./sq. in. |
| Tungsten | W | Hard (sheet) | 360 | 400 | ...... | ...... | 0.697 | 19.3 | 6092 | 0.034 | 2.4 | 1390 | 33.08 | 53.0 |
| | | Annealed | ...... | ...... | ...... | 290 | | | | | | | | |
| | | Hard (wire) | 540 | 600 | 0–8 | ...... | 0.697 | 19.3 | 6092 | 0.034 | 2.4 | 1390 | 33.08 | 53.0 |
| Zinc | Zn bal., Pb 0.08 | Hot-rolled (long.) | ...... | 19.5 | 65 | 38 | 0.258 | 7.14 | 786 | 0.094 | 18 | 746.07 | 36.56 | |
| | | Hot-rolled (transv.) | ...... | 23 | 50 | ...... | ...... | ...... | ...... | ...... | 12.8 | | | |
| | | Cold-rolled (long.) | | 21 | 50 | | | | | | | | | |
| | | Cold-rolled (transv.) | | 27 | 40 | | | | | | | | | |
| Zilloy-15 | Zn bal., Mg 0.010, Cu 1.00 | Hot-rolled (long.) | ...... | 29 | 20 | 61 | 0.259 | 7.18 | 792 | 0.0957 | 19.3 | 725.76 | 38 | |
| | | Hot-rolled (transv.) | | 40 | 10 | ...... | ...... | ...... | | | 11.7 | | | |
| | | Cold-rolled (long.) | | 36 | 25 | 80 | | | | | | | | |
| | | Cold-rolled (transv.) | | 46 | 10 | | | | | | | | | |
| Zilloy-40 | Zn bal., Cu 1.00 | Hot-rolled (long.) | ...... | 24 | 50 | 52 | 0.259 | 7.18 | 792 | 0.0957 | 16.6 | ...... | 37.5 | |
| | | Hot-rolled (transv.) | | 30 | 35 | | | | | | | | | |
| | | Cold-rolled (long.) | | 31 | 40 | 60 | | | | | | | | |
| | | Cold-rolled (transv.) | | 40 | 30 | | | | | | | | | |

| | | | | | | | | | | | | | | |
|---|---|---|---|---|---|---|---|---|---|---|---|---|---|---|
| Zinc-aluminum alloy | Zn (99.99% pure remainder), Al 3.5–4.3, Mg 0.03–0.08, Cu 0.25 max. | Die-cast | ..... | 41 | 10 | 82 | 0.24 | 6.6 | 727.9 | 0.10 | 15.2 | 783.31 | 38.2 | |
| Zinc-aluminum-copper alloy | Zn (99.99% pure remainder), Al 3.5–4.3, Mg 0.03–0.08, Cu 0.75–1.25 | Die-cast | ..... | 47.6 | 7 | 91 | 0.24 | 6.7 | 727.0 | 0.10 | 15.2 | 754.78 | 39.2 | |
| Zirconium, commercial | $O_2$ 0.07, C 0.15, Hf 1.90, Zr bal. | Annealed | 40 | 65 | 27 | B80 (Rockwell) | 0.245 | 6.5 | 3380 | 0.118 | 2.9 | 95 | 246 | 11 |
| Zircaloy 2 | Hf 0.02, Sn 1.46, Fe 0.12, Ni 0.05, Zr bal., other 0.25 | Annealed | 50 | 75 | 22 | B90 (Rockwell) | 0.237 | ..... | ..... | ..... | 3.6 | 95 | | |
| Zircaloy 3 | Hf 0.02, Sn 0.25, Fe 0.25, Ni 0.05, Zr bal., other 0.20 | Annealed | 45 | 70 | 25 | B85 (Rockwell) | | | | | | | | |

source: Perry and Chilton (eds.), *Chemical Engineers' Handbook*, 5th ed., pp. 23-38 to 23-53, McGraw-Hill Book Company, New York, 1973.

## Table 3-42. Compositions of Typical High-temperature Alloys

(See also Fig. 3-15)

| Alloy | C | Cr | Ni | Mo | Co | W | Cb | Ti | Al | Fe | Other |
|---|---|---|---|---|---|---|---|---|---|---|---|
| **Ferritic Steels** | | | | | | | | | | | |
| 1.25 Cr, Mo | .10 | 1.25 | .... | 0.50 | .... | ... | .... | .... | .... | Balance | |
| 5 Cr, Mo | .20 | 5.00 | .... | 0.50 | .... | ... | .... | .... | .... | Balance | |
| "17-22-A" S | .30 | 1.25 | .... | 0.50 | .... | ... | .... | .... | .... | Balance | |
| 410 | .10 | 12.0 | .... | ..... | .... | ... | .... | .... | .... | Balance | |
| **Austenitic Steels** | | | | | | | | | | | |
| 316 | .08 | 17.0 | 12.0 | 2.50 | .... | ... | .... | .... | .... | Balance | |
| 347 | .06 | 18.0 | 12.0 | ..... | .... | ... | 0.70 | .... | .... | Balance | |
| 16-25-6 | .10 | 16.0 | 25.0 | 6.00 | .... | ... | .... | .... | .... | Balance | |
| A-286 | .05 | 15.0 | 26.0 | 1.25 | .... | ... | .... | 1.95 | 0.20 | Balance | |
| **Nickel-base Alloys** | | | | | | | | | | | |
| Inconel | .04 | 15.5 | 76.0 | ..... | .... | ... | .... | .... | .... | 7.0 | |
| Inconel X | .04 | 15.0 | 75.0 | ..... | .... | ... | .... | 2.5 | 0.6 | 7.0 | |
| Nimonic 90 | .08 | 20.0 | 58.0 | ..... | 16 | ... | .... | 2.3 | 1.4 | 0.5 | |
| Hastelloy B | .10 | 1.0 | 65.0 | 28 | .... | ... | .... | .... | .... | 5.0 | |
| René 41 | .10 | 19.0 | 53.0 | 10 | 11 | ... | .... | 3.2 | 1.6 | 2.0 | |
| Udimet 500 | .10 | 19.4 | 55.6 | 4 | 14 | ... | .... | 2.9 | 2.9 | 0.6 | |
| **Cobalt-base Alloys** | | | | | | | | | | | |
| Vitallium (HS-21) | .25 | 27.0 | 3.0 | 5 | 62 | ... | .... | .... | .... | 1.0 | |
| X-40 (HS-31) | .40 | 25.0 | 10.0 | ..... | 55 | 8 | .... | .... | .... | 1.0 | |
| **Complex Superalloys** | | | | | | | | | | | |
| N-155 (Multimelt) | .15 | 21.0 | 20.0 | 3 | 20 | 2.5 | 1.0 | .... | .... | Balance | 0.15N |
| S-590 | .40 | 20.0 | 20.0 | 4 | 20 | 4.0 | 4.0 | .... | .... | Balance | |
| S-816 | .40 | 20.0 | 20.0 | 4 | Bal. | 4.0 | 4.0 | .... | .... | 3.0 | |
| K 42 B | .05 | 18.0 | 43.0 | ..... | 22 | ... | .... | 2.5 | 0.2 | 13 | |
| Refractaloy 26 | .05 | 18.0 | 37.0 | 3 | 20 | ... | .... | 2.8 | 0.2 | 18 | |

SOURCE: "Mechanical Metallurgy" by Dieter. Copyright © 1961, McGraw-Hill, Inc. Used by permission of McGraw-Hill Book Company.

Stress-rupture properties give the stress to rupture a test bar in a specified number of hours and at a given temperature. Stress-rupture data do not indicate the creep rate, although, in general, a metal with good stress-rupture properties also has good creep resistance.

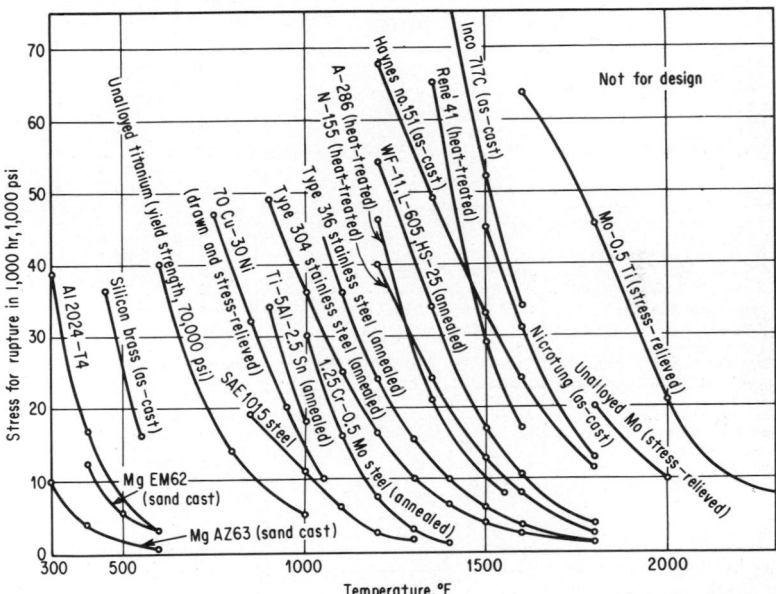

FIG. 3-15. Typical curves of stress for rupture in 1,000 hr versus temperature for selected engineering alloys. (Based on *D. P. Moon and W. F. Simmons, DMIC Memo 92, Battelle Memorial Institute, Mar. 23, 1961* in "Mechanical Metallurgy" by George Dieter, Jr. Copyright © 1961, McGraw-Hill, Inc. Used by permission.)

## Table 3-43. Mechanical and Phys

| Material | Specific gravity | Thermal conductivity, Btu/(hr)(ft²)(°F)(ft) | Coefficient of thermal expansion, $10^{-5}/°F$ | Specific heat, Btu/(lb)(°F) | Flammability, in./min | Modulus of elasticity in tension, $10^6$ psi |
|---|---|---|---|---|---|---|
| **Acrylonitrile butadiene styrene (ABS):** | | | | | | |
| High-impact................ | 1.04–1.06 | 0.08–0.12 | 4.7 | 0.35–0.38 | 1.3 | 2.6–2.9 |
| Extra-high impact........... | 1.01–1.06 | 0.08–0.12 | 4.7–5.6 | 0.35–0.38 | 1.3 | 2.1–2.6 |
| Low-temperature impact...... | 1.02 | 0.08–0.12 | 4.7–5.6 | 0.35–0.38 | 1.3 | 1.0 |
| Acetal polymer................ | 1.425 | 0.13 | 4.5 | 0.35 | 1.1 | 4.1 |
| **Acrylics, cast:** | | | | | | |
| General-purpose, type 1...... | 1.17–1.19 | 0.12 | 4.5 | 0.35 | 0.5–2.2 | 3.5–4.5 |
| General-purpose, type 2...... | 1.18–1.20 | 0.12 | 4.5 | 0.35 | 0.5–1.8 | 4.0–5.0 |
| Alkyds, molded................ | 2.22–2.24 | 0.35–0.60 | 1–3 | ........ | Self-ext. | ........ |
| **Cellulose acetate:** | | | | | | |
| Medium, type 1............. | 1.23–1.34 | 0.10–0.19 | 4.4–9.0 | 0.3–0.4 | 0.5–2.0 | ........ |
| Hard, type 2................ | 1.29–1.34 | 0.10–0.19 | 4.4–9.0 | 0.3–0.4 | 0.5–2.0 | ........ |
| Soft, type 3................ | 1.27–1.34 | 0.10–0.19 | 4.4–9.0 | 0.3–0.4 | ....... | ........ |
| **Cellulose acetate butyrate (CAB):** | | | | | | |
| Medium, type 1............. | 1.16–1.24 | 0.10–0.19 | 6–9 | 0.3–0.4 | 0.5–1.5 | ........ |
| Hard, type 2................ | 1.19–1.25 | 0.10–0.19 | 6–9 | 0.3–0.4 | 0.5–1.5 | ........ |
| Soft, type 3................ | 1.15–1.22 | 0.10–0.19 | 6–9 | 0.3–0.4 | 0.5–1.5 | ........ |
| Chlorinated polyether (Penton) | 1.4 | 11.0 | 6.6 | ........ | Self-ext. | ........ |
| **Epoxies, cast:** | | | | | | |
| General-purpose............ | 1.12–2.4 | 0.1–0.8 | 1.7–5.0 | ........ | ⎰0.3 to | ........ |
| Heat-resistant.............. | 1.15–3.2 | 0.1–0.8 | 2.8–3.3 | ........ | ⎱ self-ext. | ........ |
| **Fluorocarbons:** | | | | | | |
| Polytrifluorochlorethylene..... | 2.15 | 0.145 | 3.88 | 0.22 | Non | 1.9–3.0 |
| Polytetrafluoroethylene....... | 2.1–2.3 | 0.14 | 5.5 | 0.25 | Non | 0.38–0.65 |
| Fluorinated ethylene-propylene | 2.14–2.17 | 0.11 | 8.3–10.5 | 0.28 | Non | 0.5–0.7 |
| Melamines, unfilled........... | 1.48 | ........ | ........ | ........ | Self-ext. | ........ |
| **Polyamides, molded:** | | | | | | |
| Nylon 66 (0.2% water)....... | 1.14 | 0.14 | 5.5 | 0.3–0.5 | Self-ext. | 4.1 |
| Nylon 6.................... | 1.14 | 0.10–0.14 | 4.6–5.4 | 0.4 | Self-ext. | 2.5–3.4 |
| Nylon 11................... | 1.1 | ........ | 5.5 | 0.58 | Self-ext. | 1.8–1.9 |
| Polycarbonates............... | 1.20 | 0.11 | 3.9 | ........ | Self-ext. | 3.2 |
| Phenolics, molded (no filler)..... | 1.24–1.90 | ........ | 2.4 | ........ | Self-ext. | 7–15 |
| **Phenolics, cast:** | | | | | | |
| Type 1, mechanical and chemical.................. | 1.31 | ........ | 3.3–4.4 | ........ | Self-ext. | 4–5 |
| **Polyester, cast:** | | | | | | |
| Allyl type................... | 1.30–1.45 | 0.12 | 2.8–5.6 | 0.26–0.55 | ....... | 2–3 |
| Styrene type, rigid........... | 1.12–1.46 | 0.10–0.12 | 3.9–5.6 | 0.30–0.55 | ....... | 1.5–6.5 |
| **Silicones, molded:** | | | | | | |
| Mineral filler................ | 1.8–2.0 | 0.09–0.97 | 2.78–3.23 | ........ | 0–78 | ........ |

## ical Properties of Plastics (73.4°F)

| Tensile strength, 1,000 psi | Elongation (in 2 in.), % | Hardness, Rockwell | Impact strength (Izod notched), ft-lb/in. notch | Modulus of elasticity in flex, 10⁵ psi | Flexural strength, 1,000 psi | Compression strength (0.1% offset), 1,000 psi | Maximum recommended service temperature, °F | Heat distortion temperature, °F (264 psi) |
|---|---|---|---|---|---|---|---|---|
| 4.5–8.5<br>-8<br>_ -5 | 5–100<br>20–50<br>30–200 | R85–118<br>R85–100<br>R30–65 | 3–6<br>5–9<br>6–10 | 37–45<br>.......<br>....... | 7.5–11<br>6.8–8.0<br>3–4 | .......<br>.......<br>....... | 150<br>.......<br>....... | 185–215<br>185<br>175–185 |
| 10.0 | 15<br>(total) | M94, R120 | 1.4 | 4.1 | 14.1 | 5.2 | 185 | 212 |
| 6–9<br>8–10 | 2–7<br>2–7 | M8C–90<br>M96–102 | 0.4<br>0.4 | 3.5–4.5<br>4.0–5.0 | 12–14<br>15–17 | 12–14<br>14–18 | 140–160<br>180–200 | 150–180<br>190–225 |
| 3–4 | ....... | ........... | 0.30–0.35 | ....... | 7–10 | 16–20 | 350 | 350–400 |
| 2.7–6.5<br>6–8.5<br>4.6–7.5 | 18–54<br>6–31<br>17–40 | R68–115<br>R112–123<br>R106–121 | 1.1–4.0<br>0.4–1.9<br>0.6–2.3 | 1.1–3.5<br>2.6–4.0<br>1.9–3.4 | ........<br>........<br>........ | 14–25<br>25–36<br>22–33 | | |
| 2.9–5.7<br>5.0–6.8<br>1.9–3.8 | 47–66<br>38–54<br>60–74 | R79–112<br>R108–114<br>R59–95 | 1.0–4.3<br>0.6–2.4<br>2.5–5.4 | 0.93–1.7<br>1.5–2.0<br>0.74–1.3 | No break<br>No break<br>No break | .......<br>.......<br>....... | .......<br>.......<br>....... | 130–172<br>158–210<br>121–137 |
| 6 | 130<br>(total) | R100 | 0.4 | 1.3 | 5 | ....... | 255 | 185 |
| 2–12<br>5–14 | 2–6<br>2–5 | M75–110<br>M90–110 | 0.2–0.7<br>0.2–1.5 | 0.4–1.5<br>0.4–1.5 | 8–20<br>8–20 | 20–40<br>25–40 | 175<br>400 | 250<br>500 |
| 4.6–5.7<br>2.5–3.5<br>2.5–3.5 | 125–175<br>250–350<br>300–900 | R110–115<br>J75–95<br>D55 | 3.5–3.6<br>2.5–4.0<br>No break | 2–2.5<br>0.6<br>0.8 | 3.5<br>1.6<br>........ | 2.0<br>0.7–1.8<br>1.6 | 380<br>500<br>400 | 150–178 |
| ........ | ....... | ........... | ........ | 13 | 11–14 | 40–45 | 210 | 295 |
| 11.8<br>10.2–12<br>8.5 | 60<br>300<br>100–120 | M79, R118<br>R105–118<br>A50 | 0.9<br>1.2–3.0<br>3.3–3.6 | 4.1<br>.......<br>....... | 13.8<br>........<br>........ | .......<br>7–9.7<br>....... | 275–300<br>225–250<br>212–250 | 145 |
| 9–10.5 | 60–100<br>(total) | M70, R118 | 12–16 | 3.8 | 11–13 | 11 | 250–300 | 280–290 |
| 4.5–7.5 | ....... | M105–120 | 0.2–0.6 | 7–15 | 7–12 | 18–32 | 300–425 | 300–350 |
| 6–9 | ....... | M93–120 | 0.30–0.45 | 3–5 | 11–17 | 14–18 | ....... | 170–195 |
| 4.5–7<br>4–10 | .......<br><5 | M92–118<br>M65–115 | 0.18–0.32<br>0.18–0.40 | 3–8<br>3–9 | 6–14<br>7–19 | 20–26<br>12–37 | 300<br>250–300 | 120–320<br>120–420 |
| 4–4.3 | ....... | M89 | 0.25–0.30 | 10–13 | 6.8–7.5 | 16–20 | >700 | >900 |

## Table 3-43. Mechanical and Physical

| Material | Specific gravity | Thermal conductivity, Btu/(hr)(ft²)(°F)(ft) | Coefficient of thermal expansion, $10^{-5}/°F$ | Specific heat, Btu/(lb)(°F) | Flammability, in./min | Modulus of elasticity in tension, $10^5$ psi |
|---|---|---|---|---|---|---|
| Polystyrenes: | | | | | | |
|   General-purpose.............. | 1.04–1.07 | 0.058–0.09 | 3.3–4.8 | 0.30 | 1–1.5 | 4–5 |
|   Heat, chemical resistant...... | 1.05–1.11 | 0.046–0.09 | 3.6–3.8 | 0.30 | 0.4–1.0 | 4–6 |
| Polyethylene: | | | | | | |
|   Type I, low density.......... | 0.91–0.925 | 0.19 | 8.9–11.0 | 0.55 | 1.0 | 0.21–0.27 |
|   Type II, medium density..... | 0.926–0.940 | 0.19 | 8.3–16.7 | 0.55 | 1.0 | ........ |
|   Type III, high density....... | 0.942–0.960 | 0.19 | 8.3–16.7 | 0.46–0.55 | 1.0 | ........ |
| Polypropylene................. | 0.89–0.91 | 0.08 | 6.2 | 0.46 | ....... | 1.4–1.7 |
| Polyvinyl butyral: | | | | | | |
|   Rigid...................... | 1.08–1.12 | ........ | 4.4–12.7 | 0.4 | Slow | 3.5–4.0 |
|   Flexible.................... | 1.05 | ........ | ........ | ........ | ....... | ........ |
| Polyvinyl dichloride............ | 1.5 | ........ | ........ | ........ | ....... | 0.40 |
| Polyvinyl chloride: | | | | | | |
|   Type I, rigid............... | 1.32–1.44 | 0.07–0.10 | 2.8–3.3 | ........ | Self-ext. | 3.5–4.0 |
|   Type II, flexible............ | 1.20–1.55 | 0.07–0.10 | ........ | ........ | Self-ext. | 0.004–0.03 |
| Polyvinylidene chloride (saran).. | 1.68–1.75 | 0.053 | 8.78 | 0.32 | Self-ext. | 0.4–0.8 |
| Ureas, molded: | | | | | | |
|   Cellulose-filled.............. | 1.52 | ........ | ........ | ........ | Self-ext. | 13–16 |
| Reinforced | | | | | | |
| Phenolic | | | | | | |
|   Glass fabric filled........... | 1.80–1.95 | 0.15 | 5.0–6.0 | 0.23 | Self-ext. | 34 |
|   Asbestos fiber.............. | 1.908 | 0.17 | ........ | 0.30 | Self-ext. | 55 |
| Polyester: | | | | | | |
|   Glass fiber reinforced........ | 1.6–2.0 | 0.15 | 1.2–4 | 0.3 | Self-ext. | ........ |
| Silicones: | | | | | | |
|   Glass fabric reinforced....... | 1.6–1.93 | ........ | ........ | ........ | ....... | ........ |
| Epoxies: | | | | | | |
|   Woven-glass filled........... | 1.6–1.85 | ........ | ........ | ........ | ....... | ........ |
|   Filament-wound............. | 1.7–2.2 | 20–60 | ........ | ........ | Self-ext. | ........ |

NOTE: ASTM test results referring to thermoplastics: sp gr, D792; thermal conductivity, C177; coefficient of thermal expansion, D696; flammability, D635; modulus of elasticity, D638, D790; tensile strength, D638, D651; elongation, D638; hardness, D785; impact strength, D256; flexibility strength, D790; compression strength, D695; heat distortion temperature, D648.

Other works on the properties of plastics include S. Gross (ed.), "Modern Plastics Encyclopedia," McGraw-Hill Book Company, New York, 998 pp., 1970.

SOURCE: Perry and Chilton (eds.), "Chemical Engineers' Handbook," 5th ed., McGraw-Hill Book Company, New York, 1973.

## Properties of Plastics (73.4°F) (Continued)

| Tensile strength, 1,000 psi | Elongation (in 2 in.), % | Hardness, Rockwell | Impact strength (Izod notched), ft-lb/in. notch | Modulus of elasticity in flex, 10^5 psi | Flexural strength, 1,000 psi | Compression strength (0.1% offset), 1,000 psi | Maximum recommended service temperature, °F | Heat distortion temperature, °F (264 psi) |
|---|---|---|---|---|---|---|---|---|
| 5–8<br>10–11 | 1.5–2.5<br>1–4 | M68–80<br>M78–88 | 0.25–0.35<br>0.25–0.50 | 4–5<br>4–6 | 8–15<br>11–17 | 11.5–16<br>12–17 | 140–160<br>175–190 | 165–190<br>200–220 |
| 1.4–2.5<br><br>2.0<br><br>2.9–4.0 | 500–725<br><br>200<br><br>25–400 | C73<br>(Shore)<br>D55<br>(Shore)<br>D60<br>(Shore) | ........<br><br>........<br><br>0.4–6.0 | 13–27<br><br>43<br><br>90–125 | ........<br><br>........<br><br>........ | .......<br><br>.......<br><br>....... | 250<br><br>250<br><br>250 | 175–200<br>(soft pt.)<br>215<br>(soft pt.)<br>250<br>(soft pt.) |
| 5.0 | 500–700<br>(total) | R85–95 | 1.02 | 1.4–1.7 | 8.1 | ....... | 275–320 | 130–140 |
| 4–8.5<br>0.5–3 | 5–60<br>150–450 | L95<br>10–100<br>(Shore) | 1.2 | ....... | 10 | ....... | 115 | 61.5 |
| 7.5–9.0 | 4.5 | R117 | 0.8–5.0 | ....... | 14–17 | ....... | 210 | |
| 5.5–9.0<br>1–3.5 | 5–25<br>200–450 | R117<br>.......... | 0.25–1.2<br>Variable | 3.8–5.4<br>....... | 12–16<br>........ | 11–12<br>....... | 150–165<br>150–220 | 140–170 |
| 3–5 | 15–25 | M50–65 | 0.3–1.0 | ....... | 4–7 | 7.5–8.5 | 150–212 | 130–150 |
| 5–10 | 1.0 | E94 | 0.24–0.35 | ....... | 10–18 | 25–38 | ....... | 266–280 |
| Plastics | | | | | | | | |
| 58<br>46.1 | ........<br>........ | ..........<br>.......... | 15 | 48<br>49.8 | 87<br>52.5 | 47.5<br>20.8 | 525<br>>350 | 500–600<br>500–600 |
| 25–55 | ....... | M100 | 13–18 | 20–38 | 40–75 | 25–45 | 250–400 | 390–550 |
| 20–40 | ....... | .......... | ........ | 18–32 | 23–47 | 9–24 | 450–500 | |
| 40–85<br>80–250 | ........<br>........ | M100<br>M98–120 | 12–18<br>40–60 | 30–46<br>50–70 | 65–120<br>100–270 | 45–52<br>45–70 | 250–400<br>500 | 350–400 |

## Table 3-44. Standard Pipe Tables

### Chemical and Tensile-strength Requirements for Several ASTM and API Specificities

Continuous-weld (Furnace-welded) Pipe

| Specification | Steel | Specified chemical requirements | | | | | Specified tensile requirements | |
|---|---|---|---|---|---|---|---|---|
| | | % manganese | | % phosphorus | | % sulfur | Yield point or yield strength (min. psi) | Tensile strength (min. psi) |
| | | Min. | Max. | Min. | Max. | Max. | | |
| ASTM A120 | | ASTM A120 does not specify chemical or tensile requirements; it specifies mill hydrostatic tests. | | | | | | |
| ASTM A53 | Open hearth | | | | 0.08 | ..... | 25,000 | 45,000 |
| API 5 L | Open hearth Class I | 0.30 | 0.60 | | .045 | .060 | 25,000 | 45,000 |
| | Class II | .30 | .60 | 0.045 | .080 | .060 | 28,000 | 48,000 |

Electric Resistance-welded Pipe

| Specification | Grade | Specified chemical requirements | | | | Specified tensile requirements | |
|---|---|---|---|---|---|---|---|
| | | % carbon (max.) | % manganese (max.) | % phosphorus (max.) | % sulfur (max.) | Yield point or yield strength (min. psi) | Tensile strength (min. psi) |
| ASTM A120 | | ASTM A120 does not specify chemical or tensile requirements; it specifies mill hydrostatic tests. | | | | | |
| ASTM A53 | A | | | 0.050 | ..... | 30,000 | 48,000 |
| | B | | | .050 | ..... | 35,000 | 60,000 |
| ASTM A135 | A | | | .050 | 0.060 | 30,000 | 48,000 |
| | B | | | .050 | .060 | 35,000 | 60,000 |
| ASTM A252 | 1 | | | | ..... | 30,000 | 50,000 |
| | 2 | | | | ..... | 35,000 | 60,000 |
| API 5 L | A | 0.21 | 0.90 | .04 | .05 | 30,000 | 48,000 |
| | B | .26 | 1.15 | .04 | .05 | 35,000 | 60,000 |
| API 5 LX* nonexpanded | X-42 | .28 | 1.25 | .04 | .05 | 42,000 | 60,000 |
| | X-46 | .31 | 1.35 | .04 | .05 | 46,000 | 63,000 |
| | X-52 | .31 | 1.35 | .04 | .05 | 52,000 | 66,000 |
| API 5 LX* expanded | X-42 | .28 | 1.25 | .04 | .05 | 42,000 | 60,000 |
| | X-46 | .28 | 1.25 | .04 | .05 | 46,000 | 63,000 |
| | X-52 | .28 | 1.25 | .04 | .05 | 52,000 | 66,000 |

* All grades listed in API 5 LX are shown for information purposes only.

# Table 3-44. Standard Pipe Tables (Continued)

Seamless Pipe

| Specification | Grade | Specified chemical requirements | | | | | | Specified tensile requirements | |
|---|---|---|---|---|---|---|---|---|---|
| | | % carbon (max.) | % manganese (min.) | (max.) | % phosphorus (max.) | % sulfur (max.) | % silicon (min.) | Yield point or yield strength (min. psi) | Tensile strength (min. psi) |
| ASTM A120 | | ASTM A120 does not specify chemical or tensile requirements; it specifies mill hydrostatic tests. | | | | | | | |
| ASTM A53 | A | | | | .048 | | | 30,000 | 48,000 |
| | B | | | | .048 | | | 35,000 | 60,000 |
| ASTM A106 | A | 0.25 | 0.27 | 0.93 | .048 | .058 | .10 | 30,000 | 48,000 |
| | B | .30 | .29 | 1.06 | .048 | .058 | .10 | 35,000 | 60,000 |
| | C | .35 | .29 | 1.06 | .048 | .058 | .10 | 40,000 | 70,000 |
| ASTM A252 | 1 | | | | | | | 30,000 | 50,000 |
| | 2 | | | | | | | 35,000 | 60,000 |
| | 3 | | | | | | | 45,000 | 66,000 |
| API 5 L | A | .22 | | 0.90 | .04 | .05 | | 30,000 | 48,000 |
| | B | .27 | | 1.15 | .04 | .05 | | 35,000 | 60,000 |
| API 5 LX* | X-42 | .29 | | 1.25 | .04 | .05 | | 42,000 | 60,000 |
| | X-46 | .32 | | 1.35 | .04 | .05 | | 46,000 | 63,000 |
| | X-52 | .32 | | 1.35 | .04 | .05 | | 52,000 | 66,000 |

* All grades listed in API 5 LX are shown for information purposes only.
SOURCE: Adapted from "Youngstown Engineering Data Oil Country and Line Pipe," sec. G, Youngstown Sheet and Tube Co., Youngstown, Ohio, 1964.

# Table 3-44. Standard Pipe Tables (Continued)

Standard-weight Pipe

Dimensions, Weights, and Test Pressures—Plain Ends and Threads and Couplings

| Size: nominal, in. | Weight per foot | | Wall thickness, in. | Diameter | | No. of threads per inch | Coupling | | Test pressure, psi | | |
|---|---|---|---|---|---|---|---|---|---|---|---|
| | Nom., thds. and cplg., lb | Calculated plain ends, lb | | Outside, in. | Inside, in. | | Length, in. | Outside diameter, in. | Butt-welded | Grade A | Grade B |
| 1/8 | 0.24 | 0.24 | 0.068 | 0.405 | 0.269 | 27 | 1 3/16 | 0.563 | 700 | 700 | 700 |
| 1/4 | 0.42 | 0.42 | .088 | 0.540 | 0.364 | 18 | 1 3/16 | 0.719 | 700 | 700 | 700 |
| 3/8 | 0.57 | 0.57 | .091 | 0.675 | 0.493 | 18 | 1 3/16 | 0.875 | 700 | 700 | 700 |
| 1/2 | 0.85 | 0.85 | .109 | 0.840 | 0.622 | 14 | 1 9/16 | 1.063 | 700 | 700 | 700 |
| 3/4 | 1.13 | 1.13 | .113 | 1.050 | 0.824 | 14 | 1 5/8 | 1.313 | 700 | 700 | 700 |
| 1 | 1.68 | 1.68 | .133 | 1.315 | 1.049 | 11 1/2 | 2 | 1.576 | 700 | 700 | 700 |
| 1 1/4 | 2.28 | 2.27 | .140 | 1.660 | 1.380 | 11 1/2 | 2 1/16 | 1.900 | 1000 | 1000 | 1100 |
| 1 1/2 | 2.73 | 2.72 | .145 | 1.900 | 1.610 | 11 1/2 | 2 1/16 | 2.200 | 1000 | 1000 | 1100 |
| 2 | 3.68 | 3.65 | .154 | 2.375 | 2.067 | 11 1/2 | 2 1/8 | 2.750 | 1000 | 1000 | 1100 |
| 2 1/2 | 5.82 | 5.79 | .203 | 2.875 | 2.469 | 8 | 3 1/8 | 3.250 | 1000 | 1000 | 1100 |
| 3 | 7.62 | 7.58 | .216 | 3.500 | 3.068 | 8 | 3 1/4 | 4.000 | 1000 | 1000 | 1100 |
| 3 1/2 | 9.20 | 9.11 | .226 | 4.000 | 3.548 | 8 | 3 3/8 | 4.625 | 1200 | 1200 | 1300 |
| 4 | 10.89 | 10.79 | .237 | 4.500 | 4.026 | 8 | 3 1/2 | 5.000 | 1200 | 1200 | 1300 |
| 5 | 14.81 | 14.62 | .258 | 5.563 | 5.047 | 8 | 3 3/4 | 6.296 | .... | 1200 | 1300 |
| 6 | 19.18 | 18.97 | .280 | 6.625 | 6.065 | 8 | 4 | 7.390 | .... | 1200 | 1300 |

NOTES: The customary weight tolerance is ±5 per cent.
Taper of threads on all sizes of pipe and in couplings in sizes 2½ in. and over is ¾ in./ft on diameter. Couplings 2 in. and smaller are straight-tapped.

## Table 3-44. Standard Pipe Tables (Continued)

Extra-strong Pipe

Dimensions, Weights, and Test Pressures—Plain Ends and Threads and Couplings

| Size: nominal, in. | Weight per foot calculated plain ends, lb | Wall thickness, in. | Diameter | | Test pressure, psi | | |
|---|---|---|---|---|---|---|---|
| | | | Outside, in. | Inside, in. | Butt-welded | Grade A | Grade B |
| ⅛ | 0.31 | 0.095 | 0.405 | 0.215 | 850 | 850 | 850 |
| ¼ | 0.54 | .119 | 0.540 | 0.302 | 850 | 850 | 850 |
| ⅜ | 0.74 | .126 | 0.675 | 0.423 | 850 | 850 | 850 |
| ½ | 1.09 | .147 | 0.840 | 0.546 | 850 | 850 | 850 |
| ¾ | 1.47 | .154 | 1.050 | 0.742 | 850 | 850 | 850 |
| 1 | 2.17 | .179 | 1.315 | 0.957 | 850 | 850 | 850 |
| 1¼ | 3.00 | .191 | 1.660 | 1.278 | 1300 | 1500 | 1600 |
| 1½ | 3.63 | .200 | 1.900 | 1.500 | 1300 | 1500 | 1600 |
| 2 | 5.02 | .218 | 2.375 | 1.939 | 1300 | 1500 | 1600 |
| 2½ | 7.66 | .276 | 2.875 | 2.323 | 1300 | 1500 | 1600 |
| 3 | 10.25 | .300 | 3.500 | 2.900 | 1300 | 1500 | 1600 |
| 3½ | 12.51 | .318 | 4.000 | 3.364 | 1700 | 1700 | 1800 |
| 4 | 14.98 | .337 | 4.500 | 3.826 | 1700 | 1700 | 1800 |
| 5 | 20.78 | .375 | 5.563 | 4.813 | ⋯ | 1700 | 1800 |
| 6 | 28.57 | .432 | 6.625 | 5.761 | ⋯ | 1700 | 1800 |
| 8 | 43.39 | .500 | 8.625 | 7.625 | ⋯ | 1700 | 2400 |
| 10 | 54.74 | .500 | 10.750 | 9.750 | ⋯ | 1600 | 1900 |
| 12 | 65.42 | .500 | 12.750 | 11.750 | ⋯ | 1600 | 1900 |

NOTES: The customary weight tolerance is ±5 per cent.
Taper of threads is ¾ in./ft on diameter for all sizes.

# Table 3-44. Standard Pipe Tables (Continued)

### Double Extra-strong Pipe
Dimensions, Weights, and Test Pressures—Plain Ends and Threads and Couplings

| Size: nominal, in. | Weight per foot calculated plain ends, lb | Wall thickness, in. | Diameter | | Test pressure, psi | | |
|---|---|---|---|---|---|---|---|
| | | | Outside, in. | Inside, in. | Butt-welded | Grade A | Grade B |
| ½ | 1.71 | 0.294 | 0.840 | 0.252 | 1,000 | 1,000 | 1,000 |
| ¾ | 2.44 | .308 | 1.050 | 0.434 | 1,000 | 1,000 | 1,000 |
| 1 | 3.66 | .358 | 1.315 | 0.599 | 1,000 | 1,000 | 1,000 |
| 1¼ | 5.21 | .382 | 1.660 | 0.896 | 1,400 | 1,800 | 1,900 |
| 1½ | 6.41 | .400 | 1.900 | 1.100 | 1,400 | 1,800 | 1,900 |
| 2 | 9.03 | .436 | 2.375 | 1.503 | .... | 1,800 | 1,900 |
| 2½ | 13.70 | .552 | 2.875 | 1.771 | .... | 1,800 | 1,900 |
| 3 | 18.58 | .600 | 3.500 | 2.300 | .... | 1,800 | 1,900 |
| 4 | 27.54 | .674 | 4.500 | 3.152 | .... | 2,000 | 2,100 |
| 5 | 38.55 | .750 | 5.563 | 4.063 | .... | 2,000 | 2,100 |
| 6 | 53.16 | .864 | 6.625 | 4.897 | .... | 2,000 | 2,100 |
| 8 | 72.42 | .875 | 8.625 | 6.875 | .... | 2,800 | 2,800 |

NOTES: The customary weight tolerance is ±10 per cent.
Taper of threads is ¾ in./ft on diameter for all sizes.
SOURCE: Adapted from "Youngstown Engineering Data Oil Country and Line Pipe," sec. G, Youngstown Sheet and Tube Co., Youngstown, Ohio, 1964.

SECTION 4

# BUILDING SYSTEMS ENGINEERING

**Gregory E. Brooks, B.C.E., M.C.E., P.E.;** Chief Structural Engineer, Haines Lundberg & Waehler; Fellow, American Society of Civil Engineers; Member, American Concrete Institute, American Welding Society, Building Research Institute, National Society of Professional Engineers

**Leander Economides, B.S.M.E., P.E., R.A.;** Economides & Goldberg, Consulting Engineers; Member, American Society of Heating, Air Conditioning and Refrigeration Engineers, Building Research Institute, American Society of Mechanical Engineers, American Institute of Architects

**Bronislaus F. Winckowski, B.E.E., P.E.;** Chief Electrical Engineer, Haines Lundberg & Waehler; Senior Member, The Institute of Electrical and Electronics Engineers; Member, Illuminating Engineering Society, Building Research Institute, National Society of Professional Engineers

## CONTENTS

## STRUCTURAL

### 4-1. Loads

**Structural Design.** All structures are designed to sustain safely the weight of all permanent stationary construction (dead load) entering into a structure and the greatest loads induced by the intended occupancy (live load) or other uses.

**Design Loads.** The weights of materials most commonly used in building construction are tabulated and are classified as dead loads. Live loads are generally considered to be uniformly distributed and are classified according to occupancy.

**Dead Load.** Dead load is usually determined by the use of the weights given in Table 4-1.

## Table 4-1

| Building materials | Wt., lb/ft³ | Building materials | Wt., lb/ft³ |
|---|---|---|---|
| Aluminum | 175 | Fir | 32 |
| Ash, white | 40 | Granite, limestone, marble | 165 |
| Ashes, cinders | 45 | Iron, cast | 450 |
| Brass | 534 | Iron, wrought | 485 |
| Brick | 120 | Lead | 710 |
| Bronze | 509 | Maple | 43 |
| Cedar | 22 | Oak | 59 |
| Clay, dry | 63 | Paper | 58 |
| Clay, damp | 110 | Pine, white | 26 |
| Clay and gravel | 100 | Pine, yellow, long leaf | 44 |
| Concrete, lightweight aggregates | 90 | Poplar | 30 |
| Concrete, normal aggregates | 150 | Redwood | 26 |
| Concrete block, hollow, lightweight aggregates | 65 | Sand, gravel, dry, loose | 90–105 |
| | | Sand, gravel, dry, packed | 100–120 |
| Concrete block, hollow, normal aggregates | 85 | Sandstone, bluestone | 140 |
| | | Spruce | 27 |
| Copper | 556 | Steel | 490 |
| Earth, loose | 76 | Tin | 459 |
| Earth, packed | 95 | Zinc | 440 |
| Elm | 45 | | |

## Table 4-1 (Continued)

| Partitions and walls | Thick., in. | Wt., psf | Partitions and walls | Thick., in. | Wt., psf |
|---|---|---|---|---|---|
| Partitions:* | | | Interior walls (continued): | | |
| Hollow plaster partition........ | 4 | 22 | Concrete block, hollow-normal | 12 | 97 |
| Plaster on metal lath.......... | ¾ | 7 | Concrete block, hollow-cinder | 3 | 17 |
| Steel studs—metal lath and | | | | 4 | 24 |
| plaster (2 sides)............. | 4¾ | 15 | | 6 | 33 |
| Wood studs—wood lath | | | | 8 | 39 |
| and plaster (2 sides)........ | 5⅜ | 14 | | 12 | 63 |
| Wood studs—metal lath | | | Gypsum block, solid......... | 2 | 10 |
| and plaster (2 sides)........ | 5⅜ | 16 | | 3 | 13 |
| Wood studs—sheetrock........ | 4⅜ | 5 | | | |
| | | | Exterior walls:* | | |
| Interior walls:* | | | Brick...................... | 4 | 40 |
| Brick...................... | 4 | 40 | | 8 | 80 |
| Clay tile, hollow block....... | 3 | 17 | | 12 | 120 |
| | 4 | 18 | Block, cinder, hollow......... | 8 | 38 |
| | 6 | 25 | | 12 | 61 |
| | 8 | 31 | Block, cinder, solid.......... | 8 | 48 |
| Concrete block, hollow-normal.. | 3 | 26 | | 12 | 72 |
| | 4 | 35 | Block, normal aggregate...... | 8 | 59 |
| | 6 | 50 | | 12 | 97 |
| | 8 | 59 | | | |

| Floors, ceilings, & roofing | Thick., in. | Wt., psf | Floors, ceilings, & roofing | Thick., in. | Wt., psf |
|---|---|---|---|---|---|
| Floors: | | | Roofing (continued): | | |
| Asphalt—mastic.............. | 1 | 12 | Cooper sheet................ | ... | 2 |
| Asphalt—tile................ | ⅛ | 1 | Concrete plank.............. | 2 | 13 |
| Cement or terr. finish......... | 1 | 13 | | 2¾ | 18 |
| Cinder concrete fill............ | 2 | 10 | Felt, 4 layers............... | ... | 1 |
| Cinder concrete plank........ | 2 | 15 | Foamglass (insulation)....... | 1 | 1 |
| Concrete, lightweight.......... | 1 | 8 | Gypsum (fill)............... | 1 | 3 |
| Concrete, normal............. | 1 | 12 | Gypsum slab, precast........ | 2 | 12 |
| Floor plate................. | ⅜ | 16 | | 3 | 14 |
| Grating (1–1¼ × ³⁄₁₆)........ | 1¼ | 9 | Lead...................... | ⅛ | 8 |
| Cellular metal flooring........ | 1½ | 5 | 3-ply roofing................ | ... | 1 |
| | 3 | 7 | 4-ply felt and gravel......... | ... | 6 |
| | | | 5-ply felt and gravel......... | ... | 7 |
| Ceilings: | | | Sheathing.................. | 1 | 3 |
| Insulation.................. | 1 | 2 | Shingle, asbestos............ | ... | 4 |
| Plaster on concrete........... | ½ | 3 | Shingle, wood............... | ... | 2 |
| Plaster on metal lath......... | ¾ | 7 | Slag roofing................. | ... | 5 |
| Plaster on suspended metal lath | ... | 10 | Slate...................... | ¼ | 10 |
| Plaster on wood lath.......... | ⅞ | 6 | Steel (No. 20 gauge)......... | ... | 4 |
| Pressed steel (No. 18 gauge).... | ... | 3 | Tile, flat................... | ... | 18 |
| Sheetrock.................. | ½ | 2 | Tile, Spanish................ | ... | 8 |
| | | | Tin....................... | ... | 1 |
| Roofing: | | | Transite.................... | ... | 4 |
| Cinder fill.................. | 1 | 5 | Wood roofers (av).......... | ... | 3 |
| Concrete channel slab, light- | | | | | |
| weight, precast............. | 3½ | 14 | | | |

* Weights given do not include plaster on any surface.  Add 5 psf for plaster applied directly on each face. Add 7 psf for metal lath and plaster applied on each face.

**Live Load.** Live loads generally used for buildings are shown in Table 4-2.

**Table 4-2**

| Occupancy | Live load, psf | Occupancy | Live load, psf |
|---|---|---|---|
| Public buildings: | | Hospitals, etc. (*Continued*): | |
| Armories........................... | 150 | Operating rooms................... | 60 |
| Auditoriums, churches, etc.: | | Corridors, laboratories............ | 100 |
| Fixed seats........................ | 60 | Residential buildings: | |
| Movable seats...................... | 100 | Living areas........................ | 40 |
| Exhibition buildings: | | Corridors........................... | 100 |
| Restaurants, etc................... | 100 | Business buildings: | |
| Schools: | | Office buildings..................... | 80 |
| Classrooms........................ | 40 | Light manufacturing................ | 125 |
| Corridors.......................... | 100 | Heavy manufacturing.............. | 175 min |
| Other public buildings............... | 80 | Storage buildings: | |
| Theaters (stage floor)............... | 150 | Garages............................ | 100 |
| Institutional buildings: | | Light warehouses................... | 125 |
| Hospitals, etc.: | | Heavy warehouses................. | 250 min |
| Private rooms and wards.......... | 40 | | |

**Wind Load.** Vertical walls should be designed to resist a wind load, acting either inward or outward, as shown in Table 4-3.

**Table 4-3**

| Height, ft | Wind press., psf | Height, ft | Wind press., psf |
|---|---|---|---|
| Less than 50.......................... | 20 | 100–199............................. | 28 |
| 50–99................................ | 24 | 200 and above....................... | 30 |

**Roof Wind and Snow Loads.** Roofs should be designed for a combined wind and snow load of from 25 to 45 psf. The lesser value would be used for flat roofs in no-snow areas, and the greater values for flat roofs in heavy-snow areas. Intermediate values would be used for sloping roofs in all areas.

**Other Loads.** Other types of loading that should be considered in particular cases include earthquake loads, excessive wind loads, and impact loads.

## 4-2. Structural Framing Systems

The most economical structural framing system for a particular building is predicated on many variable conditions: use, locale, availability of material, fire resistivity, and magnitude of construction, to mention a few. The cost per square foot of floor area of an interior bay cannot be assumed as the average cost of the whole area, since the exterior-wall spandrels, elevator shafts, and wind-bracing and foundation conditions are not reflected in this single area, and these factors assume greater proportions of total building cost for smaller buildings than for larger buildings. Cost of various modes of design, however, can be determined by the design of an interior bay, which will indicate to some degree the most economical system to be chosen for a particular use and occupancy.

The unit prices used in this comparison of necessity reflect average prices for a certain locality. Any one unit price might be subject to variation. However, it is believed that the relationship between systems will not be materially affected. For the purpose of such a comparison, eight different designs of an interior bay are shown and described in Fig. 4-1 as schemes 1 to 8.

The designs are made for an office load of 80 + 20 psf, the 20 psf being an allowance for lightweight movable partitions, as required by most building codes. Weight in pounds per square foot includes floor fills, floor arches, beams, girders, and the average column weight for a 10-story structure, plus 85 per cent of the live load.

Indicated cost includes only the separately listed parts, and does not include contractor's profit, sales tax, and overhead charges.

The table of relative cost (Table 4-4) is so arranged to show:

## Table 4-4. Construction Cost Analysis

| Construction | Structural steel framing | | | | | Reinforced concrete | | |
|---|---|---|---|---|---|---|---|---|
| | Scheme 1 | Scheme 2 | Scheme 3 | Scheme 4 | Scheme 5 | Scheme 6 | Scheme 7 | Scheme 8 |
| Live + dead load, psf*....... | 160 | 163 | 182 | 203 | 150 | 192 | 222 | 208 |
| Depth of construction........ | 1'11" | 2'0¼" | 2'2½" | 2'2⅝" | 2'7½" | 2'1½" | 2'1½" | 1'2" |
| 1 Concrete................ | 0.58 | 0.54 | 0.95 | 0.91 | 0.47 | 1.38 | 1.63 | 1.30 |
| 2 Reinforcing steel......... | 0.18 | ...... | 0.37 | 0.44 | 0.07 | 1.65 | 1.38 | 1.70 |
| 3 Structural steel.......... | 2.38 | 2.61 | 2.57 | 2.99 | 1.87 | ...... | ...... | ...... |
| 4 Open web joist........... | ...... | ...... | ...... | ...... | 1.47 | ...... | ...... | ...... |
| 5 Metal cellular deck...... | 1.12 | 2.28 | ...... | ...... | ...... | ...... | ...... | ...... |
| 6 Forms................... | ...... | ...... | 2.50 | 2.50 | 0.26 | 2.50 | 2.75 | 1.96 |
| 7 Monolithic fl. finish...... | 0.23 | 0.23 | 0.23 | 0.23 | 0.23 | 0.23 | 0.23 | 0.23 |
| 8 Subtotal................ | 4.49 | 5.66 | 6.62 | 7.07 | 4.37 | 5.76 | 5.99 | 5.19 |
| 9 Cost ratio.............. | 100 | 126 | 147 | 157 | 97 | 128 | 133 | 116 |
| 10 Fire retardant........... | 0.96 | 0.91 | 0.18 | 0.18 | 1.93 | ...... | ...... | ...... |
| 11 Subtotal................ | 5.45 | 6.57 | 6.80 | 7.25 | 6.30 | 5.76 | 5.99 | 5.19 |
| 12 Cost ratio.............. | 100 | 121 | 125 | 133 | 116 | 106 | 110 | 95 |
| 13 Elect. service ducts...... | 1.03 | 0.70 | 4.62 | 4.62 | 4.62 | 4.62 | 4.62 | 4.62 |
| 14 Hung ceiling............ | 1.49 | 1.49 | 1.49 | 1.49 | 1.49 | 1.49 | 1.49 | 1.49 |
| 15 Total.................. | 7.97 | 8.76 | 12.91 | 13.36 | 12.41 | 11.87 | 12.10 | 11.30 |
| 16 Cost ratio.............. | 100 | 110 | 162 | 168 | 156 | 149 | 152 | 142 |

* Load includes floor arches, beams, girders, and columns, plus 85 per cent of live load. Depth of construction is from finished floor to underside of F. P. of girders.

1. A complete structural floor capable of supporting the design loads, but not containing underfloor electrical ducts or hung ceiling and with no fire protection rating (lines 8 and 9).

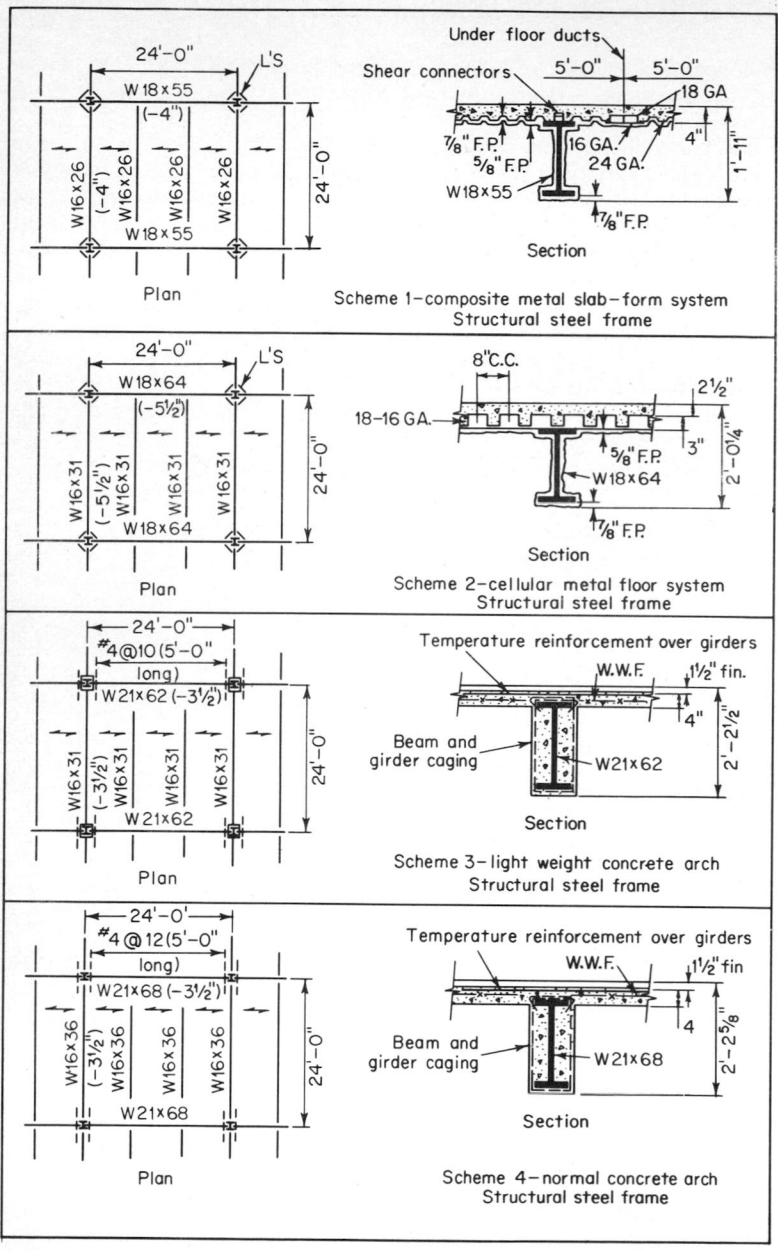

Fig. 4-1. Alternative construction schemes for interior bays.

**4**-6

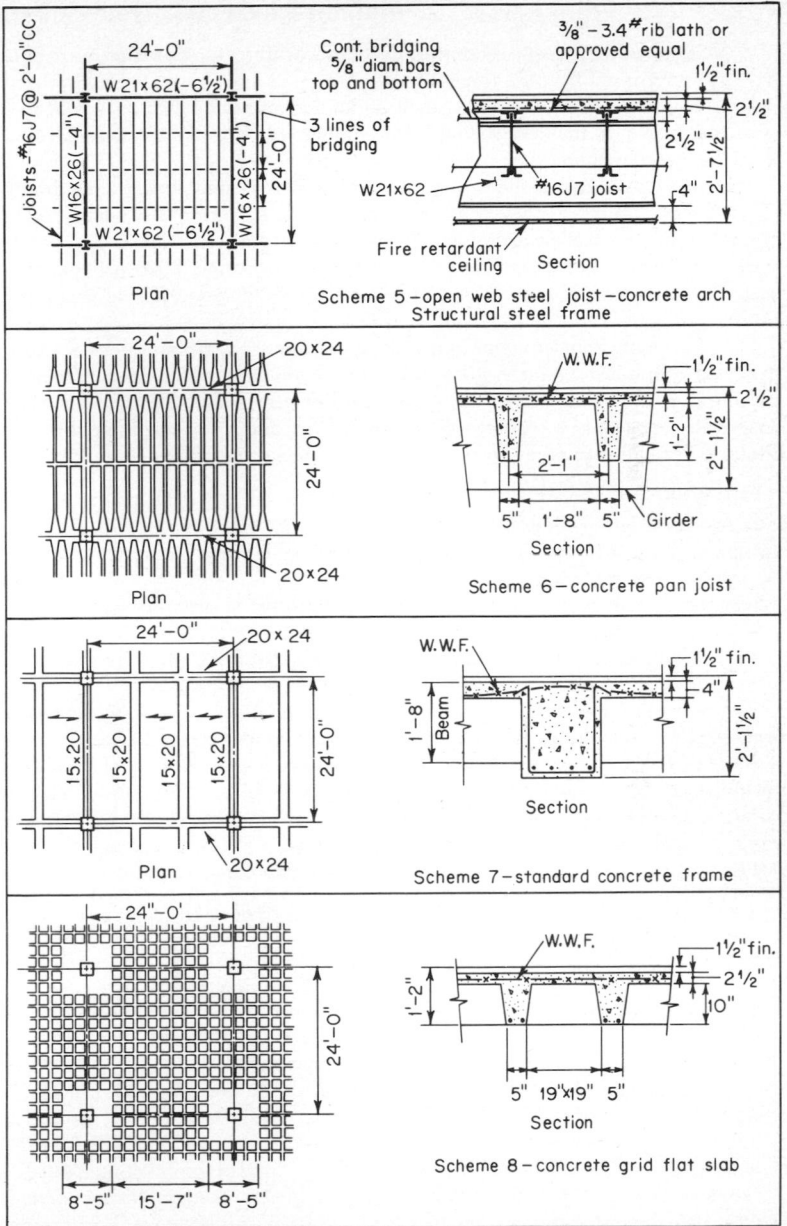

### Plan

24'-0"
W 21x62 (-6½")
W16x26 (-4')
24'-0"
Joists−#16J7 @ 2'-0"CC
W16x26 (-4')
W 21x62 (-6½")

### Section

Cont. bridging
⅝"diam.bars
top and bottom
3 lines of bridging
W21x62
#16J7 joist
Fire retardant ceiling

⅜"−3.4# rib lath or approved equal
1½"fin.
2½"
2½"
2'-7½"
4"

**Scheme 5 −open web steel joist−concrete arch Structural steel frame**

### Plan

24'-0"
20x24
24'-0"
20x24

### Section

W.W.F.
1½"fin.
2½"
1'-2"
2'-11½"
2'-1"
5" 1'-8" 5"
Girder

**Scheme 6 −concrete pan joist**

### Plan

24'-0"
20x24
24'-0"
15x20  15x20  15x20  15x20
20x24

### Section

W.W.F.
1½" fin.
4"
2'-11½"
1'-8" Beam

**Scheme 7 −standard concrete frame**

### Plan

24'-0'
24'-0"
8'-5"   15'-7"   8'-5"

### Section

W.W.F.
1½" fin.
2½"
10"
1'-2"
5" 19"x19" 5"

**Scheme 8 −concrete grid flat slab**

Fig. 4-1. Alternative construction schemes for interior bays.   (*Continued*)

2. Cost of adding required materials (if necessary) to give the maximum fire rating required by code (lines 11 and 12).

3. Cost for electrical flexibility, including additional ducts (if required) and cross headers. A finished ceiling is also included, but no mechanical duct work has been included.

Schemes 1 to 5 show structural steel framing of A36 steel, using a maximum unit stress of 24,000 psi, except for schemes 3 and 4, in which beam caging is used to allow a unit stress of 27,000 psi (AISC spec. sec. 1.11.2).

Schemes 6 to 8 are designs in reinforced concrete, using 3,000 psi concrete, with the exception of scheme 8, which requires 4,000 psi concrete to satisfy shear conditions.

For all eight floor systems without underfloor electrical ducts a steel-troweled monolithic cement floor finish has been included. In the case of schemes 3 to 8, the underfloor ducts are installed above the structural concrete floor slab, requiring a concrete fill of $1\frac{1}{2}$ in. in depth to accommodate the ducts. This fill is charged to the cost of electric service ducts (line 13).

### Fire Retardant Rating

A fire retardant rating of 3 hr is attained in the floor systems of all eight designs, with a 4-hr rating for column protection (shown on lines 10 to 12). Sprayed-on fire retardant is used in schemes 1 and 2.

Scheme 5, using open-web joist, is protected by a fire retardant ceiling attached to the underside of the joist and girders.

All columns of structural steel are covered with metal lath and fire retardant plaster or other approved fire protection giving a 4-hr rating.

The bar reinforcement or structural steel of schemes 3, 4, 6, 7, and 8 is protected by the minimum thickness of concrete cover required by code.

Scheme 1 is a design of composite construction, using shear connectors on both beams and girders, utilizing a light-gauge-steel form as reinforcement for the concrete slab. Where underfloor electrical ducts are required, cellular units are utilized. These units are usually placed 5 ft center to center. The cost for electrification is estimated as follows:

$0.33 per ft² cellular units (differential increase in cost due cellular section)
 0.70 per ft² for cross-header ducts
$1.03 per ft² (scheme 1, line 13)

Scheme 2 floor system uses metal cellular flooring with cells 8 in. center to center. No shear connectors are used in this type of floor; therefore the design is not composite in nature, requiring heavier beams and girders than are used in scheme 1. If underfloor electrical ducts are required, scheme 2 offers the greatest flexibility in the layout of underfloor ducts, having a cellular section every 8 in. The only extra cost involved for electrification of the floor system will be $0.70 per square foot for the cross-header ducts.

Schemes 3 and 4 are the usual beam and girder designs with concrete arches. Scheme 3 was included to show a comparison between lightweight and normal-weight concrete. An underfloor electrical-duct system for schemes 3 to 8 requires $1\frac{1}{2}$ in. of concrete fill to accommodate the ducts.

The estimated cost for underfloor ducts will be:

$0.39 for concrete fill (the difference in unit cost)
4.23 for grid system of ducts (including headers)
$4.62 (schemes 3 to 8, line 13)

The details of these floor systems of construction are clearly shown in the plans and sections of their particular reference.

Table 4-4 clearly demonstrates that a given framing system can change its relative economic position, depending on the design criteria established for the particular building being analyzed.

Whereas scheme 5 is the most economical for a building requiring no fire protection, it drops to fifth place when fire protection is required. Scheme 8 at no extra cost now becomes the most economical. However, when electrification of the floor becomes a design criterion, scheme 1 or 2 becomes the economical choice.

## MECHANICAL

This section includes design criteria, data, and physical laws and formulas applicable to the design of building heating, ventilating, air-conditioning, and plumbing systems.

### 4-3. Heating

**Heating Load**

$$Q_t = Q_{tr} + Q_{inf} + Q_{vent} \tag{4-1}$$

where

$Q_t$ = total heating load, Btu/hr

$Q_{tr}$ = transmission load = $AU(t_i - t_o)$    Btu/hr    (4-2)

$Q_{inf}$ = infiltration load = 1.08 (cfm)$(t_i - t_o)$    Btu/hr    (4-3)
(Tables 4-10 to 4-12)

$Q_{vent}$ = ventilation load = 1.08 (cfm)$(t_i - t_o)$    Btu/hr    (4-4)
(Tables 4-13 to 4-14)

where

$A$ = area through which heat flow occurs, ft$^2$
$U$ = over-all heat transfer coefficient = $1/R_t$    Btu/(hr)(ft$^2$)(°F)    (4-5)
(Tables 4-6 to 4-8)
$t_o$ = outside-air design dry-bulb temperature, °F (Fig. 4-2)
$t_i$ = inside design dry-bulb temperature, °F (Table 4-5)
cfm = cubic feet per minute, air
$R_t$ = thermal resistance, °F/(Btu)(hr)(ft$^2$)

$$R_t = R_1 + R_2 + R_3 + \cdots + R_n \tag{4-6}$$

where $R_1, R_2, \ldots$, are the resistances to heat flow of the individual components of a composite construction (see Table 4-9 for values of $R$ for concrete floors).

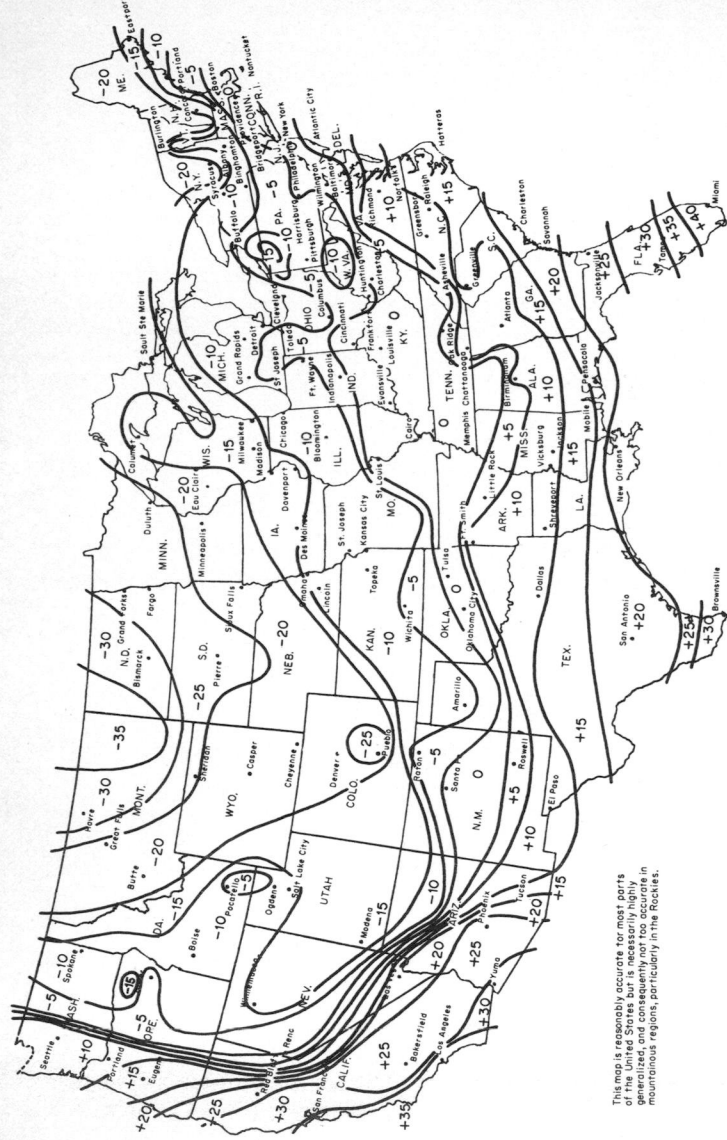

FIG. 4-2. Isotherms of winter outdoor design temperature. (*Strock and Koral,* "*Handbook of Air Conditioning, Heating, and Ventilating,*" 2d ed., 1965.)

## Table 4-5. Recommended Inside Design Conditions—Winter

| Type of application | Winter | | | | |
| --- | --- | --- | --- | --- | --- |
| | With humidification | | | Without humidification | |
| | Dry-bulb, F | Rel. hum., % | Temp. swing,* F | Dry-bulb, F | Temp. swing,* F |
| General Comfort<br>Apartment, house, hotel, office, hospital, school, etc.......................... | 74–76 | 35–30 | −3 to −4 | 75–77 | −4 |
| Retail Shops<br>(Short-term occupancy)<br>Bank, barber or beauty shop, department store, supermarket, etc........ | 72–74 | 35–30† | −3 to −4 | 73–75 | −4 |
| Low Sensible Heat Factor Applications<br>(High latent load)<br>Auditorium, church, bar, restaurant, kitchen, etc......................... | 72–74 | 40–35 | −2 to −3 | 74–76 | −4 |
| Factory Comfort<br>Assembly areas, machining rooms etc.. | 68–72 | 35–30 | −4 to −6 | 70–74 | −6 |

SOURCE: "Carrier Corporation System Design Manual," Part I, Load Estimating, 1970.
* Temperature swing is below the thermostat setting at peak winter load conditions (no lights, people, or solar heat gain).
† Winter humidification in retail clothing shops is recommended to maintain the quality texture of goods.

Transmission Load

## Table 4-6. Over-all Heat Transfer Coefficient *U*

Air-to-air heat transfer, Btu/(hr)(ft²)(°F)
Outside air 15-mph wind, inside still air

| Example | Construction | *U* |
|---|---|---|
| Frame walls............... | Wood siding, building paper, air space, gypsum lath, plaster | 0.24 |
| | Wood siding, insulation board, air space, gypsum board | 0.19 |
| Frame partition............ | Gypsum board, air space, gypsum board | 0.34 |
| Frame construction ceilings and floors | Linoleum or tile, felt, plywood, wood subfloor, air space, metal lath, plaster | 0.23 |
| Pitched roofs.............. | Asphalt shingles, building paper, wood sheathing, air space, gypsum lath, plaster | 0.28 |
| Masonry wall.............. | Face brick 4″, common brick 4″ | 0.48 |
| | Face brick 4″, common brick 4″, air space, gypsum lath, plaster | 0.29 |
| | Face brick 4″, concrete block 4″, air space, gypsum lath, plaster | 0.26 |
| Masonry partition.......... | Cement block (cinder aggregate), plaster on both sides | 0.31 |
| Concrete floor and ceiling.... | Tile, felt, plywood ⅝″, air space, metal lath, plaster | 0.23 |
| Flat masonry roof.......... | Built-up roofing, roof insulation 1″, concrete slab 4″, air space, metal lath, plaster | 0.18 |

## Table 4-7. Coefficient of Heat Transmission *U* for Windows and Skylights

Air-to-air heat transfer, Btu/(hr)(ft²)(°F)
Outside air 0°F, 15-mph wind, no solar radiation; inside still air

| Construction | Vertical glass sheets | | Horizontal glass sheets | |
|---|---|---|---|---|
| | Outdoor exposure | Indoor exposure | Outdoor exposure | Indoor exposure |
| Common window glass, single sheet............... | 1.13 | 0.75 | 1.22 | 0.96 |
| Common window glass, two sheets, 1″ air space....... | 0.53 | 0.45 | 0.63 | 0.56 |

## Table 4-8. Coefficient of Heat Transmission *U* for Wood Doors

Air-to-air heat transfer, Btu/(hr)(ft²)(°F)
Outside air 0°F, 15-mph wind, no solar radiation; inside still air

| Construction | Outdoor exposure | |
|---|---|---|
| | Single | With glass storm door |
| 1″-thick solid door (²⁵⁄₃₂″)............... | 0.64 | 0.37 |
| 2″-thick solid door (1⅝″)............... | .43 | .28 |
| Door containing wood or glass panels...... | .85 | .39 |

Table 4-9. Heat Loss of Concrete Floors at or Near Grade Level per Foot of Exposed Slab Edge, Btuh*

| Outdoor Design Temperature, F | Total Width of Insulation, In. | Value of F for Unheated Slab[a] | | | Value of F for Heated Slab[b] | | |
|---|---|---|---|---|---|---|---|
| | | R = 5.0 | R = 3.75 | R = 2.50 | R = 5.0 | R = 3.33 | R = 2.50 |
| −30 and colder | 24 | 34 | 51 | 67 | 46 | 69 | 92 |
| −25 to −29 | 24 | 32 | 48 | 64 | 44 | 66 | 88 |
| −20 to −24 | 24 | 30 | 45 | 60 | 41 | 61 | 82 |
| −15 to −19 | 24 | 28 | 43 | 57 | 39 | 59 | 78 |
| −10 to −14 | 24 | 27 | 40 | 54 | 37 | 55 | 74 |
| −5 to −9 | 24 | 25 | 38 | 51 | 35 | 52 | 70 |
| 0 to −4 | 24 | 24 | 36 | 48 | 32 | 48 | 64 |
| +5 to +1 | 24 | 22 | 33 | 44 | 30 | 45 | 60 |
| +10 to +6 | 18 | 21 | 31 | 42 | 25 | 38 | 50 |
| +15 to +11 | 12 | 21 | 31 | 42 | 25 | 38 | 50 |
| +20 to +16 | Edge only | 21 | 31 | 42 | 25 | 38 | 50 |

* Reprinted by permission from "ASHRAE Handbook of Fundamentals," ASHRAE, New York, 1972.
F = Heat loss coefficient, Btuh (per linear foot of exposed edge.)
R = Thermal resistance of insulation, 1/C.
[a] Where perimeter insulation is not required, use F = 50 for unheated slabs or F = 75 for heated slabs.
[b] Slab floors having heating pipes or ducts under the slab shall be considered as *heated slabs*.

## Table 4-10. Infiltration Through Double-Hung Wood Windows

Expressed in cubic feet per (hour) (foot of crack)

| Type of window | Pressure difference (inches of water) | | | | |
|---|---|---|---|---|---|
| | 0.10 | 0.20 | 0.30 | 0.40 | 0.50 |
| **Wood double-hung window (Locked)** (Leakage expressed as cubic feet per (hour) (foot of sash crack); only leakage around sash and through frame given) | | | | | |
| Nonweatherstripped, loose fit* | 77 | 122 | 150 | 194 | 225 |
| Nonweatherstripped, average fit† | 27 | 43 | 57 | 69 | 80 |
| Weatherstripped, loose fit | 28 | 44 | 58 | 70 | 81 |
| Weatherstripped, average fit | 14 | 23 | 30 | 36 | 42 |
| **Frame-wall leakage‡** (Leakage is that passing between the frame of a wood double-hung window and the wall) | | | | | |
| Around frame in masonry wall, not caulked | 17 | 26 | 34 | 41 | 48 |
| Around frame in masonry wall, caulked | 3 | 5 | 6 | 7 | 8 |
| Around frame in wood frame wall | 13 | 21 | 29 | 35 | 42 |

* A $\frac{3}{32}$-in. crack and clearance represent a poorly fitted window, much poorer than average.

† The fit of the average double-hung wood window was determined as $\frac{1}{16}$-in. crack and $\frac{3}{64}$-in. clearance by measurements on approximately 600 windows under heating season conditions.

‡ The values given for frame leakage are per foot of sash perimeter, as determined for double-hung wood windows. Some of the frame leakage in masonry walls originates in the brick wall itself, and cannot be prevented by caulking. For the additional reason that caulking is not done perfectly and deteriorates with time, it is considered advisable to choose the masonry frame leakage values for caulked frames as the average determined by the caulked and noncaulked tests.

Reprinted by permission from "ASHRAE Handbook of Fundamentals," ASHRAE, New York, 1972.

## Table 4-11. Infiltration through Walls

Expressed in cubic feet per (hour) (square foot)

| Type of wall | Pressure difference, inches of water | | | | |
|---|---|---|---|---|---|
| | 0.05 | 0.10 | 0.20 | 0.30 | 0.40 |
| Brick Wall* | | | | | |
| 8½ in. plain.................... | 5 | 9 | 16 | 24 | 28 |
| plastered†............... | 0.05 | 0.08 | 0.14 | 0.20 | 0.27 |
| 13 in. plain..................... | 5 | 8 | 14 | 20 | 24 |
| plastered†................. | 0.01 | 0.04 | 0.05 | 0.09 | 0.11 |
| plastered‡................. | 0.03 | 0.24 | 0.46 | 0.66 | 0.84 |
| Frame wall, lath and plaster §........ | 0.09 | 0.15 | 0.22 | 0.29 | 0.32 |

\* Constructed of porous brick and lime mortar—workmanship poor.
† Two coats prepared gypsum plaster on brick.
‡ Furring, lath, and two coats prepared gypsum plaster on brick.
§ Wall construction: bevel siding painted or cedar shingles, sheathing, building paper, wood lath, and three coats gypsum plaster.
Reprinted by permission from "ASHRAE Handbook of Fundamentals," ASHRAE, New York, 1972.

## Table 4-12. Infiltration through Doors—Winter*
15 Mph Wind Velocity†
Doors on One or Adjacent Windward Sides‡

| Description | Cfm/ft² area§ | | | | |
|---|---|---|---|---|---|
| | Infrequent use | Average use | | | |
| | | 1- and 2-story building | Tall buildings, ft | | |
| | | | 50 | 100 | 200 |
| Revolving door...................... | 1.6 | 10.5 | 12.6 | 14.2 | 17.3 |
| Glass door (³⁄₁₆″ crack).............. | 9.0 | 30.0 | 36.0 | 40.5 | 49.5 |
| Wood door 3′ × 7′.................... | 2.0 | 13.0 | 15.5 | 17.5 | 21.5 |
| Small factory door................... | 1.5 | 3.0 | | | |
| Garage and shipping-room door........ | 4.0 | 9.0 | | | |
| Ramp garage door................... | 4.0 | 13.5 | | | |

\* All values are based on the wind blowing directly at the window or door. When the prevailing wind direction is oblique to the window or doors, multiply the values by 0.60 and use the total window and door area on the windward side(s).
† Based on a wind velocity of 15 mph. For design wind velocities different from the base, multiply the table values by the ratio of velocities.
‡ Stack effect in tall buildings may also cause infiltration on the leeward side. To evaluate this, determine the equivalent velocity ($V_e$) and subtract the design velocity ($V$). The equivalent velocity is

$$V_e = \sqrt{V^2 - 1.75a} \text{ (upper section)}$$
$$= \sqrt{V^2 + 1.75b} \text{ (lower section)}$$

where $a$ and $b$ are the distances above and below the mid-height of the building, respectively, in feet.
Multiply the table values by the ratio $(V_e - V)/15$ for one-half of the windows and doors on the leeward side of the building. (Use values under one- and two-story building for doors on leeward side of tall buildings.)
§ Doors on opposite sides increase values 25 per cent.
SOURCE: "Carrier Corporation System Design Manual," Part I, Load Estimating, 1970.

**Ventilation Load**

### Table 4-13. Minimum Outdoor Air Requirements to Remove Objectionable Body Odors under Laboratory Conditions

| Type of occupants | Air space per person, ft³ | Outdoor air supply, cfm per person |
|---|---|---|
| Heating season with or without recirculation.　Air not conditioned. | | |
| Sedentary adults of average socioeconomic status....... | 100 | 25 |
| | 200 | 16 |
| | 300 | 12 |
| | 500 | 7 |
| Laborers......................................... | 200 | 23 |
| Grade school children of average socioeconomic status... | 100 | 29 |
| | 200 | 21 |
| | 300 | 17 |
| | 500 | 11 |
| Grade school children of lower socioeconomic status..... | 200 | 38 |
| Children attending private grade schools.............. | 100 | 22 |
| Heating season.　Air humidified by means of centrifugal humidifier.　Water atomization rate 8 to 10 gph.　Total air circulation 30 cfm per person. | | |
| Sedentary adults..................................... | 200 | 12 |
| Summer season.　Air cooled and dehumidified by means of a spray dehumidifier.　Spray water changed daily.　Total air circulation 30 cfm per person. | | |
| Sedentary adults..................................... | 200 | <4 |

Reprinted by permission from "ASHRAE Handbook of Fundamentals," ASHRAE, New York, 1972.

## Table 4-14. Outdoor Air Requirements

| Application | Smoking | Cfm per person | | Cfm/ft² of floor, min. |
|---|---|---|---|---|
| | | Recommended | Min. | |
| **Apartment:** | | | | |
| Average | Some | 20 | 10 | |
| Deluxe | Some | 20 | 10 | |
| Banking space | Occasional | 10 | 7½ | |
| Barber shops | Considerable | 15 | 10 | |
| Beauty parlors | Occasional | 10 | 7½ | |
| Brokers' board rooms | Very heavy | 50 | 20 | |
| Cocktail bars | .... | 40 | 25 | |
| Corridors (supply or exhaust) | .... | .... | .... | 0.25 |
| Department stores | None | 7½ | 5 | 0.05 |
| Directors' rooms | Extreme | 50 | 30 | |
| Drugstores | Considerable | 10 | 7½ | |
| Factories | None | 10 | 7½ | 0.10 |
| Five and ten cent stores | None | 7½ | 5 | |
| Funeral parlors | None | 10 | 7½ | |
| Garages | .... | .... | .... | 1.0 |
| **Hospitals:** | | | | |
| Operating rooms | None | .... | .... | 2.0 |
| Private rooms | None | 30 | 25 | 0.33 |
| Wards | None | 20 | 10 | |
| Hotel rooms | Heavy | 30 | 25 | 0.33 |
| **Kitchens:** | | | | |
| Restaurant | .... | .... | .... | 4.0 |
| Residence | .... | .... | .... | 2.0 |
| Laboratories | Some | 20 | 15 | |
| Meeting rooms | Very heavy | 50 | 30 | 1.25 |
| **Offices:** | | | | |
| General | Some | 15 | 10 | |
| Private | None | 25 | 15 | 0.25 |
| | Considerable | 30 | 25 | 0.25 |
| **Restaurants:** | | | | |
| Cafeteria | Considerable | 12 | 10 | |
| Dining-room | Considerable | 15 | 12 | |
| Schoolrooms | None | | | |
| Shop, retail | None | 10 | 7½ | |
| Theater | None | 7½ | 5 | |
| | Some | 15 | 10 | |
| Toilets (exhaust) | .... | .... | .... | 2.0 |

**Radiator and Convector Ratings.** To determine the rating of a radiator or a convector for a given space, divide the heat loss of the space by the proper factor from Table 4-15 and select the radiator or convector having an equivalent Btu per hour rating. Thus, for a 75°F room with 1000 Btu heat loss, a convector fed with 1 psig steam should be selected for a 1110-Btu rating.

## Table 4-15. Correction Factors for Various Types of Heating Units

| Steam pressure, (approx.) Gage vacuum, in. Hg | Psia | Temp. of heating medium steam or water | Cast-iron radiators Room temp., °F | | | | | Convectors Inlet air temp., °F | | | | | Finned tube Inlet air temp., °F | | | | | Baseboard Inlet air temp., °F | | | | |
|---|---|---|---|---|---|---|---|---|---|---|---|---|---|---|---|---|---|---|---|---|---|---|
| | | | 80 | 75 | 70 | 65 | 60 | 75 | 70 | 65 | 60 | 55 | 75 | 70 | 65 | 60 | 55 | 75 | 70 | 65 | 60 | 55 |
| 22.4 | 3.7 | 150 | 0.39 | 0.42 | 0.46 | 0.50 | 0.54 | 0.35 | 0.39 | 0.43 | 0.46 | 0.50 | 0.36 | 0.42 | 0.46 | 0.51 | 0.57 | 0.38 | 0.42 | 0.45 | 0.49 | 0.53 |
| 20.3 | 4.7 | 160 | 0.46 | 0.50 | 0.54 | 0.58 | 0.62 | 0.43 | 0.47 | 0.51 | 0.54 | 0.58 | 0.45 | 0.49 | 0.53 | 0.59 | 0.64 | 0.45 | 0.49 | 0.53 | 0.57 | 0.61 |
| 17.7 | 6.0 | 170 | 0.54 | 0.58 | 0.62 | 0.66 | 0.69 | 0.51 | 0.54 | 0.58 | 0.63 | 0.67 | 0.53 | 0.57 | 0.61 | 0.67 | 0.72 | 0.53 | 0.57 | 0.61 | 0.65 | 0.69 |
| 14.6 | 7.5 | 180 | 0.62 | 0.66 | 0.69 | 0.74 | 0.78 | 0.58 | 0.63 | 0.67 | 0.71 | 0.76 | 0.61 | 0.65 | 0.69 | 0.75 | 0.81 | 0.61 | 0.65 | 0.69 | 0.72 | 0.78 |
| 10.9 | 9.3 | 190 | 0.69 | 0.74 | 0.78 | 0.83 | 0.87 | 0.67 | 0.71 | 0.76 | 0.81 | 0.85 | 0.69 | 0.73 | 0.78 | 0.84 | 0.89 | 0.69 | 0.73 | 0.78 | 0.82 | 0.86 |
| 6.5 | 11.5 | 200 | 0.78 | 0.83 | 0.87 | 0.91 | 0.95 | 0.76 | 0.81 | 0.85 | 0.90 | 0.95 | 0.77 | 0.81 | 0.86 | 0.92 | 0.97 | 0.81 | 0.86 | 0.92 | 0.95 | 1.00 |
| **Psig** | | | | | | | | | | | | | | | | | | | | | | |
| 1 | 15.6 | 215 | 0.91 | 0.95 | 1.00 | 1.04 | 1.09 | 0.90 | 0.95 | 1.00 | 1.05 | 1.10 | 0.91 | 0.94 | 1.00 | 1.06 | 1.11 | 0.91 | 0.95 | 1.00 | 1.05 | 1.09 |
| 6 | 21 | 230 | 1.04 | 1.09 | 1.14 | 1.18 | 1.23 | 1.05 | 1.10 | 1.15 | 1.20 | 1.26 | 1.03 | 1.08 | 1.14 | 1.19 | 1.24 | 1.04 | 1.09 | 1.14 | 1.19 | 1.25 |
| 15 | 30 | 250 | 1.23 | 1.28 | 1.32 | 1.37 | 1.43 | 1.27 | 1.32 | 1.37 | 1.43 | 1.47 | 1.20 | 1.26 | 1.31 | 1.37 | 1.43 | 1.22 | 1.27 | 1.32 | 1.37 | 1.43 |
| 27 | 42 | 270 | 1.43 | 1.47 | 1.52 | 1.56 | 1.61 | 1.47 | 1.54 | 1.59 | 1.67 | 1.72 | 1.38 | 1.44 | 1.50 | 1.56 | 1.62 | 1.43 | 1.47 | 1.52 | 1.59 | 1.64 |
| 52 | 67 | 300 | 1.72 | 1.75 | 1.82 | 1.89 | 1.92 | 1.85 | 1.89 | 1.96 | 2.04 | 2.08 | 1.67 | 1.73 | 1.79 | 1.86 | 1.92 | 1.75 | 1.82 | 1.89 | 1.92 | 1.96 |

source: Reprinted by permission from "ASHRAE Handbook and Product Directory," ASHRAE, New York, 1975.

### Example of Use of Basic and Velocity Multiplier Charts

GIVEN:

Weight flow rate = 6,700 lb/hr

Initial steam pressure = 100 psig

Pressure drop = 11 psi/100 ft

FIND: Size of Schedule 40 pipe required and velocity of steam in pipe.

SOLUTION: The following steps are illustrated by the broken line in Fig. 4-3.

*Step* 1.  Enter diagram at a weight flow rate of 6,700 lb/hr and move vertically to the horizontal line at 100 psig.

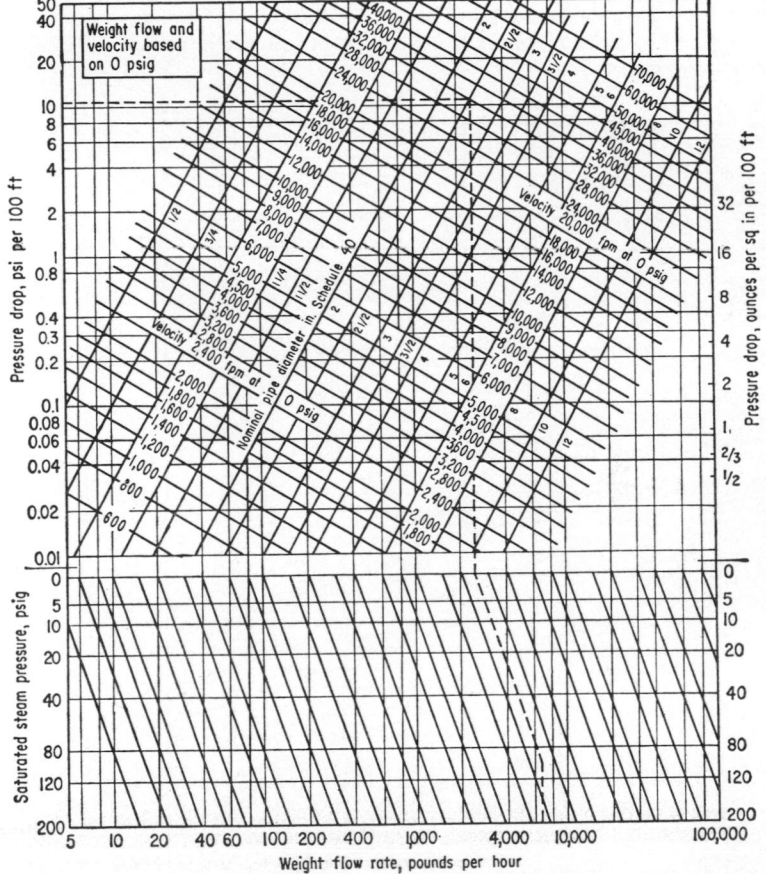

FIG. 4-3. Basic chart for weight flow rate and velocity of steam in Schedule 40 pipe based on saturation pressure of 0 psig. (*Reprinted by permission from "ASHRAE Handbook of Fundamentals," ASHRAE, New York,* 1972.)

*Step* 2.  Follow along inclined multiplier line (upward and to the left) to horizontal 0-psig line.  The equivalent weight flow at 0 psig is about 2,500 lb/hr.

*Step* 3.  Follow the 2,500-lb/hr line vertically until it intersects the horizontal line at 11 psi/100 ft pressure drop.  The nominal pipe size is 2½ in. The equivalent steam velocity at 0 psig is about 32,700 fpm.

*Step* 4.  To find the steam velocity at 100 psig, locate the value of 32,700 fpm on the ordinate of the velocity multiplier chart at 0 psig (Fig. 4-4).

*Step* 5.  Move along the inclined multiplier line (downward and to the right) until it intersects the vertical 100-psig pressure line.  The velocity as read from the right (or left) scale is about 13,000 fpm.

*Note*: The preceding steps 1 to 5 would be rearranged or reversed if different data were given.

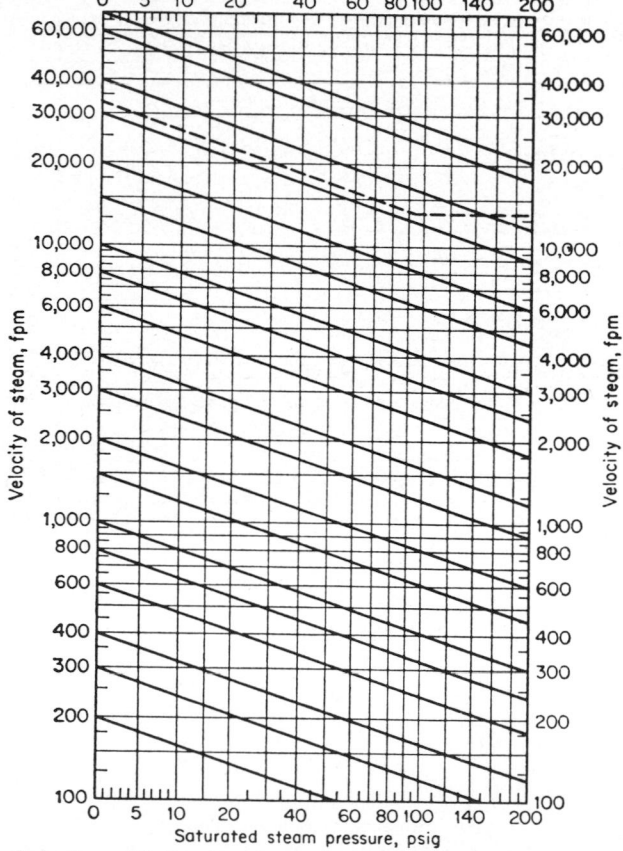

Fig. 4-4.  Velocity multiplier chart for use with Fig. 4-3.  (*Reprinted by permission from "ASHRAE Handbook of Fundamentals," ASHRAE, New York, 1972.*)

## Table 4-16. Weight Flow Rate of Steam in Schedule 40 Pipe[a] at Initial Saturation Pressures of 3.5 and 12 Psig[b,c]

Weight Flow Rate in Pounds per Hour

Pressure drop, psi per 100 ft in length

| Nom. pipe size, in. | 1/16 psi (1 oz) | | 1/8 psi (2 oz) | | 1/4 psi (4 oz) | | 1/2 psi (8 oz) | | 3/4 psi (12 oz) | | 1 psi | | 2 psi | |
|---|---|---|---|---|---|---|---|---|---|---|---|---|---|---|
| | — Saturation pressure, psig | | | | | | | | | | | | | |
| | 3.5 | 12 | 3.5 | 12 | 3.5 | 12 | 3.5 | 12 | 3.5 | 12 | 3.5 | 12 | 3.5 | 12 |
| ¾ | 9 | 11 | 14 | 16 | 20 | 24 | 29 | 35 | 36 | 43 | 42 | 50 | 60 | 73 |
| 1 | 17 | 21 | 26 | 31 | 37 | 46 | 54 | 66 | 68 | 82 | 81 | 95 | 114 | 137 |
| 1¼ | 36 | 45 | 53 | 66 | 78 | 96 | 111 | 138 | 140 | 170 | 162 | 200 | 232 | 280 |
| 1½ | 56 | 70 | 84 | 100 | 120 | 147 | 174 | 210 | 218 | 260 | 246 | 304 | 360 | 430 |
| 2 | 108 | 134 | 162 | 194 | 234 | 285 | 336 | 410 | 420 | 510 | 480 | 590 | 710 | 850 |
| 2½ | 174 | 215 | 258 | 310 | 378 | 460 | 540 | 660 | 680 | 820 | 780 | 950 | 1,150 | 1,370 |
| 3 | 318 | 380 | 465 | 550 | 660 | 810 | 960 | 1,160 | 1,190 | 1,430 | 1,380 | 1,670 | 1,950 | 2,400 |
| 3½ | 462 | 550 | 670 | 800 | 990 | 1,218 | 1,410 | 1,700 | 1,740 | 2,100 | 2,000 | 2,420 | 2,950 | 3,450 |
| 4 | 640 | 800 | 950 | 1,160 | 1,410 | 1,690 | 1,980 | 2,400 | 2,450 | 3,000 | 2,880 | 3,460 | 4,200 | |
| 5 | 1,200 | 1,430 | 1,680 | 2,100 | 2,440 | 3,000 | 3,570 | 4,250 | 4,380 | 5,250 | 5,100 | 6,100 | 7,500 | |
| 6 | 1,920 | 2,300 | 2,820 | 3,350 | 3,960 | 4,850 | 5,700 | 7,000 | 7,200 | 8,600 | 8,400 | 10,000 | 11,90_ | |
| 8 | 3,900 | 4,800 | 5,570 | 7,000 | 8,100 | 10,000 | 11,400 | 14,300 | 14,500 | 17,700 | 16,500 | 20,500 | 24,0_ | |
| 10 | 7,200 | 8,800 | 10,200 | 12,600 | 15,000 | 18,200 | 21,000 | 26,000 | 26,200 | 32,000 | 30,000 | 37,000 | 42 | |
| 12 | 11,400 | 13,700 | 16,500 | 19,500 | 23,000 | 28,400 | 33,000 | 40,000 | 41,000 | 49,500 | 48,000 | 57,500 | 67 | |

[a] Based on Moody friction factor, where flow of condensate does not inhibit the flow of steam.

[b] The weight flow rates at 3.5 psig can be used to cover saturation pressure from 1 to 6 psig, and the rates at 12 psig can be used to cover saturation pressure with an error not exceeding 8 per cent.

[c] The steam velocities corresponding to the weight flow rates given in this table can be found from the basic chart and velocity multiplier chart, Fig. 4-3.

Reprinted by permission from "ASHRAE Handbook of Fundamentals," ASHRAE, New York, 1972.

## Table 4-17. Return Main and Riser Capacities for Low-pressure Systems, Pounds per Hour

This table is based on pipe size data developed through the research investigations of The American Society of Heating, Refrigerating and Air-Conditioning Engineers.

| Pipe size, in. | 1/32 psi or 1/2 oz drop per 100 ft | | | 1/24 psi or 2/3 oz drop per 100 ft | | | 1/16 psi or 1 oz drop per 100 ft | | | 1/8 psi or 2 oz drop per 100 ft | | | 1/4 psi or 4 oz drop per 100 ft | | | 1/2 psi or 8 oz drop per 100 ft | | |
|---|---|---|---|---|---|---|---|---|---|---|---|---|---|---|---|---|---|---|
| | Wet | Dry | Vac. | Wet | Dry | Vac. | Wet | Dry | Vac. | Wet | Dry | Vac. | Wet | Dry | Vac. | Wet | Dry | Vac. |
| G | H | I | J | K | L | M | N | O | P | Q | R | S | T | U | V | W | X | Y |
| **Mains** | | | | | | | | | | | | | | | | | | |
| 3/4 | | | | | | 42 | | | 100 | | | 142 | | | 200 | | | 283 |
| 1 | 125 | 62 | | 145 | 71 | 143 | 175 | 80 | 175 | 250 | 103 | 249 | 350 | 115 | 350 | | | 494 |
| 1¼ | 213 | 130 | | 248 | 149 | 244 | 300 | 168 | 300 | 425 | 217 | 426 | 600 | 241 | 600 | | | 848 |
| 1½ | 338 | 206 | | 393 | 236 | 388 | 475 | 265 | 475 | 675 | 340 | 674 | 950 | 378 | 950 | | | 1,340 |
| 2 | 700 | 470 | | 810 | 535 | 815 | 1,000 | 575 | 1,000 | 1,400 | 740 | 1,420 | 2,000 | 825 | 2,030 | | | 2,830 |
| 2½ | 1,180 | 760 | | 1,580 | 868 | 1,360 | 1,680 | 950 | 1,680 | 2,350 | 1,230 | 2,380 | 3,350 | 1,360 | 3,350 | | | 4,730 |
| 3 | 1,880 | 1,460 | | 2,130 | 1,560 | 2,180 | 2,680 | 1,750 | 2,680 | 3,750 | 2,250 | 3,800 | 5,350 | 2,500 | 5,350 | | | 7,560 |
| 3½ | 2,750 | 1,970 | | 3,300 | 2,200 | 3,250 | 4,000 | 2,500 | 4,000 | 5,500 | 3,230 | 5,680 | 8,000 | 3,580 | 8,000 | | | 11,300 |
| 4 | 3,880 | 2,930 | | 4,580 | 3,350 | 4,500 | 5,500 | 3,750 | 5,500 | 7,750 | 4,830 | 7,810 | 11,000 | 5,580 | 11,000 | | | 15,500 |
| 5 | | | | | | 7,880 | | | 9,680 | | | 13,700 | | | 19,400 | | | 27,300 |
| 6 | | | | | | 12,600 | | | 15,500 | | | 22,000 | | | 31,000 | | | 43,800 |
| **Risers** | | | | | | | | | | | | | | | | | | |
| 3/4 | | 48 | | | 48 | 143 | | 48 | 175 | | 48 | 249 | | 48 | 350 | | | 494 |
| 1 | | 113 | | | 113 | 244 | | 113 | 300 | | 113 | 426 | | 113 | 600 | | | 848 |
| 1¼ | | 248 | | | 248 | 388 | | 248 | 475 | | 248 | 674 | | 248 | 950 | | | 1,340 |
| 1½ | | 375 | | | 375 | 815 | | 375 | 1,000 | | 375 | 1,420 | | 375 | 2,000 | | | 2,830 |
| 2 | | 750 | | | 750 | 1,360 | | 750 | 1,680 | | 750 | 2,380 | | 750 | 3,350 | | | 4,730 |
| 2½ | | | | | | 2,180 | | | 2,680 | | | 3,800 | | | 5,350 | | | 7,560 |
| 3 | | | | | | 3,250 | | | 4,000 | | | 5,680 | | | 8,000 | | | 11,300 |
| 3½ | | | | | | 4,480 | | | 5,500 | | | 7,810 | | | 11,000 | | | 15,500 |
| 4 | | | | | | 7,880 | | | 9,680 | | | 13,700 | | | 19,400 | | | 27,300 |
| 5 | | | | | | 12,600 | | | 15,500 | | | 22,000 | | | 31,000 | | | 43,800 |

Reprinted by permission from "ASHRAE Handbook of Fundamentals," ASHRAE, New York, 1972.

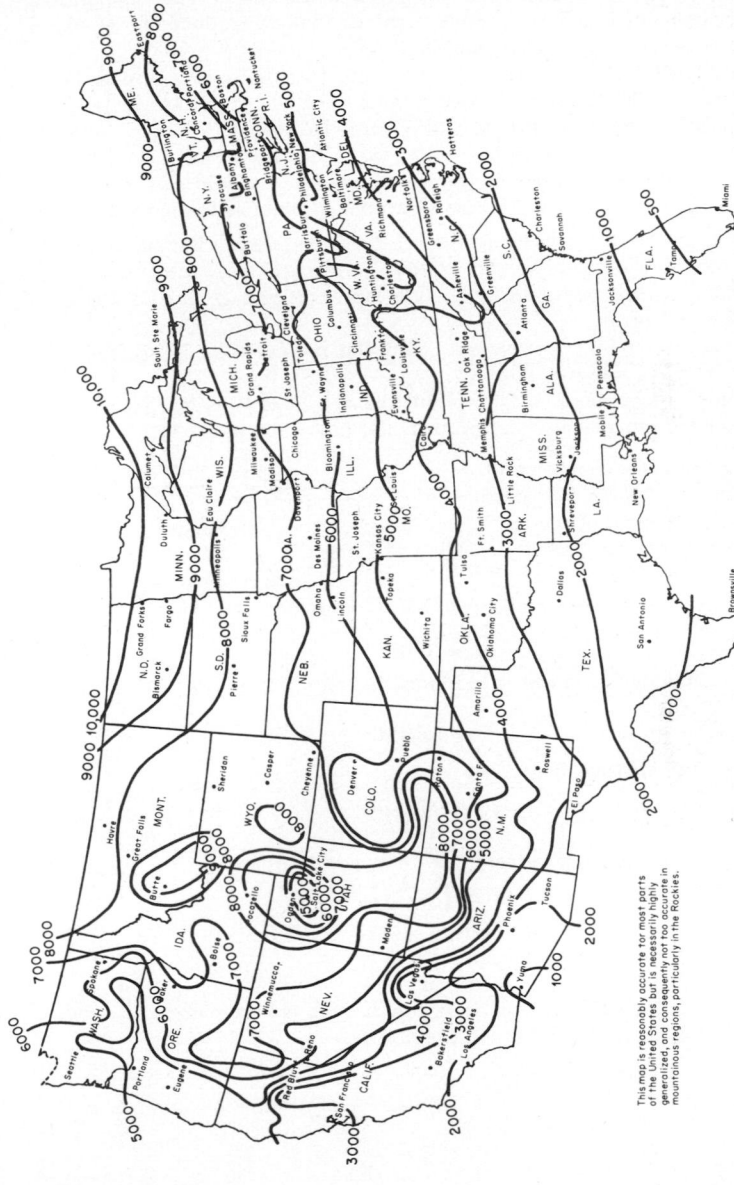

This map is reasonably accurate for most parts of the United States but is necessarily highly generalized, and consequently not too accurate in mountainous regions, particularly in the Rockies.

Fig. 4-5. Number of degree-days in a normal heating season. (*Strock and Koral, "Handbook of Air Conditioning, Heating, and Ventilating," The Industrial Press, New York, 1965.*)

**Boiler Load.** *Net Load* is the sum of direct-connected load components. These include direct radiation, infiltration, air tempering, humidification, hot water, process steam, and snow melting.

*Design Load* is the sum of the *net load* and the *piping tax*. Piping tax is the estimated heat emission in Btu per hour of the piping connecting the radiation and other apparatus to the boiler. In average heating systems it is common practice to consider the piping tax to be 20 per cent of the net load.

*Gross, or Maximum, Load* is the sum of the *design load* and the *pickup allowance*. Pickup allowance is the estimated increase in the normal load in Btu per hour, caused by the heating up of the cold system. For automatically fired boilers the sum of the piping tax and pickup allowance varies from 33.3 to 28.8 per cent of the *net load*. The larger percentage should be applied to smaller boilers.

Information on *boiler performance* and the *heating value of various fuels* may be found in Sec. 9.

**Chimneys.** Equations (4-7) to (4-10) are simplified equations for chimney sizes if the following typical values for boiler plants are assumed:

| | |
|---|---|
| Average chimney gas temperature: | $T_c = 500°F$ (960°F abs) |
| Average atmospheric temperature: | $T_0 = 62°F$ (522°F abs) |
| Average coefficient of friction: | $f = 0.016$ |
| Average chimney gas density at 0°F and 1 atm: | $0.09$ lb/ft$^3$ |
| Barometer reading, sea level: | $B_0 = 29.92$ in. Hg |

Required height of chimney above inlet, in feet:

$$H = 190D_r \tag{4-7}$$

Required minimum diameter of chimney, in feet:

$$d = 1.5W^{\frac{2}{5}} \tag{4-8}$$

Chimney gas velocity, in feet per second:

$$V_c = 13.7W^{\frac{1}{5}} \tag{4-9}$$

Stack draft, in inches of water:

$$D_r = 0.256HB_0 \left( \frac{1}{T_0 \text{ abs}} - \frac{1}{T_c \text{ abs}} \right) \tag{4-10}$$

where $D_r$ = total required draft, in. of water
    $W$ = flue gas flow rate, lb/sec
Total required draft is the sum of draft loss through the breeching and through the boiler and the required draft in the firebox.

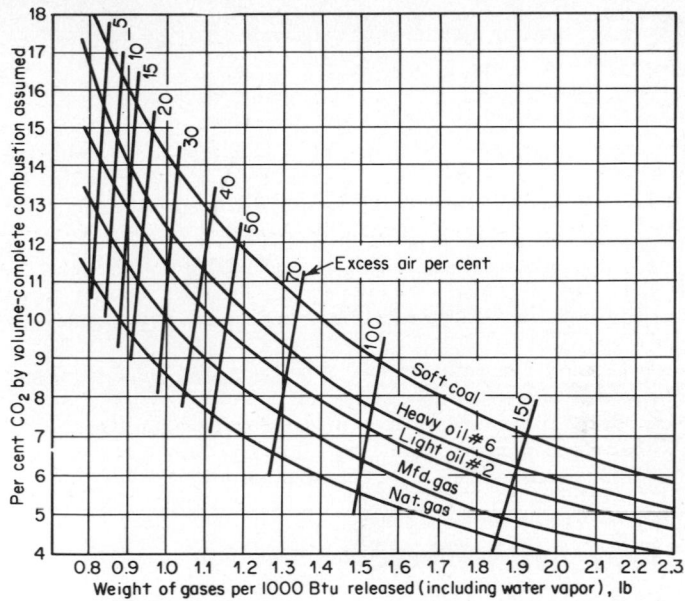

FIG. 4-6. Graphical evaluation and rate of flue gas flow from per cent $CO_2$ and fuel rate. (*Reprinted by permission from "ASHRAE Handbook & Product Directory," ASHRAE, New York, 1975.*)

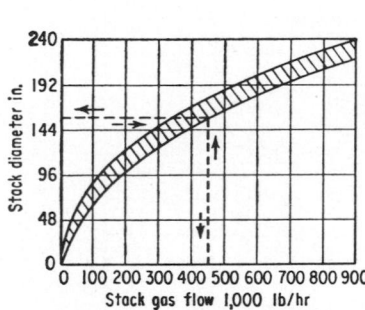

FIG. 4-7. Economical stack size based on approximately 5 per cent draft loss.

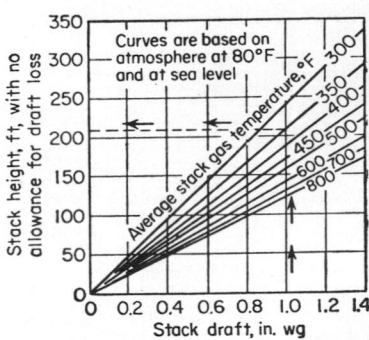

FIG. 4-8. Stack height as a function of stack draft.

## 4-4. Moisture

The moisture entering a building as water vapor may be expressed as

$$W_t = W_{trans} + W_{inf} + W_{vent} \qquad (4\text{-}11)$$

where $W_t$ = total weight of vapor, grains

$$W_{trans} = \text{transmitted vapor} = MA\theta\,\Delta\rho \qquad \text{grains} \qquad (4\text{-}12)$$

$$W_{inf} = \text{air infiltrated vapor} = W(M_o - M_i) \qquad \text{grains} \qquad (4\text{-}13)$$

$$W_{vent} = \text{ventilation air vapor} = W(M_o - M_i) \qquad \text{grains} \qquad (4\text{-}14)$$

where $M$ = permeance coefficient, perms

$$= \bar{\mu}/l,\ \text{g}/(\text{ft}^2)(\text{hr})(\text{Hg}\,\Delta\rho) \quad (\text{Table 4-18}) \qquad (4\text{-}15)$$

$A$ = area of flow path, ft$^2$
$\theta$ = time of transmission, hr
$\Delta\rho$ = vapor-pressure difference through flow path, in. Hg
$W$ = weight of air, lb
$M_o$ = moisture content of outside air, g/lb

$M_i$ = moisture content of inside air, g/lb

where 1 perm = 1 g/(ft$^2$)(hr)(in. Hg $\Delta\rho$)
$\bar{\mu}$ = permeability, perm-in., g-in./(ft$^2$)(hr)(Hg $\Delta\rho$)
$l$ = length of flow path, in.

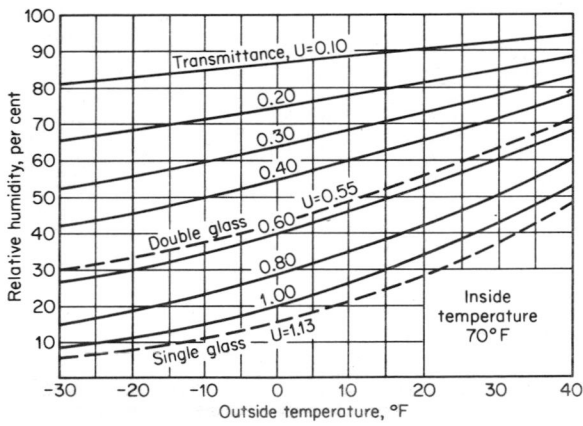

Fig. 4-9. Relative humidity at which visible condensation will appear on inside surface. (*Reprinted by permission from "ASHRAE Handbook of Fundamentals," ASHRAE, New York, 1972.*)

## Table 4-18. Permeance and Permeability of Materials to Water Vapor[a]

| Material | Permeance, perms | | | Permeability, perm-in. | | |
|---|---|---|---|---|---|---|
| | Dry cup | Wet cup | Other | Dry cup | Wet cup | Other |
| **Materials used in construction:** | | | | | | |
| Concrete (1:2:4 mix) | ...... | | ...... | ......... | 3.2 | |
| Brick masonry (4 in. thick) | ...... | ...... | 0.8 | | | |
| Concrete block (8-in. cored, limestone aggregate) | ...... | ...... | 2.4 | | | |
| Tile masonry, glazed (4 in. thick) | ...... | ...... | 0.12 | | | |
| Asbestos-cement board (0.2 in. thick) | 0.54 | | | | | |
| Plaster on metal lath (¾ in.) | ...... | ...... | 15 | | | |
| Plaster on wood lath | ...... | 11 | | | | |
| Plaster on plain gypsum lath (with studs) | ...... | ...... | 20 | | | |
| Gypsum wallboard (⅜ in. plain) | ...... | ...... | 50 | | | |
| Gypsum sheathing (½-in. asphalt impreg) | ...... | ...... | | 20 | | |
| Structural insulating board, sheathing quality | ...... | ...... | ...... | ......... | ... | 20-50 |
| interior, uncoated, ½ in | ...... | ...... | 50-90 | | | |
| Hardboard, ⅛ in. standard | ...... | ...... | 11 | | | |
| ⅛ in. tempered | | | 5 | | | |
| Built-up roofing (hot-mopped) | 0.0 | | | | | |
| Wood, sugar pine | ...... | | | ......... | ... | 0.4-5.4[b] |
| Plywood, douglas-fir, exterior glue, ¼ in. thick | ...... | ...... | 0.7 | | | |
| Plywood, douglas-fir, interior, glue, ¼ in. thick | ...... | ...... | 1.9 | | | |
| Acrylic, glass-fiber-reinforced sheet, 56 mil | 0.12 | | | | | |
| Polyester, glass-fiber-reinforced sheet, 48 mil | 0.05 | | | | | |
| **Thermal insulations:** | | | | | | |
| Air (still) | ...... | ...... | ...... | ......... | ... | 120 |
| Cellular glass | ...... | ...... | ...... | 0.0 | | |
| Corkboard | ...... | ...... | ...... | 2.1-2.6 | 9.5 | |
| Mineral wool (unprotected) | ...... | | | ......... | 116 | |
| Expanded polyurethane (R-11 blown) board stock | ...... | ...... | ...... | 0.4-1.6 | | |
| Expanded polystyrene, extruded | ...... | ...... | ...... | 1.2 | | |
| bead | ...... | ...... | ...... | 2.0-5.8 | | |
| Unicellular synthetic flexible rubber foam | ...... | ...... | ...... | 0.02-0.15 | | |
| **Plastic and metal foils and films[c]:** | | | | | | |
| Aluminum foil, 1 mil | 0.0 | | | | | |
| 0.35 mil | 0.05 | | | | | |
| Polyethylene, 2 mil | 0.16 | | | | | |
| 4 mil | 0.08 | | | | | |
| 6 mil | 0.06 | | | | | |
| 8 mil | 0.04 | | | | | |
| 10 mil | 0.03 | | | | | |
| Polyester, 1 mil | 0.7 | | | | | |
| Cellulose acetate, 125 mil | 0.4 | | | | | |
| Polyvinylchloride, unplasticized, 2 mil | 0.68 | | | | | |
| plasticized, 4 mil | 0.8-1.4 | | | | | |
| **Building papers, felts, roofing papers[d]:** | | | | | | |
| Duplex sheet, asphalt laminated, aluminum foil one side (43)[e] | 0.002 | 0.176 | | | | |
| Saturated and coated roll roofing (326)[e] | 0.05 | 0.24 | | | | |
| Kraft paper and asphalt laminated, reinforced 30-120-30 (34)[e] | 0.3 | 1.8 | | | | |
| Blanket thermal insulation back-up paper, asphalt coated (31)[e] | 0.4 | 0.6-4.2 | | | | |
| Asphalt-saturated and coated vapor-barrier paper (43)[e] | 0.2-0.3 | 0.6 | | | | |
| Asphalt-saturated but not coated sheathing paper (22)[e] | 3.3 | 20.2 | | | | |
| 15-lb asphalt felt (70)[e] | 1.0 | 5.6 | | | | |
| 15-lb tar felt (70)[e] | 4.0 | 18.2 | | | | |
| Single-kraft, double infused (16)[e] | 31 | 42 | | | | |

## Table 4-18. Permeance and Permeability of Materials to Water Vapor (Continued)

| Material | Permeance, perms | | | Permeability, perm-in. | | |
|---|---|---|---|---|---|---|
| | Dry cup | Wet cup | Other | Dry cup | Wet cup | Other |
| Liquid-applied coating materials: | | | | | | |
| Paint—2 coats: | | | | | | |
|   Asphalt paint on plywood.......... | ...... | 0.4 | | | | |
|   Aluminum varnish on wood........ | 0.3–0.5 | | | | | |
|   Enamels on smooth plaster......... | ....... | ........ | 0.5–1.5 | | | |
|   Primers and sealers on interior insulation board............... | ....... | ....... | 0.9–2.1 | | | |
|   Various primers plus 1 coat flat oil paint on plaster................... | ....... | ....... | 1.6–3.0 | | | |
|   Flat paint on interior insulation board......................... | ....... | ....... | 4 | | | |
|   Water emulsion on interior insulation board...................... | ....... | ....... | 30–85 | | | |
| Paint—exterior, 3 coats: | | | | | | |
|   White lead and oil on wood siding.. | 0.3–1.0 | | | | | |
|   White lead-zinc oxide and oil on wood.......................... | 0.9 | | | | | |
| Styrene-butadiene latex coating, 2 oz/ft² ........................ | 11 | | | | | |
| Polyvinyl acetate latex coating, 4 oz/ft²...................... | 5.5 | | | | | |
| Chloro-sulfonated polyethylene mastic, | | | | | | |
|   3.5 oz/ft²....................... | 1.7 | | | | | |
|   7.0 oz/ft²....................... | 0.06 | | | | | |
| Asphalt cut-back mastic, | | | | | | |
|   $\frac{1}{16}$ in. dry...................... | 0.14 | | | | | |
|   $\frac{3}{16}$ in. dry..................... | 0.0 | | | | | |
| Hot melt asphalt, | | | | | | |
|   2 oz/ft²......................... | 0.5 | | | | | |
|   3.5 oz/ft²...................... | 0.1 | | | | | |

[a] Table 4-18 gives the water vapor transmission rates of some representative materials. Data are provided to permit comparisons of materials; but in the selection of vapor barrier materials, exact values for permeance or permeability should be obtained from the manufacturer of the materials under consideration or secured as a result of laboratory tests. A range of values shown in the table indicated variations among mean values for materials that are similar but of different density, orientation, lot or source. The values are intended for design guidance and should not be used as design or specification data. The compilation is from a number of sources; values from dry-cup and wet-cup methods were usually obtained from investigations using ASTM E96 and C355; values shown under *Other* were obtained from investigations using such techniques as *two-temperature, special cell,* and *air-velocity*.

[b] Depending on construction and direction of vapor flow.

[c] Usually installed as vapor barriers, although sometimes used as exterior finish and elsewhere near cold side where special considerations are then required for warm-side barrier effectiveness.

[d] Low permeance sheets used as vapor barriers. High permeance used elsewhere in construction.

[e] Basis weight in lb/500 ft².

Reprinted by permission from "ASHRAE Handbook of Fundamentals," ASHRAE, 1972.

## Table 4-19. Grains of Moisture per Pound of Dry Air vs. Dew-point Temperature, °F

| DP | Grains | DP | Grains | DP | Grains | DP | Grains | DP | Grains |
|----|--------|----|--------|----|--------|----|--------|----|--------|
| 0 | 5.50 | 16 | 12.36 | 32 | 26.40 | 48 | 49.50 | 64 | 89.18 |
| 1 | 5.79 | 17 | 12.99 | 33 | 27.52 | 49 | 51.42 | 65 | 92.40 |
| 2 | 6.10 | 18 | 13.63 | 34 | 28.66 | 50 | 53.38 | 66 | 95.76 |
| 3 | 6.43 | 19 | 14.30 | 35 | 29.83 | 51 | 55.45 | 67 | 99.19 |
| 4 | 6.77 | 20 | 15.01 | 36 | 31.07 | 52 | 57.58 | 68 | 102.8 |
| 5 | 7.12 | 21 | 15.75 | 37 | 32.33 | 53 | 59.74 | 69 | 106.4 |
| 6 | 7.50 | 22 | 16.53 | 38 | 33.62 | 54 | 61.99 | 70 | 110.2 |
| 7 | 7.89 | 23 | 17.33 | 39 | 34.97 | 55 | 69.34 | 71 | 114.2 |
| 8 | 8.30 | 24 | 18.17 | 40 | 36.36 | 56 | 66.75 | 72 | 118.2 |
| 9 | 8.73 | 25 | 19.05 | 41 | 37.80 | 57 | 69.23 | 73 | 122.4 |
| 10 | 9.18 | 26 | 19.97 | 42 | 39.31 | 58 | 71.82 | 74 | 126.6 |
| 11 | 9.65 | 27 | 20.94 | 43 | 40.88 | 59 | 74.48 | 75 | 131.1 |
| 12 | 10.15 | 28 | 21.93 | 44 | 42.48 | 60 | 77.21 | 76 | 135.7 |
| 13 | 10.66 | 29 | 22.99 | 45 | 44.14 | 61 | 80.08 | 77 | 140.4 |
| 14 | 11.20 | 30 | 24.07 | 46 | 45.87 | 62 | 83.02 | 78 | 145.3 |
| 15 | 11.77 | 31 | 25.21 | 47 | 47.66 | 63 | 86.03 | 79 | 150.3 |

## Table 4-19A. Vapor Pressure of Saturated Air, Inches of Hg, vs. Dry-bulb Temperature, °F

| $T$ | $P_{sat}$ | $T$ | $P_{sat}$ | $T$ | $P_{sat}$ | $T$ | $P_{sat}$ | $T$ | $P_{sat}$ |
|----|--------|----|--------|----|--------|----|--------|----|--------|
| 0 | .03764 | 16 | .08461 | 32 | .18035 | 48 | .33629 | 64 | .60073 |
| 1 | .03966 | 17 | .08884 | 33 | .18778 | 49 | .34913 | 65 | .62209 |
| 2 | .04178 | 18 | .09326 | 34 | .19546 | 50 | .36240 | 66 | .64411 |
| 3 | .04400 | 19 | .09789 | 35 | .20342 | 51 | .37611 | 67 | .66681 |
| 4 | .04633 | 20 | .10272 | 36 | .21166 | 52 | .39028 | 68 | .69019 |
| 5 | .04877 | 21 | .10777 | 37 | .22020 | 53 | .40492 | 69 | .71430 |
| 6 | .05133 | 22 | .11305 | 38 | .22904 | 54 | .42004 | 70 | .73915 |
| 7 | .05402 | 23 | .11856 | 39 | .23819 | 55 | .43565 | 71 | .76475 |
| 8 | .05683 | 24 | .12431 | 40 | .24767 | 56 | .45176 | 72 | .79112 |
| 9 | .05977 | 25 | .13032 | 41 | .25748 | 57 | .46480 | 73 | .81828 |
| 10 | .06285 | 26 | .13659 | 42 | .26763 | 58 | .48558 | 74 | .84624 |
| 11 | .06608 | 27 | .14313 | 43 | .27813 | 59 | .50330 | 75 | .87504 |
| 12 | .06946 | 28 | .14966 | 44 | .28889 | 60 | .52159 | 76 | .90470 |
| 13 | .07299 | 29 | .15707 | 45 | .30023 | 61 | .54047 | 77 | .93523 |
| 14 | .07669 | 30 | .16452 | 46 | .31185 | 62 | .55994 | 78 | .96665 |
| 15 | .08056 | 31 | .17227 | 47 | .32386 | 63 | .58002 | 79 | .99899 |

SOURCE: "ASHRAE Guide and Data Book," chap. 3, table 2, ASHRAE, New York, 1963.

## 4-5. Cooling

**Cooling Load.**   $Q_t$, the total simultaneous cooling load, Btu/hr.

$$Q_t = Q_{ext} + Q_{int} + Q_{outside\ air} \tag{4-16}$$

where

$$Q_{ext} = \text{external heat gains} = Q_{transmission} + Q_{solar} \tag{4-17}$$

$$Q_{int} = \text{internal heat gains} = Q_{lights} + Q_{people}$$
$$+ Q_{equipment} + Q_{transmission} \tag{4-18}$$

$$Q_{tr\text{-}sol} = AU(sa\ \Delta t) \qquad \text{for walls and roofs, Btu/hr} \tag{4-19}$$

where $sa\ \Delta t$ = sol-air equivalent temperature differential, °F

$$Q_{\text{glass}} = Q_{\text{solar}} + Q_{\text{tr}} \qquad (4\text{-}20)$$

$$= \text{SHGF (SF)}A_1 + A_2U(t_o - t_i) \qquad (4\text{-}21)$$

where SHGF = solar heat gain factor (Tables 4-24 to 4-28)
       SF = shading factor (Tables 4-29 to 4-33)
      $A_1$ = area of sunlit glass, ft²
      $A_2$ = area of total glass, ft²
      $U$ = over-all coefficient (Tables 4-22 and 4-23)
      $t_o$ = outside design temp, °F (Fig. 4-12)
      $t_i$ = inside design temp, °F (Table 4-40)

$$Q_{\text{lights}} = 3.41 \times \text{wattage input to conditioned space, Btu/hr} \qquad (4\text{-}22)$$
(Figs. 4-10 and 4-11)

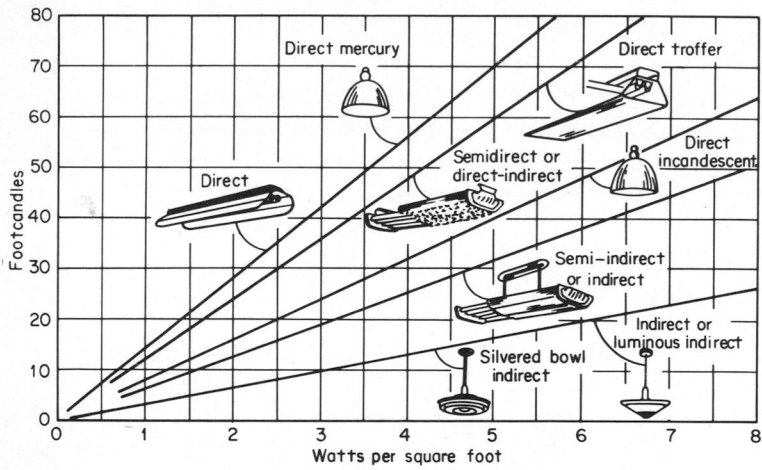

Fig. 4-10. Typical heat-gain lighting fixtures. On the curve above, follow the horizontal line, beginning at the maintained foot-candle value selected in (1), until it intersects the curve corresponding to the fixture type to be installed.

$$Q_{\text{people}} = \text{number of people } (q_{\text{sensible}} + q_{\text{latent}}), \text{ Btu/hr} \qquad (4\text{-}23)$$
(Table 4-35)

$$Q_{\text{equipment}} = q_{\text{sensible}} + q_{\text{latent}} \qquad \text{Btu/hr (Tables 4-36 to 4-39)} \qquad (4\text{-}24)$$

$$Q_{\text{outside air}} = q_{\text{sensible}} + q_{\text{latent}} \qquad \text{Btu/hr}$$
$$= 1.08(t_o - t_i) \text{ cfm} + 0.68(M_o - M_i) \text{ cfm} \qquad \text{Btu/hr} \qquad (4\text{-}25)$$

where $M_o$ = outside moisture content at design wet bulb °F, g/lb
      $M_i$ = inside moisture content at design relative humidity, g/lb
      $Q_{\text{tr}}$ = $AU(t_o - t_i)$ for partitions, floors, ceilings, Btu/hr $\qquad (4\text{-}26)$

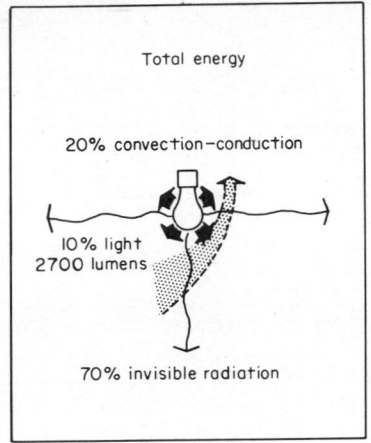

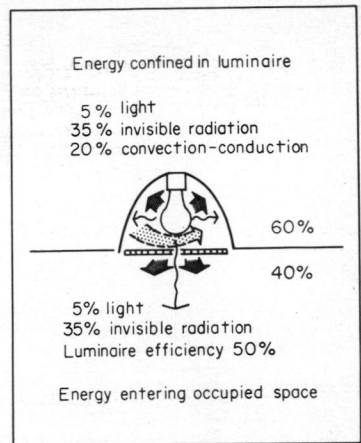

Energy output for 150-watt Incandescent lamp

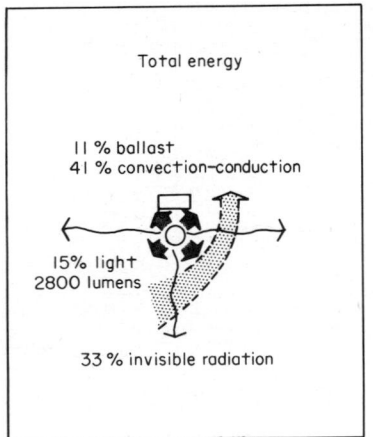

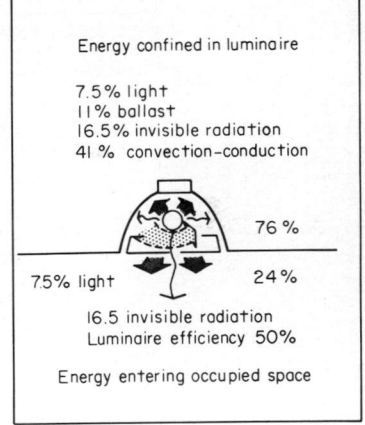

Energy output for 40-watt Fluorescent lamp and ballast

FIG. 4-11. Distribution of energy output of incandescent and fluorescent lamps. (*Light Magazine*, vol. 29, fig. 1.)

# Table 4-20A. Total Equivalent Temperature Differentials for Calculating Heat Gain through Sunlit Walls

| North Latitude Wall Facing | A.M. 8 D | A.M. 8 L | A.M. 10 D | A.M. 10 L | A.M. 12 D | A.M. 12 L | P.M. 2 D | P.M. 2 L | P.M. 4 D | P.M. 4 L | P.M. 6 D | P.M. 6 L | P.M. 8 D | P.M. 8 L | P.M. 10 D | P.M. 10 L | P.M. 12 D | P.M. 12 L | Amplitude Decrement Factor, λ | Time Lag, δ hr | South Latitude Wall Facing |
|---|---|---|---|---|---|---|---|---|---|---|---|---|---|---|---|---|---|---|---|---|---|
| **Group A** | | | | | | | | | | | | | | | | | | | λ = 0.34 | δ = 2 | |
| NE | 27 | 16 | 31 | 18 | 26 | 17 | 24 | 17 | 24 | 18 | 23 | 17 | 20 | 15 | 17 | 13 | 15 | 11 | | | SE |
| E | 32 | 18 | 41 | 24 | 37 | 22 | 29 | 20 | 28 | 20 | 26 | 19 | 23 | 16 | 20 | 14 | 18 | 13 | | | E |
| SE | 25 | 15 | 36 | 21 | 38 | 23 | 33 | 21 | 28 | 20 | 26 | 19 | 22 | 16 | 19 | 14 | 18 | 13 | | | NE |
| S | 14 | 9 | 20 | 13 | 28 | 18 | 33 | 22 | 31 | 21 | 25 | 18 | 20 | 15 | 17 | 13 | 15 | 11 | | | N |
| SW | 17 | 11 | 20 | 13 | 24 | 16 | 34 | 22 | 42 | 27 | 41 | 26 | 28 | 19 | 20 | 14 | 18 | 12 | | | NW |
| W | 17 | 11 | 20 | 13 | 24 | 16 | 30 | 20 | 42 | 27 | 48 | 30 | 33 | 22 | 22 | 15 | 19 | 13 | | | W |
| NW | 14 | 9 | 17 | 11 | 21 | 14 | 23 | 17 | 31 | 21 | 38 | 25 | 28 | 21 | 18 | 13 | 16 | 13 | | | SW |
| N | 14 | 9 | 15 | 10 | 17 | 12 | 20 | 15 | 21 | 16 | 21 | 16 | 18 | 14 | 14 | 11 | 12 | 9 | | | S |
| **Group B** | | | | | | | | | | | | | | | | | | | λ = 0.51 | δ = 3 | |
| NE | 12 | 7 | 27 | 14 | 31 | 17 | 30 | 19 | 31 | 21 | 30 | 22 | 27 | 20 | 21 | 17 | 16 | 13 | | | SE |
| E | 14 | 8 | 34 | 18 | 45 | 24 | 43 | 25 | 39 | 25 | 35 | 24 | 30 | 22 | 23 | 18 | 17 | 14 | | | E |
| SE | 9 | 5 | 25 | 13 | 39 | 21 | 44 | 26 | 41 | 26 | 37 | 25 | 31 | 23 | 24 | 18 | 17 | 14 | | | NE |
| S | 4 | 3 | 7 | 4 | 18 | 11 | 32 | 19 | 41 | 26 | 37 | 27 | 33 | 24 | 25 | 19 | 18 | 15 | | | N |
| SW | 5 | 3 | 7 | 4 | 11 | 7 | 23 | 15 | 41 | 26 | 54 | 34 | 51 | 33 | 38 | 25 | 26 | 19 | | | NW |
| W | 6 | 4 | 7 | 4 | 11 | 7 | 18 | 12 | 35 | 23 | 55 | 34 | 59 | 37 | 43 | 28 | 30 | 20 | | | W |
| NW | 5 | 3 | 6 | 4 | 11 | 7 | 17 | 12 | 26 | 18 | 41 | 27 | 47 | 31 | 36 | 24 | 25 | 18 | | | SW |
| N | 6 | 4 | 9 | 5 | 12 | 8 | 18 | 12 | 22 | 17 | 25 | 20 | 27 | 21 | 22 | 17 | 16 | 14 | | | S |
| **Group C** | | | | | | | | | | | | | | | | | | | λ = 0.40 | δ = 4 | |
| NE | 9 | 6 | 19 | 10 | 26 | 15 | 28 | 17 | 29 | 18 | 29 | 20 | 28 | 20 | 24 | 19 | 20 | 16 | | | SE |
| E | 10 | 7 | 22 | 12 | 36 | 19 | 40 | 23 | 39 | 23 | 36 | 20 | 33 | 23 | 28 | 20 | 22 | 17 | | | E |
| SE | 8 | 6 | 16 | 9 | 29 | 16 | 38 | 21 | 39 | 24 | 37 | 24 | 33 | 23 | 28 | 21 | 21 | 17 | | | NE |
| S | 7 | 5 | 7 | 4 | 12 | 7 | 22 | 14 | 32 | 20 | 36 | 24 | 34 | 24 | 29 | 21 | 23 | 17 | | | N |
| SW | 9 | 6 | 8 | 5 | 10 | 6 | 16 | 10 | 28 | 18 | 42 | 25 | 48 | 30 | 42 | 28 | 33 | 22 | | | NW |
| W | 10 | 7 | 9 | 5 | 10 | 6 | 13 | 9 | 24 | 16 | 40 | 25 | 52 | 32 | 47 | 30 | 37 | 24 | | | W |
| NW | 8 | 6 | 8 | 5 | 9 | 6 | 13 | 9 | 19 | 14 | 30 | 20 | 40 | 27 | 38 | 26 | 32 | 21 | | | SW |
| N | 7 | 5 | 8 | 5 | 10 | 7 | 14 | 9 | 18 | 13 | 22 | 16 | 25 | 19 | 23 | 18 | 19 | 16 | | | S |

Exterior color of wall—D = dark, L = light

## Group D

Column headers: SE, E, NE, N | NW, W, SW, S  
Ratio: 0.45  Value: 4

| (from) | SE | E | NE | N | NW | W | SW | S |
|---|---|---|---|---|---|---|---|---|
| NE | 16 | 17 | 17 |  | 22 | 25 | 21 | 16 |
| E | 19 | 21 | 22 | 23 | 33 | 37 | 31 | 19 |
| SE | 19 | 20 | 21 | 21 | 28 | 31 | 27 | 19 |
| S | 24 | 27 | 28 | 29 | 43 | 49 | 40 | 24 |
| SW | 21 | 23 | 24 | 24 | 32 | 34 | 28 | 19 |
| W | 28 | 33 | 34 | 35 | 51 | 55 | 42 | 25 |
| NW | 21 | 24 | 25 | 25 | 28 | 27 | 21 | 17 |
| N | 30 | 37 | 38 | 38 | 44 | 42 | 31 | 23 |
|  | 19 | 24 | 25 | 21 | 16 | 16 | 14 | 14 |
|  | 30 | 40 | 41 | 34 | 30 | 25 | 20 | 19 |
|  | 17 | 24 | 22 | 14 | 10 | 9 | 9 | 10 |
|  | 29 | 42 | 40 | 23 | 16 | 14 | 13 | 14 |
|  | 15 | 20 | 16 | 7 | 6 | 6 | 6 | 6 |
|  | 28 | 38 | 30 | 12 | 9 | 9 | 9 | 10 |
|  | 10 | 12 | 9 | 4 | 4 | 5 | 4 | 5 |
|  | 19 | 23 | 16 | 6 | 7 | 7 | 7 | 8 |
|  | 5 | 6 | 5 | 4 | 5 | 6 | 5 | 4 |
|  | 8 | 9 | 7 | 5 | 8 | 8 | 7 | 6 |

Row labels (left): NE, E, SE, S, SW, W, NW, N

## Group E

Column headers: SE, E, NE, N | NW, W, SW, S  
Ratio: 0.48  Value: 4

| (from) | SE | E | NE | N | NW | W | SW | S |
|---|---|---|---|---|---|---|---|---|
| NE | 14 | 15 | 16 | 16 | 21 | 23 | 20 | 15 |
| E | 18 | 19 | 20 | 20 | 30 | 34 | 28 | 18 |
| SE | 18 | 19 | 20 | 20 | 27 | 30 | 18 |  |
| S | 23 | 25 | 26 | 27 | 41 | 47 | 38 | 23 |
| SW | 21 | 23 | 23 | 24 | 33 | 36 | 30 | 20 |
| W | 28 | 32 | 33 | 34 | 52 | 57 | 45 | 26 |
| NW | 21 | 24 | 25 | 26 | 31 | 31 | 24 | 18 |
| N | 30 | 36 | 38 | 39 | 49 | 48 | 36 | 24 |
|  | 20 | 24 | 25 | 24 | 22 | 20 | 16 | 15 |
|  | 30 | 39 | 41 | 38 | 35 | 30 | 23 | 21 |
|  | 18 | 24 | 24 | 17 | 12 | 11 | 10 | 11 |
|  | 30 | 43 | 42 | 28 | 19 | 16 | 15 | 16 |
|  | 16 | 22 | 19 | 9 | 6 | 6 | 6 | 7 |
|  | 30 | 42 | 35 | 15 | 10 | 10 | 10 | 11 |
|  | 12 | 15 | 11 | 4 | 4 | 4 | 4 | 5 |
|  | 23 | 28 | 20 | 6 | 7 | 7 | 6 | 8 |
|  | 6 | 6 | 5 | 3 | 4 | 5 | 4 | 4 |
|  | 10 | 8 | 8 | 4 | 6 | 7 | 6 | 6 |

Row labels (left): NE, E, SE, S, SW, W, NW, N

## Group F

Column headers: SE, E, NE, N | NW, W, SW, S  
Ratio: 0.32  Value: 6

| (from) | SE | E | NE | N | NW | W | SW | S |
|---|---|---|---|---|---|---|---|---|
| NE | 17 | 19 | 19 | 19 | 25 | 27 | 23 | 17 |
| E | 23 | 26 | 27 | 27 | 37 | 41 | 33 | 21 |
| SE | 19 | 22 | 22 | 22 | 27 | 29 | 24 | 18 |
| S | 26 | 31 | 32 | 31 | 42 | 46 | 36 | 23 |
| SW | 20 | 23 | 23 | 22 | 27 | 27 | 22 | 17 |
| W | 28 | 35 | 35 | 35 | 42 | 43 | 33 | 22 |
| NW | 21 | 20 | 16 | 14 |  |  |  |  |
| N | 29 | 37 | 37 | 32 | 33 | 31 | 24 | 19 |
|  | 17 | 22 | 21 | 16 | 22 | 19 | 16 | 15 |
|  | 27 | 37 | 36 | 26 | 22 | 19 | 16 | 15 |
|  | 15 | 19 | 17 | 7 | 8 | 8 | 8 | 8 |
|  | 25 | 35 | 31 | 17 | 13 | 12 | 11 | 12 |
|  | 12 | 15 | 12 | 6 | 6 | 6 | 6 | 6 |
|  | 21 | 28 | 22 | 10 | 9 | 10 | 9 | 9 |
|  | 9 | 10 | 8 | 5 | 6 | 7 | 6 | 6 |
|  | 14 | 17 | 13 | 7 | 10 | 11 | 9 | 8 |
|  | 7 | 8 | 7 | 7 | 9 | 9 | 8 | 7 |
|  | 9 | 10 | 10 | 9 | 12 | 14 | 12 | 8 |

Row labels (left): NE, E, SE, S, SW, W, NW, N

Exterior color of wall — D = dark, L = light

| North Latitude Wall Facing | A.M. 8 D | 8 L | 10 D | 10 L | 12 D | 12 L | P.M. 2 D | 2 L | 4 D | 4 L | 6 D | 6 L | 8 D | 8 L | 10 D | 10 L | 12 D | 12 L | λ | δ hr | South Latitude Wall Facing |
|---|---|---|---|---|---|---|---|---|---|---|---|---|---|---|---|---|---|---|---|---|---|
| **Group G** | | | | | | | | | | | | | | | | | | | | | |
| NE | 11 | 9 | 10 | 15 | 20 | 12 | 24 | 14 | 25 | 16 | 26 | 17 | 27 | 18 | 26 | 18 | 23 | 17 | | | SE |
| E | 13 | 9 | 17 | 11 | 26 | 15 | 32 | 18 | 34 | 20 | 34 | 21 | 33 | 22 | 31 | 21 | 27 | 19 | | | E |
| SE | 13 | 9 | 14 | 9 | 21 | 12 | 28 | 16 | 33 | 19 | 34 | 21 | 33 | 22 | 31 | 21 | 27 | 19 | 0.25 | 6 | NE |
| S | 12 | 9 | 10 | 7 | 11 | 8 | 16 | 10 | 23 | 15 | 29 | 18 | 30 | 20 | 29 | 20 | 26 | 16 | | | N |
| SW | 16 | 11 | 13 | 9 | 13 | 8 | 14 | 9 | 20 | 13 | 27 | 19 | 37 | 24 | 39 | 25 | 35 | 23 | | | NW |
| W | 18 | 12 | 15 | 10 | 14 | 9 | 14 | 9 | 18 | 12 | 19 | 15 | 38 | 26 | 42 | 26 | 38 | 25 | | | W |
| NW | 14 | 10 | 12 | 8 | 12 | 8 | 13 | 8 | 16 | 11 | 15 | 10 | 24 | 18 | 33 | 22 | 31 | 21 | | | SW |
| N | 10 | 8 | 10 | 7 | 10 | 7 | 15 | 10 | 23 | 15 | 29 | 18 | 21 | 16 | 21 | 16 | 20 | 16 | | | S |
| **Group H** | | | | | | | | | | | | | | | | | | | | | |
| NE | 15 | 11 | 16 | 11 | 18 | 12 | 20 | 13 | 22 | 14 | 24 | 15 | 25 | 16 | 25 | 17 | 24 | 17 | | | SE |
| E | 18 | 13 | 18 | 12 | 22 | 13 | 26 | 16 | 29 | 17 | 30 | 18 | 31 | 20 | 30 | 20 | 29 | 19 | | | E |
| SE | 18 | 13 | 17 | 12 | 19 | 12 | 23 | 15 | 27 | 17 | 29 | 18 | 29 | 19 | 28 | 19 | 28 | 19 | 0.14 | 8 | NE |
| S | 16 | 12 | 14 | 10 | 14 | 10 | 15 | 10 | 19 | 12 | 23 | 15 | 25 | 17 | 26 | 18 | 26 | 18 | | | N |
| SW | 22 | 14 | 19 | 13 | 18 | 13 | 18 | 11 | 19 | 12 | 23 | 15 | 29 | 18 | 32 | 21 | 32 | 21 | | | NW |
| W | 23 | 15 | 20 | 13 | 19 | 14 | 17 | 12 | 18 | 12 | 22 | 15 | 29 | 18 | 33 | 21 | 34 | 22 | | | W |
| NW | 19 | 13 | 17 | 11 | 15 | 11 | 15 | 11 | 16 | 11 | 18 | 14 | 23 | 17 | 26 | 18 | 27 | 19 | | | SW |
| N | 13 | 10 | 14 | 9 | 11 | 9 | 13 | 10 | 15 | 10 | 16 | 11 | 17 | 13 | 18 | 14 | 18 | 16 | | | S |
| **Group I** | | | | | | | | | | | | | | | | | | | | | |
| NE | 16 | 11 | 18 | 12 | 20 | 13 | 22 | 14 | 23 | 15 | 24 | 16 | 24 | 16 | 23 | 16 | 22 | 16 | | | SE |
| E | 19 | 13 | 21 | 14 | 25 | 16 | 29 | 17 | 30 | 18 | 30 | 18 | 29 | 19 | 28 | 19 | 26 | 18 | | | E |
| SE | 18 | 13 | 19 | 12 | 22 | 14 | 27 | 16 | 28 | 18 | 29 | 18 | 29 | 19 | 28 | 18 | 26 | 18 | 0.13 | 6 | NE |
| S | 16 | 12 | 15 | 11 | 16 | 11 | 18 | 11 | 21 | 14 | 24 | 16 | 25 | 17 | 25 | 17 | 23 | 16 | | | N |
| SW | 20 | 14 | 19 | 13 | 19 | 13 | 19 | 13 | 22 | 14 | 27 | 17 | 31 | 20 | 32 | 20 | 30 | 20 | | | NW |
| W | 22 | 14 | 20 | 13 | 20 | 13 | 20 | 13 | 22 | 14 | 26 | 17 | 31 | 20 | 33 | 21 | 32 | 21 | | | W |
| NW | 18 | 12 | 16 | 11 | 16 | 11 | 17 | 12 | 18 | 13 | 21 | 14 | 25 | 18 | 27 | 18 | 26 | 18 | | | SW |
| N | 13 | 10 | 12 | 9 | 13 | 9 | 15 | 10 | 15 | 11 | 16 | 12 | 18 | 13 | 18 | 14 | 14 | 14 | | | S |

**Group J[a]**

| | | | | | | | | | | | | value | | |
|---|---|---|---|---|---|---|---|---|---|---|---|---|---|---|
| | | | | | | | | | | | | SE | | |
| | | | | | | | | | | | | E | | |
| | | | | | | | | | | | | NE | 0.10 | 9 |
| | | | | | | | | | | | | N | | |
| | | | | | | | | | | | | NW | | |
| | | | | | | | | | | | | W | | |
| | | | | | | | | | | | | SW | | |
| | | | | | | | | | | | | S | | |

| Dir | | | | | | | | | | | |
|---|---|---|---|---|---|---|---|---|---|---|---|
| NE | 18 | 13 | 17 | 12 | 18 | 12 | 19 | 13 | 21 | 14 | 16 |
| E | 22 | 15 | 20 | 14 | 21 | 14 | 24 | 15 | 26 | 17 | 19 |
| SE | 21 | 15 | 20 | 14 | 20 | 13 | 21 | 14 | 24 | 16 | 18 |
| S | 19 | 14 | 17 | 12 | 16 | 11 | 16 | 11 | 17 | 13 | 16 |
| SW | 24 | 16 | 22 | 15 | 20 | 13 | 19 | 13 | 19 | 14 | 19 |
| W | 26 | 17 | 24 | 16 | 22 | 14 | 20 | 13 | 20 | 14 | 20 |
| NW | 21 | 15 | 19 | 13 | 18 | 12 | 17 | 11 | 17 | 12 | 17 |
| N | 15 | 11 | 14 | 11 | 13 | 10 | 13 | 9 | 13 | 10 | 13 |

**Group K[a]**

| | | | | | | | | | | | | value | | |
|---|---|---|---|---|---|---|---|---|---|---|---|---|---|---|
| | | | | | | | | | | | | SE | | |
| | | | | | | | | | | | | E | | |
| | | | | | | | | | | | | NE | 0.08 | 11 |
| | | | | | | | | | | | | N | | |
| | | | | | | | | | | | | NW | | |
| | | | | | | | | | | | | W | | |
| | | | | | | | | | | | | SW | | |
| | | | | | | | | | | | | S | | |

| Dir | | | | | | | | | | | |
|---|---|---|---|---|---|---|---|---|---|---|---|
| NE | 19 | 14 | 19 | 13 | 20 | 14 | 20 | 13 | 22 | 15 | 21 |
| E | 23 | 16 | 22 | 15 | 24 | 15 | 26 | 17 | 27 | 17 | 23 |
| SE | 23 | 16 | 22 | 15 | 22 | 15 | 24 | 16 | 27 | 17 | 25 |
| S | 20 | 14 | 19 | 13 | 18 | 13 | 18 | 13 | 21 | 14 | 20 |
| SW | 25 | 16 | 23 | 15 | 22 | 14 | 21 | 14 | 24 | 16 | 24 |
| W | 26 | 17 | 24 | 16 | 23 | 15 | 22 | 14 | 24 | 16 | 24 |
| NW | 21 | 15 | 20 | 13 | 18 | 13 | 18 | 12 | 20 | 14 | 20 |
| N | 15 | 11 | 14 | 10 | 14 | 10 | 14 | 11 | 15 | 12 | 15 |

**Group L[a]**

| | | | | | | | | | | | | value | | |
|---|---|---|---|---|---|---|---|---|---|---|---|---|---|---|
| | | | | | | | | | | | | SE | | |
| | | | | | | | | | | | | E | | |
| | | | | | | | | | | | | NE | 0.08 | 8 |
| | | | | | | | | | | | | N | | |
| | | | | | | | | | | | | NW | | |
| | | | | | | | | | | | | W | | |
| | | | | | | | | | | | | SW | | |
| | | | | | | | | | | | | S | | |

| Dir | | | | | | | | | | | |
|---|---|---|---|---|---|---|---|---|---|---|---|
| NE | 18 | 13 | 18 | 13 | 19 | 14 | 21 | 14 | 22 | 15 | 21 |
| E | 22 | 15 | 22 | 14 | 23 | 14 | 25 | 15 | 27 | 16 | 22 |
| SE | 21 | 14 | 21 | 14 | 21 | 14 | 24 | 15 | 26 | 15 | 24 |
| S | 19 | 13 | 17 | 13 | 17 | 12 | 16 | 12 | 19 | 13 | 17 |
| SW | 23 | 14 | 22 | 14 | 21 | 14 | 20 | 13 | 23 | 15 | 23 |
| W | 25 | 15 | 23 | 15 | 22 | 14 | 21 | 14 | 24 | 15 | 24 |
| NW | 20 | 14 | 19 | 12 | 18 | 12 | 18 | 12 | 19 | 13 | 19 |
| N | 14 | 11 | 14 | 10 | 13 | 10 | 14 | 10 | 17 | 11 | 16 |

# Table 4-20A. Total Equivalent Temperature Differentials for Calculating Heat Gain through Sunlit Walls (Continued)

| North Latitude Wall Facing | Sun Time A.M. 8 D | 8 L | 10 D | 10 L | 12 D | 12 L | P.M. 2 D | 2 L | 4 D | 4 L | 6 D | 6 L | 8 D | 8 L | 10 D | 10 L | 12 D | 12 L | λ | Time Lag 6 hr | South Latitude Wall Facing |
|---|---|---|---|---|---|---|---|---|---|---|---|---|---|---|---|---|---|---|---|---|---|
| | | | | | | | Group Mª (D = dark, L = light) | | | | | | | | | | | | | | |
| NE | 20 | 14 | 20 | 14 | 19 | 13 | 20 | 13 | 20 | 14 | 20 | 14 | 21 | 14 | 21 | 14 | 22 | 15 | | 6 | SE |
| E | 25 | 16 | 24 | 16 | 24 | 16 | 24 | 16 | 25 | 16 | 25 | 16 | 26 | 16 | 27 | 17 | 27 | 17 | | | E |
| SE | 24 | 16 | 24 | 15 | 23 | 15 | 23 | 15 | 24 | 16 | 24 | 16 | 25 | 16 | 25 | 16 | 26 | 17 | | | NE |
| S | 21 | 14 | 20 | 14 | 19 | 13 | 19 | 13 | 19 | 13 | 19 | 13 | 20 | 14 | 20 | 14 | 21 | 15 | | | N |
| SW | 25 | 17 | 25 | 16 | 24 | 16 | 23 | 15 | 22 | 15 | 22 | 15 | 23 | 15 | 24 | 16 | 25 | 17 | 0.05 | 12 | NW |
| W | 27 | 17 | 26 | 17 | 25 | 16 | 24 | 16 | 23 | 15 | 23 | 15 | 24 | 15 | 25 | 16 | 26 | 17 | | | W |
| NW | 22 | 15 | 21 | 15 | 20 | 14 | 20 | 13 | 19 | 13 | 19 | 13 | 19 | 13 | 20 | 14 | 21 | 15 | | | SW |
| N | 15 | 12 | 15 | 11 | 14 | 11 | 14 | 11 | 14 | 11 | 14 | 11 | 14 | 11 | 15 | 11 | 16 | 12 | | | S |

ª See Table 4-20B for details of each wall grouping.

*Explanation:*

$$\left\{ \begin{array}{l} \text{Total heat transmission from solar radiation and} \\ \text{temperature difference between outside and} \\ \text{room air, Btu/(hr)/(ft}^2 \text{ wall area)} \end{array} \right\} = \left\{ \begin{array}{l} \text{Equivalent temperature} \\ \text{differential from above} \\ \text{table} \end{array} \right\} \times \left\{ \begin{array}{l} \text{Heat transmission coefficient} \\ \text{for wall, Btu/(hr)(ft}^2)(°F) \end{array} \right\}$$

1. *Application.* These values may be used for all normal air-conditioning estimates; usually without correction (except as noted below) when the load is calculated for the hottest weather.

2. *Corrections.* The values in the table were calculated for an inside temperature of 75°F and an outdoor maximum temperature of 95°F with an outdoor daily range of 21°F. The table remains approximately correct for other outdoor maximums (93 to 102°F) and other outdoor daily ranges (16 to 34°F) provided the outdoor daily average temperature remains approximately 85°F. If the room temperature is different from 75°F and/or the outdoor daily average temperature is different from 85°F, Equation 43 can be used for computing new values on the following rules can be applied:

a. For room air temperature less than 75°F, add the difference between 75°F and room air temperature; if greater than 75°F, subtract the difference.
b. For outdoor daily average temperature less than 85°F, subtract the difference between 85°F and the daily average temperature; if greater than 85°F, add the difference.
The table values will be approximately correct for the east or west wall in any latitude (0 to 50° North or South) during the hottest weather. Equation 43 should be used for obtaining values for the north or south wall in latitudes other than 40°.

3. *Color of exterior surface of wall.* Use temperature differentials for light walls only when the permanence of the light wall is established by experience. For cream color use the values for light walls. For medium colors interpolate halfway between the dark and light values. Medium colors are medium blue, medium green, bright red, light brown, unpainted wood, natural color concrete, etc. Dark blue, red, brown, green, etc., are considered dark colors.
Reprinted by permission from "ASHRAE Handbook of Fundamentals," ASHRAE, New York, 1972.

## Table 4-20B. Description of Wall Constructions

| Group | Components | Wt, lb per sq ft | U Value |
|-------|-----------|------------------|---------|
| A | 1" stucco +4" l.w. concrete block +air space | 28.6 | 0.267 |
|   | 1" stucco +air space +2" insulation | 16.3 | 0.106 |
| B | 1" stucco +4" common brick | 55.9 | 0.393 |
|   | 1" stucco +4" h.w. concrete | 62.5 | 0.481 |
| C | 4" face brick +4" l.w. concrete block +1" insulation | 62.5 | 0.158 |
|   | 1" stucco +4" h.w. concrete +2" insulation | 62.9 | 0.114 |
| D | 1" stucco +8" l.w. concrete block +1" insulation | 41.4 | 0.141 |
|   | 1" stucco +2" insulation +4" h.w. concrete block | 36.6 | 0.111 |
| E | 4" face brick +4" l.w. concrete block | 62.2 | 0.333 |
|   | 1" stucco +8" h.w. concrete block | 56.6 | 0.349 |
| F | 4" face brick +4" common brick | 89.5 | 0.360 |
|   | 4" face brick +2" insulation +4" l.w. concrete block | 62.5 | 0.103 |
| G | 1" stucco +8" clay tile +1" insulation | 62.8 | 0.141 |
|   | 1" stucco +2" insulation +4" common brick | 56.2 | 0.108 |
| H | 4" face brick +8" clay tile +1" insulation | 96.4 | 0.137 |
|   | 4" face brick +8" common brick | 129.6 | 0.280 |
|   | 1" stucco +12" h.w. concrete | 155.9 | 0.365 |
|   | 4" face brick +2" insulation +4" common brick | 89.8 | 0.106 |
|   | 4" face brick +2" insulation +4" h.w. concrete | 96.5 | 0.111 |
|   | 4" face brick +2" insulation +8" h.w. concrete block | 90.6 | 0.102 |
| I | 1" stucco +8" clay tile +air space | 62.6 | 0.209 |
|   | 4" face brick +air space +4" h.w. concrete block | 69.9 | 0.282 |
| J | face brick +8" common brick +1" insulation | 129.8 | 0.145 |
|   | 4" face brick +2" insulation +8" clay tile | 96.5 | 0.094 |
|   | 1" stucco +2" insulation +8" common brick | 96.3 | 0.100 |
| K | 4" face brick +air space +8" clay tile | 96.2 | 0.200 |
|   | 4" face brick +2" insulation +8" common brick | 129.9 | 0.098 |
|   | 4" face brick +2" insulation +8" h.w. concrete | 143.3 | 0.107 |
| L | 4" face brick +8" clay tile +air space | 96.2 | 0.200 |
|   | 4" face brick +air space +4" common brick | 89.5 | 0.265 |
|   | 4" face brick +air space +4" h.w. concrete | 96.2 | 0.301 |
|   | 4" face brick +air space +8" h.w. concrete block | 90.2 | 0.246 |
|   | 1" stucco +2" insulation +12" h.w. concrete | 156.3 | 0.106 |
| M | 4" face brick +air space +8" common brick | 129.6 | 0.218 |
|   | 4" face brick +air space +12" h.w. concrete | 189.5 | 0.251 |
|   | 4" face brick +2" insulation +12" h.w. concrete | 189.9 | 0.104 |

Reprinted by permission from "ASHRAE Handbook of Fundamentals," ASHRAE, New York, 1972.

# Table 4-21. Total Equivalent Temperature Differentials for Calculating Heat Gain through Flat Roofs

Sun Time columns — A.M.: 8, 10, 12; P.M.: 2, 4, 6, 8, 10, 12. Each time has two sub-columns, **D** and **L**.

| Description of Roof Construction[a,b] | Wt, lb per sq ft | U value Btu/(hr)(ft²)(F°) | 8 D | 8 L | 10 D | 10 L | 12 D | 12 L | 2 D | 2 L | 4 D | 4 L | 6 D | 6 L | 8 D | 8 L | 10 D | 10 L | 12 D | 12 L | λ | 6 |
|---|---|---|---|---|---|---|---|---|---|---|---|---|---|---|---|---|---|---|---|---|---|---|
| *Light Construction Roofs—Exposed to Sun* | | | | | | | | | | | | | | | | | | | | | |
| 1" insulation+steel siding | 7.4 | 0.213 | 28 | 11 | 65 | 31 | 90 | 48 | 95 | 53 | 78 | 45 | 43 | 27 | 8 | 6 | 1 | 1 | −3 | −3 | 1.0 | 0 |
| 2" insulation+steel siding | 7.8 | 0.125 | 24 | 8 | 61 | 29 | 88 | 46 | 96 | 50 | 81 | 46 | 48 | 30 | 10 | 8 | 2 | 2 | −3 | −3 | 0.99 | 1 |
| 1" insulation+1" wood[c] | 8.4 | 0.206 | 12 | 2 | 47 | 21 | 72 | 39 | 92 | 52 | 86 | 48 | 61 | 36 | 25 | 16 | 5 | 5 | 0 | 0 | 0.93 | 2 |
| 2" insulation+1" wood[c] | 8.5 | 0.122 | 8 | 0 | 41 | 18 | 72 | 36 | 90 | 48 | 88 | 48 | 65 | 38 | 30 | 19 | 7 | 7 | −1 | −1 | 0.93 | 2 |
| 1" insulation+2.5" wood[c] | 12.7 | 0.193 | 2 | −2 | 23 | 8 | 48 | 23 | 70 | 36 | 79 | 42 | 71 | 40 | 50 | 29 | 29 | 17 | 9 | 9 | 0.73 | 3 |
| 2" insulation+2.5" wood[c] | 13.1 | 0.117 | 1 | −2 | 19 | 6 | 43 | 20 | 65 | 33 | 76 | 41 | 72 | 40 | 53 | 31 | 33 | 20 | 18 | 11 | 0.68 | 4 |
| *Medium Construction Roofs—Exposed to Sun* | | | | | | | | | | | | | | | | | | | | | |
| 1" insulation+4" wood[c] | 17.3 | 0.183 | 5 | 0 | 14 | 5 | 31 | 14 | 49 | 24 | 62 | 32 | 65 | 35 | 56 | 31 | 41 | 24 | 29 | 17 | 0.51 | 5 |
| 2" insulation+4" wood[c] | 17.8 | 0.113 | 6 | 1 | 13 | 4 | 28 | 12 | 45 | 22 | 58 | 30 | 63 | 34 | 56 | 31 | 43 | 25 | 32 | 18 | 0.48 | 5 |
| 1" insulation+2" h.w. concrete | 28.3 | 0.206 | 4 | −1 | 27 | 13 | 54 | 26 | 74 | 39 | 81 | 44 | 70 | 40 | 47 | 27 | 24 | 15 | 12 | 7 | 0.75 | 5 |
| 2" insulation+2" h.w. concrete | 28.8 | 0.122 | 2 | −3 | 23 | 9 | 49 | 23 | 70 | 36 | 79 | 43 | 71 | 41 | 49 | 28 | 28 | 17 | 9 | 9 | 0.73 | 4 |
| 4" l.w. concrete | 17.6 | 0.213 | −2 | −4 | 11 | 0 | 28 | 13 | 55 | 27 | 72 | 38 | 74 | 44 | 64 | 36 | 42 | 25 | 25 | 15 | 0.87 | 4 |
| 6" l.w. concrete | 24.5 | 0.157 | 6 | 2 | 9 | 2 | 16 | 6 | 32 | 14 | 49 | 24 | 61 | 32 | 63 | 34 | 55 | 31 | 41 | 24 | 0.67 | 5 |
| 8" l.w. concrete | 31.2 | 0.125 | | | | | | | | | 49 | 24 | 61 | 32 | 63 | 34 | 55 | 31 | 41 | 24 | 0.50 | 6 |
| *Heavy Construction Roofs—Exposed to Sun* | | | | | | | | | | | | | | | | | | | | | |
| 1" insulation+4" h.w. concrete | 51.6 | 0.199 | 7 | 1 | 17 | 6 | 33 | 15 | 50 | 25 | 61 | 32 | 63 | 34 | 53 | 30 | 40 | 23 | 28 | 16 | 0.48 | 5 |
| 2" insulation+4" h.w. concrete | 52.1 | 0.120 | 7 | 2 | 15 | 6 | 30 | 13 | 46 | 23 | 58 | 30 | 61 | 33 | 54 | 30 | 41 | 23 | 31 | 17 | 0.45 | 5 |
| 1" insulation+6" h.w. concrete | 75.0 | 0.193 | 13 | 6 | 17 | 7 | 26 | 12 | 38 | 18 | 48 | 25 | 53 | 28 | 51 | 27 | 43 | 24 | 35 | 19 | 0.33 | 4 |
| 2" insulation+6" h.w. concrete | 75.4 | 0.117 | 15 | 7 | 17 | 7 | 25 | 11 | 36 | 17 | 46 | 23 | 51 | 27 | 50 | 27 | 43 | 24 | 36 | 20 | 0.30 | 6 |

*Roofs Covered with Water—Exposed to Sun*

| Outside Air Dew Point (F) | Water Layer Thickness (in.) |
|---|---|
| | |

| | | | | | | | | | | | | | | |
|---|---|---|---|---|---|---|---|---|---|---|---|---|---|---|
| **Light Construction** | 60 | 6<br>1<br>0 | | −12<br>−12 | −6<br>−8<br>−4 | −1<br>7 | 6<br>15<br>17 | 13<br>21<br>23 | 17<br>22<br>22 | 17<br>17<br>16 | 13<br>8<br>5 | 7<br>0<br>−3 | |
| | 70 | 6<br>1<br>0 | | −1<br>−5 | 0<br>2 | 4<br>10<br>12 | 11<br>19<br>21 | 18<br>25<br>26 | 21<br>26<br>26 | 21<br>21<br>19 | 17<br>12<br>9 | 12<br>5<br>2 | |
| **Heavy Construction** | 60 | 6<br>1<br>0 | | −3<br>−8<br>−9 | −4<br>−6<br>−5 | −1<br>1<br>2 | 4<br>8<br>10 | 9<br>15<br>16 | 13<br>18<br>19 | 15<br>17<br>16 | 13<br>11<br>10 | 10<br>6<br>4 | |
| | 70 | 6<br>1<br>0 | | −2<br>−2 | 2<br>0<br>1 | 4<br>6<br>8 | 9<br>14<br>16 | 14<br>20<br>21 | 18<br>23<br>23 | 20<br>21<br>21 | 18<br>16<br>15 | 15<br>11<br>9 | |

Reprinted by permission from "ASHRAE Handbook of Fundamentals," ASHRAE, New York, 1972.

a Includes outside surface resistance, 1/2-in. slag, membrane and 3/8-in. felt on the top and inside surface resistance on the bottom.

b Dark roof, $\alpha/h_o = 0.30$; light roof, $\alpha/h_o = 0.15$.

c Nominal thickness of wood.

*Explanation:* $\left\{ \begin{array}{l} \text{Total heat transmission from solar radiation and} \\ \text{temperature difference between outdoor and} \\ \text{room air, Btu/(hr)/(ft}^2\text{) of roof area} \end{array} \right\} = \left\{ \begin{array}{l} \text{Equivalent temperature} \\ \text{differential from above} \\ \text{table} \\ (°F) \end{array} \right\} \times \left\{ \begin{array}{l} \text{Heat transmission coefficient} \\ \text{for summer, Btu/(hr)/(ft}^2\text{)} \end{array} \right\}$

1. *Application.* These values may be used for all normal air conditioning estimated; usually without correction (except as noted below) in latitude 0 to 50° North or South when the load is calculated for the hottest weather.

2. *Corrections.* The values in the table were calculated for an inside temperature of 75°F and an outdoor maximum temperature of 95°F with an outdoor daily range of 21°F. The table remains approximately correct for other outdoor maximums (93 to 102°F) and other outdoor daily ranges (16 to 34°F) provided the outdoor daily average temperature remains approximately 85°F. If the room air temperature is different from 75°F and/or the outdoor daily average temperature is different from 85°F, Equation 43 can be used for computing new values or the following rules can be applied:

   a. For room air temperature less than 75°F, add the difference between 75°F and room air temperature; if greater than 75°F, subtract the difference.

   b. For outdoor daily average temperature less than 85°F, subtract the difference between 85°F and the daily average temperature; if greater than 85°F, add the difference.

3. *Attics or other spaces between the roof and ceiling.* If the ceiling is insulated and a fan is used for positive ventilation in the space between the ceiling and roof, the total temperature differential for calculating the room load may be decreased by 25 per cent.

   If the attic space contains a return duct or other air plenum, care should be taken in determining the portion of the heat gain that reaches the ceiling.

4. *Light colors.* Credit should not be taken for light colored roofs except where the permanence of light colors is established by experience, as in rural areas or where there is little smoke.

5. *For solar transmission in other months.* The table values of temperature differentials that were calculated for July 21 will be approximately correct for a roof in the following months:

### North Latitude

| Latitude (deg) | Months |
|---|---|
| 0 | All Months |
| 10 | All Months |
| 20 | All Months except Nov., Dec., Jan. |
| 30 | Mar., Apr., May, June, July, Aug., Sept. |
| 40 | April, May, June, July, Aug. |
| 50 | May, June, July |

### South Latitude

| Latitude (deg) | Months |
|---|---|
| 0 | All Months |
| 10 | All Months |
| 20 | All Months except May, June, July |
| 30 | Sept., Oct., Nov., Dec., Jan., Feb., March |
| 40 | Oct., Nov., Dec., Jan., Feb. |
| 50 | Nov., Dec., Jan. |

## Table 4-22A. Coefficients of Transmission $U$ of Flat Masonry Roofs with Built-up Roofing, with and without Suspended Ceilings* (Winter Conditions, Upward Flow)

*These Coefficients are expressed in Btu per (hour) (square foot) (Fahrenheit degree difference in temperature between the air on the two sides), and are based upon an outside wind velocity of 15 mph*

| Example | | Example of Substitution |
|---|---|---|
| Construction (heat flow up) | Resistance ($R$) | For addition of roof insulation, see Table 4-23B |
| 1. Outside surface (15 mph wind) | 0.17 | |
| 2. Built-up roofing—$\frac{3}{8}$ in. | 0.33 | |
| 3. Roof insulation (none) | — | |
| 4. Concrete slab (lt. wt. agg.) (2 in.) | 2.22 | |
| 5. Corrugated metal | 0 | |
| 6. Air space* | 0.85 | |
| 7. Metal lath and $\frac{3}{4}$ in. plas. (lt. wt. agg.) | 0.47 | |
| 8. Inside surface (still air) | 0.61 | |
| Total resistance | 4.65 | |
| $U = 1/R = 1/4.65 =$ | 0.22 | |

*To adjust $U$ values for the effect of added insulation between framing members see Table 4-23. Reprinted from "ASHRAE Handbook of Fundamentals," ASHRAE, New York, 1972.

## Table 4-22B. Coefficients of Transmission $U$ of Wood Construction Flat Roofs and Ceilings (Winter Conditions, Upward Flow)

*Coefficients are expressed in Btu per (hour) (square foot) (Fahrenheit degree difference in temperature between the air on the two sides), and are based upon an outside wind velocity of 15 mph*

| Example | Resistance ($R$) | Example of Substitution | | |
|---|---|---|---|---|
| Construction (Heat flow up) | | Delete item 3 and furnish R-19 insulation batt in lieu of item 5 air space. | | |
| 1. Outside surface (15 mph wind) | 0.17 | | | |
| 2. Built-up roofing ($\frac{3}{8}$ in.) | 0.33 | Total resistance | | 5.83 |
| 3. Roof insulation ($C = 0.72$) | 1.39 | Deduct: 3. Roof insulation | 1.39 | |
| 4. Plywood deck ($\frac{5}{8}$ in.) | 0.78 | 5. Air space | 0.85 | 2.24 |
| 5. Air space | 0.85 | | | |
| 6. Gypsum wallboard ($\frac{1}{2}$ in.) | 0.45 | Difference | | 3.59 |
| 7. Acoustical tile ($\frac{1}{2}$ in.)—glued | 1.25 | Add: 5. R-19 insulation batt | | 19.00 |
| 8. Inside surface (still air) | 0.61 | | | |
| | | Total resistance | | 22.59 |
| Total resistance | 5.83 | $U_i = 1/R = 1/22.59 =$ | | 0.04 |
| $U = 1/R = 1/5.83$ | 0.17 | | | |

Note: Adjustment for heat flow through framing members depends upon dimensions and spacing of framing.

Reprinted from "ASHRAE Handbook of Fundamentals," ASHRAE, New York, 1972.

## Table 4-22C. Coefficients of Transmission $U$ of Metal Construction Flat Roofs and Ceilings
### (Winter Conditions, Upward Flow)

*Coefficients are expressed in Btu per (hour) (square foot) (Fahrenheit degree difference in temperature between the air on the two sides), and are based on upon outside wind velocity of 15 mph*

| Example | Resistance ($R$) | Example of Substitution | |
|---|---|---|---|
| Construction | | Replace item 3 with roof insulation ($C = 0.36$) and items 6 and 7 with metal lath and $\frac{3}{4}$ in. plas. (lt. wt. agg.) | |
| 1. Outside surface (15 mph wind) | 0.17 | | |
| 2. Built-up roofing ($\frac{3}{8}$ in.) | 0.33 | Total resistance | 6.71 |
| 3. Roof insulation ($C = 0.24$) | 4.17 | Deduct 3. Roof insulation ($C = 0.24$) | 4.17 |
| 4. Metal deck | 0.00 | 6. Metal lath and | |
| 5. Air space *† | 0.99 | 7. $\frac{3}{4}$ in. plas. (sand agg.) } | 0.13 4.30 |
| 6. Metal lath and | | | |
| 7. $\frac{3}{4}$ in. plas. (sand agg.) } | 0.13 | Difference | 2.41 |
| 8. Inside surface (still air) | 0.92 | Add 3. Roof insulation ($C = 0.36$) | 2.78 |
| | | 6. Metal lath and | |
| Total resistance | 6.71 | 7. $\frac{3}{4}$ in. plas. (lt. wt. agg.) } | 0.47 3.25 |
| $U = 1/R = 1/6.71$ | 0.15 | | |
| Note: Adjustment for heat flow through metal framing members depends upon dimensions and spacing. | | Total resistance | 5.66 |
| | | $U = 1/R = 1/5.66$ | 0.18 |

\* If a vapor barrier is used beneath roof insulation it will have a negligible effect on the $U$ value.
† To adjust $U$ values for the effect of added insulation between framing members, see Table 4-23B.
Reprinted by permission from "ASHRAE Handbook of Fundamentals," ASHRAE, New York, 1972.

# Table 4-22D. Coefficients of Transmission $U$ of Pitched Roofs*

Coefficients are expressed in Btu per (hour) (square foot) (Fahrenheit degree difference in temperature between the air on the two sides), and are based on an outside wind velocity of 15 mph for heat flow upward and 7.5 mph for heat flow downward

| Example | | Example of Substitution | |
|---|---|---|---|
| Construction (Heat flow up) | Resistance $(R)$ | Find $U$ value for same construction with heat flow down (summer conditions) | |
| 1. Outside surface (15 mph wind) | 0.17 | | |
| 2. Asphalt shingle roofing | 0.44 | | |
| 3. Building paper | 0.06 | | |
| 4. Plywood deck ($\frac{5}{8}$ in.) | 0.78 | | |
| 5. Air space †(3.5 in., reflective surface) | 2.06 | Total resistance | 4.58 |
| 6. Gypsum wallboard ($\frac{1}{2}$ in.) | 0.45 | Deduct 5. Air space (3.5 in. reflective)  2.06 | |
| 7. Inside surface (still air) | 0.62 | 8. Inside surface (still air)  0.62 | 2.68 |
| | | | |
| Total resistance | 4.58 | Difference | 1.90 |
| $U = 1/R = 1/4.58$. | 0.22 | Add 5. Air space (3.5 in., reflective)  8.08 | |
| Note: Correction for heat flow through framing members depends upon dimensions and spacing. | | 8. Inside surface (still air)  0.76 | 8.84 |
| | | | |
| | | Total resistance | 10.74 |
| | | $U = 1/R = 1/10.74$. | 0.09 |
| | | Note: Correction for heat flow through framing members depends upon dimensions and spacing. | |

4-43

## Table 4-22D. Coefficients of Transmission $U$ of Pitched Roofs* (Continued)

| Example | Resistance ($R$) |
|---|---|
| Construction (Heat flow up) | |
| 1. Outside surface (15 mph wind) | 0.17 |
| 2. Asphalt shingle roofing | 0.44 |
| 3. Building paper | 0.06 |
| 4. $\frac{5}{8}$ in. plywood deck | 0.78 |
| 5. Air space | 0.90 |
| 6. $\frac{1}{2}$ in. gypsum wallboard | 0.45 |
| 7. Inside surface (still air) | 0.62 |
| Total resistance | 3.42 |
| $U_i = 1/R = 1/3.42 =$ | 0.29 |
| Note: Correction for heat flow through framing members depends upon dimensions and spacing. | |

| Example of Substitution | | |
|---|---|---|
| Find $U$-value for same construction for summer conditions (heat flow down). | | |
| Total resistance | | 3.42 |
| Deduct: 5. Air space | 0.90 | |
| 8. Inside surface (still air) | 0.62 | 1.52 |
| Difference | | 1.90 |
| Add: 5. Air space | 0.89 | |
| 8. Inside surface (still air) | 0.76 | 1.65 |
| Total resistance | | 3.55 |
| $U_i = 1/R = 1/3.55 =$ | | 0.28 |

* Pitch of roof = 45°.

† To adjust $U$ values for the effect of added insulation between framing members, see Table 4-23B.

Reprinted by permission from "ASHRAE Handbook of Fundamentals," ASHRAE, New York, 1972.

## Table 4-23A. Determination of $U$ Values Resulting from Addition of Insulation to Any Given Building Section

| Given Building Section Property[a,b] | | Added R[c,d] | | | | | | |
|---|---|---|---|---|---|---|---|---|
| | | $R=4$ | $R=6$ | $R=8$ | $R=12$ | $R=16$ | $R=20$ | $R=24$ |
| U | R | U | U | U | U | U | U | U |
| 1.00 | 1.00 | 0.20 | 0.14 | 0.11 | 0.08 | 0.06 | 0.05 | 0.04 |
| 0.90 | 1.11 | 0.20 | 0.14 | 0.11 | 0.08 | 0.06 | 0.05 | 0.04 |
| 0.80 | 1.25 | 0.19 | 0.14 | 0.11 | 0.08 | 0.06 | 0.05 | 0.04 |
| 0.70 | 1.43 | 0.19 | 0.13 | 0.11 | 0.07 | 0.06 | 0.05 | 0.04 |
| 0.60 | 1.67 | 0.19 | 0.13 | 0.10 | 0.07 | 0.06 | 0.05 | 0.04 |
| 0.50 | 2.00 | 0.18 | 0.13 | 0.10 | 0.07 | 0.06 | 0.05 | 0.04 |
| 0.40 | 2.50 | 0.16 | 0.12 | 0.10 | 0.07 | 0.05 | 0.05 | 0.04 |
| 0.30 | 3.33 | 0.14 | 0.11 | 0.09 | 0.07 | 0.05 | 0.04 | 0.04 |
| 0.20 | 5.00 | 0.11 | 0.09 | 0.08 | 0.06 | 0.05 | 0.04 | 0.03 |
| 0.10 | 10.00 | 0.06 | 0.06 | 0.06 | 0.05 | 0.04 | 0.04 | 0.03 |
| 0.08 | 12.50 | 0.06 | 0.06 | 0.05 | 0.04 | 0.04 | 0.03 | 0.03 |

[a] For $U$ or $R$-values not shown in the table, interpolate as necessary.
[b] Enter column 1 with $U$ or $R$ of the design building section.
[c] Under appropriate column heading for Added $R$, find $U$-value of resulting design section.
[d] If the insulation occupies a previously considered air space, an adjustment must be made in the given building section $R$-value.

## Table 4-23B. Determination of $U$ Value Resulting from Addition of Insulation to Uninsulated Building Sections

| U Value of Roof without Roof-Deck Insulation[a] | Conductance C of Roof-Deck Insulation | | | | | |
|---|---|---|---|---|---|---|
| | 0.12 | 0.15 | 0.19 | 0.24 | 0.36 | 0.72 |
| | U | U | U | U | U | U |
| 0.10 | 0.05 | 0.06 | 0.07 | 0.07 | 0.08 | 0.09 |
| 0.15 | 0.07 | 0.08 | 0.08 | 0.09 | 0.11 | 0.12 |
| 0.20 | 0.08 | 0.09 | 0.10 | 0.11 | 0.13 | 0.16 |
| 0.25 | 0.08 | 0.09 | 0.11 | 0.12 | 0.15 | 0.19 |
| 0.30 | 0.09 | 0.10 | 0.12 | 0.13 | 0.16 | 0.21 |
| 0.35 | 0.09 | 0.10 | 0.12 | 0.14 | 0.18 | 0.24 |
| 0.40 | 0.09 | 0.11 | 0.13 | 0.15 | 0.19 | 0.26 |
| 0.50 | 0.10 | 0.12 | 0.14 | 0.16 | 0.21 | 0.29 |
| 0.60 | 0.10 | 0.12 | 0.14 | 0.17 | 0.22 | 0.33 |
| 0.70 | 0.10 | 0.12 | 0.15 | 0.18 | 0.24 | 0.35 |

[a] Interpolation or mild extrapolation may be used.

## Table 4-24. Solar Position and Intensity; Solar Heat Gain Factor* for 24° North Latitude

| Date | Solar Time A.M. | Solar Position Alt. | Solar Position Azimuth | Direct Normal Irradiation, Btu/sq ft | N | NE | E | SE | S | SW | W | NW | Hor. | Solar Time P.M. |
|---|---|---|---|---|---|---|---|---|---|---|---|---|---|---|
| Jan 21 | 7 | 4.8 | 65.6 | 70 | 2 | 20 | 61 | 63 | 25 | 2 | 2 | 2 | 4 | 5 |
| | 8 | 16.9 | 58.3 | 239 | 11 | 41 | 190 | 218 | 114 | 11 | 11 | 11 | 55 | 4 |
| | 9 | 27.9 | 48.8 | 287 | 18 | 22 | 190 | 253 | 166 | 19 | 18 | 18 | 120 | 3 |
| | 10 | 37.2 | 36.1 | 308 | 23 | 23 | 144 | 245 | 200 | 37 | 23 | 23 | 172 | 2 |
| | 11 | 43.6 | 19.6 | 317 | 26 | 26 | 72 | 211 | 220 | 94 | 26 | 26 | 204 | 1 |
| | 12 | 46.0 | 0.0 | 320 | 27 | 27 | 28 | 160 | 227 | 160 | 28 | 27 | 215 | 12 |
| | | | | Half Day Totals | 93 | 142 | 662 | 1064 | 833 | 238 | 93 | 93 | 660 | |
| Feb 21 | 7 | 9.0 | 73.9 | 153 | 6 | 67 | 141 | 128 | 33 | 6 | 6 | 6 | 16 | 5 |
| | 8 | 21.9 | 66.4 | 261 | 14 | 80 | 220 | 224 | 89 | 14 | 14 | 14 | 83 | 4 |
| | 9 | 33.5 | 56.8 | 297 | 21 | 45 | 208 | 243 | 133 | 22 | 21 | 21 | 153 | 3 |
| | 10 | 44.5 | 43.5 | 313 | 26 | 27 | 157 | 229 | 165 | 28 | 26 | 26 | 205 | 2 |
| | 11 | 52.2 | 24.5 | 321 | 29 | 29 | 80 | 191 | 185 | 67 | 29 | 29 | 238 | 1 |
| | 12 | 55.2 | 0.0 | 323 | 30 | 30 | 31 | 133 | 192 | 133 | 31 | 30 | 249 | 12 |
| | | | | Half Day Totals | 111 | 269 | 833 | 1095 | 701 | 199 | 111 | 111 | 817 | |
| Mar 21 | 7 | 13.7 | 83.8 | 194 | 10 | 115 | 186 | 145 | 17 | 9 | 9 | 9 | 36 | 5 |
| | 8 | 27.2 | 76.8 | 267 | 18 | 124 | 234 | 204 | 48 | 18 | 18 | 18 | 111 | 4 |
| | 9 | 40.2 | 67.9 | 295 | 24 | 85 | 215 | 214 | 82 | 24 | 24 | 24 | 180 | 3 |
| | 10 | 52.3 | 54.8 | 308 | 29 | 41 | 162 | 195 | 111 | 30 | 29 | 29 | 232 | 2 |
| | 11 | 61.9 | 33.4 | 315 | 32 | 33 | 84 | 154 | 130 | 42 | 32 | 32 | 264 | 1 |
| | 12 | 66.0 | 0.0 | 317 | 33 | 33 | 35 | 95 | 137 | 95 | 35 | 33 | 275 | 12 |
| | | | | Half Day Totals | 130 | 431 | 922 | 1095 | 457 | 163 | 130 | 130 | 960 | |
| Apr 21 | 6 | 4.7 | 100.6 | 40 | 5 | 33 | 39 | 22 | 2 | 2 | 2 | 5 | 3 | 6 |
| | 7 | 18.3 | 94.9 | 203 | 19 | 151 | 198 | 127 | 15 | 14 | 14 | 14 | 58 | 5 |
| | 8 | 32.0 | 89.0 | 257 | 24 | 159 | 229 | 165 | 24 | 22 | 22 | 22 | 132 | 4 |
| | 9 | 45.6 | 81.9 | 281 | 29 | 126 | 209 | 169 | 39 | 28 | 28 | 28 | 196 | 3 |
| | 10 | 59.0 | 71.8 | 293 | 34 | 75 | 158 | 148 | 56 | 33 | 33 | 33 | 245 | 2 |
| | 11 | 71.1 | 51.6 | 298 | 36 | 39 | 85 | 107 | 70 | 38 | 36 | 36 | 275 | 1 |
| | 12 | 77.6 | 0.0 | 300 | 37 | 37 | 39 | 58 | 75 | 58 | 39 | 37 | 284 | 12 |
| | | | | Half Day Totals | 166 | 519 | 950 | 981 | 245 | 164 | 166 | 166 | 960 | |
| May 21 | 6 | 8.0 | 108.4 | 85 | 25 | 79 | 83 | 38 | 5 | 5 | 5 | 5 | 12 | 6 |
| | 7 | 21.2 | 103.2 | 203 | 43 | 171 | 196 | 105 | 17 | 17 | 17 | 17 | 72 | 5 |
| | 8 | 34.6 | 98.5 | 248 | 38 | 178 | 218 | 132 | 26 | 24 | 24 | 24 | 142 | 4 |
| | 9 | 48.3 | 93.6 | 269 | 35 | 150 | 198 | 132 | 33 | 31 | 31 | 31 | 201 | 3 |
| | 10 | 62.0 | 87.7 | 280 | 37 | 102 | 150 | 111 | 38 | 35 | 35 | 35 | 247 | 2 |
| | 11 | 75.5 | 76.9 | 286 | 40 | 55 | 83 | 74 | 44 | 39 | 38 | 38 | 274 | 1 |
| | 12 | 86.0 | 0.0 | 287 | 41 | 41 | 41 | 44 | 46 | 44 | 41 | 41 | 282 | 12 |
| | | | | Half Day Totals | 237 | 611 | 953 | 780 | 245 | 173 | 155 | 154 | 1053 | |
| June 21 | 6 | 9.3 | 111.8 | 97 | 35 | 93 | 94 | 38 | 7 | 7 | 7 | 7 | 17 | 6 |
| | 7 | 22.3 | 106.8 | 200 | 55 | 176 | 192 | 94 | 18 | 18 | 18 | 18 | 77 | 5 |
| | 8 | 35.5 | 102.6 | 242 | 49 | 184 | 212 | 117 | 26 | 26 | 26 | 26 | 144 | 4 |
| | 9 | 49.0 | 98.7 | 262 | 43 | 158 | 192 | 116 | 33 | 31 | 31 | 31 | 201 | 3 |
| | 10 | 62.6 | 95.0 | 273 | 41 | 113 | 146 | 95 | 38 | 36 | 36 | 36 | 245 | 2 |
| | 11 | 76.3 | 90.7 | 279 | 41 | 64 | 82 | 63 | 41 | 39 | 38 | 39 | 271 | 1 |
| | | | | Half Day Totals | 237 | 754 | 950 | 616 | 186 | 173 | 171 | 171 | 1090 | |

| | | | | | | | | | | | | | | |
|---|---|---|---|---|---|---|---|---|---|---|---|---|---|---|
| | 12 | 89.5 | 0.0 | 280 | 42 | 42 | 42 | 42 | 43 | 42 | 42 | 42 | 279 | 12 |
| | | | Half Day Totals | | 282 | 804 | 936 | 544 | 184 | 177 | 176 | 177 | 1095 | |
| July 21 | 6 | 8.2 | 109.0 | 81 | 25 | 76 | 80 | 36 | 5 | 5 | 5 | 5 | 13 | 6 |
| | 7 | 21.4 | 103.8 | 195 | 45 | 168 | 190 | 101 | 17 | 17 | 17 | 17 | 72 | 5 |
| | 8 | 34.8 | 99.2 | 239 | 40 | 176 | 214 | 128 | 27 | 25 | 25 | 25 | 141 | 4 |
| | 9 | 48.4 | 94.5 | 261 | 37 | 150 | 195 | 128 | 34 | 31 | 31 | 31 | 199 | 3 |
| | 10 | 62.1 | 89.0 | 272 | 39 | 104 | 149 | 108 | 39 | 36 | 36 | 36 | 243 | 2 |
| | 11 | 75.7 | 79.2 | 278 | 41 | 57 | 83 | 73 | 44 | 40 | 38 | 38 | 270 | 1 |
| | 12 | 86.6 | 0.0 | 279 | 42 | 42 | 42 | 44 | 46 | 44 | 42 | 42 | 278 | 12 |
| | | | Half Day Totals | | 247 | 751 | 933 | 598 | 189 | 176 | 172 | 172 | 1078 | |
| Aug 21 | 6 | 5.0 | 101.3 | 34 | 5 | 29 | 34 | 19 | 2 | 2 | 2 | 2 | 4 | 6 |
| | 7 | 18.5 | 95.6 | 186 | 21 | 144 | 186 | 118 | 16 | 15 | 15 | 15 | 58 | 5 |
| | 8 | 32.2 | 89.7 | 240 | 25 | 155 | 220 | 157 | 26 | 23 | 23 | 23 | 129 | 4 |
| | 9 | 45.9 | 82.9 | 265 | 31 | 126 | 202 | 162 | 39 | 30 | 30 | 30 | 191 | 3 |
| | 10 | 59.3 | 73.0 | 277 | 35 | 77 | 154 | 143 | 55 | 34 | 34 | 34 | 238 | 2 |
| | 11 | 71.6 | 53.2 | 283 | 38 | 42 | 85 | 103 | 67 | 39 | 37 | 37 | 267 | 1 |
| | 12 | 78.3 | 0.0 | 285 | 38 | 39 | 41 | 58 | 72 | 58 | 41 | 39 | 276 | 12 |
| | | | Half Day Totals | | 176 | 603 | 916 | 742 | 242 | 170 | 162 | 161 | 1027 | |
| Sept 21 | 7 | 13.7 | 83.8 | 172 | 11 | 105 | 169 | 132 | 17 | 10 | 10 | 10 | 35 | 5 |
| | 8 | 27.2 | 76.8 | 248 | 19 | 119 | 222 | 194 | 48 | 19 | 19 | 19 | 108 | 4 |
| | 9 | 40.2 | 67.9 | 277 | 26 | 84 | 207 | 206 | 81 | 26 | 26 | 26 | 174 | 3 |
| | 10 | 52.3 | 54.8 | 292 | 30 | 42 | 158 | 190 | 110 | 32 | 30 | 30 | 224 | 2 |
| | 11 | 61.9 | 33.4 | 298 | 33 | 35 | 84 | 151 | 128 | 44 | 33 | 33 | 256 | 1 |
| | 12 | 66.0 | 0.0 | 301 | 34 | 34 | 37 | 94 | 134 | 94 | 37 | 34 | 267 | 12 |
| | | | Half Day Totals | | 137 | 417 | 879 | 939 | 451 | 172 | 137 | 136 | 930 | |
| Oct 21 | 7 | 9.1 | 74.1 | 137 | 6 | 62 | 129 | 117 | 30 | 6 | 6 | 6 | 16 | 5 |
| | 8 | 22.0 | 66.7 | 246 | 15 | 78 | 211 | 213 | 85 | 15 | 15 | 15 | 82 | 4 |
| | 9 | 34.1 | 57.1 | 284 | 22 | 46 | 202 | 235 | 128 | 22 | 22 | 22 | 150 | 3 |
| | 10 | 44.7 | 43.8 | 300 | 27 | 28 | 153 | 222 | 160 | 29 | 27 | 27 | 201 | 2 |
| | 11 | 52.5 | 24.7 | 308 | 30 | 30 | 79 | 186 | 180 | 66 | 30 | 30 | 233 | 1 |
| | 12 | 55.5 | 0.0 | 311 | 31 | 31 | 32 | 130 | 186 | 130 | 32 | 31 | 244 | 12 |
| | | | Half Day Totals | | 115 | 265 | 800 | 1050 | 676 | 198 | 116 | 115 | 802 | |
| Nov 21 | 7 | 4.9 | 65.8 | 66 | 2 | 20 | 58 | 60 | 24 | 2 | 2 | 2 | 4 | 5 |
| | 8 | 17.0 | 58.4 | 232 | 12 | 41 | 186 | 213 | 111 | 12 | 12 | 12 | 54 | 4 |
| | 9 | 28.0 | 48.9 | 281 | 18 | 23 | 187 | 249 | 162 | 19 | 18 | 18 | 120 | 3 |
| | 10 | 37.3 | 36.3 | 302 | 23 | 24 | 142 | 241 | 197 | 37 | 23 | 23 | 171 | 2 |
| | 11 | 43.8 | 19.7 | 311 | 26 | 26 | 72 | 209 | 217 | 93 | 26 | 26 | 202 | 1 |
| | 12 | 46.2 | 0.0 | 314 | 27 | 27 | 29 | 158 | 224 | 158 | 29 | 27 | 213 | 12 |
| | | | Half Day Totals | | 93 | 144 | 651 | 1046 | 817 | 237 | 94 | 93 | 655 | |
| Dec 21 | 7 | 3.2 | 62.6 | 29 | 0 | 7 | 25 | 27 | 12 | 0 | 0 | 0 | 1 | 5 |
| | 8 | 14.9 | 55.3 | 225 | 10 | 29 | 174 | 209 | 118 | 10 | 10 | 10 | 44 | 4 |
| | 9 | 25.5 | 46.0 | 281 | 17 | 18 | 180 | 252 | 174 | 18 | 17 | 17 | 106 | 3 |
| | 10 | 34.3 | 33.7 | 304 | 21 | 22 | 137 | 247 | 209 | 44 | 21 | 21 | 157 | 2 |
| | 11 | 40.4 | 18.2 | 314 | 24 | 24 | 69 | 216 | 230 | 104 | 24 | 24 | 188 | 1 |
| | 12 | 42.6 | 0.0 | 317 | 25 | 25 | 27 | 167 | 237 | 167 | 27 | 25 | 199 | 12 |
| | | | Half Day Totals | | 83 | 107 | 581 | 1019 | 851 | 254 | 84 | 83 | 593 | ↑ |
| | | | | | N | NW | W | SW | S | SE | E | NE | HOR. | ←P.M. |

Reprinted by permission from "ASHRAE Handbook of Fundamentals," ASHRAE, New York, 1972.
* Total solar heat gains for DS (1/8 in.) sheet glass. Based on a ground reflectance of 0.20.

# Table 4-25. Solar Position and Intensity; Solar Heat Gain Factors* for 32° North Latitude

| Date | Solar Time A.M. | Solar Position Alt. | Solar Position Azimuth | Direct Normal Irradiation, Btuh/sq ft | N | NE | E | SE | S | SW | W | NW | Hor. | Solar Time P.M. |
|---|---|---|---|---|---|---|---|---|---|---|---|---|---|---|
| Jan 21 | 7 | 1.4 | 65.2 | 1 | 0 | 0 | 1 | 1 | 0 | 0 | 0 | 0 | 0 | 5 |
| | 8 | 12.5 | 56.5 | 202 | 8 | 29 | 160 | 189 | 103 | 9 | 8 | 8 | 32 | 4 |
| | 9 | 22.5 | 46.0 | 269 | 15 | 16 | 175 | 246 | 169 | 16 | 15 | 15 | 88 | 3 |
| | 10 | 30.6 | 33.1 | 295 | 19 | 20 | 135 | 249 | 212 | 45 | 19 | 19 | 136 | 2 |
| | 11 | 36.1 | 17.5 | 306 | 22 | 22 | 67 | 221 | 238 | 110 | 22 | 22 | 166 | 1 |
| | 12 | 38.0 | 0.0 | 309 | 23 | 23 | 25 | 174 | 246 | 174 | 25 | 23 | 176 | 12 |
| | | Half Day Totals | | | 75 | 91 | 529 | 974 | 834 | 262 | 75 | 75 | 509 | |
| Feb 21 | 7 | 6.7 | 72.8 | 111 | 4 | 47 | 102 | 95 | 26 | 4 | 4 | 4 | 9 | 5 |
| | 8 | 18.5 | 63.8 | 244 | 12 | 64 | 205 | 217 | 95 | 12 | 12 | 12 | 63 | 4 |
| | 9 | 29.3 | 52.8 | 287 | 19 | 32 | 199 | 248 | 149 | 19 | 19 | 19 | 127 | 3 |
| | 10 | 38.5 | 38.9 | 305 | 23 | 24 | 151 | 241 | 189 | 31 | 23 | 23 | 176 | 2 |
| | 11 | 44.9 | 21.0 | 314 | 26 | 26 | 76 | 208 | 213 | 87 | 26 | 26 | 207 | 1 |
| | 12 | 47.2 | 0.0 | 316 | 27 | 27 | 29 | 155 | 221 | 155 | 29 | 27 | 217 | 12 |
| | | Half Day Totals | | | 97 | 207 | 749 | 1091 | 780 | 227 | 98 | 97 | 689 | |
| Mar 21 | 7 | 12.7 | 81.9 | 184 | 9 | 105 | 176 | 142 | 19 | 9 | 9 | 9 | 31 | 5 |
| | 8 | 25.1 | 73.0 | 260 | 17 | 107 | 227 | 209 | 62 | 17 | 17 | 17 | 99 | 4 |
| | 9 | 36.8 | 62.1 | 289 | 23 | 64 | 210 | 227 | 107 | 23 | 23 | 23 | 163 | 3 |
| | 10 | 47.3 | 47.5 | 304 | 27 | 30 | 158 | 215 | 144 | 29 | 27 | 27 | 211 | 2 |
| | 11 | 55.0 | 26.8 | 310 | 30 | 31 | 82 | 179 | 168 | 58 | 33 | 30 | 242 | 1 |
| | 12 | 58.0 | 0.0 | 312 | 31 | 31 | 33 | 122 | 176 | 122 | 33 | 31 | 252 | 12 |
| | | Half Day Totals | | | 122 | 368 | 891 | 1054 | 588 | 191 | 123 | 122 | 872 | |
| Apr 21 | 6 | 6.1 | 99.2 | 66 | 9 | 54 | 65 | 37 | 3 | 3 | 3 | 3 | 7 | 6 |
| | 7 | 18.8 | 92.2 | 206 | 17 | 147 | 201 | 136 | 15 | 14 | 14 | 14 | 61 | 5 |
| | 8 | 31.5 | 84.0 | 256 | 23 | 144 | 228 | 178 | 30 | 22 | 22 | 22 | 130 | 4 |
| | 9 | 43.9 | 74.2 | 278 | 28 | 103 | 206 | 188 | 58 | 27 | 27 | 27 | 189 | 3 |
| | 10 | 55.7 | 60.3 | 290 | 32 | 52 | 156 | 173 | 87 | 33 | 32 | 32 | 234 | 2 |
| | 11 | 65.4 | 37.5 | 296 | 34 | 36 | 83 | 135 | 108 | 40 | 38 | 34 | 263 | 1 |
| | 12 | 69.6 | 0.0 | 298 | 35 | 35 | 38 | 82 | 115 | 82 | 38 | 35 | 272 | 12 |
| | | Half Day Totals | | | 159 | 559 | 965 | 898 | 359 | 174 | 150 | 149 | 1022 | |
| May 21 | 6 | 10.4 | 107.2 | 118 | 32 | 108 | 116 | 55 | 18 | 8 | 8 | 8 | 21 | 6 |
| | 7 | 22.8 | 100.1 | 211 | 35 | 170 | 204 | 118 | 27 | 18 | 18 | 18 | 81 | 5 |
| | 8 | 35.4 | 92.9 | 249 | 29 | 165 | 220 | 149 | 37 | 25 | 25 | 25 | 146 | 4 |
| | 9 | 48.1 | 84.7 | 269 | 32 | 128 | 198 | 155 | 54 | 30 | 30 | 30 | 201 | 3 |
| | 10 | 60.6 | 73.3 | 279 | 36 | 76 | 150 | 138 | 68 | 35 | 35 | 35 | 243 | 2 |
| | 11 | 72.0 | 51.9 | 285 | 38 | 41 | 82 | 102 | 74 | 39 | 37 | 37 | 269 | 1 |
| | 12 | 78.0 | 0.0 | 286 | 38 | 39 | 41 | 59 | 74 | 59 | 41 | 39 | 277 | 12 |
| | | Half Day Totals | | | 217 | 697 | 983 | 747 | 248 | 181 | 172 | 171 | 1100 | |
| June 21 | 6 | 12.2 | 110.2 | 130 | 44 | 123 | 127 | 55 | 10 | 10 | 10 | 10 | 28 | 6 |
| | 7 | 24.3 | 103.4 | 209 | 46 | 176 | 201 | 108 | 19 | 19 | 19 | 19 | 88 | 5 |
| | 8 | 36.9 | 96.8 | 244 | 36 | 171 | 214 | 135 | 28 | 26 | 26 | 26 | 151 | 4 |
| | 9 | 49.6 | 89.4 | 263 | 34 | 136 | 193 | 139 | 35 | 32 | 32 | 32 | 203 | 3 |
| | 10 | 62.2 | 79.7 | 273 | 38 | 86 | 146 | 122 | 45 | 36 | 36 | 36 | 244 | 2 |
| | 11 | 74.2 | 60.9 | 278 | 40 | 46 | 81 | 88 | 56 | 40 | 38 | 38 | 268 | 1 |

Solar Position and Solar Heat Gain Factors[*] for Sunlit Glass (32° N. Latitude)

| Date | Solar Time (A.M.) | Alt. | Azm. | 280 | N 40 | NE 41 | E 42 | SE 52 | S 60 | SW 52 | W 42 | NW 41 | HOR 276 | Solar Time (P.M.) |
|---|---|---|---|---|---|---|---|---|---|---|---|---|---|---|
| (noon ref.) | 12 | 81.5 | 0.0 | 280 | 40 | 41 | 42 | 52 | 60 | 52 | 42 | 41 | 276 | 12 |
| Half Day Totals | | | | | 252 | 744 | 972 | 672 | 222 | 186 | 180 | 180 | 1119 | |
| **July 21** | 6 | 10.7 | 107.7 | 113 | 33 | 105 | 112 | 53 | 8 | 8 | 8 | 8 | 22 | 6 |
| | 7 | 23.1 | 100.6 | 203 | 37 | 167 | 198 | 114 | 19 | 18 | 18 | 18 | 81 | 5 |
| | 8 | 35.7 | 93.6 | 241 | 31 | 163 | 216 | 145 | 28 | 26 | 26 | 26 | 145 | 4 |
| | 9 | 48.4 | 85.5 | 261 | 34 | 128 | 195 | 150 | 37 | 31 | 31 | 31 | 199 | 3 |
| | 10 | 60.9 | 74.3 | 271 | 37 | 78 | 148 | 134 | 53 | 35 | 35 | 35 | 240 | 2 |
| | 11 | 72.4 | 53.3 | 277 | 39 | 43 | 82 | 99 | 66 | 58 | 40 | 37 | 265 | 1 |
| | 12 | 78.6 | 0.0 | 278 | 40 | 40 | 42 | 58 | 71 | 58 | 42 | 40 | 273 | 12 |
| Half Day Totals | | | | | 227 | 694 | 964 | 724 | 245 | 184 | 176 | 175 | 1089 | |
| **Aug 21** | 6 | 6.5 | 100.5 | 59 | 9 | 50 | 59 | 34 | 16 | 13 | 9 | 9 | 7 | 6 |
| | 7 | 19.1 | 92.8 | 189 | 18 | 140 | 189 | 127 | 30 | 15 | 18 | 18 | 61 | 5 |
| | 8 | 31.8 | 84.7 | 239 | 25 | 141 | 219 | 170 | 56 | 23 | 20 | 20 | 127 | 4 |
| | 9 | 44.3 | 75.0 | 263 | 30 | 104 | 200 | 180 | 84 | 29 | 22 | 22 | 185 | 3 |
| | 10 | 56.1 | 61.3 | 275 | 33 | 55 | 152 | 167 | 104 | 34 | 25 | 23 | 229 | 2 |
| | 11 | 66.0 | 38.4 | 281 | 36 | 38 | 83 | 131 | 111 | 41 | 38 | 36 | 256 | 1 |
| | 12 | 70.3 | 0.0 | 283 | 37 | 37 | 40 | 80 | 111 | 80 | 40 | 37 | 265 | 12 |
| Half Day Totals | | | | | 169 | 552 | 929 | 858 | 349 | 180 | 158 | 159 | 999 | |
| **Sep 21** | 7 | 12.7 | 81.9 | 163 | 10 | 95 | 159 | 128 | 19 | 9 | 10 | 10 | 30 | 5 |
| | 8 | 25.1 | 73.0 | 240 | 18 | 103 | 215 | 199 | 60 | 18 | 18 | 18 | 96 | 4 |
| | 9 | 36.8 | 62.1 | 272 | 24 | 64 | 202 | 218 | 105 | 24 | 24 | 24 | 158 | 3 |
| | 10 | 47.3 | 47.5 | 287 | 29 | 32 | 154 | 208 | 141 | 30 | 29 | 29 | 204 | 2 |
| | 11 | 55.0 | 26.8 | 294 | 31 | 32 | 81 | 174 | 164 | 59 | 31 | 31 | 234 | 1 |
| | 12 | 58.0 | 0.0 | 296 | 32 | 32 | 34 | 120 | 171 | 120 | 34 | 32 | 244 | 12 |
| Half Day Totals | | | | | 128 | 355 | 846 | 1004 | 575 | 194 | 127 | 128 | 844 | |
| **Oct 21** | 7 | 6.8 | 73.1 | 98 | 4 | 43 | 92 | 85 | 23 | 13 | 4 | 4 | 9 | 5 |
| | 8 | 18.7 | 64.0 | 229 | 13 | 63 | 195 | 205 | 90 | 20 | 13 | 13 | 62 | 4 |
| | 9 | 29.5 | 53.0 | 273 | 19 | 33 | 193 | 239 | 144 | 32 | 19 | 19 | 125 | 3 |
| | 10 | 38.7 | 39.1 | 292 | 24 | 25 | 147 | 234 | 183 | 85 | 24 | 24 | 173 | 2 |
| | 11 | 45.1 | 21.1 | 301 | 27 | 27 | 75 | 202 | 206 | 151 | 27 | 27 | 203 | 1 |
| | 12 | 47.5 | 0.0 | 304 | 28 | 28 | 30 | 151 | 214 | 151 | 30 | 28 | 213 | 12 |
| Half Day Totals | | | | | 100 | 205 | 718 | 1044 | 750 | 225 | 101 | 100 | 677 | |
| **Nov 21** | 7 | 1.5 | 65.4 | 1 | 0 | 0 | 1 | 1 | 0 | 0 | 0 | 0 | 0 | 5 |
| | 8 | 12.7 | 56.6 | 196 | 9 | 29 | 156 | 183 | 100 | 29 | 9 | 9 | 32 | 4 |
| | 9 | 22.6 | 46.1 | 262 | 15 | 17 | 172 | 241 | 166 | 15 | 15 | 15 | 87 | 3 |
| | 10 | 30.8 | 33.2 | 288 | 20 | 20 | 134 | 244 | 209 | 45 | 20 | 20 | 135 | 2 |
| | 11 | 36.2 | 17.6 | 300 | 22 | 22 | 67 | 218 | 234 | 108 | 22 | 22 | 165 | 1 |
| | 12 | 38.2 | 0.0 | 303 | 23 | 23 | 25 | 171 | 243 | 171 | 25 | 23 | 175 | 12 |
| Half Day Totals | | | | | 76 | 92 | 521 | 955 | 820 | 258 | 77 | 76 | 505 | |
| **Dec 21** | 8 | 10.3 | 53.6 | 176 | 7 | 18 | 135 | 166 | 96 | 22 | 7 | 7 | 22 | 4 |
| | 9 | 19.8 | 43.6 | 257 | 13 | 14 | 162 | 238 | 171 | 15 | 13 | 13 | 72 | 3 |
| | 10 | 27.6 | 31.2 | 287 | 18 | 18 | 127 | 246 | 217 | 52 | 18 | 18 | 119 | 2 |
| | 11 | 32.7 | 16.4 | 300 | 20 | 20 | 63 | 222 | 243 | 116 | 20 | 20 | 148 | 1 |
| | 12 | 34.6 | 0.0 | 304 | 21 | 21 | 23 | 177 | 252 | 177 | 23 | 21 | 158 | 12 |
| Half Day Totals | | | | | 67 | 76 | 482 | 947 | 844 | 273 | 68 | 67 | 440 | ← P.M. |

Reprinted by permission from "ASHRAE Handbook of Fundamentals," ASHRAE, New York, 1972.
[*] Total solar heat gains for DS (⅛ in.) sheet glass. Based on a ground reflectance of 0.20.

## Table 4-26. Solar Position and Intensity; Solar Heat Gain Factors* for 40° North Latitude

Solar Position columns (Alt., Azimuth); Direct Normal Irradiation, Btuh/sq ft; Solar Heat Gain Factors, Btuh/sq ft (N, NE, E, SE, S, SW, W, NW, Hor.)

| Date | Solar Time A.M. | Alt. | Azimuth | Direct Normal Irradiation, Btuh/sq ft | N | NE | E | SE | S | SW | W | NW | Hor. | Solar Time P.M. |
|---|---|---|---|---|---|---|---|---|---|---|---|---|---|---|
| Jan 21 | 8 | 8.1 | 55.3 | 141 | 5 | 17 | 111 | 133 | 75 | 5 | 5 | 5 | 13 | 4 |
| | 9 | 16.8 | 44.0 | 238 | 11 | 12 | 154 | 224 | 160 | 13 | 11 | 11 | 54 | 3 |
| | 10 | 23.8 | 30.9 | 274 | 16 | 16 | 123 | 241 | 213 | 51 | 16 | 16 | 96 | 2 |
| | 11 | 28.4 | 16.0 | 289 | 18 | 18 | 61 | 222 | 244 | 118 | 18 | 18 | 123 | 1 |
| | 12 | 30.0 | 0.0 | 293 | 19 | 19 | 20 | 179 | 254 | 179 | 20 | 19 | 133 | 12 |
| | Half Day Totals | | | | 59 | 68 | 449 | 903 | 815 | 271 | 59 | 59 | 353 | |
| Feb 21 | 7 | 4.3 | 72.1 | 55 | 1 | 22 | 50 | 47 | 13 | 10 | 1 | 1 | 3 | 5 |
| | 8 | 14.8 | 61.6 | 219 | 10 | 50 | 183 | 199 | 94 | 10 | 10 | 10 | 43 | 4 |
| | 9 | 24.3 | 49.7 | 271 | 16 | 22 | 186 | 245 | 157 | 17 | 16 | 16 | 98 | 3 |
| | 10 | 32.1 | 35.4 | 293 | 20 | 21 | 142 | 247 | 203 | 38 | 20 | 20 | 143 | 2 |
| | 11 | 37.3 | 18.6 | 303 | 23 | 23 | 71 | 219 | 231 | 103 | 23 | 23 | 171 | 1 |
| | 12 | 39.2 | 0.0 | 306 | 24 | 24 | 25 | 170 | 241 | 170 | 25 | 24 | 180 | 12 |
| | Half Day Totals | | | | 81 | 144 | 634 | 1035 | 813 | 250 | 81 | 81 | 546 | |
| Mar 21 | 7 | 11.4 | 80.2 | 171 | 8 | 93 | 163 | 135 | 21 | 8 | 8 | 8 | 26 | 5 |
| | 8 | 22.5 | 69.6 | 250 | 15 | 91 | 218 | 211 | 73 | 15 | 15 | 15 | 85 | 4 |
| | 9 | 32.8 | 57.3 | 281 | 21 | 46 | 203 | 236 | 128 | 21 | 21 | 21 | 143 | 3 |
| | 10 | 41.6 | 41.9 | 297 | 25 | 26 | 153 | 229 | 171 | 28 | 25 | 25 | 186 | 2 |
| | 11 | 47.7 | 22.6 | 304 | 28 | 28 | 78 | 198 | 197 | 77 | 28 | 28 | 213 | 1 |
| | 12 | 50.0 | 0.0 | 306 | 28 | 28 | 30 | 145 | 206 | 145 | 30 | 28 | 223 | 12 |
| | Half Day Totals | | | | 112 | 310 | 849 | 1100 | 692 | 218 | 112 | 112 | 764 | |
| Apr 21 | 6 | 7.4 | 98.9 | 89 | 11 | 72 | 88 | 52 | 5 | 4 | 4 | 4 | 11 | 6 |
| | 7 | 18.9 | 89.5 | 207 | 16 | 141 | 201 | 143 | 16 | 14 | 14 | 14 | 61 | 5 |
| | 8 | 30.3 | 79.3 | 253 | 22 | 128 | 225 | 189 | 41 | 21 | 21 | 21 | 124 | 4 |
| | 9 | 41.3 | 67.2 | 275 | 26 | 80 | 203 | 204 | 83 | 26 | 26 | 26 | 177 | 3 |
| | 10 | 51.4 | 51.4 | 286 | 30 | 37 | 153 | 194 | 121 | 32 | 30 | 30 | 218 | 2 |
| | 11 | 58.7 | 29.2 | 292 | 33 | 34 | 81 | 161 | 146 | 52 | 33 | 33 | 244 | 1 |
| | 12 | 61.6 | 0.0 | 294 | 33 | 33 | 36 | 108 | 155 | 108 | 36 | 33 | 253 | 12 |
| | Half Day Totals | | | | 153 | 509 | 969 | 1003 | 489 | 196 | 146 | 145 | 962 | |
| May 21 | 5 | 1.9 | 114.7 | | | | | | | | | | | 7 |
| | 6 | 12.7 | 105.6 | 143 | 35 | 128 | 141 | 71 | 10 | 10 | 10 | 10 | 30 | 6 |
| | 7 | 24.0 | 96.6 | 216 | 28 | 165 | 209 | 131 | 20 | 18 | 18 | 18 | 87 | 5 |
| | 8 | 35.4 | 87.2 | 249 | 27 | 149 | 220 | 164 | 29 | 25 | 25 | 25 | 146 | 4 |
| | 9 | 46.8 | 76.0 | 267 | 31 | 105 | 197 | 175 | 53 | 30 | 30 | 30 | 196 | 3 |
| | 10 | 57.5 | 60.9 | 277 | 34 | 54 | 148 | 163 | 83 | 35 | 34 | 34 | 234 | 2 |
| | 11 | 66.2 | 37.1 | 282 | 36 | 38 | 81 | 130 | 105 | 42 | 36 | 36 | 258 | 1 |
| | 12 | 70.0 | 0.0 | 284 | 37 | 37 | 40 | 82 | 112 | 82 | 40 | 37 | 265 | 12 |
| | Half Day Totals | | | | 203 | 643 | 1002 | 874 | 356 | 194 | 171 | 170 | 1083 | |
| June 21 | 5 | 4.2 | 117.3 | 21 | 10 | 21 | 20 | 6 | | 1 | 1 | 1 | 2 | 7 |
| | 6 | 14.8 | 108.4 | 154 | 47 | 142 | 151 | 70 | 12 | 12 | 12 | 12 | 39 | 6 |
| | 7 | 26.0 | 99.7 | 215 | 37 | 172 | 207 | 122 | 21 | 20 | 20 | 20 | 97 | 5 |
| | 8 | 37.4 | 90.7 | 246 | 29 | 156 | 215 | 152 | 29 | 26 | 26 | 26 | 153 | 4 |
| | 9 | 48.8 | 80.2 | 262 | 33 | 113 | 192 | 161 | 45 | 31 | 31 | 31 | 201 | 3 |
| | 10 | 59.8 | 65.8 | 272 | 35 | 62 | 145 | 148 | 69 | 36 | 35 | 35 | 237 | 2 |

| | | | | | N | NW | W | SW | S | SE | E | NE | HOR. | ←P.M. |
|---|---|---|---|---|---|---|---|---|---|---|---|---|---|---|
| | 11 | 69.2 | 41.9 | 276 | 37 | 40 | 80 | 116 | 88 | 41 | 37 | 37 | 260 | 1 |
| | 12 | 73.5 | 0.0 | 278 | 38 | 38 | 41 | 71 | 95 | 71 | 41 | 38 | 267 | 12 |
| | | | Half Day Totals | | 242 | 714 | 1019 | 810 | 311 | 197 | 181 | 180 | 1121 | |
| July 21 | 5 | 2.3 | 115.2 | 2 | 0 | 2 | 1 | 0 | 0 | 0 | 0 | 0 | 0 | 7 |
| | 6 | 13.1 | 106.1 | 137 | 37 | 125 | 137 | 68 | 10 | 10 | 10 | 10 | 31 | 6 |
| | 7 | 24.3 | 97.2 | 208 | 30 | 163 | 204 | 127 | 20 | 19 | 19 | 19 | 88 | 5 |
| | 8 | 35.8 | 87.8 | 241 | 28 | 148 | 216 | 160 | 29 | 26 | 26 | 26 | 145 | 4 |
| | 9 | 47.2 | 76.7 | 259 | 32 | 106 | 194 | 170 | 52 | 31 | 31 | 31 | 194 | 3 |
| | 10 | 57.9 | 61.7 | 269 | 35 | 56 | 146 | 159 | 80 | 36 | 35 | 35 | 231 | 2 |
| | 11 | 66.7 | 37.9 | 274 | 37 | 39 | 81 | 127 | 102 | 42 | 37 | 37 | 255 | 1 |
| | 12 | 70.6 | 0.0 | 276 | 38 | 38 | 41 | 80 | 109 | 80 | 41 | 38 | 262 | 12 |
| | | | Half Day Totals | | 211 | 645 | 986 | 850 | 347 | 197 | 177 | 176 | 1074 | |
| Aug 21 | 6 | 7.9 | 99.5 | 80 | 12 | 67 | 82 | 48 | 5 | 5 | 5 | 5 | 11 | 6 |
| | 7 | 19.3 | 90.0 | 191 | 17 | 135 | 191 | 135 | 17 | 15 | 15 | 15 | 62 | 5 |
| | 8 | 30.7 | 79.9 | 236 | 23 | 126 | 216 | 180 | 40 | 22 | 22 | 22 | 122 | 4 |
| | 9 | 41.8 | 67.9 | 259 | 28 | 82 | 197 | 196 | 79 | 28 | 28 | 28 | 174 | 3 |
| | 10 | 51.7 | 52.1 | 271 | 32 | 40 | 149 | 187 | 116 | 34 | 32 | 32 | 213 | 2 |
| | 11 | 59.3 | 29.7 | 277 | 34 | 35 | 81 | 156 | 140 | 52 | 34 | 34 | 238 | 1 |
| | 12 | 62.3 | 0.0 | 279 | 35 | 35 | 38 | 105 | 149 | 105 | 38 | 35 | 247 | 12 |
| | | | Half Day Totals | | 161 | 503 | 936 | 961 | 471 | 202 | 154 | 153 | 945 | |
| Sep 21 | 7 | 11.4 | 80.2 | 149 | 8 | 84 | 146 | 121 | 21 | 8 | 8 | 8 | 25 | 5 |
| | 8 | 22.5 | 69.6 | 230 | 16 | 87 | 205 | 199 | 71 | 16 | 16 | 16 | 82 | 4 |
| | 9 | 32.8 | 57.3 | 263 | 22 | 47 | 195 | 226 | 124 | 23 | 22 | 22 | 138 | 3 |
| | 10 | 41.6 | 41.9 | 279 | 26 | 28 | 148 | 221 | 165 | 30 | 26 | 26 | 180 | 2 |
| | 11 | 47.7 | 22.6 | 287 | 29 | 29 | 77 | 192 | 191 | 77 | 29 | 29 | 206 | 1 |
| | 12 | 50.0 | 0.0 | 290 | 30 | 30 | 32 | 141 | 200 | 141 | 32 | 30 | 215 | 12 |
| | | | Half Day Totals | | 116 | 300 | 803 | 1045 | 672 | 221 | 117 | 116 | 738 | |
| Oct 21 | 7 | 4.5 | 72.3 | 48 | 1 | 20 | 45 | 41 | 12 | 1 | 1 | 1 | 3 | 5 |
| | 8 | 15.0 | 61.9 | 203 | 10 | 49 | 173 | 187 | 88 | 10 | 10 | 10 | 43 | 4 |
| | 9 | 24.5 | 49.8 | 257 | 17 | 23 | 180 | 235 | 151 | 18 | 17 | 17 | 96 | 3 |
| | 10 | 32.4 | 35.6 | 280 | 21 | 22 | 139 | 238 | 196 | 38 | 21 | 21 | 140 | 2 |
| | 11 | 37.6 | 18.7 | 290 | 23 | 23 | 70 | 212 | 224 | 100 | 23 | 23 | 167 | 1 |
| | 12 | 39.5 | 0.0 | 293 | 24 | 24 | 26 | 165 | 234 | 165 | 26 | 24 | 177 | 12 |
| | | | Half Day Totals | | 83 | 143 | 610 | 989 | 783 | 245 | 84 | 83 | 535 | |
| Nov 21 | 8 | 8.2 | 55.4 | 136 | 5 | 17 | 107 | 128 | 72 | 5 | 5 | 5 | 14 | 4 |
| | 9 | 17.0 | 44.1 | 232 | 12 | 13 | 151 | 219 | 156 | 13 | 12 | 12 | 54 | 3 |
| | 10 | 24.0 | 31.0 | 267 | 16 | 16 | 122 | 237 | 209 | 50 | 16 | 16 | 96 | 2 |
| | 11 | 28.6 | 16.1 | 283 | 19 | 19 | 61 | 218 | 240 | 116 | 19 | 19 | 123 | 1 |
| | 12 | 30.2 | 0.0 | 287 | 19 | 19 | 21 | 176 | 250 | 176 | 21 | 19 | 132 | 12 |
| | | | Half Day Totals | | 61 | 71 | 442 | 884 | 798 | 267 | 62 | 61 | 353 | |
| Dec 21 | 8 | 5.5 | 53.0 | 88 | 2 | 7 | 67 | 83 | 49 | 3 | 2 | 2 | 6 | 4 |
| | 9 | 14.0 | 41.9 | 217 | 9 | 10 | 135 | 205 | 151 | 12 | 9 | 9 | 39 | 3 |
| | 10 | 20.7 | 29.4 | 261 | 14 | 14 | 113 | 232 | 210 | 55 | 14 | 14 | 77 | 2 |
| | 11 | 25.0 | 15.2 | 279 | 16 | 16 | 56 | 217 | 242 | 120 | 16 | 16 | 103 | 1 |
| | 12 | 26.6 | 0.0 | 284 | 17 | 17 | 18 | 177 | 253 | 177 | 18 | 17 | 113 | 12 |
| | | | Half Day Totals | | 49 | 54 | 380 | 831 | 781 | 273 | 50 | 49 | 282 | ↑ |

Reprinted by permission from "ASHRAE Handbook of Fundamentals," ASHRAE, New York, 1972.

* Total solar heat gains for DS (1/8 in.) sheet glass. Based on a ground reflectance of 0.20.

## Table 4-27. Solar Position and Intensity; Solar Heat Gain Factors* for 48° North Latitude

| Date | Solar Time A.M. | Solar Position Alt. | Solar Position Azimuth | Direct Normal Irradiation, Btuh/sq ft | N | NE | E | SE | S | SW | W | NW | Hor. | Solar Time P.M. |
|---|---|---|---|---|---|---|---|---|---|---|---|---|---|---|
| Jan 21 | 8 | 3.5 | 54.6 | 36 | 1 | 4 | 28 | 34 | 19 | 1 | 1 | 1 | 2 | 4 |
| | 9 | 11.0 | 42.6 | 185 | 7 | 8 | 117 | 176 | 128 | 9 | 7 | 7 | 25 | 3 |
| | 10 | 16.9 | 29.4 | 239 | 11 | 11 | 105 | 216 | 195 | 50 | 11 | 11 | 55 | 2 |
| | 11 | 20.7 | 15.1 | 260 | 14 | 14 | 52 | 208 | 233 | 115 | 14 | 14 | 77 | 1 |
| | 12 | 22.0 | 0.0 | 267 | 15 | 15 | 16 | 171 | 245 | 171 | 16 | 15 | 85 | 12 |
| | | | | Half Day Totals | 41 | 44 | 319 | 735 | 705 | 256 | 41 | 41 | 202 | |
| Feb 21 | 7 | 1.8 | 71.7 | 3 | 0 | 1 | 3 | 3 | 0 | 0 | 0 | 0 | 0 | 5 |
| | 8 | 10.9 | 60.0 | 180 | 7 | 36 | 149 | 166 | 82 | 7 | 7 | 7 | 24 | 4 |
| | 9 | 19.0 | 47.3 | 247 | 13 | 15 | 168 | 230 | 155 | 14 | 13 | 13 | 66 | 3 |
| | 10 | 25.5 | 33.0 | 275 | 17 | 17 | 131 | 242 | 207 | 43 | 17 | 17 | 105 | 2 |
| | 11 | 29.7 | 17.0 | 288 | 19 | 19 | 65 | 221 | 240 | 112 | 19 | 19 | 129 | 1 |
| | 12 | 31.2 | 0.0 | 291 | 20 | 20 | 21 | 176 | 251 | 176 | 21 | 20 | 138 | 12 |
| | | | | Half Day Totals | 65 | 88 | 508 | 936 | 803 | 260 | 65 | 65 | 392 | |
| Mar 21 | 7 | 10.0 | 78.7 | 152 | 7 | 80 | 145 | 123 | 22 | 7 | 7 | 7 | 20 | 5 |
| | 8 | 19.5 | 66.8 | 235 | 13 | 75 | 204 | 206 | 81 | 13 | 13 | 13 | 67 | 4 |
| | 9 | 28.4 | 53.4 | 270 | 19 | 33 | 193 | 239 | 142 | 19 | 19 | 19 | 117 | 3 |
| | 10 | 35.4 | 37.8 | 287 | 22 | 23 | 146 | 237 | 189 | 33 | 22 | 22 | 156 | 2 |
| | 11 | 40.3 | 19.8 | 295 | 24 | 24 | 73 | 210 | 218 | 94 | 24 | 24 | 180 | 1 |
| | 12 | 42.0 | 0.0 | 297 | 25 | 25 | 27 | 161 | 228 | 161 | 27 | 25 | 188 | 12 |
| | | | | Half Day Totals | 98 | 256 | 790 | 1111 | 765 | 244 | 99 | 98 | 634 | |
| Apr 21 | 6 | 8.6 | 97.8 | 108 | 12 | 86 | 107 | 64 | 9 | 6 | 6 | 6 | 14 | 6 |
| | 7 | 18.6 | 86.7 | 205 | 15 | 133 | 200 | 149 | 17 | 14 | 14 | 14 | 60 | 5 |
| | 8 | 28.5 | 74.9 | 247 | 20 | 111 | 219 | 197 | 55 | 20 | 20 | 20 | 114 | 4 |
| | 9 | 37.8 | 61.2 | 269 | 25 | 60 | 198 | 216 | 107 | 25 | 25 | 25 | 161 | 3 |
| | 10 | 45.8 | 44.6 | 281 | 28 | 30 | 148 | 210 | 150 | 30 | 28 | 28 | 197 | 2 |
| | 11 | 51.5 | 24.0 | 287 | 30 | 30 | 78 | 181 | 177 | 69 | 30 | 30 | 219 | 1 |
| | 12 | 53.6 | 0.0 | 289 | 31 | 31 | 33 | 132 | 187 | 132 | 33 | 31 | 227 | 12 |
| | | | | Half Day Totals | 143 | 459 | 961 | 1086 | 604 | 225 | 139 | 138 | 879 | |
| May 21 | 5 | 5.2 | 114.3 | 41 | 16 | 39 | 38 | 13 | 2 | 2 | 2 | 2 | 4 | 7 |
| | 6 | 14.7 | 103.7 | 162 | 35 | 140 | 160 | 84 | 12 | 12 | 12 | 12 | 39 | 6 |
| | 7 | 24.6 | 93.0 | 218 | 23 | 158 | 212 | 142 | 21 | 19 | 19 | 19 | 91 | 5 |
| | 8 | 34.6 | 81.6 | 248 | 26 | 132 | 218 | 178 | 38 | 24 | 24 | 24 | 142 | 4 |
| | 9 | 44.3 | 68.3 | 264 | 29 | 82 | 194 | 192 | 77 | 29 | 29 | 29 | 186 | 3 |
| | 10 | 53.0 | 51.3 | 274 | 32 | 39 | 145 | 184 | 116 | 34 | 32 | 32 | 219 | 2 |
| | 11 | 59.5 | 28.6 | 279 | 34 | 35 | 79 | 155 | 142 | 54 | 38 | 34 | 240 | 1 |
| | 12 | 62.0 | 0.0 | 280 | 35 | 35 | 38 | 106 | 150 | 106 | 38 | 35 | 247 | 12 |
| | | | | Half Day Totals | 209 | 637 | 1058 | 1002 | 483 | 220 | 170 | 169 | 1043 | |
| June 21 | 5 | 7.9 | 116.5 | 77 | 35 | 76 | 72 | 23 | 14 | 5 | 5 | 5 | 51 | 7 |
| | 6 | 17.2 | 106.2 | 172 | 46 | 154 | 169 | 84 | 22 | 14 | 14 | 14 | 102 | 6 |
| | 7 | 27.0 | 95.8 | 219 | 29 | 165 | 211 | 135 | 34 | 21 | 21 | 21 | 152 | 5 |
| | 8 | 37.1 | 84.6 | 245 | 28 | 139 | 215 | 167 | 66 | 26 | 26 | 26 | 193 | 4 |
| | 9 | 46.9 | 71.6 | 260 | 31 | 90 | 190 | 179 | 101 | 31 | 31 | 31 | 225 | 3 |
| | 10 | 55.8 | 54.8 | 269 | 34 | 45 | 142 | 171 | | 36 | 34 | 34 | | 2 |

Table — Total Solar Heat Gain Factors (continued), latitude ≈ 64°N

| A.M. | Alt. | Azm. | | N | NW | W | SW | S | SE | E | NE | HOR. | P.M. |
|---|---|---|---|---|---|---|---|---|---|---|---|---|---|
| 11 | 62.7 | 31.2 | 273 | 36 | 37 | 78 | 142 | 125 | 49 | 36 | 36 | 245 | 1 |
| 12 | 65.4 | 0.0 | 275 | 36 | 36 | 39 | 96 | 134 | 96 | 39 | 36 | 252 | 12 |
| | | | Half Day Totals | 258 | 728 | 1098 | 952 | 434 | 224 | 186 | 185 | 1105 | |
| **July 21** | | | | | | | | | | | | | |
| 5 | 5.7 | 114.7 | 42 | 18 | 42 | 40 | 14 | 2 | 2 | 2 | 2 | 5 | 7 |
| 6 | 15.2 | 104.1 | 155 | 36 | 138 | 156 | 82 | 12 | 12 | 12 | 12 | 41 | 6 |
| 7 | 25.1 | 93.5 | 211 | 24 | 156 | 207 | 138 | 21 | 20 | 20 | 20 | 92 | 5 |
| 8 | 35.1 | 82.1 | 240 | 27 | 132 | 214 | 174 | 38 | 25 | 25 | 23 | 142 | 4 |
| 9 | 44.8 | 68.8 | 256 | 30 | 83 | 191 | 187 | 75 | 35 | 30 | 30 | 184 | 3 |
| 10 | 53.5 | 51.9 | 266 | 33 | 41 | 143 | 180 | 113 | 53 | 33 | 32 | 217 | 2 |
| 11 | 60.1 | 29.0 | 271 | 35 | 37 | 79 | 151 | 138 | 68 | 35 | 35 | 238 | 1 |
| 12 | 62.6 | 0.0 | 272 | 36 | 36 | 39 | 104 | 146 | 104 | 39 | 36 | 245 | 12 |
| | | | Half Day Totals | 218 | 643 | 1044 | 978 | 472 | 227 | 147 | 146 | 1040 | |
| **Aug 21** | | | | | | | | | | | | | |
| 6 | 9.1 | 98.3 | 98 | 13 | 81 | 100 | 60 | 7 | 6 | 7 | 6 | 16 | 6 |
| 7 | 19.1 | 87.2 | 189 | 16 | 127 | 189 | 140 | 18 | 15 | 15 | 15 | 61 | 5 |
| 8 | 29.0 | 75.4 | 231 | 21 | 110 | 211 | 188 | 53 | 21 | 21 | 21 | 113 | 4 |
| 9 | 38.4 | 61.8 | 253 | 26 | 62 | 192 | 208 | 102 | 26 | 26 | 26 | 159 | 3 |
| 10 | 46.4 | 45.1 | 265 | 30 | 32 | 145 | 202 | 144 | 68 | 30 | 30 | 193 | 2 |
| 11 | 52.2 | 24.3 | 271 | 32 | 32 | 78 | 175 | 171 | 128 | 32 | 32 | 215 | 1 |
| 12 | 54.3 | 0.0 | 273 | 33 | 33 | 35 | 128 | 180 | 128 | 35 | 33 | 222 | 12 |
| | | | Half Day Totals | 152 | 454 | 927 | 1040 | 584 | 227 | 147 | 146 | 868 | |
| **Sep 21** | | | | | | | | | | | | | |
| 7 | 10.0 | 78.7 | 131 | 7 | 71 | 128 | 108 | 21 | 14 | 14 | 14 | 19 | 5 |
| 8 | 19.5 | 66.8 | 215 | 14 | 72 | 191 | 193 | 77 | 20 | 20 | 20 | 65 | 4 |
| 9 | 28.2 | 53.4 | 251 | 20 | 33 | 184 | 227 | 136 | 34 | 23 | 23 | 113 | 3 |
| 10 | 35.4 | 37.8 | 269 | 23 | 25 | 141 | 228 | 182 | 92 | 28 | 26 | 151 | 2 |
| 11 | 40.3 | 19.8 | 277 | 26 | 26 | 73 | 203 | 211 | 156 | 35 | 26 | 174 | 1 |
| 12 | 42.0 | 0.0 | 280 | 26 | 26 | 28 | 156 | 220 | 156 | 28 | 26 | 182 | 12 |
| | | | Half Day Totals | 104 | 246 | 744 | 1050 | 736 | 242 | 147 | 104 | 612 | |
| **Oct 21** | | | | | | | | | | | | | |
| 7 | 2.0 | 71.9 | 3 | 0 | 1 | 3 | 3 | 3 | 0 | 0 | 0 | 0 | 5 |
| 8 | 11.2 | 60.2 | 165 | 8 | 35 | 139 | 155 | 76 | 15 | 13 | 13 | 24 | 4 |
| 9 | 19.3 | 47.4 | 232 | 13 | 16 | 161 | 219 | 147 | 43 | 17 | 17 | 66 | 3 |
| 10 | 25.7 | 33.1 | 261 | 17 | 18 | 127 | 233 | 199 | 109 | 20 | 20 | 103 | 2 |
| 11 | 30.0 | 17.1 | 274 | 20 | 20 | 64 | 213 | 231 | 170 | 22 | 20 | 127 | 1 |
| 12 | 31.5 | 0.0 | 278 | 20 | 20 | 22 | 170 | 242 | 170 | 22 | 20 | 136 | 12 |
| | | | Half Day Totals | 67 | 91 | 488 | 895 | 768 | 256 | 68 | 67 | 387 | |
| **Nov 21** | | | | | | | | | | | | | |
| 8 | 3.6 | 54.7 | 36 | 7 | 4 | 28 | 34 | 19 | 1 | 1 | 1 | 2 | 4 |
| 9 | 11.2 | 42.7 | 178 | 12 | 8 | 115 | 171 | 125 | 9 | 7 | 7 | 25 | 3 |
| 10 | 17.1 | 29.5 | 232 | 14 | 12 | 104 | 211 | 191 | 49 | 12 | 12 | 55 | 2 |
| 11 | 20.9 | 15.1 | 254 | 15 | 14 | 52 | 204 | 228 | 113 | 14 | 14 | 77 | 1 |
| 12 | 22.2 | 0.0 | 260 | 15 | 15 | 16 | 168 | 240 | 168 | 16 | 15 | 85 | 12 |
| | | | Half Day Totals | 41 | 45 | 316 | 719 | 690 | 252 | 42 | 41 | 202 | |
| **Dec 21** | | | | | | | | | | | | | |
| 9 | 8.0 | 40.9 | 140 | 5 | 9 | 86 | 133 | 100 | 7 | 5 | 5 | 13 | 3 |
| 10 | 13.6 | 28.4 | 214 | 9 | 12 | 91 | 194 | 179 | 49 | 9 | 9 | 37 | 2 |
| 11 | 17.3 | 17.3 | 242 | 12 | 12 | 46 | 195 | 220 | 111 | 12 | 12 | 57 | 1 |
| 12 | 18.6 | 18.6 | 250 | 12 | 12 | 14 | 163 | 233 | 163 | 14 | 12 | 64 | 12 |
| | | | Half Day Totals | 32 | 32 | 241 | 621 | 623 | 244 | 33 | 32 | 139 | |

\* Total solar heat gains for DS (⅛ in.) sheet glass. Based on a ground reflectance of 0.20.
Reprinted by permission from "ASHRAE Handbook of Fundamentals," ASHRAE, New York, 1972.

# Table 4-28. Solar Position and Intensity; Solar Heat Gain Factors* for 56° North Latitude

| Date | Solar Time A.M. | Solar Position Alt. | Solar Position Azimuth | Direct Normal Irradiation, Btuh/sq ft | N | NE | E | SE | S | SW | W | NW | Hor. | Solar Time P.M. |
|---|---|---|---|---|---|---|---|---|---|---|---|---|---|---|
| Jan 21 | 9 | 5.0 | 41.8 | 77 | 2 | 2 | 48 | 74 | 54 | 3 | 2 | 2 | 5 | 3 |
| | 10 | 9.9 | 28.5 | 170 | 6 | 6 | 73 | 156 | 143 | 38 | 6 | 6 | 20 | 2 |
| | 11 | 12.9 | 14.5 | 206 | 9 | 9 | 39 | 169 | 190 | 96 | 9 | 9 | 34 | 1 |
| | 12 | 14.0 | 0.0 | 216 | 9 | 9 | 10 | 143 | 205 | 143 | 10 | 9 | 39 | 12 |
| | | | | Half Day Totals | 21 | 21 | 168 | 475 | 489 | 205 | 22 | 21 | 78 | |
| Feb 21 | 8 | 6.9 | 59.0 | 115 | 4 | 20 | 94 | 107 | 54 | 4 | 4 | 4 | 9 | 4 |
| | 9 | 13.5 | 45.6 | 207 | 9 | 10 | 139 | 197 | 136 | 10 | 9 | 9 | 36 | 3 |
| | 10 | 18.7 | 31.2 | 245 | 13 | 13 | 115 | 223 | 196 | 45 | 13 | 13 | 64 | 2 |
| | 11 | 22.0 | 15.9 | 262 | 15 | 15 | 56 | 210 | 232 | 112 | 15 | 15 | 84 | 1 |
| | 12 | 23.2 | 0.0 | 267 | 15 | 15 | 17 | 171 | 244 | 171 | 17 | 15 | 91 | 12 |
| | | | | Half Day Totals | 48 | 60 | 405 | 819 | 738 | 252 | 49 | 48 | 239 | |
| Mar 21 | 7 | 8.3 | 77.5 | 127 | 5 | 64 | 121 | 104 | 21 | 5 | 5 | 5 | 14 | 5 |
| | 8 | 16.2 | 64.4 | 215 | 11 | 61 | 185 | 194 | 83 | 11 | 11 | 11 | 49 | 4 |
| | 9 | 23.3 | 50.3 | 253 | 16 | 23 | 179 | 233 | 148 | 17 | 16 | 16 | 89 | 3 |
| | 10 | 29.0 | 34.9 | 272 | 19 | 20 | 136 | 238 | 198 | 38 | 19 | 19 | 122 | 2 |
| | 11 | 32.7 | 17.9 | 281 | 21 | 21 | 68 | 215 | 230 | 106 | 21 | 21 | 142 | 1 |
| | 12 | 34.0 | 0.0 | 284 | 22 | 22 | 23 | 170 | 241 | 170 | 23 | 22 | 149 | 12 |
| | | | | Half Day Totals | 83 | 205 | 712 | 1080 | 799 | 260 | 83 | 83 | 490 | |
| Apr 21 | 5 | 1.4 | 108.8 | 0 | 0 | 0 | 0 | 0 | 0 | 0 | 0 | 0 | 0 | 7 |
| | 6 | 9.6 | 96.5 | 122 | 12 | 96 | 121 | 75 | 7 | 7 | 7 | 7 | 18 | 6 |
| | 7 | 18.0 | 84.1 | 201 | 14 | 123 | 196 | 152 | 21 | 13 | 13 | 13 | 56 | 5 |
| | 8 | 26.1 | 70.9 | 240 | 19 | 95 | 212 | 202 | 68 | 19 | 19 | 19 | 101 | 4 |
| | 9 | 33.6 | 56.3 | 261 | 23 | 44 | 190 | 224 | 126 | 23 | 23 | 23 | 141 | 3 |
| | 10 | 39.9 | 39.7 | 273 | 26 | 27 | 142 | 221 | 172 | 32 | 26 | 26 | 171 | 2 |
| | 11 | 44.1 | 20.7 | 279 | 27 | 27 | 74 | 196 | 201 | 86 | 30 | 27 | 189 | 1 |
| | 12 | 45.6 | 0.0 | 280 | 28 | 28 | 30 | 149 | 211 | 149 | 30 | 28 | 196 | 12 |
| | | | | Half Day Totals | 132 | 413 | 940 | 1144 | 699 | 251 | 128 | 128 | 773 | |
| May 21 | 4 | 1.2 | 125.5 | 0 | 0 | 0 | 0 | 0 | 0 | 0 | 0 | 0 | 0 | 8 |
| | 5 | 8.5 | 113.4 | 92 | 36 | 89 | 87 | 33 | 6 | 6 | 6 | 6 | 14 | 7 |
| | 6 | 16.5 | 101.5 | 175 | 32 | 148 | 173 | 97 | 14 | 13 | 13 | 13 | 48 | 6 |
| | 7 | 24.8 | 89.3 | 219 | 21 | 149 | 212 | 152 | 21 | 19 | 19 | 19 | 92 | 5 |
| | 8 | 33.1 | 76.3 | 244 | 24 | 115 | 215 | 189 | 52 | 24 | 24 | 24 | 135 | 4 |
| | 9 | 40.9 | 61.6 | 259 | 27 | 62 | 189 | 206 | 102 | 27 | 27 | 27 | 171 | 3 |
| | 10 | 47.6 | 44.2 | 268 | 30 | 33 | 141 | 200 | 145 | 33 | 30 | 30 | 199 | 2 |
| | 11 | 52.3 | 23.4 | 273 | 32 | 32 | 75 | 174 | 172 | 70 | 32 | 32 | 216 | 1 |
| | 12 | 54.0 | 0.0 | 275 | 33 | 33 | 35 | 129 | 181 | 129 | 35 | 33 | 222 | 12 |
| | | | | Half Day Totals | 223 | 651 | 1115 | 1120 | 602 | 252 | 168 | 167 | 986 | |
| June 21 | 4 | 4.2 | 127.2 | 21 | 13 | 21 | 17 | 2 | | | | | 2 | 8 |
| | 5 | 11.4 | 115.3 | 121 | 52 | 119 | 114 | 40 | 9 | 9 | 9 | 9 | 24 | 7 |
| | 6 | 19.3 | 103.6 | 185 | 42 | 160 | 182 | 97 | 16 | 16 | 16 | 16 | 61 | 6 |
| | 7 | 27.6 | 91.7 | 221 | 24 | 156 | 213 | 147 | 23 | 21 | 21 | 21 | 105 | 5 |
| | 8 | 35.9 | 78.8 | 243 | 27 | 121 | 212 | 181 | 46 | 26 | 26 | 26 | 146 | 4 |
| | 9 | 43.8 | 64.1 | 256 | 29 | 69 | 186 | 195 | 91 | 29 | 29 | 29 | 181 | 3 |
| | 10 | 50.7 | 46.4 | 264 | 32 | 35 | 139 | 189 | 132 | 35 | 32 | 32 | 208 | 2 |

Solar position and total solar heat gain table (rotated landscape table). Printed direction column labels (left→right): N, NW, W, SW, S, SE, E, NE, HOR. The first three data columns (unlabeled) are solar altitude, solar azimuth, and direct-normal heat gain. Values given for AM hours on the left; the same rows serve the PM hours on the right (→ P.M.).

**June 21 (continued, last rows)**

| AM | Alt | Azm | — | N | NW | W | SW | S | SE | E | NE | HOR | PM |
|---|---|---|---|---|---|---|---|---|---|---|---|---|---|
| 11 | 55.6 | 24.9 | 268 | 34 | 35 | 76 | 164 | 158 | 64 | 34 | 34 | 225 | 1 |
| 12 | 57.4 | 0.0 | 270 | 34 | 34 | 37 | 119 | 167 | 119 | 37 | 34 | 230 | 12 |
| Half Day Totals | | | | | | | | | | | | 1067 | |

**July 21**

| AM | Alt | Azm | — | N | NW | W | SW | S | SE | E | NE | HOR | PM |
|---|---|---|---|---|---|---|---|---|---|---|---|---|---|
| 4 | 1.7 | 125.8 | 0 | 0 | 0 | 0 | 0 | 0 | 0 | 0 | 0 | 0 | 8 |
| 5 | 9.0 | 113.7 | 91 | 37 | 89 | 87 | 33 | 6 | 6 | 6 | 6 | 15 | 7 |
| 6 | 17.0 | 101.9 | 169 | 34 | 145 | 170 | 94 | 14 | 14 | 14 | 14 | 50 | 6 |
| 7 | 25.3 | 89.7 | 212 | 22 | 147 | 208 | 148 | 22 | 20 | 20 | 20 | 93 | 5 |
| 8 | 33.6 | 76.7 | 236 | 25 | 115 | 211 | 185 | 51 | 25 | 25 | 25 | 135 | 4 |
| 9 | 41.4 | 62.0 | 251 | 28 | 63 | 186 | 201 | 99 | 34 | 28 | 28 | 171 | 3 |
| 10 | 48.2 | 44.6 | 260 | 31 | 34 | 139 | 196 | 141 | 70 | 31 | 31 | 198 | 2 |
| 11 | 52.9 | 23.7 | 265 | 33 | 33 | 76 | 171 | 168 | 126 | 33 | 33 | 215 | 1 |
| 12 | 54.6 | 0.0 | 267 | 34 | 34 | 37 | 126 | 177 | 126 | 37 | 34 | 221 | 12 |
| Half Day Totals | | | | 231 | 650 | 1101 | 1096 | 589 | 256 | 175 | 174 | 987 | |

**Aug 21**

| AM | Alt | Azm | — | N | NW | W | SW | S | SE | E | NE | HOR | PM |
|---|---|---|---|---|---|---|---|---|---|---|---|---|---|
| 5 | 2.0 | 109.2 | 1 | 0 | 1 | 1 | 0 | 0 | 0 | 0 | 0 | 0 | 7 |
| 6 | 10.2 | 97.0 | 112 | 13 | 91 | 114 | 70 | 8 | 8 | 8 | 7 | 19 | 6 |
| 7 | 18.5 | 84.5 | 186 | 16 | 119 | 186 | 144 | 21 | 21 | 15 | 15 | 58 | 5 |
| 8 | 26.7 | 71.3 | 224 | 20 | 94 | 203 | 192 | 66 | 39 | 20 | 20 | 101 | 4 |
| 9 | 34.3 | 56.7 | 245 | 24 | 46 | 184 | 215 | 121 | 84 | 24 | 24 | 139 | 3 |
| 10 | 40.5 | 40.9 | 257 | 27 | 29 | 139 | 213 | 165 | 102 | 29 | 29 | 168 | 2 |
| 11 | 44.8 | 20.9 | 263 | 29 | 29 | 74 | 189 | 193 | 163 | 32 | 30 | 187 | 1 |
| 12 | 46.3 | 0.0 | 265 | 30 | 30 | 32 | 145 | 203 | 145 | 32 | 30 | 193 | 12 |
| Half Day Totals | | | | 142 | 412 | 908 | 1095 | 673 | 254 | 137 | 136 | 768 | |

**Sep 21**

| AM | Alt | Azm | — | N | NW | W | SW | S | SE | E | NE | HOR | PM |
|---|---|---|---|---|---|---|---|---|---|---|---|---|---|
| 7 | 8.3 | 77.5 | 107 | 5 | 56 | 104 | 90 | 19 | 19 | 19 | 5 | 13 | 5 |
| 8 | 16.2 | 64.4 | 194 | 12 | 57 | 171 | 179 | 78 | 52 | 12 | 12 | 48 | 4 |
| 9 | 23.3 | 50.3 | 233 | 17 | 24 | 170 | 220 | 140 | 108 | 18 | 17 | 86 | 3 |
| 10 | 29.0 | 34.9 | 253 | 20 | 21 | 131 | 227 | 189 | 164 | 22 | 20 | 117 | 2 |
| 11 | 32.7 | 17.9 | 263 | 22 | 22 | 67 | 206 | 220 | 163 | 25 | 23 | 137 | 1 |
| 12 | 34.0 | 0.0 | 266 | 23 | 23 | 25 | 163 | 231 | 163 | 25 | 23 | 144 | 12 |
| Half Day Totals | | | | 87 | 195 | 664 | 1013 | 760 | 255 | 88 | 87 | 472 | |

**Oct 21**

| AM | Alt | Azm | — | N | NW | W | SW | S | SE | E | NE | HOR | PM |
|---|---|---|---|---|---|---|---|---|---|---|---|---|---|
| 8 | 7.1 | 59.1 | 103 | 4 | 19 | 36 | 98 | 50 | 11 | 4 | 4 | 10 | 4 |
| 9 | 13.8 | 45.7 | 192 | 10 | 11 | 132 | 185 | 128 | 43 | 10 | 10 | 36 | 3 |
| 10 | 19.0 | 31.3 | 230 | 13 | 13 | 110 | 212 | 186 | 108 | 13 | 13 | 64 | 2 |
| 11 | 22.3 | 16.0 | 247 | 15 | 15 | 55 | 201 | 222 | 164 | 18 | 16 | 83 | 1 |
| 12 | 23.5 | 0.0 | 252 | 16 | 16 | 18 | 164 | 234 | 164 | 18 | 16 | 90 | 12 |
| Half Day Totals | | | | 50 | 61 | 386 | 776 | 702 | 244 | 88 | 50 | 238 | |

**Nov 21**

| AM | Alt | Azm | — | N | NW | W | SW | S | SE | E | NE | HOR | PM |
|---|---|---|---|---|---|---|---|---|---|---|---|---|---|
| 9 | 5.2 | 41.9 | 75 | 2 | 5 | 47 | 98 | 53 | 11 | 2 | 2 | 5 | 3 |
| 10 | 10.1 | 28.5 | 164 | 7 | 7 | 72 | 152 | 139 | 37 | 7 | 7 | 21 | 2 |
| 11 | 13.1 | 14.5 | 200 | 9 | 9 | 39 | 165 | 185 | 93 | 11 | 9 | 34 | 1 |
| 12 | 14.2 | 0.0 | 210 | 10 | 10 | 11 | 140 | 200 | 140 | 11 | 10 | 40 | 12 |
| Half Day Totals | | | | 22 | 22 | 166 | 464 | 476 | 199 | 79 | 50 | 79 | |

**Dec 21**

| AM | Alt | Azm | — | N | NW | W | SW | S | SE | E | NE | HOR | PM |
|---|---|---|---|---|---|---|---|---|---|---|---|---|---|
| 9 | 1.9 | 40.5 | 5 | 0 | 0 | 3 | 46 | 9 | 0 | 0 | 0 | 0 | 3 |
| 10 | 6.6 | 27.5 | 113 | 3 | 3 | 29 | 103 | 96 | 27 | 3 | 3 | 9 | 2 |
| 11 | 9.5 | 13.9 | 165 | 6 | 6 | 8 | 135 | 154 | 78 | 6 | 6 | 19 | 1 |
| 12 | 10.6 | 0.0 | 180 | 7 | 7 | 8 | 120 | 171 | 120 | 8 | 7 | 23 | 12 |
| Half Day Totals | | | | 12 | 12 | 76 | 294 | 330 | 162 | 23 | 12 | 39 | |

| N | NW | W | SW | S | SE | E | NE | HOR | ← P.M. |
|---|---|---|---|---|---|---|---|---|---|

*Total solar heat gains for DS (⅛ in.) sheet glass. Based on a ground reflectance of 0.20.
Reprinted by permission from "ASHRAE Handbook of Fundamentals", ASHRAE, New York 1972.

## Table 4-29. Shading Coefficients for Single Glass and Insulating Glass[a]

### A. Single Glass

| Type of Glass | Nominal Thickness[b] | Solar Trans.[b] | Shading Coefficient | |
|---|---|---|---|---|
| | | | $h_0 = 4.0$ | $h_0 = 3.0$ |
| Regular Sheet | $\frac{3}{32}, \frac{1}{8}$ | 0.87 | 1.00 | 1.00 |
| Regular Plate/ | $\frac{1}{4}$ | 0.80 | 0.95 | 0.97 |
| Float | $\frac{3}{8}$ | 0.75 | 0.91 | 0.93 |
| | $\frac{1}{2}$ | 0.71 | 0.88 | 0.91 |
| Grey Sheet | $\frac{1}{8}$ | 0.59 | 0.78 | 0.80 |
| | $\frac{3}{16}$ | 0.74 | 0.90 | 0.92 |
| | $\frac{7}{32}$ | 0.45 | 0.66 | 0.70 |
| | $\frac{7}{32}$ | 0.71 | 0.88 | 0.90 |
| | $\frac{1}{4}$ | 0.67 | 0.86 | 0.88 |
| Heat-Absorbing | $\frac{3}{16}$ | 0.52 | 0.72 | 0.75 |
| Plate/Float[d] | $\frac{1}{4}$ | 0.47 | 0.70 | 0.74 |
| | $\frac{3}{8}$ | 0.33 | 0.56 | 0.61 |
| | $\frac{1}{2}$ | 0.24 | 0.50 | 0.57 |

### B. Insulating Glass[a]

| Type of Glass | Nominal Thickness[c] | Solar Trans.[b] | | Shading Coefficient | |
|---|---|---|---|---|---|
| | | Outer Pane | Inner Pane | $h_0 = 4.0$ | $h_0 = 3.0$ |
| Regular Sheet Out, Regular Sheet In | $\frac{3}{32}, \frac{1}{8}$ | 0.87 | 0.87 | 0.90 | 0.90 |
| Regular Plate/Float Out, Regular Plate/Float In | $\frac{1}{4}$ | 0.80 | 0.80 | 0.83 | 0.83 |
| Heat-Abs Plate/Float Out, Regular Plate/Float In | $\frac{1}{4}$ | 0.46 | 0.80 | 0.56 | 0.58 |

[a] Refers to factory-fabricated units with $\frac{3}{16}$, $\frac{1}{4}$, or $\frac{1}{2}$ in. air space or to prime windows plus storm windows.
[b] Refer to manufacturer's literature for values.
[c] Thickness of each pane of glass, not thickness of assembled unit.
[d] Refers to grey, bronze, and green tinted heat-absorbing plate/float glass.

Reprinted by permission from "ASHRAE Handbook of Fundamentals," ASHRAE, New York, 1972.

Table 4-30. Shading Coefficients for Single Glass with Indoor Shading by Venetian Blinds and Roller Shades

| Type of Glass | Nominal Thickness[a] | Solar Trans.[b] | Type of Shading | | | | |
|---|---|---|---|---|---|---|---|
| | | | Venetian Blinds | | Roller Shade | | |
| | | | | | Opaque | | Translucent |
| | | | Medium | Light | Dark | White | Light |
| Regular Sheet | 3/32 to 1/4 | 0.87–0.80 | 0.64 | 0.55 | 0.59 | 0.25 | 0.39 |
| Regular Plate/Float | 1/4 to 1/2 | 0.80–0.71 | | | | | |
| Regular Pattern | 1/8 to 3/32 | 0.87–0.79 | | | | | |
| Heat-Absorbing Pattern | | — | | | | | |
| Grey Sheet | 3/16, 7/32 | 0.74, 0.71 | | | | | |
| Heat-Absorbing Plate/Float[d] | 1/8, 1/4 | 0.46 | 0.57 | 0.53 | 0.45 | 0.30 | 0.36 |
| Heat-Absorbing Pattern | 3/16, 1/4 | — | | | | | |
| Grey Sheet | 1/8, 7/32 | 0.59, 0.45 | | | | | |
| Heat-Absorbing Plate/Float or Pattern | 3/8 | 0.44–0.30 | 0.54 | 0.52 | 0.40 | 0.28 | 0.32 |
| Heat-Absorbing Plate/Float[d] | | 0.34 | | | | | |
| Heat-Absorbing Plate or Pattern | — | 0.29–0.15, 0.24 | 0.42 | 0.40 | 0.36 | 0.28 | 0.31 |
| Reflective Coated Glass S.C.[c] = 0.30 | | | 0.25 | 0.23 | | | |
| 0.40 | | | 0.33 | 0.29 | | | |
| 0.50 | | | 0.42 | 0.38 | | | |
| 0.60 | | | 0.50 | 0.44 | | | |

[a] Refer to manufacturer's literature for values.
[b] For vertical blinds with opaque white and beige louvers in the tightly closed position, SC is 0.25 and 0.29 when used with glass of 0.71 to 0.80 transmittance.
[c] Shading Coefficient for glass with no shading device.
[d] Refers to grey, bronze, and green tinted heat-absorbing plate/float glass.

Reprinted by permission from "ASHRAE Handbook of Fundamentals" ASHRAE, New York, 1972.

# Table 4-31. Shading Coefficients for Insulating Glass with Indoor Shading by Venetian Blinds and Roller Shades

| Type of Glass | Nominal Thickness, each light | Solar Trans.[b] | | Type of Shading | | | | |
|---|---|---|---|---|---|---|---|---|
| | | | | Venetian Blinds[c] | | Roller Shade | | |
| | | | | | | Opaque | | Translucent |
| | | Outer Pane | Inner Pane | Medium | Light | Dark | White | Light |
| Regular Sheet Out<br>Regular Sheet In | $\frac{3}{32}, \frac{1}{8}$ | 0.87 | 0.87 | 0.57 | 0.51 | 0.60 | 0.25 | 0.37 |
| Regular Plate/Float Out<br>Regular Plate/Float In | $\frac{1}{4}$ | 0.80 | 0.80 | | | | | |
| Heat-Absorbing Plate/Float[d] Out<br>Regular Plate/Float In | $\frac{1}{4}$ | 0.46 | 0.80 | 0.39 | 0.36 | 0.40 | 0.22 | 0.30 |
| Reflective Coated Glass<br>SC[e] = 0.20<br>0.30<br>0.40 | | | | 0.19<br>0.27<br>0.34 | 0.18<br>0.26<br>0.33 | | | |

[a] Refers to factory-fabricated units with $\frac{3}{16}$, $\frac{1}{4}$, or $\frac{1}{2}$ in. air space, or to prime windows plus storm windows.
[b] Refer to manufacturer's literature for exact values.
[c] For vertical blinds with opaque white or beige louvers, tightly closed, SC is approximately the same as for opaque white roller shades.
[d] Refers to bronze or green tinted heat-absorbing plate/float glass.
[e] Shading Coefficient for glass with no shading device.

Reprinted by permission from "ASHRAE Handbook of Fundamentals," ASHRAE, New York, 1972.

4–58

Table 4-32. Shading Coefficients for Double Glazing with Between-glass Shading

| Type of Glass | Nominal Thickness, each pane | Solar Trans.[a] | | Description of Air Space | Type of Shading | | |
|---|---|---|---|---|---|---|---|
| | | Outer Pane | Inner Pane | | Venetian Blinds | | Louvered Sun Screen |
| | | | | | Light | Medium | |
| Regular Sheet Out | $\frac{3}{32}, \frac{1}{8}$ | 0.87 | 0.87 | Shade in contact with glass or shade separated from glass by air space. | 0.33 | 0.36 | 0.43 |
| Regular Sheet In | | | | Shade in contact with glass-voids filled with plastic. | — | — | 0.49 |
| Regular Plate Out | $\frac{1}{4}$ | 0.80 | 0.80 | | | | |
| Regular Plate In | | | | | | | |
| Heat-Abs. Plate/Float[b] Out | $\frac{1}{4}$ | 0.46 | 0.80 | Shade in contact with glass or shade separated from glass by air space. | 0.28 | 0.30 | 0.37 |
| Regular Plate In | | | | Shade in contact with glass-voids filled with plastic. | — | — | 0.41 |

[a] Refer to manufacturer's literature for exact values.
[b] Refers to grey, bronze, and green tinted heat-absorbing plate/float glass.
Reprinted by permission from "ASHRAE Handbook of Fundamentals," ASHRAE, New York 1972.

4-59

## Table 4-33. Shading Coefficients for Hollow Glass Block Wall Panels[a]

| Type of Glass Block[b] | Description of Glass Block | Shading Coefficient[c] | |
|---|---|---|---|
| | | Panels[d] in the Sun | Panels[e] in the Shade (N, NW, W, SW) |
| Type I | Glass Colorless or Aqua<br>Smooth Face<br>A, D: Smooth<br>B, C: Smooth or wide ribs, or flutes horizontal or vertical, or shallow configuration.<br>E: None | 0.65 | 0.40 |
| Type IA | Same as Type I except<br>A: Ceramic Enamel on exterior face. | 0.27 | 0.20 |
| Type II | Same as Type I except<br>E: Glass fiber screen. | 0.44 | 0.34 |
| Type III | Glass Colorless or Aqua<br>A, D: Narrow vertical ribs or flutes.<br>B, C: Horizontal light-diffusing prisms, or horizontal light-directing prisms.<br>E: Glass fiber screen. | 0.33 | 0.27 |
| Type IIIA | Same as Type III except<br>E: Glass fiber screen with green ceramic spray coating, or glass fiber screen and gray glass, or glass fiber screen with light-selecting prisms. | 0.25 | 0.18 |

[a] For glass block used in horizontal skylights see Tables 28 and 29, Chapter 26 of the 1963 ASHRAE Guide and Data Book.

[b] All values are for $7\frac{1}{2} \times 7\frac{1}{2} \times 3\frac{7}{8}$ in. block, set in light-colored mortar. For $11\frac{1}{4} \times 11\frac{1}{4} \times 3\frac{7}{8}$ in. block increase coefficients by 15 percent, and for $5\frac{1}{4} \times 5\frac{1}{4} \times 3\frac{7}{8}$ in. blocks reduce coefficients by 15 percent.

[c] Shading coefficients are to be applied to Heat Gain Factors for one hour earlier than the time for which the load calculation is made to allow for heat storage in the panel.

[d] Shading coefficients are for peak load condition, but provide a close approximation for other conditions. For more precise values for other conditions, see Reference 20.

[e] For NE, E, and SE panels in the shade add 50 percent to the values listed for panels in the shade.

Reprinted by permission from "ASHRAE Handbook of Fundamentals," ASHRAE, New York, 1972.

Table 4-34. Overall Coefficients of Heat Transmission ($U$ Values) for Fenestration under Summer Conditions (7.5 mph Wind Outdoors, Still Air Indoors)

| Type of Glass | U-Value | |
|---|---|---|
| | No Shading | Internal Shading[a] |
| Any Uncoated Single Glass[c] | 1.06 | 0.81 |
| Insulating Glass,[c] $\frac{3}{16}$ in. Air Space | 0.66 | 0.54 |
| uncoated $\frac{1}{4}$ in. Air Space | 0.65 | 0.52 |
| $\frac{3}{8}$ in. Air Space | 0.61 | 0.50 |
| $\frac{1}{2}$ in. Air Space | 0.59 | 0.48 |
| Prime Window Plus Storm Window, Air Space 1 in. or more | 0.54[b] | 0.47[b] |

| | No Supplementary Shading |
|---|---|
| Double Glazing with Between-Glass Shading | |
| Louvered Sun Screen Separated by Air Space | 0.63 |
| Venetian Blinds, Closed, in Air Space | 0.44 |
| Glass Block Panels[d] | |
| Types I and II | 0.56 |
| Types II, III, and IIIA | 0.48 |

[a] Values apply to tightly closed Venetian and vertical blinds, draperies, and roller shades.
[b] Values apply to storm sash with a tight air space. Air leakage present in virtually all storm windows will, in effect, increase this value.
[c] $U$-values can be substantially reduced by low-emittance coatings applied to the inner surface of single or double glazing and to an air-space surface of insulating glass. Consult manufacturers for applicable $U$-values.
[d] Values listed are for $7\frac{3}{4} \times 7\frac{3}{4} \times 3\frac{7}{8}$ in. block. For $11\frac{3}{4} \times 11\frac{3}{4} \times 3\frac{7}{8}$ in. block, reduce the listed value by 0.04, and for $5\frac{3}{4} \times 5\frac{3}{4} \times 3\frac{7}{8}$ in. block, increase the listed value by 0.04. See Table 4-33 for definition of types.

Reprinted by permission from "ASHRAE Handbook of Fundamentals," ASHRAE, New York, 1972.

## Table 4-35. Rates of Heat Gain from Occupants of Conditioned Spaces[a]

| Degree of Activity | Typical Application | Total Heat Adults, Male, Btu/Hr | Total Heat Adjusted,[b] Btu/Hr | Sensible Heat, Btu/Hr | Latent Heat, Btu/Hr |
|---|---|---|---|---|---|
| Seated at rest | Theater—Matinee | 390 | 330 | 225 | 105 |
|  | Theater—Evening | 390 | 350 | 245 | 105 |
| Seated, very light work | Offices, hotels, apartments | 450 | 400 | 245 | 155 |
| Moderately active office work | Offices, hotels, apartments | 475 | 450 | 250 | 200 |
| Standing, light work; or walking slowly | Department store, retail store, dime store | 550 | 450 | 250 | 200 |
| Walking; seated<br>Standing; walking slowly | Drug store, Bank | 550 | 500 | 250 | 250 |
| Sedentary work | Restaurant[c] | 490 | 550 | 275 | 275 |
| Light bench work | Factory | 800 | 750 | 275 | 475 |
| Moderate dancing | Dance hall | 900 | 850 | 305 | 545 |
| Walking 3 mph; moderately heavy work | Factory | 1000 | 1000 | 375 | 625 |
| Bowling[d]<br>Heavy work | Bowling alley<br>Factory | 1500 | 1450 | 580 | 870 |

[a] *Note:* Tabulated values are based on 75 F room dry-bulb temperature. For 80 F room dry-bulb, the total heat remains the same, but the sensible heat values should be decreased by approximately 20 percent, and the latent heat values increased accordingly.

[b] *Adjusted total heat gain* is based on normal percentage of men, women, and children for the application listed, with the postulate that the gain from an adult female is 85 percent of that for an adult male, and that the gain from a child is 75 percent of that for an adult male.

[c] Adjusted total heat value for *sedentary work, restaurant,* includes 60 Btu per hour for food per individual (30 Btu sensible and 30 Btu latent).

[d] For *bowling* figure one person per alley actually bowling, and all others as sitting (400 Btu per hour) or standing (550 Btu per hour).

Reprinted by permission from ASHRAE "Handbook of Fundamentals," ASHRAE, New York, 1972.

## Table 4-36. Internal Heat Gain from Miscellaneous Appliances

| Appliance | Manufacturer's rating | | Recommended rate of heat gain, Btu/hr | | |
|---|---|---|---|---|---|
| | Watts | Btu/hr | Sensible | Latent | Total |
| *Electrical Appliances* | | | | | |
| Hair dryer: | | | | | |
| Blower type................................ | 1,580 | 5,400 | 2,300 | 400 | 2,700 |
| Helmet type................................ | 705 | 2,400 | 1,870 | 330 | 2,200 |
| Permanent-wave machine, 60 heaters at 25 w, 36 in normal use........................... | 1,500 | 5,000 | 850 | 150 | 1,000 |
| Neon sign, per linear foot of tube: | | | | | |
| ½″ diam................................. | ..... | ..... | 30 | ..... | 30 |
| ⅜″ diam................................. | ..... | ..... | 60 | ..... | 60 |
| Sterilizer, instrument...................... | 1,100 | 3,750 | 650 | 1,200 | 1,850 |
| *Gas-burning Appliances* | | | | | |
| Lab burners: | | | | | |
| Bunsen, ⅞₆″ barrel......................... | ..... | 3,000 | 1,680 | 420 | 2,100 |
| Fishtail, 1½″ wide......................... | ..... | 5,000 | 2,800 | 700 | 3,500 |
| Meeker, 1″ diam........................... | ..... | 6,000 | 3,360 | 840 | 4,200 |
| Gaslight, per burner, mantle type........... | ..... | 2,000 | 1,800 | 200 | 2,000 |
| Cigar lighter, continuous flame............. | ..... | 2,500 | 900 | 100 | 1,000 |

## Table 4-37. Heat Gain from Electric Motors

| Nameplate rating of motor, hp | Average motor efficiency in continuous operation | Btu/hr to room air per rated hp of motor | | |
|---|---|---|---|---|
| | | Motor outside of room, driven device inside room | Motor in room, driven device outside of room | Motor and driven device both inside room |
| ⅛–½ | 0.60 | 2546 | 1700 | 4246 |
| ½–3 | .69 | 2546 | 1100 | 3646 |
| 3–20 | .85 | 2546 | 400 | 2946 |

General rule for motors: if $H_m$ = Btu/hr of motor input,

$$H_m = \frac{2{,}546 \times \text{hp (connected load)}}{\text{motor efficiency}}$$

NOTE: Where possible obtain actual value of motor efficiency. Where not possible: for motors use average efficiencies as listed above; for motor generators use average efficiency sets up to 3 hp. as 0.55; for larger sets use average efficiency as 0.80.

SOURCE: Strock and Koral, "Handbook of Air Conditioning, Heating, and Ventilating," The Industrial Press, New York, 1965.

## Table 4-38. Electric Motor-driven Appliances

(Motor and Driven Appliance Both in Same Room)

| Fans (blade diameters, in.) | Btu/hr | Appliances | Btu/hr |
|---|---|---|---|
| Ceiling 32 | 340 | Clock | 7 |
| 52 | 410 | Hair dryer | 1900 |
| 56 | 600 | Drink mixer | 240 |
| Desk or wall  8 | 120 | Sewing machine (domestic) | 220 |
| 10 | 140 | Vacuum cleaner (domestic) | 250 |
| 12 | 200 | Hair clipper | 78 |
| 16 | 300 | Vibrator (beauty) | 11 |

NOTE: Figures are thermal equivalents of nameplate rating, corrected for motor efficiency.

NOTE: Figures are thermal equivalents of nameplate ratings.

SOURCE: Strock and Koral, "Handbook of Air Conditioning, Heating, and Ventilating," The Industrial Press, New York, 1965.

## Table 4-39. Electric Refrigerators

With well insulated cabinet in 80F air and 40F inside cabinet

| Electric Refrigerators | | | |
|---|---|---|---|
| Electric motor and air-cooled condenser in cabinet in room air | | Cold Cabinet in Room (compressor and condenser remote) With well-insulated cabinet in 80°F air and 40°F inside cabinet allow a cooling effect as follows: | |
| Cabinet volume, ft³ | Btu/hr (thermal equivalent of motor input) | Cabinet volume, ft³ | Btu/(hr)(ft³) of cabinet volume |
| 2–4 5 6–10 12–18 | 530 710 850 1060 | 2–4 5–6 7–10 12, 14, 16 20, 25, 30 | 100–75 70–65 60–55 55–50 50, 45, 40 |

SOURCE: Strock and Koral, "Handbook of Air Conditioning, Heating, and Ventilating," The Industrial Press, New York, 1965.

## Table 4-40. Recommended Inside Design Conditions—Summer

| Type of application | Summer | | | | |
|---|---|---|---|---|---|
| | Deluxe | | Commercial practice | | |
| | Dry-bulb, °F | Rel. hum., % | Dry-bulb, °F | Rel. hum., % | Temp. swing,* °F |
| General Comfort Apartment, house, hotel, office, hospital, school, etc. | 74–76 | 50–45 | 77–79 | 50-45 | 2–4 |
| Retail Shops (Short-term occupancy) Bank, barber, or beauty shop. department store, supermarket, etc................................ | 76–78 | 50–45 | 78–80 | 50–45 | 2–4 |
| Low Sensible Heat Factor Applications (High latent load) Auditorium, church, bar, restaurant, kitchen, etc... | 76–78 | 55–50 | 78–80 | 60–50 | 1-2 |
| Factory Comfort Assembly areas, machining rooms, etc............. | 77–80 | 55–45 | 80–85 | 60–50 | 3–6 |

* Temperature swing is above the thermostat setting at peak summer load conditions.
SOURCE: "Carrier Corporation System Design Manual," Part I, Load Estimating, 1970.

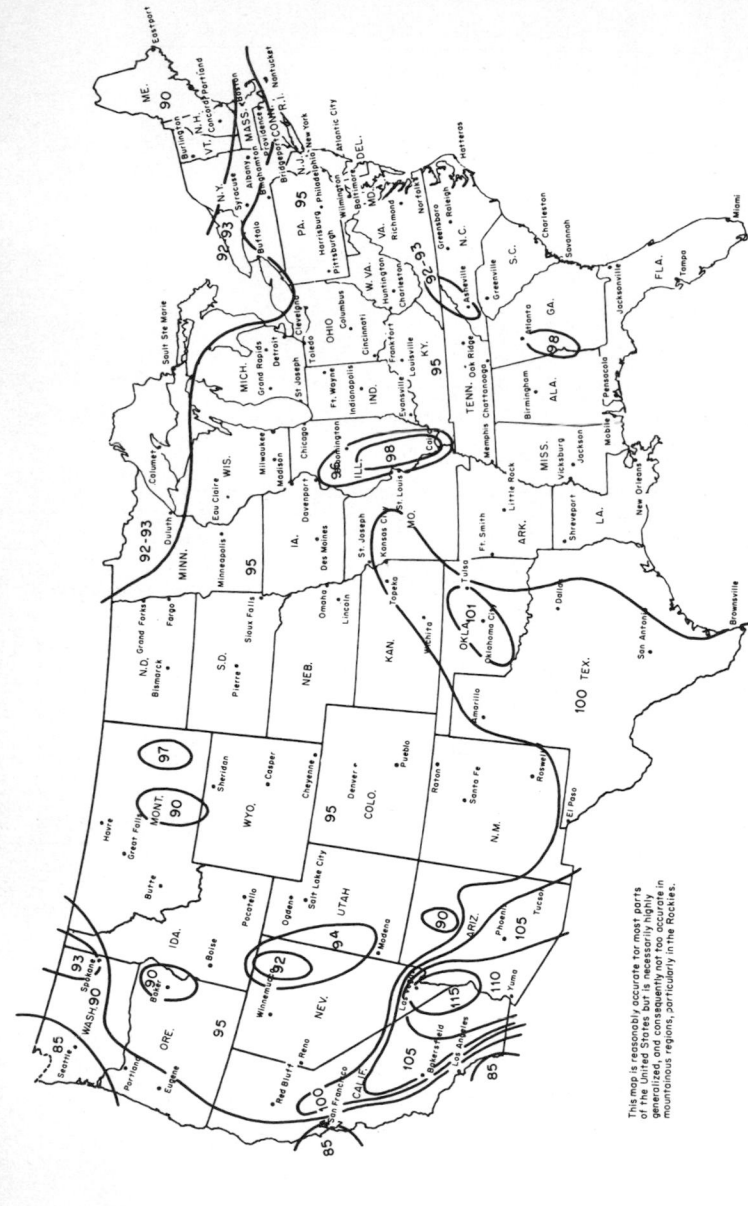

This map is reasonably accurate for most parts of the United States but is necessarily highly generalized, and consequently not too accurate in mountainous regions, particularly in the Rockies.

Fig. 4-12. Summer outside dry-bulb design temperature, °F. (Strock and Koral, "Handbook of Air Conditioning, Heating, and Ventilating," The Industrial Press, New York, 1965.)

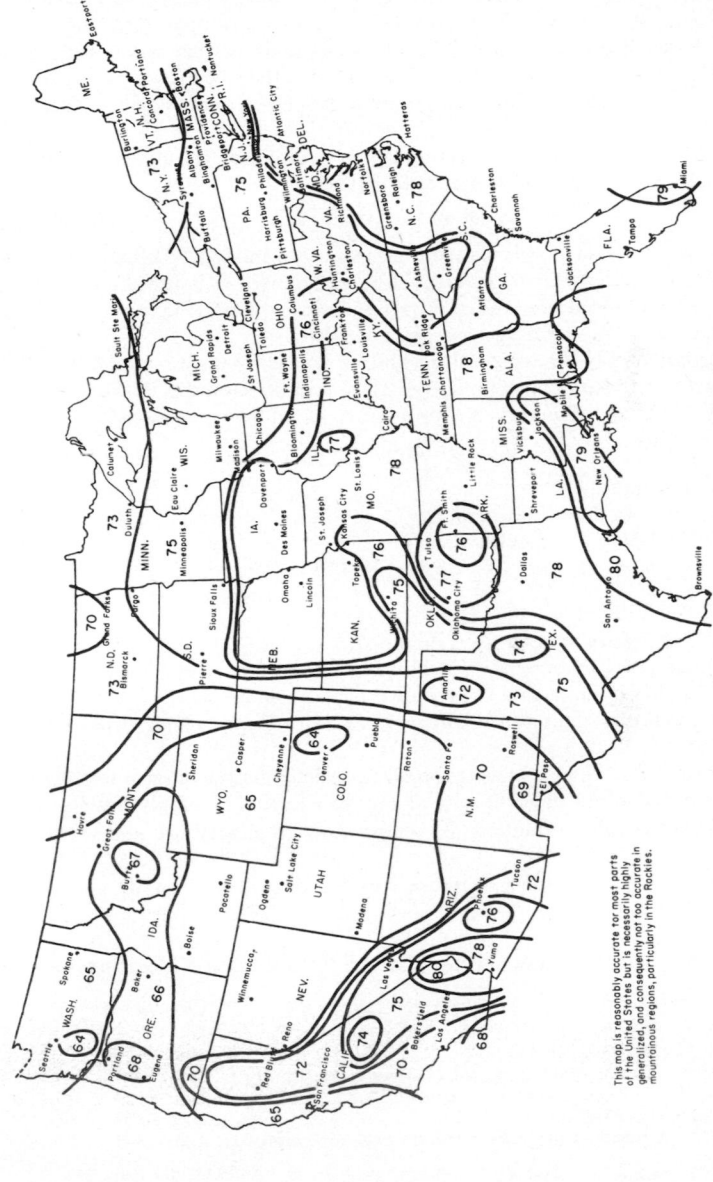

Fig. 4-13. Summer outside wet-bulb design temperature, °F. (Strock and Koral, "Handbook of Air Conditioning, Heating, and Ventilating," 2d ed., 1965.)

This map is reasonably accurate for most parts of the United States but is necessarily highly generalized, and consequently not too accurate in mountainous regions, particularly in the Rockies.

**Supply Air Temperature.**   The room cooling load is the sum of external and internal sensible and latent heat gains plus the difference in enthalpy between outside and room air for that portion of outside air that does not contact the cooling-coil surfaces.   The percentage of air that passes through a cooling coil untreated is the numerical value of the *coil bypass factor;* e.g., a bypass factor of 20 per cent represents a cooling-coil saturation efficiency of 80 per cent.

The ratio of room sensible heat gains to total room sensible and latent heat gains is the room *sensible heat ratio* (RSHR).

$$\text{RSHR} = Q_{rs}/(Q_{rs} + Q_{ri}) \qquad (4\text{-}27)$$

It represents the ratio of sensible cooling capacity to the total cooling capacity required of the supply air to satisfy room conditions.   It is used to plot the slope of the *room-condition line* on a psychrometric chart (Fig. 4-14) for the determination of the *apparatus dew point* (ADP).

The actual supply air temperature and off coil wet-bulb temperature will depend on the bypass characteristic of the selected cooling coil (Fig. 4-15).

**Supply Air Rate.**   The rate of supply air required is expressed by

$$Q_{sa} = Q_{rs}/1.08(t_r - t_s) \qquad (4\text{-}28)$$

where $Q_{sa}$ = supply air, cfm

$t_r$ = room design temperature, °F

$t_s$ = supply air temperature, °F

$1.08 = (60 \text{ min})[0.244 \text{ Btu}/(\text{lb})(°F)](0.075 \text{ lb/ft}^3) \qquad (4\text{-}29)$

## 4-6. Air Distribution

**Outlets.**   **Purpose.**   Outlets are designed:

1. To control air motion, noise level, and temperature gradients caused by the introduction of air to and the removal of air from a space

2. To counteract the natural convection and radiation effects within the room

**Supply Outlets.**   Supply outlets should be selected on the basis of manufacturers' data.   Factors which usually affect the selection of supply outlets are (1) noise, (2) location of outlet, (3) temperature of supply air, and (4) area of diffusion.

**Return Outlets.**   Selection of return registers or grilles is usually governed by face velocity.

### Table 4-41.  Recommended Return Intake Face Velocities

| Intake Location | Velocity over Gross Area, FPM |
|---|---|
| Above occupied zone | 800 up |
| Within occupied zone, not near seats | 600–800 |
| Within occupied zone, near seats | 400–600 |
| Door or wall louvers | 200–300 |
| Undercutting of doors (through undercut area) | 200–300 |

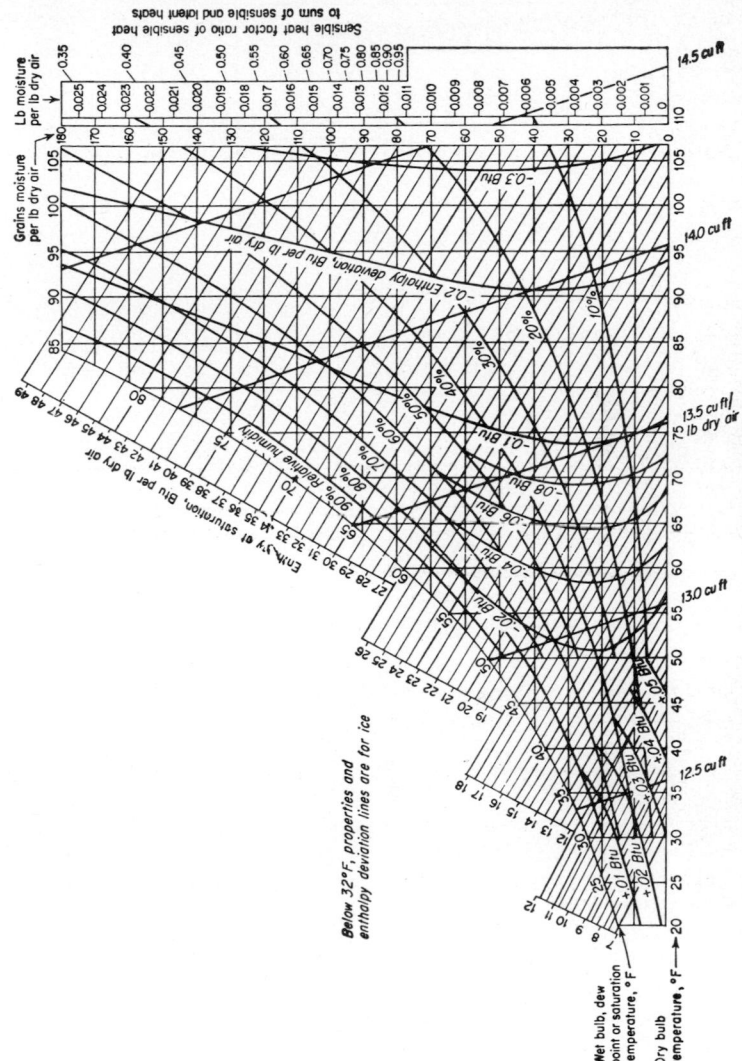

Fig. 4-14. Psychrometric chart—normal temperatures.

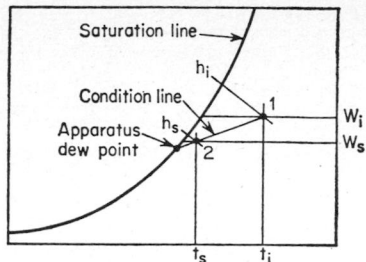

Fig. 4-15. Apparatus dew-point and condition line.

**Ductwork.** *Air Velocity.* Supply and return air ducts and apparatus are sized on the basis of air quantity, within the limitations of allowable friction losses, velocity, and noise (Table 4-42).

### Table 4-42. Recommended and Maximum Duct Velocities for Conventional Systems

| Designation | Residences | Schools, theaters, public buildings | Industrial buildings |
|---|---|---|---|
| Recommended velocities, fpm | | | |
| Outdoor air intakes*......... | 500 | 500 | 500 |
| Filters*.................... | 250 | 300 | 350 |
| Heating coils*.............. | 450 | 500 | 600 |
| Air washers................. | 500 | 500 | 500 |
| Fan outlets................. | 1,000–1,600 | 1,300–2,000 | 1,600–2,400 |
| Main ducts................. | 700–900 | 1,000–1,300 | 1,200–1,800 |
| Branch ducts............... | 600 | 600–900 | 800–1,000 |
| Branch risers.............. | 500 | 600–700 | 800 |
| Maximum velocities, fpm | | | |
| Outdoor air intakes*......... | 800 | 900 | 1,200 |
| Filters*.................... | 300 | 350 | 350 |
| Heating coils*.............. | 500 | 600 | 700 |
| Air washers................. | 500 | 500 | 500 |
| Fan outlets................. | 1,700 | 1,500–2,200 | 1,700–2,800 |
| Main ducts................. | 800–1,200 | 1,100–1,600 | 1,300–2,200 |
| Branch ducts............... | 700–1,000 | 800–1,300 | 1,000–1,800 |
| Branch risers.............. | 650–800 | 800–1,200 | 1,000–1,600 |

* These velocities are for total face area, not the net free area; other velocities in the table are for net free area.

source: "ASHRAE Guide and Data Book," chap. 12, table 6, ASHRAE, New York, 1963.

High-velocity air distribution (2,000 to 6,000 fpm) using much smaller ducts and operating at greater pressures is used when space is critical.

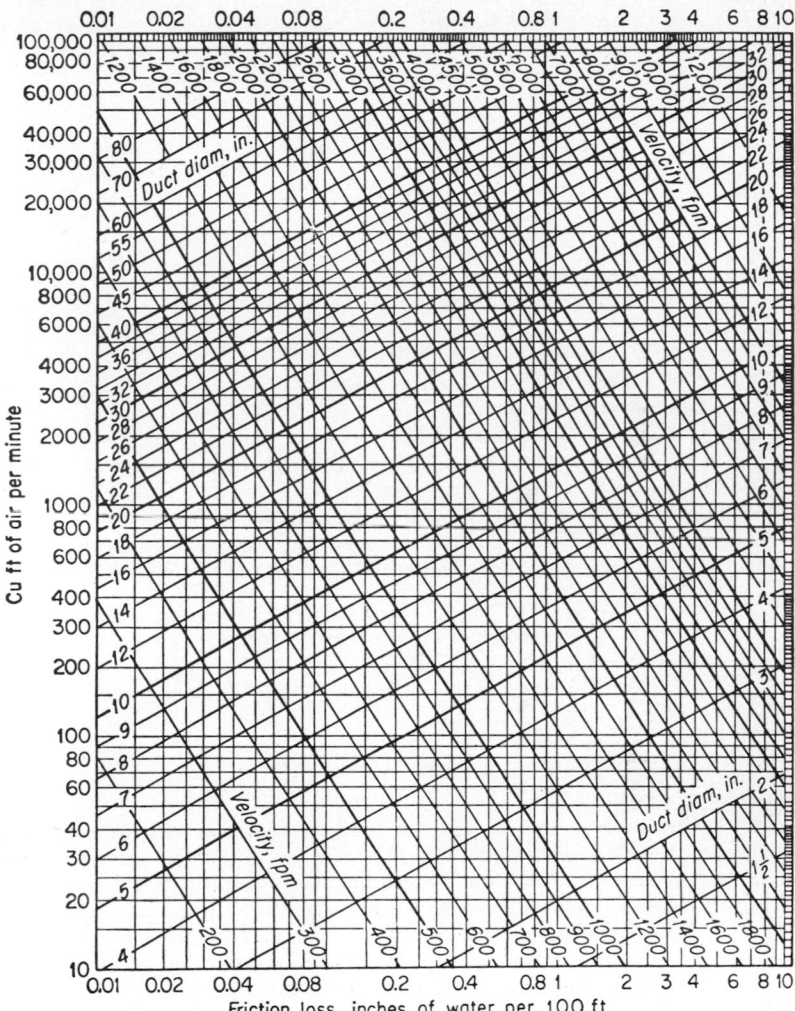

Fig. 4-16. Friction loss for usual air conditions. This chart applies to smooth round galvanized-iron ducts, and is based on air at 70°F and 29.96 in. Hg abs pressure. For air of different density the friction may be assumed to vary directly with the density.

*Pressure Losses in Duct Systems.*   Pressure losses in duct systems are due to friction of the air in contact with the sides of the duct and dynamic losses caused by changes of duct shape or direction and by obstructions to flow.

Friction:

$$H_f = f \frac{L}{D} \left( \frac{V}{4005} \right)^2 \tag{4-30}$$

where $H_f$ = head loss due to friction, in. $H_2O$
$L$ = length of duct, ft
$D$ = diameter of duct, ft
$V$ = velocity of air, fpm
$f$ = nondimensional friction coefficient

Dynamic losses:          $H_v = CV^2/4{,}005$

where $H_v$ = velocity-head loss, in. $H_2O$
$C$ = experimentally determined constant
$V$ = air velocity, fpm

*Design Methods.*   The *equal-friction method* is applicable primarily to systems using low or moderate velocities where the velocity head is not an important factor.   A friction drop per 100 ft of length is chosen, and the duct mains and branches are all sized on the basis of this friction drop.   This will invariably result in higher velocities in the mains, where they can be tolerated, and low velocities in the branches, where they are desirable.

### Table 4-43. Friction Drops

| Application | Friction Drop, In. $H_2O$/100 Ft |
|---|---|
| Noise critical, low velocity | 0.05–0.07 |
| Average application | 0.08–0.1 |
| Equipment rooms, industrial applications | 0.11–0.13 |

The *static-regain method* is used for both conventional and high-velocity systems.   It is especially applicable in the latter, where the velocity head may be appreciable.   In the static-regain method, the static pressure required to give proper air flow through the system outlets is determined, and this pressure is maintained by reducing the velocity at each branch or takeoff, so that the recovery in pressure due to reduction of velocity balances the friction loss in the preceding section of duct.   This is possible because of the convertibility of static and velocity pressures.   For practical applications it is usually assumed that 50 per cent of the velocity pressure available will be converted to static pressure.

$$H_R = 0.5 \left( \frac{V_1}{4005} \right)^2 - \left( \frac{V_2}{4005} \right)^2 \tag{4-31}$$

where $H_R$ = head recovered, in. $H_2O$
$V_1$ = system inlet velocity, fpm
$V_2$ = system outlet velocity, fpm

## Fans. *Fan Laws*

| Quantity required | Cfm | Total head delivered by wheel | Rpm | Hp | Wheel diam* |
|---|---|---|---|---|---|
| Cfm.............................. | .... | $H_t^{1/2}$ | rpm† | $hp^{1/3}$ | $D$ |
| Total head delivered by wheel..... | $cfm^2$ | ..... | $rpm^2$ | $hp^{2/3}$ | $D^2$ |
| Rpm............................. | cfm | $H_t^{1/2}$ | ..... | $hp^{1/3}$ | |
| Hp.............................. | $cfm^3$ | $H_t^{3/2}$ | $rpm^3$ | .... | $D^3$ |

\* Constant speed.
† Constant head.

### Equations

$$\text{Mechanical efficiency} = \frac{0.0001575 \times \text{cfm} \times \text{total pressure, in. } H_2O}{\text{horsepower input}} \quad (4\text{-}32)$$

Equation (4-32) is applicable to fans operating with high outlet velocity pressure relative to static pressure.

$$\text{Static efficiency} = \frac{0.0001573 \times \text{cfm} \times \text{static pressure, in. } H_2O}{\text{horsepower input}} \quad (4\text{-}33)$$

Equation (4-33) is more applicable to fans with high static pressure relative to velocity pressure.

### Characteristics

### Table 4-44. Relative Characteristics of Centrifugal Fans

| Characteristic | Backward | Radial | Forward |
|---|---|---|---|
| First cost................. | High | Medium | Low |
| Efficiency............... | High | Medium | Poor |
| Stability of operation...... | Good | Good | Poor |
| Space required........... | Medium | Medium | Small |
| Tip speed............... | High | Medium | Low |
| Resistance to abrasion..... | Medium | Good | Poor |

## Table 4-45. Outlet Velocities for Optimum Performance of Typical Ventilating Fans

| Static pressure, in. water | Centrifugal fans— outlet velocity, fpm | Tube-axial and vane-axial fans— outlet velocity at wheel diam., fpm |
|---|---|---|
| ¼ | 400–1,100 | 950–1,500 |
| ½ | 550–1,300 | 1,350–1,900 |
| ¾ | 700–1,500 | 1,650–2,350 |
| 1 | 800–1,750 | 1,900–2,700 |
| 1½ | 1,000–2,450 | 2,350–3,300 |
| 2 | 1,150–2,800 | 2,700–3,800 |
| 2½ | 1,250–3,200 | 3,000–4,300 |
| 3 | 1,400–3,500 | 3,300–4,700 |
| 4 | 1,600–4,000 | |
| 6 | 2,000–4,900 | |
| 8 | 2,300–5,650 | |
| 10 | 2,500–6,300 | |

SOURCE: "ASHRAE Guide and Data Book," chap. 40, fig. 1, ASHRAE, New York, 1963.

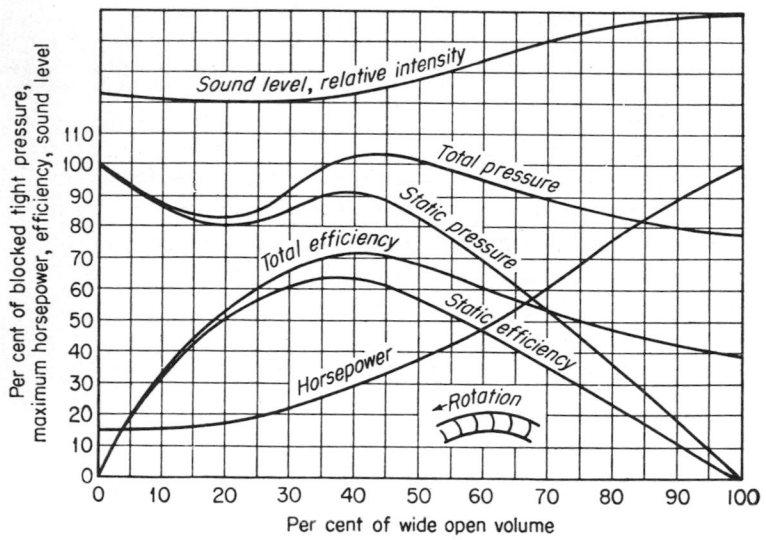

FIG. 4-17. Percentage performance curves of a forward-blade centrifugal fan.

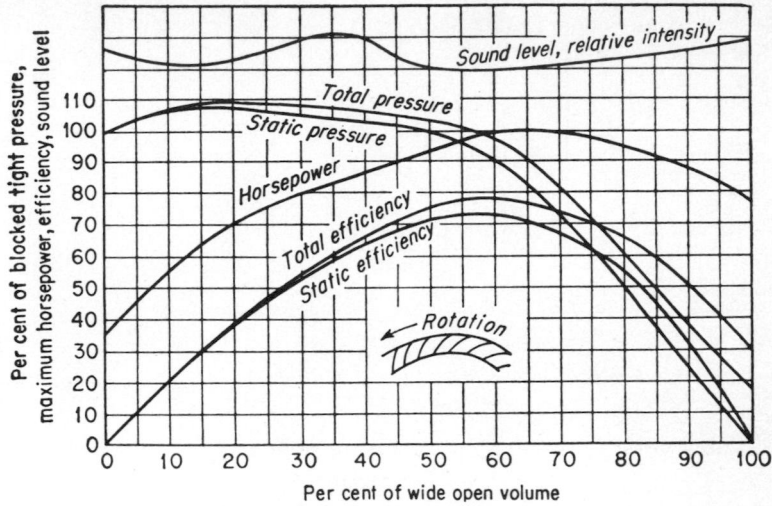

FIG. 4-18. Percentage performance curves of a backward-curved-blade centrifugal fan.

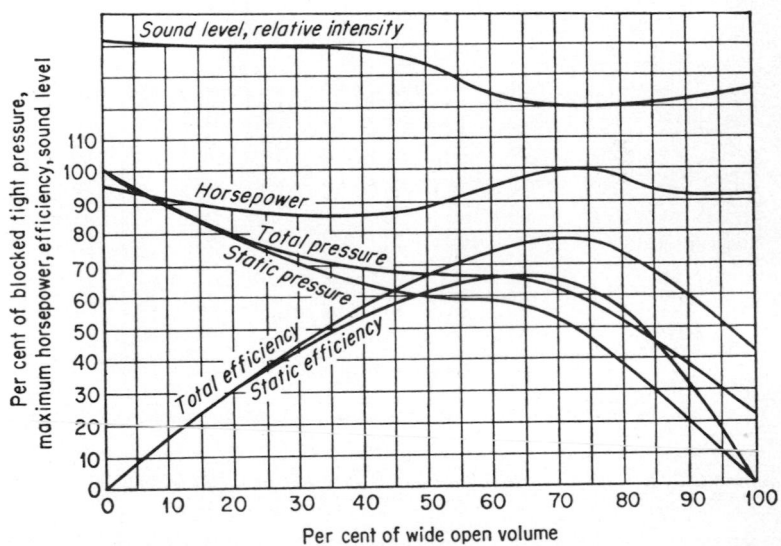

FIG. 4-19. Percentage performance curves of an axial-flow fan.

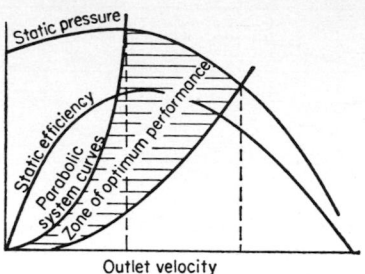

Fɪɢ. 4-20. Zone of optimum performance for fans.

*Correction Factors for Temperature and Altitude*

## Table 4-46. Correction Factor for Altitude and Temperature to Air Volume

| Altitude, ft above sea level.... | 0 | 1,000 | 2,000 | 3,000 | 4,000 | 5,000 | 6,000 | 7,000 | 8,000 |
|---|---|---|---|---|---|---|---|---|---|
| Barometric pressure, in. Hg... | 29.92 | 28.86 | 27.82 | 26.81 | 25.84 | 24.89 | 23.98 | 23.09 | 22.22 |
| Air temp, °F | Correction factors | | | | | | | | |
| 70 | 1.040 | 1.003 | 0.967 | 0.932 | 0.898 | 0.865 | 0.833 | 0.803 | 0.772 |
| 100 | 0.984 | 0.948 | .915 | .882 | .850 | .818 | .788 | .759 | .731 |
| 150 | .904 | .872 | .840 | .801 | .781 | .752 | .724 | .698 | .672 |
| 200 | .835 | .805 | .777 | .749 | .722 | .694 | .668 | .645 | .620 |
| 250 | .777 | .749 | .722 | .696 | .671 | .647 | .622 | .599 | .577 |
| 300 | .725 | .699 | .674 | .649 | .628 | .603 | .580 | .560 | .538 |
| 350 | .680 | .656 | .632 | .609 | .588 | .566 | .545 | .525 | .505 |
| 400 | .641 | .618 | .596 | .574 | .553 | .533 | .512 | .495 | .476 |
| 450 | .605 | .583 | .564 | .543 | .523 | .503 | .485 | .467 | .450 |
| 500 | .574 | .553 | .534 | .515 | .496 | .477 | .460 | .443 | .426 |
| 550 | .546 | .526 | .508 | .490 | .472 | .454 | .438 | .421 | .406 |
| 600 | .520 | .501 | .484 | .466 | .449 | .433 | .416 | .401 | .387 |
| 650 | .496 | .478 | .462 | .444 | .428 | .413 | .397 | .383 | .368 |
| 700 | .475 | .458 | .442 | .426 | .411 | .395 | .381 | .367 | .354 |

NOTE: Equivalent cfm = $\dfrac{\text{cfm at actual conditions}}{\text{correction factor}}$

SOURCE: "Bulletin 3576-B, Correction Factors for Temperature and Altitude," Buffalo Forge Co., Buffalo, N.Y.

## 4-7. Refrigeration

Refrigeration cycles are discussed in Sec. 8.   Thermodynamic data for typical refrigerants are given in Sec. 3.

## 4-8. Water Distribution

### Chilled-water Systems

Temperature differential

| Application | Temperature rise, °F |
|---|---|
| Close-coupled system on one floor | 5–8 |
| Two- or three-story building | 8–11 |
| Multistory building | 12–20 |

$$\text{Gpm} = \frac{\text{total load Btu/hr} + \text{piping heat gains} + \text{pump heat}}{500 \times \text{temperature differential}} \quad (4\text{-}34)$$

**Condenser Water Systems.** For electrically driven refrigeration compressors a temperature differential of 10°F may be assumed, and for steam-driven equipment a temperature differential of 20°F is usual. In the latter case the refrigeration and steam condensers are piped in series with a temperature rise of approximately 10°F each.

### Table 4-47. Heat Rejection of Typical Processes

| Equipment | Btu/min/ton | Btu/kwhr | Btu/bhp-hr |
|---|---|---|---|
| Refrigeration compressors, open drive | 250 | | |
| Refrigeration compressors, hermetic | 300 | | |
| Refrigeration absorption system | 550 | | |
| Steam jet refrigerating system | 550 | | |
| Steam electric power plant, kw: | | | |
| 500 | ... | 11,210 | |
| 1,000 | ... | 10,750 | |
| 5,000 | ... | 8,150 | |
| 7,500 | ... | 7,700 | |
| 10,000 | ... | 7,020 | |
| Diesel engine jacket and lube oil: | | | |
| Four-cycle, supercharged | ... | ..... | 2,600 |
| Four-cycle, nonsupercharged | ... | ..... | 3,000 |
| Two-cycle, crank-case compressor | ... | ..... | 2,000 |
| Two-cycle, pump-scavenging (large unit) | ... | ..... | 2,500 |
| Two-cycle, pump-scavenging (high-speed) | ... | ..... | 2,200 |
| Natural-gas engine: | | | |
| Four-cycle | ... | ..... | 4,500 |
| Two-cycle | ... | ..... | 4,000 |

SOURCE: "ASHRAE Guide and Data Book," chap. 37, table 4, ASHRAE, New York, 1961.

**Atmospheric Cooling Equipment.** The lowest temperature to which water can be cooled in atmospheric cooling equipment is the wet-bulb temperature of the ambient air.

**Water Cooling Effectiveness in Per Cent**

$$E = \frac{(\text{hot-water temperature—cold-water temperature}) \times 100}{\text{hot-water temperature—wet-bulb temperature of entering air}} \quad (4\text{-}35)$$

The cold-water temperature must be chosen to place the requirement within the effectiveness range of the equipment used.

### Table 4-48. Effectiveness of Water Cooling Equipment

| Cooling equipment | Water cooling effectiveness, % | | |
|---|---|---|---|
| | Minimum | Typical | Maximum |
| Spray ponds.................... | 30 | 40–50 | 68 |
| Spray-filled atmospheric towers...... | 40 | 45–55 | 60 |
| Atmospheric deck towers........... | 50 | 50–60 | 90 |
| Mechanical draft towers............ | 50 | 55–75 | 93 |

SOURCE: "Heating, Ventilating, and Air Conditioning Guide," chap. 34, table 3, ASHRAE, New York, 1958

**Makeup Water.** Makeup water is introduced to replace losses due to evaporation, drift, and blowdown.

If all water were cooled by evaporation, the loss by evaporation for the usual 10°F cooling range would be

$$\text{Evaporation } \% = \frac{Q \times 100}{8.3 \times \text{gpm} \times h_{fg}} \qquad (4\text{-}36)$$

where $Q$ = total heat rejected, Btu/hr

gpm = total condenser water circulated, gpm

$h_{fg}$ = evaporation heat of water, Btu/lb, at ambient design temperature

In practice, the loss of circulating water by evaporation due to additional cooling by sensible heat transfer will vary from about 0.64 per cent in winter to 0.88 per cent in the summer for a water-cooling range of 10°F.

Drift losses depend on the tower design, but generally, from the cooling tower, they are limited to 0.2 per cent of the circulated rate.

The makeup water replacing losses due to evaporation, drift, and blowdown introduces dissolved solids into the system.

To prevent excessive concentration, a portion of the circulating water is wasted. The quantity of blowdown depends on the original quantity of dissolved solids in the makeup water and the permissible concentration.

For larger installation, chemical water-treatment processes are used, which also require a controlled blowdown rate.

### 4-9. Pumps

The performance of pumps is discussed in Sec. 8. Methods for calculating pressure drop in pipe and fittings are presented in Sec. 2.

## 4.10. Drainage

### Sanitary Load

### Table 4-49. Drainage Fixture Unit Values for
### Various Plumbing Fixtures

| Type of Fixture or Group of Fixtures | Drainage Fixture Unit Value (d.f.u.) |
|---|---|
| Sinks: | |
| Surgeon's | 3 |
| Flushing rim (with valve) | 6 |
| Service (trap standard) | 3 |
| Service (P trap) | 2 |
| Pot, scullery, etc.* | 4 |
| Urinal, pedestal, syphon jet blowout | 6 |
| wall lip | 4 |
| stall, washout | 4 |
| Urinal trough (each 6-ft section) | 2 |
| Wash sink (circular or multiple) each set of faucets | 2 |
| Water closet, tank-operated | 4 |
| valve-operated | 6 |
| Fixtures not listed above: | |
| Trap Size 1¼ in. or less | 1 |
| 1½ in. | 2 |
| 2 in. | 3 |
| 2½ in. | 4 |
| 3 in. | 5 |
| 4 in. | 6 |

* See Sec. 11.4.2 for method of computing equivalent fixture unit values for devices or equipment which discharge continuous or semicontinuous flows into sanitary drainage systems.
SOURCE: "National Standard Plumbing Code," Table 11.4.1, 1973.

## Table 4-50. Size of Nonintegral Traps for Different Plumbing Fixtures

| Plumbing fixture | Trap size, in. |
|---|---|
| Dental lavatory | 1¼ |
| Drinking fountain | 1¼ |
| Dishwasher, commercial | 2 |
|  domestic (nonintegral trap) | 1½ |
| Floor drain | 2 |
| Food waste grinder, commercial use | 2 |
|  domestic use | 1½ |
| Kitchen sink, domestic, with food waste grinder unit | 1½ |
|  domestic | 1½ |
|  domestic, with dishwasher | 1½ |
| Lavatory, common | 1¼ |
|  barber shop, beauty parlor, or surgeon's | 1½ |
|  multiple type (wash fountain or wash sink) | 1½ |
| Laundry tray (1 or 2 compartments) | 1½ |
| Shower stall or drain | 2 |
| Sinks: | |
|  Surgeon's | 1½ |
|  Flushing rim type, flush valve supplied | 3 |
|  Service type with floor outlet trap standard | 3 |
|  Service trap with P trap | 2 |
|  Commercial (pot, scullery, or similar type) | 2 |
|  Commercial (with food grinder unit) | 2 |

* Separate trap required for wash tray and separate trap required for sink compartment with food waste grinder unit.

SOURCE: "National Standard Plumbing Code," Table 5.2, 1973.

## Table 4-51. Size and Length of Vents

| Size of soil or waste stack, in. | Fixture units connected | Diameter of vent required, in. | | | | | | | | |
|---|---|---|---|---|---|---|---|---|---|---|
| | | 1¼ | 1½ | 2 | 2½ | 3 | 4 | 5 | 6 | 8 |
| | | Maximum length of ven, ft | | | | | | | | |
| 1½ | 8 | 50 | 150 | | | | | | | |
| 1½ | 10 | 30 | 100 | | | | | | | |
| 2 | 12 | 30 | 75 | 200 | | | | | | |
| 2 | 20 | 26 | 50 | 150 | | | | | | |
| 2½ | 42 | ... | 30 | 100 | 300 | | | | | |
| 3 | 10 | ... | 30 | 100 | 100 | 600 | | | | |
| 3 | 30 | ... | ... | 60 | 200 | 500 | | | | |
| 3 | 60 | ... | ... | 50 | 80 | 400 | | | | |
| 4 | 100 | ... | ... | 35 | 100 | 260 | 1,000 | | | |
| 4 | 200 | ... | ... | 30 | 90 | 250 | 900 | | | |
| 4 | 500 | ... | ... | 20 | 70 | 180 | 700 | | | |
| 5 | 200 | ... | ... | ... | 35 | 80 | 350 | 1,000 | | |
| 5 | 500 | ... | ... | ... | 30 | 70 | 300 | 900 | | |
| 5 | 1,100 | ... | ... | ... | 20 | 50 | 200 | 700 | | |
| 6 | 350 | ... | ... | ... | 25 | 50 | 200 | 400 | 1,300 | |
| 6 | 620 | ... | ... | ... | 15 | 30 | 125 | 300 | 1,100 | |
| 6 | 960 | ... | ... | ... | ... | 24 | 100 | 250 | 1,000 | |
| 6 | 1,900 | ... | ... | ... | ... | 20 | 70 | 200 | 700 | |
| 8 | 600 | ... | ... | ... | ... | ... | 50 | 150 | 500 | 1,300 |
| 8 | 1,400 | ... | ... | ... | ... | ... | 40 | 100 | 400 | 1,200 |
| 8 | 2,200 | ... | ... | ... | ... | ... | 30 | 80 | 350 | 1,100 |
| 8 | 3,600 | ... | ... | ... | ... | ... | 25 | 60 | 250 | 800 |
| 10 | 1,000 | ... | ... | ... | ... | ... | ..... | 75 | 125 | 1,000 |
| 10 | 2,500 | ... | ... | ... | ... | ... | ..... | 50 | 100 | 500 |
| 10 | 3,800 | ... | ... | ... | ... | ... | ..... | 30 | 80 | 350 |
| 10 | 5,600 | ... | ... | ... | ... | ... | ..... | 25 | 60 | 250 |

SOURCE: "National Standard Plumbing Code," Table 12.16.6, 1973.

## Table 4-52. Maximum Length of Trap Arm

| Size of fixture drain, in. | Distance—trap to vent |
|---|---|
| 1¼ | 2 ft 6 in |
| 1½ | 3 ft 6 in |
| 2 | 5 ft |
| 3 | 6 ft |
| 4 | 10 ft |

SOURCE: 'National Standard Plumbing Code," Table 12.8.1, 1973.

**Storm-water Load**

$$S = ARC/96 \qquad\qquad (4\text{-}37)$$

where $S$ = storm-water quantity, gpm
$A$ = area being drained, ft$^2$
$R$ = design rate of rainfall, in./hr
$C$ = ratio of runoff to rainfall
Design rate of rainfall varies with locality but is usually between 3 and 6 in./hr.

### Table 4-53. Runoff Coefficients for Rational Formula

| Type of area | Flat: slope <2% | Rolling: slope 2–10% | Hilly: slope >10% |
|---|---|---|---|
| Pavements, roofs, etc.......... | 0.90 | 0.90 | 0.90 |
| City business areas............ | .80 | .85 | .85 |
| Suburban residential areas..... | .45 | .50 | .55 |
| Dense residential areas........ | .60 | .65 | .70 |
| Grassed areas................. | .25 | .30 | .30 |
| Earth areas.................. | .60 | .65 | .70 |
| Cultivated land: | | | |
|   Impermeable (clay, loam).... | .50 | .55 | .60 |
|   Permeable (sand)........... | .25 | .30 | .35 |
| Meadows and pasture lands.... | .25 | .30 | .35 |
| Forests and wooded areas...... | .10 | .15 | .20 |

*Pipe Sizing*

### Table 4-54. Building Drains and Sewers

| Pipe diameter, in. | Maximum number of fixture units that may be connected to any portion of the building drain or the building sewer including branches of the building drain | | | |
|---|---|---|---|---|
| | Fall per foot | | | |
| | $\frac{1}{16}$ in. | $\frac{1}{8}$ in. | $\frac{1}{4}$ in. | $\frac{1}{2}$ in. |
| 2 | ..... | ..... | 21 | 26 |
| 2½ | ..... | ..... | 24 | 31 |
| 3 | ..... | 36* | 42* | 50* |
| 4 | ..... | 180 | 216 | 250 |
| 5 | ..... | 390 | 480 | 575 |
| 6 | ..... | 700 | 840 | 1,000 |
| 8 | 1,400 | 1,600 | 1,920 | 2,300 |
| 10 | 2,500 | 2,900 | 3,500 | 4,200 |
| 12 | 3,900 | 4,600 | 5,600 | 6,700 |
| 15 | 7,000 | 8,300 | 10,000 | 12,000 |

* Not over two water closets or two bathroom groups.
NOTE: On-site sewers that serve more than one building may be sized according to the current standards and specifications of the Administrative Authority for public sewers.
SOURCE: "National Standard Plumbing Code," Table 11.5.1A, 1973.

*Size of Combined Drains and Sewers.* For combined storm and sanitary systems, drain sizing is based on fixture units and the storm drainage area is converted to equivalent fixture units.

Where the total fixture-unit load on the combined drain is less than 256 fixture units, the equivalent drainage area in horizontal projection is taken as 1,000 ft².

When the total fixture-unit load exceeds 256 fixture units, each fixture unit is considered the equivalent of 3.9 ft² of drainage area.

If the rainfall to be provided for is more or less than 4 in./hr, the 1,000-ft² equivalent and the 3.9 ft² are adjusted by multiplying by 4 and dividing by the rainfall in inches per hour to be provided for.

## Table 4-55. Size of Roof Gutters*

| Diameter of gutter, in.† | Maximum projected roof area for gutters, $\frac{1}{16}$-in. slope‡ | |
|---|---|---|
| | ft² | gpm |
| 3 | 170 | 7 |
| 4 | 360 | 15 |
| 5 | 625 | 26 |
| 6 | 960 | 40 |
| 7 | 1,380 | 57 |
| 8 | 1,990 | 83 |
| 10 | 3,600 | 150 |

* Table 4-55 is based on a maximum rate of rainfall of 4 in./hr for a 5-min duration and 10-year return period. Where maximum rates are more or less than 4 in./hr, the figures for drainage area shall be adjusted by multiplying by 4 and dividing by the local rate in inches per hour.
† Gutters other than semicircular may be used provided they have an equivalent cross-sectional area.
‡ Capacities given for slope of $\frac{1}{16}$ in./ft shall be used when designing for greater slopes.
SOURCE: "National Standard Plumbing Code." Table 13.6.3, 1973.

## Table 4-56. Size of Vertical Conductors and Leaders*

| Size of leader or conductor, in.† | Maximum projected roof area | |
|---|---|---|
| | ft² | gpm |
| 2 | 544 | 23 |
| 2½ | 987 | 41 |
| 3 | 1,610 | 67 |
| 4 | 3,460 | 144 |
| 5 | 6,280 | 261 |
| 6 | 10,200 | 424 |
| 8 | 22,000 | 913 |

* Table 4-56 is based on a maximum rate of rainfall of 4 in./hr and on the hydraulic capacities of vertical circular pipes flowing between one-third and one-half full at terminal velocity, computed by the method of NBS Mono. 31. Where maximum rates are more or less than 4 in./hr, the figures for drainage area shall be adjusted by multiplying by 4 and dividing by the local rate in inches per hour.
† The area of rectangular leaders shall be equivalent to that of the circular leader or conductor required. The ratio of width to depth of rectangular leaders shall not exceed 3 to 1.
SOURCE: "National Standard Plumbing Code," Table 13.6.1, 1973.

## Table 4-57. Horizontal Fixture Branches and Stacks

| Pipe diameter, in. | Maximum number of fixture units that may be connected to | | | |
| | Any horizontal fixture branch* | Stack sizing for 3 stories or 3 intervals | Stack sizing for more than 3 stories | |
| | | | Total for stack | Total at story or branch interval |
|---|---|---|---|---|
| 1½ | 3 | 4 | 8 | 2 |
| 2 | 6 | 10 | 24 | 6 |
| 2½ | 12 | 20 | 42 | 9 |
| 3 | 20† | 48† | 72† | 20† |
| 4 | 160 | 240 | 500 | 90 |
| 5 | 360 | 540 | 1,100 | 200 |
| 6 | 620 | 960 | 1,900 | 350 |
| 8 | 1,400 | 2,200 | 3,600 | 600 |
| 10 | 2.500 | 3,800 | 5,600 | 1,000 |
| 12 | 3,900 | 6,000 | 8,400 | 1,500 |
| 15 | 7,000 | | | |

* Does not include branches of the building drain.
† Not more than two water closets or bathroom groups within each branch interval nor more than six water closets or bathroom groups on the stack.
‡ Stacks shall be sized according to the total accumulated connected load at each story or branch interval and may be reduced in size as this load decreases to a minimum diameter of half of the largest size required.
SOURCE: "National Standard Plumbing Code," Table 11.5.1B, 1973.

## Table 4-58. Size of Horizontal Storm Drains*

| Drain diameter, in. | Maximum projected area for drains of various slopes | | | | | |
| | ⅛-in. slope | | ¼-in. slope | | ½-in. slope | |
| | ft² | gpm | ft² | gpm | ft² | gpm |
|---|---|---|---|---|---|---|
| 3 | 822 | 34 | 1,160 | 48 | 1,644 | 68 |
| 4 | 1,880 | 78 | 2,650 | 110 | 3,760 | 156 |
| 5 | 3,340 | 139 | 4,720 | 196 | 6,680 | 278 |
| 6 | 5,350 | 222 | 7,550 | 314 | 10,700 | 445 |
| 8 | 11,500 | 478 | 16,300 | 677 | 23,000 | 956 |
| 10 | 20,700 | 860 | 29,200 | 1,214 | 41,400 | 1,721 |
| 12 | 33,300 | 1,384 | 47,000 | 1,953 | 66,600 | 2,768 |
| 15 | 59,500 | 2,473 | 84,000 | 3,491 | 119,000 | 4,946 |

* Table 4-58 is based on a maximum rate of rainfall of 4 in./hr. Where maximum rates are more or less than 4 in./hr, the figures for drainage area shall be adjusted by multiplying by 4 and dividing by the local rate in inches per hour.
SOURCE: "National Standard Plumbing Code," Table 13.6.2, 1973.

## 4-11. Cold Water

### Table 4-59. Estimating Demand

| Supply systems predominantly for flush tanks | | Supply systems predominantly for flush valves | |
|---|---|---|---|
| Load, water supply fixture units | Demand gpm | Load, water supply fixture units | Demand gpm |
| 6 | 5 | | |
| 8 | 6.5 | | |
| 10 | 8 | 10 | 27 |
| 12 | 9.2 | 12 | 28.6 |
| 14 | 10.4 | 14 | 30.2 |
| 16 | 11.6 | 16 | 31.8 |
| 18 | 12.8 | 18 | 33.4 |
| 20 | 14 | 20 | 35 |
| 25 | 17 | 25 | 38 |
| 30 | 20 | 30 | 41 |
| 35 | 22.5 | 35 | 43.8 |
| 40 | 24.8 | 40 | 46.5 |
| 45 | 27 | 45 | 49 |
| 50 | 29 | 50 | 51.5 |
| 60 | 32 | 60 | 55 |
| 70 | 35 | 70 | 58.5 |
| 80 | 38 | 80 | 62 |
| 90 | 41 | 90 | 64.8 |
| 100 | 43.5 | 100 | 67.5 |
| 120 | 48 | 120 | 72.5 |
| 140 | 52.5 | 140 | 77.5 |
| 160 | 57 | 160 | 82.5 |
| 180 | 61 | 180 | 87 |
| 200 | 65 | 200 | 91.5 |
| 225 | 70 | 225 | 97 |
| 250 | 75 | 250 | 101 |
| 275 | 80 | 275 | 105.5 |
| 300 | 85 | 300 | 110 |
| 400 | 105 | 400 | 126 |
| 500 | 125 | 500 | 142 |
| 750 | 170 | 750 | 178 |
| 1,000 | 208 | 1,000 | 208 |
| 1,250 | 240 | 1,250 | 240 |
| 1,500 | 267 | 1,500 | 267 |
| 1,750 | 294 | 1,750 | 294 |
| 2,000 | 321 | 2,000 | 321 |
| 2,250 | 348 | 2,250 | 348 |
| 2,500 | 375 | 2,500 | 375 |
| 2,750 | 402 | 2,750 | 402 |
| 3,000 | 432 | 3,000 | 432 |
| 4,000 | 525 | 4,000 | 525 |
| 5,000 | 593 | 5,000 | 593 |
| 6,000 | 643 | 6,000 | 643 |
| 7,000 | 685 | 7,000 | 685 |
| 9,000 | 718 | 8,000 | 718 |
| 8,000 | 745 | 9,000 | 745 |
| 10,000 | 769 | 10,000 | 769 |

SOURCE: "National Standard Plumbing Code," Table 10.13.2B, 1973.

## Table 4-60. Water Consumption per Capita

| Occupancy | Gal as stated or gpcpd | Occupancy | Gal as stated or gpcpd |
|---|---|---|---|
| Office buildings............. | 27–45 | Laundries, per pound....... | 3–5.7 |
| Grade schools.............. | 5–10 | Hotels, per room.......... | 300–525 |
| High schools............... | 15–20 | Hospitals, per bed......... | 125–350 |
| Restaurants, per meal....... | 0.5–4 | | |

## Table 4-61. Sizing the Water Supply System*

| Fixture | Occupancy | Type of supply control | Load in fixture units† |
|---|---|---|---|
| Bathroom group‡........ | Private | Flush valve for closet | 8 |
| Bathroom group‡........ | Private | Flush tank for closet | 6 |
| Bathtub................ | Private | Faucet | 2 |
| Bathtub................ | Public | Faucet | 4 |
| Clothes washer.......... | Private | Faucet | 2 |
| Clothes washer.......... | Public | Faucet | 4 |
| Combination fixture..... | Private | Faucet | 3 |
| Kitchen sink............ | Private | Faucet | 2 |
| Kitchen sink............ | Hotel, restaurant | Faucet | 4 |
| Laundry trays (1 to 3)... | Private | Faucet | 3 |
| Lavatory............... | Private | Faucet | 1 |
| Lavatory............... | Public | Faucet | 2 |
| Separate shower........ | Private | Mixing valve | 2 |
| Service sink............ | Office, etc. | Faucet | 3 |
| Shower head........... | Private | Mixing valve | 2 |
| Shower head........... | Public | Mixing valve | 4 |
| Urinal, pedestal | Public | Flush valve | 10 |
| stall or wall...... | Public | Flush valve | 5 |
| stall or wall...... | Public | Flush tank | 3 |
| Water closet............ | Private | Flush valve | 6 |
| Water closet............ | Private | Flush tank | 3 |
| Water closet............ | Public | Flush valve | 10 |
| Water closet............ | Public | Flush tank | 5 |

Water supply outlets for items not listed above shall be computed at their maximum demand, but in no case less than:

| Fixture, in. | Number of fixture units | |
|---|---|---|
| | Private use | Public use |
| ⅜ | 1 | 2 |
| ½ | 2 | 4 |
| ¾ | 3 | 6 |
| 1 | 6 | 10 |

* For supply outlets likely to impose continuous demands, estimate continuous supply separately and add to total demand for fixtures.

† The given weights are for total demand. For fixtures with both hot and cold water supplies, the weights for maximum separate demands may be taken as ¾ the listed demand for the supply.

‡ A bathroom group for the purposes of this table consists of not more than one water closet, one lavatory, one bathtub, one shower stall or not more than one water closet, two lavatories, one bathtub or one separate shower stall.

SOURCE: "National Standard Plumbing Code," Table 10.13.2.A, 1973.

## Table 4-62. Proper Flow and Pressure Required During Flow for Different Fixtures

| Fixture | Flow pressure* | Flow, gpm |
|---|---|---|
| Ordinary basin faucet.................. | 8 | 3.0 |
| Self-closing basin faucet............... | 12 | 2.5 |
| Sink faucet, ⅜ in...................... | 10 | 3.5 |
| Sink faucet, ½ in...................... | 5 | 4.5 |
| Dishwasher........................... | 15–25 | † |
| Bathtub faucet........................ | 5 | 6.0 |
| Laundry tub cock, ¼ in............... | 5 | 5.0 |
| Shower............................... | 12 | 3–10 |
| Ball-cock for closet.................... | 15 | 3.0 |
| Flush valve for closet................. | 10–20 | 15–40‡ |
| Flush valve for urinal................. | 15 | 15.0 |
| Garden hose, 50 ft, and sill cock........ | 30 | 5.0 |

* Flow pressure is the pressure psig in the pipe at the entrance to the particular fixture considered.
† Varies, see manufacturers' data.
‡ Wide range due to variation in design and type of flush-valve closets.
Reprinted by permission from "ASHRAE Handbook of Fundamentals," ASHRAE, New York, 1972.

## 4-12. Hot Water

### Table 4-63. Maximum Daily (24-hr) Requirements for Hot Water in Gallons

| Apartments and private homes with no. of rooms | Number of bathrooms | | | | |
|---|---|---|---|---|---|
| | 1 | 2 | 3 | 4 | 5 |
| 1 | 60 | | | | |
| 2 | 70 | | | | |
| 3 | 80 | | | | |
| 4 | 90 | 120 | | | |
| 5 | 100 | 140 | | | |
| 6 | 120 | 160 | 200 | | |
| 7 | 140 | 180 | 220 | | |
| 8 | 160 | 200 | 240 | 250 | |
| 9 | 180 | 220 | 260 | 275 | |
| 10 | 200 | 240 | 280 | 300 | |
| 11 | ... | 260 | 300 | 340 | |
| 12 | ... | 280 | 325 | 380 | 450 |
| 13 | ... | 300 | 350 | 420 | 500 |
| 14 | ... | ... | 375 | 460 | 550 |
| 15 | ... | ... | 400 | 500 | 600 |
| 16 | ... | ... | ... | 540 | 650 |
| 17 | ... | ... | ... | 580 | 700 |
| 18 | ... | ... | ... | 620 | 750 |
| 19 | ... | ... | ... | ... | 800 |
| 20 | ... | ... | ... | ... | 850 |

Hotels:
  Room with basin................................. 10
  Room with bath—transient..................... 50
  Room with bath—resident...................... 60
  Two rooms with bath........................... 80
  Three rooms with bath......................... 100
  Public shower................................. 200
  Public basins................................. 150
  Slop sink..................................... 30
Office buildings:
  White-collar worker (per person)*............ 2–3
  Other workers (per person).................... 4.0
  Cleaning per 10,000 sq ft..................... 30.0
Hospitals:
  Per bed...................................... 80–100

\* The value for white-collar workers is for office occupancy only, not including allowance for employees lunch rooms, dining rooms, etc. Requirements for these areas should be calculated separately.

Reprinted by permission from "ASHRAE Handbook and Product Directory—Systems, ASHRAE, New York, 1973.

## Table 4-64. Estimated Hot Water Demand Characteristics for Various Types of Buildings

| Type of building | Hot water required per person | Max. hourly demand in relation to day's use | Duration of peak load, hr | Storage capacity in relation to day's use | Heating capacity in relation to day's use |
|---|---|---|---|---|---|
| Residences, apartments, hotels, etc.†·‡ | 20–40 gpd* | ⅐ | 4 | ⅕ | ⅐ |
| Office buildings. . . . . . . . . . . . . . . . . . . . . | 2–3 gpd* | ⅕ | 2 | ⅕ | ⅙ |
| Factory buildings. . . . . . . . . . . . . . . . . | 5 gpd* | ⅓ | 1 | ⅖ | ⅛ |

\* At 140°F.
† Daily hot water requirements and demand characteristics vary with the type of hotel. The better class hotel has a relatively high daily consumption with a low peak load. The commercial hotel has a lower daily consumption but a high peak load.
‡ The increasing use of dishwashers and laundry washing machines in residences and apartments requires additional allowances of 15 gal per dishwasher and 40 gal per laundry washer.
Reprinted by permission from "ASHRAE Handbook and Product Directory—Systems," ASHRAE, New York, 1973.

**Hot Water for Kitchens.** Although, in private dwellings, a water temperature of 140°F is reasonable for dishwashing, in public places sanitation regulations call for 180°F water. Most of the dishwashing machines now available on the market require 180°F water. The amount of 180°F water needed in restaurants per day may be determined according to the American Gas Association method outlined in the following paragraphs:

1. Multiply the number of meals per day by the number of dishes per meal (6 for low-price restaurants, 8 for medium-price restaurants, and 10 for high-price restaurants) to determine the total number of dishes per day.

2. Divide the total number of dishes per day by the average number of dishes per rack to find the number of racks per day.

3. Multiply the number of racks per day by the gallons of 180°F water (using 1.5 gal for single-tank machines and 0.75 gal for two-tank machines). This product will give the gallons of 180°F water per day for rinse sprays.

4. Multiply the number of meal periods per day (one, two, or three) by the dishwashing tank capacity in gallons, giving the gallons of 180°F water per day necessary to fill the tanks.

5. Add values from (3) and (4) to obtain the total number of gallons of 180°F water required per day.

For purposes other than dishwashing, a considerable amount of 140°F water is used. To find the daily 140°F water requirement in a restaurant, multiply the total number of meals served per day by the gallons of 140°F water per meal. Low-price restaurants on the average utilize 0.9 gal of 140°F water per meal; medium- and high-price restaurants use 1.2 and 1.5 gal per meal, respectively.

## Table 4-65. Hot Water Demand per Fixtures for Various Types of Buildings

Gallons of water per hour per fixture, calculated at a final temperature of 140 F

| | Apartment House | Club | Gym-nasium | Hospital | Hotel | Industrial Plant | Office Building | Private Residence | School | Y.M.C.A. |
|---|---|---|---|---|---|---|---|---|---|---|
| 1. Basins, private lavatory | 2 | 2 | 2 | 2 | 2 | 2 | 2 | 2 | 2 | 2 |
| 2. Basins, public lavatory | 4 | 6 | 8 | 6 | 8 | 12 | 6 | .... | 15 | 8 |
| 3. Bathtubs | 20 | 20 | 30 | 20 | 20 | .... | .... | 20 | .... | 30 |
| 4. Dishwashers[a] | 15 | 50-150 | .... | 50-150 | 50-200 | 20-100 | .... | 15 | 20-100 | 20-100 |
| 5. Foot basins | 3 | 3 | 12 | 3 | 3 | 12 | .... | 3 | 3 | 12 |
| 6. Kitchen sink | 10 | 20 | .... | 20 | 30 | 20 | 20 | 10 | 20 | 20 |
| 7. Laundry, stationary tubs | 20 | 28 | .... | 28 | 28 | .... | .... | 20 | .... | 28 |
| 8. Pantry sink | 5 | 10 | .... | 10 | 10 | .... | 10 | 5 | 10 | 10 |
| 9. Showers | 30 | 150 | 225 | 75 | 75 | 225 | 30 | 30 | 225 | 225 |
| 10. Slop sink | 20 | 20 | .... | 20 | 30 | 20 | 20 | 15 | 20 | 20 |
| 11. Hydro-therapeutic showers | | | | 400 | | | | | | |
| 12. Hubbard baths | | | | 600 | | | | | | |
| 13. Leg baths | | | | 100 | | | | | | |
| 14. Arm baths | | | | 35 | | | | | | |
| 15. Sitz baths | | | | 30 | | | | | | |
| 16. Continuous-flow baths | | | | 165 | | | | | | |
| 17. Circular wash sinks | | | | 20 | 20 | 30 | 20 | | 30 | |
| 18. Semi-circular wash sinks | | | | 10 | 10 | 15 | 10 | | 15 | |
| 19. Demand factor | 0.30 | 0.30 | 0.40 | 0.25 | 0.25 | 0.40 | 0.30 | 0.30 | 0.40 | 0.40 |
| 20. Storage capacity factor[b] | 1.25 | 0.90 | 1.00 | 0.60 | 0.80 | 1.00 | 2.00 | 0.70 | 1.00 | 1.00 |

[a] Dishwasher requirements should be taken from Table 13 or from manufacturers' data for the model to be used, if this is known.

[b] Ratio of storage tank capacity to probable maximum demand per hour. Storage capacity may be reduced where an unlimited supply of steam is available from a central street steam system or large boiler plant.

Reprinted by permission from "ASHRAE Handbook & Product Directory," ASHRAE, New York, 1973.

## 4.13. Gas Piping

### Table 4-66. Capacity of Gas Piping, ft³/hr

At pressure drop of 0.3 in. water.  Specific gravity = 0.60

| Pipe length, ft | Iron pipe size (IPS), in. | | | | | | | | |
|---|---|---|---|---|---|---|---|---|---|
| | ½ | ¾ | 1 | 1¼ | 1½ | 2 | 2½ | 3 | 4 |
| 10 | 132 | 278 | 520 | 1,050 | 1,600 | 3,050 | 4,800 | 8,500 | 17,500 |
| 20 | 92 | 190 | 350 | 730 | 1,100 | 2,100 | 3,300 | 5,900 | 12,000 |
| 30 | 73 | 152 | 285 | 590 | 890 | 1,650 | 2,700 | 4,700 | 9,700 |
| 40 | 63 | 130 | 245 | 500 | 760 | 1,450 | 2,300 | 4,100 | 8,300 |
| 50 | 56 | 115 | 215 | 440 | 670 | 1,270 | 2,000 | 3,600 | 7,400 |
| 60 | 50 | 105 | 195 | 400 | 610 | 1,150 | 1,850 | 3,250 | 6,800 |
| 70 | 46 | 96 | 180 | 370 | 560 | 1,050 | 1,700 | 3,000 | 6,200 |
| 80 | 43 | 90 | 170 | 350 | 530 | 990 | 1,600 | 2,800 | 5,800 |
| 90 | 40 | 84 | 160 | 320 | 490 | 930 | 1,500 | 2,600 | 5,400 |
| 100 | 38 | 79 | 150 | 305 | 460 | 870 | 1,400 | 2,500 | 5,100 |
| 125 | 34 | 72 | 130 | 275 | 410 | 780 | 1,250 | 2,200 | 4,500 |
| 150 | 31 | 64 | 120 | 250 | 380 | 710 | 1,130 | 2,000 | 4,100 |
| 175 | 28 | 59 | 110 | 225 | 350 | 650 | 1,050 | 1,850 | 3,800 |
| 200 | 26 | 55 | 100 | 210 | 320 | 610 | 980 | 1,700 | 3,500 |

From ANSI *Standard Installation of Gas Appliances and Gas Piping*, ANSI Z21.30-1959.  Reprinted by permission from "ASHRAE Handbook of Fundamentals," ASHRAE, New York, 1972.

### Table 4-67. Multipliers for Various Specific Gravities

| Sp gr | Multiplier | Sp gr | Multiplier | Sp gr | Multiplier |
|---|---|---|---|---|---|
| 0.35 | 1.31 | 0.75 | 0.895 | 1.40 | 0.655 |
| .40 | 1.23 | 0.80 | .867 | 1.50 | .633 |
| .45 | 1.16 | 0.85 | .841 | 1.60 | .612 |
| .50 | 1.10 | 0.90 | .817 | 1.70 | .594 |
| .55 | 1.04 | 1.00 | .775 | 1.80 | .577 |
| .60 | 1.00 | 1.10 | .740 | 1.90 | .565 |
| .65 | 0.962 | 1.20 | .707 | 2.00 | .547 |
| .70 | .926 | 1.30 | .680 | 2.10 | .535 |

Reprinted by permission from "ASHRAE Handbook of Fundamentals," ASHRAE, New York, 1972.

### Table 4-68. Common Gas Appliances

Maximum Gas Consumption in Ft³/Hour

| Appliance | Natural gas 1050 Btu/ft³ | Mixed gas 800 Btu/ft³ | Manufactured gas 550 Btu/ft³ |
|---|---|---|---|
| Range, domestic, 4 top, 1 oven burners............. | 60 | 80 | 115 |
| Range, domestic, 6 top, 2 oven burners............. | 100 | 135 | 200 |
| Hot plate, domestic or laundry stove per burner...... | 8.5 | 11 | 16 |
| Room heater, radiant type, single, domestic......... | 2 | 2.5 | 4 |
| Water heater, instantaneous, automatic, per 1 gpm capacity..................................... | 36 | 47 | 68 |
| Refrigerator.................................... | 2.6 | 3.1 | 4.5 |

## ELECTRICAL

### 4-14. Power Systems

The typical circuit arrangements of power systems found in buildings may be classified as follows: radial, secondary selective, secondary (spot) network, and primary selective.

The radial arrangement employs a single power source and one circuit to each load. An equipment failure will result in a power outage until difficulty is corrected. The high quality of modern distribution equipment provides the service reliability which justifies the use of the radial arrangement for a majority of applications.

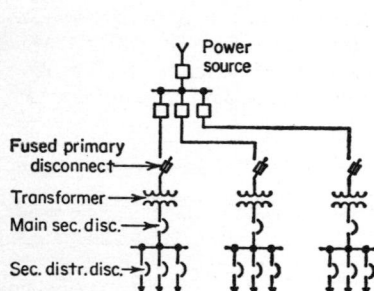

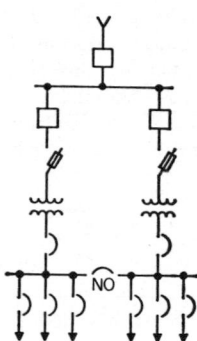

Fig. 4-21. Radial-circuit arrangement.

Fig. 4-22. Secondary-selective-circuit arrangement.

The secondary selective arrangement is in effect two radial systems with a secondary tie between them. It is provided in buildings where a greater degree of reliability is desired. This arrangement permits any secondary bus to be energized from either of two sources.

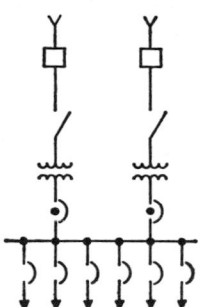

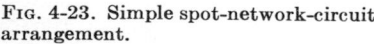

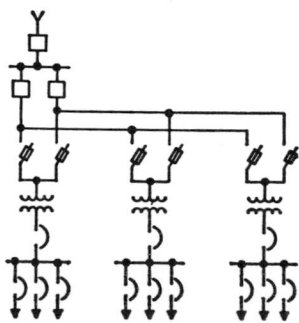

Fig. 4-23. Simple spot-network-circuit arrangement.

Fig. 4-24. Primary-selective-circuit arrangement.

The secondary-network arrangement is one where a high degree of service continuity is desired, as in large institutional buildings. The arrangement consists of two or more transformers energized by separate primary circuits, with the respective secondaries joined together.

The primary-selective arrangement provides an alternative power source to the substation transformers, but does not provide an alternative source of power to the secondary loads in event of a transformer outage.

The local prevailing rules of the Electric Service Company will usually determine the type and voltage of service available, regardless of the building size. This service may be from the secondary-network system in the street or for buildings of large magnitude, a spot network being instituted for the specific building load. The service voltage may be either 208Y/120 or 480Y/277 volts. In some of the current taller buildings, spot-network vaults are established by the Electric Service Company on intermediate floors in addition to the basement. In areas where buildings can be served at voltages greater than the utilization voltage, in the range of 2,400 to 13,800 volts, greater flexibility is available for circuit arrangements and in the selection and establishment of the utilization voltage.

## 4-15. System Voltages

Distribution systems may be classified according to voltages, levels used to carry the power directly to the branch circuits, or to load-center transformers or substations at which feeders to branch circuits originate.

The nominal system voltages listed in the left-hand columns of Table 4-69 are officially designated as standard nominal voltages in the United States by ANSI C84.1-1970. For the low-voltage systems the associated nominal system voltages in the right-hand column are obsolete and should not be used. For primary distribution voltage systems, the numbers in the right-hand column may designate an older system in which the voltage tolerance limits are maintained at a different level than the standard nominal system voltage.

Typical voltage-level utilization and application are outlined in Table 4-70.

## Table 4-69. Nominal System Voltages

| Standard nominal system voltages | Associated nominal system voltages |
|---|---|
| Low-voltage Systems | |
| 120 | 110, 115, 125 |
| 120/240* | 110/220, 115/230, 125/250 |
| 208Y/120* | 216Y/125 |
| 240/120* | |
| 240 | 230, 250 |
| 480Y/277* | 416Y/240, 460Y/265 |
| 480* | 440, 460 |
| 600 | 550, 575 |
| Primary Distribution Voltage Systems | |
| 2,400 | 2,200, 2,300 |
| 4,160Y/2,400 | |
| 4,160* | 4,000 |
| 4,800 | 4,600 |
| 6,900 | 6,600,   7,200 |
| 8,320Y/4,800 | |
| 12,000Y/6,930 | 11,000, 11,500 |
| 12,470Y/7,200* | |
| 13,200Y/7,620* | |
| 13,800Y/7,970 | |
| 13,800* | 14,400 |
| 20,780Y/12,000 | |
| 22,860Y/13,200 | |
| 23,000 | |
| 24,940Y/14,400* | |
| 34,500Y/19,920* | |
| 34,500 | 33,000 |

* Preferred standard nominal system voltages.
SOURCE: "IEEE Recommended Practice for Electric Power Systems in Commercial Buildings," IEEE Std. 241-1974, p. 56.

## Table 4-70. Voltage Levels—Utilization and Application

| Typical nominal voltage levels | Utilization | Application |
|---|---|---|
| 120/240............. | Light and power (light at 120 volts, power at 120 and 240 volts) | Small loads such as individual homes, multifamily dwellings, and small commercial occupancies |
| 208Y/120, 3 phase..... | Light and power (light at 120 volts, power at 120 volts and 208 volts, 1 phase, and 208 volts, 3 phase) | Commercial buildings and small industrial shops with limited electrical load |
| 240................. | Power | Commercial and industrial buildings |
| 480................. | Power | Commercial and industrial buildings with substantial motor loads |
| 600 | | |
| 480Y/277, 3 phase..... | Light and power (light at 277 volts, 1 phase, and power at 480 volts, 3 phase) | Commercial and industrial buildings |
| 2,400................ | Distribution | Industrial, heavy motor loads directly and lighting through transformation |
| 4,163................ | Distribution | Large-area, spread-out commercial institutional buildings such as shopping centers, schools, and motels; supply load centers and transformers for lighting and power |
| 4,800................ | Distribution | Industrial, with substations for stepping voltage to lower levels for lighting and power |
| 12,470Y/7,200........ 13,200Y/7,620 13,800 | Distribution | Large industrial plans with substations for stepping voltage to lower levels for lighting and power |

## 4-16. Building Loads

This subsection contains tables which permit the establishment of the anticipated electrical load for the building.  With the area and the knowledge of the building utilization, the building-load density can be formulated.  Total building load can be estimated by application of pertinent factors in Table 4-71.  Individual building-load densities are obtained by application of pertinent items and factors in Tables 4-72 to 4-76.  The demand load is obtained by application of items and factors contained in Table 4-77.

### Table 4-71. Load Density in Representative Plants and Buildings

| Type of industry | Light and power, volt-amp demand/ft² | Type of industry | Light and power, volt-amp demand/ft² |
|---|---|---|---|
| Commercial | | Manufacturing | |
| Bank.. | 6–8 | Appliance | 7–12 |
| Department store | 8–11 | Automotive | 7.5–12 |
| Hotel | 6–9 | Beet sugar refinery | 19 |
| Office building | 6–14 | Cigarette manufacture | 11 |
| Restaurant | 12–18 | Chemical | 10–15 |
| Small store | 5–8 | Electronics, industrial | 6–10 |
| Shopping center | 7–10 | Foundry* | 11–15 |
| School | 4–7 | Glass | 1.5–8.5 |
| | | Heavy machinery | 7–13 |
| | | Light machinery | 11–15 |
| | | Metal fabricating and assembly | 3–8 |
| | | Small device, industrial | 4.5–10 |
| | | Textile | 12 |

* Large electric furnace loads are not included.  They should be considered separately.
SOURCE: "Electrical Equipment Specifications Manual," Book III, Load Estimating Data Table 5.3, p. 2, General Electric Co., 1959.

## Table 4-72. Approximate Electric Load, in Watts per Square Foot, for Various Footcandle Levels*

| Foot-can-dle level | Coefficient of utilization | | | | | | | | | | | | | | | |
|---|---|---|---|---|---|---|---|---|---|---|---|---|---|---|---|---|
| | 0.20 | 0.24 | 0.28 | 0.32 | 0.36 | 0.40 | 0.44 | 0.48 | 0.52 | 0.56 | 0.60 | 0.64 | 0.68 | 0.72 | 0.76 | 0.80 |
| 10 | 1.4 | 1.2 | 1.0 | 0.9 | 0.8 | 0.7 | 0.6 | 0.6 | 0.5 | 0.5 | 0.5 | 0.4 | 0.4 | 0.4 | 0.4 | 0.4 |
| 20 | 2.9 | 2.4 | 2.0 | 1.8 | 1.6 | 1.4 | 1.3 | 1.2 | 1.1 | 1.0 | 0.9 | 0.9 | 0.8 | 0.8 | 0.7 | 0.7 |
| 30 | 4.3 | 3.6 | 3.1 | 2.7 | 2.5 | 2.3 | 2.0 | 1.8 | 1.7 | 1.5 | 1.4 | 1.3 | 1.3 | 1.2 | 1.1 | 1.1 |
| 50 | 7.2 | 6.0 | 5.1 | 4.5 | 4.0 | 3.6 | 3.2 | 3.0 | 2.8 | 2.6 | 2.4 | 2.2 | 2.1 | 2.0 | 1.9 | 1.8 |
| 80 | 11.4 | 9.5 | 8.2 | 7.2 | 6.4 | 5.7 | 5.2 | 4.8 | 4.4 | 4.1 | 3.8 | 3.6 | 3.4 | 3.2 | 3.0 | 2.9 |
| 100 | 14.3 | 11.9 | 10.2 | 8.9 | 8.0 | 7.2 | 6.5 | 6.0 | 5.5 | 5.1 | 4.8 | 4.5 | 4.2 | 4.0 | 3 8 | 3.6 |

NOTE: Apply correction factor of Table 4-65 for specific light source.
* 50 lm/W; 0.70 in light-loss factor.
SOURCE: "IEEE Recommended Practice for Electric Power Systems in Commercial Buildings," IEEE Std. 241-1974, p. 286.

## Table 4-73. Approximate Correction Factors for Some Typical Lamps

| Lamp type | Cool white and warm white | Deluxe cool white and warm white |
|---|---|---|
| Fluorescent lamps: | | |
| 40 W T12, 430 mA................ | 0.72 | 1.02 |
| 8-ft slimline, 430 mA............. | 0.70 | 1.02 |
| 8-ft high output, 800 mA.......... | 0.67 | 0.95 |
| 8-ft extra high output, 1.5 A........ | 0.71–0.78 | 1.02 |
| Incandescent lamps: | | |
| 100 W.......................... | | 2.86 |
| 150 W.......................... | | 2.74 |
| 200 W.......................... | | 2.60 |
| 300 W.......................... | | 2.50 |
| 500 W.......................... | | 2.33 |
| 750 W.......................... | | 2.25 |
| 1000 W......................... | | 2.14 |
| Mercury lamps, deluxe white: | | |
| 100 W.......................... | | 1.39 |
| 175 W.......................... | | 1.20 |
| 250 W.......................... | | 1.14 |
| 400 W.......................... | | 0.97 |
| 700 W.......................... | | 0.89 |
| 1000 W......................... | | 0.85 |
| Metal halide lamps: | | |
| 400 W.......................... | | 0.64 |
| High-pressure sodium lamps: | | |
| 400 W.......................... | | 0.48 |

NOTE: To find the correction factor for lamps not listed here, divide 50 by the lumens per watt of the lamp (including ballast losses). Lamp lumen data can be found in the catalogs of lamp manufacturers.
SOURCE: "IEEE Recommended Practice for Electric Power Systems in Commercial Buildings," IEEE Std. 241-1974, p. 286.

## Table 4-74. Air-conditioning Load Density

Based on 1.5 Kva/Ton, Air-cooled Units*

| *Application* | *Demand* | |
|---|---|---|
| Banks | 4.5–6 | va/ft² |
| Barber shops | 5–6 | va/ft² |
| Bars and taverns | 165–210 | va/seat |
| Beauty parlors | 750 | va/booth |
| | | |
| Department stores: | | |
|    Main floor | 7.5–10 | va/ft² |
|    Upper floors | 4.5–6 | va/ft² |
|    Top floors | 5–7.5 | va/ft² |
|    Bargain basement | 6–10 | va/ft² |
|    Normal basement | 4.5–6 | va/ft² |
| | | |
| Dress shops | 4.5–10 | va/ft² |
| Drugstores | 4.5–10 | va/ft² |
| Funeral parlors | 3.75–5 | va/ft² |
| Grocery stores | 3.75–5 | va/ft² |
| | | |
| Night clubs: | | |
|    Convention type | 190–210 | va/seat |
|    Week-end peak | 165–190 | va/seat |
| | | |
| Offices: | | |
|    Multistory | 3–3.75 | va/ft² |
|    Single floor | 3.75–4.5 | va/ft² |
|    Top floor | 5–6 | va/ft² |
| | | |
| Restaurants: | | |
|    Cafeterias | 165–210 | va/seat |
|    Hotel dining-rooms | 130 | va/seat |
|    Family restaurants | 125–150 | va/seat |
| Shoe shops | 4.5–9.0 | va/ft² |
| Supermarkets | 3.75–5.25 | va/ft² |
| | | |
| Theaters: | | |
|    Continuous performances | 82.5–100 | va/seat |
|    Neighborhood | 75–82.5 | va/seat |

* For water-cooled units multiply load by 0.75.

## Table 4-75. Air-conditioning Equipment

Kva Demand (Air-cooled)

| Type of drive | Tons | Btu | Equipment kva demand |
|---|---|---|---|
| Induction-motor drive | 1 | 12,000 | 1.5 |
| 0.8 PF synchronous-motor drive | 1 | 12,000 | 1.65 |
| 1.0 PF synchronous-motor drive | 1 | 12,000 | 1.3 |

SOURCE: "Electrical Equipment Specifications Manual," Book III, Load Estimating Data Tables 5.6 and 5.7, p. 3, General Electric Co., 1959.

### Table 4-76. Kva-demand Material-handling Loads

| Load | Kva demand* | Load | Kva demand* |
|---|---|---|---|
| Conveyors............... | 1–15 | Escalators.............. | 10–40 |
| Cranes: | | Hoists: | |
|   Gantry................. | 25–200 |   Ash and cinder........ | 1–5 |
|   Traveling bridge....... | 5–200 |   Tramrail 1-ton........ | 1.5–3 |
| Dumbwaiters............ | 1/2–5 |   Tramrail 5-ton........ | 6–10 |
| Elevators: | |   Warehouse loading.... | 1–3 |
|   1-ton freight........... | 3–20 | | |
|   5-ton freight........... | 7.5–20 | | |
|   10-passenger........... | 7.5–30 | | |
|   20-passenger........... | 7.5–50 | | |
|   27-passenger........... | 10–60 | | |

\* Demand depends upon rate of travel as well as size of load.
SOURCE: "Electrical Equipment Specifications Manual," Book III, Load Estimating Data Table 5.4, p. 2, General Electric Co., 1959.

### Table 4-77. Demand Factors of Utilization Equipment

| Equipment | Range, per unit | Equipment | Range, per unit |
|---|---|---|---|
| Arc furnaces.................... | 0.90–0.100 | Hand tools................... | 0.20–0.40 |
| Arc welders.................... | .20–.50 | Induction furnaces and heating | |
| Compressors................... | .20–.50 |   equipment.................... | .80–1.0 |
| Conveyors..................... | .90–.100 | Lighting...................... | .75–1.0 |
| Cranes........................ | | Paper mills................... | .50–.70 |
| | | Resistance ovens, heaters, | |
| Elevators (quantity) | |   furnaces.................... | .80–1.0 |
| | | Resistance welders.............. | .05–.400 |
|   1–2 | 1.0 | Rubber mills.................. | .50–.70 |
|   3 | 0.9 | Pumps........................ | .20–.50 |
|   4 | .775 | Rolling mills.................. | .20–.50 |
|   6 | .55 | Refineries.................... | .50–.70 |
|   10 | .48 | Textile mills.................. | .70–1.0 |
|   20 | .44 | Ventilation, blower motors....... | .20–.50 |

SOURCE: "Electrical Equipment Specifications Manual," Book III, Load Estimating Data Table 5.5, p. 2, General Electric Co., 1959.

## 4-17. Distribution

The purpose of any electric system is to provide a continuous supply of energy to the utilization equipment at reasonable cost. A typical power distribution system consisting of transformer, switchboard, motor control center, panelboards, feeders, lighting and power arrangements, and their relationship in an integrated system is outlined in Fig. 4-25.

Wire and cables in conduit and busways are used as feeders and capable of carrying large and small blocks of power from main switchboard to load centers to loads.

The feeder-conductor volt-drop limitations to building loads are outlined in Table 4-78.

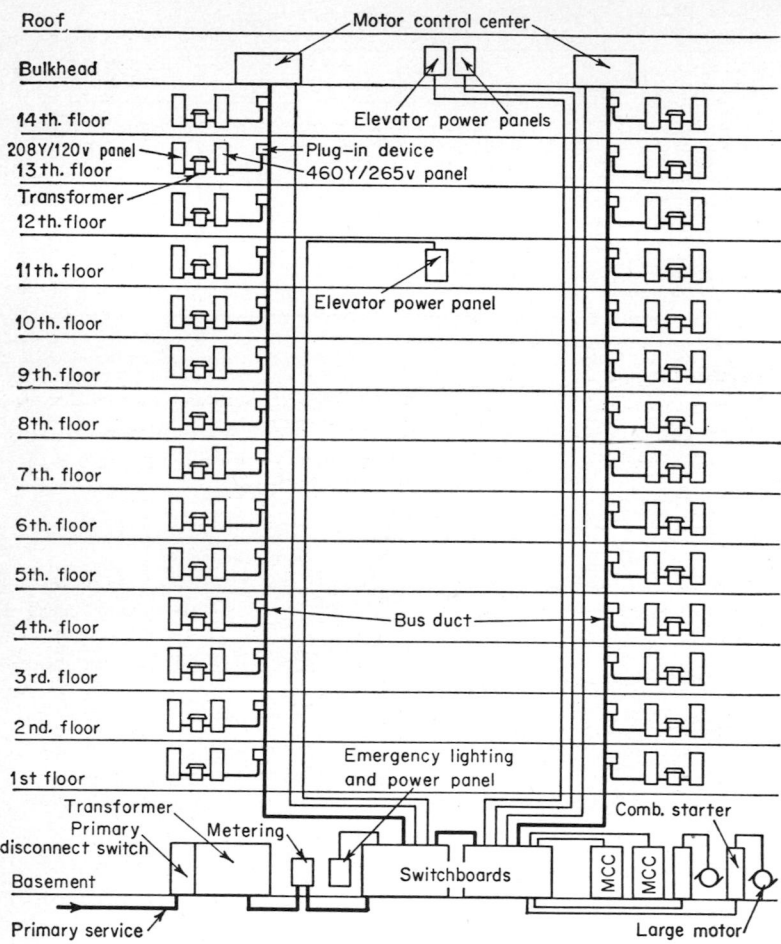

Fig. 4-25. Typical power-riser diagram.

## Table 4-78. Feeder-conductor Volt-drop Limitations

| Load | Limitations,* % |
|---|---|
| Power, heating or lighting or combination thereof.................. | 3 |
| Max. total drop for conductors for feeders and branch circuits...... | 5 |

\* Recommended by National Electrical Code 1975.

The IPCEA has published ampere ratings of cables insulated with oil-impregnated paper, varnished cambric, and rubber compounds. The National

Electrical Code publishes current ratings of low-voltage cables for most applications in commercial buildings. Wiring in insured buildings must be installed at rated values which do not exceed those in the NEC or other local codes which are more restrictive than IPCEA ratings.

Because of the versatility, flexibility, and economic feasibility for the method of electrical distribution in large commercial and institutional buildings, feeder and plug-in busways are being widely accepted. Various bus ducts are listed in Table 4-79.

## Table 4-79. Bus Ducts

| Type | Volt rating | Ampere rating | Conductor |
|------|-------------|---------------|-----------|
| Plug-in | 600 ac | 100 | Copper |
| Plug-in | 600 ac-dc | 225, 400, 600, 800, 1000 | Copper or aluminum |
| Plug-in or feeder, low impedance | 600 ac-dc | 600, 800, 1000, 1350, 1600, 2000, 2500, 3000, 4000, 5000 | Copper or aluminum |
| Plug-in or feeder, high frequency, 120 to 10,000 cycles/sec | 800 ac | 400, 500, 700 | Copper or aluminum |
| Feeder | 600 dc | 225, 400, 600, 800, 1000, 1350, 1600, 2000, 2500, 3000, 4000, 5000 | Copper or aluminum |
| Feeder, current limiting | 600 ac | 1000, 1350, 1600, 2000, 2500, 3000, 4000 | Copper |

Circuit breakers, fuses, safety switches, and combinations of these devices provide protection in a building distribution system against short circuits, overloads, and undervoltage by controlling the flow of current up to their respective rating. These devices may be contained in switchboards, panelboards, control centers, or in individual enclosures. Typical devices are listed in Table 4-80.

## Table 4-80. Ampere Rating of Standard Fuses. Safety Switches, Pressure Switches, and Circuit Breakers (600 Volts or Less)

### Single-element Fuses*

| | | | |
|---|---|---|---|
| 15 | 70 | 225 | 1,000 |
| 20 | 80 | 250 | 1,200 |
| 25 | 90 | 300 | 1,600 |
| 30 | 100 | 350 | 2,000 |
| 35 | 110 | 400 | 2,500 |
| 40 | 125 | 450 | 3,000 |
| 45 | 150 | 500 | 4,000 |
| 50 | 175 | 600 | 5,000 |
| 60 | 200 | 800 | 6,000 |

### Enclosed General-purpose Safety Switches

| | | | |
|---|---|---|---|
| 30 | 100 | 400 | 800 |
| 60 | 200 | 600 | 1,200 |

### Enclosed Pressure Switches

| | | | |
|---|---|---|---|
| 800 | 1,600 | 2,500 | 4,000 |
| 1,200 | 2,000 | 3,000 | 5,000 |

### Nonadjustable Trip Circuit Breakers

| | | | |
|---|---|---|---|
| 15 | 70 | 200 | 400 |
| 20 | 100 | 225 | 500 |
| 30 | 125 | 250 | 600 |
| 40 | 150 | 300 | 700 |
| 50 | 175 | 350 | 800 |

* Dual-element fuses that provide both motor-running protection and short-circuit protection are available in a much greater range of sizes.

Table 4-81 outlines a basis for selection of panelboards.

## Table 4-81. Panelboards

| Usage | Type | Max. circuits per panel | Remarks |
|---|---|---|---|
| Lighting........ | Switch and fuse or circuit breaker | 42* | Includes lighting and appliance panelboards † |
| Power.......... | Switch and fuse or circuit breaker | None | Physical size is limiting factor |

* Where more than 42 circuits originate at one location, use two panels. Not more than 42 overcurrent devices shall be installed in a lighting and/or lighting and appliance panelboard, or cabinet.
† Lighting and appliance panelboard is defined as having more than 10 per cent of its overcurrent devices rated 30 amp or less, for which neutral connections are provided.

## 4-18. Motors and Controls

The general requirements for motor provisions are outlined in the National Electrical Code (NEC). The design of motor installations must conform to the code requirements and should include considerations for adequacy, flexibility, voltage drop, and safety.

Motors can be classified as outlined in Table 4-83.

The full-load currents (amperes) of representative motors at typical voltages are outlined in Tables 4-82 and 4-84. These data are useful in determining the wiring and setting of protective devices.

## Table 4-82. Full-load Currents in Amperes Single-phase Alternating-current Motors

| Horsepower | 115V | 230V |
|:----------:|:----:|:----:|
| 1/6 | 4.4 | 2.2 |
| 1/4 | 5.8 | 2.9 |
| 1/3 | 7.2 | 3.6 |
| 1/2 | 9.8 | 4.9 |
| 3/4 | 13.8 | 6.9 |
| 1 | 16 | 8 |
| 1½ | 20 | 10 |
| 2 | 24 | 12 |
| 3 | 34 | 17 |
| 5 | 56 | 28 |
| 7½ | 80 | 40 |
| 10 | 100 | 50 |

The values of full-load currents are for motors running at usual speeds and motors with normal torque characteristics. Motors built for especially low speeds or high torques may have higher full-load currents, and multispeed motors will have full load current varying with speed, in which case the nameplate current rating shall be used.

To obtain full-load currents of 208- and 200-volt motors, increase corresponding 230-volt motor full-load currents by 10 and 15 percent, respectively.

The voltages listed are rated motor voltages. Corresponding nominal system voltages are 110 to 120 and 220 to 240.

Reproduced by permission from the 1975 National Electrical Code, copyright National Fire Protection Association, Boston, Mass.

## Table 4-83. Classification of Motors

| | Type | Speed characteristics | Full voltage | | Hp range | Application— see footnotes (a) to (e) |
|---|---|---|---|---|---|---|
| | | | Starting torque | Starting current | | |
| Constant-speed Drive | | | | | | |
| Polyphase a-c | Squirrel-cage general-purpose Design A | Constant | Normal 1-2.5 times$^f$ | High 6-8 times | All | (a) Fans and (c) centrifugal pumps and centrifugal compressors |
| | Squirrel-cage Design B | Constant | Normal 1-2.5 times$^f$ | Normal 5-6 times | Medium small | (a) Fans and centrifugal pumps and centrifugal compressors |
| | Squirrel-cage Design C | Constant | High 2-2.5 times$^f$ | Normal 5-6 times | Medium small | (b) Reciprocating pumps and compressors (e) started loaded |
| | Squirrel-cage Design F | Constant | Low 1.25 | Low 4 times | Medium large | Fans, centrifugal pumps, and compressors |
| | Wound rotor | Constant or variable | High 1-2.5 times (with secondary control) | Low 1-3 times (with secondary control) | All | (a) Hoists (b) reciprocating pumps and compressors (c) and frequent (e) or hard start |
| | Synchronous high speed | Exactly constant | Normal 0.75-1.75 times | Normal 5-7 times | Medium large | (a) Fans and centrifugal pumps and centrifugal compressors |
| | Synchronous low speed | Exactly constant | Low 0.3-0.4 times | Low 3-4 times | Medium large | (a) Reciprocating compressors starting unloaded |
| | Two-value capacitor | Constant | High | Normal | Small | (b) Pumps and compressors |
| | Permanent split capacitor | Constant | Low | Normal | Fractional | (a) Fans, blowers |

| | | | | | | |
|---|---|---|---|---|---|---|
| Single-phase a-c | Capacitor start | Constant | Moderate | Normal | Small fractional | (a) Fans and pumps |
| | Repulsion induction | Constant | High | Normal | Medium small | (a) Fans<br>(b) pumps and compressors |
| | Split phase | Constant and adjustable | Normal | Normal | Fractional | (a) Fans<br>(b) pumps and compressors<br>(d) fans—direct |
| | Adjustable-speed Drive | | | | | |
| Polyphase a-c | Squirrel-cage high slip; transformer adjustment | Variable | Normal | Normal | Medium small | (a) Fans |
| | Squirrel-cage separate winding or regrouped poles | Constant multispeed | Normal or high | Normal or low | All | (a) Fans,<br>(b) pumps, and<br>(c) compressors |
| | Wound rotor | Variable | High (with secondary control) | Low (with secondary control) | All | (a) Fans<br>(b) centrifugal pumps and compressors |
| Single-phase a-c | Repulsion | Variable | High | Normal | Low and fractional | (a) Fans, centrifugal pumps<br>(b) compressors |
| | Capacitor low-torque tapped winding | Variable two-speed | Low | Normal | Fractional | (d) Fans, direct |
| | Capacitor low-torque transformer adjustment | Variable | Low | Low | Fractional | (d) Fans |
| | Split-phase regrouped poles | Constant | Normal | Normal | Fractional | (d) Fans |

a Drives having medium or low starting torque and inertia $WR^2$ such as fans and centrifugal pumps or reciprocating pumps and compressors started unloaded.
b Drives having high starting torques, such as reciprocating pumps and compressors starting loaded.
c Similar to (a) except where frequent or hard starting (large $WR^2$) requires a higher starting and accelerating torque.
d Fans direct-connected.
e Stoker drives.
f Torque depends on hp rating and speed. See *NEMA Standard* MG1-4.10, Motors and Generators.
SOURCE: "Heating, Ventilating, Air Conditioning Guide," vol. 37, pp. 642 and 643, ASHRAE, New York, 1959.

### Table 4-84. Full-load-current* Three-phase A-C Motors

| Hp | Induction-type squirrel-cage and wound-rotor, A | | | | | Synchronous-type unity power-factor,† A | | | |
|---|---|---|---|---|---|---|---|---|---|
|    | 115V | 230V | 460V | 575V | 2,300V | 220V | 440V | 550V | 2,300V |
| ½  | 4    | 2    | 1    | 0.3  |      |      |      |      |      |
| ¾  | 5.6  | 2.8  | 1.4  | 1.1  |      |      |      |      |      |
| 1  | 7.2  | 3.6  | 1.8  | 1.4  |      |      |      |      |      |
| 1½ | 10.4 | 5.2  | 2.6  | 2.1  |      |      |      |      |      |
| 2  | 13.6 | 6.8  | 3.4  | 2.7  |      |      |      |      |      |
| 3  | .... | 9.6  | 4.8  | 3.9  |      |      |      |      |      |
| 5  | .... | 15.2 | 7.6  | 6.1  |      |      |      |      |      |
| 7½ | .... | 22   | 11   | 9    |      |      |      |      |      |
| 10 | .... | 28   | 14   | 11   |      |      |      |      |      |
| 15 | .... | 42   | 21   | 17   |      |      |      |      |      |
| 20 | .... | 54   | 27   | 22   |      |      |      |      |      |
| 25 | .... | 68   | 34   | 27   | ...  | 54   | 27   | 22   |      |
| 30 | .... | 80   | 40   | 32   | ...  | 65   | 33   | 26   |      |
| 40 | .... | 104  | 52   | 41   | ...  | 86   | 43   | 35   |      |
| 50 | .... | 130  | 65   | 52   | ...  | 108  | 54   | 44   |      |
| 60 | .... | 154  | 77   | 62   | 16   | 128  | 64   | 51   | 12   |
| 75 | .... | 192  | 96   | 77   | 20   | 161  | 81   | 65   | 15   |
| 100| .... | 248  | 124  | 99   | 26   | 211  | 106  | 85   | 20   |
| 125| .... | 312  | 156  | 125  | 31   | 264  | 132  | 106  | 25   |
| 150| .... | 360  | 180  | 144  | 37   | ...  | 158  | 127  | 30   |
| 200| .... | 480  | 240  | 192  | 49   | ...  | 210  | 168  | 40   |

For full-load currents of 208- and 200-volt motors, increase the corresponding 230-volt motor full-load current by 10 and 15 per cent, respectively.

* These values of full-load current are for motors running at speeds usual for belted motors and motors with normal torque characteristics. Motors built for especially low speeds or high torques may require more running current, and multispeed motors will have full-load current varying with speed, in which case the nameplate current rating shall be used.

† For 90 and 80 per cent power factor the above figures shall be multiplied by 1.1 and 1.25 respectively. The voltages listed are rated motor voltages. Corresponding nominal system voltages are 110 to 120, 220 to 240, 440 to 480 and 550 to 600 volts.

Reproduced by permission from the 1975 National Electrical Code, copyright National Fire Protection Association, Boston, Mass.

## 4-19. Telephones

Well-planned communication facilities for both present and future needs incorporated into buildings during initial construction or major alterations will be beneficial throughout the life of the building.

Modern buildings may require teletypewriter service, data-transmission service—connections between data-processing machines over telephone facilities, centrex service—permitting the dialing of outgoing calls as well as receiving incoming calls without attendant, and public telephone service.

Communication needs of the building include:

**Raceway.** Underfloor ducts or cellular floor systems to serve as a telephone cable distribution facility.

**Apparatus Closets.** To house relay cabinets and auxiliary apparatus of modern key telephone systems.

**PBX Equipment Rooms.** To house the large equipment required for PBX service.

**Cable Riser Systems.** For bringing cables from the main terminal room to the various building floors.

**Main Terminal Room.** Connecting point between building and outside facilities.

**Service from the Street.** Aerial or underground service into the building.

A desirable distribution arrangement for commercial buildings is to divide the floor space into zones of no more than 10,000 ft², and preferably from 4,000 to 6,000 ft², to handle the distribution cables. The size of the raceway in each zone should be one square inch for every 100 ft² of office area, which is predicated upon the average allocation of one desk and telephone per 100 ft² of floor area. The raceway for the distribution cables can be provided as follows:

**Underfloor Ducts.** Spaced $4\frac{1}{2}$ to 6 ft between parallel runs or feeds with cross runs and junction boxes located every 40 ft or less.

**Cellular Floor.** Appropriate cell utilization with header ducts connected to cell area at intervals for maximum coverage, usually no greater than 50 ft.

The communications equipment in each zone is connected to relay cabinets and other apparatus in a central closet in each zone. The type and size closet is outlined in Table 4-85.

## Table 4-85. Telephone Zone Closets

| Specification | Walk-in closet | Shallow closet |
|---|---|---|
| Depth: | | |
|    Minimum................... | 3 ft | 1½ ft |
|    Maximum................... | None | 2½ ft |
| Width: | | |
|    Minimum................... | 5 ft | 3 ft |
|    Maximum................... | None | None |
| Floor area per 100 ft² served...... | 4 ft² | None |
| Length of walls per 1,000 ft² served | 2½ ft | 2½ ft |
| Minimum height of doors........ | 6 ft 8 in. | 6 ft 8 in.* |
| Minimum width of doors........ | 3 ft | 3 ft† |

* When shallow closets are used, the center post between double doors should be eliminated, if possible.
† Minimum for single door, 2½ ft for double doors.
SOURCE: Bell Telephone System, Telephone-planning Fact File, AIA File 31-i-5.

## 4-20. Signal and Communications Systems

Different types of buildings require a variety of signal and communications systems.

**Industrial Buildings.** *Burglar Alarm.* Burglar alarm system may be used to protect all doors, windows, elevator openings, skylights, etc.

*Clock and Program Systems.* These are used for indicating the time of day and operating signal devices such as bells or horns at predetermined times, such as starting and stopping work, rest periods, lunch periods, etc.

**Door Alarm.**  Door alarm system is used to signal the guard room when certain restricted areas have been entered or vacated by individuals.

**Fire Alarm.**  Fire alarm systems should be of the closed-circuit supervised type.  Generally, noncoded systems are limited to small plants, since they only transmit a general alarm and do not indicate the location of the operated station.  Coded systems are preferable.

**Fire Detection.**  Automatic fire detection system may be used separately or combined with the manual fire alarm systems.

**Intercom.**  Intercommunicating system may be provided in various forms.

**P.A.**  Public address sound system may be used throughout the plant for paging, radio programs, recordings, announcements, and entertainment.

**Paging.**  Paging system is used to call and locate individuals.

**Smoke Detection.**  This system is used to detect smoke in ventilating, air-conditioning, and dust-collecting ducts.

**Sprinkler System.**  This alarm system is used to signal when sprinkler heads open, when noticeable leaks occur, when water flow valves operate in either dry or wet systems, when post indicator valves operate or are left open, or when the shutoff valves are placed in any subnormal position.

**Watchman's System.**  Watchman's supervisory system should be of a type which will require the watchmen on the various tours to produce a record in the superintendent's or chief guard's quarters at the start and at the finish of each tour.

**Commercial Buildings.  Fire Systems.**  Fire alarm system of the manual type should be provided for the protection of the general public and the employees within the building.

A fire alarm system of the automatic type should be provided where records or files are kept or stored.  A fire-line signal system is used exclusively by members of the fire department to transmit signals to the pump room.

A fire-line telephone system consists of a master station telephone located in the pump room, submaster station telephones located in the auxiliary pump room and at the building entrance, and outlying telephones located on each of the other floors.  This system is of the common-talking type.

**Schools.  Clock and Program System.**  Clock and program system provides the means of showing correct time throughout the premises and to denote the different periods in a day's schedule.

**Fire Alarms.**  Any fire signal should be distinctive from all other signals and should be audible to everyone in the building.

A fire alarm system for use in schools is one of four types, having a common characteristic: they are all closed-circuit, electrically supervised.

In small schools, either the noncoded or master-coded type is frequently used.

In large schools and colleges, the coded types of system are used.

**Intercom System.**  Telephones are used for intercommunication between the principal's office and the main office and the classrooms.

**Sound and Radio Distribution System.**  A sound and radio distribution system enables the distribution of radio programs, recordings, lectures, and announcements.

**Hospitals.** *Clock System.* A clock system is important in hospitals, both for keeping time and for administering anesthesia.

*Emergency Call.* An emergency feature may be added to any nurse-call system. This is used by the nurse to call assistance to a patient's room when the occasion requires.

*Fire Alarm System.* A fire alarm system for use in hospitals is usually of the General Alarm type.

*Nurse Call.* A nurse-call system is used by patients to call a nurse to the bedside. There are two general types of such systems, the visual and the audio.

*Paging Systems.* Paging systems are used to locate doctors and other members of the staff throughout the building. The visual system uses lamp annunciators throughout and also incorporates an audible signal such as a buzzer or chime.

The sound system consists of loudspeakers throughout one or more hospital buildings.

"In" and "out" systems are used by the doctors and other members of the staff to designate whether or not they are in the building.

*Sound and Radio System.* Sound and radio systems enable patients to listen in on one or more channels of radio programs, TV, recordings, announcements, etc.

## 4-21. Grounding

The purpose of grounding is to provide protection of personnel, equipment, and circuits by eliminating the possibility of dangerous or excessive voltage. The considerations of grounding may be subdivided as follows:

**System Grounding.** In an electric power distribution system system grounding is concerned with the nature and location of an intentional electric interconnection between the electric system conductors and ground (earth). Specific methods of power system grounding include the underground system, solid neutral grounding, low-resistance grounding, high-resistance grounding, high-resistance grounding with traceable signal at fault, corner-of-the-delta grounding, mid-phase grounding and low-reactance grounding.

From a practical standpoint, two methods of low-voltage system grounding, namely the solidly grounded neutral and high-resistance grounded neutral methods, will best fulfill the majority of application requirements. Principal characteristics of major methods of grounding low-voltage systems are outlined in Table 4-86.

**Equipment Grounding.** This designation pertains to that system of electric conductors by which all metallic structures, through which energized conductors run, will be interconnected. The equipment grounding ensures freedom from electric shock to personnel by maintaining a low potential difference between nearby metallic members and provides an adequate and effective electric conductor over which short-circuit currents involving ground can flow without sparking or other thermal distress so as to avoid a fire hazard to combustible material.

**Static Grounding.** Static grounding is concerned with the connection to ground of static accumulation on equipment, materials being handled, or even on operating personnel in a manner to eliminate a potentially hazardous operating condition, where the discharge of a static charge to ground or other equipment in the presence of flammable or explosive materials can cause fires and explosions.

**Lightning Protection Gounding.** The conduction to earth of current discharges in the atmosphere originating from electric charges in cloud formations is lightning protection grounding. The function of the lightning grounding system is to convey the lightning discharge currents safely to earth without incurring damaging potential differences across electrical insulation in the power system, without overheating the lighting grounding conductors, and without the disrupting breakdown of air between the lightning rod conductors and other metallic members of the structure.

**Connection to Earth.** This is one of the most important parts of the grounding system and should be one of low resistance. For large substations, the earth resistance should not exceed 1 ohm. For smaller substations and industrial plants, an earth resistance of less than 5 ohms should be obtained if practicable. The National Electrical Code (1975) states that the maximum resistance shall not exceed 25 ohms. Ground electrodes may be underground metallic piping systems, metal building, framework, well casings, steel piling, and other underground metal structures installed for purposes other than grounding, or they may be made electrodes specifically designed for grounding purposes. Made electrodes may be driven rods, buried strips or cables, grids, buried plates, and counterpoises.

## Table 4-86. Principal Characteristics of Major Methods of Grounding Low-voltage Systems

| System property | Type of system grounding | | |
| --- | --- | --- | --- |
| | Solid* | High resistance† | Ungrounded |
| Immediate shutdown of faulty circuit on occurrence of first ground fault | Yes | No | No |
| Control of transient overvoltages due to arcing ground faults | Yes | Yes | No |
| Control of impressed or self-generated steady-state overvoltages | Yes | No | No |
| Flash hazard to personnel during ground fault (no escalation of fault) | Severe | Essentially zero | Essentially zero |
| Arcing fault damage to equipment during ground fault (no escalation) | May be severe unless fault is promptly removed | Usually minor unless fault removal is so prolonged as to cause fault escalation | Usually minor but transient over-voltages may cause fault escalation or multiple insulation failures |
| Shock hazard, unfaulted phases to ground, during ground fault | Line-to-neutral voltage | Approximately line-to-line voltage | May be several times line-to-neutral voltage |
| Shock hazard, equipment frame to ground during solid internal line-to-ground fault | Moderate | Minimum | Small |
| Detection of arcing faults | L-L or L-G arcing faults readily detected, esp. with ground fault relaying | Ground detectors and fault locating equipment required for L-G arcing faults. L-L faults readily detected by phase overcurrent devices unless fault current is severely limited | Ground detectors and fault locating equipment required for L-G arcing faults. Transient over-voltages may meanwhile cause additional insulation breakdowns. L-L faults readily detected by phase overcurrent devices unless fault current is severely limited |
| Suitable for four-wire, three-phase service | Yes | No | No |

\* For optimum results, use of solid grounding method should include sensitive ground fault relaying.
† For optimum results, use of high-resistance grounding method should include equipment and procedures for alarming, tracing, and removing the ground fault promptly.
SOURCE: General Electric Electrical Equipment Specifiers' Guide, 1972 11-AP-3, p. 37.

SECTION 5

# CHEMICAL ENGINEERING

**Robert E. White, D.Ch.E.;** Professor and Chairman, Department of Chemical Engineering, Villanova University; Fellow, American Institute of Chemical Engineers; Member, American Society for Engineering Education

**C. Michael Kelly, Ph.D.;** Associate Professor, Air Products and Chemicals Professor of Chemical Engineering, Department of Chemical Engineering, Villanova University; Member, American Institute of Chemical Engineers, American Society for Engineering Education

## CONTENTS

# INTRODUCTION

**Basic Dimensions.** Basic dimensions of parameters are given in terms of the following symbols, and may be found with various compatible combinations of the units shown.

$$L = \text{length, ft, cm}$$
$$\theta = \text{time, sec, hr}$$
$$T = \text{temperature, }^\circ\text{R or K}$$
$$M = \text{mass, lb mass, grams, tons}$$
$$\text{moles} = \text{mass/molecular weight, gram-moles, lb-moles}$$
$$F = \text{force, lb force, dynes}$$
$$\text{conc.} = \text{concentration, moles per unit volume}$$
$$E = \text{energy, cal, Btu, joules}$$

## Nomenclature

$A$ = filter area, $L^2$

    = collector plate area, $L^2$

    = heat-transfer surface area, $L^2$

    = membrane area, $L^2$

    = chemical reactant species

    = frequency factor, $\text{moles}/(\theta \cdot \text{conc.}^{n-1})$

$a$ = surface area of packing per unit volume of contactor, $L^2/L^3$

    = stoichiometric coefficient for reactant $A$

$B$ = chemical reactant species

$b$ = stoichiometric coefficient for reactant $B$

$C$ = chemical reactant species

    = concentration, when subscripted, $\text{moles}/L^3$

$C_D$ = drag coefficient

$C_F$ = heat capacity of feed, $E/(M \cdot T)$

$c$ = stoichiometric coefficient of reactant $C$

    = controlled variable

$D$ = distillate product rate, $\text{moles}/\theta$

    = chemical reactant species

$D_a$ = impeller diameter, $L$

$D_p$ = particle diameter, $L$

    = diameter of ion-exchange resin bead, $L$

$D_i$ = fluid-phase diffusion coefficient, $L^2/\theta$

$d$ = indicates differentiation

$d_F$ = packing characteristic length, $L$

$E$ = energy of activation, $E$

    = extract flow rate, $M/\theta$

$E_c$ = column plate efficiency

$e$ = error signal

exp $=$ exponential function (natural antilogarithm)

$F$ $=$ feed flow rate, moles/$\theta$, $M/\theta$, $L^3/\theta$

$f$ $=$ amount of vapor introduced to flash still or feed tray in rectifying column, moles/mole

$=$ collected feed stream variables in process control

$G$ $=$ gas volumetric flow rate, $L^3/\theta$

$=$ rate of solute-free solvent removed in solvent separator, $M/\theta$

$G_M$ $=$ gas molar flow rate, moles/$\theta$

$g$ $=$ gravitational acceleration, $L/\theta^2$

$g_c$ $=$ gravitational constant, $LM/(\theta^2 F)$

$H$ $=$ velocity head of agitated fluid, $E/M$

$H_{OG}$ $=$ height of transfer unit (HTU) based on over-all (gas-phase) driving force, $L$

$H_{OL}$ $=$ height of transfer unit (HTU) based on over-all (liquid-phase) driving force, $L$

$K'$ $=$ distribution coefficient for liquid-liquid equilibrium

$K_e$ $=$ equilibrium constant

$=$ optimal controller gain

$K_G$ $=$ over-all gas-phase mass-transfer coefficient, moles/$(L^2\theta$ conc.$)$

$K_L$ $=$ over-all liquid-phase mass-transfer coefficient, moles/$(\theta L^2$ conc.$)$

$K_u$ $=$ ultimate controller gain

$k$ $=$ reaction rate constant, forward reaction

$k'$ $=$ reaction rate constant, reverse reaction

$k_G$ $=$ gas-phase mass-transfer coefficient, moles/$(\theta L^2$ conc.$)$

$k_L$ $=$ liquid-phase mass-transfer coefficient, moles/$(\theta L^2$ conc.$)$

$L$ $=$ membrane thickness, $L$

$=$ liquid molar flow rate in distillation column, moles/$\theta$

$L_M$ $=$ liquid molar flow rate, moles/$\theta$

$L_e$ $=$ extract-layer flow rate from last stage, $M/\theta$

$L_r$ $=$ raffinate-layer flow rate from last stage, $M/\theta$

$l$ $=$ nominal pore length in catalyst particle (radius of sphere or cylinder, half-thickness of slab), $L$

$m$ $=$ controller output signal

$=$ slope of vapor-liquid equilibrium line

$N$ $=$ rotation speed of impeller, rpm

$=$ number of equilibrium stages

$=$ number of moles of subscripted species

$N_{OG}$ $=$ number of transfer units, based on over-all (gas-phase) driving force

$N_{OL}$ $=$ number of transfer units, based on over-all (liquid-phase) driving force

$N_M$ $=$ molar mass-transfer rate, moles/$\theta$

$N_P$ $=$ impeller power number

$N_Q$ $=$ impeller discharge coefficient

$N_T$ $=$ number of theoretical stages at total reflux

$N_{\text{act}}$ $=$ actual number of stages in column

$N_{\text{theor}}$ = number of equilibrium stages (theoretical stages) in column

$n$ = order of chemical reaction

$p$ = permeability of menbrane

$P$ = pressure of liquid, $F/L^2$

    = agitator shaft power, $E/\theta$

    = chemical reactant species

    = vapor pressure of subscripted component, $F/L^2$

    = number of moles in still-pot (batch distillation)

$P_u$ = ultimate period, $\theta$

$p$ = partial pressure of subscripted component, $F/L^2$

    = stoichiometric coefficient of reactant $P$

$Q$ = impeller discharge rate, $L^3/\theta$

    = rate of feed of solute-free solvent to raffinate end of extractor, $M/\theta$

    = chemical reactant species

$q$ = rate of heat transfer, $E/\theta$

    = stoichiometric coefficient of reactant $Q$

$R$ = ideal gas constant, $E/(T \cdot \text{moles})$

    = raffinate-layer flow rate, $M/\theta$

    = reflux ratio

$r$ = specific reaction rate, or specific reaction rate of subscripted species (forward reaction), $\text{moles}/(\theta L^3)$

$r'$ = specific reaction rate (reverse reaction), $\text{moles}/(\theta L^3)$

$\bar{r}$ = average pore radius, $L$

$r_{\text{net}}$ = net forward rate of a reversible reaction, $\text{moles}/(\theta L^3)$

$S$ = amount of material distilled in Rayleigh distillation, moles

    = solvent content of extract layer, $M_{\text{solvent}}/M_{\text{dissolved material}}$

$s$ = controller set point

    = empirical constant for filter-cake compressibility

    = average of mutual solubilities of solute-free contacted liquids

    = solvent content of raffinate layer, $M_{\text{solvent}}/M_{\text{dissolved solids}}$

$T$ = absolute temperature, °R or K

$t$ = temperature, °C or °F

$U$ = over-all heat transfer coefficient, $E/(L^2\theta T)$

$U_i$ = dialysis coefficient in subscripted phase, $\text{moles}/(L^2\theta)$ conc.

$V$ = total volume upstream of a given point in a plug flow reactor, $L^3$

    = volume of contactor or reactor, $L^3$

    = molar vapor rate in column, $M/\theta$

$V_C$ = superficial velocity of continuous phase, $L^3/(L^2\theta)$

$V_D$ = superficial velocity of dispersed phase, $L^3/(L^2\theta)$

$V_e$ = extract-layer flow rate from last stage, $M/\theta$

$V_r$ = raffinate-layer flow rate from last stage, $M/\theta$

$v$ = total volume of feed to ion-exchange process, $L^3$

$v_t$ = terminal settling velocity, $L/\theta$

$W$ = mass of solids retained in filter cake, $M$

    = bottom product rate, $\text{moles}/\theta$

$w$ = equilibrium composition of raffinate phase, $M_{\text{solute}}/M_{\text{solute-free phase}}$

$w'$ = equilibrium composition of extract phase, $M_{\text{solute}}/M_{\text{solute-free phase}}$

$X$ = mass fraction solute in raffinate, solvent-free basis

= moles of subscripted component converted per mole of feed to continuous reactor

$x$ = mole fraction (of subscripted component) in liquid phase

$x^*$ = mole fraction in liquid phase which would be in equilibrium with actual gas phase at that point in contactor.

$Y$ = mass fraction of solute in extract, solvent-free basis

$y$ = mole fraction (of subscripted component) in vapor or gas phase

$y^*$ = mole fraction in vapor phase which would be in equilibrium with actual liquid phase at that point in contactor

$Z$ = total column height, $L$

$Z_t$ = tray spacing, $L$

$Z_f$ = solvent content of feed, $M_{\text{solvent}}/M_{\text{solvent-free stream}}$

$\alpha, \alpha_{12}$ = relative volatility

$\alpha$ = reaction order with respect to reactant $A$

= specific cake resistance, $L/M$

= empirical constant for wet scrubber efficiencies

$\alpha'$ = empirical constant for filter cake resistance

$\beta$ = reaction order with respect to reactant $B$

$\gamma$ = activity coefficient

= empirical constant for wet scrubber efficiencies

= reaction order with respect to reactant $C$

$\Delta$ = difference or change

$\Delta P_c$ = pressure drop across filter cake, $F/L^2$

$\theta$ = time

$\theta_{1/2}$ = reaction half-life

$\lambda_m$ = molar latent heat of vaporization, $E/\text{moles}$

$\lambda$ = latent heat of vaporization, $E/M$

$\mu$ = viscosity $M/(L\theta)$

$\Pi$ = total pressure, $F/L^2$

$\rho$ = density, $M/L^3$

= reaction order with respect to reactant $P$

$\rho_f$ = density of fluid, $M/L^3$

$\rho_p$ = density of particle, $M/L^3$

$\sigma$ = reaction order with respect to reactant $Q$

$\sigma$ = interfacial tension, lb mass/hr$^2$

$\sigma'$ = interfacial tension, dynes/cm

$\phi(\ )$ = arbitrary function of $(\ )$

$\psi(\ )$ = arbitrary function of $(\ )$

$\tau_I$ = reset time, $\theta$

$\tau_D$ = rate action derivative time, $\theta$

### Subscripts

$A$ = reactant $A$

$B$ = reactant $B$

$C$ = reactant $C$
    = continuous phase
$D$ = disperse phase
$d$ = distillate product
$F$ = feed stream
$i$ = component $i$
    = at interface
$L_1$ = liquid phase 1
$L_2$ = liquid phase 2
$lm$ = logarithmic mean
$M$ = molar property
$m$ = molar property
$O$ = initial value
$p$ = in still pot
$P$ = reactant $P$
$Q$ = reactant $Q$
$w$ = bottom product
in = entering
out = leaving
1 = component 1
    = at position 1 (upstream) or initial time
2 = component 2
    = at position 2 (downstream) or final time

## DIFFUSIONAL OPERATIONS

### 5-1. Mass-transfer Fundamentals

The transfer of material from one phase to another is a primary means of separating multicomponent solutions.  In general, two equilibrium phases of a multicomponent mixture will have different chemical compositions, and this difference offers a means for separating a mixture into its individual components.  Repetitive phase changes can provide increasingly pure solutions, and in the limiting case can produce pure individual components. Analysis of these phase-change separation processes depends on three factors: thermodynamic equilibrium, mass-transfer rates, and pattern of contact between phases.

**Equilibria.**  When two phases are brought into contact, their temperatures, pressures, and compositions will adjust until thermodynamic equilibrium is reached.  The compositions of the phases are functions of temperature, pressure, and over-all composition.  The number of such variables which may be specified independently is fixed according to the Gibbs phase rule of thermodynamics.  Data relating equilibrium phase compositions are ultimately dependent on experimental measurement.  Phase equilibrium data for many systems have been tabulated in the literature (e.g., "International Critical Tables," "Chemical Engineers' Handbook," *Journal of Chemical and Engineering Data*).  Estimating techniques are available for predicting phase equilibria, as are equations correlating experimental data. Since proper design is dependent on the accuracy of the equilibrium relationships used, the use of accurate experimental data is preferable.

**Mass-transfer Rates.** The amount of contacting required to bring two phases into equilibrium is dependent on the rate of mass transfer. The rate at which mass is transferred between phases is controlled by the *driving force* for mass transfer, the *resistance* to mass transfer and the *interfacial area* between phases, according to

$$N_{M_i} = K_L a V (x_i - x_i{}^*) = K_G a V (y_i - y_i{}^*) \tag{5-1}$$

The over-all mass-transfer coefficients are dependent on resistance to mass transfer in interfacial films (in a manner analogous to film resistances in convective heat transfer), which depend on molecular parameters, fluid turbulence near the interface, and the equilibrium relationship between phases. For equilibrium relationships which are essentially the straight lines $y = mx$, mass-transfer coefficients are given by

$$\frac{1}{K_G} = \frac{1}{k_G} + \frac{m}{k_L} \tag{5-2}$$

$$\frac{1}{K_L} = \frac{1}{k_L} + \frac{1}{mk_G} \tag{5-3}$$

The coefficients are ordinarily determined experimentally as volumetric coefficients $K_G a$ and $K_L a$, since interfacial areas are difficult to determine.

**Continuous Contacting.** Continuous contact processes may be run with cocurrent, crosscurrent, or countercurrent flow patterns. In terms of efficient mass-transfer countercurrent contacting, which allows the greatest amount of mass transfer between phases for a given initial composition difference between feed streams, is preferred. Other flow patterns are generally used only in special cases where countercurrent flow is impractical.

The *transfer unit* is a standard degree of separation used to describe the performance of a contacting device. The greater the number of transfer units, the more thorough the separation. The number of transfer units required to accomplish a specified degree of separation between components is dependent on the product compositions and relative flow rates of the two phases, and may be based on either phase, according to

$$N_{OG} = \int_{y_2}^{y_1} \frac{(1 - y)_{lm}}{(1 - y)(y - y^*)} \, dy \cong \int_{y_2}^{y_1} \frac{dy}{y - y^*} \tag{5-4}$$

$$N_{OL} = \int_{x_1}^{x_2} \frac{(1 - x)_{lm}}{(1 - x)(x - x^*)} \, dx \cong \int_{x_2}^{x_1} \frac{dx}{x - x^*} \tag{5-5}$$

The approximation holds for dilute solutions, where transfer of material does not change the over-all stream molar flow rates, and is a true equality for equimolar counterdiffusion.

The *height of a transfer unit* (*HTU*) is an indication of the amount of contacting required to accomplish the standard separation of one transfer unit. It is dependent on mass-transfer coefficients, packing type and specific surface area, flow patterns of the contacting phases, and flow rates. The heights of the over-all transfer units defined by Eqs. (5-4) and (5-5) are

$$H_{OG} = \frac{G_M}{K_G a (1 - y)_{lm}} \tag{5-6}$$

$$H_{OL} = \frac{L_M}{K_L a (1 - x)_{lm}} \tag{5-7}$$

The height of a transfer unit is often much more constant than the mass-transfer coefficient for a given column and packing at various flow rates and is therefore more commonly used. HTU's are determined empirically by measuring mass transfer in a given column, calculating $N_{OG}$ or $N_{OL}$ by Eq. (5-4) or (5-5) and dividing into column height, that is,

$$H_{OG} = \frac{Z}{N_{OG}} \tag{5-8}$$

$$H_{OL} = \frac{Z}{N_{OL}} \tag{5-9}$$

In design applications, the number of transfer units required for a specified separation is calculated by Eq. (5-4) or (5-5). This is combined with HTU data for similar (or pilot plant) operations to obtain total column height.

Nomenclature for Eqs. (5-1) through (5-9) corresponds to that used for gas absorption. The equivalent equations for other continuous contacting operations have slightly different forms appropriate to the parameters commonly encountered. A thorough treatment of mass transfer is found in Treybal's comprehensive text (Treybal, Robert E., "Mass Transfer Operations," 2d ed., McGraw-Hill Book Company, New York, 1968).

**Staged Operations.** In staged contacting two phases initially not in equilibrium are held in contact for a length of time assumed sufficient to attain equilibrium. The two phases are then separated and each is fed to an adjacent stage where it is again held in contact with a nonequilibrium mixture of the opposite phase. The two phases flow from stage to stage in opposite directions. As one phase advances through the contactor it becomes progressively more concentrated in a particular component or group of components. As the other phase advances in the opposite direction it becomes progressively less concentrated in that same component or group of components.

Design of staged contacting systems (or performance analysis of existing systems) requires repetitive computation using equations based on three basic principles: material balances on the various independent components, equilibrium relationships between streams, and degree of approach to equilibrium (efficiency) of each contacting stage. The approach to equilibrium depends on the same factors as control mass-transfer rates in continuous contacting.  For practical calculations, efficiencies are always estimated on the basis of experience with similar contacting equipment and multiphase systems.

## 5-2. Distillation

**Definitions.**  *Simple distillation* is the partial vaporization of a solution of liquid components with separate recovery of vapor and liquid residue.  The concentration of the more volatile components (sometimes termed the "lighter" components) is greatest in the condensed vapor, while the concentration of the less volatile components ("heavier" components) is greatest in the liquid residue.  The degree of separation in a simple distillation process depends on the relative volatility of the components, which, in turn, depends on the thermodynamic properties of the mixture.  Ordinarily single-stage distillation does not provide adequate separation of components. (The major exception is evaporation, which is treated separately in Sec. 5-7.) Simple distillation may be carried out as a batch operation, in which case the instantaneous compositions of vapor product and liquid residue change continuously.  Alternatively, it may be carried out as a continuous operation, with liquid feed to the still pot to replenish the vapor removed.  In this case the vapor product composition remains constant once steady state is achieved.

*Rectification* is continuous countercurrent contact of the vapor resulting from a simple distillation with a condensed portion of the vapor product. This countercurrent contact results in greater enrichment of the vapor (often termed "overhead") product with the more volatile components than is possible with a single stage of simple distillation.  The condensed vapor returned to accomplish this is termed *reflux*.  In rectification the feed is to the simple distillation stage (bottom) and the more important product is removed as a vapor (top).

*Stripping* is continuous countercurrent contacting of the liquid feed with the vapor resulting from a simple distillation.  This countercurrent contacting results in a more complete removal of the more volatile components from the liquid product than could be accomplished by a single stage of simple distillation.  In stripping, the liquid feed is introduced at the top of the column and the vapor is supplied by partial vaporization of the liquid stream at the bottom of the column in a *reboiler*.

Commonly rectification is carried out on the vapor product leaving a stripping operation.  This is termed *fractional distillation*, or *fractionation*. In a fractional distillation, column feed is introduced at a *feed stage* within the column.  Above the feed stage is the rectification section and below it the stripping section.  Overhead product is condensed and a portion returned as reflux.  Bottom product is partially vaporized, the vapor returned to the column as *boilup*.  The stripping section is for recovery of overhead product and purification of bottom product, while the rectification section is for purification of top product and recovery of bottom product.  The important factors controlling separation are the number of stages in each section, the vapor/liquid flow ratios and the relative volatilities of the components.

**Equilibrium Data.** Vapor-liquid equilibrium data are required for the design of stills. In general, these must be observed experimentally. For binary systems, they are reported usually as tables or graphs of corresponding $x$ and $y$ values. Many such data are summarized by Hala, Wichterle, et al. ("Vapour Liquid Equilibrium Data at Normal Pressures," Pergamon Press, New York, 1968), by Smith, Block, and Hickman [in Perry and Chilton (eds.), "Chemical Engineers' Handbook," 5th ed., sec. 13, McGraw-Hill Book Company, New York, 1973], and by Hala, Pick, Fried, and Vilim ("Vapour-liquid Equilibrium," 2d ed., Pergamon Press, New York, 1967).

If equilibrium data are not available, they may be estimated for a binary system by a modification of Raoult's law:

$$y_1 = \frac{P_1}{\Pi} = \frac{\gamma_1 x_1 P_1^0}{\Pi} = \frac{\alpha_{12} x_1}{1 + (\alpha_{12} - 1)x_1} \tag{5-10}$$

For ideal mixtures $\gamma_1 = \gamma_2 = 1.0$; as a system departs from ideality, use of Eq. (5-10) becomes less reliable. Values of $\gamma$ may be estimated for a number of binaries by a method summarized by Smith et al. (*loc. cit.*).

**Simple Batch Distillation (Rayleigh or Differential Distillation).** In this case a batch of material is charged to a still pot, boiling is initiated, and the vapors are continuously removed, condensed, and collected until their average composition has reached a desired value. As the distillation proceeds, the concentration of less volatile components in the vapor continually increases. Sometimes successive portions of the vapor product (termed "cuts") are collected separately; these decrease in purity (with respect to the more volatile component) as the distillation proceeds. Ordinarily simple distillation is used only when there are large differences in relative volatility among the components.

Simple distillation is analyzed by assuming that at any instant the vapors are in equilibrium with the average liquid composition in the still. Though not strictly correct, this is usually a good assumption. In this case the composition of the liquid in the still is related to the total amount vaporized by

$$\ln \frac{S_2}{S_1} = \int_{x_1}^{x_2} \frac{dx}{y - x} \tag{5-11}$$

where $y$ is related to $x$ by the equilibrium relationship $y = y(x, \Pi, t)$. The average composition of the vapors collected during this period is related to the initial and final liquid compositions and quantities by a simple material balance:

$$x_{\text{ave}} = \frac{x_1 S_1 - x_2 S_2}{S_1 - S_2} \tag{5-12}$$

**Equilibrium Flash Distillation.** Liquid at an elevated temperature and pressure is throttled into a *flash chamber* maintained at a pressure below the vapor pressure of the liquid. A certain fraction of the feed vaporizes, reducing the system temperature until the vapor pressure of the remaining liquid at the new temperature matches the flash chamber pressure. Vapor and liquid fractions are led separately from the still. If vapor and liquid streams are in equilibrium, this is equivalent to a single-stage simple distillation.

Compositions and flow rates of vapor and liquid streams are obtained by simultaneous solution of material balances

$$(1 - f)x_i + fy_i = x_{Fi} \tag{5-13}$$

and energy balance

$$(t_F - t)C_F = F\lambda_m \tag{5-14}$$

together with the equilibrium relationship

$$y_i = y(x_i, \Pi, t) \tag{5-15}$$

and the relationship between temperature, vapor pressure, and total pressure

$$P_i = P_i(t) \tag{5-16}$$

$$\Pi = \Sigma P_i(t) \tag{5-17}$$

where $f$ is the fraction vaporized, and $C_F$ and $\lambda_m$ are the heat capacity and latent heat of vaporization of the feed (assumed constant). Equations (5-13) through (5-17) are solved simultaneously by trial and error.

**Continuous Binary Rectification.** *Plate Columns.* *Plate-to-plate Calculations.* The simultaneous solution of material balance, energy balance, and equilibrium relationships between each successive two stages in a column permits the exact computation of the column behavior from one terminal stream to another. Until recently a generally impractical procedure, it now may be utilized economically if the need for repeated designs is sufficient to justify its being programmed for a digital computer. This method is beyond the scope of this book (see Smith et al., *loc. cit.*).

*McCabe-Thiele Graphical Method.* If the molar latent heat of vaporization is relatively independent of composition, and if heat losses are negligible, liquid and vapor molar flow rates within the column will be constant and the following graphical procedure is acceptable:

1. Plot the vapor-liquid equilibrium data as $y$ vs. $x$.

2. Write and plot the operating-line (material balance) equations for each section of the column which relate passing streams within the column to feed or product streams. With reference to Fig. 5-1, the equations are, for section II (enriching section),

$$y_n = x_{n+1}\frac{L_{n+1}}{V_n} + x_d\frac{D}{V_n} = \frac{R}{R+1}x_{n+1} + \frac{x_d}{R+1} \tag{5-18}$$

where $R = L_E/D = $ *reflux ratio* and for section III (stripping section),

$$y_m = x_{m+1} \frac{L_{m+1}}{V_m} - x_w \frac{W}{V_m} \qquad (5\text{-}19)$$

The liquid rates in the two sections of the column are constant, but generally different because of the introduction of the feed, which may be liquid, vapor, or a mixture (see below).

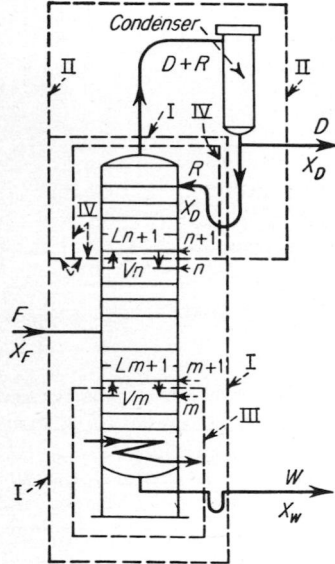

FIG. 5-1. Schematic of continuous distillation column.

3. Determine the number of stages by stepping from operating line vertically to equilibrium line, then horizontally to operating line. Begin with bottom product and continue in this fashion until the feed composition is reached, then use upper operating line until upper product concentration is reached.

4. Correct the theoretical number of stages to the actual number of stages using stage efficiencies (estimated from previous experience) according to

$$N_{\text{act}} = N_{\text{theor}}/\text{efficiency} \qquad (5\text{-}20)$$

The upper and lower operating lines intersect on a *feed line* described by Eq. (5-13), where $f$ is the moles of vapor introduced to the rectifying section per mole of feed introduced to the feed stage. Note that $f$ is fractional between zero and unity if the feed is introduced as a mixture of liquid and vapor; $f = 0$ if feed is a saturated liquid; $f < 0$ if feed is a cold liquid; $f = 1$ if feed is saturated vapor; $f > 1$ if feed is a superheated vapor. In Fig. 5-2a, five different feed lines are shown with the same upper operating line and five lower operating lines leading to the same bottom product composition. Intersections between operating lines for feeds of different qualities are represented by $C_1$ through $C_5$:

$$C_1(f < 0), \ C_2(f = 0), \ C_3(0 < f < 1), \ C_4(f = 1), \text{ and } C_5(f > 1).$$

Three sets of operating lines are shown on Fig. 5-2b. Lines $a$ and $b$, intersecting the feed line on the equilibrium curve, represent minimum reflux and require an infinite number of theoretical stages to separate the feed with composition $x_F$ into products with compositions $x_d$ and $x_w$. Line $e$, where both operating lines coincide with the diagonal, represents total reflux, which requires the smallest number of stages to achieve the desired separation,

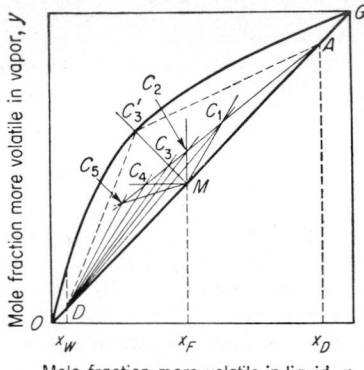

FIG. 5-2a. Effect of thermal condition of feed on operating lines and minimum reflux ratio.

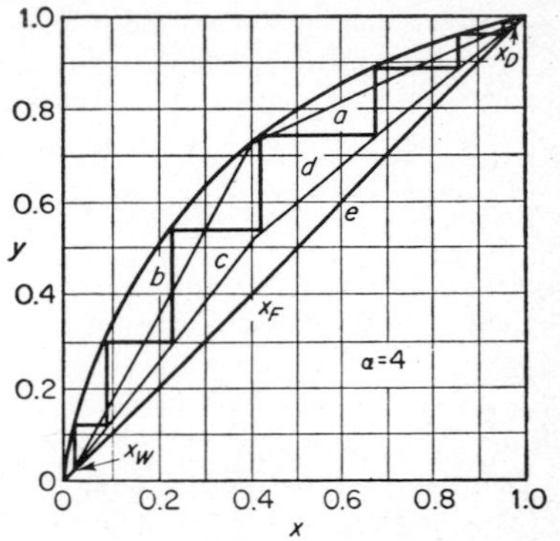

FIG. 5-2b. McCabe-Thiele graphical method for determining number of theoretical stages.

but produces no product. Lines $c$ and $d$ represent practical operating lines between the limits of minimum and total reflux.

### Analytical Method for Mixtures of Constant Relative Volatility

1. Solve for the number of theoretical plates necessary at "total" reflux (the condition when vapor and liquid rates within the column are infinitely large compared to feed, overhead, and bottom drawoff rates), using the Fenske-Underwood equation

$$N_T + 1 = \log (x_1/x_2)_d (x_2/x_1)_w / \log \alpha \quad (5\text{-}21)$$

2. Estimate the minimum reflux ratio (the ratio of liquid to distillate rates if the column were infinitely tall) by using

$$\frac{R}{R_{\min}} = \frac{x_d[1 + (\alpha - 1)x_F] - \alpha x_F}{(\alpha - 1)x_F(1 - x_F)} \quad (5\text{-}22)$$

3. By use of Fig. 5-3, estimate the number of theoretical trays necessary for the reflux ratio to be employed.

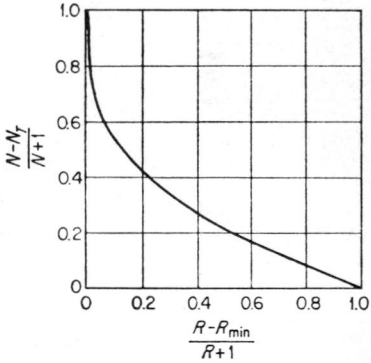

FIG. 5-3. Correlation of Gilliland for number of theoretical stages.

**Batch Binary Rectification.** In batch distillations, generally three "products" are withdrawn from the still. These are an initial product high in purity with regard to the more volatile or light-key component, an intermediate product which will usually be recycled for redistillation, and finally a product high in purity with regard to the heavy or less volatile component. Obviously a desirable separation is one which will minimize the middle or recycle cut. For most batch distillations the following rule of thumb will hold and is useful in setting the reflux to be used.

$$(L/D)(\alpha - 1) \geq 10 \qquad (5\text{-}23)$$

If a constant-reflux ratio is maintained, the product purity of the more volatile component will drop off as the distillation proceeds. The speed at which this decline in purity occurs will be a function of the particular reflux ratio employed, the relative volatility, and the amount of volatile component originally present.

It must be remembered that the practicing engineer is most often confronted with a different problem in designing for continuous operation than for batch distillation. In the former a column is usually designed and built for a given separation; in the latter the usual problem is how a given piece of equipment should be operated to effect a desired separation. We shall therefore assume that the equipment and the number of plates are specified.

Of importance in planning for the operation of any batch distillation apparatus are answers to the following questions:

1. What is the overhead-product composition as a function of still-pot composition?

2. How many moles of steam (as a heating medium) will be required to effect the separation?

For the constant-reflux case the vapor requirement, and thus the steam requirement, is obtained from

$$V = (L/D + 1)D \qquad (5\text{-}24)$$

The relation between still-pot and overhead compositions is obtained by plotting on a $y - x$ diagram (as shown in Fig. 5-4) lines of constant slope equal to

$$L/V = (L/D)/(L/D + 1) \qquad (5\text{-}25)$$

and stepping off the number of theoretical plates in the column. To obtain the relation between the amount distilled and the still-pot composition, plot $x_p$ vs. $1/(x_p - x_d)$. The area under the curve is equal then to $P/D$, where $P$ represents the amount of liquid in the still pot.

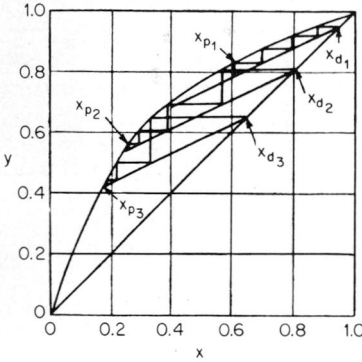

Fig. 5-4. Diagram for batch distillation at constant reflux ratio.

A second method of operating batch-distillation columns is to maintain product purity over a period of time by constantly increasing the reflux ratio. The relation between amount distilled and still-pot composition is now found by a simple material balance. The steam requirement is obtained by finding the area under a curve of $\dfrac{P_0(x_0 - x_p)}{(x_p - x_d)^2(1 - L/V)}$ versus $x_p$. Appropriate values of $x_p$ are read as a function of $L/V$ from a $y - x$ plot.

**Multicomponent Rectification.** This subject is too complicated for treatment here. The interested reader should refer to a more comprehensive source (e.g., Smith et al., *op. cit.*).

**Plate Efficiency.** The estimation of plate efficiencies is empirical. In a properly designed column, the value should exceed 0.60 and may exceed 0.95. Fair et al. present estimation methods and typical values [in Perry and Chilton (eds.), "Chemical Engineers' Handbook," 5th ed., sec. 18, pp. 13–19, McGraw-Hill Book Company, New York, 1973].

**Limiting Factors.** Proper operation of a column requires gas velocities within a narrow range of values. Velocities must be high enough to give good gas-liquid contacting, yet low enough to prevent entrainment or excessive pressure drop with resultant flooding. These factors control column diameter for a given vapor rate.

Tray spacing must be large enough to prevent carryover of liquid from the tray below, either by entrainment or because of froth height. Some liquid-vapor systems cause severe operating problems because of formation of a relatively stable foam with considerable liquid carryover to the tray above.

A detailed discussion of design procedures with regard to these problems is beyond the scope of this work. The interested reader is referred to Treybal (*loc. cit.*, pp. 126–150).

**Design Optimization.** When designing a distillation process the designer usually must meet specified product purities for both products, or one specified purity plus a percentage recovery of a component. Choice of a given reflux ratio specified the number of stages required to meet product specifications. Higher reflux ratios lead to smaller numbers of stages (hence lower column heights and column costs). Normal optimal design procedure is to determine minimum reflux ratio, then incrementally increase reflux ratio toward total reflux. For each reflux ratio capital and operating costs are calculated, and an acceptable design is determined by the economic balance. Final details of design can then be determined using this reflux ratio. In practice the economical reflux ratio is often between 1.2 and 2.0 times the minimum.

## 5-3. Solvent Extraction

**Definitions.** Solvent extraction consists of the transfer of a component dissolved in a liquid (called the *feed solution*) to a second liquid (called the *solvent*) to form an *extract* solution of the transferred component and to leave a *raffinate* solution relatively lean in the transferred component. Solvent extraction is used when distillation is impractical, as with close-boiling or temperature-sensitive mixtures. There is a strong analogy between extraction and distillation, solubility being the counterpart of volatility and the solvent, that of heat (Fig. 5-5).

**Equilibrium Data.** Phase equilibria for liquids are so specific that it is best to refer to laboratory data for the system in question. Maddox [in Perry and Chilton (eds.), "Chemical Engineers' Handbook," 5th ed., sec. 14, McGraw-Hill Book Company, New York, 1973] cites many such data.

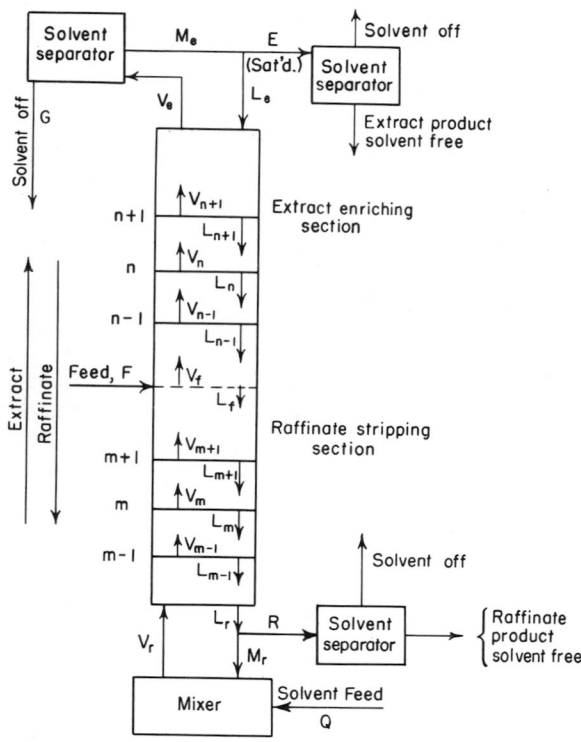

FIG. 5-5. Flow diagram for countercurrent multistage extraction with reflux.

For some systems the equilibrium is well approximated by the ideal-distribution law

$$K' = w/w' \tag{5-26}$$

with consequent simplification of design procedures. In solvent extraction fewer theoretical plates and much lower plate efficiencies are encountered than in distillation or absorption, with corresponding aggravation of inaccuracies implicit in simplified methods. For this reason, shortcut approximations should be used with caution.

**Countercurrent Extraction.** A feed mixture of two completely miscible components $A$ and $B$ is to be separated into its components by extraction with a solvent. If the solvent is partially miscible with each component of the feed, countercurrent extraction with reflux may be used to separate the feed components completely (in the limiting case). In this case the method of operation is presented schematically in Fig. 5-5. If the solvent is partially miscible with one feed component $A$ and totally miscible with the other component $B$, complete separation cannot be achieved. In this case, essentially pure component $A$ can be obtained at the solvent-feed end of the column, but the ratio of components $A$ and $B$ in the overhead product is limited by the shape of the equilibrium diagram. Where purity of component $B$ is limited, the extract enriching section is not employed; i.e., the feed is at the top of the column, there is no extract reflux, and the top product after solvent recovery is a solution of $A$ and $B$, although richer in component $B$ than the feed stream.

Assume that the solvent is only partially miscible with each feed component, and the equipment is as shown in Fig. 5-5. Reflux is furnished to the top of the column by removing sufficient solvent from the extract phase (e.g., by distillation) to make it miscible with the raffinate phase. (Note that this does not change the ratio of $A$ to $B$.) A portion of this raffinate phase is returned to the column as reflux, the remainder is purified further to remove any remaining solvent and removed as product.

Raffinate phase at the bottom of the column is withdrawn and purified to remove any solvent present, e.g., by distillation. Raffinate reflux is sometimes supplied at the bottom of the column by adding a portion of the raffinate before purification to the solvent feed in a mixer. This is done primarily to ensure adequate dispersion of solvent in raffinate for good interphase contact, and does not increase recovery of component $A$ as extract reflux increases recovery of component $B$.

All recovered solvent is recycled to the solvent feed end of the column, and supplemented by makeup solvent if product purification is not complete.

Contacting may be continuous, with one phase dispersed in the other, using a packed column or spray column, or it may be stagewise, using a tray column as in distillation or a countercurrent system of mixer-settler stages. Design procedures are presented for staged extraction systems. These are applicable for continuous systems by determining empirically the column height equivalent to a theoretical stage (HETS).

A simplified design procedure similar to McCabe-Thiele graphical method (see subsection under 5.2) may be used, assuming that extract-layer and raffinate-layer flow rates are constant throughout the column.

1. Equilibrium data are plotted as mass fraction of $A$ in the extract layer (*ordinate*) against mass fraction of $A$ in the raffinate layer (*abscissa*), *both fractions being on a solvent-free basis.*

2. Extract and raffinate products are located on the $Y = X$ line.

3. Operating lines through these points and of slope $L_e/V_e$ and $L_r/V_r$ are drawn.

4. Plates are stepped off as in the McCabe-Thiele method previously described.

A more precise design procedure is illustrated in Fig. 5-6 and is outlined below.  All flow rates and concentrations are on a solvent-free basis, unless otherwise noted.

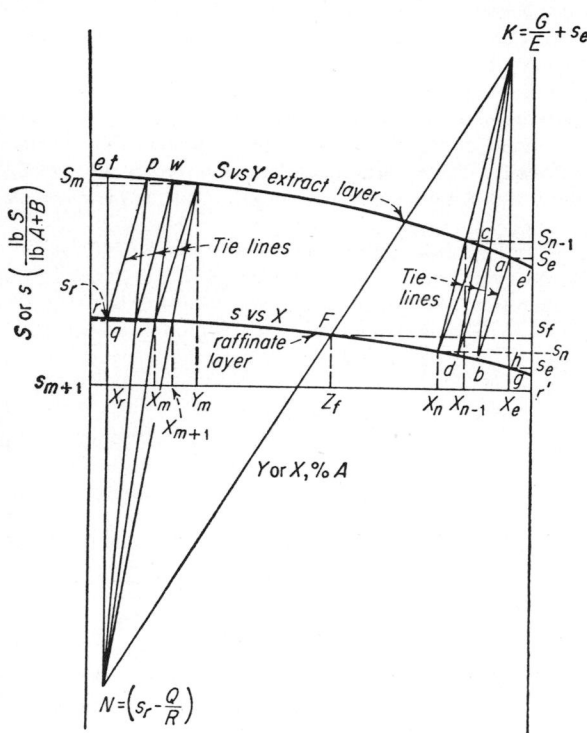

Fig. 5-6.   Graphical stepwise calculation of equilibrium stages on a solvent content-concentration diagram for operation with reflux.

1. From known equilibrium relationships construct the $S$ vs. $Y$ (extract layer) and $s$ vs. $X$ (raffinate layer) lines on working diagram.

2. Locate the operating point $K$ at an abscissa of $X_e$ (extract-product composition) and an ordinate of $G/E + S_e$.

3. Locate the operating point $N$ at an abscissa of $X_r$ (the $A$ content of the raffinate product) and an ordinate of $(s_r - Q/R)$.  A line joining $K$ and $N$ will intersect the $s$ vs. $X$ line at $Z_f$ (solvent content of the feed).

It is now possible to "walk" across the diagram to determine the number of theoretical stages necessary to effect the separation.   A line is drawn from

$K$ to $X_e$ intersecting the $S$ vs. $Y$ curve at $a$. Line $ab$ is an equilibrium tie line wherein the composition represented by $b$ is that in equilibrium with the composition represented by $a$. Another line from $K$ can then be drawn to the point $b$ so established, and another equilibrium tie line $cd$ is drawn. The procedure is repeated until a ray from $K$ coincides with the line joining $K$ and $N$. To the left of this dividing line the same procedure is followed, using point $N$ where point $K$ was used before. The number of theoretical stages is then obtained by counting the total number of rays drawn from the two operating points.

Minimum reflux on this type of diagram is obtained by moving $K$ vertically downward and $N$ vertically upward (at such a relative rate that $F$ always lies on a line joining them) until the line between them coincides with a tie line through $F$. The ordinate of $K$ then corresponds to a point of minimum reflux. Economic balances of column size vs. heat loads in solvent recovery of course determine the optimum degree of departure from this minimum-reflux point in actual operation.

**Column Efficiency.** For perforated-plate columns, the over-all efficiency may be estimated as the fraction $E_c$:

$$E_c = \frac{89,500 Z_t{}^{0.5}}{\sigma g_c}\left(\frac{V_D}{V_C}\right)^{0.42} = \frac{0.9 Z'_t{}^{0.5}}{\sigma'}\left(\frac{V_D}{V_C}\right)^{0.42} \tag{5-27}$$

where $Z_t$ is tray spacing in ft and $Z'_t$ is tray spacing in in.

For packed columns, the efficiency is expressed in the height assigned to a transfer unit (HTU) or theoretical stage (HETS). Ellis [*Ind. Chem.*, **28**, 483 (1952)] shows that, for *rough estimates*, the following empirical relationships are useful for towers packed with Raschig rings larger than $\frac{3}{8}$ in.

1. Transfer of solute from continuous aqueous to dispersed organic phase.

$$\text{HETS} = \frac{94.5 \mu_C (12 d_F)^b (V_C/V_D)^{0.5}}{10^{0.0683s}\,\Delta\rho} \tag{5-28}$$

2. Transfer of solute from dispersed organic to continuous aqueous phase,

$$\text{HETS} = \frac{69 \mu_C (12 d_F)^b}{10^{0.0535s}\,\Delta\rho} \tag{5-29}$$

where $b = 2.15/10^{0.096s}$, $d_F$ is in inches, $\mu$ is in lb/ft sec, and $\rho$ is in lb/ft³. Here $s$ is the average of the mutual solubilities of the solute-free contacted liquids in each other, expressed as weight per cent, and provides a rough measure of interfacial tension. For liquid pairs as insoluble as toluene and water, $s$ may be taken as zero.

**Extraction-tower Diameter.** Limiting flows, and hence minimum allowable diameters, for liquid-liquid extraction columns may be calculated using Fig. 5-7. While strictly applicable only to packed towers, the figure may be used with extreme caution for tray columns in the absence of better data. Column diameters must be larger (i.e., liquid velocities lower) than those specified in the flooding correlation.

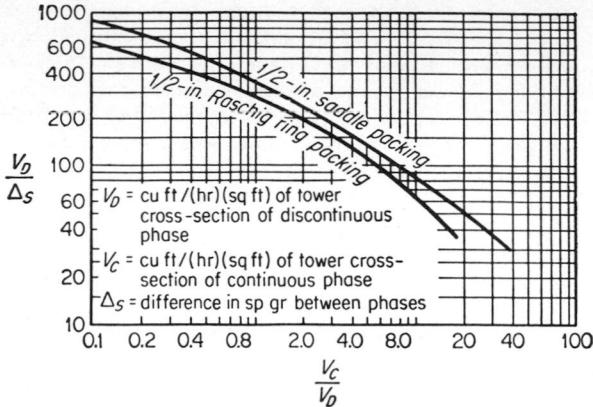

FIG. 5-7.  Colburn correlation of flooding data for packed extraction columns

## 5-4. Gas Absorption

**Definitions.**  Gas absorption consists of the transfer of a component from a gas phase to a liquid phase.  The liquid phase is called the *solvent,* or *absorbent;* the transferred gas is called the *solute,* or *absorbate.*  Usually, the solute is selectively absorbed from a carrier gas.  Fundamental considerations and design methods that apply to absorption are useful generally for the reverse operation of *desorption,* or *stripping.*

**Equilibrium Data.**  Gas solubility in a liquid is measured as a function of partial pressure or concentration of the gas in the equilibrium vapor phase. Solubilities sometimes are reported in the form of Henry's law constants. Equilibrium data for many systems may be found in standard reference sources (e.g., "International Critical Tables;" also Sherwood and Pigford, "Absorption and Extraction," McGraw-Hill Book Company, New York, 1952).

**Equipment.**  Gas absorption or stripping is accomplished in three principal types of equipment: *absorption columns,* packed or plate; *spray chambers* or towers; *bubble-sparged tanks,* frequently agitated.  Only absorption columns, by far the most important, will be treated here.

**Column Height.**  The height of a *packed column* is determined by the degree of separation to be achieved and by a characteristic contacting effectiveness of the packing.  The former may be expressed by stream compositions or by number of transfer units [Eq. (5-4)]; the latter, by the appropriate transfer coefficient [Eq. (5-2)] or HTU [Eq. (5-6)].  The height of a transfer unit varies with application and should be determined experimentally for the gas-liquid system, column packing, and column loadings employed.

The number of transfer units required may be calculated from Eq. (5-4) or, if operating and equilibrium lines are approximately straight, by Fig. 5-8, use of which requires knowledge of the slope of the equilibrium line $m$ and stipulation of $G_M/L_M$. The latter should be an economic selection. For most columns, 0.7 is an acceptable value for $mG_M/L_M$, but something less may be used if the solute is of low economic value.

The number of ideal plates in a *plate column* may be determined by the McCabe-Thiele procedure (Sec. 5-2). Once the values of $K_L$ and $K_G$ are

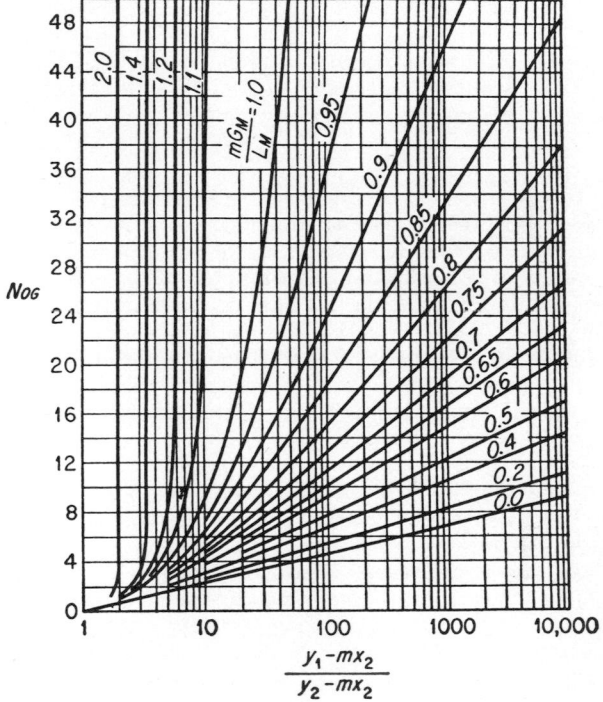

FIG. 5-8. Number of transfer units in an absorption column. Subscripts 1 and 2 refer to the concentrated and dilute ends, respectively.

estimated for the particular conditions under consideration and then combined to give a value for $H_{OG}$ by use of Eqs. (5-6) and (57),

$$Z = H_{OG}N_{OG} \tag{5-30}$$

The estimation of $N_{OG}$ may be obtained from Fig. 5-8, assuming that the operating and equilibrium lines are both straight or, at worst, only slightly curved.

## 5-5. Humidification

The most frequently encountered gas-vapor system is that of air and water vapor. A distinct terminology has been developed for this system, but the principles and mechanisms involved apply to any gas-vapor system. Humidification is the process by which the moisture content of air is increased; dehumidification is the reverse process.

If dry air and liquid water are held in intimate contact, some of the water will vaporize into the air in an effort to establish equilibrium between the phases. The concentration and the partial pressure of water vapor in the air will increase until the air becomes *saturated;* that is, until the partial pressure of the water vapor equals the vapor pressure of water at the equilibrium temperature.

The concentration of water vapor in the air is expressed as the *absolute humidity* (pounds of water vapor per pound of dry air), the *molal humidity* (mols water vapor per mol dry air), *relative humidity* (ratio of the actual partial pressure of vapor to the partial pressure if saturated at the existing temperature) or *percentage humidity* (ratio of actual humidity to the humidity if saturated at the existing temperature). The moisture level in air is indicated by the *dew point* (the temperature to which an air sample must be cooled to reach saturation) and by the difference between the air temperature and the *wet-bulb temperature* (the temperature assumed by a water-wet body held in a fast-flowing stream of the air). The above properties for the air–water vapor system are commonly presented graphically on a humidity or psychrometric chart as shown in Fig. 5-9.

Humidification or dehumidification are often the unavoidable secondary results of other operations such as drying, compression, absorption, and water cooling. The evaporative cooling of water is accomplished by adiabatically contacting it with a relatively large amount of unsaturated air; the resultant vaporization of some of the water cools the remainder, which approaches wet-bulb or adiabatic saturation temperature corresponding to the condition of the air.

The contacting of air and water is accomplished with spray ponds or cooling towers. Spray ponds depend on natural forces for air movement but power must be supplied to atomize and spray the water into the air. Spray ponds require large land areas, and water loss through entrainment is often high. Spray nozzle performance is critical in the design since it controls power demand, air-water contact surface, and water loss by drift. Table 5-1 presents guidelines for the design of spray ponds.

Cooling towers are preferred for most large industrial installations because they conserve land area, are less sensitive to atmospheric changes, and offer greater flexibility in design and operation. They are of two types: atmospheric, which depends on the wind for air cross-flow, and mechanical, in which fans pull the air upward through the tower. In both the water is pumped to the top of the structure where it is distributed uniformly across the tower and then cascades from grid deck to grid deck.

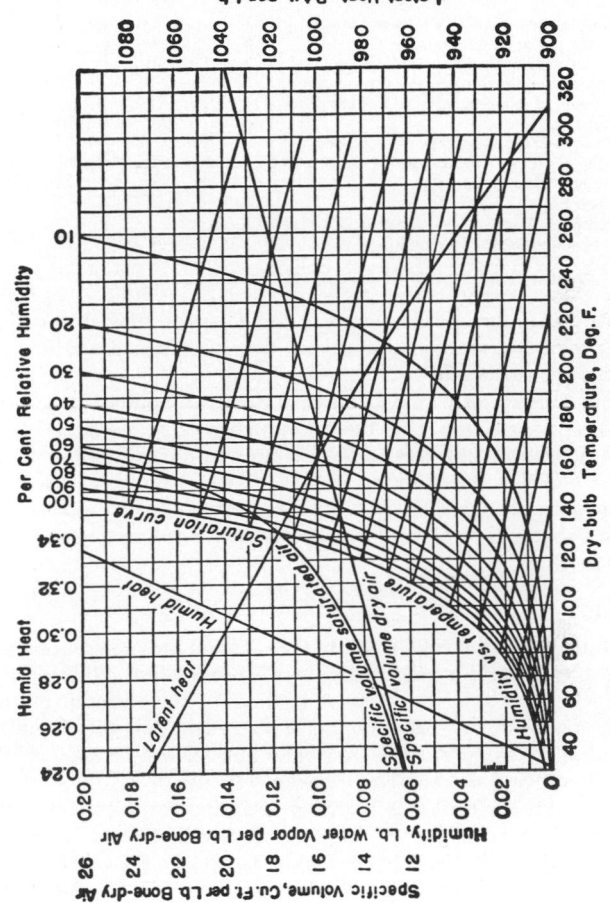

Fig. 5-9. Humidity chart for air–water vapor mixtures.

## Table 5-1. Spray-pond Engineering Data and Design*

| Recommendations | Usual | Minimum | Maximum |
|---|---|---|---|
| Nozzle capacity, gal./min. each . . . . . | 35–50 | 10 | 60 |
| Nozzles per 12-ft. length of pipe . . . . . | 5–6 | 4 | 8 |
| Height of nozzles above sides of basin, ft. . | 7–8 | 2 | 10 |
| Nozzle pressure, lb./sq. in. . . . . . . . . | 5–7 | 4 | 10 |
| Size of nozzles and nozzle arms, in. . . . . | 2 | $1\frac{1}{4}$ | $2\frac{1}{2}$ |
| Distance between spray lateral piping, ft. . | 25 | 13 | 38 |
| Distance of nozzles from side of pond, un- | | | |
|     fenced, ft. . . . . . . . . . . . . . . . | 25–35 | 20 | 50 |
| Distance of nozzles from side of pond, | | | |
|     fenced, ft. . . . . . . . . . . . . . . | 12–18 | 10 | 25 |
| Height of louver fence, ft. . . . . . . . . | 12 | 6 | 18 |
| Depth of pond basin, ft. . . . . . . . . . | 4–5 | 2 | 7 |
| Friction loss per 100 ft. pipe, in. of water | 1–3 | . . . . | 6 |
| Design wind velocity, m.p.h. . . . . . . . | 5 | 3 | 10 |

*From *Spray Pond Bull.* SP-51, p. 3, Marley Co.

The design of induced-draft towers is based on the allowable air velocity, the cooling load and range, the approach to wet-bulb temperature, and the transfer coefficients attained.

Norris [in Perry and Chilton (eds.), "Chemical Engineers' Handbook," 5th ed., sec. 12, McGraw-Hill Book Company, New York, 1973] gives a design procedure, but the services of a reputable tower supplier are recommended. As a very rough estimate, 1 ft of tower height will yield 1°F of cooling when the water rate is 2 to 3 gpm/ft² of tower cross section, the upward air velocity is 300 ft/min and the approach to the wet-bulb temperature is 10 to 15°F.

## 5-6. Drying

The unit operation, drying, refers to the removal by vaporization of liquid from a solid; the liquid usually constitutes a relatively small fraction of the wet solid. Beyond this, the term is applied to the removal of traces of vapor from a gas and of small amounts of water from another liquid. Only drying of a solid is discussed here and water and air will be used as examples of wetting liquid and surrounding gas, respectively.

The equipment for drying is classified according to the means by which the necessary heat is brought to the evaporating liquid and also according to the form and disposition of the wet solid. *Direct dryers* deliver heat to the liquid by direct contact with the hot gas stream into which the liquid vaporizes; *indirect dryers* supply the heat through a wall which separates the wet solid from a heat source, such as condensing steam.

Simultaneous heat transfer and mass transfer occur at equivalent rates during drying; the wet solid assumes the temperature needed to maintain this balance. The rate of drying depends on (1) the temperature of the heat source, (2) the resistance to transfer of heat to the vaporization site, (3) the resistance to transfer of mass from the vaporization site, and (4) the concentration (or partial pressure) of vapor in the gas in contact with the wet solid.  In turn, (2) and (3) depend on the physical form and characteristics of the solid and its wetness (liquid concentration); thus, the rate of drying can vary greatly as the process proceeds.  It is very common for some of the liquid wetting a solid to be dispersed in such a way that it does not exert its normal vapor pressure; in such cases, the final extent of drying is limited by the temperature and vapor content of air in contact with the wet solid.

Typical of indirect dryers are drum dryers (for slurries), can dryers (for textiles), and cylinder dryers (for paper), all of which are rotating cylindrical vessels usually internally heated with steam.  The material being dried is continuously applied to the outside surface, carried around for a part of a revolution, and then removed.  Sufficient drying may be accomplished in one pass, as on a slow-turning drum dryer, or it may require repeated passes over a series of similar dryers as in paper making.  Generally, the drying rates can only be established experimentally; rates can range from the equivalent of 1000 to 4000 Btu/(hr)(ft$^2$) of dryer surface.

A wide variety of direct dryers are available, which differ primarily in the method of contacting the wet solid with the drying gas.  The simplest and lowest cost is the batch tray drier in which the wet solid is spread on trays held in an enclosure and hot air is blown across the surface of the solid until it is sufficiently dry.  The drying rate is usually low; depends greatly on the temperature, humidity, and velocity of the air; and can decrease substantially as the drying proceeds.  If the wet solid can be granulated or pelletized and held on a perforated tray, improved rates can be obtained by passing the gas through the solid bed rather than across it.

A modification of the tray dryer is the truck dryer in which the trays are loaded on trucks and rolled into the dryer compartment.  Since the loading and unloading are done outside of the dryer proper, the dryer is used much more efficiently.

Rotary dryers offer continuous operation and are particularly adaptable to finely divided, nonsticking and nonagglomerating solids.  The wet solid is fed into the upper end of a slightly sloped, rotating cylinder through which heated air or hot combustion gas is passed.  Internal flights lift the solid and shower it through the hot gas while also advancing it to the discharge end.  If the gas velocity is high, fine particles may be carried out with the gas and must be recovered in cyclones or bag filters.

The steam-tube dryer is a variation of the rotary dryer in which a number of tubes are supported longitudinally within the rotating cylinder and supply the required heat.  Only a small flow of air through the cylinder is needed to sweep out the vapor and the loss of fines is minimized.

Spray dryers are particularly applicable when the feed is a solution or a very dilute suspension that can be atomized into fine droplets.  The dryer consists of an atomizing device, a hot-gas source, and a drying chamber in which the droplets contact the hot gases.  The atomizing device may be a pressure spray nozzle or a rotating disk.  The hot gases rise vertically counter to the falling droplets.  The product of the spray dryer is usually a spherical particle, often hollow, which is advantageous in certain cases.

In pneumatic conveyor or flash dryers the wet solid is dispersed in a high-velocity stream of hot gas and drying occurs as the gas carries the solids to a separating device. Solids disintegration and solids size classification are often carried out in conjunction with flash drying.

Empirical relationships for estimating drying rates under very restricted conditions and performance data for several types of dryers are given by McCormick [in Perry and Chilton (eds.), "Chemical Engineers' Handbook," 5th ed., sec. 20, pp. 16–63, McGraw-Hill Book Company, New York, 1973]. However, experimentally determined rates for the specific material and particular dryer type are much preferred when preparing dryer specifications.

## 5-7. Evaporation

Evaporation is the operation by which a volatile liquid is separated from a solution or suspension by vaporization. The separation is usually not complete and either or both the vapor and concentrated liquor may be the desired product.

Direct-fired or steam-jacketed kettles may serve as evaporators, but the most common types use a tubular heating surface and are steam heated. The tubes may be horizontal or vertical and the heating steam may be inside or surrounding the tubes. The tube bundle is placed within or external to the body of the evaporator and so arranged that the boiling liquor can be circulated through or around the tubes. The body also serves to allow disengagement of the vapor from the liquor as it is formed. Figure 5-10 illustrates several types of tubular evaporators.

An evaporator is basically a heat-transfer device. Its evaporative capacity is related to the temperature difference between heating steam and the boiling solution, the area of the tubular heat-transfer surface, the heat transfer coefficient and the latent heat of vaporization of the evaporating liquid:

$$\text{Capacity} = \frac{q}{\lambda} = \frac{(UA)(\Delta t)}{\lambda} \tag{5-31}$$

The coefficient $U$ is chiefly a function of the velocity of the solution past the heating surface and its viscosity, the extent of fouling of the heating surface, the temperature difference between steam and liquor, and the height of the liquor level relative to the tubes. The fouling of the tubes can become significant or even the controlling factor when scale-forming materials are present in the liquor. The values of $U$ are nearly always derived from experimentally determined rates of evaporation as a function of $\Delta t$. The temperature difference observed, however, is more often apparent than real because it is inferred from pressure measurements, which correspond to temperature values that do not reflect boiling-point rise of the liquid due to dissolved solute, or temperature shifts on the steam side due to vapor superheat or condensate subcooling. Such $\Delta t$ values are known as *apparent*

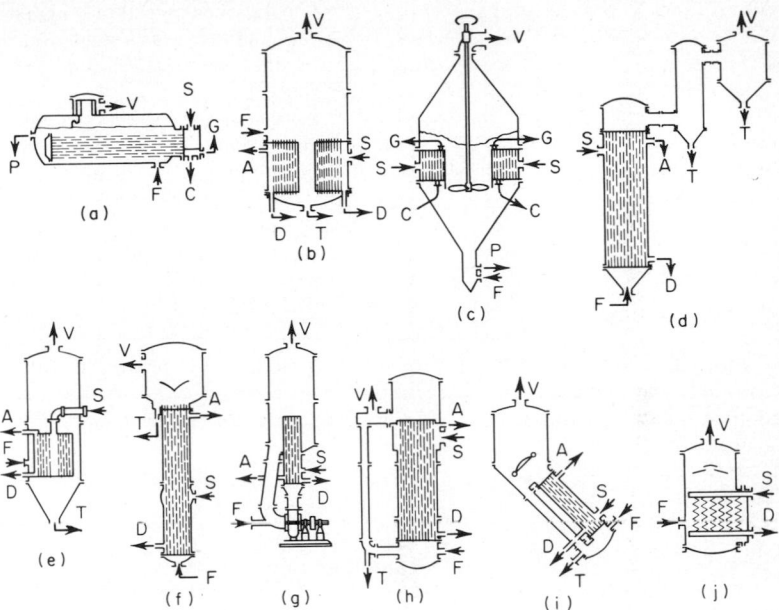

Fig. 5-10.  Typical evaporator designs.  (*a*) Horizontal tube.  (*b*) Short tube vertical.  (*c*) Propeller calandria.  (*d*) Long tube vertical without vapor head. (*e*) Basket type.  (*f*) Long tube vertical.  (*g*) Forced circulation.  (*h*) Long tube vertical with downtake.  (*i*) Buflovac inclined tube.  (*j*) Coiled tube.

*temperature differences,* and the values of $U$ corresponding to them are known as *apparent coefficients.*  The error on the steam side usually is small.  When the boiling-point rise is known and can be used to adjust the temperature difference, nearly correct values of $\Delta t$ and $U$ result; such values are known as *temperature difference and coefficient corrected for boiling-point rise.*  When not otherwise stipulated, the values reported for evaporators usually are these.

Illustrative over-all coefficients are given in Figs. 5-11 to 5-14.  These coefficients are intended only for preliminary estimation; they are unreliable for accurate design.

**Multiple-effect Evaporation.**  A multiple-effect evaporator is merely a series of similar evaporators so connected that the vapor from one body is the heating medium for the next.  Passing from single to multiple effect does not alter the major features of body construction; it merely affects the interconnecting piping and the operation.

The purpose of multiple-effect evaporation is to improve thermal efficiency.  One pound of steam supplied to the first effect will evaporate approximately one pound of water in that effect.  This pound of water

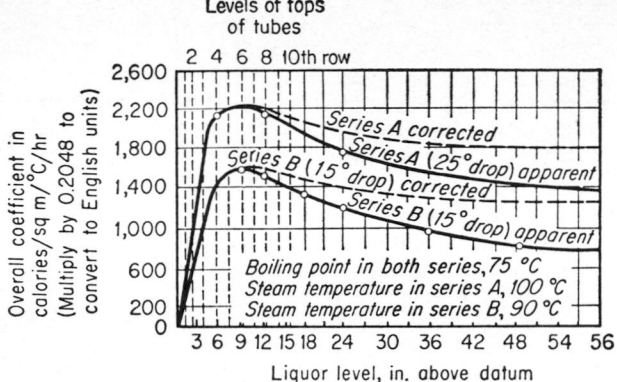

FIG. 5-11.   Heat transfer coefficients in a horizontal-tube evaporator.

vapor will then pass to the steam space of the second effect and, in condensing, will evaporate approximately another pound of water, and so on, so that in $N$ effects, 1 lb of steam will evaporate approximately (but somewhat less than) $N$ lb of water.   The pressure must be progressively reduced from effect to effect in order to produce a temperature difference between the boiling liquid of that effect and the condensing vapor from the preceding effect.

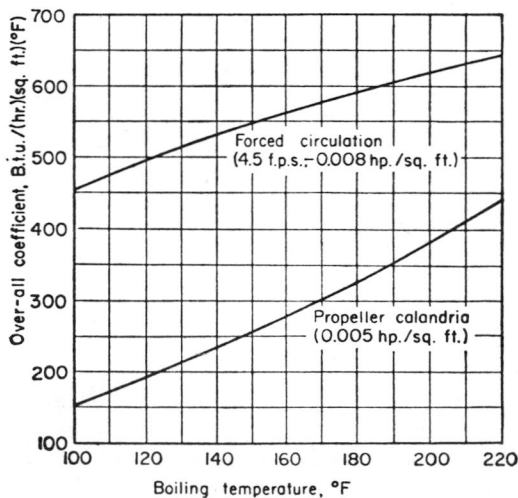

FIG. 5-12.   Heat transfer coefficients in salt evaporators.

If it is assumed that the terminal temperatures (temperature of heating steam available and temperature corresponding to the vacuum that can be produced in the condenser) are fixed, then passing from a single to a multiple effect does not increase the capacity of an evaporator. If a single-effect evaporator is operating between these terminal conditions and requires $A$ ft² heating surface to accomplish the desired evaporation, an $N$-effect evaporator to be used between the same terminal conditions for the same weight of water evaporated will require $N$ bodies of approximately $A$ ft² each to accomplish the same result. In short, passing from single- to multiple-effect operation decreases steam cost but increases apparatus cost.

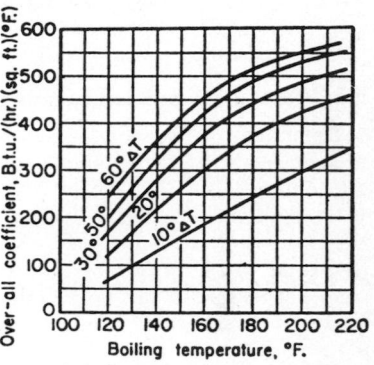

Fig. 5-13. Heat transfer coefficients for water in short tube.

The effects are commonly made the same size. The total area required may be calculated from the simultaneous solution of heat and material balance statements written for the several effects and the entire evaporator (e.g., see McCabe and Smith, "Unit Operations of Chemical Engineering," 3rd ed., pp. 454–459, McGraw-Hill Book Company, New York, 1976).

**Thermocompression.** The simplest, though not the least expensive, means of reducing the thermal energy requirements of evaporation is to

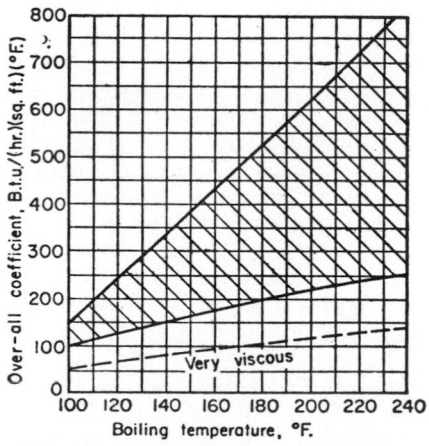

Fig. 5-14. General range of LTV coefficients.

compress the vapor from a single-effect evaporator so that the vapor can be used as the heating medium in the same evaporator. The compression may be accomplished by mechanical means or by a steam jet. In order to keep the compressor cost and power requirements within reason, the evaporator must work with a fairly narrow temperature difference, usually from about 10 to 20°F. This means that a large evaporator heating surface is needed, partially offsetting the advantages of thermocompression.

## 5-8. Adsorption and Ion Exchange

**Definitions.** *Adsorption* is separation of the components of a fluid phase brought about by contacting the fluid phase with a fixed solid phase. The solid phase, termed the *sorbent*, consists generally of highly porous particles with greatly extended surface area, most of which is internal. One or more components of the fluid phase, termed *sorbates*, enter the sorbent particles and bond to the surface of the solid. In *physical adsorption* weak forces such as ionic or Van der Waals forces bond sorbate molecules to the surface. In *chemisorption* true chemical bonds form between sorbate and sorbent surface. *Molecular sieves* possess a large number of adsorption sites of uniform type and essentially molecular dimensions and can be extremely selective toward a given component or class of components.

*Ion exchange* is transfer of ions from a fluid phase to a solid phase with simultaneous transfer of other similarly charged ions in the reverse direction. The solid medium is a synthetic organic resin with functional groups having a given charge bonded to the resin matrix and hence immobile. Mobile ions of opposite charge are distributed throughout the matrix to maintain electrical neutrality. *Cation-exchange resins* have negatively charged functional groups (such as sulfonic groups) bound to the matrix, while *anion-exchange resins* have positively charged bound groups (quaternary ammonium groups in *strongly basic resins* and other amine groups in *weakly basic resins*). If the mobile ion $X$ is predominant, the resin is said to be in the $X$ *form* (e.g., if $Cl^-$ ions are the predominant mobile ion, the resin is in *chloride form*).

Once the sorbent has become saturated, the sorbate is stripped from the sorbent and recovered, returning the sorbent to its original conditions for reuse. This is termed *regeneration* and may be accomplished by heating, chemical reaction, or by contacting with a *regenerant*.

**Mechanisms.** Design or operation of an ion-exchange or adsorption system depends on two factors: sorption equilibrium and rate of mass transfer. The relationship between solute concentration in the fluid phase and equilibrium amounts of solute retained in the solid phase is generally a strong function of temperature, and the curve describing this relationship is known as an *isotherm*. Many theoretical descriptions of sorption equilibrium have been developed, and are summarized by Vermeulen et al. [in Perry and Chilton (eds.), "Chemical Engineers' Handbook," 5th ed., sec. 16, McGraw-Hill Book Company, New York, 1973]. Isotherms for a particular sorbent-sorbate system must be determined experimentally.

The important rate processes for adsorption are as follows: turbulent eddy transport within the fluid phase to the boundary of the sorbent particle, molecular diffusion through pores to the interior of the particle, reaction at the phase boundary to establish equilibrium between fluid and solid phases, diffusion in the sorbed state (especially important for molecular sieves). In ion exchange, there is also equimolar counterdiffusion of the exchanged ion. Ordinarily transport to the particle and reaction at the surface are relatively rapid, thus diffusion within pores or in the sorbed state is the rate-controlling step.

**Continuous Operation.** For simplification, assume adsorption (or ion exchange) of a single solute (or ion) on a fixed bed, packed initially with adsorbent (or resin) free of the desired sorbate. The fluid phase is assumed to move slowly relative to the rate at which equilibrium is attained on the particle surface. At the outset of flow through the bed, all the solute is adsorbed on the packing within a short distance of the inlet end. As the sorbed-phase concentration approaches equilibrium with the inlet solute concentration, solute must flow further through the bed before being adsorbed. During this time the effluent concentration of sorbate is essentially zero, but eventually some of the solute passes through the entire depth of the bed without being adsorbed and appears in the effluent. This is termed *breakthrough*, and thereafter the effluent concentration rises until finally it equals the inlet concentration. At this time the entire bed has a sorbed-phase concentration in equilibrium with the effluent concentration, and the bed is said to be *exhausted*. Concentration vs. time profiles illustrating this sequence are presented in Fig. 5-15.

At some point in the operation, usually when the effluent reaches some specified maximum allowable concentration, the bed must be taken off stream and regenerated to restore its adsorptive capacity. The maximum allowable effluent concentration is based on product purity requirements, emissions limitations to protect environmental quality, or economic balances between the value of recovered solute and cost of regeneration.

Regeneration usually is accomplished by contacting the exhausted bed with a regenerant, often at elevated temperature. The solute is desorbed into a much smaller volume of regenerant than in the initial carrier stream, thus giving a much increased concentration of solute than initially fed to the system. (If regeneration is by heat alone, the regeneration yields pure sorbate.) In ion exchange, the sorbate is removed in a relatively small quantity of a regenerant solution which is highly concentrated in the alternate ion (e.g., in a concentrated NaCl or HCl solution if the resin is regenerated in the chloride form). After repeated regenerations, adsorbents usually lose adsorptive capacity, requiring addition of fresh make-up adsorbent.

Because bed-type adsorption is basically a batch operation, multiple beds in series must be used to attain continuous operation. When the first bed in the series is exhausted the feed is switched by a valving system to the second bed, thus making that the first bed, the exhausted bed is regenerated

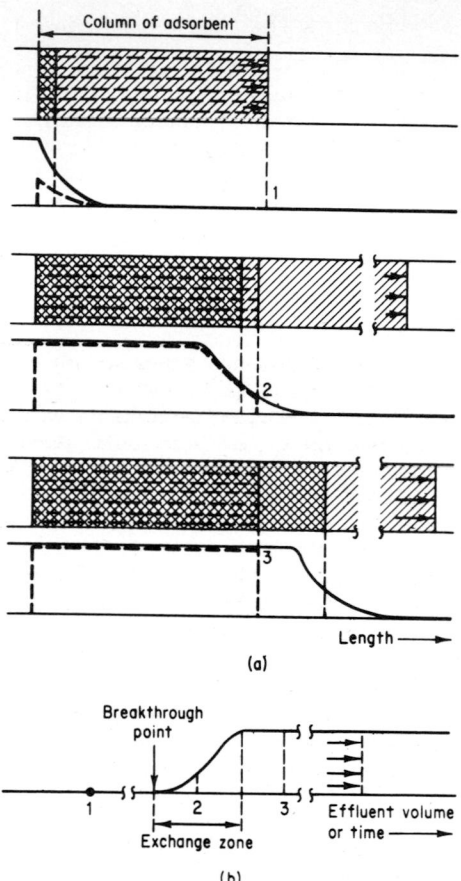

FIG. 5-15. (a) Typical column-concentration profiles. (b) Effluent-concentration history for fixed beds. In (a), dashes show span of packed bed. Crosshatching, single, indicates penetration by carrier fluid; double, by solute. Length scale to right of bed is contracted to show entire exchange zone.

and, again by means of valves, put on-stream as the last bed of the series. Thus, each bed in the series successively becomes the first bed.

**Design.** Design for continuous operation generally depends on scale-up from an experimentally determined breakthrough curve for a particular sorbent-sorbate fluid system. It is generally found that for a given system the ratio of effluent concentration is a function of the quantity of liquid treated (expressed as volumes of liquid per volume of solid, dimensionless), residence time, and effective mass-transfer rate. For a given system at a given set of equilibrium conditions the same functional relationship must hold:

$$X = f\left(\frac{F^a V}{D_p^b F'}\right)\left(\frac{v}{V}\right) \tag{5-32}$$

where $F^a/D_p^b$ is a dimensional group controlling mass-transfer rate, $V/F'$ is the residence time and $(v/V)$ is the total throughput in bed volumes. If mass transfer is rapid enough to ensure instantaneous equilibrium at all points, the effects of the mass-transfer parameter and residence time are not significant, and

$$X = f\left(\frac{v}{V}\right) \tag{5-33}$$

The shape of this curve is dependent on equilibrium considerations only (i.e., the feed concentration and the adsorption isotherm).

The shape of the breakthrough curve can be used to calculate the length of bed which has reached equilibrium with the feed (i.e., exhausted) and the bed loading, or amount of solute adsorbed. [Collins, J. J., *Chem. Eng. Prog. Symp. Ser.*, **63**(74), 31 (1967), also in Perry and Chilton (eds.), "Chemical Engineers' Handbook," 5th ed., sec. 16, McGraw-Hill Book Company, New York, 1973]. From this information one can calculate the required time between regenerations as a function of bed volume and feed flow rate, and also the appropriate number of bed volumes treated to ensure complete exhaustion before regeneration for multiple beds in series.

### 5-9. Membrane Processes

Several diffusional separation processes involve transfer of one or more components from one fluid phase through a porous solid medium or membrane to a second fluid phase. The porous medium, generally very thin, does not permit flow of the fluid but does allow the transfer of material between phases by molecular diffusion through pores. Membrane processes are classified according to the driving force across the membrane causing this diffusion; in *membrane permeation* the driving force is a difference in partial pressure; in *dialysis* the driving force is a concentration difference; in *ultrafiltration* and *reverse osmosis* the driving force is a difference in pressure; and in *electrodialysis* the driving force is a voltage gradient.

Due to its physical, chemical, or electrochemical properties, a particular membrane material offers different resistances to the diffusional flow of different materials. This allows the separation of the components of solutions from one another.

**Membrane Permeation.** For steady-state processes the rate of permeation of a component through the membrane is given by

$$N_M = \frac{P}{L}(\Delta p) \tag{5-34}$$

where $\Delta p$ is the difference in partial pressure of that component across the membrane, and $P$ is the *permeability*. Permeability is a function of the membrane composition and its structure, diffusing component, and system temperature and pressure. Correlations are available to predict permeabilities for various permeate-membrane combinations and the effects of temperature and pressure [Perry and Chilton (eds.), "Chemical Engineers' Handbook," 5th ed., sec. 17, McGraw-Hill Book Company, New York, 1973].

Membrane permeation is most useful for those separations where a more conventional process is impractical, and where the product can be recovered as a vapor. Some examples are separation of isomeric mixtures, thermally unstable compounds, and fluorides of different isotopes of uranium. Process design considerations include trade-offs between membrane selectivity, permeation rates, membrane strength, and durability. Membrane separation is a capital-intensive process, rather than energy intensive, and is only practical for difficult separations involving components with considerable economic value.

**Dialysis and Osmosis.** If two liquid solutions with different compositions are separated by a permeable membrane, the concentration gradient causes molecular diffusion through the membrane. For relatively dilute solutions, where no significant volume change occurs as the diffusion proceeds, mass-transfer rates are described by

$$N_{M_i} = U_i A \Delta C_{i.lm} \tag{5-35}$$

The over-all dialysis coefficient $U_i$ is determined by three film coefficients, analogous to heat transfer

$$\frac{1}{U_i} = \frac{1}{U_{i_{L1}}} + \frac{1}{U_{i_m}} + \frac{1}{U_{i_{L2}}} \tag{5-36}$$

where the individual coefficients are for the membrane and the two liquid films. If adequate mixing takes place in the two liquids, the membrane coefficient is controlling, $U_i = U_{i_m}$.

If the membrane resistance to diffusion differs significantly among solute components, mass transfer proceeds at different rates for the different components. In such cases membranes may be used to recover a single component preferentially from a complex solution. In such cases the complex solution is separated from a solvent by a selective membrane through which the desired component diffuses at a higher rate than other components. The resulting dilute solution of recovered solute is then concentrated by addition of make-up solute and returned to the plant for reuse. This is the basis for caustic recovery processes in the viscose-rayon industry and acid recovery from metallurgical liquors.

Design of a dialysis process depends on choice of membrane, concentration driving forces employed, and physical arrangement of the contacting streams. It is desirable to use a membrane which is thin and which has a high selectivity for the desired components. Thin membranes offer higher transfer rates but greater problems with mechanical stability. When large transfer areas are involved, the most effective and preferred arrangement is countercurrent staging of modules, with crosscurrent contacting within modules. Typical module designs have membranes separated by corrugated spacers, with corrugation at right angles on alternating layers, providing mechanical support for the membranes.

Design trade-offs are total membrane area (capital cost) versus value of recovered solute, and membrane type and thickness (maintenance costs) versus transfer rate (size and thus capital costs).

**Reverse Osmosis and Ultrafiltration.** Reverse osmosis and ultrafiltration processes recover the pure solvent from a solution. If a liquid solution is separated from the pure solvent by a membrane permeable to the solvent only, the osmotic pressure causes the solvent to diffuse from the pure liquid side to the solution side. A mechanical pressure applied to the solution side slows the diffusion process. If sufficient pressure is applied, the flow is reversed; i.e., solvent is transferred from the solution to the pure side, counter to the concentration gradient.

For dilute solutions, large solute molecules or high-pressure differentials, Eq. (5-34) applies for reverse osmosis processes. For concentrated solutions of solutes with low molecular weight, a correction for osmotic pressure is required.

In most applications the limiting factor is concentration polarization of the membrane. As solvent is removed from the solution, solutes greatly increase in concentration in a thin layer near the membrane. This causes an osmotic pressure which opposes the applied pressure gradient and reduces the rate of transfer. Another significant problem is buildup of deposits on the membrane surface.

Normal design trade-offs involve membrane thickness, pressure differential, membrane surface area, and percentage of solvent recovered.

**Electrodialysis.** If an electric field is imposed on an electrolytic solution, negatively charged ions are attracted toward the positive pole of the field and positively charged ions to the negative pole. If the solution is partitioned by semipermeable membranes (alternately anion-permeable and cation-permeable) the flow of ions under the influence of the electric field will create alternately dilute and concentrated compartments (see Fig. 5-16). The feed solution is fed to the diluting compartments and a salt-recovery solution is fed to the concentrating compartments. Electrodialysis may be carried out on either a continuous or a batch basis.

Important parameters in electrodialysis are membrane selectivity, electric field strength, current efficiency, and desired degree of deionization of the feed. The current efficiency, or amount of solute transferred per coulomb of electric charge, and the total voltage drop required to maintain a constant field strength both increase with the number of compartments. The rate of transfer increases with increasing field strength, but for a given membrane, selectivity decreases as electric field strength is increased.

Operating problems include concentration polarization of membranes,

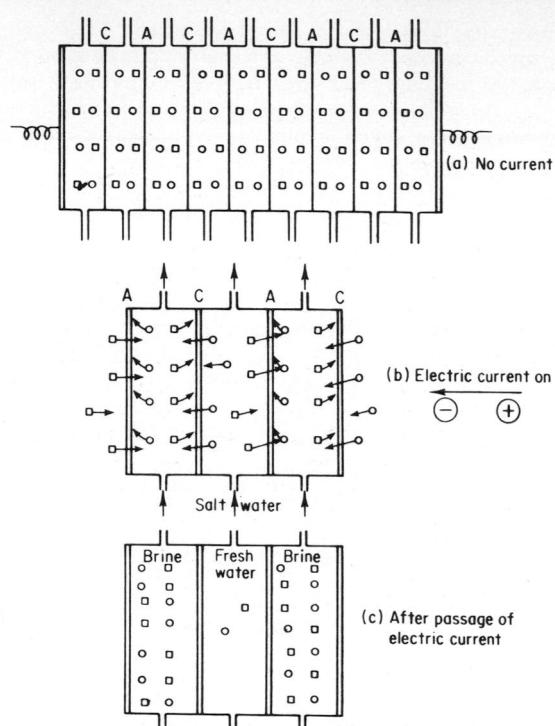

FIG. 5-16. Principles of electrodialysis. ○ Positive ion (e.g., sodium). □ Negative ion (e.g., chloride). *C* and *A*, cation and anion permeable membrane, respectively. Ion migration under action of electric current causes salt depletion in alternate compartments and salt enrichment in adjacent ones. (*Speigler, "Salt Water Purification," John Wiley & Sons, Inc., New York*, 1962.)

scaling or fouling of membrane surfaces, mechanical instability of membranes and current control. Typical design trade-offs are: larger number of parallel cells (increased current efficiency) vs. higher membrane costs and higher operating voltages, larger membrane area vs. electric power costs, higher concentrate cell concentration vs. greater concentration polarization and membrane scaling, lower diluent cell concentrations vs. increased electrical resistance and power costs.

## MULTIPHASE CONTACTING AND PHASE DISTRIBUTION

### 5-10. Agitation

Agitation is motion imparted to material to promote heat or mass transfer to, from, or within the material, or to distribute another phase through the material.  Examples include mixing of miscible liquids, formation of uniform solids suspensions, promotion of heat transfer between a process fluid and a heat-exchange surface, promoting dissolution of a gas in a liquid.  The most important process applications of agitation involve a freely fluid liquid as the primary phase.  The key equipment is a rotating element (sometimes more than one) called the *agitator* or *impeller*.  Turbines, paddles, propellers, and special shapes are used, the choice depending on the properties of the agitated material and the type of agitation desired.

**Agitation Power.**  Over a wide range of operating conditions the parameter which most completely determines the performance of an agitator is the power delivered to the fluid—the greater the power delivered, the more effective the agitation.  The power required by a rotating impeller cannot be computed directly; rather it must be measured for a geometrically similar model of the impeller and its surroundings, then scaled up or down.  The most useful scaling correlations involve the impeller Reynolds number $D_a^2 N \rho / \mu$, the power number $P g_c / D_a^5 \rho N^3$, and the Froude number $g / N^2 D_a$.  Examples of this correlation are presented in Fig. 5-17.  Additional power data are available in Holland and Chapman, "Liquid Mixing and Processing in Stirred Tanks," Reinhold Publishing Company, New York, 1966; Sterbacek and Tausk, "Mixing in the Chemical Industry," Pergamon Press, London, 1965; and Uhl and Gray, "Mixing Theory and Practice," Academic Press, Inc., New York, 1967.  Since agitator geometry (including impeller

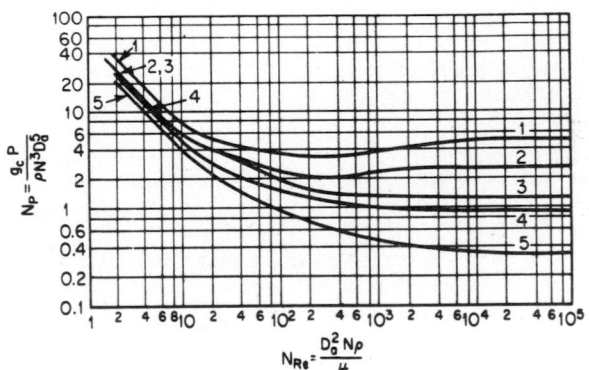

Fig. 5-17.  Impeller power correlations.  (1) Six-blade turbine (curved blades), $D_a/W_i = 5$, with four baffles of height equal to $\frac{1}{12}$ agitator tank diameter.  (2) Six-blade turbine (straight vertical blades), $D_a/W_i = 8$; same baffle arrangement as (1).  (3) Six-blade turbine (straight blades with 45° pitch), $D_a/W_i = 8$; same baffle arrangement as (1).  (4) Propeller, pitch equal to $2D_a$; four baffles of height equal to $\frac{1}{10}$ agitator tank diameter.  (5) Propeller, pitch equal to $D_a$; same baffle arrangement as (4).  (*Curves 4 and 5 from Rushton, Costich, and Everett, Chem. Eng. Prog.*, **46**, 395, 467 (1950), *by permission; curves 2 and 3 from Bates, Fondy, and Corpstein, Ind. Eng. Chem. Process Design Dev.*, **2**, 310 (1963), *by permission of the copyright owner, the American Chemical Society.*)

dimensions, impeller position, vessel dimensions, baffles, etc.) affects the relationship between Reynolds number and power number, it is preferable to use experimental data rather than published correlations.

**Functional Performance of Agitators.** The degree of effectiveness of agitation is related to the intensity of shear, the level of turbulence, and the circulation rate produced by the impeller. For an impeller of given design, these quantities are determined by the agitator speed and power delivery. The following equations relate velocity head, power, and circulation rate for geometrically similar impellers under turbulent flow conditions:

$$Q = N_Q N D_a{}^3 \tag{5-37}$$

$$H = \frac{N_P N^2 D_a{}^2}{N_Q g_c} \tag{5-38}$$

$$P = \rho Q H = N_P \rho N^3 \frac{D_a{}^5}{g_c} \tag{5-39}$$

where $Q$ is impeller discharge rate, ft$^3$/sec; $H$ is velocity head, ft-lb force/lb mass; $N_P$ is power number; $N_Q$ is discharge coefficient, dimensionless.

The intensity of turbulence is primarily a function of the velocity head induced by the impeller. To achieve a higher discharge velocity head at a given rate of power consumption requires smaller impeller diameter and higher rotational speeds. The high-intensity turbulence of such an impeller is appropriate for: (1) production of large interfacial area and small droplets in gas-liquid and immiscible liquid systems, (2) promotion of mass transfer between phases, (3) solids deagglomeration by shear forces, (4) rapid mixing in the impeller discharge system, and (5) suspension of particles with relatively high settling rates.

To achieve a larger impeller discharge rate at constant power consumption requires larger impeller diameters and lower rotational speeds. A large-diameter impeller with its associated high discharge rate is appropriate for: (1) short times to complete mixing of a miscible liquid throughout the entire agitator volume, (2) promotion of heat transfer at a heat-exchange surface within the vessel, (3) maintaining uniform temperatures and concentrations throughout a backmix reactor, and (4) suspension of particles with relatively low settling rates.

## 5-11. Spray Generation

A spray is a mechanically produced, unstable suspension of liquid droplets in a gas. Sprays are generated from a continuous liquid by nozzles, of which there are three principal types: *pressure nozzles, rotating nozzles,* and *gas-atomizing nozzles* (or two-fluid nozzles). The nozzle converts mechanical energy into the surface free energy required to form liquid droplets, either by means of a pressure head loss (pressure nozzles and gas-atomizing nozzles) or a spinning shaft (rotating nozzles). The smaller the average droplet diameter, the more energy is required per pound of liquid dispersed.

### Table 5-2. Discharge Rates and Included Angle of Spray of Typical Pressure Nozzles

| Nozzle type | Orifice diam., in. | Discharge, gpm, and included angle of spray | | | | | | | |
|---|---|---|---|---|---|---|---|---|---|
| | | 10 psi | | 25 psi | | 50 psi | | 100 psi | |
| | | Dis-charge | Angle, deg | Dis-charge | Angle, deg | Dis-charge | Angle, deg | Dis-charge | Angle, deg |
| Hollow cone...... | 0.046 | ..... | .. | 0.10 | 65 | 0.135 | 68 | 0.183 | 75 |
| | .140 | 0.535 | 82 | 0.81 | 88 | 1.10 | 90 | 1.50 | 93 |
| | .218 | 1.25 | 83 | 1.88 | 86 | 2.55 | 89 | 3.45 | 92 |
| | .375 | 7.2 | 62 | 11.8 | 70 | 16.5 | 70 | | |
| Solid cone........ | .047 | ..... | .. | 0.167 | 65 | 0.235 | 70 | 0.34 | 70 |
| | .188 | 1.60 | 55 | 2.46 | 58 | 3.42 | 60 | 4.78 | 60 |
| | .250 | 3.35 | 65 | 5.40 | 70 | 7.50 | 70 | 10.4 | 75 |
| | .500 | 17.5 | 86 | 27.5 | 84 | 38.7 | 73 | | |
| Fan.............. | .031 | 0.085 | 40 | 0.132 | 90 | 0.182 | 110 | 0.252 | 110 |
| | .093 | 0.70 | 70 | 1.12 | 76 | 1.57 | 80 | 2.25 | 80 |
| | .187 | 2.25 | 50 | 3.70 | 59 | 5.35 | 65 | 7.70 | 65 |
| | .375 | 9.50 | 66 | 15.40 | 74 | 22.10 | 75 | 30.75 | 75 |

SOURCE: Data furnished through the courtesy of the Spray Engineering Co., Cambridge, Mass.

**Pressure Nozzles.** In pressure nozzles, classified as *hollow-cone, solid-cone, fan,* or *impact* types depending on the shape of the spray, the fluid is throttled through small openings, thereby attaining high velocities. Droplets are formed because of shear forces at the nozzle exit, by its instability at such high levels of inertia, or by impact with another jet or a solid surface. Discharge rates and droplet sizes are functions of orifice opening and applied pressure drop. Typical values are given by Tables 5-2 and 5-3.

### Table 5-3. Drop-size Distributions Produced by Three Hollow-cone Nozzles of the Same Design

| Nominal drop diam., $\mu$ | Number of drops in each size group | | | | | |
|---|---|---|---|---|---|---|
| | 0.063-in. orifice diam. | | | 0.086-in. orifice diam. | | 0.128-in. orifice diam., 200 psi |
| | 50 psi | 100 psi | 200 psi | 100 psi | 200 psi | |
| 10 | 375 | 800 | 1700 | 100 | 300 | 100 |
| 25 | 200 | 280 | 580 | 60 | 150 | 50 |
| 50 | 160 | 180 | 260 | 41 | 100 | 45 |
| 100 | 50 | 60 | 70 | 26 | 34 | 27 |
| 150 | 27 | 31 | 35 | 14 | 18 | 15 |
| 200 | 19 | 23 | 27 | 9 | 12 | 11 |
| 300 | 8 | 9 | 11 | 5 | 8 | 6 |
| 400 | 2 | 4 | 4 | 4 | 7 | 3 |
| 500 | 1 | 1 | .... | 2 | 1 | 2 |
| 600 | 1 | ... | .... | 1 | ... | 1 |

NOTE: $1\mu = 10^{-4}$ cm = 0.0000394 in. The nominal diameter is the mid-diameter of a drop group which includes a finite range of sizes. The 25 group includes drops from 17.5 to 37.5 $\mu$; the 50 group contains drops rom 37.5 to 75$\mu$, etc. The number of drops has been adjusted in each case so that the total amount of fluid sprayed is the same for each size distribution.

**Gas-atomizing Nozzles.** These nozzles disintegrate a stream of liquid by contact with a high-velocity stream of gas. The liquid may be preatomized by a pressure nozzle or injected as a generally continuous sheet. In either case the primary energy source for atomization is the gas pressure drop, not the liquid pressure drop. Two-fluid nozzles produce very fine droplets at the price of high energy consumption. Typical drop-size distribution for a gas-atomizing nozzle is given in Table 5-4.

### Table 5-4. Drop-size Distribution of a Small Atomizing Nozzle

| Drop diam., $\mu$ | Number of drops | Drop diam., $\mu$ | Number of drops |
|---|---|---|---|
| 2 | 390,000 | 35 | 1,730 |
| 5 | 340,000 | 40 | 1,080 |
| 10 | 165,000 | 45 | 650 |
| 15 | 40,200 | 50 | 430 |
| 20 | 11,680 | 60 | 350 |
| 25 | 4,970 | 70 | 220 |
| 30 | 2,160 | | |

NOTE: The fluid pressure and the gas pressure were each 15 psi. The total quantity of fluid represented by this size distribution is the same as that in Table 5-8, so that the numbers of drops are directly comparable.

## 5-12. Gas Sparging

A sparger is a distributor that disperses gas into the body of a liquid by emitting bubbles or jets of gas through an individual orifice, an array of orifices, or a porous structure. Spargers are used to promote gas-liquid mass transfer or to produce dispersions; sometimes they are used as gentle agitators.

**Simple Bubblers.** Open-end pipes or perforated tubes or plates with orifices $\frac{1}{8}$ to $\frac{1}{2}$ in. in diameter are used as spargers. A perforated tube should be designed so that the pressure drop across the individual orifice is large compared to the pressure drop for flow through the tube. At practical operating rates, simple bubblers produce jets rather than bubbles; the jets disintegrate, but the resulting cloud of bubbles may include some as large as 0.5 in. Their effectiveness as mass-transfer promoters is orders of magnitude below that of vigorously agitated tanks or packed columns, and they are used only for very easy transport operations (e.g., air humidification) or for gentle mechanical agitation.

**Porous Septa.** Porous plates, tubes, or disks are made by bonding or sintering carefully sized particles of carbon, ceramic, metal, or polymer. The resulting septa may be used as spargers to produce much smaller bubbles than will result from a simple bubbler.

The gas flux through a porous septum is limited on the lower side by the requirement that, for good performance, the whole sparger surface should bubble uniformly, and on the higher side by the onset of serious bubble coalescence. In a practical range of fluxes, the size of bubbles produced is a direct function of both pore size and pressure drop.   Figure 5-18 shows the recommended limit of flux density for a typical porous medium.   The working pressure drop across typical porous media is larger than the dry permeability.   Wet-permeability values should be used in all design calculation.

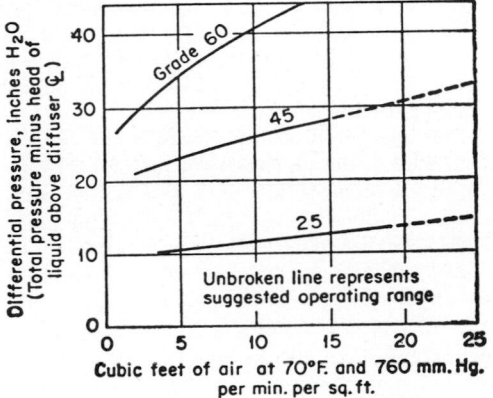

Fig. 5-18.   Pressure drop across porous carbon diffusers submerged in water at 70°F.   (*National Carbon Co.*)

Porous spargers are used generally to promote gas absorption.   They are of the same order of effectiveness as packed or tray columns or agitated vessels, but no generalized data or methods are available for their specification as mass-transfer devices.   Their advantages of simplicity and inexpensiveness are balanced by their susceptibility to plugging and their awkwardness for countercurrent operations.

## 5-13. Fluidization

**Definitions.**   If a gas is passed upward through an unrestrained and unconsolidated bed of granular solids with ever-increasing velocity, the pressure drop across the bed due to friction will increase until it becomes equivalent to the weight of the bed plus the friction between fluid and walls. With further increase in the gas velocity, the bed tends to rise as a unit, but its unconsolidated character causes it instead to expand until the increased porosity allows the friction again just to balance the pressure drop.   As the bed becomes more expanded, individual particles achieve freedom to interchange position, and the bed can circulate.   As the gas velocity is further increased the drag force upon individual particles eventually becomes large enough to balance gravitational forces, and individual particles can be entrained and carried out of the bed.   The latter conditions are termed *particulate* or *continuous fluidization,* the former termed *aggregative* or *batch* *fluidization.* The designation *fluidized bed* usually refers to aggregative fluidization.

**Continuous Fluidization.** One primary application of continuous fluidization is *pneumatic conveying* of a dispersed solid by means of compressed air. The solid phase is introduced into the carrier gas, e.g., by aspiration or by a screw conveyor, is immediately fluidized, and is transported to the destination where it is removed from the carrier gas by means of a cyclone or other collection device. The gas-pressure drop is greater than would occur if no solids were present, increasing with increased solids concentration. Empirical correlations between pressure drop and solids loading are imprecise, and experimental data should be used where possible.

Two operating problems associated with pneumatic conveying are *saltation*, or deposition of particles in the pipeline with restriction of gas flow to the unblocked section, and *erosion* of the pipe wall caused primarily by inertial impact of particles at elbows, etc. Saltation can be prevented by employing superficial gas velocities in excess of a certain minimum velocity, termed the *saltation velocity*, which is a function of particle type and mass loading in the gas, and which must generally be determined experimentally. Erosion is best combatted by protection at the inertial impact point, either by an erosion plate, reinforced pipe walls, or a saltation layer resulting from the flow pattern.

**Batch Fluidization.** Fluidized beds sometimes are called *boiling beds*. Indeed, the expanded suspended mass of the bed does resemble a boiling liquid. This mass has a zero angle of repose, seeks its own level, and assumes the shape of the containing vessel. Just as in a vessel designed for boiling a liquid, space must be provided for vertical expansion of the solids and for disengaging splashed and entrained material.

**Conditions for Fluidization.** The size of solid particles which can be fluidized varies greatly, from less than 1 micron to $2\frac{1}{2}$ in. It is generally concluded that particles distributed in size between 65 mesh and 10 microns are the best for smooth fluidization (least formation of large bubbles). Large particles cause instability and result in slugging or massive surges. Small particles (less than 10 microns) frequently, even though dry, act as if damp, forming agglomerates or fissures in the bed, or spouting. Adding finer-sized particles to a coarse bed or coarser-sized particles to a bed of fines often results in better fluidization.

The upward velocity of the gas is usually between 0.5 and 10 fps. This velocity is based on the flow through the empty vessel and is frequently referred to as the superficial velocity. Its upper limit is fixed by the terminal free-settling velocity of the smallest particles in the bed that should not be carried over. The velocity used is best determined by test in equipment where visual observations of the action of the bed can be made. The flow required to maintain a completely homogeneous bed of solids, whereby coarse or heavy particles will not segregate from the fluidized portion, is very different from the minimum fluidizing velocity discussed in most studies of fluidization.

Bed height is determined by a number of factors, either individually or collectively, such as:

1. Space-time yield
2. Gas-contact time
3. $L/D$ ratio required to provide staging
4. Space required for internal heat exchangers
5. Solids-retention time

Generally, bed heights are not less than 12 in. or more than 50 ft.

For details beyond the scope of this section, reference should be made to: Leva, "Fluidization," McGraw-Hill Book Company, New York, 1959; Othmer and Zenz, "Fluidization and Fluid Particle Systems," Reinhold Publishing Corporation, New York, 1960; and Wells, in Perry and Chilton (eds.), "Chemical Engineers' Handbook," 5th ed., sec. 20, McGraw-Hill Book Company, New York, 1973.

**Heat Transfer and Mixing in Fluidized Beds.** Heat-exchange surfaces have been used to provide means of removing or adding heat to fluidized beds. Usually, these surfaces are provided in the form of vertical tubes manifolded at top and bottom. Other shapes have been used such as horizontal bayonets. In any such installations adequate provision must be made for abrasion of the exchanger surface by the bed. Normally, the transfer rate is 5 to 25 times that for solids-free gas.

Heat transfer from solids to gas and gas to solids usually results in a coefficient of about 3 to 10 Btu/(hr)(ft$^2$)(°F). However, the large area of the solids per cubic foot of bed (15,000 ft$^2$/ft$^3$ for 60-micron particles of 40 lb/ft$^3$ bulk density) results in the rapid approach of gas and solids temperatures. With a fairly good distributor, essential equalization of temperatures occurs within 1 to 3 in. of the top of the distributor.

Bed thermal conductivities in the vertical direction have been measured in the laboratory in the range of 20,000 to 30,000 Btu/(hr)(ft$^2$)(°F/ft). Horizontal conductivities for $\frac{1}{8}$-in. particles in the range of 1000 Btu/(hr)(ft$^2$)(°F/ft) have been measured in large-scale experiments.

Except at large $L/D$ ratios, the temperature in the fluidized bed is uniform, the temperature at any point being, generally, within 10°F of any other point. The solids, too, will be well mixed. For all practical purposes, beds with $L/D$ ratios of from 1.0 to 4 can be considered to be completely mixed continuous-reaction vessels as far as the solids are concerned.

**Equipment.** The use of the fluidization technique requires in almost all cases the employment of a fluidized-bed system rather than an isolated piece of equipment. Figure 5-19 illustrates the arrangement of components of a system used in cases where the flow of solids is small, such as is generally encountered in noncatalytic usages of the fluidized beds or in catalytic units where there is little or no deactivation of the catalyst. Figure 5-20 illustrates a catalytic-type unit such as is used for petroleum cracking where large quantities of solids flow into and out of the reactor, and to and from the catalyst regenerator, which also is usually a fluidized bed. It is obvious that, in the simplified form, the only difference between a fluidized catalytic-

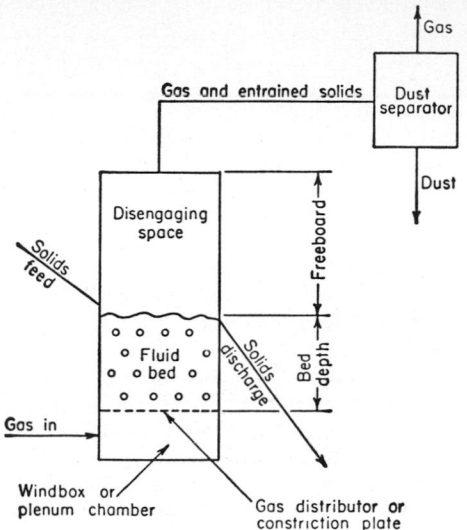

FIG. 5-19.   Noncatalytic fluidized-bed system.

cracking unit and fluidized-bed units used in most other cases is the method and point of solids feed.

The major parts of a fluidized-bed system can be listed as follows:
1. Reaction vessel
    a. Fluidized-bed portion
    b. Disengaging space or freeboard
    c. Gas distributor
2. Solids feeder or flow control
3. Solids discharge
4. Dust separator for the exit gases
5. Instrumentation
6. Gas supply

The reactor is usually a vertical cylinder; however, there is no real limitation on shape.   The specific design features vary with operating conditions, available space, and use.   The lack of moving parts tends toward simple, clean design.

The freeboard or disengaging height is frequently chosen rather arbitrarily or based on experience.   It has been established that carry-over of solids entrained by the gases is reduced as the vertical distance between the top of the dense-phase fluidized bed and gas-outlet port is increased.   Small-scale experiments have also shown that the size distribution of the solids entrained by the gases is reduced as the freeboard height or cross-sectional area is increased.   However, for some distance (from a few inches to a number of

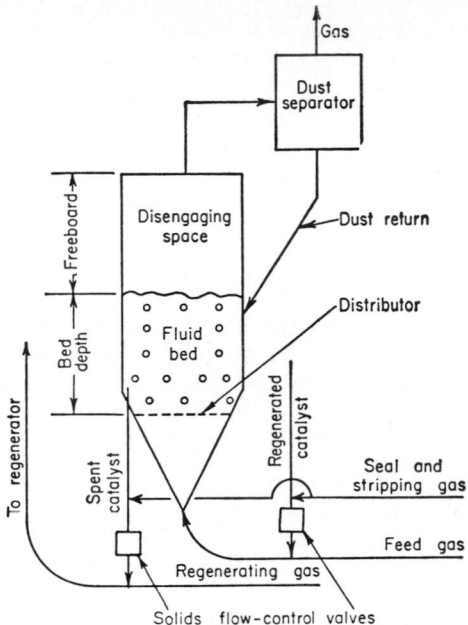

FIG. 5-20.   Catalytic fluidized-bed system.

feet) the size distribution of the solids in the dilute suspension just above the fluid bed is the same as the size distribution of the solids in the fluid bed.

The gas distributor has a considerable effect on proper operation of the fluidized bed.   Basically, there are two types: (1) for use where the inlet gas contains solids, (2) for use where the inlet gas is clean.   In most cases, the distributor is designed to prevent backflow of solids during normal operation, and in many cases it is designed to prevent backflow during shutdown.   In order to provide distribution, it is necessary to restrict the gas or gas and solids flow so that pressure drops across the restriction amount to from a few inches of water to a few pounds per square inch.   As a general rule, pressure drops in excess of 2 psi are not used.

In cases where both solids and gases pass through the distributor, such as in catalytic-cracking units, a number of variations are or have been used, such as concentric rings in the same plane, with the annuli open, concentric rings in the form of a cone, grids of T bars or other structural shapes, flat metal perforated plates supported or reinforced with structural members, and dished and perforated plates concave both upward and downward.   The last two forms are generally more economical.

Fluidized-bed reactors usually are designed by scaling up a laboratory or pilot unit. Considerable difficulty has been encountered in such scale-up because of the staging effect achieved in high $L/D$ ratio units used in the laboratory or semiworks as compared with the lower $L/D$ ratios used in commercial units.

## MECHANICAL SEPARATIONS AND PHASE COLLECTION

### 5-14. Filtration

Filtration is the mechanical phase separation of a fluid-solid suspension or slurry by passage of the liquid through a porous septum, or filter medium, which retains the solids. The clarified liquid product is called the *filtrate* and the retained solids, the *filter cake*. As the filtration proceeds, the liquid in the feed slurry must flow through the cake that is being formed as well as the filter medium, therefore, the resistance to flow continually increases. The driving force for this flow can be gravity, a pump-developed pressure on the upstream side of the filter medium, or a vacuum applied to the downstream side. Filtering devices are categorized on the basis of the method for creating the driving force and the mode of operation.

**Types of Filters.** Gravity filters are the simplest and are always intermittent in operation. They are usually in the form of a false-bottomed tank; the false bottom is perforated and supports the filter medium. The slurry is pumped into the tank and the liquid drains by gravity through the medium into the lower portion of the tank. The rate of filtration decreases continually as the filter cake builds up and the unit must eventually be shut down and the cake removed.

The *plate-and-frame filter press* is an intermittent pressure filter. The plates are solid with channelled or ribbed faces; the filter medium (e.g., filter cloth) is laid over the faces of each plate. The frames are hollow and of the same outer dimensions as the plates but are made in a variety of thicknesses. The plates carrying the medium and the frames are arranged alternately in a clamping device which, when tightened, forms a leak-tight assembly of cavities into which the slurry is pumped and paths for the escape of the filtrate which has passed through the filter medium. Channels and ports in each plate and frame permit the simultaneous flow of slurry into all frames, the flow of filtrate from all plates, and a flow for washing of the filter cake collected in each frame. When the frames become full, or when the resistance to flow becomes excessive, the clamping device is opened and the cake removed from each frame.

*Leaf filters* are also intermittent pressure filters. The leaf is a hollow frame covered with a wire support screen and the filter medium. A series of leaves are mounted in a pressure housing into which the slurry is pumped. The cake builds up on the leaves as the filtrate passes through the medium into the interior of the leaf and then through a conduit to the exterior of the housing. When the cake has built up to a predetermined thickness or the back pressure becomes excessive, the housing is drained and opened and the filter cake removed.

There are several *continuous vacuum filter* designs but all are similar in basic principle. The filter medium is carried on a rotating drum or disk mounted on a hollow horizontal shaft. Vacuum is applied to the plenum behind the medium and the cake is formed during that part of the revolution when the medium is immersed in the slurry. The vacuum is maintained and the cake is dewatered after emerging from the slurry. At some later point in the revolution the vacuum is released and the cake is scraped off, leaving the medium ready for reimmersion in the slurry. The filtrate drains from the plenum through the central shaft to vacuum receivers. Provision can also be made for washing and air drying the cake before removing it.

**Rate of Filtration.** The rate of filtration is obviously the rate at which the filtrate can be forced through the cake and filter medium. The flow resistance of the cake is a function of its particle-size distribution, compressibility, thickness, and the viscosity of the filtrate. The volumetric flow rate through a hard, granular, noncompressible cake can be related to the operating conditions and cake properties:

$$\frac{dV}{d\theta} = \frac{A\Delta P_c}{\mu\alpha g_c(W/A)} \tag{5-41}$$

where $\alpha$ is the specific cake resistance (constant). For compressible cakes $\alpha$ is a function of pressure drop, often correlated by

$$\alpha = \alpha' \, \Delta P_c{}^S \tag{5-42}$$

where $\alpha'$ and $S$ are empirical constants depending on the particular solids being filtered. The compressibility $S$ varies from zero for a hard, granular, noncompressible cake to unity for a soft, readily deformed, and compressible cake; for most industrial slurry solids, $S$ ranges from 0.1 to 0.8. For compressible cakes the filtration rate is given by

$$\frac{dV}{d\theta} = \frac{A\Delta P_c{}^{(1-S)}}{\mu\alpha' g_c(W/A)} \tag{5-43}$$

The above rate expressions do not consider the flow resistance offered by the filter medium. This can be significant and even controlling particularly if the medium becomes blinded by deposition of solids in its interstices. A wide variety of filter-medium materials and weaves are available and the choice involves a trade-off between the resistance to flow and the clarity of the filtrate but must also consider resistance to wear and chemical attack, tendency to blind, mechanical strength, and cost.

The sizing of a filter for a particular slurry starts with laboratory or pilot-plant-scale tests followed by scale-up to the desired capacity. The choice of the type of filter involves a variety of factors and is best made with the advice of filter manufacturers. Table 5-5 may be used as a preliminary guide to filter selection.

## Table 5-5. Filter Selection: Slurry Characteristics*

| | Fast filtering | Medium filtering | Slow filtering | Dilute | Very dilute |
|---|---|---|---|---|---|
| **Slurry characteristics** | | | | | |
| Cake-formation rate | in./sec. | in./min. | 0.05–0.25 in./min. | <0.05 in./min. | No cake |
| Usual solids concentration | >20% | 10–20% | 1–10% | <5% | <0.1% |
| Settling rate | Very rapid | Rapid | Slow | Slow | |
| Leaf test rate, lb./(hr.)/(sq. ft.) | >500 | 50–500 | 5–50 | <5 | |
| Filtrate rate, gal./(min.)/(sq. ft.) | >5 | 0.2–5 | 0.01–0.02 | 0.01–2 | 0.01–2 |
| **Filters:** | | | | | |
| Continuous vacuum filters: | | | | | |
|   multicompartment drum | X | X | X | | |
|   single-compartment drum | X | | | | |
|   Dorrco | X | | | | |
|   top-feed | X | | | | |
|   horizontal-table | X | X | | | |
|   tilting-pan | X | X | | | |
|   horizontal-belt | X | X | | | |
|   disk | | X | X | | |
|   precoat | | | | | |
| Batch vacuum leaf | | | | X | X |
| Batch nutsche | X | X | X | X | X |
| Batch pressure filters: | | | | | |
|   plate-and-frame | | X | X | X | X |
|   vertical leaf | | X | X | X | X |
|   tubular | | X | X | X | X |
|   horizontal plate | X | X | X | X | X |
|   cartridge, edge | | | | | X |
| Continuous pressure filters: | | | | | |
|   drum | | X | X | X | X |
|   fest | | X | X | X | X |
|   precoat | | | | | X |

* Adapted from Porter et al., Chem. Eng., 78(4), 40 (1971).

## 5-15. Settling and Sedimentation

The separation of suspended solids from a fluid by gravitational forces is termed settling or sedimentation; differentiation between the two terms is not precise. *Settling* usually refers to very dilute suspensions of discrete rigid particles or discrete liquid droplets. *Sedimentation* is applied to higher concentrations of solids and to solids that *flocculate* (i.e., agglomerate to form rigid lattices).

In the free or unhindered settling of discrete particles, the particle will accelerate until drag forces offered by the liquid exactly balance the net gravitational forces (i.e., particle weight minus buoyant forces), and thereafter will settle at a constant *terminal velocity* expressed by

$$v_t = \sqrt{\frac{4gD_p(\rho_p - \rho_f)}{3\rho_f C_D}} \tag{5-44}$$

Where the drag coefficient $C_D$ is a function of particle shape and the Reynolds number based on particle diameter, $N_{Re} = D_p v_t \rho_f / \mu$. At very low $N_{Re}$ (<0.3), Stokes' law is valid for spherical or near-spherical rigid particles, and $C_D = 24/N_{Re}$; at higher $N_{Re}$, the coefficient decreases with increasing $N_{Re}$, but becomes less dependent on $N_{Re}$, becoming essentially constant; as the sphericity of a particle decreases, the drag coefficient increases. Nonrigid particles exhibit similar behavior, but droplet deformation and internal circulation affect the relationship between $C_D$ and $N_{Re}$. [For more thorough discussion see Boucher and Alves, in Perry and Chilton (eds.), "Chemical Engineers' Handbook," 5th ed., sec. 5, pp. 61-65, McGraw-Hill Book Company, New York, 1973].

If the concentration of particles is sufficiently high that the particles collide or interact in any way, *hindered settling* results. In some cases Eq. (5-44) can be modified to give approximate values for hindered-settling velocities, but normally data must be obtained experimentally for each specific case. Since particle interactions increase drag forces, hindered-settling velocities are always lower than free-settling velocities. As the concentration of particles increases, interactions increase, leading to reduced terminal velocities.

When the suspended particles are flocculent, either naturally or by the addition of promoters termed *flocculating agents*, the settling behavior can vary greatly, depending on the concentration. At very low solids concentrations flocs can settle freely and unhindered in a manner similar to single rigid particles. At higher concentrations the flocs continue to grow as they settle, either by coalescence of flocs or by continued flocculation of fine particles. As a result the settling velocity changes as the flocs settle. At still higher concentrations, the flocs coalesce completely and tend to settle as a single porous mass. In this case fluid flow phenomena more closely resemble flow through a porous bed of solids than flow around a submerged particle.

In typical settling operations, the solids concentration (after a period of settling) becomes a function of depth. Distinct zones appear in which different mechanisms control settling velocities (i.e., free settling, hindered settling, flow in a porous bed), leading to the term *zone settling*. Since no mathematical analysis of settling is available other than for free settling of discrete particles, laboratory settling data must be obtained to evaluate design parameters.

### 5-16. Centrifugation

Centrifugal force can be applied to enhance the separation of liquid-solid or liquid-liquid suspensions by either settling or filtration.

Centrifugal force is exerted on a mass that is following a curved path; the extent of the force depends on the radius of curvature and the angular velocity, increasing as velocity is increased or curvature is decreased. Thus, a centrifugal force field thousands of times stronger than normal gravity can be generated in properly designed devices called *centrifuges*. A centrifugal force field can produce the same effects as a gravitational field but at rates proportional to the relative strengths of the fields. For example, the separation of a liquid-solids suspension by settling, which may be very slow when brought about by gravity, can be speeded up greatly by placing the suspension in a strong centrifugal field.

A centrifuge for the separation of phases consists of a rotatable vessel (the bowl or basket) in which the centrifugal force is generated, a drive mechanism for rotating the vessel, and means for introducing the separable mixture and removing, individually, the separated phases. The treatment vessel may be a bowl with solid walls or a basket with perforated walls. The operation may be batchwise or continuous; the former usually gives a more complete separation but the latter requires less operating labor. A summary of types and characteristics of centrifuges is given in Table 5-6.

Centrifuges may be classified by the operation carried out; namely, sedimentation or filtration. *Sedimentation centrifuges* are usually continuous-operation, solid-bowl machines yielding a dense phase of thickened sludge and a light phase of clear liquid. The concentration of solids in the sludge can be controlled by the throughput rate (or residence time) but often is limited by the flow (by the method of discharge) characteristics of the sludge. In the case of liquid-liquid suspensions, the degree of separation of the light and heavy phases is limited only by the residence time at any particular force field.

*Filtration centrifuges* usually have a perforated-wall cylindrical basket lined with a suitable filter medium, such as fine woven cloth or wire screen, which will retain the solids and allow the liquid to pass through. The centrifugal force acts directly on the liquid to push it through a cake of solids during buildup of the cake and also to drive it out of the pores of the cake during the dry spin. The latter action is much more effective than forcing air through the cake as in many standard filters. Both batch and continuous designs are available. The method of discharging the finished cake is a principal distinction among specific designs of both types of centrifuge.

## Table 5-6. Characteristics of Commercial Centrifuges

| Method of separation | Rotor type | Centrifuge type | Manner of liquid discharge | Manner of solids discharge or removal | Centrifuge speed for solids discharge | Capacity |
|---|---|---|---|---|---|---|
| Sedimentation | Batch | Ultracentrifuge | ......... | ......... | ......... | 1 ml. |
| | | Laboratory, clinical | Batch | Batch manual | Zero | To 6 liters |
| | Tubular | Supercentrifuge | Continuous° | Batch manual | Zero | To 1200 gal./hr. |
| | | Multipass clarifier | Continuous° | Batch manual | Zero | To 3000 gal./hr. |
| | Disk | Solid wall | Continuous° | Batch manual | Zero | To 30,000 gal./hr. |
| | | Light-phase skimmer | Continuous | Continuous for light-phase solids | Full | To 1200 gal./hr. |
| | | Peripheral nozzles | Continuous | Continuous | Full | To 24,000 gal./hr. |
| | | Peripheral valves | Continuous | Intermittent | Full | To 3000 gal./hr. |
| | | Peripheral annulus | Continuous | Intermittent | Full | To 12,000 gal./hr. |
| | Solid bowl | Constant-speed horizontal | Continuous° | Cyclic | Full (usually) | To 60 cu. ft. |
| | | Variable-speed vertical | Continuous° | Cyclic | Zero or reduced | To 16 cu. ft. |
| | | Continuous decanter | Continuous | Continuous screw conveyor | Full | To 18,000 gal./hr. To 75 tons/hr. solids |
| Sedimentation and filtration | .......... | Screen bowl decanter | Continuous | Continuous | Full | To 16,000 gal./hr. To 75 tons/hr. solids |
| Filtration | Conical screen | Wide-angle screen | Continuous | Continuous | Full | To 40 tons/hr. solids |
| | | Differential conveyor | Continuous | Continuous | Full | To 40 tons/hr. solids |
| | | Vibrating, oscillating, and tumbling screens | Continuous | Essentially continuous | Full | To 100 tons/hr. solids |
| | | Reciprocating pusher | Continuous | Essentially continuous | Full | Limited data |
| | Cylindrical screen | Reciprocating pusher, single and multistage | Continuous | Essentially continuous | Full | To 30 tons/hr. solids |
| | | Horizontal | Cyclic | Intermittent, automatic | Full (usually) | To 25 tons/hr. solids |
| | | Vertical, underdriven | Cyclic | Intermittent, automatic or manual | Zero or reduced | To 6 tons/hr. solids |
| | | Vertical, suspended | Cyclic | Intermittent, automatic or manual | Zero or reduced | To 10 tons/hr. solids |

° Interrupted during solids unloading.

The operating mode of batch filtering centrifuges can be varied readily to provide the best conditions for the particular liquid-solids system. Continuous machines are much less flexible and must be much more closely specified for a particular application.

The design or specification of centrifuges, depending on the operation carried out, calls for knowledge of the type and rate of settling, volume of liquid retained in the bowl and the throughput rate, filtration or drainage rate, compressibility and porosity of the cake, and the particle-size range of the solids. Generally predictions based on theory are very risky and the specification of centrifuges should be based on scale-up from tests on laboratory machines of similar type and geometry. Most centrifuge manufacturers can provide testing services and are well versed in scale-up methods.

### 5-17. Screening

Screening is the mechanical separation of a mixture of particles into two or more fractions by means of a surface with multiple uniform perforations, termed a *screen*. Material retained on the first screen in a series (i.e., largest openings) is termed *oversize*, that passing through the last in a series (i.e., smallest openings) is termed *undersize*, or *fines*. A screen may consist of cloth woven from various fibers, a perforated plate or uniformly spaced parallel bars. Screens are specified by the number of openings per linear inch (*mesh count*) or by the dimension of the openings, measured between and perpendicular to adjacent wires or bars (*aperture or clear opening*).

Commercial screening machines vary in specific design but have common principles. The material to be screened is fed to a screen surface which slants downward from inlet to outlet. The screen surface is rotated, vibrated, shaken, oscillated, or otherwise mechanically energized to bring the material into contact with the screen and to help impel oversize particles through to the outlet. Undersize particles coming in contact with the screen fall through and are conveyed to a different outlet.

The efficiency of a screening device is dependent on the thickness of the particle bed above the screen, the length of time the particles are in contact with the screen, the amount of mechanical vibration (or rotation, etc.) to bring undersize particles into contact with the screen and the characteristics of the particle mixture (agglomeration tendency, shape, etc.). Efficiency is defined in terms of per cent of desirable fraction retained by the screening device divided by per cent of that fraction in the feed. For a given screen, increased throughput generally results in reduced efficiency. The primary design variable is screen area required, which depends on the throughput to be handled and the desired efficiency.

### 5-18. Wet Classification

Wet classification is the separation of a mixture of particles into two or more fractions according to particle size or particle density by contact with a fluidizing medium, often water. It is used as a unit operation in the chemical process industry primarily for raw materials treatment (i.e., ore beneficiation, coal washing, etc.).   Most types of wet classification utilize the different settling velocities of the different particles to remove coarse or dense particles (termed *sand*) from fine particles.   Zones of settling are created (either hindered or free settling) in which the sands are retained behind a weir and removed in the underflow, and fines are carried along with the overflow. Sequential settling zones may be used to give several fractions as products. Mechanical agitation is often used to create zones of increased liquid velocities and to keep coarser particles in suspension, as well as water jets. Wet classification seldom produces sharp divisions of particle size or density between fractions, and it is most useful when there is a large difference in settling velocities between desired product and wastes.   The application and operating characteristics of several types of wet classifiers are given in Table 5-7.

**Jigging.**   Jigging is the separation of materials of different specific gravities by the pulsation of a stream of liquid flowing through a bed of the materials.   The liquid pulsates, or "jigs," up and down, causing the heavy material to work down to the bottom of the bed and the lighter material to rise to the top.   Each product is then drawn off separately.

The throughout capacity and power requirements of jigs depend on the character of the feed, the separation required, and the type of equipment used.   The water consumption is high, 1,200 to 2,500 gal water/ton of solids processed.

**Tabling.**   Tabling is the classification of particulate solids by means of an inclined, riffled, shaking surface (called a table) across which water or air is flowed.   The particles are classified principally on the basis of density difference.

Wet tables require finer feed (dense ore, 6 to 150 mesh; light material, such as coal, <1 in.) than air tables (which handle ore up to $\frac{1}{4}$ in. and coal up to 3 in.).

**Froth Flotation.**   Froth flotation is the fractionation of particulate solids based on differences in interfacial tensions between the solids, water, and air.   It has been an important process in the beneficiation of ore.   The ore particles are suspended in a liquid at a pulp density of 15 to 35 per cent solids by mechanical or air agitation.   The slurry is treated with chemicals, called *promoters*, which render the surfaces of specific minerals air-avid and water repellent.   Air bubbles are then introduced by direct aeration (see Sec. 5-12, Gas Sparging), agitation or injection of water saturated with air at a much higher pressure.   The air bubbles adhere to the treated particles, and carry them to the surface froth which is skimmed off.

The valuable concentrates from froth flotation may be either the froth product which collects at the top or the underflow product.   In the case of metallic sulfide ores of copper, lead, zinc, nickel, mercury, and molybdenum, and native gold and silver, the values collect in the froth.   In glass-sand flotation, iron-bearing minerals are floated off in the froth, while high-grade silica values appear as underflow.

## Table 5-7. Sizes, Limitations, and Major Applications of Wet Classification Machines

| Type of classifier | Normal size range, ft: Width | Normal size range, ft: Diam. | Normal size range, ft: Max. length | Normal mesh of separation range* | Normal feed tonnage range | Max. oversize in feed | Normal overflow, % solids range | Normal sand product, % solids range | Motor range, hp | Typical applications |
|---|---|---|---|---|---|---|---|---|---|---|
| Nonmechanical: | | | | | | | | | | |
| Cone classifier | ...... | 2–12 | ....... | 28–325 | 2–100 tons/hr | ¼ in. | 5–30 | 35–60 | None | For desliming and primary dewatering |
| Liquid cyclone | ...... | 10 mm to 4 ft | 9 | 48 mesh to 5 μ | ½–1500 gpm | 14–325 mesh | 5–30 | 55–70 | Power for pressure head 5–60 psi | For medium or fine separations and closed-circuit grinding |
| Mechanical: | | | | | | | | | | |
| Drag classifier | 1–10 | ........ | Not critical | 28–200 | 5–350 tons/hr | 1½ in. | 5–30 | 70–83 | 1–10 | For desliming, conveying, and closed-circuit grinding |
| Rake and spiral classifiers | 1–20 | ........ | 40 | 20–200 | 5–350 tons/hr | 1 in. | 5–30 | 75–83 | ½–25 | Closed-circuit grinding, washing and dewatering, desliming, process feed control |
| Bowl classifier | 1½–20 | 4–28 | 40 | 100–325 | 5–200 tons/hr | ½ in. | 5–25 | 75–80 | Bowl: 1–7½<br>Rake: 1–25 | Closed-circuit grinding usually in secondary circuits |
| Bowl desiltor | 4–16 | 20–50 | 40 | 100–325 | 5–250 tons/hr | ½ in. | 1–15 | 75–83 | Bowl: 1–10<br>Rake: 5–25 | Recovery of fine sand, limestone, coal, and fine phosphate rock from large flow volumes |
| Hydroseparator | ...... | 10–150 | ....... | 100–325 | 5–700 tons/hr | ¼ in. | 1–20 | 30–50 | 1–15 | For fine separation where large feed volumes are involved and drainage not critical |
| Solid-bowl centrifuge | ...... | 18–54 in. | 70 in. | 200 mesh to 1 μ | 10–600 gpm | ¼ in. | 1–40 | 10–70 | 15–150 | For fine-size fractionating |
| Sand washer | ...... | 7–12 | ....... | 28–65 | 25–125 tons/hr | 1 in. | 5–15 | 75–80 | 5–10 | For desliming and dewatering large tonnages of solids |
| Countercurrent classifier | ...... | 1½–10 | 40 | 35–100 | 1–600 tons/hr | 3 in. | 5–30 | 75–83 | ¼–25 | Sand-slime separations, washing, closed-circuit grinding |
| Hydraulic: | | | | | | | | | | |
| Sizer | 1½–20 | ........ | 5–20 | 8–150 | 2–100 tons/hr | 3/16 in. | 1–10 | 40–60 | 1–2 for air pressure | Multiproduct unit for exceptionally clean sands fractionated into narrow size ranges; min. 3 tons hydraulic water per ton sand |
| Super Sorter† | 6 | ........ | 40 | 8–150 | 40–150 tons/hr | ⅜ in. | 1–10 | 40–60 | 1 to operate pincer valves | Multiproduct unit for exceptionally clean sands fractionated into narrow size ranges; min. 3 tons hydraulic water per ton sand |
| Siphon Sizer‡ | ...... | 3–30 | ....... | 14–150 | 1–100 tons/hr | 1 in. | 1–10 | 40–60 | None | Two-product unit efficient for desliming and exceptionally clean sands, washing, closed-circuit grinding; min. 2 tons hydraulic water per ton sand |
| Hydroscillator‡ | 4–12 | 4–14 | 40 | 20–150 | 5–250 tons/hr | ½ in. | 5–30 | 75–83 | Oscillator: 3–10<br>Rakes: 5–20 | Two-product unit for exceptionally clean sand having low moisture content; closed-circuit grinding, washing; min. 0.5 ton hydraulic water per ton sand |

* Size of screen retaining 1½ per cent of the overflow solids. † Trademark of Deister Concentrator Co., Inc. ‡ Trademark of Dorr-Oliver Inc.

## 5-19. Crystallization

Crystallization is the production and recovery of solid material from a solution brought about by reduction of solubility due to temperature change or by evaporation of solvent.  A saturated solution is fed to a vessel, often an agitated vessel, and heat is added or removed so that the concentration in the solution is above the solubility at the operating conditions, i.e., supersaturated.  The circulating slurry of crystals in supersaturated liquor is continuously withdrawn, and crystals are removed, e.g., by settling, centrifugation, or filtration.  The mother liquor, saturated at the operating conditions, is returned to the process.

In the crystallization vessel, two processes take place simultaneously: crystal nucleation and crystal growth.  For a given solvent/solute system the rate of nucleation depends almost entirely on the degree of supersaturation in the crystallizer, increasing rapidly as supersaturation increases. The rate of crystal growth depends on number of growth sites (i.e., total surface area), resistance to diffusional transport to growth site, resistance to incorporation into crystal, and degree of supersaturation.  In general, the greater the degree of supersaturation maintained in the crystallizer, the greater the amount of nucleation, the faster the rate of crystallization, and the smaller the product crystals.  Analytical techniques which are available to predict crystallization rates, crystal sizes, and nucleation are beyond the scope of this treatment; see Bennett [in Perry and Chilton (eds.), "Chemical Engineers' Handbook," 5th ed., sec. 17, pp. 8–18, McGraw-Hill Book Company, New York, 1973].

## CHEMICAL KINETICS AND REACTOR DESIGN

## 5-20. Introduction

Nearly all industrial chemical processes involve one or more steps where the process stream undergoes a chemical transformation.  Analysis and design of *chemical reactors* to bring about such transformations are a most important facet of chemical engineering practice, requiring understanding of many other phases of chemical engineering, notably fluid flow, heat and mass transfer, thermodynamics and process control, in addition to a fundamental understanding of the chemical reactions involved.

The principles governing the mechanisms of chemical reactions and the rates at which they proceed comprise the field of chemical kinetics.  Although the theory of chemical kinetics is imperfect, it is a useful guide for analyzing the results of experimental investigations, and is thus the basis for design of practical industrial reactor systems.

The rate of a chemical reaction is best described in terms of number of moles of reactant converted (or moles of product produced) per unit of reactor volume per unit time.  Thus for the reaction with the stoichiometric equation

$$aA + bB + cC \rightleftarrows pP + qQ \tag{5-45}$$

the rate of reaction may be described by

$$rV = \frac{-1}{a}\frac{dN_A}{d\theta} = \frac{-1}{b}\frac{dN_B}{d\theta} = \frac{-1}{c}\frac{dN_C}{d\theta} = \frac{1}{p}\frac{dN_P}{d\theta} = \frac{1}{q}\frac{dN_Q}{d\theta} \quad (5\text{-}46)$$

The rate of appearance (or disappearance) of any product (or reactant) can be obtained in terms of the rate of change of any other participant by multiplication of Eq. (5-46) by the appropriate stoichiometric coefficient. If the reaction volume is constant, the concentration of reactants may be introduced

$$r = -\frac{1}{a}\frac{d(N_A/V)}{d\theta} = -\frac{1}{a}\frac{dC_A}{d\theta} \quad (5\text{-}47)$$

This is approximately true for most liquid-phase reactions taking place in a tank where material is neither added nor removed (*batch reaction*), or where no significant change in liquid level occurs in a flow reaction. It is exact for gas-phase reactions (flow or batch) confined in a rigid vessel.

For steady-state flow reactions with no longitudinal mixing, the composition at any point is constant with time, and the reaction rate may be defined in terms of change in composition with position as

$$r = \frac{F}{a}\frac{dX_A}{dV} \quad (5\text{-}48)$$

where $F$ is volumetric feed rate, $X_A$ is moles of $A$ converted per unit volume of total feed, and $V$ is the reactor volume upstream of that point in the reactor.

Reactions are classified as homogeneous or heterogeneous depending on whether they occur in a single phase or involve contact of several phases, as exothermic or endothermic according to whether they liberate energy or absorb energy from the surroundings, as simple or complex depending on whether or not the rate law follows the stoichiometric coefficients directly.

## 5-21. Homogeneous Reactions

Reactions in which both reactants and products are in the same phase throughout the reaction are termed *homogeneous reactions*. Some reactions involving phase change may be treated as homogeneous even if there is a phase change between the initial reactant state and final product state provided the phase change occurs rapidly enough so as not to affect the overall reaction rate. (A reaction which results in a precipitated product, for example, may be treated as a homogeneous reaction during which the product concentration is constant at its solubility level.) Practically, homogeneous reactions must occur within a liquid or gas phase.

**Reaction Order.** In general, the rates of reactions whose mechanisms are simple have been found to be proportional to integral powers of the concentrations of some or all of the reacting components. That is, for a reaction involving reactants $A$, $B$, and $C$.

$$r = kC_A{}^\alpha C_B{}^\beta C_C{}^\gamma \quad (5\text{-}49)$$

The exponents are experimentally determined and are not necessarily equal to the stoichiometric coefficients of the reaction equation. In general they have a value between 0 and 3. For the special case of single-step reactions involving a small number of reacting molecules, the exponents $\alpha$, $\beta$, and $\gamma$ will be integers.

Rate laws for more complex reactions having several steps, involving catalyzed reactions, chain reactions, etc., will be much more complicated than Eq. (5-49). Such reactions may sometimes be described by Eq. (5-49) in order to obtain a correlation between rates and reactant concentrations. If this is done, the exponents $\alpha$, $\beta$, and $\gamma$ will often have noninteger values. When the reaction mechanism is unknown, this is often the only available procedure. It should be used with caution, however, and usually the values of $\alpha$, $\beta$, and $\gamma$ which are experimentally determined apply only to the conditions (reactant concentration, catalyst concentration, temperature, etc.) for which the data were taken.

The reaction kinetically described by Eq. (5-49) is said to be $\alpha$th order with respect to $A$, $\beta$th order with respect to $B$, and $\gamma$th order with respect to $C$; as a whole, its order is $(\alpha + \beta + \gamma)$th. Theory suggests that order may be related to molecularity of the reaction mechanism; if so, the order of simple homogeneous reactions with respect to each component should be finite and represented by an integer. Most, in fact, are of first, second, or third order, the latter being rare. Under certain conditions, however, reactions appear to be of zeroth order with respect to others, especially in complex reaction mechanisms.

To postulate a reaction mechanism and order without reference to actual rate data is speculative and dangerous if a reactor design is to be based on the postulate. The order assigned a reaction must rationalize reliable experimental kinetic data. On the other hand, data that indicate a homogeneous reaction to be of exotic order should be critically examined, or the method of their treatment should be questioned, or both.

Equation (5-49) gives the rate of a reaction proceeding irreversibly among three components. If the reaction of interest were, instead, a reversible one (as, strictly speaking, all reactions are), such as

$$A + B + C \leftrightarrows P + Q \tag{5-50}$$

Eq. (5-49) would describe only the rate of the forward half-reaction. The reverse might be expected to exhibit a rate proportional to simple powers of the concentrations of the products $P$ and $Q$:

$$r' = k'C_P{}^\rho C_Q{}^\sigma \tag{5-51}$$

Thus the reverse half-reaction would be $\rho$th order with respect to $P$, $\sigma$th order with respect to $Q$, and $(\rho + \sigma)$th order over-all.

It should be noted that the net rate of a reversible reaction is the algebraic sum of its forward and reverse rates. For the reaction described by Eqs. (5-49) and (5-51),

$$r_{\text{net}} = r - r' \tag{5-52}$$

The coefficients $k$ and $k'$ of Eqs. (5-49) and (5-51) are known as specific rate constants, peculiar to a particular reaction and temperature but independent of concentrations of reactants.

**Integrated Rate Equations.** If one substitutes the appropriate rate law [e.g., Eq. (5-49)] into Eq. (5-47) and relates the concentrations by means of a material balance, it is possible to integrate Eq. (5-47) to give reactant concentrations as a function of time. For irreversible reactions of integral order, these integrated equations have a simple form. Table 5-8 presents integrated equations for simple irreversible, constant-volume reactions.

Reaction rate constants [$k$ in Eq. (5-49)] may be determined from experimental data by plotting concentration vs. time in a manner determined by the form of the integrated rate law [e.g., for a first-order reaction plot $\ln(C_{A0}/C_A)$ vs. time; the reaction rate constant is the slope of the line].

Reversible reactions, consecutive reactions $(A \rightarrow B \rightarrow D)$, and parallel reactions $\left( A \underset{C}{\overset{B}{\diagup\diagdown}} \right)$ require much more complicated rate equations for their description. Whereas the formulation of an appropriate rate statement is often simple, the solution of the resulting differential equations is likely to be difficult and is beyond the scope of this section. Typical solutions are presented in monographs on applied kinetics (e.g., Walas. "Reaction Kinetics for Chemical Engineers," McGraw-Hill Book Company, New York, 1959). Many rate equations previously considered too difficult to solve because of their demand for awkward or tedious numerical approximation methods can now be solved by means of electronic analog or digital computers.

**Equilibrium and Kinetics.** Inasmuch as all chemical reactions are limited by a chemical equilibrium, reaction kinetics really describes the rate of approach to that equilibrium rather than to a stoichiometric completeness of the reaction. At equilibrium, the net rate of reaction Eq. (5-52) is zero; whence it follows that for a reaction whose rate law is described by the molecularity (i.e., stoichiometric equation) the equilibrium constant for the reaction is related to the forward and reverse specific rate constants, thus:

$$\frac{C_A{}^a C_B{}^l C_C{}^c}{C_P{}^p C_Q{}^q} = K_c = \frac{k}{k'} \tag{5-53}$$

It is clear that the larger the value of $K_c$, the larger the magnitude of the forward rate constant relative to the reverse and the closer to stoichiometric completeness the equilibrium conversion. Also, the greater the concentration of reactant above the equilibrium concentration, the faster the progress of the reaction toward equilibrium. All other things being equal, conditions that increase the value of the equilibrium constant are favorable to the net kinetics of the reaction. Equilibrium constants are discussed in Sec. 2.

**Effect of Temperature.** Homogeneous reactions are strongly temperature-dependent. Their specific reaction rate always increases with increas-

# Table 5-8. Rate Equations for Simple Order

| Order | Differential equation | Constant-volume process |
|---|---|---|
| Zero | $-\dfrac{dN_A}{Vd\theta} = k$ | $k(\theta - \theta_0) = C_A^0 - C_A$ |
| One-half | $-\dfrac{dN_A}{Vd\theta} = kC_A^{1/2}$ | $k(\theta - \theta_0) = 2(C_A^{01/2} - C_A^{1/2})$ |
| First | $-\dfrac{dN_A}{Vd\theta} = kC_A$ | $k(\theta - \theta_0) = \ln \dfrac{C_A^0}{C_A}$ |
| Second | $-\dfrac{dN_A}{Vd\theta} = kC_A^2$ | $k(\theta - \theta_0) = \dfrac{1}{C_A} - \dfrac{1}{C_A^0}$ |
|  | $-\dfrac{dN_A}{Vd\theta} = kC_AC_B$ | $k(\theta - \theta_0) = \dfrac{1}{C_B^0 - C_A^0} \ln \dfrac{C_AC_A^0 + C_A^0C_B^0 - C_A^{02}}{C_AC_B^0} \qquad C_A^0 \neq C_B^{0*}$ |
| Third | $-\dfrac{dN_A}{Vd\theta} = kC_A^3$ | $2k(\theta - \theta_0) = \dfrac{1}{C_A^2} - \dfrac{1}{C_A^{02}}$ |
|  | $-\dfrac{dN_A}{Vd\theta} = kC_AC_BC_C$ | $k(\theta - \theta_0) = \dfrac{1}{(C_B^0 - C_A^0)(C_C^0 - C_A^0)} \ln \dfrac{C_A^0}{C_A} + \dfrac{1}{(C_B^0 - C_C^0)(C_B^0 - C_A^0)} \ln \left(\dfrac{C_B^0}{C_A + C_B^0 - C_A^0}\right)$ $+ \dfrac{1}{(C_C^0 - C_B^0)(C_C^0 - C_A^0)} \ln \left(\dfrac{C_C^0}{C_A + C_C^0 - C_A^0}\right) \qquad C_B^0 \neq C_C^0 \neq C_A^0†$ |

NOTE: $C^0$ and $\theta_0$ are initial conditions for time and concentration, respectively.
* If $C_A^0 = C_B^0$, use expression for $-dN_A/Vd\theta = kC_A^2$.
† If $C_A^0 = C_B^0 = C_C^0$, use expression for $-dN_A/Vd\theta = kC_A^3$.

ing temperature. The effect of temperature is described by the semitheoretical relation of Arrhenius:

$$k = Ae^{-E/RT} \tag{5-54}$$

The coefficient $A$ (called the frequency factor) and the exponent $E$ (called the energy of activation) have theoretical interpretations, but they are best regarded by the process designer as empirical quantities peculiar to a particular chemical reaction and evaluable from experimental rate data. Thus a plot of ln $k$ against $1/T$ should be linear and should have a slope of $-E/R$ and an intercept of ln $A$, provided $A$ and $E$ are independent of temperature. In fact, both the frequency factor and energy of activation vary slightly with temperature, but over the temperature ranges normally encountered, they may be assigned constant average values without serious error. Figure 5-21 shows an Arrhenius plot for a second-order reaction. The activa-

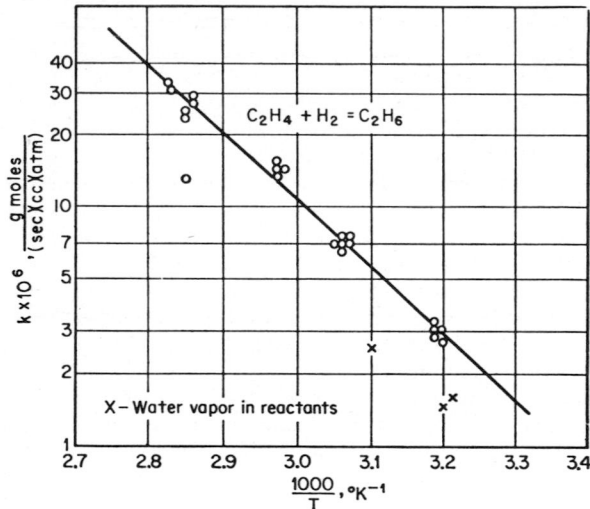

Fig. 5-21. Arrhenius plot for hydrogenation of ethylene. (*Smith, "Chemical Engineering Kinetics," p. 74, McGraw-Hill Book Company, New York, 1956.*)

tion energy represented by the line drawn through the data is 10,000 cal/g mole.

The failure of rate data to fit Eq. (5-54) may be accepted as evidence that:

1. A reversible reaction has been treated as if it were irreversible, and the effect of temperature on the equilibrium is significant.

2. An otherwise incorrect mechanism has been assigned to the reaction.

3. The specific rate constant has been evaluated from the experimental data incorrectly.

4. The reaction is heterogeneous, and its rate is influenced by adsorption or by some other physical process.

Energies of activation range from less than 1,000 to greater than 100,000 cal/g mole. For most reactions, the value will be between 10,000 and 70,000 cal/g mole.

A long popular rule of thumb states that the rate of a reaction approximately doubles for each 10°C rise in temperature. Inspection of Eq. (5-54) shows that this can be true only for particular combinations of activation energy and temperature. At high temperatures especially, typical activation energies are such that considerably more than 10°C is required to double the reaction rate.

**Effect of Concentration.** At constant temperature, the specific rate constant is assumed to be independent of the concentration of reactants and products, so that equations like (5-49) or (5-51) show explicitly the effect of concentration on the progress of the reaction. In general, this is a valid assumption for homogeneous uncatalyzed reactions. If there seems to be a dependency of $k$ on concentration, the most likely reasons are that the wrong order (or mechanism) has been postulated, that a catalyst is influencing the reaction, or that the temperature has not remained constant. Heterogeneous reactions may yield apparent rate constants that reflect complex combinations of physical and chemical processes, and hence may vary with concentration.

**Homogeneous Catalyzed Reactions.** A catalyst is a substance which affects the rate of a chemical reaction without entering the reaction in any stoichiometric sense. The catalyst may undergo net physical or chemical change in the course of the reaction, but often it does neither. Trace amounts of a catalyst can greatly influence the reaction rate and mechanism. A catalyst can be positive (increase the rate) or negative (decrease the rate); if not otherwise stipulated, a positive effect is implied. Negative catalysts are called inhibitors.

Some homogeneous gas-phase and liquid-phase (most commonly the latter) reactions can be catalyzed by materials that are soluble in the reacting mass and therefore do not destroy the homogeneity of the system. Sometimes homogeneous catalysis occurs without the prior knowledge of the kinetic experimenter. In such a case, grossly incorrect conclusions can be drawn about the kinetics of the reaction. Usually, catalytic behavior will be signaled by one or more of the following phenomena:

1. Irrationally rapid or accelerating rate of reaction
2. Irrationally slow rate of reaction
3. Apparent zero or fractional order with respect to known reactants
4. Abnormal temperature dependency of the rate

Although homogeneous catalysts are likely to be effective in very small amounts, most catalyzed reactions will exhibit a definite order with respect to the catalyst below a particular concentration. This order should be determined experimentally for the most reliable statement of the kinetics of the reaction. In some instances, however, the catalyst concentration is kept constant, and the reaction may be described satisfactorily for design purposes by the assignment of apparent orders to the reactants and products, with no explicit ordering of the catalyst.

## 5-22. Heterogeneous Reactions

A chemical reaction is said to be heterogeneous if more than one phase is an active participant and if transfer of materials to phase boundaries has an effect on the rate of reaction. Heterogeneous reactions commonly involve fluid-solid mass transfer (e.g., catalysis of a fluid reaction mixture on the surface of solid catalyst pellets, combustion of a solid fuel in air, acid leaching of metals from ores) or mass transfer between two fluid phases (e.g., absorption of gaseous sulfur dioxide by weak aqueous sodium hydroxide, nitration of toluene by nitric acid). In the limiting cases of extremely high reaction rates heterogeneous reactions are analyzed as mass-transfer problems. Reactions where mass-transfer rates are sufficiently rapid can often be analyzed as homogeneous reactions taking place in one of the fluid phases.

**Uncatalyzed Heterogeneous Reactions.** In an uncatalyzed heterogeneous reaction, chemical action occurs among components that are simultaneously being transferred physically from phase to phase. The apparent rate of the reaction is in fact the rate of a more complicated process. It will be influenced not only by factors affecting chemical kinetics, but also by those affecting the rate of interphase mass transfer. Among the latter are:

1. Amount of interfacial surface
2. Concentration of reactants in each phase
3. Concentration of products in each phase
4. Relative velocity at the interface
5. Temperature (in its effects on both phase equilibrium and diffusivities)
6. Presence of a solid resistance at the interface (e.g., an ash layer formed on a reacting solid)

Sometimes the conditions of a heterogeneous process are such that the chemical reaction is relatively rapid, whence the rate of physical transport becomes effectively that of the over-all process. Sometimes the reverse is true. More generally, the rates of physical transfer and chemical reaction are of the same order of magnitude, in which case each contributes significantly to the kinetics of the over-all process.

Inasmuch as temperature effects in chemical and physical processes are quite different, the selection of operating temperature in a heterogeneous reaction may determine which component process is controlling. Figure 5-22 shows the rate of a gas-solid reaction, in which an ash film is formed, as a function of temperature. Over section $AB$ the process rate is essentially that of the chemical reaction; over section $CD$, that of the diffusion of reactant through the gas film; and over $EF$, that of diffusion through the ash layer. Over sections $BC$ and $DE$ more than one phenomenon controls. *It should be noted that the over-all rate never can be greater than that of the slowest component process.*

Whenever more than one phenomenon determines the effective reaction rate, the over-all rate equation becomes very difficult to solve. Design equations combining reaction effects with mass-transfer effects have been derived for a number of special cases. These are still largely not supported by extensive experimental data. Such systems are often treated as modified mass-transfer problems [e.g., Maddox, in Perry and Chilton (eds.), "Chemical Engineers' Handbook," 5th ed., sec. 14, pp. 6–8, McGraw-Hill Book Company, New York, 1973].

A number of techniques exist which may be used to scale-up reactor systems from empirical data. These must be applied carefully and generally are restricted by the conditions of the experiment. Such scale-up methods are beyond the scope of this text. See Lin, in Perry and Chilton (eds.), "Chemical Engineers' Handbook," 5th ed., sec. 4, pp. 16–20, McGraw-Hill Book Company, New York, 1973.

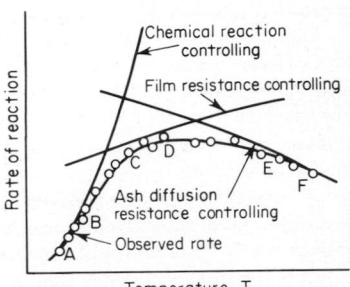

Fig. 5-22.    Rate of reaction as affected by combined resistances. (*Levenspiel, "Chemical Reaction Engineering," p. 355, John Wiley & Sons, Inc., New York, 1962.*)

**Catalyzed Heterogeneous Reactions.**    Although heterogeneous reactions responsive to catalysis may involve any combination of phases, the examples most common and industrially most important are solid-fluid systems in which the catalyst is the solid phase. The reactants and products may be gaseous, liquid, or both. The solid catalyst may be a container wall, a metal gauze, or a granular mass. Usually it is the latter, in either fixed- or fluidized-bed form, with particles seldom larger than 0.25 in. A reaction catalyzed by a solid is believed to take place at the surface of the solid (the surface may be internal, i.e., interstitial within a porous particle) and to involve activated adsorption or chemisorption on that surface.

The mechanism of a fluid-phase reaction catalyzed by a solid is extremely complex and may comprise as many as seven sequential steps:

1. Diffusion of reactants to the outside catalyst surface from the body of the fluid phase

2. Diffusion of reactants into the catalyst pores (or through an inert deposit to active surface regions)

3. Adsorption of reactants

4. Chemical reaction in the adsorbed state

5. Desorption of products

6. Diffusion of products from the catalyst pores

7. Diffusion of products from the outside catalyst surface into the body of the fluid phase

Any one or more of these steps may be slow enough to control the rate of the entire sequence. Often control can be ascribed to a single step, and an adequate design procedure can be based on this premise.

Analysis of a catalyzed heterogeneous process, then, consists in examining the data for evidence of the rate-controlling step (the experimental program must have been planned to yield such evidence) and applying the data to evaluate coefficients and indices of whatever equations appropriately describe the rate.

The procedure is outlined as follows:

1. If the degree of conversion depends on the linear velocity of reacting fluid with respect to the catalyst (at constant space velocity), mass transfer between the body of the fluid and the external surface of the catalyst is controlling. The rate of the process is then dependent on the rate of diffusion to or from the catalyst, and the transfer coefficient may be calculated from established correlations for mass-transfer coefficients. The estimated value of the mass-transfer coefficient in the fluid phase can be used to calculate the concentration of reactants at the surface of the catalyst particle.

2. If the degree of conversion is independent of fluid velocity but depends on size of the catalyst pellets, diffusion of reactants or products within the pores of the catalyst is a rate-controlling step. In such a case, the data should be treated to evaluate an effectiveness factor, defined as the ratio of observed reaction rate to that which would obtain if pore diffusional resistance were negligible.

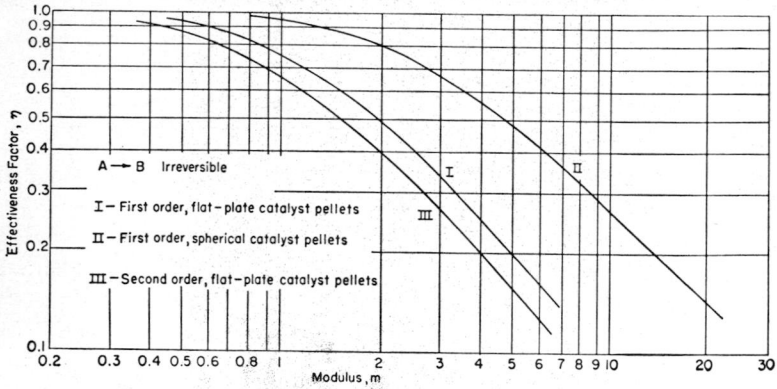

Fig. 5-23. Effectiveness factor for equations of simple order.

Figure 5-23 is a typical plot of effectiveness factor against a modulus

$$m = 1 \sqrt{2k(C_i{}^0)^{n-1}(\bar{r}D_i)}$$

in which $C^0$ is the concentration of reactant at the external catalyst surface, and $\bar{r}$, the average pore radius, is calculated as $2V_g/S_g$ (Examples are given in Table 5-9).

## Table 5-9. Values of Internal Surface Area, Pore Volume, and Average Pore Radius for Typical Catalysts

| Catalyst | $S_g$, m²/g | $V_g$, cc/g | $\bar{r} = 2V_g/S_g$, A |
|---|---|---|---|
| Activated carbons................................. | 500–1500 | 0.6–0.8 | 10–20 |
| Silica gels...................................... | 200–700 | 0.4 | 15–100 |
| Silica-alumina cracking catalysts ~10–20% Al₂O₃......... | 200–700 | 0.2–0.7 | 15–150 |
| Silica-alumina (steam-deactivated)..................... | 67 | 0.519 | 155 |
| Silica-magnesia microsphere: | | | |
|   Nalco, 25% MgO............................. | 630 | 0.451 | 14.3 |
|   Nalco, steam treated, 621°C, 400 psig for 24 hr......... | 322 | 0.283 | 17.6 |
| Da-5 silica-magnesia................................ | 656 | 0.365 | 11.1 |
| Activated clays................................... | 150–225 | 0.4–0.52 | ~100 |
| TCC clay pellets (MgO, CaO, Fe₂O₃, SO₄) = ~10%...... | 276 | 0.363 | 26.3 |
| Clays: | | | |
|   Montmorillonite (raw)............................ | 214 | 0.297–0.306 | ~28 |
|   Montmorillonite (heated 550°C)..................... | 212 | 0.268 | 25.2 |
|   Vermiculite.................................... | 35 | 0.063–0.057 | ~314 |
| Activated alumina (Alorico)........................... | 175 | 0.388 | 45 |
| CoMo on alumina.................................. | 168–251 | 0.261–0.331 | 20–40 |
| Kieselguhr (Celite 296)............................. | 4.2 | 1.14 | 11,000 |
| Fe-synthetic NH₃ catalyst........................... | 4–13 | 0.12 | 200–1000 |
| Co-ThO₂-Kieselguhr 100:18:100 (reduced) pellets......... | 42.3 | 0.73 | 345 |
| Co-ThO₂-MgO (100:6:12) (reduced) granular.............. | 84.1 | 0.80 | 190 |
| Co-Kieselguhr 100:200 (reduced) granular............... | 22.8 | 2.31 | 2030 |
| Porous plate (Coors No. 760)......................... | 1.6 | 0.172 | 2150 |
| Pumice.......................................... | 0.38 | | |
| Fused copper catalyst............................... | 0.23 | | |
| Ni film.......................................... | 8.4 | | |
| Ni on pumice, 91.8% pumice.......................... | 1.27 | | |

$S_g$ = catalyst surface area
$V_c$ = catalyst pore volume
$\bar{r}$ = average radius of pore
A = angstrom unit = $1 \times 10^{-8}$ cm

3. If absorption-desorption or chemical reaction at the catalyst surface is the controlling process, a mechanism must be found that will identify which of the possibilities are rate-controlling and which are equilibrium steps. For example, for the stoichiometric reaction

$$A \leftrightarrows R \qquad (5\text{-}55)$$

the following mechanistic steps involving the participants $A$ and $R$ and a catalyst site $s$ may be postulated:

$$A + s \rightleftarrows As \qquad (5\text{-}55a)$$

$$As \rightleftarrows Rs \qquad (5\text{-}55b)$$

$$Rs \rightleftarrows R + s \qquad (5\text{-}55c)$$

## Table 5-10. Mechanisms and Their Corresponding Rate Equations

| Chemical equation | Catalytic steps | Rate equation* |
|---|---|---|
| $A \rightleftharpoons R$ | $A + s \rightleftharpoons As$ | $r = \dfrac{k(C_A - C_R/K)}{1 + K_R C_R}$ |
| | $As \rightleftharpoons Rs$ | $r = \dfrac{k(C_A - C_R/K)}{1 + K_A C_A + K_R C_R}$ |
| | $Rs \rightleftharpoons R + s$ | $r = \dfrac{k(C_A - C_R/K)}{1 + K_A C_A}$ |
| $A \rightleftharpoons R$ | $2A + s \rightleftharpoons A_2 s$ | $r = \dfrac{k(C_A^2 - C_R^2/K^2)}{1 + K_R C_R + K_R C_R^2}$ |
| | $A_2 s + s \rightleftharpoons 2As$ | $r = \dfrac{k(C_A^2 - C_R^2/K^2)}{(1 + K_R C_R + K_A C_A^2)^2}$ |
| | $As \rightleftharpoons Rs$ | $r = \dfrac{k(C_A - C_R/K)}{1 + K_A C_A^2 + K_A' C_A + K_R C_R}$ |
| | $Rs \rightleftharpoons R + s$ | $r = \dfrac{k(C_A - C_R/K)}{1 + K_A C_A^2 + K_A' C_A}$ |
| $A \rightleftharpoons R$ | $A + 2s \rightleftharpoons 2A_{\frac{1}{2}} s$ | $r = \dfrac{k(C_A - C_R/K)}{(1 + \sqrt{K_R C_R} + K_R' C_R)^2}$ |
| | $2A_{\frac{1}{2}} s \rightleftharpoons Rs + s$ | $r = \dfrac{k(C_A - C_R/K)}{(1 + \sqrt{K_A C_A} + K_R C_R)^2}$ |
| | $Rs \rightleftharpoons R + S$ | $r = \dfrac{k(C_A - C_R/K)}{1 + \sqrt{K_A C_A} + K_A' C_A}$ |
| $A \rightleftharpoons R + S$ | $A + s \rightleftharpoons As$ | $r = \dfrac{k(C_A - C_R C_S/K)}{1 + K_{RS} C_R C_S + K_R C_R + K_S C_S}$ |
| | $As + s \rightleftharpoons Rs + Ss$ | $r = \dfrac{k(C_A - C_R C_S/K)}{(1 + K_A C_A + K_R C_R + K_S C_S)^2}$ |
| | $\left.\begin{array}{l} Rs \rightleftharpoons R + s \\ Ss \rightleftharpoons S + s \end{array}\right\}$ | $r = \dfrac{k(C_A - C_R C_S/K)}{C_S(1 + K_A C_A + (K_{AS} C_A/C_S) + K_S C_S)}$ |
| $A \rightleftharpoons R + S$ | $A + s \rightleftharpoons As$ | $r = \dfrac{k(C_A - C_R C_S/K)}{1 + K_R C_R + K_{RS} C_R C_S}$ |
| | $As \rightleftharpoons Rs + s$ | $r = \dfrac{k(C_A - C_R C_S/K)}{1 + K_A C_A + K_R C_R}$ |
| | $Rs \rightleftharpoons R + s$ | $r = \dfrac{k(C_A - C_R C_S/K)}{C_S(1 + K_A C_A + K_{AS} C_A/C_S)}$ |
| $A + B \rightleftharpoons R$ | $A + s \rightleftharpoons As$ | $r = \dfrac{k(C_A - C_R/K C_B)}{1 + (K_{RB} C_R/C_B) + K_B C_B + K_R C_R}$ |
| | $B + s \rightleftharpoons Bs$ | $r = \dfrac{k(C_B - C_R/K C_A)}{1 + K_A C_A + (K_{RA} C_R/C_A) + K_R C_R}$ |
| | $As + Bs \rightleftharpoons Rs + s$ | $r = \dfrac{k(C_A C_B - C_R/K)}{(1 + K_A C_A + K_B C_B + K_R C_R)^2}$ |
| | $Rs \rightleftharpoons R + s$ | $r = \dfrac{k(C_A C_B - C_R/K)}{1 + K_A C_A + K_B C_B + K_{AB} C_A C_R}$ |
| $A + B \rightleftharpoons R + S$ | $A + s \rightleftharpoons As$ | $r = \dfrac{k(C_A - C_R C_S/K C_B)}{1 + (K_{RS} C_R C_S/C_B) + K_B C_B + K_R C_R + K_S C_S}$ |
| | $B + s \rightleftharpoons Bs$ | $r = \dfrac{k(C_B - C_R C_S/K C_A)}{1 + (K_{RS} C_R C_S/C_A)^{\frac{1}{2}} + K_A C_A + K_R C_R + K_S C_S}$ |
| | $As + Bs \rightleftharpoons Rs + Ss$ | $r = \dfrac{k(C_A C_B - C_R C_S/K)}{(1 + K_A C_A + K_B C_B + K_R C_R + K_S C_S)^2}$ |
| | $\left.\begin{array}{l} Rs \rightleftharpoons R + s \\ Ss \rightleftharpoons S + s \end{array}\right\}$ | $r = \dfrac{k[(C_A C_B/C_S) - C_R/K]}{1 + K_A C_A + K_B C_B + K_S C_S + K_{AB} C_A C_B/C_S}$ |

### Table 5-10. Mechanisms and Their Corresponding Rate Equations (Continued)

| Chemical equation | Catalytic steps | Rate equation* |
|---|---|---|
| $A + B \rightleftharpoons R + S$ | $A + 2s \rightleftharpoons 2A_{\frac{1}{2}}s$ | $r = \dfrac{k(C_A - C_R C_S/KC_B)}{[1 + K_{RS}C_R C_S/C_B + K_B C_B + K_R C_R + K_S C_S]^2}$ |
| | $B + s \rightleftharpoons Bs$ | $r = \dfrac{b(C_B - C_R C_S/KC_A)}{1 + \sqrt{K_A C_A} + (K_{RS}C_R C_S/C_A) + K_R C_R + K_S C_S}$ |
| | $2A_{\frac{1}{2}}s + Bs \rightleftharpoons Rs + Ss + s$ | $r = \dfrac{k(C_A C_B - C_R C_S/K)}{(1 + \sqrt{K_A C_A} + K_B C_B + K_R C_R + K_S C_S)^3}$ |
| | $Rs \rightleftharpoons R + s$ | $r = \dfrac{k(C_A C_B/C_S - C_R/K)}{1 + K_A\sqrt{C_A} + K_B C_B + (K_{AB}C_A C_B/C_S) + K_S C_S}$ |
| | $Ss \rightleftharpoons S + s$ | $r = \dfrac{k(C_A C_B/C_R - C_S/K)}{1 + \sqrt{K_A C_A} + K_B C_B + K_R C_R + K_{AB}C_A C_B/C_R}$ |
| $A + B \rightleftharpoons R + S$ | $B + s \rightleftharpoons Bs$ | $r = \dfrac{k(C_B - C_S C_R/KC_A)}{1 + K_R C_R + K_{RS}C_R C_S/C_A}$ |
| | $A + Bs \rightleftharpoons Rs + S$ | $r = \dfrac{k(C_A C_B - C_R C_S/K)}{1 + K_R C_R + K_B C_B}$ |
| | $Rs \rightleftharpoons R + s$ | $r = \dfrac{k[(C_A C_B/C_S) - C_R/K]}{1 + (K_{AB}C_A C_D/C_S) + K_B C_B}$ |

NOTE: $K_{AB}\ldots$ = combined equilibrium constants; $K$ = over-all equilibrium constant for the chemical equation; $k$ = constant.
* The rate equation is opposite the catalytic step assumed to be rate-controlling.

Assumption of each of these in order as the controlling step results in a different rate equation, which may be validated against experimental kinetic data. If none meet the test, a new mechanism must be tried. Lin (*op. cit.*, pp. 7–8) describes the procedure for obtaining rate equations based on postulated mechanisms. The resulting rate equations for a number of simple examples are summarized in Table 5-10.

The discovery of a suitable rate equation by the methods described does not constitute establishment of the true mechanism by which the heterogeneous catalytic reaction is occurring. Nevertheless, whenever the true mechanism is unknown—and it usually is—a rate-equation formulation by the kind of semitheoretical approach outlined offers the most reliable device for rationalizing kinetic data and extending them to conditions not exactly covered in the experiment that yielded them.

## 5-23. Interpretation of Kinetic Data

The design of a chemical reactor should be based on properly collected and analyzed laboratory data. These data will, in general, be one of three types:

1. Measurements of composition as a function of time in a batch reactor of constant volume operated at various temperatures and pressures

2. Measurements of outlet composition as a function of feed rate to a flow reactor of constant volume operated at various pressure and temperature levels

3. Measurements of composition as a function of time in a variable-

volume batch reactor operated at constant temperature and substantially constant pressure.

The third type of data is much less common than the other two, and the experimental technique is more difficult. Data of the second type are generally the most dependable and simple to obtain. This method has the advantage of direct applicability to flow-type reactors. Data of the first type should not be used for the design of flow reactors unless it is certain that the extent of mixing is the same in both the batch and flow systems. In all cases it is important that the temperature does not vary with time in the batch reactor or with position in the flow reactor.

Data should be taken and analyzed to determine reaction order with respect to all participants, reaction rate constants, and temperature dependence. The first step is to determine the reaction order. This treatment consists, generally, of testing the validity of the rate equations, differential or integrated, suspected to be appropriate, until one is found that fits the data. Many techniques exist for testing the validity of proposed rate equations. The best use some integrated form of the rate equation, though preliminary assessment can often be made by determining the differential reaction rate as a function of concentration. The use of differential rate equations requires the direct measurement of reaction rates in the laboratory (usually not feasible) or the differentiation of composition-time data (a procedure highly susceptible to error). Differential equations are useful for a quick, tentative assessment of a reaction, but integrated equations are preferred for final evaluation.

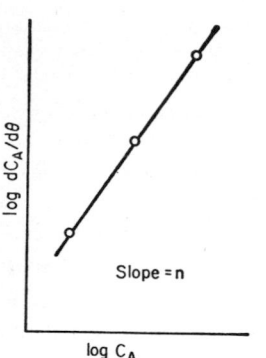

Fig. 5-24. Determining reaction order: differentiation. (*Walas, "Reaction Kinetics for Chemical Engineers," Fig. 2-1, McGraw-Hill Book Company, New York, 1959.*)

For simple reactions involving more than one reactant, the order with respect to any one component can be determined by holding all other reactant concentrations constant. This may be accomplished by using large excesses of these other reactants. For the specific reaction $A + B \rightarrow C$, with reactant $B$ present in great excess, assumed to be substantially irreversible, several analysis techniques are presented below, as examples of the general approach.

1. *Differential rate equation.* A logarithmic plot of $dC_A/d\theta$ (or, by approximation, $\Delta C_A/\Delta\theta$) against $C_A$ yields a straight line, of which the slope is $n$, the order of the reaction (Fig. 5-24).

2. *Integrated rate equation, first order.* A plot of $-\log (C_A/C_{Ao})$ against $\theta$ yields a straight line through the origin (Fig. 5-25).

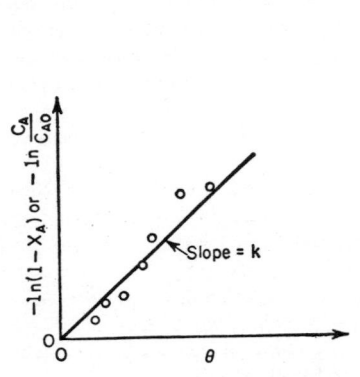

FIG. 5-25.   Test for the first-order reaction. (*Levenspiel, op. cit., Fig. 3, p. 48.*)

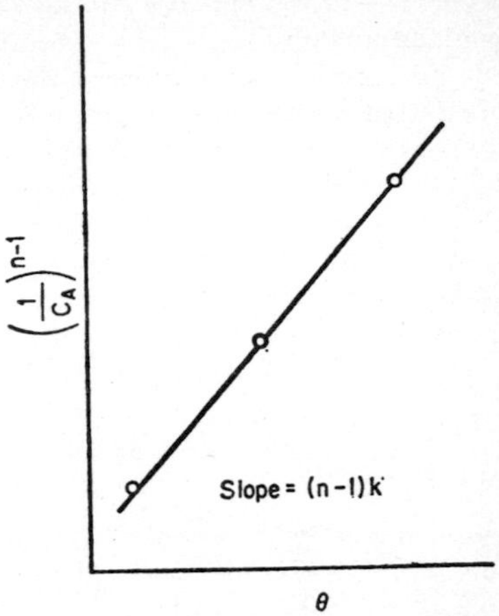

FIG. 5-26.   Determining order of reaction: integrated equation. (*Walas, op. cit., Fig. 2-2.*)

3. *Integrated rate equation, other than first order.*   When $n > 1$, the general rate equation

$$-\frac{dC_A}{d\theta} = kC_A{}^n \tag{5-56}$$

is integrated between the limits $C_{AO}$ and $C_A$ to give

$$\left(\frac{1}{C_A}\right)^{n-1} - \left(\frac{1}{C_{AO}}\right)^{n-1} = (n-1)k\theta \tag{5-57}$$

From Eq. (5-57), a plot of $(1/C_A)^{n-1}$ against $\theta$ yields a straight line with the intercept $(1/C_{AO})^{n-1}$ and the slope $k(n-1)$ (Fig. 5-26).

4. *Integrated rate equation: method of half-life.*   The half-life, or time for 50 per cent conversion, is a useful criterion for order.   Integration of Eq. (5-56) between the limits $C_{AO}$ and $0.5C_{AO}$ yields the following values for half-life, $\theta_{1/2}$:

$$\theta_{1/2} = \begin{cases} \dfrac{1}{2k}\,C_{AO} & n = 0 \\[2ex] \dfrac{0.69}{k} & n = 1 \\[2ex] \dfrac{1}{kC_{AO}} & n = 2 \\[2ex] \dfrac{2^{n-1}-1}{k(n-1)(C_{AO})^{n-1}} & n = 3 \end{cases}$$

Hence, if experiments are run at different initial concentrations and log $\theta_{1/2}$ is plotted against log $C_{AO}$, a straight line of slope $1 - n$ results (Fig. 5-27).

5. *Integrated rate equation: method of reference curves.* Inspection of the integrated rate equations indicates that the ratio of times required for any two degrees of conversion is dependent only on those conversion fractions and on the reaction order. Walas (*op. cit.*, p. 35) has calculated such ratios, taking 90 per cent conversion as the arbitrary convenient reference, to give the useful curves of Fig. 5-28. If one plots per cent conversion, $100C_A/C_{AO}$ against $\theta/\theta_{0.9}$, to the same scale as in Fig. 5-28, comparison of the two graphs will reveal the order of the reaction.

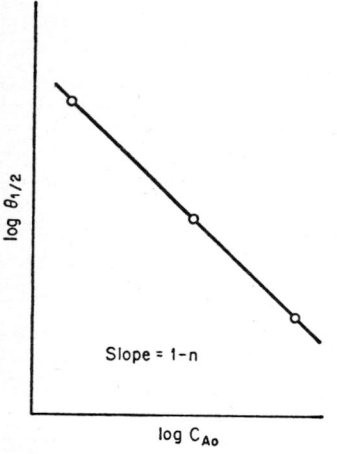

Fig. 5-27. Determining order of reaction: half-lives. (*Walas, op. cit.,* Fig. 2-3.)

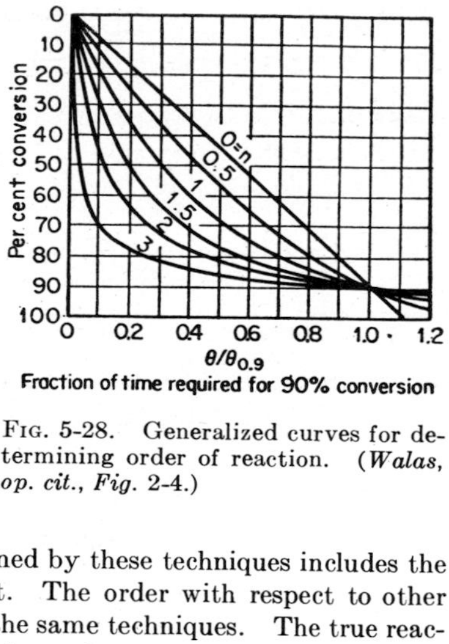

Fig. 5-28. Generalized curves for determining order of reaction. (*Walas, op. cit., Fig. 2-4.*)

The reaction rate constant $k$ determined by these techniques includes the concentration of the excess component. The order with respect to other components must be determined using the same techniques. The true reaction rate constant can be determined from the pseudo rate constants by dividing out the (constant) concentrations of reactants raised to the appropriate order.

Failure of the experimental data to fit the form of the equations indicated in steps 1 through 4 above is evidence that the postulated simple rate equation is in error. For example, the reaction may be reversible, may involve chain steps, etc.

Sometimes complicated reactions can be simplified in their analysis by approximations permitted in the way experimental data are taken. For example, the early stages of reversible reaction among initially pure reactants will act substantially as if the reaction were irreversible, and the data may be so treated. Again the order of each of several reactants sometimes can be determined individually from experiments in which all but the reactant of interest is present in large stoichiometric excess. Extreme care must be taken in such approximations, however, inasmuch as sampling and analysis problems can induce large experimental error.

## 5-24. Reactor Design

Once a suitable rate equation that fits the experimental kinetic data has been discovered, design of the plant reactor can proceed. Five steps are involved:

1. Selection of the type of reactor
2. Selection of the shape or proportions of reactor
3. Sizing the reactor
4. Selection of materials of construction
5. Design of reactor auxiliaries

**Type of Reactor.** Reactors generally are of four basic types: batch, semicontinuous stirred tank, continuous stirred tank, and tubular (plug-flow). The choice of type depends on the state of the reactants and prod-

## Table 5-11. Types of Reactors and Their Applications

| Type | Conditions Suitable |
|---|---|
| Batch | Intermittent operation<br>Holdup of charge for testing required<br>Individual fine adjustment necessary<br>Small production rate<br>Liquid-liquid or liquid-solid reactions<br>Long induction period involved |
| Semicontinuous | Gas-liquid reactions<br>Large excess of one reagent desired, for liquid-liquid or liquid-solid reactions |
| Continuous stirred tank | Homogeneous liquid-phase reactions best<br>Steady availability of all reactants<br>Steady demand for product<br>Liquid-solid reactions satisfactory if solids are easily suspended<br>Gas-solid reactions satisfactory if fluidization is feasible |
| Tubular (plug-flow) | Solid-catalyzed gas-phase reactions appropriate<br>Homogeneous fluid-phase reactions best<br>Steady availability of reactants<br>Steady demand for product |

ucts, the nature of the reaction, the rate of production, and the character of the rest of the process of which the reaction is a part. The choice is often determined by economic factors. Table 5-11 indicates some of the conditions for which each type may be suitable.

**Shape of Reactor.** Except for the simplest of reactions, it is desirable that a plant reactor be of the same type and generally of the same shape as the laboratory unit from which the kinetic data for the reaction were obtained. In extremely complicated cases a pilot prototype of the plant unit should be operated. Even in the latter instance, however, rational scale-up of the pilot unit will be required, and may lead to a plant reactor of different proportions from the prototype.

For perfect scale-up, complete similarity (geometrical, kinematic, dynamic, thermodynamic, and chemical) should be preserved between small and large models. If one is to operate with the same process stream in both models, as one must, complete similarity is impossible. A compromise must then be made, frequently requiring longer reaction times or lower reactor productivity (rate of production per unit volume) in the large unit than in the small, but meeting the most critical demands of the system— good mixing, catalyst distribution, or temperature control, for example.

Walas (*op. cit.*, chap. 10) and Lin (*op. cit.*, pp. 16–20) give excellent discussions of reactor scale-up.

**Sizing the Reactor.** The volume of a reactor is calculated directly from the rate equation, such as Eq. (5-49), which may be rewritten as

$$\int_0^\theta V\,d\theta = \int_{N_{A_2}}^{N_{A_1}} \frac{dN_A}{r_A} \tag{5-58}$$

the correct-order expression being inserted for $r_A$. For flow reactors, a more useful form of Eq. (5-58) is

$$\int_0^V \frac{dV}{F} = \int_0^{X_A} \frac{dX_A}{r_A} \tag{5-59}$$

written in terms of the molal flow rate $F$ and degree of conversion $X_A$, moles of $A$ converted per mole of $F$. For steady-state continuous stirred tanks, the rate expression is simply

$$r_A = (F/V)(C_{A1} - C_{A2}) \tag{5-60}$$

where the subscripts 1 and 2 refer to the concentration of reactant $A$ in the entering and leaving streams, respectively.

A plug-flow reactor (no longitudinal mixing) and a batch reactor (no concentration or temperature gradients) require the shortest residence time possible for a given reaction of finite order to proceed to the desired extent. The residence time is identical for these two types. The productivity (average rate of product availability per unit of reactor volume) is reduced for the batch reactor by the outage time (time required for emptying, cleaning, and refilling between batches).

Single-stage continuous stirred-tank reactors, which provide a reaction environment of the constant composition of the effluent or completed-reaction stream, have the lowest productivity of all reactors. As the total volume of the continuous stirred-tank reactor is subdivided into a series of equal-volume stages, the productivity increases, approaching that of a plug-flow reactor as the number of stages approaches infinity. Figure 5-29

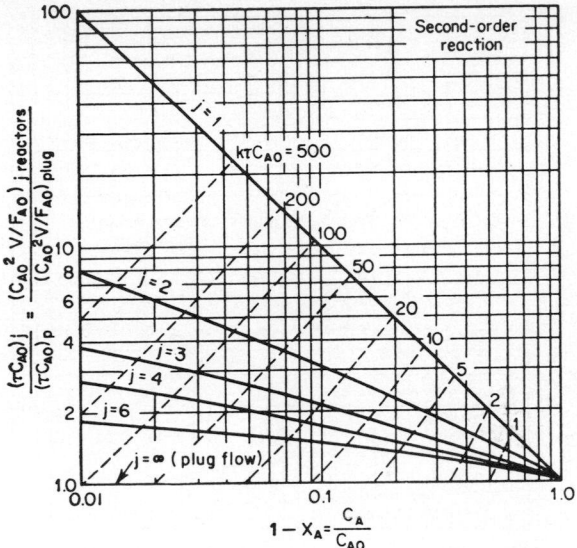

FIG. 5-29. Comparison of plug-flow and a series of $j$ equal-size backmix reactors.

$$2A \rightarrow R$$

$$A + B \rightarrow R \qquad C_{A0} = C_{B0}$$

with negligible expansion. For the same processing rate of identical feed, the ordinate measures the volume ratio $V_j/V_p$ or space-time ratio $\tau_j/\tau_p$ directly. (*Levenspiel, op. cit., Fig. 7, p. 141.*)

shows a comparison of residence times in stirred-tank reactors of $j$ equal stages and in plug-flow reactors for second-order irreversible reactions of various specific reaction constants and initial concentrations.

In actual practice, the assumptions of perfect mixing in continuous stirred tanks and of no longitudinal mixing in tubular reactors are only approximations. For stirred tanks of proper design with vigorous agitation and for small-diameter, high-velocity tubular reactors, the approximations are well within the limits of design accuracy. For large-diameter tubes and for packed beds, the effect of longitudinal mixing can be appreciable, resulting in a larger requirement of reactor volume. Figure 5-30 gives an idea of the effect of longitudinal mixing on the required reactor volume for irreversible second-order reactions. The axial dispersion coefficient $D$ must be determined experimentally for a given reactor.

**Materials of Construction.** Materials for the fabrication of all process equipment are selected, first for their ability to withstand chemical attack and thus avoid process-stream contamination, and second for their economic

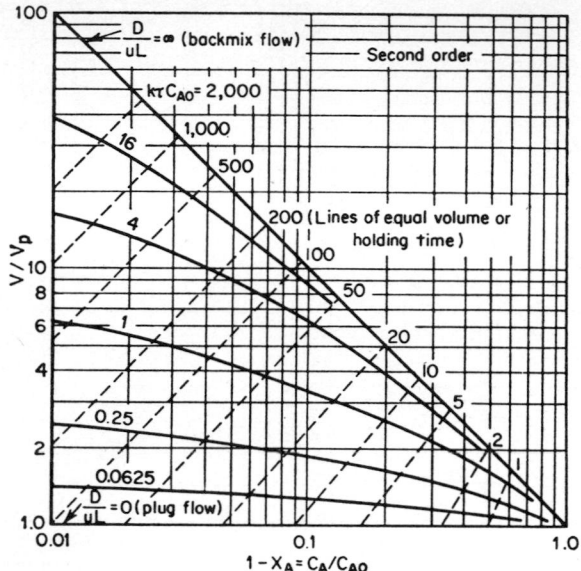

FIG. 5-30.  Comparison of real and ideal (plug-flow) reactors.  (*Levenspiel, op. cit., Fig.* 24, *p.* 280.)

life.  Resistance to corrosion is especially important in reactors because (1) the combination of composition, temperature, and mechanical conditions is likely to be more severe there than in most other pieces of process equipment, and (2) trace contamination due to dissolved metal can be disastrous in its catalytic effect.  The high pressures, high or low temperatures, and rapid temperature changes that obtain in many reactors also make mechanical and structural integrity difficult to obtain.

Detailed tables summarizing the chemical and mechanical characteristics of materials are available in a number of review sources [e.g., Norden, in Perry and Chilton (eds.), "Chemical Engineers' Handbook," 5th ed., sec. 23, McGraw-Hill Book Company, New York, 1973] and in Sec. 3 of this manual.

**Optimal Design.**  The optimization of reaction conditions and of the reactor design is highly complicated, because of the many variables that are involved, and is seldom achieved or even attempted.  Certain aspects of the optimization must be considered, however, to arrive at a reasonable, if not optimal, operation.

The temperature of the reaction is chosen with regard to the following considerations: (1) reaction rates increase with temperature, and high temperatures favor low residence time in a reactor; (2) undesirable side reactions may be minimized by the proper choice of temperature; (3) equilibrium of exothermic reactions is less favorable, the higher the temperature; (4) maintaining high reactor temperatures may require high thermal costs; (5) temperature must be chosen with regard to its effect on catalysts; (6) high temperatures are identified with high corrosion rates and with costly materials of construction. To achieve a compromise between kinetic and equilibrium effects in an exothermic reaction, a programmed temperature change during the course of the reaction may be used, as in sulfuric acid converters. Figure 5-31 shows how such programming can achieve the maximum average rate of reaction.

Reactant concentrations are selected in such a way as to allow maximum conversion of the most expensive or critical reactant, with due regard to product isolation and reactant-recovery costs. Pressure is generally the equivalent of concentration in a gas-phase reaction, and must be selected with additional regard for the equipment costs associated with high-pressure reactors. Pressure does not affect liquid-phase reactions.

Degree of conversion must be chosen, keeping separation, recycle, and subsequent processing steps and their costs in mind.

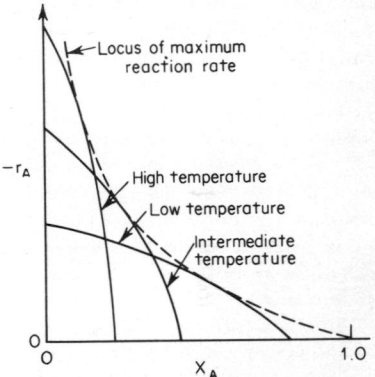

FIG. 5-31. Reaction rate as a function of conversion and temperature for reversible exothermic reactions, using a given feed material. Dashed line shows the temperature to use at each composition for optimum operations. (*Levenspiel, op. cit., Fig.* 6, *p.* 217.)

In general, these costs are balanced against the cost of the reactor and its operation; the former are high for low conversion, whereas the latter may be high for high conversion.

The optimization of reactor type, proportion, size, and materials of construction is extremely complicated and cannot be treated here. The interested reader is referred to special literature on the subject (e.g., Aris, "The Optimal Design of Chemical Reactors," Academic Press Inc., New York, 1961).

# PROCESS CONTROL

## 5-25. Introduction

Chemical processing of materials is generally carried out as a sequence of unit operations, either on a batch or continuous-flow basis. The output of any one process step is usually the input to one or more subsequent steps. For each subprocess there are a number of independently adjustable variables which must be specified by the engineer or process operator. The values specified for these *process variables* determine how the subprocess operates on the feed to produce the output, and determine such factors as yield, purity, reaction time, etc. Examples of such independently adjustable variables are process-stream flow rates, energy input rates, and cooling-water flow rates.

Design procedures generally specify approximate values of these variables for a given input to the subprocess and a given desired output. However, a process must generally be "fine tuned" to determine the appropriate relationship between input parameters and process variables required to achieve a given output or product.

The field of process control deals with the use of automatic devices to set values of process variables in order to achieve a given process output. Inputs to a given subprocess are often variable, depending on previous process steps, feed material compositions, temperatures of available cooling water, etc. Since changes in inputs to a process generally require changes in process variables in order to maintain a given output condition, the dynamic response of processes and their control systems is of primary importance.

**Simple Control Systems.** In the simplest control system, no information about either process input or output is used to determine the setting of process variables. Upsets in process inputs lead to changes in output, with no compensation by the control system to restore output to the desired condition. Obviously this type of control is useful only for processes where the output response to input changes is either small or tolerable within its range of variance.

**Feedback Control.** Information about some aspect of the output, called the *controlled variable,* is used to determine the adjustment to be made in

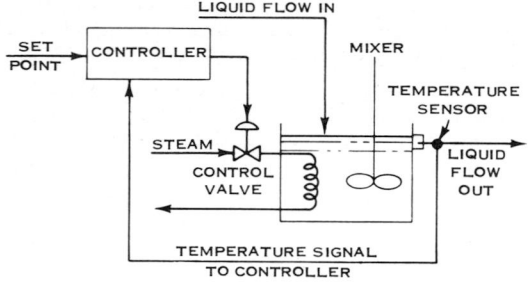

Fig. 5-32. Automatic feedback control of heat-exchange process.

some process variables. Such control systems, in which measured values of a controlled variable are used to adjust a process variable upstream of the measurement point, are termed *closed-loop systems.* A simple feedback control process is illustrated by Fig. 5-32. Systems which do not make use of such feedback from downstream points to upstream points are termed *open-loop systems.*

**Feedforward Control.** An input variable to the process is measured and transmitted to a feedforward controller. The controller uses a system model to determine values of process variables which will produce the desired value of the output (controlled variable). The effectiveness of pure feedforward control depends on whether the system model accurately predicts the response of the process to input and process variable changes. Feedforward control is most useful for those situations where there is a long time lag between upsets in input variables and the associated changes in the controlled variable. A simple feedforward control system is illustrated in Fig. 5-33.

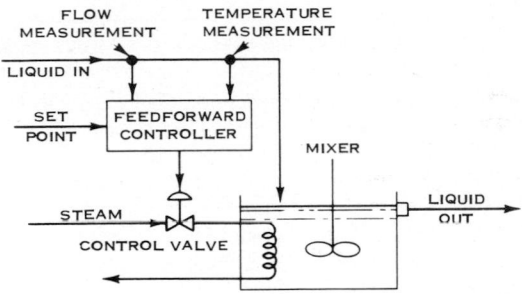

Fig. 5-33.  Feedforward control of heat-exchange process.

One significant drawback to feedforward control is that few system models are accurate in predicting controlled variable responses. This is overcome by utilizing a feedback controller based on measurement of the controlled variable to modify the signal from the feedforward controller. This is most often accomplished by simply adding the signal from the feedback controller to that of the feedforward controller. Such control schemes (termed *combined feedforward and feedback systems*) are often the only effective method for controlling processes with long time lags between changes in process variables and the response of the controlled variable. The feedforward control adjusts the process to these changes, and the feedback signal "trims" the signal of the feedforward controller to compensate for inadequacies of the system model. An example of combined feedforward and feedback control is shown in Fig. 5-34.

**Cascade Control.** Another approach to controlling processes with a long time lag between input variable change and controlled variable response is *cascade control*. The primary control loop measures the controlled variable and adjusts the set point on a secondary control loop. The secondary control loop measures some intermediate process variable and compares this to the output of the primary loop controller to determine adjustments in the manipulated variable. The secondary loop can be either feedforward or feedback, but must have a shorter time constant than the primary loop.

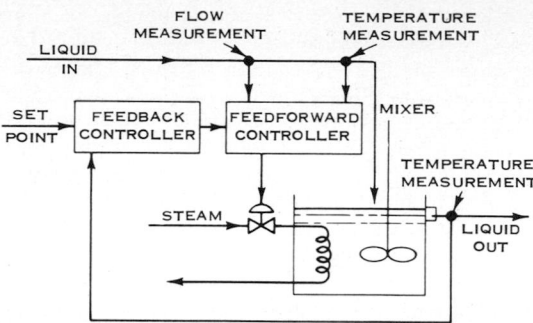

Fig. 5-34.   Combined feedforward and feedback control of heat exchange.

Cascade control is valuable when the manipulated variable is subject to fluctuations outside the process being controlled, for example the pressure of a steam line being used to supply heat to a reboiler.   The secondary loop, with its short time lag, prevents fluctuations in the input condition of the manipulated variable from affecting the process.   The primary loop ensures that the adjustments made in the manipulated variable will give the desired output of the controlled variable.   A simple cascade control system is illustrated in Fig. 5-35.

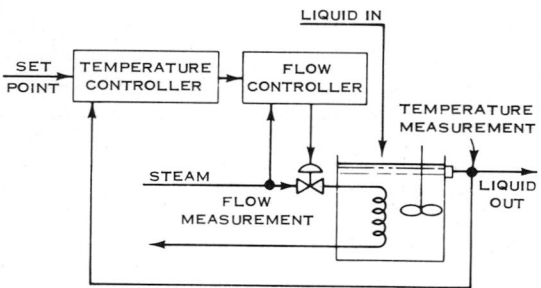

Fig. 5-35.   Cascade control system in which disturbances originating in the steam supply are prevented from entering the heat-exchanger process.

## 5-26. Control System Analysis

Analysis of a process (either controlled or uncontrolled) may be carried out by deriving and solving the differential equations describing the dynamic (time-dependent) behavior of the system.   Each independent element of the system is analyzed in terms of its governing physical, chemical, and electrical principles to obtain a series of simultaneous algebraic and time-dependent differential equations.

A generalized block diagram for such a system is shown in Fig. 5-36.   The

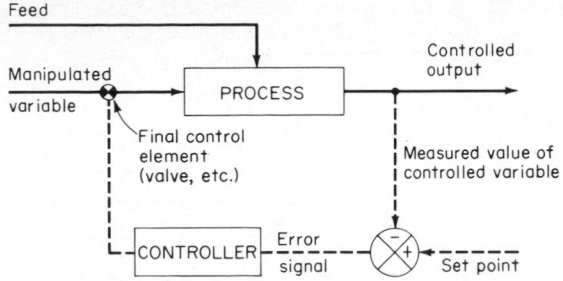

Fig. 5-36.   Generalized block diagram of feedback control system.

output *signal* of the controller is a function of the *error*, which is the difference between the measured value of the controlled variable and the desired value which is represented by the controller *set point*.   Equations describing this system are

$$f = f(\theta) \tag{5-61}$$

$$c = \phi(f, m, \theta) \tag{5-62}$$

$$m = \psi(e) \tag{5-63}$$

$$e = c - s \tag{5-64}$$

where $c$ is the signal from an instrument measuring the controlled variable, $s$ is the controller set point, $m$ is the controller output signal, and $f$ is the feed to the system.   The function represents the physical, chemical, and thermodynamic equations describing the process.   The form of the function $\psi$ depends on the nature of the controller.

More sophisticated system models include additional equations describing the dynamic behavior of measuring and transmission elements, as well as the characteristics of the control valve.

To design or analyze a control system for a given process, the parameters of the controller function $\psi$ are adjusted, and the responses of the controlled process to various changes in $f$ are evaluated.   There are several approaches to this procedure.   The first is to linearize all equations in terms of a very narrow range of values near the normal steady-state operating conditions, and then solve these linear simultaneous equations by means of Laplace transforms.   This procedure is widely used and quite useful, since elements of the more standard controllers have simple Laplace transforms; the response of many systems has been analyzed in terms of transfer functions in the Laplace domain.   Detailed derivation and explanation of the theory of Laplace transforms are beyond the scope of this work.   (See, for example, Buckley, P. S., "Techniques of Process Control," John Wiley & Sons, Inc., New York, 1964.)

One major disadvantage of Laplace transforms is the approximation introduced by linearization of the process model. This is acceptable for small variations about the steady-state operating values, but is likely to be significantly in error for large-scale variations such as at start-up, in major process upsets, or in deliberate shifts to significantly different steady-state operating conditions.

An alternate procedure to the use of Laplace transforms is to solve the system model equations on either an analog or digital computer. This has the advantage of simulating system response to any input function, both simple pulse, step change, or sine wave inputs (such as are used in Laplace transform analysis) and more complex inputs which are not amenable to Laplace transforms. Furthermore, it may not be necessary to linearize the system model equations, making them applicable for large-scale process changes. This procedure has the disadvantage of requiring far more effort to design and tune those simple control systems where Laplace transforms are applicable.

Yet another procedure for analysis of control systems is the frequency-response approach. Deliberate sine-wave manipulations are made in the process input variables, and the associated response of the process output variable is measured. The relationships between amplitudes of input variable changes and output responses for different frequencies of sine-wave input are used to design controller systems. One method of frequency-response analysis (the Bode plot) is described in Sec. 5-28.

## 5-27. Feedback Control Systems

**On-Off Control.** The simplest and most common control system is the on-off controller. This is exemplified by the thermostat used in home heating. The controller puts out either a maximum or a minimum signal depending on the sign (but not the magnitude) of the error signal.

On-off control is simple, inexpensive, and stable control. If the controller is a perfect pure relay, instantaneously assuming either the maximum or minimum output condition as the error becomes different from zero, and if there is no time lag in the system or controller transmission lines, then on-off control is optimal control, responding to system upsets in the minimum time. In practice, time lags invariably occur in both the control system elements (transmission lines, control valves) and the process itself. Because of this, the control system tends to overcompensate, delivering a corrective control signal longer than necessary to drive the error to zero. This causes successive overshoot and undershoot, with the controlled variable oscillating around the desired value, as shown in Fig. 5-37. Typically, response dynamics of the controlled system will be such that the time-averaged system output will differ from the set point, even though the oscillation is around the set point. This difference is termed offset and generally is undesirable.

The magnitude and period of the oscillations are of importance in on-off control, and depend on both the process system dynamics and the controller characteristics. It is desirable to reduce the amplitude of the oscillations and eliminate offset, while keeping the period of the oscillations as long as possible. Rapid oscillations can induce maintenance problems with the mechanical elements of the controller system, for example, through too rapid cycling of valves. High-amplitude oscillations can cause poor product quality or unstable operations.

Rapid cycling of the controller is often reduced by using a controller with hysteresis or dead band, which eliminates chattering of the controller system when at steady state but increases the amplitude of oscillations. The magnitude of oscillations can be reduced by using a smaller adjustment in the manipulated variable. This, however, slows the rate of return to steady state after a process upset.

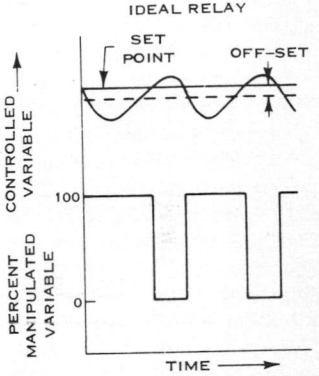

FIG. 5-37. Relationship between controlled variables and the manipulated variable for an ideal relay.

For processes where small-scale deviations from steady-state conditions can be tolerated and time lags are relatively small, on-off control (usually with hysteresis or dead band) is often suitable. It is the least sophisticated and therefore least expensive control system. In practice, feasibility of on-off control is usually determined by whether the process can stand the rapid and frequent transition from fully-on to fully-off conditions. For example, the water-hammer effect resulting from rapid closure of a valve may not be acceptable. Additionally, rapid fluctuation of the final control element (valve, etc.) may cause frequent failure and maintenance problems.

**Proportional Control.** One approach to eliminating overshooting which occurs in on-off control is to reduce the adjustment to the manipulated variable as the magnitude of the error signal approaches zero. In *proportional control* the controller output is proportional to the magnitude of the error signal, up to the maximum output of the controller, after which point it is a constant value. The proportionality constant by which the error signal is multiplied to obtain controller output is the *controller gain*. The inverse of the controller gain (times 100 per cent) is the *proportional band*, which is similar to the dead band or hystersis band in on-off control. The proportional controller is tuned to the process by adjusting the gain (or the width of the proportional band.)

The proportional controller suffers some of the disadvantages of the on-off controller, with certain exceptions. When the gain is high a system with lag exhibits overshoot, oscillating around the set point. However, the amplitude of the oscillations can be made to decrease with time by using a low enough gain. As a result the system can be made to seek a stable steady state, with no oscillations. The higher the gain, the greater will be the initial overshoot as the system responds to an upset, the more rapid will be the oscillations, and the slower the decay to a new steady state. In the limiting case, a proportional control with very high gain becomes almost equivalent to an on-off controller, exhibiting a limit cycle around a steady-state value which does not decay.

If the gain is lowered to achieve stability, the amplitude of fluctuations will die out to produce a steady-state value. However, the new steady-state value may not be the desired output value. This offset is one of the most serious deficiencies of simple proportional control. It arises because a gain setting sufficiently low to achieve stability may not provide enough adjustment to the manipulated variable to correct for a changed process input.

For systems with no lag (or small lag), response to a process upset is slower than an on-off controller, becoming faster as gain is increased. For this type of system on-off control is preferable, except for technical considerations previously mentioned. For systems with lag, a controller having proportional gain is preferable, often mandatory.

**Proportional Plus Integral Control.** (*PI or Two-mode Control*). The most significant drawback of proportional control is the offset which occurs in optimally tuned stable control systems. This problem can be reduced by adding *integral* or *reset action* to the controller. The output of a two-mode controller is proportional to the sum of the error (as in proportional control) plus the integral of the error with respect to time, up to the maximum controller output. A small error signal integrated over a long enough period of time can provide enough controller output to drive the process to the set point, thus eliminating offset.

The PI controller has two degrees of freedom which are adjusted to tune the controller to the process: proportional band or gain and reset time. *Reset time* is the time required for the integral mode output of the controller to equal the proportional mode output, assuming a constant error signal. Some controllers are calibrated in resets per minute, which is the inverse of reset time. Increasing the reset time decreases the rate at which the controller drives a small error to zero. In the limiting case, long reset time gives straight proportional control.

**Proportional Plus Integral Plus Rate Control.** (*PID or Three-mode Control*). The main disadvantages of PI control are slow response to a sudden change and a tendency to overshoot the desired value when tuned to give optimal response. Both of these disadvantages can be reduced by addition of *derivative* or *rate action* to the controller. In three-mode control, the controller output is proportional to the sum of error signal, the integral of the error signal and the time-derivative of the error signal. This allows the controller to handle a rapidly changing error signal, since the faster the error increases, the greater the controller output to correct that error. On the other hand, as the corrective action of the controller catches up to the process disturbance and the magnitude of the error becomes smaller, the time derivative of the error becomes negative. This reduces the corrective action of the controller and thus limits the magnitude of the overshoot. This allows the use of somewhat higher gain for rapid controller action plus shorter reset times for rapid controller action and elimination of offset. Generally, high rate action is desirable; practically, this is limited by noise in the system. Noisy signals involve high time derivatives and create rapid fluctuation of the rate action of the controller, which reduces controller stability.

Three-mode controllers are the most sophisticated and expensive of simple feedback control systems. They are best suited for systems where precise control is required and no special problems, such as large-scale process time lags, occur. They are especially useful for systems where inertial effects are large.

## 5-28. Tuning Control Systems

**On-Off Systems.** Tuning on-off control systems consists primarily of adjusting the relationship between amplitude and frequency of oscillations around the set point. This is done primarily by adjusting the dead band or hysteresis band of the controller. It is generally a trade-off (based on experience) between allowable variation in the controlled variable and increased maintenance cost due to wear in the control equipment. Offset is reduced by adjusting the relationship between maximum and minimum manipulated variable corresponding to maximum and minimum controller output. If the offset is such that more corrective action is needed, the ratio of maximum value to minimum value is increased; if the offset indicates less corrective action, the ratio is reduced.

**PID Controllers.** The criteria for tuning three-mode controllers vary but generally include stability of the closed-loop system, the rate at which overshoot decays, and minimization of the error integral. It is obvious that the process must be designed so that the manipulated variables have sufficient range to maintain the desired steady-state output condition over the anticipated range of inputs to the system. This point is sometimes overlooked and cannot be compensated for by any control system.

Specific procedures for tuning controllers are beyond the scope of this text. [See Perry and Chilton (eds.), "Chemical Engineers' Handbook," 5th ed., sec. 22, McGraw-Hill Book Company, New York, 1973.] General procedures are presented below. The first step consists of determining the response of the open-loop (uncontrolled) system to a sine-wave variation of the manipulated variable. This may be done experimentally by manipulating the system, or analytically by using Laplace transform techniques on the system model equations. In either case a plot is made of the phase lag between the system response and the manipulated variable vs. the logarithm of the frequency of oscillation of the manipulated variable. A second plot is made, on log-log coordinates, of amplitude ratio (ratio of system response to maximum input value of manipulated variable) vs. frequency.

If the frequency response characteristics of the control system are superimposed on these plots (note that the shape of the control system plots depends on the controller gain, reset time, and rate-action time constants of the controller), the characteristics of the closed-loop system can be obtained by graphical addition of the two plots. The tuning procedure then involves adjusting the controller parameters to ensure that the controlled system is stable and to minimize the error integral.

Instability occurs when the phase lag approaches 180° and the amplitude ratio approaches unity. In this situation the controller continually adjusts the process in the wrong direction, and the system gain is high enough to cause increasingly large oscillations around the set point. This is analogous to pushing a pendulum at the limit of its swing. If the controlled system has zero phase lag, the controller corrects the process at the time its effect is strongest. This is analogous to retarding a pendulum at the center of its swing.

In some cases disturbances enter the system at some point other than the manipulated input variable. If the major sources of disturbance are not in the manipulated variable, the formalized technique above will not give adequate control. In such cases the control system parameters obtained above are used as a starting point to tune the controller using a trial-and-error search technique. The controller parameters are adjusted slightly and changes in the total system performance are analyzed. This is a difficult procedure and calls for experience and good judgment.

There are many other methods for tuning controllers. One frequently used for tuning existing controllers is to determine the *ultimate gain* and *ultimate period*. The integral and derivative modes are turned off and the proportional gain is adjusted to obtain a metastable oscillation in the controlled system (i.e., an oscillation whose amplitude neither decays nor increases with time). The proportional gain at which this occurs is the ultimate gain, and the period of the oscillation is the ultimate period. The settings of the three controller parameters are then determined from the ultimate gain and period by one of various tuning-criteria rules. For example if $K_u$ is the ultimate gain and $P_u$ is the ultimate period, then

$$K_c = 0.6K_u \tag{5-65}$$

$$\tau_I = 0.5P_u \tag{5-66}$$

$$\tau_D = 0.125P_u \tag{5-67}$$

will produce a decay of successive oscillations to one-fourth the amplitude of the previous oscillation, where $K_c$ is the controller gain, $\tau_I$ is the reset time, and $\tau_D$ is the rate-action derivative time.

### 5-29. Design of Control Systems

Design of control systems is a complex subject which cannot be presented in great detail in any condensed text. The best procedure is generally to design the control system along with the process, with close interaction between design teams. Control systems "added on" after the fact seldom perform as well as those designed for the system from initial stages. Several critically important principles, often overlooked in the design of a control system, are presented below.

One independent process input variable must be manipulated for each controlled variable. It is best to select combinations of measured variable/manipulated variable that give the shortest possible time lag between input adjustment and response; this usually means as close together in the process flowsheet as possible. The paired variables should interact strongly rather than weakly.

For control loops where the controlled variable can be allowed to oscillate within a certain range, on-off controllers are the best choice, provided that there are no restrictions due to mechanical limitations of the control system or dynamic limitations on allowable rate of manipulation of input variables. If on-off control is unacceptable, the next choice would be proportional control. If the resulting offset is also unacceptable, it can be eliminated by adding integral mode (i.e., PI controller). Derivative control action is added to eliminate overshoot (PID control). If there is a significant time lag in the system, especially "dead time," some form of feedforward control is indicated. The measuring device used must be adequately sensitive to changes in the controlled variable and preferably fast acting.

One problem which is less obvious is control system incompatibility. It is possible that manipulation of one input variable to control a given output variable will cause upsets in a second output variable. It is desirable to reduce interaction between control loops whenever possible. The subject of interacting loops is complex and beyond the scope of this text.

Proper design of a process control system can eliminate many later operating problems and can improve the process, increasing yields, improving product purity, and shortening processing times.

SECTION 6

# CIVIL ENGINEERING

**Austin E. Brant, Jr., B.C.E., M.S., P.E.;** Partner, Tippetts-Abbett-McCarthy-Stratton, Engineers and Architects; Fellow, American Society of Civil Engineers, Institute of Traffic Engineers; Member, Transportation Research Board, Operations Research Society of America

**Leonard S. Oberman, B.S.C.E., P.E.;** Associate, Tippetts-Abbett-McCarthy-Stratton, Engineers and Architects; Fellow, American Society of Civil Engineers; Member, Permanent International Association of Navigation Congresses, American Road Builders Association

## CONTENTS

# SURVEYING

## 6-1. Measurement of Distance

**Units of Measurement.** Distances are usually measured in feet and tenths, hundredths, and (for accurate work) thousandths of feet. For many older surveys, distances were measured in chains and links. A chain is 66 ft in length and is divided into 100 links, each 7.92 in. long. The metric system, in which the unit of distance is the meter and its decimal fractions or multiples, is in use in most countries and is expected to replace the English system of measurement in the United States. See Tables 1-1 and 1-2 as well as the inside covers for conversion factors.

Four methods used for the direct measurement of distance are pacing, stadia reading, taping, and electronic distance recording.

**Pacing.** Pacing is a rapid means of checking more accurate measurements of distance. The precision of pacing under average conditions is from 1:100 to 1:200.

**Stadia Reading.** The use of stadia furnishes a rapid method of determining distances with a fair degree of accuracy. Under average conditions, a precision of from 1:300 to 1:1,000 can be obtained (Sec. 6-4).

**Measurement with Tape.** The most commonly used method of determining distance is by measurement with a tape. Steel tapes, ranging in length from 50 to 300 ft, are generally used, but tapes of other materials may be used where accuracy is not essential. The precision of a tape measurement depends on the degree of refinement with which the measurement is made. The precision of taping ordinarily used in surveys is from 1:3,000 to 1:5,000.

For ordinary taping, a tape accurate to 0.01 ft should be used. The tension of the tape should be about 15 lb. The temperature should be determined within 10°F, and the slope of the ground within 2 per cent, and the proper corrections applied. The correction to be applied for temperature when using a steel tape is

$$C_t = 0.0000065s(T - T_0) \tag{6-1}$$

The correction to be made to measurements on a slope is

$$C_h = s(1 - \cos\theta) \qquad \text{exact} \tag{6-2}$$

$$\text{or} \qquad = 0.00015s\theta^2 \qquad \text{approximate} \tag{6-2a}$$

$$\text{or} \qquad = h^2/2s \qquad \text{approximate} \tag{6-2b}$$

where $C_t$ = temperature correction to measured length, ft
$C_h$ = correction to be subtracted from slope distance, ft
$s$ = measured length, ft
$T$ = temperature at which measurements are made, °F

$T_0$ = temperature at which tape is standardized, °F

$h$ = difference in elevation at ends of measured length, ft

$\theta$ = slope angle, deg

In more accurate taping, using a tape standardized when fully supported throughout, corrections should also be made for tension and for support conditions. The correction for tension is

$$C_p = \frac{(P_m - P_s)s}{SE} \tag{6-3}$$

The correction for sag when not fully supported is

$$C_s = \frac{w^2 L^3}{24 P_m{}^2} \tag{6-4}$$

where $C_p$ = tension correction to measured length, ft

$C_s$ = sag correction to measured length for each section of unsupported tape, ft

$P_m$ = actual tension, lb

$P_s$ = tension at which tape is standardized, lb (usually 10 lb)

$S$ = cross-sectional area of tape, in.²

$E$ = modulus of elasticity of tape, psi (29 million psi for steel)

$w$ = weight of tape, lb/ft

$L$ = unsupported length, ft

**Electronic Distance Measurement (EDM).** Electronic measuring devices, utilizing principles similar to radar, are now in general use. Distances are measured by determination of the time required for a wave (infrared, radio, or laser beam) to travel at the speed of light to and from a point. For some types, readings of distance are obtained directly from a counter, eliminating the need for calculations. Advantages include ability to take readings across bodies of water, rugged terrain, and brush much more rapidly than with tape measurement. The precision of EDM ranges up to 1 : 300,000.

## 6-2. Measurement of Difference in Elevation

Difference in elevation may be measured by three methods: barometric leveling, stadia leveling, and direct leveling. Barometric methods are used for rough or preliminary work. Stadia reading is a rapid method and will give results having an error, in feet, of 1.0 $\sqrt{\text{distance in miles}}$ (Sec. 6-4). Direct leveling is the most accurate and most commonly used method for determining difference in elevation. EDM instruments (Sec. 6-1) make it possible to determine differences in elevation by measurement of slope distances and vertical angles.

**Rough Leveling.** Rough leveling is practiced on preliminary or reconnaissance surveys. Sights are permitted up to 1,000 ft in length, and rod readings are made to 0.1 ft. Precision in feet is 0.4 $\sqrt{\text{distance in miles}}$.

**Ordinary Leveling.** Ordinary leveling is used in the construction and location of highways, railroads, and the like. Sights are permitted up to 500 ft in length, and rod readings are made to 0.01 ft. Precision in feet is 0.1 $\sqrt{\text{distance in miles}}$.

**Accurate Leveling.** Accurate leveling is used for establishing important bench marks. Sights are limited to 300 ft in length, and rod readings are made to 0.001 ft. Precision in feet is 0.05 $\sqrt{\text{distance in miles}}$.

**Precise Leveling.** Precise leveling is used for establishing bench marks at widely separated locations. Sights are limited to 300 ft in length, and rod readings are made to 0.001 ft. Special equipment and extreme care are used, and several runs are usually made. Precision in feet is 0.02 $\sqrt{\text{distance in miles}}$.

## 6-3. Measurement of Angles

Angles may be measured with either a compass or a transit. The precision of compass measurement is from 30′ to 1°. The precision of transit measurements is from 1″ to 2′, depending on the type of instrument used and the care exercised. Angle measurements should have a precision consistent with distance measurements. Surveys in which distances are measured to 0.01 ft should have angles measured to 15″, with a resulting accuracy of better than 1 : 10,000, and surveys with distances measured to 0.1 ft and angles measured to 1′ should have an accuracy of about 1 : 5,000.

## 6-4. Stadia Surveying

In stadia surveying, a transit having horizontal stadia cross hairs above and below the central horizontal cross hair is used. The difference in the rod readings at the stadia cross hairs is termed the rod intercept. The intercept may be converted to the horizontal and vertical distances between the instrument and the rod by the following formulas:

$$H = Ki(\cos a)^2 + (f + c) \cos a \tag{6-5}$$

$$V = \tfrac{1}{2}Ki(\sin 2a) + (f + c) \sin a \tag{6-6}$$

where $H$ = horizontal distance between center of transit and rod, ft
$V$ = vertical distance between center of transit and point on rod intersected by middle horizontal cross hair, ft
$K$ = stadia factor (usually 100)
$i$ = rod intercept, ft
$a$ = vertical inclination of line of sight, measured from the horizontal, deg
$f + c$ = instrument constant, ft (usually taken as 1 ft)

In the use of these formulas, distances are usually calculated to feet and differences in elevation to tenths of feet.

## 6-5. Latitudes and Departures

The latitude of a line is the projection of the line upon a true or assumed north-south meridian. The latitude of a line of length $s$ is $s \cos \beta$ where $\beta$ is the bearing of the line (the angle between the direction of the line and the direction of the north-south axis). If the line is considered as running from the southerly end to the northerly end, the latitude is positive; if from the northerly end to the southerly end, the latitude is negative.

The departure of a line is the projection of the line upon a parallel at right angles to the meridian. The departure of a line of length $s$ is $s \sin \beta$. If the line is considered as running from the westerly end to the easterly end, the departure is positive; if from the easterly end to the westerly end, the departure is negative.

### 6-6. Balancing a Closed Traverse

The geometry of a closed traverse requires that the algebraic sum of the latitudes and of the departures be zero. If these sums are not zero, but are small enough to indicate that the discrepancy is not the result of an actual error, the traverse may be balanced by adjusting each line. The compass rule is usually used in balancing. The rule states: The correction to be applied to the latitude (or departure) of any line in the traverse is to the total error in latitude (or departure) as the length of the line is to the length of the traverse.

### 6-7. Calculation of Areas of Land

Areas of land within the limits of a balanced traverse are usually calculated by the method of double meridian distances. For convenience in the use of this method, a meridian is assumed to pass through one corner (usually the most westerly corner) of the traverse. The meridian distance of any point in the traverse is then the departure of the point, departures to the east of the meridian being considered positive and departures to the west, negative. The double meridian distance of a line in the traverse is the sum of the departures (or meridian distances) of the two ends of the line. Double meridian distances (DMD) may be calculated by the following rules:

1. The DMD of the first line in the traverse (the line one end of which is on the reference meridian) is equal to the departure of that line.

2. The DMD of any other line is equal to the DMD of the preceding line, plus the departure of the preceding line, plus the departure of the line itself.

3. The DMD of the last line is the same as the departure of that course with opposite sign.

Algebraic values should be used with due regard for signs. The area of the traverse is equal to one-half the algebraic sum of the products of the DMD and the latitude of each line.

The area of irregular tracts of land may be determined by the trapezoidal method or by Simpson's method. Both of these methods require that a straight line enclose one side of the area and that offsets from this line be measured at regular intervals to the irregular boundary. In addition, Simpson's method requires that the number of offsets be odd (or the number of regular intervals be even). The trapezoidal method states: The area is equal to the product of the interval between offsets and the sum of the intermediate offsets and one-half each end offset. Simpson's method states: The area is equal to one-third the product of the interval between offsets and the sum of the end offsets, twice each odd intermediate offset, and four times each even intermediate offset.

## 6-8. Circular Curves

Circular curves are the most common type of horizontal curve used to connect intersecting tangent (or straight) sections of highways or railroads. In the United States, two methods of defining circular curves are in use: the first, in general use in railroad work, defines the degree of curve as the central angle subtended by a *chord* of 100 ft in length; the second, used in highway work, defines the degree of curve as the central angle subtended by an *arc* of 100 ft in length. In the metric system, the degree of curve is sometimes expressed as the number of degrees subtended by an arc or chord 20 m long.

The terms and symbols generally used in reference to circular curves are listed below and shown in Figs. 6-1 and 6-2.

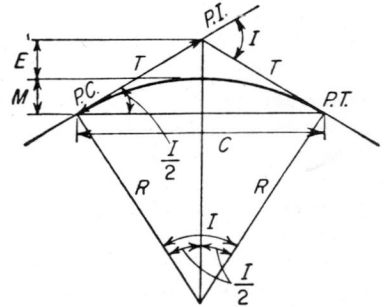

Fig. 6-1. Circular curve.

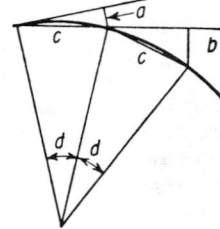

Fig. 6-2. Offsets to circular curve.

PC = point of curvature, beginning of curve
PI = point of intersection of tangents
PT = point of tangency, end of curve
$R$ = radius of curve, ft
$D$ = degree of curve (see above)
$I$ = deflection angle between tangents at PI, also central angle of curve
$T$ = tangent distance, distance from PI to PC or PT, ft
$L$ = length of curve from PC to PT measured on 100-ft chord for chord definition, on arc for arc definition, ft
$C$ = length of long chord from PC to PT, ft
$E$ = external distance, distance from PI to mid-point of curve, ft
$M$ = mid-ordinate, distance from mid-point of curve to mid-point of long chord, ft

$d$ = central angle for portion of curve ($d < D$)
$l$ = length of curve (arc) determined by central angle $d$, ft
$c$ = length of curve (chord) determined by central angle $d$, ft
$a$ = tangent offset for chord of length $c$, ft
$b$ = chord offset for chord of length $c$, ft

### Equations of Circular Curves

$$R = 5{,}729.578/D \qquad \text{exact for arc definition, approximate for} \qquad (6\text{-}7)$$
$$\text{chord definition}$$
$$= 50/\sin \tfrac{1}{2}D \qquad \text{exact for chord definition} \qquad (6\text{-}8)$$
$$T = R \tan \tfrac{1}{2}I \qquad \text{exact} \qquad (6\text{-}9)$$
$$E = R \operatorname{exsec} \tfrac{1}{2}I = R(\sec \tfrac{1}{2}I - 1) \qquad \text{exact} \qquad (6\text{-}10)$$
$$M = R \operatorname{vers} \tfrac{1}{2}I = R(1 - \cos \tfrac{1}{2}I) \qquad \text{exact} \qquad (6\text{-}11)$$
$$C = 2R \sin \tfrac{1}{2}I \qquad \text{exact} \qquad (6\text{-}12)$$
$$L = 100I/D \qquad \text{exact} \qquad (6\text{-}13)$$
$$L - C = L^3/24R^2 = C^3/24R^2 \qquad \text{approximate} \qquad (6\text{-}14)$$
$$d = Dl/100 \qquad \text{exact for arc definition} \qquad (6\text{-}15)$$
$$Dc/100 \qquad \text{approximate for chord definition} \qquad (6\text{-}16)$$
$$\sin \tfrac{1}{2}d = c/2R \qquad \text{exact for chord definition} \qquad (6\text{-}17)$$
$$a = c^2/2R \qquad \text{approximate} \qquad (6\text{-}18)$$
$$b = c^2/R \qquad \text{approximate} \qquad (6\text{-}19)$$

**Layout of Circular Curve.** The field layout of a circular curve depends on the geometric property of a circle that the angle between a tangent and a chord is one-half the included angle. The procedure is shown in Fig. 6-3, where the length of the first chord (or arc) is so chosen that point 1 is at an even 100-ft station. Point 1 is located by measurement of the chord distance $c$ from the PC and by the deflection angle $\tfrac{1}{2}d$ from the tangent. Point 2 is then located by measurement of the 100-ft chord (or the chord corresponding to the 100-ft arc) from point 1 and by the total deflection angle ($\tfrac{1}{2}D + \tfrac{1}{2}d$) from the tangent. Succeeding points

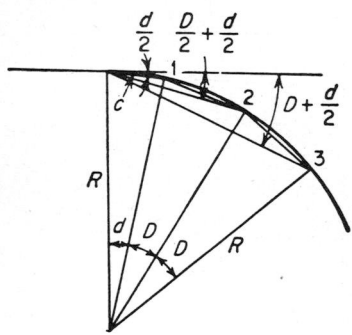

FIG. 6-3. Layout of a circular curve.

are similarly located. The entire curve can be laid out with the transit set at the PC.

## 6-9. Parabolic Curves

Parabolic curves are used to connect sections of highways or railroads of differing gradient. The use of a parabolic curve provides a gradual change in direction along the curve. The terms and symbols generally used in reference to parabolic curves are listed below and shown on Fig. 6-4.

PVC = point of vertical curvature, beginning of curve

PVI = point of vertical intersection of grades on either side of curve

PVT = point of vertical tangency, end of curve

$G_1$ = grade at beginning of curve, ft/ft

$G_2$ = grade at end of curve, ft/ft

$L$ = length of curve, ft

$R$ = rate of change of grade, ft/ft$^2$

$V$ = elevation of PVI, ft

$E_0$ = elevation of PVC, ft

$E_t$ = elevation of PVT, ft

$x$ = distance of any point on the curve from the PVC, ft

$E_x$ = elevation of point $x$ distant from PVC, ft

$x_s$ = distance from PVC to lowest point on a sag curve or highest point on a summit curve, ft

$E_s$ = elevation of lowest point on a sag curve or highest point on a summit curve, ft

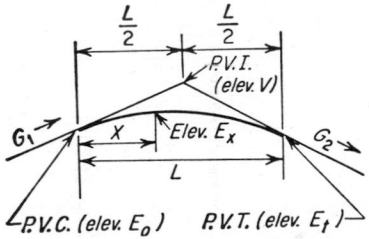

FIG. 6-4. Vertical parabolic curve (summit curve).

**Equations of Parabolic Curves.** In the parabolic-curve equations given below, algebraic quantities should always be used. Upward grades are positive and downward grades are negative.

$$R = (G_2 - G_1)/L \qquad \text{(6-20)}$$
Note: $K$ as used on Figs. 6-10 and 6-11 is equal to $1/100R$.

$$E_0 = V - \tfrac{1}{2}LG_1 \qquad \text{(6-21)}$$

$$E_x = E_0 + G_1x + \tfrac{1}{2}Rx^2 \qquad \text{(6-22)}$$

$$x_s = -G_1/R \qquad \text{(6-23)}$$
Note: If $x_s$ is negative or if $x_s > L$, the curve does not have a high point or a low point.

$$E_s = E_0 - G_1{}^2/2R \qquad \text{(6-24)}$$

## 6-10. Photogrammetry

Photogrammetry is a method of obtaining measurements through use of ground or aerial photography. For large-scale projects, aerial photogrammetry techniques permit substantial savings in mapping time, but establishment of the location and elevation of control points by conventional ground survey are necessary. Through photoanalysis and interpretation, a trained interpreter can obtain reliable qualitative information concerning the type and characteristics of the soils, surface and ground waters, and manufactured features such as roads and bridges. Matched groups of overlapping photographs are viewed through a stereoscope and deductive and inductive methods used for evaluation. A selective field check is an important part of the photo-interpretative methodology.

The scale of aerial photography is given as a ratio, such as 1:6,000, equivalent to 1 in. = 500 ft. For a standard camera with a 6-in. focal length lens and using 9-in. square film, the following relationships apply:

| Flight height above ground, ft | Photographic scale | Coverage per single photograph, mi² | Map scale (5 × enlargement of photographs) | Contour interval |
|---|---|---|---|---|
| 1,200 | 1: 2,400 | 0.11 | 1″ = 40′ | 1 ft |
| 3,000 | 1: 6,000 | 0.72 | 1″ = 100′ | 2.5 ft |
| 4,000 | 1: 8,000 | 1.29 | 1″ = 153′ | 1 m |
| 6,000 | 1:12,000 | 2.90 | 1″ = 200′ | 5 ft |
| 8,000 | 1:16,000 | 5.16 | 1″ = 267′ | 2 m |
| 12,000 | 1:24,000 | 11.62 | 1″ = 400′ | 10 ft |
| 18,000 | 1:36,000 | 26.15 | 1″ = 600′ | 15 ft |
| 20,000 | 1:40,000 | 32.28 | 1″ = 667′ | 5 m |

For mapping, the photographed area is usually covered in series of parallel strips, with photographs of the same strip overlapping 60 per cent and photographs of adjacent strips overlapping 30 per cent. Horizontal and vertical ground control is required for the preparation of maps.

## SOIL MECHANICS AND FOUNDATIONS

### 6-11. Grain Size

The grain size classification of soils used by the U.S. Department of Agriculture is given as follows.

| Soil type | Particle diam, mm | Soil type | Particle diam, mm |
|---|---|---|---|
| Gravel | >2.0 | Sand, very fine | 0.10–0.05 |
| Gravel, fine | 2.0–1.0 | Silt | 0.05–0.005 |
| Sand, coarse | 1.0–0.5 | Clay | 0.005–0.0002 |
| Sand, medium | 0.5–0.25 | Colloids | <0.0002 |
| Sand, fine | 0.25–0.10 | | |

### 6-12. Bureau of Public Roads Soil Classification

The U.S. Bureau of Public Roads (now the Federal Highway Administration of the U.S. Department of Transportation) developed a detailed method for classifying soils for use as highway subgrades which was subsequently expanded and adopted by the American Association of State Highway and Transportation Officials (AASHTO). Soils are classified in seven major groups as shown in Table 6-1 and described below.

#### Granular Materials

*Group A-1.* This group includes granular materials with or without non-

# Table 6-1. Classification of Highway Subgrade Materials*

| General classification | Granular materials (35% or less passing No. 200 sieve—0.075 mm) | | | | | | | Silt-clay materials (more than 35% passing No. 200 sieve—0.075 mm) | | | |
|---|---|---|---|---|---|---|---|---|---|---|---|
| | A-1 | | A-3 | A-2 | | | | A-4 | A-5 | A-6 | A-7** |
| Group classification | A-1-a | A-1-b | | A-2-4 | A-2-5 | A-2-6 | A-2-7 | | | | A-7-5, A-7-6 |
| Sieve analysis, % passing: | | | | | | | | | | | |
| No. 10 (2.00 mm) | 50 max | | | | | | | | | | |
| No. 40 (0.425 mm) | 30 max | 50 max | 51 min | | | | | | | | |
| No. 200 (0.075 mm) | 15 max | 25 max | 10 max | 35 max | 35 max | 35 max | 35 max | 36 min | 36 min | 36 min | 36 min |
| Characteristics of fraction passing No. 40: | | | | | | | | | | | |
| Liquid limit | | | NP | 40 max | 41 min | 40 max | 41 min | 40 max | 41 min | 40 max | 41 min |
| Plasticity index | 6 max | 6 max | | 10 max | 10 max | 11 min | 11 min | 10 max | 10 max | 11 min | 11 min |
| Group index | 0 | 0 | 0 | 0 | 0 | 4 max | 4 max | 8 max | 12 max | 16 max | 20 max |
| Usual types of significant constituent materials | Stone fragments, gravel, and sand | | Fine sand | Silty or clayey gravel and sand | | | | Silty soils | | Clayey soils | |
| General rating as subgrade | Excellent to good | | | | | | | Fair to poor | | | |

* Classification procedure: With required test data available, proceed from left to right on above chart and correct group will be found by process of elimination. The first group from the left into which the test data will fit is the correct classification.  **Plasticity index of A-7-5 subgroup is equal to or less than LL minus 30. Plasticity index of A-7-6 subgroup is greater than LL minus 30.

plastic or feebly plastic soil binders.   Subgroup A-1-a includes materials consisting predominantly of stone fragments or gravel, either with or without a well-graded binder of fine material.   Subgroup A-1-b includes materials consisting predominantly of coarse sand either with or without a well-graded soil binder.

*Group A-3.*   This group includes fine beach sand or fine desert blow sand without silty or clay fines or with a very small amount of nonplastic silt and stream-deposited mixtures of poorly graded fine sand with limited amounts of coarse sand and gravel.

*Group A-2.*   This group includes a wide variety of "granular" materials which are at the border line between materials falling in groups A-1 and A-3 and the silt-clay materials of groups A-4, A-5, A-6, and A-7.   Subgroups A-2-4 and A-2-5 include such materials as gravel and coarse sand with silt content or plasticity index in excess of the limitations of group A-1 and fine sand with nonplastic silt content in excess of the limitations of group A-3. Subgroups A-2-6 and A-2-7 include materials similar to those described under subgroups A-2-4 and A-2-5, except that the fine portion contains plastic clay having the characteristics of the A-6 or A-7 group.

### Silt-Clay Materials

*Group A-4.*   The typical material of this group is a nonplastic or moderately plastic silty soil.   The group also includes mixtures of fine silty soil and sand and gravel.

*Group A-5.*   The typical material of this group is similar to that described under group A-4, except that it is usually of diatomaceous or micaceous character and may be highly elastic as indicated by the high liquid limit.

*Group A-6.*   The typical material of this group is a plastic clay soil.   The group also includes mixtures of fine clayey soil and sand and gravel.   Materials of this group usually have a high volume change between wet and dry states.

*Group A-7.*   The typical material of this group is similar to that described under group A-6, except that it has the high-liquid-limit characteristics of the A-5 group and may be elastic as well as subject to high volume change. Subgroup A-7-5 includes those materials with moderate plasticity indexes which may be highly elastic as well as subject to considerable volume change. Subgroup A-7-6 includes those materials with high plasticity indexes in relation to liquid limit which are subject to extremely high volume change.

### Group Index

The group index is used as an approximate within-group evaluation of the materials of the A-2-6, A-2-7, A-4, A-5, A-6, and A-7 groups.

$$\text{Group index} = 0.2a + 0.005ac + 0.01bd$$

where $a$ = that portion of the percentage passing the No. 200 sieve greater
  than 35 and not exceeding 75 per cent, expressed as a positive
  whole number (1 to 40)

  $b$ = that portion of the percentage passing the No. 200 sieve greater than 15 and not exceeding 55 per cent, expressed as a positive whole number (1 to 40)

$c$ = that portion of the numerical liquid limit greater than 40 and not exceeding 60, expressed as a positive whole number (1 to 20)

$d$ = that portion of the numerical plasticity index greater than 10 and not exceeding 30, expressed as a positive whole number (1 to 20)

Under average conditions of good drainage and thorough compaction, the supporting value of a material as a subgrade is in inverse ratio to its group index; that is, a group index of 0 indicates a good subgrade material and a group index of 20 indicates a very poor subgrade material.

## 6-13. Relationship among Soil Classifications

Other important soil classifications and measures of supporting strength include the following:

1. California Bearing Ratio, the ratio (expressed as a percentage) of the load required to cause a specified penetration in a given soil to the load required to cause the same penetration in a compacted gravel
2. Casagrande soil classification
3. Civil Aeronautics Administration soil classification
4. Resistance value $R$
5. Bearing value

The approximate relationships among these classifications are shown in Fig. 6-5.

In the Casagrande soil classification, the following symbols are used:

G = gravel, gravelly soil
S = sand, sandy soil
O = organic silt or clay
C = clay
M = silt or very fine sand
F = fine
P = poorly graded
W = well graded
L = low to medium compressibility
H = high compressibility

## 6-14. Relationship of Weights and Volumes in Soil

The unit weight of soil varies, depending on the amount of water contained in the soil. Three unit weights are in general use: the saturated unit weight $\gamma_{sat}$, the dry unit weight $\gamma_{dry}$, and the buoyant unit weight $\gamma_b$.

$$\gamma_{sat} = (G + e)\gamma_0/(1 + e) = (1 + w)G\gamma_0/(1 + e) \qquad S = 100\% \qquad (6\text{-}25)$$
$$\gamma_{dry} = G\gamma_0/(1 + e) \qquad S = 0\% \qquad (6\text{-}26)$$
$$\gamma_b = (G - 1)\gamma_0/(1 + e) \qquad S = 100\% \qquad (6\text{-}27)$$

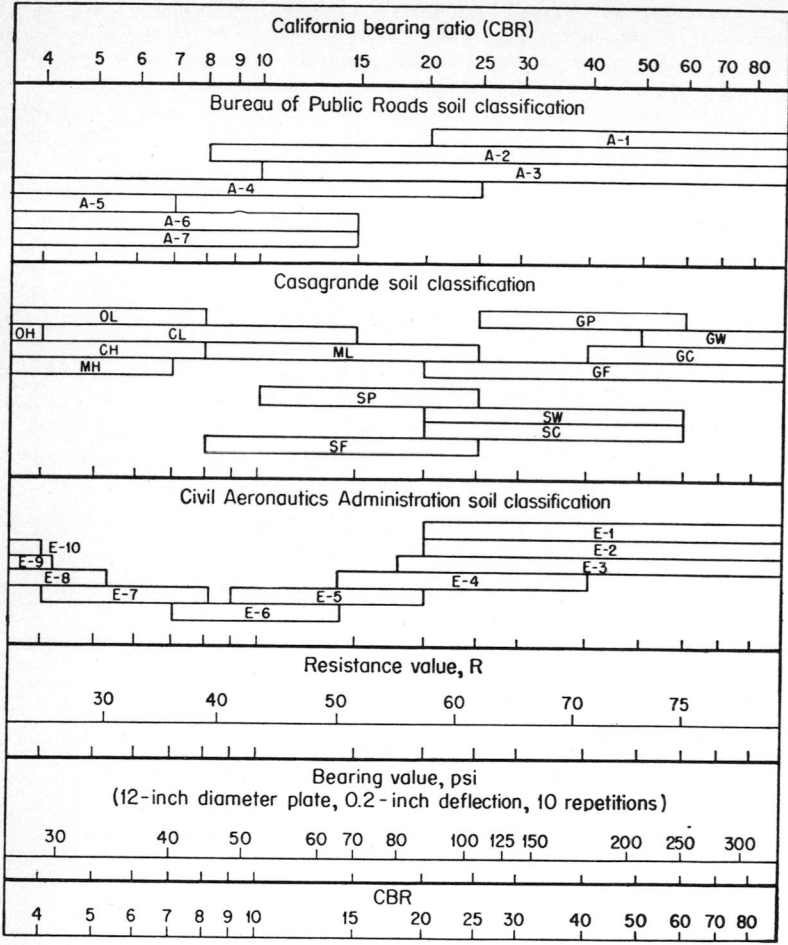

Fig. 6-5. Approximate relationship among soil classifications.

Unit weights are generally expressed in pounds per cubic foot or grams per cubic centimeter. Representative values of unit weights for a soil with a specific gravity of 2.73 and a void ratio of 0.80 are

$$\gamma_{sat} = 122 \text{ lb/ft}^3 = 1.96 \text{ g/cm}^3$$
$$\gamma_{dry} = 95 \text{ lb/ft}^3 = 1.52 \text{ g/cm}^3$$
$$\gamma_b = 60 \text{ lb/ft}^3 = 0.96 \text{ g/cm}^3$$

The symbols used in Eqs. (6-25) to (6-27) and in Fig. 6-6 are

$G$ = specific gravity of soil solids (specific gravity of quartz is 2.67; for majority of soils specific gravity ranges between 2.65 and 2.85; organic soils would have lower specific gravities)

$\gamma_0$ = unit weight of water (62.4 lb/ft³ or 1.0 g/cm³)

$e$ = voids ratio, volume of voids in mass of soil divided by volume of solids in same mass [also equal to $n/(1-n)$, where $n$ is porosity—volume of voids in mass of soil divided by total volume of same mass]

$S$ = degree of saturation, volume of water in mass of soil divided by volume of voids in same mass

$w$ = water content, weight of water in mass of soil divided by weight of solids in same mass (also equal to $Se/G$)

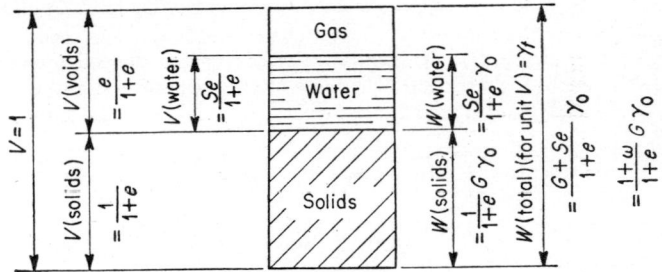

Total volume (solids + water + gas) = 1

Fig. 6-6. Relationship of weights and volumes in soil.

## 6-15. Atterberg Limits

The Atterberg limits are used to define the change in the strength properties of fine-grained soils with a change in water content. The *liquid limit* $w_l$ is the highest water content at which the soil has a small but definite shear resistance. At the liquid limit, the cohesion of the soil is practically zero. The *plastic limit* $w_p$ is the lowest water content at which the soil is plastic. The *shrinkage limit* $w_s$ is the lowest water content that can occur in a soil when it is completely saturated.

The *plasticity index* $I_p$ is the liquid limit minus the plastic limit and is the range of water content throughout which the soil is plastic. When the plastic limit is equal to the liquid limit, the plasticity index is zero and the soil is entirely lacking in plasticity.

## 6-16. Permeability

The coefficient of permeability of a soil is the volume of water which would be forced through a mass of soil having a unit cross-sectional area and a unit length by a unit head of water. The permeability of sand usually ranges from $20 \times 10^{-4}$ to $3{,}000 \times 10^{-4}$ cm/sec (5 to 850 ft/day). The permeability of clays is usually less than $10 \times 10^{-4}$ cm/sec (2.8 ft/day).

Natural soils occurring in stratified formations have a permeability in the direction of stratification much greater than in the direction perpendicular to the stratification.

## 6-17. Internal Friction and Cohesion

The angle of *internal friction* for a soil is expressed by

$$\tan \phi = \tau / \sigma \qquad (6\text{-}28)$$

where $\phi$ = angle of internal friction
$\tan \phi$ = coefficient of internal friction
$\sigma$ = normal force on given plane in cohesionless soil mass
$\tau$ = shearing force on same plane when sliding on plane is impending

For medium and coarse sands, the angle of internal friction is about 30 to 35°. The angle of internal friction for clays ranges from practically 0 to 20°.

The *cohesion* of a soil is the shearing strength which the soil possesses by virtue of its intrinsic pressure. The value of the ultimate cohesive resistance of a soil is usually designated by $c$. Average values for $c$ are given below.

| General soil type | Cohesion $c$, psf | General soil type | Cohesion $c$, psf |
|---|---|---|---|
| Almost liquid clay........ | 100 | Medium clay.......... | 1,000 |
| Very soft clay............ | 200 | Damp, muddy sand...... | 400 |
| Soft clay................ | 400 | | |

## 6-18. Vertical Pressures in Soils

The vertical stress in a soil caused by a vertical, concentrated surface load may be determined with a fair degree of accuracy by the use of elastic theory. Two equations are in common use, the Boussinesq and the Westergaard. The Boussinesq equation applies to an elastic, isotropic, homogeneous mass which extends infinitely in all directions from a level surface. The vertical stress at a point in the mass is

$$\sigma_z = 3P / 2\pi z^2 [1 + (r/z)^2]^{5/2} \qquad (6\text{-}29)$$

The Westergaard equation applies to an elastic material laterally reinforced with horizontal sheets of negligible thickness and infinite rigidity, which prevent the mass from undergoing lateral strain. The vertical stress at a point in the mass, assuming a Poisson's ratio of zero, is

$$\sigma_z = P / \pi z^2 [1 + 2(r/z)^2]^{3/2} \qquad (6\text{-}30)$$

where $\sigma_z$ = vertical stress at a point, psf
$P$ = total concentrated surface load, lb
$z$ = depth of point at which $\sigma_z$ acts, measured vertically downward from surface, ft
$r$ = horizontal distance from projection of surface load $P$ to point at which $\sigma_z$ acts, ft

For values of $r/z$ between 0 and 1, the Westergaard equation gives stresses appreciably lower than those given by the Boussinesq equation. For values of $r/z$ greater than 2.2, both equations give stresses less than $P/100z^2$.

The Westergaard equation is somewhat preferable for use in analyses in sedimentary soils because the assumptions on which it is based are probably nearer to the conditions existing in stratified soils.

Equations (6-29) and (6-30) may be used for loads spread over an area, provided that the area of loading has a maximum dimension less than one-third the depth $z$ at which the stress is to be computed. Areas having greater dimensions should be subdivided for purposes of the computation, and the resulting stresses added.

## 6-19. Lateral Pressures in Soils, Forces on Retaining Walls

The Rankine theory of lateral earth pressures, used for estimating approximate values for lateral pressures on retaining walls, assumes that the pressure on the back of a vertical wall is the same as the pressure that would exist on a vertical plane in an infinite soil mass. Friction between the wall and the soil is neglected. The pressure on a wall consists of (1) the lateral pressure of the soil held by the wall, (2) the pressure of the water, if any, behind the wall, and (3) the lateral pressure from any surcharge on the soil behind the wall.

Symbols used in this section are as follows:

$\gamma$ = unit weight of soil, lb/ft³ (saturated unit weight, dry unit weight, or buoyant unit weight, depending on conditions)

$P$ = total thrust of soil, lb/linear ft of wall

$H$ = total height of wall, ft

$\phi$ = angle of internal friction of soil, deg

$i$ = angle of inclination of ground surface behind wall with horizontal; also angle of inclination of line of action of total thrust $P$ and pressures on wall with horizontal

$K_A$ = coefficient of active pressure

$K_P$ = coefficient of passive pressure

$c$ = cohesion, psf

**Lateral Pressure of Cohesionless Soils.** For walls that retain cohesionless soils and are free to move an appreciable amount, the total thrust from the soil is

$$P = \frac{1}{2} \gamma H^2 \cos i \, \frac{\cos i - \sqrt{(\cos i)^2 - (\cos \phi)^2}}{\cos i + \sqrt{(\cos i)^2 - (\cos \phi)^2}} \tag{6-31}$$

When the surface behind the wall is level, the thrust is

$$P = \frac{1}{2}\gamma H^2 K_A \tag{6-32}$$

$$K_A = [\tan (45° - \phi/2)]^2 \tag{6-33}$$

The thrust is applied at a point $H/3$ above the bottom of the wall, and the pressure distribution is triangular, with the maximum pressure of $2P/H$ occurring at the bottom of the wall.

**For walls that** retain cohesionless soils and are free to move only a slight amount, the total thrust is $1.12P$, where $P$ is as given above. The thrust is applied at the mid-point of the wall and the pressure distribution is trapezoidal with the maximum pressure of $1.4P/H$ extending over the middle six-tenth of the height of the wall.

For walls that retain cohesionless soils and are completely restrained (very rare), the total thrust from the soil is

$$P = \frac{1}{2} \gamma H^2 \cos i \frac{\cos i + \sqrt{(\cos i)^2 - (\cos \phi)^2}}{\cos i - \sqrt{(\cos i)^2 - (\cos \phi)^2}} \qquad (6\text{-}34)$$

When the surface behind the wall is level, the thrust is

$$P = \frac{1}{2}\gamma H^2 K_P \qquad (6\text{-}35)$$
$$K_P = [\tan (45° + \phi/2)]^2 \qquad (6\text{-}36)$$

The thrust is applied at a point $H/3$ above the bottom of the wall, and the pressure distribution is triangular, with the maximum pressure of $2P/H$ occurring at the bottom of the wall.

**Lateral Pressure of Cohesive Soils.** For walls that retain cohesive soils and are free to move a considerable amount over a long period of time, the total thrust from the soil (assuming a level surface) is

$$P = \frac{1}{2}\gamma H^2 K_A - 2cH \sqrt{K_A} \qquad (6\text{-}37)$$

or, since highly cohesive soils generally have small angles of internal friction,

$$P = \frac{1}{2}\gamma H^2 - 2cH \qquad (6\text{-}38)$$

The thrust is applied at a point somewhat below $H/3$ from the bottom of the wall, and the pressure distribution is approximately triangular.

For walls that retain cohesive soils and are free to move only a small amount or not at all, the total thrust from the soil is

$$P = \frac{1}{2}\gamma H^2 K_P \qquad (6\text{-}39)$$

since the cohesion would be lost through plastic flow.

**Water Pressure.** The total thrust from water retained behind a wall is

$$P = \frac{1}{2}\gamma_0 H^2 \qquad (6\text{-}40)$$

where $H$ = height of water above bottom of wall, ft
$\gamma_0$ = unit weight of water, lb/ft³ (62.4 lb/ft³ for fresh water and 64 lb/ft³ for salt)

The thrust is applied at a point $H/3$ above the bottom of the wall, and the pressure distribution is triangular, with the maximum pressure of $2P/H$ occurring at the bottom of the wall. Regardless of the slope of the surface behind the wall, the thrust from water is always horizontal.

**Lateral Pressure from Surcharge.** The effect of a surcharge on a wall retaining a cohesionless soil or an unsaturated cohesive soil can be accounted for by applying a uniform horizontal load of magnitude $K_A p$ over the entire height of the wall, where $p$ is the surcharge in pounds per square foot. For saturated cohesive soils the full value of the surcharge $p$ should be considered as acting over the entire height of the wall as a uniform horizontal load. $K_A$ is defined in list of nomenclature, above.

## 6-20. Stability of Slopes

**Cohesionless Soils.** A slope in a cohesionless soil without seepage of water is stable if

$$i < \phi \qquad (6\text{-}41)$$

With seepage of water parallel to the slope, and assuming the soil to be saturated, an infinite slope in a cohesionless soil is stable if

$$\tan i < (\gamma_b/\gamma_{\text{sat}}) \tan \phi \qquad (6\text{-}42)$$

where $i$ = slope of ground surface
$\phi$ = angle of internal friction of soil
$\gamma_b, \gamma_{\text{sat}}$ = unit weights, lb/ft³ (Sec. 6-14)

**Cohesive Soils.** A slope in a cohesive soil is stable if

$$H < C/\gamma N \qquad (6\text{-}43)$$

where $H$ = height of slope, ft
$C$ = cohesion, lb/ft²
$\gamma$ = unit weight, lb/ft³
$N$ = stability number, dimensionless

For failure on the slope itself, without seepage water,

$$N = (\cos i)^2(\tan i - \tan \phi) \qquad (6\text{-}44)$$

Similarly, with seepage of water,

$$N = (\cos i)^2[\tan i - (\gamma_b/\gamma_{\text{sat}}) \tan \phi] \qquad (6\text{-}44a)$$

where terms are as defined for Eq. (6-42).

For failure encompassing all or part of the slope, together with soil at the top or toe of the slope, approximate values of the stability number $N$ are given in Table 6-2. In the use of formula (6-44a) and Table 6-2, appropriate values must be used for $\phi$ and $\gamma$. When the slope is submerged, $\phi$ is the angle

### Table 6-2. Stability Numbers for Simple Slopes

| $i$ | $N$ for various values of $\phi$ | | | |
|---|---|---|---|---|
| | 0° | 5° | 15° | 25° |
| 90° | 0.261 | 0.239 | 0.199 | 0.165 |
| 75° | .219 | .196 | .154 | .118 |
| 60° | .191 | .165 | .120 | .082 |
| 45° | .170 | .141 | .085 | .048 |
| 30° | .156 | .114 | .048 | .012 |
| 15° | .145 | .072 | | |

of internal friction of the soil and $\gamma$ is equal to $\gamma_b$.  When the surrounding water is removed from a submerged slope in a short time (sudden drawdown), $\phi$ is the weighted angle of internal friction [equal to $(\gamma_b/\gamma_{sat})\phi$] and $\gamma$ is equal to $\gamma_{sat}$.

## 6-21. Bearing Capacity of Soils

The approximate ultimate bearing capacity under a long footing at the surface of a soil is given by Prandtl's equation as

$$q_u = (c/\tan \phi + \tfrac{1}{2}\gamma_{dry}b\sqrt{K_p})(K_p e^{\pi \tan \phi} - 1) \qquad (6\text{-}45)$$

where $q_u$ = ultimate bearing capacity of soil, $lb/ft^2$
   $c$ = cohesion, $lb/ft^2$
   $\phi$ = angle of internal friction, deg
   $\gamma_{dry}$ = unit weight of dry soil, $lb/ft^3$ (Sec. 6-14)
   $b$ = width of footing, ft
   $d$ = depth of footing below surface, ft
   $K_p$ = coefficient of passive pressure = $[\tan (45 + \phi/2)]^2$
   $e$ = 2.718 . . .

For footings below the surface, the ultimate bearing capacity of the soil may be modified by the factor $1 + Cd/b$.  The coefficient $C$ is about 2 for cohesionless soils and about 0.3 for cohesive soils.  The increase in bearing capacity with depth for cohesive soils is often neglected.

Typical values of the allowable bearing capacity of various soils as given in the National Building Code of the National Board of Fire Underwriters are shown in Table 6-3.  These values represent the ultimate bearing capacity divided by an appropriate safety factor.

### Table 6-3. Allowable Bearing Capacity of Soils

| Soil | Allowable Bearing Capacity, Tons/Ft² |
|------|-------------------------------------:|
| Medium soft clay | 1.5 |
| Medium stiff clay | 2.5 |
| Sand, fine, loose | 2 |
| Sand, coarse, loose; compact fine sand; loose sand-gravel mixture | 3 |
| Gravel, loose; compact coarse sand | 4 |
| Sand-gravel mixture, compact | 6 |
| Hardpan and exceptionally compacted or partially cemented gravels or sands | 10 |
| Sedimentary rocks, such as hard shales, sandstones, limestones, and silt stones, in sound condition | 15 |
| Foliated rocks, such as schist or slate, in sound condition | 40 |
| Massive bedrock, such as granite, diorite, gneiss, and trap rock, in sound condition | 100 |

## 6-22. Settlement under Foundations

The approximate relationship between loads on foundations and settlement is

$$\frac{q}{P} = C_1\left(1 + \frac{2d}{b}\right) + \frac{C_2}{b} \qquad (6\text{-}46)$$

where $q$ = load intensity, $lb/ft^2$
   $P$ = settlement, in.
   $d$ = depth of foundation below ground surface, ft

$b$ = width of foundation, ft
$C_1$ = coefficient dependent on internal friction
$C_2$ = coefficient dependent on cohesion

The coefficients $C_1$ and $C_2$ are usually determined by bearing-plate loading tests.

## 6-23. Allowable Loads on Piles

(See also Sec. 6-54, Pile Driving.)

A dynamic formula extensively used in the United States to determine the allowable static load on a pile is the *Engineering News* formula. For piles driven by a drop hammer, the allowable load is

$$P_a = 2WH/(p + 1) \tag{6-47}$$

For piles driven by a single-acting hammer, the allowable load is

$$P_a = 2WH/(p + 0.1) \tag{6-48}$$

For piles driven by a double-acting hammer, the allowable load is

$$P_a = 2E/(p + 0.1) \tag{6-48a}$$

where $P_a$ = allowable pile load, lb
$W$ = weight of hammer, lb
$H$ = height of drop or stroke, ft
$E$ = actual energy delivered per blow, ft-lb
$p$ = penetration of pile per blow, in.

For a group of piles penetrating a soil stratum of good bearing characteristics and transferring their loads to the soil by point bearing on the ends of the piles, the total allowable load would be the sum of the individual allowable loads for each pile. For piles transferring their loads to the soil by skin friction on the sides of the piles, the total allowable load would be less than the sum of the individual allowable loads for each pile, because of the interaction of the shearing stresses and strains caused in the soil by each pile.

## 6-24. Types of Piles

Foundation piles used to carry structure loads may be timber, concrete, composite (timber with concrete upper section), or steel. Piles which distribute the load throughout their length to the soil are called friction piles. Those which carry the load to firm substrata are end-bearing piles. Sheet piles used to retain soil or water may be wood planking, steel sheeting, or precast concrete sheets. Waterproofing is obtained by interlocking or overlapping of sections.

## HIGHWAY AND TRAFFIC ENGINEERING

(See also Sec. 6-55, Pavements.)

For detailed data on the geometric design of highways, reference should be made to "A Policy on Geometric Design of Rural Highways" and "A Policy on Arterial Highways in Urban Areas."[1]   Information on the capacity of highways will be found in the "Highway Capacity Manual."[2]   Much of the material herein is taken from these publications.

## 6-25. Highway Design Controls

**Vehicle Characteristics.**   Dimensions of the four design vehicles recommended for use by AASHTO[1] as controls for geometric design are shown in Table 6-4.   Minimum turning paths for these vehicles are shown in Fig. 6-7. The vehicle which should be used in design is the largest one which represents a significant percentage of the traffic.   For design of most highways accommodating truck traffic, one of the design semitrailer combinations should be used.   A design check should be made for the largest vehicle expected, in order to ensure that such a vehicle can negotiate the designated turns, particularly if pavements are curbed.

**Design Speed.**   The design speed of a highway is the maximum safe speed that can be maintained over a specified section when conditions are favorable, so that the design features of the highway govern the speed.

**Traffic.**   The principal measures of traffic volume and character and the relationship between the various elements for rural highways are shown in Table 6-5.   Determination of the relationship between the traffic elements for urban highways usually requires special study.

**Types of Arterial Highways.**   A major street is an arterial highway with intersections at grade and direct access to abutting property and on which geometric design and traffic control measures are used to expedite the safe movement of through traffic.   An expressway is a divided arterial highway with full or partial control of access and generally with grade separations at intersections.   A freeway is an expressway with full control of access.   A parkway is a type of arterial highway provided for noncommercial traffic, with full or partial control of access and usually located within a park or ribbon of parklike development.

## 6-26. Elements of Geometric Design

**Stopping Sight Distance.**   Design stopping sight distance is the minimum distance required for a vehicle traveling at or near the design speed to stop before reaching an object in its path.   It is the sum of the distances traveled during perception and brake reaction time and the distance traveled while braking to a stop.   Stopping sight distance is measured from a point 3.75 ft above the road surface to a point 6 in. above the road surface.   The sight

[1] American Association of State Highway and Transportation Officials (AASHTO).

[2] Transportation Research Board, National Academy of Sciences—National Research Council.

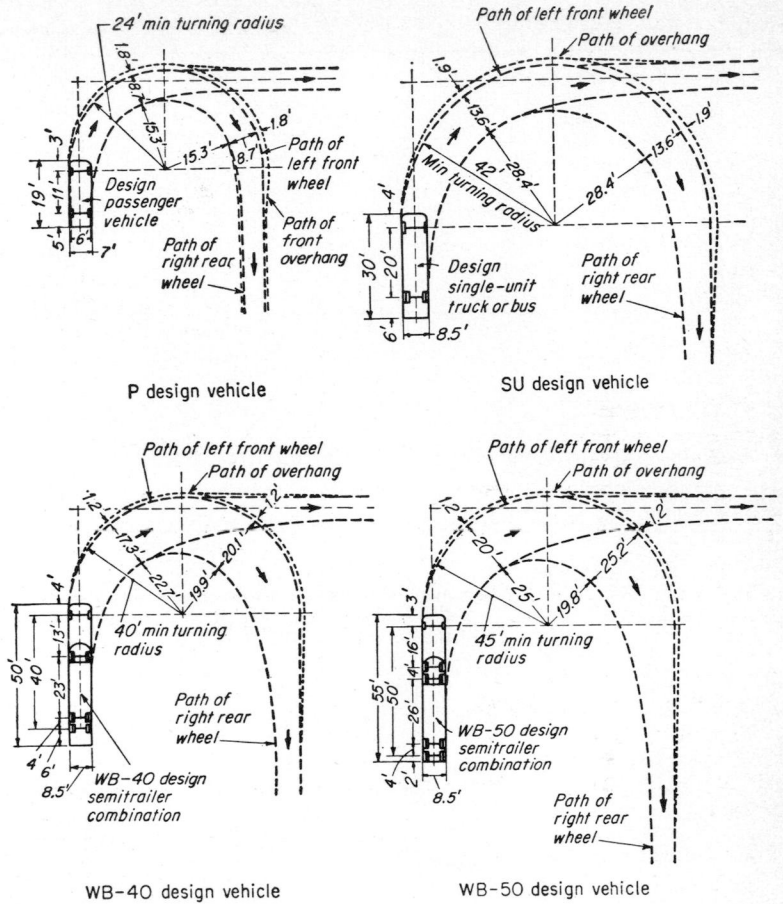

P design vehicle

SU design vehicle

WB-40 design vehicle

WB-50 design vehicle

Fig. 6-7. Minimum turning paths for design vehicles.

### Table 6-4. Dimensions of Design Vehicles

| Design vehicle type and symbol | Dimensions. ft | | | | | |
|---|---|---|---|---|---|---|
| | Wheelbase | Overhang | | Over-all length | Over-all width | Height |
| | | Front | Rear | | | |
| Passenger car, P............................ | 11 | 3 | 5 | 19 | 7 | |
| Single-unit truck, SU...................... | 20 | 4 | 6 | 30 | 8.5 | 13.5 |
| Semitrailer combination, intermediate, WB-40.. | 13 + 27 = 40 | 4 | 6 | 50 | 8.5 | 13.5 |
| Semitrailer combination, large, WB-50........ | 20 + 30 = 50 | 3 | 2 | 55 | 8.5 | 13.5 |

### Table 6-5. Traffic Elements and Their Relation for Rural Highways

*Traffic Element*        *Explanation and Nationwide Percentage or Factor*

Average daily traffic, ADT... Average 24-hr volume for a given year; total for both directions of travel, unless otherwise specified

DHV.................... Design hour volume (two-way unless otherwise specified), usually the thirtieth highest hourly volume of the design year (30HV)

K........................ DHV expressed as a percentage of ADT, both two-way: normal range 12 to 18%

D........................ Directional distribution of DHV, one-way volume in predominant direction of travel expressed as a percentage of two-way DHV: general range 55 to 80%, average 67%

T........................ Trucks (exclusive of light delivery trucks) expressed as a percentage of DHV: normal range 5 to 12%, average 8%.

distance at every point on a highway should be at least as great as the minimum distances shown in Table 6-6.

## Table 6-6. Stopping Sight Distance

| Design Speed, Mph | Min Stopping Sight Distance, Ft | Design Speed, Mph | Min Stopping Sight Distance, Ft |
|---|---|---|---|
| 15 | 80 | 40 | 275 |
| 20 | 120 | 50 | 350 |
| 25 | 160 | 60 | 475 |
| 30 | 200 | 70 | 600 |
|  |  | 80 | 750 |

**Passing Sight Distance.** Design passing sight distance is the minimum distance required to make safely a normal passing maneuver on two- and three-lane highways at passing speeds representative of nearly all drivers, commensurate with design speed. Passing sight distance is measured from a point 3.75 ft above the road surface to a second point 4.5 ft above the road surface. The minimum passing sight distance is shown in Table 6-7.

**Maximum Horizontal Curvature and Superelevation.** The maximum horizontal curvature for a given design speed is limited by the maximum rate of superelevation and the allowable side friction. The maximum superelevation that is considered generally desirable is 0.10 ft/ft pavement width. Values from 0.06 to 0.12 are used for maximum superelevation rates, depending on local conditions such as ice formation, frequency of intersection, and similar factors. For a maximum superelevation rate of 0.10 ft/ft, the design

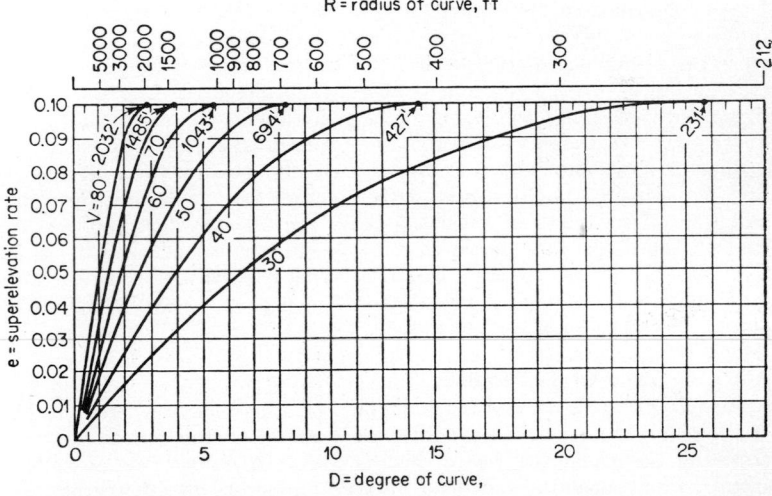

FIG. 6-8. Design superelevation rates for maximum superelevation rate of 0.10 ft/ft.

## Table 6-7. Passing Sight Distance
## for Two-lane Highways

| Design Speed, Mph | Min Passing Sight Distance, Ft |
|---|---|
| 30 | 1100 |
| 40 | 1500 |
| 50 | 1800 |
| 60 | 2100 |
| 70 | 2500 |
| 80 | 2700 |

superelevation rates recommended by AASHTO for various speeds are shown on Fig. 6-8. The figure also indicates the maximum curvature for various design speeds at the maximum superelevation rate of 0.10.

**Maximum Grades.** The maximum grades recommended by AASHTO for main highways are shown in Table 6-8. Maximum grades for secondary highways may be about 2% steeper than those shown in the table.

## Table 6-8. Maximum Grades, Per Cent

| Topography | Design speed, mph | | | | | |
|---|---|---|---|---|---|---|
| | 30 | 40 | 50 | 60 | 70 | 80 |
| Flat............... | 6 | 5 | 4 | 3 | 3 | 3 |
| Rolling............. | 7 | 6 | 5 | 4 | 4 | 4 |
| Mountainous........ | 9 | 8 | 7 | 6 | 5 | |

**Sight Distance on Horizontal Curves.** Stopping sight distance must be provided on all horizontal curves. Figure 6-9 shows the required clearance from the center line of the inside lane to provide the minimum stopping sight distance. Design of two-lane highways for passing sight distance must in general be confined to tangent or very flat alignment conditions because of the excessive clearances that would be required on curves.

**Sight Distance on Vertical Curves.** Vertical curves must be designed to provide stopping sight distance. Other factors that enter into the determination of the length of a vertical curve are rider comfort and drainage control. It is generally impractical to design vertical curves for passing sight distance. The minimum length of vertical curve $L$ for various algebraic differences in grade $A$ is shown in Fig. 6-10 for crest vertical curves and in Fig. 6-11 for sag vertical curves.

## 6-27. Highway Cross Sections

**Pavement Type and Cross Slope.** The type of pavement is determined by the volume and composition of traffic, the availability of materials, the initial cost, and the extent and cost of maintenance. *High-type* pavements have smooth riding qualities and good antiskid properties in all weather, and should support adequately the expected volume and weight of vehicles without fatigue. *Intermediate-type* pavements vary from those only slightly less

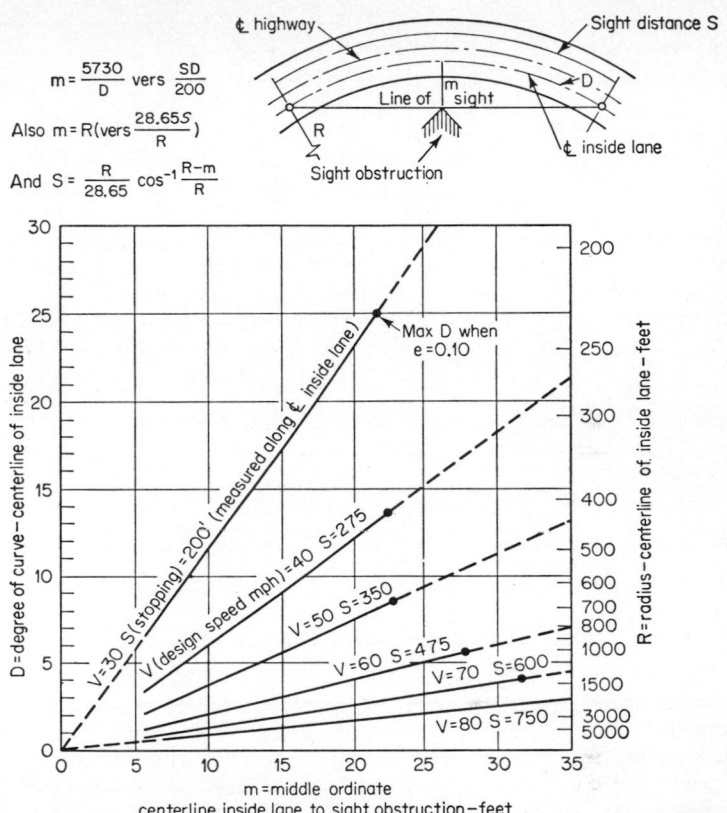

$$m = \frac{5730}{D} \text{ vers } \frac{SD}{200}$$

Also $m = R(\text{vers} \frac{28.65S}{R})$

And $S = \frac{R}{28.65} \cos^{-1} \frac{R-m}{R}$

Fig. 6-9. Stopping sight distance on horizontal curves (open road conditions).

costly than the high type to surface treatments. *Low-type* surfaces range from surface-treated earth to loose surfaces such as earth, shell, or gravel.

The range of cross slopes applicable to each type of pavement for adequate drainage is as follows:

| Surface Type | Cross Slope, Ft/Ft |
|---|---|
| High | 0.01 –0.02 |
| Intermediate | 0.015–0.03 |
| Low | 0.02 –0.04 |

When curbs are located at the pavement edge, the above values should be increased slightly.

**Vertical Clearance.** Clear heights of 14 ft should be provided over all highways, except for routes limited to noncommercial traffic, where 12.5 ft is adequate. In many states, 16 ft is required on interstate highway routes.

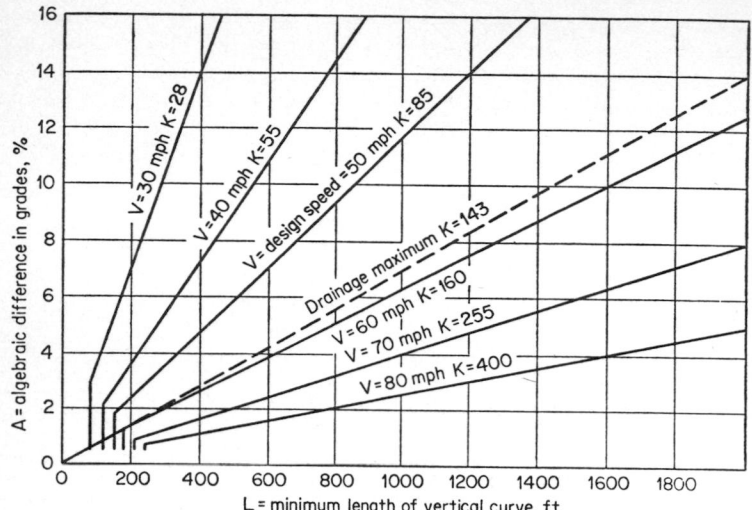

FIG. 6-10. Design controls for crest vertical curves (stopping sight distance).

These clearances are often increased by 4 to 6 in. to provide for future resurfacing.

**Pavement Width.** The desirable width of pavement on a highway is 12 ft per lane. This width may be reduced where traffic is light or speed is low.

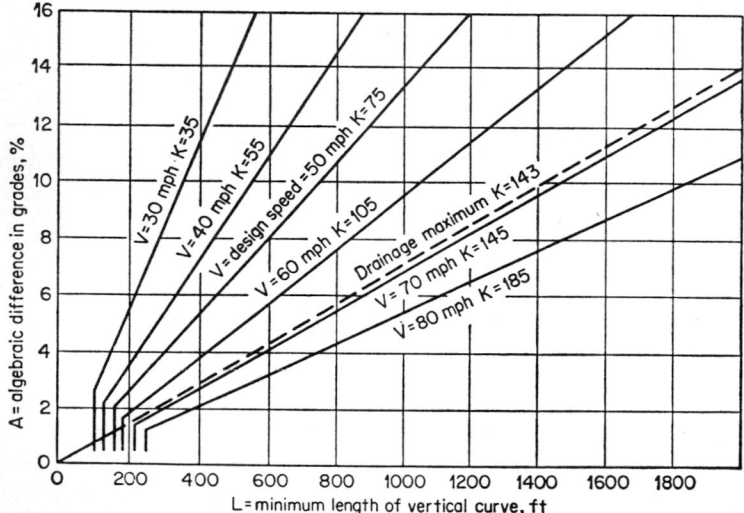

FIG. 6-11. Design controls for sag vertical curves.

**Shoulder Width.** The usable width of shoulder is that which can be used when a driver makes an emergency stop. The over-all width of shoulder, the dimension between the edge of pavement and the intersection of the shoulder and side-slope planes, is 1 to 3 ft greater than the usable shoulder width, except where the side slopes are 4 : 1 or flatter. Usable shoulders 10 to 12 ft wide are desirable on all highways, but narrower shoulders may be used in low-volume highways.

### Table 6-9. Design Guides for Two-lane Rural Highways

| Cross-section element | Element dimension, ft | | |
|---|---|---|---|
| | Low | Intermediate | High |
| Surfacing............ | 18–20 | 20–24 | 24 |
| Usable shoulder...... | 4–8 | 8 | 10 |
| Roadway............. | 26–36 | 36–40 | 44 |
| Border. each side..... | 18–25* | 20– 30* | 25– 35* |
| Right of way........ | 66–80* | 80–100* | 100–120* |

* Preferably more.

**Design Guides.** Design guides for cross-section elements for two-lane rural highways are shown in Table 6-9.

Four-lane highways should, if possible, be designed as divided highways. Design guides for cross-section elements for various types of four-lane divided rural highways are shown in Table 6-10. These cross-section elements should provide a balanced total section. Where restrictions are necessary, the border width should be reduced before decreasing the median, and both should be cut to a minimum before considering a reduction in shoulder or lane width.

### Table 6-10. Design Guides for Four-lane Divided Rural Highways

| Cross-section element | Element dimension, ft | | |
|---|---|---|---|
| | Restricted | Intermediate | Desirable |
| Pavement. each...... | 24 | 24 | 24 |
| Usable shoulder...... | 8– 10 | 10 | 10– 12 |
| Median............. | 4– 15 | 20* | 40* |
| Border. each side..... | 12– 15* | 25– 40* | 50– 80* |
| Right of way........ | 90–110* | 140–180* | 210–310* |

* Preferably more.

Design guides for major streets in urban areas are shown in Table 6-11. Design speeds for major streets are 30 mph in built-up districts and as high as 50 mph in outlying areas.

## Table 6-11. Design Guides for Major Urban Streets

| Section | Type of urban area | Through traffic lanes | | Median— width, ft | | Shoulders, pavement widening at curbs, or parking lanes— width, ft | | Border— width, ft | | Right of way— width, ft | |
|---|---|---|---|---|---|---|---|---|---|---|---|
| | | No. | Width, ft* | | | | | | | | |
| | | | A    B | A | B | A | B | A | B | A | B |
| Shoulders— no curbs | Res. | 2 | 11   12 | 0 | 0 | 10 | 10 | 12 | 20 | 66 | 84 |
| | Res. | 4 | 11   12 | 0 | 14 | 10 | 10 | 8 | 12 | 80 | 106 |
| Curbed— no parking | Com. | 4 | 11   12 | 0 | 4 | 1 | 2 | 8 | 12 | 62 | 80 |
| | Res. | 4 | 11   12 | 0 | 4 | 1 | 2 | 12 | 16 | 70 | 88 |
| | Com. | 6 | 11   12 | 0 | 4 | 1 | 2 | 8 | 12 | 84 | 104 |
| | Res. | 6 | 11   12 | 0 | 4 | 1 | 2 | 12 | 16 | 92 | 112 |
| Curbed with parking lanes | Com. | 4 | 11   12 | 0 | 4 | 10 | 11 | 8 | 12 | 80 | 98 |
| | Res. | 4 | 11   12 | 0 | 4 | 10 | 10 | 12 | 16 | 88 | 104 |
| | Com. | 6 | 11   12 | 0 | 4 | 10 | 11 | 8 | 12 | 102 | 122 |
| | Res. | 6 | 11   12 | 0 | 4 | 10 | 10 | 12 | 16 | 110 | 128 |
| Divided with parking lanes† | Com. | 4 | 11   12 | 4 | 14 | 10 | 12 | 8 | 12 | 84 | 110 |
| | Res. | 4 | 11   12 | 4 | 14 | 10 | 11 | 12 | 16 | 92 | 116 |
| | Com. | 6 | 11   12 | 4 | 14 | 10 | 12 | 8 | 12 | 106 | 134 |
| | Res. | 6 | 11   12 | 4 | 14 | 10 | 11 | 12 | 16 | 114 | 140 |

NOTE: A = acceptable minimum, B = desirable minimum, Res. = residential, Com. = commercial.
* Ten-foot widths may be considered in special cases, but not on two-lane streets.
† Without parking lanes, deduct 20 ft from right of way.

## Table 6-12. Design Guides for Expressways-at-grade

| Cross-section element | Element dimension, ft | | |
|---|---|---|---|
| | Restricted | Intermediate | Desirable |
| Pavement, each: | | | |
| Four-lane.......... | 24 | 24 | 24 |
| Six-lane........... | 36 | 36 | 36 |
| Shoulder............ | 10 | 10 | 10 |
| Median............. | 4 | 14–25 | 40* |
| Border, each side..... | 12 | 20 | 30* |
| Right of way: | | | |
| Four-lane.......... | 96 | 120–130 | 170* |
| Six-lane........... | 120 | 145–155 | 195* |

* Preferably more.

An expressway-at-grade is intermediate between a major street and a freeway with respect to design features. The expressway-at-grade is a surface facility practically free from roadside interference and on which crossing or entering traffic from minor streets is eliminated. Design speeds range from 40 mph in built-up districts to 60 mph in outlying areas. Design guides for expressways-at-grade are shown in Table 6-12.

Design guides for freeways are similar to those for expressways-at-grade. Design speeds range from 50 mph in built-up districts to 60 or 70 mph in outlying areas. Frontage roads are often required for depressed freeways to provide continuity in the local street system. Grade separations for cross-streets may occur at intervals of one to two blocks in downtown areas, three to five blocks in intermediate areas, and at greater distances in outlying areas. Interchanges are generally located about 2 miles apart in urban areas, 4 miles apart in suburban areas, and 8 miles apart in rural areas.

## 6-28. Highway Capacity and Levels of Service

The capacity of a highway is the maximum number of vehicles which has a reasonable expectation of passing over a given section of a lane or a roadway in one direction (or in both directions for a two-lane or a three-lane highway) during a given time period under prevailing roadway and traffic conditions.

### Table 6-13. Operating Speeds and Service Volumes, Two-lane Highway

(Uninterrupted Flow Conditions; Rural)

| Level of service | Operating speed, mph | % of length with passing sight distance of 1,500 ft or more | Maximum service volume under ideal conditions (total passenger cars per hour, both directions) | | | |
|---|---|---|---|---|---|---|
| | | | For 70-mph design speed | For 60-mph design speed | For 50-mph design speed | For 40-mph design speed |
| A | 60 or more | 100 | 400 | 1 | 1 | 1 |
| | | 50 | 270 | 1 | 1 | 1 |
| | | 0 | 80 | 1 | 1 | 1 |
| B | 50 or more | 100 | 900 | 800 | 1 | 1 |
| | | 50 | 720 | 540 | 1 | 1 |
| | | 0 | 480 | 240 | 1 | 1 |
| C | 40 or more | 100 | 1,400 | 1,320 | 1,120 | 1 |
| | | 50 | 1,270 | 1,070 | 850 | 1 |
| | | 0 | 1,080 | 760 | 360 | 1 |
| D | 35 or more | 100 | 1,700 | 1,660 | 1,500 | 1,160 |
| | | 50 | 1,650 | 1,550 | 1,350 | 960 |
| | | 0 | 1,600 | 1,320 | 1,020 | 380 |
| E[2] | 30± | n.p.[3] | 2,000 | 2,000 | 2,000 | 2,000 |
| F | Less than 30 | n.p.[3] | Variable | Variable | Variable | Variable |

[1] This level of service not attainable at this design speed.
[2] Capacity.
[3] No passing at this level.

## Table 6-14. Operating Speeds and Service Volumes, Freeways and Expressways

(Uninterrupted Flow Conditions; Rural or Small Metropolitan Areas)

| Level of service | Operating speed, mph | Maximum service volume under ideal conditions (total passenger cars per hour, one direction) | | | |
|---|---|---|---|---|---|
| | | For 70-mph design speed | | For 60-mph design speed per lane in one direction | For 50-mph design speed per lane in one direction |
| | | For two lanes in one direction | Each additional lane in one direction | | |
| $A$ | 60 or more | 1,400 | 1,000 | 1 | 1 |
| $B$ | 55 or more | 2,000 | 1,500 | 500 | 1 |
| $C$ | 50 or more | 2,300 | 1,400 | 700 | 1 |
| $D$ | 40 or more | 2,800 | 1,400 | 1,200 | 700 |
| $E^2$ | 30 to 35 | 4,000 | 2,000 | 2,000 | 2.000 |
| $F$ | Less than 30 | Variable | Variable | Variable | Variable |

[1] This level of service not attainable at this design speed.
[2] Capacity.

## Table 6-15. Effects of Narrow Traffic Lanes and Restricted Lateral Clearances

| Clearance from pavement edge to obstruction | Percentage of service volume* | | | | | | | |
|---|---|---|---|---|---|---|---|---|
| | Lanes with obstruction on one side | | | | Lanes with obstruction on both sides | | | |
| | 12-ft | 11-ft | 10-ft | 9-ft | 12-ft | 11-ft | 10-ft | 9-ft |
| Two-lane Highways | | | | | | | | |
| 6 | 100 | 86 | 77 | 70 | 100 | 86 | 77 | 70 |
| 4 | 96 | 83 | 74 | 68 | 92 | 79 | 71 | 65 |
| 2 | 91 | 78 | 70 | 64 | 81 | 70 | 63 | 57 |
| 0 | 85 | 73 | 66 | 60 | 70 | 60 | 54 | 49 |
| Freeways and Expressways, Two Lanes Each Direction | | | | | | | | |
| 6 | 100 | 97 | 91 | 81 | 100 | 97 | 91 | 81 |
| 4 | 99 | 96 | 90 | 80 | 98 | 95 | 89 | 79 |
| 2 | 97 | 94 | 88 | 79 | 94 | 91 | 86 | 76 |
| 0 | 90 | 87 | 82 | 73 | 81 | 79 | 74 | 66 |

* Applies to level of service $B$ for two-lane highways and to all levels for freeways and expressways.

Capacity is equivalent to level of service $E$, defined below with other levels of service:

A—free flow, low volumes, high speeds, little or no restriction in maneuverability

*B*—stable flow, operating speeds somewhat restricted by traffic conditions (level *B* suitable for design of rural highways)

*C*—stable flow, operating speeds satisfactory but closely controlled by traffic conditions (level *C* suitable for urban design)

*D*—approaching unstable flow, tolerable operating speed, little freedom to maneuver

*E*—unstable flow, momentary stoppages (level *E* is capacity)

*F*—forced flow, congestion (volumes below capacity)

Service volumes for two-lane highways and for freeways and expressways under uninterrupted flow conditions are shown in Tables 6-13 and 6-14. These service volumes must be adjusted for roadway and traffic conditions. Adjustments for lane widths and lateral clearances are given in Table 6-15. Adjustments for trucks may be made by converting trucks to equivalent passenger cars using the factors from Table 6-16. The service volumes shown do not apply at intersections or in the vicinity of ramp termini.

### Table 6-16. Average Passenger Car Equivalent of Trucks over Extended Lengths of Highways

| Type of route | Level of service | Level terrain | Rolling terrain | Mountainous terrain |
|---|---|---|---|---|
| Freeway and expressways....... | All | 2 | 4 | 8 |
| Two-lane highways........... | *A* | 3 | 4 | 7 |
| | *B* and *C* | 2.5 | 5 | 10 |
| | *D* and *E* | 2 | 5 | 12 |

Typical design service volumes per lane for urban arterial routes with allowances for the factors discussed above and for roadside and intersection interferences are shown in Fig. 6-12.

## 6-29. Intersection Capacity and Levels of Service

The capacity of a signalized intersection approach is the maximum number of vehicles that the approach can reasonably accommodate under the existing geometric, environmental, and traffic characteristics and controls. Capacity is equivalent to level of service *E*, defined below with other levels of service:

*A*—free operation, no vehicle waits longer than one red indication: load factor = 0.0

*B*—stable operation, occasional approach cycle fully utilized, many drivers somewhat restricted by traffic conditions: load factor = 0.1 or less (level *B* suitable for design of rural intersections)

*C*—stable operation, intermittent loading, most drivers somewhat restricted by traffic conditions: load factor = 0.3 or less (level *C* suitable for design of urban intersections)

*D*—approaching instability, substantial delays during short peaks within peak period: load factor = 0.7 or less

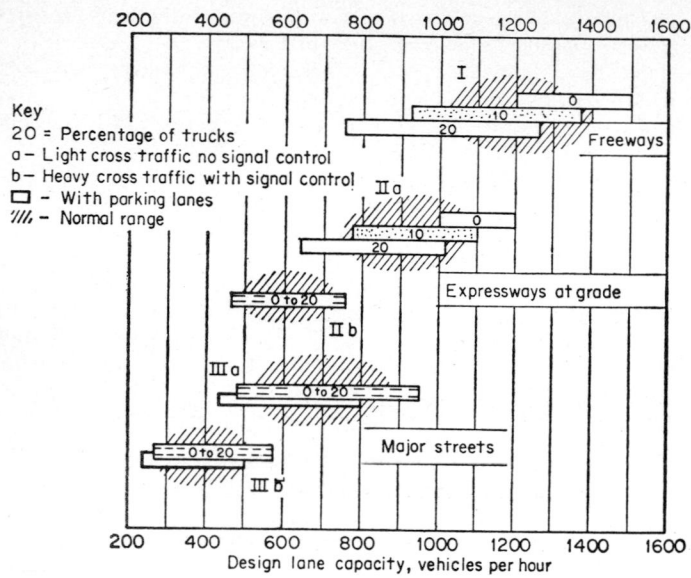

Fig. 6-12. Lane capacities for urban routes.

*E*—delays of several cycles, queues developing: load factor = 0.7 to 1.0, depending on conditions, with an average of 0.85 (level *E* is capacity)

*F*—jammed conditions, traffic flow controlled by downstream conditions, volumes unpredictable

Load factor is the proportion of green-signal intervals that are fully utilized.

Service volumes at intersections on two-way and one-way urban streets with parking are shown on Figs. 6-13 and 6-14. Service volumes of intersections are expressed as vehicles per hour of green signal time, and the service volume for an approach must be adjusted for the proportion of total time allocated to the approach. Adjustments must also be made for metropolitan-area size, the peak-hour factor (for intersections, the ratio of the volume during the peak hour to four times the volume during the peak 15 minutes: range 0.25 to 1.00), the location within the metropolitan area, and the effects of commercial-vehicle traffic, turning movements, and parking prohibitions.

Peak-hour factors of 1.00 are rarely found. Where long lines of waiting vehicles are typically present, a peak-hour factor of 0.90 or 0.95 may be used. The usual conditions in a metropolitan area are equivalent to a peak-hour factor of 0.85, but where a high rate of flow occurs over a period shorter than an hour, factors of 0.75, 0.70, or less should be used. Adjustments to service volumes for various peak hours are combined on Figs. 6-13 and 6-14 with adjustments for metropolitan-area size. The figures also indicate adjustments for the location of the intersection within the metropolitan area.

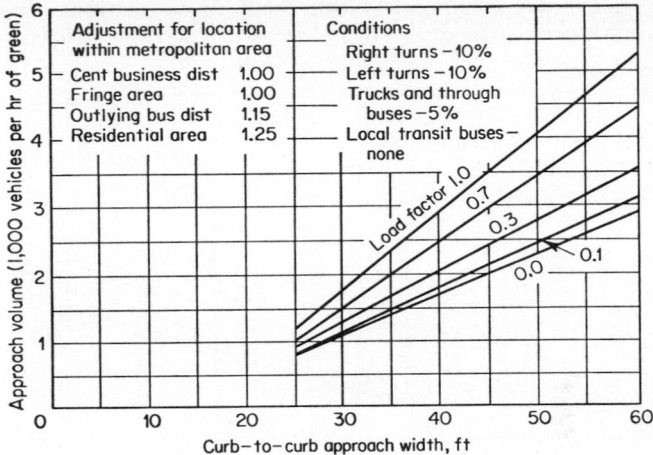

Adjustment for Peak-hour Factor and Metropolitan-area Size

| Metropolitan-area pop. (1,000's) | Peak-hour factor | | | | | | |
|---|---|---|---|---|---|---|---|
| | 0.70 | 0.75 | 0.80 | 0.85 | 0.90 | 0.95 | 1.00 |
| Over 1,000 | 1.00 | 1.05 | 1.09 | 1.14 | 1.19 | 1.24 | 1.29 |
| 1,000 | 0.97 | 1.02 | 1.07 | 1.11 | 1.16 | 1.21 | 1.26 |
| 750 | 0.94 | 0.99 | 1.04 | 1.09 | 1.14 | 1.18 | 1.23 |
| 500 | 0.91 | 0.96 | 1.01 | 1.06 | 1.11 | 1.16 | 1.21 |
| 375 | 0.88 | 0.93 | 0.98 | 1.03 | 1.08 | 1.13 | 1.18 |
| 250 | 0.85 | 0.90 | 0.95 | 1.00 | 1.05 | 1.10 | 1.15 |
| 175 | 0.82 | 0.87 | 0.92 | 0.97 | 1.02 | 1.07 | 1.12 |
| 100 | 0.80 | 0.85 | 0.89 | 0.94 | 0.99 | 1.04 | 1.09 |
| 75 | 0.77 | 0.82 | 0.87 | 0.92 | 0.96 | 1.01 | 1.06 |

Fig. 6-13. Urban intersection approach service volume, in vehicles per hour of green-signal time, for one-way streets with parking both sides.

For streets with approach widths of 21 to 29 ft where turning movements differ from 10 per cent right and 10 per cent left, multiply the service volume by

$[1.00 - (0.005)(R - 10)]$ $[1.00 - (0.010)(L - 10)]$ [for two-way streets]
$[1.00 - (0.005)(R - 10)]$ $[1.00 - (0.005)(L - 10)]$ [for one-way streets]

where $R$ = percentage of right turns (maximum $R = 30$)
$L$ = percentage of left turns (maximum $L = 30$)

Special adjustments must be made for intersections having separate lanes and/or separate signal indications for turning movements.

For intersections where the percentage of trucks and through buses differs from 5 per cent, multiply the service volume by $[1.00 - (0.010)(T - 5)]$ where $T$ is the percentage of trucks and through buses. Adjustments for buses which stop in the vicinity of the intersection require special procedures.

Prohibition of parking can result in increases of 30 to 50 per cent over the service volumes shown on Fig. 6-14 for two-way streets and 20 per cent to as much as 100 per cent over the volumes shown on Fig. 6-13 for one-way streets.

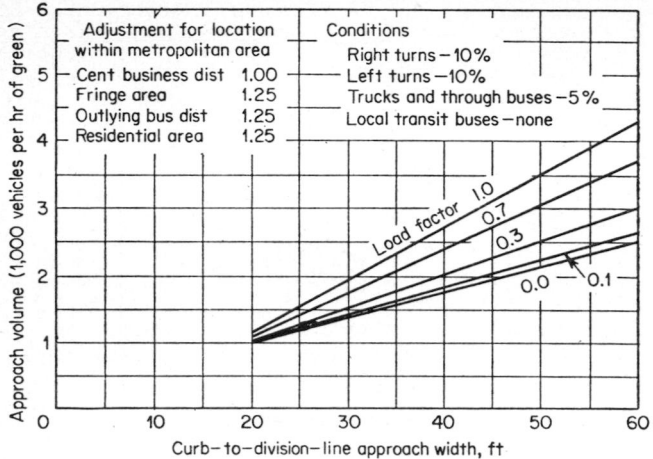

Adjustment for Peak-hour Factor and Metropolitan-area Size

| Metropolitan-area pop. (1,000's) | Peak-hour factor | | | | | | |
|---|---|---|---|---|---|---|---|
| | 0.70 | 0.75 | 0.80 | 0.85 | 0.90 | 0.95 | 1.00 |
| Over 1,000 | 1.00 | 1.05 | 1.10 | 1.14 | 1.19 | 1.24 | 1.29 |
| 1,000 | 0.97 | 1.02 | 1.07 | 1.11 | 1.16 | 1.21 | 1.27 |
| 750 | 0.94 | 0.99 | 1.04 | 1.09 | 1.13 | 1.18 | 1.23 |
| 500 | 0.91 | 0.96 | 1.01 | 1.06 | 1.11 | 1.15 | 1.20 |
| 375 | 0.89 | 0.93 | 0.98 | 1.03 | 1.08 | 1.12 | 1.17 |
| 250 | 0.86 | 0.91 | 0.95 | 1.00 | 1.05 | 1.10 | 1.14 |
| 175 | 0.83 | 0.88 | 0.92 | 0.97 | 1.02 | 1.07 | 1.11 |
| 100 | 0.80 | 0.85 | 0.90 | 0.94 | 0.99 | 1.04 | 1.09 |
| 75 | 0.77 | 0.82 | 0.87 | 0.91 | 0.96 | 1.01 | 1.06 |

Fig. 6-14. Urban intersection approach service volume, in vehicles per hour of green-signal time, for two-way streets with parking.

## Table 6-17. Street Space Used for Parking

| Angle of parking at curb | Width of street used when parked, ft | Width needed for parking plus maneuvering, ft | Length of curb per car, ft |
|---|---|---|---|
| Parallel.... | 7 | 19 | 22.0 |
| 45° | 17 | 29 | 11.3 |
| 60° | 18 | 36 | 9.2 |
| 90° | 17 | 40 | 8.0 |

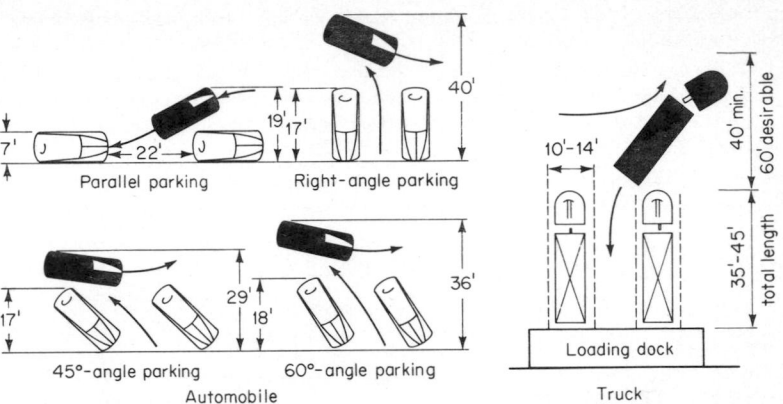

Fig. 6-15. Storage and maneuvering space used for various parking positions.

### 6-30. Parking Requirements

**Automobile Parking.** The space used for automobile parking is shown in Fig. 6-15 and Table 6-17.

**Truck Parking.** Truck loading/unloading platform heights range from 48 to 52 in. to match truck floors. Parking stall should be 10 to 14 ft in width. Other dimensions are shown on Fig. 6-15.

## RAILROADS

### 6-31. Track Gage

Standard railroad gage in North America is 4 ft 8½ in. between inside of rails, measured ⅝ in. below top of rail. The gage may be varied from 4 ft 8⅜ in. on some high-speed tangent track to 4 ft 9⅛ in. on curves of small radius. Table 6-18 lists major railroad gages in use throughout the world.

### 6-32. Track Materials

Ties are usually oak, pine, or fir, but concrete ties are in use. Ties are generally 8 ft 6 in. or 9 ft (recommended) in length for standard gage and range from 6 in. deep and 6 in. wide for yard and sidetracks to 7 in. deep and 10 in. wide for principal main lines. Ties spacing varies from 19½ in. on centers for main lines to 24 in. for secondary tracks.

Rail is designated by weight per yard and section. Weights up to 155 lb/yd are used for high-speed track, but rail ranging from 100 to 130 lb/yd is in more common use. Rail is usually rolled in 39-ft lengths, but 78-ft lengths are often provided to decrease maintenance costs. The use of continuous welded rail with lengths of 5,000 to 6,000 ft is increasing. Length is governed by turnout and signal locations since ends of signal circuits require insulated joints.

Rails are connected by means of joint bars which hold adjoining rails in horizontal and vertical alignment. Tie plates are placed under rails to distribute rail loads and reduce wear. Both rail and tie plates are held to the

## Table 6-18. Major Railway Gages of the World

| 36 in. (914 mm) | 39⅜ in. (1,000 mm) | 42 in. (1,067 mm) | 56½ in. (1,435 mm) | 60 in. (1,524 mm) |
|---|---|---|---|---|
| Colombia | Algeria* | Angola | Algeria* | Czecho- |
| El Salvador | Argentina* | Australia* | Australia* | slovakia* |
| Guatemala | Austria* | Costa Rica | Austria* | Finland |
| Haiti | Bangladesh | Ecuador | Argentina* | Panama* |
| Honduras* | Belgium* | Gabon | Belgium* | USSR* |
| Panama* | Bolivia | Honduras* | Brazil* | |
| Venezuela* | Brazil* | Indonesia | Bulgaria | |
| | Chile* | Malawi | Canada | **63 in. (1,600 mm)** |
| | Czechoslovakia* | Mozambique | China | |
| | Egypt* | New Zealand | Cuba | Brazil* |
| | Ethiopia | Nicaragua | Czechoslovakia* | |
| | France* | Nigeria | Denmark | |
| | Germany, West* | Norway* | Egypt* | |
| | Germany, East* | Philippines | France* | |
| | Kenya | Rhodesia | Germany, West* | |
| | Malaysia | Sudan | Germany, East* | **65⅝ in. (1,668 mm)** |
| | Peru* | Union of | Great Britain | |
| | Puerto Rico | South Africa | Greece | Spain* |
| | Spain* | Venezuela* | Hungary | |
| | Switzerland* | Zaire* | Iran | |
| | Tanzania | Zambia | Iraq | |
| | Thailand | | Ireland | |
| | Tunisia | | Italy | |
| | USSR* | | Japan | **66 in. (1,676 mm)** |
| | Vietnam | | Mexico | |
| | Zaire* | | Morocco | Argentina* |
| | | | Netherlands | Chile* |
| | | | Norway* | India |
| | | | Paraguay | Pakistan |
| | | | Peru* | Portugal |
| | | | Poland | Spain* |
| | | | Romania | |
| | | | Sweden | |
| | | | Switzerland* | |
| | | | Syria | |
| | | | Turkey | |
| | | | United States | |
| | | | Uruguay | |

\* Countries having more than one standard gage.

ties by means of spikes. Longitudinal rail movement is prevented by rail anchors.

## 6-33. Curvature and Superelevation

Maximum curvature for railroads is usually determined by train operating speeds and allowable superelevation. Maximum superelevation is 6 to 7 in., but may be less on a particular railroad. The AREA (American Railway Engineering Association) formula for equilibrium superelevation is

(6-49)

$$e = 0.0007DV^2$$

$D$ = degree of curve, deg
$V$ = speed, mph

For high-speed trains, as much as 3 in. of unbalanced superelevation may be permitted, so that

$$e = 0.0007DV^2 - 3 \qquad (6\text{-}49a)$$

Curvature may also be limited because of coupling difficulties on curves of more than 6°. Some railroads have established an absolute maximum curvature of 16° because of dimensions of the rigid wheelbase of cars and engines and permissible swing of couplers. Reverse curves should be separated by at least two car lengths.

Vertical curves should be of sufficient length to limit gradient changes to 0.05%/100 ft for sags and 0.10%/100 ft for crests on main lines; secondary lines may have vertical curves one-half the lengths required for main lines.

## 6-34. Clearances

Track clearances are established from center line of track for horizontal dimensions and from top of rail for vertical dimensions. Track spacings vary among railroads, but the following are common centerline spacings:

| | |
|---|---|
| Main track–main track | 13 ft 6 in. |
| Main track–yard or passing tracks | 15 ft 0 in. |
| Yard track–yard track | 13 ft 6 in. |
| Ladder track–yard track | 18 ft 0 in. |
| Ladder track–ladder track | 19 ft 0 in. |

Track spacings are increased on curves at a rate of 2 in./deg. Where adjacent tracks have different superelevation, additional clearance is provided of 3 in./in. of difference in superelevation.

Clearances from track to fixed structures vary by railroad and by state. Generally, vertical clearances of 22 ft are required, except at building entrances, where 17 ft may be permitted. Horizontal clearances of 8 ft and 8 ft 6 in. are standard for tangent track, with additional allowances of 1 in./deg of curve and 3 in./in. of superelevation for curved track. The clearance recommendations of the AREA are shown on Fig. 6-16a.

## 6-35. Turnouts and Crossovers

Turnouts, used to divert trains from one track to another, consist of a switch, a frog to carry the wheel flanges over crossing rails, closure rails

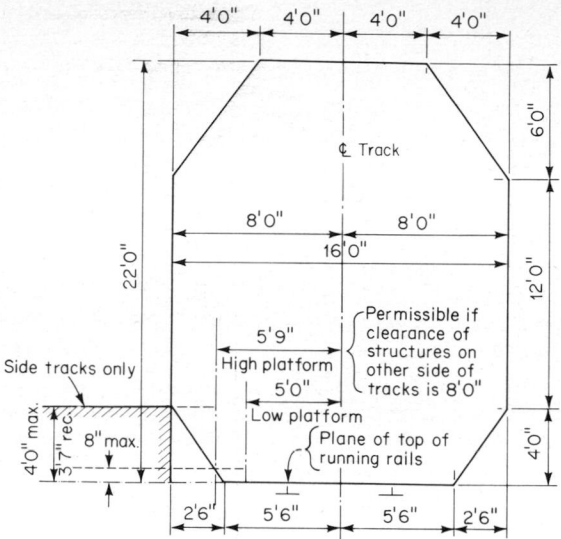

Fɪɢ. 6-16a. AREA clearance recommendations.

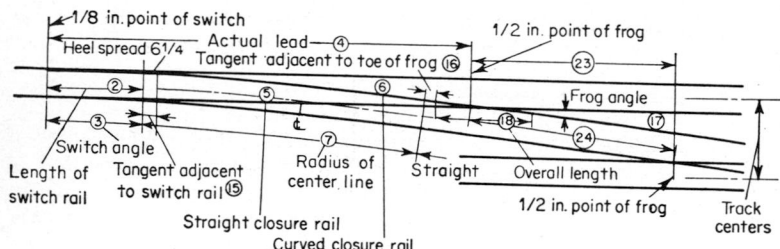

Fɪɢ. 6-16b. Standard turnouts and crossovers. (Refer to table on pages 6-42 and 6-43 for dimensional data)

connecting the switch rails and the frog, and guard rails to guide the flanges at the frog. Control points for turnouts are the actual or $\frac{1}{8}$-in. point of the switch (ground to a width of $\frac{1}{8}$ in.) and the actual or $\frac{1}{2}$-in. point of frog. Standard turnouts, shown on Fig. 6-16b, are identified by the frog number. Turnouts with No. 16 to No. 20 frogs are used for high-speed main-line movements, No. 10 to No. 12 for slow-speed main-line movements, and No. 8 for yards and sidings.

Crossovers, used to transfer trains between parallel tracks, consist of two turnouts and connecting rails. Data for crossovers are also shown on Fig. 6-16b.

## WATER SUPPLY, SEWERAGE, AND DRAINAGE

### 6-36. Water Supply and Treatment

**Quantity of Water.** Average annual water requirements in metropolitan areas with metered systems generally range from 100 gallons per capita per day (gpcpd) to 200 gpcpd, with a median value of about 150 gpcpd. Unmetered supply systems have considerably higher consumption, and large water-using industries require special determination of demand.

Seasonal variations in water demand occur largely because of irrigation, lawn sprinkling, and air-conditioning loads, and maximum monthly consumption is generally about 125 per cent of average annual demand but may range up to 200 per cent of average annual demand. Maximum daily demands of 150 per cent of average annual demand and maximum hourly demands of from 200 to 250 per cent of annual average demand are commonly used for design.

Fire demand is often the determining factor in the design of mains, distribution storage tanks, and pumps, even though the total quantity of water required for fire fighting is small during a long period. For communities of less than 200,000 population, the fire demand is given by the National Board of Fire Underwriters as

$$Q = 1,020 \sqrt{P}(1 - 0.01 \sqrt{P}) \tag{6-50}$$

where $Q$ = fire demand, gpm
$P$ = population in thousands
The fire demand is added to the normal demand on the maximum day to determine the total maximum demand.

**Design Period.** Pipes less than 12 in. in diameter are generally designed to be adequate for the full development of the area served; pipes more than 12 in. in diameter and wells, distribution systems, and filtration and treatment plants are generally designed for the flow expected 15 to 25 years in the future; and large dams and conduits are generally designed to be adequate for 25 to 50 years.

## Turnout and Crossover Data

| (1) Frog number | Properties of switches | | (4) Actual lead | Closure distance | | Lead curve | |
|---|---|---|---|---|---|---|---|
| | (2) Length of switch rail | (3) Switch angle | | (5) Straight closure rail | (6) Curved closure rail | (7) Radius of center line | (8) Degree of curve |
| | Ft In. | Deg Min Sec | Ft In. | Ft In. | Ft In. | Ft | Deg Min Sec |
| 5 | 11-0 | 2-39-34 | 42-6½ | 28-0 | 28-4 | 177.80 | 32-39-56 |
| 6 | 11-0 | 2-39-34 | 47-6 | 32-9 | 33-0 | 258.57 | 22-17-58 |
| 7 | 16-6 | 1-46-22 | 62-1 | 40-10½ | 41-1¼ | 365.59 | 15-43-16 |
| 8 | 16-6 | 1-46-22 | 68-0 | 46-5 | 46-7½ | 487.28 | 11-46-44 |
| 9 | 16-6 | 1-46-22 | 72-3½ | 49-5 | 49-7¼ | 615.12 | 9-19-30 |
| 10 | 16-6 | 1-46-22 | 78-9 | 55-10 | 56-0 | 779.39 | 7-21-24 |
| 11 | 22-0 | 1-19-46 | 91-10¼ | 62-10¼ | 63-0 | 927.27 | 6-10-56 |
| 12 | 22-0 | 1-19-46 | 96-8 | 66-10½ | 67-0 | 1104.63 | 5-11-20 |
| 14 | 22-0 | 1-19-46 | 107-0¾ | 76-5¼ | 76-6¾ | 1581.20 | 3-37-28 |
| 15 | 30-0 | 0-58-30 | 126-4½ | 86-11½ | 87-0¾ | 1720.77 | 3-19-48 |
| 16 | 30-0 | 0-58-30 | 131-4 | 91-11 | 92-0 | 2007.12 | 2-51-18 |
| 18 | 30-0 | 0-58-30 | 140-11½ | 99-11 | 100-0 | 2578.79 | 2-13-20 |
| 20 | 30-0 | 0-58-30 | 151-11½ | 110-11 | 111-0 | 3289.29 | 1-44-32 |

## (for Use with Fig. 6-16b)

| | | Properties of frogs | | Data for crossovers | | | |
|---|---|---|---|---|---|---|---|
| | | | | 13′ 0″ track centers | | For change of 1′ 0″ of track centers | |
| (15) Tangent adjacent to switch rail | (16) Tangent adjacent to toe frog | (17) Frog angle | (18) Over-all length | (23) Straight track | (24) Crossover track | Straight track | Crossover track |
| Ft | Ft | Deg Min Sec | Ft In. | Ft In. | Ft In. | Ft In. | Ft In. |
| 0.00 | 0.78 | 11-25-16 | 9-0 | 16-10 5/16 | 18-1 7/8 | 4-11 7/16 | 5-0 5/8 |
| 0.00 | 1.75 | 9-31-38 | 10-0 | 20-5 1/2 | 21-6 1/2 | 5-11 1/2 | 6-0 1/2 |
| 0.01 | 0.00 | 8-10-16 | 12-0 | 24-0 3/8 | 24-11 5/8 | 6-11 9/16 | 7-0 7/16 |
| 0.64 | 0.00 | 7-09-10 | 13-0 | 27-7 1/8 | 28-4 7/8 | 7-11 5/8 | 8-0 3/8 |
| 0.00 | 0.17 | 6-21-35 | 16-0 | 31-1 5/8 | 31-10 3/8 | 8-11 11/16 | 9-0 5/16 |
| 2.08 | 0.00 | 5-43-29 | 16-6 | 34-8 1/8 | 35-3 7/8 | 9-11 11/16 | 10-0 5/16 |
| 0.00 | 0.13 | 5-12-18 | 18-8 1/2 | 38-2 1/2 | 38-9 1/2 | 10-11 3/4 | 11-0 1/4 |
| 0.00 | 0.50 | 4-46-19 | 20-4 | 41-8 3/4 | 42-3 1/4 | 11-11 3/4 | 12-0 1/4 |
| 0.24 | 0.00 | 4-05-27 | 23-7 | 48-9 1/4 | 49-2 13/16 | 13-11 13/16 | 14-0 1/4 |
| 1.56 | 0.00 | 3-49-06 | 24-4 1/2 | 52-3 7/16 | 52-8 5/8 | 14-11 13/16 | 15-0 3/16 |
| 0.66 | 0.00 | 3-34-47 | 26-0 | 55-9 5/8 | 56-2 1/2 | 15-11 13/16 | 16-0 3/16 |
| 0.57 | 0.00 | 3-10-56 | 29-3 | 62-9 7/8 | 63-2 3/16 | 17-11 13/16 | 18-0 3/16 |
| 2.47 | 0.00 | 2-51-51 | 30-10 1/2 | 69-10 | 70-2 | 19-11 7/8 | 20-0 1/8 |

**Quality of Water.** The outstanding requirement for a domestic water supply is freedom from pathogenic bacteria. In addition, there are reasonable limits for certain impurities, as listed below:

| Impurity | Limit, ppm | Impurity | Limit, ppm |
|---|---|---|---|
| Turbidity | 10 | Iron plus manganese | 0.3 |
| Color | 20 | Magnesium | 125 |
| Lead | 0.1 | Total solids | 500 |
| Fluoride | 1.5 | Total hardness (calcium | |
| Copper | 3.0 | plus magnesium salts) | 100 |

**Water Treatment.** Water treatment usually consists of filtration through either a slow or a rapid sand filter and disinfection with chlorine. In addition, water may be softened to remove hardness and aerated to remove iron and manganese.

The slow sand filter operates at a rate of 2 to 10 million gal/(acre)(day) and is effective in removing tastes and odors from raw water. About 99 per cent of the bacterial content is also removed.

The rapid sand filter operates at a rate of 125 to 250 million gal/(acre)(day). However, preliminary treatment of the raw water is required, including chemical coagulation and sedimentation. The entire treatment process is effective in removing about 99.98 per cent of the bacterial content, but removal of the color and turbidity is less dependable than for the slow sand filter and requires particular attention to the coagulation process.

Softening is accomplished by the addition of lime, or lime and soda ash, and sedimentation. The addition of lime and passage through a zeolite softener is also used. Iron and manganese may be removed by aeration and sedimentation.

Disinfection by the addition of chlorine is the final stage of any treatment process. Common practice is to add sufficient chlorine so that a small free chlorine residual is maintained.

## 6-37. Water Distribution

Transmission mains connecting the source of supply to the distribution system must be large enough to supply at least the maximum daily demand plus fire flow. If the distribution system does not include storage, supply mains must also be adequate to deliver maximum hourly demands.

Both transmission and distribution mains are usually designed by using the Hazen-Williams formula. This equation with factors necessary for its use and a nomograph to facilitate its application are presented in Sec. 2.

## 6-38. Sewage Collection and Treatment

**Quantity of Sewage.** The average flow of sewage from a metropolitan area is about 100 gpcpd (gallons per capita per day). This rate may vary from

240 gpcpd in a maximum hour, 160 gpcpd on a maximum day, 70 gpcpd on a minimum day, and 40 gpcpd in a minimum hour.   In addition, infiltration of ground water into sewers may be taken at about 600 gal/(day)(in. diam)(mile).

**Design Period.**   Laterals and submains less than 15 in. in diameter are generally designed to be adequate for the full development of the area served; main sewers, outfalls, and intercepter sewers are generally designed for the flow expected from 40 to 50 years in the future; and treatment works are generally designed for the flow expected from 10 to 25 years in the future.

**Sewer Design.**   Sewers should be at least 8 in. in diameter and should be laid on a grade sufficient to produce a velocity of 2 fps when flowing full, to prevent the deposition of suspended solids.   Sewer lines should be designed with straight alignment and uniform grade between manholes, which should have a maximum spacing of 400 ft.

**Quality of Sewage.**   Sewage is approximately 99.92 per cent water, with the remaining 0.08 per cent (800 ppm by weight) composed of organic and mineral matter, as shown below:

|  | Organic matter, ppm | Mineral matter, ppm |
|---|---|---|
| Suspended solids.... | 100 | 50 |
| Colloidal solids...... | 140 | 60 |
| Dissolved solids..... | 160 | 290 |

The "biochemical oxygen demand" of sewage, BOD, is the quantity of oxygen which must be supplied during the aerobic stabilization of sewage, and thus is a direct measure of the pollutional effect.   Residential sewage has an average BOD of 0.24 lb oxygen per capita.

**Sewage Treatment.**   The degree of sewage treatment required should be based on the size, characteristics, and usage of the receiving body of water and upon the amount and quality of sewage to be treated.   Complete sewage treatment might include preliminary treatment, such as screening to remove large suspended solids, grit removal, and grease removal; primary treatment, such as plain sedimentation or chemical precipitation; secondary treatment of a biological nature, such as the trickling filter or the activated sludge process; final treatment by chlorination; and finally disposal by dilution in a body of water.   Approximate values of the BOD and suspended-solids removal of primary and secondary treatment are shown below:

| Treatment process | Percentage removal | | Treatment process | Percentage removal | |
|---|---|---|---|---|---|
| | BOD | Suspended solids | | BOD | Suspended solids |
| Plain sedimentation......... | 25–40 | 40–70 | Trickling filter............. | 80–95 | 80–90 |
| Chemical precipitation....... | 50–75 | 70–90 | Activated sludge.......... | 85–95 | 85–95 |

## 6-39. Sizes and Slopes of Sewers

Sewer sizes and slopes are usually designed by using the Manning formula

$$v = \frac{1.486}{n} R^{2/3} S^{1/2} \tag{6-51}$$

where $v$ = average velocity of flow, fps
   $n$ = coefficient of roughness
   $R$ = hydraulic radius, ft = $A/P = D/4$ for circular conduit flowing full
   $S$ = hydraulic gradient, ft head loss/ft length
   $A$ = cross-sectional area of flow, ft$^2$
   $P$ = wetted perimeter, ft
   $D$ = diameter of circular conduit, ft

For circular sewers flowing full, the Manning formula can be written as

$$Q = (0.4632/n)D^{2/3}S^{1/2} = \text{conveyance factor} \times S^{1/2} \tag{6-52}$$

where $Q$ = quantity of flow, cfs.

The factors necessary for the use of this equation and nomographs to facilitate its application are presented in Sec. 2.

## 6-40. Quantity of Runoff

The rational method for the determination of the quantity of storm water which appears as runoff involves the use of

$$Q = ciA \tag{6-53}$$

where $Q$ = runoff from rainfall, cfs
   $c$ = coefficient of runoff, dimensionless
   $i$ = rainfall intensity, expressed as a rate, in. rain/hr
   $A$ = tributary area, acres

These factors are discussed below.

**Coefficient of Runoff.** The coefficient of runoff for a particular area depends on the character of the surface, the type and extent of vegetation, the slope of the surface, and other less important factors. Approximate values of the coefficient of runoff $c$ are given in Table 6-19.

### Table 6-19. Runoff Coefficients for Rational Formula

| Type of area | Flat: slope $<2\%$ | Rolling: slope $2–10\%$ | Hilly: slope $>10\%$ |
|---|---|---|---|
| Pavements, roofs, etc.......... | 0.90 | 0.90 | 0.90 |
| City business areas........... | .80 | .85 | .85 |
| Suburban residential areas..... | .45 | .50 | .55 |
| Dense residential areas........ | .60 | .65 | .70 |
| Grassed areas................ | .25 | .30 | .30 |
| Earth areas.................. | .60 | .65 | .70 |
| Cultivated land: | | | |
|   Impermeable (clay, loam).... | .50 | .55 | .60 |
|   Permeable (sand).......... | .25 | .30 | .35 |
| Meadows and pasture lands.... | .25 | .30 | .35 |
| Forests and wooded areas..... | .10 | .15 | .20 |

Rainfall Intensity. The rainfall intensity is dependent on the recurrence interval and the time of concentration. The recurrence interval is the period of time within which, on the average, a rainfall of a given intensity will be equaled or exceeded only once. Recurrence intervals of from 5 to 25 years are generally used, but for important structures periods of 100 years have been used.

For a particular area and a given recurrence interval, a study of rainfall records will permit the determination of an intensity-duration curve, which gives the rainfall intensity (in inches per hour) as a function of the duration of rainfall. The rainfall intensity is greatest for short periods and decreases sharply as the duration of rainfall becomes greater. The intensity to use for a particular design is that for which the duration is equal to the time of concentration.

Time of Concentration. The time of concentration for a particular inlet to a drainage system is the time required for rainfall falling on the most remote part of the tributary area drained by the inlet to reach the inlet. At this time, the entire area tributary to the inlet will be contributing to the runoff and the total runoff will be a maximum. The time for water to flow overland from the most remote part of the tributary area to the inlet may be approximated by

$$t = C(L/Si^2)^{1/3} \qquad (6\text{-}54)$$

where $t$ = time of overland flow, min
$\quad L$ = distance of overland flow, ft
$\quad S$ = slope of land, ft/ft
$\quad i$ = rainfall intensity, in./hr
$\quad C$ = coefficient: 0.5 for paved areas, 1.0 for bare earth, 2.5 for turf
For any portions of the flow carried in ditches, the time of flow to the inlet may be computed by means of the Manning formula [Eq. (6-51)].

## 6-41. Flow in Drainage Channels

Drainage channels are usually of such lengths that head losses other than those due to friction are negligible. Design of drainage channels is generally by the Manning formula:

$$v = (1.486/n)R^{2/3}S^{1/2} \qquad (6\text{-}55)$$

where $S = h_f/l$ = hydraulic gradient, ft head loss/ft length
$\quad n$ = coefficient of roughness, dimensionless
$\quad R$ = hydraulic radius, ft = $A/P$
$\quad A$ = cross-sectional area of flow, ft$^2$
$\quad P$ = wetted perimeter, ft
$\quad h_f$ = head loss due to friction, ft
$\quad l$ = length of channel or conduit, ft
$\quad v$ = average velocity of flow, fps
Factors necessary for the use of this equation and nomographs to facilitate its application are presented in Sec. 2.

## STRUCTURAL ANALYSIS AND DESIGN[1]

The four basic types of load-carrying members are *ties*, which carry axial tension; *columns*, which carry axial compression; *beams*, which carry transverse loads; and *shafts*, which carry torsional loads. Many structural members have as their primary loading a combination of the basic types of loading, and most have secondary loadings of a type other than their primary loading.

### 6-42. Ties

The stress in a tie carrying axial tension or tension applied at the centroid of the cross section of the tie is

$$f_a = P/A \tag{6-56}$$

and the total strain, or total elongation, of a tie under load is

$$e = Pl/AE = f_a l/E \tag{6-57}$$

where $f_a$ = tensile unit stress, psi
$P$ = total axial load, lb
$A$ = cross-sectional area of tie, in.$^2$
$e$ = total elongation, in. or ft
$l$ = total length of member, in. or ft
$E$ = modulus of elasticity, psi

### 6-43. Columns

The stress and strain in a short column, sometimes called a strut, carrying an axial compressive load is the same as given above for a tie, except that the stress is compressive and the strain is a shortening. Long, slender columns usually fail by buckling, and the strength of these columns is not determined by the strength of the material of which the column is made. The critical elastic buckling load for a long, slender column axially loaded is given by Euler's formula as

$$P_c = \pi^2 EI/l^2 = \pi^2 AE/(l/r)^2 \tag{6-58}$$

The critical buckling load for a column intermediate between the short column and the long, slender column is given by the secant formula as

$$P_c = \frac{f_y A}{1 + \dfrac{ec}{r^2} \sec\left(\dfrac{l}{2r}\sqrt{\dfrac{P_c}{AE}}\right)} \tag{6-59}$$

where $P_c$ = critical buckling load, lb
$I$ = least moment of inertia of cross-sectional area, in.$^4$
$A$ = cross-sectional area, in.$^2$
$r$ = least radius of gyration of cross-sectional area $(r = \sqrt{I/A})$, in.
$f_y$ = yield-point stress of material, psi

[1] The properties of geometric shapes, such as moment of inertia, section modulus, and radius of gyration, are given in Sec. 8. Properties for structural shapes are given in Table 6-20 (Lumber) and Table 6-28a (Steel).

$ec/r^2$ = factor introducing effect of column crookedness and unintentional eccentricity of loading (generally taken as 0.25)

$l$ = effective length of column (see below), in.

$E$ = modulus of elasticity, psi

The ratio of the effective length of columns to their actual length is approximately as follows:

| End Conditions | Effective Length as Percentage of Actual Length |
|---|---|
| One end fixed, one end free | 200 |
| One end fixed, one end free to turn but not move | 70 |
| Both ends free to turn but not move | 100 |
| Both ends fixed | 50 |
| Pin-ended columns | $87\frac{1}{2}$ |
| Riveted-end columns | 75 |

## 6-44. Beams

**Shear and Moment.** The vertical *shear* for a section of a beam is the algebraic sum of the external forces acting perpendicularly to the beam's longitudinal axis to either side of the section. For convenience, the forces that lie to the left of the section are usually used. When the sum of these forces is upward, the shear at the section is positive. The bending *moment* for a section of a beam is the algebraic sum of the moments, about the centroid of the section, of the external forces and moments that act on the beam to either side of the section. For convenience, the forces and moments that lie to the left of the section are usually used. When the sum of these moments is clockwise, the upper fibers of the beam are in compression and the bending moment at the section is positive. Shears and bending moments for beams with loading conditions of frequent occurrence are shown in Fig. 6-17.

### List of Symbols Used for Fig. 6-17

$R_1, R_2$ = reactions at left and right ends of beam, respectively, lb

$V$ = shear, lb

$M$ = moment, lb-in.

$\Delta$ = deflection, in.

$w$ = intensity of distributed load on beam, lb/in.

$l$ = length of beam, in. ($x$, $a$, and $b$ in same units)

$W$ = total load on beam, lb

$P$ = concentrated load on beam, lb

$E$ = modulus of elasticity, psi

$I$ = moment of inertia of beam cross section, in.$^4$

**Statically Determinate Beams.** The reactions, shears, and moments in statically determinate beams (those having only three unknown reaction components) can be determined by the three equations of statics which can be applied to a beam:

$$\Sigma H = 0 \qquad (6\text{-}60)$$

$$\Sigma V = 0 \qquad (6\text{-}61)$$
$$\Sigma M = 0 \qquad (6\text{-}62)$$

where $\Sigma H$ = algebraic sum of all horizontal forces acting on beam, including horizontal reaction forces

$\Sigma V$ = algebraic sum of all vertical forces acting on beam, including vertical reaction forces

$\Sigma M$ = algebraic sum of all moments acting on beam and of moments of all forces, including reactions, acting on beam, about any point in beam (generally chosen as at one of reactions)

The reactions, shears, and moments for beams with loading conditions of frequent occurrence are shown in Fig. 6-17.

**Statically Indeterminate Beams.**   The moments at the points of support of statically indeterminate beams may be determined by either the three-moment equation or the slope-deflection equations.   The three-moment equation for a beam with a constant moment of inertia between supports is

$$M_{AB}\frac{l_{AB}}{I_{AB}} + 2M_{BA}\left(\frac{l_{AB}}{I_{AB}} + \frac{l_{CB}}{I_{CB}}\right) + M_{CB}\frac{l_{CB}}{I_{CB}} = -\left(\frac{6A_{AB}a_{AB}}{I_{AB}l_{AB}} + \frac{6A_{CB}a_{CB}}{I_{CB}l_{CB}}\right)$$
$$+ \left(\frac{6Eh_{AB}}{l_{AB}} + \frac{6Eh_{CB}}{l_{CB}}\right) \qquad (6\text{-}63)$$

The symbols used in the three-moment equation are shown in Fig. 6-18 and listed below.

$M_{AB}, M_{BA}, M_{CB}$ = moments in statically indeterminate beam at points $A$, $B$, and $C$, respectively, lb-in.

$l_{AB}, l_{CB}$ = lengths of spans $AB$ and $CB$, in.

$I_{AB}, I_{CB}$ = moments of inertia of beam cross section between $A$ and $B$ and between $C$ and $B$, in.[4]

$A_{AB}, A_{CB}$ = areas of moment diagrams, considering sections of beam between supports to be simply supported, between $A$ and $B$ and between $C$ and $B$, lb-in.[2]

$a_{AB}, a_{CB}$ = distance from $A$ and $C$, respectively, to the centroids of areas $A_{AB}$ and $A_{CB}$, in.

$h_{AB}, h_{CB}$ = deflection of $A$ and $C$ above $B$, in.

$E$ = modulus of elasticity of beam material, psi

In using the three-moment equation, moments are considered positive when they cause a compressive stress in the upper fibers of the beam.   Beams of more than two spans may be analyzed by this method by writing the three-moment equation for successive, overlapping two-span sections.

The slope-deflection equations for a beam with a constant moment of inertia between supports are

$$M_{AB} = M_{FAB} + (EI_{AB}/l_{AB})(4\theta_A + 2\theta_B - 6R) \qquad (6\text{-}64)$$
$$M_{BA} = M_{FBA} + (EI_{AB}/l_{AB})(4\theta_B + 2\theta_A - 6R) \qquad (6\text{-}65)$$

The terminology is the same as that used for the three-moment equation, with the following additions:

## 1. Simple Beam—Uniformly Distributed Load

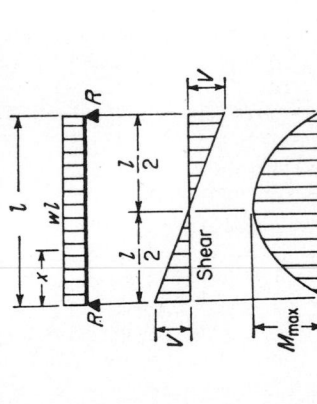

$$R = V = \frac{wl}{2}$$

$$V_x = w\left(\frac{l}{2} - x\right)$$

$$M_{max} \text{ (at center)} = \frac{wl^2}{8}$$

$$M_x = \frac{wx}{2}(l - x)$$

$$\Delta_{max} \text{ (at center)} = \frac{5wl^4}{384EI}$$

$$\Delta_x = \frac{wx}{24EI}(l^3 - 2lx^2 + x^3)$$

## 2. Simple Beam—Load Increasing Uniformly to One End

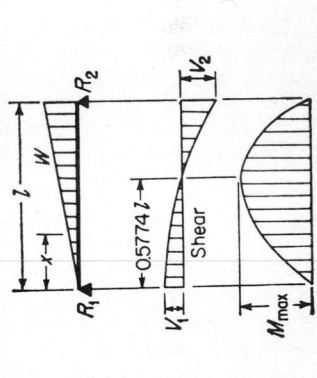

$$R_1 = V_1 = \frac{W}{3}$$

$$R_2 = V_{2max} = \frac{2W}{3}$$

$$V_x = \frac{W}{3} - \frac{Wx^2}{l^2}$$

$$M_{max}\left(\text{at } x = \frac{l}{\sqrt{3}} = .5774l\right) = \frac{2Wl}{9\sqrt{3}} = .1283Wl$$

$$M_x = \frac{Wx}{3l^2}(l^2 - x^2)$$

$$\Delta_{max}\left(\text{at } x = l\sqrt{1 - \sqrt{\frac{8}{15}}} = .5193l\right) = .01304\frac{Wl^3}{EI}$$

$$\Delta_x = \frac{Wx}{180EIl^2}(3x^4 - 10l^2x^2 + 7l^4)$$

FIG. 6-17. Beam diagrams and formulas for various static loading conditions.

## 3. Simple Beam—Load Increasing Uniformly to Center

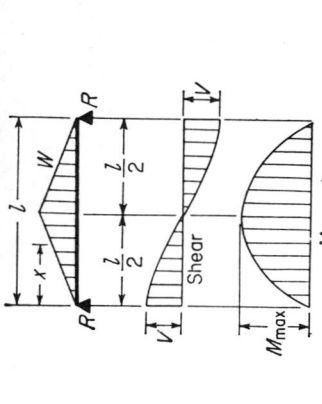

$$R = V = \frac{W}{2}$$

$$V_x \left(\text{when } x < \frac{l}{2}\right) = \frac{W}{2l^2}(l^2 - 4x^2)$$

$$M_{\max} \text{ (at center)} = \frac{Wl}{6}$$

$$M_x \left(\text{when } x < \frac{l}{2}\right) = Wx\left(\frac{1}{2} - \frac{2x^2}{3l^2}\right)$$

$$\Delta_{\max} \text{ (at center)} = \frac{Wl^3}{60EI}$$

$$\Delta_x = \frac{Wx}{480EIl^2}(5l^2 - 4x^2)^2$$

## 4. Simple Beam—Uniform Load Partially Distributed

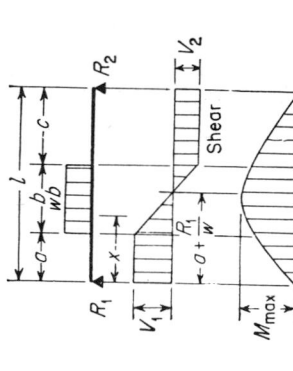

$$R_1 = V_1 \text{ (max when } a < c) = \frac{wb}{2l}(2c + b)$$

$$R_2 = V_2 \text{ (max when } a > c) = \frac{wb}{2l}(2a + b)$$

$$V_x \text{ [when } x < a \text{ and } > (a + b)] = R_1 - w(x - a)$$

$$M_{\max} \left(\text{at } x = a + \frac{R_1}{w}\right) = R_1\left(a + \frac{R_1}{2w}\right)$$

$$M_x \text{ (when } x < a) = R_1 x$$

$$M_x \text{ [when } x > a \text{ and } < (a + b)] = R_1 x - \frac{w}{2}(x - a)^2$$

$$M_x \text{ (when } x > (a + b)) = R_2(l - x)$$

Fig. 6-17. (*Continued*)

## 5. Simple Beam—Uniform Load Partially Distributed at One End

$R_1 = V_{1\,\max} \quad = \dfrac{wa}{2l}(2l - a)$

$R_2 = V_2 \quad = \dfrac{wa^2}{2l}$

$V \text{ (when } x < a) \quad = R_1 - wx$

$M_{\max} \left(\text{at } x = \dfrac{R_1}{w}\right) = \dfrac{R_1^2}{2w}$

$M_x \text{ (when } x < a) \quad = R_1 x - \dfrac{wx^2}{2}$

$M_x \text{ (when } x > a) \quad = R_2(l - x)$

$\Delta_x \text{ (when } x < a) \quad = \dfrac{wx}{24EIl}\left[a^2(2l - a)^2 - 2ax^2(2l - a) + lx^3\right]$

$\Delta_x \text{ (when } x > a) \quad = \dfrac{wa^2(l - x)}{24EIl}(4xl - 2x^2 - a^2)$

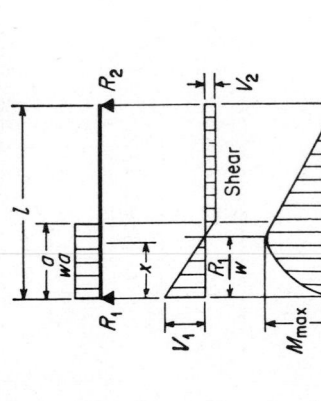

## 6. Simple Beam—Concentrated Load at Center

$R = V \quad = \dfrac{P}{2}$

$M_{\max} \text{ (at point of load)} = \dfrac{Pl}{4}$

$M_x \left(\text{when } x < \dfrac{l}{2}\right) = \dfrac{Px}{2}$

$\Delta_{\max} \text{ (at point of load)} = \dfrac{Pl^3}{48EI}$

$\Delta_x \left(\text{when } x < \dfrac{l}{2}\right) = \dfrac{Px}{48EI}(3l^2 - 4x^2)$

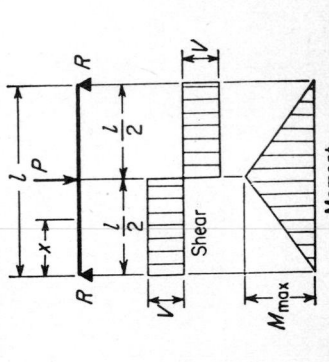

Fig. 6-17. (Continued)

## 7. Simple Beam—Concentrated Load at Any Point

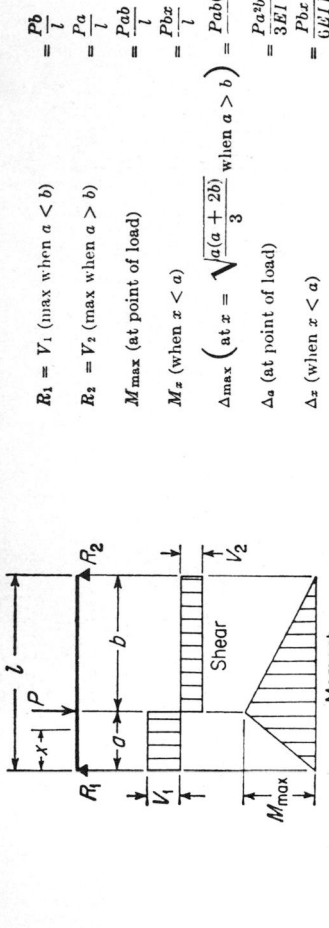

$$R_1 = V_1 \text{ (max when } a < b\text{)} = \frac{Pb}{l}$$

$$R_2 = V_2 \text{ (max when } a > b\text{)} = \frac{Pa}{l}$$

$$M_{\max} \text{ (at point of load)} = \frac{Pab}{l}$$

$$M_x \text{ (when } x < a\text{)} = \frac{Pbx}{l}$$

$$\Delta_{\max} \left( \text{at } x = \sqrt{\frac{a(a+2b)}{3}} \text{ when } a > b \right) = \frac{Pab(a+2b)\sqrt{3a(a+2b)}}{27EIl}$$

$$\Delta_a \text{ (at point of load)} = \frac{Pa^2b^2}{3EIl}$$

$$\Delta_x \text{ (when } x < a\text{)} = \frac{Pbx}{6EIl}(l^2 - b^2 - x^2)$$

## 8. Cantilever Beam—Load Increasing Uniformly to Fixed End

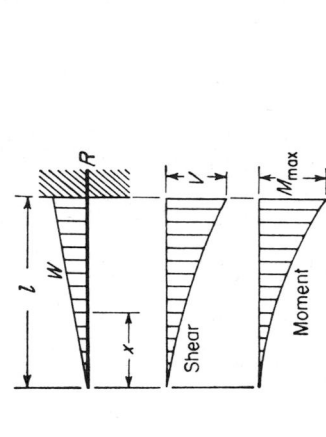

$$R = V = W$$

$$V_x = W\frac{x^2}{l^2}$$

$$M_{\max} \text{ (at fixed end)} = \frac{Wl}{3}$$

$$M_x = \frac{Wx^3}{3l^2}$$

$$\Delta_{\max} \text{ (at free end)} = \frac{Wl^3}{15\,EI}$$

$$\Delta_x = \frac{W}{60EIl^2}(x^5 - 5l^4x + 4l^5)$$

Fig. 6-17. (*Continued*)

6–54

## 9. Cantilever Beam—Uniformly Distributed Load

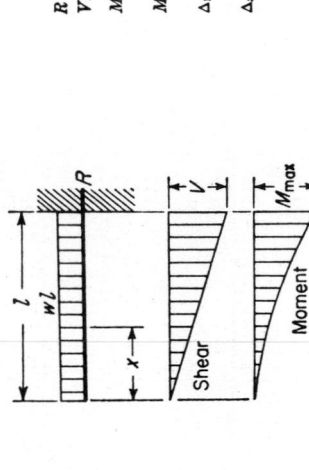

$$R = V = wl$$
$$V_x = wx$$
$$M_{max} \text{ (at fixed end)} = \frac{wl^2}{2}$$
$$M_x = \frac{wx^2}{2}$$
$$\Delta_{max} \text{ (at free end)} = \frac{wl^4}{8EI}$$
$$\Delta_x = \frac{w}{24EI}(x^4 - 4l^3x + 3l^4)$$

## 10. Cantilever Beam—Concentrated Load at Any Point

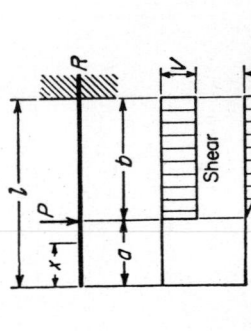

$$R = V \text{ (when } x > a) = P$$
$$M_{max} \text{ (at fixed end)} = Pb$$
$$M_x \text{ (when } x > a) = P(x - a)$$
$$\Delta_{max} \text{ (at free end)} = \frac{Pb^2}{6EI}(3l - b)$$
$$\Delta_a \text{ (at point of load)} = \frac{Pb^3}{3EI}$$
$$\Delta_x \text{ (when } x < a) = \frac{Pb^2}{6EI}(3l - 3x - b)$$
$$\Delta_x \text{ (when } x > a) = \frac{P(l - x)^2}{6EI}(3b - l + x)$$

Fig. 6-17. (*Continued*)

## 11. Cantilever Beam—Concentrated Load at Free End

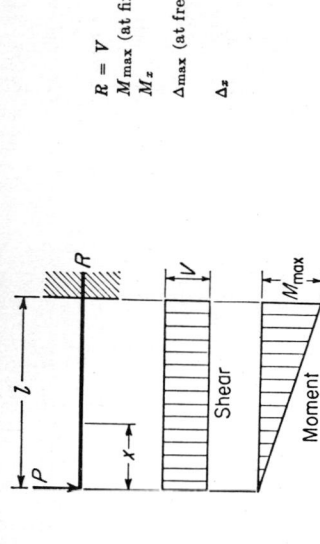

$$R = V = P$$
$$M_{max} \text{ (at fixed end)} = Pl$$
$$M_x = Px$$
$$\Delta_{max} \text{ (at free end)} = \frac{Pl^3}{3EI}$$
$$\Delta_x = \frac{P}{6EI}(2l^3 - 3l^2x + x^3)$$

## 12. Beam Fixed at Both Ends—Uniformly Distributed Loads

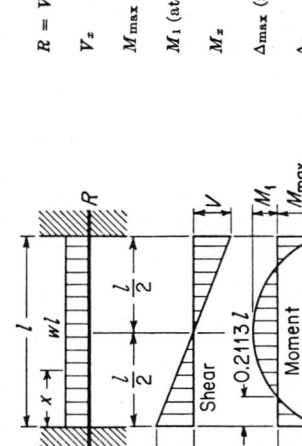

$$R = V = \frac{wl}{2}$$
$$V_x = w\left(\frac{l}{2} - x\right)$$
$$M_{max} \text{ (at ends)} = \frac{wl^2}{12}$$
$$M_1 \text{ (at center)} = \frac{wl^2}{24}$$
$$M_x = \frac{w}{12}(6lx - l^2 - 6x^2)$$
$$\Delta_{max} \text{ (at center)} = \frac{wl^4}{384EI}$$
$$\Delta_x = \frac{wx^2}{24EI}(l - x)^2$$

Fig. 6-17. (*Continued*)

## 13. Beam Fixed at Both Ends—Concentrated Load at Center

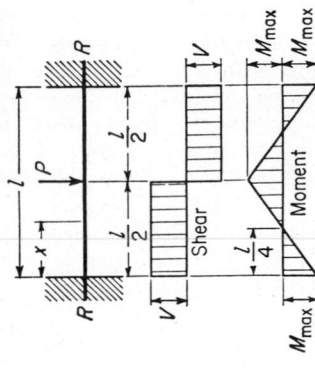

$$R = V = \frac{P}{2}$$

$$M_{max} \text{ (at center and ends)} = \frac{Pl}{8}$$

$$M_x \left( \text{when } x < \frac{l}{2} \right) = \frac{P}{8}(4x - l)$$

$$\Delta_{max} \text{ (at center)} = \frac{Pl^3}{192EI}$$

$$\Delta_x = \frac{Px^2}{48EI}(3l - 4x)$$

## 14. Beam Fixed at Both Ends—Concentrated Load at Any Point

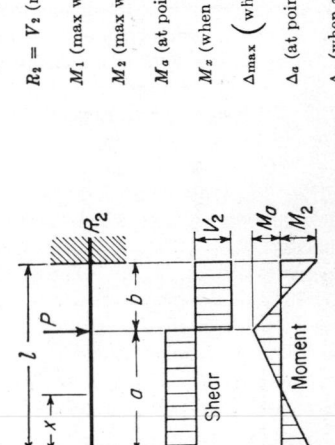

$$R_1 = V_1 \text{ (max when } a < b) = \frac{Pb^2}{l^3}(3a + b)$$

$$R_2 = V_2 \text{ (max when } a > b) = \frac{Pa^2}{l^3}(a + 3b)$$

$$M_1 \text{ (max when } a < b) = \frac{Pab^2}{l^2}$$

$$M_2 \text{ (max when } a > b) = \frac{Pa^2b}{l^2}$$

$$M_a \text{ (at point of load)} = \frac{2Pa^2b^2}{l^3}$$

$$M_x \text{ (when } x < a) = R_1x - \frac{Pab^2}{l^2}$$

$$\Delta_{max} \left( \text{when } a > b \text{ at } x = \frac{2al}{3a + b} \right) = \frac{2Pa^3b^2}{3EI(3a + b)^2}$$

$$\Delta_a \text{ (at point of load)} = \frac{Pa^3b^3}{3EIl^3}$$

$$\Delta_x \text{ (when } x < a) = \frac{Pb^2x^2}{6EIl^3}(3al - 3ax - bx)$$

Fig. 6-17. (*Continued*)

6–57

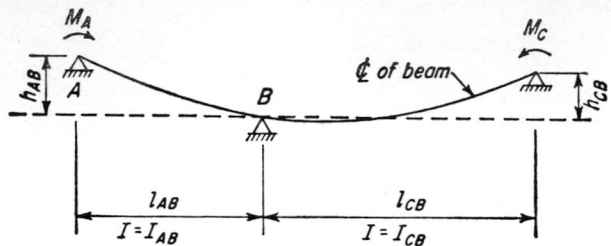

Diagram of statically indeterminate beam

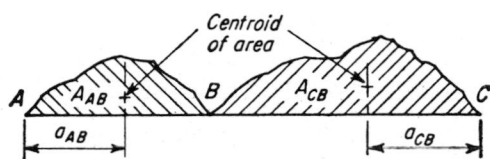

Moment diagram for each section of beam
considered to be simply supported

Fig. 6-18. Statically indeterminate beam.

$M_{FAB}$, $M_{FBA}$ = moments in member $AB$ at end $A$ and end $B$, respectively, with both ends of member considered to be fixed, lb-in.

$\theta_A$, $\theta_B$ = slopes of statically indeterminate beam at end $A$ and end $B$, radians

$R$ = ratio of deflection of one end of beam, with respect to the other end, to length of beam, with both measured in same units, radians

In the use of the slope-deflection equations, all moments and slopes should be considered positive when clockwise ($R$ is considered to be a slope). The equations are written for each section of the beam and are solved simultaneously for the slopes. These slopes are then substituted in the original equations to determine the moments at the points of support. After the signs of these moments are revised to conform to the normal beam convention (positive moment causes compression in the upper fibers of the beam) shears and reactions may be determined.

**Stresses.** The stress in a beam due to moment is

$$f = My/I \qquad f_{\max} = Mc/I \qquad (6\text{-}66) \quad (6\text{-}67)$$

and the stress due to shear is

$$s = VQ/It \qquad (6\text{-}68)$$

where $f$ = fiber stress, psi

$y$ = distance from neutral axis to point at which stress is $f$, in.

$c$ = distance from neutral axis to outermost surface of beam, in.

$s$ = horizontal (and vertical) shearing stress, psi

$Q$ = statical moment, or first moment, about neutral axis of cross-sectional area of beam between plane on which $s$ occurs and outer face of beam, in.[3]

$t$ = thickness of beam at plane on which $s$ occurs, in.

$M$ = moment, lb-in.

$I$ = moment of inertia of beam cross section, in.[4]

$V$ = shear, lb

When the bending moment is positive, the fiber stress is compression at the top of the beam and tension at the bottom.

**Deflection.** The general equation for the deflection of a beam is

$$\Delta = \int_0^l \int_0^l \frac{M}{EI}\, dx\, dx + \int_0^l \frac{s}{G}\, dx \qquad (6\text{-}69)$$

where $\Delta$ = deflection of beam from its longitudinal axis, in.

$x$ = distance measured parallel to beam axis, in.

$M$ = moment, expressed as a function of $x$, lb-in.

$s$ = shearing unit stress, expressed as a function of $x$, psi

$G$ = shearing modulus of elasticity, psi

$E$ = modulus of elasticity, psi

$l$ = length of beam, in.

The first term of Eq. (6-69) expresses the deflection due to moment; the second term, the deflection due to shear.

The deflection of a beam due to moment is shown in Fig. 6-17 for beams with loading conditions of frequent occurrence. The deflection of a beam due to shear is negligible in the usual case where the length of the beam is great in comparison with the depth of the beam. In short deep beams, however, the deflection due to shear may require consideration.

## 6-45. Shafts and Torsion Members

The shearing stress in a cylindrical shaft carrying a twisting moment is

$$s = T\rho/J \qquad (6\text{-}70)$$

Since the maximum stress occurs at the outer edge of the shaft and $J$ for a solid cylindrical shaft is $\pi d^4/32$,

$$s = 16T/\pi d^3 \qquad (6\text{-}71)$$

The angle of twist in a cylindrical shaft is

$$\theta = Tl/JG \qquad (6\text{-}72)$$

where $s$ = shearing stress, psi

$T$ = twisting moment or torque, lb-in.

$J$ = polar moment of inertia of cross-sectional area of shaft with respect to center, in.[4]

hp = horsepower transmitted by shaft

$n$ = shaft speed, rpm

$d$ = shaft diameter, in.

$l$ = shaft length, in.

$\theta$ = angle of twist in length $l$, radians
$G$ = shearing modulus of elasticity, psi
$\rho$ = distance from center to point in shaft, in.

## 6-46. Stress Due to Temperature Change

Material subjected to a temperature increase expands in length a total of

$$e = l\epsilon \, \Delta t \qquad (6\text{-}73)$$

For a straight member so restrained that the ends are unable to move, the compressive stress caused by a temperature increase is

$$f_t = E\epsilon \, \Delta t \qquad (6\text{-}74)$$

where $e$ = total change in length, in.
$l$ = length of member, in.
$\epsilon$ = coefficient of thermal expansion, dimensionless
$\Delta t$ = change in temperature, °F
$f_t$ = stress due to temperature change, psi
$E$ = modulus of elasticity, psi

For a decrease in temperature, a tensile stress would be set up in the member. Average values of $\epsilon$ for several structural materials, expressed as the change in length per unit of length and related to the Fahrenheit scale, are as follows:

| | | | |
|---|---|---|---|
| Aluminum... | 0.0000125 | Concrete... | 0.0000062 |
| Brick....... | .0000050 | Steel....... | .0000065 |
| Cast iron.... | .0000062 | Timber..... | .0000025 |

## 6-47. Combined Stresses

When an element of a body is subjected to stresses, either direct or shearing, on a given plane, the stresses on other planes may be determined by the following equations. (For nomenclature see Fig. 6-19; all stresses have units of pounds per square inch.)

$$f_n = \tfrac{1}{2}(f_1 + f_2) + \tfrac{1}{2}(f_1 - f_2)\cos 2\theta + s_s \sin 2\theta \quad (6\text{-}75)$$

$$s_t = \tfrac{1}{2}(f_1 - f_2)\sin 2\theta - s_s \cos 2\theta \qquad (6\text{-}76)$$

The maximum and minimum normal stresses are

$$f_n = \tfrac{1}{2}(f_1 + f_2) \pm \tfrac{1}{2}[(f_1 - f_2)^2 + 4s_s^2]^{1/2} \quad (6\text{-}77)$$

These maximum stresses occur when

$$\tan 2\theta = 2s_s/(f_1 - f_2) \quad (6\text{-}78)$$

The maximum shearing stress is

$$s_t = \tfrac{1}{2}[(f_1 - f_2)^2 + 4s_s^2]^{1/2} \quad (6\text{-}79)$$

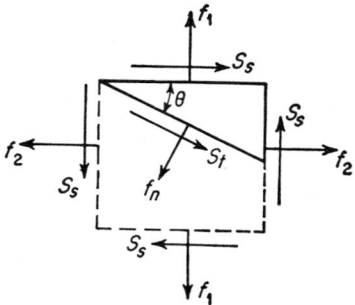

Fig. 6-19. Combined stresses.

The maximum shear occurs on planes at 45° to the planes of the maximum and minimum normal stresses. There is no shearing stress on the planes of maximum and minimum normal stress. However, there are usually normal stresses on the planes of maximum shear.

## 6-48. Timber Design

**Lumber Sizes.** Lumber is designated by a nominal size. The size of unfinished lumber is the same as the nominal size, but the dimensions of dressed or finished lumber are from $\frac{3}{8}$ to $\frac{1}{2}$ in. smaller. Dressed sizes and properties of standard lumber cross sections are given in Table 6-20.

**Lumber Grades.** Stress-grade lumber consists of three classifications:

1. *Beams and stringers.* Lumber of rectangular cross section, 5 in. or more thick and 8 in. or more wide, graded with respect to its strength in bending when loaded on the narrow face.

2. *Joists and planks.* Lumber of rectangular cross section, 2 in. to, but not including, 5 in. thick and 4 in. or more wide, graded with respect to its strength in bending when loaded either on the narrow face as a joist or on the wide face as a plank.

3. *Posts and timbers.* Lumber of square, or approximately square, cross section 5 by 5 in. or larger, graded primarily for use as posts or columns carrying longitudinal load but adapted for miscellaneous uses in which the strength in bending is not especially important.

Allowable unit stresses apply only for loading for which lumber is graded.

**Lumber Groups.** Lumber species are separated into four groupings, based on specific gravities, for determining allowable loads for nails, spikes, screws, and bolts. Typical listings for the groups are given in Table 6-24.

**Working Stresses.** The allowable unit stresses and the modulus of elasticity for some of the commonly used species of lumber, as recommended by the National Lumber Manufacturers Association, are shown in Table 6-21. These stresses are for normal conditions of loading, and should be decreased by 10 per cent for lumber fully stressed to working load for many years. For short-term loading, stress should be increased as follows:

| *Loading Conditions* | *Percentage Increase in Allowable Unit Stress* |
|---|---|
| Loading of 2 months' duration (snow) | 15 |
| Loading of 7 days' duration | 25 |
| Wind or earthquake loading | $33\frac{1}{3}$ |
| Impact loading | 100 |

The allowable unit stresses in Table 6-21 apply to lumber used under conditions continuously dry. These stresses also apply to lumber used under conditions where the moisture content of the wood is at or above the fiber saturation point, as when continuously submerged, except that, under such conditions of use, the allowable unit stresses in compression parallel to grain is reduced one-tenth, in compression perpendicular to the grain is reduced one-third, and the values of the modulus of elasticity are reduced one-eleventh. Allowable unit stresses for lumber pressure-impregnated with preservative are the same as shown in Table 6-21, but these stresses should be reduced by 10 per cent for lumber pressure-impregnated with fire-retardant chemicals.

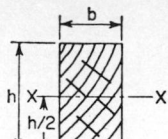

# Table 6-20. Properties of Sections for Standard Lumber Sizes

### Dressed (S4S) Sizes

Moment of inertia and section modulus are given with respect to $xx$ axis, with dimensions $b$ and $h$ as shown on sketch.

| Nominal size $b$ $h$ | Standard dressed size S4S $b$ $h$ | Area of section $A = bh$ | Moment of inertia $I = \dfrac{bh^3}{12}$ | Section modulus $S = \dfrac{bh^2}{6}$ | Board feet per linear foot of piece |
|---|---|---|---|---|---|
| 2 × 4 | 1⅝ × 3⅝ | 5.89 | 6.45 | 3.56 | ⅔ |
| 2 × 6 | 1⅝ × 5½ | 8.93 | 22.53 | 8.19 | 1 |
| 2 × 8 | 1⅝ × 7½ | 12.19 | 57.13 | 15.23 | 1⅓ |
| 2 × 10 | 1⅝ × 9½ | 15.44 | 116.10 | 24.44 | 1⅔ |
| 2 × 12 | 1⅝ × 11½ | 18.69 | 205.95 | 35.82 | 2 |
| 3 × 4 | 2⅝ × 3⅝ | 9.52 | 10.42 | 5.75 | 1 |
| 3 × 6 | 2⅝ × 5½ | 14.43 | 36.40 | 13.23 | 1½ |
| 3 × 8 | 2⅝ × 7½ | 19.69 | 92.29 | 24.61 | 2 |
| 3 × 10 | 2⅝ × 9½ | 24.94 | 187.55 | 39.48 | 2½ |
| 3 × 12 | 2⅝ × 11½ | 30.19 | 332.69 | 57.86 | 3 |
| 4 × 6 | 3⅝ × 5½ | 19.95 | 50.25 | 18.28 | 2 |
| 4 × 8 | 3⅝ × 7½ | 27.19 | 127.44 | 33.98 | 2⅔ |
| 4 × 10 | 3⅝ × 9½ | 34.44 | 259.00 | 54.43 | 3⅓ |
| 4 × 12 | 3⅝ × 11½ | 41.69 | 459.43 | 79.90 | 4 |
| 4 × 14 | 3⅝ × 13½ | 48.94 | 743.24 | 110.11 | 4⅔ |
| 4 × 16 | 3⅝ × 15½ | 56.19 | 1,124.92 | 145.15 | 5⅓ |
| 6 × 6 | 5½ × 5½ | 30.25 | 76.26 | 27.73 | 3 |
| 6 × 8 | 5½ × 7½ | 41.25 | 193.36 | 51.56 | 4 |
| 6 × 10 | 5½ × 9½ | 52.25 | 392.96 | 82.73 | 5 |
| 6 × 12 | 5½ × 11½ | 63.25 | 697.07 | 121.23 | 6 |
| 6 × 14 | 5½ × 13½ | 74.25 | 1,127.67 | 167.06 | 7 |
| 6 × 16 | 5½ × 15½ | 85.25 | 1,706.78 | 220.23 | 8 |
| 6 × 18 | 5½ × 17½ | 96.25 | 2,456.38 | 280.73 | 9 |
| 8 × 8 | 7½ × 7½ | 56.25 | 263.67 | 70.31 | 5⅓ |
| 8 × 10 | 7½ × 9½ | 71.25 | 535.86 | 112.81 | 6⅔ |
| 8 × 12 | 7½ × 11½ | 86.25 | 950.55 | 165.31 | 8 |
| 8 × 14 | 7½ × 13½ | 101.25 | 1,537.73 | 227.81 | 9½ |
| 8 × 16 | 7½ × 15½ | 116.25 | 2,327.42 | 300.31 | 10⅔ |
| 8 × 18 | 7½ × 17½ | 131.25 | 3,349.61 | 382.81 | 12 |
| 8 × 20 | 7½ × 19½ | 146.25 | 4,634.30 | 475.31 | 13⅓ |
| 10 × 10 | 9½ × 9½ | 90.25 | 678.76 | 142.90 | 8⅓ |
| 10 × 12 | 9½ × 11½ | 109.25 | 1,204.03 | 209.40 | 10 |
| 10 × 14 | 9½ × 13½ | 128.25 | 1,947.80 | 288.56 | 11⅔ |
| 10 × 16 | 9½ × 15½ | 147.25 | 2,948.07 | 380.40 | 13⅓ |
| 10 × 18 | 9½ × 17½ | 166.25 | 4,242.84 | 484.90 | 15 |
| 10 × 20 | 9½ × 19½ | 185.25 | 5,870.11 | 602.06 | 16⅔ |
| 12 × 12 | 11½ × 11½ | 132.25 | 1,457.51 | 253.48 | 12 |
| 12 × 14 | 11½ × 13½ | 155.25 | 2,357.86 | 349.31 | 14 |
| 12 × 16 | 11½ × 15½ | 178.25 | 3,568.71 | 460.48 | 16 |
| 12 × 18 | 11½ × 17½ | 201.25 | 5,136.07 | 586.98 | 18 |
| 12 × 20 | 11½ × 19½ | 224.25 | 7,105.92 | 728.81 | 20 |
| 12 × 22 | 11½ × 21½ | 247.25 | 9,524.28 | 885.98 | 22 |
| 12 × 24 | 11½ × 23½ | 270.25 | 12,437.13 | 1,058.48 | 24 |

Decking (Based on Strip 1 Ft Wide and of Thickness Indicated)

| | | | | | |
|---|---|---|---|---|---|
| 1′0 × 2 | 12 × 1⅝ | 19.50 | 4.29 | 5.28 | 2 |
| 1′0 × 3 | 12 × 2⅝ | 31.50 | 18.00 | 13.76 | 3 |
| 1′0 × 4 | 12 × 3½ | 42.00 | 42.88 | 24.50 | 4 |

SOURCE: National Lumber Manufacturers Association.

The stresses indicated in Table 6-21 are reduced for Douglas fir and western hemlock when the lumber contains splits parallel to the grain.

The deflection of wood beams may be calculated by the standard methods of mechanics discussed in Sec. 6-44. Under long-term loading, lumber acquires a permanent set about equal to the deflection, but the strength is not reduced.

**Bearing.** The allowable unit stresses given for compression perpendicular to the grain apply to bearings of any length at the ends of beams, and to all bearings 6 in. or more in length at other locations. When calculating the required bearing area at the ends of beams, no allowance should be made for the fact that, as the beam bends, the pressure upon the inner edge of the bearing is greater than at the end of the beam. For bearings of less than 6 in. in length and not nearer than 3 in. to the end of the member, the allowable stress for compression perpendicular to the grain should be modified by multiplying by the factor $(l + \frac{3}{8})/l$, where $l$ is the length of the bearing in inches measured along the grain of the wood.

**Beams.** The extreme fiber stress in bending for a rectangular beam is

$$f = 6M/bh^2 = M/S \tag{6-80}$$

A beam of circular cross section is assumed to have the same strength in bending as a square beam having the same cross-sectional area.

The horizontal shearing stress in a rectangular beam is

$$H = 3V/2bh \tag{6-81}$$

For a rectangular beam with a notch in the lower face at the end, the horizontal shearing stress is

$$H = (3V/2bd_1)(h/d_1) \tag{6-82}$$

A gradual change in cross section rather than a square notch decreases the shearing stress nearly to that computed for the actual depth above the notch.

### Nomenclature for Eqs. (6-80) to (6-84)

$f$ = maximum fiber stress, psi
$M$ = bending moment, lb-in.
$h$ = depth of beam, in.
$b$ = width of beam, in.
$S$ = section modulus ($= bh^2/6$ for rectangular section), in.$^3$
$H$ = horizontal shearing stress, psi
$V$ = total shear, lb
$d_1$ = depth of beam above notch, in.
$l$ = span of beam, in.
$P$ = concentrated load, lb
$V^1$ = modified total end shear, lb
$W$ = total uniformly distributed load, lb
$x$ = distance from reaction to concentrated load, in.

For simple beams, the span should be taken as the distance from face to face of supports plus one-half the required length of bearing at each end, and

### Table 6-21. Allowable Unit Stresses, Stress-grade Lumber for Normal Loading Conditions

| Species, commercial grade, and modulus of elasticity | Allowable unit stresses, psi | | | | | |
|---|---|---|---|---|---|---|
| | Extreme fiber in bending, $f$, and tension parallel to grain, $t$ | | Horizontal shear, $H$ | Compression perpendicular to grain, $c\perp$ | Compression parallel to grain, $c$ | |
| | J and P B and S | P and T | | | J and P P and T | B and S |
| *Douglas fir* | | | | | | |
| $E = 1,760,000$ psi | | | | | | |
| Dense select structural......... | 2,050 | 1,900 | 120 | 455 | 1,650 | 1,500 |
| Select structural.... ......... | 1,900 | 1,750 | 120 | 415 | 1,500 | 1,400 |
| Dense construction............. | 1,750 | 1,500 | 120 | 455 | 1,400 | 1,200 |
| Construction................... | 1,500 | 1,200 | 120 | 390 | 1,200 | 1,000 |
| Standard (J and P only)........ | 1,200 | ..... | 95 | 390 | 1,000 | |
| *Pine, southern, 5 in. thick and up* | | | | | | |
| $E = 1,760,000$ psi | | | | | | |
| Dense structural 86............ | 2,400 | 2,400 | 150 | 455 | 1,800 | 1,800 |
| Dense structural 72............ | 2,000 | 2,000 | 135 | 455 | 1,550 | 1,550 |
| Dense structural 65............ | 1,800 | 1,800 | 120 | 455 | 1,400 | 1,400 |
| Dense structural 58............ | 1,600 | 1,600 | 105 | 455 | 1,300 | 1,300 |
| No. 1 dense SR................ | 1,600 | 1,600 | 120 | 455 | 1,500 | 1,500 |
| No. 1 SR..................... | 1,400 | 1,400 | 120 | 390 | 1,300 | 1,300 |
| No. 2 dense SR................ | 1,400 | 1,400 | 105 | 455 | 1,050 | 1,050 |
| No. 2 SR..................... | 1,200 | 1,200 | 105 | 390 | 900 | 900 |
| *Pine, Norway* (J and P only) | | | | | | |
| $E = 1,320,000$ | | | | | | |
| Prime structural............... | 1,200 | ..... | 75 | 360 | 900 | |
| Common structural............. | 1,100 | ..... | 75 | 360 | 775 | |
| Utility structural.............. | 950 | ..... | 75 | 360 | 650 | |
| *Spruce, eastern* (J and P only) | | | | | | |
| $E = 1,320,000$ | | | | | | |
| 1450 f structural grade......... | 1,450 | ..... | 110 | 300 | 1,050 | |
| 1300 f structural grade......... | 1,300 | ..... | 95 | 300 | 975 | |
| 1200 f structural grade......... | 1,200 | ..... | 95 | 300 | 900 | |
| *Redwood* | | | | | | |
| $E = 1,320,000$ psi | | | | | | |
| Dense structural............... | 1,700 | ..... | 110 | 320 | 1,450 | 1,450 |
| Heart structural................ | 1,300 | ..... | 95 | 320 | 1,100 | 1,100 |
| *Hemlock, eastern* | | | | | | |
| $E = 1,210,000$ | | | | | | |
| Select structural............... | 1,300 | ..... | 85 | 360 | 850 | 850 |
| Prime structural (J and P only).. | 1,200 | ..... | 60 | 360 | 775 | |
| Common structural (J and P only) | 1,100 | ..... | 60 | 360 | 650 | |
| Utility structural (J and P only). | 950 | ..... | 60 | 360 | 600 | |
| *Hemlock, western* | | | | | | |
| $E = 1,540,000$ | | | | | | |
| Select structural (J and P only).. | 1,600 | ..... | 100 | 365 | 1,200 | |
| Construction................... | 1,500 | 1,200 | 100 | 365 | 1,100 | 1,000 |
| Standard (J and P only)........ | 1,200 | ..... | 80 | 365 | 1,000 | |

NOTE: J and P = joists and planks, B and S = beams and stringers, P and T = posts and timbers.

for continuous beams the span should be taken as the distance between the centers of bearing on supports.

When determining $V$, neglect all loads within a distance from either support equal to the depth of the beam.

In the stress grade of solid-sawn beams, allowances for checks, end splits, and shakes have been made in the assigned unit stresses. For such members,

Eq. (6-81) does not indicate the actual shear resistance because of the redistribution of shear stress that occurs in checked beams. For a solid-sawn beam which does not qualify using Eq. (6-81) and the $H$ values in Table 6-21, the modified reaction $V^1$ should be determined as follows:

For concentrated loads:
$$V^1 = \frac{10P(l - x)(x/h)^2}{9l[2 + (x/h)^2]} \tag{6-83}$$

For uniform loading:
$$V^1 = \frac{W}{2}\left(1 - \frac{2h}{l}\right) \tag{6-84}$$

The sum of the $V^1$ values from Eqs. (6-83) and (6-84) should be substituted for $V$ in Eq. (6-81), and the resulting $H$ values checked against those given in Table 6-22. Shear values in Table 6-22 should be adjusted for duration of loading as described under working stresses in this section.

### Table 6-22. Modified Allowable Horizontal Shear

$H$ in Pounds per Square Inch
For Use When Total Shear Is Determined Using Eqs. (6-83) and (6-84)

```
Douglas fir...........................145
Pine, southern.......................175
Pine, Norway........................130
Spruce, eastern......................130
Redwood.............................110
Hemlock, eastern....................110
Hemlock, western...................120
```

**Columns.** The allowable unit stress on timber columns consisting of a single piece of lumber or a group of pieces glued together to form a single member is

$$\frac{P}{A} = \frac{3.619E}{(l/r)^2} \tag{6-85}$$

For columns of square or rectangular cross section, this formula becomes

$$\frac{P}{A} = \frac{0.30E}{(l/d)^2} \tag{6-86}$$

For columns of circular cross section, the formula becomes

$$\frac{P}{A} = \frac{0.22E}{(l/d)^2} \tag{6-87}$$

The allowable unit stress $P/A$ may not exceed the allowable compressive stress $c$. The ratio $l/d$ must not exceed 50. Values of $P/A$ are subject to the duration of loading adjustment given previously.

#### Nomenclature for Eqs. (6-85) to (6-87)

$P$ = total allowable load, lb
$A$ = area of column cross section, in.$^2$
$c$ = allowable unit stress in compression parallel to grain, psi
$d$ = dimension of least side of column, in.
$l$ = unsupported length of column between points of lateral support, in.

$E$ = modulus of elasticity, psi

$r$ = least radius of gyration of column, in.

For members loaded as columns, the allowable unit stresses for bearing on end grain (parallel to grain) are given in Table 6-23. These allowable stresses

**Table 6-23. Allowable Unit Stresses for End Grain in Bearing, Psi**

| Species | Sawn lumber 4 in. and less in thickness and glued laminated lumber | Sawn lumber more than 4 in. in thickness |
|---|---|---|
| Douglas fir, dense........ | 2,600 | 2,050 |
| Douglas fir.............. | 2,200 | 1,750 |
| Pine, dense............. | 2,600 | 2,050 |
| Pine, southern......... | 2,200 | 1,750 |
| Pine, Norway........... | 1,600 | 1,250 |
| Spruce, eastern........ | 1,600 | 1,250 |
| Redwood.............. | 2,050 | 1,650 |
| Hemlock, eastern....... | 1,450 | 1,150 |
| Hemlock, western....... | 1,800 | 1,450 |

apply provided there is adequate lateral support and end cuts are accurately squared and parallel. When stresses exceed 75 per cent of values given, bearing must be on a snug-fitting metal plate. These stresses apply under conditions continuously dry, and must be reduced by 27 per cent for glued laminated lumber and lumber 4 in. or less in thickness and by 9 per cent for sawn lumber more than 4 in. in thickness, for lumber exposed to weather.

**Combined Bending and Axial Load.** Members under combined bending and axial load should be so proportioned that the quantity

$$P_a/P + M_a/M < 1 \qquad (6\text{-}88)$$

where $P_a$ = total axial load on member, lb

$P$ = total allowable axial load, lb

$M_a$ = total bending moment on member, lb-in.

$M$ = total allowable bending moment, lb-in.

**Compression at Angle to Grain.** The allowable unit compressive stress when the load is at an angle to the grain is

$$c' = c(c\perp)/[c(\sin \theta)^2 + (c\perp)(\cos \theta)^2] \qquad (6\text{-}89)$$

where $c'$ = allowable unit stress at angle to grain, psi

$c$ = allowable unit stress parallel to grain, psi

$c\perp$ = allowable unit stress perpendicular to grain, psi

$\theta$ = angle between direction of load and direction of grain

**Timber Connections.** *Split Rings and Bolts.* Split rings are available in two sizes: 2½ in. diameter with a ½-in. bolt and 4 in. diameter with a ¾-in. bolt. The allowable single-shear load for a split ring and bolt connector varies widely depending on the type of wood, the thickness of material, the edge distance, and other factors. With adequate edge distances, the allowable load on the 2½-in. ring ranges up to 3,160 lb and on the 4-in. ring to

6,140 lb. Detailed data on loads for these and other timber connectors can be found in the "National Design Specification for Stress-Grade Lumber and its Fastenings," recommended by the National Lumber Manufacturers Association.

**Toothed Rings and Bolts.** Toothed rings are available in four sizes: 2 in. diameter with a ½-in. bolt, 2⅝ in. diameter with a ⅝-in. bolt, 3⅜ in. diameter with a ¾-in. bolt, and 4 in. diameter with a ¾-in. bolt. The allowable single-shear load for a 2-in. ring ranges up to 1,330 lb, for the 2⅝-in. ring to 2,270 lb, for the 3⅜-in. ring to 3,180 lb, and for the 4-in. ring to 3,700 lb. The reference cited above should be consulted for further data.

**Shear Plates and Bolts.** Shear plates are available in two sizes: 2⅝ in. diameter with a ¾-in. bolt and 4 in. diameter with either a ¾- or a ⅞-in. bolt. The allowable single-shear load for a 2⅝-in. plate ranges up to 2,900 lb and for the 4-in. plate up to 5,090 lb. The reference cited above should be consulted for further data.

**Bolts.** The allowable double-shear load for a bolt loaded at both ends depends on the factors discussed under split rings and bolts and also on the ratio of the length of the bolt in the wood member to the bolt diameter. Allowable loads in Douglas fir and southern Pine range from a maximum of 1,290 lb for a ½-in. bolt to a maximum of 8,040 lb for a 1¼-in. bolt.

**Nails and Spikes.** The allowable withdrawal load per inch of penetration of a common nail or spike driven into side grain (perpendicular to fibers) of seasoned wood, or unseasoned wood which will remain wet, is

$$p = 1{,}380G^{\frac{5}{2}}D \qquad (6\text{-}90)$$

where $p$ = allowable load per inch of penetration into member receiving point, lb

$D$ = diameter of nail or spike, in.

$G$ = specific gravity of wood, oven-dry (see Table 6-24)

When nails or spikes are driven into side grain in unseasoned wood which will season subsequently under load, or in wood pressure-impregnated with fire-retardant chemicals, the allowable withdrawal load is one-fourth that given. Nails and spikes should not be loaded in withdrawal from end grain of wood.

The total allowable lateral load for a nail or spike driven into side grain of seasoned wood is

$$p = CD^{\frac{3}{2}} \qquad (6\text{-}91)$$

where $p$ = allowable load per nail or spike, lb

$D$ = diameter of nail or spike, in.

$C$ = coefficient dependent on group number of wood (see Table 6-24)

For nails or spikes, values of $C$ for the four groups into which stress-grade lumber is classified are given in Table 6-24.

The loads apply where the nail or spike penetrates into the member receiving

### Table 6-24. Group Number, Specific Gravity, and Coefficient for Common Lumber Species

| Species | Group number | Specific gravity $G$ | $G^2$ | $G^{5/2}$ | $C$ Nails or spikes | $C$ Wood screws |
|---|---|---|---|---|---|---|
| Oak, Red................ | I | 0.66 | 0.436 | 0.354 | 2,040 | 4,800 |
| Ash.................... | I | 0.64 | 0.410 | 0.328 | 2,040 | 4,800 |
| Pine, southern........... | II | 0.59 | 0.348 | 0.267 | 1,650 | 3,960 |
| Douglas fir.............. | II | 0.51 | 0.260 | 0.186 | 1,650 | 3,960 |
| Pine, Norway............ | III | 0.47 | 0.221 | 0.151 | 1,350 | 3,240 |
| Hemlock, western........ | III | 0.44 | 0.194 | 0.128 | 1,350 | 3,240 |
| Redwood................ | III | 0.42 | 0.176 | 0.114 | 1,350 | 3,240 |
| Hemlock, eastern........ | IV | 0.43 | 0.185 | 0.121 | 1,080 | 2,520 |
| Spruce................. | IV | 0.41 | 0.168 | 0.108 | 1,080 | 2,520 |

its point at least 10 diameters for Group I species, 11 diameters for Group II species, 13 diameters for Group III species, and 14 diameters for Group IV species. Allowable loads for lesser penetrations are directly proportional to the penetration, but the penetration must be at least one-third that specified. When nails or spikes are driven into side grain in unseasoned wood which will remain wet or will be loaded before seasoning, or in wood pressure-impregnated with fire-retardant chemicals, the allowable lateral load per nail or spike is three-fourths that given above. The allowable lateral load for nails or spikes driven into end grain is two-thirds that given above for side grain.

Nails and spikes are usually designated by pennyweight sizes. The diameters and lengths of common sizes are shown in Table 6-25.

*Wood Screws.* The allowable withdrawal load per inch of penetration of the threaded portion of a wood screw into side grain of seasoned wood which remains dry is

$$p = 2,850G^2D \tag{6-92}$$

where $p$ = allowable load per inch of penetration of threaded portion into member receiving point, lb

$D$ = diameter of wood screw, in.

$G$ = specific gravity of wood, oven-dry (see Table 6-24)

Wood screws should not be loaded in withdrawal from end grain.

The total allowable lateral load for wood screws driven into the side grain of seasoned wood which remains dry is

$$p = CD^2 \tag{6-93}$$

where $p$ = allowable load per wood screw, lb

$D$ = diameter of wood screw, in.

$C$ = coefficient dependent on group number of wood (Table 6-24)

## Table 6-25. Nail and Spike Sizes

| Size (pennyweight) | Length, in. | Diam $D$, in. | $D^{3/2}$ |
|---|---|---|---|
| | | Nails | |
| 6 | 2 | 0.113 | 0.038 |
| 8 | 2½ | .131 | .047 |
| 10 | 3 | .148 | .057 |
| 12 | 3¼ | .148 | .057 |
| 16 | 3½ | .162 | .065 |
| 20 | 4 | .192 | .084 |
| 30 | 4½ | .207 | .094 |
| 40 | 5 | .225 | .107 |
| 50 | 5½ | .244 | .122 |
| 60 | 6 | .263 | .135 |
| | | Spikes | |
| 10 | 3 | .192 | .084 |
| 12 | 3¼ | .192 | .084 |
| 16 | 3½ | .207 | .094 |
| 20 | 4 | .225 | .107 |
| 30 | 4½ | .244 | .122 |
| 40 | 5 | .263 | .135 |
| 50 | 5½ | .283 | .150 |
| 60 | 6 | .283 | .150 |
| 5⁄16 in. | 7 | .312 | .175 |
| ⅜ in. | 8–12 | .375 | .230 |

For screws, values of $C$ for the four groups into which stress-grade lumber is classified are given in Table 6-24.

The allowable lateral load for wood screws driven into end grain is two-thirds that given for side grain.

Allowable loads for withdrawal and lateral resistance apply where the wood screw penetrates at least seven times its diameter. For penetration between seven and four times the diameter, the allowable load should be reduced proportionately; a screw should not be used with a penetration of less than 4 diam. When the screw is driven into wood which is exposed to the weather, or wood pressure-impregnated with fire-retardant chemicals, 75 per cent, and when always wet, 67 per cent of the load determined above should be used.

Wood screws are usually designated by a gauge number. The diameter of common sizes is shown in Table 6-26.

## 6-49. Steel Design

**Working Stresses.** Structural steel has a weight of 490 lb/ft³, a modulus of elasticity of 29 million psi, and a shearing modulus of 12 million psi. Structural steels used in buildings and the minimum yield points are listed in Table 6-27.

### Table 6-26. Wood-screw Sizes

| Gauge no. | Diam $D$, in. | $D^2$ | Gauge no. | Diam $D$, in. | $D^2$ |
|---|---|---|---|---|---|
| 0 | 0.060 | 0.0036 | 10 | 0.190 | 0.0361 |
| 1 | .073 | .0053 | 11 | .203 | .0412 |
| 2 | .086 | .0074 | 12 | .216 | .0467 |
| 3 | .099 | .0098 | 14 | .242 | .0586 |
| 4 | .112 | .0125 | 16 | .268 | .0718 |
| 5 | .125 | .0156 | 18 | .294 | .0864 |
| 6 | .138 | .0190 | 20 | .320 | .1024 |
| 7 | .151 | .0228 | 24 | .372 | .1384 |
| 8 | .164 | .0269 | | | |
| 9 | .177 | .0313 | | | |

### Table 6-27. Designations and Yield Points for Structural Steels

| ASTM Specification | Designation | Specified minimum yield point $F_y$, psi* |
|---|---|---|
| "Structural Steel". . . . . . . . . . . . . . . . . . . . . . . . . . . . . . . . . . . . | A36 | 36,000 |
| "Structural Steel". . . . . . . . . . . . . . . . . . . . . . . . . . . . . . . . . . . . | A529 | 42,000 |
| "High-strength Low-alloy Structural Steel". . . . . . . . . . . . . . . | A242 | 42,000 |
| "High-strength Structural Steel". . . . . . . . . . . . . . . . . . . . . . . . | A440 | 42,000 |
| "High-strength Low-alloy Structural Magnesium Vanadium Steel". . . . . . . . . . . . . . . . . . . . . . . . . . . . . . . . . . . . | A441 | 42,000 |
| "High-strength Low-alloy Structural Columbium Vanadium Steel, Grade 42". . . . . . . . . . . . . . . . . . . . . . . . | A572 | 42,000 |
| "High-strength Low-alloy Structural Steel with 50,000 psi Minimum Yield Point". . . . . . . . . . . . . . . . . . . . . . . . | A588 | 50,000† |
| "High-Yield Strength, Quenched and Tempered Alloy Steel Plate". . . . . . . . . . . . . . . . . . . . . . . . . . . . . . . . | A514 | 90,000 |

* Values given are for heavy sections and for plates 1½ to 4 in. thick; higher values may be allowed for light sections and thinner plates and lower values for thicker plates.
† For sections weighing more than 600 lb/ft, $F_y = 42,000$ psi.

The allowable unit working stresses for structural steel as given in the "Specification for the Design, Fabrication and Erection of Structural Steel for Buildings" of the American Institute of Steel Construction are shown below. Somewhat lower stresses than those shown are used for bridges, in recognition of the more severe service and the greater possibility of overloading such structures.

### Tension

Tension on net section, except at pinholes: $F_t = 0.60F_y$        (6-94)

Tension on net section at pinholes:       $F_t = 0.45F_y$        (6-95)

where $F_t$ = allowable tensile stress, psi.

The slenderness ratio $Kl/r$ [defined following Eq. (6-101)] preferably should not exceed 240 for main members or 300 for bracing or other secondary members, other than rods.

*Shear*

Shear on gross section: $\qquad F_v = 0.40F_y \qquad$ (6-96)

where $F_v$ = allowable shear stress, psi.

*Compression*

Compression on the gross section of axially loaded compression members when $Kl/r$ is less than $C_c$:

$$F_a = \frac{\left[1 - \dfrac{(Kl/r)^2}{2C_c{}^2}\right]F_y}{SF} \qquad (6\text{-}97)$$

Compression on the gross section of axially loaded columns when $Kl/r$ exceeds $C_c$:

$$F_a = \frac{149{,}000{,}000}{(Kl/r)^2} \qquad (6\text{-}98)$$

Compression on the gross section of axially loaded bracing and secondary members when $l/r$ exceeds 120:

$$F_{as} = \frac{F_a \text{ [from Eq. (6-97) or (6-98), depending on } C_c]}{1.6 - l/200r} \qquad (6\text{-}99)$$

Compression on the gross area of plate girder stiffeners:

$$F_a = 0.60F_y \qquad (6\text{-}100)$$

Compression on the web of rolled shapes at the toe of the fillet:

$$F_a = 0.75F_y \qquad (6\text{-}101)$$

where $F_a$ = allowable comprehensive stress permitted in absence of bending moment, psi

$\quad F_{as}$ = allowable comprehensive stress permitted in absence of bending moment for bracing and other secondary members, psi

$\quad F_y$ = minimum yield point, psi

$\quad K$ = effective-length factor (suggested design values shown in Fig. 6-20)

$\quad l$ = actual unbraced length, in.

$\quad r$ = radius of gyration corresponding to $K$ and $l$, in. $(= \sqrt{I/A})$

$\quad I$ = moment of inertia, in.[4]

$\quad A$ = gross cross-sectional area, in.[2]

$\quad C_c$ = slenderness ratio separating elastic and inelastic buckling

$$C_c = \sqrt{\frac{2\pi^2 E}{F_y}} \qquad (6\text{-}102)$$

$SF$ = factor of safety

$$SF = 1.67 + \frac{3(Kl/r)}{8C_c} - \frac{(Kl/r)^3}{8C_c{}^3} \qquad (6\text{-}103)$$

The slenderness ratio $Kl/r$ of compression members must not exceed 200.

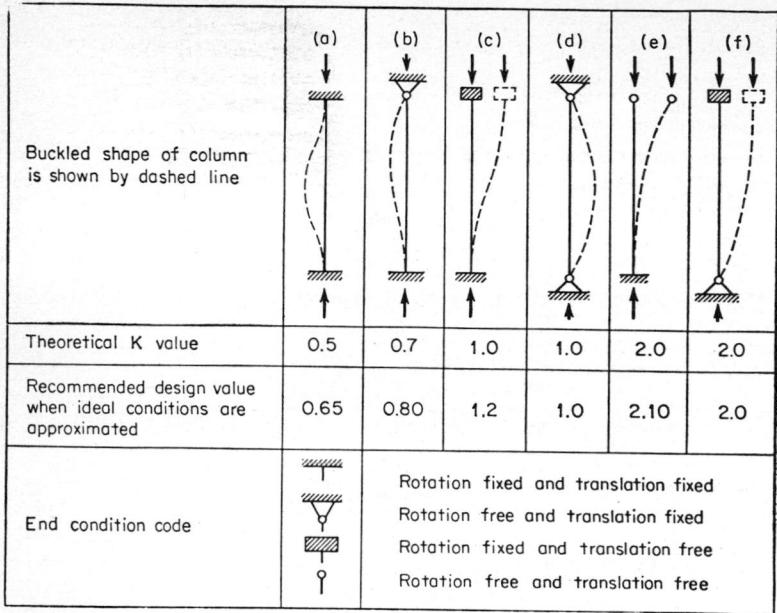

| | (a) | (b) | (c) | (d) | (e) | (f) |
|---|---|---|---|---|---|---|
| Buckled shape of column is shown by dashed line | | | | | | |
| Theoretical K value | 0.5 | 0.7 | 1.0 | 1.0 | 2.0 | 2.0 |
| Recommended design value when ideal conditions are approximated | 0.65 | 0.80 | 1.2 | 1.0 | 2.10 | 2.0 |
| End condition code | | Rotation fixed and translation fixed | | | | |
| | | Rotation free and translation fixed | | | | |
| | | Rotation fixed and translation free | | | | |
| | | Rotation free and translation free | | | | |

FIG. 6-20. Effective-length factors for members subject to axial load.

*Bending*

Tension and compression on extreme fibers of laterally supported compact shapes[1] having an axis of symmetry in the plane of loading:

$$F_b = 0.66F_y \qquad (6\text{-}104)$$

where $F_b$ = allowable bending stress in absence of axial load, psi
Laterally supported members have transverse movement of the compression flange prevented at points of support not more than $2,400b_f/\sqrt{F_y}$, or $20,000,000A_f/dF_y$ in. apart, where

$b_f$ = compression flange width, in.
$A_f$ = cross-sectional area of compression flange, in.[2]
$d$ = depth of member, in.

[1] A compact shape has the flanges continuously connected to the web or webs; the width of unstiffened projecting elements of the compression flange does not exceed $2,050/\sqrt{F_y}$ times the flange thickness; the width of flange plates does not exceed $6,000/\sqrt{F_y}$ times the flange-plate thickness; and the depth of the web does not exceed $20,200[1 - 3.74(f_a/F_a)]/\sqrt{F_y}$ times the web thickness where $(f_a/F_a)$ is the ratio of computed axial stress to allowable axial stress in the absence of bending moment, except that it need not be less than $8,100/\sqrt{F_y}$.

Tension and compression on extreme fibers of laterally supported unsymmetrical members (except channels) or box-type members, and tension on other rolled shapes or built-up members:

$$F_b = 0.60F_y \tag{6-105}$$

Compression on extreme fibers of other rolled shapes and built-up members (except box-type members), the larger value from Eqs. (6-106) and (6-107), but not more than $0.60F_y$:

$$F_b = \left[ 1.0 - \frac{(l/r)^2}{2C_c^2} \right] 0.60F_y \tag{6-106}$$

$$F_b = \frac{12,000,000}{ld/A_f} \tag{6-107}$$

where $l$ = unsupported length of compression flange, in.
$\quad r$ = radius of gyration of compression flange plus one-sixth web about an axis in plane of web, in.
$\quad d$ = depth of member, in.
$\quad A_f$ = cross-sectional area of compression flange, in.²
$\quad C_c = \sqrt{2\pi^2 E/F_y}$

Equation (6-106) may be further modified in certain cases by consideration of the moments at each end of the unsupported length.

Compression on extreme fibers of channels, the value from Eq. (6-107) but not more than $0.60F_y$.

Tension and compression on extreme fibers of pins:

$$F_b = 0.90F_y \tag{6-108}$$

Tension and compression on extreme fibers of rectangular bearing plates:

$$F_b = 0.75F_y \tag{6-109}$$

**Bearing**

Bearing on milled surfaces and pins in reamed, drilled, or bored holes:

$$F_p = 0.90F_y \tag{6-110}$$

Bearing on bolts or rivets:

$$F_p = 1.35F_y \tag{6-111}$$

where $F_p$ = allowable bearing stress, psi.

**Columns and Tension Members.** Columns are designed on the basis of the gross area of the section used. Tension members, however, are designed on the basis of net area, with deductions made for rivet and other holes. In determining net area, net width is obtained by deducting from the gross width the sum of the diameters of all the holes in any chain of holes in any diagonal or zigzag direction and adding for each gauge space in the chain the quantity $s^2/4g$, where $s$ is the longitudinal spacing (pitch) in inches of any two successive holes, and $g$ is the transverse spacing (gauge) in inches of the same two holes. Several chains of holes should be tried until the one giving the least net width is found. The net area of a tension member taken through a hole is limited to 85 per cent of the gross area. The diameter of a rivet or bolt hole is taken as $\frac{1}{8}$ in. greater than the nominal diameter of the rivet or bolt.

**Beams.** The extreme fiber stress in bending for a steel beam is computed as

$$f_b = M/S \qquad (6\text{-}112)$$

where $f_b$ = maximum fiber stress, psi
$M$ = bending moment, lb-in.
$S$ = section modulus ($= I/c$), in.$^3$
$I$ = moment of inertia of cross-sectional area, in.$^4$
$c$ = distance from extreme fiber to neutral axis, in. ($c$ = one-half the depth for a symmetrical cross section)

The section moduli for standard rolled-steel sections are shown in Table 6-28A. For beams requiring greater section moduli, built-up sections or plate girders are generally used.

The shearing stress in a steel beam with flanges is relatively constant over the depth of the web, and may be computed as

$$f_v = V/d_w t \qquad (6\text{-}113)$$

where $f_v$ = shearing stress in web, psi
$V$ = total shear at section, lb
$d_w$ = depth of web, in.
$t$ = thickness of web, in.

Built-up sections will usually require stiffeners to prevent web buckling.

## 6-50. Structural Shapes

Designations for structural shapes are related to the profiles of the shapes. **W** shapes have essentially parallel flange surfaces. **HP** bearing pile shapes have essentially parallel flange surfaces and equal web and flange thicknesses. **S** shapes and American Standard channels (**C**) have a slope of approximately 2 in 12 on their inner flange surfaces. The letter **M** designates shapes not classified as **W**, **HP**, or **S**. **MC** designates channels not American Standard. Current and former designations for structural shapes are shown in Table 6-28B.

## 6-51. Steel Connections

**Rivets.** Rivets vary in size from $\frac{3}{8}$ to $1\frac{1}{4}$ in. in diameter, with the $\frac{3}{4}$- and $\frac{7}{8}$-in. sizes most commonly used. The standard sizes and cross-sectional areas of rivets are shown in Table 6-29. Rivet holes are considered to be $\frac{1}{8}$ in. larger in diameter than the rivet.

**Bolts.** Both turned and unfinished bolts are available in the same sizes as rivets. Larger sizes are also available. Cross-sectional areas are as shown in Table 6-29.

**Welds.** The allowable loads on butt welds of the same size as the connected members are the same as for the members. The allowable load per

## Table 6-28A.

# $S_x$ ELASTIC SECTION MODULUS TABLE
## For shapes used as beams

| $S_x$ In.³ | Shape | $S_x$ In.³ | Shape | $S_x$ In.³ | Shape |
|---|---|---|---|---|---|
| 1110 | W 36 × 300 | 252 | S 24 × 120 | 110 | W 21 × 55 |
|      |            | 250 | W 24 × 100 | 108 | W 18 × 60 |
| 1030 | W 36 × 280 | 250 | W 21 × 112 | 107 | W 12 × 79 |
|      |            |     |            | 104 | W 16 × 64 |
| 952  | W 36 × 260 | 243 | W 27 × 94  | 103 | S 18 × 70 |
|      |            | 236 | S 24 × 105.9 | 103 | W 14 × 68 |
| 894  | W 36 × 245 |     |            |     |           |
|      |            | 221 | W 24 × 94  | 98.4 | W 18 × 55 |
| 837  | W 36 × 230 | 220 | W 18 × 114 | 97.5 | W 12 × 72 |
| 813  | W 33 × 240 |     |            | 94.4 | W 16 × 58 |
| 742  | W 33 × 220 | 212 | W 27 × 84  |      |           |
|      |            | 202 | W 18 × 105 | 93.3 | W 21 × 49 |
| 671  | W 33 × 200 | 199 | S 24 × 100 | 92.2 | W 14 × 61 |
|      |            | 198 | W 21 × 96  | 89.4 | S 18 × 54.7 |
| 665  | W 36 × 194 |     |            | 89.1 | W 18 × 50 |
| 651  | W 30 × 210 | 197 | W 24 × 84  | 88.0 | W 12 × 65 |
|      |            | 189 | W 14 × 119 |      |           |
| 622  | W 36 × 182 | 187 | S 24 × 90  |      |           |
| 587  | W 30 × 190 | 185 | W 18 × 96  | 81.6 | W 21 × 44 |
|      |            |     |            | 80.8 | W 16 × 50 |
| 580  | W 36 × 170 |     |            | 79.0 | W 18 × 45 |
|      |            | 176 | W 24 × 76  | 78.1 | W 12 × 58 |
| 542  | W 36 × 160 | 176 | W 14 × 111 | 77.8 | W 14 × 53 |
| 530  | W 30 × 172 | 175 | S 24 × 79.9 | 75.1 | MC 18 × 58 |
|      |            | 169 | W 21 × 82  | 73.7 | W 10 × 66 |
| 504  | W 36 × 150 | 166 | W 16 × 96  | 72.5 | W 16 × 45 |
| 494  | W 27 × 177 | 164 | W 14 × 103 | 70.7 | W 12 × 53 |
| 487  | W 33 × 152 | 161 | S 20 × 95  | 70.2 | W 14 × 48 |
|      |            | 157 | W 18 × 85  | 69.7 | MC 18 × 51.9 |
| 448  | W 33 × 141 |     |            |      |           |
| 446  | W 27 × 160 | 153 | W 24 × 68  | 68.4 | W 18 × 40 |
|      |            | 152 | S 20 × 85  | 67.1 | W 10 × 60 |
| 440  | W 36 × 135 | 151 | W 21 × 73  | 64.8 | S 15 × 50 |
| 414  | W 24 × 160 | 151 | W 16 × 88  | 64.7 | W 12 × 50 |
|      |            | 151 | W 14 × 95  |      |           |
| 406  | W 33 × 130 | 142 | W 18 × 77  |      |           |
| 404  | W 27 × 145 |     |            | 64.6 | W 16 × 40 |
| 380  | W 30 × 132 |     |            | 64.3 | MC 18 × 45.8 |
| 373  | W 24 × 145 | 140 | W 21 × 68  | 62.7 | W 14 × 43 |
|      |            | 138 | W 14 × 87  | 61.6 | MC 18 × 42.7 |
| 359  | W 33 × 118 | 131 | W 14 × 84  | 60.4 | W 10 × 54 |
| 355  | W 30 × 124 |     |            | 59.6 | S 15 × 42.9 |
| 332  | W 24 × 130 | 130 | W 24 × 61  | 58.2 | W 12 × 45 |
|      |            | 129 | W 18 × 70  |      |           |
| 329  | W 30 × 116 | 128 | S 20 × 75  |      |           |
| 317  | W 21 × 142 | 128 | W 16 × 78  | 57.9 | W 18 × 35 |
|      |            | 127 | W 21 × 62  | 56.5 | W 16 × 36 |
| 300  | W 30 × 108 | 121 | W 14 × 78  | 54.7 | W 14 × 38 |
| 300  | W 27 × 114 | 118 | S 20 × 65.4 | 54.6 | W 10 × 49 |
| 300  | W 24 × 120 | 118 | W 18 × 64  | 53.8 | C 15 × 50 |
| 284  | W 21 × 127 | 116 | W 16 × 71  | 51.9 | W 12 × 40 |
| 276  | W 24 × 110 | 116 | W 12 × 85  | 50.8 | S 12 × 50 |
|      |            |     |            | 49.1 | W 10 × 45 |
| 270  | W 30 × 99  | 114 | W 24 × 55  |      |           |
| 267  | W 27 × 102 | 112 | W 14 × 74  | 48.6 | W 14 × 34 |

SOURCE: American Institute of Steel Construction Manual of Steel Construction, 7th edition, 1970, with 1974 Supplements.

## Table 6-28A. (Continued)

# ELASTIC SECTION MODULUS TABLE
## For shapes used as beams
$$S_x$$

| $S_x$ In.³ | Shape | $S_x$ In.³ | Shape | $S_x$ In.³ | Shape |
|---|---|---|---|---|---|
| 47.2 | W 16 × 31 | 21.1 | M 14 × 17.2 | 7.76 | M 10 × 9 |
| 46.5 | C 15 × 40 | 20.8 | W 8 × 24 | 7.80 | W 8 × 10 |
| 46.0 | W 12 × 36 | 20.7 | C 10 × 30 | 7.78 | C 7 × 14.75 |
| 45.4 | S 12 × 40.8 | | | 7.37 | S 6 × 12.5 |
| 42.2 | W 10 × 39 | 18.8 | W 10 × 19 | 7.25 | W 6 × 12 |
| 42.0 | C 15 × 33.9 | 18.2 | C 10 × 25 | 6.93 | C 7 × 12.25 |
| 41.9 | W 14 × 30 | 17.6 | W 12 × 16.5 | | |
| 39.5 | W 12 × 31 | 17.0 | W 8 × 20 | 6.40 | MC 10 × 8.4 |
| | | 16.7 | W 6 × 25 | 6.08 | C 7 × 9.8 |
| 38.3 | W 16 × 26 | 16.2 | W 10 × 17 | 6.09 | S 5 × 14.75 |
| 38.2 | S 12 × 35 | 16.2 | S 8 × 23 | 5.80 | C 6 × 13 |
| 36.4 | S 12 × 31.8 | 15.8 | C 10 × 20 | 5.08 | W 6 × 8.5 |
| | | | | 5.06 | C 6 × 10.5 |
| 35.1 | W 14 × 26 | 14.8 | W 12 × 14 | 4.92 | S 5 × 10 |
| 35.0 | W 10 × 33 | 14.4 | S 8 × 18.4 | | |
| 34.2 | W 12 × 27 | 14.1 | W 8 × 17 | | |
| 31.1 | W 8 × 35 | 13.8 | W 10 × 15 | 4.62 | M 8 × 6.5 |
| 30.8 | W 10 × 29 | 13.5 | C 10 × 15.3 | 4.42 | MC 10 × 6.5 |
| 29.4 | S 10 × 35 | 13.5 | C 9 × 20 | 4.38 | C 6 × 8.2 |
| | | 13.4 | W 6 × 20 | 3.56 | C 5 × 9 |
| | | 12.1 | S 7 × 20 | | |
| 28.9 | W 14 × 22 | | | | |
| 27.4 | W 8 × 31 | 12.0 | M 12 × 11.8 | 3.44 | M 7 × 5.5 |
| 27.0 | C 12 × 30 | 11.8 | W 8 × 15 | 3.39 | S 4 × 9.5 |
| 26.6 | M 10 × 29.1 | 11.3 | C 9 × 15 | 3.00 | C 5 × 6.7 |
| 26.5 | W 10 × 25 | 11.0 | C 8 × 18.75 | 3.04 | S 4 × 7.7 |
| | | 10.6 | C 9 × 13.4 | | |
| 25.3 | W 12 × 22 | | | | |
| 24.7 | S 10 × 25.4 | 10.5 | W 10 × 11.5 | 2.40 | M 6 × 4.4 |
| 24.3 | W 8 × 28 | 10.5 | S 7 × 15.3 | 2.29 | C 4 × 7.25 |
| 24.1 | C 12 × 25 | 10.2 | W 6 × 16 | 1.95 | S 3 × 7.5 |
| 23.6 | M 10 × 22.9 | 10.0 | W 6 × 15.5 | 1.93 | C 4 × 5.4 |
| | | 9.90 | W 8 × 13 | 1.68 | S 3 × 5.7 |
| | | | | 1.38 | C 3 × 6 |
| 21.5 | C 12 × 20.7 | 9.23 | MC 12 × 10.6 | 1.24 | C 3 × 5 |
| 21.5 | W 10 × 21 | 9.03 | C 8 × 13.75 | | |
| 21.3 | W 12 × 19 | 8.77 | S 6 × 17.25 | | |
| | | 8.14 | C 8 × 11.5 | 1.10 | C 3 × 4.1 |

inch of fillet weld is determined on the minimum cross section; for an equal leg weld, the minimum section at the throat is 0.707 times the dimension of the weld leg.

**Working Stresses.** *Rivets.* Allowable stresses for A502, Grade 1 hot-driven rivets are 20,000 psi in tension and 15,000 psi in shear; for A502, Grade 2 hot-driven rivets, stresses are 27,000 psi in tension and 20,000 psi in shear.

*Bolts.* Allowable stresses for A307 bolts are 20,000 psi in tension and 10,000 psi in shear; for other threaded parts of other steels, stresses are $0.60F_y$ in tension and $0.30F_y$ in shear.

### Table 6-28B. Hot-rolled Structural Steel Shape Designations

| Current designation | Type of shape | Former designation |
|---|---|---|
| W 24 × 76<br>W 14 × 26 | W shape | 24 WF 76<br>14 B 26 |
| S 24 × 100 | S shape | 24 I 100 |
| M 8 × 18.5<br>M 10 × 9<br>M 8 × 34.3 | M shape | 8 M 18.5<br>10 JR 9.0<br>8 × 8 M 34.3 |
| C 12 × 20.7 | American Standard Channel | 12 C 20.7 |
| MC 12 × 45<br>MC 12 × 10.6 | Miscellaneous Channel | 12 × 4 C 45.0<br>12 JR C 10.6 |
| HP 14 × 73 | HP shape | 14 BP 73 |
| L 6 × 6 × ¾<br>L 6 × 4 × ⅝ | Equal Leg Angle<br>Unequal Leg Angle | ∠ 6 × 6 × ¾<br>∠ 6 × 4 × ⅝ |
| WT 12 × 38<br>WT 7 × 13 | Structural Tee cut from W shape | ST 12 WF 38<br>ST 7 B 13 |
| ST 12 × 50 | Structural Tee cut from S shape | ST 12 I 50 |
| MT 4 × 9.25<br>MT 5 × 4.5<br>MT 4 × 17.15 | Structural Tee cut from M shape | ST 4 M 9.25<br>ST 5 JR 4.5<br>ST 4 M 17.15 |
| PL ½ × 18 | Plate | PL 18 × ½ |
| Bar 1 ◻<br>Bar 1¼ φ<br>Bar 2½ × ½ | Square Bar<br>Round Bar<br>Flat Bar | Bar 1 ◻<br>Bar 1¼ φ<br>Bar 2½ × ½ |
| Pipe 4 Std.<br>Pipe 4 X - Strong<br>Pipe 4 XX - Strong | Pipe | Pipe 4 Std.<br>Pipe 4 X-Strong<br>Pipe 4 XX-Strong |
| TS 4 × 4 × .375<br>TS 5 × 3 × .375<br>TS 3 OD × .250 | Structural Tubing:  Square<br>Structural Tubing:  Rectangular<br>Structural Tubing:  Circular | Tube 4 × 4 × .375<br>Tube 5 × 3 × .375<br>Tube 3 OD × .250 |

SOURCE: Amercian Institute of Steel Construction Manual of Steel Construction, 7th edition, 1970, with 1974 Supplements.

### Table 6-29. Rivet and Bolt Diameters and Areas

| Rivets and bolts | | Bolts only | | Rivets and bolts | | Bolts only | |
|---|---|---|---|---|---|---|---|
| Nominal diam, in. | Cross-sec area, in.² | Cross-sec area, thread root, in.² | Threads/ in. | Nominal diam, in. | Cross-sec area, in.² | Cross-sec area, thread root, in.² | Threads/ in. |
| ⅜ | 0.110 | 0.068 | 16 | ⅞ | 0.601 | 0.419 | 9 |
| ½ | .196 | .126 | 13 | 1 | 0.785 | .551 | 8 |
| ⅝ | .307 | .202 | 11 | 1⅛ | 0.994 | .693 | 7 |
| ¾ | .442 | .302 | 10 | 1¼ | 1.227 | .890 | 7 |

Allowable stresses for A325 and A490 bolts are shown below:

### Allowable Bolt Stresses, Psi (Buildings)

| Bolts | Tension | Shear | | |
|---|---|---|---|---|
| | | Friction-type connections | Bearing-type connections | |
| | | | Threading excluded from shear planes | Threading not excluded from shear planes |
| A325............... | 40,000 | 15,000 | 22,000 | 15,000 |
| A490............... | 54,000 | 20,000 | 32,000 | 22,500 |

*Welds.* The allowable stress for welds on A36, A242, and A441 steels is 21,000 psi; except that complete-penetration groove welds with any type of loading and partial-penetration groove welds loaded in compression, bearing, or tension parallel to the axis of the weld may be stressed to the full allowable stress of the connected material.

## 6-52. Reinforced-concrete Design

**Concrete Mixes.** The proportioning of concrete ingredients is by weight or volume. Weight measures are considered more reliable. Concrete mixes are designated by the proportion of each ingredient in the order: cement, sand, coarse aggregate. For example, a 1:2:3 mix is one part cement, two parts sand, and three parts stone or gravel. A bag of cement (94 lb) is equivalent to 1 ft³. Water for concrete should be free of injurious amounts of oils, acids, alkalis, salts, or organic matter.

**Strength and Durability of Concrete.** The most important factor affecting the strength and durability of concrete is the water-cement ratio. For concrete made from average materials, compressive strengths to be used for design are shown in Table 6-30. Strengths greater than those shown may be used, based on compressive-strength tests. Water-cement ratios for various types of construction and exposure conditions are shown in Table 6-31, but any concrete subject to freezing temperatures while wet should have a water-cement ratio not more than 6 gal per bag and should contain entrained air.

**Design Methods.** Reinforced concrete may be designed by either one of two methods: working-stress design or ultimate-strength design. Both methods are permitted under current codes, and the selection of a method is left to the designer. Both methods are discussed herein; reference should be made to the American Concrete Institute (ACI) Building Codes 318-63 and 318-71 with 1973 Supplement which are the bases for the following discussion.

## Table 6-30. Maximum Permissible Water-cement Ratios for Concrete

| Specified compressive strength at 28 days, psi $f'_c$ | Maximum permissible water-cement ratio* | | | |
|---|---|---|---|---|
| | Non-air-entrained concrete | | Air-entrained concrete | |
| | U.S. gal per 94-lb bag of cement | Absolute ratio by weight | U.S. gal per 94-lb bag of cement | Absolute ratio by weight |
| 2.500 | 7.3 | 0.65 | 6.1 | 0.54 |
| 3,000 | 6.6 | 0.58 | 5.2 | 0.46 |
| 3.500 | 5.8 | 0.51 | 4.5 | 0.40 |
| 4.000 | 5.0 | 0.44 | 4.0 | 0.35 |

* Including free surface moisture on aggregates.
SOURCE: "Building Code Requirements for Reinforced Concrete (ACI 318-71)" American Concrete Institute, with 1973 Supplement.

## Table 6-31. Maximum Permissible Water-cement Ratios (Gal per Bag) for Different Types of Structures and Degrees of Exposure

| Type of structure | Exposure conditions* | | | | | |
|---|---|---|---|---|---|---|
| | Severe wide range in temperature or frequent alternations of freezing and thawing (air-entrained concrete only) | | | Mild temperature, rarely below freezing, or rainy, or arid | | |
| | In air | At the water line or within the range of fluctuating water level or spray | | In air | At the water line or within the range of fluctuating water level or spray | |
| | | In fresh water | In sea water or in contact with sulfates† | | In fresh water | In sea water or in contact with sulfates† |
| Thin sections, such as railings, curbs, sills, ledges, ornamental or architectural concrete, reinforced piles, pipe, and all sections with less than 1 in. concrete cover over reinforcing............. | 5.5 | 5.0 | 4.5‡ | 6 | 5.5 | 4.5‡ |
| Moderate sections, such as retaining walls, abutments, piers, girders, beams....... | 6.0 | 5.5 | 5.0‡ | § | 6.0 | 5.0‡ |
| Exterior portions of heavy (mass) sections | 6.5 | 5.5 | 5.0‡ | § | 6.0 | 5.0‡ |
| Concrete deposited by tremie under water | ... | 5.0 | 5.0 | ... | 5.0 | 5.0 |
| Concrete slabs laid on the ground........ | 6.0 | ... | .... | § | | |
| Concrete protected from the weather, interiors of buildings, concrete below ground........................ | § | ... | .... | § | | |
| Concrete which will later be protected by enclosure or backfill but which may be exposed to freezing and thawing for several years before such protection is offered........................ | 6.0 | ... | .... | § | | |

* Air-entrained concrete should be used under all conditions involving severe exposure, and may be used under mild exposure conditions to improve workability of the mixture.
† Soil or ground water containing sulfate concentrations of more than 0.2 per cent.
‡ When sulfate-resisting cement is used, maximum water-cement ratio may be increased by 0.5 gal per bag.
§ Water-cement ratio should be selected on the basis of strength and workability requirements.
SOURCE: "Recommended Practice for Selecting Proportions for Concrete (ACI 614-59)," American Concrete Institute, 1959.

**Reinforcing Bars.**  Steel bars for concrete reinforcement are available in the sizes shown in Table 6-32.

### Table 6-32. ASTM Standard Reinforcing Bars

| Bar size no.) | Wt, lb/ft | Diam, in. | Cross-sectional area, in.[6] | Perimeter, in. |
|---|---|---|---|---|
| 3 | 0.376 | 0.375 | 0.11 | 1.178 |
| 4 | 0.668 | 0.500 | 0.20 | 1.571 |
| 5 | 1.043 | 0.625 | 0.31 | 1.963 |
| 6 | 1.502 | 0.750 | 0.44 | 2.356 |
| 7 | 2.044 | 0.875 | 0.60 | 2.749 |
| 8 | 2.670 | 1.000 | 0.79 | 3.142 |
| 9 | 3.400 | 1.128 | 1.00 | 3.544 |
| 10 | 4.303 | 1.270 | 1.27 | 3.990 |
| 11 | 5.313 | 1.410 | 1.56 | 4.430 |
| 14 | 7.650 | 1.693 | 2.25 | 5.319 |
| 18 | 13.600 | 2.257 | 4.00 | 7.091 |

### Working-stress Design

*Design Loadings.*  In working-stress design, members should be designed to withstand actual service loads, consisting of dead loads, live loads, wind loads, and earthquake loads in any combination.  Members subject to stress produced by wind or earthquake may be proportioned for stresses one-third greater than those given in Tables 6-33 and 6-34, provided that the section thus required is not less than required for dead plus live loads.

*Working Stresses.*  Allowable unit stresses for concrete and reinforcing steel are given in Tables 6-33 and 6-34.

**Beams.**  Concrete beams may be considered to be of three principal types: rectangular beams with tensile reinforcing only, T beams with tensile reinforcing only, and beams with tensile and compressive reinforcing.

*Rectangular Beams with Tensile Reinforcing Only.*  This type of beam includes slabs (for which $b = 12$ in. when the moment and shear are expressed per foot of width).  The stresses in the concrete and steel are

$$f_c = 2M/kjbd^2 \tag{6-114}$$

$$f_s = M/A_s jd = M/pjbd^2 \tag{6-115}$$

where $b$ = width of beam (equals 12 in. for slab), in.

$d$ = effective depth of beam, measured from compressive face of beam to centroid of tensile reinforcing (Fig. 6-21), in.

$M$ = bending moment, lb-in.

$f_c$ = compressive stress in extreme fiber of concrete, psi

$f_s$ = stress in reinforcement, psi

$A_s$ = cross-sectional area of tensile reinforcing, in.$^2$

$j$ = ratio of distance between centroid of compression and centroid of tension to depth $d$

## Table 6-33. Allowable Stresses in Concrete

| Description | | For any strength of concrete | Allowable stresses — For strength of concrete shown below | | | |
|---|---|---|---|---|---|---|
| | | | $f'_c = 2,500$ psi | $f'_c = 3,000$ psi | $f'_c = 4,000$ psi | $f'_c = 5,000$ psi |
| Modulus of elasticity ratio: $n$ ......... | | $\dfrac{29{,}000{,}000}{w^{1.5}33\sqrt{f'_c}}$ | | | | |
| For concrete weighing 145 lb per cu ft | $n$ | ......... | 10 | 9 | 8 | 7 |
| Flexure: $f_c$ | | | | | | |
| Extreme fiber stress in compression... | $f_c$ | $0.45f'_c$ | 1,125 | 1,350 | 1,800 | 2,250 |
| Extreme fiber stress, in tension in plain concrete footings and walls........ | $f_c$ | $1.6\sqrt{f'_c}$ | 80 | 88 | 102 | 113 |
| Shear: $v$ (as a measure of diagonal tension at a distance $d$ from the face of the support) | | | | | | |
| Beams with no web reinforcement.... | $v_c$ | $1.1\sqrt{f'_c}$ | 55 | 60 | 70 | 78 |
| Joists with no web reinforcement..... | $v_c$ | $1.2\sqrt{f'_c}$ | 61 | 66 | 77 | 86 |
| Members with vertical or inclined web reinforcement or properly combined bent bars and vertical stirrups..... | $v$ | $5\sqrt{f'_c}$ | 250 | 274 | 316 | 354 |
| Slabs and footings (peripheral shear).. | $v_c$ | $2\sqrt{f'_c}$ | 100 | 110 | 126 | 141 |
| Bearing: $f_c$ | | | | | | |
| On full area..................... | | $0.25f'_c$ | 625 | 750 | 1,000 | 1,250 |
| On one-third area or less*.......... | | $0.375f'_c$ | 938 | 1,125 | 1,500 | 1,875 |

* This increase is permitted only when the least distance between the edges of the loaded and unloaded areas is a minimum of one-fourth of the parallel side dimension of the loaded area. The allowable bearing stress on a reasonably concentric area greater than one-third but less than the full area is to be interpolated between the values given.

NOTE: $f'_c$ = compressive strength of concrete, psi; $n$ = ratio of modulus of elasticity of steel to that of concrete; $w$ = weight of concrete, lb/ft³.

SOURCE: "Building Code Requirements for Reinforced Concrete (ACI 318-63)," American Concrete Institute, 1963.

## Table 6-34. Allowable Stresses in Steel for Concrete Reinforcement

| | Psi |
|---|---|
| *In tension* | |
| For billet-steel or axle-steel concrete-reinforcing bars of structural grade..................... | 18,000 |
| For main reinforcement, ⅜ in. or less in diameter, in one-way slabs of not more than 12-ft span, 50% of the minimum yield strength specified by the American Society for Testing Materials for the reinforcement used, but not to exceed............... | 30,000 |
| For deformed bars with a yield strength of 60,000 psi or more and in sizes No. 11 and smaller....... | 24,000 |
| For all other reinforcement............................................ | 20,000 |
| *In compression, vertical column reinforcement* | |
| Spiral columns, 40% of the minimum yield strength, but not to exceed........................... | 30,000 |
| Tied columns, 85% of the value for spiral columns, but not to exceed........................... | 25,500 |
| Composite and combination columns: | |
| Structural steel sections: | |
| For ASTM A36 Steel............................................ | 18,000 |
| Cast-iron sections............................................ | 10,000 |
| *Spirals [yield strength for use in Eq. (6-129)]* | |
| Hot-rolled rods, intermediate grade............................................ | 40,000 |
| Hot-rolled rods, hard grade............................................ | 50,000 |
| Hot-rolled rods, ASTM A432 grade and cold-drawn wire..................................... | 60,000 |

SOURCE: "Building Code Requirements for Reinforced Concrete (ACI 318-63)," American Concrete Institute, 1963.

$k$ = ratio of depth of compression area to depth $d$

$p$ = ratio of cross-sectional area of tensile reinforcing to area of the beam ($= A_s/bd$)

For approximate design purposes, $j$ may be assumed to be ⅞ and $k$ ⅜. For average structures, the following guides to the depth $d$ of a reinforced concrete beam may be used.

| Member | $d$ |
|---|---|
| Roof and floor slabs | $l/25$ |
| Light beams | $l/15$ |
| Heavy beams and girders | $l/12$–$l/10$ |

where $l$ is the span of the beam or slab in inches.   The width of a beam should be at least $l/32$.

Cross-section of beam            Stress diagram

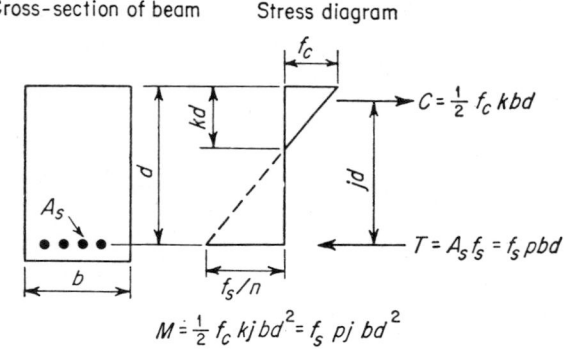

$$M = \tfrac{1}{2} f_c \, kj \, bd^2 = f_s \, pj \, bd^2$$

FIG. 6-21. Rectangular concrete beam with tensile reinforcing only.

For a balanced design, one in which both the concrete and the steel are stressed to the maximum allowable stress, the following formulas may be used.

$$bd^2 = M/K \qquad\qquad (6\text{-}116)$$

$$K = \tfrac{1}{2} f_c k j = p f_s j \qquad\qquad (6\text{-}117)$$

Values of $K$, $k$, $j$, and $p$ for commonly used stresses are given in Table 6-35.

**T Beams with Tensile Reinforcing Only.**   When a concrete slab is constructed monolithically with the supporting concrete beams, a portion of the slab acts as the upper flange of the beam.   The effective flange width should not exceed (1) one-fourth the span of the beam, (2) the width of the web portion of the beam plus 16 times the thickness of the slab, or (3) the center-to-center distance between beams.   T beams where the upper flange is not a portion of a slab should have a flange thickness not less than one-half the width of the web and a flange width not more than 4 times the width of the web.   For preliminary designs, the formulas given above for rectangular beams with tensile reinforcing only can be used, since the neutral axis is usually in or near the flange. The area of tensile reinforcing will usually be critical.

**Beams with Tensile and Compressive Reinforcing.** Beams with compressive reinforcing are generally used when the size of the beam is limited. The allowable beam dimensions are used in the formulas given above to determine the moment which could be carried by a beam without compressive reinforcement. The reinforcing requirements may then be approximately determined from

$$A_s = 8M/7f_s d \tag{6-118}$$

$$A_{sc} = (M - M')/nf_c d \tag{6-119}$$

where $A_s$ = total cross-sectional area of tensile reinforcing, in.$^2$
  $A_{sc}$ = cross-sectional area of compressive reinforcing, in.$^2$
  $M$ = total bending moment, lb-in.
  $M'$ = bending moment which would be carried by beam of balanced design and same dimensions with tensile reinforcing only, lb-in.
  $n$ = ratio of modulus of elasticity of steel to that of concrete

**Check of Stresses in Beam.** Beams designed by the above approximate formulas should be checked to ensure that the actual stresses do not exceed the allowable, and that the reinforcing is not excessive. This can be accomplished by determining the moment of inertia of the beam. In this determination, the concrete below the neutral axis should not be considered as stressed, while the reinforcing steel should be transformed into an equivalent concrete section. For tensile reinforcing, this transformation is made

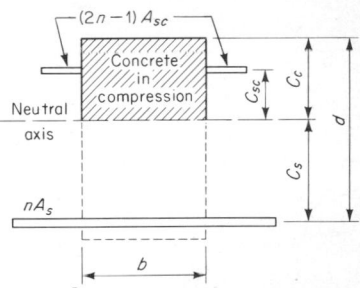

FIG. 6-22. Transformed section of concrete beam.

by multiplying the area $A_s$ by $n$, the ratio of the modulus of elasticity of steel to that of concrete. For compressive reinforcing, the area $A_{sc}$ is multiplied by $(2n - 1)$. This factor includes allowances for the concrete in compression replaced by the compressive reinforcing and for the plastic flow of concrete. The neutral axis is then located by solving

$$\tfrac{1}{2}bc_c^2 + (2n - 1)A_{sc}c_{sc} = nA_s c_s \tag{6-120}$$

for the unknowns $c_c$, $c_{sc}$, and $c_s$ (Fig. 6-22). The moment of inertia of the transformed beam section is

$$I = \tfrac{1}{3}bc_c^3 + (2n - 1)A_{sc}c_{sc}^2 + nA_s c_s^2 \tag{6-121}$$

and the stresses are

$$f_c = Mc_c/I \tag{6-122}$$
$$f_{sc} = 2nMc_{sc}/I \tag{6-123}$$
$$f_s = nMc_s/I \tag{6-124}$$

where $f_c, f_{sc}, f_s$ = actual unit stresses in extreme fiber of concrete, in compressive reinforcing steel, and in tensile reinforcing steel, respectively, psi

$c_c, c_{sc}, c_s$ = distances from neutral axis to face of concrete, to compressive reinforcing steel, and to tensile reinforcing steel, respectively, in.

$I$ = moment of inertia of transformed beam section, in.[4]

$b$ = beam width, in.

and $A_s$, $A_{sc}$, $M$, and $n$ are as defined for Eqs. (6-118) and (6-119).

### Table 6-35. Values of $K, k, j, p$ for Rectangular Sections (Balanced Conditions)

| $f'_c$ | $f_c$ | $K$ | $k$ | $j$ | $p$ |
|---|---|---|---|---|---|
| | | $f_s = 18,000$ psi | | | |
| 2,500 | 1,125 | 207 | 0.429 | 0.857 | 0.0134 |
| 3,000 | 1,350 | 248 | .429 | .857 | .0161 |
| 4,000 | 1,800 | 340 | .444 | .852 | .0222 |
| 5,000 | 2,250 | 444 | .467 | .845 | .0292 |
| | | $f_s = 20,000$ psi | | | |
| 2,500 | 1,125 | 196 | .403 | .866 | .0113 |
| 3,000 | 1,350 | 236 | .403 | .866 | .0136 |
| 4,000 | 1,800 | 325 | .419 | .861 | .0188 |
| 5,000 | 2,250 | 422 | .440 | .853 | .0247 |
| | | $f_s = 24,000$ psi | | | |
| 2,500 | 1,125 | 177 | .359 | .880 | .0084 |
| 3,000 | 1,350 | 213 | .359 | .880 | .0101 |
| 4,000 | 1,800 | 295 | .375 | .875 | .0141 |
| 5,000 | 2,250 | 387 | .396 | .868 | .0186 |

**Shear and Diagonal Tension in Beams.** The shearing unit stress, as a measure of diagonal tension, in a reinforced concrete beam is

$$v = V/bd \tag{6-125}$$

where $v$ = shearing unit stress, psi

$V$ = total shear, lb

$b$ = width of beam (for T beam use width of stem), in.

$d$ = effective depth of beam

If the value of the shearing unit stress as computed above exceeds the allowable shearing unit stress ($v_c$ in Table 6-33) web reinforcement should be provided. Such reinforcement will usually consist of stirrups. The cross-sectional area required for a stirrup placed perpendicular to the longitudinal reinforcement is

$$A_v = (V - V')s/f_v d \tag{6-126}$$

where $A_v$ = cross-sectional area of web reinforcement in distance $s$ (measured parallel to longitudinal reinforcement), in.[2]

$f_v$ = allowable unit stress in web reinforcement, psi
$V$ = total shear, lb
$V'$ = shear which concrete alone could carry ($= v_c bd$), lb
$s$ = spacing of stirrups in direction parallel to that of longitudinal reinforcing, in.
$d$ = effective depth, in.

Stirrups should be so spaced that every 45° line extending from the middepth of the beam to the longitudinal tension bars is crossed by at least one stirrup. If the total shearing unit stress is in excess of 3 $\sqrt{f'_c}$ psi, every such line should be crossed by at least two stirrups. The shear stress at any section should not exceed 5 $\sqrt{f'_c}$ psi.

**Bond and Anchorage for Reinforcing Bars.** In beams in which the tensile reinforcing is parallel to the compression face, the bond stress on the bars is

$$u = \frac{V}{jd\Sigma_0} \qquad (6\text{-}127)$$

where $u$ = bond stress on surface of bar, psi
$V$ = total shear, lb
$d$ = effective depth of beam, in.
$\Sigma_0$ = sum of perimeters of tensile reinforcing bars, in.

For preliminary design, the ratio $j$ may be assumed to be $\frac{7}{8}$. Bond stresses may not exceed the values shown in Table 6-36. To provide sufficient anchor-

### Table 6-36. Allowable Bond Stresses, Psi

| | Horizontal bars with more than 12 in. of concrete cast below the bar | Other bars |
|---|---|---|
| Tension bars with sizes and deformations conforming to ASTM A305... | $\dfrac{3.4\sqrt{f'_c}}{D}$ or 350, whichever is less | $\dfrac{4.8\sqrt{f'_c}}{D}$ or 500, whichever is less |
| Tension bars with sizes and deformations conforming to ASTM A408.... | $2.1\sqrt{f'_c}$ | $3\sqrt{f'_c}$ |
| Deformed compression bars.......... | $6.5\sqrt{f'_c}$ or 400, whichever is less | $6.5\sqrt{f'_c}$ or 400, whichever is less |
| Plain bars....................... | $1.7\sqrt{f'_c}$ or 160, whichever is less | $2.4\sqrt{f'_c}$ or 160, whichever is less |

NOTE: $f'_c$ = compressive strength of concrete, psi; $D$ = nominal diameter of bar, in.

age to develop the strength of reinforcing steel, tensile bars should be extended beyond the point at which they are needed to resist stress and should be terminated in a compression region, over a support, or with a hook.

**Columns.** The principal columns in a structure should have a minimum diameter of 10 in. or, for rectangular columns, a minimum thickness of 8 in. and a minimum gross cross-sectional area of 96 in.[2]

*Short Columns, Spiral Reinforcing.* For short columns with closely spaced spiral reinforcing enclosing a circular concrete core reinforced with vertical **bars,** the maximum allowable load is

$$P = A_g(0.25f'_c + f_s p_g) \qquad (6\text{-}128)$$

where $P$ = total allowable axial load, lb

$A_g$ = gross cross-sectional area of column, in.$^2$

$f'_c$ = compressive strength of concrete, psi

$f_s$ = allowable stress in vertical concrete reinforcing, psi, equal to 40 per cent of the minimum yield strength, but not to exceed 30,000 psi

$p_g$ = ratio of cross-sectional area of vertical reinforcing steel to gross area of column $A_g$

The ratio $p_g$ should not be less than 0.01 nor more than 0.08. The minimum number of bars to be used is six, and the minimum size is no. 5. The spiral reinforcing to be used in a spirally reinforced column is

$$p_s = 0.45(A_g/A_c - 1)f'_c/f_y \qquad (6\text{-}129)$$

where $p_s$ = ratio of spiral volume to concrete-core volume (out-to-out spiral)

$A_c$ = cross-sectional area of column core (out-to-out spiral), in.$^2$

$f_y$ = yield strength of spiral reinforcement, psi, but not to exceed 60,000 psi

The center-to-center spacing of the spirals should not exceed one-sixth of the core diameter. The clear spacing between spirals should not exceed one-sixth the core diameter, or 3 in., nor be less than $1\frac{3}{8}$ in., or $1\frac{1}{2}$ times the maximum size of coarse aggregate used.

*Short Columns with Ties.* The maximum allowable load on short columns reinforced with longitudinal bars and separate lateral ties is 85 per cent of that given in Eq. (6-128) for spirally reinforced columns. The ratio $p_g$ for a tied column should not be less than 0.01 nor more than 0.08. The longitudinal reinforcing should consist of at least four bars, and the minimum size is no. 5.

Ties should be at least $\frac{1}{4}$ in. in diameter, and should be spaced apart not over 16 bar diameters, 48 tie diameters, or the least dimension of the column.

*Long Columns.* Allowable column loads where compression governs design must be adjusted for column length, as follows:

1. If the ends of the column are fixed so that a point of contraflexure occurs between the ends, applied axial loads and moments should be divided by $R$ from Eq. (6-130) ($R$ cannot exceed 1.0).

$$R = 1.32 - 0.006h/r \qquad (6\text{-}130)$$

2. If relative lateral displacement of the ends of the column is prevented and the member is bent in single curvature, applied axial loads and moments should be divided by $R$ from Eq. (6-131) ($R$ cannot exceed 1.0).

$$R = 1.07 - 0.008h/r \qquad (6\text{-}131)$$

where $h$ = unsupported length of column, in.

$r$ = radius of gyration of gross concrete area, in.

= 0.30 times depth for rectangular column

= 0.25 times diameter for circular column

$R$ = long-column load reduction factor

Applied axial load and moment when tension governs design should be similarly adjusted, except that the factor $R$ varies linearly with the axial load from the values given by Eqs. (6-130) and (6-131) at the balanced condition, as defined by Eq. (6-137) to a value of 1.0 when the axial load is 0.

**Combined Bending and Compression.** The strength of a symmetrical column is controlled by compression if the equivalent axial load $N$ has an eccentricity $e$ in each principal direction no greater than given by Eq. (6-132) or (6-133) and by tension if $e$ exceeds these values in either principal direction.

For spiral columns: $\qquad e_b = 0.43p_g m D_s + 0.14t$ $\qquad\qquad$ (6-132)

For tied columns: $\qquad e_b = (0.67p_g m + 0.17)d$ $\qquad\qquad$ (6-133)

where $e$ = eccentricity, in.

$\quad e_b$ = maximum permissible eccentricity, in.

$\quad N$ = eccentric load normal to cross section of column

$\quad p_g$ = ratio of area of vertical reinforcement to gross concrete area

$\quad m = f_y/0.85f'_c$

$\quad D_s$ = diameter of circle through centers of longitudinal reinforcement, in.

$\quad t$ = diameter of column or over-all depth of column, in.

$\quad d$ = distance from extreme compression fiber to centroid of tension reinforcement, in.

$\quad f_y$ = yield point of reinforcement, psi

Design of columns controlled by compression is based on Eq. (6-134), except that the allowable load $N$ may not exceed the allowable load $P$ [Eq. (6-128)] permitted when the column supports axial load only.

$$\frac{f_a}{F_a} + \frac{f_{bx}}{F_b} + \frac{f_{by}}{F_b} \le 1.0 \qquad\qquad (6\text{-}134)$$

where $f_a$ = axial load divided by gross concrete area, psi

$\quad f_{bx}, f_{by}$ = bending moment about $x$ and $y$ axes, divided by section modulus of corresponding transformed uncracked section, psi

$\quad F_b$ = allowable bending stress permitted for bending alone, psi

$\quad F_a = 0.34(1 + p_g m)f'_c$

The allowable bending moment on columns controlled by tension varies linearly with the axial load from $M_0$ when the section is in pure bending to $M_b$ when the axial load is $N_b$.

For spiral columns: $\qquad M_0 = 0.12A_{st}f_y D_s$ $\qquad\qquad$ (6-135)

For tied columns: $\qquad M_0 = 0.40A_s f_y(d - d')$ $\qquad\qquad$ (6-136)

where $A_{st}$ = total area of longitudinal reinforcement, in.²

$\quad f_y$ = yield strength of reinforcement, psi

$\quad D_s$ = diameter of circle through centers of longitudinal reinforcement, in.

$\quad A_s$ = area of tension reinforcement, in.²

$\quad d$ = distance from extreme compression fiber to centroid of tension reinforcement, in.

$d'$ = distance from extreme compression fiber to centroid of compression reinforcement, in.

$N_b$ and $M_b$ are the axial load and moment at the balanced condition, i.e., when the eccentricity $e$ equals $e_b$ as determined from Eq. (6-132) or (6-133). At this condition, $N_b$ and $M_b$ should be determined from Eq. (6-134) so that

$$M_b = N_b e_b \qquad (6\text{-}137)$$

When bending is about two axes,

$$\frac{M_x}{M_{0x}} + \frac{M_y}{M_{0y}} \leq 1 \qquad (6\text{-}138)$$

where $M_x$ and $M_y$ are bending moments about the $x$ and $y$ axes, and $M_{0x}$ and $M_{0y}$ are the values of $M_0$ for bending about these axes.

### Ultimate-Strength Design

*Loadings.*  In ultimate-strength design, proportioning of members is based upon design loads determined from appropriate combinations of dead loads, live loads, wind loads, and earthquake loads.  The design loads are determined by multiplying the actual loads by various safety factors. Design loads are

$$U = 1.4D + 1.7L \qquad (6\text{-}139)$$

or, with wind load a factor,

$$U = 0.75(1.4D + 1.7L + 1.7W) \qquad (6\text{-}140)$$

or, with $L$ absent,

$$U = 0.9D + 1.3W \qquad (6\text{-}141)$$

where $U$ = design load, lb  (use maximum value from above equations)
 $D$ = dead loads, lb
 $L$ = live loads plus impact, lb
 $W$ = wind loads, lb

For earthquake loading, substitute $1.1E$ for $W$ in Eqs. (6-140) and (6-141), where $E$ = earthquake loads, lb.

*Capacity Reduction Factors.*  To compensate for variations in dimensions, workmanship, and materials, the ultimate-strength equations include a capacity reduction factor $\phi$.  Typical values of $\phi$ are shown in Table 6-37 and should be applied to computed resisting forces and moments.

*Assumptions in Design.*  In ultimate-strength design, basic assumptions include: (1) maximum strain occurs at extreme compression fiber and is 0.003 in./in., (2) stress in reinforcing bars below the yield strength, $f_y$, is 29,000,000 psi times the steel strain, (3) strain in concrete is directly proportional to the distance from neutral axis, (4) tensile strength of concrete is neglected in flexural calculations, and (5) concrete stress intensity of $0.85f'_c$ is uniformly distributed over a depth $a = kc$, where $c$ is the distance

## Table 6-37. Capacity Reduction Factors—Reinforced Concrete

| Condition | $\phi$ Value |
|---|---|
| Bending, with or without axial tension | 0.90 |
| Axial tension | 0.90 |
| Axial compression, with or without combined bending | 0.70* |
| Shear and torsion | 0.85 |
| Bearing on concrete | 0.70 |

* May be increased to 0.75 when spiral reinforcement given by Eq. (6-129) is provided; values up to 0.90 are allowable under special conditions when axial compressive loading is small.

from the extreme compression fiber to the neutral axis. For $f'_c \leq 4,000$ psi, use $k = 0.85$; for each 1,000 psi in excess of 4,000 psi, $k$ is reduced by 0.05.

**Design Values.** For conventional structures, values normally used are compressive strength of concrete $f'_c = 4,000$ psi and yield strength of reinforcement $f_y = 60,000$ psi.

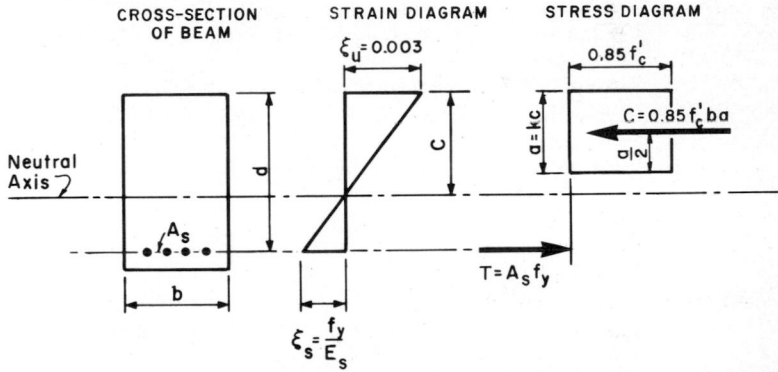

Fig. 6-23. Assumptions in ultimate-strength design.

**Rectangular Beams with Tensile Reinforcing Only.** This type of beam includes slabs (for which $b = 12$ in. when the moment and shear are expressed per foot of width). The ultimate resisting moment is

$$M_u = \phi[bd^2f'_c q(1 - 0.59q)] = \phi\left[ A_s f_y \left( d - \frac{a}{2} \right) \right] \qquad (6\text{-}142)$$

where $A_s$ = area of tensile reinforcement, in.$^2$

$a$ = depth of rectangular stress block = $A_s f_y/0.85f'_c b$, in.

$b$ = width of compressive face of flexural member, in.

$d$ = distance from extreme compression fiber to centroid of tension reinforcement, in.

$f'_c$ = compressive strength of concrete, psi

$f_y$ = yield strength of reinforcement, psi

$M_u$ = ultimate resisting moment, in.-lb

$p$ = $A_s/bd$, $<0.75p_b$

$p_b$ = reinforcement ratio producing balanced conditions[1]

$q = A_s f_y/bdf'_c = pf_y/f'_c$

$\phi$ = capacity reduction factor, Table 6-37

$k$ = concrete stress intensity factor = 0.85 when $f'_c = \leq 4{,}000$ psi

**T Beams with Tensile Reinforcing Only.** For preliminary designs, the formulas given above for rectangular beams with tensile reinforcing only can be used, since the neutral axis is usually in or near the flange. The area of tensile reinforcing will usually be critical.

**Beams with Tensile and Compressive Reinforcing.** Beams with compression reinforcing are generally used when the size of the beam is limited, or where $p > 0.75p_b$. The ultimate resisting moment is then

$$M_u = \phi\left[ (A_s - A_{sc})f_y\left(d - \frac{a}{2}\right) + A_{sc}f_y(d - d') \right] \qquad (6\text{-}143)$$

where $A_s$ = area of tensile reinforcement, in.[2]

$A_{sc}$ = area of compressive reinforcement, in.[2]

$a$ = depth of equivalent rectangular stress block = $(A_s - A_{sc})f_y/0.85f'_cb$, in.

$b$ = width of compressive face of flexural member, in.

$d$ = distance from extreme compression fiber to centroid of tension reinforcement, in.

$d'$ = distance from extreme compression fiber to centroid of compression steel, in.

$f_y$ = yield strength of reinforcement, psi

$M_u$ = ultimate resisting moment, in.-lb

$p = A_s/bd$ ⎫
$p' = A_{sc}/bd$ ⎬ $p - p' < 0.75p_b$

$p_b$ = reinforcement ratio producing balanced conditions, Eq. (6-142a)

$\phi$ = capacity reduction factor, Table 6-37

**Shear and Diagonal Tension in Beams.** The ultimate shearing unit stress, as a measure of diagonal tension, in a reinforced concrete beam is

$$v_u = V_u/bd \qquad (6\text{-}144)$$

where $v_u$ = ultimate shearing unit stress, psi

$V_u$ = ultimate total shear, lb

$b$ = width of beam (for T beam use width of stem), in.

$d$ = effective depth of beam, in.

For design, the maximum shear is considered to occur at a distance $d$ from the face of the support. The shear stress carried by concrete $v_c$ should not exceed $2\phi \sqrt{f'_c}$. Wherever the ultimate shear stress $v_u$ exceeds the shear stress $v_c$, web reinforcement is mandatory. Such reinforcement

---

[1] Balanced conditions occur when the tension reinforcement is at its yield strength $f_y$ and the concrete in compression is at its assumed ultimate strain of 0.003; balanced conditions exist when

$$p_b = \left(\frac{0.85kf'_c}{f_y}\right)\left(\frac{87{,}000}{87{,}000 + f_y}\right) \qquad (6\text{-}142a)$$

will usually consist of stirrups. The cross-sectional area required for a stirrup placed perpendicular to the longitudinal reinforcement is

$$A_u = V_u s / \phi f_y d \qquad (6\text{-}145)$$

where $A_u$ = total area of web reinforcement in tension within a distance $s$ measured in a direction parallel to the longitudinal reinforcements, in.²

    $d$ = effective depth, in.

    $f_y$ = yield strength of reinforcement, psi

    $s$ = spacing of stirrups, in.

    $V_u$ = total ultimate shear, lb

    $\phi$ = capacity reduction factor, Table 6-37

The shear stress $v_u$ should not exceed $10\phi \sqrt{f'_c}$ and $f_y$ should not exceed 60,000 psi. Stirrups should be anchored at both ends to be considered effective and so spaced that every 45° line extending from the middepth of the beam to the longitudinal tension bars is crossed by at least one stirrup.

***Bond and Anchorage for Reinforcing Bars.*** In beams in which the tensile reinforcing is parallel to the compression face, the bond stress on the bars is

$$u_u = \frac{V_u \phi}{j d \Sigma_0} \qquad (6\text{-}146)$$

where $u_u$ = ultimate bond stress on surface of bar, psi

    $V_u$ = total ultimate shear, lb

    $jd$ = distance between centroid of compression and centroid of tension, in.

    $\Sigma_0$ = sum of perimeters of tensile reinforcing bars, in.

    $\phi$ = capacity reduction factor, Table 6-37

    $d$ = effective depth of beam, in.

For preliminary design, the ratio $j$ may be assumed to be ⅞. Bond stresses should not exceed the values shown in Table 6-38. To provide sufficient anchorage to develop the strength of reinforcing steel, tensile bars should be extended beyond the point at which they are needed to resist stress and should be terminated in a compression region, over a support, or with a hook.

### Table 6-38. Maximum Bond Stresses, Psi

| | Horizontal bars with more than 12 in. of concrete cast below the bar | Other bars |
|---|---|---|
| Tension bars with sizes and deformations conforming to ASTM A305......... | $\dfrac{6.7 \sqrt{f'_c}}{D}$ or 560 psi* | $\dfrac{9.5 \sqrt{f'_c}}{D}$ or 800 psi* |
| Tension bars with sizes and deformations conforming to ASTM A408......... | $4.2 \sqrt{f'_c}$ | $6 \sqrt{f'_c}$ |
| Deformed compression bars......... | $13 \sqrt{f'_c}$ or 800 psi* | $13 \sqrt{f'_c}$ or 800 psi* |
| Plain bars......... | 250 psi | 250 psi |

\* Use lower of two values.

NOTE: $f'_c$ = compressive strength of concrete, psi; $D$ = nominal diameter of bar, in.

*Columns—General.* Columns should be designed for the axial load computed by using Eqs. (6-139), (6-140), and (6-141) and for the actual eccentricity $e$ of the applied loading which should not be less than $0.05t$ for spirally reinforced columns or $0.10t$ for tied columns where $t$ is the overall depth of a rectangular section or diameter of a circular section. The maximum load capacities given by Eqs. (6-147) through (6-152) apply only to short columns; adjustment factors for long columns are given by Eqs. (6-130) and (6-131).

*Short Columns—Rectangular.* The ultimate strength of short rectangular columns where the reinforcement is in one or two faces, each parallel to the axis of bending and all reinforcement in any one face is located at approximately the same distance from the axis of bending, is computed by the empirical formulas

$$P_u = \phi(0.85f'_c ba + A_{sc}f_y - A_s f_s) \tag{6-147}$$

$$P_u e' = \phi\left[0.85f'_c ba\left(d - \frac{a}{2}\right) + A_{sc}f_y(d - d')\right] \tag{6-148}$$

where $a$ = depth of equivalent rectangular stress block, in.
    $A_g$ = gross area of section, in.²
    $A_s$ = area of tension reinforcement, in.²
   $A_{sc}$ = area of compression reinforcement, in.²
   $A_{st}$ = total area of longitudinal reinforcement, in.²
     $b$ = width of compression face of flexural member, in.
     $d$ = distance from extreme compression fiber to centroid of tension reinforcement, in.
    $d'$ = distance from extreme compression fiber to centroid of compression reinforcement, in.
    $D$ = overall diameter of circular section, in.
    $D_s$ = diameter of the circle through centers of reinforcement arranged in a circular pattern, in.
     $e$ = eccentricity of axial load at end of member measured from plastic centroid* of the section, calculated by conventional methods of frame analysis, in.
    $e'$ = eccentricity of axial load at end of member measured from the centroid of the tension reinforcement, calculated by conventional methods of frame analysis, in.
   $f'_c$ = compressive strength of concrete, psi
    $f_s$ = calculated stress in reinforcement when less than the yield strength $f_y$, psi
    $f_y$ = yield strength of reinforcement, psi
    $m$ = $f_y/0.85f'_c$
    $P_u$ = axial load capacity under combined axial load and bending, lb
    $p_t$ = $A_{st}/A_g$
     $t$ = overall depth of a rectangular section or diameter of a circular section, in.
    $\phi$ = capacity reduction factor, Table 6-37

---

\* Centroid of the resistance to load computed for assumptions that concrete is stressed uniformly to $0.25f'_c$ and steel is stressed uniformly to $f_y$.

**Short Columns—Circular.** The ultimate strength of short circular columns with reinforcing bars circularly arranged is computed by the following empirical formulas.

When tension controls:

$$P_u = \phi \left\{ 0.85f'_c D^2 \left[ \sqrt{\left(\frac{0.85e}{D} - 0.38\right)^2 + \frac{p_t m D_s}{2.5D}} - \left(\frac{0.85e}{D} - 0.38\right) \right] \right\} \quad (6\text{-}149)$$

When compression controls:

$$P_u = \phi \left[ \frac{A_{st}f_y}{\dfrac{3e}{D_s} + 1} + \frac{A_g f'_c}{\dfrac{9.6De}{(0.8D + 0.67D_s)^2} + 1.18} \right] \quad (6\text{-}150)$$

**Short Columns—Square.** The ultimate strength of short square columns with reinforcing bars circularly arranged is computed by the following empirical formulas.

When tension controls:

$$P_u = \phi \left\{ 0.85btf'_c \left[ \sqrt{\left(\frac{e}{t} - 0.5\right)^2 + 0.67\frac{D_s}{t} p_t m} - \left(\frac{e}{t} - 0.5\right) \right] \right\} \quad (6\text{-}151)$$

When compression controls

$$P_u = \phi \left[ \frac{A_{st}f_y}{\dfrac{3e}{D_s} + 1} + \frac{A_g f'_c}{\dfrac{12te}{(t + 0.67D_s)^2} + 1.18} \right] \quad (6\text{-}152)$$

Equations (6-147) through (6-152) for the ultimate strength of columns subjected to combined axial compression and bending require assumption of a concrete cross section and reinforcement, computation of an allowable ultimate load and eccentricity, and comparison of the allowable loading with the design loading. This procedure usually requires iteration and may involve several assumptions regarding member size and amount of reinforcement. In practice, columns are usually designed by reference to extensive design tables, such as appear in "Concrete Reinforcing Steel Handbook," Concrete Reinforcing Steel Institute (1975), which lists allowable axial loads and eccentricities for a large selection of cross sections and reinforcement arrangements. An alternative method involves the development of a strain diagram for an assumed member with the concrete strain in the extreme compressive fiber at 0.003 in./in.; use of the compressive stress in concrete as $0.85f'_c$ over an area determined by using the $k$ factor described under Assumptions in Design;* computation of the tensile and compressive stresses in the reinforcement using strains compatible with the strains in the concrete; converting reinforcement strains to stresses using

* See page 6-88

$E_s$ = 29,000,000 psi; and determination of resisting forces using appropriate stresses and areas. The axial load and eccentricity determined from the summation of the resisting forces represent the allowable ultimate loading for the assumed member and strain diagram.

**Concrete Protection for Reinforcing Steel.** The concrete protection for reinforcing steel should not be less than that given in Table 6-39.

### Table 6-39. Minimum Concrete Protection for Reinforcement

|  | Minimum Concrete Protection, In. |
|---|---|
| Concrete placed directly against the ground | 3 |
| Concrete exposed to weather: | |
|   Bars less than no. 5 in size | 1½ |
|   Bars no. 5 in size or larger | 2 |
| Concrete not exposed to ground or weather: | |
|   Slabs and walls | ¾ |
|   Beams, girders, and columns | 1½ |
| Concrete exposed to sea water | 4 |

## CONSTRUCTION ENGINEERING

The construction of civil engineering projects involves the scheduling and planning of personnel, materials, and equipment. Consideration is given to means of access to the project area, delivery schedules for materials, availability of skilled and unskilled labor, and construction procedures suited to the project schedule and climate.

### 6-53. Earthmoving

The volume and density of earth changes when it is excavated, hauled, and compacted as fill. Payment for earthmoving may be based on bank-measure volume (in a borrow pit or cut prior to excavation) or fill volume (after compaction). Loose measure volume (after excavation but before compaction) is used to determine haul payloads. The change from the bank-measure volume to the volume in fill after compaction is termed *shrinkage;* the change from the bank-measure volume to the loose volume before compaction is termed *swell.* Moisture content of soils directly affects shrinkage and swell allowances. Typical volume relationships for various materials are shown below:

| Typical soil type | Volume factors | | |
|---|---|---|---|
|  | Bank measure | Loose | Compacted fill |
| Sand | 1.00 | 1.10 | 0.95 |
| Loam | 1.00 | 1.25 | 0.90 |
| Clay | 1.00 | 1.45 | 0.90 |
| Rock (blasted) | 1.00 | 1.50 | 1.30 |

## 6-54. Pile Driving

There are three major types of pile hammers, classified by power source and action:

1. Single-acting hammers, raised by compressed air, steam, or other means, and dropped

2. Double-acting hammers, similar to single-acting but with power-assisted downward stroke

3. Vibratory drivers, transmitting a vibration to the pile through clamps

Variables governing the selection of a pile hammer include type and weight of pile, ground conditions, penetration requirements, and soil characteristics. For pile driving, soils can be separated into two groups: cohesive and noncohesive. Table 6-40 is a guide to pile hammer selection.

### Table 6-40. Guide to Selection of Pile Hammers

| Pile type | Cohesive soils | | Noncohesive soils | |
|---|---|---|---|---|
| | Soft | Stiff | Soft | Stiff |
| Wood............ | DA | SA | DA | SA |
| Open pipe......... | V-DA | DA-SA | V-DA | V-DA-SA |
| Closed pipe....... | DA-SA | SA | DA | SA |
| H-pile............ | V-DA | DA-SA | V-DA | V-DA-SA |
| Sheet............. | V-DA | DA-SA | V | V-DA |
| Concrete......... | DA-SA | SA | DA | SA |

V = vibratory, DA = double-acting, SA = single-acting.

## 6-55. Pavements

The two basic types of pavements are rigid and flexible. The principal rigid type is concrete with or without a bituminous wearing surface. Flexible types consist of a variable depth of bituminous material mixed with aggregate with or without a separate bituminous surface course. Both types require a base course of granular material. Soil-cement mixtures may be used to improve the bearing value of marginal or substandard granular base or subbase materials.

**Rigid Types.** Concrete slabs are designed to distribute wheel loads over large areas of subgrade. Typical concrete thicknesses for primary roads range from 8 to 12 in. Transverse joints are provided to relieve temperature stresses, and longitudinal joints are formed between lanes to control irregular cracking and mark traffic lanes. Pavement concrete is usually reinforced with wire mesh. A properly placed concrete pavement has a probable life of 25 years or more.

**Flexible Types** Types of bituminous construction most common for pavements are: surface treatment, penetration macadam, mixed-in-place, and plant mix. The usual thicknesses of bituminous surface courses and their probable life expectancies are:

| Type | Thickness, in. | Probable life, years |
|---|---|---|
| Surface treatment...................... | ½–1 | 5–8 |
| Penetration macadam.................. | 2½–3 | 15–17 |
| Mixed-in-place........................ | 1½–3 | 10–12 |
| Plant mix (asphaltic concrete)........... | 2–3 per course | 15–20 |

Bituminous paving materials are solutions of asphalt cement in a solvent (rapid-curing, RC, and medium-curing, MC) or straight-run distillations or blends (slow-curing, SC). Within these three grades, additional classifications are made on the basis of viscosity. The relative hardness of the asphalt cement varies from high for the RC grades, to medium for MC, to low for the SC types. The type of bituminous construction selected is influenced by thickness of the required surface course, by the climate, materials, and equipment available, and by the relative costs.

## ECONOMIC, SOCIAL, AND ENVIRONMENTAL CONSIDERATIONS

The scope of civil engineering includes many types of capital projects where the expenditure of funds must be justified by economic and financial analyses. Major civil engineering projects usually require analysis to determine their effect on community and regional social structures and on the environment.

### 6-56. Economic Analyses

Economic analyses compare the economic benefits of a project with its economic costs, while financial analyses compare the monetary return from a project with its financial costs. Economic analyses may be summarized in terms of a benefit cost ratio using the formula

$$B/C = \frac{\Delta t + \Delta u + \Delta a}{\Delta c + \Delta m} \qquad (6\text{-}153)$$

where $B/C$ = benefit cost ratio

$\Delta t$ = present worth of reduction in user time costs (computed on annual basis over the life of a project, discounted to present worth)

$\Delta u$ = present worth of reduction in user operating costs (computed on annual basis over the life of a project, discounted to present worth)

$\Delta a$ = present worth of reduction in user accident and damage costs (computed on annual basis over the life of a project, discounted to present worth)

$\Delta c$ = present worth of incremental costs of initial and recurring capital investments over the life of a project

$\Delta m$ = present worth of increase in maintenance and operating costs (computed on annual basis over the life of a project, discounted to present worth)

In some cases the benefits of an engineering project are determined as the net value of increased farm or industrial output attributable to the project. In Eq. (6-153) the interest cost to be used is the opportunity cost of capital or market rate of interest; the opportunity cost of capital generally ranges from 8 per cent to 15 per cent. The life of a project is determined from consideration of the inherent durability of project components but should not exceed the period in which technological obsolescence may occur. For most construction projects, the life will range up to 30 years, but longer lives may be used for major projects. The benefit cost ratio $(B/C)$ indicates the economic justification of a project. A project with a benefit cost ratio of one is marginal; values higher than one indicate greater justification.

An internal rate of return calculation is often used instead of a benefit cost ratio. The internal rate of return is determined by setting $B/C$ in Eq. (6-153) equal to unity and determining the interest rate used for discounting at which the numerator equals the denominator. The interest rate is then compared with the opportunity cost of capital.

Benefit cost or internal rate of return calculations should be made on an incremental basis for comparison among alternatives. One alternative which must be included is the "do nothing" alternative. Incremental analyses among alternatives permit the determination of the incremental return obtained from successively greater capital investments.

Economic analyses usually involve comparison of benefits which accrue to the public with costs which are incurred by governments. Financial analyses involve the same type of computation but consideration is limited to revenues and costs received by and incurred by a private or quasi-public organization.

## 6-57. Social and Environmental Factors

Civil engineering projects may have significant effects on the social structure of a community, a region, or a nation. Projects may affect population growth, quality of life, community relationships, and distribution of income. Analyses of project effects involve consideration of changes in demographic or cultural patterns to determine the positive or negative benefits of project development.

Civil engineering projects normally cause some environmental changes. Project planning should maximize positive environmental effects and minimize negative effects. Such effects may be short term, generally limited to the construction period, or long term, occurring through the life of the project.

For social and environmental analyses, the impact of a project normally involves consideration of the following:

1. Biological resources— wildlife, vegetation, fish, etc.
2. Air quality and pollution[1]
3. Noise control

[1] See Sec. 10.

  4. Water quality and flow[1]
    *a.* Ground water
    *b.* Streams and rivers
    *c.* Lakes, beaches and shores, and estuaries and other larger bodies of
        water
  5. Commitment of natural resources—types, approximate amount, and
probable source of needed construction materials
  6. Aesthetics
    *a.* Visual quality of the facility
    *b.* Special architectural or engineering features
  7. Open space and recreational opportunities
  8. Archaeological or paleontological resources, with consideration of sal-
vage operations during construction
  9. Historical areas or sites
  10. Future land uses of surrounding area, including neighborhood
compatibility
  11. Potential for joint development of public or private facilities
  12. Comprehensive plans for the development of the region or area
  13. Programs of public agencies
  14. Displacements of persons, businesses, and farms and proposed reloca-
tion treatment
  15. Residential and business property values
  16. Local area tax base
  17. Local churches, clubs, and other social activities
  18. Community services, such as police and fire protection, health services,
public and private utilities, mail
  19. Educational institutions, including modification of attendance boundaries or the actual school environment
  20. Area employment in terms of number and location of jobs
Consideration should be given to the interactions among these factors within
a complete system.
  Environmental and social considerations are weighed in conjunction with
economic and financial factors in the assessment of project justification and
selection.

  [1] See Sec. 10.

# ELECTRICAL ENGINEERING

**D. L. Whitehead, M.S.**; Manager, Engineering Laboratories, High Voltage Section, Westinghouse Electric Corporation; Fellow, Institute of Electrical and Electronics Engineers; Committee Member, National Electrical Manufacturers Association, ASA

## CONTENTS

## UNITS

Three distinct sets of units are used in the electrical engineering field: the centimeter-gram-second electrostatic units, cgs esu; the centimeter-gram-second electromagnetic units, cgs emu; and the meter-kilogram-second, mks or practical, units. The units of the International System (SI) are of the same magnitude as in the mks system, i.e., the conversion factors listed in Table 7-1A for mks units can be used for SI conversions. However, nomenclature and symbols for SI units vary and are given in Table 7-1B. From geometrical relationships the factor $4\pi$ inherently appears in some of the basic equations. If $4\pi$ is arbitrarily placed in the denominator of Coulomb's law expressions, the more commonly used formulas in engineering do not contain the $4\pi$ term. The three systems of units are then called *rationalized* in contrast to the *unrationalized* systems that carry the $4\pi$ term separately. The cgs esu system of units is commonly used when working electrostatic-field problems, the cgs emu units when working magnetic-field problems and related problems in physics, and the mks units when working practical-circuit problems. All of the systems have some units that are of

## Table 7-1A. Conversion Factors between MKS (Practical), CGS Electrostatic (ESU), and CGS Electromagnetic (EMU) Systems of Units

| Quantity | Symbol | Mks unit | Conversion factors | | Cgs (esu) unit | Conversion factors | | Cgs (emu) unit | Conversion factors | |
|---|---|---|---|---|---|---|---|---|---|---|
| | | | To cgs (esu) | To cgs (emu) | | To cgs (emu) | To mks | | To cgs (esu) | To mks |
| Acceleration | $a$ | Meter per second per second | $10^2$ | $10^2$ | Centimeter per second per second | $1$ | $10^{-2}$ | Centimeter per second per second | $1$ | $10^{-2}$ |
| Area | $A$ | Square meter | $10^4$ | $10^4$ | Square centimeter | $1$ | $10^{-4}$ | Square centimeter | $1$ | $10^{-4}$ |
| Capacitance | $C$ | Farad | $9 \times 10^{11}$ | $10^{-9}$ | Statfarad | $\frac{1}{9} \times 10^{-20}$ | $\frac{1}{9} \times 10^{-11}$ | Abfarad | $9 \times 10^{20}$ | $10^9$ |
| Charge | $Q$ | Coulomb | $3 \times 10^9$ | $10^{-1}$ | Statcoulomb | $\frac{1}{3} \times 10^{-10}$ | $\frac{1}{3} \times 10^{-9}$ | Abcoulomb | $3 \times 10^{10}$ | $10$ |
| Charge density: | | | | | | | | | | |
| Linear | $q$ | Coulomb per meter | $3 \times 10^7$ | $10^{-3}$ | Statcoulomb per centimeter | $\frac{1}{3} \times 10^{-10}$ | $\frac{1}{3} \times 10^{-7}$ | Abcoulomb per centimeter | $3 \times 10^{10}$ | $10^3$ |
| Area | $\sigma$ | Coulomb per square meter | $3 \times 10^5$ | $10^{-5}$ | Statcoulomb per square centimeter | $\frac{1}{3} \times 10^{-10}$ | $\frac{1}{3} \times 10^{-5}$ | Abcoulomb per square centimeter | $3 \times 10^{10}$ | $10^5$ |
| Volume | $\rho$ | Coulomb per cubic meter | $3 \times 10^3$ | $10^{-7}$ | Statcoulomb per cubic centimeter | $\frac{1}{3} \times 10^{-10}$ | $\frac{1}{3} \times 10^{-3}$ | Abcoulomb per cubic centimeter | $3 \times 10^{10}$ | $10^7$ |
| Conductance | $G$ | Mho | $9 \times 10^{11}$ | $10^{-9}$ | Statmho | $\frac{1}{9} \times 10^{-20}$ | $\frac{1}{9} \times 10^{-11}$ | Abmho | $9 \times 10^{20}$ | $10^9$ |
| Conductivity | $\gamma$ | Mho per meter | $9 \times 10^9$ | $10^{-11}$ | Statmho per centimeter | $\frac{1}{9} \times 10^{-20}$ | $\frac{1}{9} \times 10^{-9}$ | Abmho per centimeter | $9 \times 10^{20}$ | $10^{11}$ |
| Current | $I$ | Ampere | $3 \times 10^9$ | $10^{-1}$ | Statampere | $\frac{1}{3} \times 10^{-10}$ | $\frac{1}{3} \times 10^{-9}$ | Abampere | $3 \times 10^{10}$ | $10$ |
| Current density | $g$ | Ampere per square meter | $3 \times 10^5$ | $10^{-5}$ | Statampere per square centimeter | $\frac{1}{3} \times 10^{-10}$ | $\frac{1}{3} \times 10^{-5}$ | Abampere per square centimeter | $3 \times 10^{10}$ | $10^5$ |
| Elastance | $S$ | Daraf | $\frac{1}{9} \times 10^{-11}$ | $10^9$ | Statdaraf | $9 \times 10^{20}$ | $9 \times 10^{11}$ | Abdaraf | $\frac{1}{9} \times 10^{-20}$ | $10^{-9}$ |
| Electric intensity | $E$ | Volt per meter | $\frac{1}{3} \times 10^{-4}$ | $10^6$ | Statvolt per centimeter | $3 \times 10^{10}$ | $3 \times 10^4$ | Abvolt | $\frac{1}{3} \times 10^{-10}$ | $10^{-6}$ |
| Electrostatic flux | $X$ | .......... | $3 \times 10^9$ | $10^{-1}$ | .......... | $\frac{1}{3} \times 10^{-10}$ | $\frac{1}{3} \times 10^{-9}$ | .......... | $3 \times 10^{10}$ | $10$ |
| Electrostatic flux density | $D$ | .......... | $3 \times 10^5$ | $10^{-5}$ | .......... | $\frac{1}{3} \times 10^{-10}$ | $\frac{1}{3} \times 10^{-5}$ | .......... | $3 \times 10^{10}$ | $10^5$ |
| Energy | $W$ | Joule | $10^7$ | $10^7$ | Erg | $1$ | $10^{-7}$ | Erg | $1$ | $10^{-7}$ |
| Force | $F$ | Newton | $10^5$ | $10^5$ | Dyne | $1$ | $10^{-5}$ | Dyne | $1$ | $10^{-5}$ |
| Inductance | $L$ | Henry | $\frac{1}{9} \times 10^{-11}$ | $10^9$ | Stathenry | $9 \times 10^{20}$ | $9 \times 10^{11}$ | Abhenry | $\frac{1}{9} \times 10^{-20}$ | $10^{-9}$ |
| Length | $l$ | Meter | $10^2$ | $10^2$ | Centimeter | $1$ | $10^{-2}$ | Centimeter | $1$ | $10^{-2}$ |
| Magnetic flux | $\Phi$ | Weber | $\frac{1}{3} \times 10^{-2}$ | $10^8$ | .......... | $3 \times 10^{10}$ | $3 \times 10^2$ | Maxwell | $\frac{1}{3} \times 10^{-10}$ | $10^{-8}$ |
| Magnetic flux density | $B$ | Weber per square meter | $\frac{1}{3} \times 10^{-6}$ | $10^4$ | .......... | $3 \times 10^{10}$ | $3 \times 10^6$ | Gauss | $\frac{1}{3} \times 10^{-10}$ | $10^{-4}$ |
| Magnetic intensity | $H$ | Praoersted | $3 \times 10^7$ | $10^{-3}$ | .......... | $\frac{1}{3} \times 10^{-10}$ | $\frac{1}{3} \times 10^{-7}$ | Oersted | $3 \times 10^{10}$ | $10^3$ |
| Magnetic linkages | $\lambda$ | Weber-turn | $\frac{1}{3} \times 10^{-2}$ | $10^8$ | .......... | $3 \times 10^{10}$ | $3 \times 10^2$ | Maxwell-turn | $\frac{1}{3} \times 10^{-10}$ | $10^{-8}$ |
| Magnetomotive force | $F$ | Pragilbert | $3 \times 10^9$ | $10^{-1}$ | .......... | $\frac{1}{3} \times 10^{-10}$ | $\frac{1}{3} \times 10^{-9}$ | Gilbert | $3 \times 10^{10}$ | $10$ |

# Table 7-1A. Conversion Factors between MKS (Practical), CGS Electrostatic (ESU), and CGS Electromagnetic (EMU) Systems of Units (Continued)

| Quantity | Symbol | Mks unit | Conversion Factors To cgs (esu) | To cgs (emu) | Cgs (esu) unit | Conversion factors To cgs (emu) | To mks | Cgs (emu) unit | Conversion factors: To cgs (esu) | To mks |
|---|---|---|---|---|---|---|---|---|---|---|
| Mass | $m$ | Kilogram | $10^3$ | $10^3$ | Gram | $1$ | $10^{-3}$ | Gram | $1$ | $10^{-3}$ |
| Permeability | $\mu$ | | $\tfrac{1}{9} \times 10^{-13}$ | $10^7$ | | $9 \times 10^{20}$ | $9 \times 10^{13}$ | Gauss per oersted | $\tfrac{1}{9} \times 10^{-20}$ | $10^{-7}$ |
| Permeability of free space | $\mu_0$ | $10^{-7}$ | $\tfrac{1}{9} \times 10^{-13}$ | $10^7$ | | $9 \times 10^{20}$ | $9 \times 10^{13}$ | $1$ | $\tfrac{1}{9} \times 10^{-20}$ | $10^{-7}$ |
| Permeance | $P$ | Weber per pragilbert | $\tfrac{1}{9} \times 10^{-11}$ | $10^9$ | | $9 \times 10^{20}$ | $9 \times 10^{11}$ | Maxwell per gilbert | $\tfrac{1}{9} \times 10^{-20}$ | $10^{-9}$ |
| Permittivity | $\epsilon$ | Farads per meter | $9 \times 10^9$ | $10^{-11}$ | | $\tfrac{1}{9} \times 10^{-20}$ | $\tfrac{1}{9} \times 10^{-9}$ | | $9 \times 10^{20}$ | $10^{11}$ |
| Permittivity of free space | $\epsilon_0$ | $\tfrac{1}{9} \times 10^{-9}$ | $9 \times 10^9$ | $10^{-11}$ | $\tfrac{1}{9} \times 10^{-20}$ | $\tfrac{1}{9} \times 10^{-20}$ | $\tfrac{1}{9} \times 10^{-9}$ | $\tfrac{1}{9} \times 10^{-20}$ | $9 \times 10^{20}$ | $10^{11}$ |
| Potential difference | $v$ | Volt | $\tfrac{1}{3} \times 10^{-2}$ | $10^8$ | Statvolt | $3 \times 10^{10}$ | $3 \times 10^2$ | Abvolt | $\tfrac{1}{3} \times 10^{-10}$ | $10^{-8}$ |
| Power | $P$ | Watt | $10^7$ | $10^7$ | Erg per second | $1$ | $10^{-7}$ | Erg per second | $1$ | $10^{-7}$ |
| Reluctance | $\mathscr{R}$ | Pragilbert weber | $9 \times 10^{11}$ | $10^{-9}$ | | $\tfrac{1}{9} \times 10^{-20}$ | $\tfrac{1}{9} \times 10^{-11}$ | Gilbert per maxwell | $9 \times 10^{20}$ | $10^9$ |
| Reluctivity | $\nu$ | | $9 \times 10^{13}$ | $10^{-7}$ | | $\tfrac{1}{9} \times 10^{-20}$ | $\tfrac{1}{9} \times 10^{-13}$ | Oersted per gauss | $9 \times 10^{20}$ | $10^7$ |
| Resistance | $R$ | Ohm | $\tfrac{1}{9} \times 10^{-11}$ | $10^9$ | Statohm | $9 \times 10^{20}$ | $9 \times 10^{11}$ | Abohm | $\tfrac{1}{9} \times 10^{-20}$ | $10^{-9}$ |
| Resistivity | $\rho$ | Ohm-meter | $\tfrac{1}{9} \times 10^{-9}$ | $10^{11}$ | Statohm centimeter | $9 \times 10^{20}$ | $9 \times 10^9$ | Abohm centimeter | $\tfrac{1}{9} \times 10^{-20}$ | $10^{-11}$ |
| Time | $t$ | Second | $1$ | $1$ | Second | $1$ | $10^{-7}$ | Second | $1$ | $1$ |

## Table 7-1B. Nomenclature and Symbols for SI Units

| | | |
|---|---|---|
| Acceleration | $m/s^2$ | meter per second per second |
| Area | $m^2$ | square meter |
| Capacitance | F | farad |
| Charge | C | coulomb |
| Charge density: | | |
| Linear | C/m | coulomb per meter |
| Area | $C/m^2$ | coulomb per square meter |
| Volume | $C/m^3$ | coulomb per cubic meter |
| Conductance | S | siemens |
| Conductivity | S/m | siemens per meter |
| Current | A | ampere |
| Current density | $A/m^2$ | ampere per square meter |
| Electric intensity | V/m | volt per meter |
| Force | N | newton |
| Inductance | H | henry |
| Length | m | meter |
| Magnetic flux | Wb | weber |
| Magnetic flux density | T | tesla |
| Magnetomotive force | A | ampere (or ampere-turn) |
| Mass | kg | kilogram |
| Potential difference | V | volt |
| Power | W | watt |
| Resistance | Ω | ohm |
| Time | s | second |

SOURCE: "Letter Symbols for Units Used in Science and Technology," ANSI Y10.19-1969, New York, American National Standards Institute, 1969.

inconvenient size; however, the mks system has the advantage of using units of practical size in the expressions most commonly used in engineering. Table 7-1A gives the conversion factors between the three systems of units. the three systems of units.

## ELECTRIC-CIRCUIT THEORY

### 7-1. D-C Circuits

The flow of direct current in any circuit is determined only by the resistance of that circuit for a given applied direct voltage other than initial transients when the voltage is applied or changed. The resistance of a pure inductance is zero, and the resistance of a pure capacitance is infinite.

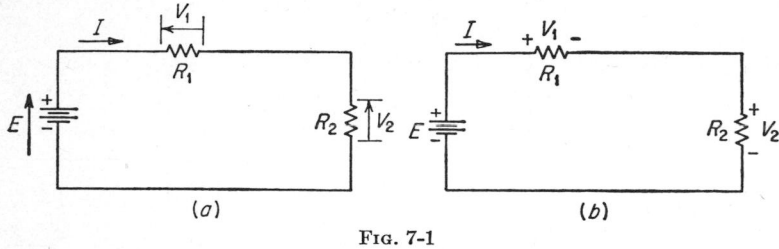

FIG. 7-1

Applied voltages are indicated by $E$, current by $I$, and resistance by $R$. In contrast to applied voltages, voltage drops are indicated by $V$. Arrow directions are assigned to the voltages and currents. Although arbitrary, one convenient system for designating voltage and current directions is the use of closed heads on the arrows for currents and open heads for voltage rises, with the head of the arrow the point of higher potential. Plus and minus symbols can also be used, the heads of arrows being $+$ and the tails $-$ (Fig. 7-1$a$ and $b$).

*Ohm's law* states that, for a steady current, the current through a given circuit is directly proportional to the total electromotive force in the circuit and inversely proportional to the total resistance in the circuit. Expressed in equation form,

$$I = E/R \quad \text{or} \quad E = IR \tag{7-1}$$

where $I$ is in amperes, $E$ is in volts, and $R$ is in ohms.

*Resistors in series* (Fig. 7-2) are added to obtain the total resistance

$$R = R_1 + R_2 + R_3 + \cdots \quad \text{ohms} \tag{7-2}$$

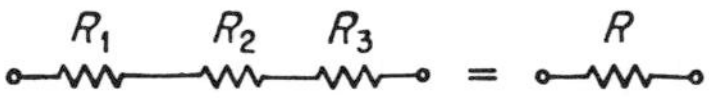

$$ R_1 \quad R_2 \quad R_3 \qquad R $$

FIG. 7-2

*Resistors in parallel* (Fig. 7-3) can be lumped to a single equivalent resistor

by
$$1/R = 1/R_1 + 1/R_2 + 1/R_3 + \cdots \qquad \text{ohms}^{-1} \qquad (7\text{-}3)$$
or
$$G = G_1 + G_2 + G_3 + \cdots \qquad \text{mhos} \qquad (7\text{-}3a)$$

where $G$ is the reciprocal of resistance, or conductance.

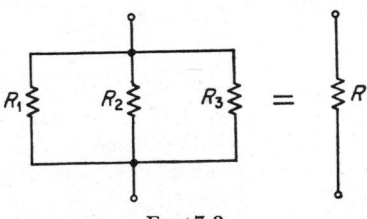

FIG. 7-3

*Two resistors in parallel* (Fig. 7-4) are equivalent to

$$R = R_1R_2/(R_1 + R_2) \qquad \text{ohms} \qquad (7\text{-}4)$$

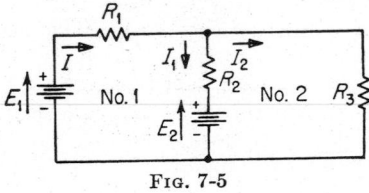

FIG. 7-4

*Kirchhoff's voltage law* states that the sum of the voltage drops around any closed loop in a circuit is equal to the sum of the emfs or driving voltages around the same loop (Fig. 7-5).

For loop 1
$$R_1I + R_2I_1 = E_1 - E_2 \qquad \text{volts}$$
and for loop 2
$$-R_2I_1 + R_3I_2 = E_2 \qquad \text{volts} \qquad (7\text{-}5)$$

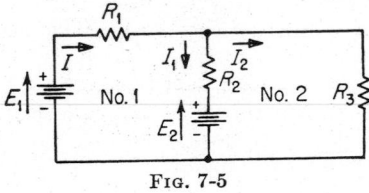

FIG. 7-5

*Kirchhoff's current law* states that the sum of all the currents flowing into or out of a junction is equal to zero. In Fig. 7-6

$$I - I_1 - I_2 = 0 \qquad \text{amp} \qquad (7\text{-}6)$$

Current $I$ flowing into two resistors in parallel (Fig. 7-7) divides in the proportion

$$I_1 = I R_2/(R_1 + R_2) \; (7) \qquad \text{and} \qquad I_2 = I R_1/(R_1 + R_2) \qquad (7\text{-}7)$$

where $I = E(R_1 + R_2)/R_1 R_2$ amp, $E$ is in volts, and $R$ is in ohms.

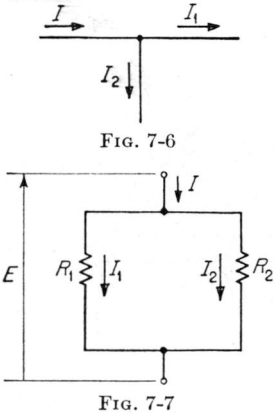

FIG. 7-6

FIG. 7-7

*Resistance R of a homogeneous material* of uniform cross section varies directly as its length and inversely as its cross section

$$R = \rho \, l/A \qquad \text{ohms} \qquad (7\text{-}8)$$

where $\rho$ = specific resistance of conductor
   $l$ = conductor length
   $A$ = conductor cross section

$l$ and $A$ must be in consistent units, feet and square feet, centimeters and square centimeters, etc.  $\rho$ must correspondingly be in ohms per cubic foot, per cubic centimeter, etc.  A convenient consistent set of units for calculating the resistance of wires is to express $\rho$ in ohms per circular mil–foot, $l$ in feet, and $A$ in circular mils.

*Resistivity $\rho$* at 20°C for 100 per cent conductivity copper and 61 per cent conductivity aluminum is as follows:

| $\rho$ | | Unit |
|---|---|---|
| Copper | Aluminum | |
| $1.724 \times 10^{-6}$ | $2.828 \times 10^{-6}$ | ohms/cm³ |
| $0.6788 \times 10^{-6}$ | $1.113 \times 10^{-6}$ | ohms/in.³ |
| $10.37$ | $17.01$ | ohms/(cir mil)(ft) |

A *circular mil* is the area of a circle whose diameter is 1 mil (0.001 in.).

To obtain the number of circular mils in a solid cylindrical wire, express the diameter in mils and then square it, or cir mils = $(1,000 \times$ in. diam$)^2$.

*Temperature change* causes change in the resistance of a conductor

$$R_t = R_0(1 + \alpha t) \qquad \text{ohms} \qquad (7\text{-}9)$$

where $R_t$ = resistance at desired temperature $t$, °C

$\quad R_0$ = resistance at 0°C

$\quad \alpha$ = temperature coefficient of resistance at 0°C

For copper $\alpha = 0.00427$, for aluminum $\alpha = 0.0039$ at 0°C. For most other pure metals $\alpha = 0.004$ approximately. Generally the resistance of a conductor will be known at one temperature and it is desired to find its resistance at some other temperature. The following general formula may be used:

$$R_{t2}/R_{t1} = (M + t_2)/(M + t_1) \qquad (7\text{-}10)$$

where $R_{t2}$ = d-c resistance at $t_2$°C, ohms

$\quad R_{t1}$ = d-c resistance at $t_1$°C, ohms

$\quad M$ = constant for type of conductor material: 234.5 for annealed 100 per cent conductivity copper, 241.5 for hard-drawn 97.3 per cent conductivity copper, 228.1 for aluminum

*Skin effect* will cause an increase of the resistance of a conductor if current other than direct is passed through it. For any alternating current of frequency $f$ the resistance can be expressed by

$$R_f = KR_{dc} \qquad \text{ohms} \qquad (7\text{-}11)$$

where $R_f$ = a-c resistance at desired frequency $f$ cps, ohms

$\quad R_{dc}$ = d-c resistance at any known temperature, ohms

$\quad K$ = a value taken from Table 7-2

In Table 7-2 $K$ is shown as a function of $X$ and

$$X = 0.063598 \sqrt{\mu f/R_{dc}} \qquad (7\text{-}12)$$

where $\mu$ = permeability of conductor (1.0 for nonmagnetic materials)

$\quad f$ = frequency of alternating current, cps

## Table 7-2. Skin-effect Table

| $X$ | $K$ | $X$ | $K$ | $X$ | $K$ | $X$ | $K$ | $X$ | $K$ | $X$ | $K$ | $X$ | $K$ | $X$ | $K$ |
|---|---|---|---|---|---|---|---|---|---|---|---|---|---|---|---|
| 0.0 | 1.00000 | 0.5 | 1.00032 | 1.0 | 1.00519 | 1.5 | 1.02582 | 2.0 | 1.07816 | 2.5 | 1.17538 | 3.0 | 1.31809 | 3.5 | 1.49202 |
| 0.1 | 1.00000 | 0.6 | 1.00067 | 1.1 | 1.00758 | 1.6 | 1.03323 | 2.1 | 1.09375 | 2.6 | 1.20056 | 3.1 | 1.35102 | 3.6 | 1.52879 |
| 0.2 | 1.00001 | 0.7 | 1.00124 | 1.2 | 1.01071 | 1.7 | 1.04205 | 2.2 | 1.11126 | 2.7 | 1.22753 | 3.2 | 1.38504 | 3.7 | 1.56587 |
| 0.3 | 1.00004 | 0.8 | 1.00212 | 1.3 | 1.01470 | 1.8 | 1.05240 | 2.3 | 1.13069 | 2.8 | 1.25620 | 3.3 | 1.41999 | 3.8 | 1.60314 |
| 0.4 | 1.00013 | 0.9 | 1.00340 | 1.4 | 1.01969 | 1.9 | 1.06440 | 2.4 | 1.15207 | 2.9 | 1.28644 | 3.4 | 1.45570 | 3.9 | 1.64051 |

*Network reduction* can be accomplished in d-c circuits by combining in series and in parallel various branches. When delta or star branches are present (Fig. 7-8), it is necessary to be able to convert from delta to star and from star to delta. The following equations are used, with all resistances in ohms.

From delta to star:

$$R_a = \frac{R_{ab}R_{ca}}{R_{ab} + R_{bc} + R_{ca}} \qquad R_b = \frac{R_{ab}R_{bc}}{R_{ab} + R_{bc} + R_{ca}}$$

$$R_c = \frac{R_{bc}R_{ca}}{R_{ab} + R_{bc} + R_{ca}} \qquad (7\text{-}13)$$

From star to delta:

$$R_{ab} = \frac{R_aR_b + R_bR_c + R_cR_a}{R_c} \qquad R_{bc} = \frac{R_aR_b + R_bR_c + R_cR_a}{R_a}$$

$$R_{ca} = \frac{R_aR_b + R_bR_c + R_cR_a}{R_b} \qquad (7\text{-}14)$$

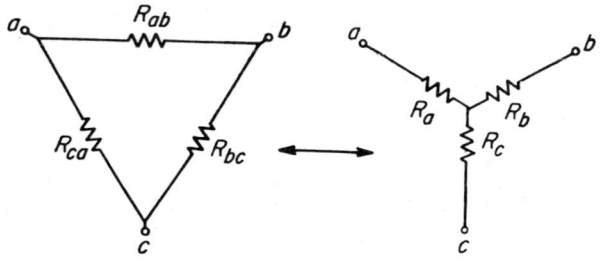

Fig. 7-8

*Network theorems* used in d-c circuits are similar to those used in a-c circuits (Sec. 7-2) with resistance substituted for impedance. Circuits are usually solved by reduction, series and parallel combinations, star-delta and delta-star transformations, or by mesh current calculations (Sec. 7-2). If there is more than one voltage source and the solution is to be made by reduction, it is desirable to use the *superposition theorem*, which states: In any network consisting of voltage sources and linear resistances (or impedances for a-c circuits) the current flowing at any point is the sum of the currents which would flow if each voltage source were considered separately, all other sources being replaced at the time by resistances (impedances) equal to their internal resistances (impedances). For example (Fig. 7-9),

$$I_1 = I'_1 + I''_1 \qquad I_2 = I'_2 + I''_2 \qquad I_3 = I'_3 + I''_3 \qquad \text{amp} \quad (7\text{-}15)$$

It is assumed that the internal resistances of $E_1$ and $E_2$ have been combined with $R_1$ and $R_2$. The theorem is valid for any number of sources.

The *d-c power through a conductor* is given by

$$P = EI \quad \text{watts} \qquad (7\text{-}16)$$

where $E$ = voltage across terminals of circuit, volts

$I$ = current through conductor, amp

The *energy delivered to a conductor* carrying direct current during a time $T$ is

$$W = PT = EIT \quad \text{watt-sec or joules} \quad (7\text{-}17)$$

where $P$ is expressed in watts, $E$ in volts, $I$ in amperes, and $T$ in seconds. If $E$ is expressed in kilovolts, $I$ in amperes, and $T$ in hours, the power is expressed in kilowatts and the energy in kilowatthours.

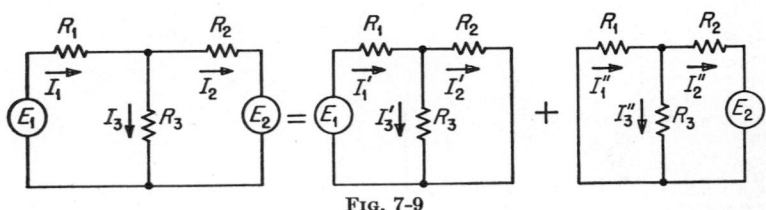

Fig. 7-9

## 7-2. A-C Circuits

*Kirchhoff's and Ohm's laws* apply to a-c as well as d-c circuits. Impedance must be used instead of resistance in the defining equations, and the numerical work is accomplished by complex numbers. Voltages and currents are represented by planars (vectors lying in one plane) which can be plotted on a single plane, commonly called the complex plane, with real portions of the complex number plotted along the abscissa and the $j$ or imaginary numbers plotted as ordinates. Root mean square (rms) values of current and voltage are plotted as the magnitude of the planars; the planar positions are located by the real and imaginary components. The origin of *imaginary numbers* stems from equations of the form $X^2 + K^2 = 0$. The solution takes the form

$$X = \pm \sqrt{-K^2} = \pm K \sqrt{-1}$$
$$= \pm jK$$

where $j = \sqrt{-1}$ and is called an imaginary number. Some of the unique properties of $j$ are

$$j = \sqrt{-1} \quad j^3 = -j \quad j^5 = j$$
$$j^2 = -1 \quad j^4 = +1 \quad j^6 = -1$$

If we consider an equation of the form $X^2 - 2X + 10 = 0$, we get

$$X = (2 \pm \sqrt{4 - 40})/2 = 1 \pm j3 \quad \text{or} \quad X = a + jb$$

which is known as a *complex number* and can be plotted on the complex plane, $a$ units along the real axis and $b$ units along the imaginary axis, as shown in

**Fig. 7-10.** A complex number can be represented in four common forms:

| | | |
|---|---|---|
| Orthogonal: | $A = a + jb$ | (7-18) |
| Vectorial: | $A = \bar{A}\,\lfloor\underline{\theta}$ | (7-19) |
| where | $\bar{A} = \sqrt{a^2 + b^2}$ | |
| | $\theta = \tan^{-1}(b/a)$ | |
| Trigonometric: | $A = \bar{A}(\cos\theta + j\sin\theta)$ | (7-20) |
| Exponential: | $A = \bar{A}\epsilon^{j\theta}$ | (7-21) |

Alternating voltages and currents, being sinusoidal in nature, can be represented in any of the four forms. For example, 100 volts is applied to

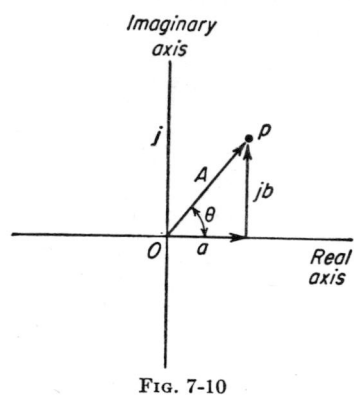

FIG. 7-10

a circuit which takes a current at 80 per cent power factor ($\theta = \cos^{-1} 0.80$). The voltage can be represented in each of the four forms, using current as a reference.

1. $a + jb = 100(\cos\theta + j\sin\theta) = 80.0 + j\,60.0$ volts
2. $\bar{A}\,\lfloor\underline{\theta} = 100\lfloor\underline{36.9°}$ volts
3. $\bar{A}(\cos\theta + j\sin\theta) = 100(\cos 36.9° + j\sin 36.9°)$ volts
4. $\bar{A}\epsilon^{j\theta} = 100\epsilon^{j36.9°}$

Addition and subtraction can best be done in the orthogonal form. Multiplication, division, powers, and roots can best be done in the exponential form or vector form.

Orthogonal addition and subtraction:

$$(a + jb) + (c + jd) = (a + c) + j(b + d) \tag{7-22}$$
$$(a + jb) - (c + jd) = (a - c) + j(b - d) \tag{7-23}$$

Exponential multiplication, division, powers and roots:

$$\bar{A}\epsilon^{j\theta_1} \cdot \bar{B}\epsilon^{j\theta_2} = \overline{AB}\epsilon^{j(\theta_1 + \theta_2)} = \overline{AB}\,\lfloor\underline{\theta_1 + \theta_2} \tag{7-24}$$

$$\frac{\bar{A}\epsilon^{j\theta_1}}{\bar{B}\epsilon^{j\theta_2}} = \frac{\bar{A}}{\bar{B}}\,\epsilon^{j(\theta_1-\theta_2)} = \frac{\bar{A}}{\bar{B}}\,\underline{|\theta_1 - \theta_2} \qquad (7\text{-}25)$$

$$(\bar{A}\epsilon^{j\theta})^n = (\bar{A})^n\epsilon^{jn\theta} = \bar{A}^n\underline{|n\theta} \qquad (7\text{-}26)$$

$$\sqrt[n]{\bar{A}\epsilon^{j\theta}} = \sqrt[n]{\bar{A}}\,\epsilon^{j\theta/n} = \sqrt[n]{\bar{A}}\,\underline{|\theta/n} \qquad (7\text{-}27)$$

**Circuit Parameters.** A-c circuits are made up of resistance, self-inductance, mutual inductance, and capacitance. The units of impedance in ohms are $R$, $+j2\pi fL$, $\pm j2\pi fM$, and $-j/2\pi fC$, where $R$ is the resistance in ohms, $L$ is the self-inductance in henrys, $M$ is the mutual inductance in henrys, $C$ is the capacitance in farads, and $f$ is the frequency in cycles per second. A circuit may contain any or all of the above parameters with a total impedance of

$$\begin{aligned} Z &= R + j2\pi fL \pm j2\pi fM - j/2\pi fC \qquad \text{ohms} \\ &= R + jX_L \pm jX_M - jX_c \qquad \text{ohms} \end{aligned} \qquad (7\text{-}28)$$

*Impedances* can be added directly in series. For parallel circuits the reciprocals of the impedance elements can be added to give the total *admittance*

$$Y = 1/Z = g - jb \qquad \text{mhos} \qquad (7\text{-}29)$$

where $g$ is the *conductance* and $b$ is the *susceptance* in mho units.

The *sign of the mutual M* is determined by the direction of the mutual flux relative to the direction of the flux of the self-inductance of the circuit into

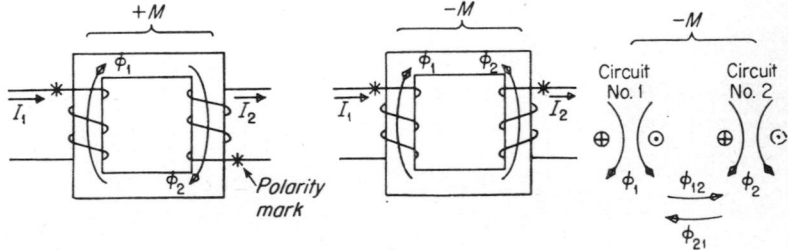

FIG. 7-11

which the mutual is being considered. If the mutual flux *adds* to the flux of self-inductance, the sign of the mutual is *plus*. If the mutual flux is in *opposition* to or *subtracts* from the self-inductance flux, the sign of the mutual is *minus*. The *right-hand rule* is useful in determining the direction of self and mutual fluxes. Grasp the conductor in the right hand with the thumb in the direction of the current and the fingers will surround the conductor in the direction of the flux (Fig. 7-11).

The *impedance of a circuit* is expressed in ohms, per cent, or per unit (p.u.). When working in ohms, all the impedance values must be on the same *voltage* basis. When working in per cent or per unit, all impedance values must be on the same kva base.   To convert from per cent to ohms use

$$\text{Ohms} = \frac{10 \times \text{per cent} \times \text{kv}^2{}_{L-L}}{\text{kva}} \qquad \text{per phase} \qquad (7\text{-}30)$$

$$\text{Per cent} = \frac{\text{ohms} \times \text{kva}}{10 \times \text{kv}^2{}_{L-L}} \qquad \text{per phase} \qquad (7\text{-}31)$$

$$\text{Per unit (pu)} = \text{per cent}/100 \qquad\qquad\qquad (7\text{-}32)$$

Per cent is expressed in numerical form, i.e., one hundred per cent = 100 per cent = 1.00 per unit.  $\text{kv}_{L-L}$ is line-to-line voltage in kilovolts and kva is the three-phase kilovolt-ampere base.   The equations are equally valid if line-to-neutral voltage in kilovolts and single-phase kilovolt amperage is used.

*Transformer turns ratio* are squared and multiplied by the ohms when transferring ohms from one side of a transformer to the other.   Per cent impedance is moved from one side of a transformer to the other unchanged (Fig. 7-12).   Unequal transformer turns ratio in circuits involving step-up and step-down transformers can be readily solved by first converting to ohms on a common kv base as viewed from the point of greatest interest in the circuit and then converting to per cent (Fig. 7-13).   The per-phase impedance diagram in ohms on $n$ kv base as viewed from $F$ is shown in Fig. 7-14, the per-phase diagram in per unit in Fig. 7-15.

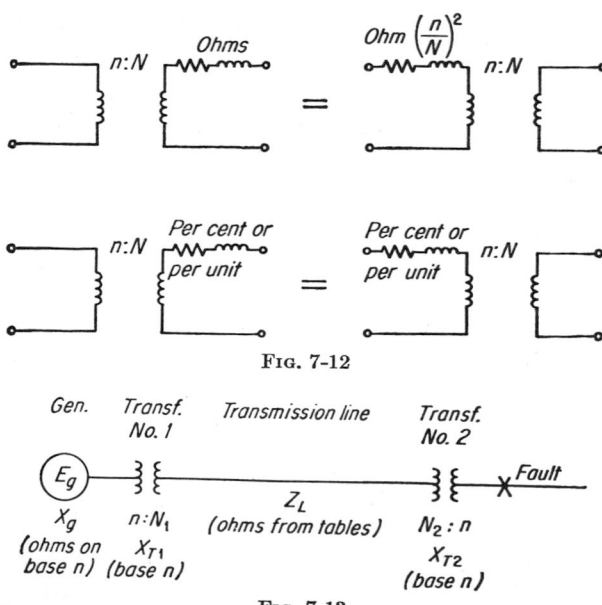

Fig. 7-12

Fig. 7-13

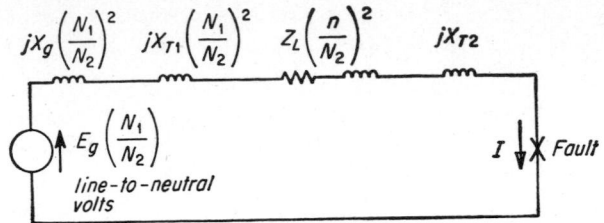

Fig. 7-14

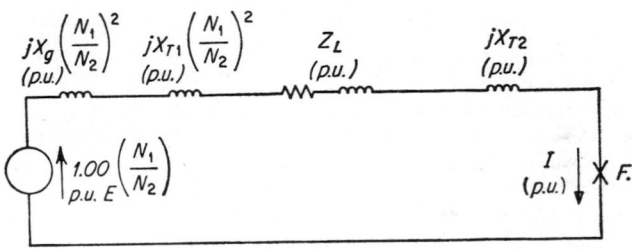

Fig. 7-15

*Per unit current* $I_{pu}$ is given by

$$I_{pu} = E_{pu}/Z_{pu} \qquad \text{per unit} \qquad (7\text{-}33)$$

where $Z_{pu} = jX_{g\ pu}(N_1/N_2)^2 + jX_{T1\ pu}(N_1/N_2)^2 + Z_{L\ pu} + jX_{T2\ pu}$

per unit, ohms

The individual per unit reactances and impedances must be on the same arbitrary kva base.

*Normal current* $I_n$ is defined as

$$I_n = \frac{\text{three-phase base kva}}{\sqrt{3}\ \text{line-to-line kv}} \qquad \text{amp} \qquad (7\text{-}34)$$

$$I = I_{pu} \times I_n \qquad \text{amp} \qquad (7\text{-}35)$$

*Name-plate data* for heavy generators and transmission equipment give nominal kv, kva, and per cent impedance based on nominal ratings from which equivalent ohms at any desired kv base can be obtained by using Eq. (7-30).

*Capacitors* are usually built with an average $+5$ per cent kva tolerance, and the equivalent ohms can be found from

$$X_c = 1{,}000\ \text{kv}^2/1.05\ \text{kva} \qquad \text{ohms} \qquad (7\text{-}36)$$

where kv = rated voltage of capacitor, line-to-line for three-phase kva and line-to-neutral for one-phase kva

*Shunt reactors* have 100 per cent reactance, and the equivalent ohms are given by

$$X_L = 1,000 \text{ kv}^2/\text{kva} \qquad \text{ohms} \qquad (7\text{-}37)$$

*Series reactors* have a name-plate per cent reactance based on the through-circuit kva. The kva of reactor parts is also given. For example, a 5 per cent, 1,000-kva, series reactor will have 5 per cent reactance based on 20,000 kva circuit kva (1,000 kva is 5 per cent of 20,000 kva).

$$X_{L \text{ series}} = (10 \times \text{per cent} \times \text{kv}^2{}_{L-L})/\text{circuit kva} \qquad \text{ohms} \qquad (7\text{-}38)$$

Loads can be converted to equivalent ohms by

$$Z = 1,000 \text{ kv}^2/(P - jQ) = [1,000 \text{ kv}^2/(P^2 + Q)](P + jQ) \qquad \text{ohms} \qquad (7\text{-}39)$$

where kv = the line-to-line voltages, kv
    $P$ = three-phase kw = kva $\times$ cos $\theta$
    $Q$ = reactive kva lagging (three-phase) kvar
       = kva $\times$ sin $\theta$
    $\theta$ = power-factor angle

**Network Solutions.** The solution of electrical circuits is based on the application of Ohm's and Kirchhoff's laws. For circuits of any complexity the solution can be systematized by the use of mesh currents, voltages, and impedances. The following steps are suggested in setting up any given circuit:

1. Label individual impedances in any consistent manner.
2. Set up arrow direction of voltages (establish direction of voltage rise).
3. Number the meshes and set up arrow directions for the branch and mesh currents.

*Mesh impedances* are defined in general as: $Z_{pq}$ is the voltage drop in volts in the reference direction in mesh $q$ due to 1 amp of current in the reference direction in mesh $p$. Curved current arrows show the reference direction in each mesh. In linear circuits the current is directly proportional to the voltage, and in bilateral circuits the impedance is independent of the direction of current flow. In such circuits $Z_{pq} = Z_{qp}$.

*Self-impedance* is defined as $Z_{pp}$, which is equal to the sum of all the individual impedances around mesh $p$, and $p$ is any one of the meshes in a given circuit. For example in the three-mesh circuit of Fig. 7-16

$$Z_{11} = Z_a + Z_c + Z_d$$
$$Z_{22} = Z_b + Z_c + Z_e$$
$$Z_{33} = Z_d + Z_e + Z_f$$

*Mutual impedance* is defined as $Z_{pq}$, which is equal to the sum of the individual impedances that are common to meshes $p$ and $q$ where $p$ and $q$ are any two meshes of a given circuit. The signs of the mutuals are determined by the arbitrary current direction; when mesh currents are in the same direction in a common branch, the mutual is $+$, when in opposite directions, the sign is $-$. In Fig. 7-16 the mutuals are given by

$$Z_{12} = -Z_c \qquad Z_{13} = -Z_d \qquad Z_{23} = -Z_e$$

*Mesh emf* is defined as the sum of the voltages around a given mesh in the reference direction:

$$E_{11} = E_a - E_b \qquad E_{22} = E_b - E_c \qquad E_{33} = E_d$$

*Mesh-voltage equations* are obtained by applying Kirchhoff's voltage law around the various meshes:

$$Z_{11}I_1 + Z_{21}I_2 + Z_{31}I_3 = E_{11}$$
$$Z_{12}I_1 + Z_{22}I_2 + Z_{32}I_3 = E_{22}$$
$$Z_{13}I_1 + Z_{23}I_2 + Z_{33}I_3 = E_{33}$$

These three independent equations can be solved in any desired manner for

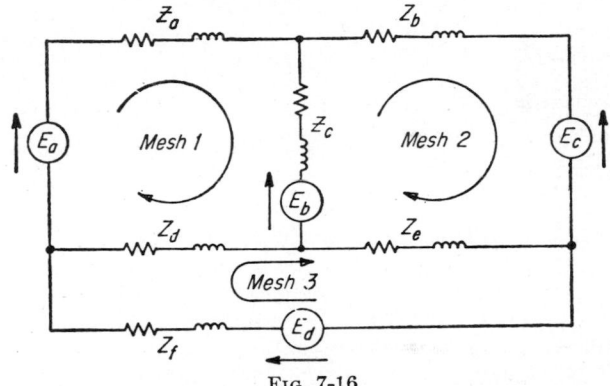

Fig. 7-16

the currents; one of the best methods when two or more meshes are involved is by means of determinants. For example

$$I_1 = E_{11}A_{11}/D + E_{22}A_{12}/D + E_{33}A_{13}/D \qquad \text{amp} \qquad (7\text{-}40)$$

where

$$D = \begin{vmatrix} Z_{11} & Z_{21} & Z_{31} \\ Z_{12} & Z_{22} & Z_{32} \\ Z_{13} & Z_{23} & Z_{33} \end{vmatrix} = Z_{11}Z_{22}Z_{33} - Z_{23}{}^2 Z_{11} + 2Z_{12}Z_{23}Z_{13} - Z_{13}{}^2 Z_{22}$$

and

$$A_{11} = + \begin{vmatrix} Z_{22} & Z_{32} \\ Z_{23} & Z_{33} \end{vmatrix} = Z_{22}Z_{33} - Z_{23}{}^2$$

$$A_{12} = - \begin{vmatrix} Z_{21} & Z_{31} \\ Z_{23} & Z_{33} \end{vmatrix} = - Z_{21}Z_{33} + Z_{31}Z_{23}$$

$$A_{13} = + \begin{vmatrix} Z_{21} & Z_{31} \\ Z_{22} & Z_{32} \end{vmatrix} = Z_{21}Z_{32} - Z_{31}Z_{22}$$

The *general solution for the current in any mesh p* of a network consisting of $n$ meshes is given by

$$I_p = E_{11}A_{p1}/D + E_{22}A_{p2}/D + \cdots + E_{nn}A_{pn}/D \quad \text{amp} \quad (7\text{-}41)$$

where $D$ is given by

$$D = \begin{vmatrix} Z_{11} & Z_{21} & \cdots & Z_{n1} \\ Z_{12} & Z_{22} & \cdots & Z_{n2} \\ \cdots & \cdots & \cdots & \cdots \\ Z_{1n} & Z_{2n} & \cdots & Z_{nn} \end{vmatrix}$$

and the $A$'s are the cofactors of $D$ formed from $D$ by drawing intersecting lines through the row and column intersecting on the element whose cofactor is being found. For example, in a three-mesh circuit

$$D = \begin{vmatrix} Z_{11} & Z_{21} & Z_{31} \\ Z_{12} & Z_{22} & Z_{32} \\ Z_{13} & Z_{23} & Z_{33} \end{vmatrix}$$

From which cofactor $A_{11}$ corresponding to element $Z_{11}$ is seen to be

$$A_{11} = \begin{vmatrix} Z_{22} & Z_{32} \\ Z_{23} & Z_{33} \end{vmatrix}$$

The *sign of the cofactor* is determined from its position in the original determinant. If the sum of the row and the column is even, the sign is $+$, if odd, the sign is $-$. The sign can also be determined by multiplying the cofactor by $(-1)^{r+c}$, where $r$ is the number of the row and $c$ the number of the column from which the cofactor is being formed. For example, the second row and first column would give $(-1)^{2+1} = -1$.

*Driving-point impedance* is the ratio of the voltage applied to two terminals of a network to the current that flows through those two terminals with all other voltage sources in the network replaced by their internal impedance. The driving-point admittance is the reciprocal of the driving-point impedance. The driving point-impedance, $Z'_{nn}$ and the driving-point admittance $Y'_{nn}$ can be readily found by using determinants

$$Z'_{nn} = D/A_{nn} \text{ ohms} \quad \text{and} \quad Y'_{nn} = A_{nn}/D \quad \text{mhos} \quad (7\text{-}42)$$

where $n$ denotes the driving-point mesh.

*Transfer impedance* $Z'_{pq}$ is the ratio of the voltage in one mesh, say mesh $p$, to the current in a second mesh $q$.

$$Z'_{pq} = D/A_{pq} \quad \text{ohms} \quad (7\text{-}43)$$

The transfer admittance is given by

$$Y'_{pq} = A_{pq}/D \quad \text{mhos} \quad (7\text{-}44)$$

*Solution of a-c networks by reduction* is accomplished by taking a single source of voltage and then reducing the network to a single impedance by means of series and parallel combinations as shown for direct current, Sec.

**7-1**, with impedance substituted for resistance. Star-delta impedance conversions are made as shown on page 7-8 and in Fig. 7-8. Other voltage sources are similarly treated, and the final currents are obtained by superposing or adding the results of the individual solutions to obtain the total solution. Some star-connected impedances have mutuals between star branches which must be eliminated before further reduction can be made.

A *star circuit with mutuals* can be converted to a star without mutuals, as shown in Fig. 7-17, where

$$Z_A = Z_a + Z_{bc} - Z_{ab} - Z_{ca} \quad \text{ohms}$$
$$Z_B = Z_b + Z_{ca} - Z_{bc} - Z_{ab} \quad \text{ohms}$$
$$Z_C = Z_c + Z_{ab} - Z_{ca} - Z_{bc} \quad \text{ohms}$$

*Polarity marks* as shown require that, with all reference directions from the center outward as shown, all self and mutual drops are from center out-

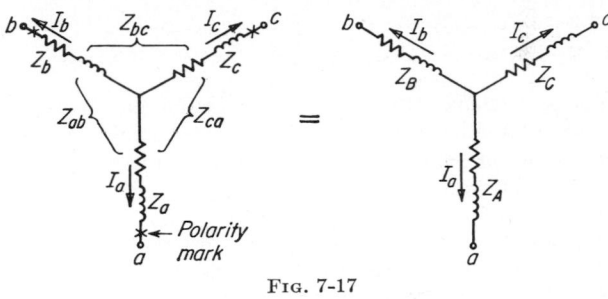

Fɪɢ. 7-17

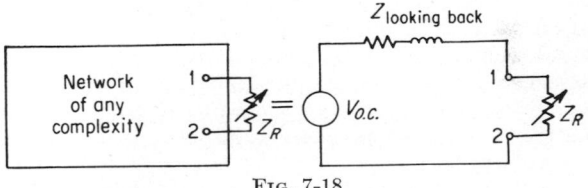

Fɪɢ. 7-18

ward. Polarity marks on a two-winding transformer indicate a $-M$ mutual for current into one polarity mark and out the other as shown in Fig. 7-11.

**Network Theorems.** *Thévenin's Theorem.* The current in any impedance (including short-circuit) connected to any two terminals of a network is the same as if that impedance were connected to a simple generator whose generated voltage is the open-circuit voltage of the terminals in question, and whose impedance is the impedance of the network looking back from the terminals with all generators replaced by impedances equal to their internal impedance. See Fig. 7-18, where

$V_{o-c}$ = open-circuit voltage at terminals 1-2

$Z_{l.b.}$ = impedance looking back from terminals 1-2

*Norton's Theorem.* The current in any impedance (including short-circuit) connected to any two terminals of a network is the same as if the impedance were connected to a constant-current generator whose current is equal to the current which flows through the two terminals when these terminals are short-circuited, the constant-current generator being in shunt with an impedance equal to the impedance looking back into the network from the terminals in question. See Fig. 7-19, where

$I_{s-c}$ = short circuit current at terminals 1-2

$Z_{l.b.}$ = looking-back impedance from terminals 1-2

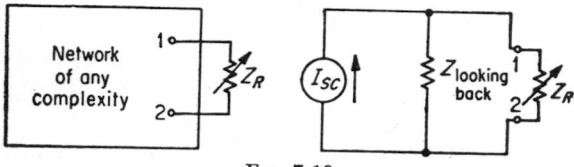

FIG. 7-19

*Maximum-power-transfer Theorem.* Maximum power will be absorbed by one network from another connected to it at two terminals if the impedances looking into the two networks from the junction are conjugates of each other $(a + jb$ and $a - jb)$.

*Reciprocal Theorem.* If a voltage $E$ is applied between any two terminals of a passive linear network and the current $I$ is measured in any branch, the ratio $E/I$ will be unchanged if the location of the voltage source and measuring point of the current are interchanged.

**Equivalent Circuits.** Equivalent circuits at all frequencies can be found for some of the simple configurations. These circuits are equivalent at the input terminals only and are enclosed in a dotted box to indicate terminal equivalence only. They are useful in designing circuits such as filters and computer elements to have a given response with a choice of elements used to make up the circuit. All resistance is in ohms, inductance in henrys, and capacitance in farads.

*Resistance and Inductance Circuits.* The circuits of Fig. 7-20 are equivalent when

$$R_1 = R_B{}^2/(R_A + R_B) \qquad\qquad R_B = R_1 + R_2$$
$$L_1 = L_A/(1 + R_A/R_B)^2 \qquad\quad R_A = R_2 + R_2{}^2/R_1$$
$$R_2 = R_A R_B/(R_A + R_B) \qquad\quad L_A = L_1(1 + R_2/R_1)^2$$

The circuits of Figs. 7-21 are equivalent when

$$L_1 = L_B{}^2/(L_A + L_B) \qquad\qquad L_A = (L_2/L_1)(L_1 + L_2)$$
$$R_1 = R_A/(1 + L_A/L_B)^2 \qquad\quad L_B = L_1 + L_2$$
$$L_2 = L_A L_B/(L_A + L_B) \qquad\qquad R_A = R_1(1 + L_2/L_1)^2$$

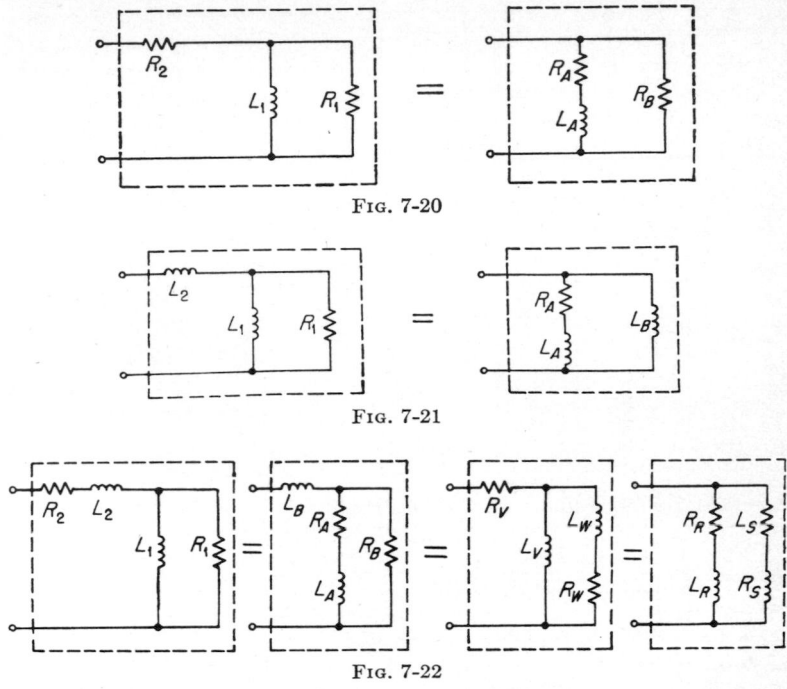

FIG. 7-20

FIG. 7-21

FIG. 7-22

All four of the circuits of Fig. 7-22 are equivalent when

$$R_1 = \frac{R_B{}^2}{R_A + R_B} = \frac{R_w}{(1 + L_w/L_v)^2} = \frac{(L_R R_s - L_s R_R)^2}{(L_R + L_s)^2 (R_R + R_s)}$$

$$R_2 = R_A R_B/(R_A + R_B) = R_v = R_R R_s/(R_R + R_s)$$

$$L_2 = L_B = L_v L_w/(L_v + L_w) = L_R L_s/(L_R + L_s)$$

$$L_1 = L_A/(1 + R_A/R_B)^2 = L_v{}^2/(L_v + L_w)$$
$$= (L_R R_s - L_s R_R)^2/(R_R + R_s)^2 (L_R + L_s)$$

$$R_A = R_2 + R_2{}^2/R_1 \qquad\qquad R_B = R_1 + R_2$$

$$L_B = L_2 \qquad\qquad L_A = L_1(1 + R_2/R_1)^2$$

$$R_v = R_2 \qquad\qquad R_w = R_1(1 + L_2/L_1)^2$$

$$L_v = L_1 + L_2 \qquad\qquad L_w = \frac{L_2}{L_1}(L_1 + L_2)$$

$$L_s = (R_2{}^2/2L_1)[K + \sqrt{K^2 - 4L_1 L_2 K/R_2{}^2}]$$

$$K = (L_1/R_1 + L_1/R_2 + L_2/R_2)^2 - 4L_1 L_2/R_1 R_2$$

$$L_R = L_s L_2/(L_s - L_2) \qquad R_s = \frac{(L_s - L_R)}{(L_1/R_1 + L_1/R_2 + L_2/R_2 - L_R/R_2)}$$

$$R_R = R_2 R_s/(R_s - R_2)$$

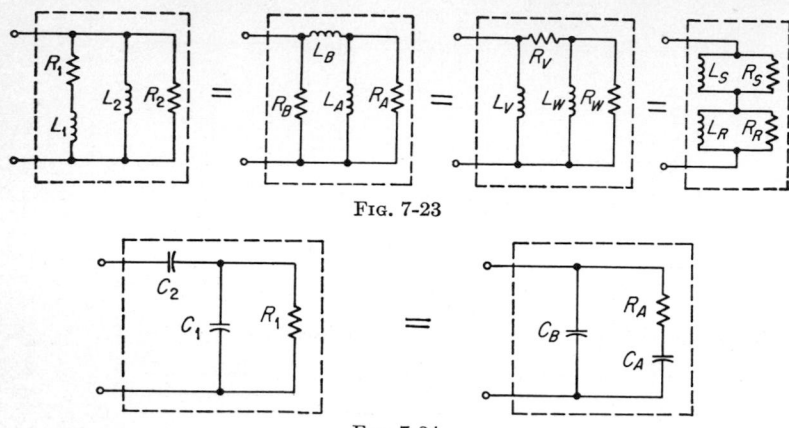

FIG. 7-23

FIG. 7-24

The four-element circuits of Fig. 7-23 are equivalent when

$$L_1 = (L_B/L_A)(L_A + L_B) = L_w(1 + R_v/R_w)^2$$
$$= L_A L_s(L_R + L_s)(R_R + R_s)^2/(L_R R_s - L_s R_R)^2$$
$$L_2 = L_A + L_B = L_v = L_R + L_s$$
$$R_1 = R_A(1 + L_B/L_A)^2 = R_v + R_v^2/R_w$$
$$= R_R R_s(R_R + R_s)(L_R + L_s)^2/(L_R R_s - L_s R_R)^2$$
$$R_2 = R_B = R_v + R_w = R_R + R_s$$

$$L_A = L_2^2/(L_1 + L_2) \qquad\qquad L_B = L_1 L_2/(L_1 + L_2)$$
$$R_A = R_1/(1 + L_1/L_2)^2 \qquad\qquad R_B = R_2$$
$$L_v = L_2 \qquad\qquad L_w = L_1/(1 + R_1/R_2)^2$$
$$R_v = R_1 R_2/(R_1 + R_2) \qquad\qquad R_w = R_2^2/(R_1 + R_2)$$
$$R_s = \frac{2L_2^2}{R_1} \frac{1}{K + \sqrt{K^2 - 4L_2^2 K/R_1 R_2}}$$
$$K = (L_1/R_1 + L_2/R_1 + L_2/R_2)^2 - 4L_1 L_2/R_1 R_2$$
$$R_R = R_2 - R_s$$
$$L_s = (L_1/R_1 + L_2/R_1 + L_2/R_2 - L_2/R_R) R_R R_s/(R_R - R_s)$$
$$L_R = L_2 - L_s$$

*Equivalent Resistance and Capacitance Circuits.* The circuits of Fig. 7-24 are equivalent when

$$C_2 = C_A + C_B \qquad\qquad C_B = C_1 C_2/(C_1 + C_2)$$
$$R_1 = R_A/(1 + C_B/C_A)^2 \qquad\qquad C_A = C_2^2/(C_1 + C_2)$$
$$C_1 = (C_B/C_A)(C_A + C_B) \qquad\qquad R_A = R_1(1 + C_1/C_2)^2$$

The circuits of Fig. 7-25 are equivalent when

$$R_1 = R_B^2/(R_A + R_B) \qquad\qquad R_B = R_1 + R_2$$
$$C_1 = C_A(1 + R_A/R_B)^2 \qquad\qquad R_A = (R_2/R_1)(R_1 + R_2)$$
$$R_2 = R_A R_B/(R_A + R_B) \qquad\qquad C_A = C_1/(1 + R_2/R_1)^2$$

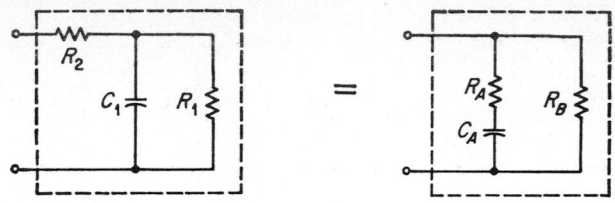

FIG. 7-25

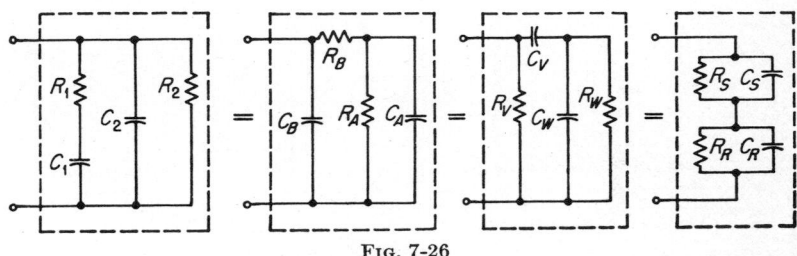

FIG. 7-26

The circuits of Fig. 7-26 are equivalent when

$R_1 = (R_B/R_A)(R_A + R_B) = R_w(1 + C_w/C_v)^2$
$\qquad\qquad = R_R R_s(R_R + R_s)(C_R + C_s)^2/(R_R C_R - R_s C_s)^2$
$R_2 = R_A + R_B = R_v = R_R + R_s$
$C_1 = C_A/(1 + R_B/R_A)^2 = C_v^2/(C_v + C_w)$
$\qquad\qquad = (R_R C_R - R_s C_s)^2/(R_R + R_s)^2(C_R + C_s)$
$C_2 = C_B = C_v C_w/(C_v + C_w) = C_R C_s/(C_R + C_s)$

$R_A = R_2^2/R_1 + R_2 \qquad\qquad R_B = R_1 R_2/(R_1 + R_2)$
$C_A = C_1(1 + R_1/R_2)^2 \qquad\quad C_B = C_2$
$R_v = R_2 \qquad\qquad\qquad\qquad R_w = R_1/(1 + C_2/C_1)^2$
$C_v = C_1 + C_2 \qquad\qquad\qquad C_w = (C_2/C_1)(C_1 + C_2)$
$C_s = (K + \sqrt{K^2 - 4R_2^2 C_1 C_2 K}/2R_2^2 C_1$
$K = (R_1 C_1 + R_2 C_1 + R_2 C_2)^2 - 4R_1 C_1 R_2 C_2$
$C_R = C_s C_2/(C_s - C_2) \qquad R_s = (R_1 C_1 + R_2 C_1 + R_2 C_2 - R_2 C_R)/(C_s - C_R)$
$R_R = R_2 - R_s$

Each of the circuits of Fig. 7-27 will be equivalent at all frequencies when

$C_1 = (C_B/C_A)(C_A + C_B) = C_w(1 + R_w/R_v)^2$
$\qquad\qquad = C_R C_s(C_R + C_s)(R_R + R_s)^2/(R_R C_R - L_s C_s)^2$
$C_2 = C_A + C_B = C_v = C_R + C_s$
$R_1 = R_A/(1 + C_B/C_A)^2 = R_v^2/(R_v + R_w)$
$\qquad\qquad = (R_R C_R - R_s C_s)^2/(C_R + C_s)^2(R_R + R_s)$
$R_2 = R_B = (R_v R_w/(R_v + R_w) = R_R R_s/(R_R + R_s)$

$C_A = C_2{}^2/(C_1 + C_2)$ $\qquad\qquad$ $C_B = C_1 C_2/(C_1 + C_2)$

$R_A = R_1(1 + C_1/C_2)^2$ $\qquad\qquad$ $R_B = R_2$

$C_v = C_2$ $\qquad\qquad$ $C_w = C_1/(1 + R^2/R_1)^2$

$R_v = R_1 + R_2$ $\qquad\qquad$ $R_w = (R_2/R_1)(R_1 + R_2)$

$R_s = (K + \sqrt{K^2 - 4R_1 R_2 C_2{}^2 K})/2R_1 C_2{}^2$

$K = (R_1 C_1 + R_1 C_2 + R_2 C_2)^2 - 4R_1 R_2 C_1 C_2$

$R_R = R_s R_2/(R_s - R_2)$ $\qquad$ $C_s = R_1 C_1 + R_1 C_2 + R_2 C_2 - R_R C_2/(R_s - R_R)$

$C_R = C_2 - C_s$

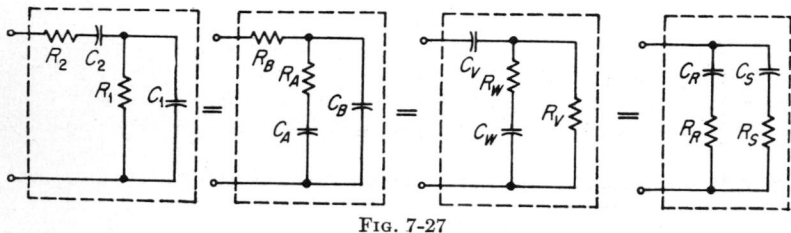

Fig. 7-27

**Power Transmission.** *Transmission-line Representation.* Three-phase power circuits are usually solved on a per-phase basis, since in a balanced system the only difference between the voltages and currents in the different

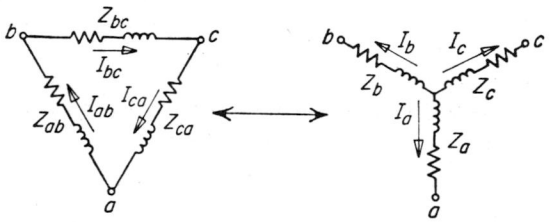

Fig. 7-28

phases is that the $b$ and $c$ phases are displaced 120° lagging and leading from the $a$ phase quantities. Delta (pi) systems are solved by first converting to an equivalent wye (T).

*Delta to star* (pi to T) transformation of impedances are made by the rule: Star impedances are the product of adjacent delta impedances divided by the sum of all the delta impedances (Fig. 7-28).

$Z_a = Z_{ca} Z_{ab}/D$ $\qquad\qquad$ $I_{ab} = -(Z_{ca}/D)I_a + (Z_{bc}/D)I_b$

$Z_b = Z_{ab} Z_{bc}/D$ $\qquad\qquad$ $I_{bc} = -(Z_{ab}/D)I_b + (Z_{ca}/D)I_c$

$Z_c = Z_{bc} Z_{ca}/D$ $\qquad\qquad$ $I_{ca} = -(Z_{bc}/D)I_c + (Z_{ab}/D)I_a$

$D = Z_{ab} + Z_{bc} + Z_{ca}$

*Star to delta* transformation of impedances can be made as follows:

$$Z_{ab} = D'Z_aZ_b \qquad\qquad I_a = I_{ca} - I_{ab}$$
$$Z_{bc} = D'Z_bZ_c \qquad\qquad I_b = I_{ab} - I_{bc}$$
$$Z_{ca} = D'Z_cZ_a \qquad\qquad I_c = I_{bc} - I_{ca}$$
$$D' = 1/Z_a + 1/Z_b + 1/Z_c$$

*ABCD constants* of a linear bilateral circuit having distinct input (sending end) and output (receiving end) terminals can be defined as:

$A$ = voltage impressed at sending end per volt at open-circuited receiving end; a dimensionless voltage ratio

$B$ = voltage impressed at sending end per ampere in short-circuited receiving end, same as transfer impedance; also, equal to voltage impressed at receiving end per ampere in short-circuited sending-end terminals

$C$ = current in amperes into sending end per volt on open-circuited receiving end; has dimensions of admittance

$D$ = current in amperes into sending end per ampere in short-circuited receiving end; a dimensionless current ratio

For all passive networks

$$AD - BC = 1 \tag{7-45}$$

*Voltage and current relationships* in terms of *ABCD* constants for transmission-type network are

$$E_s = AE_R + BI_R \quad \text{volts} \qquad E_R = DE_s - BI_s \quad \text{volts}$$
$$I_s = CE_R + DI_R \quad \text{amp} \qquad I_R = AI_s - CE_s \quad \text{amp}$$

The values of $A$, $B$, $C$, and $D$ can be determined for typical networks by referring to Tables 7-3 and 7-4.

**Power-circle Diagrams.** Power-circle diagrams can be drawn by using the equations of Table 7-5. *Real and reactive* power is expressed by

$$P + jQ = E\hat{I} \quad \text{va} \tag{7-46}$$

where $P$ = real power

$Q$ = reactive power

$E$ = vector voltage

$\hat{I}$ = conjugate current in vector form (if $I = a + jb$, then $\hat{I} = a - jb$)

*Two-station systems* can be analyzed by means of circle diagrams obtained from equations in Table 7-5. A step-by-step procedure is as follows:

1. Calculate vector to center, and locate $C_s$ or $C_R$.

2. Calculate radius vector $R_{SO}$ or $R_{RO}$ for $\theta = 0$ ($\epsilon^{j\theta} = 1$).

3. Add (1) and (2) to obtain real and reactive power for sending and receiving voltages in phase. Plot this as power for $\theta = 0$.

4. Draw circle using center from (1) and passing through power for $\theta = 0$. Draw reference radius vector from center to power for $\theta = 0$ point to serve as reference from which angles are measured.

5. Corresponding sending and receiving conditions are found at the same angle on the corresponding circles.

| Type of network | | $A$ |
|---|---|---|
| Series impedance | $E_S$ —⌇⌇— $Z$ —⌇⌇— $E_R$ | $1$ |
| Shunt admittance | $E_S$ ⫴ $Y$ $E_R$ | $1$ |
| Transformer | $E_S$ $\frac{Z_T}{2}$ $Y_T$ $\frac{Z_T}{2}$ $E_R$ | $1 + \dfrac{Z_T Y_T}{2}$ |
| Transformer ratio | $E_S$ $1{:}N$ $E_R$ | $\dfrac{1}{N}$ |
| Transmission line* | $E_S$ $Z$ $y$ $Z$ $E_R$ | $\cosh \sqrt{ZY} =$ $\left(1 + \dfrac{ZY}{2} + \dfrac{Z^2 Y^2}{24} + \cdots\right)$ |
| $A_1 B_1 C_1 D_1$ in series with impedance | $E_S$ \|$A_1\ B_1\ C_1\ D_1$\| $Z$ $E_R$ | $A_1$ |
| Impedance in series with $A_1 B_1 C_1 D_1$ | $E_S$ $Z$ \|$A_1\ B_1\ C_1\ D_1$\| $E_R$ | $A_1 + C_1 Z$ |
| Sending-end impedance in series with $A_1 B_1 C_1 D_1$ and receiver impedance | $E_S$ $Z_S$ \|$A_1\ B_1\ C_1\ D_1$\| $Z_R$ $E_R$ | $A_1 + C_1 Z_S$ |
| $A_1 B_1 C_1 D_1$ and shunt admittance at receiver | $E_S$ \|$A_1\ B_1\ C_1\ D_1$\| $Y_R$ $E_R$ | $A_1 + B_1 Y_R$ |
| $A_1 B_1 C_1 D_1$ and shunt admittance at sending end | $E_S$ $Y_S$ \|$A_1\ B_1\ C_1\ D_1$\| $E_R$ | $A_1$ |

*ABCD* constants for different types of networks

| B | C | D |
|---|---|---|
| $Z$ | 0 | 1 |
| 0 | $Y$ | 1 |
| $Z_T\left(1 + \dfrac{Z_T Y_T}{4}\right)$ | $Y_T$ | $1 + \dfrac{Z_T Y_T}{2}$ |
| 0 | 0 | $N$ |
| $\sqrt{Z/Y}\sinh\sqrt{ZY} =$ $Z\left(1 + \dfrac{ZY}{6} + \dfrac{Z^2 Y^2}{120} + \cdots\right)$ | $\sqrt{Y/Z}\sinh\sqrt{ZY} =$ $Y\left(1 + \dfrac{ZY}{6} + \dfrac{Z^2 Y^2}{120} + \cdots\right)$ | Same as $A$ |
| $B_1 + A_1 Z$ | $C_1$ | $D_1 + C_1 Z$ |
| $B_1 + D_1 Z$ | $C_1$ | $D_1$ |
| $B_1 + A_1 Z_R + D_1 Z_S + C_1 Z_S Z_R$ | $C_1$ | $D_1 + C_1 Z_R$ |
| $B_1$ | $C_1 + D_1 Y_R$ | $D_1$ |
| $B_1$ | $C_1 + A_1 Y_S$ | $D_1 + B_1 Y_S$ |

Table 7-3

| Type of network | | A |
|---|---|---|
| $A_1B_1C_1D_1$ and shunt admittance at both ends | | $A_1 + B_1Y_R$ |
| $A_1B_1C_1D_1$ in series with $A_2B_2C_2D_2$ | | $A_1A_2 + C_1B_2$ |
| Two general networks with series impedance | | $A_1A_2 + C_1B_2 + C_1A_2Z$ |
| Two general networks in parallel | | $\dfrac{A_1B_2 + B_1A_2}{B_1 + B_2}$ |

\* $z$ = ohms/mile; $Z = sz$, where $s$ = miles; $Y = sy$, total susceptance.

*ABCD* constants for different types of networks

| B | C | D |
|---|---|---|
| $B_1$ | $C_1 + A_1Y_S + D_1Y_R + B_1Y_SY_R$ | $D_1 + B_1Y_S$ |
| $B_1A_2 + D_1B_2$ | $A_1C_2 + C_1D_2$ | $B_1C_2 + D_1D_2$ |
| $B_1A_2 + D_1B_2 + D_1A_2Z$ | $A_1C_2 + C_1D_2 + C_1C_2Z$ | $B_1C_2 + D_1D_2 + D_1C_2Z$ |
| $\dfrac{B_1B_2}{B_1 + B_2}$ | $\dfrac{C_1 + C_2 + (A_1 - A_2)(D_2 - D_1)}{B_1 + B_2}$ | $\dfrac{B_1D_2 + D_1B_2}{B_1 + B_2}$ |

**Table 7-4. Conversion Formulas for Transmission-type Networks**

To Convert from

| To Convert to | | ABCD | Impedance | Equivalent pi | Equivalent T | Nomenclature |
|---|---|---|---|---|---|---|
| **ABCD** | $A =$ | $ABCD$ constants<br>$E_s = AE_R + BI_R$<br>$I_s = CE_R + DI_R$<br>$E_R = DE_s - BI_s$<br>$I_R = -CE_s + AI_s$ | $Z_{22}/Z_{12}$ | $1 + ZY_R$ | $1 + Z_sY$ | |
| | $B =$ | | $(Z_{11}Z_{22} - Z_{12}^2)/Z_{12}$ | $Z$ | $Z_R + Z_s + YZ_RZ_s$ | |
| | $C =$ | | $1/Z_{12}$ | $Y_R + Y_s + ZY_RY_s$ | $Y$ | |
| | $D =$ | | $Z_{11}/Z_{12}$ | $1 + ZY_s$ | $1 + Z_RY$ | |
| **Impedance** | $Z_{11} =$ | $\dfrac{D}{C}$ | Impedance constants<br>$E_1 = Z_{11}I_1 + Z_{12}I_2$<br>$E_2 = Z_{12}I_1 + Z_{22}I_2$ | $\dfrac{1 + ZY_s}{Y_R + Y_s + ZY_RY_s}$ | $Z_R + \dfrac{1}{Y}$ | |
| | $Z_{12} =$ | $\dfrac{1}{C}$ | | $\dfrac{1}{Y_R + Y_s + ZY_RY_s}$ | $\dfrac{1}{Y}$ | |
| | $Z_{22} =$ | $\dfrac{A}{C}$ | | $\dfrac{1 + ZY_R}{Y_R + Y_s + ZY_RY_s}$ | $Z_s + \dfrac{1}{Y}$ | |
| **Equiv pi** | $Y_R =$ | $\dfrac{A - 1}{B}$ | $\dfrac{Z_{22} - Z_{12}}{Z_{11}Z_{22} - Z_{12}^2}$ | Equivalent pi | $\dfrac{YZ_s}{Z_R + Z_s + YZ_RZ_s}$ | |
| | $Z =$ | $B$ | $\dfrac{Z_{11}Z_{22} - Z_{12}^2}{Z_{12}}$ | | $Z_R + Z_s + YZ_RZ_s$ | |
| | $Y_s =$ | $\dfrac{D - 1}{B}$ | $\dfrac{Z_{11} - Z_{12}}{Z_{11}Z_{22} - Z_{12}^2}$ | | $\dfrac{YZ_R}{Z_R + Z_s + YZ_RZ_s}$ | |
| **Equiv T** | $Z_R =$ | $\dfrac{D - 1}{C}$ | $Z_{11} - Z_{12}$ | $\dfrac{ZY_s}{Y_R + Y_s + ZY_RY_s}$ | Equivalent T | |
| | $Y =$ | $C$ | $\dfrac{1}{Z_{12}}$ | $Y_R + Y_s + ZY_RY_s$ | | |
| | $Z_s =$ | $\dfrac{A - 1}{C}$ | $Z_{22} - Z_{12}$ | $\dfrac{ZY_R}{Y_R + Y_s + ZY_RY_s}$ | | |

## Table 7-5. Equations for Plotting Circle Diagrams

| Type of network | Sending end circle | | | Receiving end circle | | |
|---|---|---|---|---|---|---|
| | Power input* | Vector to center, $C_s$† | Radius vector, $R_{s0}$ | Power output | Vector to center, $C_R$ | Radius vector, $R_{R0}$ |
| ABCD | $P_s + jQ_s$ | $\dfrac{\hat{D}}{\hat{B}}\,\bar{E}_s^2$ | $-\dfrac{\bar{E}_R\bar{E}_s\epsilon^{+j\theta}}{\hat{B}}$ | $P_R + jQ_R$ | $-\dfrac{\hat{A}}{\hat{B}}\,\bar{E}_R^2$ | $+\dfrac{\bar{E}_R\bar{E}_s\epsilon^{-j\theta}}{\hat{B}}$ |
| Equivalent pi | $P_s + jQ_s$ | $\left(\dfrac{1}{\hat{Z}}+\hat{Y}_s\right)\bar{E}_s^2$ | $-\dfrac{\bar{E}_R\bar{E}_s\epsilon^{+j\theta}}{\hat{Z}}$ | $P_R + jQ_R$ | $-\left(\dfrac{1}{\hat{Z}}+\hat{Y}_R\right)\bar{E}_R^2$ | $+\dfrac{\bar{E}_R\bar{E}_s\epsilon^{-j\theta}}{\hat{Z}}$ |
| Equivalent T | $P_s + jQ_s$ | $\dfrac{(1+\hat{Z}_R\hat{Y})\bar{E}_s^2}{\hat{Z}_R+\hat{Z}_s+\hat{Y}\hat{Z}_R\hat{Z}_s}$ | $-\dfrac{\bar{E}_R\bar{E}_s\epsilon^{+j\theta}}{\hat{Z}_R+\hat{Z}_s+\hat{Y}\hat{Z}_R\hat{Z}_s}$ | $P_R + jQ_R$ | $-\dfrac{(1+\hat{Z}_s\hat{Y})\bar{E}_R^2}{\hat{Z}_R+\hat{Z}_s+\hat{Y}\hat{Z}_R\hat{Z}_s}$ | $+\dfrac{\bar{E}_R\bar{E}_s\epsilon^{-j\theta}}{\hat{Z}_R+\hat{Z}_s+\hat{Y}\hat{Z}_R\hat{Z}_s}$ |
| Mesh impedance | $P_2 + jQ_2$ | $\dfrac{\hat{Z}_{11}\bar{E}_2^2}{\hat{Z}_{11}\hat{Z}_{22}-\hat{Z}_{12}^2}$ | $-\dfrac{\hat{Z}_{12}\bar{E}_1\bar{E}_2\epsilon^{+j\theta}}{\hat{Z}_{11}\hat{Z}_{22}-\hat{Z}_{12}^2}$ | $P_1 + jQ_1$ | $\dfrac{\hat{Z}_{22}\bar{E}_1^2}{\hat{Z}_{11}\hat{Z}_{22}-\hat{Z}_{12}^2}$ | $-\dfrac{\hat{Z}_{12}\bar{E}_1\bar{E}_2\epsilon^{-j\theta}}{\hat{Z}_{11}\hat{Z}_{22}-\hat{Z}_{12}^2}$ |

* For $P$ and $Q$ in megawatts and megavolt amperes, use $E_s$ and $E_R$ in kilovolts line-to-line.
† Bar over $E_s$ and $E_R$ indicates scalar values; ^ over constants indicates conjugates.

*Multistation systems* can be treated as two-station systems when there is a single path for power flow. For example, suppose circle diagrams are to be constructed for interconnected stations $A$, $B$, and $C$. First consider stations $A$ and $B$ as a two-station system with power flow to or from $C$ remaining constant. Then, with known conditions at $B$, treat $B$ to $C$ as a two-station system.

**Steady-state Stability.** Steady-state stability limit is defined as the maximum power flow possible through some point in the system when the entire system or the part of the system to which the stability limit refers is operating with stability. For a simple two-machine system, neglecting losses, the steady-state limit is reached when

$$P = \bar{E}_g \bar{E}_m / X \qquad \text{watts} \tag{7-47}$$

where $P$ = three-phase power, watts

$E_g$ = line-to-line internal volts of generator

$E_m$ = line-to-line internal volts of motor

$X$ = reactance between generator and motor internal voltages, ohms/phase

A single generator connected to an infinite bus has a per-unit pull-out torque (POT) (assuming constant speed) of

$$POT = E_i E_s / (X_g + X_s) \qquad \text{per unit} \tag{7-48}$$

where $E_i$ = internal voltage of generator, volts

$E_t$ = generator terminal voltage, volts

$E_s$ = infinite system voltage per unit

$X_g$ = per-unit generator reactance

$X_s$ = per-unit infinite system reactance

Pull-out torque of single machine and infinite bus by the *air-gap voltage method* is given by

$$POT = E_{ag} E_s / (X_d + X_s) \qquad \text{per unit} \tag{7-49}$$

where $E_{ag}$ = per-unit air-gap voltage of generator

$X_d$ = per-unit synchronous reactance of generator

Pull-out torque by *synchronous-reactance method*

$$POT = E_d E_s / (X_d + X_s) \qquad \text{per unit} \tag{7-50}$$

where $E_d$ is the p-u voltage behind synchronous reactance, $E_d = E_t + IX_d$.

*Short-circuit ratio* (SCR) is the ratio of field current required to produce rated voltage on the no-load saturation curve to the field current required to produce rated armature current with a three-phase short circuit at the generator terminals.

$$POT = E_i E_s / (1/SCR + X_s) \qquad \text{per unit} \tag{7-51}$$

where $E_i = E_t + I(1/SCR) \qquad$ per unit

### Table 7-6. Reactance Constants of Synchronous Machines, Typical Normal Values, Per Cent

| Type of machine | Synchronous* direct axis $X_d$ | Transient† $X'_d$ | Subtransient‡ $X''_d$ | Negative-¶ sequence $X_2$ |
|---|---|---|---|---|
| Turbogenerator: | | | | |
| 3,600 rpm.............. | 110 | 15 | 9 | 9 |
| 1,800 rpm.............. | 110 | 23 | 15 | 15 |
| Salient-pole generator: | | | | |
| With dampers.......... | 115 | 37 | 24 | 24 |
| Without dampers....... | 115 | 35 | 32 | 55 |
| Synchronous condenser.... | 180 | 40 | 25 | 24 |

\* Unsaturated-current values.
† Rated-current values, use stability calculations.
‡ Short-circuit-current values, use three-phase and unbalanced-fault calculations.
¶ Short-circuit-current values, use unbalanced-fault calculations.

***Transient Stability.*** Transient stability limit refers to the maximum power that can be transmitted over a given system during a system disturbance and maintain synchronism.

The *natural frequency* of synchronous machines connected to an infinite bus and shaft-connected to reciprocating machinery is given by

$$f_n = (35,200/n) \sqrt{P_r f / WR^2} \qquad \text{cpm} \qquad (7\text{-}52)$$

where $f_n$ = natural frequency, cycles/min
$n$ = speed of machine, rpm
$P_r$ = synchronizing power = shaft power in kilowatts divided by angular displacement of rotor in electrical radians, kw/δ
$f$ = frequency of circuit, cps
$WR^2$ = moment of inertia of synchronous machine and shaft-connected prime mover or load, lb-ft²

$$\delta = \tan^{-1} (IX_q \cos \phi / E_t + \bar{I} X_q \sin \phi) \qquad \text{radians} \qquad (7\text{-}53)$$

where δ = rotor displacement angle, electrical radians
$\bar{I}$ = per-unit armature current
$E_t$ = per-unit armature terminal voltage
$\phi$ = power-factor angle
$X_q$ = per-unit quadrature-axis synchronous reactance

*Inertia constant* of synchronous machines is given by

$$H = \text{kw-sec/kva} = 0.231(WR^2 \text{ rpm}^2 \times 10^{-6})/\text{kva} \qquad (7\text{-}54)$$

where $WR^2$ = moment of inertia, lb-ft²
rpm = speed in revolutions per minute

The *equivalent inertia constant* for reducing a two-machine system to a single machine and an infinite bus is

$$H_{eq\,a} = H_a/(1 + H_a \text{ kva}_a/H_b \text{ kva}_b) \qquad (7\text{-}55)$$

where $H_a$ = inertia constant of machine $a$
        $H_b$ = inertia constant of machine $b$
        $\text{kva}_a$ = capacity machine $a$, kva
        $\text{kva}_b$ = capacity machine $b$, kva
        $H_{eq\,a}$ = equivalent inertia constant for machine $a$

The *acceleration of a synchronous generator* when subjected to an accelerating or decelerating power $\Delta P$ is given by

$$\alpha = 180f\ \Delta P/H\ \text{kva} \qquad \text{deg/sec}^2 \qquad (7\text{-}56)$$

where $\alpha$ = acceleration or deceleration, electrical deg/sec$^2$
      $f$ = system frequency, cps
      $\Delta P$ = accelerating or decelerating power, kw

**Symmetrical Components.** Vector $a$ is of unit length and is oriented 120° in a positive (counterclockwise) direction from the reference direction. A vector operated upon by $a$ is not changed in magnitude, but simply rotated 120° in the positive direction.

Properties of the vector operator $a$

$$1 + j0 = \epsilon^{j0}$$
$$a = -\tfrac{1}{2} + j\,\tfrac{1}{2}\sqrt{3} = \epsilon^{j120}$$
$$a^2 = -\tfrac{1}{2} - j\,\tfrac{1}{2}\sqrt{3} = \epsilon^{j240}$$
$$a^3 = 1 + j0 = \epsilon^{j0}$$
$$a^4 = a \qquad a^5 = a^2$$
$$1 + a + a^2 = 0$$
$$a + a^2 = -1 = \epsilon^{j180}$$
$$a - a^2 = j\sqrt{3} = \sqrt{3}\,\epsilon^{j90}$$
$$a^2 - a = -j\sqrt{3} = \sqrt{3}\,\epsilon^{j270}$$
$$1 - a = \tfrac{3}{2} - j\,\tfrac{1}{2}\sqrt{3} = \sqrt{3}\,\epsilon^{j330}$$
$$1 - a^2 = \tfrac{3}{2} + j\,\tfrac{1}{2}\sqrt{3} = \sqrt{3}\,\epsilon^{j30}$$

$$a - 1 = -\tfrac{3}{2} + j\,\tfrac{1}{2}\sqrt{3} = \sqrt{3}\,\epsilon^{j150}$$
$$a^2 - 1 = -\tfrac{3}{2} - j\,\tfrac{1}{2}\sqrt{3} = \sqrt{3}\,\epsilon^{j210}$$
$$1 + a = \tfrac{1}{2} + j\,\tfrac{1}{2}\sqrt{3} = \epsilon^{\,j60}$$
$$1 + a^2 = \tfrac{1}{2} - j\,\tfrac{1}{2}\sqrt{3} = \epsilon^{j300}$$
$$(1 + a)(1 + a)^2 = 1 = \epsilon^{j0}$$
$$(1 - a)(1 - a^2) = 3 = 3\epsilon^{j0}$$
$$(1 + a)/(1 + a^2) = a = \epsilon^{j120}$$
$$(1 - a)/(1 - a^2) = -a = \epsilon^{j300}$$
$$(1 + a)^2 = a = \epsilon^{j120}$$
$$(1 + a^2)^2 = a^2 = \epsilon^{j240}$$

Resolution of *three unbalanced vectors* into balanced symmetrical components, zero, positive, and negative sequence, can be done as follows:

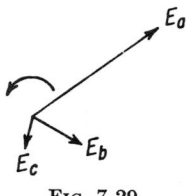

Fig. 7-29

$$E_0 = \tfrac{1}{3}(E_a + E_b + E_c) \qquad \text{zero sequence, volts}$$
$$E_1 = \tfrac{1}{3}(E_a + aE_b + a^2E_c) \qquad \text{positive sequence, volts}$$
$$E_2 = \tfrac{1}{3}(E_a + a^2E_b + aE_c) \qquad \text{negative sequence, volts}$$

This resolution applies to both voltages and currents in three-phase systems.
The unbalanced vectors can be expressed as functions of the balanced

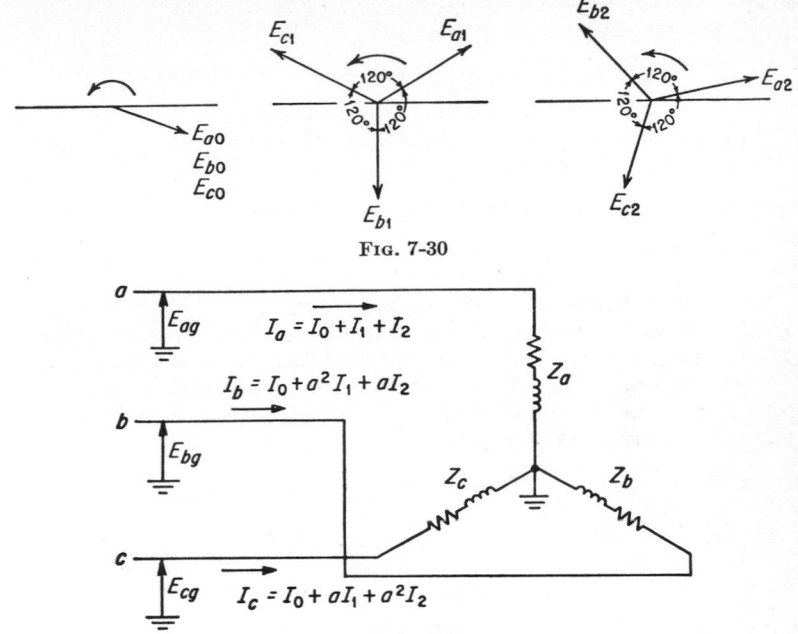

FIG. 7-30

FIG. 7-31

**components**

$$E_a = E_{a0} + E_{a1} + E_{a2} = E_0 + E_1 + E_2 \quad \text{volts}$$
$$E_b = E_{b0} + E_{b1} + E_{b2} = E_0 + a^2E_1 + aE_2 \quad \text{volts}$$
$$E_c = E_{c0} + E_{c1} + E_{c2} = E_0 + aE_1 + a^2E_2 \quad \text{volts}$$

*Unbalanced impedances* as shown in Fig. 7-31 can also be broken into symmetrical components.

$$Z_0 = \tfrac{1}{3}(Z_a + Z_b + Z_c) \quad \text{ohms} \quad E_0 = I_0Z_0 + I_1Z_2 + I_2Z_1 \quad \text{volts}$$
$$Z_1 = \tfrac{1}{3}(Z_a + aZ_b + a^2Z_c) \quad \text{ohms} \quad E_1 = I_0Z_1 + I_1Z_0 + I_2Z_2 \quad \text{volts}$$
$$Z_2 = \tfrac{1}{3}(Z_a + a^2Z_b + aZ_c) \quad \text{ohms} \quad E_2 = I_0Z_2 + I_1Z_1 + I_2Z_0 \quad \text{volts}$$

If $Z_a = Z_b = Z_c$, then $Z_1 = Z_2 = 0$ and $Z_0 = Z_a$, so that

$$E_0 = I_0Z_0 \qquad E_1 = I_1Z_0 \qquad E_2 = I_2Z_0 \quad \text{volts}$$

*Mutual impedances* between phases (Fig. 7-32) can also be resolved into symmetrical components.

$$Z_{m0} = \tfrac{1}{3}(Z_{mbc} + Z_{mca} + Z_{mab}) \quad \text{ohms}$$
$$Z_{m1} = \tfrac{1}{3}(Z_{mbc} + aZ_{mca} + a^2Z_{mab}) \quad \text{ohms}$$
$$Z_{m2} = \tfrac{1}{3}(Z_{mbc} + a^2Z_{mca} + aZ_{mab}) \quad \text{ohms}$$

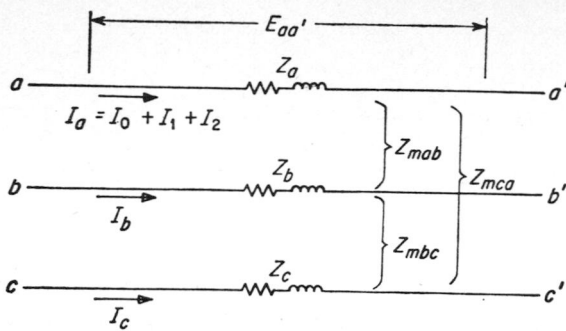

FIG. 7-32

The associated voltage drops are

$$E_0 = \tfrac{1}{3}(E_{aa'} + E_{bb'} + E_{cc'}) \qquad \text{volts}$$
$$\quad = I_0(Z_0 + 2Z_{m0}) + I_1(Z_2 - Z_{m2}) + I_2(Z_1 - Z_{m1}) \qquad \text{volts}$$
$$E_1 = \tfrac{1}{3}(E_{aa'} + aE_{bb'} + a^2E_{cc'}) \qquad \text{volts}$$
$$\quad = I_0(Z_1 - Z_{m1}) + I_1(Z_0 - Z_{m0}) + I_2(Z_2 + 2Z_{m2}) \qquad \text{volts}$$
$$E_2 = \tfrac{1}{3}(E_{aa'} + a^2E_{bb'} + aE_{cc'}) \qquad \text{volts}$$
$$\quad = I_0(Z_2 - Z_{m2}) + I_1(Z_1 + 2Z_{m1}) + I_2(Z_0 - Z_{m0}) \qquad \text{volts}$$

For symmetrical impedance values

$$E_0 = I_0(Z_0 + 2Z_{m0}) = I_0Z_0 \qquad \text{volts}$$
$$E_1 = I_1(Z_0 - Z_{m0}) = I_1Z_1 \qquad \text{volts}$$
$$E_2 = I_2(Z_0 - Z_{m0}) = I_2Z_2 \qquad \text{volts}$$

*Three-phase-power* expressed in terms of symmetrical components:

$$P = 3(E_0I_0 \cos\theta_0 + E_1I_1 \cos\theta_1 + E_2I_2 \cos\theta_2) \qquad \text{watts} \qquad (7\text{-}57)$$

where $\theta_0$, $\theta_1$, and $\theta_2$ are the angles between $E_0$ and $I_0$, $E_1$ and $I_1$, and $E_2$ and $I_2$ respectively.

**Sequence Networks.**  Sequence networks used for solving fault problems on three-phase systems consist of positive-, negative-, and zero-sequence networks each set up as viewed from the point of fault.  The values of positive-, negative-, and zero-sequence impedance to be inserted in each network can be obtained by passing unit current of that sequence through the original three-phase network and calculating or measuring the corresponding voltage drop.  In all cases positive current direction must be the same in each sequence network.  In many cases complex circuits are lumped to a single equivalent generator and impedance.  All quantities are in volts, amperes, and ohms, or all are in per unit.

*Three-phase short-circuit* sequence network connections (Fig. 7-33)

$$I_{1F} = I_F = E_{a1}/Z_1$$

where $Z_1 = jX_1$ if system resistance is neglected

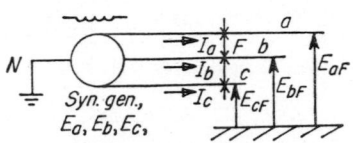

(*a*) Equivalent system for three-phase short circuit

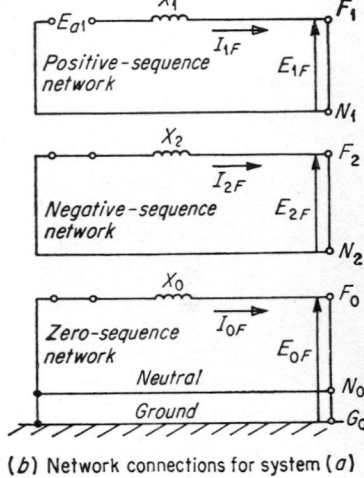

(*b*) Network connections for system (*a*)

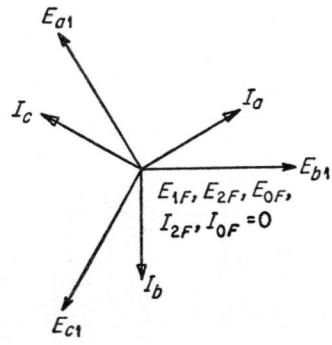

(*c*) Vector diagram for system (*a*)

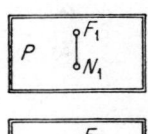

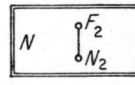

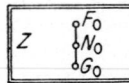

(*d*) Condensed representation of (*b*)

Fig. 7-33

*Single-line-to-ground fault* on ungrounded generator (Fig. 7-34)

$$I_F = I_{1F} = I_{2F} = I_{0F} = 0$$
$$E_{aF} = 0 \qquad E_{bF} = \sqrt{3}\, E_{b1}\epsilon^{-j30} \qquad E_{cF} = \sqrt{3}\, E_{c1}\epsilon^{j30}$$

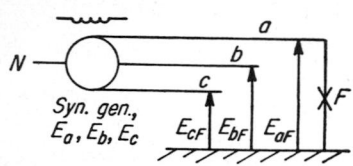

(*a*) Equivalent system for
single-line-to-ground fault
ungrounded system

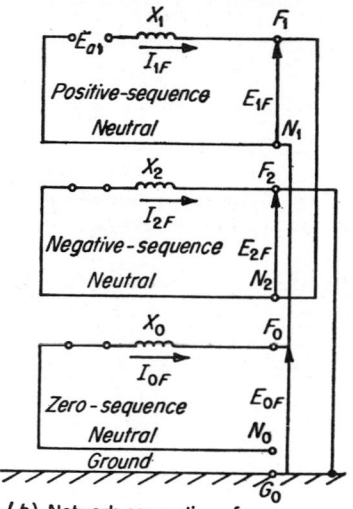

(*b*) Network connections for
system (*a*)

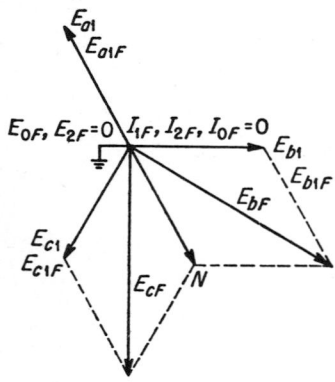

(*c*) Vector diagram for system (*a*)

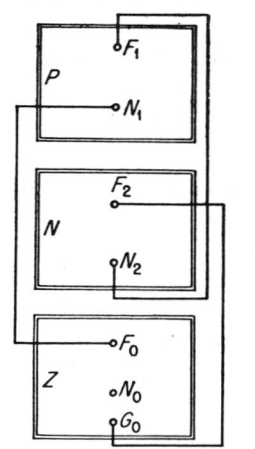

(*d*) Condensed representation
of (*b*)

Fig. 7-34

*Single-line-to-ground* fault on generator grounded through a neutral reactor (Fig. 7-35)

$$I_{1F} = I_{2F} = I_{0F} = E_{a1}/(Z_1 + Z_2 + Z_0 + j3X_N)$$
$$I_F = I_{1F} + I_{2F} + I_{0F} = 3I_{0F}$$
$$E_{1F} = E_{a1}(Z_2 + Z_0 + j3X_N)/(Z_1 + Z_2 + Z_0 + j3X_N)$$
$$E_{2F} = -E_{a1}Z_2/(Z_1 + Z_2 + Z_0 + j3X_N)$$
$$E_{0F} = -E_{a1}(Z_0 + j3X_N)/(Z_1 + Z_2 + Z_0 + j3X_N)$$

where $Z_1 = jX_1$, $Z_2 = jX_2$, and $Z_0 = jX_0$ if system resistance is neglected.

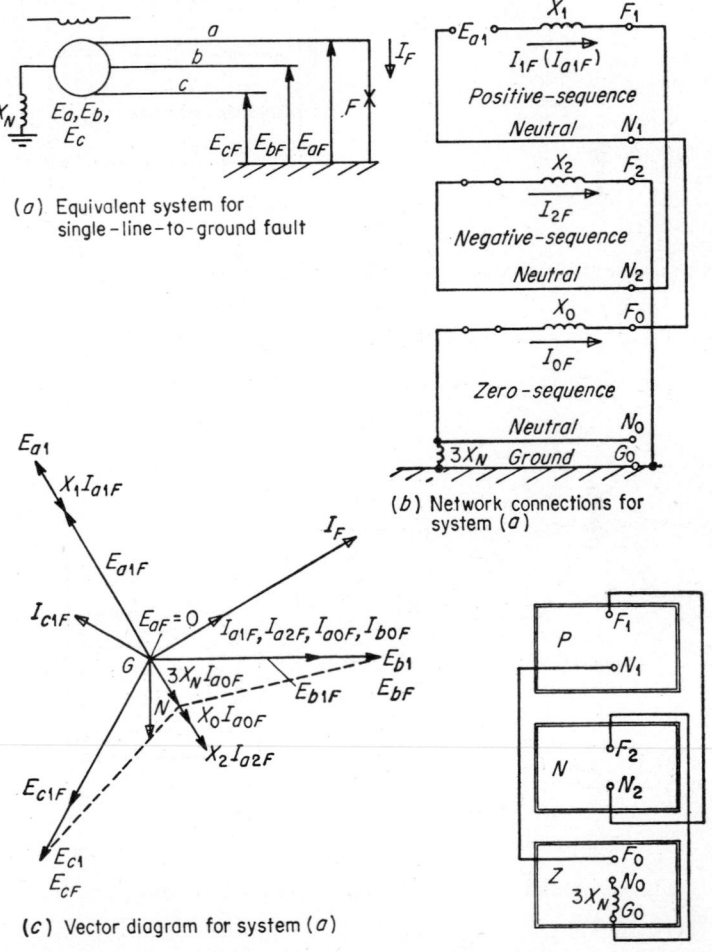

(*a*) Equivalent system for single-line-to-ground fault

(*b*) Network connections for system (*a*)

(*c*) Vector diagram for system (*a*)

(*d*) Condensed representation of (*b*)

FIG. 7-35

*Line-to-line fault* on grounded or ungrounded generator (Fig. 7-36)

$$I_{1F} = -I_{2F} = E_{a1}/(Z_1 + Z_2)$$
$$I_F = \sqrt{3}\, I_{1F} \qquad E_{1F} = E_{2F} = E_{a1}Z_2/(Z_1 + Z_2)$$

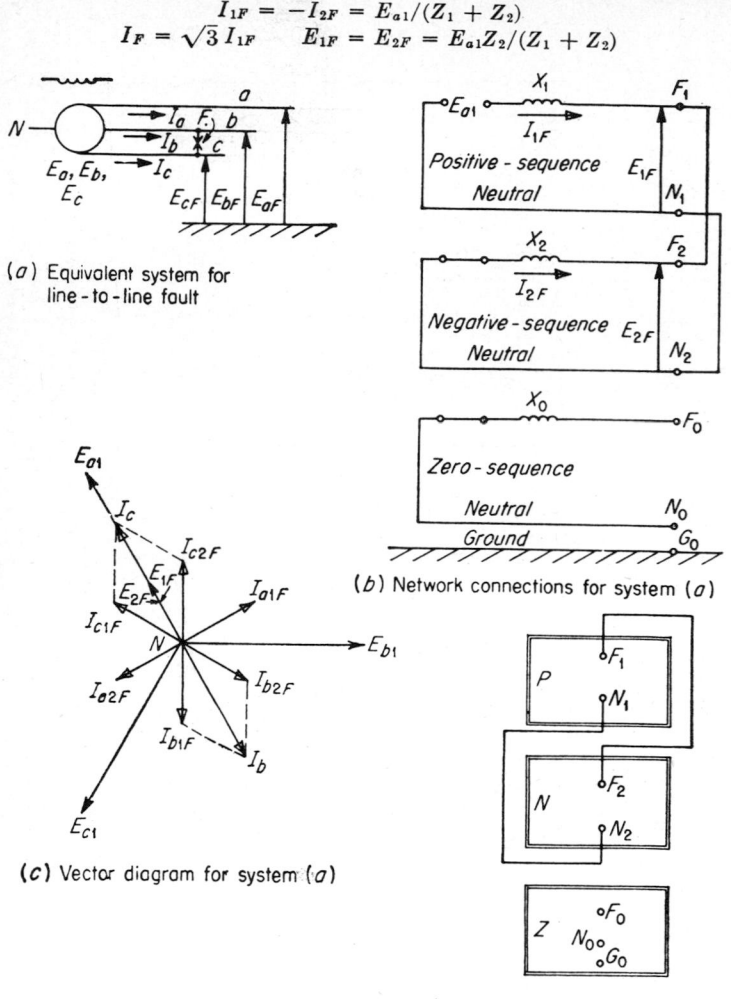

(*a*) Equivalent system for line-to-line fault

(*b*) Network connections for system (*a*)

(*c*) Vector diagram for system (*a*)

(*d*) Condensed representation of (*b*)

Fig. 7-36

*Double-line-to-ground fault* on generator grounded through a neutral reactor (Fig. 7-37)

$$I_F = I_{1F} + I_{2F} + I_{0F}$$
$$I_{1F} = E_{a1}(Z_2 + Z_0 + j3X_N)/[Z_1Z_2 + Z_1(Z_0 + j3X_N) + Z_2(Z_0 + j3X_N)]$$
$$I_{2F} = -(Z_0 + j3X_N)E_{a1}/[Z_1Z_2 + Z_1(Z_0 + j3X_N) + Z_2(Z_0 + j3X_N)]$$
$$I_{0F} = -Z_2E_{a1}/[Z_1Z_2 + Z_1(Z_0 + j3X_N) + Z_2(Z_0 + j3X_N)]$$
$$E_{1F} = E_{2F} = E_{0F} = \frac{Z_2(Z_0 + j3X_N)E_{a1}}{[Z_1Z_2 + Z_1(Z_0 + j3X_N) + Z_2(Z_0 + j3X_N)]}$$

(a) Equivalent system for double-
line-to-ground fault

(b) Network connections for
system (a)

(c) Vector diagram for system (a)

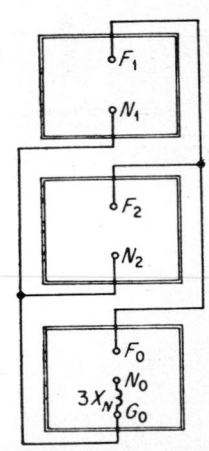

(d) Condensed representation
of (b)

Fig. 7-37

*One line open* (Fig. 7-38)

$$I_{1F} = E_{a1}(Z_2 + Z_0)/(Z_1Z_2 + Z_1Z_0 + Z_2Z_0)$$
$$I_{2F} = -Z_0E_{a1}/(Z_1Z_2 + Z_1Z_0 + Z_2Z_0)$$
$$I_{0F} = -Z_2E_{a1}/(Z_1Z_2 + Z_1Z_0 + Z_2Z_0)$$

$$E_{1x} - E_{1y} = E_{2x} - E_{2y} = E_{0x} - E_{0y} = \frac{Z_2Z_0E_{a1}}{Z_1Z_2 + Z_1Z_0 + Z_2Z_0}$$

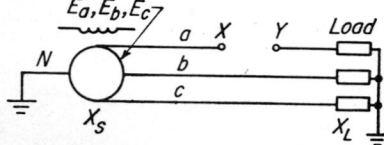

(*a*) Equivalent system for one line open

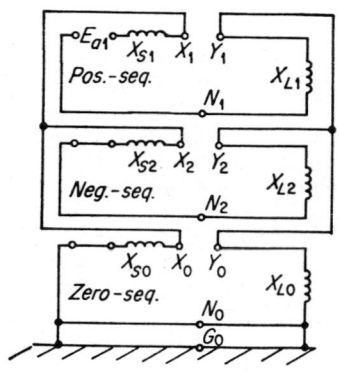

(*b*) Network connections for system (*a*)

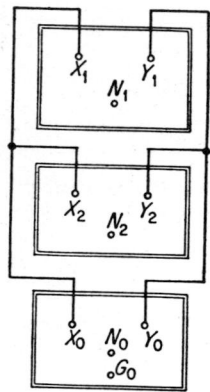

(*c*) Condensed representation of (*b*)

Fɪɢ. 7-38

*Two lines open* (Fig. 7-39)

$$I_{1F} = I_{2F} = I_{0F} = E_{a1}/(Z_1 + Z_2 + Z_0)$$
$$I_F = I_a = 3I_{0F}$$
$$E_{1x} - E_{1y} = E_{a1}(Z_2 + Z_0)/(Z_1 + Z_2 + Z_0)$$
$$E_{2x} - E_{2y} = -E_{a1}Z_2/(Z_1 + Z_2 + Z_0)$$
$$E_{0x} - E_{0y} = -E_{a1}Z_0/(Z_1 + Z_2 + Z_0)$$

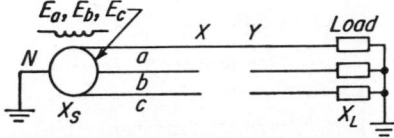

(*a*) Equivalent system for two lines open

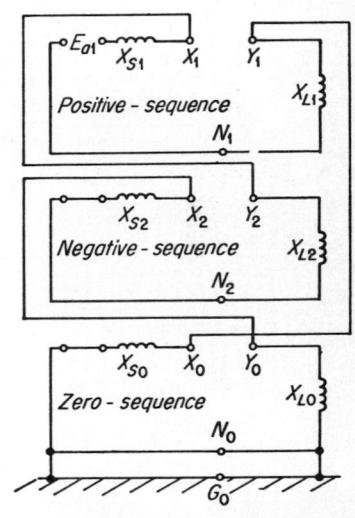

(*b*) Net work connections for system (*a*)

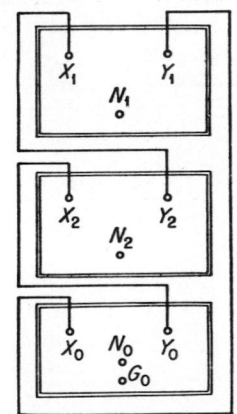

(*c*) Condensed representation of (*b*)

FIG. 7-39

*Impedance in one line* (Fig. 7-40)

$$I_{1F} = E_{a1}(ZZ_0 + ZZ_2 + 3Z_0Z_2)/(ZZ_1Z_0 + ZZ_1Z_2 + 3Z_1Z_2Z_0 + ZZ_2Z_0)$$
$$I_{2F} = -E_{a1}ZZ_0/(ZZ_1Z_0 + ZZ_1Z_2 + 3Z_1Z_2Z_0 + ZZ_2Z_0)$$
$$I_{0F} = -E_{a1}ZZ_2/(ZZ_1Z_0 + ZZ_1Z_2 + 3Z_1Z_2Z_0 + ZZ_2Z_0)$$

$$E_{1x} - E_{1y} = E_{2x} - E_{2y} = E_{0x} - E_{0y}$$

$$= \frac{E_{a1}ZZ_2Z_0}{ZZ_1Z_0 + ZZ_1Z_2 + 3Z_1Z_2Z_0 + ZZ_2Z_0}$$

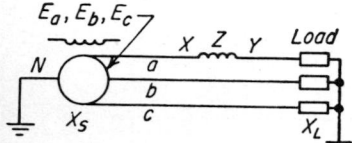

(a) Equivalent system for impedance in one line

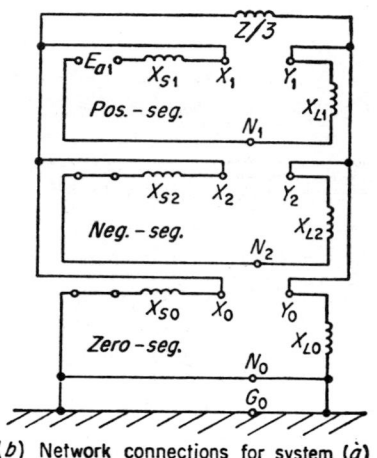

(b) Network connections for system (a)

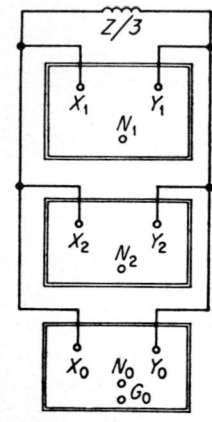

(c) Condensed representation of (b)

Fig. 7-40

## Table 7-7. Transformer Impedances, Typical Normal Values, Per Cent Full-load KVA Base

| Voltage class, kv | Single-phase kva rating | | | | | |
| --- | --- | --- | --- | --- | --- | --- |
| | 3 | 10 | 25 | 50 | 100 | 500 |
| 2.5 | 2.2 | 2.2 | 2.5 | 2.4 | 3.3 | 4.8 |
| 15 | 2.8 | 2.4 | 2.3 | 2.5 | 3.2 | 5.0 |
| 25 | ... | 5.2 | 5.2 | 5.2 | 5.2 | 5.2 |
| 69 | ... | ... | ... | 6.5 | 6.5 | 6.5 |
| 138 | | | | | | |
| 161 | | | | | | |
| 230 | | | | | | |

| Voltage class, kv | Single-phase kva rating | | | | |
| --- | --- | --- | --- | --- | --- |
| | 1,000 | 5,000 | 10,000 | 25,000 | 50,000 |
| 2.5 | | | | | |
| 15 | 4.5–8.0 | 4.5–8 | 4.5–8.0 | | |
| 25 | 5.5–9.0 | 5.5–9 | 5.5–9.0 | 5 5–9 0 | |
| 69 | 7.0–11 | 7.0–11 | 7.0–11 | 7.0–11 | |
| 138 | 8.5–17 | 8.5–17 | 8.5–17 | 8.5–17 | 8.5–17 |
| 161 | ....... | 9.5–18 | 9.5–18 | 9.5–18 | 9.5–18 |
| 230 | ....... | ...... | 11–20 | 11–20 | 11–20 |

*Positive- and negative-sequence impedances* of three-phase transmission lines can be obtained from Tables 7-8 through 7-11

$$x_1 = x_2 = x_a + x_d \text{ ohms/(phase)(mile)}$$

where $x_d$ = one-third of sum of $x_d$'s from tables for three spacings between line conductors

*Zero-sequence impedance* of aerial lines is dependent upon a number of factors including the type of grounding, circuit configuration, and the number and type of ground wires. The fundamental equations are quite involved, but can be simplified by defining

$$r_e = 0.00477f \text{ ohms/(phase)(mile)}$$
$$x_e = 0.006985f \log 4.6655 \times 10^6 \rho/f \text{ ohms/(phase)(mile)}$$

where $f$ = frequency, cps
　　　$\rho$ = earth resistivity, ohms/m$^3$

*Single three-phase circuit* with earth return but without ground wires, zero-sequence impedance

$$z_0 = r_a + r_e + j(x_e + x_a - 2x_d) \qquad \text{ohms/(phase)(mile)}$$

where $r_a$, $r_e$, $x_e$, and $x_a$ are obtained directly from the conductor tables and

$$x_d = \tfrac{1}{3}(x_{d(ab)} + x_{d(bc)} + x_{d(ca)}) \qquad \text{ohms/(phase)(mile)}$$

where $x_{d(ab)} = x_d$ from tables for conductor spacing $a$ to $b$ etc.

*Mutual zero-sequence impedance* between two three-phase circuits with earth return but without ground wires

$$z_{0(m)} = r_e + j(x_e - 3x_d) \qquad \text{ohms/(phase)(mile)}$$

where $x_d = \tfrac{1}{9}(x_{d(aa')} + x_{d(ab')} + x_{d(ac')} + x_{d(ba')} + x_{d(bb')} + x_{d(bc')}$
$$+ x_{d(ca')} + x_{d(cb')} + x_{d(cc')})$$

$x_{d(aa')} = x_d$ from tables for conductor spacing of $a$ to $a'$, etc.

*Single three-phase circuit* with ground wire and earth return, zero-sequence impedance

$$z_{0(g)} = 3r_a + r_e + j(x_e + 3x_a) \qquad \text{ohms/(phase)(mile)}$$

*Single three-phase circuit* with two ground wires and earth return, zero-sequence impedance

$$z_{0(g)} = \tfrac{3}{2}r_a + r_e + j(x_e + \tfrac{3}{2}x_a - \tfrac{3}{2}x_d)$$

where $x_d = x_d$ from table for spacing between ground wires.

*Zero-sequence impedance* of a three-phase circuit with $n$ ground wires and earth return

$$z_{0(g)} = \frac{3}{n}r_a + r_e + j\left[x_e + \frac{3x_a}{n} - \frac{3(n-1)}{n}x_d\right] \qquad \text{ohms/(phase)(mile)}$$

where $x_d = (1/n)(n-1)$ (sum of $x_d$'s for all possible distances between all ground wires).

*Zero-sequence mutual impedance* between one circuit with earth return and $n$ ground wires with earth return

$$z_{0(m)} = r_e + j(x_e - 3x_d) \qquad \text{ohms/(phase)(mile)}$$

where $x_d = (1/3n)(x_{d(ag1)} + x_{d(bg1)} + x_{d(cg1)} + \cdots + x_{d(agn)}$
$$+ x_{d(bgn)} + x_{d(cgn)})$$

*Zero-sequence impedance,* one three-phase circuit with $n$ ground wires and earth return

$$z_0 = z_{0(a)} - z_0{}^2{}_{(ag)}/z_{0(g)}$$

where $z_{0(a)}$ = zero-sequence impedance of three-phase circuit

$z_{0(g)}$ = zero-sequence impedance of $n$ ground wires

$z_{0(ag)}$ = zero-sequence mutual impedance between three-phase circuit as one group of conductors and ground wires as other conductor group

*Shunt capacitive reactance of three-phase circuits* can be obtained from Tables 7-8 through 7-11 for positive and negative sequence where

$$x'_1 = x'_2 = x'_a + x'_d \quad \text{megohms/(phase)(mile)}$$

*Divide* by number of miles of line to get total reactance.

$$x'_d = \tfrac{1}{3} \times \text{sum of } x_d\text{'s for three distances between line conductors}$$
$$= \tfrac{1}{3}(x'_{d(ab)} + x'_{d(ac)} + x'_{d(bc)})$$

*Zero-sequence shunt capacitive reactance* of a single three-phase circuit and earth

$$x'_{0(a)} = x'_a + x'_e - 2x'_d \quad \text{megohms/(conductor)(mile)}$$

*Divide* by number of miles of line to get total reactance.

*Zero-sequence shunt capacitive reactance* of one ground wire and earth

$$x'_{0(g)} = 3x'_{a(g)} + x'_{e(g)} \quad \text{megohms/(conductor)(mile)}$$
$$x'_e = \frac{12.30}{f} \log 2h \quad \text{megohms/(conductor)(mile)}$$

where $f$ = frequency, cps
$h$ = height above ground, ft

*Zero-sequence capacitive reactance* of *two* ground wires and earth

$$x'_{0(g)} = \tfrac{3}{2}x'_{a(g)} + x'_{e(g)} - \tfrac{3}{2}x'_d \quad \text{megohms/(conductor)(mile)}$$

where $x'_d = x'_d$ for distance between ground wires

*Zero-sequence capacitive reactonce* of $n$ ground wires and earth

$$x'_{0(g)} = x'_e + (3/n)x'_a - [3(n-1)/n]x'_d \quad \text{megohms/(conductor)(mile)}$$

where $x'_d = \dfrac{2}{n(n-1)}$ sum of all $x'_d$'s for all possible distances between all possible *pairs* of ground wires

or $x'_d = \dfrac{1}{n(n-1)}$ sum of all $x'_d$'s for all possible distances between *all* ground wires

*Zero-sequence capacitive reactance* between one circuit and earth and $n$ ground wires and earth

$$x'_{0(ag)} = x'_e - 3x'_d \quad \text{megohms/(conductor)(mile)}$$
$$x'_d = (1/3n)(x'_{d(ag1)} + x'_{d(bg1)} + x'_{d(cg1)} + \cdots + x'_{d(agn)} + x'_{d(bgn)} + x'_{d(cgn)})$$

*Zero-sequence capacitive reactance* of one circuit with $n$ ground wires

$$x'_0 = x'_{0(a)} - x'_{0(ag)}{}^2/x'_{0(g)} \quad \text{megohms/(conductor)(mile)}$$

*Shunt capacitive reactance* of single-phase circuit with identical conductors $a$ and $b$

$$x' = 2(x'_a + x'_d) \quad \text{megohms/(mile)}(circuit)$$
$$x'_d = x'_d \quad \text{for spacing } a \text{ to } b$$

when $a$ and $b$ conductors are not identical

$$x' = x'_{a(a)} + x'_{a(b)} + 2x'_d \qquad \text{megohms/(mile)}(circuit)$$

*Shunt capacitive reactance* of one conductor and earth

$$x' = x'_a + \tfrac{1}{3}x'_e \qquad \text{megohms/mile}$$

### Table 7-8. Characteristics of Copper Conductors
Hard-drawn, 97.3 Per Cent Conductivity

| Conductor size | | OD, in. | Wt, lb/mile | Capacity,* amp | $x'_a$† | $r_a$‡ | $x_a$¶ |
|---|---|---|---|---|---|---|---|
| Cir mils | Awg or B.&S. | | | | | | |
| 1,000,000 | ... | 1.152 | 16,300 | 1,300 | 0.0901 | 0.0685 | 0.400 |
| 900,000 | ... | 1.092 | 14,670 | 1,220 | .0916 | 0.0752 | .406 |
| 800,000 | ... | 1.029 | 13,040 | 1,130 | .0934 | 0.0837 | .413 |
| 750,000 | ... | 0.997 | 12,230 | 1,090 | .0943 | 0.0888 | .417 |
| 700,000 | ... | .963 | 11,410 | 1,040 | .0954 | 0.0947 | .422 |
| 600,000 | ... | .891 | 9,781 | 940 | .0977 | 0.109 | .432 |
| 500,000 | ... | .814 | 8,151 | 840 | .1004 | 0.130 | .443 |
| 450,000 | ... | .770 | 7,336 | 780 | .1020 | 0.144 | .451 |
| 400,000 | ... | .726 | 6,521 | 730 | .1038 | 0.162 | .458 |
| 350,000 | ... | .679 | 5,706 | 670 | .1058 | 0.184 | .466 |
| 300,000 | ... | .629 | 4,891 | 610 | .1080 | 0.215 | .476 |
| 250,000 | ... | .574 | 4,076 | 540 | .1108 | 0.257 | .487 |
| 211,600 | 4/0 | .522 | 3,450 | 480 | .1136 | 0.303 | .503 |
| 167,800 | 3/0 | .464 | 2,736 | 420 | .1171 | 0.382 | .518 |
| 133,100 | 2/0 | .414 | 2.170 | 360 | .1205 | 0.481 | .532 |
| 105,500 | 1/0 | .368 | 1,720 | 310 | .1240 | 0.607 | .546 |
| 83,690 | 1 | .328 | 1,364 | 270 | .1274 | 0.765 | .560 |
| 66,370 | 2 | .320 | 1,071 | 240 | .1281 | 0.955 | .571 |
| 52,630 | 3 | .285 | 850 | 200 | .1315 | 1.20 | .585 |
| 41,740 | 4 | .254 | 674 | 180 | .1349 | 1.52 | .599 |
| 33,100 | 5 | .226 | 534 | 150 | .1384 | 1.91 | .613 |
| 26,250 | 6 | .162 | 420 | 120 | .1483 | 2.39 | .637 |
| 20,800 | 7 | .144 | 333 | 110 | .1517 | 3.01 | .651 |
| 16,510 | 8 | .129 | 264 | 90 | .1552 | 3.80 | .665 |

* Approximate current-carrying capacity for conductor at 75°C, air at 25°C, wind 1.4 mph (2 fps), 60 cycles.
† $x'_a$ = shunt capacitive reactance at 1 ft, megohms/mile.
‡ $r_a$ = resistance at 50°C, 60 cycles, ohms/(conductor)(mile).
¶ $x_a$ = reactance at 1-ft spacing, 60 cycles, ohms/(conductor)(mile).

## Table 7-9. Characteristics of Copperweld and Copperweld-Copper Conductors

| Conductor | Cu equiv, cir mils or awg | OD, in. | Wt, lb/mile | Capacity, amp* | $x'_a$† | $r_a$‡ | $x_a$¶ |
|---|---|---|---|---|---|---|---|
| | | | Copperweld-Copper | | | | |
| 350 E | 350,000 | 0.788 | 7,409 | 660 | 0.1012 | 0.204 | 0.456 |
| 250 E | 250,000 | .666 | 5,292 | 540 | .1064 | 0.278 | .476 |
| 4/0 E | 4/0 | .613 | 4,479 | 490 | .1088 | 0.326 | .486 |
| 3/0 E | 3/0 | .545 | 3,552 | 420 | .1123 | 0.406 | .501 |
| 350 EK | 350,000 | .735 | 6,536 | 680 | .1034 | 0.188 | .452 |
| 250 EK | 250,000 | .621 | 4,669 | 540 | .1084 | 0.261 | .472 |
| 4/0 EK | 4/0 | .571 | 3,951 | 490 | .1109 | 0.308 | .483 |
| 4/0 S | 4/0 | .633 | 4,210 | 490 | .1079 | 0.330 | .477 |
| 2/0 S | 2/0 | .502 | 2,658 | 360 | .1148 | 0.513 | .506 |
| 250 V | 250,000 | .637 | 4,699 | 530 | .1077 | 0.278 | .480 |
| 4/0 V | 4/0 | .586 | 3,977 | 480 | .1101 | 0.325 | .490 |
| 2/0 V | 2/0 | .465 | 2,502 | 360 | .1170 | 0.505 | .518 |
| 4/0 F | 4/0 | .550 | 3,750 | 470 | .1120 | 0.320 | .505 |
| 1/0 F | 1/0 | .388 | 1,870 | 310 | .1224 | 0.627 | .547 |
| 1 F | 1 | .346 | 1,483 | 270 | .1258 | 0.785 | .561 |
| 2 F | 2 | .308 | 1,176 | 230 | .1293 | 0.985 | .575 |
| 2 A | 2 | .366 | 1,356 | 240 | .1241 | 0.978 | .591 |
| 4 A | 4 | .290 | 853 | 180 | .1310 | 1.544 | .620 |
| 6 A | 6 | .230 | 536 | 140 | .1379 | 2.44 | .648 |
| 8 A | 8 | .199 | 392 | 100 | .1422 | 3.87 | .666 |
| 8 C | 8 | .179 | 320 | 100 | .1460 | 3.87 | .678 |
| 9½ D | 9½ | .174 | 298 | 85 | .1462 | 5.43 | .709 |
| | | | Copperweld, 30 per cent Conductivity | | | | |
| 19 no. 9 | 76,000 | .572 | 3,696 | 370 | .1109 | 0.792 | .637 |
| 7 no. 4 | 89,300 | .613 | 4,324 | 410 | .1088 | 0.672 | .629 |
| 7 no. 5 | 70,800 | .546 | 3,429 | 350 | .1122 | 0.848 | .643 |
| 7 no. 6 | 56,100 | .486 | 2,719 | 300 | .1157 | 1.069 | .657 |
| 7 no. 7 | 44,500 | .433 | 2,157 | 260 | .1191 | 1.348 | .671 |
| 7 no. 8 | 35,300 | .385 | 1,710 | 230 | .1226 | 1.699 | .685 |
| 7 no. 9 | 28,000 | .343 | 1,356 | 200 | .1260 | 2.14 | .699 |
| 7 no. 10 | 22,200 | .306 | 1,076 | 170 | .1294 | 2.70 | .713 |
| 3 no. 5 | 30,350 | .392 | 1,467 | 220 | .1221 | 1.963 | .683 |
| 3 no. 6 | 24,100 | .349 | 1,163 | 190 | .1255 | 2.47 | .697 |
| 3 no. 7 | 19,100 | .311 | 922 | 160 | .1289 | 3.12 | .711 |
| 3 no. 8 | 15,150 | .277 | 732 | 140 | .1324 | 3.93 | .725 |
| 3 no. 9 | 12,010 | .247 | 580 | 120 | .1358 | 4.96 | .739 |
| 3 no. 10 | 9,528 | .220 | 460 | 100 | .1392 | 6.26 | .753 |

* Approximate current-carrying capacity; copperweld at 125°C, copperweld-copper at 75°C, air at 25°C, wind 1.4 mph (2 fps), 60 cycles.
† Shunt capacitive reactance at 1 ft, megohms/mile.
‡ Resistance of copperweld conductors at 25°C, copperweld-copper at 50°C, ohms/(conductor)(mile) at 60 cycles.
¶ Reactance at 1-ft spacing, 60 cycles, ohms/(conductor)(mile).

## Table 7-10. Characteristics of Aluminum Cable, Steel-reinforced

| Conductor Size, cir mils or Awg | Cu equiv, cir mils or Awg* | OD, in | Wt, lb/mile | Capacity amp† | $x'_a$‡ | $r_a$§ | $z_a$¶ |
|---|---|---|---|---|---|---|---|
| 1,590,000 | 1000,000 | 1.545 | 10,777 | 1,380 | 0.0814 | 0.0684 | 0.359 |
| 1,510,000 | 950,000 | 1.506 | 10,237 | 1,340 | .0821 | 0.0720 | .362 |
| 1,431,000 | 900,000 | 1.465 | 9,699 | 1,300 | .0830 | 0.0760 | .365 |
| 1,351,000 | 850,000 | 1.424 | 9,160 | 1,250 | .0838 | 0.0803 | .369 |
| 1,272,000 | 800,000 | 1.382 | 8,621 | 1,200 | .0847 | 0.0851 | .372 |
| 1,192,500 | 750,000 | 1.338 | 8,082 | 1,160 | .0857 | 0.0906 | .376 |
| 1,113,000 | 700,000 | 1.293 | 7,544 | 1,110 | .0867 | 0.0969 | .380 |
| 1,033,500 | 650,000 | 1.246 | 7,019 | 1,060 | .0878 | 0.104 | .385 |
| 954,000 | 600,000 | 1.196 | 6,479 | 1,010 | .0890 | 0.113 | .390 |
| 900,000 | 566,000 | 1.162 | 6,112 | 970 | .0898 | 0.119 | .393 |
| 874,500 | 550,000 | 1.146 | 5,940 | 950 | .0903 | 0.123 | .395 |
| 795,000 | 500,000 | 1.093 | 5,399 | 900 | .0917 | 0.138 | .401 |
| 666,000 | 419,000 | 1.000 | 4,527 | 800 | .0943 | 0.160 | .412 |
| 636,000 | 400,000 | 0.977 | 4,319 | 770 | .0950 | 0.169 | .414 |
| 605,000 | 380,500 | .953 | 4,109 | 750 | .0957 | 0.178 | .417 |
| 556,500 | 350,000 | .927 | 4,039 | 730 | .0965 | 0.186 | .420 |
| 477,000 | 300,000 | .858 | 3,462 | 670 | .0988 | 0.216 | .430 |
| 397,500 | 250,000 | .783 | 2,885 | 590 | .1015 | 0.259 | .441 |
| 336,400 | 4/0 | .721 | 2,442 | 530 | .1039 | 0.306 | .451 |
| 266,800 | 3/0 | .642 | 1,936 | 460 | .1074 | 0.385 | .465 |
| 4/0 | 2/0 | .563 | 1,542 | 340 | .1113 | 0.592 | .581 |
| 3/0 | 1/0 | .502 | 1,223 | 300 | .1147 | 0.723 | .621 |
| 2/0 | 1 | .447 | 970 | 270 | .1182 | 0.895 | .641 |
| 1/0 | 2 | .398 | 769 | 230 | .1216 | 1.12 | .656 |
| 1 | 3 | .355 | 610 | 200 | .1250 | 1.38 | .665 |
| 2 | 4 | .316 | 484 | 180 | .1285 | 1.69 | .665 |
| 4 | 6 | .250 | 304 | 140 | .1355 | 2.57 | .659 |

* Based on copper 97 per cent; aluminum 61 per cent.
† Approximate current-carrying capacity for conductor at 75°C, air at 25°C, wind 1.4 mph (2 fps), 60 cycles.
‡ $x'_a$ = capacitive reactance at 1 ft, megohms/mile.
§ Resistance, ohms/(conductor)(mile) at 50°C, 60 cycles.
¶ Reactance at 1-ft spacing, 60 cycles, ohms/(conductor)(mile).

## Table 7-11. Reactance Spacing Factors

| | $x_d$, Separation in Feet | | | | | | | | | |
|---|---|---|---|---|---|---|---|---|---|---|
| Feet | 0 | 1 | 2 | 3 | 4 | 5 | 6 | 7 | 8 | 9 |
| 0 | ..... | 0.000 | 0.084 | 0.133 | 0.168 | 0.195 | 0.217 | 0.236 | 0.252 | ·0.267 |
| 10 | 0.279 | .291 | .302 | .311 | .320 | .329 | .336 | .344 | .351 | .357 |
| 20 | .364 | .369 | .375 | .380 | .386 | .391 | .395 | .400 | .404 | .409 |
| 30 | .413 | .417 | .421 | .424 | .428 | .431 | .435 | .438 | .441 | .445 |

| | $x_d$, Separation in Inches* | | | | | | | | | |
|---|---|---|---|---|---|---|---|---|---|---|
| In. | 0 | 1 | 2 | 3 | 4 | 5 | 6 | 7 | 8 | 9 |
| 0 | ..... | $\overline{.302}$ | $\overline{.217}$ | $\overline{.169}$ | $\overline{.134}$ | $\overline{.107}$ | $\overline{.085}$ | $\overline{.066}$ | $\overline{.050}$ | $\overline{.035}$ |
| 10 | $\overline{.023}$ | $\overline{.011}$ | .000 | .010 | .019 | .027 | .035 | .042 | .049 | .056 |
| 20 | .062 | .068 | .074 | .079 | .084 | .089 | .094 | .098 | .103 | .107 |
| 30 | .111 | .115 | .119 | .123 | .126 | .130 | .133 | .137 | .140 | .143 |
| 40 | .146 | .149 | .152 | .155 | .158 | .160 | .163 | .166 | .168 | .171 |

| | $x'_d$, Separation in Feet | | | | | | | | | |
|---|---|---|---|---|---|---|---|---|---|---|
| Feet | 0 | 1 | 2 | 3 | 4 | 5 | 6 | 7 | 8 | 9 |
| 0 | ...... | .0000 | .0206 | .0326 | .0411 | .0478 | .0532 | .0577 | .0617 | .0652 |
| 10 | .0683 | .0711 | .0737 | .0761 | .0783 | .0804 | .0823 | .0841 | .0858 | .0874 |
| 20 | .0889 | .0903 | .0917 | .0930 | .0943 | .0955 | .0967 | .0978 | .0989 | .0999 |
| 30 | .1010 | .1020 | .1030 | .1040 | .1050 | .1060 | .1060 | .1070 | .1080 | .1090 |

| | $x'_d$, Separation in Inches* | | | | | | | | | |
|---|---|---|---|---|---|---|---|---|---|---|
| In. | 0 | 1 | 2 | 3 | 4 | 5 | 6 | 7 | 8 | 9 |
| 0 | ...... | $\overline{.0737}$ | $\overline{.0532}$ | $\overline{.0411}$ | $\overline{.0326}$ | $\overline{.0260}$ | $\overline{.0206}$ | $\overline{.0160}$ | $\overline{.0120}$ | $\overline{.009}$ |
| 10 | $\overline{.0050}$ | $\overline{.0030}$ | .0000 | .0023 | .0045 | .0066 | .0085 | .0103 | .0120 | .0136 |
| 20 | .0151 | .0166 | .0180 | .0193 | .0206 | .0218 | .0229 | .0240 | .0251 | .0262 |
| 30 | .0272 | .0281 | .0291 | .0300 | .0309 | .0317 | .0326 | .0334 | .0342 | .0349 |
| 40 | .0357 | .0364 | .0371 | .0378 | .0385 | .0392 | .0398 | .0405 | .0411 | .0417 |

| | Values of $x_e$ | | | | | | | | |
|---|---|---|---|---|---|---|---|---|---|
| $\rho$† | 1 | 5 | 10 | 50 | 100 | 500 | 1,000 | 5,000 | 10,000 |
| $x_e$ | 2.05 | 2.35 | 2.47 | 2.77 | 2.89 | 3.19 | 3.31 | 3.61 | 3.73 |

* Bar over number indicates negative value.
† $\rho$ = earth resistivity, meter-ohms.

## Table 7-12. 60-Cycle Three-conductor Belted-paper-insulated Cables

| Voltage class, kv | Insulation thickness, mils | | Awg (B.&S.) or MCM* | Approx max current-carrying capacity† | Positive and negative sequence | | Zero sequence | | |
|---|---|---|---|---|---|---|---|---|---|
| | Conductor | Belt | | | Series reactance, ohms/mile‡ | Shunt capacitive reactance, ohms/mile¶ | Resistance, ohms/mile§ | Series reactance, ohms/mile§ | Shunt capacitive reactance, ohms/mile¶ |
| 3 | 70 | 40 | 6 | 68 | 0.192 | 6,700 | 9.67 | 0.322 | 12,500 |
| | 70 | 40 | 4 | 89 | .181 | 5,800 | 8.06 | .298 | 11,200 |
| | 70 | 40 | 2 | 115 | .171 | 5,100 | 6.39 | .278 | 9,800 |
| | 70 | 40 | 0 | 149 | .156 | 4,400 | 5.06 | .256 | 8,600 |
| | 70 | 40 | 00 | 170 | .142 | 3,500 | 5.69 | .259 | 6,700 |
| | 70 | 40 | 000 | 193 | .138 | 2,700 | 5.28 | .246 | 5,100 |
| | 70 | 40 | 0000 | 218 | .135 | 2,400 | 4.57 | .237 | 4,600 |
| | 70 | 40 | 350 | 288 | .129 | 1,800 | 3.61 | .219 | 3,700 |
| | 70 | 40 | 500 | 348 | .126 | 1,500 | 2.89 | .214 | 3,000 |
| | 75 | 40 | 750 | 427 | .123 | 1,300 | 2.37 | .204 | 2,500 |
| 15 | 170 | 85 | 2 | 106 | .217 | 8,600 | 4.20 | .323 | 15,000 |
| | 160 | 75 | 0 | 138 | .193 | 7,100 | 3.62 | .288 | 12,800 |
| | 155 | 75 | 00 | 156 | .185 | 6,500 | 3.25 | .280 | 12,000 |
| | 155 | 75 | 000 | 178 | .180 | 6,000 | 2.99 | .272 | 11,300 |
| | 155 | 75 | 0000 | 202 | .174 | 5,600 | 2.64 | .263 | 10,600 |
| | 155 | 75 | 250 | 221 | .168 | 5,300 | 2.50 | .256 | 10,200 |
| | 155 | 75 | 350 | 267 | .152 | 5,100 | 2.54 | .250 | 7,200 |
| | 155 | 75 | 500 | 321 | .145 | 4,600 | 2.26 | .239 | 6,200 |
| | 155 | 75 | 600 | 352 | .142 | 4,300 | 1.97 | .231 | 5,700 |
| | 155 | 75 | 750 | 393 | .139 | 4,000 | 1.77 | .226 | 5,100 |

* All cables no. 0 Awg and larger have sector-shaped conductors.
† Three similar, loaded, nonshielded cables in a duct bank assumed; earth temperature 20°C, 100 per cent load factor.
‡ For approximate resistance use $r_a$ of conductors of same Awg or MCM from Table 7-8.
¶ For specific inductive capacity of 3.7.
§ Based upon all return current in the sheath, none in ground.

## Table 7-13. Dimensions, Weight, and Resistance of Pure Copper Wire

| AWG | Diam, in. | Area, $d^2$, cir mils | Lb/1,000 ft (bare wire) | Ft length/lb | Resistance, 77°F, ohms/1,000 ft (bare wire) |
|---|---|---|---|---|---|
| | 1.152 | 1,000,000 | 3,088 | 0.3238 | 0.0108 |
| | 1.031 | 800,000 | 2,470 | 0.4048 | 0.0135 |
| | 0.964 | 700,000 | 2,161 | 0.4627 | 0.0154 |
| | .893 | 600,000 | 1,853 | 0.5397 | 0.0180 |
| | .813 | 500,000 | 1,544 | 0.6477 | 0.0216 |
| | .728 | 400,000 | 1,235 | 0.8 97 | 0.0270 |
| | .575 | 250,000 | 772 | 1.30 | 0.0431 |
| 0000 | .4600 | 211,600 | 653.3 | 1.53 | 0.0509 |
| 000 | .4096 | 167,800 | 518.1 | 1.93 | 0.0642 |
| 00 | .3648 | 133,100 | 410.9 | 2.43 | 0.0811 |
| 0 | .3248 | 105,500 | 325.8 | 3.07 | 0.102 |
| 1 | .2893 | 83,690 | 258.9 | 3.87 | 0.129 |
| 2 | .2576 | 66,370 | 204.9 | 4.88 | 0.162 |
| 3 | .2294 | 52,640 | 162.5 | 6.15 | 0.205 |
| 4 | .2043 | 41,740 | 128.9 | 7.76 | 0.259 |
| 6 | .1620 | 26,250 | 81.05 | 12.34 | 0.410 |
| 8 | .1284 | 16,510 | 49.98 | 20.01 | 0.641 |
| 10 | .1018 | 10,380 | 31.43 | 31.82 | 1.018 |
| 12 | .0808 | 6,530 | 19.77 | 50.59 | 1.619 |
| 14 | .0640 | 4,107 | 12.43 | 80.44 | 2.575 |
| 16 | .0508 | 2,583 | 7.82 | 127.90 | 4.094 |
| 18 | .0403 | 1,624 | 4.92 | 203.40 | 6.510 |
| 20 | .0319 | 1,022 | 3.09 | 323.4 | 10.35 |
| 22 | .0254 | 642 | 1.95 | 514.2 | 16.46 |
| 24 | .0201 | 404 | 1.22 | 817.7 | 26.17 |
| 26 | .0159 | 254 | 0.77 | 1,300 | 41.62 |
| 28 | .0126 | 159.8 | .48 | 2,067 | 66.17 |
| 30 | .0100 | 100.5 | .30 | 3,287 | 105.2 |
| 32 | .0080 | 63.2 | .19 | 5,227 | 167.3 |
| 34 | .0063 | 39.7 | .12 | 8,310 | 266.0 |
| 36 | .0050 | 25.0 | .076 | 13,210 | 423.0 |
| 38 | .0040 | 15.7 | .047 | 21,010 | 672.6 |
| 40 | .0031 | 9.89 | .030 | 33,410 | 1,069 |
| 42 | .0025 | 6.22 | .019 | 52,800 | 1,701 |
| 44 | .0020 | 3.91 | .012 | 82,500 | 2,703 |
| 46 | .0016 | 2.46 | .008 | 128,800 | 4,299 |
| 48 | .0012 | 1.55 | .004 | 229,600 | 6,836 |
| 50 | .0010 | 0.97 | .003 | 330,000 | 10,870 |

*Characteristics of Conductors.* Tables 7-8 to 7-13 give resistance and reactance constants of conductors at 60 cycles.

The compactness of the tables is secured by arranging the constants for the different conductors for 1-ft spacing and using additional tables of spacing factors to take care of other spacings. The formulas relating to these constants follow.

*Three-phase circuit,* impedance to neutral

$$z = r_a + j(x_a + x_d) \qquad \text{ohms/mile}$$

*Note:* For unsymmetrical spacings, use an effective spacing equal to the cube root of the product of the three spacings.

*Example:* Determine the impedance to neutral of a 60-cycle line with 795,000 cir mil ACSR conductor, 54 aluminum strands, with conductor separation = 26 ft. From table of ACSR:

$$r_a = 0.138 \qquad \text{and} \qquad x_a = 0.401 \text{ ohms/mile}$$

From reactance-spacing-factor tables:

$$x_d = 0.395 \text{ ohms/mile}$$
$$z = r_a + j(x_a + x_d)$$
$$= 0.138 + j\,0.796 \qquad \text{ohms/mile}$$

*Shunt Capacitive Reactance*

$$x' = x'_a + x'_d \qquad \text{megohms/mile}$$

*Example:* Determine shunt capacitive reactance of the above transmission line.

From tables,   $x'_a = 0.0917$    $x'_d = 0.0967$ megohms/mile
$$x' = x'_a + x'_d = 0.1884 \text{ megohms/mile}$$

For total line *divide* by number of miles.

*Single-phase circuit,* without earth return

$$\text{Total impedance of circuit} = 2[r_a + j(x_a + x_d)] \qquad \text{ohms/mile}$$

With line and neutral wires, the latter grounded:

$$\text{Total impedance} = z_A - m^2/z_N \qquad \text{ohms/mile}$$

where $z_A$ = line wire $z$    ohms/mile
     $z_N$ = neutral wire $z$    ohms/mile
     $z = (r_a + r_e/3) + j(x_e/3 + x_a)$    ohms/mile
     $m = r_e/3 + j(x_e/3 - x_d)$    ohms/mile

*Current-carrying Capacity of Conductors.* The current-carrying capacity of conductors can be determined approximately from the $I^2R$ losses and the convection and radiation characteristics of the conductors.

$$I^2R = (W_c + W_r)A \qquad \text{watts} \tag{7-58}$$
or   $$I = \sqrt{(A/R)(W_c + W_r)} \qquad \text{amp}$$

where $I$ = conductor current, amp
$R$ = conductor resistance, ohms/ft
$W_c$ = watts/in.$^2$ dissipated by convection
$W_r$ = watts/in.$^2$ dissipated by radiation
$A$ = conductor surface area, in.$^2$/ft length

and $$W_c = (0.0128 \sqrt{pv}/T_a{}^{0.123} \sqrt{d}) \Delta t \qquad \text{watts/in.}^2 \qquad (7\text{-}59)$$

where $p$ = pressure, atm ($p = 1.0$ for atmospheric pressure)
$v$ = air velocity, fps
$T_a$ = average of absolute temperatures of conductor and air, °K
$d$ = outside diameter of conductor, in.
$\Delta t$ = temperature rise, °C

and $$W_r = 36.8E[(T/1,000)^4 - (T_0/1,000)^4] \qquad \text{watts/in.}^2 \qquad (7\text{-}60)$$

where $E$ = relative emissivity of conductor surface (1.0 for black body, 0.5 for average oxidized copper)
$T$ = absolute temperature of conductor, °K
$T_0$ = absolute temperature of surroundings, °K

## 7-3. Transients in Electric Circuits

**D-C Circuits.** When a d-c voltage $E$ is impressed on a circuit consisting of a *resistance* of $R$ ohms and *inductance* of $L$ henrys, the transient current $i$ which flows is a function of time and is given by

$$i = (E/R)(1 - \epsilon^{-Rt/L}) + I_0\epsilon^{-Rt/L} \qquad \text{amp} \qquad (7\text{-}61)$$

where $I_0$ is the current flowing in the circuit at the instant before the voltage $E$ is impressed at $t = 0$. $I_0$ is positive if flowing in the positive direction of $E$ and negative if flowing in opposition to $E$. The current $I_0$ flowing in the circuit at $t = 0$ is the instantaneous value at that time and is consequently a constant and independent of its past time variation. If the source voltage is short-circuited, the current will decay to zero, its value at any time being

$$i = I_0\epsilon^{-Rt/L} \qquad \text{amp} \qquad (7\text{-}62)$$

where $I_0$ = current flowing at $t = 0$, time at which short-circuit is applied
The *time constant* $T$ of a $RL$ circuit is that value of $t$ required for the current to build up to 63.2 per cent of its final value when the circuit is being energized with zero initial current, or to decay to 36.8 per cent of its initial value when short-circuited.

$$T = L/R \qquad \text{sec} \qquad (7\text{-}63)$$

where $R$ is in ohms and $L$ in henrys.
If a *capacitor* of $C$ farads and a *resistor* of $R$ ohms are connected in series to a d-c voltage $E$, the transient current $i$ which flows is given by

$$i = [(E - E_c)/R]\epsilon^{-t/RC} \qquad \text{amp} \qquad (7\text{-}64)$$

where $E_c$ is the voltage on the capacitor the instant before the voltage $E$ is applied and may be either plus or minus, depending upon whether it acts in opposition to or conjunction with the impressed voltage. $E - E_c$ represents the net voltage in the direction of positive $i$.

$$E_c = Q_0/C \quad \text{volts} \tag{7-65}$$

where $Q_0$ = initial charge on capacitor $C$

The *capacitor charge* $q$ is the integral of the current $i$.

$$q = CE(1 - \epsilon^{-t/RC}) + CE_c\epsilon^{-t/RC} \quad \text{coulombs} \tag{7-66}$$

The voltage across the capacitor at any time $t$ after the voltage $E$ is impressed is

$$e_c = q/C = E(1 - \epsilon^{-t/RC}) + E_c\epsilon^{-t/RC} \quad \text{volts} \tag{7-67}$$

The *current* $i$ flowing in an $RC$ circuit when the source voltage is short-circuited and having an initial voltage of $E_c$ is

$$i = (E_c/R)\epsilon^{-t/RC} \quad \text{amp} \tag{7-68}$$

The *charge* $q$ on $C$ for a short-circuited $RC$ circuit at any time $t$ is

$$q = CE_c\epsilon^{-t/RC} \quad \text{coulombs} \tag{7-69}$$

The *voltage* across the capacitor under the same conditions is

$$e_c = E_c\epsilon^{-t/RC} \quad \text{volts} \tag{7-70}$$

The *time constant* $T$ of an $RC$ circuit is that value of $t$ required for the charge to build up to 63.2 per cent of its final value when the circuit is being energized with zero initial charge on the capacitor, or to decay to 36.8 per cent of its initial value when short-circuited.

$$T = RC \quad \text{sec} \tag{7-71}$$

where $R$ is in ohms and $C$ in farads.

When a circuit consisting of series-connected *resistance, inductance,* and *capacitance RLC* is energized from a d-c source, the transient current is dependent upon the relative magnitudes of $R$, $L$, and $C$. It may take one of three forms:

1. A damped exponential wave when $R^2/4L^2 > 1/LC$
2. A critically damped exponential wave when $R^2/4L^2 = 1/LC$
3. An exponentially damped sine wave when $R^2/4L^2 < 1/LC$

**Case 1.** $R^2/4L^2 > 1/LC$

$$i = \left[\frac{E - E_c - (a - b)I_0}{2bL}\right]\epsilon^{-(a-b)t} - \left[\frac{E - E_c - (a + b)I_0}{2bL}\right]\epsilon^{-(a+b)t} \quad \text{amp}$$

$$(7\text{-}72)$$

where $a = R/2L$
$b = \sqrt{R^2/4L^2 - 1/LC}$
$E_c = Q_0/C$
$Q_0 = $ initial charge on $C$
$I_0 = $ initial current in circuit before $t = 0$

If at the time of application of voltage there is no initial current and no initial charge on the capacitor, the expression for the transient current becomes

$$i = (E/2bL)\epsilon^{-at}(\epsilon^{bt} - \epsilon^{-bt}) \quad \text{amp}$$
$$= (E/bL)\epsilon^{-at}\sinh bt \quad \text{amp} \quad (7\text{-}73)$$

The time for the current to rise to maximum value is

$$t_m = (1/b)\tanh^{-1}(b/a) \quad \text{sec} \quad (7\text{-}74)$$

**Case 2.** $R^2/4L^2 = 1/LC$

$$i = \left[I_0 + \frac{2(E - E_c) - RI_0}{2L}t\right]\epsilon^{-(R/2L)t} \quad \text{amp} \quad (7\text{-}75)$$

If no initial current or charge is present

$$i = (E/L)t\epsilon^{-(R/2L)t} \quad \text{amp} \quad (7\text{-}76)$$

Time required for current to reach maximum value:

$$t_m = 2L/R \text{ sec} \quad (7\text{-}77)$$

Maximum current is

$$i_{\max} = 0.736E/R \quad \text{amp} \quad (7\text{-}78)$$

**Case 3.** $R^2/4L^2 < 1/LC$

$$i = \left\{\frac{[2(E - E_c) - RI_0]}{2\beta L}\sin \beta t + I_0 \cos \beta t\right\}\epsilon^{-(R/2L)t} \quad \text{amp} \quad (7\text{-}79)$$

where $\beta = \sqrt{1/LC - R^2/4L^2}$

If no initial current or charge is present, the transient current becomes

$$i = (E/\beta L)\epsilon^{-at}\sin \beta t \quad \text{amp} \quad (7\text{-}80)$$

where $a = R/2L$

The frequency of oscillation

$$f = (1/2\pi)\sqrt{1/LC - R^2/4L^2} \quad \text{cps} \quad (7\text{-}81)$$

If $R^2/4L^2$ is negligible, the undamped or natural frequency is

$$f = 1/(2\pi\sqrt{LC}) \quad \text{cps} \quad (7\text{-}82)$$

Time at which first current maximum occurs

$$t_m = \sigma/\beta \qquad (7\text{-}83)$$

where $\sigma = \tan^{-1}(\beta/a)$

Any of the positive maximum values of current can be found from

$$I_n = (E\sqrt{LC}/L)\epsilon^{-at_n} \qquad (7\text{-}84)$$

where $t_n$ = time at which $n$th current maximum occurs

*Numerical decrement* is defined as the difference between any two current peaks separated by a complete cycle divided by the larger of the two.

$$\text{Numerical decrement} = 1 - \epsilon^{-R/2Lf} \qquad (7\text{-}85)$$

*Logarithmic decrement* is defined as the logarithm of the ratio of two consecutive positive current maxima.

$$\text{Logarithmic decrement} = \log_\epsilon (I_n/I_{n+1}) = R/2Lf \qquad (7\text{-}86)$$

where $f$ = frequency of oscillation

The *discharge current* which flows when the source voltage is short-circuited is obtained in each of the three cases by setting $E = 0$.

**A-C Circuits.** If an alternating voltage of the form $e = E_m \sin(\omega t + \lambda)$ is applied to a *RL series circuit*, a current which is made up of two components, steady-state and transient, will flow.

$$
\begin{aligned}
i &= i_s + i_t \\
&= (E_m/Z)\sin(\omega t + \lambda - \theta) - (E_m/Z)\sin(\lambda - \theta)\epsilon^{-(R/L)t} \quad (7\text{-}87)
\end{aligned}
$$

where $E_m = \sqrt{2}\,E$

$\quad E$ = rms value of applied a-c voltage

$\quad Z = \sqrt{R^2 + (\omega L)^2}$

$\quad \omega = 2\pi f$

$\quad \lambda$ = angular displacement between $e = 0$ and $t = 0$ measured positively from $e = 0$, the zero point for $e$ being at the intersection where $e$ is rising in the positive direction, deg or radians

$\quad \theta = \tan^{-1} \omega L/R$, the steady-state power-factor angle

If an *alternating voltage* is applied to an *RC* circuit, the current which flows is

$$i = \frac{E_m}{Z}\sin(\omega t + \lambda - \theta) + [(E_m/R)\sin\theta\cos(\lambda - \theta) + Q_0/RC]\,\epsilon^{-t/RC}$$

$$\qquad (7\text{-}88)$$

where $Z = \sqrt{R^2 + (-1/\omega C)^2}$

$\quad \theta = \tan^{-1}(-1/\omega CR)$

$\quad Q_0$ = initial charge on $C$

If an alternating voltage is applied to a *RLC* circuit, the current which flows is dependent upon whether $R^2/4L^2$ is greater than, equal to, or less than $1/LC$.

**Case 1.**  $R^2/4L^2 > 1/LC$

$$i = \frac{E_m}{Z} \sin (\omega t + \lambda - \theta) + \epsilon^{-at}[(E_d/bL) \sinh bt - (E_m/Z) \sin (\lambda - \theta) \\ \cos bt] \quad (7\text{-}89)$$

where $Z = \sqrt{R^2 + (\omega L - 1/\omega c)^2}$
$\quad\quad\ \theta = \tan^{-1} [(\omega L - 1/\omega C)/R]$
$\quad\quad\ a = R/2L$
$\quad\quad\ b = \sqrt{R^2/4L^2 - 1/LC}$

$$E_d = \left[ E_m \sin \lambda - (E_m \omega L/Z) \cos (\lambda - \theta) - \frac{Q_0}{C} - \frac{E_m R}{2Z} \sin (\lambda - \theta) \right] \\ (7\text{-}90)$$

**Case 2.**  $R^2/4L^2 = 1/LC$

$$i = (E_m/Z) \sin (\omega t + \lambda - \theta) + \epsilon^{-at}[(E_d/L)t - (E_m/Z) \sin (\lambda - \theta)] \quad (7\text{-}91)$$

**Case 3.**  $R^2/4L^2 < 1/LC$

$$i = (E_m/Z) \sin (\omega t + \lambda - \theta) + \epsilon^{-at}[(E_d/\beta L) \sin \beta t \\ - (E_m/Z) \sin (\lambda - \theta) \cos \beta t] \quad (7\text{-}92)$$

where $\quad\quad\quad\quad\quad\quad\quad \beta = \sqrt{1/LC - R^2/4L^2}$

The nature of the *transient response* of a circuit is indicated by the roots of the *determinantal equation* $D(p)$ for that circuit. $D(p)$ for a simple series circuit can be found by first writing the impedance of the circuit in terms of $j\omega$ and then substituting $p$ for $j\omega$ and equating to zero. For example,

$$Z = R + j\omega L + 1/j\omega C \quad (7\text{-}93)$$
$$D(p) = R + pL + 1/pC = 0 \quad (7\text{-}94)$$

For complex networks $D(p)$ can be found from the general expression for $D$ as defined on page 7-16 for the generalized impedance network by substituting $p$ for $j\omega$ in the impedance elements, and then setting $D(p) = 0$ to solve for the values of $p$.

## 7-4. Traveling Waves on Transmission Lines

The *surge impedance* of a conductor is equal to

$$Z = \sqrt{L/C} \quad \text{ohms} \quad (7\text{-}95)$$

where $L$ = inductance, henrys/unit length conductor
$\quad\quad\ C$ = capacitance, farads/unit length conductor
For a single aerial conductor parallel to the earth and with zero earth resistivity,

$$L = 7.410 \times 10^{-4} \log 2h/r \quad \text{henrys/mile} \quad (7\text{-}96)$$
and $\quad\quad\quad C = (3.882 \times 10^{-8})/\log (2h/r) \quad \text{farads/mile} \quad (7\text{-}97)$

where $h$ = height of conductor above ground
$\quad\quad\ r$ = radius of conductor in same units

For cables

$$L = 7.410 \times 10^{-4} \log (r_2/r_1) \quad \text{henrys/mile} \quad (7\text{-}98)$$
$$C = (3.882 \times 10^{-8} \epsilon)/\log (r_2/r_1) \quad \text{farads/mile} \quad (7\text{-}99)$$

where $r_1$ = radius of conductor

$r_2$ = inner radius of sheath

$\epsilon$ = permittivity

The surge impedance of a typical aerial line is approximately 500 ohms and of a typical cable 50 ohms. The relationship between a traveling voltage wave and current wave is

$$e = iZ \qquad (7\text{-}100)$$

The *velocity of propagation* of a surge or traveling wave is

$$v = 1/\sqrt{LC} = 984 \text{ ft}/\mu \text{ sec for aerial conductor} \qquad (7\text{-}101)$$

For cables

$$v = 984/\sqrt{\epsilon} \text{ ft}/\mu\text{sec} \qquad (7\text{-}102)$$

where $\epsilon$ has a range of about 2.5 to 4.0.

The *propagation constant* of a conductor when energized with alternating current is

$$\alpha = \sqrt{zy} = \alpha_1 + j\beta \qquad (7\text{-}103)$$

where $z$ = series impedance/(unit length)(phase)

$y$ = shunt admittance/(unit length)(phase to neutral)

$\alpha_1$ = attenuation constant, nepers/mile

$\beta$ = wavelength constant, radians/mile

*Wavelength*

$$\lambda = 2\pi/\beta \qquad \text{miles} \qquad (7\text{-}104)$$

*Velocity of propagation* of an alternating voltage is

$$v = f\lambda = \omega/\beta \qquad (7\text{-}105)$$

## 7-5. Nonsinusoidal Periodic Waves

The solution of circuit problems when the applied voltages are *nonsinusoidal periodic waves* can be accomplished by resolving the wave to sinusoidal components by use of *Fourier's series* and applying the superposition theorem. Fourier's theorem states that any function $f(x)$ which within an interval is finite, single-valued, and continuous or has only a finite number of discontinuities may be represented by a series of the form

$$f(x) = A_0/2 + A_1 \cos x + A_2 \cos 2x + \cdots + A_n \cos nx + \cdots$$
$$+ B_1 \sin x + B_2 \sin 2x + \cdots + B_n \sin nx + \cdots$$

$$= A_0/2 + \sum_{n=1}^{n=\infty} A_n \cos nx + \sum_{n=1}^{n=\infty} B_n \sin nx \qquad (7\text{-}106)$$

where $A_n = 1/\pi \int_0^{2\pi} f(x) \cos nx\, dx$

$\qquad B_n = 1/\pi \int_0^{2\pi} f(x) \sin nx\, dx$

$\qquad A_0/2 = 1/2\pi \int_0^{2\pi} f(x)\, dx$

## ELECTRICAL MACHINERY

### 7-6. Rectifiers

For the condition of no grid delay and no overlap, the d-c no-load voltage of a rectifier is given by

$$E_{d0} = \sqrt{2}\, Ep/\pi \sin\,(\pi/p) \qquad \text{volts} \qquad\qquad (7\text{-}107)$$

where $E_{d0}$ = d-c voltage at no load
$\qquad E$ = rms line-to-neutral secondary or anode voltage
$\qquad p$ = number of secondary phases

The *output voltage* of a $p$-phase rectifier with grid delay but no load is

$$E_d = E_{d0} \cos \alpha \qquad \text{volts} \qquad\qquad (7\text{-}108)$$

where $\alpha$ = grid-delay angle

The output voltage of a $p$-phase rectifier with grid delay and operating under load is

$$E_d = E_{d0} - E_{d0}(1 - \cos \alpha) - (p/2\pi)XI_{dc} \qquad \text{volts} \qquad (7\text{-}109)$$

where $X$ = commutating reactance; reactance to neutral of one anode circuit which includes anode reactors, rectifier transformer, and supply circuit using generator subtransient reactance, ohms
$\qquad I_{dc}$ = load current, amp

The *regulation formula* for a $p$-phase grid-controlled rectifier:

$$E_d = \begin{matrix}\text{no-load}\\\text{voltage}\end{matrix} - \begin{matrix}\text{reduction by}\\\text{grid delay}\end{matrix} - \begin{matrix}\text{reduction}\\\text{by overlap}\end{matrix} - \begin{matrix}\text{resistance}\\\text{drop}\end{matrix} - \begin{matrix}\text{arc}\\\text{drop}\end{matrix}$$

$$= \quad E_{d0} \quad - E_{d0}(1 - \cos \alpha) - (p/2\pi)XI_{dc} - \quad W/I_{dc} \quad - \quad A \qquad \begin{matrix}\text{volts}\\(7\text{-}110)\end{matrix}$$

where $W$ = total copper losses, watts
$\qquad A$ = arc drop, volts

The *angle of overlap u* without grid control.

$$\cos u = 1 - XI_{dc}/\sqrt{2}\, E \sin\,(\pi/p) \qquad\qquad (7\text{-}111)$$

When grid control is present

$$\cos\,(u + \alpha) = \cos \alpha - XI_{dc}/\sqrt{2}\, E \sin\,(\pi/p) \qquad\qquad (7\text{-}112)$$

## 7-7. D-C Motors and Generators

The *emf equation* for a d-c generator is

$$E = Z\phi_a \frac{\text{rpm}}{60} \frac{\text{poles}}{\text{paths}} \times 10^{-8} \quad \text{volts} \quad (7\text{-}113)$$

where $E$ = generated voltage between terminals
$Z$ = total number of active conductors
$\phi_a$ = flux per pole which crosses air gap and is cut by armature conductors
rpm = armature speed in revolutions per minute
paths = number of parallel circuits through armature
The *terminal voltage* of a d-c generator $V$

$$V = E - IR \quad (7\text{-}114)$$

where $I$ = armature current
$R$ = resistance between brushes
$E$ = generated voltage between terminals
The *armature torque of a generator*

$$T = 0.1175 Z I \phi_a \frac{\text{poles}}{\text{paths}} \times 10^{-8} \quad \text{lb-ft} \quad (7\text{-}115)$$

*Power input $P_i$ to a generator*

$$\begin{aligned} P_i &= 1.903 T \text{ rpm} \times 10^{-4} \quad \text{hp} \\ &= 0.1420 T \text{ rpm} \quad \text{watts} \\ &= \text{power output of a d-c motor} \end{aligned} \quad (7\text{-}116)$$

*Power output $P_0$ of a generator*

$$\begin{aligned} P_0 &= VI \quad \text{watts} \\ &= \text{power input to a d-c motor} \end{aligned} \quad (7\text{-}117)$$

## 7-8. A-C Motors and Generators

*Frequency $f$ of the voltage generated in a synchronous generator.*

$$f = p \times \text{rpm}/120 \quad \text{cps} \quad (7\text{-}118)$$

where $p$ = number of poles
The internal generated voltage for a full-pitch winding is approximately

$$E = 2.1 f \phi N \times 10^{-8} \quad \text{volts} \quad (7\text{-}119)$$

where $\phi$ = flux per pole, maxwells
$f$ = frequency, cps
$N$ = number of series-connected active conductors per phase
The *transient open-circuit time constant $T'_{d0}$* of a machine is the time in seconds for the field current to build up to 0.632 times its final value after the application of field voltage with the armature open-circuited. The

mathematical expression for this relation is

$$I_f = (E_{dc}/R_f)(1 - \epsilon^{-t/T'_{d0}}) \qquad \text{amp} \qquad (7\text{-}120)$$

where $E_{dc}$ = exciter voltage, volts
$R_f$ = resistance of field winding, ohms
$t$ = time, sec
$T'_{d0}$ can be readily found from this expression if an oscillogram of $I_f$ is available.

The *transient short-circuit time constant* $T'_d$ which determines the rate of decay of the transient component of the armature current during short circuit is given by

$$T'_d = (X'_d/X_d)T'_{d0} \qquad \text{sec} \qquad (7\text{-}121)$$

where $X'_d$ = transient reactance
$X_d$ = synchronous reactance
The *subtransient short-circuit time constant* $T''_d$, which determines the rate of decay of the subtransient component of the armature current during short-circuit, is expressed in

$$\Delta i'' = (i''_d - i'_d)\epsilon^{-t/T''_d} \qquad (7\text{-}122)$$

where $\Delta i''$ can be obtained from short-circuit oscillograms that show the subtransient and transient currents, $i''_d$ and $i'_d$.

The *armature short-circuit time constant* $T_a$ is given by

$$T_a = X_2/2\pi f R_a \qquad \text{sec} \qquad (7\text{-}123)$$

where $X_2$ = negative-sequence reactance, ohms
$f$ = frequency, cps.
$R_a$ = armature resistance, ohms

## 7-9. Induction Motors

The induction-motor equivalent circuit is shown in Fig. 7-41, where
$r_s$ = stator resistance, ohms
$X_s$ = stator leakage reactance at rated frequency, ohms
$r_r$ = rotor resistance, ohms
$X_r$ = rotor leakage reactance at rated frequency, ohms
$Z_m$ = shunt impedance to include the effect of magnetizing current and no-load losses, ohms
$E_s$ = applied voltage, volts
$i_s$ = stator current, amp
$i_r$ = rotor current, amp
$S$ = slip
The slip $S$ of an induction machine is given by

$$S = 1 - \text{rpm}_r/\text{rpm}_{syn} \qquad \text{per unit} \qquad (7\text{-}124)$$

where $\text{rpm}_r$ = rotor speed
$\quad\text{rpm}_{syn}$ = synchronous speed
*Total shaft power* of induction motor is

$$P = [(1 - S)/S]3r_r i_r^2 \quad \text{watts}$$
$$= (1/746)[(1 - S)/S]3r_r i_r^2 \quad \text{hp} \tag{7-125}$$

*Induction-motor efficiency,* neglecting losses other than rotor copper loss

$$\text{efficiency} = 100(1 - S) \quad \text{per cent} \tag{7-126}$$

*Shaft torque* is given by

$$T = \frac{7.04}{(\text{rpm})_{syn}} \frac{(3r_r i_r^2)_{\text{watts}}}{(S)_{\text{per unit}}} \quad \text{lb-ft} \tag{7-127}$$

or approximately

$$T = \frac{21.12}{\text{rpm}_{syn}} \frac{r_r/S}{(r_s + r_r/S)^2 + (X_s + X_r)^2} \quad \text{lb ft} \tag{7-128}$$

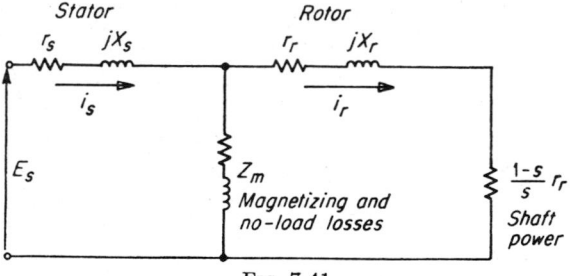

FIG. 7-41

## 7-10. Transformers

The voltage induced in a transformer winding is given by

$$E = 4.44fnAB_{\max} \times 10^{-8} \quad \text{volts} \tag{7-129}$$

where $f$ = frequency, cps
$\quad n$ = number of turns in winding
$\quad A$ = cross-sectional area of uniform magnetic circuits, cm$^2$
$\quad B_{\max}$ = maximum flux density in core, lines/cm$^2$

A *two-winding transformer* with a primary winding $P$ of $n_1$ turns and a secondary winding $S$ of $n_2$ turns has an equivalent circuit as shown in Fig. 7-42. The corresponding vector diagram is shown in Fig. 7-43, where
$\quad N = n_2/n_1$
$\quad Z_m$ = magnetizing impedance, ohms
$\quad Z_p = R_p + j\omega[L_p - (n_1/n_2)M] \quad \text{ohms}$
$\quad Z_s = R_s + j\omega[L_s - (n_2/n_1)M] \quad \text{ohms}$
$R_p, R_s$ = primary and secondary effective winding resistances, ohms
$\quad \omega = 2\pi f$

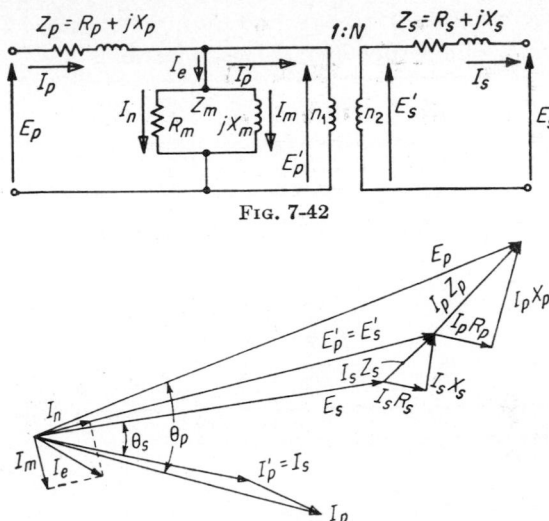

FIG. 7-42

FIG. 7-43

$L_p$, $L_s$ = primary and secondary winding self-inductance, henrys

$M$ = mutual inductance between windings, henrys

$Z_{ps}$ = $Z_p + (1/N^2)Z_s$   ohms leakage impedance between $p$ and $s$ windings, measured on $p$ winding with $s$ winding short-circuited

In many calculations it is customary to make

$$Z_p = (1/N^2)Z_s = \tfrac{1}{2}Z_{ps}, \quad \text{ohms} \tag{7-130}$$

## 7-11. Decibels

In communications work, it is convenient to consider the transmission characteristic of a system in terms of attenuation, or the decrease in power along the transmission system. The ratio between the voltages, currents, and powers at any two points on such a system is a measure of the attenuation of the circuit between these two points. It is not usually convenient to express these transmission losses or gains in terms of the voltage and current or power ratios directly. The losses so expressed cannot be added to obtain the total loss, but must be *multiplied*. Consequently, these ratios are usually expressed in decibels (db) which can be added directly and can be defined as

$$db = 10 \log (P_1/P_2) \quad db = 20 \log (E_1/E_2) \quad db = 20 \log (I_1/I_2)$$

The last two formulas are valid only if the impedance levels of the circuits upon which the two currents or voltages are based are the same.

Various power and voltage or current ratios and the corresponding decibels and efficiencies are shown in the Table 7-14.

## Table 7-14. Power, Voltage, and Current Ratios and Their Corresponding Values in Decibels

| Power ratio | Voltage or current ratio | Decibels (db) | Efficiency, % |
|---|---|---|---|
| 1.26 | 1.12 | 1.0 | 79.5 |
| 1.58 | 1.26 | 2.0 | 63.4 |
| 2.0 | 1.41 | 3.0 | 50.0 |
| 3.16 | 1.78 | 5.0 | 31.6 |
| 5.01 | 2.24 | 7.0 | 20.0 |
| 10.0 | 3.16 | 10.0 | 10.0 |
| 50.12 | 7.08 | 17.0 | 1.99 |
| 100.0 | 10.0 | 20.0 | 1.0 |
| 1,000.0 | 31.6 | 30.0 | 0.1 |
| $10^5$ | 316.2 | 50.0 | .001 |
| $10^8$ | 10,000.0 | 80.0 | .000001 |
| $10^{10}$ | 100,000.0 | 100.0 | .00000001 |

## 7-12. Resistor and Capacitor Color Codes

### Table 7-15. Color Code for Fixed Resistors
Values in Ohms

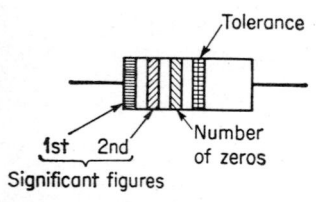

1st   2nd
Significant figures

Tolerance

Number of zeros

Resistor with axial wire leads

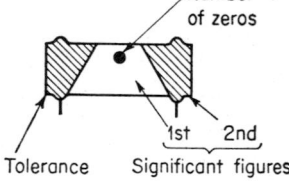

Number of zeros

Tolerance   1st   2nd   Significant figures

Resistor with radial wire leads

| Resistor with radial wire leads.... | Body | End | Dot or band | End | | |
|---|---|---|---|---|---|---|
| Resistor with axial wire leads..... | 1st band | 2nd band | 3rd band | End band | | |
| Color | Value | Value | Value | Color | Tolerance, % | |
| Black......................... | 0 | 0 | None | Gold........... | ± 5 | |
| Brown......................... | 1 | 1 | 0 | Silver | ± 10 | |
| Red........................... | 2 | 2 | 00 | None........... | ± 20 | |
| Orange......................... | 3 | 3 | 000 | | | |
| Yellow......................... | 4 | 4 | 0000 | | | |
| Green......................... | 5 | 5 | 00000 | | | |
| Blue.. ......................... | 6 | 6 | 000000 | | | |
| Violet......................... | 7 | 7 | 0000000 | | | |
| Gray......................... | 8 | 8 | 00000000 | | | |
| White......................... | 9 | 9 | 000000000 | | | |

## Table 7-16. Color Code for JAN Fixed Mica Capacitors

Color-code scheme for JAN standard fixed mica capacitors.  The significance of the letters denoting characteristic will be found in Specification JAN C-5.

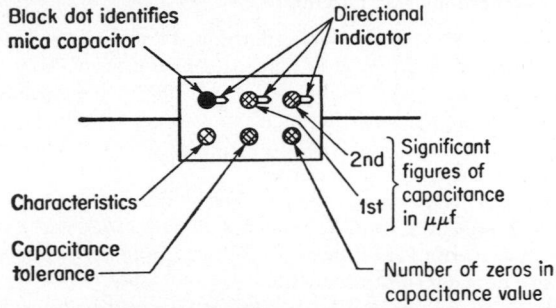

| Color | Capacitance, μμf (significant fig) | Decimal multiplier | Tolerance, % | Characteristic |
|---|---|---|---|---|
| Black...... | 0 | 1 | 20 (M)* | A |
| Brown..... | 1 | 10 | ...... | B |
| Red....... | 2 | 100 | 2 (G)* | C |
| Orange.... | 3 | 1,000 | ...... | D |
| Yellow..... | 4 | ........ | ...... | E |
| Green..... | 5 | ........ | ...... | F |
| Blue....... | 6 | ........ | ...... | G |
| Violet..... | 7 | | | |
| Gray...... | 8 | | | |
| White | 9 | | | |
| Gold...... | ... | 0.1 | 5 (J)* | |
| Silver..... | ... | .01 | 10 (K)* | |

* Code letter for indicated % tolerance.

## 7-13. High-voltage Flashover

### Flashover Definitions and Formulas

*Flashover* denotes an electrical breakdown or disruptive discharge over the surface of an insulating material in a gaseous or liquid medium.

*Sparkover* denotes an electrical breakdown or disruptive discharge through or in a gaseous or liquid insulating medium.

*Puncture* denotes a failure of a solid insulating material when a disruptive discharge occurs through that material.

*Partial discharge,* sometimes called *corona,* is an electrical discharge that only partially bridges the insulation between conductors.  It occurs at any point where the electric field intensity (voltage gradient) exceeds the breakdown strength of the insulating medium.

*Breakdown strength* of insulation systems is determined by a multitude of factors.  The more common factors are types of insulation (solid, liquid, or gaseous), voltage gradients as influenced by insulation geometry and homogeneity of insulation, contamination, and in gaseous systems the type of gas, pressure, and humidity.

*Wave shapes* of the impressed voltage influences the magnitude of the breakdown voltage.  Three types of voltage waves are used for determining breakdown strength.  They are lightning impulse, switching impulse, and power frequency (60 or 50 Hz) voltages.

*Lightning impulse* is an aperiodic transient voltage or current that rises rapidly to a peak value and then falls more slowly to zero.  The standard impulse voltage wave rises to a crest in 1.2 μsec (microseconds) and falls to 0.5 crest in 50 μsec.  Designated 1.2/50.

*Switching impulses* are similar to lightning impulses except for longer fronts and tails.  The standard switching impulse is 250/2,500 μsec.

*Time-lag characteristics* of insulation are demonstrated by a curve obtained by plotting the *crest* value of the applied voltage wave against the time to flashover.

*Insulation coordination* is attained by selecting system component characteristics, clearance values, and protective devices characteristics to achieve the desired operating characteristics.

*Critical sparkover* voltage is the magnitude of voltage at which there is a 50 per cent probability of disruptive discharge.

*Withstand voltage* is a specified value of test voltage which, when applied to a test piece under specified conditions, results in no flashovers.  In some cases a single flashover can be ignored if a specified larger number of tests result in no further flashover.

**Flashover Correction Factors.**  All electrical apparatus is designed to meet specified performance under standard conditions of test.  For flashover tests in open air (rarely at standard conditions of atmospheric pressure, humidity, and temperature), correction factors must be applied to correct the observed flashover values to standard conditions.

*Standard Atmospheric Conditions.* These are atmospheric pressure $b_0$, 1,013 mbar (760 mm Hg at 0°C); humidity $h_0$, 11 g water/m³ (0.44 in. Hg); and ambient temperature $t_0$, 20°C (68°F).

If the observed flashover or sparkover voltage is $V$ at ambient conditions of barometric pressure $b$, humidity $h$, and temperature $t$, the flashover voltage corrected to standard conditions is given by

$$V_0 = V k_h / k_d \tag{7-131}$$

where $V_0$ = flashover voltage at standard conditions

$V$ = flashover voltage under ambient conditions

$k_h$ = humidity correction factor = $k^w$ where $k$ is determined from Fig. 7-44. For gaps with uniform fields such as sphere gaps, $w = 0$ and $k_h$ equals 1. For nonuniform fields such as insulator strings and rod-gaps, $w = 1$ for $\pm$ d-c and positive impulse voltages. For negative impulse waves, $w = 0.8$. For alternating voltages and positive switching surges determine $w$ from Fig. 7-45

$k_d$ = air density correction factor

$$k_d = \left(\frac{b}{b_0}\right)^m \left(\frac{273 + t_0}{273 + t}\right)^n \tag{7-132}$$

where $b$ = atmospheric pressure, mbar

= $1.333 H (1.8 \cdot 10^{-4} t)$

$H$ = barometric pressure, mm Hg at $t$°C

$t$ = ambient temperature, °C

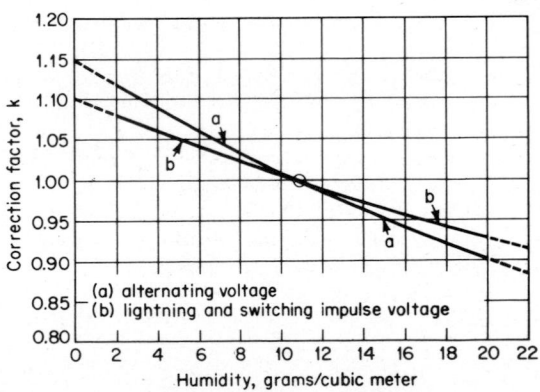

FIG. 7-44. Humidity correction.

$m, n$ = exponents depending on type and polarity of applied voltage and
gap electrode configuration. For d-c and impulse voltages of
both polarities and all field configuration, $m$ and $n = 1$. For
uniform gap fields and alternating and switching impulses of
both polarity, $m$ and $n = 1$. For nonuniform gap fields, use
Fig. 7-45. For values of $m$ and $n = 1$, $k_d$ = relative air density
(RAD) = $0.289\, b/(273 + t)$

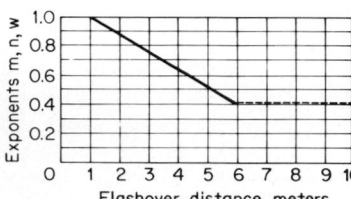

Fig. 7-45

***Electrical Clearance Requirements.*** For conductors in air, the electrical
clearance requirements can be estimated by the use of rod-gap flashover data
as given in Table 7-17. Considerable accuracy can be achieved if the
conductors are reasonably symmetrical and have no sharp projections that
would create points of high-voltage gradients. The reduction in clearance
that can be achieved by careful shielding and conductor geometry to produce
a uniform electric field can be evaluated by comparing the rod-gap data from
Table 7-17 to the sphere gap data in Tables 7-18 and 7-19. For example, a
rod gap with a 10-cm gap will flashover at 79-kv crest 60-Hz voltage.
Spheres of 10-cm diameter require a gap of only 2.8 cm to withstand the same
voltage.

## Table 7-17. Rod-gap Sparkover Crest Voltages

| Gap spacing | | | Critical (50%) sparkover, kv crest | | | |
|---|---|---|---|---|---|---|
| | | | (1.2 × 5) wave | | (1.2 × 50) wave | |
| cm. | in. | 60 Hz | Positive | Negative | Positive | Negative |
| 2 | 0.8 | 25 | 32 | 32 | 32 | 32 |
| 3 | 1.2 | 36 | 41 | 42 | 41 | 42 |
| 4 | 1.6 | 46 | 50 | 51 | 50 | 51 |
| 5 | 2.0 | 54 | 59 | 62 | 59 | 62 |
| 6 | 2.4 | 60 | 64 | 70 | 64 | 70 |
| 8 | 3.1 | 70 | 79 | 86 | 77 | 86 |
| 10 | 3.9 | 79 | 93 | 101 | 90 | 100 |
| 12 | 4.7 | 87 | 112 | 110 | 104 | 117 |
| 14 | 5.5 | 96 | 130 | 135 | 121 | 134 |
| 16 | 6.3 | 104 | 148 | 151 | 137 | 149 |
| 18 | 7.1 | 112 | 165 | 167 | 147 | 163 |
| 20 | 7.9 | 121 | 183 | 184 | 159 | 179 |
| 25 | 9.8 | 143 | 227 | 227 | 186 | 221 |
| 30 | 11.8 | 168 | 269 | 268 | 219 | 259 |
| 35 | 13.8 | 193 | 311 | 309 | 256 | 300 |
| 40 | 15.7 | 230 | 352 | 350 | 284 | 336 |
| 45 | 17.7 | 245 | 391 | 394 | 312 | 367 |
| 50 | 19.7 | 271 | 431 | 438 | 342 | 393 |
| 60 | 23.6 | 324 | 509 | 522 | 395 | 462 |
| 70 | 27.6 | 373 | 588 | 607 | 454 | 532 |
| 80 | 31.5 | 424 | 667 | 691 | 514 | 597 |
| 90 | 35.4 | 476 | 741 | 771 | 573 | 661 |
| 100 | 39.4 | 524 | 820 | 860 | 632 | 726 |
| 120 | 47.2 | 626 | 963 | 1,020 | 741 | 850 |
| 140 | 55.1 | 726 | 1,111 | 1,189 | 859 | 980 |
| 160 | 63.0 | 826 | 1,270 | 1,358 | 973 | 1,109 |
| 180 | 70.9 | 927 | 1,442 | 1,547 | 1,110 | 1,258 |
| 200 | 78.7 | 1,028 | 1,566 | 1,686 | 1,205 | 1,363 |
| 220 | 86.6 | 1,135 | 1,719 | 1,855 | 1,324 | 1,492 |
| 240 | 94.5 | .... | 1,877 | 2,034 | 1,442 | 1,631 |

NOTE: Standard atmospheric conditions: 20°C, 1,013 mbar pressure and 11 g water/m³.

## Table 7-18. Sphere Gap with One Sphere Grounded

Peak values of disruptive-discharge voltages in kilovolts (50% values for impulse tests). Valid for alternating voltages, full negative standard impulses and impulses with longer tails and direct voltages of either polarity. Atmospheric reference conditions: 20°C and 1,013 mbar.

| Gap spacing, cm | Sphere diameter, cm | | | | | | | | |
|---|---|---|---|---|---|---|---|---|---|
| | 2 | 5 | 6.25 | 10 | 12.5 | 15 | 25 | 50 | 75 |
| 0.30 | 11.2 | 11.2 | | | | | | | |
| 0.40 | 14.4 | 14.3 | 14.2 | | | | | | |
| 0.50 | 17.4 | 17.4 | 17.2 | 16.8 | 16.8 | 16.8 | | | |
| 0.60 | 20.4 | 20.4 | 20.2 | 19.9 | 19.9 | 19.9 | | | |
| 0.70 | 23.2 | 23.4 | 23.2 | 23.0 | 23.0 | 23.0 | | | |
| 0.80 | 25.8 | 26.3 | 26.2 | 26.0 | 26.0 | 26.0 | | | |
| 0.90 | 28.3 | 29.2 | 29.1 | 28.9 | 28.9 | 28.9 | | | |
| 1.0 | 30.7 | 32.0 | 31.9 | 31.7 | 31.7 | 31.7 | 31.7 | | |
| 1.2 | (35.1)* | 37.6 | 37.5 | 37.4 | 37.4 | 37.4 | 37.4 | | |
| 1.4 | (38.5) | 42.9 | 42.9 | 42.9 | 42.9 | 42.9 | 42.9 | | |
| 1.5 | (40.0) | 45.5 | 45.5 | 45.5 | 45.5 | 45.5 | 45.5 | | |
| 1.6 | ...... | 48.1 | 48.1 | 48.1 | 48.1 | 48.1 | 48.1 | | |
| 1.8 | ...... | 53.0 | 53.5 | 53.5 | 53.5 | 53.5 | 53.5 | | |
| 2.0 | ...... | 57.5 | 58.5 | 59.0 | 59.0 | 59.0 | 59.0 | 59.0 | 59.0 |
| 2.2 | ...... | 61.5 | 63.0 | 64.5 | 64.5 | 64.5 | 64.5 | 64.5 | 64.5 |
| 2.4 | ...... | 65.5 | 67.5 | 69.5 | 70.0 | 70.0 | 70.0 | 70.0 | 70.0 |
| 2.6 | ...... | (69.0) | 72.0 | 74.5 | 75.0 | 75.0 | 75.5 | 75.5 | 75.5 |
| 2.8 | ...... | (72.5) | 76.0 | 79.5 | 80.0 | 80.5 | 81.0 | 81.0 | 81.0 |
| 3.0 | ...... | (75.5) | 79.5 | 84.0 | 85.0 | 85.5 | 86.0 | 86.0 | 86.0 |
| 3.5 | ...... | (82.5) | (87.5) | 95.0 | 97.0 | 98.0 | 99.0 | 99.0 | 99.0 |
| 4.0 | ...... | (88.5) | (95.0) | 105 | 108 | 110 | 112 | 112 | 112 |
| 4.5 | ...... | ...... | (101) | 115 | 119 | 122 | 125 | 125 | 125 |
| 5.0 | ...... | ...... | (107) | 123 | 129 | 133 | 137 | 138 | 138 |
| 5.5 | ...... | ...... | ...... | (131) | 138 | 143 | 149 | 151 | 151 |
| 6.0 | ...... | ...... | ...... | (138) | 146 | 152 | 161 | 164 | 164 |

| Gap spacing, cm | Sphere diameter, cm | | | | | | | | |
|---|---|---|---|---|---|---|---|---|---|
| | 10 | 12.5 | 15 | 25 | 50 | 75 | 100 | 150 | 200 |
| 7.0 | (150)* | (161) | 169 | 184 | 189 | 190 | 190 | 190 | |
| 8.0 | ...... | (174) | (185) | 206 | 214 | 215 | 215 | 215 | |
| 9.0 | ...... | (185) | (198) | 226 | 239 | 240 | 241 | 241 | |
| 10 | ...... | (195) | (209) | 244 | 263 | 265 | 266 | 266 | 266 |
| 12 | ...... | ...... | (229) | 275 | 309 | 315 | 318 | 318 | 318 |
| 14 | ...... | ...... | ...... | (302) | 353 | 363 | 366 | 366 | 366 |
| 16 | ...... | ...... | ...... | (326) | 392 | 410 | 414 | 414 | 414 |
| 18 | ...... | ...... | ...... | (347) | 429 | 453 | 462 | 462 | 462 |
| 20 | ...... | ...... | ...... | (366) | 460 | 492 | 510 | 510 | 510 |
| 22 | ...... | ...... | ...... | ...... | 489 | 530 | 555 | 560 | 560 |
| 24 | ...... | ...... | ...... | ...... | 515 | 565 | 595 | 610 | 610 |
| 26 | ...... | ...... | ...... | ...... | (540) | 600 | 635 | 655 | 660 |
| 28 | ...... | ...... | ...... | ...... | (565) | 635 | 675 | 700 | 705 |
| 30 | ...... | ...... | ...... | ...... | (585) | 665 | 710 | 745 | 750 |
| 32 | ...... | ...... | ...... | ...... | (605) | 695 | 745 | 790 | 795 |
| 34 | ...... | ...... | ...... | ...... | (625) | 725 | 780 | 835 | 840 |
| 36 | ...... | ...... | ...... | ...... | (640) | 750 | 815 | 875 | 885 |
| 38 | ...... | ...... | ...... | ...... | (665) | (775) | 845 | 915 | 930 |
| 40 | ...... | ...... | ...... | ...... | (670) | (800) | 875 | 955 | 975 |
| 45 | ...... | ...... | ...... | ...... | ...... | (850) | 945 | 1 050 | 1 080 |

## Table 7-18. Sphere Gap with One Sphere Grounded (Continued)

| Gap spacing, cm | Sphere diameter, cm | | | | | | | | |
|---|---|---|---|---|---|---|---|---|---|
| | 10 | 12.5 | 15 | 25 | 50 | 75 | 100 | 150 | 200 |
| 50 | ...... | ...... | ...... | ...... | ...... | (895) | 1 010 | 1 130 | 1 180 |
| 55 | ...... | ...... | ...... | ...... | ...... | (935) | (1 060) | 1 210 | 1 260 |
| 60 | ...... | ...... | ...... | ...... | ...... | (970) | (1 110) | 1 280 | 1 340 |
| 65 | ...... | ...... | ...... | ...... | ...... | ...... | (1 160) | 1 340 | 1 410 |
| 70 | ...... | ...... | ...... | ...... | ...... | ...... | (1 200) | 1 390 | 1 480 |
| 75 | ...... | ...... | ...... | ...... | ...... | ...... | (1 230) | 1 440 | 1 540 |
| 80 | ...... | ..... | ...... | ...... | ...... | ...... | ...... | (1 490) | 1 600 |
| 85 | ...... | ...... | ...... | ...... | ...... | ...... | ...... | (1 540) | 1 660 |
| 90 | ...... | ...... | ...... | ...... | ...... | ...... | ...... | (1 580) | 1 720 |
| 100 | ...... | ...... | ...... | ...... | ...... | ...... | ...... | (1 660) | 1 840 |
| 110 | ...... | ...... | ...... | ...... | ...... | ...... | (1 730) | (1 940) | |
| 120 | ...... | ...... | ...... | ...... | ...... | ...... | (1 800) | (2 020) | |
| 130 | ...... | ...... | ...... | ...... | ...... | ...... | ...... | (2 100) | |
| 140 | ...... | ...... | ...... | ...... | ...... | ...... | ...... | (2 180) | |
| 150 | ...... | ...... | ...... | ...... | ...... | ...... | ...... | (2 250) | |

* The figures in parentheses are of doubtful accuracy.

## Table 7-19. Sphere Gap with One Sphere Grounded

Peak values of disruptive-discharge voltages in kilovolts (50% values). Valid for full positive standard impulses, and impulses with longer tails.
Atmospheric reference conditions: 20°C and 1,013 mbar.

| Gap spacing, cm | Sphere diameter, cm | | | | | | | | |
|---|---|---|---|---|---|---|---|---|---|
| | 2 | 5 | 6.25 | 10 | 12.5 | 15 | 25 | 50 | 75 |
| 0.30 | 11.2 | 11.2 | | | | | | | |
| 0.40 | 14.4 | 14.3 | 14.2 | | | | | | |
| 0.50 | 17.4 | 17.4 | 17.2 | 16.8 | 16.8 | 16.8 | | | |
| 0.60 | 20.4 | 20.4 | 20.2 | 19.9 | 19.9 | 19.9 | | | |
| 0.70 | 23.2 | 23.4 | 23.2 | 23.0 | 23.0 | 23.0 | | | |
| 0.80 | 25.8 | 26.3 | 26.2 | 26.0 | 26.0 | 26.0 | | | |
| 0.90 | 28.3 | 29.2 | 29.1 | 28.9 | 28.9 | 28.9 | | | |
| 1.0 | 30.7 | 32.0 | 31.9 | 31.7 | 31.7 | 31.7 | 31.7 | | |
| 1.2 | (35.1)* | 37.8 | 37.6 | 37.4 | 37.4 | 37.4 | 37.4 | | |
| 1.4 | (38.5) | 43.3 | 43.2 | 42.9 | 42.9 | 42.9 | 42.9 | | |
| 1.5 | (40.0) | 46.2 | 45.9 | 45.5 | 45.5 | 45.5 | 45.5 | | |
| 1.6 | ...... | 49.0 | 48.6 | 48.1 | 48.1 | 48.1 | 48.1 | | |
| 1.8 | ...... | 54.5 | 54.0 | 53.5 | 53.5 | 53.5 | 53.5 | | |
| 2.0 | ...... | 59.5 | 59.0 | 59.0 | 59.0 | 59.0 | 59.0 | 59.0 | 59.0 |
| 2.2 | ...... | 64.0 | 64.0 | 64.5 | 64.5 | 64.5 | 64.5 | 64.5 | 64.5 |
| 2.4 | ...... | 69.0 | 69.0 | 70.0 | 70.0 | 70.0 | 70.0 | 70.0 | 70.0 |
| 2.6 | ...... | (73.0) | 73.5 | 75.5 | 75.5 | 75.5 | 75.5 | 75.5 | 75.5 |
| 2.8 | ...... | (77.0) | 78.0 | 80.5 | 80.5 | 80.5 | 81.0 | 81.0 | 81.0 |
| 3.0 | ...... | (81.0) | 82.0 | 85.5 | 85.5 | 85.0 | 86.0 | 86.0 | 86.0 |
| 3.5 | ...... | (90.0) | (91.5) | 97.5 | 98.0 | 98.5 | 99.0 | 99.0 | 99.0 |
| 4.0 | ...... | (97.5) | (101) | 109 | 110 | 111 | 112 | 112 | 112 |
| 4.5 | ...... | ...... | (108) | 120 | 122 | 124 | 125 | 125 | 125 |
| 5.0 | ...... | ...... | (115) | 130 | 134 | 136 | 138 | 138 | 138 |
| 5.5 | ...... | ...... | ...... | (139) | 145 | 147 | 151 | 151 | 151 |
| 6.0 | ...... | ...... | ...... | (148) | 155 | 158 | 163 | 164 | 164 |

| Gap spacing, cm | Sphere diameter, cm | | | | | | | | |
|---|---|---|---|---|---|---|---|---|---|
| | 10 | 12.5 | 15 | 25 | 50 | 75 | 100 | 150 | 200 |
| 7.0 | ...... | ...... | ...... | (163)* | (173) | 178 | 187 | 189 | 190 |
| 8.0 | ...... | ...... | ...... | ...... | (189) | (196) | 211 | 214 | 215 |
| 9.0 | ...... | ...... | ...... | ...... | (203) | (212) | 233 | 239 | 240 |
| 10 | ...... | (215) | (226) | 254 | 263 | 265 | 266 | 266 | 266 |
| 12 | ...... | ...... | (249) | 291 | 311 | 315 | 318 | 318 | 318 |
| 14 | ...... | ...... | ...... | (323) | 357 | 363 | 366 | 366 | 366 |
| 16 | ...... | ...... | ...... | (350) | 402 | 411 | 414 | 414 | 414 |
| 18 | ...... | ...... | ...... | (374) | 442 | 458 | 462 | 462 | 462 |
| 20 | ...... | ...... | ...... | (395) | 480 | 505 | 510 | 510 | 510 |
| 22 | ...... | ...... | ...... | ...... | 510 | 545 | 555 | 560 | 560 |
| 24 | ...... | ...... | ...... | ...... | 540 | 585 | 600 | 610 | 610 |
| 26 | ...... | ...... | ...... | ...... | 570 | 620 | 645 | 655 | 660 |
| 28 | ...... | ...... | ...... | ...... | (595) | 660 | 685 | 700 | 705 |
| 30 | ...... | ...... | ...... | ...... | (620) | 695 | 725 | 745 | 750 |
| 32 | ...... | ...... | ...... | ...... | (640) | 725 | 760 | 790 | 795 |
| 34 | ...... | ...... | ...... | ...... | (660) | 755 | 795 | 835 | 840 |
| 36 | ...... | ...... | ...... | ...... | (680) | 785 | 830 | 880 | 885 |
| 38 | ...... | ...... | ...... | ...... | (700) | (810) | 865 | 925 | 935 |
| 40 | ...... | ...... | ...... | ...... | (715) | (835) | 900 | 965 | 980 |
| 45 | ...... | ...... | ...... | ...... | ...... | (890) | 980 | 1 060 | 1 090 |

## Table 7-19. Sphere Gap with One Sphere Grounded (Continued)

| Gap spac-ing, cm | Sphere diameter, cm | | | | | | | | |
|---|---|---|---|---|---|---|---|---|---|
| | 10 | 12.5 | 15 | 25 | 50 | 75 | 100 | 150 | 200 |
| 50 | ...... | ...... | ...... | ...... | ...... | (940) | 1 040 | 1 150 | 1 190 |
| 55 | ...... | ...... | ...... | ...... | ...... | (985) | (1 100) | 1 240 | 1 290 |
| 60 | ...... | ...... | ...... | ...... | ...... | (1 020) | (1 150) | 1 310 | 1 380 |
| 65 | ...... | ...... | ...... | ...... | ...... | ...... | (1 200) | 1 380 | 1 470 |
| 70 | ...... | ...... | ...... | ...... | ...... | ...... | (1 240) | 1 430 | 1 550 |
| 75 | ...... | ...... | ...... | ...... | ...... | ...... | (1 280) | 1 480 | 1 620 |
| 80 | ...... | ...... | ...... | ...... | ...... | ...... | ...... | (1 530) | 1 690 |
| 85 | ...... | ...... | ...... | ...... | ...... | ...... | ...... | (1 580) | 1 760 |
| 90 | ...... | ...... | ...... | ...... | ...... | ...... | ...... | (1 630) | 1 820 |
| 100 | ...... | ...... | ...... | ...... | ...... | ...... | ...... | (1 720) | 1 930 |
| 110 | ...... | ...... | ...... | ...... | ...... | ...... | ...... | (1 790) | (2 030) |
| 120 | ...... | ...... | ...... | ...... | ...... | ...... | ...... | (1 860) | (2 120) |
| 130 | ...... | ...... | ...... | ...... | ...... | ...... | ...... | ...... | (2 200) |
| 140 | ...... | ...... | ...... | ...... | ...... | ...... | ...... | ...... | (2 280) |
| 150 | ...... | ...... | ...... | ...... | ...... | ...... | ...... | ...... | (2 350) |

* The figures in parentheses are of doubtful accuracy.

## ILLUMINATION

### Definition of Terms

*Brightness* (*B*) is that property of a light source that specifies the ability of an element of the source to produce luminous effects. It may be expressed in two ways: candles per unit area, such as candles/sq in. (c/in.²), or lumens per unit area.

*Footcandle* (fc) is the illumination at a point on a surface which is one foot from and perpendicular to a uniform point source of one candle.

*Footlambert* (fl) is the brightness of a surface emitting or reflecting one lumen per square foot.

Footlambert (fl) = footcandles (fc) $\times$ reflection factor

$$= \frac{\text{lumens (incident)} \times \text{reflection factor}}{\text{area (sq ft) of surface}} \quad (7\text{-}133)$$

*Illumination* (*E*) is the density of luminous flux on a given surface. The unit of measure is the footcandle.

*Lambert* (*B'*) is the brightness of a surface emitting or reflecting one lumen per square centimeter.

*Lumen* (lm) is the quantity of light flux falling on a surface of one square foot from a uniform point source of one candle. A one-square-foot section from a sphere of one-foot radius with a one-candle source at its center would be such a surface. The lumen differs from the candle in that it is a measure of light flux, irrespective of direction.

$$\text{Incident lumens} = \text{footcandles} \times \text{area (sq ft)} \quad (7\text{-}134)$$

*Luminous flux* (*F*) is the time rate of flow of light. The unit of measure is the lumen.

*Luminous intensity* (*I*) (candlepower) is that property of a light source which specifies its ability as a whole to produce luminous effects. The standard unit of intensity in a given direction is the International Candle. An ordinary wax candle has a luminous intensity of approximately one candle.

$$\text{Candlepower (cp)} = \text{footcandles (fc)} \times \text{distance squared } (D^2) \quad (7\text{-}135)$$

where $D$ = distance in feet from light source to illuminated surface.

*Mean spherical candlepower* (MSCP) is the average candlepower of a source in all directions.

$$\text{MSCP} = \frac{\text{lumens}}{12.57} \quad (7\text{-}136)$$

**Inverse-square Law.** Illumination decreases inversely as the square of the distance. When the light rays are perpendicular to the surface:

$$\text{Illumination } E = \frac{\text{luminous intensity}}{\text{distance squared}} = \frac{I}{D^2} \quad \text{footcandles} \quad (7\text{-}137)$$

where $I$ is the candlepower and $D$ is the distance in feet.

When the light rays are not perpendicular to the surface, the horizontal illumination is

$$E_h = \frac{I \times \cos \theta}{D^2} \quad \text{footcandles} \qquad (7\text{-}138)$$

and the vertical illumination is

$$E_v = \frac{I \times \sin \theta}{D^2} \quad \text{footcandles} \qquad (7\text{-}139)$$

In both cases $\theta$ is measured from the vertical.

### Table 7-20. Illumination Conversion Factors

| Quantity | Multiply number of | By | To obtain |
|---|---|---|---|
| Brightness........... | blondels | 0.0002054 | Candles/sq in. |
| Brightness........... | Candles/sq cm | 6.45 | Candles/sq in. |
| Brightness........... | footlamberts | 0.002210 | Candles/sq in. |
| Brightness........... | lamberts | 2.054 | Candles/sq in. |
| Brightness........... | blondels | 0.00003183 | Stilbs |
| Brightness........... | Candles/sq cm | 1 | Stilbs |
| Brightness........... | Candles/sq in. | 0.1550 | Stilbs |
| Brightness........... | footlamberts | 0.0003425 | Stilbs |
| Brightness........... | lamberts | 0.3183 | Stilbs |
| Brightness........... | Candles/sq cm | 31,416 | blondels |
| Brightness........... | Candles/sq in. | 4,870 | blondels |
| Brightness........... | footlamberts | 10.76 | blondels |
| Brightness........... | lamberts | 10,000 | blondels |
| Brightness........... | Stilbs | 31,416 | blondels |
| Brightness........... | blondels | 0.0929 | footlamberts |
| Brightness........... | Candles/sq cm | 2,919 | footlamberts |
| Brightness........... | Candles/sq in. | 452 | footlamberts |
| Brightness........... | lamberts | 929 | footlamberts |
| Brightness........... | Stilbs | 2,919 | footlamberts |
| Brightness........... | blondels | 0.0001 | lamberts |
| Brightness........... | Candles/sq cm | 3.1416 | lamberts |
| Brightness........... | Candles/sq in. | 0.487 | lamberts |
| Brightness........... | footlamberts | 0.001076 | lamberts |
| Illumination........ | lumens/sq cm | 929 | footcandles |
| Illumination........ | lumens/sq meter | 0.0929 | footcandles |
| Illumination........ | lumens/sq ft | 1 | footcandles |
| Illumination........ | lux | 0.0929 | footcandles |
| Illumination........ | phot | 929 | footcandles |
| Illumination........ | footcandles | 10.76 | lux |
| Illumination........ | lumens/sq cm | 10,000 | lux |
| Illumination........ | lumens/sq meter | 1 | lux |
| Illumination........ | lumens/sq ft | 10.76 | lux |
| Illumination........ | phot | 10,000 | lux |
| Illumination........ | foot-candles | 0.001076 | phot |
| Illumination........ | lumens/sq cm | 1 | phot |
| Illumination........ | lumens/sq meter | 0.0001 | phot |
| Illumination........ | lumens/sq ft | 0.001076 | phot |
| Illumination........ | lux | 0.0001 | phot |
| Luminous flux....... | light-watts | 680 | lumens |
| Luminous flux....... | youngs | 680 | lumens |

**Application.**   Generally acceptable lighting levels in terms of footcandles for various types of installations are listed in Table 7-21.   This is the average illumination at the work level.   The number of lamps required to produce a required level of illumination is given by

Number of lamps

$$= \frac{\text{footcandles} \times \text{area}}{\text{lumens per lamp} \times \text{coefficient of utilization} \times \text{maintenance factor}} \quad (7\text{-}140)$$

The lumens per lamp for a number of standard bulbs of incandescent, mercury, and fluorescent types are listed in Tables 7-22 to 7-24.   The coefficient of utilization and maintenance factor (MF) are selected from Table 7-25, corresponding to the luminaire that is to be used.   A luminaire usually consists of a number of lamps.   The total number of luminaires required is

$$\text{Number of luminaires} = \frac{\text{number of lamps}}{\text{lamps per luminaire}} \quad (7\text{-}141)$$

To use Table 7-25 it is first necessary to determine a room index from Table 7-26, which covers a wide variety of room dimensions and light-mounting heights.   Table 7-27 gives average data on diffuse-reflection ratios, and Table 7-28 lists maximum-brightness ratios that are acceptable.

*Brightness ratios* are determined as follows: Determine the reflection values of the task, desk, floor, walls, and ceilings from Table 7-27.   The brightness of the task and desk is determined by multiplying the average foot-candles by the reflection factor.   Brightness of walls, ceilings, and floor is determined from Table 7-30.

## Table 7-21. Illumination Levels, Interior Lighting

| | Foot-candles Maintained in Service (Not Initial Values) |
|---|---|
| Assembly (manufacturing): | |
|   Rough | 20 |
|   Medium | 50 |
|   Fine | 100 |
|   Extra fine | 300* |
| Auditoriums: | |
|   Assembly only | 10 |
|   Exhibitions | 30 |
| Banks: | |
|   Lobby | 20 |
|   Cages and offices | 50 |
| Barber shops and beauty parlors | 50 |
| Bathrooms: | |
|   General lighting | 5 |
|   At mirror (on face) | 40 |
| Bedrooms: | |
|   General lighting | 5 |
|   At mirror (on face) | 20 |
| Churches: | |
|   Auditorium | 10 |
|   Sunday-school rooms | 20 |
|   Pulpit | 20 |
| Classrooms, on desks and chalkboards: | |
|   Typical | 30 |
|   Sight-saving or special | 50 |
| Depots and stations: | |
|   Waiting room | 20 |
|   Ticket rack and counter | 50 |
|   Concourse | 5 |
|   Platforms | 5 |
| Dining rooms: | |
|   Homes (general lighting) | 5 |
|   Hotels and restaurants | 10 |
| Drafting rooms | 50 |
| Elevators | 10 |
| Garages: | |
|   Storage | 10 |
|   Repair and servicing | 50 |
| Gymnasiums: | |
|   Exhibitions and matches | 30 |
|   General exercise | 20 |
|   Assemblies | 10 |
|   Dances | 5 |
|   Lockers and shower rooms | 10 |
| Halls and corridors | 5 |
| Homes (see specific rooms) | |
| Hospitals: | |
|   Private rooms and wards: | |
|     General lighting | 5 |
|     Supplementary for reading | 20 |
|   Surgery: | |
|     General lighting | 50 |
|     Operating table | 1800 |
|   Obstetrical: | |
|     Delivery room | 50 |
|     Delivery table | 200 |
|     Examination table | 50 |
| Inspection: | |
|   Rough | 20 |
|   Medium | 50 |
|   Fine | 100 |
|   Extra fine | 200 or more* |
| Ironing | 40 |
| Kitchens: | |
|   General lighting | 10 |
|   Supplementary (at task) | 40 |

## Table 7-21. Illumination Levels, Interior Lighting (Continued)

|  | Foot-candles Maintained in Service (Not Initial Values) |
|---|---|
| Laboratories: |  |
| General lighting | 30 |
| Work tables | 50 |
| Close work | 100 |
| Living rooms (see also specific visual task): |  |
| General lighting | 5 |
| Lobbies | 20 |
| Machine shops: |  |
| Rough bench and machine work | 20 |
| Medium bench and machine work | 50 |
| Fine bench and machine work | 100 |
| Extra fine bench and machine work | 200 or more* |
| Mail rooms | 30 |
| Museums and art galleries: |  |
| General lighting | 10 |
| On displays | 50 |
| Offices: |  |
| Casual visual tasks: inactive file rooms, reception rooms, stairways, washrooms, and other service areas | 10 |
| Ordinary visual tasks: general office work (except for work classified as "difficult visual tasks"), private office work, general correspondence, conference rooms, active file rooms, mail rooms | 30 |
| Difficult visual tasks: auditing and accounting, business-machine operation, transcribing and tabulation, bookkeeping, drafting, designing | 50 |
| Reading: |  |
| Short periods, material of reasonably good visibility | 20 |
| Prolonged periods or smaller type | 40 |
| Proofreading | 100 |
| Schools (see specific rooms) |  |
| Sewing: |  |
| Coarse work, high contrast between thread and fabric | 20 |
| Light fabrics, occasional periods | 40 |
| Light to medium fabrics, prolonged periods | 80 |
| Dark fabrics, fine detail, low contrast | 150 or more* |
| Show windows: |  |
| Low surrounding brightness: |  |
| General displays | 50 |
| Feature displays | 100 |
| Medium surrounding brightness: |  |
| General displays | 100 |
| Feature displays | 200 |
| High surrounding brightness: |  |
| General displays | 200 |
| Feature displays | 500 |
| Stairways | 10 |
| Storage and stock rooms: |  |
| Rough bulky material | 5 |
| Medium material | 10 |
| Fine material requiring care | 20 |
| Store interiors: |  |
| Circulation areas | 20 |
| General merchandising areas | 50 |
| Showcases, wall cases, and open-counter displays | 100* |
| Feature displays | 200* |
| Theaters and motion-picture houses: |  |
| Auditorium during intermission | 5 |
| Auditorium during picture | 0.1 |
| Foyer | 5 |
| Lobby | 20 |
| Toilets and washrooms | 10 |
| Waiting rooms | 20 |
| Woodworking: |  |
| Rough sawing and bench work | 30 |
| Sizing, planing, rough sanding, veneering, medium machine and bench work | 50 |
| Fine bench and machine work, fine sanding and finishing | 100 |
| Writing | 20 |

\* Usually obtained by supplementary luminaires in combination with general lighting systems providing not less than one-tenth of the recommended value for the task.

SOURCE: "Foot Candle Tables," Bulletin A-4981, p. 3, Westinghouse Electric Corporation, Bloomfield, N.J.

## Table 7-22. Incandescent-lamp Data

| Watts | Bulb | Base | Finish | Rated avg life, hr | Initial lumens |
|---|---|---|---|---|---|
| General-service Lamps | | | | | |
| 100 | A-21 | Med. | I.f | 750 | 1,620 |
| 150 | A-23 | Med. | I.f-cl. | 750 | 2,600 |
| 200 | PS-30 | Med. | I.f.-cl. | 750 | 3,700 |
| 300 | PS-30 | Med. | I.f.-cl. | 750 | 5,900 |
| 300 | PS-35 | Mogul | I.f.-cl. | 1,000 | 5,650 |
| 500 | PS-40 | Mogul | I.f.-cl. | 1,000 | 9,900 |
| 750 | PS-52 | Mogul | I.f.-cl. | 1,000 | 15,600 |
| 1,000 | PS-52 | Mogul | I.f.-cl. | 1,000 | 21,500 |
| 1,500 | PS-52 | Mogul | I.f.-cl. | 1,000 | 33,000 |
| Projector and Reflector Lamps | | | | | |
| 75 | PAR-38 | Med. skt. | Projector spot | 1,000 | 450 (0–15°) |
| 75 | PAR-38 | Med. skt. | Projector flood | 1,000 | 550 (0–30°) |
| 150 | PAR-38 | Med. skt. | Projector spot | 1,000 | 1,150 (0–15°) |
| 150 | PAR-38 | Med. skt. | Projector flood | 1,000 | 1,400 (0–30°) |
| 75 | R-30 | Med. | Reflector spot | 1,000 | 220 (0–15°) |
| 75 | R-30 | Med. | Reflector flood | 1,000 | 300 (0–30°) |
| 150 | R-40 | Med. | Reflector spot | 1,000 | 600 (0–15°) |
| 150 | R-40 | Med. | Reflector flood | 1,000 | 800 (0–30°) |
| 300 | R-40 | Med. | Reflector spot | 1,000 | 1,350 (0–15°) |
| 300 | R-40 | Med. | Reflector flood | 1,000 | 1,600 (0–30°) |

SOURCE: "Illumination Design Data for Interiors," p. 6, Westinghouse Electric Corporation, Bloomfield, N.J.

## Table 7-23. Mercury-lamp Data

| Designation | Watts | Bulb | Base | Ballast loss/lamp, watts | Rated avg life, hr* | Initial lumens |
|---|---|---|---|---|---|---|
| A-H1 | 400 | T-16 | Mogul | 40† | 4,000 | 16,000 |
| A-H12 | 1,000 | T-28 | Mogul | 85† | 3,000 | 60,000 |
| A-H9 | 3,000 | T-9½ | S.C. term | 165‡ | 5,000 | 120,000 |

* Rated average life under specified test conditions at 5 hr per start. At 10 hr/start, rated average life is 6,000 hr.
† Single lamp high PF 110 to 125-volt ballasts. Losses for two-lamp ballasts are generally lower.
‡ Single lamp high PF, 230-volt ballast.
SOURCE: "Illumination Design Data for Interiors," p. 6, Westinghouse Electric Corporation, Bloomfield, N.J

## Table 7-24. Fluorescent-lamp Data

| Bulb | Watts | Base | Rated avg life, hr | Rated initial lumens* | | |
|---|---|---|---|---|---|---|
| | | | | White | Std cool white | Std warm white |
| Preheat Lamps | | | | | | |
| 33″ T-12 | 25 | Med. bipin | 7,500† | 1,430 | 1,370 | 1,440 |
| 48″ T-12 | 40 | Med. bipin | 7,500† | 2,480 | 2,370 | 2,500 |
| 60″ T-17 | 90 | Mog. bipin | 7,500† | 4,860 | 4,650 | 4,900 |
| Instant-start Lamps | | | | | | |
| 48″ T-12 | 40 | Med. bipin | 6,000‡ | 2,480 | 2,370 | 2,500 |
| 60″ T-17 | 40 | Mog. bipin | 6,000‡ | ..... | 2,300 | |
| Slimline Lamps¶ | | | | | | |
| 48″ T-12 | 38§ | Single pin | 6,000‡ | 2,320 | 2,200 | 2,340 |
| | 52 | | | 3,020 | 2,870 | 3,050 |
| 72″ T-12 | 59 | Single pin | 6,000‡ | 3,660 | 3,500 | 3,700 |
| | 72 | | | 4,300 | 4,100 | 4,340 |
| 96″ T-12 | 75 | Single pin | 6,000‡ | 4,800 | 4,575 | 4,850 |
| | 96 | | | 5,800 | 5,540 | 5,860 |
| 96″ T-8 | 34 | Single pin | 6,000‡ | 2,280 | 2,180 | 2,300 |
| | 51 | | | 3,300 | 3,150 | 3,330 |
| | 69 | | | 4,350 | 4,150 | 4,390 |

* Lumens measured after 100 hr burning at 80°F ambient and under specified test conditions. The lumen outputs of the de luxe cool white and de luxe warm white lamps are approximately 40 per cent less than those of the corresponding standard cool white and standard warm white. The lumen values of daylight and soft white lamps are 85 and 73 per cent, respectively, of the white values.
† Life under specified test conditions at 3 burning hours per start. Lamp life is slightly longer for more burning hours per start.
‡ Life (tentative) under specified test conditions at 12 burning hours per start. Lamp life is somewhat shorter for fewer burning hours per start.
¶ Slimlines may be operated at any current density within their design range. The figures listed for the 96″ T-8 Slimline are for 120, 200, and 300 ma. The data listed for the T-12 Slimlines are for 425 and 600 ma.
§ Operates on a standard 40-watt instant-start ballast at 420 ma.
SOURCE: "Illumination Design Data for Interiors," p. 6, Westinghouse Electric Corporation, Bloomfield, N.J.

## Table 7-25. Coefficients of Utilization
For Explanation of Symbols, See Notes on Page 7-80

| Luminaire | Ceiling.. | 75% | | | 50% | | | 30% | |
|---|---|---|---|---|---|---|---|---|---|
| | Walls... | 50% | 30% | 10% | 50% | 30% | 10% | 30% | 10% |
| | Room index | Coefficient of utilization | | | | | | | |

**MF**
G −.75 ↑0
M −.65
P −.55 ↓79
Direct, RLM dome reflector, MS = 1.0 x MH

| | | | | | | | | | |
|---|---|---|---|---|---|---|---|---|---|
| | J | .37 | .31 | .27 | .36 | .31 | .27 | .31 | .27 |
| | I | .45 | .41 | .38 | .45 | .40 | .37 | .40 | .37 |
| | H | .49 | .45 | .42 | .49 | .45 | .42 | .45 | .42 |
| | G | .53 | .49 | .46 | .53 | .49 | .46 | .48 | .46 |
| | F | .56 | .53 | .49 | .55 | .52 | .49 | .51 | .49 |
| | E | .61 | .58 | .55 | .60 | .57 | .55 | .56 | .55 |
| | D | .66 | .63 | .60 | .64 | .62 | .60 | .61 | .60 |
| | C | .67 | .65 | .62 | .66 | .64 | .62 | .63 | .61 |
| | B | .71 | .68 | .66 | .69 | .67 | .65 | .66 | .64 |
| | A | .72 | .70 | .67 | .71 | .68 | .67 | .67 | .66 |

**MF**
G −.75 ↑0
M −.65
P −.55 ↓70
Direct, RLM deep-bowl reflector, MS = 1.0 x MH

| | | | | | | | | | |
|---|---|---|---|---|---|---|---|---|---|
| | J | .35 | .31 | .28 | .34 | .31 | .28 | .30 | .28 |
| | I | .43 | .39 | .37 | .42 | .39 | .37 | .39 | .37 |
| | H | .46 | .44 | .42 | .46 | .44 | .42 | .43 | .42 |
| | G | .50 | .47 | .45 | .49 | .47 | .45 | .46 | .45 |
| | F | .53 | .50 | .47 | .51 | .49 | .47 | .49 | .47 |
| | E | .56 | .54 | .51 | .56 | .54 | .51 | .53 | .51 |
| | D | .61 | .58 | .56 | .59 | .57 | .56 | .56 | .56 |
| | C | .62 | .60 | .57 | .61 | .58 | .57 | .58 | .56 |
| | B | .64 | .62 | .61 | .63 | .61 | .60 | .60 | .59 |
| | A | .65 | .63 | .61 | .64 | .62 | .61 | .61 | .60 |

**MF**
G −.75 ↑0
M −.60
P −.40 ↓75
Direct, high bay, narrow spread, MS =.6 x MH

| | | | | | | | | | |
|---|---|---|---|---|---|---|---|---|---|
| | J | .43 | .40 | .39 | .42 | .40 | .39 | .40 | .38 |
| | I | .51 | .50 | .49 | .50 | .49 | .48 | .49 | .46 |
| | H | .55 | .54 | .53 | .54 | .53 | .52 | .53 | .52 |
| | G | .59 | .58 | .57 | .58 | .56 | .55 | .56 | .55 |
| | F | .61 | .60 | .58 | .59 | .58 | .58 | .58 | .57 |
| | E | .64 | .63 | .62 | .63 | .62 | .61 | .61 | .60 |
| | D | .68 | .65 | .64 | .66 | .65 | .64 | .64 | .63 |
| | C | .69 | .67 | .66 | .67 | .66 | .64 | .64 | .64 |
| | B | .70 | .68 | .67 | .68 | .67 | .66 | .66 | .65 |
| | A | .71 | .70 | .68 | .69 | .67 | .67 | .67 | .66 |

**MF**
G −.75 ↑0
M −.65
P −.50 ↓75
Direct, high bay, medium or wide spread, MS = 1.0 x MH

| | | | | | | | | | |
|---|---|---|---|---|---|---|---|---|---|
| | J | .40 | .36 | .34 | .39 | .36 | .34 | .36 | .33 |
| | I | .48 | .45 | .43 | .47 | .44 | .43 | .44 | .42 |
| | H | .52 | .50 | .48 | .51 | .49 | .47 | .49 | .47 |
| | G | .55 | .53 | .52 | .55 | .52 | .51 | .52 | .51 |
| | F | .58 | .56 | .53 | .56 | .55 | .53 | .55 | .53 |
| | E | .62 | .60 | .56 | .61 | .59 | .57 | .58 | .57 |
| | D | .66 | .63 | .61 | .64 | .62 | .61 | .62 | .61 |
| | C | .67 | .65 | .62 | .66 | .64 | −62 | .63 | .62 |
| | B | .69 | .67 | .66 | .67 | .65 | .64 | .65 | .64 |
| | A | .70 | .68 | .67 | .69 | .67 | .65 | .66 | .64 |

**MF**
G −.80 ↑0
M −.72
P −.65 ↓70
Direct, heavy duty
narrow spread, MS =.5 x MH
medium spread, MS =.8 x MH

| | | | | | | | | | |
|---|---|---|---|---|---|---|---|---|---|
| | J | .40 | .38 | .36 | .39 | .38 | .36 | .38 | .36 |
| | I | .48 | .46 | .45 | .47 | .46 | .45 | .45 | .43 |
| | H | .52 | .51 | .50 | .51 | .50 | .49 | .50 | .48 |
| | G | .55 | .54 | .53 | .54 | .53 | .52 | .53 | .51 |
| | F | .57 | .56 | .55 | .56 | .55 | .54 | .55 | .53 |
| | E | .60 | .59 | .58 | .59 | .58 | .57 | .57 | .56 |
| | D | .64 | .61 | .60 | .62 | .60 | .59 | .60 | .59 |
| | C | .64 | .63 | .61 | .63 | .62 | .60 | .60 | .60 |
| | B | .65 | .64 | .63 | .64 | .63 | .62 | .62 | .61 |
| | A | .66 | .65 | .64 | .64 | .63 | .62 | .62 | .62 |

## Table 7-25. Coefficients of Utilization (Continued)

| Luminaire | Ceiling.. | 75% | | | 50% | | | 30% | |
|---|---|---|---|---|---|---|---|---|---|
| | Walls... | 50% | 30% | 10% | 50% | 30% | 10% | 30% | 10% |
| | Room index | Coefficient of utilization | | | | | | | |
| **MF** G –.80 ↑0  M –.72  P –.65 ↓70 — Direct, heavy duty, wide spread, MS = 1.1 x MH | J | .37 | .34 | .31 | .36 | .34 | .31 | .34 | .31 |
| | I | .45 | .42 | .41 | .44 | .41 | .40 | .41 | .39 |
| | H | .48 | .46 | .45 | .49 | .45 | .44 | .45 | .44 |
| | G | .52 | .50 | .48 | .51 | .49 | .48 | .49 | .48 |
| | F | .55 | .52 | .51 | .50 | .51 | .50 | .51 | .50 |
| | E | .57 | .56 | .54 | .57 | .55 | .53 | .55 | .53 |
| | D | .62 | .59 | .57 | .60 | .58 | .57 | .57 | .57 |
| | C | .63 | .61 | .58 | .62 | .59 | .58 | .59 | .57 |
| | B | .64 | .62 | .61 | .63 | .61 | .60 | .60 | .59 |
| | A | .66 | .64 | .62 | .64 | .62 | .61 | .62 | .60 |
| **MF** G –.70 ↑5  M –.60  P –.45 ↓58 — Direct, RLM Glassteel diffuser, MS = 1.0 x MH | J | .27 | .23 | .20 | .26 | .23 | .20 | .22 | .20 |
| | I | .34 | .30 | .28 | .33 | .29 | .27 | .29 | .27 |
| | H | .37 | .34 | .31 | .36 | .33 | .31 | .32 | .30 |
| | G | .40 | .37 | .34 | .39 | .36 | .34 | .35 | .33 |
| | F | .42 | .39 | .37 | .40 | .38 | .36 | .37 | .36 |
| | E | .46 | .43 | .41 | .45 | .42 | .40 | .41 | .40 |
| | D | .49 | .47 | .44 | .48 | .46 | .44 | .44 | .43 |
| | C | .51 | .49 | .46 | .49 | .47 | .46 | .46 | .44 |
| | B | .53 | .51 | .49 | .51 | .49 | .48 | .48 | .47 |
| | A | .54 | .53 | .51 | .52 | .51 | .49 | .49 | .48 |
| **MF** G –.60 ↑0  M –.50  P –.40 ↓67 — Direct, RLM silvered-bowl diffuser, MS =.8 x MH | J | .38 | .36 | .35 | .38 | .36 | .35 | .36 | .35 |
| | I | .46 | .45 | .44 | .45 | .44 | .43 | .44 | .42 |
| | H | .49 | .49 | .48 | .49 | .48 | .47 | .48 | .47 |
| | G | .53 | .52 | .51 | .52 | .51 | .50 | .51 | .49 |
| | F | .55 | .54 | .53 | .53 | .53 | .52 | .53 | .51 |
| | E | .57 | .57 | .56 | .57 | .56 | .55 | .55 | .54 |
| | D | .61 | .59 | .58 | .59 | .58 | .57 | .57 | .56 |
| | C | .62 | .61 | .59 | .60 | .59 | .58 | .58 | .57 |
| | B | .63 | .62 | .61 | .61 | .60 | .59 | .59 | .58 |
| | A | .64 | .63 | .62 | .62 | .61 | .60 | .60 | .59 |
| **MF** G –.75 ↑0  M –.65  P –.55 ↓65 — Direct, vapor-tight, wide spread, MS = 1.0 x MH | J | .31 | .26 | .23 | .30 | .26 | .23 | .26 | .23 |
| | I | .38 | .34 | .31 | .37 | .33 | .31 | .33 | .31 |
| | H | .41 | .38 | .34 | .41 | .38 | .34 | .37 | .34 |
| | G | .45 | .41 | .39 | .44 | .41 | .39 | .40 | .39 |
| | F | .47 | .44 | .41 | .46 | .43 | .41 | .43 | .41 |
| | E | .51 | .48 | .46 | .50 | .48 | .46 | .47 | .46 |
| | D | .55 | .52 | .50 | .54 | .52 | .50 | .51 | .50 |
| | C | .56 | .54 | .52 | .55 | .53 | .52 | .52 | .51 |
| | B | .59 | .57 | .55 | .58 | .56 | .54 | .55 | .54 |
| | A | .60 | .58 | .56 | .59 | .57 | .56 | .56 | .55 |
| **MF** G –.70 ↑0  M –.60  P –.50 ↓53 — Direct, prismatic lens, medium spread, MS =.8 x MH | J | .25 | .22 | .20 | .24 | .22 | .20 | .22 | .20 |
| | I | .31 | .28 | .26 | .29 | .28 | .26 | .28 | .26 |
| | H | .34 | .31 | .29 | .32 | .31 | .29 | .30 | .28 |
| | G | .36 | .33 | .32 | .34 | .33 | .31 | .32 | .30 |
| | F | .38 | .35 | .34 | .36 | .34 | .33 | .34 | .32 |
| | E | .40 | .39 | .38 | .39 | .37 | .36 | .37 | .35 |
| | D | .43 | .41 | .40 | .42 | .40 | .39 | .39 | .38 |
| | C | .45 | .43 | .42 | .44 | .41 | .40 | .40 | .40 |
| | B | .48 | .45 | .44 | .47 | .43 | .42 | .42 | .41 |
| | A | .50 | .47 | .46 | .48 | .46 | .45 | .45 | .42 |

## Table 7-25. Coefficients of Utilization (Continued)

| Luminaire | Ceiling.. | 75% | | | 50% | | | 30% | |
|---|---|---|---|---|---|---|---|---|---|
| | Walls... | 50% | 30% | 10% | 50% | 30% | 10% | 30% | 10% |
| | Room index | Coefficient of utilization | | | | | | | |

**MF -.75 ↑0 ↓62** — Direct, PAR-38, 150-watt shielded to 45°, total lamp lumens = 1850, MS = .5 x MH

| Room index | 50% | 30% | 10% | 50% | 30% | 10% | 30% | 10% |
|---|---|---|---|---|---|---|---|---|
| J | .52 | .49 | .47 | .51 | .49 | .47 | .48 | .47 |
| I | .55 | .53 | .51 | .54 | .52 | .51 | .51 | .50 |
| H | .57 | .55 | .53 | .56 | .54 | .53 | .53 | .53 |
| G | .58 | .57 | .55 | .57 | .56 | .55 | .55 | .54 |
| F | .59 | .58 | .57 | .58 | .57 | .56 | .56 | .56 |
| E | .61 | .60 | .59 | .60 | .59 | .58 | .58 | .57 |
| D | .63 | .62 | .61 | .61 | .61 | .60 | .60 | .59 |
| C | .64 | .64 | .63 | .63 | .63 | .62 | .62 | .61 |
| B | .65 | .65 | .64 | .64 | .64 | .63 | .63 | .62 |
| A | .66 | .66 | .65 | .65 | .65 | .64 | .64 | .63 |

**MF G-.65 ↑0 M-.55 P-.45 ↓79** — Direct, RLM, 2 40-watt lamps, MS = 1.0 x MH

| Room index | 50% | 30% | 10% | 50% | 30% | 10% | 30% | 10% |
|---|---|---|---|---|---|---|---|---|
| J | .38 | .32 | .28 | .37 | .32 | .28 | .31 | .28 |
| I | .47 | .42 | .39 | .46 | .41 | .38 | .40 | .37 |
| H | .51 | .47 | .44 | .50 | .47 | .43 | .46 | .43 |
| G | .55 | .51 | .48 | .54 | .51 | .47 | .50 | .47 |
| F | .58 | .54 | .51 | .57 | .53 | .51 | .52 | .50 |
| E | .63 | .60 | .57 | .62 | .59 | .56 | .58 | .55 |
| D | .68 | .64 | .61 | .66 | .64 | .61 | .63 | .60 |
| C | .70 | .67 | .63 | .68 | .65 | .64 | .64 | .62 |
| B | .73 | .70 | .68 | .71 | .68 | .67 | .67 | .66 |
| A | .74 | .72 | .70 | .72 | .70 | .68 | .69 | .67 |

**MF G-.65 ↑0 M-.55 P-.45 ↓72** — Direct, RLM, 3 40-watt lamps, MS = 1.0 x MH

| Room index | 50% | 30% | 10% | 50% | 30% | 10% | 30% | 10% |
|---|---|---|---|---|---|---|---|---|
| J | .34 | .29 | .25 | .33 | .29 | .25 | .28 | .25 |
| I | .42 | .38 | .35 | .41 | .37 | .34 | .37 | .34 |
| H | .46 | .42 | .39 | .44 | .42 | .39 | .41 | .39 |
| G | .50 | .46 | .43 | .48 | .45 | .41 | .44 | .41 |
| F | .53 | .49 | .46 | .51 | .47 | .44 | .47 | .44 |
| E | .57 | .54 | .51 | .56 | .52 | .50 | .52 | .50 |
| D | .61 | .58 | .55 | .59 | .56 | .54 | .56 | .54 |
| C | .63 | .60 | .57 | .61 | .58 | .56 | .58 | .56 |
| B | .66 | .64 | .61 | .64 | .60 | .59 | .60 | .59 |
| A | .67 | .65 | .62 | .66 | .62 | .61 | .62 | .61 |

**MF G-.60 ↑0 M-.50 P-.45 ↓71** — Direct, RLM, 2·85-watt lamps, MS = 1.0 x MH

| Room index | 50% | 30% | 10% | 50% | 30% | 10% | 30% | 10% |
|---|---|---|---|---|---|---|---|---|
| J | .33 | .28 | .25 | .33 | .28 | .25 | .28 | .25 |
| I | .41 | .37 | .34 | .40 | .36 | .33 | .36 | .33 |
| H | .45 | .41 | .38 | .44 | .41 | .38 | .40 | .38 |
| G | .48 | .45 | .42 | .48 | .45 | .42 | .43 | .42 |
| F | .51 | .48 | .45 | .50 | .47 | .45 | .46 | .45 |
| E | .55 | .53 | .50 | .55 | .52 | .50 | .51 | .50 |
| D | .60 | .57 | .54 | .58 | .56 | .54 | .55 | .54 |
| C | .61 | .59 | .56 | .60 | .57 | .56 | .57 | .55 |
| B | .64 | .62 | .60 | .62 | .60 | .59 | .60 | .58 |
| A | .65 | .63 | .61 | .64 | .62 | .60 | .61 | .60 |

**MF G-.70 ↑0 M-.65 P-.55 ↓60** — Direct, dust and vapor-tight, MS = 1.0 x MH

| Room index | 50% | 30% | 10% | 50% | 30% | 10% | 30% | 10% |
|---|---|---|---|---|---|---|---|---|
| J | .29 | .26 | .23 | .28 | .26 | .23 | .25 | .23 |
| I | .35 | .32 | .31 | .35 | .32 | .30 | .32 | .30 |
| D | .38 | .36 | .34 | .38 | .36 | .34 | .35 | .34 |
| G | .41 | .39 | .37 | .41 | .39 | .37 | .38 | .37 |
| F | .44 | .41 | .39 | .42 | .41 | .39 | .40 | .39 |
| E | .46 | .45 | .42 | .46 | .44 | .42 | .44 | .42 |
| D | .50 | .48 | .46 | .49 | .47 | .46 | .46 | .46 |
| C | .51 | .49 | .47 | .50 | .48 | .47 | .48 | .46 |
| B | .53 | .51 | .50 | .52 | .50 | .49 | .49 | .49 |
| A | .54 | .52 | .50 | .53 | .51 | .50 | .50 | .49 |

## Table 7-25. Coefficients of Utilization (Continued)

| Luminaire | Ceiling... | 75% | | | 50% | | | 30% | |
|---|---|---|---|---|---|---|---|---|---|
| | Walls... | 50% | 30% | 10% | 50% | 30% | 10% | 30% | 10% |
| | Room index | Coefficient of utilization | | | | | | | |
| **MF** G-.70 ↑0 M-.60 P-.50 ↓80 — Direct, 3-kw mercury, MS = 1.0 x MH | J | .38 | .32 | .28 | .37 | .32 | .28 | .31 | .28 |
| | I | .47 | .42 | .39 | .46 | .41 | .38 | .41 | .38 |
| | H | .51 | .47 | .43 | .50 | .47 | .43 | .46 | .43 |
| | G | .55 | .51 | .47 | .54 | .51 | .47 | .49 | .47 |
| | F | .58 | .54 | .51 | .56 | .53 | .81 | .52 | .51 |
| | E | .63 | .59 | .56 | .62 | .59 | .56 | .58 | .56 |
| | D | .67 | .64 | .61 | .66 | .63 | .61 | .63 | .61 |
| | C | .69 | .67 | .64 | .67 | .65 | .63 | .64 | .63 |
| | B | .72 | .70 | .67 | .71 | .68 | .67 | .67 | .66 |
| | A | .74 | .71 | .69 | .72 | .70 | .68 | .69 | .67 |
| **MF** G-.65 ↑0 M-.55 P-.45 ↓64 — Direct, RLM with louvers, MS = .9 x MH | J | .33 | .28 | .26 | .32 | .28 | .26 | .28 | .26 |
| | I | .39 | .36 | .34 | .39 | .35 | .34 | .35 | .34 |
| | H | .43 | .40 | .38 | .42 | .40 | .38 | .39 | .38 |
| | G | .46 | .43 | .41 | .45 | .43 | .41 | .42 | .41 |
| | F | .48 | .46 | .43 | .47 | .45 | .43 | .45 | .43 |
| | E | .52 | .50 | .47 | .51 | .49 | .47 | .48 | .47 |
| | D | .55 | .53 | .51 | .54 | .52 | .51 | .52 | .51 |
| | C | .57 | .55 | .52 | .56 | .53 | .52 | .53 | .52 |
| | B | .59 | .57 | .56 | .57 | .56 | .55 | .55 | .54 |
| | A | .60 | .58 | .56 | .59 | .57 | .56 | .56 | .55 |
| **MF** G-.70 ↑0 M-.60 P-.50 ↓50 — Direct, Troffer, glass, MS = 1.0 x MH | J | .28 | .27 | .26 | .28 | .27 | .26 | .28 | .26 |
| | I | .34 | .33 | .32 | .34 | .32 | .32 | .33 | .31 |
| | H | .36 | .36 | .36 | .36 | .36 | .35 | .35 | .35 |
| | G | .39 | .38 | .38 | .38 | .38 | .37 | .37 | .36 |
| | F | .41 | .40 | .39 | .40 | .39 | .38 | .39 | .38 |
| | E | .43 | .42 | .42 | .42 | .42 | .40 | .41 | .40 |
| | D | .46 | .44 | .43 | .44 | .43 | .43 | .42 | .42 |
| | C | .46 | .45 | .44 | .45 | .44 | .43 | .43 | .43 |
| | B | .47 | .45 | .45 | .46 | .44 | .44 | .44 | .44 |
| | A | .47 | .46 | .46 | .46 | .45 | .45 | .45 | .44 |
| **MF** G-.70 ↑0 M-.60 P-.50 ↓47 — Direct, Troffer, glass, MS = 1.0 x MH | J | .27 | .25 | .24 | .26 | .25 | .24 | .26 | .24 |
| | I | .32 | .31 | .30 | .31 | .30 | .30 | .30 | .29 |
| | H | .34 | .34 | .33 | .34 | .33 | .33 | .33 | .32 |
| | G | .36 | .35 | .35 | .36 | .35 | .35 | .35 | .34 |
| | F | .39 | .38 | .37 | .37 | .37 | .36 | .37 | .36 |
| | E | .40 | .40 | .38 | .39 | .39 | .38 | .39 | .38 |
| | D | .43 | .41 | .40 | .41 | .40 | .40 | .40 | .39 |
| | C | .43 | .42 | .41 | .42 | .41 | .40 | .40 | .40 |
| | B | .44 | .43 | .42 | .43 | .42 | .41 | .41 | .41 |
| | A | .45 | .44 | .43 | .43 | .42 | .42 | .42 | .41 |
| **MF** G-.70 ↑0 M-.60 P-.55 ↓61 — Direct, Troffer, louvers, MS = .8 x MH | J | .33 | .31 | .30 | .33 | .31 | .30 | .30 | .29 |
| | I | .40 | .38 | .38 | .39 | .38 | .37 | .38 | .36 |
| | H | .43 | .42 | .41 | .42 | .41 | .41 | .41 | .40 |
| | G | .46 | .45 | .44 | .46 | .44 | .43 | .44 | .43 |
| | F | .49 | .47 | .46 | .47 | .46 | .45 | .46 | .45 |
| | E | .51 | .50 | .49 | .50 | .49 | .48 | .49 | .47 |
| | D | .55 | .52 | .51 | .53 | .52 | .50 | .51 | .50 |
| | C | .55 | .54 | .52 | .54 | .53 | .52 | .52 | .51 |
| | B | .56 | .55 | .54 | .55 | .53 | .53 | .53 | .52 |
| | A | .57 | .56 | .55 | .56 | .55 | .53 | .54 | .53 |

## Table 7-25. Coefficients of Utilization (Continued)

| Luminaire | Ceiling.. | 75% | | | 50% | | | 30% | |
|---|---|---|---|---|---|---|---|---|---|
| | Walls... | 50% | 30% | 10% | 50% | 30% | 10% | 30% | 10% |
| | Room index | Coefficient of utilization | | | | | | | |
| **MF** G −.75 ↑8 M −.65 ↓50 P −.55 Semidirect, surface−mounted, MS = 1.0 x MH | J | .21 | .17 | .14 | .20 | .16 | .14 | .16 | .14 |
| | I | .26 | .22 | .20 | .25 | .21 | .19 | .21 | .19 |
| | H | .29 | .25 | .23 | .28 | .25 | .22 | .24 | .22 |
| | G | .32 | .28 | .25 | .30 | .27 | .25 | .26 | .24 |
| | F | .34 | .30 | .27 | | | .30 | .27 | .29 | .27 |
| | E | .38 | .34 | .31 | .36 | .33 | .31 | .32 | .30 |
| | D | .41 | .37 | .34 | .39 | .36 | .34 | .35 | .33 |
| | C | .42 | .39 | .36 | .41 | .38 | .36 | .37 | .35 |
| | B | .45 | .42 | .39 | .42 | .40 | .39 | .39 | .38 |
| | A | .47 | .44 | .41 | .45 | .42 | .40 | .41 | .39 |
| **MF** G −.75 ↑9 M −.65 ↓55 P −.55 Semidirect, surface− mounted, MS = 1.0 x MH | J | .24 | .20 | .19 | .23 | .20 | .17 | .19 | .17 |
| | I | .30 | .26 | .23 | .29 | .25 | .23 | .25 | .23 |
| | H | .33 | .29 | .27 | .32 | .29 | .26 | .28 | .26 |
| | G | .36 | .32 | .30 | .34 | .32 | .29 | .30 | .29 |
| | F | .39 | .35 | .32 | .37 | .34 | .31 | .33 | .31 |
| | E | .42 | .39 | .35 | .41 | .38 | .35 | .36 | .34 |
| | D | .45 | .41 | .39 | .44 | .41 | .38 | .40 | .38 |
| | C | .47 | .44 | .41 | .45 | .42 | .40 | .41 | .39 |
| | B | .50 | .47 | .44 | .48 | .45 | .43 | .44 | .42 |
| | A | .52 | .49 | .46 | .50 | .47 | .45 | .45 | .44 |
| **MF** G −.75 ↑18 M −.65 ↓53 P −.55 Semidirect, surface − mounted MS = 1.0 x MH | J | .23 | .19 | .17 | .23 | .18 | .16 | .17 | .16 |
| | I | .29 | .25 | .22 | .28 | .24 | .21 | .23 | .21 |
| | H | .32 | .26 | .25 | .31 | .28 | .25 | .26 | .24 |
| | G | .36 | .32 | .29 | .34 | .30 | .27 | .29 | .26 |
| | F | .40 | .35 | .31 | .37 | .33 | .30 | .31 | .29 |
| | E | .43 | .39 | .35 | .41 | .37 | .34 | .35 | .32 |
| | D | .47 | .42 | .39 | .44 | .40 | .37 | .38 | .36 |
| | C | .49 | .45 | .41 | .46 | .42 | .39 | .40 | .38 |
| | B | .52 | .48 | .45 | .49 | .45 | .43 | .43 | .41 |
| | A | .54 | .51 | .47 | .51 | .47 | .45 | .44 | .43 |
| **MF** G −.75 ↑24 M −.65 ↓66 P −.55 Semidirect, surface − mounted, MS = 1.0 x MH | J | .29 | .24 | .22 | .29 | .23 | .20 | .22 | .20 |
| | I | .37 | .32 | .28 | .36 | .30 | .27 | .29 | .27 |
| | H | .41 | .35 | .32 | .39 | .34 | .32 | .33 | .30 |
| | G | .46 | .41 | .37 | .43 | .38 | .34 | .37 | .33 |
| | F | .51 | .44 | .39 | .47 | .42 | .39 | .39 | .37 |
| | E | .55 | .49 | .44 | .52 | .47 | .43 | .44 | .41 |
| | D | .60 | .53 | .49 | .56 | .51 | .47 | .48 | .46 |
| | C | .62 | .57 | .52 | .58 | .53 | .50 | .51 | .48 |
| | B | .66 | .61 | .57 | .62 | .57 | .54 | .54 | .52 |
| | A | .68 | .64 | .60 | .65 | .60 | .57 | .56 | .55 |
| **MF** G −.75 ↑39 M −.70 ↓45 P −.65 General diffuse, enclosing globe, MS = 1.2 x MH | J | .24 | .20 | .16 | .22 | .18 | .16 | .17 | .15 |
| | I | .30 | .25 | .23 | .27 | .23 | .21 | .22 | .19 |
| | H | .33 | .29 | .26 | .31 | .27 | .24 | .25 | .22 |
| | G | .37 | .33 | .30 | .34 | .30 | .27 | .27 | .25 |
| | F | .41 | .36 | .32 | .36 | .33 | .31 | .31 | .27 |
| | E | .45 | .41 | .37 | .41 | .37 | .33 | .33 | .30 |
| | D | .49 | .44 | .40 | .44 | .40 | .37 | .36 | .33 |
| | C | .51 | .47 | .43 | .46 | .42 | .39 | .38 | .35 |
| | B | .55 | .51 | .47 | .49 | .45 | .43 | .40 | .38 |
| | A | .57 | .53 | .50 | .51 | .47 | .45 | .42 | .40 |

## Table 7-25. Coefficients of Utilization (Continued)

| Luminaire | Room index | Ceiling 75% Walls 50% | 30% | 10% | Ceiling 50% Walls 50% | 30% | 10% | Ceiling 30% Walls 30% | 10% |
|---|---|---|---|---|---|---|---|---|---|
| | | Coefficient of utilization | | | | | | | |
| **MF** G −.75 ↑30 M −.65 P −.55 ↓59 Semidirect, ceiling − mounted,* MS = 1.0 x MH | J | .30 | .25 | .21 | .28 | .23 | .20 | .22 | .19 |
| | I | .38 | .33 | .29 | .35 | .30 | .27 | .29 | .26 |
| | H | .42 | .37 | .34 | .39 | .35 | .32 | .33 | .30 |
| | G | .46 | .41 | .37 | .42 | .38 | .35 | .35 | .33 |
| | F | .50 | .45 | .41 | .49 | .41 | .38 | .38 | .36 |
| | E | .55 | .50 | .46 | .50 | .46 | .43 | .43 | .40 |
| | D | .60 | .55 | .51 | .54 | .50 | .47 | .47 | .45 |
| | C | .62 | .58 | .54 | .56 | .52 | .50 | .49 | .47 |
| | B | .66 | .62 | .59 | .60 | .56 | .54 | .52 | .50 |
| | A | .68 | .65 | .61 | .62 | .58 | .56 | .54 | .52 |
| **MF** G −.70 ↑19 M −.65 P −.60 ↓49 Semidirect, ceiling − mounted,* 2 or 4 lamps, MS = .9 x MH | J | .28 | .25 | .23 | .23 | .21 | .19 | .18 | .16 |
| | I | .34 | .31 | .29 | .28 | .26 | .25 | .22 | .21 |
| | H | .37 | .34 | .33 | .31 | .29 | .28 | .25 | .24 |
| | G | .41 | .38 | .36 | .34 | .32 | .30 | .27 | .26 |
| | F | .43 | .41 | .38 | .36 | .33 | .32 | .29 | .27 |
| | E | .46 | .44 | .42 | .38 | .37 | .35 | .31 | .30 |
| | D | .50 | .47 | .45 | .41 | .39 | .37 | .33 | .32 |
| | C | .52 | .49 | .46 | .42 | .40 | .39 | .34 | .33 |
| | B | .54 | .51 | .50 | .44 | .42 | .41 | .36 | .35 |
| | A | .56 | .53 | .51 | .46 | .43 | .42 | .37 | .36 |
| **MF** G − .70 ↑46 M − .65 P − .60 ↓33 Direct − indirect, suspension − mounted, 2 or 4 lamps, MS = 1.2 x MH | J | .26 | .23 | .20 | .23 | .21 | .19 | .19 | .17 |
| | I | .31 | .28 | .27 | .28 | .26 | .24 | .23 | .20 |
| | H | .35 | .32 | .30 | .31 | .28 | .27 | .26 | .24 |
| | G | .38 | .35 | .33 | .34 | .31 | .30 | .28 | .27 |
| | F | .41 | .38 | .35 | .36 | .34 | .32 | .30 | .28 |
| | E | .44 | .42 | .39 | .39 | .37 | .35 | .32 | .31 |
| | D | .48 | .45 | .42 | .42 | .39 | .38 | .34 | .33 |
| | C | .50 | .49 | .44 | .45 | .41 | .39 | .35 | .34 |
| | B | .53 | .50 | .48 | .46 | .43 | .42 | .37 | .36 |
| | A | .54 | .52 | .50 | .47 | .45 | .43 | .39 | .37 |
| **MF** G − .65 ↑20 M − .55 P − .50 ↓47 Semidirect, ceiling − mounted,* glass bottom, MS = .9 x MH | J | .28 | .23 | .21 | .23 | .20 | .18 | .17 | .16 |
| | I | .33 | .30 | .28 | .28 | .25 | .23 | .22 | .20 |
| | H | .36 | .33 | .31 | .30 | .28 | .26 | .24 | .23 |
| | G | .39 | .36 | .34 | .33 | .30 | .29 | .26 | .25 |
| | F | .42 | .39 | .36 | .35 | .32 | .31 | .28 | .26 |
| | E | .45 | .42 | .40 | .38 | .36 | .34 | .30 | .29 |
| | D | .48 | .45 | .44 | .40 | .38 | .36 | .32 | .31 |
| | C | .50 | .47 | .44 | .42 | .39 | .38 | .33 | .32 |
| | B | .53 | .50 | .48 | .43 | .41 | .40 | .35 | .34 |
| | A | .54 | .52 | .49 | .45 | .43 | .41 | .36 | .35 |
| **MF** G − .65 ↑49 M − .55 P − .50 ↓33 Direct − indirect, suspension − mounted, glass bottom, MS = 1.2 x MH | J | .27 | .24 | .22 | .24 | .22 | .21 | .21 | .19 |
| | I | .33 | .30 | .29 | .29 | .27 | .26 | .25 | .23 |
| | H | .36 | .34 | .32 | .32 | .30 | .29 | .28 | .26 |
| | G | .39 | .37 | .35 | .36 | .33 | .32 | .30 | .28 |
| | F | .43 | .40 | .37 | .39 | .35 | .34 | .31 | .30 |
| | E | .46 | .43 | .41 | .41 | .38 | .37 | .34 | .32 |
| | D | .50 | .46 | .44 | .43 | .41 | .39 | .36 | .35 |
| | C | .52 | .49 | .46 | .45 | .43 | .41 | .37 | .36 |
| | B | .55 | .52 | .50 | .47 | .45 | .44 | .38 | .37 |
| | A | .56 | .54 | .52 | .48 | .47 | .45 | .40 | .38 |
| **MF** G − .60 ↑56 M − .50 P − .40 ↓20 Semi − indirect, suspension − mounted, 2 or 4 lamps, MS = 1.2 x MH | J | .18 | .14 | .13 | .14 | .12 | .10 | .09 | .08 |
| | I | .22 | .19 | .17 | .18 | .15 | .14 | .12 | .11 |
| | H | .25 | .22 | .20 | .20 | .18 | .16 | .14 | .13 |
| | G | .28 | .25 | .22 | .22 | .20 | .18 | .16 | .15 |
| | F | .30 | .27 | .24 | .24 | .22 | .20 | .17 | .16 |
| | E | .34 | .30 | .25 | .27 | .24 | .22 | .19 | .18 |
| | D | .37 | .33 | .31 | .29 | .26 | .25 | .21 | .20 |
| | C | .39 | .36 | .33 | .30 | .28 | .26 | .22 | .21 |
| | B | .42 | .39 | .37 | .34 | .30 | .28 | .24 | .23 |
| | A | .44 | .41 | .39 | .34 | .32 | .30 | .25 | .24 |

\* Data based upon photometric curve run with false ceiling plate installed above luminaire in accordance with standard test procedure.

## Table 7-25. Coefficients of Utilization (Continued)

| Luminaire | Ceiling.. | 75 | | | 50% | | | 30% | |
|---|---|---|---|---|---|---|---|---|---|
| | Walls... | 50% | 30% | 10% | 50% | 30% | 10% | 30% | 10% |
| | Room index | Coefficient of utilization | | | | | | | |
| **MF** G-.70↑79 M-.60 P-.50↓3 <br> Indirect, glass, plastic, or metal, MS = 1.2 x MH | J | .16 | .13 | .11 | .12 | .10 | .08 | .06 | .05 |
| | I | .20 | .16 | .15 | .15 | .13 | .11 | .08 | .07 |
| | H | .23 | .20 | .17 | .17 | .14 | .13 | .10 | .08 |
| | G | .26 | .23 | .20 | .20 | .17 | .15 | .11 | .10 |
| | F | .29 | .26 | .23 | .22 | .19 | .17 | .12 | .11 |
| | E | .32 | .29 | .26 | .24 | .21 | .19 | .13 | .12 |
| | D | .36 | .32 | .30 | .26 | .24 | .22 | .15 | .14 |
| | C | .38 | .35 | .32 | .28 | .25 | .24 | .16 | .15 |
| | B | .42 | .39 | .36 | .30 | .29 | .27 | .18 | .17 |
| | A | .44 | .41 | .39 | .33 | .30 | .29 | .19 | .18 |
| **MF** G-.65↑85 M-.60 P-.55↓0 <br> Indirect, silvered bowl, MS = 1.2 x MH | J | .17 | .14 | .12 | .13 | .11 | .09 | .07 | .06 |
| | I | .21 | .17 | .16 | .16 | .14 | .12 | .09 | .08 |
| | H | .24 | .21 | .18 | .18 | .15 | .14 | .11 | .09 |
| | G | .27 | .24 | .21 | .21 | .18 | .16 | .12 | .11 |
| | F | .30 | .27 | .23 | .23 | .20 | .18 | .13 | .12 |
| | E | .33 | .30 | .27 | .25 | .22 | .20 | .14 | .13 |
| | D | .37 | .33 | .31 | .27 | .25 | .23 | .16 | .15 |
| | C | .39 | .36 | .33 | .29 | .26 | .25 | .17 | .16 |
| | B | .43 | .40 | .37 | .31 | .30 | .28 | .19 | .18 |
| | A | .45 | .42 | .40 | .34 | .31 | .30 | .20 | .19 |

| Typical luminaire | Estimated maintenance factors* | Distribution and max spacing | Room index | Ceiling | | | | | | | |
|---|---|---|---|---|---|---|---|---|---|---|---|
| | | | | 75% | | | 50% | | | 30% | |
| | | | | Walls | | | | | | | |
| | | | | 50% | 30% | 10% | 50% | 30% | 10% | 30% | 10% |
| Luminous ceiling using thin corrugated plastic diffuser having a reflectance of .40 and transmittance of .50 | G-.65↑0 M-.65 P-.45↓68 | Direct | J | .22 | .16 | .12 | Estimates based on calculations with cavity reflectance = 75%, cavity efficiency = 60%, apparent ceiling reflectance = 60%, floor reflectance = 14% | | | | |
| | | | I | .27 | .22 | .19 | | | | | |
| | | | H | .33 | .28 | .24 | | | | | |
| | | | G | .38 | .32 | .29 | | | | | |
| | | | F | .41 | .37 | .33 | | | | | |
| | | | E | .46 | .42 | .39 | | | | | |
| | | | D | .49 | .46 | .43 | | | | | |
| | | | C | .52 | .49 | .46 | | | | | |
| | | | B | .55 | .52 | .50 | | | | | |
| | | | A | .57 | .55 | .53 | | | | | |
| 45° plastic louverall below 2-lamp 40-watt industrial type fluorescent units and bare lamps | G-.70↑0 M-.65 P-.55↓60 | Direct | | with reflectors shallow cavity 75% | | | without reflectors shallow cavity 75% | | | | |
| | | | J | .28 | .25 | .23 | .25 | .20 | .19 | | |
| | | | I | .31 | .29 | .27 | .29 | .25 | .23 | | |
| | | | H | .34 | .32 | .30 | .32 | .28 | .26 | | |
| | | | G | .37 | .35 | .33 | .35 | .32 | .30 | | |
| | | | F | .40 | .37 | .35 | .38 | .34 | .32 | | |
| | | | E | .43 | .41 | .38 | .41 | .38 | .36 | | |
| | | | D | .45 | .43 | .40 | .43 | .40 | .39 | | |
| | | | C | .46 | .44 | .42 | .45 | .42 | .41 | | |
| | | | B | .48 | .45 | .43 | .47 | .44 | .43 | | |
| | | | A | .48 | .46 | .44 | .48 | .46 | .44 | | |
| 45° white metal louverall | G-.70↑0 M-.65 P-.55↓50 | Direct | | with reflectors shallow cavity 75% | | | without reflectors shallow cavity 75% | | | | |
| | | | J | .23 | .20 | .19 | .23 | .19 | .18 | | |
| | | | I | .27 | .24 | .22 | .26 | .23 | .21 | | |
| | | | H | .30 | .27 | .25 | .29 | .26 | .24 | | |
| | | | G | .32 | .29 | .28 | .32 | .29 | .27 | | |
| | | | F | .34 | .31 | .30 | .34 | .31 | .29 | | |
| | | | E | .36 | .33 | .32 | .36 | .33 | .32 | | |
| | | | D | .38 | .35 | .34 | .38 | .35 | .34 | | |
| | | | C | .39 | .37 | .36 | .39 | .37 | .36 | | |
| | | | B | .41 | .39 | .38 | .41 | .38 | .38 | | |
| | | | A | .42 | .40 | .39 | .42 | .40 | .39 | | |

## Table 7-25. Coefficients of Utilization (Continued)

**Luminaire 1**

MF
G – .70 ↑0
M – .60
P – .55 ↓53

Direct, Troffer, louvers, MS = .8 x MH

| Ceiling | 75% | | | 50% | | | 30% | |
|---|---|---|---|---|---|---|---|---|
| Walls | 50% | 30% | 10% | 50% | 30% | 10% | 30% | 10% |
| Room index | | | | Coefficient of utilization | | | | |
| J | .29 | .27 | .26 | .29 | .27 | .26 | .27 | .26 |
| I | .35 | .34 | .33 | .35 | .33 | .33 | .33 | .31 |
| H | .38 | .37 | .36 | .37 | .36 | .36 | .36 | .35 |
| G | .40 | .39 | .39 | .40 | .39 | .38 | .38 | .38 |
| F | .43 | .42 | .49 | .41 | .40 | .39 | .40 | .39 |
| E | .45 | .44 | .43 | .44 | .43 | .42 | .43 | .42 |
| D | .48 | .46 | .45 | .46 | .45 | .44 | .45 | .44 |
| C | .48 | .47 | .45 | .47 | .46 | .45 | .45 | .45 |
| B | .50 | .48 | .47 | .48 | .47 | .46 | .46 | .46 |
| A | .50 | .49 | .48 | .49 | .48 | .47 | .48 | .46 |

**Luminaire 2**

MF
G –.70 ↑0
M –.65   Metal | Plastic
P –.55 ↓50     ↓59

Direct, louverall ceiling, shielded to 45°, cavity reflectance 75%

Coefficients for plastic louvers based on the use of bare lamps without reflectors.

| Room index | METAL Cavity—75% | | | PLASTIC Cavity—75% | | |
|---|---|---|---|---|---|---|
| J | .23 | .20 | .19 | .25 | .20 | .19 |
| I | .27 | .24 | .22 | .29 | .25 | .23 |
| H | .30 | .27 | .25 | .32 | .28 | .26 |
| G | .32 | .29 | .28 | .35 | .32 | .30 |
| F | .34 | .31 | .30 | .38 | .35 | .32 |
| E | .36 | .33 | .32 | .41 | .38 | .36 |
| D | .38 | .35 | .34 | .43 | .40 | .39 |
| C | .39 | .37 | .36 | .45 | .42 | .41 |
| B | .41 | .39 | .38 | .47 | .44 | .43 |
| A | .42 | .40 | .39 | .48 | .46 | .45 |

**Luminaire 3**

MF
G –.70 ↑0
M –.60
P –.55 ↓60

Direct, surface-mounted, 2 or 4 lamps, MS = 1.0 x MH

| Room index | 75% | | | 50% | | | 30% | |
|---|---|---|---|---|---|---|---|---|
| | 50% | 30% | 10% | 50% | 30% | 10% | 30% | 10% |
| J | .29 | .26 | .23 | .28 | .26 | .23 | .25 | .23 |
| I | .35 | .32 | .31 | .35 | .32 | .30 | .32 | .30 |
| H | .38 | .36 | .34 | .38 | .36 | .34 | .35 | .34 |
| G | .41 | .39 | .37 | .41 | .39 | .37 | .38 | .37 |
| F | .44 | .41 | .39 | .42 | .41 | .39 | .40 | .39 |
| E | .46 | .45 | .42 | .46 | .44 | .42 | .44 | .42 |
| D | .50 | .48 | .46 | .49 | .47 | .46 | .46 | .46 |
| C | .51 | .49 | .47 | .50 | .48 | .47 | .48 | .46 |
| B | .53 | .51 | .50 | .52 | .50 | .49 | .49 | .49 |
| A | .54 | .52 | .50 | .53 | .51 | .50 | .50 | .49 |

**Luminaire 4**

MF
G –.70 ↑5
M –.65
P –.60 ↓47

Semidirect, surface–mounted, MS = 1.0 x MH

| Room index | 75% | | | 50% | | | 30% | |
|---|---|---|---|---|---|---|---|---|
| | 50% | 30% | 10% | 50% | 30% | 10% | 30% | 10% |
| J | .26 | .23 | .22 | .25 | .23 | .22 | .23 | .21 |
| I | .31 | .29 | .28 | .30 | .28 | .27 | .28 | .26 |
| H | .34 | .32 | .31 | .32 | .31 | .30 | .30 | .29 |
| G | .36 | .34 | .34 | .35 | .33 | .32 | .33 | .32 |
| F | .38 | .36 | .35 | .36 | .35 | .34 | .35 | .33 |
| E | .40 | .39 | .37 | .39 | .38 | .36 | .37 | .35 |
| D | .43 | .41 | .39 | .41 | .40 | .38 | .39 | .38 |
| C | .45 | .42 | .40 | .42 | .41 | .39 | .40 | .39 |
| B | .46 | .44 | .42 | .44 | .42 | .41 | .41 | .40 |
| A | .46 | .45 | .43 | .45 | .43 | .42 | .42 | .41 |

NOTES: *Symbols*
MF   maintenance factor
G   good maintenance factor
M   medium maintenance factor
P   poor maintenance factor
MS   maximum spacing
MH   mounting height

*Distribution.* Curves represent shape and not quantity of light. Fluorescent curves are taken normal to lamp axis.

↑ 48 per cent up

↓ 36 per cent down

48 + 36 = 84% luminaire efficiency

## Table 7-26. Room Index

| Ceiling height in feet for semi-indirect and indirect lighting | | 9, 9½ | 10–11½ | 12–13½ | 14–16½ | 17–20 | 21–24 | 25–30 | 31–36 | 37–50 | | |
|---|---|---|---|---|---|---|---|---|---|---|---|---|
| Mounting height above floor in feet for direct and semidirect lighting | | 7, 7½ | 8, 8½ | 9, 9½ | 10–11½ | 12–13½ | 14–16½ | 17–20 | 21–24 | 25–30 | 31–36 | 37–50 |

Room index*

| Room width, ft | Room length, ft | 7, 7½ | 8, 8½ | 9, 9½ | 10–11½ | 12–13½ | 14–16½ | 17–20 | 21–24 | 25–30 | 31–36 | 37–50 |
|---|---|---|---|---|---|---|---|---|---|---|---|---|
| **9** (8½–9) | 8–10 | H | I | J | J | | | | | | | |
| | 10–14 | H | I | I | J | | | | | | | |
| | 14–20 | G | H | I | J | J | | | | | | |
| | 20–30 | G | G | H | I | J | J | | | | | |
| | 30–42 | F | G | H | I | J | J | J | | | | |
| | 42–up | E | F | G | H | I | J | J | | | | |
| **10** (9½–10½) | 10–14 | G | H | I | J | J | | | | | | |
| | 14–20 | G | H | I | J | J | J | | | | | |
| | 20–30 | F | G | H | I | J | J | | | | | |
| | 30–42 | F | G | G | H | I | J | J | | | | |
| | 42–60 | E | F | G | G | H | I | J | | | | |
| | 60–up | E | F | F | G | H | I | J | | | | |
| **12** (11–12½) | 10–14 | G | H | I | I | J | J | | | | | |
| | 14–20 | F | G | H | I | J | J | | | | | |
| | 20–30 | F | G | G | H | I | J | J | | | | |
| | 30–42 | E | F | G | G | H | I | J | | | | |
| | 42–60 | E | F | F | F | H | H | I | | | | |
| | 60–up | E | E | E | F | G | H | I | | | | |
| **14** (13–15½) | 14–20 | F | G | H | H | I | J | J | | | | |
| | 20–30 | E | F | G | H | I | J | J | | | | |
| | 30–42 | E | F | F | F | H | I | J | J | | | |
| | 42–60 | E | E | E | F | F | H | I | J | J | | |
| | 60–90 | D | E | E | E | F | G | H | I | J | J | |
| | 90–up | D | D | E | E | F | F | G | I | J | J | |
| **17** (16–18½) | 14–20 | E | F | G | H | I | J | J | | | | |
| | 20–30 | E | F | F | G | H | I | J | | | | |
| | 30–42 | D | E | F | F | G | H | I | J | J | | |
| | 42–60 | D | E | E | E | F | G | H | I | J | J | |
| | 60–110 | D | E | E | E | F | F | G | H | I | J | |
| | 110–up | C | D | E | E | E | F | G | H | I | J | |
| **20** (19–21½) | 20–30 | D | E | F | G | H | I | J | J | | | |
| | 30–42 | D | E | E | F | G | H | I | J | J | | |
| | 42–60 | D | E | E | E | F | G | H | I | J | J | |
| | 60–90 | C | D | D | E | E | F | G | H | I | J | |
| | 90–140 | C | D | D | E | E | F | F | H | I | I | J |
| | 140–up | C | D | D | E | E | F | F | H | H | I | J |
| **24** (22–26) | 20–30 | D | E | E | F | G | H | I | J | J | | |
| | 30–42 | C | D | D | E | F | G | H | I | J | J | |
| | 42–60 | C | D | D | E | E | F | G | H | I | J | |
| | 60–90 | C | C | D | D | E | F | G | H | I | J | |
| | 90–140 | C | C | D | D | E | E | F | G | H | I | J |
| | 140–up | C | C | D | D | E | E | F | F | H | I | J |
| **30** (27–33) | 30–42 | C | D | D | E | F | G | H | I | J | J | |
| | 42–60 | C | C | D | D | E | F | G | H | I | J | J |
| | 60–90 | B | C | C | D | D | E | F | G | H | I | J |
| | 90–140 | B | C | C | C | D | D | E | F | G | H | I |
| | 140–180 | B | B | C | C | D | D | E | E | F | G | H |
| | 180–up | B | B | C | C | D | D | E | E | F | G | H |

## Table 7-26. Room Index (Continued)

| Ceiling height in feet for semi-indirect and indirect lighting | | 9, 9½ | 10–11½ | 12–13½ | 14–16½ | 17–20 | 21–24 | 25–30 | 31–36 | 37–50 | | |
|---|---|---|---|---|---|---|---|---|---|---|---|---|
| Mounting height above floor in feet for direct and semidirect lighting | | 7, 7½ | 8, 8½ | 9, 9½ | 10–11½ | 12–13½ | 14–16½ | 17–20 | 21–24 | 25–30 | 31–36 | 37–50 |
| **Room width, ft** | **Room length, ft** | Room index [*] | | | | | | | | | | |
| 36 (34–39) | 30–42 | B | C | D | E | F | F | H | I | I | J | J |
| | 42–60 | B | C | C | D | E | F | G | H | I | J | J |
| | 60–90 | A | C | C | C | E | E | F | H | H | I | J |
| | 90–140 | A | B | C | C | D | E | F | G | H | I | J |
| | 140–200 | A | B | C | C | D | D | F | F | G | H | I |
| | 200–up | A | B | C | C | D | D | E | F | F | G | H |
| 42 (40–45) | 42–60 | A | B | C | C | E | F | G | H | I | I | J |
| | 60–90 | A | B | B | C | D | E | F | G | H | I | J |
| | 90–140 | A | B | B | B | D | D | E | F | G | H | I |
| | 140–200 | A | A | B | B | C | D | D | E | F | G | H |
| | 200–up | A | A | A | B | C | D | D | E | F | F | G |
| 50 (46–55) | 42–60 | A | A | B | C | D | E | F | G | H | I | J |
| | 60–90 | A | A | B | C | C | D | F | F | G | H | I |
| | 90–140 | A | A | A | C | C | D | E | F | F | G | I |
| | 140–200 | A | A | A | C | C | C | D | E | E | F | G |
| | 200–up | A | A | A | C | C | D | E | E | F | G | H |
| 60 (56–67) | 60–90 | A | A | A | B | C | D | E | F | G | H | I |
| | 90–140 | A | A | A | B | C | C | D | E | F | G | H |
| | 140–200 | A | A | A | B | C | C | C | D | E | F | H |
| | 200–up | A | A | A | B | C | C | D | D | E | F | H |
| 75 (68–90) | 60–90 | A | A | A | A | B | C | D | E | F | G | I |
| | 90–140 | A | A | A | A | B | B | C | D | E | F | H |
| | 140–200 | A | A | A | A | B | B | C | C | D | E | G |
| | 200–up | A | A | A | A | B | B | B | C | D | E | G |

[*] "Room index" is the classification of a room according to its proportions; large and small rooms of the same proportion have the same index. Hence, for large rooms of dimensions greater than those shown, divide each dimension by the same number and use the index determined for the smaller room.

EXAMPLE: A room 200 by 600 by 40 ft would have the same room index as a room 50 by 150 by 10 ft.

SOURCE: "Essential Data for Lighting Design," p. 2, General Electric Company, Cleveland, Ohio, November, 1951.

## Table 7-27. Diffuse-reflection Factors

| Color | Average reflection factor | Color | Average reflection factor |
|---|---|---|---|
| White............ | 0.88 | Medium: | |
| Very Light: | | Blue green.... | 0.54 |
| Blue green..... | .76 | Yellow........ | .65 |
| Cream........ | .81 | Buff.......... | .63 |
| Blue.......... | .65 | Gray......... | .61 |
| Buff.......... | .76 | Dark: | |
| Gray.......... | .83 | Blue.......... | .08 |
| Light: | | Yellow........ | .50 |
| Blue green..... | .72 | Brown........ | .10 |
| Cream........ | .79 | Gray......... | .25 |
| Blue.......... | .55 | Green........ | .07 |
| Buff .......... | .70 | Black......... | .03 |
| Gray.......... | .73 | Wood Finishes: | |
| | | Maple........ | .42 |
| | | Walnut....... | .16 |
| | | Mahogany.... | .12 |

SOURCE: "Illumination Design Data for Interiors," pp. 2, 6, Westinghouse Electric Corporation, Bloomfield, N.J.

## Table 7-28. Brightness Ratios

| Area | Max Ratio |
|---|---|
| Between task and surroundings............. | 3:1 |
| Between task and remote surfaces (walls).... | 10:1 |
| Between fixtures and adjacent surface....... | 20:1 |
| Anywhere in normal field of view........... | 40:1 |

SOURCE: "IES Handbook," Sec. 9. Illuminating Engineering Society, New York, 1952.

## Table 7-29. Approximate Ballast Loss per Lamp, Watts

| Bulb | Watts | Starter switch no. or current, ma | 110–125 volt | | | | 220–250 volt, high PF | |
|---|---|---|---|---|---|---|---|---|
| | | | Single lamp | | Two-lamp high PF | | | |
| | | | Low PF | High PF | Series | Lead lag | Single lamp | Two lamp |
| Preheat Lamps | | | | | | | | |
| 48″ T-12 | 40 | FS-4 | 8.5 | 11 | ... | 7.8 | 8.5 | 9.3 |
| 60″ T-17 | 90 | FS-85 | .... | 25 | ... | 19.5 | ... | 16 |
| Rapid-start Lamps | | | | | | | | |
| 48″ T-12 | 40 | 430 | 12 | 12 | ... | 8.5 | | |
| Instant-start Lamps | | | | | | | | |
| 48″ and 60″ | 40 | 415 | .... | 20 | 11 | 12 | | |
| Slimline Lamps | | | | | | | | |
| 72″ T-12 | 55 | 425 | .... | 25 | 16 | 15 | | |
| | 67 | 600 | .... | ... | ... | 17.5 | | |
| 96″ T-12 | 74 | 425 | .... | 29 | 16 | 17.5 | | |
| | 95 | 600 | .... | ... | ... | 29 | | |
| 96″ T-8 | 50 | 200 | .... | 20 | ... | 16 | | |
| | 69 | 300 | .... | 25 | ... | 23 | | |

SOURCE: "Illumination Design Data for Interiors," pp. 2, 6, Westinghouse Electric Corporation, Bloomfield, N.J.

## Table 7-30. Brightness Ratios for Direct, Uniformly Diffusing, Indirect, and Luminous-ceiling Systems

$$A = \frac{\text{average wall brightness (midway between floor and ceiling)}}{\text{average illumination at work plane}}$$

| Ceiling reflectance | 0.80 | | | | 0.70 | | | 0.50 | | |
|---|---|---|---|---|---|---|---|---|---|---|
| Wall reflectance... | 0.80 | 0.50 | 0.30 | 0.10 | 0.50 | 0.30 | 0.10 | 0.50 | 0.30 | 0.10 |
| Room coef* | Luminous Ceilings or Indirect Luminaires (Floor Reflectance: 0.30) | | | | | | | | | |
| 0.0 | 0.520 | 0.325 | 0.195 | 0.0650 | 0.325 | 0.195 | 0.0650 | 0.325 | 0.195 | 0.0650 |
| 0.1 | .536 | .332 | .198 | .0657 | .332 | .198 | .0657 | .332 | .198 | .0656 |
| 0.2 | .551 | .340 | .202 | .0667 | .340 | .202 | .0667 | .340 | .202 | .0667 |
| 0.3 | .567 | .348 | .206 | .0680 | .348 | .206 | .0680 | .348 | .206 | .0680 |
| 0.4 | .583 | .357 | .212 | .0697 | .357 | .212 | .0697 | .357 | .212 | .0696 |
| 0.5 | .598 | .367 | .218 | .0717 | .367 | .218 | .0717 | .367 | .218 | .0716 |
| 0.7 | .631 | .389 | .231 | .0765 | .388 | .231 | .0765 | .388 | .231 | .0762 |
| 1.0 | .681 | .426 | .256 | .0856 | .426 | .256 | .0856 | .426 | .256 | .0851 |
| | Luminous Ceilings or Indirect Luminaires (Floor Reflectance: 0.10) | | | | | | | | | |
| 0.0 | .440 | .275 | .165 | .0550 | .275 | .165 | .0550 | .275 | .165 | .0550 |
| 0.1 | .463 | .288 | .172 | .0570 | .287 | .172 | .0570 | .288 | .172 | .0570 |
| 0.2 | .486 | .300 | .179 | .0592 | .300 | .179 | .0592 | .300 | .179 | .0592 |
| 0.3 | .508 | .313 | .186 | .0616 | .313 | .186 | .0615 | .313 | .186 | .0615 |
| 0.4 | .530 | .326 | .194 | .0641 | .327 | .194 | .0641 | .327 | .194 | .0641 |
| 0.5 | .552 | .340 | .202 | .0668 | .340 | .202 | .0668 | .340 | .202 | .0668 |
| 0.7 | .594 | .368 | .220 | .0728 | .368 | .220 | .0728 | .368 | .220 | .0728 |
| 1.0 | .655 | .413 | .249 | .0833 | .413 | .249 | .0833 | .413 | .249 | .0833 |
| | Direct Luminaires (Floor Reflectance: 0.30) | | | | | | | | | |
| 0.0 | .218 | .137 | .082 | .0273 | .128 | .077 | .0255 | .113 | .068 | .0225 |
| 0.1 | .329 | .193 | .115 | .0372 | .189 | .112 | .3640 | .182 | .100 | .0350 |
| 0.2 | .342 | .200 | .115 | .0367 | .197 | .113 | .3610 | .189 | .109 | .0350 |
| 0.3 | .363 | .205 | .117 | .0366 | .202 | .113 | .3630 | .194 | .112 | .0353 |
| 0.4 | .384 | .216 | .119 | .0370 | .211 | .118 | .3670 | .205 | .115 | .0361 |
| 0.5 | .409 | .224 | .124 | .0384 | .220 | .123 | .3820 | .217 | .120 | .0375 |
| 0.7 | .415 | .253 | .139 | .0429 | .249 | .139 | .4280 | .247 | .137 | .0424 |
| 1.0 | .594 | .331 | .192 | .0602 | .328 | .192 | .6020 | .327 | .191 | .0598 |
| | Direct Luminaires (Floor Reflectance: 0.10) | | | | | | | | | |
| 0.0 | .072 | .045 | .027 | .0090 | .042 | .026 | .0085 | .038 | .023 | .0075 |
| 0.1 | .211 | .127 | .074 | .0242 | .125 | .073 | .0238 | .062 | .071 | .0234 |
| 0.2 | .239 | .140 | .080 | .0255 | .138 | .079 | .0253 | .134 | .077 | .0249 |
| 0.3 | .266 | .151 | .086 | .0269 | .150 | .085 | .0268 | .146 | .083 | .0265 |
| 0.4 | .298 | .165 | .092 | .0287 | .164 | .089 | .0286 | .160 | .090 | .0282 |
| 0.5 | .329 | .181 | .100 | .0310 | .179 | .100 | .0309 | .176 | .098 | .0308 |
| 0.7 | .402 | .217 | .121 | .0372 | .217 | .120 | .0372 | .214 | .119 | .0370 |
| 1.0 | .552 | .308 | .179 | .0565 | .307 | .179 | .0564 | .304 | .179 | .0562 |
| | Uniformly Diffusing Luminaires (Floor Reflectance: 0.30) | | | | | | | | | |
| 0.0 | .351 | .220 | .132 | .0439 | .209 | .125 | .0418 | .183 | .110 | .0367 |
| 0.1 | .696 | .442 | .268 | .0903 | .453 | .274 | .0925 | .478 | .289 | .0977 |
| 0.2 | .705 | .455 | .278 | .0945 | .464 | .284 | .0967 | .490 | .298 | .1020 |
| 0.3 | .714 | .466 | .287 | .0987 | .476 | .294 | .1009 | .500 | .306 | .1063 |
| 0.4 | .725 | .478 | .297 | .1029 | .488 | .304 | .1052 | .512 | .313 | .1107 |
| 0.5 | .736 | .490 | .307 | .1074 | .500 | .314 | .1096 | .523 | .320 | .1151 |
| 0.7 | .757 | .514 | .328 | .1165 | .523 | .334 | .1188 | .545 | .334 | .1245 |
| 1.0 | .790 | .551 | .359 | .1308 | .559 | .364 | .1329 | .576 | .354 | .1377 |
| | Uniformly Diffusing Luminaires (Floor Reflectance: 0.10) | | | | | | | | | |
| 0.0 | .596 | .372 | .223 | .0744 | .363 | .218 | .0727 | .342 | .205 | .0683 |
| 0.1 | .653 | .415 | .252 | .0848 | .424 | .257 | .0867 | .447 | .271 | .0913 |
| 0.2 | .663 | .427 | .269 | .0853 | .437 | .268 | .0912 | .460 | .382 | .0960 |
| 0.3 | .675 | .441 | .272 | .0906 | .450 | .278 | .0957 | .473 | .293 | .1007 |
| 0.4 | .688 | .454 | .283 | .0959 | .464 | .289 | .1003 | .487 | .304 | .1055 |
| 0.5 | .701 | .468 | .294 | .1011 | .477 | .300 | .1051 | .500 | .315 | .1103 |
| 0.7 | .729 | .496 | .316 | .1115 | .505 | .322 | .1148 | .525 | .336 | .1201 |
| 1.0 | .769 | .538 | .350 | .1272 | .545 | .356 | .1300 | .562 | .368 | .1348 |

## Table 7-30. Brightness Ratios for Direct, Uniformly Diffusing, Indirect, and Luminous-ceiling Systems (Continued)

$$B = \frac{\text{average ceiling brightness}}{\text{average illumination at work plane}}$$

| Ceiling reflectance | 0.80 | | | | 0.70 | | | 0.50 | | |
|---|---|---|---|---|---|---|---|---|---|---|
| Wall reflectance... | 0.80 | 0.50 | 0.30 | 0.10 | 0.50 | 0.30 | 0.10 | 0.50 | 0.30 | 0.10 |
| Room coef* | Luminous Ceilings or Indirect Luminaires (Floor Reflectance: 0.30) | | | | | | | | | |
| 0.0 | 1.000 | 1.000 | 1.000 | 1.0000 | 1.000 | 1.000 | 1.0000 | 1.000 | 1.000 | 1.0000 |
| 0.1 | 1.074 | 1.108 | 1.129 | 1.1510 | 1.108 | 1.129 | 1.1510 | 1.108 | 1.129 | 1.1510 |
| 0.2 | 1.153 | 1.228 | 1.277 | 1.3260 | 1.228 | 1.277 | 1.3260 | 1.228 | 1.277 | 1.3260 |
| 0.3 | 1.236 | 1.363 | 1.446 | 1.5280 | 1.363 | 1.446 | 1.5280 | 1.363 | 1.446 | 1.5280 |
| 0.4 | 1.324 | 1.514 | 1.638 | 1.7610 | 1.514 | 1.638 | 1.7610 | 1.514 | 1.638 | 1.7590 |
| 0.5 | 1.418 | 1.682 | 1.856 | 2.0300 | 1.682 | 1.856 | 2.0300 | 1.682 | 1.856 | 2.0270 |
| 0.7 | 1.625 | 2.078 | 2.386 | 2.6990 | 2.078 | 2.386 | 2.6990 | 2.078 | 2.386 | 2.6910 |
| 1.0 | 1.989 | 2.861 | 3.481 | 4.1380 | 2.861 | 3.481 | 4.1380 | 2.861 | 3.481 | 4.1120 |
| | Luminous Ceilings or Indirect Luminaires (Floor Reflectance: 0.10) | | | | | | | | | |
| 0.0 | 1.000 | 1.000 | 1.000 | 1.0000 | 1.000 | 1.000 | 1.0000 | 1.000 | 1.000 | 1.0000 |
| 0.1 | 1.086 | 1.114 | 1.134 | 1.1530 | 1.114 | 1.134 | 1.1520 | 1.114 | 1.134 | 1.1520 |
| 0.2 | 1.174 | 1.241 | 1.285 | 1.3280 | 1.241 | 1.285 | 1.3280 | 1.241 | 1.285 | 1.3280 |
| 0.3 | 1.267 | 1.381 | 1.457 | 1.5310 | 1.381 | 1.457 | 1.5310 | 1.381 | 1.457 | 1.5310 |
| 0.4 | 1.364 | 1.537 | 1.651 | 1.7650 | 1.537 | 1.651 | 1.7650 | 1.537 | 1.651 | 1.7650 |
| 0.5 | 1.466 | 1.710 | 1.872 | 2.0350 | 1.710 | 1.872 | 2.0350 | 1.710 | 1.872 | 2.0350 |
| 0.7 | 1.686 | 2.113 | 2.406 | 2.7050 | 2.113 | 2.406 | 2.7050 | 2.113 | 2.406 | 2.7050 |
| 1.0 | 2.067 | 2.906 | 3.506 | 4.1460 | 2.906 | 3.506 | 4.1460 | 2.905 | 3.506 | 4.1460 |
| | Direct Luminaires (Floor Reflectance: 0.30) | | | | | | | | | |
| 0.0 | 0.234 | 0.234 | 0.234 | 0.2340 | 0.210 | 0.210 | 0.2100 | 0.150 | 0.150 | 0.1500 |
| 0.1 | .245 | 0.232 | 0.212 | 0.2020 | 0.196 | 0.188 | 0.1760 | .140 | .132 | 0.1260 |
| 0.2 | .252 | 0.214 | 0.190 | 0.1700 | 0.186 | 0.166 | 0.1480 | .133 | .118 | 0.1060 |
| 0.3 | .264 | 0.206 | 0.173 | 0.1440 | 0.179 | 0.151 | 0.1260 | .127 | .108 | 0.0897 |
| 0.4 | .278 | 0.202 | 0.160 | 0.1230 | 0.175 | 0.140 | 0.1080 | .124 | .099 | 0.0769 |
| 0.5 | .297 | 0.202 | 0.151 | 0.1070 | 0.175 | 0.131 | 0.0938 | .124 | .093 | 0.0668 |
| 0.7 | .338 | 0.211 | 0.143 | 0.0847 | 0.183 | 0.125 | 0.0732 | .129 | .088 | 0.0528 |
| 1.0 | .440 | 0.357 | 0.162 | 0.0737 | 0.224 | 0.143 | 0.0477 | .158 | .102 | 0.0460 |
| | Direct Luminaires (Floor Reflectance: 0.10) | | | | | | | | | |
| 0.0 | .080 | 0.080 | 0.080 | 0.0800 | 0.070 | 0.070 | 0.0700 | .050 | .050 | 0.0500 |
| 0.1 | .096 | 0.084 | 0.076 | 0.0689 | 0.073 | 0.067 | 0.0592 | .052 | .047 | 0.0430 |
| 0.2 | .116 | 0.091 | 0.075 | 0.0602 | 0.079 | 0.065 | 0.0526 | .056 | .046 | 0.0386 |
| 0.3 | .140 | 0.099 | 0.075 | 0.0534 | 0.085 | 0.065 | 0.0470 | .061 | .046 | 0.0334 |
| 0.5 | .195 | 0.121 | 0.080 | 0.0451 | 0.105 | 0.070 | 0.0396 | .074 | .049 | 0.0285 |
| 0.7 | .234 | 0.151 | 0.092 | 0.0419 | 0.131 | 0.080 | 0.0368 | .094 | .057 | 0.0262 |
| 1.0 | .394 | 0.223 | 0.134 | 0.0489 | 0.194 | 0.117 | 0.0436 | .136 | .083 | 0.0310 |
| | Uniformly Diffusing Luminaires (Floor Reflectance: 0.30) | | | | | | | | | |
| 0.0 | .578 | 0.578 | 0.578 | 0.5780 | 0.535 | 0.535 | 0.5350 | .433 | .433 | 0.4330 |
| 0.1 | .637 | 0.634 | 0.633 | 0.6310 | 0.589 | 0.587 | 0.5860 | .479 | .477 | 0.4760 |
| 0.2 | .689 | 0.694 | 0.693 | 0.6950 | 0.642 | 0.643 | 0.6450 | .522 | .520 | 0.5240 |
| 0.3 | .735 | 0.748 | 0.758 | 0.7670 | 0.694 | 0.732 | 0.7620 | .561 | .563 | 0.5770 |
| 0.4 | .777 | 0.805 | 0.826 | 0.8480 | 0.745 | 0.764 | 0.8490 | .602 | .606 | 0.6350 |
| 0.5 | .817 | 0.864 | 0.898 | 0.9360 | 0.797 | 0.830 | 0.9420 | .640 | .650 | 0.6970 |
| 0.7 | .889 | 0.982 | 1.050 | 1.1340 | 0.902 | 0.969 | 1.1400 | .715 | .740 | 0.8330 |
| 1.0 | .988 | 1.167 | 1.310 | 1.4830 | 1.062 | 1.195 | 1.4850 | .825 | .878 | 1.0610 |
| | Uniformly Diffusing Luminaires (Floor Reflectance: 0.10) | | | | | | | | | |
| 0.0 | .889 | 0.889 | 0.889 | 0.8890 | 0.824 | 0.824 | 0.8240 | .667 | .667 | 0.6670 |
| 0.1 | .565 | 0.564 | 0.563 | 0.5610 | 0.523 | 0.522 | 0.5210 | .425 | .424 | 0.4230 |
| 0.2 | .631 | 0.634 | 0.637 | 0.6120 | 0.589 | 0.591 | 0.5930 | .478 | .480 | 0.4810 |
| 0.3 | .689 | 0.703 | 0.712 | 0.6990 | 0.651 | 0.660 | 0.6690 | .527 | .535 | 0.5430 |
| 0.4 | .741 | 0.769 | 0.789 | 0.7920 | 0.711 | 0.731 | 0.7510 | .574 | .590 | 0.6370 |
| 0.5 | .790 | 0.835 | 0.869 | 0.8910 | 0.771 | 0.803 | 0.8380 | .619 | .645 | 0.6750 |
| 0.7 | .878 | 0.967 | 1.040 | 1.1000 | 0.888 | 0.953 | 1.0260 | .704 | .757 | 0.8190 |
| 1.0 | .993 | 1.166 | 1.305 | 1.4650 | 1.061 | 1.190 | 1.3450 | .824 | .929 | 1.0550 |

## Table 7-30. Brightness Ratios for Direct, Uniformly Diffusing, Indirect, and Luminous-ceiling Systems (Continued)

$$C = \frac{\text{average floor brightness}}{\text{average illumination at work plane}}$$

| Ceiling reflectance | 0.80 | | | | 0.70 | | | 0.50 | | |
|---|---|---|---|---|---|---|---|---|---|---|
| Wall reflectance... | 0.80 | 0.50 | 0.30 | 0.10 | 0.50 | 0.30 | 0.10 | 0.50 | 0.30 | 0.10 |
| **Room coef*** | | | | | | | | | | |
| Luminous Ceilings or Indirect Luminaires (Floor Reflectance: 0.30) | | | | | | | | | | |
| 0.0 | 0.300 | 0.300 | 0.300 | 0.3000 | 0.300 | 0.300 | 0.3000 | 0.300 | 0.300 | 0.3000 |
| 0.1 | .293 | .290 | .288 | .2860 | .290 | .288 | .2860 | .290 | .288 | .2860 |
| 0.2 | .286 | .280 | .277 | .2730 | .280 | .277 | .2730 | .280 | .277 | .2730 |
| 0.3 | .279 | .271 | .266 | .2610 | .271 | .266 | .2610 | .271 | .266 | .2610 |
| 0.4 | .273 | .262 | .255 | .2490 | .262 | .255 | .2490 | .262 | .255 | .2480 |
| 0.5 | .266 | .253 | .245 | .2370 | .253 | .245 | .2370 | .253 | .245 | .2370 |
| 0.7 | .254 | .236 | .226 | .2160 | .236 | .226 | .2160 | .236 | .226 | .2150 |
| 1.0 | .237 | .213 | .200 | .1880 | .213 | .200 | .1880 | .213 | .200 | .1860 |
| Luminous Ceilings or Indirect Luminaires (Floor Reflectance: 0.10) | | | | | | | | | | |
| 0.0 | .100 | .100 | .100 | .1000 | .100 | .100 | .1000 | .100 | .100 | .1000 |
| 0.1 | .097 | .096 | .096 | .0954 | .096 | .096 | .0954 | .096 | .096 | .0954 |
| 0.2 | .095 | .093 | .092 | .0910 | .093 | .092 | .0910 | .093 | .092 | .0910 |
| 0.3 | .092 | .090 | .088 | .0868 | .090 | .088 | .0868 | .090 | .088 | .0868 |
| 0.4 | .090 | .087 | .085 | .0828 | .087 | .085 | .0828 | .087 | .085 | .0828 |
| 0.5 | .087 | .084 | .081 | .0790 | .084 | .081 | .0789 | .084 | .081 | .0789 |
| 0.7 | .083 | .078 | .075 | .0718 | .078 | .075 | .0718 | .078 | .075 | .0718 |
| 1.0 | .077 | .070 | .066 | .0623 | .070 | .066 | .0623 | .070 | .066 | .0623 |
| Direct Luminaires (Floor Reflectance: 0.30) | | | | | | | | | | |
| 0.0 | .300 | .300 | .300 | .3000 | .300 | .300 | .3000 | .300 | .300 | .3000 |
| 0.1 | .302 | .300 | .299 | .2980 | .301 | .300 | .2990 | .301 | .300 | .2990 |
| 0.2 | .302 | .300 | .299 | .2990 | .301 | .300 | .2980 | .301 | .299 | .2980 |
| 0.3 | .303 | .300 | .299 | .2980 | .301 | .299 | .2970 | .302 | .300 | .2980 |
| 0.4 | .303 | .301 | .298 | .2970 | .301 | .299 | .2970 | .303 | .300 | .2980 |
| 0.5 | .303 | .302 | .300 | .2990 | .302 | .300 | .2990 | .303 | .302 | .2990 |
| 0.7 | .303 | .302 | .301 | .3000 | .302 | .301 | .2990 | .305 | .303 | .3000 |
| 1.0 | .302 | .302 | .302 | .3010 | .303 | .302 | .3010 | .306 | .304 | .3010 |
| Direct Luminaires (Floor Reflectance: 0.10) | | | | | | | | | | |
| 0.0 | .100 | .100 | .100 | .1000 | .100 | .100 | .1000 | .100 | .100 | .1000 |
| 0.1 | .101 | .100 | .100 | .1000 | .101 | .100 | .1000 | .101 | .100 | .1000 |
| 0.2 | .101 | .101 | .100 | .1000 | .101 | .100 | .1000 | .101 | .100 | .1000 |
| 0.3 | .101 | .101 | .100 | .1000 | .101 | .100 | .1000 | .101 | .100 | .1000 |
| 0.4 | .101 | .101 | .100 | .1000 | .101 | .100 | .1000 | .101 | .100 | .1000 |
| 0.5 | .101 | .101 | .100 | .1000 | .101 | .100 | .1000 | .101 | .100 | .1000 |
| 0.7 | .100 | .100 | .100 | .1000 | .101 | .100 | .1000 | .101 | .100 | .1000 |
| 1.0 | .100 | .100 | .100 | .1000 | .100 | .100 | .1000 | .101 | .101 | .1000 |
| Uniformly Diffusing Luminaires (Floor Reflectance: 0.30) | | | | | | | | | | |
| 0.0 | .300 | .300 | .300 | .3000 | .300 | .300 | .3000 | .300 | .300 | .2000 |
| 0.1 | .300 | .298 | .295 | .2930 | .298 | .296 | .2940 | .300 | .297 | .2950 |
| 0.2 | .300 | .295 | .291 | .2870 | .296 | .292 | .2880 | .299 | .293 | .2910 |
| 0.3 | .298 | .292 | .287 | .2820 | .293 | .289 | .2840 | .298 | .289 | .2870 |
| 0.4 | .296 | .289 | .283 | .2770 | .291 | .285 | .2800 | .296 | .285 | .2840 |
| 0.5 | .293 | .286 | .280 | .2740 | .288 | .282 | .2760 | .295 | .281 | .2810 |
| 0.7 | .288 | .281 | .275 | .2680 | .284 | .278 | .2710 | .292 | .274 | .2770 |
| 1.0 | .280 | .275 | .270 | .2630 | .279 | .274 | .2660 | .288 | .265 | .2740 |
| Uniformly Diffusing Luminaires (Floor Reflectance: 0.10) | | | | | | | | | | |
| 0.0 | .100 | .100 | .100 | .1000 | .100 | .100 | .1000 | .100 | .100 | .1000 |
| 0.1 | .100 | .099 | .099 | .0980 | .100 | .099 | .0982 | .100 | .099 | .0986 |
| 0.2 | .100 | .098 | .097 | .0920 | .099 | .098 | .0964 | .100 | .099 | .0973 |
| 0.3 | .099 | .097 | .096 | .0914 | .098 | .096 | .0948 | .099 | .098 | .0960 |
| 0.4 | .098 | .096 | .095 | .0906 | .097 | .095 | .0934 | .099 | .097 | .0949 |
| 0.5 | .097 | .095 | .093 | .0898 | .096 | .094 | .0922 | .098 | .096 | .0940 |
| 0.7 | .095 | .093 | .091 | .0884 | .094 | .092 | .0902 | .097 | .095 | .0925 |
| 1.0 | .092 | .091 | .090 | .0871 | .092 | .091 | .0887 | .095 | .094 | .0914 |

* Room coef $= \dfrac{\text{room height} \times (\text{room length} + \text{width})}{2 \times \text{room length} \times \text{width}}$

SOURCE: "IES Handbook," p. 9-35. Illuminating Engineering Society, New York, 1952.

*Visual comfort* is determined by first calculating the glare factor, which is given by

$$\text{Glare factor} = AB^2/D^2\alpha^2 S^{0.6} \qquad (7\text{-}142)$$

where $A$ = apparent area of source, in.$^2$

$B$ = brightness of source, fl, divided by 1,000

$D$ = distance from source to eye, ft, divided by 10

$\alpha$ = angle above horizontal, deg, divided by 10

$s$ = surrounding brightness, fl, divided by 10

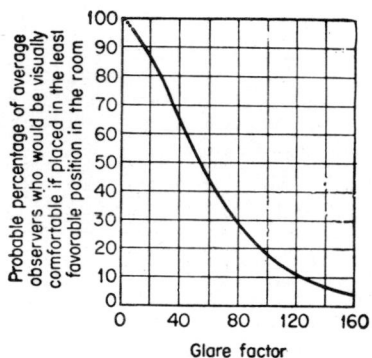

Fig. 7-46. The practical relationship between glare factor and the percentage of occupants who will be visually comfortable.

The total glare factor equals the sum of the glare factors for each fixture in a given reference direction. The total glare factor is entered in the curve of Fig. 7-46 to determine the visual comfort. The values in the equation are found from the physical size of the fixtures (area) and brightness curves. Correct lamp application for color is determined from Tables 7-31 and 7-32. Table 7-29 gives typical losses for fluorescent-lamp ballast, which serves as an aid in the determination of branch-circuit loading.

## Table 7-31. Color of Lamps

| Lamp type | Basic features | Use |
|---|---|---|
| Incandescent.... | Oranges emphasized | General lighting, highlighting special areas, lighting meats and meat products |
| Standard cool white | Deficient in red | Displaying woods and furs, general lighting where the effect of color is not critical |
| De luxe cool white | Same appearance as cool white with red added | General lighting |
| Standard warm white | Beige tint emphasizes yellow greens | Home lighting (favorable to complexions) |
| De luxe warm white | Same as standard warm white with red added | Home lighting (favorable to complexions), use alone where warm red is needed in abundance |
| White.......... | Emphasizes yellows, yellow greens, and oranges | Food-store lighting except for meats, general lighting of schools, offices, etc. |
| Daylight........ | Emphasizes blues and greens, tends to gray red, oranges, and yellows | General lighting, displaying iced foods |

## Table 7-32. Effect of Colored Light on Colored Objects

| Object color | Red light | Blue light | Green light | Yellow light |
|---|---|---|---|---|
| White....... | Light pink | Very light blue | Very light green | Very light yellow |
| Black....... | Reddish black | Blue black | Greenish black | Orange black |
| Red......... | Brilliant red | Dark bluish red | Yellowish red | Bright red |
| Light blue... | Reddish blue | Bright blue | Greenish blue | Light reddish blue |
| Dark blue... | Dark reddish purple | Brilliant blue | Dark greenish blue | Light reddish purple |
| Green....... | Olive green | Green blue | Brilliant green | Yellow green |
| Yellow...... | Red orange | Light reddish brown | Light greenish yellow | Brilliant light orange |
| Brown....... | Brown red | Bluish brown | Dark olive brown | Brownish orange |

SOURCE: "Westinghouse Lighting Handbook," p. 4-16, Westinghouse Electric Corporation, Bloomfield, N.J., 1954.

# SECTION 8

# MECHANICAL ENGINEERING

**Theodore Baumeister, B.S., M.E., P.E.;** Stevens Professor Emeritus of Mechanical Engineering, Columbia University; Fellow, American Society of Mechanical Engineers

**Eugene A. Avallone, B.M.E., M.S., M.E., P.E.;** Professor of Mechanical Engineering, The City College of the City University of New York; Member, American Society of Mechanical Engineers

## CONTENTS*

* For stationary engineering, and energy conversion in general, see Sec. 9.

**8-1**

# MECHANICS

## 8-1. Formulas of Motion

### Nomenclature

$t$ = time, sec
$s$ = linear displacement, ft
$v$ = linear velocity, fps
$V_0$ = linear velocity at time zero, fps
$a$ = linear acceleration, ft/sec$^2$
$\theta$ = angular displacement, radians
$\omega$ = angular velocity, radians/sec
$\omega_0$ = angular velocity at time zero, radians/sec
$\alpha$ = angular acceleration, radians/sec$^2$
$w$ = weight of body, lb mass
$f$ = force of acceleration, lb force
$g_c$ = conversion factor = 32.2 (lb mass)(ft)/(lb force)(sec$^2$)

| $v$ = constant | $\omega$ = constant | $v$ = variable | $\omega$ = variable |
|---|---|---|---|
| $v = s/t$ | $\omega = \theta/t$ | $v = ds/dt$ | $\omega = d\theta/dt$ |

| $a$ = constant | $\alpha$ = constant | $a$ = variable | $\alpha$ = variable |
|---|---|---|---|
| $v = V_0 + at$ | $\omega = \omega_0 + \alpha t$ | $a = \dfrac{dv}{dt} = \dfrac{d^2s}{dt^2}$ | $\alpha = \dfrac{d\omega}{dt} = \dfrac{d^2\theta}{dt^2}$ |
| $s = V_0 t + \frac{1}{2}at^2$ | $\theta = \omega_0 t + \frac{1}{2}\alpha t^2$ | $v = \int a\, dt$ | $\omega = \int \alpha\, dt$ |
| $v = \sqrt{V_0^2 + 2as}$ | $\omega = \sqrt{\omega_0^2 + 2\alpha\theta}$ | $s = \int v\, dt$ | $\theta = \int \omega\, dt$ |

For uniform acceleration

$$f = (w/g_c)a \qquad (8\text{-}1)$$

## 8-2. Dynamic Similarity (Theory of Models)

Two or more systems are dynamically similar when:
1. They are geometrically similar.
2. The paths of motion in the two systems are similar.

3. The centers of gravity are in the same position.
4. The radii of gyration are proportional.

The times for performing the same motion need not be the same.

## 8-3. Statics

Any force system in space will be in equilibrium if the resultant force and resultant moment are both equal to zero. This can be expressed by

$$\Sigma F_x = \Sigma F_y = \Sigma F_z = 0 \qquad (8\text{-}2a)$$
$$\Sigma M_x = \Sigma M_y = \Sigma M_z = 0 \qquad (8\text{-}2b)$$

where $F$ = force, lb
$M$ = moment, ft-lb
$x, y, z$ = orthogonal axes

**Center of Gravity.** The center of gravity of a body is that point through which the resultant of all gravity forces acting on the body passes. This point is fixed in the body and is independent of the position of the body.

**Law of Motion of the Center of Gravity.** The center of gravity of a system of particles or rigid bodies, under the action of several impressed forces, moves exactly as if the whole weight of the system were concentrated there and as if it were acted on by a force equal in magnitude and direction to the vector sum of all impressed forces. The coordinates of the center of gravity of the system being $\bar{x}$, $\bar{y}$, $\bar{z}$, its motion may be determined by the equations:

$$\frac{W}{g_c}\frac{d^2\bar{x}}{dt^2} = \Sigma F_{x_i} \qquad \frac{W}{g_c}\frac{d^2\bar{y}}{dt^2} = \Sigma F_{y_i} \qquad \frac{W}{g_c}\frac{d^2\bar{z}}{dt^2} = \Sigma F_{z_i} \qquad (8\text{-}3)$$

where $W$ = total weight of system, lb mass
$F_{x_i}, F_{y_i}, F_{z_i}$ = projections of impressed forces on coordinate axes, lb force
$t$ = time, sec
$g_c$ = (lb mass)(ft)/(lb force)(sec²)

### Centers of Gravity of Lines[1,2]

Straight line. The center of gravity is at its midpoint.
Circular arc $AB$, Fig. 8-1. $x_0 = r \sin c/\text{rad } c$ and

$$y_0 = 2r \sin^2 \tfrac{1}{2}c/\text{rad } c$$

Circular arc $AC$, Fig. 8-2. $x_0 = r \sin c/\text{rad } c$, $y_0 = 0$.
Quadrant $AB$, Fig. 8-3. $x_0 = y_0 = 2r/\pi = 0.6366r$.
Semicircumference $AC$ Fig. 8-3. $y_0 = 2r/\pi = 0.6366r$, $x_0 = 0$.
Combination of arcs and straight line (Fig. 8-4). $AD$ and $BC$ are two quadrants of radius $r$.

$$y_0 = \{ABr + 2[0.5\pi r(r - 0.6366r)]\}/[AB + 2(0.5\pi r)], \quad x_0 = 0$$

[1] In this discussion, rad $c$ = angle $c$ measured in radians.
[2] Source: T. Baumeister (ed.), "Standard Handbook for Mechanical Engineers," 7th ed., pp. 3-10, McGraw-Hill Book Company, New York, 1967.

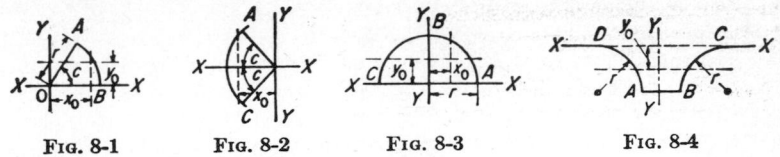

FIG. 8-1      FIG. 8-2      FIG. 8-3      FIG. 8-4

### Centers of Gravity of Plane Areas[1,2]

Triangle. Center of gravity lies at the intersection of the lines joining the vertices with the mid-points of the sides and at a distance from any side equal to one-third of the corresponding altitude.

Parallelogram. Center of gravity lies at the point of intersection of the diagonals.

Trapezoid (Fig. 8-5). Center of gravity lies on the line joining the mid-points $m$ and $n$ of the parallel sides. The distances $h_a$ and $h_b$ are

$$h_a = h(a + 2b)/3(a + b) \qquad h_b = h(2a + b)/3(a + b)$$

Draw $BE = a$ and $CF = b$; $EF$ will then intersect $mn$ at the center of gravity.

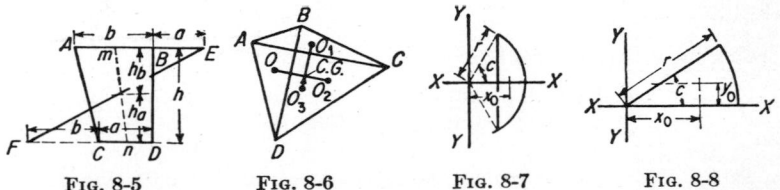

FIG. 8-5      FIG. 8-6      FIG. 8-7      FIG. 8-8

Any quadrilateral. The center of gravity of any quadrilateral may be determined by the general rule for areas, or graphically by dividing it into two sets of triangles by means of the diagonals. Find the center of gravity of each of the four triangles and connect the centers of gravity of the triangles belonging to the same set. The intersection of these lines will be the center of gravity of area. Thus in Fig. 8-6, $O$, $O_1$, $O_2$, and $O_3$ are, respectively, the centers of gravity of the triangles $ABD$, $ABC$, $BDC$, and $ACD$. The intersection of $O_1O_3$ with $OO_2$ is the center of gravity.

Segment of a circle (Fig. 8-7). $x_0 = \frac{2}{3}r \sin^3 c/(\text{rad } c - \cos c \sin c)$. A segment may be considered to be a sector from which a triangle is subtracted, and the general rule applied.

Sector of a circle (Fig. 8-8). $x_0 = \frac{2}{3}r \sin c/\text{rad } c, y_0 = \frac{4}{3}r \sin^2 \frac{1}{2}c/\text{rad } c$.

Semicircle. $x_0 = \frac{4}{3}r/\pi = 0.4244r$, $y_0 = 0$.

Quadrant (90° sector). $x_0 = y_0 = \frac{4}{3}r/\pi = 0.4244r$.

Parabolic half segment, area $ABO$, Fig. 8-9. $x_0 = \frac{3}{5}x_1$, $y_0 = \frac{3}{8}y_1$.

[1] In this discussion rad $c$ = angle $c$ measured in radians.

[2] Source: T. Baumeister (ed.), "Standard Handbook for Mechanical Engineers," 7th ed., pp. 3-10, McGraw-Hill Book Company, New York, 1967.

Parabolic spandrel, area $AOC$, Fig. 8-9.   $x'_0 = \frac{3}{10}x_1,\ y'_0 = \frac{3}{4}y_1$.

Quadrant of an ellipse, area $OAB$, Fig. 8-10.   $x_0 = \frac{4}{3}A/\pi,\ y_0 = \frac{4}{3}b/\pi$.

The center of gravity of a figure such as that shown in Fig. 8-11 may be determined as follows: Divide the arm $OABC$ into a number of parts by lines drawn perpendicular to the axis $XX$, for example, 11, 22, 33, etc.   These parts will be approximately triangles, rectangles, or trapezoids.   The area

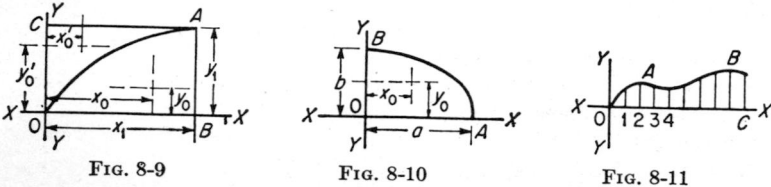

FIG. 8-9               FIG. 8-10                     FIG. 8-11

of each division may be obtained by taking the product of its mean height and its base.   The center of gravity of each area may be obtained as previously shown.   The sum of the moments of all the areas about $XX$ and $YY$, respectively, divided by the sum of the areas will give approximately the distances from the center of gravity of the whole area to the axes $XX$ and $YY$.   The greater the number of areas taken the more nearly exact the result.

### Centers of Gravity of Solids[1]

Prism or cylinder with parallel bases.   The center of gravity lies in the center of the line connecting the centers of gravity of the bases.

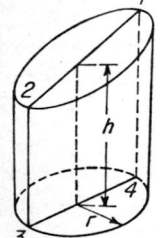

FIG. 8-12

Oblique frustum of a right circular cylinder (Fig. 8-12). Let 1-2-3-4 be the plane of symmetry.   The distance from the base to the center of gravity is $\frac{1}{2}h + (r^2 \tan^2 c)/8h$, where $c$ is the angle of inclination of the oblique section to the base.   The distance of the center of gravity from the axis of the cylinder is $r^2 \tan c/4h$.

Pyramid or cone.   The center of gravity lies in the line connecting the center of gravity of the base with the vertex and at a distance of one-fourth the altitude above the base.

Truncated pyramid.   If $h$ is the height of the truncated pyramid and $A$ and $B$ the area of its bases, the distance of the center of gravity from the surface of $A$ is $h(A + 2\sqrt{AB} + 3B)/4(A + \sqrt{AB} + B)$.

Truncated circular cone.   If $h$ is the height of the frustum and $R$ and $r$ the radii of the bases, the distance from the surface of the base whose radius is $R$ to the center of gravity is $h(R^2 + 2Rr + 3r^2)/4(R^2 + Rr + r^2)$.

Segment of a sphere, volume $ABC$, Fig. 8-13.   $x_0 = 3(2r - h)^2/4(3r - h)$.

Hemisphere.   $x_0 = 3r/8$.

[1] Source: T. Baumeister (ed.), "Standard Handbook for Mechanical Engineers," 7th ed., pp. 3-11, McGraw-Hill Book Company, New York, 1967.

Hollow hemisphere. $x_0 = 3(R^4 - r^4)/8(R^3 - r^3)$, where $R$ and $r$ are the outer and inner radii, respectively.

Sector of a sphere, volume $OABCO$, Fig. 8-13. $x'_0 = \frac{3}{8}(2r - h)$.

Ellipsoid with semiaxes $a$, $b$, and $c$. For each octant, distance from center of gravity to each of the bounding planes equals $\frac{3}{8}$ times the length of semiaxis perpendicular to the plane considered.

The motion of the center of gravity of a system does not change when the internal forces of the system vary. Its motion is not affected when internal forces are created or disappear, as occurs when parts of the system collide or explode.

**Moment of Inertia.** *Rectangular Moment of Inertia.* The rectangular moment of inertia $I$ (Table 8-1) with respect to a given axis is defined by the equations

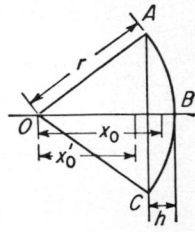

FIG. 8-13

$$I = \int y^2 \, dm \qquad \text{lb-ft}^2 \qquad \text{for solid body} \qquad (8\text{-}4)$$
and
$$I = \int y^2 \, dA \qquad \text{ft}^4 \qquad \text{for plane area} \qquad (8\text{-}5)$$

where $y$ = distance from elements of mass or area to reference axis, ft
$dm$ = element of mass, lb
$dA$ = element of area, ft$^2$

*Radius of Gyration.* The radius of gyration is a length $K$ feet such that

$$I = \int y^2 \, dm = K^2 m \qquad \text{for solid body} \qquad (8\text{-}6)$$
and
$$I = \int y^2 \, dA = K^2 A \qquad \text{for plane area} \qquad (8\text{-}7)$$

where $m$ = total mass, lb
$A$ = area, ft$^2$

*Rectangular Moments of Inertia and Radii of Gyration about Parallel Axes.* Moment of inertia about any reference axis is equal to

$$I = I_{CG} + a^2 m \qquad \text{for solid body} \qquad (8\text{-}8)$$
and
$$I = I_{CG} + a^2 A \qquad \text{for plane area} \qquad (8\text{-}9)$$

where $I_{CG}$ = moment of inertia of solid, lb-ft$^2$, or area, ft$^4$, about axis parallel to reference axis and passing through center of gravity

$a$ = distance between reference axis and axis passing through center of gravity, ft

Likewise, $$K^2 = K^2_{cg} + a^2 \qquad (8\text{-}10)$$

where $K_{cg}$ = radius of gyration through center of gravity, ft

*Polar Moment of Inertia.* The polar moment of inertia $J$ of an area is taken about an axis perpendicular to the area and is equal to

$$J = I_1 + I_2 \qquad \text{ft}^4 \qquad (8\text{-}11)$$

where $I_1$ and $I_2$ = moments of inertia about any two mutually perpendicular axes lying in plane of area and intersecting axis perpendicular to plane

## Table 8-1. Properties of Various Cross Sections

$I$ = moment of inertia; $I/c$ = section modulus; $r = \sqrt{I/A}$ = radius of gyration

| Section | Moment of inertia | Section modulus | Radius of gyration |
|---|---|---|---|

$I = \dfrac{bh^3}{12}$

$\dfrac{I}{c} = \dfrac{bh^2}{6}$

$r = \dfrac{h}{\sqrt{12}} = 0.289h$

$\dfrac{bh^3}{3}$

$\dfrac{bh^2}{3}$

$\dfrac{h}{\sqrt{3}} = 0.577h$

$\dfrac{b^3h^3}{6(b^2 + h^2)}$

$\dfrac{b^2h^2}{6\sqrt{b^2 + h^2}}$

$\dfrac{bh}{\sqrt{6(b^2 + h^2)}}$

$\dfrac{bh}{12}(h^2 \cos^2 a + b^2 \sin^2 a)$

$\dfrac{bh}{6}\left(\dfrac{h^2 \cos^2 a + b^2 \sin^2 a}{h \cos a + b \sin a}\right)$

$\sqrt{\dfrac{h^2 \cos^2 a + b^2 \sin^2 a}{12}}$

---

$I = \dfrac{b}{12}(H^3 - h^3)$

$\dfrac{I}{c} = \dfrac{b}{6}\dfrac{H^3 - h^3}{H}$

$r = \sqrt{\dfrac{H^3 - h^3}{12(H - h)}}$

$\dfrac{H^4 - h^4}{12}$

$\dfrac{1}{6}\dfrac{H^4 - h^4}{H}$

$\sqrt{\dfrac{H^2 + h^2}{12}}$

$\dfrac{H^4 - h^4}{12}$

$\dfrac{\sqrt{2}}{12}\dfrac{H^4 - h^4}{H}$

$\sqrt{\dfrac{H^2 + h^2}{12}}$

$\dfrac{bh^3}{36}; c = \dfrac{2}{3}h$

$\dfrac{bh^2}{24}$

$\dfrac{h}{\sqrt{18}}$

---

$I = \dfrac{bh^3}{12}$

$\dfrac{I}{c} = \dfrac{bh^2}{12}$

$r = \dfrac{h}{\sqrt{6}}$

$\dfrac{5\sqrt{3}}{16}R^4$

$\frac{5}{8}R^3$

$\sqrt{\dfrac{5}{24}}\,R$

$\dfrac{5\sqrt{3}}{16}R^4$

$\dfrac{5\sqrt{3}}{16}R^3$

$\dfrac{1 + 2\sqrt{2}}{6}R^4$

$0.6906R^3$

$0.475R$

## Table 8-1. Properties of Various Cross Sections (Continued)

| Section | Moment of inertia | Section modulus | Radius of gyration |
|---|---|---|---|
| Equilateral polygon<br>$A$ = area<br>$R$ = radius circumscribed circle<br>$r$ = radius inscribed circle<br>$n$ = no. sides<br>$a$ = length of side<br>Axis as in preceding section of octagon | $I = \dfrac{A}{24}(6R^2 - a^2)$<br><br>$= \dfrac{A}{48}(12r^2 + a^2)$<br><br>$= \dfrac{AR^2}{4}$ (approx) | $\dfrac{I}{c} = \dfrac{I}{r}$<br><br>$= \dfrac{I}{R\cos\dfrac{180°}{n}}$<br><br>$= \dfrac{AR}{4}$ (approx) | $\sqrt{\dfrac{6R^2 - a^2}{24}} \approx \dfrac{R}{2}$<br><br>$\sqrt{\dfrac{12r^2 + a^2}{48}}$ |

| | | | |
|---|---|---|---|
| | $I = \dfrac{6b^2 + 6bb_1 + b_1{}^2}{36(2b + b_1)}h^3$<br><br>$c = \dfrac{1}{3}\dfrac{3b + 2b_1}{2b + b_1}h$ | $\dfrac{I}{c} = \dfrac{6b^2 + 6bb_1 + b_1{}^2}{12(3b + 2b_1)}h^2$ | $\dfrac{h\sqrt{12b^2 + 12bb_1 + 2b_1{}^2}}{6(2b + b_1)}$ |

| | | | |
|---|---|---|---|
| | | $I = \dfrac{BH^3 + bh^3}{12}$<br><br>$\dfrac{I}{c} = \dfrac{BH^3 + bh^3}{6H}$ | $\sqrt{\dfrac{BH^3 + bh^3}{12(BH + bh)}}$ |

| | | | |
|---|---|---|---|
| | | $I = \dfrac{BH^3 - bh^3}{12}$<br><br>$\dfrac{I}{c} = \dfrac{BH^3 - bh^3}{6H}$ | $\sqrt{\dfrac{BH^3 - bh^3}{12(BH - bh)}}$ |

| | | | |
|---|---|---|---|
| | $I = \frac{1}{3}(Bc_1{}^3 - B_1h^3 + bc_2{}^3 - b_1h_1{}^3)$<br><br>$c_1 = \dfrac{1}{2}\dfrac{aH^2 + B_1d^2 + b_1d_1(2H - d_1)}{aH + B_1d + b_1d_1}$ | | $\sqrt{\dfrac{I}{(Bd + bd_1) + a(h + h_1)}}$ |

| | | | |
|---|---|---|---|
| | | $I = \frac{1}{3}(Bc_1{}^3 - bh^3 + ac_2{}^3)$<br><br>$c_1 = \dfrac{1}{2}\dfrac{aH^2 + bd^2}{aH + bd}$<br><br>$c_2 = H - c_1$<br><br>$r = \sqrt{\dfrac{I}{[Bd + a(H - d)]}}$ | |

| | | | |
|---|---|---|---|
| | $I = \dfrac{\pi d^4}{64} = \dfrac{\pi r^4}{4} = \dfrac{A}{4}r^2$<br><br>$= 0.05d^4$ (approx) | $\dfrac{I}{c} = \dfrac{\pi d^3}{32} = \dfrac{\pi r^3}{4} = \dfrac{A}{4}r$<br><br>$= 0.1d^3$ (approx) | $\dfrac{r}{2} = \dfrac{d}{4}$ |

## Table 8-1. Properties of Various Cross Sections (Continued)

| Section | Moment of inertia | Section modulus | Radius of gyration |
|---|---|---|---|
| $dm = \frac{1}{2}(D + d)$ $s = \frac{1}{2}(D - d)$ | $I = \frac{\pi}{64}(D^4 - d^4)$ $= \frac{\pi}{4}(R^4 - r^4)$ $= \frac{1}{4}A(R^2 + r^2)$ $= 0.05(D^4 - d^4)$ (approx) | $\frac{I}{c} = \frac{\pi}{32}\frac{D^4 - d^4}{D}$ $= \frac{\pi}{4}\frac{R^4 - r^4}{R}$ $= 0.8d_m{}^2s$ (approx) when $\frac{s}{d_m}$ is very small | $\frac{\sqrt{R^2 + r^2}}{2}$ $= \frac{\sqrt{D^2 + d^2}}{4}$ |
| | $I = r^4\left(\frac{\pi}{8} - \frac{8}{9\pi}\right)$ $= 0.1098r^4$ | $\frac{I}{c_2} = 0.1908r^3$ $\frac{I}{c_1} = 0.2587r^3$ $c_1 = 0.4244r$ | $\frac{\sqrt{9\pi^2 - 64}}{6\pi} = 0.264r$ |
| | $I = 0.1098(R^4 - r^4)$ $- \frac{0.283R^2r^2(R - r)}{R + r}$ $= 0.3tr_1{}^3$ (approx) when $\frac{t}{r_1}$ is very small | $c_1 = \frac{4}{3\pi}\frac{R^2 + Rr + r^2}{R + r}$ $c_2 = R - c_1$ | $\sqrt{\frac{2I}{\pi(R^2 - r^2)}}$ $= 0.31r_1$ (approx) |
| | $I = \frac{\pi a^3 b}{4} = 0.7854a^3b$ | $\frac{I}{c} = \frac{\pi a^2 b}{4} = 0.7854a^2b$ | $\frac{a}{2}$ |
| | $I = \frac{\pi}{4}(a^3b - a_1{}^3b_1)$ $= \frac{\pi}{4}a^2(a + 3b)t$ (approx) | $\frac{I}{c} = \frac{\pi}{4}a(a + 3b)t$ (approx) | $\sqrt{\frac{I}{(\pi ab - a_1b_1)}} =$ $\frac{a}{2}\sqrt{\frac{a + 3b}{a + b}}$ (approx) |
| | $I = \frac{1}{12}\left[\frac{3\pi}{16}d^4 + b(h^3 - d^3) + b^3(h - d)\right]$ $\frac{I}{c} = \frac{1}{6h}\left[\frac{3\pi}{16}d^4 + b(h^3 + d^3) + b^3(h - d)\right]$ | | $\sqrt{\frac{I}{\pi\frac{d^2}{4} + 2b(h - d)}}$ (approx) |

## Table 8-1. Properties of Various Cross Sections (Continued)

| Section | Moment of inertia | Section modulus | Radius of gyration |
|---|---|---|---|
| 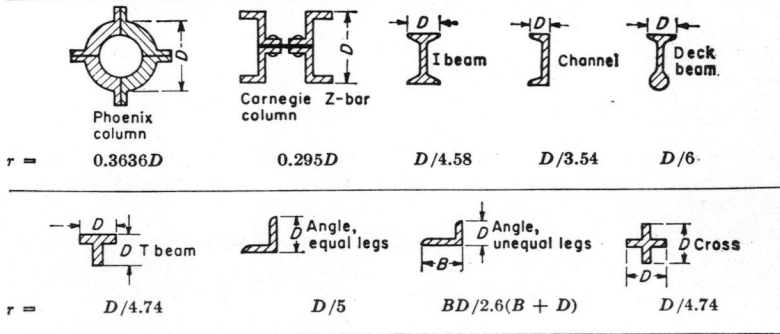 | $I = \dfrac{t}{4}\left(\dfrac{\pi B^3}{16} + B^2 h + \dfrac{\pi B h^2}{2} + \dfrac{2}{3}h^3\right)$ <br> $h = H - \tfrac{1}{2}B$ <br> $\dfrac{I}{c} = \dfrac{2I}{H+t}$ | | $\sqrt{\dfrac{I}{2\left(\dfrac{\pi B}{4} + h\right)t}}$ |

| Section | Moment of inertia and section modulus | | Radius of gyration |
|---|---|---|---|
| **Corrugated sheet iron, parabolically curved** | $I = \dfrac{64}{105}(b_1 h_1{}^3 - b_2 h_2{}^3)$, where <br><br> $h_1 = \tfrac{1}{2}(H+t)$ $\quad b_1 = \tfrac{1}{4}(B + 2.6t)$ <br> $h_2 = \tfrac{1}{2}(H-t)$ $\quad b_2 = \tfrac{1}{4}(B - 2.6t)$ <br><br> $\dfrac{I}{c} = \dfrac{2I}{H+1}$ | | $r = \sqrt{\dfrac{3I}{t(2B + 5.2H)}}$ |

### Approximate Values of Least Radius of Gyration $r$

| | Phoenix column | Cornegie Z-bar column | I beam | Channel | Deck beam |
|---|---|---|---|---|---|
| $r =$ | $0.3636D$ | $0.295D$ | $D/4.58$ | $D/3.54$ | $D/6$ |

| | T beam | Angle, equal legs | Angle, unequal legs | Cross |
|---|---|---|---|---|
| $r =$ | $D/4.74$ | $D/5$ | $BD/2.6(B+D)$ | $D/4.74$ |

NOTE: Square, axis same as first rectangle: side $= h$, $I = h^4/12$, $I/c = h^3/6$, $r = 0.289h$.
Square, diagonal taken as axis: $I = h^4/12$, $I/c = 0.1179h^3$, $r = 0.289h$.
SOURCE: T. Baumeister (ed.), "Standard Handbook for Mechanical Engineers." 7th ed., pp. 5-39. McGraw-Hill Book Company, New York, 1967.

**Product of Inertia.** The product of inertia with respect to two coordinate axes $x$ and $y$ is defined by

$$P_{xy} = \int xy \, dA \qquad \text{ft}^4 \qquad \text{for area} \qquad (8\text{-}12)$$

and

$$P_{xy} = \int xy \, dm \qquad \text{lb-ft}^2 \qquad \text{for solid body} \qquad (8\text{-}13)$$

where $x, y$ = distances from element of area or mass to axes $x$, $y$
   $dA$ = element of area, ft$^2$
   $dm$ = element of mass, lb

*Moments of Inertia of Important Solids (Homogeneous).*[1]

*Nomenclature*

$m$ = mass per unit volume of body, lb/ft³
$M$ = total mass of body, lb
$r$ = radius, ft
$I$ = moment of inertia, lb-ft²

Solid circular cylinder about its axis. $I = \pi r^4 m a/2 = M r^2/2$, where $a$ is the length of axis of cylinder.

Solid circular cylinder about an axis through the center of gravity and perpendicular to axis of cylinder. $I = M(r^2 + a^2/3)/4$.

Hollow circular cylinder about its axis. $I = \pi m a (r_1^4 - r_2^4)/2$, where $r_1$ and $r_2$ are the outer and inner radii, feet and $a$ is the length (feet).

Thin hollow circular cylinder about its axis. $I = M r^2$.

Solid sphere about a diameter. $I = 8 m \pi r^5/15 = 2 M r^2/5$.

Thin hollow sphere about a diameter. $I = 2 M r^2/3$.

Thick hollow sphere about a diameter. $I = 8 m \pi (r_1^5 - r_2^5)/15$, where $r_1$ and $r_2$ are the outer and inner radii (feet).

Rectangular prism about an axis through center of gravity and perpendicular to a face whose dimensions are $a$ and $b$ (feet). $I = M(a^2 + b^2)/12$.

Solid right circular cone about an axis through its apex and perpendicular to its axis. $I = 3M(r^2/4 + h^2)/5$, where $h$ is the altitude of the cone in feet and $r$ is the radius of the base in feet.

Solid right circular cone about its axis of revolution. $I = 3 M r^2/10$.

Ellipsoid with semiaxes $a$, $b$, and $c$.

$I$ about diameter $2c$ ($z$ axis) $= 4 m \pi a b c (a^2 + b^2)/15$

where the equation of ellipsoid is: $x^2/a^2 + y^2/b^2 + z^2/c^2 = 1$.

Ring with circular section (Fig. 8-14). $I_{yy} = \frac{1}{2} m \pi^2 R a^2 (4R^2 + 3a^2)$, $I_{zz} = m \pi^2 R a^2 [R^2 + (5a^2/4)]$.

*Approximate Moments of Inertia of Solids.*[1] In order to determine the moment of inertia of a solid, it is necessary to know all its dimensions. In the case of a rod of mass $M$ (pounds) and length $l$ (feet) with shape and size of the cross section unknown (Fig. 8-15), making the approximation that the weight is all concentrated along the axis of the rod, the moment of inertia about $YY$ will be $I_{yy} = \int_0^l (M/l)x^2 \, dx = M l^2/3$ lb-ft², where $x$ has units of feet.

A thin plate may be treated in the same way (Fig. 8-16).

$$I_{yy} = \int_0^l (M/l)x^2 \, dx$$

Here the mass of the plate is assumed to be concentrated at its middle layer.

[1] Source: T. Baumeister (ed.), "Standard Handbook for Mechanical Engineers," 7th ed., pp. 3-15, McGraw-Hill Book Company, New York, 1967.

Thin ring, or cylinder (Fig. 8-17). Assume the mass $M$ (pounds) of the ring or cylinder to be concentrated at a distance $r$ (feet) from $O$. The moment of inertia about an axis through $O$ perpendicular to plane of ring or along axis of cylinder will be $I = Mr^2$ lb-ft². This will be greater than the exact moment of inertia, and $r$ is sometimes taken as the distance from $O$ to the center of gravity of the cross section of the rim.

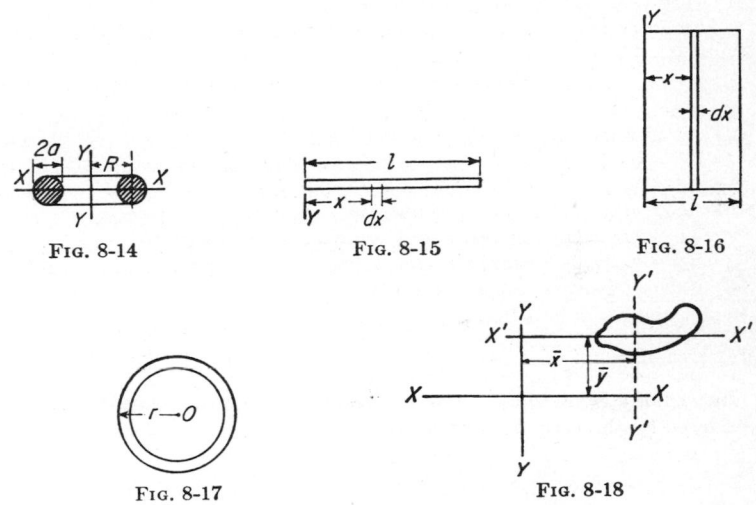

FIG. 8-14                   FIG. 8-15                  FIG. 8-16

FIG. 8-17                            FIG. 8-18

Parallel axis rule for product of inertia (Fig. 8-18).

$$P_{xy} = P_{x'y'} + A\overline{xy} \qquad \text{ft}^4 \qquad\qquad (8\text{-}14)$$

and
$$P_{xy} = P_{x'y'} + m\overline{xy} \qquad \text{lb-ft}^2 \qquad\qquad (8\text{-}15)$$

where $P_{x'y'}$ = product of inertia through axis through center of gravity

$\bar{x}, \bar{y}$ = coordinates of center of gravity with respect to reference axes

**Principal Axes.** If the rectangular coordinates axes $x$, $y$, $z$ through any point in a body are chosen in such directions that the products of inertia about that point are equal to zero, that is,

$$P_{xz} = \int xz\, dm = 0 \qquad P_{yz} = \int yz\, dm = 0 \qquad P_{xy} = \int xy\, dm = 0 \quad (8\text{-}16)$$

the axes are called *principal axes* of the body at the given point. The corresponding moments of inertia about the point are called *principal moments of inertia*. The three coordinate planes $xy$, $yz$, $xz$ are called *principal planes*.

### General Rules about Principal Axes

1. The axis perpendicular to a plane of symmetry is the principal axis.
2. For two perpendicular planes of symmetry

$$P_{xz} = P_{zy} = 0 \qquad\qquad (8\text{-}17)$$

Therefore if $z$ is a principal axis,

$$P_{zx} = P_{zy} = P_{xy} = 0 \qquad (8\text{-}18)$$

and $0x$, $0y$, $0z$ are all principal axes. The line of intersection of the planes and the two axes in the planes are all principal axes.

3. If a body is in rotation, the axis of rotation is the principal axis, and any other two axes will also be principal axes.

4. The principal axes about which the moments of inertia are maximum and minimum are axes of stable free rotation. The third axis has unstable equilibrium.

5. If a body is rotated about its principal axis, there will be no reactions in the bearings because

$$M_z = P_{xy}\omega^2 = 0 \qquad (8\text{-}19)$$

where $\omega$ = angular velocity, radians/sec

$P_{xy}$ = product of inertia about principal ($xy$) axis, lb-ft$^2$

$M_z$ = reaction normal to principal axis, lb-ft$^2$/sec$^2$

6. Unless a body is rotated about the principal axis through the center of gravity, there will be a bearing reaction due to centrifugal force.

## 8-4. Kinetics

**Energy of a Rigid Body.** The *kinetic energy* of a rigid body is the energy possessed by the body by virtue of its motion.

$$\text{Kinetic energy} = \tfrac{1}{2}mv^2 \qquad \text{translation} \qquad (8\text{-}20)$$
$$\text{Kinetic energy} = \tfrac{1}{2}I_0\omega^2 \qquad \text{rotation} \qquad (8\text{-}21)$$

where $m$ = mass, lb

$I_0$ = moment of inertia about axis of rotation, lb-ft$^2$

$v$ = velocity, fps

$\omega$ = angular velocity, radians/sec

The *potential energy* of a rigid body is the energy possessed by the body by virtue of its position, i.e., that energy which is available to do work.

**Free Harmonic Vibrations of Systems with One Degree of Freedom.** If an elastic system is disturbed from its position of equilibrium by a force, the elastic restoring forces of the system in the disturbed position will no longer be in equilibrium with the loading, and vibrations will ensue (Fig. 8-19).

### Nomenclature

$k$ = spring constant of elastic system, lb force/ft

$v = dx/dt$ = velocity, fps

$v_0$ = initial velocity, fps

$t$ = time, sec

$W$ = weight (neglecting spring weight as small compared with weight $W$), lb mass

$f$ = frequency of oscillation, sec$^{-1}$

$p$ = period of oscillation = $\sqrt{kg_c/W}$, sec$^{-1}$

$t$ = time for one complete oscillation, sec

$\omega$ = $p$, in the case of rotation, radians/sec

$g_c$ = 32.2 (lb mass)(ft)/(lb force)(sec²)

$x$ = displacement of $W$ from equilibrium position, ft

$x_0$ = initial displacement of $W$ from equilibrium position, ft

$$\frac{W}{g_c}\frac{d^2x}{dt^2} - kx = 0 \qquad (8\text{-}22)$$

$$t = 2\pi/p \qquad f = 1/t = p/2\pi \qquad p = 2\pi f \qquad p = 2\pi/t \qquad (8\text{-}23)$$

The equation of motion is

$$x = x_0 \cos pt + (v_0/p) \sin pt \qquad (8\text{-}24)$$

**Natural Frequency.** If $\partial_{ST}$ is the deflection of the spring caused by the weight $W$, then $\partial_{ST} = W/k$ and

$$\omega_n = \sqrt{g_c/\partial_{ST}} = \text{number of free oscillations per } 2\pi \text{ sec} \qquad (8\text{-}25)$$

Then the natural frequency is

$$f_n = 3.14\sqrt{1/\partial_{ST}} \qquad \text{cps} \qquad (8\text{-}26)$$

**Torsional Vibration.** If a disk is supported as shown in Fig. 8-20 and subjected to a couple in the plane of the disk which is suddenly removed, free torsional vibrations of the elastic system consisting of the shaft and disk will be produced.

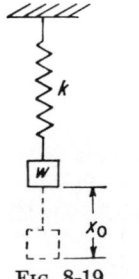
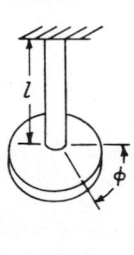
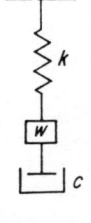
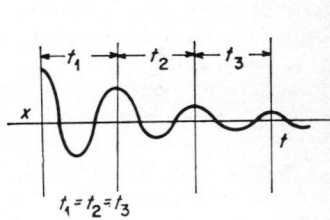

FIG. 8-19    FIG. 8-20    FIG. 8-21    FIG. 8-22

Let $\phi$ = angle of twist of shaft at any moment, radians

$k$ = torque moment necessary to produce angle of twist of 1 radian in shaft, lb force–ft

$\omega_0$ = initial angular velocity, radians/sec

$\phi_0$ = initial angle of twist of shaft, radians

$J$ = polar moment of inertia of disk (neglecting shaft $J$ as small compared to $J$ of disk), lb mass–ft

and $p$, $f$, $t$, and $g_c$ are as defined above for elastic vibration. The period of the torsional vibration is

$$p = \sqrt{k/J} \qquad (8\text{-}27)$$

The frequency

$$f = (1/2\pi) \sqrt{kg_c/J} \tag{8-28}$$

The equation of motion is

$$\phi = \phi_0 \cos pt + (\omega_0/p) \sin pt \tag{8-29}$$

***Damped Free Vibrations.*** Assuming viscous damping, i.e., damping proportional to velocity, such as may exist in dashpots (Fig. 8-21),

$$x = Ae^{(-\alpha+\beta)t} + Be^{(-\alpha-\beta)t} \tag{8-30}$$

where[1] $\alpha = cg_c/2W$ $\tag{8-31}$

$\beta = \sqrt{c^2g_c^2/4W^2 - kg_c/W}$ $\tag{8-32}$

$A, B$ = constants of integration

$c$ = damping coefficient, lb force/ft

When $c^2g_c^2/4W^2 > kg_c/W$, $\beta$ is real and positive. The result is exponential decay, in which $x \to 0$ as $t \to \infty$.

When $c^2g_c^2/4W^2 < kg_c/W$, $\beta$ is imaginary. This case is more representative of the usual case of damped vibration, the amplitude diminishing after each cycle according to the physical constants of the system. The frequency, however, does not change, that is $t_1 = t_2 = t_3$ in Fig. 8-22.

When $c^2g_c^2/4W^2 = kg_c/W$, $\beta = 0$. For this condition

$$c = c_{cr} = \sqrt{4Wk/g_c} \tag{8-33}$$

or, critical damping (see Fig. 8-23). This is a boundary case and rarely exists.

### Forced Vibrations without Damping

$P$ = impressed force, lb force, with frequency $\omega$, sec$^{-1}$

= $P_0 \cos \omega t$, where $t$ is time for one vibration, sec

$p$ = natural frequency of system, sec$^{-1}$, of weight $W$, lb mass

$g_c$ = 32.2 (lb mass)(ft)/(lb force)(sec$^2$)

$$x = A \sin pt + B \cos pt + (P_0 g_c/W)[1/(p^2 - \omega^2)] \cos \omega t \tag{8-34}$$

where $A$ and $B$ are constants of integration.

In general, the vibrations due to the first two terms die out shortly and only those remain which are due to the forcing frequency $\omega$. At such time, then,

$$x = (P_0 g_c/W)[1/(p^2 - \omega^2)] \cos \omega t \tag{8-35}$$

Let $\partial_{ST} = P_0/k$ = static deflection, in feet, resulting from $P_0$, where $k$ is the system constant, in pounds force per foot, and

$$x_0 = x_{max} = (P_0 g_c/W)[1/(p^2 - \omega^2)]. \tag{8-36}$$

Then

$$x_0/\partial_{ST} = [1/(1 - \omega^2/p^2)] = \gamma \tag{8-37}$$

[1] For nomenclature see p. 8–14.

This relation can be plotted as shown in Fig. 8-24.   It is seen that:
1. When $\omega/p = 0$, or when $\omega$ is small compared to $p$, $x = \delta_{ST}$ or nearly so.
2. When $\omega/p = \infty$, that is, when $\omega$ is very large compared to $p$, $x = 0$.
3. When $\omega/p = 1$, or $\omega = p$, $x = \infty$.   This is the case of resonance.

Figure 8-24 shows that, when the applied forced frequency becomes larger than the natural frequency of the body, the deflection of the body is opposite in direction to that of the force.

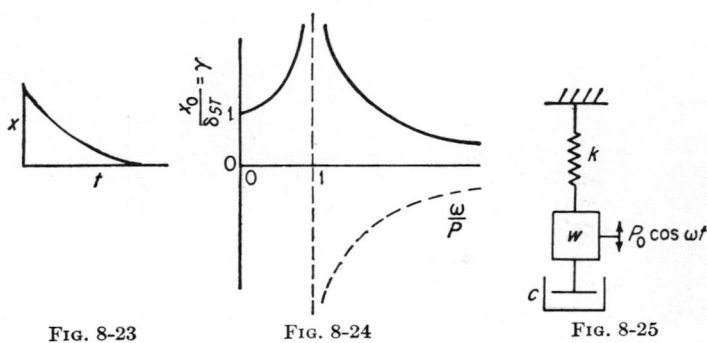

FIG. 8-23            FIG. 8-24                    FIG. 8-25

In vibration isolation, a quantity known as the transmission ratio is defined as equal to $1/(\omega^2/p^2 - 1)$.   For practical vibration isolation, $\omega/p \geq \sqrt{2}$.   This is accomplished by supplying a small value of $p$ through use of very soft springs or by increasing the mass of the machine or its foundation.

**Forced Vibrations with Damping** (Fig. 8-25).   The motion of the weight $W$ at any time is

$$x = e^{-\alpha t}(A \cos \beta t + B \sin \beta t) + C \sin \omega t + D \cos \omega t \qquad (8\text{-}38)$$

where $A$, $B$, $C$, and $D$ are constants of integration and $\alpha$, $\beta$ are as defined by Eqs. (8-31) and (8-32).

For steady-state application, the last two terms only are of interest, that is,

$$x = C \sin \omega t + D \cos \omega t \qquad (8\text{-}39)$$

where
$$C = \frac{P_0 g_c}{W} \frac{W^2 g_c}{(p^2 - \omega^2)W^2 + c^2 g_c^2 \omega^2} \qquad (8\text{-}40)$$

$$D = \frac{P_0 g_c}{W} \frac{W^2(p^2 - \omega^2)}{(p^2 - \omega^2)^2 W^2 + c^2 g_c^2 \omega^2} \qquad (8\text{-}41)$$

Let
$$R = \sqrt{C^2 + D^2} = \frac{P_0}{k} \frac{1}{\sqrt{(1 - \omega^2/p^2)^2 + (2c\omega/C_{cr}p)^2}} \qquad (8\text{-}42)$$

and
$$\theta = \tan^{-1}(C/D) = \tan^{-1}[cg_c\omega/W(p^2 - \omega^2)] \qquad (8\text{-}43)$$

Figures 8-26 and 8-27 can be drawn:

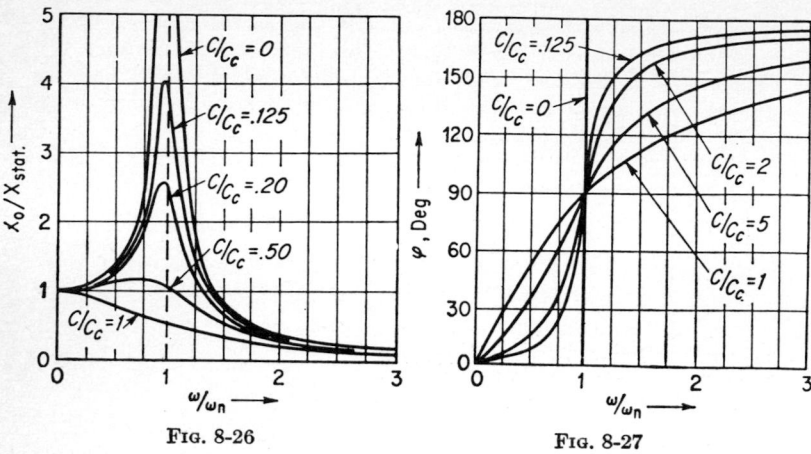

FIG. 8-26                                    FIG. 8-27

## 8-5. Torsion (See Table 8-2)

**Torsion** (Solid Circular Shafts) (See Fig. 8-28)

$$S_v = M_t c / J \qquad\qquad (8\text{-}44)$$

where $S_v$ = shear stress, psi
$\quad M_t$ = twisting moment = $Pl$, in-lb
$\quad c$ = distance from center to stressed surface of interest, in.
$\quad J$ = polar moment of inertia of cross section, in.[4]

**Combined Torsion and Bending** (Solid Circular Shafts) (Fig. 8-29)

$$\sigma_{\max} = (16/\pi d^3)(M_b + \sqrt{M_b{}^2 + M_t{}^2}) \qquad (8\text{-}45)$$

where $\sigma_{\max}$ = maximum stress, psi
$\quad M_t$ = torque, in-lb.
$\quad M_b$ = moment due to bending load, in-lb = $Wx$
$\quad d$ = diameter of bar, in.

$$M = \sigma I / c \qquad\qquad (8\text{-}46)$$

where $M$ = bending moment, lb-in.
$\quad \sigma$ = elastic stress at distance $c$ from neutral axis, psi
$\quad c$ = distance from neutral axis to plane at which stress $\sigma$ is calculated, in.
$\quad I$ = rectangular moment of inertia of cross-sectional area about neutral axis, in.[4]
$\quad I/c$ = section modulus where $c$ is distance to the outermost fiber, in.[3]

## Table 8-2. Torsion of Shafts of Various Cross Sections

$G$ = Shear Modulus of Elasticity, psi

| Cross section | Torsional resisting moment $M_t$ | Angular deflection, $a_1$ (length = 1 in., radius = 1 in.) | | Work of torsion ($V$ = volume) |
|---|---|---|---|---|
| | | In terms of torsional moment | In terms of max shear | |
| | $\dfrac{\pi}{16} d^3 S_v$ | $\dfrac{M_t}{GJ} = \dfrac{32}{\pi d^4}\dfrac{M_t}{R}$ | $2\dfrac{S_{v_{max}}}{G}\dfrac{1}{d}$ | $\dfrac{1}{4}\dfrac{S^2_{v_{max}}}{G}V$ (Note 1) |
| | $\dfrac{\pi}{16}\dfrac{D^4 - d^4}{D}S_v$ | $\dfrac{32}{\pi(D^4 - d^4)}\dfrac{M_t}{G}$ | $2\dfrac{S_{v_{max}}}{G}\dfrac{1}{D}$ | $\dfrac{1}{4}\dfrac{S^2_{v_{max}}}{G}\dfrac{D^2 + d^2}{D^2}V$ (Note 2) |
| | $\dfrac{\pi}{16}b^2hS_v$ ($h > b$) | $\dfrac{16}{\pi}\dfrac{b^2 + h^2}{b^3h^2}\dfrac{M_t}{G}$ | $\dfrac{S_{v_{max}}}{G}\dfrac{b^2 + h^2}{bh^2}$ | $\dfrac{1}{8}\dfrac{S^2_{v_{max}}}{G}\dfrac{b^2 + h^2}{h^2}V$ (Note 3) |
| | $\dfrac{2}{9}b^2hS_v$ ($h > b$) | $3.6\dfrac{b^2 + h^2}{b^3h^3}\dfrac{M_t}{G}$ * | $0.8\dfrac{S_{v_{max}}}{G}\dfrac{b^2 + h^2}{bh^2}$ * | $\dfrac{4}{45}\dfrac{S^2_{v_{max}}}{G}\dfrac{b^2 + h^2}{h^2}V$ (Note 4) |
| | $\dfrac{2}{9}h^3S_v$ | $7.2\dfrac{1}{h^4}\dfrac{M_t}{G}$ | $1.6\dfrac{S_{v_{max}}}{G}\dfrac{1}{h}$ | $\dfrac{8}{45}\dfrac{S^2_{v_{max}}}{G}V$ (Note 5) |
| | $\dfrac{b^3}{20}S_v$ | $4.62\dfrac{1}{b^4}\dfrac{M_t}{G}$ | $2.31\dfrac{S_{v_{max}}}{G}\dfrac{1}{b}$ | |
| | $\dfrac{b^3}{1.09}S_v$ | $0.967\dfrac{1}{b^4}\dfrac{M_t}{G}$ | $0.9\dfrac{S_{v_{max}}}{G}\dfrac{1}{b}$ | |

* When

| | $h/b = $ | 1 | 2 | 4 | 8 |
|---|---|---|---|---|---|
| Coefficient 3.6 becomes = | | 3.56 | 3.50 | 3.35 | 3.21 |
| Coefficient 0.8 becomes = | | 0.79 | 0.78 | 0.74 | 0.71 |

NOTES: (1) $S_{v_{max}}$ at circumference. (2) $S_{v_{max}}$ at outer circumference. (3) $S_{v_{max}}$ at $A$; $S_{v_B} = 16M_t/\pi bh^2$. (4) $S_{v_{max}}$ at middle of side $h$; in middle of $b$, $S_v = 9M_t/2bh^2$. (5) $S_{v_{max}}$ at middle of side.

SOURCE: T. Baumeister (ed.),"Standard Handbook for Mechanical Engineers," 7th ed., p. 5-53. McGraw-Hill Book Company, New York, 1967.

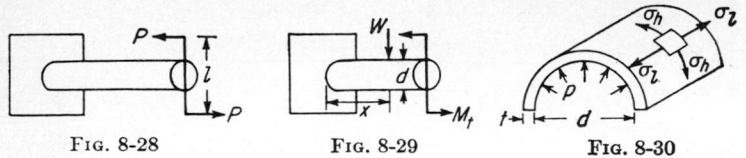

FIG. 8-28                 FIG. 8-29                 FIG. 8-30

## 8-6. Cylinder Stresses

**Stresses in Thin-walled Tubes or Cylinders** (Fig. 8-30)

$$\sigma_h = pd/2t \qquad \sigma_l = pd/4t \qquad (8\text{-}47)$$

where $\sigma_h$ = hoop stress, psi
$\quad \sigma_l$ = longitudinal stress, psi
$\quad d$ = internal diameter, in.
$\quad p$ = internal pressure, psi
$\quad t$ = thickness of tube wall, in.

**Stresses in Thick Cylinders or Tubes** (Fig. 8-31)
For internal pressure only:

$$\sigma_r = [a^2 p_i/(b^2 - a^2)](1 - b^2/r^2) \qquad (8\text{-}48)$$
$$\sigma_t = [a^2 p_i/(b^2 - a^2)](1 + b^2/r^2) \qquad (8\text{-}49)$$

For external pressure only:

$$\sigma_r = [- p_0 b^2/(b^2 - a^2)](1 - a^2/r^2) \qquad (8\text{-}50)$$
$$\sigma_t = [- p_0 b^2/(b^2 - a^2)](1 + a^2/r^2) \qquad (8\text{-}51)$$

where $\sigma_r$ = stress in radial direction, psi
$\quad \sigma_t$ = stress in tangential direction, psi
$\quad a$ = internal radius of cylinder, in.
$\quad b$ = external radius of cylinder, in.
$\quad r$ = radial measurement, in.
$\quad p_i$ = internal pressure, psi
$\quad p_0$ = external pressure, psi

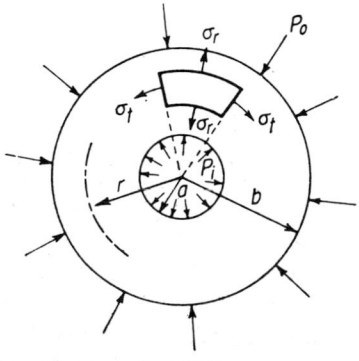

FIG. 8-31

## 8-7. Columns

**Long Columns.** Euler's formula for failure by buckling ($l/r > 120$)

$$W = n\pi^2 EI/l^2 \qquad (8\text{-}52)$$

where $n = \tfrac{1}{4}$ for one end fixed, one end free; 1 for both ends of column rounded; 4 for both ends fixed; 2 for one end rounded, one fixed

$W$ = load, lb
$E$ = modulus of elasticity, psf
$I$ = rectangular moment of inertia, ft$^4$
$l$ = length of column, ft
$r$ = least radius of gyration, ft
$l/r$ = slenderness ratio

**Short Columns.** Rankine's formula ($l/r = 20$ to $120$)

$$W = \sigma A/(1 + kl^2/r^2) \qquad (8\text{-}53)$$

where $W$ = design load, lb
$\sigma$ = design stress, psf
$A$ = area of cross section, ft$^2$
$l$ = length, ft
$r$ = least radius of gyration, ft
$k$ = values from Table 8-3, dimensionless

### Table 8-3. Values of $k$ (Merriman)

| Material | Both ends fixed | Both ends rounded | One end fixed, one end rounded | One end fixed, one end free |
|---|---|---|---|---|
| Timber.......... | 1/3,000 | 1.95/3,000 | 4/3,000 | 16/3,000 |
| Cast iron....... | 1/5,000 | 1.95/5,000 | 4/5,000 | 16/5,000 |
| Wrought iron.... | 1/36,000 | 1.95/36,000 | 4/36,000 | 16/36,000 |
| Steel........... | 1/25,000 | 1.95/25,000 | 4/25,000 | 16/25 000 |

source: L. S. Marks (ed.), "Mechanical Engineers' Handbook," 5th ed., p. 466, McGraw-Hill Book Company, Inc, New York, 1951.

## MACHINE DESIGN DATA

### 8-8. Failure

In the broadest sense, a structure or structural element experiences a failure when it can no longer satisfactorily perform its design function. Failure may be due to yielding or fracture, according as the material is ductile or brittle in nature. A ductile failure may result in eventual fracture of the member, though usually the part will have failed long before it breaks. A brittle failure, however, always results in fracture.

### 8-9. Stress Concentration

A loaded structural member whose design is such that it is subjected to sudden changes in shape and/or stress level must be investigated for the effects of stress concentrations, or stress raisers. Most failures start at the surface of the material. It is necessary to consider the surface condition of the stressed part, which gives rise to stress raisers due to surface discontinuities. Thus, the usually encountered geometric stress concentration factors arise from macrodiscontinuities in the form of notches, fillets, rough surface finish, press fits, inclusions, residual stresses, etc., and their effects are often the most important consideration in design of parts subjected to repeated loading and fatigue failure.

The total stress concentration factor, or stress multiplying factor, for a particular design depends on the material as well as type of stress raiser. The disruption in geometry results in a geometrical stress concentration factor, which, by itself, is independent of any variable save the geometry of the discontinuity. That resulting from a relatively simple notch is shown in Figs. 8-32 and 8-33. Note that the geometrical stress concentration factor multiplies the average axial stress $\sigma_n = P/A$ by a factor of 2.

Although some geometrical stress concentration factors can be obtained mathematically for simple notches, most data of this kind are supplied by experimental methods of stress analysis. (See Peterson, "Stress Concentration Design Factors," John Wiley & Sons, Inc., New York, 2d ed., 1974.) A given notch will result in different geometric stress concentration factors for different type loads. Figures 8-34 through 8-42 are useful for some of the most common design situations encountered. Common unified screw threads have a geometric stress concentration factor of between 5 and 6; stress concentration factor for keyways is shown in Fig. 8-43.

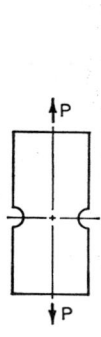

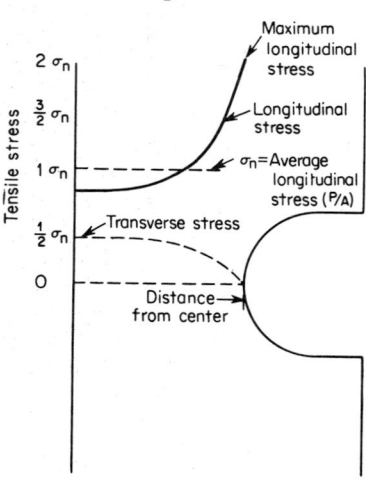

FIG. 8-32. Notched tensile specimen. The elastic stress distribution across center of this specimen is shown in Fig. 8-33.

FIG. 8-33. Elastic stress distribution across center of notched specimen shown in Fig. 8-32. $A$ = cross-sectional area at root of notch.

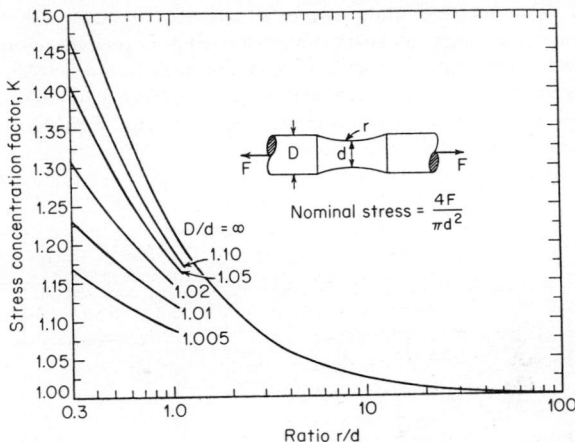

Fig. 8-34. Tensioning of a solid round bar with a shallow fillet. (*Source: H. J. Faupel, "Engineering Design," John Wiley & Sons, Inc., New York, 1964.*)

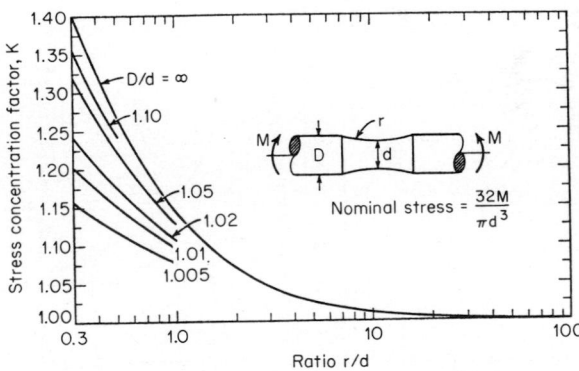

Fig. 8-35. Bending of a solid round bar with a shallow fillet. (*Source: Faupel, op. cit.*)

Sharp notching and/or high surface roughness gives rise to high stress concentration factor; accordingly, expected fatigue life is reduced. (See Sec. 8-10.) The nature of the material enters into the over-all stress concentration factor through its sensitivity to notching, or its index of sensitivity $q$, ranging between 0 and 1. For $q = 0$, the material experiences no reduction in fatigue strength due to a given notch; on the other hand, if $q = 1$, there is a full reduction in fatigue strength, as indicated by the geometrical stress concentration factor. Notch sensitivity is less for a ductile material than for a hard material (Fig. 8-44).

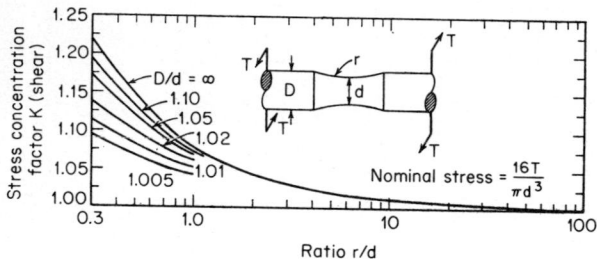

FIG. 8-36. Torsion of a solid round bar with a shallow fillet. (*Source: Faupel, op. cit.*)

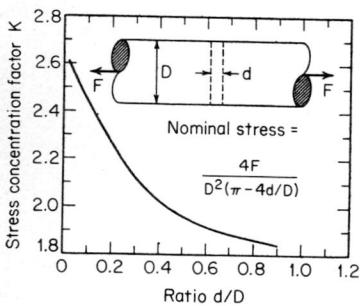

FIG. 8-37. Tensioning of a solid circular bar with a small transverse hole. (*Source: Faupel, op. cit.*)

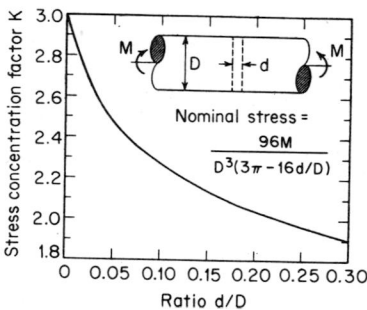

FIG. 8-38. Bending of a solid circular bar with a small transverse hole. (*Source: Faupel, op. cit.*)

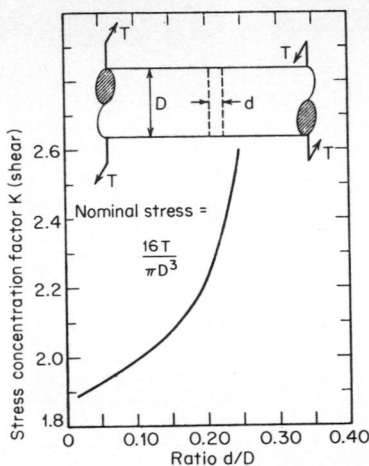

FIG. 8-39. Torsion of a solid circular bar with a small transverse hole. (*Source: Faupel, op. cit.*)

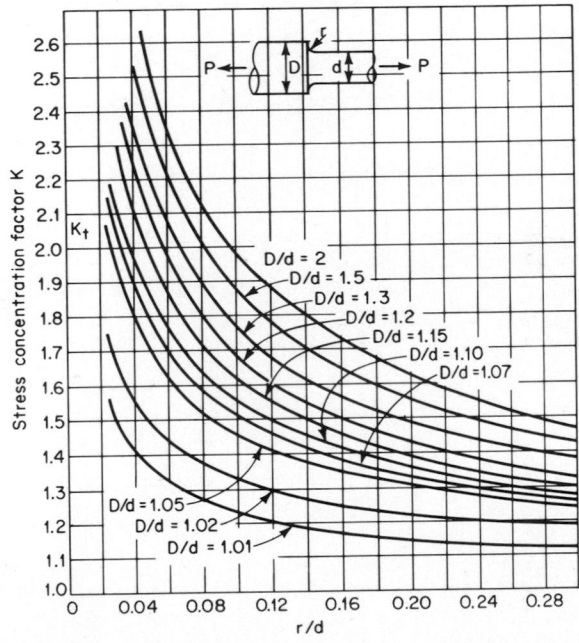

FIG. 8-40. Tensioning of a solid circular shaft with a shoulder fillet. (*Source: R. E. Peterson, "Stress Concentration Design Factors," John Wiley & Sons, Inc., New York, 2d ed., 1974.*)

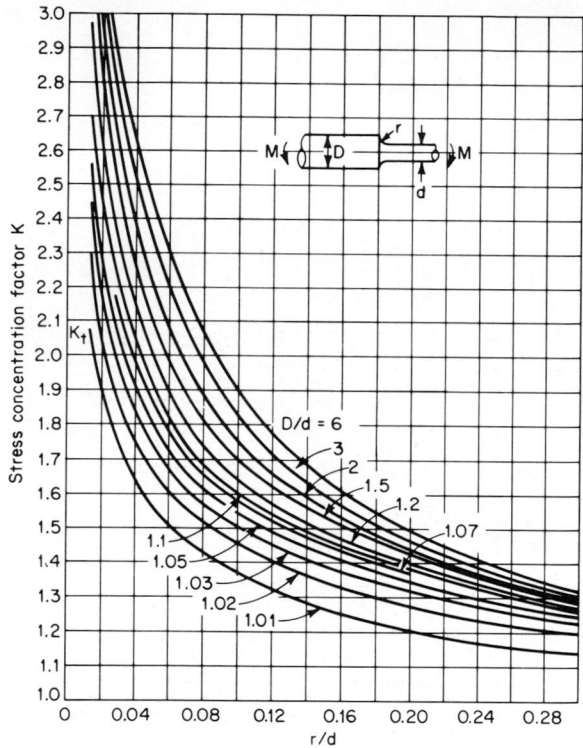

FIG. 8-41. Bending of a solid circular shaft with a shoulder fillet.   (*Source: Peterson, op. cit.*)

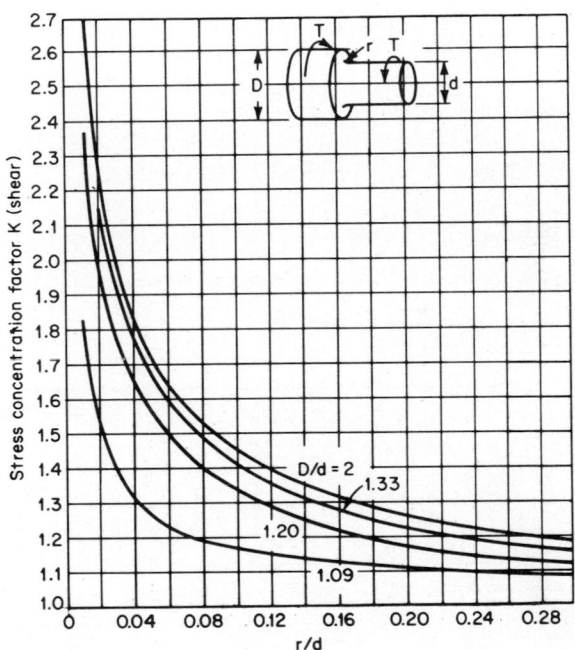

Fig. 8-42. Torsion of a solid circular shaft with a shoulder fillet. (*Source: Peterson, op. cit.*)

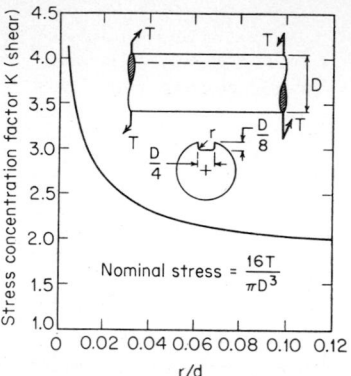

Fig. 8-43. Stress concentration factors for the straight portion of a keyway in a solid circular shaft subjected to torsion. (*Source: M. M. Leven, Proc. Soc. Exp. Stress Anal.,* **7**(2), 1949.)

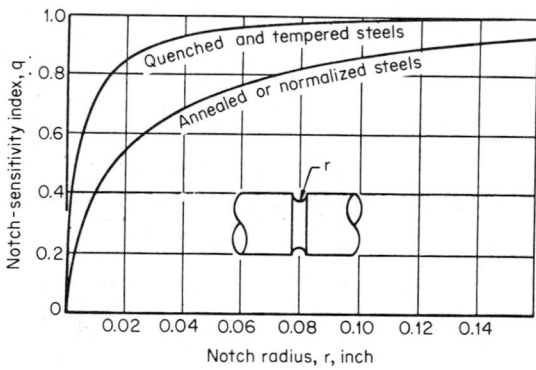

Fig. 8-44. Variation of notch sensitivity index $q$ in rotating bending tests of steels. (*Source: H. J. Grover, et al., "The Fatigue of Metals and Structures," NAVAER* 00-25-534, *Department of the Navy,* 1954.)

## 8-10. Fatigue

Structural or mechanical members loaded repetitively are subject to fatigue failure if improperly designed. The designer must know the properties of the material which are to be subjected to a particular type and level of repeated loading. Where fatigue data and design are concerned, the following definitions are suggested in the ASTM Manual of Fatigue Testing:

**Stress Cycle.** A stress cycle is the smallest section of the stress-time function which is repeated periodically and identically.

Figure 8-45 illustrates stress cycles commonly used and indicates, diagrammatically, many of the following terms.

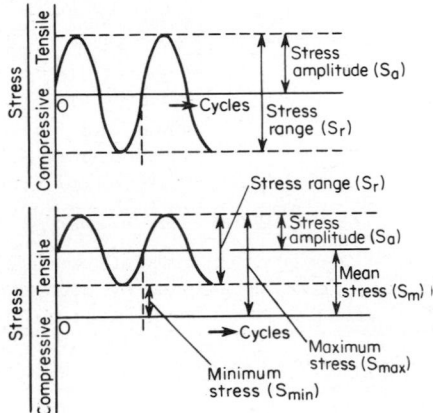

Fig. 8-45. Typical stress cycles in fatigue testing.

**Nominal Stress, $S$.** The stress calculated on the net section by simple theory such as $S = P/A$ or $S = Mc/I$ or $S_s = Tc/J$ without taking into account the variation in stress conditions caused by geometrical discontinuities such as holes, grooves, fillets, etc.

**Maximum Stress, $S_{max}$.** The highest algebraic value of the stress in the stress cycle, tensile stress being considered positive and compressive stress negative.

**Minimum Stress, $S_{min}$.** The lowest algebraic value of the stress in the stress cycle, tensile stress being considered positive and compressive stress negative.

**Range of Stress, $S_r$.** The algebraic difference between the maximum and minimum stress in one cycle, that is, $S_r = S_{max} - S_{min}$. For many cases of fatigue testing, the stress varies equally above and below zero stress, but other types of variation may be experienced.

**Alternating Stress Amplitude (or Variable Stress Component), $S_a$.** One-half the range of stress, that is, $S_a = S_r/2$.

**Mean Stress (or Steady-stress Component), $S_m$.** The algebraic mean of the maximum and minimum stress in one cycle, that is, $S = (S_{max} + S_{min})/2$.

**Stress Ratio, $R$.** The algebraic ratio of the minimum stress to the maximum stress in one cycle, that is, $R = S_{min}/S_{max}$.

**Stress Cycles Endured, $n$.** The number of cycles which a specimen has endured at any stage of a fatigue test.

**Fatigue Life, $N$.** The number of stress cycles which can be sustained for a given test condition.

*SN* **Diagram.** A plot of stress against number of cycles to failure. It is usually plotted $S$ versus log $N$, but a plot of log $S$ versus log $N$ is sometimes used (Fig. 8-46).

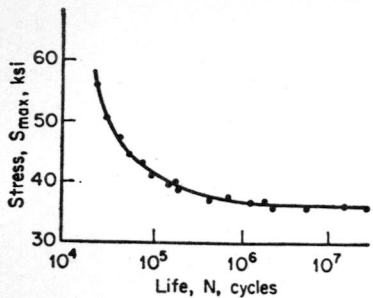

**Fig. 8-46.** *SN* curve for a steel tested in rotating beam machine. Note well-defined endurance limit at 36 ksi.

**Fatigue Limit (or Endurance Limit),** $S_e$. The limiting value of the stress below which a material can presumably endure an infinite number of stress cycles, that is, the stress at which the *SN* diagram becomes horizontal and appears to remain so. It should be noted that certain materials and environment preclude the attainment of a fatigue limit.

If the stress is not completely reversed, it is necessary to state what is meant by the fatigue limit. It may be expressed in terms of the alternating stress amplitude or the maximum stress; whichever method is used, it is also necessary to state the value of the mean stress, minimum stress, or stress ratio.

Most fatigue test data have been obtained through a rotating-beam test which subjects the material to completely reversed bending stresses. Data are usually reported on an *SN* curve (Fig. 8-46). Note that long life is bought at the expense of lowered operating stress level. The material in Fig. 8-46 shows a well-defined endurance limit. For a material whose *SN* curve shows no stress value to which it can be subjected for an infinite number of stress cycles, it is practice to report an endurance strength at a given number of stress cycles, usually between $10^8$ and $10^9$ cycles. Extensive fatigue data are found in Grover et al., "The Fatigue of Metals and Structures," U.S. Department of the Navy, NAVAER 00-25-534, 1954.

## 8-11. Gears

$$P_c = \pi D_p/N = \text{circular pitch, in./tooth} \tag{8-54}$$
$$P_d = N/D_p = \text{diametral pitch, teeth/in. of pitch diameter} \tag{8-55}$$

where $N$ = number of teeth in gear
$D_p$ = pitch diameter, in.

$$\text{Gear ratio} = \frac{N_2}{N_1} = \frac{\text{product of teeth of all driving gears}}{\text{product of teeth of all driven gears}} \tag{8-56}$$

$$\text{Pulley ratio} = \frac{N_2}{N_1} = \frac{\text{product of diameters of all driving pulleys}}{\text{product of diameters of all driven pulleys}} \tag{8-57}$$

where $N_1$ = speed of first shaft, rpm
$N_2$ = speed of last shaft, rpm

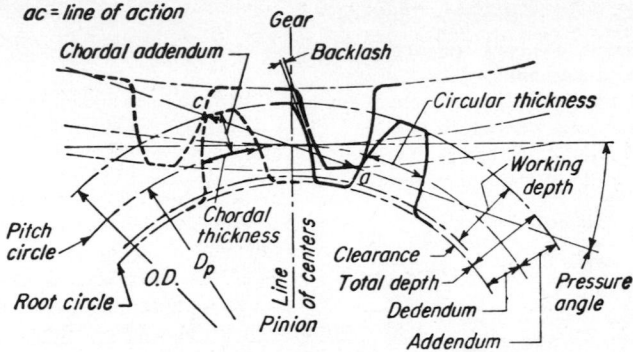

Fig. 8-47

## Table 8-4. Gear Tooth Proportions in Various Systems

| Pressure angle | $14\frac{1}{2}°$ | $20°$ | |
| --- | --- | --- | --- |
| Depth of tooth | Full | Full | Stub |
| Addendum | $1/P_d$ | $1/P_d$ | $0.8/P_d$ |
| Minimum dedendum including clearance | $1.157/P_d$ | $1.157/P_d$ | $1/P_d$ |
| Minimum clearance | $0.157/P_d$ | $0.157/P_d$ | $0.2/P_d$ |
| Minimum total depth | $2.157/P_d$ | $2.157/P_d$ | $1.8/P_d$ |
| Outside diameter | $(2+N)/P_d$ | $(2+N)/P_d$ | $(1.6+N)/P_d$ |

SOURCE: T. Baumeister (ed.), "Standard Handbook for Mechanical Engineers," 7th ed., p. 8-133, McGraw-Hill Book Company, New York, 1967.

## 8-12. Screws and Screw Threads

Threaded members are used as fasteners, for power transmission, and to provide adjustments.

Fasteners and adjusting screws generally take the form shown in Fig. 8-48, though in cases where considerable force or power is involved, square or Acme threads are used (Figs. 8-49 and 8-50, Tables 8-5 and 8-6). Power screws are generally of the square or Acme-thread form.

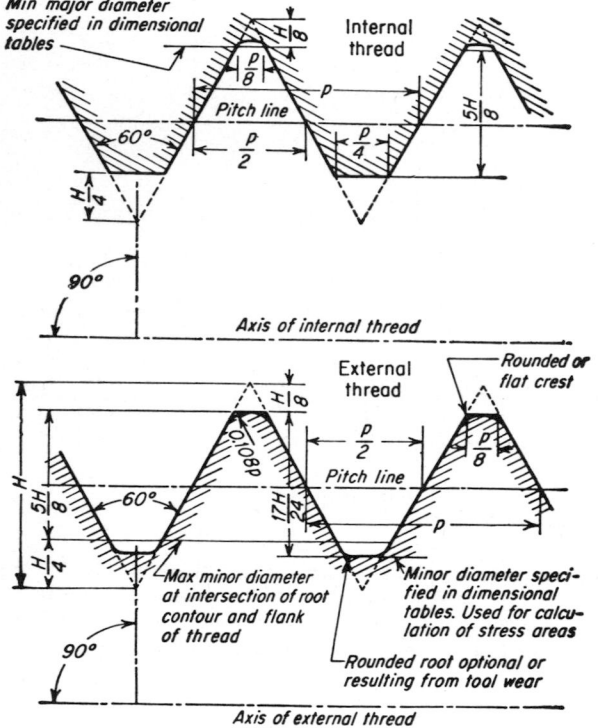

Fig. 8-48. 60° unified thread forms.

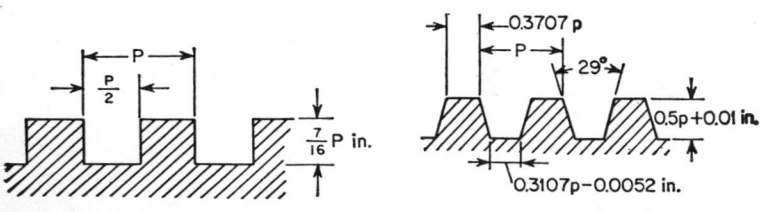

Fig. 8-49. Square thread.                    Fig. 8-50. Acme thread.

## Table 8-5. Standard Square Threads

| Bolt diam., in. | Threads per in. | Root diam., in. | Root area, in.² |
|---|---|---|---|
| 1/4 | 10 | 0.1625 | 0.0207 |
| 5/16 | 9 | 0.2153 | 0.0375 |
| 3/8 | 8 | 0.2658 | 0.0555 |
| 7/16 | 7 | 0.3125 | 0.0767 |
| 1/2 | 6 1/2 | 0.3656 | 0.1049 |
| 9/16 | 6 | 0.4167 | 0.1364 |
| 5/8 | 5 1/2 | 0.4666 | 0.1709 |
| 11/16 | 5 | 0.5125 | 0.2063 |
| 3/4 | 5 | 0.5750 | 0.2597 |
| 13/16 | 4 1/2 | 0.6181 | 0.3000 |
| 7/8 | 4 1/2 | 0.6806 | 0.3638 |
| 15/16 | 4 | 0.7188 | 0.4058 |
| 1 | 4 | 0.7813 | 0.4804 |
| 1 1/8 | 3 1/2 | 0.8750 | 0.6013 |
| 1 1/4 | 3 1/2 | 1.0000 | 0.7854 |
| 1 3/8 | 3 | 1.0834 | 0.9201 |
| 1 1/2 | 3 | 1.2084 | 1.1462 |
| 1 5/8 | 2 3/4 | 1.307 | 1.3414 |
| 1 3/4 | 2 1/2 | 1.400 | 1.5394 |
| 1 7/8 | 2 1/2 | 1.525 | 1.8265 |
| 2 | 2 1/4 | 1.612 | 2.0422 |
| 2 1/4 | 2 1/4 | 1.862 | 2.7245 |
| 2 1/2 | 2 | 2.063 | 3.3410 |
| 2 3/4 | 2 | 2.313 | 4.2000 |
| 3 | 1 3/4 | 2.500 | 4.9087 |
| 3 1/4 | 1 3/4 | 2.750 | 5.9396 |
| 3 1/2 | 1 5/8 | 2.962 | 6.8930 |
| 3 3/4 | 1 1/2 | 3.168 | 7.8853 |
| 4 | 1 1/2 | 3.418 | 9.1756 |

## Table 8-6. Acme Screw Threads

| Threads per in. | Depth of thread, in. | Thickness at root of thread, in. |
|---|---|---|
| 1 | 0.5100 | 0.6345 |
| 1 1/2 | .3850 | .4772 |
| 2 | .2600 | .3199 |
| 3 | .1767 | .2150 |
| 4 | .1350 | .1625 |
| 5 | .1100 | .1311 |
| 6 | .0933 | .1101 |
| 7 | .0814 | .0951 |
| 8 | .0725 | .0839 |
| 9 | .0655 | .0751 |
| 10 | .0600 | .0681 |

The shapes of threaded members are limitless, depending upon the application; the heads, however, are most often of the forms shown in Fig. 8-51.

Set screws prevent motion between two parts by compressive forces set up by driving the end of the set screw tightly against one of the parts. The

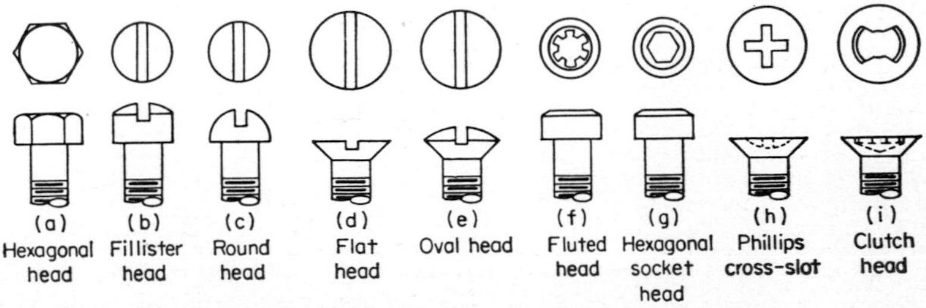
(a) Hexagonal head
(b) Fillister head
(c) Round head
(d) Flat head
(e) Oval head
(f) Fluted head
(g) Hexagonal socket head
(h) Phillips cross-slot
(i) Clutch head

FIG. 8-51. Typical head forms for threaded fasteners.

## Table 8-7. Thread Form and Formulas, Unified and American

| Thread Form | Formula |
|---|---|
| Angle of thread | $2a = 60°$ |
| Half angle of thread | $a = 30°$ |
| Number of threads per inch | $n = 1/p$ |
| Pitch of thread | $p = 1/n$ |
| Height of sharp V thread | $H = 0.86603p$ |
| | $= 0.86603/n$ |
| Height of external thread* | $h_s = 0.61343p$ |
| | $= 0.61343/n$ |
| | $= {}^{17}\!/_{24}H$ |
| Height of internal thread | $h_n = 0.54127p$ |
| | $= 0.54127/n$ |
| | $= \frac{5}{8}H$ |
| Depth of thread engagement | $h_e = 0.54127p$ |
| | $= 0.54127/n$ |
| Flat at crest of external thread | $F_{cs} = 0.125p$ |
| | $= 0.125/n$ |
| | $= p/8$ |
| Truncation of external-thread crest | $f_{cs} = 0.10825p$ |
| | $= 0.10825/n$ |
| | $= H/8$ |
| Truncation of external-thread rounded root* | $s_{rs} = 0.14434p$ |
| | $= 0.14434/n$ |
| | $= H/6$ |
| Flat at crest of internal thread | $F_{cn} = 0.25p$ |
| | $= 0.25/n$ |
| | $= p/4$ |
| Truncation of internal-thread crest | $f_{cn} = 0.21651p$ |
| | $= 0.21651/n$ |
| | $= H/4$ |
| Flat at root of internal thread | $F_{rn} = 0.125p$ |
| | $= 0.125/n$ |
| | $= p/8$ |
| Truncation of internal-thread root | $f_{rn} = 0.10825p$ |
| | $= 0.10825/n$ |
| | $= H/8$ |
| Addendum of external thread | $h_{as} = 0.32476p$ |
| | $= 0.32476/n$ |
| | $= \frac{3}{8}H$ |
| Major diameter of external thread (nominal diameter)† | $D_s$ |
| Pitch diameter of external thread† | $E_s = D - 2h_{as}$‡ |
| | $= D - 0.64952p$ |
| | $= D - 0.64952/n$ |
| Minor diameter of external thread | $K_s = D_s - 2h_s$ |
| | $= D_s - 1.22687p$ |
| | $= D_s - 1.22687/n$ |
| Major diameter of internal thread† | $D_n$ |
| Pitch diameter of internal thread† | $E_n$ |
| Minor diameter of internal thread | $K_n = D_n - 2h_n$ |
| | $= D_n - 1.08253p$ |
| | $= D_n - 1.08253/n$ |

* For calculating minor diameter and stress-area values in tables.

† As external and internal threads have the same basic major diameters and the same basic pitch diameters, hereinafter the subscripts are omitted from $D$ and $E$.

‡ $2h_{as} = h_b =$ the "basic height" $h$ of the original American National form.

source: Extracted from American National Standard "Unified Screw Threads," ANSI B1.1-1960, with the permission of the publisher, The American Society of Mechanical Engineers. An optional supplement B1.1a-1968) is available for metric translation.

most common types of set-screw points are shown in Fig. 8-52. Other variations in point type and ingenious locking features are available from the industry. Data for standard 60° thread forms are given in Tables 8-7 to 8-10.

Oval  Cup  Cone  Dog  Flat  Hanger

Fig. 8-52. Typical set-screw points.

Tapping screws have become widely used over the past forty years. Designed for pierced or drilled holes, they cut or form their own threads as they are driven. Of the types shown in Table 8-11, the drive screw (Type U) is hammered into place; all others are screwed into place. The heads of the latter usually receive a straight or Phillips-head driver or a hexagonal-socket head wrench.

### Table 8-8. Coarse-thread Series, UNC and NC, Basic Dimensions

| Sizes | Basic major diam, in. | Thds/in. | Basic pitch diam,* in. | Minor diam ext thds, in. | Minor diam int thds, in. | Lead angle at basic pitch diam | Section at minor diam, in.² | Tensile stress area, in.² |
|---|---|---|---|---|---|---|---|---|
| | $D$ | $n$ | $E$ | $K_s$ | $K_n$ | $\lambda$ | at $D - 2h_b$ | |
| 1 (0.073) | 0.0730 | 64 | 0.0629 | 0.0538 | 0.0561 | 4° 31' | 0.0022 | 0.0026 |
| 2 (0.086) | 0.0860 | 56 | 0.0744 | 0.0641 | 0.0667 | 4° 22' | 0.0031 | 0.0036 |
| 3 (0.099) | 0.0990 | 48 | 0.0855 | 0.0734 | 0.0764 | 4° 26' | 0.0041 | 0.0048 |
| 4 (0.112) | 0.1120 | 40 | 0.0958 | 0.0813 | 0.0849 | 4° 45' | 0.0050 | 0.0060 |
| 5 (0.125) | 0.1250 | 40 | 0.1088 | 0.0943 | 0.0979 | 4° 11' | 0.0067 | 0.0079 |
| 6 (0.138) | 0.1380 | 32 | 0.1177 | 0.0997 | 0.1042 | 4° 50' | 0.0075 | 0.0090 |
| 8 (0.164) | 0.1640 | 32 | 0.1437 | 0.1257 | 0.1302 | 3° 58' | 0.0120 | 0.0139 |
| 10 (0.190) | 0.1900 | 24 | 0.1629 | 0.1389 | 0.1449 | 4° 39' | 0.0145 | 0.0174 |
| 12 (0.216) | 0.2160 | 24 | 0.1889 | 0.1649 | 0.1709 | 4° 1' | 0.0206 | 0.0240 |
| 1/4 | 0.2500 | 20 | 0.2175 | 0.1887 | 0.1959 | 4° 11' | 0.0269 | 0.0317 |
| 5/16 | 0.3125 | 18 | 0.2764 | 0.2443 | 0.2524 | 3° 40' | 0.0454 | 0.0522 |
| 3/8 | 0.3750 | 16 | 0.3344 | 0.2983 | 0.3073 | 3° 24' | 0.0678 | 0.0773 |
| 7/16 | 0.4375 | 14 | 0.3911 | 0.3499 | 0.3602 | 3° 20' | 0.0933 | 0.1060 |
| 1/2 | 0.5000 | 13 | 0.4500 | 0.4056 | 0.4167 | 3° 7' | 0.1257 | 0.1416 |
| 9/16 | 0.5625 | 12 | 0.5084 | 0.4603 | 0.4723 | 2° 59' | 0.1620 | 0.1816 |
| 5/8 | 0.6250 | 11 | 0.5660 | 0.5135 | 0.5266 | 2° 56' | 0.2018 | 0.2256 |
| 3/4 | 0.7500 | 10 | 0.6850 | 0.6273 | 0.6417 | 2° 40' | 0.3020 | 0.3340 |
| 7/8 | 0.8750 | 9 | 0.8028 | 0.7387 | 0.7547 | 2° 31' | 0.4193 | 0.4612 |
| 1 | 1.0000 | 8 | 0.9188 | 0.8466 | 0.8647 | 2° 29' | 0.5510 | 0.6051 |
| 1 1/8 | 1.1250 | 7 | 1.0322 | 0.9497 | 0.9704 | 2° 31' | 0.6931 | 0.7627 |
| 1 1/4 | 1.2500 | 7 | 1.1572 | 1.0747 | 1.0954 | 2° 15' | 0.8898 | 0.9684 |
| 1 3/8 | 1.3750 | 6 | 1.2667 | 1.1705 | 1.1946 | 2° 24' | 1.0541 | 1.1538 |
| 1 1/2 | 1.5000 | 6 | 1.3917 | 1.2955 | 1.3196 | 2° 11' | 1.2938 | 1.4041 |
| 1 3/4 | 1.7500 | 5 | 1.6201 | 1.5046 | 1.5335 | 2° 15' | 1.7441 | 1.8983 |
| 2 | 2.0000 | 4 1/2 | 1.8557 | 1.7274 | 1.7594 | 2° 11' | 2.3001 | 2.4971 |
| 2 1/4 | 2.2500 | 4 1/2 | 2.1057 | 1.9774 | 2.0094 | 1° 55' | 3.0212 | 3.2464 |
| 2 1/2 | 2.5000 | 4 | 2.3376 | 2.1933 | 2.2294 | 1° 57' | 3.7161 | 3.9976 |
| 2 3/4 | 2.7500 | 4 | 2.5876 | 2.4433 | 2.4794 | 1° 46' | 4.6194 | 4.9326 |
| 3 | 3.0000 | 4 | 2.8376 | 2.6933 | 2.7294 | 1° 36' | 5.6209 | 5.9659 |
| 3 1/4 | 3.2500 | 4 | 3.0876 | 2.9433 | 2.9794 | 1° 29' | 6.7205 | 7.0992 |
| 3 1/2 | 3.5000 | 4 | 3.3376 | 3.1933 | 3.2294 | 1° 22' | 7.9183 | 8.3268 |
| 3 3/4 | 3.7500 | 4 | 3.5876 | 3.4433 | 3.4794 | 1° 16' | 9.2143 | 9.6546 |
| 4 | 4.0000 | 4 | 3.8376 | 3.6933 | 3.7294 | 1° 11' | 10.6084 | 11.0805 |

* British: effective diameter.
NOTE: Bold type below rule indicates unified threads, UNC.
SOURCE: Extracted from American National Standard "Unified Screw Threads," ANSI B1.1-1960, with the permission of the publisher, The American Society of Mechanical Engineers.

## Table 8-9. Fine-thread Series, UNF and NF, Basic Dimensions

| Sizes | Basic major diam, in. | Thds/in. | Basic pitch diam,* in. | Minor diam ext thds, in. | Minor diam int thds, in. | Lead angle at basic pitch diam | Section at minor diam, in.$^2$ | Tensile stress area, in.$^2$ |
|---|---|---|---|---|---|---|---|---|
| | $D$ | $n$ | $E$ | $K_s$ | $K_n$ | $\lambda$ | at $D - 2h_b$ | |
| 0 (*0.060*) | 0.0600 | 80 | 0.0519 | 0.0447 | 0.0465 | 4° 23′ | 0.0015 | 0.0018 |
| 1 (*0.073*) | 0.0730 | 72 | 0.0640 | 0.0560 | 0.0580 | 3° 57′ | 0.0024 | 0.0027 |
| 2 (*0.086*) | 0.0860 | 64 | 0.0759 | 0.0668 | 0.0691 | 3° 45′ | 0.0034 | 0.0039 |
| 3 (*0.099*) | 0.0990 | 56 | 0.0874 | 0.0771 | 0.0797 | 3° 43′ | 0.0045 | 0.0052 |
| 4 (*0.112*) | 0.1120 | 48 | 0.0985 | 0.0864 | 0.0864 | 3° 51′ | 0.0057 | 0.0065 |
| 5 (*0.125*) | 0.1250 | 44 | 0.1102 | 0.0971 | 0.1004 | 3° 45′ | 0.0072 | 0.0082 |
| 6 (*0.138*) | 0.1380 | 40 | 0.1218 | 0.1073 | 0.1109 | 3° 44′ | 0.0087 | 0.0101 |
| 8 (*0.164*) | 0.1640 | 36 | 0.1460 | 0.1299 | 0.1339 | 3° 28′ | 0.0128 | 0.0146 |
| 10 (*0.190*) | 0.1900 | 32 | 0.1697 | 0.1517 | 0.1562 | 3° 21′ | 0.0175 | 0.0199 |
| 12 (*0.216*) | 0.2160 | 28 | 0.1928 | 0.1722 | 0.1773 | 3° 22′ | 0.0226 | 0.0257 |
| 1/4 | 0.2500 | 28 | 0.2268 | 0.2062 | 0.2113 | 2° 52′ | 0.0326 | 0.0362 |
| 5/16 | 0.3125 | 24 | 0.2854 | 0.2614 | 0.2674 | 2° 40′ | 0.0524 | 0.0579 |
| 3/8 | 0.3750 | 24 | 0.3479 | 0.3239 | 0.3299 | 2° 11′ | 0.0809 | 0.0876 |
| 7/16 | 0.4375 | 20 | 0.4050 | 0.3762 | 0.3834 | 2° 15′ | 0.1090 | 0.1185 |
| 1/2 | 0.5000 | 20 | 0.4675 | 0.4387 | 0.4459 | 1° 57′ | 0.1486 | 0.1597 |
| 9/16 | 0.5625 | 18 | 0.5264 | 0.4943 | 0.5024 | 1° 55′ | 0.1888 | 0.2026 |
| 5/8 | 0.6250 | 18 | 0.5889 | 0.5568 | 0.5649 | 1° 43′ | 0.2400 | 0.2555 |
| 3/4 | 0.7500 | 16 | 0.7094 | 0.6733 | 0.6823 | 1° 36′ | 0.3513 | 0.3724 |
| 7/8 | 0.8750 | 14 | 0.8286 | 0.7874 | 0.7977 | 1° 34′ | 0.4805 | 0.5088 |
| 1 | 1.0000 | 14 | 0.9536 | 0.9124 | 0.9227 | 1° 22′ | 0.6464 | 0.6791 |
| 1 | 1.0000 | 12 | 0.9459 | 0.8978 | 0.9098 | 1° 36′ | 0.6245 | 0.6624 |
| 1 1/8 | 1.1250 | 12 | 1.0709 | 1.0228 | 1.0348 | 1° 25′ | 0.8118 | 0.8549 |
| 1 1/4 | 1.2500 | 12 | 1.1959 | 1.1478 | 1.1598 | 1° 16′ | 1.0237 | 1.0721 |
| 1 3/8 | 1.3750 | 12 | 1.3209 | 1.2728 | 1.2848 | 1° 9′ | 1.2602 | 1.3137 |
| 1 1/2 | 1.5000 | 12 | 1.4459 | 1.3978 | 1.4098 | 1° 3′ | 1.5212 | 1.5799 |

* British: effective diameter.
NOTE: Bold type below rules indicates unified threads, UNF.
SOURCE: Extracted from American National Standard "Unified Screw Threads," ANSI B1.1-1960, with the permission of the publisher, The American Society of Mechanical Engineers.

## Table 8-10. Extra-fine-thread Series, NEF, Basic Dimensions

| Sizes | Basic major diam, in. | Thds/in. | Basic pitch diam,* in. | Minor diam ext thds, in. | Minor diam int thds, in. | Lead angle at basic pitch diam | Section at minor diam, in.² | Tensile stress area, in. |
|---|---|---|---|---|---|---|---|---|
| | $D$ | $n$ | $E$ | $K_s$ | $K_n$ | $\lambda$ | at $D - 2h_b$ | |
| 12 (*0.216*) | 0.2160 | 32 | 0.1957 | 0.1777 | 0.1822 | 2° 55′ | 0.0242 | 0.0269 |
| ¼ | 0.2500 | 32 | 0.2297 | 0.2117 | 0.2162 | 2° 29′ | 0.0344 | 0.0377 |
| 5⁄16 | 0.3125 | 32 | 0.2922 | 0.2742 | 0.2787 | 1° 57′ | 0.0581 | 0.0622 |
| ⅜ | 0.3750 | 32 | 0.3547 | 0.3367 | 0.3412 | 1° 36′ | 0.0878 | 0.0929 |
| 7⁄16 | 0.4375 | 28 | 0.4143 | 0.3937 | 0.3988 | 1° 34′ | 0.1201 | 0.1270 |
| ½ | 0.5000 | 28 | 0.4768 | 0.4562 | 0.4613 | 1° 22′ | 0.1616 | 0.1695 |
| 9⁄16 | 0.5625 | 24 | 0.5354 | 0.5114 | 0.5174 | 1° 25′ | 0.2030 | 0.2134 |
| ⅝ | 0.6250 | 24 | 0.5979 | 0.5739 | 0.5799 | 1° 16′ | 0.2560 | 0.2676 |
| 11⁄16 | 0.6875 | 24 | 0.6604 | 0.6364 | 0.6424 | 1° 9′ | 0.3151 | 0.3280 |
| ¾ | 0.7500 | 20 | 0.7175 | 0.6887 | 0.6959 | 1° 16′ | 0.3685 | 0.3855 |
| 13⁄16 | 0.8125 | 20 | 0.7800 | 0.7512 | 0.7584 | 1° 10′ | 0.4388 | 0.4573 |
| ⅞ | 0.8750 | 20 | 0.8425 | 0.8137 | 0.8209 | 1° 5′ | 0.5153 | 0.5352 |
| 15⁄16 | 0.9375 | 20 | 0.9050 | 0.8762 | 0.8834 | 1° 0′ | 0.5979 | 0.6194 |
| 1 | 1.0000 | 20 | 0.9675 | 0.9387 | 0.9459 | 0° 57′ | 0.6866 | 0.7095 |
| 1 1⁄16 | 1.0625 | 18 | 1.0264 | 0.9943 | 1.0024 | 59′ | 0.7702 | 0.7973 |
| 1⅛ | 1.1250 | 18 | 1.0889 | 1.0568 | 1.0649 | 56′ | 0.8705 | 0.8993 |
| 1 3⁄16 | 1.1875 | 18 | 1.1514 | 1.1193 | 1.1274 | 53′ | 0.9770 | 1.0074 |
| 1¼ | 1.2500 | 18 | 1.2139 | 1.1818 | 1.1899 | 50′ | 1.0895 | 1.1216 |
| 1 5⁄16 | 1.3125 | 18 | 1.2764 | 1.2443 | 1.2524 | 48′ | 1.2082 | 1.2420 |
| 1⅜ | 1.3750 | 18 | 1.3389 | 1.3068 | 1.3149 | 45′ | 1.3330 | 1.3684 |
| 1 7⁄16 | 1.4375 | 18 | 1.4014 | 1.3693 | 1.3774 | 43′ | 1.4640 | 1:5010 |
| 1½ | 1.5000 | 18 | 1.4639 | 1.4318 | 1.4399 | 42′ | 1.6011 | 1.6397 |
| 1 9⁄16 | 1.5625 | 18 | 1.5264 | 1.4943 | 1.5024 | 40′ | 1.7444 | 1.7846 |
| 1⅝ | 1.6250 | 18 | 1.5889 | 1.5568 | 1.5649 | 38′ | 1.8937 | 1.9357 |
| 1 11⁄16 | 1.6875 | 18 | 1.6514 | 1.6193 | 1.6274 | 37′ | 2.0493 | 2.0929 |
| 1¾ | 1.7500 | 16 | 1.7094 | 1.6733 | 1.6823 | 40′ | 2.1873 | 2.2382 |
| 2 | 2.0000 | 16 | 1.9594 | 1.9233 | 1.9323 | 35′ | 2.8917 | 2.9501 |

* British: effective diameter.
SOURCE: Extracted from American National Standard "Unified Screw Threads," ANSI B1.1-1960, with the permission of the publisher, The American Society of Mechanical Engineers.

# Table 8-11. Standard Tapping Screws

## Thread forming

Type A

Spaced thread, gimlet point. Often called sheet-metal screw. Produces strongest joint in light-gauge sheet metal. Used in pierced or punched holes where sharp starting point is needed and exposed point does not matter. For sheets up to 20 gauge use No. 6 screw; larger screws may be used up to 18 gauge. Do not use on thicknesses larger than 18 gauge (0.048 in.). Fastest driving of all screw types, except for Type U drive screws. Can be used in easily deformed plastics or metals, with pilot hole less than diameter of screw, to increase joint strength.

Type B

Spaced thread, blunt point. Used for sheet metals thicker than 18 gauge. Has slight taper on front end insufficient for self-aligning, but when placed in a pilot-hole taper holds screw upright, making it easy to drive. Drives faster than any screw except Types A and U, with less driving torque than A. Can be used in nonferrous castings, plastics, or soft metals.

Type BP

Spaced thread, same as Type B, but has a cone point. Can be used where holes are slightly misaligned.

Type C

Blunt-point screw with threads same pitch as standard machine screw. Used where finer-pitch screw is desirable with chip-free assembly. More engaged thread surface increases frictional resistance to loosening. Small helix angle results in backout torque component under vibration. Obtains higher clamping forces for same applied driving torque.

Type U

Multiple thread, blunt point, metallic drive screw. Threads have high helix angle. Hammered or forced in, this screw has good holding power even when subjected to vibration. Used for permanent fastening, since it is difficult to remove. Do not use in material thinner than 1 diam of the screw. Usually does not have a driving slot or recess.

# Thread cutting

### Type D

Blunt-point screw with a single narrow flute. Approximate machine-screw thread. Used in same manner as Type C where less driving torque is needed. This screw is very good for low-strength metals and plastics; for high-strength, brittle metals such as cast iron; and for rethreading clogged pre-tapped holes. Easy starting. Gives the highest clamping force for a given torque of any tapping screw.

### Type F

Approximate machine-screw thread, blunt point. Tapered thread may be complete or incomplete at the producer's option. Recommended for same general application conditions as Type C screws but where low driving torque is needed. Because of the five evenly spaced cutting grooves and large chip cavities, this screw is used in a wide range of materials such as aluminum, zinc, die castings, carbon and stainless-steel sheet and shapes, cast iron, brass, plastics, etc. Drives faster than a machine screw and resists vibration. Chip space of the flutes is not suitable for deep penetration because of clogging.

### Type G

Blunt point with single through slot which forms two cutting edges. Approximate machine screw threads, front end having incomplete tapered threads. Recommended for same general usage as Type C but where less driving torque is required. Has higher percentage of thread and longer thread engagement than Type C, making it useful for low-strength metals and plastics.

### Type BF

Spaced thread same as Type B, blunt point, with five evenly spaced cutting grooves and chip cavities. The grooves of this thread remove only a small part of the material to maintain maximum shear strength in the threaded wall of the hole. Wall thickness should be at least 1½ times the major diameter of the screw. Chip room and cutting-groove design are helpful in brittle plastics and die castings to reduce stripping. Used for producing long thread engagement, particularly in blind holes. Permits faster driving than is possible with fine-thread types.

### Type T

Blunt point with single wide flute. Approximate machine screw threads. Usage same as Type D, except wide flute provides more chip clearance.

### Type BG

Spaced thread, blunt point. Single slot has two cutting edges. Especially useful for brittle or friable material where threads that are too close together will cause material to crumble. Since this screw removes the least amount of material from thread wall, maximum stripping strength is maintained in plastics and soft materials. Can produce long thread engagement, particularly in blind holes. Drives faster than fine threads.

### Type BT

Spaced thread, blunt point. Same as Type BG except for single wide flute. Flute provides room for twisted curly chips so that they do not cause binding or reaming of the hole.

SOURCE: "Machine Design," Sept. 29, 1960. Copyright 1960, Penton Publishing Co.

## 8-13. Bolt Preload and Gasketed Joints

Threaded fasteners develop their holding power by being put into tension as the screw is tightened. The danger of permanently deforming a screw exists if excessive torque is applied in "making up" the joint, with the possibility of breaking the threaded fastener through combined shear and tension. When one of the threaded members is softer than the other, the softer threads may shear or strip.

In practice, the average initial tensile load $F_i$ put on a tightened through-bolt or stud of major diameter $= d$ is approximately

$$F_i = 16,000d \qquad (8\text{-}58)$$

Thus a $\frac{5}{8}$-11 UNC bolt will be subjected to a 10,000-lb tensile load when tightened or, based on a root area of 0.225 in.$^2$, a direct tensile stress due to tightening alone of 45,000 psi. High-strength fasteners, generally of alloy steel, should be used if high initial tensile preloads are required in high-pressure joints and the like.

A bolted joint consists of two resilient members, the bolt and the gasket. Denote tensile preload force by $F_i$, external applied load by $F_e$, and total load by $F_t$. Then the total bolt load is

$$F_t = KF_e + F_i \qquad (8\text{-}59)$$

where $K$, a function of the relative stiffnesses of gasket and bolt, will vary between 0 and 1 (Table 8-12).

### Table 8-12. Values of K for Eq. (8-59)

| Type of Joint | K |
|---|---|
| Soft packing with studs | 0.90–1.00 |
| Soft packing with through bolts | 0.60–0.75 |
| Asbestos | 0.50–0.60 |
| Soft-copper gasket with long through bolts | 0.40–0.50 |
| Hard-copper gasket with long through bolts | 0.20–0.30 |
| Metal-to-metal joints with through bolts | 0.00 |

Obviously, should $F_e$ becomes so large that the gasket loses contact with the faces of the joint, the bolt will be loaded with the entire external load; that is, $F_t = F_e$.

## 8-14. Shafts

A shaft is the fundamental torque-transmitting machine element, and can be subjected separately or simultaneously to torsion, bending, and axial loading. The ASME Code for Design of Transmission Shafting, applicable to ductile materials with ultimate tensile strength about twice the ultimate shear strength, results in the following design equation:

$$d_0{}^3 = \frac{16}{\pi s_s} \sqrt{\left[ K_m M + \frac{\alpha F_a d_0 (1 + K^2)}{8} \right]^2 + (K_t T)^2} \; \frac{1}{1 - K^4} \qquad (8\text{-}60)$$

where $d_0$ = shaft diameter, in.

$F_a$ = axial tension or compression, lb

$K$ = ratio of inside to outside diameter of hollow shafts

$K_m$ = combined shock and fatigue factor to be applied to computed bending moment (Table 8-13)

$K_t$ = combined shock and fatigue factor to be applied to computed torsional moment (Table 8-13)

$M$ = maximum bending moment, lb-in.

$T$ = maximum torsional moment, lb-in.

$s_s$ = maximum stress permissible in shear, psi

$\alpha$ = ratio of maximum intensity of stress resulting from axial load to average axial stress

$$\alpha = \begin{cases} \dfrac{1}{1 - 0.0044 \left(\dfrac{L}{k}\right)} & \text{for } \dfrac{L}{k} < 115 \qquad (8\text{-}61) \\[3ex] \dfrac{s_y}{n\pi^2 E} \left(\dfrac{L}{k}\right)^2 & \text{for } \dfrac{L}{k} \geq 115 \qquad (8\text{-}62) \end{cases}$$

where $L$ = length between bearing, in.

$k$ = radius of gyration of shaft, in.

$s_y$ = compression yield stress, psi

$E$ = modulus of elasticity, psi

$n$ = 1 for free (simple) ends; $n$ = 2.5–3 for fixed ends (rigid bearings)

A shaft of brittle material loaded in both torsion and bending is designed based on the maximum allowable tensile stress $s_{t.\max}$:

$$d_0{}^3 = \frac{16}{\pi s_t} \left[ K_m M + \sqrt{(K_m M)^2 + (K_t T)^2} \right] \frac{1}{1 - K^4} \qquad (8\text{-}63)$$

## Table 8-13. Constants for ASME Code Equations for Shafting

| Type loading | $K_m$ | $K_t$ |
|---|---|---|
| Stationary shafts: | | |
|   Gradually applied load........ .............. | 1.0 | 1.0 |
|   Suddenly applied load....................... | 1.5–2.0 | 1.5–2.0 |
| Rotating shafts: | | |
|   Gradually applied or steady load.............. | 1.5 | 1.0 |
|   Suddenly applied loads, minor shocks only..... | 1.5–2.0 | 1.0–1.5 |
|   Suddenly applied loads, heavy shocks.......... | 2.0–3.0 | 1.5–3.0 |

## Table 8-14. Maximum Permissible Working Stresses for Shafts

| Grade of shafting | Simple bending | Simple torsion | Combined stress |
|---|---|---|---|
| "Commercial steel" shafting without allowance for keyways...................... | 16,000 | 8,000 | 8,000 |
| "Commercial steel" shafting with allowance for keyways............................ | 12,000 | 6,000 | 6,000 |
| Steel purchased under definite specifications.. | 60% of the elastic limit but not over 36% of the ultimate in tension | 30% of the elastic limit but not over 18% of the ultimate in tension | 30% of the elastic limit but not over 18% of the ultimate in tension |

## Table 8-15. Standard Diameters and Tolerances of Finished Transmission (T) and Machinery (M) Shafting

### All Dimensions Given in Inches

| Stock diam of shafting | | Diam toler-ance† | Stock diam of shafting | | Diam toler-ance† | Stock diam of shafting | | Diam toler-ance† |
| T | M | | T | M | | T | M | |
|---|---|---|---|---|---|---|---|---|
| | 1/2 | 0.002 | 1 15/16 | 1 15/16 | 0.003 | | 3 3/4 | 0.004 |
| | 9/16 | .002 | | 2 | .003 | | 3 7/8 | .004 |
| | 5/8 | .002 | | 2 1/16 | .004 | 3 15/16 | 4 | .004 |
| | 1 1/16 | .002 | | 2 1/8 | .004 | | 4 1/4 | .005 |
| | 3/4 | .002 | 2 3/16 | 2 3/16 | .004 | 4 7/16 | 4 1/2 | .005 |
| | 1 3/16 | .002 | | 2 1/4 | .004 | | 4 3/4 | .005 |
| | 7/8 | .002 | | 2 5/16 | .004 | 4 13/16 | 5 | .005 |
| 1 5/16 | 1 5/16 | .002 | | 2 3/8 | .004 | | 5 1/4 | .005 |
| | 1 | .002 | 2 7/16 | 2 7/16 | .004 | 5 7/16 | 5 1/2 | .005 |
| | 1 1/16 | .003 | | 2 1/2 | .004 | | 5 3/4 | .005 |
| | 1 1/8 | .003 | | 2 5/8 | .004 | 5 13/16 | 6 | .005 |
| 1 3/16 | 1 3/16 | .003 | 2 15/16 | 2 3/4 | .004 | | 6 1/4 | .005 |
| | 1 1/4 | .003 | | 2 7/8 | .004 | 6 1/2 | 6 1/2 | .005 |
| | 1 5/16 | .003 | | 3 | .004 | | 6 3/4 | .005 |
| | 1 3/8 | .003 | | 3 1/8 | .004 | 7 | 7 | .005 |
| 1 7/16 | 1 7/16 | .003 | 3 7/16 | 3 1/4 | .004 | | 7 1/4 | .005 |
| | 1 1/2 | .003 | | 3 3/8 | .004 | 7 1/2 | 7 1/2 | .005 |
| | 1 9/16 | .003 | | 3 1/2 | .004 | | 7 3/4 | .005 |
| | 1 5/8 | .003 | | 3 5/8 | .004 | 8 | 8 | .005 |
| 1 11/16 | 1 11/16 | .003 | | | | | | |
| | 1 3/4 | .003 | | | | | | |
| | 1 13/16 | .003 | | | | | | |
| | 1 7/8 | .003 | | | | | | |

† Shaft tolerances are *negative* and represent the maximum allowable variation below the exact nominal size. For example, the maximum diameter of the 1½-in. shaft is 1.500 in. and its minimum allowable diameter is 1.497 in.

SOURCE: L. S. Marks (ed.), "Mechanical Engineers' Handbook," 5th ed., p. 911, McGraw-Hill Book Company, Inc., New York, 1951.

## 8-15. Keys, Pins, Splines

Keys are used to fasten hubbed members to shafts, and are loaded in shear and compression when transmitting torque from one to the other (Fig. 8-53). Square and Woodruff keys are most commonly used, though many other

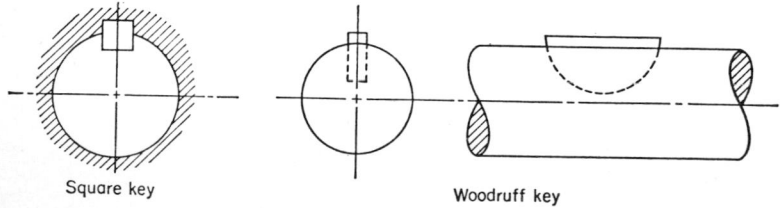

Square key        Woodruff key

FIG. 8-53

types exist.  Standard square and rectangular keys and standard keyseats are shown in Table 8-16.  Pins are keys positioned at right angles to the shaft center line and are generally round or conical (taper pins) (Fig. 8-54). Hollow spring pins (roll pins) (Fig. 8-55) have come into wide use recently.

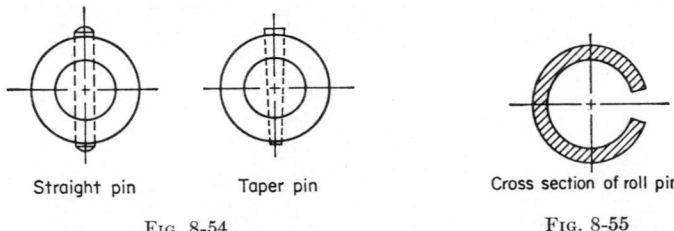

| Straight pin | Taper pin | Cross section of roll pin |

FIG. 8-54                     FIG. 8-55

They can replace both solid straight pins and tapered pins because they combine the advantages of both, i.e., simple tooling, ease of removal, ability to be driven from either side.

Splines are effectively a multiplicity of longitudinal keys, either straight or helical, whose teeth are machined integral with the male and female members, in contrast with the common key, which is a third member employed between the male and female elements.  The automotive and machine-tool industries employ them extensively in their transmission devices and have caused them to be standardized.  Straight-sided standard splines are listed in Table 8-17.

## Table 8-16. Standard Square and Rectangular Keys and Keyseats*

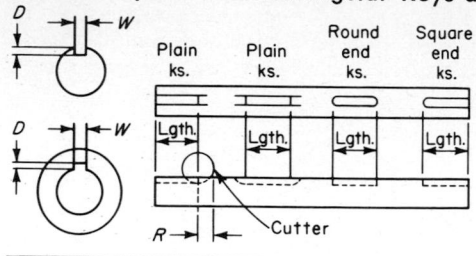

| Shaft size | W Width +.002 in. −.000 in. | D† Depth Regular | Shallow | R Max cutter runout | Max diam of cutter |
|---|---|---|---|---|---|
| 5/16–7/16 | 3/32 | 3/64 | ... | 1/2 | 3 1/4 |
| 1/2–9/16 | 1/8 | 1/16 | ... | 9/16 | 3 1/4 |
| 5/8–7/8 | 3/16 | 3/32 | ... | 1 1/16 | 3 1/4 |
| 15/16–1 1/4 | 1/4 | 1/8 | ... | 1 3/16 | 4 |
| 1 5/16–1 3/4 | 3/8 | 3/16 | ... | 1 1/16 | 5 |
| 1 13/16–2 1/4 | 1/2 | 1/4 | 1/8 | 1 3/16 | 5 |
| 2 5/16–2 3/4 | 5/8 | 5/16 | 3/16 | 1 5/16 | 5 |
| 2 13/16–3 1/4 | 3/4 | 3/8 | 3/16 | 1 9/16 | 5 1/2 |
| 3 5/16–3 3/4 | 7/8 | 7/16 | 1/4 | 1 11/16 | 5 1/2 |
| 3 13/16–4 1/2 | 1 | 1/2 | 1/4 | 1 3/4 | 5 1/2 |
| 4 9/16–5 1/2 | 1 1/4 | 5/8 | 1/4 | 1 13/16 | 5 1/2 |
| 5 9/16–6 1/2 | 1 1/2 | 3/4 | 1/4 | 2 1/8 | 5 1/2 |
| 6 9/16–7 1/2 | 1 3/4 | 3/4 | 1/4 | 2 1/8 | 5 1/2 |
| 7 9/16–9 | 2 | 3/4 | 3/8 | 2 1/8 | 5 1/2 |
| 9 1/16–11 | 2 1/2 | 7/8 | 3/8 | 2 3/16 | 6 |
| 11 1/16–13 | 3 | 1 | 3/8 | 2 7/16 | 6 |

Shaft keyseat: Always make straight and never taper. Always make regular depth even when shallow depth is used in hub.

Hub keyseat: Make straight unless taper keyseat is specified (see below). Make regular depth unless shallow depth is specified.

Taper keyseat: Should never be used in shafts. Taper is 1/8 in./ft. Depth at large end is equal to $D$.

* Square keys per "Keys and Keyseats," ANSI B17.1-1967 (R1973).

† Tolerance on Depth:

+0.010, −0.000 in. for keyseat in shaft.

+0.010, −0.000 in. (preferably +0.010 in.) for straight keyseat in hub.

+0.000, −0.010 in. (preferably −0.010 in.) for taper keyseat in hub.

SOURCE: Extracted from "Engineering Standards-Multiple V-Belt Drives," 1951. with the permission of the publisher, Multiple V-Belt Drive & Mechanical Power Transmission Association, Chicago, Illinois.

## Table 8-17. SAE Standard Splines

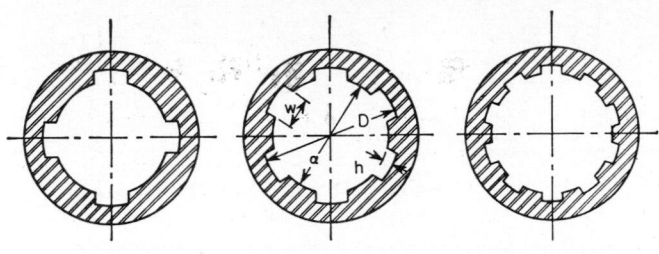

|  | 4 spline | 6 spline | 10 spline |
|---|---|---|---|
| Permanent fit | $d = 0.85D$<br>$w = 0.241D$<br>$h = 0.075D$ | $d = 0.90D$<br>$w = 0.25D$<br>$h = 0.05D$ | $d = 0.91D$<br>$w = 0.156D$<br>$h = 0.045D$ |
| To slide when not under load | $d = 0.75D$<br>$w = 0.241D$<br>$h = 0.125D$ | $d = 0.85D$<br>$w = 0.25D$<br>$h = 0.075D$ | $d = 0.86D$<br>$w = 0.156D$<br>$h = 0.07D$ |
| To slide when under load |  | $d = 0.80D$<br>$w = 0.25D$<br>$h = 0.10D$ | $d = 0.81D$<br>$w = 0.156D$<br>$h = 0.095D$ |

NOTE: Shaft dimensions 0.001 in. under nominal for small shafts and 0.002 in. for large shafts.

## 8-16. Belts and Sheaves

### Belt Lengths

Total belt length $L = 2S \cos \theta + \pi[(R + r) + (R - r)\theta/90]$    in.    (8-64)

where $S$ = center distance between pulleys, in.

$R$ = radius of large pulley (flat belt) or pitch radius of large pulley (V belt), in.

$r$ = radius of small pulley (flat belt) or pitch radius of small pulley (V belt), in.

$\theta = \sin^{-1}[(R - r)/S]$    deg

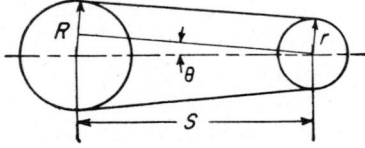

FIG. 8-56

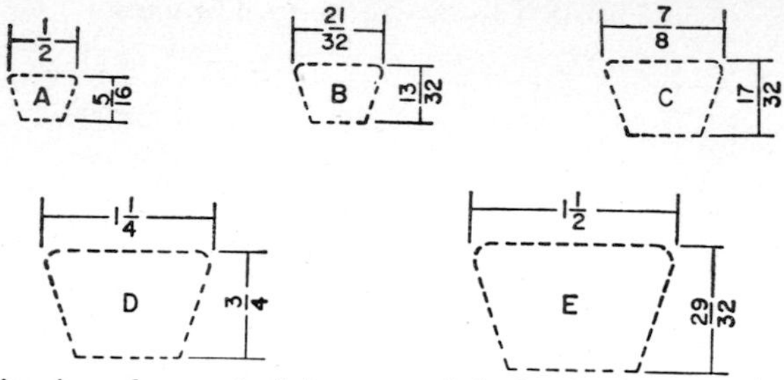

NOTE: The dimensions and cross-sectional shapes are nominal only. Actual cross-sectional dimensions of belts made by one manufacturer may differ from those of the same belt cross section made by another manufacturer. Because of different constructions and different methods of manufacturing, the cross-sectional shapes and the included angle between the side walls may be different for different manufacturers. However, all standard cross sections will operate interchangeably in standard grooves.

SOURCE: Extracted from "Engineering Standards-Multiple V-Belt Drives," 1951, with the permission of the publisher, Multiple V-Belt Drive & Mechanical Power Transmission Association, Chicago 3, Illinois.

FIG. 8-57. V-belt cross sections.

## Table 8-18. Measuring Tensions

| Belt* | Tension,† lb | Belt* | Tension,† lb |
|-------|--------------|-------|--------------|
| A | 50 | D | 300 |
| B | 65 | E | 400 |
| C | 165 | | |

* See Table 8-19.

† The tension specified is the sum of tension in the two strands of the belt.

SOURCE: Extracted from "Engineering Standard-Multiple V-Belt Drives," 1951, with the permission of the publisher, Multiple V-Belt Drive & Mechanical Power Transmission Association, Chicago, Illinois.

## Table 8-19. Standard Belt Sizes

For Dimensions and Cross-sectional Shapes See Fig. 8-57

| Std nominal length | Std pitch lengths, in. | | | | | Permissible deviations from std pitch length, in. | Matching limits for one set, in. |
|---|---|---|---|---|---|---|---|
| | A | B | C | D | E | | |
| 26 | 27.3 | ..... | ..... | ..... | ..... | +0.7−0.3 | 0.10 |
| 31 | 32.3 | ..... | ..... | ..... | ..... | +0.7−0.3 | 0.10 |
| 33 | 34.3 | ..... | ..... | ..... | ..... | +0.8−0.4 | 0.20 |
| 35 | 36.3 | 36.8 | ..... | ..... | ..... | +0.8−0.4 | 0.20 |
| 38 | 39.3 | 39.8 | ..... | ..... | ..... | +0.8−0.4 | 0.20 |
| 42 | 43.3 | 43.8 | ..... | ..... | ..... | +0.8−0.4 | 0.20 |
| 46 | 47.3 | 47.8 | ..... | ..... | ..... | +0.8−0.4 | 0.20 |
| 48 | 49.3 | 49.8 | ..... | ..... | ..... | +0.9−0.5 | 0.20 |
| 51 | 52.3 | 52.8 | 53.9 | ..... | ..... | +0.9−0.5 | 0.20 |
| 53 | 54.3 | 54.8 | ..... | ..... | ..... | +0.9−0.5 | 0.20 |
| 55 | 56.3 | 56.8 | ..... | ..... | ..... | +0.9−0.5 | 0.20 |
| 60 | 61.3 | 61.8 | 62.9 | ..... | ..... | +0.9−0.5 | 0.20 |
| 62 | 63.3 | 63.8 | ..... | ..... | ..... | +0.9−0.5 | 0.20 |
| 64 | 65.3 | 65.8 | ..... | ..... | ..... | +0.9−0.5 | 0.20 |
| 66 | 67.3 | 67.8 | ..... | ..... | ..... | +0.9−0.5 | 0.20 |
| 68 | 69.3 | 69.8 | 70.9 | ..... | ..... | +0.9−0.5 | 0.20 |
| 71 | 72.3 | 72.8 | ..... | ..... | ..... | +0.9−0.5 | 0.20 |
| 75 | 76.3 | 76.8 | 77.9 | ..... | ..... | +0.9−0.5 | 0.20 |
| 78 | 79.3 | 79.8 | ..... | ..... | ..... | +1.0−0.5 | 0.30 |
| 80 | 81.3 | ..... | ..... | ..... | ..... | +1.0−0.5 | 0.30 |
| 81 | ..... | 82.8 | 83.9 | ..... | ..... | +1.0−0.5 | 0.30 |
| 83 | ..... | 84.8 | ..... | ..... | ..... | +1.0−0.5 | 0.30 |
| 85 | 86.3 | 86.8 | 87.9 | ..... | ..... | +1.0−0.5 | 0.30 |
| 90 | 91.3 | 91.8 | 92.9 | ..... | ..... | +1.0−0.5 | 0.30 |
| 96 | 97.3 | ..... | 98.9 | ..... | ..... | +1.0−0.5 | 0.30 |
| 97 | ..... | 98.8 | ..... | ..... | ..... | +1.0−0.5 | 0.30 |
| 105 | 106.3 | 106.8 | 107.9 | ..... | ..... | +1.1−0.5 | 0.40 |
| 112 | 113.3 | 113.8 | 114.9 | ..... | ..... | +1.1−0.5 | 0.40 |
| 120 | 121.3 | 121.8 | 122.9 | 123.3 | ..... | +1.2−0.5 | 0.40 |
| 128 | 129.3 | 129.8 | 130.9 | 131.3 | ..... | +1.3−0.6 | 0.40 |
| 136 | ..... | 137.8 | 138.9 | ..... | ..... | +1.3−0.6 | 0.40 |
| 144 | ..... | 145.8 | 146.9 | 147.3 | ..... | +1.4−0.6 | 0.40 |
| 158 | ..... | 159.8 | 160.9 | 161.3 | ..... | +1.5−0.6 | 0.40 |
| 162 | ..... | ..... | 164.9 | 165.3 | ..... | +1.6−0.6 | 0.40 |
| 173 | ..... | 174.8 | 175.9 | 176.3 | ..... | +1.7−0.7 | 0.50 |
| 180 | ..... | 181.8 | 182.9 | 183.3 | 184.5 | +1.7−0.7 | 0.50 |
| 195 | ..... | 196.8 | 197.9 | 198.3 | 199.5 | +1.8−0.8 | 0.50 |
| 210 | ..... | 211.8 | 212.9 | 213.3 | 214.5 | +2.0−0.8 | 0.50 |
| 240 | ..... | 240.3 | 240.9 | 240.8 | 241.0 | +2.2−0.9 | 0.50 |
| 270 | ..... | 270.3 | 270.9 | 270.8 | 271.0 | +2.4−1.0 | 0.50 |
| 300 | ..... | 300.3 | 300.9 | 300.8 | 301.0 | +2.5−1.2 | 0.60 |
| 330 | ..... | ..... | 330.9 | 330.8 | 331.0 | +2.5−1.2 | 0.60 |
| 360 | ..... | ..... | 360.9 | 360.8 | 361.0 | +2.5−1.2 | 0.60 |
| 390 | ..... | ..... | 390.9 | 390.8 | 391.0 | +3.0−1.5 | 0.70 |
| 420 — | ..... | ..... | 420.9 | 420.8 | 421.0 | +3.5−2.0 | 0.80 |
| 480 | ..... | ..... | ..... | 480.8 | 481.0 | +4.0−2.5 | 0.90 |
| 540 | ..... | ..... | ..... | 540.8 | 541.0 | +4.5−3.0 | 1.10 |
| 600 | ..... | ..... | ..... | 600.8 | 601.0 | +5.0−3.5 | 1.30 |
| 660 | ..... | ..... | ..... | 660.8 | 661.0 | +6.0−4.0 | 1.50 |

## Table 8-20. Suggested Service Factors

| Application | Electric motors | | | | | | | | | | Engines | | | | Line shafts and clutch starting |
| | A-c | | | | | | | | D-c | | Gas and diesel | | | Steam | |
| | Squirrel cage | | | | Synchronous | | Single phase | | | | | | | | |
| | Normal torque, line start | Normal torque, compensator start | High torque | Wound rotor (slip ring) | Normal torque | High torque | Repulsion and split-phase | Capacitor | Shunt wound | Compound wound | 4 or more cyl, above 700 rpm | 4 or more cyl, below 700 rpm | 3 or less cyl, (refer to factory) | Steam | Line shafts and clutch starting |
|---|---|---|---|---|---|---|---|---|---|---|---|---|---|---|---|
| **Agitators, paddle-propeller:** | | | | | | | | | | | | | | | |
| Liquid | 1.0 | 1.0 | 1.2 | | | | | | | | | | | | |
| Semiliquid | 1.2 | 1.0 | 1.4 | 1.2 | | | | | | | | | | | |
| **Brick and clay machinery:** | | | | | | | | | | | | | | | |
| Auger machine | | 1.2 | 1.4 | 1.4 | | | | | | 1.4 | | | | | 2.0 |
| De-airing machine | | 1.2 | 1.4 | 1.4 | | | | | | 1.4 | | | | | 2.0 |
| Cutting table | | 1.2 | 1.4 | 1.4 | | | | | | | | | | | 2.0 |
| Pug mill | 1.5 | 1.3 | 1.8 | 1.5 | | | | | | | | | | | 2.0 |
| Mixer | | 1.2 | 1.6 | 1.4 | | | | | | | | | | | |
| Granulator | | 1.2 | 1.4 | 1.4 | | | | | | | | | | | |
| Dry press | | 1.2 | 1.6 | 1.4 | | | | | | | | | | | |
| Rolls | | 1.2 | 1.4 | 1.4 | | | | | | | | | | | |
| **Bakery machinery:** | | | | | | | | | | | | | | | |
| Dough mixer | 1.2 | | | | | | 1.2 | 1.0 | | | | | | | |
| **Compressors:** | | | | | | | | | | | | | | | |
| Centrifugal | 1.2 | 1.2 | | 1.4 | 1.4 | | | | | 1.2 | 1.2 | | | | |
| Rotary | 1.2 | 1.2 | | 1.4 | 1.4 | | 1.2 | 1.2 | | 1.2 | 1.2 | | | | |
| **Reciprocating:** | | | | | | | | | | | | | | | |
| 3 or more cyl | 1.2 | 1.2 | | 1.4 | 1.4 | | | | | 1.2 | | | | | |
| 1 or 2 cyl | 1.4 | 1.4 | | 1.5 | 1.5 | | | | | 1.2 | | | | | |
| **Conveyors:** | | | | | | | | | | | | | | | |
| Apron | | 1.4 | 1.6 | | | | | | | 1.4 | | | | | 1.6 |
| Belt (ore, coal, sand) | | 1.2 | 1.4 | | | | | | | 1.2 | | | | | 1.4 |
| Belt (light package) | | 1.0 | 1.1 | | | | | | | 1.0 | | | | | 1.2 |
| Oven | | 1.0 | 1.1 | | | | | | | 1.0 | | | | | 1.2 |
| Screw | | 1.6 | 1.8 | | | | | | | 1.6 | | | | | 1.8 |
| Bucket | | 1.4 | 1.6 | | | | | | | 1.4 | | | | | 1.6 |
| Pan | | 1.4 | 1.6 | | | | | | | 1.4 | | | | | 1.6 |
| Flight | | 1.6 | 1.8 | | | | | | | 1.6 | | | | | 1.8 |
| Elevator | | 1.4 | 1.6 | | | | | | | 1.4 | | | | | 1.6 |
| **Crushing machinery:** | | | | | | | | | | | | | | | |
| Jaw crusher | | 1.4 | 1.6 | 1.4 | | | | | | 1.4 | 1.4 | | | | 1.6 |
| Gyratory crusher | | 1.4 | 1.6 | 1.4 | 1.4 | 1.6 | | | | 1.4 | 1.4 | | | | 1.6 |
| Cone crusher | | 1.4 | 1.6 | 1.4 | | | | | | 1.6 | 1.4 | | | | 1.6 |
| Crushing rolls | | 1.4 | 1.6 | 1.4 | | | | | | 1.4 | 1.4 | | | | 1.6 |
| Ball and pebble mill | | 1.4 | 1.6 | 1.4 | 1.4 | 1.6 | | | | 1.4 | 1.6 | | | | 1.6 |
| Tube mill | | 1.4 | 1.6 | 1.4 | 1.4 | | | | | 1.4 | | | | | 1.6 |
| **Fans and blowers:** | | | | | | | | | | | | | | | |
| Centrifugal | 1.2 | 1.2 | | 1.4 | | | | | | 1.2 | 1.2 | | | | |
| Propeller | 1.4 | 1.4 | 2.0 | 1.6 | | 2.0 | | | | 1.4 | 1.4 | | | | |
| Induced draft | 1.2 | 1.2 | | 1.4 | | | | | | 1.4 | 1.4 | | | | |
| Mine fan | 1.6 | 1.4 | 2.0 | | | 2.0 | | | | | 1.6 | | | | |
| Positive blower | 1.6 | 1.6 | | 2.0 | 2.0 | 2.0 | | | | | 1.6 | | | | |
| Exhauster | 1.2 | 1.2 | | 1.4 | | | | | | 1.4 | | | | 1.5 | 1.5 |
| **Flour, feed, cereal-mill machinery:** | | | | | | | | | | | | | | | |
| Bolter and sifter | | 1.0 | | | | | | | | | | | | | |
| Grinder and hammermill | | 1.4 | | | | | | | | 1.6 | | | | | |
| Purifier and reel | 1.2 | 1.4 | | | | | | | | | | | | | |
| Main line-shaft drive | 1.4 | 1.4 | | 1.6 | 1.4 | 1.4 | | | | 1.8 | | | | | |
| Separator | 1.0 | 1.0 | | | | | | | | | | | | | |
| Roller mill | | 1.4 | | | | | | | | | | | | | |
| Generators and exciters | 1.2 | | | | | | | | | 1.2 | | | | 1.4 | 1.4 |

## Table 8-20. Suggested Service Factors (Continued)

| Application | Electric motors | | | | | | | | | | Engines | | | | |
|---|---|---|---|---|---|---|---|---|---|---|---|---|---|---|---|
| | A-c | | | | | | | | D-c | | Gas and diesel | | | | |
| | Squirrel cage | | | | Synchronous | | Single phase | | | | | | | | |
| | Normal torque, line start | Normal torque, compensator start | High torque] | Wound rotor (slip ring) | Normal torque | High torque | Repulsion and split-phase | Capacitor | Shunt wound | Compound wound | 4 or more cyl, above 700 rpm | 4 or more cyl, below 700 rpm | 3 or less cyl, (refer to factory) | Steam | Line ssafts and clutch starting |
| **Laundry machinery:** | | | | | | | | | | | | | | | |
| Washer | 1.2 | | | | | | | | | 1.2 | | | | | |
| Extractor | 1.2 | | | | | | | | | 1.2 | | | | | |
| Tumbler | 1.2 | | | | | | | | | 1.2 | | | | | |
| Dampener | 1.2 | | | | | | | | | 1.2 | | | | | |
| Flat-work ironer | 1.2 | | | | | | | | | 1.2 | | | | | |
| Line shafts | 1.4 | 1.4 | | 1.4 | 1.4 | 2.0 | 1.4 | 1.4 | 1.4 | 1.4 | 1.6 | | | 1.6 | 1.6 |
| **Machine tools:** | | | | | | | | | | | | | | | |
| Grinder | 1.2 | | | 1.4 | | | 1.2 | 1.0 | 1.2 | 1.2 | | | | | |
| Boring mill | 1.2 | | | 1.4 | | | | | 1.2 | 1.2 | | | | | |
| Lathe | 1.0 | | | 1.2 | | | 1.0 | 1.0 | 1.0 | 1.0 | | | | | |
| Milling machine | 1.2 | | | 1.4 | | | | | 1.2 | 1.2 | | | | | |
| Screw machine | 1.0 | | | 1.0 | | | 1.0 | 1.0 | 1.0 | 1.0 | | | | | |
| Cam cutter | 1.0 | | | 1.0 | | | | | 1.0 | 1.0 | | | | | |
| Planer | 1.2 | | | 1.4 | | | 1.2 | 1.0 | 1.2 | 1.2 | | | | | |
| Shaper | 1.0 | | | 1.0 | | | 1.0 | 1.0 | 1.0 | 1.0 | | | | | |
| Drill press | 1.0 | | | 1.0 | | | 1.0 | 1.0 | 1.0 | 1.0 | | | | | |
| Drop hammer | 1.0 | | | 1.0 | | | 1.0 | 1.0 | 1.0 | 1.0 | | | | | |
| Shears | 1.2 | | | 1.4 | | | 1.2 | 1.2 | 1.2 | 1.0 | | | | | |
| **Mills:** | | | | | | | | | | | | | | | |
| Pebble | | 1.4 | 1.6 | 1.4 | | | | | | 1.4 | | | | | 1.6 |
| Rod | | 1.4 | 1.6 | 1.4 | | | | | | 1.4 | | | | | 1.6 |
| Ball | | 1.4 | 1.6 | 1.4 | | | | | | 1.4 | | | | | 1.6 |
| Roller | | 1.4 | 1.6 | 1.4 | | | | | | 1.4 | | | | | 1.6 |
| Flaking | | 1.6 | 1.6 | 1.4 | | | | | | 1.4 | | | | | 1.6 |
| Tumbling barrel | | 1.6 | 1.6 | 1.4 | | | | | | 1.4 | | | | | 1.6 |
| **Oil-field machinery:** | | | | | | | | | | | | | | | |
| Slush pump | | | | | | | | | | 1.4 | 1.4 | 1.6 | | 1.4 | 1.4 |
| Pumping unit | 1.2 | 1.2 | 1.4 | | | | | | | 1.4 | | | | | 1.6 |
| Pipeline pump, centrifugal | 1.2 | 1.2 | 1.4 | | | | | | | 1.4 | 1.4 | | | | |
| Draw works (intermittent)* | | | | | | | | | | 1.3 | | 1.0 | 1.0 | | |
| **Paper machinery:** | | | | | | | | | | | | | | | |
| Jordan engine | 1.5 | 1.3 | 1.8 | 1.5 | 1.6 | 1.8 | | | 1.5 | 1.5 | | | | | |
| Beater | 1.4 | 1.4 | | 1.4 | | | | | 1.4 | 1.4 | | | | | 1.8 |
| Calender | 1.2 | 1.0 | | 1.2 | | | | | 1.2 | 1.2 | | | | | |
| Agitator | 1.2 | 1.2 | 1.4 | 1.2 | | | | | 1.2 | 1.2 | | | | | 1.6 |
| Dryer | 1.2 | 1.2 | | 1.2 | | | | | 1.2 | 1.2 | | | | | |
| Paper machine | 1.4 | 1.4 | | 1.5 | | | | | 1.5 | 1.5 | | | | | 1.6 |
| **Printing machinery:** | | | | | | | | | | | | | | | |
| Rotary press | 1.2 | 1.2 | | 1.2 | | | | | 1.2 | | | | | | |
| Embossing press | 1.2 | 1.2 | | 1.2 | | | | | 1.2 | 1.2 | | | | | |
| Folder | 1.2 | 1.2 | | 1.2 | | | | | 1.2 | | | | | | |
| Paper cutter | 1.2 | 1.2 | | 1.2 | | | | | 1.2 | 1.2 | | | | | |
| Linotype machine | 1.2 | 1.2 | | | | | | | 1.2 | | | | | | |
| Flat-bed press | 1.2 | 1.2 | | 1.2 | | | | | 1.2 | | | | | | |
| **Pumps:** | | | | | | | | | | | | | | | |
| Centrifugal | 1.2 | 1.2 | 1.4 | 1.4 | | | 1.2 | 1.2 | | | | | | | |
| Gear | 1.2 | 1.2 | 1.4 | 1.4 | | | 1.2 | 1.2 | | | | | | | |
| Rotary | 1.2 | 1.2 | 1.4 | 1.4 | | | 1.2 | 1.2 | 1.2 | | 1.2 | | | | |
| Reciprocating: | | | | | | | | | | | | | | | |
|   3 or more cyl | 1.2 | 1.2 | | 1.4 | 1.4 | 1.6 | | | | | 1.8 | | | 1.8 | |
|   1 or 2 cyl | 1.4 | 1.4 | | 1.6 | 1.6 | 1.8 | | | | | 2.0 | | | 2.0 | |
| Dredge pumps | 1.4 | 1.4 | | 1.4 | | | | | | | 2.0 | | | 2.0 | |

## Table 8-20. Suggested Service Factors (Continued)

| Application | Electric motors | | | | | | | | | | Engines | | | | |
|---|---|---|---|---|---|---|---|---|---|---|---|---|---|---|---|
| | A-c | | | | | | | | D-c | | Gas and diesel | | | | |
| | Squirrel cage | | | | Synchronous | | Single phase | | | | | | | | |
| | Normal torque, line start | Normal torque, compensator start | High torque | Wound rotor (slip ring) | Normal torque | High torque | Repulsion and split-phase | Capacitor | Shunt wound | Compound wound | 4 or more cyl, above 700 rpm | 4 or more cyl, below 700 rpm | 3 or less cyl, (refer to factory) | Steam | Line shafts and clutch starting |
| **Rubber-plant machinery:** | | | | | | | | | | | | | | | |
| Calender.................... | 1.4 | 1.4 | 1.4 | 1.4 | ... | 1.8 | ... | ... | ... | ... | ... | ... | ... | 2.0 | |
| Banbury mill.............. | 1.4 | 1.4 | 1.4 | 1.4 | ... | 1.8 | ... | ... | ... | ... | ... | ... | ... | 2.0 | |
| Mixer..................... | 1.4 | 1.4 | 1.4 | 1.4 | ... | 1.8 | ... | ... | ... | ... | ... | ... | ... | 2.0 | |
| **Screens:** | | | | | | | | | | | | | | | |
| Vibrating................. | 1.2 | 1.2 | 1.4 | | | | | | | | | | | | |
| Conical................... | 1.2 | 1.2 | | | | | | | | | | | | | |
| Revolving................. | 1.2 | 1.2 | | | | | | | | | | | | | |
| **Textile machinery:** | | | | | | | | | | | | | | | |
| Spinning frame.......... | 1.6 | ... | | 1.8 | | | | | | | | | | | |
| Twister................... | 1.6 | ... | | 1.8 | | | | | | | | | | | |
| Loom..................... | 1.2 | | | | | | | | | | | | | | |
| Warper................... | 1.2 | | | | | | | | | | | | | | |
| Reel...................... | 1.2 | | | | | | | | | | | | | | |

\* Hoisting service factor based on total engine horsepower (continuous rating). Electric drive factor based on continuous rating of motor.

SOURCE: Extracted from "Engineering Standards-Multiple V-Belt Drives," 1951, with the permission of the publisher, Multiple V-Belt Drive & Mechanical Power Transmission Association, Chicago, Illinois.

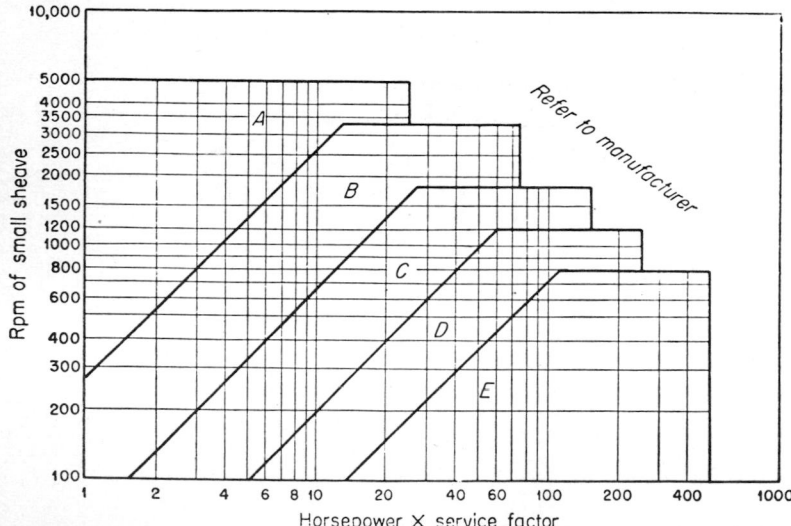

FIG. 8-58. *Source: "Engineering Standards—Multiple V-Belt Drives," 1951, with the permission of the publisher, Multiple V-Belt Drive & Mechanical Power Transmission Association, Chicago, Illinois.*

## Sheaves

### Table 8-21. Groove Dimensions for V-belt Sheaves

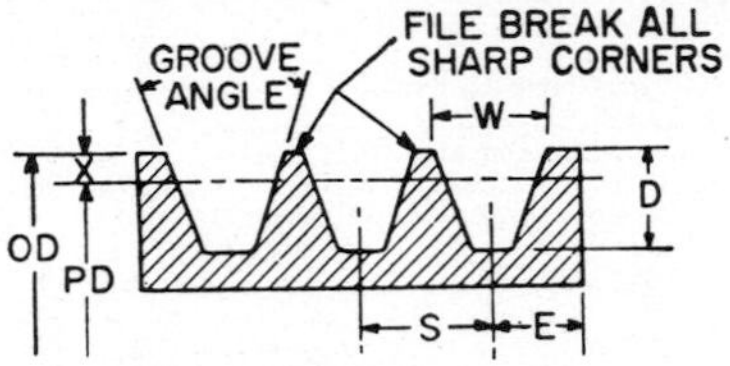


Face Width of Stock and Standard Sheaves
Face width = $S(N - 1) + 2E$

where $N$ = number of grooves

| Belt | Min recom-mended pitch diam | Pitch diam | Groove angle, deg | $W$ | $D$ | $X$ | $S$ | $E$ |
|---|---|---|---|---|---|---|---|---|
| | | | Standard-groove Dimensions | | | | | |
| *A* | 3.0 | 2.6–5.4<br>>5.4 | 34<br>38 | 0.494<br>0.504 | 0.490 | 0.125 | 5/8 | 3/8 |
| *B* | 5.4 | 4.6–7.0<br>>7.0 | 34<br>38 | 0.637<br>0.650 | 0.580 | .175 | 3/4 | 1/2 |
| *C* | 9.0 | 7.0–7.99<br>8.0–12.0<br>>12.0 | 34<br>36<br>38 | 0.879<br>0.887<br>0.895 | 0.780 | .200 | 1 | 11/16 |
| *D* | 13.0 | 12.0–12.99<br>13.0–17.0<br>>17.0 | 34<br>36<br>38 | 1.259<br>1.271<br>1.283 | 1.050 | .300 | 1 7/16 | 7/8 |
| *E* | 21.0 | 18.0–24.0<br>>24.0 | 36<br>38 | 1.527<br>1.542 | 1.300 | .400 | 1 3/4 | 1 1/8 |
| | | | Deep-groove Dimensions | | | | | |
| *A* | 3.0 | 2.6–5.4<br>>5.4 | 34<br>38 | 0.589<br>0.611 | 0.645 | .280 | 3/4 | 7/16 |
| *B* | 5.4 | 4.6–7.0<br>>7.0 | 34<br>38 | 0.747<br>0.774 | 0.760 | .355 | 7/8 | 9/16 |
| *C* | 9.0 | 7.0–7.99<br>8.0–12.0<br>>12.0 | 34<br>36<br>38 | 1.066<br>1.085<br>1.105 | 1.085 | .505 | 1 1/4 | 13/16 |
| *D* | 13.0 | 12.0–12.99<br>13.0–17.0<br>>17.0 | 34<br>36<br>38 | 1.513<br>1.541<br>1.569 | 1.465 | .715 | 1 3/4 | 1 1/16 |
| *E* | 21.0 | 18.0–24.0<br>>24.0 | 36<br>38 | 1.816<br>1.849 | 1.745 | .845 | 2 1/16 | 1 5/16 |

SOURCE: Extracted from "Engineering Standards-Multiple V-Belt Drives," 1951, with the permission of the publisher, Multiple V-Belt Drive & Mechanical Power Transmission Association, Chicago, Illinois.

### Table 8-22. Incline of Belt Conveyors

| Material Conveyed | Incline, Deg |
|---|---|
| Briquets and egg-shaped material | 7–12 |
| Wet-mixed concrete | 10–15 |
| Sized coal | 13–18 |
| Washed and screened gravel | 13–18 |
| Loose cement | 15–20 |
| Crushed and screened coke | 15–20 |
| Sand | 15–20 |
| Glass batch | 15–20 |
| Run-of-mine coal | 17–22 |
| Run-of-bank gravel | 17–22 |
| Crushed ore | 20–25 |
| Crushed stone | 20–26 |
| Tempered foundry sand | 20–25 |
| Wood chips | 22–28 |

## 8-17. Bearings

The primary function of bearings is to reduce friction between moving parts, and consequently, to reduce wear and wasted power.

**Fluid Film Bearings.** In a properly designed bearing of this type, hydrodynamic action results in a thin pressurized film of viscous lubricant which separates the moving surfaces and supplies load-carrying capacity. In rotating machinery, the viscous lubricant adheres to both the shaft, or journal, and bearing surfaces. Relative motion between the journal and bearing results in shearing the lubricant film. Energy is consumed in this fluid shearing action, but the over-all coefficient of friction and concomitant power losses are markedly less than would be the case were the moving surfaces to slide directly on each other.

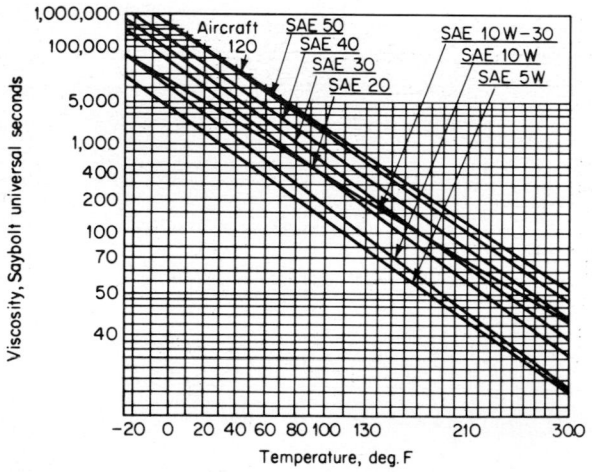

Fig. 8-59. Variation of viscosity with temperature for petroleum oils.

Variables entering into the design of fluid film bearings include diameter and length of bearing, relative speed, lubricant viscosity, etc. Viscosity is the lubricant's most important property, and is extremely temperature-sensitive (Fig. 8-59). The viscosity of gaseous lubricants, highly compressible compared to liquid ones, varies with pressure as well as temperature. A convenient index for evaluation of journal bearings is mean bearing pressure. Table 8-23 shows the range of values experienced for rotating machinery listed.

In a properly designed and operating fluid film bearing, surface wear is zero as long as the moving surfaces do not touch. Any wear which does occur will most likely take place at start-up, reversals, and under unexpected high overloads. When it is imperative to have no contact at any time between moving elements, oil lifts are employed. (See D. D. Fuller, "Theory and Practice of Lubrication for Engineers," John Wiley & Sons, Inc., New York, 1956.)

The effective coefficient of friction varies with a convenient parameter $ZN/P$, shown in Fig. 8-60. Good design calls for the $ZN/P$ value to be in the area of full fluid film lubrication; a range of 30 to 300 is recommended.

### Table 8-23. Current Practice in Mean Bearing Pressures

| Type of bearing | Permissible press., psi, of projected area | Type of bearing | Permissible press., psi, of projected area |
|---|---|---|---|
| Diesel engines, main bearings | 800–1,500 | Automotive gasoline engines, main bearings | 500– 600 |
| Crankpin | 1,000–2,000 | Crankpin | 1,500–2,000 |
| Wrist pin | 1,800–2,000 | Air compressors, main bearings | 120– 240 |
| Electric-motor bearings | 100– 200 | Crankpin | 240– 400 |
| Marine diesel engines, main bearings | 400– 600 | Crosshead pin | 400– 800 |
| Crankpin | 1,000–1,400 | Aircraft-engine crankpin | 700–2,000 |
| Marine line-shaft bearings | 25– 35 | Centrifugal pumps | 80– 100 |
| Steam engines, main bearings | 150– 500 | Generators, low or medium speed | 90– 140 |
| Crankpin | 800–1,500 | Roll-neck bearings | 1,500–2,000 |
| Crosshead pin | 1,000–1,800 | Locomotive crankpins | 1,500–1,900 |
| Flywheel bearings | 200– 250 | Railway-car-axle bearings | 300– 350 |
| Marine steam engine, main bearings | 275– 500 | Miscellaneous ordinary bearings | 80– 150 |
| Crankpin | 400– 600 | Light line shaft | 15– 25 |
| Steam turbines and reduction gears | 100– 220 | Heavy line shaft | 100– 150 |

SOURCE: T. Baumeister (ed.), "Standard Handbook for Mechanical Engineers," 7th ed., p. 8-159, McGraw-Hill Book Company, New York, 1967.

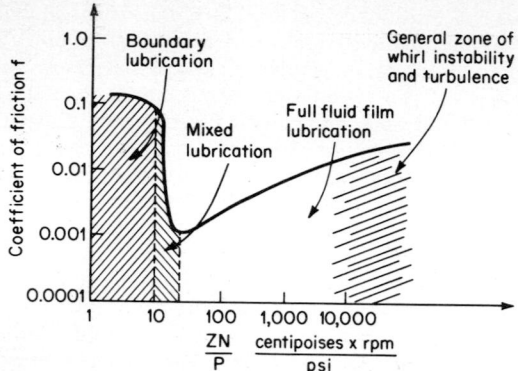

Fig. 8-60. Various zones of lubrication for a journal bearing. (*Source: T. Baumeister* (*ed.*), "*Standard Handbook for Mechanical Engineers,*" 7th ed., *McGraw-Hill Book Company, New York, 1967.*)

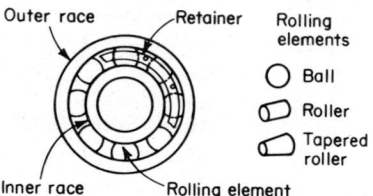

Fig. 8-61. Basic parts of rolling element bearings.

**Rolling Contact Bearings.**    In these bearings, there is solid contact between members in relative motion.    That motion is rolling rather than sliding; thus both the coefficient of friction and power loss are lower (Fig. 8-61).    Parts in rolling contact are highly loaded.    The design and application of such bearings are based on life (i.e., number of stress cycles) as well as on load.    Both the yield strength and the fatigue strength of the material are considered in design of the bearings.    Applications are always made by selection of a specific type and size, based on manufacturers' catalogs listed capacities, etc.

Common configurations of these bearings, revealing basically the type load to be resisted, are shown in Fig. 8-62.    The rolling element takes the form of a ball or sphere, a portion of a sphere, or straight or tapered rollers.

Table 8-24 is a brief guide to selection of ball-bearing types.

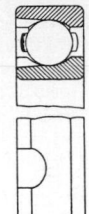

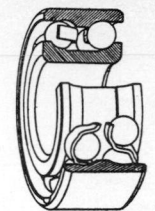

Single-row deep-
groove ball bearing
without filling slot

Single-row ball bearing
with filling slot

Single-row angular-
contact ball bearing

Double-row angular-
contact ball bearing

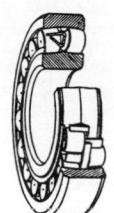
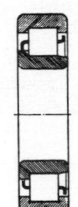
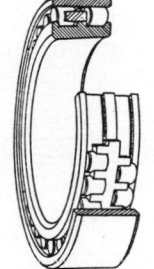
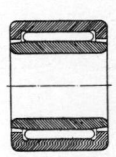

Spherical roller
bearing with separate
guide flange

Type NJ
Cylindrical-
roller bearings

Double-row cylindrical-
roller bearing

Needle bearing

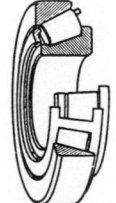
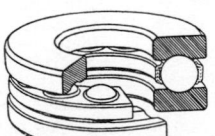
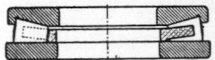

Tapered-roller bearing
with large
contact angle

One direction thrust
ball bearing

Tapered roller
thrust bearing

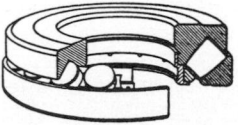
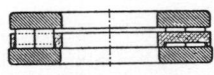

Spherical roller
thrust bearing

Cylindrical roller
thrust bearing

FIG. 8-62. Common configurations of rolling element bearings. (*Fafair Bearing Co., New Britain, Conn.*)

## Table 8-24. Guide to Selection of Ball-bearing Types

| Bearing type | Characteristics and abilities |
|---|---|
| Single-row radial *nonfilling slot type* | Designed to carry radial and combined loads and thrust in either direction. High speed. Preferred where specific characteristics of other type bearings are not required. |
| Single-row radial *filling slot type* | Higher radial capacity than single-row radial type, except at extremely low speeds; moderate thrust capacity in either direction; moderate high-speed ability. Long life. |
| Single-row radial | Heavy radial and one-direction thrust capacities. Seldom used for pure radial loads when mounted singly. Best for duplex mounting and high speeds and offer extreme rigidity when duplexed. |
| Single-row angular-contact | Well suited for duplex mountings, these bearings have heavy one-direction thrust and moderate radial capacity. Offer high axial and moderate radial rigidity when duplexed. |
| Self-aligning | Provide maximum capacity in applications where self-alignment is needed and where a larger outside diameter is permissible. |
| Double-row angular-contact | Heavy radial capacity and moderate thrust capacity in either direction with moderate speed ability. Good radial and axial rigidity. For reversing thrust loads, duplex pairs of single-row angular-contact bearings are preferred. |
| One and two shields | Same characteristics as single-row radial types. Offer effective grease retention and exclusion of coarse dirt. Types with two shields are prelubricated. (Some double-row bearings available with one shield.) |
| Felt-seal | Same features as single-row radial type, with effective grease retention and dirt exclusion. Prelubricated; moderate high-speed ability. |
| Aircraft type | For oscillatory and intermittent slow-speed service only, with high radial and thrust capacity. Made in inch dimensions. Corrosion resistant, sealed and prelubricated. Offer space and weight savings. |

SOURCE: Fafair Bearing Co., New Britain, Conn., Catalog 56, 1958.

## 8-18. Friction Clutches

### Table 8-25. Service Factors for Shock and Variable Load

| Type of Service | Factor |
|---|---|
| Driving machine: | |
| Electric motor, steady load | 1.0 |
| Fluctuating load | 1.5 |
| Gas engine, single cylinder | 1.5 |
| Multiple cylinder | 1.0 |
| Diesel engine, high speed | 1.5 |
| Large, slow speed | 2.0 |
| Driven machine: | |
| Generator, steady load | 1.0 |
| Fluctuating load | 1.5 |
| Blower | 1.0 |
| Compressor, depending on the number of cylinders | 2.0–2.5 |
| Pumps, centrifugal | 1.0 |
| Single-acting | 2.0 |
| Double-acting | 1.5 |
| Line shaft | 1.5 |
| Woodworking machinery | 1.75 |
| Hoists, elevators, cranes, shovels | 2.0 |
| Hammer mills, ball mills, crushers | 2.0 |
| Brick machinery | 3.0 |
| Rock crushers | 3.0 |

To compensate for high torques required to overcome inertia of driven members upon starting and for the reduced friction torque capacity while the friction surfaces are slipping prior to full engagement, the factors in Table 8–25 are multiplied by 1.5 to 2.0.

## 8-19. Mechanical Springs

Basically, mechanical springs are resilient structures whose properties depend upon the material used and the shape it is given. Such springs can take an unlimited variety of shapes; the materials most often used are found in Table 8-26. The shape most commonly found is the cylindrical, helical-coiled, round-wire spring, illustrated in Fig. 8-63 with 13½ coils. The treat-

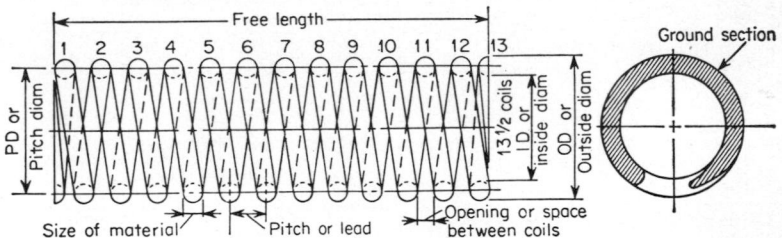

Fig. 8-63. Helical-coil spring.

ment given the ends depends on the spring's intended use; some common types are shown in Figs. 8-64 and 8-65.

Table 8-26. Physical Properties of

| | Material | Analysis | Tensile properties | | |
|---|---|---|---|---|---|
| | | | Ultimate strength, psi | Elastic limit, psi | Modulus of elasticity |
| **Flat cold-rolled spring steel** | Clock spring steel AS 100, SAE 1095 | C .90–1.05% <br> Mn .30–.50% | 180,000 to 340,000 | 150,000 to 310,000 | 30,000,000 |
| | Flat spring steel AS 101, SAE 1074 | C .70–.80% <br> Mn .50–.80% | 160,000 to 320,000 | 125,000 to 280,000 | 30,000,000 |
| | Flat spring steel AS 102, SAE 1060 | C .50–.65% <br> Mn .60–.90% <br> P and S .04% max | 160,000 to 280,000 | 120,000 to 180,000 | 30,000,000 |
| **Carbon steel wires** | High-carbon wire AS 8 | C .85–.95% <br> Mn .25–.60% | 200,000 to 250,000 | 160,000 to 210,000 | 30,000,000 |
| | Oil-tempered wire AS 10, ASTM A229-41 | C .60–.70% <br> Mn .60–.90% | 155,000 to 300,000 | 120,000 to 250,000 | 30,000,000 |
| | Music wire AS 5, ASTM A228-47 | C .70–1.00% <br> Mn .30–.60% | 250,000 to 500,000 | 150,000 to 350,000 | 30,000,000 |
| | Hard-drawn spring wire AS 20, ASTM A227-47 | C .60–.70% <br> Mn .90–1.20% | 150,000 to 300,000 | 100,000 to 200,000 | 30,000,000 |
| **Hot-rolled alloy steel** | Silico-manganese alloy steel AS 70, SAE 9260 | C .55–.65% <br> Mn .60–.90% <br> Si 1.80–2.20% | 200,000 to 250,000 | 180,000 to 230,000 | 30,000,000 |
| **Alloy and stainless spring materials** | Chrome-vanadium alloy steel AS 32, SAE 6150 | C .48–.53% Si .20–.35% <br> Mn .70–.90% Cr .80–1.10% <br> P .04 max V .15 min <br> S .04 max <br> subject to standard tolerances | 200,000 to 250,000 | 180,000 to 230,000 | 30,000,000 |
| | Chrome-silicon alloy steel AS 33, SAE 9254 | C .50–.60% <br> Mn .50–.80% <br> Si 1.20–1.60% <br> Cr .50–.80% | 250,000 to 325,000 | 220,000 to 300,000 | 30,000,000 |
| | 18-8 type stainless AS 35, SAE 30302 | Cr 17–20% <br> Ni 6–10% <br> C .08–.15% <br> Mn 2% max <br> Si .30–.75% | 160,000 to 330,000 | 60,000 to 260,000 | 28,000,000 |
| | Type 316 stainless SAE 30316 | Cr 16–18% <br> Ni 10–14% C .08% max <br> Mo 2–3% Si 1% max <br> Mn 2% max P .04% max <br> S .03% max | 170,000 to 250,000 | 130,000 to 200,000 | 28,000,000 |

## Commonly Used Spring Materials

| *Rockwell hardness | Torsional properties of wire | | | Process of manufacture, Chief uses, Special Properties |
|---|---|---|---|---|
| | Ultimate strength, psi | Elastic limit, psi | Modulus in torsion | |
| C40–52 | Not used | Not used | Not used | Cold-rolled and heat-treated before forming. Clock and motor springs, miscellaneous flat springs. |
| Annealed B70–85 Temp'd C38–50 | Not used | Not used | Not used | Cold-rolled or annealed or tempered. Miscellaneous flat springs. Most popular spring steel. |
| Annealed B70–85 Temp'd C38–50 | Not used | Not used | 11,500,000 | Use cold-rolled and annealed. Miscellaneous flat springs, static loads. |
| C44–48 | 160,000 to 200,000 | 110,000 to 150,000 | 11,500,000 | Cold-rolled or drawn. High-grade helical springs or wire forms. |
| C42–46 | 115,000 to 200,000 | 80,000 to 130,000 | 11,500,000 | Cold-drawn and heat-treated before coiling. General use. |
| | 150,000 to 300,000 | 90,000 to 180,000 | 11,500,000 to 12,000,000 depending on size | Patented and cold-drawn. Miscellaneous small springs of various types—high quality. |
| | 120,000 to 220,000 | 75,000 to 130,000 | 11,500,000 | Patented and cold-drawn. Same uses as music wire but lower-quality wire. |
| C42–52 | 140,000 to 175,000 | 100,000 to 130,000 | 11,500,000 | Hot- or cold-rolled or drawn. Better heat resistance than chrome-vanadium. |
| C42–48 | 140,000 to 175,000 | 100,000 to 130,000 | 11,500,000 | Cold-rolled or drawn. Special applications. |
| C47–51 | 160,000 to 200,000 | 130,000 to 160,000 | 11,500,000 | Hot- or cold-rolled or drawn. Used at high stresses. Resists heat well to 450°F. |
| C35–45 | 120,000 to 240,000 | 45,000 to 140,000 | 10,000,000 | Cold-rolled or drawn. Best corrosion resistance. Fair temperature resistance. |
| C35–45 | 120,000 to 220,000 | 80,000 to 130,000 | 11,000,000 | Cold-rolled or drawn. Heat-treated after forming. Resists corrosion when polished. Good temperature resistance. |

Table 8-26. Physical Properties of Commonly

Nonferrous spring materials

| Material | Analysis | | Tensile properties | | |
|---|---|---|---|---|---|
| | | | Ultimate strength, psi | Elastic limit, psi | Modulus of elasticity |
| Spring brass AS 55, AS 155 | Cu 64–72%<br>Zn remainder | | 100,000 to 130,000 | 40,000 to 60,000 | 15,000,000 |
| Nickel-silver | Cu 56%<br>Zn 25%<br>Ni 18% | | 135,000 to 150,000 | 80,000 to 110,000 | 16,000,000 |
| Phosphor-bronze AS 60, AS 160 | Cu 91–93%<br>Sn 7–9%<br>or<br>Cu 94–96%<br>Sn 4–6% | | 100,000 to 150,000 | 60,000 to 110,000 | 15,000,000 |
| Silicon-bronze AS 46, AS 146 (made under various trade names) | Si 2–3%<br>Small amounts of Sn or Mn, balance copper | | 100,000 to 150,000 | 60,000 to 110,000 | 15,000,000 |
| Monel AS 40, AS 140 | Ni (+Co) 63.0–70.0<br>Cu remainder  Mg 2.00 max<br>Fe 2.50 max  Si .50 max<br>C .30 max  S .024 max | | 120,000 to 165,000 | 85,000 to 125,000 | 26,000,000 |
| K Monel AS 40, AS 140 | Ni (+Co) 63.0–70.0<br>Cu remainder  Mg 1.50 max<br>Fe 2.00 max  Si 1.00 max<br>Al 2.0–4.0  S .01 max<br>C .25 max  Ti .25–1.00 | | 120,000 to 180,000 | 85,000 to 140,000 | 26,000,000 |
| Inconel AS 40, AS 140 | Ni (+Co) 72.00 min<br>Cu .50 max  Mg 1.00 max<br>Fe 6.0–10.0  Si .50 max<br>Cr 14.0–17.0  S .015 max<br>C .15 max | | 140,000 to 185,000 | 110,000 to 140,000 | 31,000,000 |
| Inconel X AS 40, AS 140 | Ni (+Co) 70.00 min<br>Cu .50 max  Mg 1.00 max<br>Fe 5.0–9.0  Si .50 max<br>Cr 14.0–17.0  S .01 max<br>Al .40–1.0  Ti 2.00–2.50<br>C .08 max<br>Cb (+Ti) .70–1.20 | | 130,000 to 220,000 | 90,000 to 150,000 | 31,000,000 |
| Duranickel AS 40, AS 140 | Ni (+Co) 93.00 min<br>Cu .25 max  Mg .50 max<br>Fe .60 max  Si 1.00 max<br>Al 4.00–4.75  S .01 max<br>C .30 max  Ti .25–1.00 | | 125,000 to 205,000 | 80,000 to 140,000 | 30,000,000 |
| Beryllium copper AS 45, AS 145 | Cu 98%<br>Be 2% | | 160,000 to 200,000 | 100,000 to 150,000 | 16,000,000 to 18,500,000 subject to heat-treatment |

SOURCE: Handbook of Mechanical Spring Design, Associated Spring Corp., 1955.

## Used Spring Materials (Continued)

| *Rockwell hardness | Torsional properties of wire | | | Process of manufacture, Chief uses, Special Properties |
|---|---|---|---|---|
| | Ultimate strength, psi | Elastic limit, psi | Modulus in torsion | |
| B90 | 45,000 to 90,000 | 30,000 to 60,000 | 5,500,000 | Cold-rolled or drawn. For electrical conductivity at low stresses. For corrosion resistance. |
| B95–100 | 85,000 to 100,000 | 60,000 to 70,000 | 5,500,000 | Cold-rolled or drawn. Better quality than brass. Also used for its color. Corrosion-resistant. |
| B90–100 | 80,000 to 105,000 | 50,000 to 85,000 | 6,250,000 | Cold-rolled or drawn. Used for corrosion resistance and electrical conductivity. |
| B90–100 | 80,000 to 105,000 | 50,000 to 85,000 | 6,250,000 | Cold-rolled or drawn. Used as substitute for phosphor-bronze where lower cost is necessary. |
| C23–32 | 85,000 to 110,500 | 50,000 to 70,000 | 9,500,000 | Cold-rolled or drawn. Resists corrosion. Moderate stresses to 400°F. |
| C23–35 | 85,000 to 130,000 | 50,000 to 75,000 | 9,500,000 | Same as Monel except higher operational stresses can be employed to 450°F. Precipitation-hardened by thermal treatment. |
| C25–37 | 95,000 to 130,000 | 55,000 to 80,000 | 11,000,000 | Cold-rolled or drawn. Resists corrosion. High stresses to 650°F. |
| C24–46 | 90,000 to 155,000 | 50,000 to 90,000 | 12,000,000 | Resists corrosion and oxidation. Can be used to 1000°F for prolonged periods of service; up to 1200°F for short periods of intermittent temperature exposure. |
| C25–43 | 85,000 to 145,000 | 50,000 to 85,000 | 11,000,000 | Cold-rolled or drawn. Precipitation-hardened by heat-treatment. Resists corrosion. High stresses to 600°F. |
| C35–42 | 100,000 to 130,000 | 65,000 to 95,000 | 6,000,000 to 7,000,000 subject to heat-treatment | Cold-rolled or drawn. Corrosion resistance like copper. High physicals for electrical work. Low hysteresis. |

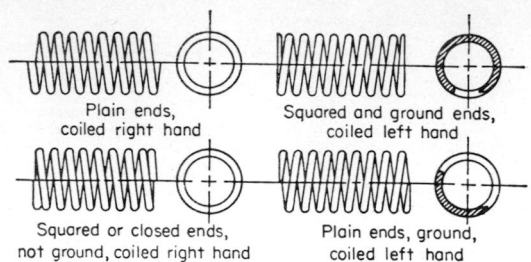

FIG. 8-64. Compression-spring ends.

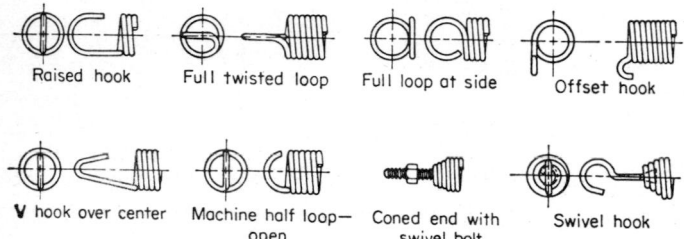

FIG. 8-65. Extension-spring ends.

A coiled spring made of round wire will be subjected to a maximum stress given by

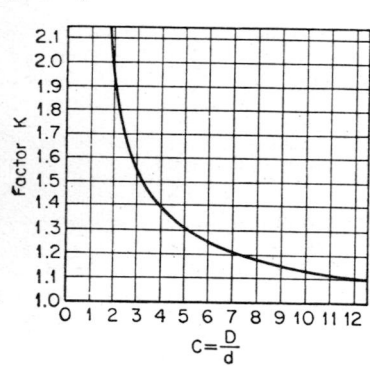

FIG. 8-66. Wahl correction factor.

$$\sigma_{max} = \frac{8PD}{\pi d^3}\left(\frac{4C-1}{4C-4} + \frac{0.615}{C}\right) \tag{8-65}$$

where $P$ = load on spring, lb
$D$ = mean diameter of coil, in. (outside diameter minus the wire diameter)
$d$ = diameter of wire, in.
$\sigma_{max}$ = torsional stress, psi
$C = D/d$

The expression in parentheses in Eq. (8-65) is the Wahl correction factor $K$, from Fig. 8-66. It accounts for the added stresses in the coils due to curvature and shear.

**Leaf Springs.** These are beams configured as cantilevers or simply supported. They have variable stresses throughout their length and are, therefore, inefficient. Simple leaf springs are shown in Fig. 8-67. The arrangement in Fig. 8-68 overcomes this objection to a large degree.

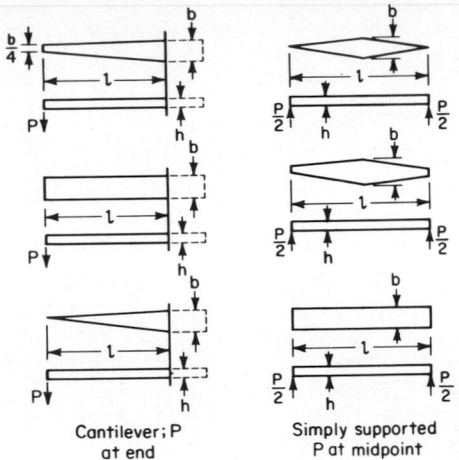

FIG. 8-67. Simple leaf springs.

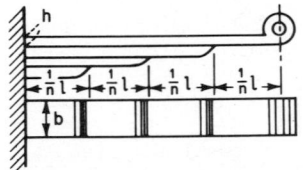

FIG. 8-68. Laminated leaf spring.

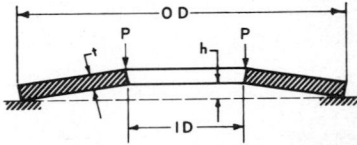

FIG. 8-69. Sectional view of Belleville spring.

**Belleville Springs.** Often called dished washers, Belleville springs occupy a very small space. They are stressed in a very complex manner, and provide unusual spring rate curves (Fig. 8-69). These springs are nonlinear, but for some proportions, they behave with approximately linear characteristics in a limited range (Fig. 8-70). Likewise, for some proportions they can be used through a spectrum of spring rates, from positive to flat and then through a negative region. The snap-through action, shown at point $A$ in Fig. 8-70, can be useful in particular applications requiring reversal of spring rates. These springs are applied for very high spring rates and to provide special spring rates. They are particularly sensitive to slight variations in their geometry.

**Springs in Combination.** Springs used in combination are configured either in series or in parallel. Pertinent characteristics of the combinations are shown in Fig. 8-71.

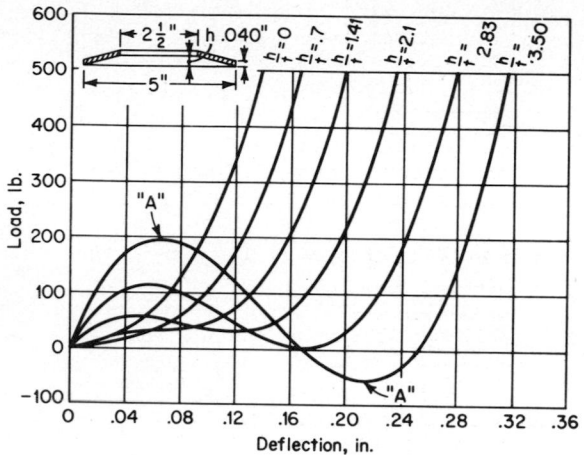

Fig. 8-70. Load-deflection curves for a family of Belleville springs.

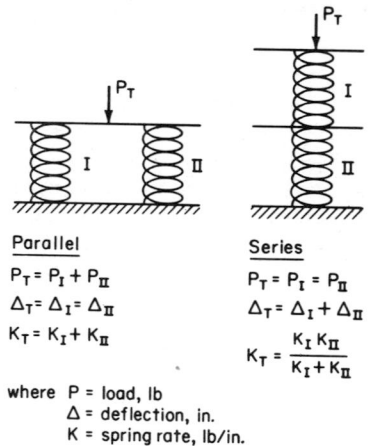

Parallel

$P_T = P_I + P_{II}$

$\Delta_T = \Delta_I = \Delta_{II}$

$K_T = K_I + K_{II}$

Series

$P_T = P_I = P_{II}$

$\Delta_T = \Delta_I + \Delta_{II}$

$K_T = \dfrac{K_I K_{II}}{K_I + K_{II}}$

where  P = load, lb
       Δ = deflection, in.
       K = spring rate, lb/in.

Fig. 8-71. Springs in parallel and in series.

## 8-20. Flywheels

A flywheel is a rotating energy-storing device used to impart more uniform rotation to a machine with fundamentally intermittent operating characteristics (reciprocating machines, punch presses, etc.). Energy is stored in the flywheel by increasing its angular momentum during the time when no external work is done, and then released by a decrease in its angular momentum when external work is done. The degree to which a flywheel succeeds in keeping speed reasonably constant throughout a cycle of operation can be measured by the coefficient of fluctuation $C_f$, representative values for which in several classes of service are found in Table 8-27.

### Table 8-27. Coefficient of Speed Fluctuation

| Type of Equipment | $C_f$ |
|---|---|
| Crushing machinery | 0.200 |
| Electrical machinery | 0.003 |
| Direct-driven | 0.002 |
| Engines with belt transmission | 0.030 |
| Flour-milling machinery | 0.020 |
| Gear-wheel transmission | 0.020 |
| Hammering machinery | 0.200 |
| Machine tools | 0.030 |
| Papermaking machinery | 0.025 |
| Pumping machinery | 0.030–0.050 |
| Shearing machinery | 0.030–0.050 |
| Spinning machinery | 0.010–0.020 |
| Textile machinery | 0.025 |

SOURCE: Kent, "Mechanical Engineer's Handbook," 12th ed., John Wiley & Sons, Inc., New York.

Let $\omega_{av}$ = average angular velocity during cycle, radians/sec

$\omega_{min}$ = minimum angular velocity during cycle, radians/sec

$\omega_{max}$ = maximum angular velocity during cycle, radians/sec

$C_f$ = coefficient of speed fluctuation

$I$ = mass moment of inertia of flywheel, lb-ft-sec²

$\Delta KE$ = energy released from flywheel during working portion of cycle (i.e., while speed changes from $\omega_{max}$ to $\omega_{min}$), ft-lb

Then

$$\omega_{av} = \frac{\omega_{max} + \omega_{min}}{2} \tag{8-66}$$

$$C_f = \frac{\omega_{max} - \omega_{min}}{\omega_{av}} \tag{8-67}$$

$$\Delta KE = \tfrac{1}{2} I (\omega_{max}{}^2 - \omega_{min}{}^2) \tag{8-68}$$

## 8-21. Drill and Tap Sizes

### Table 8-28. Diameters and Areas of Small Drills
Number, Letter, Metric, and Fractional Drills in Order of Size

| No. | Ltr | Mm | In. | Diam, in. | Area, in.² | No. | Ltr | Mm | In. | Diam, in. | Area, in.² |
|---|---|---|---|---|---|---|---|---|---|---|---|
| | | 0.100 | ... | 0.003900 | 0.0000119 | 54 | ... | ..... | ... | 0.055000 | 0.0023760 |
| | | 0.150 | ... | .005900 | .0000273 | | | 1.400 | ... | .055100 | .0023860 |
| | | 0.200 | ... | .007800 | .0000477 | | | 1.450 | ... | .057000 | .0025440 |
| | | 0.250 | ... | .009800 | .0000753 | | | 1.500 | ... | .059000 | .0027390 |
| | | 0.300 | ... | .011800 | .0001091 | 53 | ... | ..... | ... | .059500 | .0027810 |
| 80 | ... | | | .013500 | .0001429 | | | 1.550 | ... | .061000 | .0029210 |
| | | 0.350 | ... | .013700 | .0001468 | | | | 1/16 | .062500 | .0030680 |
| 79 | ... | ..... | ... | .014500 | .0001650 | | | 1.600 | ... | .062990 | .0031160 |
| | | | 1/64 | .015620 | .0001920 | 52 | ... | | | .063500 | .0031670 |
| | | 0.400 | ... | .015740 | .0001950 | | | 1.650 | ... | .064900 | .0033060 |
| 78 | ... | | | .016000 | .0002010 | | | 1.700 | ... | .066920 | .0035180 |
| | | 0.450 | ... | .017700 | .0002450 | 51 | ... | | | .067000 | .0035260 |
| 77 | ... | | | .018000 | .0002540 | | | 1.750 | ... | .068800 | .0037140 |
| | | 0.500 | ... | .019680 | .0003040 | 50 | ... | | | .070000 | .0038480 |
| 76 | ... | ..... | ... | .020000 | .0003140 | | | 1.800 | ... | .070860 | .0039440 |
| 75 | ... | | | .021000 | .0003460 | | | 1.850 | ... | .072800 | .0041620 |
| | | 0.550 | ... | .021600 | .0003650 | 49 | ... | | | .073000 | .0041850 |
| 74 | ... | | | .022500 | .0003980 | | | 1.900 | ... | .074800 | .0043940 |
| | | 0.600 | ... | .023620 | .0004380 | 48 | ... | | | .076000 | .0045360 |
| 73 | ... | ..... | ... | .024000 | .0004520 | | | 1.950 | ... | .076700 | .0046180 |
| 72 | ... | | | .025000 | .0004910 | | | | 5/64 | .078120 | .0047940 |
| | | 0.650 | ... | .025500 | .0005100 | 47 | ... | | | .078500 | .0048400 |
| 71 | ... | | | .026000 | .0005310 | | | 2.000 | ... | .078740 | .0048690 |
| | | 0.700 | ... | .027560 | .0005970 | | | 2.050 | ... | .080700 | .0051120 |
| 70 | ... | ..... | ... | .028000 | .0006160 | 46 | ... | ..... | ... | .081000 | .0051530 |
| 69 | ... | | | .029250 | .0006720 | 45 | ... | | | .082000 | .0052810 |
| | | 0.750 | ... | .029500 | .0006830 | | | 2.100 | ... | .082670 | .0053690 |
| 68 | ... | ..... | ... | .031000 | .0007550 | | | 2.150 | ... | .084600 | .0056150 |
| | | | 1/32 | .031250 | .0007670 | 44 | ... | | | .086000 | .0058090 |
| | | 0.800 | ... | .031490 | .0007790 | | | 2.200 | ... | .086610 | .0058920 |
| 67 | ... | ..... | ... | .032000 | .0008040 | | | 2.250 | ... | .088500 | .0061490 |
| 66 | ... | | | .033000 | .0008550 | 43 | ... | | | .089000 | .0062210 |
| | | 0.850 | ... | .033400 | .0008710 | | | 2.300 | ... | .090550 | .0064400 |
| 65 | ... | | | .035000 | .0009620 | | | 2.35 | ... | .092500 | .0067150 |
| | | 0.900 | ... | .035430 | .0009860 | 42 | ... | ..... | ... | .093500 | .0068600 |
| 64 | ... | ..... | ... | .036000 | .0010180 | | | | 3/32 | .093750 | .0069030 |
| 63 | ... | | | .037000 | .0010750 | | | 2.400 | ... | .094480 | .0070120 |
| | | 0.950 | ... | .037400 | .0010910 | 41 | ... | | | .096000 | .0072380 |
| 62 | ... | | | .038000 | .0011340 | | | 2.45 | ... | .096400 | .0072960 |
| 61 | ... | ..... | ... | .039000 | .0011950 | 40 | ... | ..... | ... | .098000 | .0075430 |
| | | 1.000 | ... | .039370 | .0012170 | | | 2.50 | ... | .098420 | .0076090 |
| 60 | ... | | | .040000 | .0012570 | 39 | ... | | | .099500 | .0077760 |
| 59 | ... | ..... | ... | .041000 | .0013200 | 38 | ... | ..... | ... | .101500 | .0080910 |
| | | 0.105 | ... | .041300 | .0013510 | | | 2.60 | ... | .102360 | .0082290 |
| 58 | ... | | | .042000 | .0013850 | 37 | ... | | | .104000 | .0084900 |
| 57 | ... | ..... | ... | .043000 | .0014520 | | | 2.70 | ... | .106300 | .0088750 |
| | | 1.100 | ... | .043300 | .0014730 | 36 | ... | ..... | ... | .106500 | .0089080 |
| | | 1.150 | ... | .045200 | .0016100 | | | 2.75 | ... | .108200 | .0091890 |
| 56 | ... | ..... | ... | .046500 | .0016980 | | | | 7/64 | .109370 | .0093960 |
| | | | 3/64 | .046870 | .0017260 | 35 | ... | ..... | ... | .110000 | .0095030 |
| | | 1.200 | ... | .047240 | .0017530 | | | 2.80 | ... | .110240 | .0095440 |
| | | 1.250 | ... | .049200 | .0019000 | 34 | ... | ..... | ... | .111000 | .0096770 |
| | | 1.300 | ... | .051181 | .0020570 | 33 | ... | ..... | ... | .113000 | .0100290 |
| 55 | ... | ..... | ... | .052000 | .0021240 | | | 2.90 | ... | .114170 | .0102380 |
| | | 1.350 | ... | .053100 | .0022060 | 32 | ... | ..... | ... | .116000 | .0105680 |

## Table 8-28. Diameters and Areas of Small Drills (Continued)

| No. | Ltr | Mm | In. | Diam, in. | Area, in.² | No. | Ltr | Mm | In. | Diam, in. | Area, in.² |
|---|---|---|---|---|---|---|---|---|---|---|---|
| 31 | ... | 3.00 | | 0.118110 | 0.0109590 | 7 | ... | | | 0.201000 | 0.0317310 |
| | | | | .120000 | .0113100 | | | | 13/64 | .203120 | .0324030 |
| | | 3.10 | | .122050 | .0116990 | 6 | ... | | | .204000 | .0326850 |
| | | | 1/8 | .125000 | .0122720 | | | 5.20 | | .204730 | .0329180 |
| | | 3.20 | | .125980 | .0124660 | 5 | ... | | | .205500 | .0331681 |
| 30 | ... | 3.25 | | .127900 | .0128020 | | | 5.25 | | .206600 | .0335200 |
| | | | | .128500 | .0129690 | | | 5.30 | | .208600 | .0341960 |
| | | 3.30 | | .129920 | .0132570 | 4 | ... | | | .209000 | .0343070 |
| | | 3.40 | | .133860 | .0140730 | | | 5.400 | | .212600 | .0354990 |
| 29 | ... | | | .136000 | .0145270 | 3 | ... | | | .213000 | .0356330 |
| 28 | ... | 3.50 | | .137800 | .0149130 | | | 5.500 | | .216540 | .0368250 |
| | | | | .140500 | .0155040 | | | | 7/32 | .218750 | .0375830 |
| | | | 9/64 | .140620 | .0155310 | | | 5.600 | | .220470 | .0381770 |
| | | 3.60 | | .141730 | .0157770 | 2 | ... | | | .221000 | .0383600 |
| 27 | ... | | | .144000 | .0162860 | | | 5.700 | | .224410 | .0395520 |
| 26 | ... | 3.70 | | .145670 | .0166660 | | | 5.750 | | .226300 | .0402430 |
| | | | | .147000 | .0169720 | 1 | ... | | | .228000 | .0408280 |
| | | 3.75 | | .147600 | .0171060 | | | 5.800 | | .228350 | .0409520 |
| 25 | ... | | | .149500 | .0175540 | | | 5.900 | | .232280 | .0423770 |
| | | 3.80 | | .149610 | .0175790 | | A | | | .234000 | .0430050 |
| 24 | ... | | | .152000 | .0181460 | | | | 15/64 | .234370 | .0431410 |
| | | 3.90 | | .153540 | .0185160 | | | 6.000 | | .236220 | .0438250 |
| 23 | ... | | | .154000 | .0186270 | | B | | | .238000 | .0444880 |
| | | | 5/32 | .156250 | .0191750 | | | 6.100 | | .240150 | .0452990 |
| 22 | ... | | | .157000 | .0193590 | | C | | | .242000 | .0459960 |
| 21 | ... | 4.00 | | .157480 | .0194780 | | | 6.200 | | .244100 | .0467970 |
| | | | | .159000 | .0198560 | | D | | | .246000 | .0475290 |
| 20 | ... | | | .161000 | .0203580 | | | 6.250 | | .246060 | .0475480 |
| | | 4.10 | | .161420 | .0204640 | | | 6.300 | | .248030 | .0483170 |
| | | 4.2 | | .165360 | .0214740 | | E | | 1/4 | .250000 | .0490870 |
| 19 | ... | | | .166000 | .0216420 | | | 6.400 | | .251970 | .0498630 |
| | | 4.25 | | .167300 | .0220560 | | | 6.500 | | .255910 | .0514340 |
| | | 4.30 | | .169290 | .0225050 | | F | | | .257000 | .0518750 |
| 18 | ... | | | .169500 | .0225650 | | | 6.600 | | .259800 | .0530280 |
| | | | 11/64 | .171875 | .0232020 | | G | | | .261000 | .0535020 |
| 17 | ... | | | .173000 | .0235060 | | | 6.700 | | .263700 | .0546480 |
| | | 4.40 | | .173230 | .0235680 | | | | 17/64 | .265600 | .0554120 |
| 16 | ... | | | .177000 | .0246060 | | | 6.750 | | .265700 | .0554140 |
| | | 4.50 | | .177170 | .0246520 | | H | | | .266000 | .0555720 |
| 15 | ... | | | .180000 | .0254470 | | | 6.800 | | .267720 | .0562910 |
| 14 | ... | 4.60 | | .181100 | .0257600 | | | 6.900 | | .271650 | .0579590 |
| | | | | .182000 | .0260160 | | I | | | .272000 | .0581070 |
| 13 | ... | | | .185000 | .0268800 | | | 7.000 | | .275500 | .0596510 |
| | | 4.70 | | .185040 | .0268920 | | J | | | .277000 | .0602630 |
| | | 4.75 | | .187000 | .0274570 | | | 7.100 | | .279500 | .0613670 |
| | | | 3/16 | .187500 | .0276120 | | K | | | .281000 | .0620160 |
| | | 4.80 | | .188980 | .0280480 | | | | 9/32 | .281250 | .0621260 |
| 12 | ... | | | .189000 | .0280550 | | | 7.200 | | .283470 | .0631080 |
| 11 | ... | | | .191000 | .0286520 | | | 7.250 | | .285400 | .0639700 |
| | | 4.90 | | .192910 | .0292290 | | | 7.300 | | .287400 | .0648740 |
| 10 | ... | | | .193500 | .0294070 | | L | | | .290000 | .0660520 |
| 9 | ... | | | .196000 | .0301720 | | | 7.400 | | .291330 | .0666630 |
| | | 5.00 | | .196850 | .0304340 | | M | | | .295000 | .0683490 |
| 8 | ... | | | .199000 | .0311030 | | | 7.500 | | .295200 | .0684770 |
| | | 5.10 | | .200790 | .0316640 | | | | 19/64 | .296875 | .0692180 |

## Table 8-28. Diameters and Areas of Small Drills (Continued)

| No. | Ltr | Mm | In. | Diam, in. | Area, in.² | No. | Ltr | Mm | In. | Diam, in. | Area, in.² |
|---|---|---|---|---|---|---|---|---|---|---|---|
|  |  | 7.600 | .... | 0.299220 | 0.0703150 |  |  |  | 1/2 | 0.500000 | 0.1963500 |
|  | N | ..... | .... | .302000 | .0716310 |  |  | 13.00 | .... | 0.511800 | .2050700 |
|  |  | 7.700 | .... | .303140 | .0721780 |  |  |  | 33/64 | 0.515600 | .2087500 |
|  |  | 7.750 | .... | .305100 | .0730570 |  |  |  | 17/32 | 0.531200 | .2216600 |
|  |  | 7.800 | .... | .307090 | .0740650 |  |  | 13.50 | .... | 0.531400 | .2217900 |
|  |  | 7.900 | .... | .311020 | .0759760 |  |  |  | 35/64 | 0.546800 | .2341200 |
|  |  |  | 5/16 | .312500 | .0766990 |  |  | 14.00 | .... | 0.551100 | .2384400 |
|  |  | 8.000 | .... | .314960 | .0779120 |  |  |  | 9/16 | 0.562500 | .2485000 |
|  | O | ..... | .... | .316000 | .0784270 |  |  | 14.50 | .... | 0.570800 | .2560400 |
|  |  |  |  | .318890 | .0798720 |  |  |  | 37/64 | 0.578100 | .2624800 |
|  |  | 8.100 | .... | .322830 | .0818560 |  |  | 15.00 | .... | 0.590500 | .2733900 |
|  |  | 8.200 | .... | .323000 | .0819400 |  |  |  | 19/32 | 0.593700 | .2768800 |
|  | P | 8.250 | .... | .324800 | .0827810 |  |  |  | 39/64 | 0.609300 | .2916100 |
|  |  | 8.300 | .... | .326800 | .0838650 |  |  | 15.50 | .... | 0.610200 | .2922400 |
|  |  |  | 21/64 | .328120 | .0845580 |  |  |  | 5/8 | 0.625000 | .3068000 |
|  |  | 8.400 | .... | .330700 | .0858980 |  |  | 16.00 | .... | 0.629900 | .3117200 |
|  | Q | ..... | .... | .332000 | .0865700 |  |  |  | 41/64 | 0.640600 | .3223200 |
|  |  | 8.500 | .... | .334650 | .0879550 |  |  | 16.50 | .... | 0.649600 | .3308100 |
|  |  | 8.600 | .... | .338580 | .0900370 |  |  |  | 21/32 | 0.656200 | .3382400 |
|  | R | ..... | .... | .339000 | .0902590 |  |  | 17.00 | .... | 0.669200 | .3514200 |
|  |  | 8.700 | 11/32 | .342500 | .0921430 |  |  |  | 43/64 | 0.671800 | .3645400 |
|  |  |  |  | .343700 | .0928060 |  |  |  | 11/16 | 0.687500 | .3712200 |
|  |  | 8.750 | .... | .344400 | .0929120 |  |  | 17.50 | .... | 0.688900 | .3726700 |
|  |  | 8.800 | .... | .346400 | .0942740 |  |  |  | 45/64 | 0.703100 | .3882800 |
|  | S | .... | .... | .348000 | .0951150 |  |  | 18.00 | .... | 0.708600 | .3943400 |
|  |  | 8.900 | .... | .350400 | .0964280 |  |  |  | 23/32 | 0.718700 | .4057400 |
|  | T | 9.000 | .... | .354300 | .0986070 |  |  | 18.50 | .... | 0.723300 | .4165700 |
|  |  |  |  | .358000 | .1006600 |  |  |  | 47/64 | 0.734300 | .4235600 |
|  |  | 9.100 | .... | .358300 | .1008110 |  |  | 19.00 | .... | 0.748900 | .4393500 |
|  |  |  | 23/64 | .359300 | .1014340 |  |  |  | 3/4 | 0.750000 | .4417900 |
|  |  | 9.200 | .... | .362200 | .1030390 |  |  |  | 49/64 | 0.765600 | .4587500 |
|  |  | 9.250 | .... | .364100 | .1039860 |  |  | 19.50 | .... | 0.767700 | .4623300 |
|  | U | 9.300 | .... | .366100 | .1052910 |  |  |  | 25/32 | 0.781200 | .4793700 |
|  |  |  |  | .368000 | .1063620 |  |  | 20.00 | .... | 0.787400 | .4868600 |
|  |  | 9.400 | .... | .370100 | .1075670 |  |  |  | 51/64 | 0.796800 | .4977200 |
|  |  | 9.500 | .... | .374020 | .1098680 |  |  | 20.50 | .... | 0.807000 | .5115800 |
|  | V |  | 3/8 | .375000 | .1104470 |  |  |  | 13/16 | 0.812500 | .5184900 |
|  |  |  |  | .377000 | .1116230 |  |  | 21.00 | .... | 0.826700 | .5368500 |
|  |  | 9.600 | .... | .377950 | .1121930 |  |  |  | 53/64 | 0.828100 | .5386100 |
|  |  | 9.700 | .... | .381800 | .1145430 |  |  |  | 27/32 | 0.843700 | .5591400 |
|  |  | 9.750 | .... | .383800 | .1156890 |  |  | 21.50 | .... | 0.846400 | .5626600 |
|  | W | 9.800 | .... | .385800 | .1169170 |  |  |  | 55/64 | 0.859300 | .5799200 |
|  |  |  |  | .385900 | .1170210 |  |  | 22.00 | .... | 0.866100 | .5892000 |
|  |  | 9.900 | .... | .389300 | .1193150 |  |  |  | 7/8 | 0.875000 | .6013200 |
|  |  |  | 25/64 | .390620 | .1198440 |  |  | 22.50 | .... | 0.885800 | .6162900 |
|  |  | 10.00 | .... | .393700 | .1217380 |  |  |  | 57/64 | 0.890600 | .6629700 |
|  | X | ..... | .... | .397000 | .1237860 |  |  | 23.00 | .... | 0.905500 | .6400700 |
|  | Y | ..... | .... | .404000 | .1281900 |  |  |  | 29/32 | 0.906200 | .6450400 |
|  |  |  | 13/32 | .406200 | .1295220 |  |  |  | 59/64 | 0.921800 | .6674600 |
|  | Z | ..... | .... | .413000 | .1339650 |  |  | 23.50 | .... | 0.925100 | .6722200 |
|  |  | 10.50 | .... | .413300 | .1342000 |  |  |  | 15/16 | 0.937500 | .6902900 |
|  |  |  | 27/64 | .421800 | .1397200 |  |  | 24.00 | .... | 0.944800 | .7011900 |
|  |  | 11.00 | .... | .433000 | .1471800 |  |  |  | 61/64 | 0.953100 | .7150200 |
|  |  |  | 7/16 | .437500 | .1503300 |  |  | 24.50 | .... | 0.964500 | .7297900 |
|  |  | 11.50 | .... | .452700 | .1607600 |  |  |  | 31/32 | 0.968700 | .7370800 |
|  |  |  | 29/64 | .453100 | .1612400 |  |  | 25.00 | .... | 0.984200 | .7614400 |
|  |  |  | 15/32 | .468700 | .1725700 |  |  |  | 63/64 | 0.998430 | .7634500 |
|  |  | 12.00 | .... | .472400 | .1749000 |  |  |  | 1.0 | 1.000000 | .7854000 |
|  |  |  | 31/64 | .484300 | .1842500 |  |  | 25.50 | .... | 1.003000 | .7915200 |
|  |  | 12.50 | .... | .492100 | .1901400 |  |  |  |  |  |  |

SOURCE: Colvin and Stanley, "American Machinists' Handbook," 8th ed., pp. 137–139. McGraw-Hill Book Company, Inc., New York, 1945.

Table 8-29. Tap-drill Sizes for American Standard Screw Threads*

| Size, no. or in. | Coarse-thread series Thds/in. | Drill size | Fine-thread series Thds/in. | Drill size | Size, no. or in. | Coarse-thread series Thds/in. | D-ill size | Fine-thread series Thds/in. | Drill size |
|---|---|---|---|---|---|---|---|---|---|
| 0 | ... | ...... | 80 | $3\!\!/_{64}$ | $3\!\!/_4$ | 10 | $21\!\!/_{32}$ | 16 | $11\!\!/_{16}$ |
| 1 | 64 | No. 53 | 72 | No. 53 | $7\!\!/_8$ | 9 | $49\!\!/_{64}$ | 14 | $13\!\!/_{16}$ |
| 2 | 56 | No. 50 | 64 | No. 50 | 1 | 8 | $7\!\!/_8$ | 14 | $15\!\!/_{16}$ |
| 3 | 48 | No. 47 | 56 | No. 45 | $1\!\!/_8$ | 7 | $63\!\!/_{64}$ | 12 | $1\,3\!\!/_{64}$ |
| 4 | 40 | No. 43 | 48 | No. 42 | $1\!\!/_4$ | 7 | $1\,7\!\!/_{64}$ | 12 | $1\,11\!\!/_{64}$ |
| 5 | 40 | No. 38 | 44 | No. 37 | $1\,3\!\!/_8$ | 6 | $1\,7\!\!/_{32}$ | 12 | $1\,19\!\!/_{64}$ |
| 6 | 32 | No. 36 | 40 | No. 33 | $1\,1\!\!/_2$ | 6 | $1\,23\!\!/_{64}$ | 12 | $1\,27\!\!/_{64}$ |
| 8 | 32 | No. 29 | 36 | No. 29 | $1\,3\!\!/_4$ | 5 | $1\,35\!\!/_{64}$ | | |
| 10 | 24 | No. 25 | 32 | No. 21 | 2 | $4\!\!/_2$ | $1\,25\!\!/_{32}$ | | |
| 12 | 24 | No. 16 | 28 | No. 14 | $2\!\!/_4$ | $4\!\!/_2$ | $2\!\!/_{32}$ | | |
| $\tfrac14$ | 20 | No. 7 | 28 | No. 3 | $2\!\!/_2$ | 4 | $2\!\!/_4$ | | |
| $\tfrac{5}{16}$ | 18 | F | 24 | I | $2\tfrac34$ | 4 | $2\!\!/_2$ | | |
| $\tfrac38$ | 16 | $\tfrac{5}{16}$ | 24 | Q | 3 | 4 | $2\tfrac34$ | | |
| $\tfrac{7}{16}$ | 14 | U | 20 | $29\!\!/_{64}$ | $3\!\!/_4$ | 4 | 3 | | |
| $\tfrac12$ | 13 | $27\!\!/_{64}$ | 20 | $29\!\!/_{64}$ | $3\!\!/_2$ | 4 | $3\!\!/_4$ | | |
| $\tfrac{9}{16}$ | 12 | $31\!\!/_{64}$ | 18 | $33\!\!/_{64}$ | $3\tfrac34$ | 4 | $3\!\!/_2$ | | |
| $\tfrac58$ | 11 | $17\!\!/_{32}$ | 18 | $37\!\!/_{64}$ | 4 | 4 | $3\tfrac34$ | | |

* The sizes listed are the commercial tap drills to produce approx 75 per cent full thread.
SOURCE: T. Baumeister (ed.), "Standard Handbook for Mechanical Engineers," 7th ed., p. 8-30, McGraw-Hill Book Company, New York, 1967.

## 8-22. Screw Threads for Pipes

American Standard Taper Pipe Thread, ANSI B2.1-1968 is shown in Fig. 8-72 and is made to the following specifications: The taper is 1 in. in 16 in. or 0.75 in./ft. The basic length of the effective external taper thread is determined by $L_2 = p(0.8D + 6.8)$, where $D$ is the basic outside diameter of the pipe (Table 8-30).

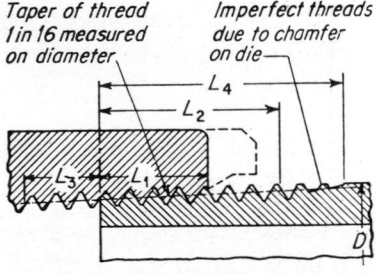

Taper of thread 1 in 16 measured on diameter

Imperfect threads due to chamfer on die

FIG. 8-72. Source: T. Baumeister (ed.), "Standard Handbook for Mechanical Engineers," 7th ed., p. 8-23, McGraw-Hill Book Company, New York, 1967.

## Table 8-30. ANSI Taper Pipe Thread

(ANSI B2.1-1968)

| Nominal pipe size | Pipe OD | Thds/in. | Thread pitch, $P$ | $L_1$* | $L_2$† | $L_3$‡ | $L_4$¶ |
|---|---|---|---|---|---|---|---|
| 1⁄16 | 0.3125 | 27 | 0.03704 | 0.160 | 0.2611 | 0.1111 | 0.3896 |
| 1⁄8 | 0.405 | 27 | .03704 | 0.180 | 0.2639 | .1111 | 0.3924 |
| 1⁄4 | 0.540 | 18 | .05556 | 0.200 | 0.4018 | .1667 | 0.5946 |
| 3⁄8 | 0.675 | 18 | .05556 | 0.240 | 0.4078 | .1667 | 0.6006 |
| 1⁄2 | 0.840 | 14 | .07143 | 0.320 | 0.5337 | .2143 | 0.7815 |
| 3⁄4 | 1.050 | 14 | .07143 | 0.339 | 0.5457 | .2143 | 0.7935 |
| 1 | 1.315 | 11½ | .08696 | 0.400 | 0.6828 | .2609 | 0.9845 |
| 1¼ | 1.660 | 11½ | .08696 | 0.420 | 0.7068 | .2609 | 1.0085 |
| 1½ | 1.900 | 11½ | .08696 | 0.420 | 0.7235 | .2609 | 1.0252 |
| 2 | 2.375 | 11½ | .08696 | 0.436 | 0.7565 | .2609 | 1.0582 |
| 2½ | 2.875 | 8 | .12500 | 0.682 | 1.1375 | .2500 | 1.5712 |
| 3 | 3.500 | 8 | .12500 | 0.766 | 1.2000 | .2500 | 1.6337 |
| 3½ | 4.000 | 8 | .12500 | 0.821 | 1.2500 | .2500 | 1.6837 |
| 4 | 4.500 | 8 | .12500 | 0.844 | 1.3000 | .2500 | 1.7337 |
| 5 | 5.563 | 8 | .12500 | 0.937 | 1.4063 | .2500 | 1.8400 |
| 6 | 6.625 | 8 | .12500 | 0.958 | 1.5125 | .2500 | 1.9462 |
| 8 | 8.625 | 8 | .12500 | 1.063 | 1.7125 | .2500 | 2.1462 |
| 10 | 10.750 | 8 | .12500 | 1.210 | 1.9250 | .2500 | 2.3587 |
| 12 | 12.750 | 8 | .12500 | 1.360 | 2.1250 | .2500 | 2.5587 |
| 14 OD | 14.000 | 8 | .12500 | 1.562 | 2.2500 | .2500 | 2.6837 |
| 16 OD | 16.000 | 8 | .12500 | 1.812 | 2.4500 | .2500 | 2.8837 |
| 18 OD | 18.000 | 8 | .12500 | 2.000 | 2.6500 | .2500 | 3.0837 |
| 20 OD | 20.000 | 8 | .12500 | 2.125 | 2.8500 | .2500 | 3.2837 |
| 24 OD | 24.000 | 8 | .12500 | 2.375 | 3.2500 | .2500 | 3.6837 |

* Hand-tight engagement length.
† Effective thread external length.
‡ Wrench make-up length for internal thread length.
¶ Over-all length external thread.

## Table 8-31. Dimensions of Standard Hose Couplings*

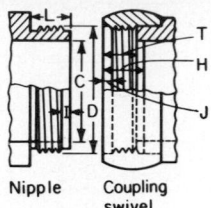

Nipple  Coupling
swivel

(All dimensions in inches.)

| Service and nominal size | Inside diam, $C$ | Diam of thread, $D$ | No. of threads per in. | $L$ | $I$ | $H$ | $T$ |
|---|---|---|---|---|---|---|---|
| Garden: <br> ½, ⅝, ¾ | $2\frac{5}{32}$ | $1\frac{1}{16}$ | $11\frac{1}{2}$ | $\frac{9}{16}$ | $\frac{1}{8}$ | $1\frac{7}{32}$ | $\frac{3}{8}$ |
| Chemical: <br> ¾, 1 | $1\frac{1}{32}$ | $1\frac{3}{8}$ | 8 | $\frac{5}{8}$ | $\frac{5}{32}$ | $1\frac{9}{32}$ | $1\frac{5}{32}$ |
| Fire: <br> 1½ | $1\frac{17}{32}$ | 2 | 9 | $\frac{5}{8}$ | $\frac{5}{32}$ | $1\frac{9}{32}$ | $1\frac{5}{32}$ |
| Other connections: | | | | | | | |
| ½ | $1\frac{7}{32}$ | $1\frac{3}{16}$ | 14 | $\frac{1}{2}$ | $\frac{1}{8}$ | $1\frac{5}{32}$ | $\frac{5}{16}$ |
| ¾ | $2\frac{5}{32}$ | $1\frac{1}{32}$ | 14 | $\frac{9}{16}$ | $\frac{1}{8}$ | $1\frac{7}{32}$ | $\frac{3}{8}$ |
| 1 | $1\frac{9}{32}$ | $1\frac{9}{32}$ | $11\frac{1}{2}$ | $\frac{9}{16}$ | $\frac{5}{32}$ | $1\frac{7}{32}$ | $\frac{3}{8}$ |
| 1¼ | $1\frac{9}{32}$ | $1\frac{5}{8}$ | $11\frac{1}{2}$ | $\frac{5}{8}$ | $\frac{5}{32}$ | $1\frac{9}{32}$ | $1\frac{5}{32}$ |
| 1½ | $1\frac{17}{32}$ | $1\frac{7}{8}$ | $11\frac{1}{2}$ | $\frac{5}{8}$ | $\frac{5}{32}$ | $1\frac{9}{32}$ | $1\frac{5}{32}$ |
| 2 | $2\frac{1}{32}$ | $2\frac{11}{32}$ | $11\frac{1}{2}$ | $\frac{3}{4}$ | $\frac{3}{16}$ | $2\frac{3}{32}$ | $1\frac{9}{32}$ |

* ANSI B2.4-1966.

## 8-23. Coefficients of Static and Sliding Friction

Coefficients of static and sliding friction are defined as the ratios of frictional force to the normal force between two bodies in contact. (See Table 8-32.)

## Table 8-32. Coefficients of Static and Sliding Friction

| Materials | Static coef | | Sliding coef | |
|---|---|---|---|---|
| | Dry | Greasy | Dry | Greasy |
| Hard steel on hard steel........... | 0.78 | 0.11 | 0.42 | 0.029 |
| | .... | .23 | .... | .081 |
| | .... | .15 | .... | .080 |
| | .... | .11 | .... | .058 |
| | .... | .0075 | .... | .084 |
| | .... | .0052 | .... | .105 |
| | .... | ...... | .... | .096 |
| | .... | ...... | .... | .108 |
| | .... | ...... | .... | .12 |
| Mild steel on mild steel........... | 0.74 | ...... | 0.57 | .09 |
| | .... | ...... | .... | .19 |
| Hard steel on graphite........... | 0.21 | .09 | .... | .... |
| Hard steel on babbitt (ASTM 1).... | 0.70 | .23 | 0.33 | .16 |
| | .... | .15 | .... | .06 |
| | .... | .08 | .... | .11 |
| | .... | .085 | | |
| Hard steel on babbitt (ASTM 8).... | 0.42 | .17 | 0.35 | .14 |
| | .... | .11 | .... | .065 |
| | .... | .09 | .... | .07 |
| | .... | .08 | .... | .08 |
| Hard steel on babbitt (ASTM 10)... | .... | .25 | .... | .13 |
| | .... | .12 | .... | .06 |
| | .... | .10 | .... | .055 |
| | .... | .11 | | |
| Mild steel on cadmium silver...... | .... | ...... | .... | .097 |
| Mild steel on phosphor bronze...... | .... | ...... | 0.34 | .173 |
| Mild steel on copper lead...... | .... | ...... | .... | .145 |
| Mild steel on cast iron............. | .... | .183 | 0.23 | .133 |
| Mild steel on lead.................. | 0.95 | .5 | 0.95 | .3 |
| Nickel on mild steel............... | .... | ...... | 0.64 | .178 |
| Aluminum on mild steel........... | 0.61 | ...... | 0.47 | .... |
| Magnesium on mild steel........... | .... | ...... | 0.42 | .... |
| Cadmium on mild steel............ | .... | ...... | 0.46 | .... |
| Copper on mild steel............. | 0.53 | ...... | 0.36 | .18 |
| Nickel on nickel.................. | 1.10 | ...... | 0.53 | .12 |
| Brass on mild steel............... | 0.51 | ...... | 0.44 | .... |
| Brass on cast iron................. | .... | ...... | 0.30 | .... |
| Zinc on cast iron.................. | 0.85 | ...... | 0.21 | .... |
| Magnesium on cast iron........... | .... | ...... | 0.25 | .... |
| Copper on cast iron................ | 1.05 | ...... | 0.29 | .... |
| Tin on cast iron.................. | .... | ...... | 0.32 | .... |
| Lead on cast iron................. | .... | ...... | 0.43 | .... |
| Aluminum on aluminum............ | 1.05 | ...... | 1.4 | .... |
| Glass on glass.................... | 0.94 | .01 | 0.40 | .09 |
| | .... | .005 | .... | .116 |
| Carbon on glass.................. | .... | ...... | .... | .18 |
| Garnet on mild steel............. | .... | ...... | .... | .39 |
| Glass on nickel.................. | 0.78 | ...... | .... | .56 |
| Copper on glass.................. | 0.68 | ...... | .... | .53 |
| Cast iron on cast iron............ | 1.10 | ...... | 0.15 | .070 |
| | .... | ...... | .... | .064 |
| Bronze on cast iron.............. | .... | ...... | .22 | .077 |
| Oak on oak (parallel to grain)..... | 0.62 | ...... | .48 | .164 |
| | .... | ...... | .... | .067 |
| Oak on oak (perpendicular to grain).. | .54 | ...... | .32 | .072 |
| Leather on oak (parallel).......... | .61 | ...... | .52 | |
| Cast iron on oak.................. | .... | ...... | .49 | .075 |
| Leather on cast iron.............. | .... | ...... | .56 | .36 |
| | .... | ...... | .... | .13 |
| Laminated plastic on steel......... | .... | ...... | .... | .35 .05 |
| Fluted rubber bearing on steel...... | .... | ...... | .... | .05 |

source: T. Baumeister (ed.), "Standard Handbook for Mechanical Engineers," 7th ed., p. 3-35, McGraw-Hill Book Company, New York, 1967.

## COMPRESSOR, FAN, AND PISTON-MACHINE PERFORMANCE

### Nomenclature

$a$ = area of piston (with deductions for piston and tail rods, when present), in.$^2$

$C$ = constant, 12 for 4-cycle engine, 20 for 2-cycle engine

$C_p$ = heat capacity at constant pressure, Btu/(lb)($^\circ$F)

$C_v$ = heat capacity at constant volume, Btu/(lb)($^\circ$F)

$c$ = clearance, decimal fraction of displacement

$D$ = piston displacement, ft$^3$/cycle

$D'$ = piston displacement, cfm

$d$ = bore, in.

$F$ = net force at arm bearing point, lb

$g_c$ = conversion factor, 32.2 (lb mass)(ft)/(lb force)(sec$^2$)

$h$ = enthalpy ($= pv/J + u$), Btu/lb

$h_s$ = static head, ft fluid

$h_t$ = total head, ft fluid

$h_v$ = velocity head, ft fluid

$h''_w$ = head, in. $H_2O$

$h_1$ = enthalpy at compressor supply, Btu/lb

$h_2$ = enthalpy at compressor delivery, Btu/lb

$J$ = mechanical equivalent of heat, 778 ft-lb/Btu

$k$ = ratio of specific heats, $C_p/C_v$

$L$ = length of stroke, ft

   = length of brake arm (shaft center line to arm bearing point), ft

$l$ = stroke length, in.

mep = mean effective pressure, psi

$N$ = shaft rpm

   = number of stages

$n$ = number cycles completed per minute

   = polytropic exponent, $pv^n$ = constant for polytropic process

$p$ = pressure, lb force/ft$^2$ abs

$p_1$ = supply pressure, lb force/in.$^2$ abs

$p_2$ = delivery pressure, lb force/in.$^2$ abs

$R_p$ = ratio of pressures = $p_2/p_1$

$R_{p1}, R_{p2}, R_{p3}$ = compression ratio in stages one, two, three, respectively

$u$ = internal or intrinsic energy, Btu/lb

$v$ = volume, ft$^3$

$\bar{v}$ = specific volume, ft$^3$/lb

$\bar{w}_f$ = density of fluid, lb/ft$^3$

$\bar{w}_w$ = density of water, 62.4 lb/ft$^3$

$\Delta W$ = work done on or by fluid, ft-lb force/lb

$_1$ = entering or initial conditions

$_2$ = leaving or final conditions

### 8-24. Adiabatic (Isentropic) Compressor Standards

The ideal compressor cycle is shown in Fig. 8-73 where there are three phases: (1) admission from $a$ to 1, (2) compression from 1 to 2, and (3) delivery from 2 to $b$. For a perfect gas with reversible adiabatic (or isentropic) compression ($pv^k$ = const) the work is given as[1]

$$\Delta W_{\text{adiabatic cycle}} = 144 p_1 v_1 [k/(k-1)](R_p^{(k-1)/k} - 1) \qquad (8\text{-}69)$$

If the compression is isentropic and for a real gas, the thermodynamic properties of which are known (as for refrigerants), then for a perfect compressor the work of the cycle is

$$\Delta W_{\text{adiabatic cycle}} = (h_2 - h_1)778 \qquad (8\text{-}70)$$

Equations (8-69) and (8-70) give identical answers for a perfect gas.

### 8-25. Isothermal Compressor Standards

If compressors are so cooled that temperature is constant during compression, the isothermal standard prevails as shown in Fig. 8-74 ($pv$ = const) and there is a saving in work over the adiabatic value of Eq. (8-69). The work is given as[1]

$$\Delta W_{\text{isothermal}} = 144(p_1 v_1) \log_e R_p \qquad (8\text{-}70a)$$

### 8-26. Multistage Compressor Standards

The work of a compressor is reduced by the use of multistage compression with intercooling between stages. If cooling is complete and the gas enters the succeeding stage at the same temperature at which it enters the machine, the intercooling is said to be perfect. Minimum work is then obtained with unique values of pressure between the stages, called best receiver pressure. It is determined by[1]

$$R_{p1} = R_{p2} = R_{p3} \cdots = R_p^{1/N} \qquad (8\text{-}71)$$

With isentropic compression in each stage, best receiver pressure, and perfect intercooling, the work of the ideal cycle is

$$\Delta W_{\text{multistage}} = 144 N p_1 v_1 [k/(k-1)](R_p^{(k-1)/Nk} - 1) \qquad (8\text{-}72)$$

The isothermal standard of Eq. (8-70a) applies equally well to single- and multiple-stage compression.

### 8-27. Capacity

Capacity is expressed on a volume basis and for air is given on the "free air" basis. This measures capacity at the ambient pressure, temperature, and humidity.

For a positive-displacement machine without clearance, the volume is represented as $v_1 - v_a$ of Fig. 8-75. This is obtained from the dimensions of the cylinders[1]

[1] For nomenclature see p. 8-73.

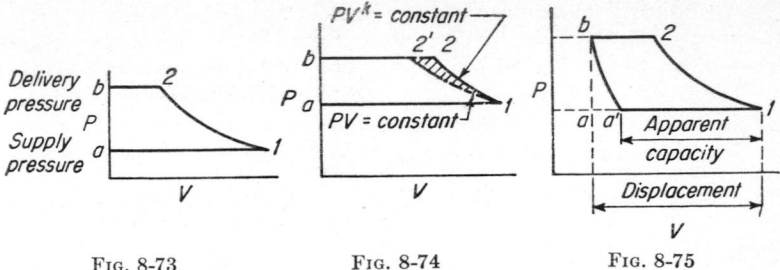

Fig. 8-73          Fig. 8-74          Fig. 8-75

$$\text{Piston displacement, ft}^3/\text{cycle} = D = (\pi d^2/4)(l/1{,}728) \qquad (8\text{-}73)$$

$$D' = \text{piston displacement, cfm} = \frac{\pi d^2}{4} nl \frac{1}{1{,}728} \qquad (8\text{-}74)$$

$$= \frac{d^2 l n}{2{,}200}$$

As the machine contains clearance which runs from 2 to 20 per cent of the displacement, there is a clearance reexpansion loss, so that point $a$ shifts to position $a'$ and the length $v_1 = v'_a$ is less than the displacement $v_1 - v_a$ in Fig. 8-75. This apparent capacity is calculable by

$$\text{Apparent capacity} = D(1 + c - cR_p{}^{1/k}) \qquad (8\text{-}75)$$

The actual capacity, as metered, for a real compressor is less than this apparent value because of suction heating, suction pressure drop, and leakage losses.

Volumetric efficiency is the ratio of capacity to displacement, and if the former is a real metered value, then

$$\text{Actual volumetric eff, } \% = \frac{\text{actual metered capacity}}{\text{piston displacement}} \times 100 \qquad (8\text{-}76)$$

If the apparent capacity is used from Eq. (8-75), then

$$\text{Apparent volumetric eff, } \% = (1 + c - cR_p{}^{1/k}) \times 100 \qquad (8\text{-}77)$$

The relation between these two is called slippage efficiency and is defined as

$$\text{slippage eff, } \% = \frac{\text{actual volumetric eff}}{\text{apparent volumetric eff}} \times 100 \qquad (8\text{-}78)$$

## 8-28. Ideal Horsepower of Compressors

The equations for ideal work, (8-68), (8-69), (8-70), and (8-72), apply equally well to compressors involving clearance, because work is independent of clearance. For a volume flow rate of 100 cfm the equations can be rewritten as[1]

[1] For nomenclature see p. 8–73.

$$\text{Isentropic or adiabatic hp/100 cfm} = \frac{k}{k-1}\frac{p_1}{2.292}(R_p^{(k-1)/k} - 1) \quad (8\text{-}79)$$

$$= \frac{(h_2 - h_i) \times \text{lb/min}}{0.4242\bar{v}_1} \quad (8\text{-}80)$$

where $\bar{v}_1$ = specific volume, ft³/lb, at supply pressure

$$\text{Isothermal hp/100 cfm} = (p_1/2.292)\log_e R_p \quad (8\text{-}81)$$

In a multistage compressor with perfect intercooling and best receiver pressure:

$$\text{Isentropic or adiabatic hp/100 cfm} = \frac{Np_1}{2.292}\frac{k}{k-1}R_p^{(k-1)/Nk} - 1) \quad (8\text{-}82)$$

These equations for work and horsepower can be conveniently solved by use of Fig. 8-76.

## 8-29. Compression Efficiency

The actual power required by a compressor can be compared to the ideal power (for the same capacity) to give

$$\text{Compression eff, \%} = \frac{\text{ideal hp}}{\text{actual hp}} \times 100 \quad (8\text{-}83)$$

The ideal value may be obtained from Eqs. (8-79) to (8-82) giving adiabatic and isothermal compression efficiencies.

The actual horsepower may be obtained from (1) the compressor-cylinder indicator card, (2) the shaft of the compressor, or (3) it may be actual power input to the motor terminals of an electrically driven unit. Care must be taken to specify the base.

## 8-30. Fan Performance

A fan is a compressor in which the change in density of the gaseous fluid, on passage through the machine, is negligibly small.

### Definitions

Standard air:   Air at 68°F, 29.92 in. Hg pressure, and 5 per cent relative humidity.   It has a density of 0.07488 lb/ft³ and a specific volume of 13.3 ft³/lb and is the basis for measuring fan performance.

Capacity:   The volume delivered by a fan, $Q$, expressed in cubic feet per minute.

Head:   The difference between the pressures on the suction and discharge sides of a fan, variously expressed as feet of fluid, inches of water, pounds per square inch, etc.   Conversion is as follows[1]

$$h_t = (h''_w/12)(\bar{w}_w/\bar{w}_f) \qquad \text{ft fluid} \qquad (8\text{-}84)$$
$$h_t = h''_w \times 69.5 \qquad \text{ft std air} \qquad (8\text{-}85)$$
$$\text{Pressure} = h''_w/27.7 \qquad \text{psi} \qquad (8\text{-}86)$$

[1] For nomenclature see p. 8-73.

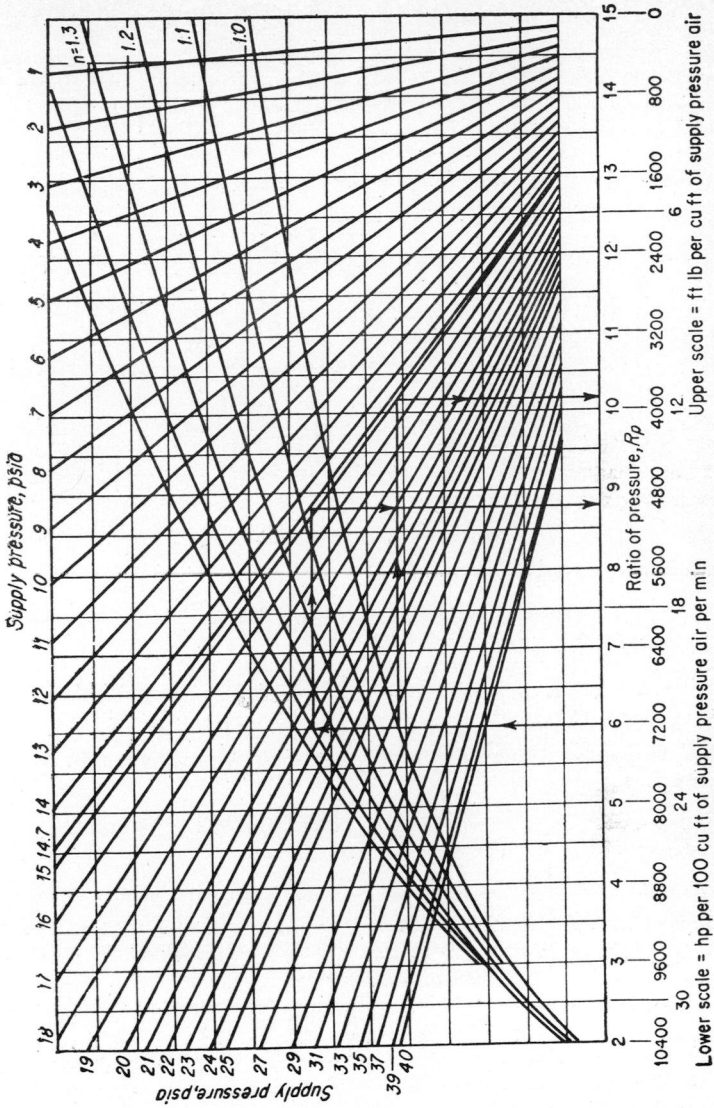

FIG. 8-76

*Static, Velocity, and Total Heads.* As shown in Fig. 8-77, three pressures can be read in a fan duct. *Static head $h_s$* is a directly obtained pressure reading; *velocity head $h_v$* is obtained from the flow rate in the duct and must be an average value obtained by a traverse. Then *total head*

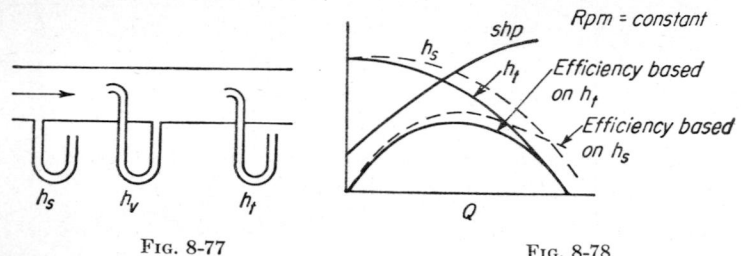

FIG. 8-77                                    FIG. 8-78

$$h_t = h_s + h_v \tag{8-87}$$

Conversion is by

$$\text{Velocity} = \sqrt{2g_c h_v} \quad \text{fps} \tag{8-88}$$

Substituting the conversion of Eq. (8-84)

$$\text{Velocity} = 1{,}096.2 \sqrt{h''_w/\bar{w}_f} \quad \text{fpm} \tag{8-89}$$

and if standard air is used

$$\text{Velocity} = 4{,}005 \sqrt{h''_w} \quad \text{fpm} \tag{8-90}$$

Fan performance is variously based on static and total head, the former being generally more realistic because it is the only form of head usable in overcoming a system resistance.

**Horsepower of Fans.** The ideal or air horsepower is given by[1]

$$\text{Air hp} = Qh''_w/6{,}355 \tag{8-91}$$

Static or total head may be used, giving two alternative values of ideal horsepower, the latter being larger.

**Shaft horsepower** (shp). Shaft-horsepower input to drive the fan is measured by a suitable dynamometer. *Fan efficiency* is defined as

$$\text{Fan eff, } \% = \frac{\text{air hp}}{\text{shp}} \times 100 \tag{8-92}$$

This value may be on the static or total basis.

**Fan Characteristics.** Fans, like other fluid-acceleration machines, operate with characteristic curves. A set of characteristics is plotted in Fig. 8-45 for a representative fan. These curves are exactly definitive, and the fan must operate at some point on the characteristic.

**Fan Laws.** See pump laws, Sec. 8-33, which are identical to the fan laws.

[1] For nomenclature see p. 8–73.

## 8-31. Performance Characteristics of Piston Machines

**Mean Effective Pressure.**  In piston and cylinder machines it is convenient to measure performance through the use of mean effective pressure. As illustrated by Fig. 8-79, the mean effective pressure is defined as the difference in pressure on the two sides of the piston, which difference tends to move the piston in an engine or resist its motion in a pump.  It can be defined as

Mean effective pressure, mep = mean forward pressure, mfp

— mean back pressure, mbp    (8-93)

Thus in the two illustrations of Fig. 8-79, for a single-acting and a double-acting mechanism, the mep is the same (100 psi) because in $A$,

$$\text{mep} = 115 - 15 = 100 \text{ psi};$$

in $B$, mep = 215 — 115 = 100 psi.

These pressures are mean values which can be considered as prevailing throughout the stroke.  They may be calculated for ideal cyclic conditions by utilizing the methods of thermodynamics and fluid dynamics.  Thus in Fig. 8-80 the area of the $P$-$V$ diagram is the work of the cycle, expressible in foot pounds.  If that area is divided by the length of the diagram, i.e., by the stroke or displacement, the result is a vertical height for a rectangle of equivalent area, and this height is the mep, as

$$\text{mep} = \frac{\text{work of cycle, ft-lb}}{\text{displacement, ft}^3 \times 144 \text{ in.}^2/\text{ft}^2} = \frac{\text{lb}}{\text{in.}^2} \qquad (8\text{-}94)$$

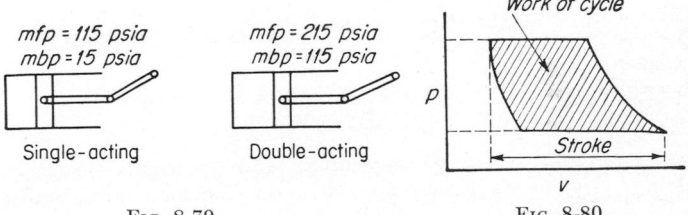

FIG. 8-79                    FIG. 8-80

Actual mean effective pressure is determined from the planimetered area of the indicator card, dividing by length, and applying spring scale.  Thus with Fig. 8-81,

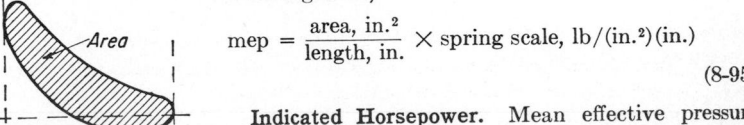

$$\text{mep} = \frac{\text{area, in.}^2}{\text{length, in.}} \times \text{spring scale, lb/(in.}^2)\text{(in.)}$$

$$(8\text{-}95)$$

**Indicated Horsepower.**  Mean effective pressure is used to calculate the indicated horsepower, thus[1]

FIG. 8-81            Indicated hp = mep $Lan$/33,000    (8-96)

[1] For nomenclature see p. 8–73.

The items $a$ and $L$ are obtained directly from the bore and stroke. The number of cycles completed per minute depends upon the mechanism construction, i.e., single- or double-acting; number of cylinders; and number of strokes or revolutions needed to complete a cycle (Table 8-33).

### Table 8-33. General Data on Engine, Pump, and Compressor Constructions

| Machine | Strokes per cycle | Single- or double-acting |
|---|---|---|
| Steam engine. | 2 | DA |
| Air compressors: | | |
|   Large. | 2 | DA |
|   Small. | 2 | SA |
| Refrigeration compressors: | | |
|   Large. | 2 | DA |
|   Small. | 2 | SA |
| Direct-acting pumps. | 2 | DA |
| Triplex pumps. | 2 | SA |
| Compressed-air engines: | | |
|   Large. | 2 | DA |
|   Small. | 2 | SA |
| Internal-combustion engines: | | |
| High-speed automotive gasoline. | 4 | SA |
| Small high-speed gasoline. | 2 or 4 | SA |
| High-speed diesel. | 2 or 4 | SA |
| Medium-speed diesel. | 2 or 4 | SA |
| Low-speed diesel. | 2 or 4 | SA or DA |
| Natural gas: | | |
|   Small. | 2 or 4 | SA |
|   Large. | 2 or 4 | DA |

**Brake or Shaft Horsepower.** As measured on a prony brake or dynamometer, brake or shaft horsepower is determined by[1]

$$hp = 2\pi \, LFN/33,000 \qquad (8\text{-}97)$$
$$= LFN/5,250 \qquad (8\text{-}98)$$

**Brake Mean Effective Pressure, or Brake Mean Pressure.** On high-speed engines and compressors it is not possible to take indicator cards, but brake-horsepower readings can be expressed as equivalent brake mean pressure by combining results of Eq. (8-97) in Eq. (8-98), or[1]

$$\text{Brake mean pressure, psi} = \frac{\text{brake hp}}{Lan} \times 33,000 \qquad (8\text{-}99)$$

**Mean Friction Pressures.** Mean friction pressure measures the losses between cylinders and shaft, or

On engines:

Mean friction pressure, psi = indicated mean effective pressure

$$- \text{ brake mean pressure} \qquad (8\text{-}100)$$

[1] For nomenclature see p. 8-73.

On compressors and pumps:

Mean friction pressure, psi = brake mean pressure
$$\quad\quad\quad\quad\quad\quad\quad\quad - \text{ indicated mean pressure} \quad\quad (8\text{-}101)$$

**Mechanical Efficiency.** Mechanical efficiency is another device for expressing the losses between cylinder and shaft, or

On engines:
$$\text{Mechanical eff} = \frac{\text{brake hp}}{\text{indicated hp}} \times 100 \quad\quad (8\text{-}102)$$

On pumps and compressors:
$$\text{Mechanical eff} = \frac{\text{indicated hp}}{\text{brake hp}} \times 100 \qu\quad\quad (8\text{-}103)$$

## PUMP PERFORMANCE

### 8-32. General

**Nomenclature**

$G$ = capacity, lb/hr
$p$ = head, psi
$Q$ = capacity, gpm
$H$ = head, ft

In American practice, pump sizes generally refer to the discharge-pipe diameter in inches.

**Standard Water.** Pump performance is predicated on the use of cold water (less than 85°F). The specific gravity of water at 60°F is usually referred to as unity, with density of 62.4 lb/ft³ and specific volume of 0.016 ft³/lb.

**Capacity.** The capacity of a pump is expressed on a volume basis, generally in gallons per minute (1 United States gallon = 251 in.³).

**Head.** The head developed $H$ is expressed in feet of water. Conversion to pressure $p$, in pounds per square inch, is computed from

$$p = \frac{H}{2.3} \quad \text{for cold water} \quad\quad (8\text{-}104)$$

The head delivered by a pump is expressed as total dynamic head, which is the sum of the static and velocity heads. The velocity head is usually a minor item in pump performance. Conversion of head to velocity $v$ is by

$$v = \sqrt{2g_cH}$$
$$\quad = 8.02 \sqrt{H} \quad \text{fps} \quad\quad (8\text{-}105)$$

**Horsepower.** The ideal or water horsepower is given by

$$\text{Water hp} = (QH \times \text{sp gr})/3{,}960 \quad\quad (8\text{-}106)$$
$$\quad\quad\quad\quad = Gp/(857{,}200 \times \text{sp gr}) \qu\quad (8\text{-}107)$$

where sp gr = specific gravity of fluid being pumped.

**Shaft Horsepower.** The horsepower input to drive the pump is measured by dynamometer.
Pump efficiency is defined as

$$\text{Pump eff, } \% = \frac{\text{water hp}}{\text{shp}} \times 100 \quad (8\text{-}107a)$$

**Pump Characteristics.** Centrifugal and axial-flow pumps, like fans, must operate at some point on the characteristic curves (Fig. 8-82). These curves are exactly definitive for the performance of a centrifugal or axial-flow pump, and operation must lie on these curves.

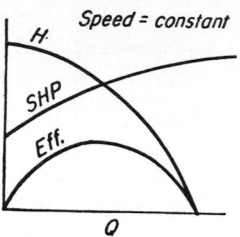

Fig. 8-82

### 8-33. Pump Laws

For a given centrifugal or axial-flow pump operating at a given point on the efficiency curve, if the speed $n$ (rpm) is changed, then

$$Q \propto n \qquad H \propto n^2 \qquad \text{shp} \propto n^3 \qquad (8\text{-}108)$$

where shp = shaft horsepower
For a series of similar pumps operating at a given point on the efficiency curve, if the size of the pump is changed as measured by the impeller diameter $D$ (feet) for convenience, but with the speed held constant,

$$Q \propto D^3 \qquad H \propto D^2 \qquad \text{shp} \propto D^5 \qquad (8\text{-}109)$$

For a series of similar pumps operating at a given point on the efficiency curve, if the size of the pump is changed as measured by the impeller diameter $D$ but with the tip speed ($u = \pi D n$) held constant

$$Q \propto D^2 \qquad H = \text{const} \qquad \text{shp} \propto D^2 \qquad (8\text{-}110)$$

Specific speed is a useful criterion for defining the adaptability of a pump to a particular service. It is the speed in rpm at which an homologous pump of the series, with suitable diameter, would run in order to deliver 1 gpm at 1 ft head. The specific speed can be plotted as a further characteristic, but the value of the parameter at maximum efficiency is most useful and is always implied. It is derived from the pump-law equations (8-108) and (8-109) by eliminating $D$ from among the four expressions for $Q$, $D$, and $H$ to give

$$N_S = \text{specific speed} = \frac{\text{rpm} \times \sqrt{\text{gpm}}}{\text{head}^{3/4}} = \frac{n\sqrt{Q}}{H^{3/4}} \qquad (8\text{-}111)$$

Specific speed is the true criterion for judging high-speed and low-speed pumps. It must be reduced to the single-stage single-inlet-impeller basis. It is related to experience on pump selection as in Fig. 8-83.

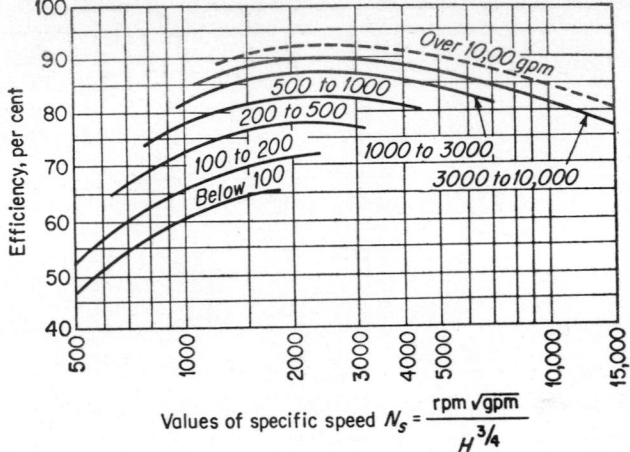

Values of specific speed $N_s = \dfrac{\text{rpm }\sqrt{\text{gpm}}}{H^{3/4}}$

Fig. 8-83. *Source: T. Baumeister (ed.), "Standard Handbook for Mechanical Engineers" 7th ed., p.* 14-32, *McGraw-Hill Book Company, New York,* 1967.

## 8-34. Positive-displacement Pumps

The positive-displacement pump may be (1) of a type which is driven through a crank-connecting rod mechanism or (2) of a type which has a steam or compressed-air piston cylinder directly connected through a piston rod to the pump cylinder. In the latter form the stroke is not rigorously fixed. Slip is defined by

$$\text{Slip} = \left(1.00 - \frac{\text{actual volume discharged}}{\text{piston displacement}}\right) \times 100 \qquad \text{per cent} \qquad (8\text{-}112)$$

Speeds are limited on positive-displacement pumps because of the inertia effects of liquid columns. Recommended practice is reflected in the data of Table 8-34.

### Table 8-34. Basic Speeds of Standard Pumps

| Stroke, in. | Simplex and duplex steam | | Duplex and triplex steam | | Stroke, in. | Simplex and duplex steam | | Duplex and triplex steam | |
|---|---|---|---|---|---|---|---|---|---|
| | Rpm | Fpm | Rpm | Fpm | | Rpm | Fpm | Rpm | Fpm |
| 3 | 74 | 37 | 105 | 52 | 10 | 45 | 75 | 57 | 95 |
| 4 | 71 | 47 | 90 | 60 | 12 | 41 | 81 | 52 | 104 |
| 5 | 64 | 53 | 80 | 66 | 15 | 36 | 90 | 47 | 117 |
| 6 | 60 | 60 | 74 | 74 | 18 | 33 | 99 | 43 | 129 |
| 8 | 51 | 68 | 64 | 85 | 24 | 27 | 108 | 37 | 148 |

NOTE: Rpm refers to revolutions per minute of driver, fpm refers to lineal displacement in feet per minute per revolution.

SOURCE: T. Baumeister (ed.), "Standard Handbook for Mechanical Engineers," 7th ed., p. 14-8, McGraw-Hill Book Company, New York, 1967.

**Suction Lift and Suction Head.** A pump can theoretically operate with either a positive or negative pressure on the suction. The latter condition makes it possible to locate a pump physically above the water level in the sump. The theoretical vertical distance is the barometric height (Table 8-35).

### Table 8-35. Barometric Pressure vs. Altitude

| Altitude, ft above sea level | Barometer, in. Hg | Altitude, ft above sea level | Barometer, in. Hg |
|---|---|---|---|
| 0 | 29.92 | 6,000 | 24.00 |
| 1,000 | 28.84 | 8,000 | 22.30 |
| 2,000 | 27.80 | 10,000 | 20.72 |
| 3,000 | 26.80 | 12,000 | 19.24 |
| 4,000 | 25.83 | 14,000 | 17.88 |
| 5,000 | 24.88 | | |

Practical vertical distance or suction lift = (barometric head)
  − (velocity head) − (friction losses in pipe fittings and valves
  + pressure of gas in solution + vapor pressure of liquid
    + margin for assuring continuity of liquid column)  (8-113)

The item of vapor pressure is widely variable and is a function of the temperature of the liquid. With water this limitation is reflected in the data of Fig. 8-84.

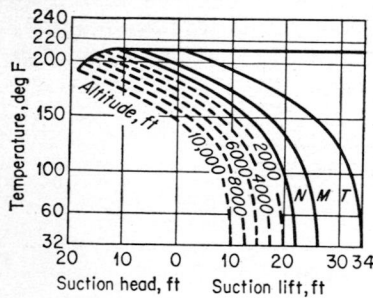

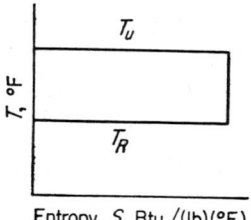

FIG. 8-84. *Source: L. S. Marks (ed.), "Mechanical Engineers' Handbook," 5th ed., p. 1837, McGraw-Hill Book Company, Inc., New York, 1951.*

FIG. 8-85

## REFRIGERATION

### 8-35. Coefficient of Performance C.O.P.

$$\text{C.O.P.} = \frac{\text{refrigeration}}{\text{work done}} = \frac{h_2 - h_1}{h_3 - h_2} \qquad \text{as cooling machine} \quad (8\text{-}114)$$

$$\text{C.O.P.} = \frac{\text{heat delivered to condenser}}{\text{work done}} = \frac{h_3 - h_6}{h_3 - h_2} \quad \text{as warming machine}$$

$$(8\text{-}115)$$

The maximum value of C.O.P. is given by the Carnot cycle (Fig. 8-85), where the absolute temperatures of the refrigerating coil $T_R$ and the condensing coil $T_H$ prevail instead of enthalpy, since the C.O.P. is independent of the properties of the refrigerant used.

$$\text{C.O.P.} = T_R/(T_H - T_R) \qquad \text{as cooling machine} \qquad (8\text{-}116)$$
$$\text{C.O.P.} = T_H/(T_H - T_R) \qquad \text{as warming machine} \qquad (8\text{-}117)$$

To facilitate calculations on refrigeration cycles it is necessary to have tables and charts showing the thermodynamic properties of the refrigerants. Representative data for several common refrigerants are shown in Fig. 8-86.

The type of refrigeration system chosen will most often depend on the system load. Typical practice in considering selection of the refrigeration system is indicated in Table 8-36.

Ideal performance of some common refrigerants and their operating temperature ranges are shown in Table 8-37. For a chosen refrigerant operating in the temperature and pressure range and in the type machine indicated, the quantities shown for piston displacement and theoretical horsepower, when adjusted for mechanical efficiency, will provide a first-order approximation of the size of compressor and input power. An extended range of such data for an ammonia refrigeration system is shown in Table 8-38. Similar data in graphical form are shown in Figs. 8-87 and 8-88 for the representative refrigerant and operating conditions cited in the headings. Note that the rotary compressor in Fig. 8-88 is a relatively small machine.

## 8-36. Refrigeration Systems

### Table 8-36. Typical Refrigeration Systems*

| System load, tons | System type | | |
|---|---|---|---|
| | Often used | Occasionally used | Rarely used |
| 0–5 | Unit system with reciprocating compressor | Central-station built-up system; reciprocating compressor | Central-station built-up units |
| 5–25 | Central-station built-up systems; reciprocating compressor | Central-station built-up system; reciprocating compressor | Absorption or adsorption units |
| 25–50 | Central-station built-up systems; reciprocating compressor | Central-station built-up system; centrifugal compressor | Absorption units |
| 50–400 | Central-station built-up systems; reciprocating compressor | Central-station built-up systems; steam-jet and centrifugal compressors | |
| 400 and up | Central system; centrifugal and/or absorption unit | Central station built-up steam-jet unit | |

* Adapted from ASHRAE data.

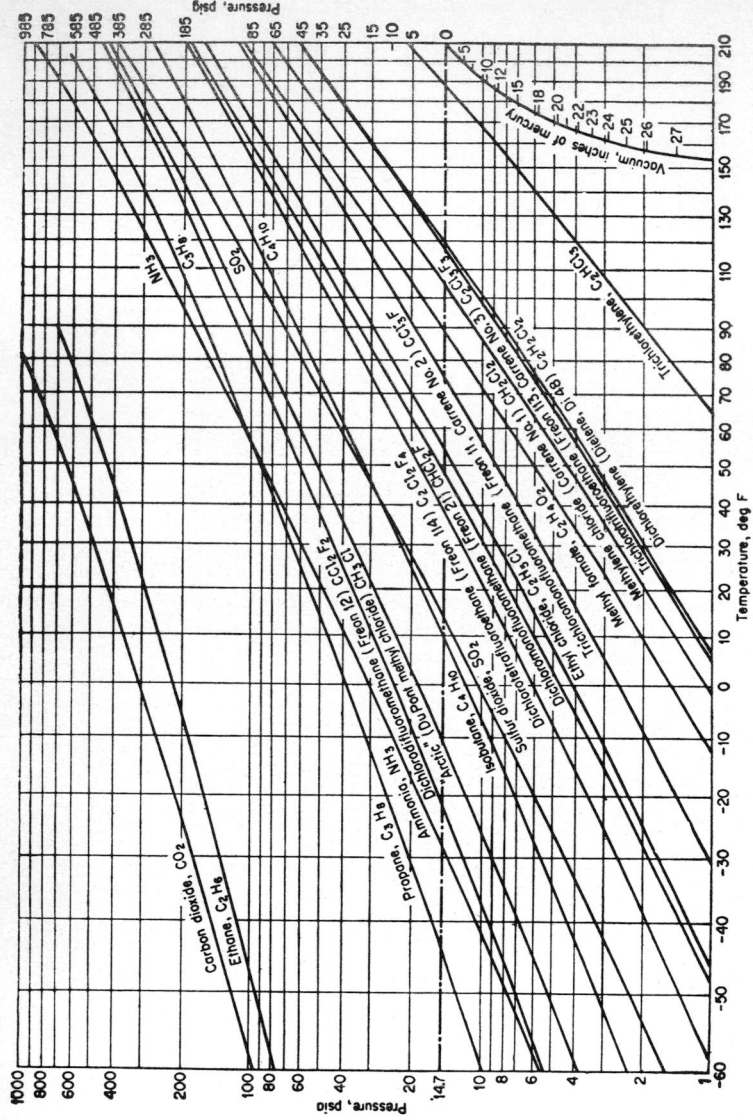

Fig. 8-86. Properties of common refrigerants. *Source: Electrochemicals Department, E. I. du Pont de Nemours & Company, Inc., Wilmington, Delaware.*

**8**–86

## Table 8-37. Ideal Performance of Refrigerants for Various Temperature Ranges

| Refrigerant and (number) | Operating temperature range, deg F | Suction pressure, psia | Head pressure, psia | Ratio of head to suction pressure | With dry and saturated suction vapor, per ton | | | Temperature at end of compression, deg F | Type of compressor |
|---|---|---|---|---|---|---|---|---|---|
| | | | | | Weight of vapor, lb per min | Piston displacement, cfm | Theoretical hp | | |
| Air (729) | 5-86 | 14.7 | 73.5 | 5.0 | 7.02 | 82.3 | 2.82 | 277.0 | Recip. and exp. cyl |
| Water (718) | 32-86 | 0.0885 | 0.6152 | 6.95 | 0.1957 | 647.0 | 0.618 | 332 | Centrif. |
| | 32-100 | 0.0885 | 0.9492 | 10.73 | 0.1985 | 656.2 | 0.819 | 420 | or |
| | 40-100 | 0.1217 | 0.9492 | 7.80 | 0.1978 | 483.4 | 0.687 | 366 | ejector |
| Carbon dioxide ($CO_2$) (744) | 5-86 | 332.0 | 1,043.0 | 3.14 | 3.61 | 0.960 | 1.827 | 160.3 | Recip. |
| Ammonia ($NH_3$) (717) | 5-86 | 34.27 | 169.2 | 4.94 | 0.421 | 3.44 | 0.99 | 209.8 | Recip. |
| | 20-100 | 48.21 | 211.9 | 4.40 | 0.421 | 2.49 | 0.94 | 212.8 | |
| | 40-100 | 73.32 | 211.9 | 2.89 | 0.427 | 1.70 | 0.65 | 176.0 | |
| Freon 11 ($CCl_3F$) (11) | 5-86 | 2.931 | 18.28 | 6.20 | 3.058 | 37.0 | 0.94 | 112.7 | Centrif. |
| | 20-100 | 4.342 | 23.60 | 5.44 | 3.086 | 26.2 | 0.89 | 122.1 | |
| | 40-100 | 7.032 | 23.60 | 3.36 | 2.945 | 16.08 | 0.63 | 114.2 | |
| Freon 12 ($CCl_2F_2$) (12) | 5-86 | 26.51 | 107.9 | 4.07 | 3.916 | 5.82 | 1.00 | 100.2 | Recip. |
| | 20-100 | 35.75 | 131.6 | 3.68 | 4.054 | 4.54 | 0.97 | 112.5 | or |
| | 40-100 | 51.68 | 131.6 | 2.55 | 3.880 | 3.07 | 0.67 | 108.0 | centrif. |
| Freon 21 ($CHCl_2F$) (21) | 5-86 | 5.243 | 31.23 | 5.96 | 2.364 | 20.87 | 0.94* | 99.0* | Rotary |
| | 20-100 | 7.699 | 40.04 | 5.20 | 2.404 | 15.61 | 0.94* | 110.4* | |
| | 40-100 | 12.32 | 40.04 | 3.25 | 2.315 | 10.19 | 0.68* | 105.9* | |
| Freon 22 ($CHClF_2$) (22) | 5-86 | 43.02 | 174.5 | 4.06 | 2.926 | 3.65 | 1.03 | 131.7 | Recip. |
| | 20-100 | 57.98 | 212.6 | 3.67 | 3.023 | 2.83 | 0.99 | 152.7 | |
| | 40-100 | 83.72 | 212.6 | 2.54 | 2.936 | 1.93 | 0.68 | 131.0 | |
| Methylene chloride ($CH_2Cl_2$) (30) | 5-86 | 1.28 | 10.07 | 8.56 | 1.485 | 74.0 | 0.96* | 205.1* | Centrif. |
| | 20-100 | 1.92 | 13.25 | 6.90 | 1.520 | 47.72 | 0.91* | 157.1* | |
| | 40-100 | 3.38 | 13.25 | 3.92 | 1.493 | 27.76 | 0.63* | 167.7* | |
| Methyl chloride ($CH_3Cl$) (40) | 5-86 | 21.15 | 94.70 | 4.48 | 1.331 | 5.95 | 0.96 | 178.1 | Recip. |
| | 20-100 | 29.16 | 116.7 | 4.00 | 1.363 | 4.51 | 0.90 | 184.4 | |
| | 40-100 | 43.25 | 116.7 | 2.69 | 1.342 | 3.07 | 0.62 | 157.0 | |
| Sulphur dioxide ($SO_2$) (764) | 5-86 | 11.81 | 66.45 | 5.63 | 1.415 | 9.08 | 0.97 | 191.4 | Recip. |
| | 20-100 | 17.18 | 84.52 | 4.92 | 1.453 | 6.52 | 0.92 | 193.4 | |
| | 40-100 | 27.10 | 84.52 | 3.12 | 1.444 | 4.17 | 0.63 | 162.6 | |
| Propane ($C_3H_8$) (290) | 5-86 | 42.1 | 155.3 | 3.69 | 1.653 | 4.10 | 1.35* | 92.9* | Recip. |
| | 20-100 | 55.5 | 187.0 | 3.37 | 1.730 | 3.29 | 1.32* | 103.9* | |
| | 40-100 | 78.0 | 187.0 | 2.40 | 1.646 | 2.26 | 0.90* | 101.5* | |
| Ethane ($C_2H_6$) (170) | 5-86 | 236.0 | 675.9 | 2.87 | 3.41 | 1.82 | 2.180 | 105 | Recip. |
| Ethyl chloride ($C_2H_5Cl$) (160) | 5-86 | 4.65 | 27.10 | 5.83 | 1.405 | 24.0 | 0.95* | 106.3* | Rotary |
| | 20-100 | 6.80 | 34.79 | 5.12 | 1.425 | 17.19 | 0.92* | 116.1* | |
| | 40-100 | 10.79 | 34.79 | 3.22 | 1.375 | 10.73 | 0.63* | 109.9* | |

* Approximate.

SOURCE: T. Baumeister "Standard Handbook for Mechanical Engineers." 7th ed., p. 18-5, McGraw-Hill Book Company, New York, 1967.

Table 8-38. Theoretical Horsepower (Hp) and Theoretical Volume (Cfm) of Dry Ammonia Gas Pumped per Minute to Produce 1 Ton of Refrigeration

| Suction pressure and temperature | | Condenser pressure, psig (temperature, deg F) | | | | | | | | | |
|---|---|---|---|---|---|---|---|---|---|---|---|
| | | 103 (65) | | 127 (75) | | 153 (85) | | 182 (95) | | 215 (105) | |
| Psig | Deg F | Hp | Cfm | Hp | Cfm | Hp | Cfm | Hp | Cfm | Hp | Cfm |
| 4 | −20 | 1.058 | 5.84 | 1.205 | 5.96 | 1.361 | 6.09 | 1.525 | 6.23 | 1.691 | 6.43 |
| 6 | −15 | 0.997 | 5.35 | 1.145 | 5.46 | 1.300 | 5.58 | 1.461 | 5.70 | 1.546 | 5.83 |
| 9 | −10 | 0.903 | 4.66 | 1.045 | 4.76 | 1.193 | 4.86 | 1.347 | 4.97 | 1.435 | 5.08 |
| 13 | −5 | 0.818 | 4.09 | 0.954 | 4.17 | 1.094 | 4.25 | 1.244 | 4.35 | 1.321 | 4.44 |
| 16 | 0 | 0.735 | 3.59 | 0.865 | 3.66 | 1.002 | 3.74 | 1.147 | 3.83 | 1.219 | 3.91 |
| 20 | 5 | 0.666 | 3.20 | 0.795 | 3.27 | 0.928 | 3.34 | 1.066 | 3.41 | 1.138 | 3.49 |
| 24 | 10 | 0.592 | 2.87 | 0.726 | 2.93 | 0.854 | 2.99 | 0.991 | 3.06 | 1.060 | 3.12 |
| 28 | 15 | 0.541 | 2.59 | 0.664 | 2.65 | 0.792 | 2.71 | 0.922 | 2.76 | 0.994 | 2.82 |
| 33 | 20 | 0.474 | 2.31 | 0.592 | 2.36 | 0.715 | 2.41 | 0.842 | 2.46 | 0.903 | 2.51 |
| 39 | 25 | 0.410 | 2.06 | 0.523 | 2.10 | 0.599 | 2.15 | 0.767 | 2.20 | 0.829 | 2.24 |
| 45 | 30 | 0.351 | 1.85 | 0.461 | 1.89 | 0.576 | 1.93 | 0.694 | 1.97 | 0.759 | 2.01 |
| 51 | 35 | 0.300 | 1.70 | 0.410 | 1.74 | 0.521 | 1.77 | 0.640 | 1.81 | 0.701 | 1.85 |

SOURCE: T. Baumeister (ed.), "Standard Handbook for Mechanical Engineers," 7th ed., p. 18-9, McGraw-Hill Book Company, New York, 1967.

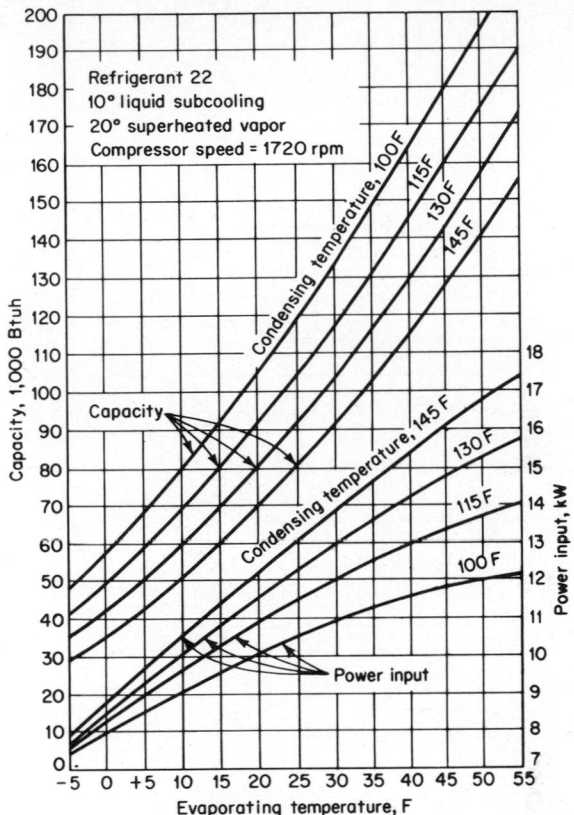

FIG. 8-87. Typical capacity and power input curves for a reciprocating compressor. (*Source: ASHRAE Guide and Data Book, ASHRAE, New York,* 1967.)

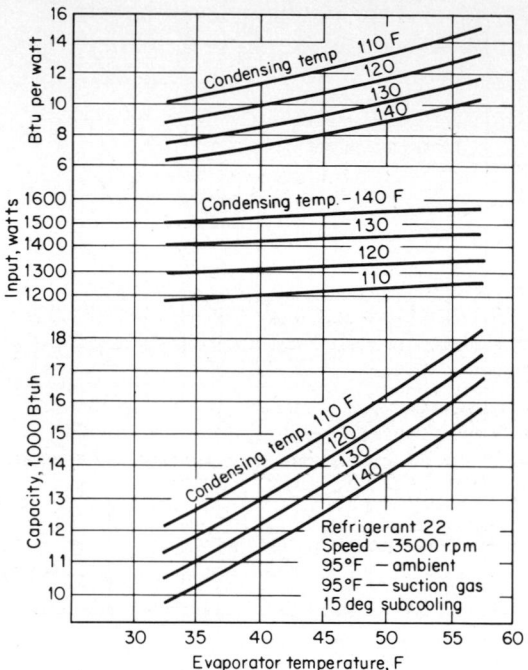

Fig. 8-88. Typical performance curves for a rotary compressor. (*Source: ASHRAE Guide and Data Book, ASHRAE, New York, 1967.*)

SECTION 9

# ENERGY CONVERSION

**John W. Bartlett, Ph.D.;** Manager, Process Evaluation, Pacific Northwest
Laboratories, Battelle Memorial Institute; Member, American Institute of
Chemical Engineers, American Nuclear Society, American Association for
the Advancement of Science

## CONTENTS

## INTRODUCTION

This chapter describes the principles and characteristics of the major energy conversion concepts in use today and expected to be used in the near future. Emphasis is placed on methods for generating large quantities of electrical energy because of the present and projected relative growth of consumption of electrical energy. Most of the effort aimed at improving existing energy conversion systems and developing new ones involves electricity.

Combustion of coal and nuclear reactors are expected to be the major suppliers of electrical energy for the next few decades. Other conversion methods are not yet competitive (economically or because of policy constraints) or not sufficiently developed. Coal combustion and nuclear power are therefore emphasized; basic principles for other energy conversion methods are outlined.

### 9-1. Energy Sources, Availability, and Consumption

**Sources.** Energy is supplied to the earth from only two basic sources: direct solar input and tidal energy from the moon and sun. Energy is available from the two supply sources (solar and tides) and natural resources. There are only five basic natural resources: falling water, wind, geothermal, fossil fuels, and nuclear fuels. A power balance for the earth based on these energy shources is shown in Fig. 9-1.

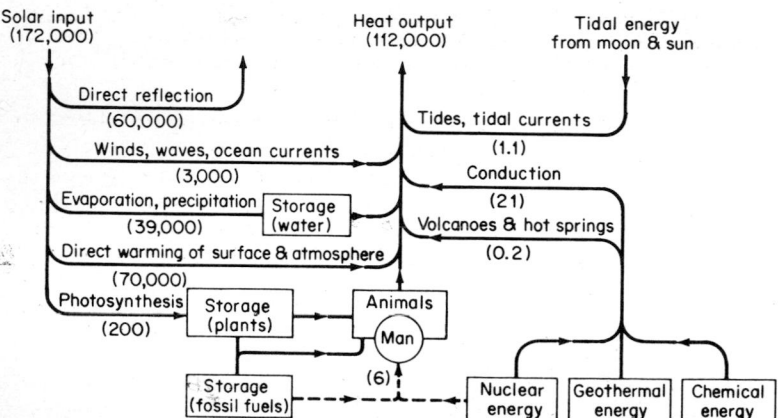

FIG. 9-1. Power balance for the earth. Numbers in parenthesis are billions of kwt. (*Source: M. K. Hubbert, 1962 Report to the National Academy of Sciences.*)

**Availability.** The practical availability of solar, tidal, and wind energy sources depends on capability to devise economic conversion methods. Availability of geothermal, fossil, and nuclear energy depends on (in addition to economic conversion systems) a capability to find the resources and extract them in useful forms. The amounts of energy available from each source therefore depend on the amount present on earth, and people's capability to use it.

Availability of solar energy and nuclear fuels for fission and fusion that are contained in the seas (i.e., not including uranium ores) is, in principle, limitless. Geothermal, fossil, ore-based nuclear, and falling water supplies are finite. Many experts predict that supplies of the petroleum-based fraction of fossil fuels will be depleted in decades, as illustrated by Table 9-1.

### Table 9-1. Depletion of Fossil Fuels

Supply for all fuels has been converted from the usual units (cubic feet, barrels, metric tons) to trillion kWh

| Fuel | Initial supply, trillion kWh | $Y_P$*, date, A.D. | $Y_{90}$‡, date, A.D. |
|------|------|------|------|
| Natural gas (U.S. less Alaska)......... | 407 | 1980 | 2015 |
| Crude oil, U.S. less Alaska............ | 275 | 1970 | 2000 |
| World.................... | 2,240 | 1990 | 2020 |
| Coal, U.S........................... | 5,920 | 2175 | 2400 |
| World........................ | 34,400 | 2110 | 2300 |

\* $Y_P$ = year of peak consumption.
‡ $Y_{90}$ = year in which 90 percent of total cumulative consumption is behind us.
SOURCE: Condensed and adapted from M. King Hubbert, in "Resources and Man," W. A. Freeman & Co., San Francisco, 1969.

Thus, maintenance of energy availability in the future is possible but will require changes in sources and conversion systems.

**Consumption.** Historical consumption patterns in the United States are shown in Fig. 9-2. The breakdown by consumption sector is illustrated by Table 9-2 and Fig. 9-3. In recent years, consumption in the United States

### Table 9-2. Breakdown of Energy and Electricity Consumption in the United States—1968

| All energy sources, % | | Electric component, % | |
|------|------|------|------|
| Electric generation for all purposes.......... | 22.5 | Household............ | 32 |
| Nonelectric household and commercial........ | 21.8 | Commercial........... | 21 |
| Nonelectric transportation.................... | 24.2 | Industrial............. | 43 |
| Nonelectric industrial....................... | 31.0 | Other................ | 4 |
| Other.................................... | 0.5 | | |

SOURCE: Statistical Abstract of the United States, 1970.

was increasing at an annual rate of more than 4 per cent. As a result of the oil embargo of 1973, consumption patterns and growth rates are expected to change.

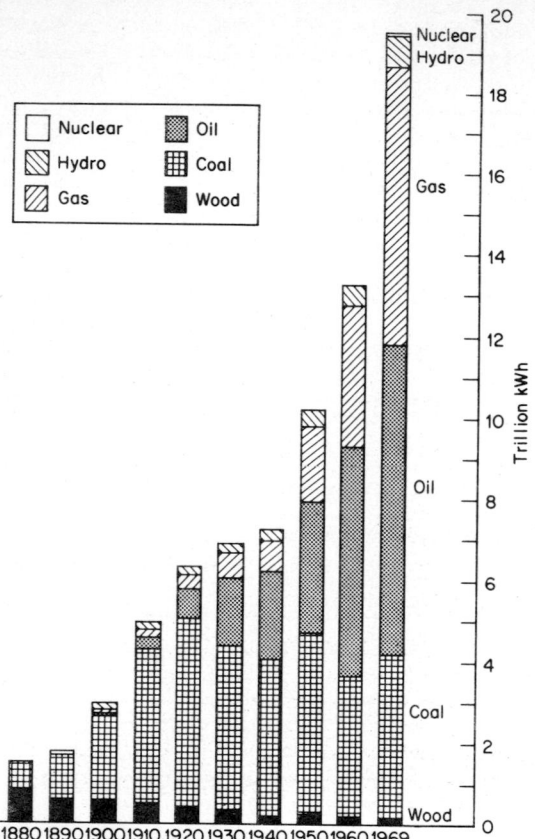

Fig. 9-2. United States energy consumption, 1880–1969. (*Source: "Energy in the United States," Statistical Abstract of the United States, 1970.*)

## 9-2. Fundamentals of Energy Conversion

The law of conservation of energy states that energy cannot be created or destroyed. All "consumptions" are therefore really only changes in form. The second law of thermodynamics states that changes in form proceed in such a way as to reduce, on the whole, the ability of energy to do work. An implication of the second law of thermodynamics is that energy used in conversion processes that do work cannot be recycled.

Conversion of energy from less useful to more useful forms degrades a fraction of the initial energy present into relatively useless low-temperature heat. This degraded energy is associated with the "waste streams," such as

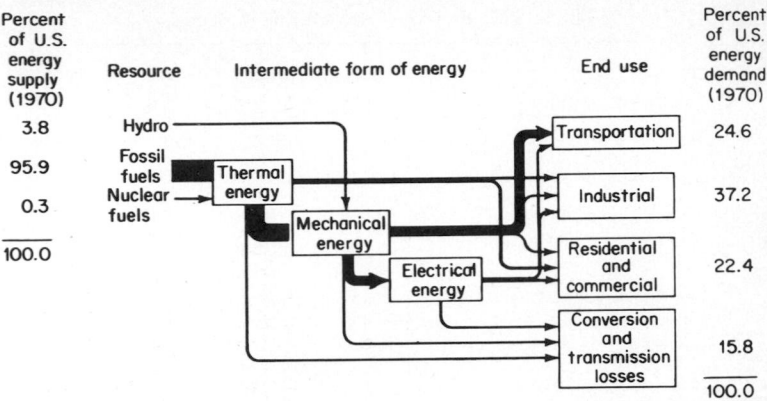

FIG. 9-3. Energy resources and uses.

the exhaust gases from internal combustion engines or the effluent cooling water from plant condensers. Unless it is somehow extracted or used, this waste energy has no utility for us. Since the energy conversion systems most widely used (steam power plants and internal combustion engines) have overall efficiencies between 25 and 40 per cent, a large fraction of the energy input to these systems comes out in the waste streams.

Most energy conversion systems in large-scale use involve a working fluid. The fluid may be gas, vapor, or liquid; when it is a gas or vapor, the performance of that portion of the system using the working fluid can be analyzed in terms of thermodynamics and power cycles (e.g., Carnot, Brayton, Rankine, etc., cycles) outlined in Sec. 2. The over-all system and its performance may, and frequently do, involve conversion processes other than that associated with a working fluid.

Basic forms of energy useful to people are: chemical, thermal, mechanical, electrical, radiant, and kinetic. Some commonly used energy conversion devices, the conversions involved, and typical efficiencies are listed in Table 9-3. Some of these devices, such as the steam power plant, use intermediate forms of energy in addition to the input and output forms listed.

The energy forms can be graded according to "quality" of usefulness which, in thermodynamic terms, is related to the temperature at which it is used. The higher the temperature, the greater the usefulness. In these terms, stored chemical or nuclear energy in the form of fuels has the highest quality; fossil fuels are burned in combustion chambers and nuclear fuels are fissioned in reactor cores. The next highest form of energy is thermal, which may be associated with a high-temperature working fluid. Thermal energy can be degraded to mechanical energy, which in turn can be degraded to electrical energy. Devices and systems that use these energy forms are shown in Fig. 9-4. Characteristics of these systems are described below.

## Table 9-3. Energy Conversion Devices

| Device | Energy forms | | Typical efficiency, % |
| --- | --- | --- | --- |
| | Input | Output | |
| Electric generator......... | M | E | 98 |
| Dry cell battery........... | C | E | 90 |
| Steam boiler.............. | C | T | 88 |
| Fuel cell.................. | C | E | 60 |
| Liquid-fuel rocket......... | T | K | 48 |
| Steam turbine............. | T | M | 45 |
| Steam power plant........ | C | E | 40 |
| Gas turbine.............. | C | M | 35 |
| Solid-state laser........... | E | R | 30 |
| Automobile engine......... | C | M | 25 |
| Fluorescent lamp.......... | E | R | 20 |
| Solar cell................. | R | E | 10 |
| Incandescent lamp........ | E | R | 5 |

C—chemical, E—electrical, K—kinetic, M—mechanical, T—thermal, R—radiant.

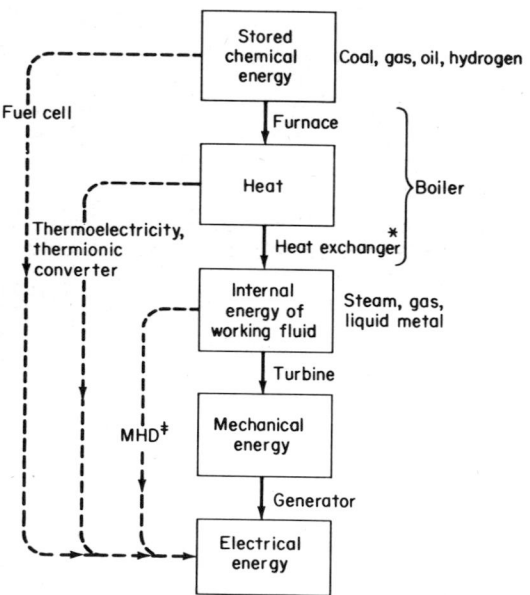

*None required if combustion gases serve as working fluid

‡MHD = magnetohydrodynamics

FIG. 9-4. Energy conversion with chemical fuels.

## FOSSIL-FIRED POWER PLANTS

Energy conversion systems using coal, oil, or natural gas as the fuel convert the chemical energy of the fuel to electrical energy via the sequence of conversions shown in Fig. 9-4. Performance of these systems depends on the quality of the fuel and the efficiency of the boiler, turbine, and generator. As shown in Table 9-3, over-all efficiency of modern plants is on the order of 40 per cent.

Concern about depletion of oil and natural gas reserves has produced large research and development programs aimed at finding effective processes for converting coal to fuel forms that can be used in power plants designed to operate with oil or gas. Thus, although the power plants themselves represent well-developed technology, major developments are occurring in processes associated with their use.

**Conversion Factors.** Engineering units used in this section can be converted to SI units as follows:

| To convert from | To | Multiply by |
|---|---|---|
| Btu (60°F) | calories (gram) | 252 |
| Btu (60°F) | joules | 1,054.7 |
| pounds | grams | 453.59 |
| °F | °C | (°F − 32)/1.8 |
| ft² | cm² | 929.0 |

For a more complete listing of conversion factors, see Tables 1-1 and 1-2 and inside covers.

### 9-3. Fuels and Combustion

The basic requirements of calculations on fuels and combustion are concerned with evaluation of air needed, the heat of reaction, and analysis of the products of combustion.

**Air Requirements.** The air requirements are given in Table 9-4 on both gravimetric and volumetric bases. In these and other calculations the air analysis is taken as

| Basis | $O_2$, % | $N_2$, % |
|---|---|---|
| Gravimetric | 23.2 | 76.8 |
| Volumetric | 21.0 | 79.0 |

$N_2$ is not entirely chemical nitrogen, but includes all other gases inert to the combustion reaction.

A useful approximate formula for estimating the air requirements for the combustion of coals and oils is

$$\text{Air required, lb/lb fuel} = \frac{\text{higher heating value of fuel, Btu/lb}}{1,300} \quad (9\text{-}1)$$

## Table 9-4. Heating Value and the Products of Combustion

| Substance | Molecular or atomic wt | Over-all combustion reaction | Heat of combustion* | | | |
|---|---|---|---|---|---|---|
| | | | High-heat value | | Low-heat value | |
| | | | Btu/lb | Btu/ft³† | Btu/lb | Btu/ft³ |
| 1 Graphite............ | 12.01 | $C + O_2 \rightarrow CO_2$ | 14,087 | ..... | 14,087 | ..... |
| 2 Carbon (coal)......... | (12.01) | $C + O_2 \rightarrow CO_2$ | 14,447 | ..... | 14,447 | ..... |
| 3 Carbon (coal)......... | (12.01) | $C + 0.5O_2 \rightarrow CO$ | 4,341 | ..... | 4,341 | ..... |
| 4 Carbon monoxide..... | 28.01 | $CO + 0.5O_2 \rightarrow CO_2$ | 4,344 | 321 | 4,344 | 321 |
| 5 Sulfur.............. | 32.06 | $S + O_2 \rightarrow SO_2$ | 3,980 | ..... | 3,980 | ..... |
| 6 Hydrogen............ | 2.016 | $H_2 + 0.5O_2 \rightarrow H_2O$ | 60,958 | 325 | 51,571 | 275 |
| 7 Hydrogen sulfide...... | 34.08 | $H_2S + 1.5O_2 \rightarrow SO_2 + H_2O$ | 7,180 | 639 | 6,620 | 589 |
| 8 Carbon disulfide...... | 76.13 | $CS_2 + 3O_2 \rightarrow CO_2 + 2SO_2$ | 12,050 | 620 | 12,050 | 620 |
| 9 Ammonia............ | 17.03 | $2NH_3 + 1.5O_2 \rightarrow N_2 + 3H_2O$ | 9,668 | 435 | 8,001 | 360 |
| 10 Methane............ | 16.04 | $CH_4 + 2O_2 \rightarrow CO_2 + 2H_2O$ | 23,861 | 1,011 | 21,502 | 911 |
| 11 Ethane............. | 30.07 | $C_2H_6 + 3.5O_2 \rightarrow 2CO_2 + 3H_2O$ | 22,304 | 1,772 | 20,416 | 1,622 |
| 12 Propane............ | 44.09 | $C_3H_8 + 5O_2 \rightarrow 3CO_2 + 4H_2O$ | 21,646 | 2,522 | 19,929 | 2,322 |
| 13 Butane............. | 58.12 | $C_4H_{10} + 6.5O_2 \rightarrow 4CO_2 + 5H_2O$ | 21,293 | 3,270 | 19,665 | 3,020 |
| 14 Hexane (vapor)...... | 86.17 | $C_6H_{14} + 9.5O_2 \rightarrow 6CO_2 + 7H_2O$ | 20,928 | 4,765 | 19,391 | 4,414 |
| 15 Octane (vapor)....... | 114.23 | $C_8H_{18} + 12.5O_2 \rightarrow 8CO_2 + 9H_2O$ | 20,747 | 6,266 | 19,256 | 5,812 |
| 16 Ethylene........... | 28.05 | $C_2H_4 + 3O_2 \rightarrow 2CO_2 + 2H_2O$ | 21,625 | 1,603 | 20,276 | 1,503 |
| 17 Propylene.......... | 42.08 | $C_3H_6 + 4.5O_2 \rightarrow 3CO_2 + 3H_2O$ | 21,032 | 2,338 | 19,683 | 2,188 |
| 18 Butylene........... | 56.10 | $C_4H_8 + 6O_2 \rightarrow 4CO_2 + 4H_2O$ | 20,833 | 3,076 | 19,484 | 2,877 |
| 19 Acetylene.......... | 26.04 | $C_2H_2 + 2.5O_2 \rightarrow 2CO_2 + H_2O$ | 21,460 | 1,473 | 20,734 | 1,423 |
| 20 Benzene (vapor)..... | 78.11 | $C_6H_6 + 7.5O_2 \rightarrow 6CO_2 + 3H_2O$ | 18,172 | 3,745 | 17,446 | 3,595 |
| 21 Toluene (vapor)..... | 92.13 | $C_7H_8 + 9O_2 \rightarrow 7CO_2 + 4H_2O$ | 18,422 | 4,490 | 17,601 | 4,285 |
| 22 Naphthalene (vapor).. | 128.16 | $C_{10}H_8 + 12O_2 \rightarrow 10CO_2 + 4H_2O$ | 17,300 | 5,854 | 16,700 | 5,654 |
| 23 Methyl alcohol (vapor) | 32.04 | $CH_3OH + 1.5O_2 \rightarrow CO_2 + 2H_2O$ | 10,270 | 855 | 9,080 | 755 |
| 24 Ethyl alcohol (vapor) | 46.07 | $C_2H_5OH + 3O_2 \rightarrow 2CO_2 + 3H_2O$ | 13,170 | 1,575 | 11,930 | 1,425 |
| 25 Lignite‡.......... | ....... | $C_{24}H_{18}O_5 + 28.5O_2 \rightarrow 24CO_2 + 9H_2O$ | 12,055 | ..... | 11,505 | ..... |
| 26 Bituminous coal‡.... | ....... | $C_{24}H_{20}O_2 + 29.0O_2 \rightarrow 24CO_2 + 10H_2O$ | 14,550 | ..... | 14,055 | ..... |
| 27 Anthracite‡........ | ....... | $C_{48}H_{18}O + 52O_2 \rightarrow 48CO_2 + 9H_2O$ | 15,230 | ..... | 14,940 | ..... |

* Data for heat of combustion for items 1 to 4 and 10 to 21 based on F. D. Rossini et al., Selected Values of Properties of Hydrocarbons, *Natl. Bur. Standards (U.S.) Circ.* C461, November, 1947. Reactants and products at 25°C (77°F).

† Based upon the standard cubic foot at 60°F and 30 in. Hg of a perfect gas. One lb mole = 379 std ft³.

**Heating Value.** High or gross heating values of fuels are distinguished from low or net heating values by the condition of the water formed in the products of combustion; if it is condensed to a liquid instead of remaining in the vapor form, the latent heat becomes available. The difference between the high and the low heating values is computed by

$$\text{Difference} = 9H_2(1,090.7 - 0.545t) \qquad \text{Btu/lb fuel} \qquad (9\text{-}2)$$

where $H_2$ = weight of hydrogen in fuel, lb/lb

$t$ = temperature of atmosphere, °F

In American practice it is customary to buy fuels, and to guarantee the performance of fuel-burning equipment, on the basis of the high heating value. Heating values are best determined by calorimetric methods, but if such data are lacking, the high heating value of a coal can be estimated by

## of Various Solid, Liquid, and Gaseous Fuels

| | | Combustion with theoretical amount of air | | | | | | | | | | | | Lb air/ 1,000 Btu | | CO₂, % by vol (dry basis) | |
|---|---|---|---|---|---|---|---|---|---|---|---|---|---|---|---|---|---|
| Lb/lb | | | | | | | Ft³/ft³ | | | | | | | | | | |
| Required | | Products of combustion | | | | | Required | | Products of combustion | | | | | | | | |
| O₂ | Air | N₂ | CO₂ | H₂O | SO₂ | Total | O₂ | Air | N₂ | CO₂ | H₂O | SO₂ | Total | Lhv | Hhv | | |
| 2.67 | 11.50 | 8.83 | 3.67 | .... | .... | 12.50 | .... | .... | .... | .... | .... | .... | .... | 0.82 | 0.82 | 21.0 | 1 |
| 2.67 | 11.50 | 8.83 | 3.67 | .... | .... | 12.50 | .... | .... | .... | .... | .... | .... | .... | 0.80 | 0.80 | 21.0 | 2 |
| 1.33 | 5.73 | 4.40 | 2.33§ | .... | .... | 6.73 | .... | .... | .... | .... | .... | .... | .... | .... | .... | .... | 3 |
| 0.57 | 2.46 | 1.89 | 1.57 | .... | .... | 3.46 | 0.50 | 2.38 | 1.88 | 1.00 | .... | .... | 2.88 | 0.57 | 0.57 | 34.6 | 4 |
| 1.00 | 4.31 | 3.31 | .... | .... | 2.00 | 5.31 | .... | .... | .... | .... | .... | .... | .... | 1.08 | 1.08 | .... | 5 |
| 7.94 | 34.34 | 26.40 | .... | 8.98 | .... | 35.38 | 0.50 | 2.38 | 1.88 | .... | 1.00 | .... | 2.88 | 0.67 | 0.56 | .... | 6 |
| 1.41 | 6.10 | 4.69 | .... | 0.53 | 1.88 | 7.10 | 1.50 | 7.14 | 5.64 | .... | 1.00 | .... | 7.64 | .92 | .85 | .... | 7 |
| 1.26 | 5.44 | 4.18 | 0.58 | .... | 1.68 | 6.44 | 3.00 | 14.28 | 11.28 | 1.00 | .... | 2.00 | 14.28 | .45 | .45 | 21.0¶ | 8 |
| 1.41 | 6.08 | 5.49 | .... | 1.59 | .... | 7.08 | 0.75 | 3.57 | 3.32 | .... | .... | 1.50 | 4.82 | .76 | .63 | .... | 9 |
| 4.00 | 17.27 | 13.27 | 2.74 | 2.25 | .... | 18.26 | 2.00 | 9.52 | 7.52 | 1.00 | 2.00 | .... | 10.52 | .80 | .72 | 11.7 | 10 |
| 3.73 | 16.12 | 12.39 | 2.92 | 1.80 | .... | 17.11 | 3.50 | 16.65 | 13.15 | 2.00 | 3.00 | .... | 18.15 | .79 | .72 | 13.2 | 11 |
| 3.64 | 15.70 | 12.06 | 2.99 | 1.64 | .... | 16.69 | 5.00 | 23.80 | 18.80 | 3.00 | 4.00 | .... | 25.80 | .79 | .72 | 13.8 | 12 |
| 3.58 | 15.44 | 11.86 | 3.03 | 1.56 | .... | 16.45 | 6.50 | 30.90 | 24.40 | 4.00 | 5.00 | .... | 33.40 | .79 | .72 | 14.1 | 13 |
| 3.53 | 15.21 | 11.68 | 3.07 | 1.46 | .... | 16.21 | 9.50 | 45.20 | 35.70 | 6.00 | 7.00 | .... | 48.70 | .79 | .78 | 14.4 | 14 |
| 3.50 | 15.10 | 11.60 | 3.08 | 1.42 | .... | 16.10 | 12.50 | 59.50 | 47.00 | 8.00 | 9.00 | .... | 64.00 | .78 | .73 | 14.5 | 15 |
| 3.42 | 14.75 | 11.33 | 3.14 | 1.29 | .... | 15.76 | 2.00 | 14.28 | 11.28 | 2.00 | 2.00 | .... | 15.28 | .73 | .68 | 15.1 | 16 |
| 3.42 | 14.75 | 11.33 | 3.14 | 1.29 | .... | 15.76 | 4.50 | 21.40 | 16.90 | 3.00 | 3.00 | .... | 22.90 | .74 | .70 | 15.1 | 17 |
| 3.42 | 14.75 | 11.33 | 3.14 | 1.29 | .... | 15.76 | 6.00 | 28.55 | 22.55 | 4.00 | 4.00 | .... | 30.55 | .76 | .71 | 15.1 | 18 |
| 3.07 | 13.23 | 10.16 | 3.38 | 0.69 | .... | 14.23 | 2.50 | 11.90 | 9.40 | 2.00 | 1.00 | .... | 12.40 | .64 | .62 | 17.5 | 19 |
| 3.07 | 13.23 | 10.16 | 3.38 | 0.69 | .... | 14.23 | 7.50 | 35.70 | 28.20 | 6.00 | 3.00 | .... | 37.20 | .76 | .73 | 17.5 | 20 |
| 3.13 | 13.50 | 10.35 | 3.35 | 0.78 | .... | 14.48 | 9.00 | 42.80 | 33.80 | 7.00 | 4.00 | .... | 44.80 | .77 | .73 | 17.2 | 21 |
| 3.00 | 12.93 | 9.93 | 3.44 | 0.56 | .... | 13.93 | 12.00 | 57.10 | 45.10 | 10.00 | 4.00 | .... | 59.10 | .77 | .75 | 18.1 | 22 |
| 1.60 | 6.90 | 5.30 | 1.37 | 1.13 | .... | 7.80 | 1.50 | 7.14 | 5.64 | 1.00 | 2.00 | .... | 8.64 | .76 | .67 | 15.1 | 23 |
| 2.08 | 8.96 | 6.89 | 1.91 | 1.17 | .... | 9.96 | 3.00 | 14.28 | 11.28 | 2.00 | 3.00 | .... | 16.28 | .75 | .68 | 15.1 | 24 |
| 2.14 | 9.22 | 7.08 | 2.71 | 0.43 | .... | 10.22 | .... | .... | .... | .... | .... | .... | .... | .80 | .76 | 19.5 | 25 |
| 2.60 | 11.21 | 8.69 | 3.08 | 0.52 | .... | 12.21 | .... | .... | .... | .... | .... | .... | .... | .80 | .77 | 18.6 | 26 |
| 2.74 | 11.83 | 9.08 | 3.46 | 0.28 | .... | 12.83 | .... | .... | .... | .... | .... | .... | .... | .79 | .78 | 19.5 | 27 |

‡ Computed from average moisture- and ash-free ultimate analysis of 17 lignites, 27 medium and high-volatile coals, and 5 anthracite coals in the United States. The formulas for lignite, bituminous, and lignitic coals, given in the third column, do not represent the true constitution of the coal molecule, which is much more complex, but are adequate for stoichiometric calculations.

¶ $CO_2 + SO_2$. § CO.

source: Kent, "Mechanical Engineers' Handbook Power," 12th ed., pp. 2–04, 2–05, John Wiley & Sons, Inc. New York, 1950.

Dulong's equation

$$\text{Btu/lb} = 14{,}544C + 62{,}028(H - O/8) + 4{,}050S \qquad (9\text{-}3)$$

where C, H, O, and S are the weight percentages of these four elements obtained from the ultimate analysis of the fuel.

The heating values of petroleum and its products can be estimated from the specific gravities, as shown in Fig. 9-5. Combustion data for representative fuels are given in Table 9-4.

**Products of Combustion.** Flue-gas analyses are made in order to determine the effectiveness of combustion operations and are ordinarily given on the dry, volumetric basis. If the nitrogen content of the fuel is small, then the excess air can be computed from

$$\text{Excess air} = \frac{3.78(O_2 - CO/2)}{N_2 - 3.78(O_2 - CO/2) \times 100} \qquad \text{per cent} \qquad (9\text{-}4)$$

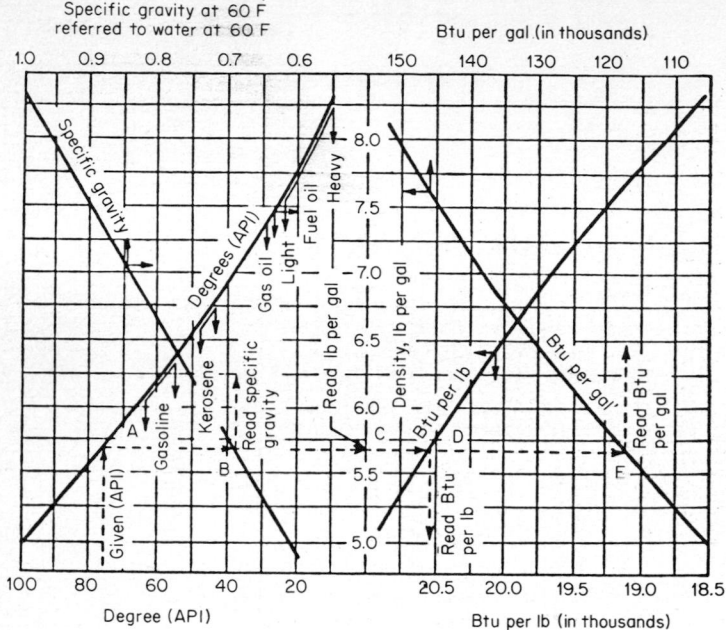

FIG. 9-5. Heating value from specific gravity. (*Source: "Combustion Engineering," p. 25.14, Combustion Engineering, Inc., New York, 1947.*)

### Table 9-5. Excess Air at Furnace Outlet

| Fuel | Excess air, % | Fuel | Excess air, % |
|------|---------------|------|---------------|
| Coal | 10–40 | Natural gas | 5–10 |
| Coke | 20–40 | Refinery gas | 8–15 |
| Wood | 25–50 | Blast-furnace gas | 15–25 |
| Bagasse | 25–45 | Coke-oven gas | 5–10 |
| Oil | 8–15 | | |

SOURCE: "Combustion Engineering," 1st ed., 1947. p. 10.3, Combustion Engineering, Inc., New York, N. Y.

where $O_2$, $N_2$, and $CO$ are percentages by volume obtained from the flue-gas analysis. Some customary excess-air values are given in Table 9-5.

## 9-4. Power-plant Performance Factors

Heat rate is defined for the over-all thermal performance as

$$\text{Heat rate, Btu/kwhr} = \frac{\text{heat supplied in fuel for period, Btu}}{\text{energy output for period, kwhr}} \quad (9\text{-}5)$$

$$\text{Thermal eff, } \% = \frac{3,412.75}{\text{heat rate}} \times 100 \tag{9-6}$$

$$\text{Capacity factor, } \% = \frac{\text{average load for period, kw}}{\text{rated capacity, kw}} \times 100 \tag{9-7}$$

$$\text{Load factor, } \% = \frac{\text{average load for period, kw}}{\text{peak load during period, kw}} \times 100 \tag{9-8}$$

## 9-5. Boiler Performance

### Nomenclature

$h_{\text{steam}}$ = enthalpy leaving boiler unit (superheater outlet), Btu/lb
$h_{\text{feed water}}$ = enthalpy of feed water entering boiler unit (economizer inlet), Btu/lb
$W_m$ = moisture content of fuel, lb/lb
$t_{\text{fuel}}$ = fuel temperature, °F
$t_{fg}$ = flue-gas temperature, °F
$H_2$ = lb hydrogen/lb fuel, from ultimate analysis
$W_{da}$ = weight dry air supplied, lb/lb fuel
$W_w$ = weight water vapor/lb dry air
$t_a$ = ambient temperature or temperature of air entering air heater, °F
$W_{dg}$ = weight dry flue gases, lb/lb fuel
C = lb carbon/lb fuel, from ultimate analysis
CO = CO in flue gas, dry volumetric basis, per cent
$CO_2$ = $CO_2$ in flue gas, dry volumetric basis, per cent
refuse = lb refuse/lb fuel, as burned in boiler furnace
ash = lb ash/lb fuel, from ultimate analysis

### Heat Added to Steam

$$\Delta Q = h_{\text{steam}} - h_{\text{feed water}} \quad \text{Btu/lb} \tag{9-9}$$

With a resuperheater, the heat added as reheat must be included and

$$h_{\text{reheat}} = h_{\text{leaving reheater}} - h_{\text{entering reheater}} \tag{9-10}$$

**Boiler Rating and Steam Output.** Older procedures used the term "developed boiler horsepower" to measure the output of the boiler, or the heat added to the steam, defining it as the evaporation of 34.5 lb water from and at 212°F. Thus,

$$\text{Developed boiler hp} = 34.5 \times 970.4 = 33,479 \text{ Btu/hr} \tag{9-11}$$

The rated boiler horsepower was defined as

$$\text{Rated boiler hp} = 10 \text{ ft}^2 \text{ heating surface} \tag{9-12}$$

Thus, $\qquad \text{Per cent rating} = \dfrac{\text{developed boiler hp}}{\text{rated boiler hp}} \times 100 \tag{9-13}$

*Factor of Evaporation, F.E.*

$$\text{F.E.} = \frac{\text{actual heat absorbed in converting water to steam}}{\text{latent heat of steam from and at 212°F}}$$

$$= (h_{\text{steam}} - h_{\text{feed water}})/970.4 \tag{9-14}$$

*Evaporation*

$$\text{Actual evaporation, A.E.,} = \frac{\text{lb steam made during period}}{\text{lb fuel fired during period}} \tag{9-15}$$

$$\begin{array}{l} \text{Equivalent evaporation, E.E., lb} \\ \text{steam/lb fuel, from and at 212°F} \end{array} = \begin{array}{l} \text{actual evaporation} \\ \times \text{ factor of evaporation} \end{array}$$

$$= \text{A.E.} \times \text{F.E.} \tag{9-16}$$

*Boiler Efficiency*

$$\text{Boiler eff} = \frac{\text{heat added to steam over period, Btu}}{\text{heat supplied in fuel over period, Btu}} \tag{9-17}$$

In American practice the heat supplied in the fuel is the high or gross heating value on the "as fired basis."

**Heat Balance and Losses.** By the first law of thermodynamics it is possible to account for all the heat supplied in the fuel by adding all the losses to the heat supplied to the steam.

$$\text{Loss due to moisture in fuel} = W_m(1{,}090.7 - t_{\text{fuel}} + 0.455 t_{fg})$$
$$\text{Btu/lb fuel} \tag{9-18}$$

$$\begin{array}{l} \text{Loss due to hydrogen burning to water vapor instead of liquid} \\ = 9 \times \text{H}_2(1{,}090.7 - t_{\text{fuel}} + 0.455 t_{fg}) \qquad \text{Btu/lb fuel} \tag{9-19} \end{array}$$

$$\text{Loss due to moisture in air} = W_{da} \times W_w \times 0.47(t_{fg} - t_a)$$
$$\text{Btu/lb fuel} \tag{9-20}$$

$$\text{Loss due to dry stack gases} = 0.24 W_{dg}(t_{fg} - t_a) \qquad \text{Btu/lb fuel} \tag{9-21}$$

$$\begin{array}{l} \text{Loss due to incomplete combustion of carbon} \\ = 10{,}160(\text{C} \times \text{CO})/(\text{CO}_2 + \text{CO}) \qquad \text{Btu/lb fuel} \tag{9-22} \end{array}$$

$$\text{Loss due to combustible in refuse} = 14{,}600(\text{refuse} - \text{ash})$$
$$\text{Btu/lb fuel} \tag{9-23}$$

$$\begin{array}{l} \text{Loss due to radiation and unaccounted for, Btu/lb fuel} \\ = \text{heating value of fuel} - \text{sum of Eqs. (9-18) to (9-23)} \tag{9-24} \end{array}$$

## 9-6. Sulfur Dioxide Emissions

Most coal now being mined for electrical power generation has a "high" sulfur content, i.e., emissions of sulfur dioxide with flue gases will exceed clear-air regulations. Releases of pollutants in the United States during 1968 are shown in Table 9-6.

## Table 9-6. Sources of Air Pollution in the United States—1968

(Percentages by weight)

| Source | Carbon monoxide (100 million tons) | Sulfur oxides (33 million tons) | Hydro-carbons (32 million tons) | Nitrogen oxides (21 million tons) | Particulates (28 million tons) |
|---|---|---|---|---|---|
| Fuel burning for transportation..... | 63.8% | 2.4% | 51.9% | 39.3% | 4.3% |
| Fuel burning in stationary sources... | 1.9 | 73.5 | 2.2 | 48.5 | 31.4 |
| Industrial processes other than fuel burning.......................... | 9.6 | 22.0 | 14.4 | 1.0 | 26.5 |
| Solid-waste disposal................ | 7.8 | 0.3 | 5.0 | 2.9 | 3.9 |
| Miscellaneous*.................... | 16.9 | 1.8 | 26.5 | 8.3 | 33.9 |

\* Includes forest fires, agricultural burning, coal waste fires, gasoline marketing.

The high-rank bituminous coals of the midwest and east typically contain more than 3 per cent sulfur. Only about 10 per cent of known domestic coal reserves contain 1 per cent or less sulfur, and these are usually held for metallurgical use. Clean-air standards will require less than 1 per cent sulfur, and some local standards would set the limit as low as 0.3 per cent. Methods to limit sulfur dioxide emissions are therefore necessary.

The sulfur is present in coal in inorganic and organic forms. The organic material, which is on average about half of the sulfur in the coal, is contained in the combustible matter and cannot be removed by crushing and cleaning. Such precombustion operations are, however, useful for removing the inorganic sulfur, which is usually present as pyrites, etc. Crushing, washing, and froth flotation can be used to remove the pyritic material.

Organic sulfur can, in principle, be removed during combustion or from the flue gas. Removal during combustion requires feeding limestone with the coal to a fluidized-bed combustion chamber. Removal from flue gases requires catalytic oxidation of $SO_2$ or removal by scrubbing with lime, limestone, double alkali, magnesium oxide, sodium sulfate, or sodium citrate. Development of processes involving these options is aimed at proving technical performance and minimizing costs of pollution control operations.

## 9-7. Coal Gasification

Coal gasification was a widely used technology early in the twentieth century. It fell into disuse as a result of cheap, abundant natural gas and petroleum. Now that these resources are becoming expensive and supplies are diminishing, coal gasification is being revived and new processes are being developed.

Gasification involves the heating of coal, as in distillation, and the subsequent reaction of the solid residue with air, oxygen, steam, or various mixtures of these reactants. Two types of products can result: a high-Btu gas, which has a heating value similar to that of natural gas, or a low-Btu gas, sometimes called producer gas, which is a suitable fuel for some boilers.

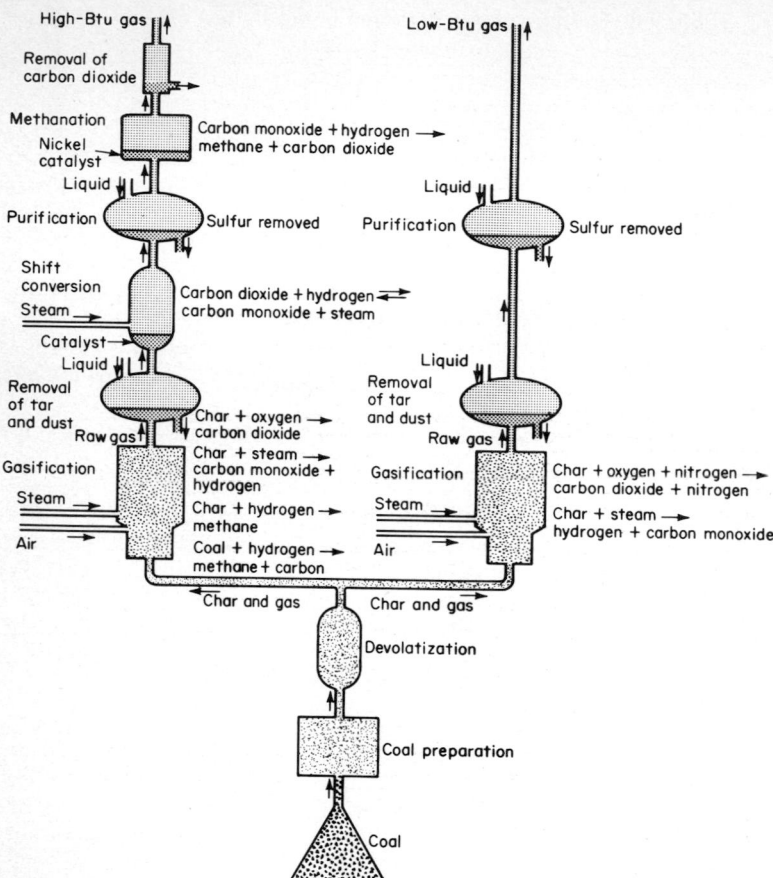

FIG. 9-6. Production of high- and low-Btu gas from coal. (*Adapted from Harry Perry, "Coal Gasification," Scientific American, March* 1974.)

The process can also be used to provide the raw materials for manufacture of synthetics such as alcohols, ketones, and other petroleum products.

Two gasification processes have previously been developed and used on a commercial scale. The *Lurgi process* reacts coal with oxygen and steam at above-atmospheric pressures to produce a high-Btu gas. The *Koppers-Totzek process* also uses steam and oxygen in the gasification but operates at atmospheric pressure. Neither process is ideal, and many alternative processing concepts are under development. Basic concepts for processing to produce high- or low-Btu gas are illustrated by Fig. 9-6.

In general, technology for coal gasification and liquefaction is known and available. The major limitation to its application is cost in comparison with costs for natural products. As costs of natural gas and petroleum increase and supplies diminish, coal conversion processes can be expected to reappear in commercial use. All will add cost penalties to energy supply because they serve only to convert fuel form.

## NUCLEAR FISSION

Consensus projections indicate that only nuclear fission can maintain adequate supplies of energy while fossil-fueled sources decrease as a result of increasing cost and exhausted supplies and new, large-scale sources (such as solar power and fusion) are still being developed. Nuclear fission is therefore expected to be the major, growing source of energy in the next few decades.

Design of fission-powered systems is based on principles of nuclear physics and engineering that are outlined here. Nuclear engineering is involved in designing the fuel-bearing region (known as the *core*), the core-containment vessel, auxiliary systems and components, and shielding. The core and its associated equipment in a nuclear power plant basically replace the boiler in a conventional system. Balance-of-plant designs (i.e., the steam generator, turbine, and electricity generator portions of the system) can follow conventional practice or differ markedly. One of the characteristics of nuclear power systems is the wide variety of designs possible in comparison with fossil-fired systems.

### 9-8. Atomic Structure and Nuclear Particles

Nuclear engineering requires knowledge and manipulation of atoms and the particles that comprise them. An atom is composed of a dense nucleus surrounded by electrons arranged in various orbits. The atom is mostly free space; the atomic radius is on the order of $10^{-8}$ cm, and the radius of the nucleus is $10^{-12}$ to $10^{-13}$ cm.

Ordinary chemical reactions involve changes in the orbital electrons of the atom; nuclear reactions such as fission involve the nucleus. There are great differences in the energy involved in these two types of reactions. For example, the combustion reaction (oxidation of carbon to $CO_2$) releases

4.1 electron volts (ev) of energy; in contrast, fission of a uranium atom releases approximately $2 \times 10^8$ ev (i.e., 200 Mev) of energy.

The physicists have demonstrated that nuclei are composed of many fundamental particles. Relatively few of these particles are, however, of interest in nuclear reactors. These include the neutron and proton (the major constituents of the nucleus, also called nucleons), the electron, and the gamma ray. Gamma rays are high-energy electromagnetic radiation originating in the nucleus; they have neither mass nor charge. Electrons are of low mass [approximately $5.49 \times 10^{-4}$ atomic mass units (amu), where one amu = $1.66 \times 10^{-24}$ g on the physical mass scale] and may have either a positive or negative charge. The positively charged electron is called a positron.

The proton has a mass of 1.00759 amu and is positively charged; the neutron is electrically neutral and has a mass of 1.00898 amu. The sum of the number of protons and neutrons in a nucleus defines the mass of an atom; the number of protons defines the atomic number, and therefore the specific element in the periodic table. Individual atoms may possess the same number of protons but different numbers of neutrons (called *isotopes* of a given element) or they may be of the same mass but different charge (*isobars*). Complete designation of an atom therefore requires definition of charge or element and mass. To illustrate the symbolism used, consider uranium, which has an atomic number of 92. The isotope of uranium which has a mass of 238 amu would be designated $_{92}U^{238}$, $U^{238}$, U 238, or uranium 238.

Physicists are still engaged in unraveling the mystery of the structure of the nucleus. The most important aspect of this mystery, from a practical point of view, is that the nucleus, composed of positively charged particles with large repulsive forces between them, manages to be held together as a dense-packed mass. Clearly, a "nuclear glue" characterized by large attractive forces is required. Details of this nuclear glue are not understood. It is characterized, however, by the "binding energy." Experiments have shown that the actual mass of a nucleus is less than the sum of the masses of the constituent nucleons; the binding energy can be shown to correspond to this mass defect.

The binding energy per nucleon is a measure of the stability of a nucleus. When the binding energy per nucleon is high, the nucleus is most stable. Figure 9-7 illustrates the variation of binding energy per nucleon as a function of atomic mass. It can be shown that energy must be released when nuclei of low binding energy per nucleon are converted to nuclei of high binding energy per nucleon. Hence, from Fig. 9-7, combination of light nuclei (fusion) or splitting of heavy nuclei (fission) should release energy.

## 9-9. Radioactivity

A radioactive nucleus contains energy in excess of that characteristic of a stable configuration. To achieve stability, the nucleus must emit the excess energy: the process is called radioactive decay. The energy emission may involve release of a nuclear particle or one or more gamma rays, or both. Several basic modes of decay have been identified, and are discussed below.

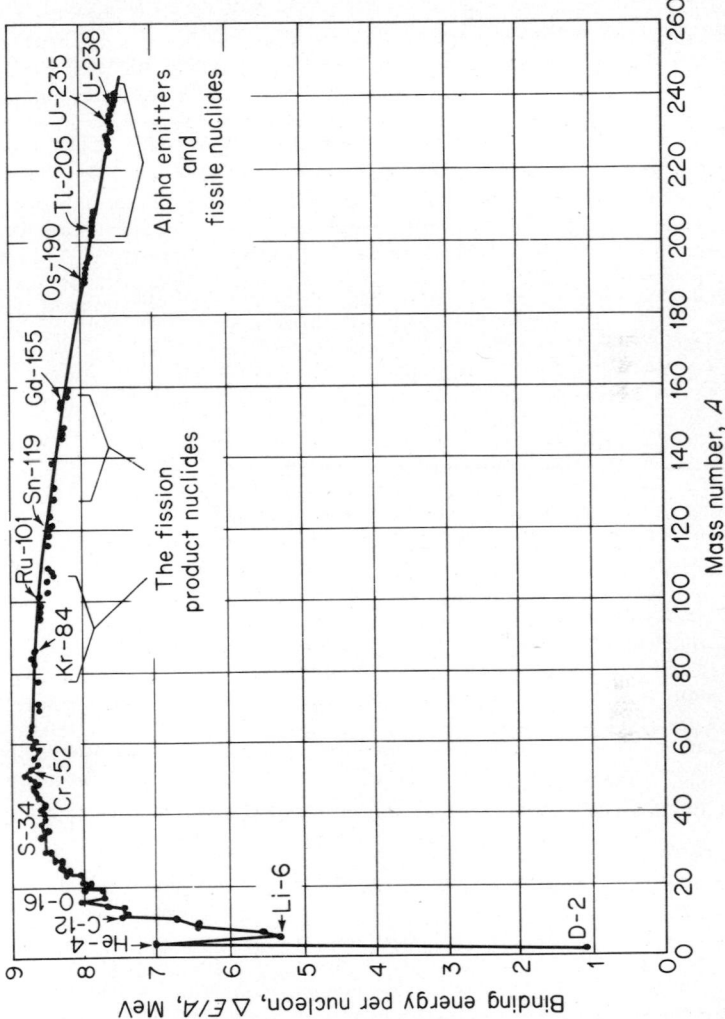

Fig. 9-7. Binding energy per nucleon for stable nuclei. (*After Glasstone and Edlund,* "*Elements of Reactor Theory,*" *D. Van Nostrand Company, Inc., Princeton, N.J., 1952.*)

The time interval between formation and decay of a single radioactive isotope cannot be predicted. It has been observed, however, that a large number of radioactive isotopes of a given kind will decay at a fixed rate characterized by a quantity known as the decay constant $\lambda$. The rate of decay, or alternatively, disintegration, is related to the decay constant by

$$dN/dt = -\lambda N \tag{9-25}$$

where $N$ is the number of radioactive atoms present at time $t$. The units of $\lambda$ are $(\text{time})^{-1}$. If at time $t = 0$ there were $N_0$ radioactive atoms present, Eq. (9-25) gives

$$N = N_0 e^{-\lambda t} \tag{9-26a}$$

or
$$\ln N/N_0 = -\lambda t \tag{9-26b}$$

which shows that $\lambda$ may be determined as the slope of the straight line obtained with a semilogarithmic plot of disintegration rate vs. time.

Values of $\lambda$ can be determined only by experiment, and show great variation. It is frequently more convenient to report the half-life $T_{1/2}$, which is defined as the time for the number of radioactive atoms present to decrease by a factor of 2, and may be derived from Eq. (9-26b) in terms of $\lambda$ as

$$T_{1/2} = \ln (2)/\lambda = 0.693/\lambda \tag{9-26c}$$

Measured values of the half-life range from $10^{-7}$ to $10^{10}$ years.

Many radioactive nuclides have been identified. Some are naturally occurring, but most have been man-made by bombarding stable nuclei with high-energy particles. Nearly all naturally occurring radioisotopes have mass numbers greater than 80. Man-made isotopes, however, cover the entire spectrum of elements, and many are commercially available.

The quantity $\lambda N$ is known as the disintegration rate, or *activity*. The basic unit of activity is the curie, defined as $3.7 \times 10^{10}$ disintegrations per second (dis/sec). One curie corresponds approximately to the activity of one gram of radium and is a large amount of activity. More frequently encountered amounts are the millicurie, $3.7 \times 10^7$ dis/sec, and the micro-curie, $3.7 \times 10^4$ dis/sec.

## 9-10. Modes of Radioactive Decay

Two major modes of radioactive decay—alpha-particle emission and beta-particle emission—are of present interest. Decay by emission of beta parti-cles (electrons) characterizes most useful radioisotopes; many of the naturally occurring radioactive nuclides, however, decay by alpha-particle emission. Both modes of decay frequently also involve gamma-ray emission. For nuclear engineering purposes, the properties of the particles and radiations emitted during radioactive decay are of major interest since they govern shielding and personnel-protection requirements.

**Alpha Decay.** The alpha particle is the nucleus of a helium atom. It is composed of two neutrons and two protons and therefore carries two positive charges. For practical purposes, the initial energy of all alpha particles emitted from a given kind of radioactive nuclide may be assumed to be constant; initial alpha-particle energies range from about 3 to 10 Mev.

Because of their large mass and charge, alpha particles rapidly lose their kinetic energy when passing through a medium by causing ionization of that medium. After sufficient energy has been lost, the slowly moving alpha particle picks up two electrons and becomes a helium atom.

The range of alpha particles from a given source in a given medium is constant since initial energies are essentially constant. This range is not great (approximately 1 in., in air, for a 4-Mev alpha particle) because of the rapid loss of energy by ionization. The range of alpha particles is inversely proportional to the density of the medium; most alpha particles are stopped by a sheet of paper and will not penetrate human skin. Hence alpha particles are not in general a serious external hazard to humans. When ingested into the body, however, they do considerable damage to tissue because of their great ionizing power.

**Beta Decay.** When a beta particle is emitted from the nucleus during radioactive decay, the transformation in the nucleus may be described by the relation

$$_0N^1 \rightarrow {}_1H^1 + \beta^- + \nu \tag{9-27}$$

i.e., a neutron is converted to a proton, the beta particle, and a neutrino, designated by $\nu$. (The neutrino has never been identified experimentally, but its existence and properties can be demonstrated theoretically.) From Eq. (9-3) it may be inferred that, although little change in total mass of the nucleus has occurred, the mass number of the nucleus has increased by one. Thus the atom is transformed to a different element by beta decay.

The neutrinos carry off different amounts of energy for each transformation, and hence the beta particles from a given radionuclide are found to have a continuous spectrum of energies, terminating in a definite maximum energy $E_0$. Values of $E_0$ are of interest for shielding purposes, and it is these values that are tabulated in the literature.

Other modes of beta-particle decay, such as electron capture and positron emission, occur. These are relatively infrequent, however, and are not discussed here.

Because of their low mass and charge, beta particles have much larger ranges than alpha particles (for example, the range of 3 Mev beta particles in air is about 43 ft), although they interact with materials in essentially the same way. In addition, ranges for beta particles are not clearly defined because of secondary interactions with the medium. It is possible, however, to specify the thickness of a given material required to reduce ionization by beta particles nearly to zero. As a first approximation, the thickness required may be assumed to be inversely proportional to the density of the medium.

**Gamma Rays.** As previously noted, radioactive decay frequently involves gamma-ray emission. These gamma rays are of great concern because their high energy and great penetrating power make them difficult to stop by shielding. Gamma rays interact with materials by several processes; they also produce ionization, but indirectly.

All disintegrations of a given kind of radioisotope do not always produce the same gamma rays. In other words, although the initial radionuclide and the decay, or "daughter," nucleus may be the same, various decay schemes and various gamma rays may be involved. It is important, again for shielding purposes, that the frequency of each mode of decay, and the gamma rays associated with each, be identified. This is done experimentally, and the data are tabulated in the literature.

## 9-11. Nuclear Reactions

Many kinds of reactions of incident particles and radiations with an atomic nucleus are possible. Relatively few, however, are of concern in nuclear reactors. The most important are those in which the neutron is the incident particle; fission is an example of these.

Nuclear reactions are of two basic types: scattering reactions, in which the identity of the incident particle is preserved, and absorption reactions, in which the incident particle is absorbed by the nucleus to form a new, highly excited compound nucleus. In absorption reactions the identity of the incident particle is lost; when the excited nucleus loses its energy, new reaction products are formed. In all nuclear reactions, total energy must be conserved either as mass or energy. The equivalence of mass and energy is given by the famous Einstein relation $E = mc^2$, where $E$ is the energy equivalent of the mass $m$, and $c$ is the velocity of light.

A shorthand notation is widely used to describe absorption reactions. Consider as an example the reaction

$$_0N^1 + Ni^{58} \rightarrow Co^{58} + {_1}H^1 \tag{9-28}$$

which indicates that absorption of a neutron in the nucleus of a $Ni^{58}$ atom produces a $Co^{58}$ atom and a proton. This reaction is written as $Ni^{58}(n,p)Co^{58}$ in the conventional notation. This form of expression has been adopted because of its simplicity and ease of identifying incident and reaction product particles.

Other types of neutron-induced reactions are $(n,n)$, $(n,\gamma)$, $(n,2n)$, and $(n,f)$ reactions. The latter designates the fission process. Fission and $(n,\gamma)$ reactions are most important in nuclear reactors. The $(n,\gamma)$ reactions are the most predominant mechanism by which radioactive species are produced because energy considerations permit them to occur with relative ease.

**Reaction Cross Sections.** A measure of the probability of nucleus-particle interaction is required to make quantitative calculations of nuclear reaction rates. The quantity which designates this probability, for a single nucleus, is the microscopic cross section $\sigma$. The term *cross section* is derived from the fact that $\sigma$ is essentially a measure of the effective cross-sectional area the nucleus presents to the incident particle.

As would be expected from the fact that atoms are mostly free-space, cross sections are extremely small. Values of cross sections are reported in the literature in terms of *barns*; one barn is defined as $10^{-24}$ cm$^2$. Measured values of $\sigma$ range from about $10^{-3}$ to $10^6$ barns. Cross sections can be determined only by experimental measurement.

Every nucleus has a specific cross section for each specific kind of nuclear reaction that can occur; i.e., the cross section is different, for a given nucleus, for neutron scattering, neutron absorption, proton absorption, etc. In addition, for each specific kind of reaction, the cross section is a function of the energy of the incident particle. Thus cross sections for a given reaction must be measured at various incident-particle energies of interest.

Of all the nuclear reactions possible, neutron reactions such as scattering, $(n,\gamma)$, and $(n,f)$ are most important. Hence extensive measurements of $\sigma$ as a function of energy have been made for these reactions.

Shorter tabulations of cross sections are also found in the literature (e.g., Table 9-7). Such values are specifically for absorption of so-called "thermal" neutrons. These are neutrons with energies of 0.025 ev (or equivalently, velocities of 2,200 m/sec). The latter values pertain to neutrons of most probable velocity in the Maxwell-Boltzmann distribution for thermal equilibrium at 20°C. If calculations of reaction rates are to be made for other temperatures and, as is the case with nuclear reactors, for environments containing neutrons with a wide spectrum of energies, appropriate corrections to the tabulated values must be made. Correction procedures are described in the literature.

Neutron absorption cross sections show great variation with neutron energy. Typically, at low energies ($< 0.1$ ev) $\sigma$ is proportional to $1/v$, where $v$ is the neutron velocity. In the intermediate range (0.1 to $10^3$ ev) sharp peaks, or "resonances," occur at specific energies. This is called the resonance region. At high energies, the cross section approaches the geometric cross section of the nucleus. Variations in cross section as a function of neutron energy are illustrated for some nuclides in Fig. 9-8.

**Nuclear Reaction Rates.** Actual rates at which nuclear reactions occur depend on the cross section for the particular reaction, the number of incident particles available, and the number of target nuclei. In general, calculations must be made for specific reactions for specific isotopes because different isotopes of a given element will have different cross sections and the total cross section for a given isotope is the sum of contributions for various types of reactions such as scattering, $(n,\gamma)$, etc. Calculation of actual reaction rates in nuclear reactors is a complex process because the reaction rates are a function of material thickness, and the number density of incident particles is spatially dependent. The following procedure, however, is typical for neutron reactions.

The volumetric rate of neutron reaction, $R$, is given by

$$R = \sigma \phi N_T \tag{9-29}$$

where $\sigma$ = cross section for the particular reaction, cm$^2$
$N_T$ = target nucleus density, atoms/cm$^3$
$\phi$ = "neutron flux," in neutrons/cm$^2$-sec

### Table 9-7. Thermal-neutron-absorption Cross Sections

| Element | Isotope | Isotopic abundance, % | Cross section, barns* |
|---|---|---|---|
| H | . . . . . | . . . . . . . | 0.33 |
|  | $H^1$ | 100 | 0.33 |
|  | $H^2$ | 0.015 | 0.46 mb |
| He | . . . . . | . . . . . . . | Variable |
|  | $He^3$ | 0.00013 | np 5,200 |
|  | $He^4$ | 100 | 0 |
| Li | . . . . . | . . . . . . . | 67 |
|  | $Li^6$ | 7.5 | nα 910 |
|  | $Li^7$ | 92.5 | 33 mb |
| Be | $Be^9$ | 100 | 9.0 mb |
| B | . . . . . | . . . . . . . | 750 |
|  | $B^{10}$ | 18.8 | nα 3,990 |
|  | $B^{11}$ | 81.2 | 50 mb |
| C | . . . . . | . . . . . . . | 4.5 mb |
|  | $C^{12}$ | 98.9 |  |
|  | $C^{13}$ | 1.1 | 1.0 mb |
| N | . . . . . | . . . . . . . | 1.78 |
|  | $N^{14}$ | 99.6 | np 1.70; nα 0.10 |
|  | $N^{15}$ | 0.37 | 0.024 mb |
| O | . . . . . | . . . . . . . | 0.2 mb |
|  | $O^{16}$ | 99.76 | Very small |
|  | $O^{17}$ | 0.037 | nα 0.5 |
|  | $O^{18}$ | 0.20 | 0.21 mb |
| F | $F^{19}$ | 10C | 10 mb |
| Ne | . . . . . | . . . . . . . | 2.8 |
| Na | $Na^{23}$ | 100 | 0.49 |
| Mg | . . . . . | . . . . . . . | 59 mb |
| Al | $Al^{27}$ | 100 | 0.22 |
| Si | . . . . . | . . . . . . . | 0.13 |
| P | $P^{31}$ | 100 | 0.19 |
| S | . . . . . | . . . . . . . | 0.49 |
| Cl | . . . . . | . . . . . . . | 31.6 |
| A | . . . . . | . . . . . . . | 0.62 |
| K | . . . . . | . . . . . . . | 1.97 |
| Ca | . . . . . | . . . . . . . | 0.43 |
| Ti | . . . . . | . . . . . . . | 5.6 |
| V | . . . . . | . . . . . . . | 4.7 |
| Cr | . . . . . | . . . . . . . | 2.9 |
| Mn | $Mn^{55}$ | 100 | 12.6 |
| Fe | . . . . . | . . . . . . . | 2.43 |
| Co | $Co^{59}$ | 100 | 34 |
| Ni | . . . . . | . . . . . . . | 4.5 |
| Cu | . . . . . | . . . . . . . | 3.59 |
| Zn | . . . . . | . . . . . . . | 1.06 |
| Zr | . . . . . | . . . . . . . | 0.18 |
| Mo | . . . . . | . . . . . . . | 2.4 |
| Cd | . . . . . | . . . . . . . | 2,400 |
| In | . . . . . | . . . . . . . | 190 |

## Table 9-7. Thermal-neutron-absorption Cross Sections (Continued)

| Element | Isotope | Isotopic abundance, % | Cross section, barns* |
|---------|---------|-----------------------|-----------------------|
| Sn      | . . . . | . . . . . . .         | 0.65                  |
| Xe      | . . . . | . . . . . . .         | 35                    |
|         | $Xe^{135}$ | 0                  | $3.5 \times 10^6$     |
| Sm      | . . . . | . . . . . . .         | 6,500                 |
|         | $Sm^{149}$ | 13.8               | 50,000                |
| Eu      | . . . . | . . . . . . .         | 4,500                 |
| Gd      | . . . . | . . . . . . .         | 44,000                |
| Hf      | . . . . | . . . . . . .         | 115                   |
| Ta      | . . . . | . . . . . . .         | 21.3                  |
| Au      | $Au^{197}$ | 100                | 94                    |
| Hg      | . . . . | . . . . . . .         | 380                   |
| Pb      | . . . . | . . . . . . .         | 0.17                  |
| Bi      | $Bi^{209}$ | 100                | 32 mb                 |
| Th      | $Th^{232}$ | 100                | 7.0                   |
|         | $Th^{233}$ | 0                  | 1,400                 |
| Pa      | $Pa^{233}$ | 0                  | 37                    |
| U       | . . . . | . . . . . . .         | nγ 3.50, nf 3.92      |
|         | $U^{235}$  | 0.714              | nγ 101, nf 549        |
|         | $U^{238}$  | 99.3               | 2.80                  |
|         | $U^{239}$  | 0                  | 22                    |
| Pu      | $Pu^{239}$ | 0                  | nγ 361, nf 664        |

* mb means millibarns, or $10^{-3}$ barns; one mb = $10^{-27}$ cm².
SOURCE: R. Stephenson, "Introduction to Nuclear Engineering," p. 375, McGraw-Hill Series in Chemical Engineering, McGraw-Hill Book Company, Inc., New York, 1954.

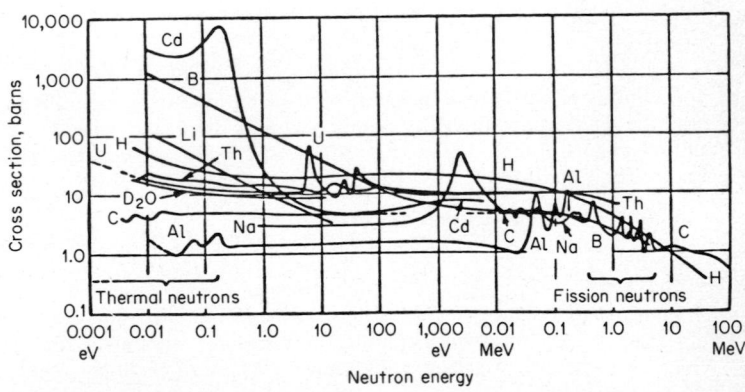

FIG. 9-8. Total neutron cross sections for some reactor materials. ("*Neutron Cross Sections,*" *AECU* 2040.)

The neutron flux is properly interpreted as the product $nv$, where $n$ is the number density of neutrons (neutrons/cm³) having the velocity $v$ in cm/sec. Equation (9-29) is frequently written

$$R = \Sigma \phi \qquad (9\text{-}30)$$

where $\Sigma$, called the *macroscopic cross section,* is defined by

$$\Sigma = \rho(N_a/A)\sigma \qquad (9\text{-}31)$$

where $\rho$ = mass density, g/cm³, of specific isotope for which microscopic cross section is $\sigma$

$N_a$ = Avogadro's number

$A$ = atomic mass of target isotope, g/g mole.

It is important to recognize potential pitfalls in the use of Eq. (9-29) or (9-30). As previously mentioned, $\sigma$ is a function of neutron energy, and, in a nuclear reactor—and in many other circumstances—neutrons with a spectrum of energies are present. Hence proper evaluation of the total reaction rate really requires evaluation of the integral of the product $\sigma(E)\phi(E)$, where the argument represents the energy dependence of $\sigma$ and $\phi$, and $\phi$ is the actual neutron flux in the material in question. An alternative procedure is to use the thermal neutron cross section for the reaction and a properly weighted *effective thermal neutron flux.* The best procedure, however, is to determine experimentally the *activation product* $\sigma\phi$ for the particular system. Even this presents difficulties because $\phi$ is frequently position-dependent.

To illustrate the use of Eq. (9-29), consider a frequently encountered problem: determination of the amount of radioactive species present as a function of time, when the radionuclide is the product of a neutron absorption reaction. The procedure will be illustrated for the $Cu^{63}(n,\gamma)Cu^{64}$ reaction, assumed to occur in pure copper exposed to an effective thermal neutron flux of $10^{14}$ neutrons/cm²-sec.

The equation describing the amount of $Cu^{64}$ present at any time is

$$dN^{64}/dt = R - \lambda N^{64} \qquad (9\text{-}32)$$

i.e., the number of $Cu^{64}$ atoms present, represented by $N^{64}$, is the difference between the amount produced by reaction $R$ and the loss by decay. The production rate is obtained from Eq. (9-5) as $R = \sigma\phi N^{63}$, and Eq. (9-32) becomes, when integrated for the condition $N^{64} = 0$ when $t = 0$,

$$N^{64}(t) = \frac{\sigma\phi N^{63}}{\lambda}(1 - e^{-\lambda t}) \qquad (9\text{-}33)$$

Numerical values are obtained as follows: Since an effective thermal neutron flux is given, the thermal neutron absorption cross section from the literature, 4.5 barns, may be used, assuming the reaction occurs at 20°C. The literature also gives the isotopic abundance of $Cu^{63}$ as 69.09 per cent and the half-life of $Cu^{64}$ as 12.9 hrs. Then

$$N^{63} = \frac{0.6909 \text{ g Cu}^{63}}{\text{g Cu}} \times \frac{8.92 \text{ g Cu}}{\text{cm}^3} \times \frac{6.023 \times 10^{23} \text{ atoms Cu}^{63}}{\text{g mole Cu}^{63}}$$

$$\times \frac{1 \text{ g mole Cu}^{63}}{63 \text{ g Cu}^{63}} = 5.9 \times 10^{22} \text{ atoms Cu}^{63}/\text{cm}^3 \quad (9\text{-}34)$$

and from Eq. (9-26c),

$$\lambda = \frac{0.693}{T_{\frac{1}{2}}} = \frac{0.693}{12.9(3,600)} = 1.49 \times 10^{-5} \text{ sec}^{-1} \quad (9\text{-}35)$$

Substituting values into Eq. (9-33),

$$N^{64}(t) = \frac{(4.5 \times 10^{-24})(10^{14})(5.9 \times 10^{22})}{1.49 \times 10^{-5}} (1 - e^{-\lambda t})$$

$$= (1.78 \times 10^{18})(1 - e^{-\lambda t}) \text{ atoms Cu}^{64}/\text{cm}^3 \quad (9\text{-}36)$$

At equilibrium $(dN^{64}/dt = 0)$, the atomic density of $Cu^{64}$ is simply $1.78 \times 10^{18}$ atoms $Cu^{64}/cm^3$. The activity $\lambda N$ of this amount of $Cu^{64}$ is

$$(1.49 \times 10^{-5})(1.78 \times 10^{18}) = 2.65 \times 10^{13} \text{ disintegrations/sec}$$

or $(2.65 \times 10^{13})/(3.7 \times 10^{10}) = 7.17 \times 10^2$ curies.

## 9-12. Nuclear Fission

In the (n,f) reaction, the target nucleus splits to form two new nuclei of lighter mass, called *fission fragments*, and, most important to sustaining the reaction in nuclear reactors, several free neutrons. Only three nuclides, $U^{235}$, $U^{233}$, and $Pu^{239}$, are for practical purposes fissionable by neutrons of all energies; they are referred to as *fissile* nuclides. Of these, only $U^{235}$ occurs in nature (its isotopic abundance in natural uranium is 0.72 per cent). The $U^{233}$ and $Pu^{239}$ are produced from $Th^{232}$ and $U^{238}$, respectively, by neutron absorption; the latter are referred to as *fertile* nuclides.

The mechanism of fission may be explained in terms of the liquid-drop model. The nucleus is viewed as analogous to a drop of liquid; the liquid drop is held together by surface tension forces, and the nucleus is held together by binding energy. When sufficient excitation energy is imparted to the liquid, surface tension forces will be overcome and the drop will split in two. Similarly, when the energy of the excited compound nucleus formed after neutron absorption exceeds the binding energy, the nucleus splits into two fragments. The excess energy is carried off primarily as kinetic energy of the fragments. As previously noted, the total energy released per fission is on the order of 200 Mev, of which about 95 per cent is available for power production. The remainder is carried off by neutrinos.

Important fission properties differ for the various fissile nuclides. For example, the total energy released per fission varies slightly. Similarly, the average number of neutrons released varies, as does the fission cross section, both absolutely and as a fraction of the total neutron absorption cross section. The latter lead to definition of the regeneration factor $\eta$, which is one of the most important physical constants related to fission chain reactors. The

## Table 9-8. Properties of the Fissile Nuclides*

|  | U²³³ | U²³⁵ | Pu²³⁹ |
|---|---|---|---|
| Useful energy per fission, Mev† | 191 | 193 | 201 |
| Total absorption cross section $\sigma_a$, barns | 578 | 683 | 1,028 |
| Fission cross section $\sigma_f$, barns | 525 | 575 | 577 |
| $\sigma_a/\sigma_f$ | 1.10 | 1.18 | 1.39 |
| Fission neutrons per fission | 2.51 | 2.44 | 2.89 |
| Regeneration factor | 2.28 | 2.07 | 2.08 |
| Delayed neutron fraction‡ | 0.0026 | 0.0065 | 0.0021 |

* From H. S. Isbin, "Introductory Nuclear Reactor Theory," p. 461, Reinhold Publishing Corporation, New York. 1963.
† From L. J. Templin (ed.), "Reactor Physics Constants," 2d ed., USAEC, ANL-5800, July, 1963.
‡ From G. R. Keepin, and T. F. Wimett, Reactor Kinetic Functions: A New Evaluation, *Nucleonics*, **16** (.10), 89 (1958).

average number of neutrons released per fission is conventionally given the symbol $\nu$. The regeneration factor is then defined in terms of $\nu$ as

$$\eta = \nu(\Sigma_f/\Sigma_a) \tag{9-37}$$

where $\Sigma_f$ is the macroscopic fission cross section for the fissile nuclide, and $\Sigma_a$ is the total neutron absorption cross section for the fuel material. In words, $\eta$ is the number of fission neutrons produced by thermal fission per thermal neutron absorbed in the fuel. This parameter is the key factor in neutron economy (Sec. 9-13). Values of $\eta$ and other properties of fissile nuclides are given in Table 9-8.

**Fission Neutrons.** Neutrons released by fission are divided into two fractions, prompt and delayed. As implied by the name, the latter are emitted some time after the fission event, apparently in conjunction with decay of certain of the fission products. Although delayed neutron fractions are very small (Table 9-8), their existence is probably the major factor permitting controlled utilization of nuclear power: they are the key to safe reactor operation (Sec. 9-15).

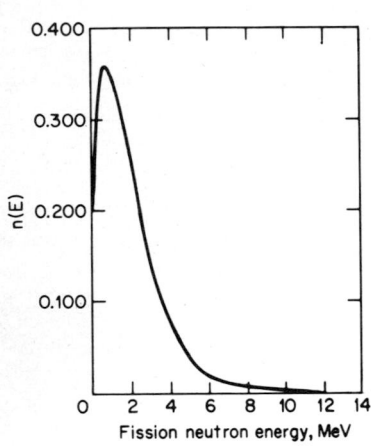

FIG. 9-9. Energy spectrum of prompt fission neutrons. (*Glasstone and Sesonske, "Nuclear Reactor Engineering," D. Van Nostrand Company, Inc., Princeton, N.J.,* 1963.)

Prompt fission neutrons are emitted with a spectrum of energies as shown in Fig. 9-9. Most have energies in the range 1 to 2 Mev, but a few have energies in excess of 10 Mev.   The latter are an important consideration in shielding.

The delayed neutrons fall into six groups, each characterized by an exponential decay rate.   The six groups are the same for the three fissile materials, but the distribution of delayed neutrons in the six groups differs.   Half-lives of the groups range between approximately 0.23 and 56 sec.   $Kr^{87}$ and $Xe^{137}$ have been identified as the neutron emitters for two of the groups.

**Fission Products.**   The fission process occurs in more than 40 ways, producing fission fragments with mass numbers ranging from about 72 to 160.   These fission fragments are highly radioactive, and decay in a succession of steps involving formation of other radionuclides.   The nuclei which result from this process—over 200 radioactive species—are known collectively as *fission products.*

The mass distribution of fission products for fissioning of $U^{235}$ by thermal and 14 Mev neutrons is shown in Fig. 9-10.   Curves for the other fissile species are similar.   It may be noted from Fig. 9-10 that the maximum yield

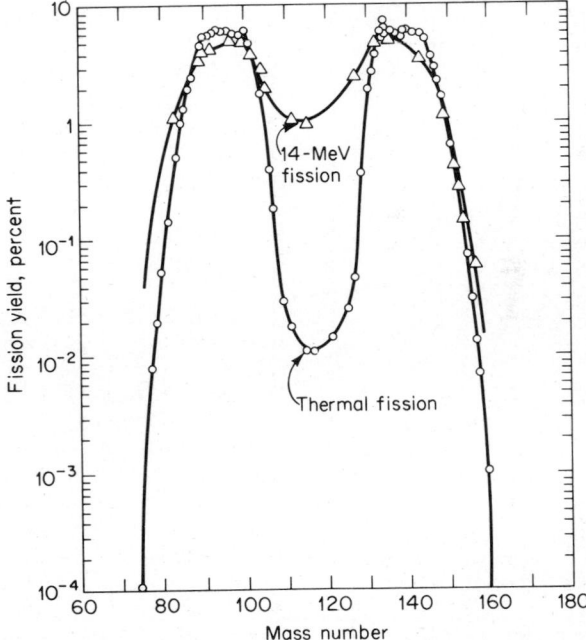

FIG. 9-10.   Mass distribution of $U^{235}$ fission products.   (*Glasstone and Sesonske, op. cit.*)

is about 6 per cent; the maxima occur at mass numbers of approximately 95 and 135.

The radioactive fission products give off energy as gamma rays and beta particles during decay. In the nuclear reactor, this energy is rapidly manifested as heat, which must be removed to prevent core meltdown. For this reason, and also for proper design of spent-fuel reprocessing facilities, it is important to know the magnitude of this *decay heat power,* as it is called, as a function of time after fission ceases.

The decay heat power can be determined as a fraction of fission power from the expression

$$\frac{P}{P_0} = 6.1 \times 10^{-3}[(\tau - T_0)^{0.2} - \tau^{-0.2}] \qquad (9\text{-}38)$$

where $P$ = decay heat power
   $P_0$ = fission or reactor power (both in arbitrary but identical units)
   $\tau$ = time in days since cessation of fission
   $T_0$ = number of days for which fission occurred (at constant rate)

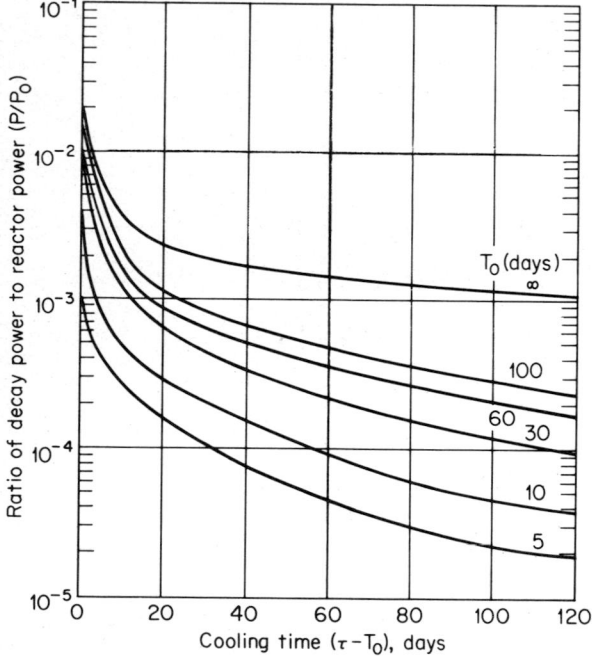

Fig. 9-11. Fission-product-decay heat power.  (*Glasstone and Sesonske, op. cit.*)

As suggested by Eq. (9-38), the ratio $P/P_0$ is a function of $T_0$ as well as cooling time. Equation (9-38) is shown graphically for several values of $T_0$ in Fig. 9-11

## 9-13. Physics of the Nuclear Reactor

From the physicist's point of view, design of a nuclear reactor is a problem of neutron economy. About 2.5 neutrons are produced per fission event (Sec. 9-12), only one of which must be absorbed to produce fission again and thereby sustain a chain reaction. However, many processes compete for neutrons in the reactor. These may be briefly summarized as (1) loss from the system at boundaries, (2) absorption in nonfissionable materials, (3) nonfission absorption in fissile materials, and (4) absorption to produce fission. The physicist's objective is to construct a balance sheet for neutrons which involves these four processes and sustains a chain reaction in the framework of engineering requirements for the reactor (Sec. 9-16).

**Classification of Reactors.** Nuclear reactors can be classified according to a variety of standards. The most fundamental, however, is the kinetic energy of the neutrons causing most of the fissions. On this basis, there are two major types: *fast* reactors, in which most fissions are caused by neutrons of high energy, and *thermal* reactors, which operate primarily on low-energy thermal neutrons. Since fission neutrons are born at high energies (about 2 Mev; Sec. 9-9), fast reactors operate with neutrons of or near fission energy. In thermal reactors, the fast-fission neutrons are slowed down (to take advantage of the larger cross sections at thermal energies); this slowing down is accomplished by materials known as *moderators* that are put into the core. Thermal reactors predominate in the spectrum of operating reactors in the world today, primarily because of greater design flexibility. In the future, however, fast reactors should become more important in order to make best use of nuclear fuel resources.

**Breeding and Conversion.** Fast reactors are expected to become important because of their potential for producing more fissionable fuel than they consume, i.e., *breeding*. If new fuel generation involves production of a fissile material different from that being consumed (e.g., a reactor operating on $U^{235}$ produces $Pu^{239}$ from the fertile $U^{238}$), the process is called *conversion*. Many combinations of types of reactors and fissile and fertile materials have been and are being considered to maximize potential for conversion and breeding. Development of fast reactors for breeding has been slow because of difficult, but apparently soluble, technological problems with heat transfer, materials, and dynamic stability.

Because known world reserves of the only naturally occurring fissile material, $U^{235}$, are relatively small, breeding and conversion involving the fertile materials $Th^{232}$ and $U^{238}$ are mandatory for effective use of all potential fission energy reserves (which could supply man's needs for a century).

## 9-14. Physics Design of Reactor Cores

The ensuing discussion is directed primarily at thermal reactors because of the importance of neutron slowing-down processes in these systems. Basic concepts such as definition of reactor parameters, method of sizing the core, etc., apply equally well, however, to fast reactors.

Processes that occur for neutrons in a reactor may be described qualitatively as follows: Immediately after birth, the fast-fission neutrons begin to move rapidly through the reactor because of their high kinetic energy. As they do so, they encounter and interact with atoms of materials in the reactor core. These encounters with other atoms may produce one of two results: the neutron may be absorbed in the nucleus of the struck atom, or it may simply suffer a collision in which some of the neutron's kinetic energy is transferred to the struck atom.

These collisions are known as scattering reactions. They are of two types, *elastic* and *inelastic*. In the elastic, or "billiard-ball," collisions, kinetic energy of the neutron-atom pair is conserved. Kinetic energy is not conserved in inelastic collisions, which are generally restricted to nuclei of fairly high mass number and neutrons of energy in excess of 0.1 Mev; the neutron is captured by the nucleus, and part or all of its kinetic energy is converted to excitation energy of the nucleus.

If the neutron is not absorbed during the above processes, its energy is gradually reduced (i.e., moderated—the moderator nuclei are targets for scattering reactions) as a result of the scattering collisions, so that the desired objective, neutrons of thermal energy, is achieved. These thermal neutrons are then available for absorption in fuel to produce fission and thereby re-initiate the above process.

Throughout the foregoing sequence of events, the neutrons are always subject to possible leakage from the system at boundaries. To reduce loss of neutrons by leakage, *reflectors* are placed at the boundaries of reactor cores. The reflectors scatter some of the leaked-out neutrons back into the core so that they remain available to cause fission.

**Reactor Parameters.** Quantitative calculations of the rate at which the above processes occur are required to size a reactor core. To make these calculations, parameters defined in terms of material properties important to these processes are used. These parameters and their symbols are as follows:

*Diffusion Coefficient $D$.* The diffusion coefficient is defined by the equation

$$J = -D \operatorname{grad} \phi \tag{9-39}$$

where $J$ is the neutron current in a particular direction, and $\phi$ is the neutron flux. A good approximation to $D$ may be obtained from the relation

$$D = \frac{1}{3\Sigma_s(1 - \bar{\mu}_0)} \tag{9-40}$$

where $\Sigma_s$ is the macroscopic scattering cross section for the medium, and $\bar{\mu}_0$ is the average cosine of the scattering angle per collision, given in terms of the mass number $A$ of the medium by $\bar{\mu}_0 = \frac{2}{3}A$. Note that $D$ has units of length.

**Diffusion Length L.** The diffusion length is defined by

$$L \equiv \sqrt{D/\Sigma_a} \qquad (9\text{-}41)$$

where $\Sigma_a$ is the macroscopic neutron absorption cross section of the medium. It is a measure of the distance a neutron travels from the point where it becomes thermal to the point where it is absorbed.

**Average Logarithmic Energy Decrement ξ.** This parameter is a measure of the average energy loss the neutron suffers in elastic collisions with nuclei. It is determined to a high degree of accuracy from the relation

$$\xi \cong \frac{2}{A + \frac{2}{3}} \qquad (9\text{-}42)$$

To minimize the size of the reactor and possibilities of nonfission neutron absorption, it is desirable that $\xi$ have as large a value as possible, i.e., that $A$ be small (hence, moderator materials should be of low atomic mass).

**Fermi Age τ.** The Fermi age was defined in conjunction with a model in which the slowing down of neutrons, which proceeds in discrete steps of energy loss for each scattering collision, is represented as a continuous process. This parameter is of great practical importance in reactor design (see below). It is a function of neutron energy $E$, and is defined by the relation

$$\tau(E) \equiv \int_{E_0}^{E} \frac{D}{\xi \Sigma_s E} \, dE \qquad (9\text{-}43)$$

where $E_0$ is the energy of the neutrons at the beginning of the slowing-down process. The Fermi age is a measure of the distance (note that the units of $\tau$ are length squared) a neutron has traveled from the point of origin to the point where its Fermi age is $\tau$. An important value of $\tau$ is $\tau_{\text{thermal}}$, corresponding to $E = E_{\text{thermal}}$. This is a measure of the distance the neutron travels to achieve thermal velocity.

**Migration Length M.** This parameter is a measure of the total distance the neutron travels from birth as a fission neutron to absorption as a thermal neutron. As would be expected, it is defined in terms of $L$ and $\tau$ by

$$M \equiv \sqrt{L^2 + \tau} \qquad (9\text{-}44)$$

The parameter that actually appears in reactor equations is the *migration area* $M^2$, where

$$M^2 = L^2 + \tau \qquad (9\text{-}45)$$

**Design Methods for the Steady State.** The basic problem in reactor design is to devise a useful mathematical model which is descriptive of the physical processes discussed above and utilizes the parameters representative of the effects of reactor geometry and materials on these processes. The ensuing discussion is an outline of the basis for development and use of such models.

The reader is cautioned that in practice, elaborate, complex computer programs are actually used for reactor design. These programs have their origin, however, in the concepts given here.

Neutron behavior in reactors can be described rigorously by Boltzmann transport theory. The complexity of reactors, however, prohibits detailed solution of the resulting equations. Hence the basis for reactor design lies in an approximation to transport theory known as *diffusion theory*. The essential feature of diffusion theory is that neutron leakage from the reactor is described as a diffusional process.

The material balance for neutrons in the reactor must take account of production (by fission) and losses (by leakage and all absorption reactions). In diffusion theory, the neutron balance for steady-state operation takes the form

$$D \nabla^2 \phi - \Sigma_a \phi + S = 0 \tag{9-46}$$

where $\nabla^2$ = Laplacian operator

$\phi$ = neutron flux

$\Sigma_a$ = effective macroscopic absorption cross section

$S$ = source term

Equation (9-46) is known as the *diffusion equation*, and is the basis of reactor design. It is strictly applicable only in systems containing monoenergetic neutrons, and also at points more than two or three neutron mean free paths from boundaries and strong sources and absorbers. As will be seen, however, these restrictions can be obviated.

To solve Eq. (9-46), it is essential to obtain a representation for the source term $S$ and define reactor geometry and boundary conditions. As the first step in the solution, it is convenient to take as reference a fictitious reactor, infinite in size, so that no neutron leakage occurs. For this system a parameter known as the *infinite multiplication factor* $k_\infty$ is defined. It is given by

$$k_\infty = \frac{\text{no. of fission neutrons produced in a given generation}}{\text{no. of neutrons absorbed in the preceding generation}} \tag{9-47}$$

where a "neutron generation" is the fission-birth, slowing-down absorption cycle previously described.

A similar parameter, the *effective multiplication factor* $k_{\text{eff}}$, may now be defined for a finite reactor. It is given in terms of $k_\infty$ and a factor which corrects $k_\infty$ for leakage losses in the finite reactor:

$$k_{\text{eff}} = k_\infty P \tag{9-48}$$

where $P$ is defined as the *nonleakage probability*. It should be apparent from these definitions that the reactor is operating at steady state when $k_{\text{eff}}$ has a value of exactly unity. When $k_{\text{eff}} = 1$, the reactor is said to be *critical;* when $k_{\text{eff}}$ is less than or greater than unity, the reactor is subcritical and supercritical, respectively.

The infinite multiplication factor may be defined in terms of parameters representative of the effect of material properties on physical processes of scattering and absorption that occur during a neutron generation. The relationship is

$$k_\infty = \epsilon p f \eta \qquad (9\text{-}49)$$

where $\eta$ is the regeneration factor previously defined (Sec. 9-9), and the other parameters are as defined below. Equation (9-49) is known as the *four-factor equation*. Because of the definitions of $\epsilon$, $p$, $f$, and $\eta$, it is a powerful means for determining the effect of changes in reactor materials on criticality. These parameters are defined as:

**Fast-fission Factor $\epsilon$.** This factor accounts for neutron production during the slowing-down process by fissions at high energies. It may be defined as the ratio of the total number of fission neutrons produced by fast and thermal fission to the number produced by thermal fission.

**Resonance Escape Probability $p$.** This parameter is the ratio of the number of neutrons leaving the resonance region (Sec. 9-9) at low energies to the number entering at high energies. It is a complex function of the macroscopic absorption and scattering cross sections of the materials in the core.

**Thermal Utilization $f$.** This factor is defined as the fraction of all thermal neutrons absorbed that are absorbed in fuel material (which may include nonfissionable material such as $U^{238}$). The exact definition of $f$ depends on whether the system is homogeneous or heterogeneous; an acceptable general expression for a reactor of volume $V$ is

$$f = \frac{V_F \Sigma_{aF} \phi_F}{V_F \Sigma_{aF} \phi_F + V_m \Sigma_{am} \phi_m + V_i \Sigma_{ai} \phi_i} \qquad (9\text{-}50)$$

where the subscripts $F$, $m$, and $i$ represent fuel, moderator, and impurity (e.g., structural) materials, respectively.

In general, $\epsilon$ does not differ much from unity and, as shown in Table 9-8, values of $\eta$ for the fissile materials are similar. The infinite multiplication factor for a given reactor therefore depends strongly on values of $p$ and $f$, both of which are dependent on the amount, dispersion, and properties of materials in the reactor.

**Solution of the Diffusion Equation.** One may take the viewpoint that the objective in solving Eq. (9-46), the diffusion equation, is to obtain an expression for the nonleakage probability $P$ [Eq. (9-48)], in terms of reactor materials and geometry. The actual expression obtained for $P$ depends basically on two factors: the expression used for the source term $S$ and the method of treating the neutron energy spectrum in the core. In practice, the neutron energy spectrum is considered to consist of several groups, each containing monoenergetic neutrons. This leads to a diffusion equation for each group; the source term for each equation is then the neutrons entering that group from the group of next-highest energy neutrons.

The above approach leads to quite complex representations of neutron behavior. The general method by which expressions for $P$ are developed may be illustrated, however, by the following simple model. This model is for an unreflected homogeneous reactor in which all neutrons are considered to have the same energy (i.e., "one-group" theory). The derivation will be illustrated for the steady state, which presupposes that $k_{\text{eff}}$ is unity.

For the situation assumed, all neutrons are absorbed at the same energy and at a total rate $\Sigma_a\phi$. Since $k_\infty$ fission neutrons are produced per absorption, the source term is simply $k_\infty\Sigma_a\phi$. Equation (9-46) then becomes

$$D\nabla^2\phi - \Sigma_a\phi + k_\infty\Sigma_a\phi = 0 \tag{9-51}$$

or upon rearrangement and introduction of the diffusion length [Eq. (9-41)],

$$\nabla^2\phi + \frac{k_\infty - 1}{L^2}\phi = 0 \tag{9-52}$$

Equation (9-52) indicates only the effect of reactor materials on neutron behavior. It is now necessary to consider the effect of neutron leakage and reactor geometry on the spatial distribution of the neutron flux.

The neutron flux distribution is represented by the relationship

$$\nabla^2\phi + B^2\phi = 0 \tag{9-53}$$

which is subject to boundary conditions imposed by the shape of the reactor (e.g., spherical, cylindrical, etc.). The constant $B^2$ is known as the "buckling," because it measures the bending (i.e., buckling) of the neutron flux.

At this point two operations are possible. First, Eq. (9-53) may be solved independently, with appropriate boundary conditions, to determine the flux distribution on the basis of purely geometrical considerations. Second, it may be noted that Eqs. (9-52) and (9-53) may be satisfied simultaneously if the coefficients of $\phi$ are identical. Equating coefficients,

$$\frac{k_\infty - 1}{L^2} = B_c^2 \tag{9-54}$$

where the subscript on $B^2$ indicates that the value of $B^2$ that satisfies Eq. (9-54) is the one for which the reactor is critical. If this value is now made equal to the value obtained by independent solution of Eq. (9-53), the reactor will actually be critical.

Two bucklings may therefore be distinguished. That arising from solution of Eq. (9-53) is known as the *geometric buckling*, and that given by Eq. (9-54) is the *material buckling*. When the reactor is critical, the material and geometric bucklings are identical. Expressions for the geometric buckling for various reactor geometries are given in Table 9-9. It will be noted that these expressions indicate the dimensions of the reactor. When the two bucklings are equal, these are the dimensions for criticality.

Equation (9-54) may be rearranged to

$$\frac{k_\infty}{1 + L^2B_c^2} = 1 \tag{9-55}$$

and it will be recalled that since steady state was assumed, $k_{\text{eff}}$ must be unity. Hence

$$\frac{k_\infty}{1 + L^2B_c^2} = 1 = k_{\text{eff}} = k_\infty P \tag{9-56}$$

### Table 9-9. Geometric Bucklings for Various Reactor Shapes

| Geometry | Buckling | Minimum critical volume |
|---|---|---|
| Sphere | $\dfrac{\pi^2}{R}$ | $\dfrac{130}{B_c{}^3}$ |
| Rectangular parallelepiped | $\dfrac{\pi^2}{a}+\dfrac{\pi^2}{b}+\dfrac{\pi^2}{c}$ | $\dfrac{161}{B_c{}^3}\ (a = b = c)$ |
| Finite cylinder | $\dfrac{(2.405)^2}{R}+\dfrac{\pi^2}{H}$ | $\dfrac{148}{B_c{}^3}\ (H = 1.847R)$ |

$R$ = radius; $a,b,c$ = length of sides; $H$ = height.

and therefore the nonleakage probability is given by $1/(1 + L^2B_c{}^2)$ according to one-group theory.

Equation (9-55) is known as the *critical equation* for one-group theory. It is, as noted, the result of a very simple model; its predictions of critical size are therefore, at best, approximate. To obtain more reliable estimates of critical dimensions, models that more accurately describe physical processes for the neutrons are required.

Critical equations for two other, more accurate models are as follows: The *age-diffusion model*, which utilizes the Fermi continuous slowing-down approximation mentioned above, gives

$$\frac{k_\infty e^{-B_c{}^2\sigma_{\text{thermal}}}}{1 + L^2B_c{}^2} = 1 \tag{9-57}$$

which for a large reactor reduces to

$$\frac{k_\infty}{1 + M^2B_c{}^2} = 1 \tag{9-58}$$

*Two-group theory*, in which the neutron energy spectrum is divided into a fast group and a thermal group, gives

$$\frac{k_\infty}{(1 + \tau B_c{}^2)(1 + L^2B_c{}^2)} = 1 \tag{9-59}$$

Many *multigroup* methods for determining critical size are also available. These can be quite accurate, but they are also quite elaborate, and require iterative solution on computers. Many of the computer programs, or "codes," used to determine the critical size of reactors are available in the literature.

The above critical equations were derived assuming the reactor is homogeneous. In practice, of course, fuel, moderator, and structural materials are distinct (homogeneous reactors are in development). It is therefore frequently desirable to subdivide the reactor into small "unit cells," each of which may be treated as a homogeneous entity. Such a procedure adds considerably to the complexity—but also the accuracy—of the calculations.

## 9-15. Reactor Kinetics and Control

The power output of a nuclear reactor is varied by controlling the neutron flux. Many methods of flux control are available, but the most common is to insert in the core materials which have very large neutron absorption cross sections, called *poisons*. To sustain operation for long periods of time without refueling, the reactor is built with fuel in excess of that required for criticality. The poison materials provide "negative fuel" that compensates for the excess fuel (as fuel is consumed, the poison must gradually be removed) and, in conjunction with the control system, prevent the reactor from becoming supercritical during transient operations such as startup.

Poisons are usually inserted as an array of metallic control rods dispersed throughout the core (see Sec. 9-18 for a discussion of poison materials). The control rods are connected mechanically to drive motors that are actuated as a result of signals received from neutron-detection instruments. Much of the operation of the control system is automatic. An important safety feature is that operator actions which tend to increase the neutron flux are subject to automatic controls built into the system.

Poisons may also be inserted as "burnable" (i.e., gradually depleted) poisons added to the fuel matrix or the coolant. Methods of control other than poisons include addition or removal of fuel, variation of the amount of moderator in the core, and movement of sections of the core or reflector.

**Reactor Control Parameters.** A fundamental concept in reactor control is the *neutron lifetime* $\ell$, which is defined as the average time between successive generations for an infinite reactor. The effective lifetime, defined for a finite reactor, is the neutron lifetime multiplied by the nonleakage probability $P$. The prompt neutron generation time $\ell^*$ is defined by

$$\ell^* \equiv \ell/k_\infty \tag{9-60}$$

and characterizes the lifetime of prompt neutrons in the reactor.

Another fundamental concept in reactor control is the reactivity $\rho$, defined by

$$\rho \equiv \frac{k_{\text{eff}} - 1}{k_{\text{eff}}} = \frac{k_{\text{ex}}}{k_{\text{eff}}} = \frac{\Delta k}{k_{\text{eff}}} \cong \delta k \tag{9-61}$$

The reactivity is frequently taken to be equivalent to $\Delta k$, or alternatively, $\delta k$.

The significance of these parameters and delayed neutrons in reactor control may be illustrated by considering changes in neutron density in a reactor not operating at steady state. The change in neutron density $n$ with time is given by

$$\frac{dn}{dt} = n\frac{\delta k}{\ell} \tag{9-62}$$

which gives, with $n = n_0$ at $t = 0$,

$$n(t) = n_0 e^{(\delta k/\ell)t} = n_0 e^{t/\phi} \tag{9-63}$$

where $\theta$ is the reactor period. The effect of the delayed neutrons is to increase the neutron lifetime, and therefore the reactor period, from about $10^{-3}$ sec, characteristic of the prompt neutrons, to $10^{-1}$ sec. Thus, by inspection of Eq. (9-63), in a given period of time the neutron density changes by a much smaller amount when delayed neutrons control the reactor period. The neutron density would change at a rate too fast for the electromechanical systems to control if the delayed neutrons did not control the period.

**Kinetic-analysis Fundamentals.** Because of their powerful influence on reactor dynamics, the delayed-neutron contribution to neutron economy is clearly distinguished in kinetic studies. The one-group diffusion equation for a bare, homogeneous reactor [Eq. (9-46)] becomes

$$D \nabla^2 \phi - \Sigma_a \phi + k_\infty \Sigma_a \phi (1 - \beta) + \sum_{i=1}^{6} \lambda_i C_i = \frac{dn}{dt} \tag{9-64}$$

where $\beta$ is the delayed-neutron fraction (Table 9-8), and $(1 - \beta)$ is therefore the prompt-neutron fraction. The contribution of the six delayed-neutron groups, each characterized by a concentration $C_i$ and decay constant $\lambda_i$, is indicated by the summation.

Equation (9-64) may be written

$$\frac{dn}{dt} = \frac{\rho - \beta}{\ell^*} n + \sum_{i=1}^{6} \lambda_i C_i \tag{9-65}$$

with which are associated the equations descriptive of delayed-neutron behavior,

$$\frac{dC_i}{dt} = \frac{\beta_i}{\ell^*} n - \lambda_i C_i \qquad i = 1, 2, \ldots, 6 \tag{9-66}$$

which, it will be noted, is similar to Eq. (9-32) for neutron reactions.

Equations (9-65) and (9-66) are fundamental to reactor kinetics; they are basic to development of transfer functions descriptive of reactor dynamic response. However, as for steady-state design, relationships used in practice are considerably more complex than those given here. A detailed discussion of methods in use is given in many texts.

An important aspect of unsteady-state operation that can be extremely dangerous is the "prompt-critical condition," for which $\delta k = \rho = \beta$. The reactor is critical on prompt neutrons alone, and the period is therefore extremely short. The power level could rise at such a rate that the core would melt before corrective action could be taken. Control systems are carefully designed to prevent achievement of the prompt-critical condition.

**Temperature Effects.** Changes in temperature exert great influence on reactor kinetics because they affect materials density, core dimensions, neutron energy, and cross sections. Temperature effects are basically determined by differentiating the critical equation with respect to temperature (i.e., determining $dk_{\text{eff}}/dT$) and investigating the change with temperature in the range of interest for each resulting term.

Safety considerations require that the reactor temperature coefficient be negative (i.e., if temperature increases, power decreases). It should be noted that a negative coefficient can be achieved only by proper design. Some contributions to the coefficient are positive, and some are negative; the magnitude and sign of each must be determined. All reactors are designed to have negative temperature coefficients, generally on the order of $10^{-5}$ to $10^{-4}$ per degree Fahrenheit, at operating temperatures.

A major contributor to the over-all temperature coefficient is the *Doppler coefficient*, which describes the effect of temperature changes on neutron absorption in the resonance region. In general, as the temperature increases, the resonance peaks broaden, and increased neutron absorption occurs. In fissile material, the Doppler coefficient is therefore positive; in other materials it is negative.

**Fission-product Poisoning.** Two of the fission products—$Xe^{135}$ and $Sm^{149}$—have very large thermal neutron absorption cross sections ($2.7 \times 10^6$ and $4.2 \times 10^4$ barns, respectively). The magnitudes of these cross sections and the amounts in which the isotopes are formed are sufficient to have an effect on the multiplication factor.

These *fission-product poisons* influence the reactivity of the reactor. The amount of poisons present depends on reactor operating history and the thermal neutron flux. It can be shown, however, that a definite maximum equilibrium *poisoning*, defined as the ratio of the number of neutrons absorbed by the poison to the number absorbed by fuel, exists for operating reactors, as shown in Fig. 9-12.

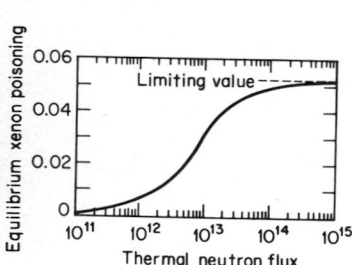

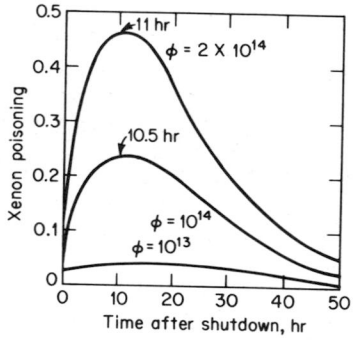

Fig. 9-12. Equilibrium xenon poisoning during reactor operation. (*Glasstone and Sesonske, op. cit.*)

Fig. 9-13. Xenon poisoning after shutdown. (*Glasstone and Sesonske, op. cit.*)

Figure 9-12 shows that the maximum poisoning during operation is relatively small. The poisoning can achieve very large values, however, after reactor shutdown. This phenomenon occurs because the precursor of $Xe^{135}$, which is $I^{135}$, has a relatively long half-life (6.7 hr) and hence continues to produce $Xe^{135}$ from its decay after shutdown.

The poisoning achieved after shutdown is a strong function of neutron flux, as shown in Fig. 9-13. The very high value of poisoning achieved for fluxes of $10^{14}$ and greater is in some cases the limiting factor in operating neutron flux and core life. To restart the reactor when the poisoning is a maximum, the core must have available sufficient excess reactivity (as excess fuel) to "override peak xenon"; near the end of core life this capacity is limited. In ship propulsion reactors, where startup at any time is essential, this limitation becomes quite important.

## 9-16. Engineering Design of Nuclear Reactors

A unique feature of nuclear power generation is the wide variety of design concepts that may be utilized successfully. Many core configurations, component designs, and materials combinations have been used in the past, and more innovations may be expected for the future. All power reactors have certain common components, however, as outlined below.

**Reactor Coolants.** The coolant, which removes heat generated by fission and radiation heating of core structures, may be a gas (air, helium, $CO_2$), molten salt, water (light or heavy), liquid metal, or organic liquid. In some reactor designs, coolant flow through the core is orificed to match the heat-generation distribution.

**Reactor Vessel.** The core and associated components such as control-rod assemblies, support structures, and reflectors are housed in the reactor vessel. In pressurized, water-cooled reactors the vessel must be able to withstand high operating pressures (up to about 2,000 psig). A major problem in reactor-vessel design is thermal stress; other design problems arise from the need to provide fuel-handling facilities and control-rod drives. Pressure vessels for water-cooled reactors are built in accordance with sec. III of the ASME Code.

**Fuel-element Cladding.** Individual fuel elements in heterogeneous reactors are sheathed in a cladding which acts as a barrier, preventing escape of fission products from the fuel to the coolant. Cladding materials and properties are detailed in Sec. 9-18.

**Thermal Shields.** The radiations emitted by nuclear reactions in the core cause extensive heating of adjacent structural materials, including the reactor vessel. To prevent excessive thermal stresses in the vessel as a result of radiation-induced internal heat generation in the region of the core, thermal shields are placed between the core and the vessel. The thermal shields must be cooled (heat generation in the shields is about 3 per cent of the total output of the reactor). The shields must have high absorption coefficients for neutrons and electromagnetic radiation and high thermal conductivities to prevent overheating. Steels are commonly used as shield materials.

**Fuel-handling Systems.** A system must be provided for loading and unloading fuel in the reactor. The system may also be capable of moving fuel from one position to another in the core. Refueling of water-cooled reactors is usually done with the reactor shut down and the vessel closure head removed. In other systems, however, refueling may be accomplished without costly reactor shutdown.

**Containment Vessels.** All nuclear power reactors are required to be housed in containment structures designed to retain fission products and gases that might be released as a result of the maximum credible accident (see discussion of hazard analysis below). In many installations, this structure is a low-leakage vessel totally enclosing the reactor vessel and associated external components such as reactor coolant piping and steam generators. Such vessels are usually cylindrical or spherical and fabricated from steels not subject to brittle failure.

### 9-17. Thermal Design of Reactor Cores

Core design objectives for heat transfer and physics are basically at odds. To optimize neutron economy, the core should be small; to ease heat removal, the core should be large. Every operating core represents a compromise of these different objectives.

Nuclear reactors operate with high heat fluxes (on the order of 100,000 and 500,000 $Btu/(hr)(ft^2)$ in water-cooled and liquid-metal-cooled systems, respectively). To prevent fuel melting and release of fission products, a large number of fuel elements of small cross section is therefore required (for example, the Yankee reactor in Rowe, Mass., contains over 23,000 individual fuel elements; the nuclear ship Savannah contains over 32,000). Power densities in reactors vary considerably for different coolants, as shown in Table 9-10.

### Table 9-10. Power Densities for Various Types of Nuclear Reactors

| Reactor Type | Power Density, Kw (Thermal)/ft³ |
|---|---|
| Gas-cooled—natural U | 15 |
| High-temperature gas | 220 |
| Sodium graphite | 290 |
| Organic-cooled | 390 |
| Heavy water—natural U | 510 |
| Boiling water | 820 |
| Pressurized water | 1,550 |
| Sodium-cooled fast breeder | 21,500 |
| Nuclear rocket reactor | 280,000 |
| Conventional forced-convection boiler | 280 |

SOURCE: Abstracted from S. Glasstone and A. Sesonske, "Nuclear Reactor Engineering," D. Van Nostrand Company, Inc., Princeton. N.J., 1963.

Standard techniques appropriate for the coolant and heat transfer regime are used to determine heat transfer coefficients (Sec. 5). A detailed, point-by-point heat transfer analysis is required, however, to prevent fuel-element failure. The general procedure is to develop a heat transfer correlation for determining the peak central temperature (PCT) of a fuel element and apply this correlation to individual portions of the core. The correlation must account for the thermal resistance of the fuel, cladding, corrosion-product deposits on the cladding, coolant film, and bulk coolant.

In every reactor core there will be one point in one fuel element which operates at the highest temperature. The thermal analysis must locate this point and assure that fuel-element failure (burnout) will not occur. To achieve this goal, hot-spot analysis involving the so-called *hot-channel factors* is used. The hot-channel factors are of two basic types, nuclear and engineering. They account for variations in neutron flux and design parameters such as fuel-element dimensions, uncertainty in coolant flow rate, uncertainty in the heat transfer film coefficient, etc.

A large number of hot-channel factors may be defined (in general, for a given parameter as the ratio of the maximum-worst-deviation value to the nominal value) and quantitatively determined for a given core. However, they are combined into three basic, over-all factors used in the PCT correlation: (1) the factor for coolant temperature rise, $F_{\Delta T}$; (2) the film temperature-drop factor, $F_{\theta}$; and (3) the heat generation factor, $F_q$.

The method by which individual hot-channel factors are combined to determine the over-all factors is a subject of considerable debate. Two basic methods are available: statistical, in which probabilities of occurrence are assigned to each factor, and the "factor-product" method, in which all contributors to each over-all factor are multiplied together. The latter is of course extremely conservative.

### 9-18. Nuclear Reactor Materials

Conventional engineering materials are used in reactors except where special nuclear requirements must be met. Reactor materials must satisfy applicable, conventional criteria such as tensile strength, ductility, corrosion resistance, etc., and those used in the core must also have nuclear properties appropriate to their function.

A choice of materials for each major reactor component is available. There are, however, two general restrictions: materials specifications for reactors are generally more stringent than for nonnuclear applications because interactions with neutrons can affect suitability, and materials that are mutually compatible must be selected.

The combined engineering, nuclear, and compatibility requirements for nuclear materials have led to extensive research and development programs designed to produce new alloys and materials with improved physical properties for reactor use. Results of this work are reported in *Reactor Materials*, a quarterly publication of the U.S. government.

**Fuel Materials.** Nuclear fuels in common use include natural uranium, slightly enriched uranium (0.95 to ~6.0 per cent $U^{235}$), and fully enriched uranium (93 per cent $U^{235}$). The fuel materials also include large amounts of the fertile species $U^{238}$ and $Th^{232}$. As more and more reactors operate to produce fissile nuclides from these fertile species, it may be expected that $Pu^{239}$ and $U^{233}$ will become important reactor fuels.

Natural uranium is generally used in both heavy-water and in gas-cooled reactors. Reactors cooled and moderated with light water use slightly enriched uranium as fuel; the actual enrichment is dictated by criticality and core-endurance requirements. Fully enriched uranium is used extensively in research and other special-purpose reactors, but thus far has received limited use in nuclear power stations.

Uranium has very poor metallurgical and physical properties. It is dimensionally unstable after irradiation and thermal cycling and fails catastrophically with short exposure to water at 100°C or more. Its apparent physical properties are also extremely sensitive to purity, state of cold work, grain size, and orientation of grains. For these reasons, it is usually desirable to use uranium alloys or compounds as fuel materials; however, in the gas-cooled reactors that are so prevalent in Europe and England it is found economical to use natural uranium metal as fuel.

Reactors that burn slightly or fully enriched uranium use $UO_2$ or uranium alloys as the fuel material. Common alloying materials include zirconium, aluminum, molybdenum, and stainless steels. $UO_2$, in spite of low thermal conductivity, is widely used in water-cooled power reactors.

Current nuclear fuel development programs have as their major objectives development of improved fuel-fabrication techniques and better fuel materials. Promising examples of the latter are ceramics, such as $UC$, $UC_2$, and $UN$. These materials show much better resistance to adverse effects of irradiation than uranium metal, and are capable of withstanding long periods of exposure in the core. They have high melting points, reasonably good thermal conductivities, and are easily fabricated but react with water. Successful development of these materials is expected to result in much cheaper production of nuclear power as a result of reduced core-fabrication costs and longer periods of exposure between shutdowns for refueling.

Fuel exposure, or *burnup*, between refuelings is an extremely important factor in determining nuclear power costs. Burnup is usually expressed in megawatt-days of heat energy produced per metric ton (or tonne; equivalent to 2,200 lb) of uranium, abbreviated as Mwt-days/tonne U, or MWD/T. Thermal reactors currently in operation in the United States achieve burnups of about 10,000 MWD/T, with values between 20,000 and 30,000 or more anticipated for the future. The natural uranium-fueled, gas-cooled reactors in England and Europe achieve much lower burnups—on the order of 3,000 MWD/T—primarily because of loss of reactivity in natural uranium.

Radiation effects that limit fuel burnup may be summarized as dimensional instability (elongation or swelling) as a result of accumulation of fission products and gases, adverse changes in mechanical properties, relatively increased parasitic capture of neutrons, and loss of reactivity due to fuel consumption. It is expected that new fuel materials and fabrication techniques (too numerous to detail here) and programmed movement of fuel from one position in the core to another will greatly reduce these adverse effects, and thereby improve burnup and reduce fuel costs in the future.

**Moderator Materials.** Moderator materials in common use today include light water, heavy water, and graphite. Beryllium, BeO, and lithium 7 are also good moderator materials, but are toxic and highly reactive. Development programs for these materials are in progress.

Graphite is the most commonly used moderator in power reactors. In United States reactors, however, light water is most commonly used; it is cheap, readily available, and serves also as the coolant. Heavy water is actually a better moderator because deuterium has a much lower absorption cross section for parasitic capture of neutrons than hydrogen. Its high present cost ($24.50 per pound), however, has limited its use in power reactors, except where natural uranium can be economically used as fuel. Heavy water is present in natural water to the extent of 140 to 150 ppm; the two species may be separated by distillation, chemical exchange, or electrolysis. The United States maintains large separation plants in Savannah River, Ga.

The graphite used in reactors is of high density (to limit reactor size) and high purity ($\sim$20 ppm total impurities). A highly pure material is required to minimize parasitic capture of neutrons. The product resulting from the graphite-manufacturing process may be easily machined or cast in desired shapes.

**Cladding Materials.** The cladding must be a high-integrity structural material with good corrosion resistance, high thermal conductivity, and a low neutron absorption cross section. Stainless steels and aluminum are currently the most commonly used cladding materials; however, zirconium alloys are actually superior to these because of their very low neutron absorption cross section ($\sim$0.2 vs. $\sim$3 barns for steel) and excellent corrosion resistance.

Other cladding materials used in thermal reactors include Magnox (a magnesium alloy, used primarily in the United Kingdom gas-cooled reactors), magnesium-zirconium alloys, Incaloy, ceramic coatings, and graphite. Claddings used in fast reactors include stainless steels, zirconium, niobium, and other refractory metals and ceramics. Properties of these and other materials discussed in this section may be found in Sec. 3 of this manual.

**Reactor Control Materials.** Materials used to control thermal reactors by neutron absorption have as their most important characteristic large neutron absorption cross sections; the elements that contain isotopes suitable for this application include boron, cadmium, europium, and hafnium.

Control materials are used in the reactor in two principal ways. The most common is by control rods, a method that requires the control materials to withstand shock, vibration, and wear. The other method is to use a burnable poison which can be evenly distributed in the fuel, placed in discrete positions in the core, or dissolved in the coolant. Use of burnable poisons requires careful calculation of control-material concentrations; poison burnup and fuel-depletion rates must be comparable.

The control materials may be utilized in various forms. Boron, for example, is used in boron–stainless-steel alloys, as $B_4C$ particles dispersed in zirconium or fuel alloys, and as boric acid added to the coolant. The latter is a commonly used backup safety method in water-cooled reactors. Cadmium is used as a major constituent of silver-indium-cadmium alloys. These alloys have relatively low melting points, however, and poor corrosion resistance in water containing even small amounts ($\sim$5 ppm) of oxygen. They also have poor strength properties.

Hafnium has been found to be an excellent control material. In addition to its good nuclear properties, it has high strength and good corrosion resistance. It is readily shaped and welded as the pure metal. Hafnium is expensive, however. It is obtained only as a by-product of zirconium ores, and processing costs to separate the two elements are high. The separation is essential, because hafnium impurities in zirconium used in reactor cores would greatly increase parasitic capture of neutrons. Hence hafnium is readily available for limited use; but fabrication costs are also quite high because of the need to avoid impurities which adversely affect physical properties of the material.

Most operating power station reactors use boron as the control material. The Shippingport, Pa., reactor and many mobile reactors use hafnium. The Yankee reactor in Rowe, Mass., has used a silver-indium-cadmium alloy, as do many research reactors. The very high cost and metallurgical problems with europium and other lanthanon poisons have prohibited their use except as burnable poisons within the fuel.

**Reactor Coolants.** The basic types of materials available as reactor coolants (water, gases, liquid metals, molten salts, and organic liquids) have previously been mentioned briefly. Of all of these, light water is the most commonly used in the United States because physical-property data are readily available and it is economical. Reactors using the other types of coolants have also been built, however, and several alternatives in each category are available, as outlined below.

*Gases.* Air, hydrogen, helium, and carbon dioxide are in principle good reactor coolants. Each, however, has certain specific deficiencies: the oxygen in air reacts with reactor materials, hydrogen is extremely hazardous and reacts with many metals, helium is quite expensive, and $CO_2$ is a relatively poor heat-transfer medium. In general, all gaseous coolants also have the disadvantage of being poor heat-transfer agents compared with liquids.

Carbon dioxide is the most commonly used gas coolant because of the problems with the other gases. Some advanced-concept reactors, however, use helium as the coolant because of superior heat-transfer and physical properties.

*Liquid Metals.* These materials are virtually a necessity for fast reactors in which coolants with good moderating properties are undesirable. In general, the excellent thermal properties of the liquid metals provide the advantage of operation at high temperatures with high power densities (Table 9-10) and compact cores. On the debit side, however, technological knowledge of the liquid metals is relatively sparse, their cost is high, and their generally high chemical reactivity at elevated temperatures makes selection of compatible materials a problem.

The most meaningful criterion for selection of liquid metals as reactor coolants is the neutron absorption cross section, which must be small to minimize parasitic capture. On this basis, the liquid metals listed in Table 9-11 are found to be suitable reactor coolants. Of these, liquid sodium and the NaK eutectic (22 weight per cent Na), which is liquid at ordinary temperatures, are at present most attractive because they have fewer disadvantages than the others. A major disadvantage of the use of sodium as a reactor coolant is formation of $Na^{24}$ from $Na^{23}$. The $Na^{24}$ has a significantly long half-life (15 hr) and emits two high-energy gamma rays with decay. Shielding of reactor components is therefore required, and access to the reactor compartment for maintenance may be restricted because of high radiation levels for two or three days.

### Table 9-11. Properties of Liquid-metal Reactor Coolants

| Metal | $\sigma_a$, barns | Melting point, °F |
|---|---|---|
| Bismuth | 0.032 | 520 |
| Lithium 7 | 0.033 | 367 |
| Lead | 0.17 | 621 |
| Sodium | 0.50 | 208 |
| Tin | 0.65 | 450 |
| Potassium | 2.0 | 145 |
| Gallium | 2.7 | 86 |
| Thallium | 3.3 | 576 |

*Molten Salts.* These materials have been considered as reactor coolants because they permit economic utilization of all grades of reactor fuels. They are used in a fluid-fueled reactor in which the coolant and fuel are a homogeneous slurry pumped throughout the system. The molten salts are advantageous for fluid-fueled systems because, unlike aqueous fuels, they can be used at low pressures and high temperatures.

Only two molten-salt reactor systems have been built and operated; both had demonstration of feasibility as the major objective. The first was the Aircraft Reactor Experiment, now terminated, and the most recent is the Molten Salt Reactor Experiment (MSRE) located in Oak Ridge, Tenn. The MSRE utilizes a fuel salt which has a composition of 65 mole per cent $Li^7F$, 29.1 per cent $BF_2$, 5 per cent $ZrF_4$ and 0.9 per cent $UF_4$; its melting point is 842°F. This salt, and other structural materials, were specially developed for the MSRE reactor.

*Organic Liquids.* Organic liquids have many potential advantages as reactor coolants: the systems may be operated at low pressure; the organics are not highly corrosive; they are good moderators; induced radioactivity is low, and physical properties are well known. It has been found, however, that the organic coolant tends to decompose when exposed to high temperatures and radiation. This characteristic could result in serious operational problems, such as plugging of flow channels and fouling of heat-transfer surfaces. To eliminate these problems decomposed coolant must be removed by purification and replaced with fresh coolant.

## 9-19. Fission Power Systems

Feasible fission power concepts number in the thousands. For economic reasons, however, practical systems are limited to a few basic concepts and variations on these.

The type of reactor selected depends on local economic conditions. In Great Britain, where capital finance charges are small, gas-cooled reactors fueled with natural uranium are preferred. Canada is committed to reactors cooled with heavy water. In the United States, light water reactors fueled with slightly enriched uranium have been economical.

Reactor types are usually defined by the type of coolant used. Types currently in use and those projected for use in the United States are described below.

**Light-water Reactors.** There are two types of light-water reactor (LWR) systems: the boiling-water reactor (BWR), and the pressurized-water reactor (PWR). Both use ordinary (light) water as coolant and moderator, and both are fueled with slightly enriched uranium (about 3.5 per cent $U^{235}$) in the form of oxide pellets contained in zirconium tubing.

In a BWR system (Fig. 9-14) the core coolant is allowed to boil. The steam is dried and sent to the turbines; the water phase is recirculated in the reactor vessel. In a PWR system (Fig. 9-15) the core coolant is kept in a liquid state; a heat exchanger external to the reactor vessel is used to generate steam for the turbines.

The BWR and PWR systems are competitive because of design and economic trade-offs. The PWR provides higher thermal efficiencies: the PWR reactor outlet temperature is about 600°F (320°C); the BWR outlet temperature is about 550°F (290°C). The cost of steam generators and the pressurizer for a PWR are offset by higher costs for reactor vessel internal components in the BWR.

**Gas-cooled Reactors.** The high-temperature gas-cooled reactor (HTGR) (Fig. 9-16) uses helium coolant, graphite moderation, thorium and highly

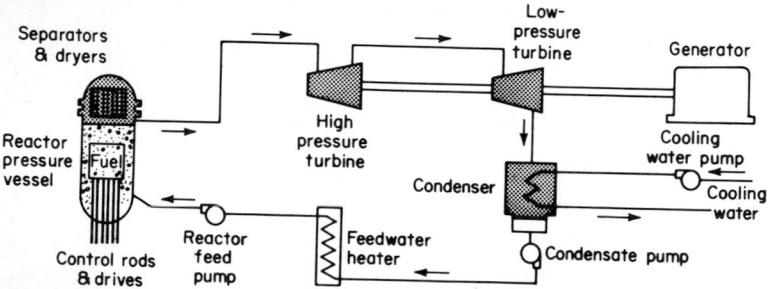

FIG. 9-14. Schematic diagram of a boiling-water reactor system. (*Source: "The Nuclear Industry 1973," publication of the U.S. Atomic Energy Commission.*)

enriched uranium as fuel materials, and a prestressed-concrete reactor vessel (PCRV). Coolant outlet temperatures are about 1400°F (780°C); coolant temperature rise through the core is about 330°C (in LWRs it is restricted to about 16°C).

The HTGR core consists of vertical columns of hexagonal graphite elements. Coolant flows in channels separated by graphite from the fuel pins which contain small spheres of coated fuel material. Pressures in the primary system are low (700 psia) in comparison with LWR pressures.

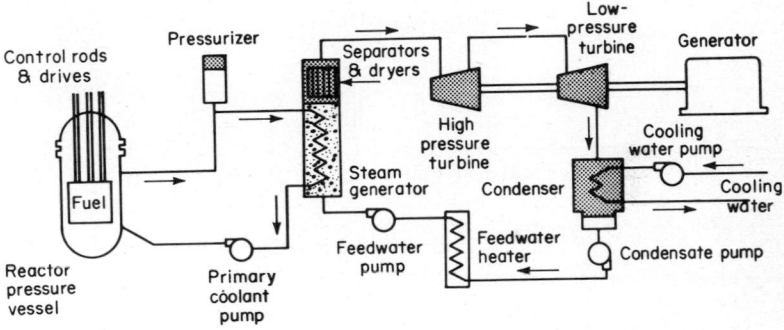

Fig. 9-15. Schematic diagram of a pressurized water reactor system. (*Source:* "*The Nuclear Industry* 1973," *publication of the U.S. Atomic Energy Commission.*)

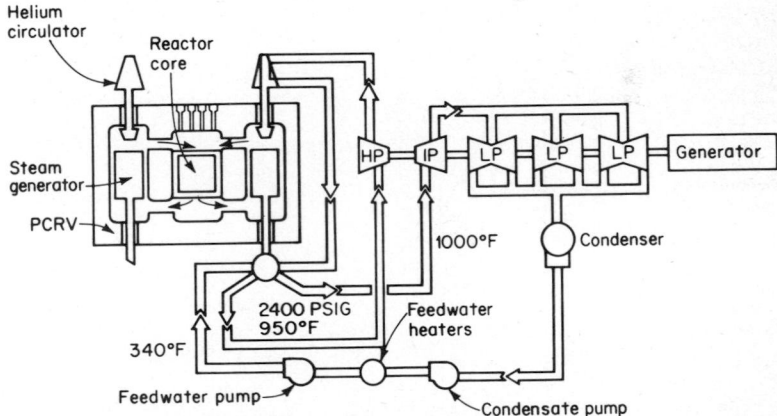

Fig. 9-16. Schematic diagram of a high-temperature gas-cooled reactor system. (*Source:* "*The Nuclear Industry* 1973," *publication of the U.S. Atomic Energy Commission.*)

**Liquid-metal-cooled Reactors.** Since liquid metals do not moderate neutrons, they offer potential for good breeding and conversion, i.e., efficient use of fission fuel materials. The liquid-metal fast breeder reactor (LMFBR) is simple in concept (Fig. 9-17) but complex in practice because of design and operating problems associated with use of sodium or sodium-potassium as the coolant.

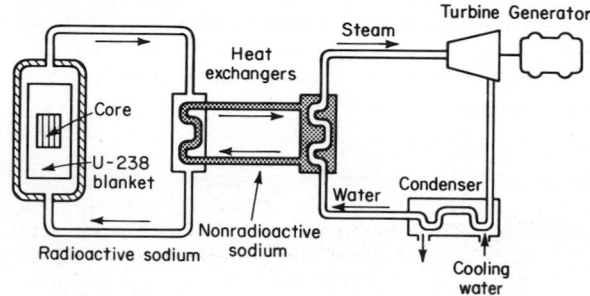

Fig. 9-17. Schematic diagram of a fast breeder reactor system.

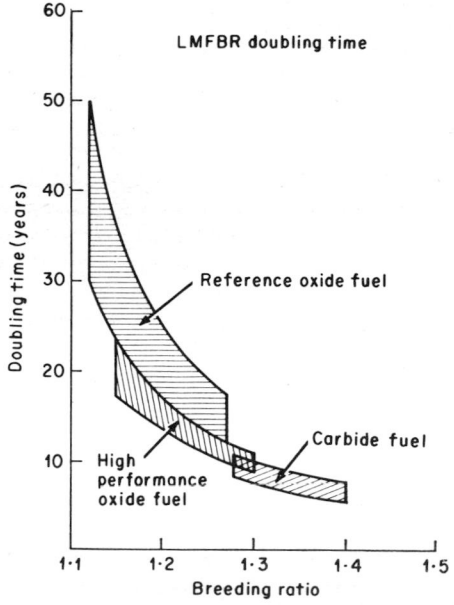

Fig. 9-18. LMFBR doubling time as a function of breeding ratio.

Efficient use of fission fuels in LMFBRs depends on the *breeding ratio*, which can be defined as the rate of production of fissile material compared to the rate of consumption, and the *doubling time*, which is approximately the length of time to produce sufficient excess fissile plutonium to inventory the in-core and ex-core fissile requirements for a similar new reactor.

The breeding ratio is a complex function of core design. It is related to the doubling time, as shown in Fig. 9-18, by the types of fuel used. Current LMFBR development work is aimed at improving economics and fuel utilization by developing new fuels that can reduce the doubling time. It also is seeking to upgrade operational performance through design development work aimed at attaining higher operating temperatures.

## 9-20. Nuclear Fuel Cycle

Nuclear fission is distinguished from all other energy conversion concepts by the need to reprocess spent fuel. The need arises because nuclear and mechanical features of core designs prevent the complete burnup of a fuel loading. Nuclear limitations are associated with loss of capability to sustain criticality as a result of fuel depletion and buildup of fission products. Mechanical limitations arise from radiation damage that causes distortion, loss of strength, and loss of thermal conductivity in the fuel elements. The net result is that fuel elements must be removed from the reactor when only 10 to 30 per cent of the fissile material has been consumed.

Reprocessing of the spent fuel has three objectives: (1) recovery of fissile uranium and reconversion into fresh fuel; (2) extraction and utilization of fissile plutonium created by neutron capture in $U^{238}$, and (3) isolation of fission products and other wastes. Since the fissile uranium and plutonium have value as fuels, the first two steps are potentially revenue-producing. The third is an overhead cost for the fission power economy.

Operations in the nuclear fuel cycle are shown in Fig. 9-19. The conversion, enrichment, fuel-element fabrication, and reprocessing steps have no parallel in other energy conversion systems. Nuclear power is economically competitive despite the costs of these operations because of the low unit cost of fissile fuels.

The conversion step changes solid uranium oxide into gaseous uranium hexafluoride, a form suitable for the enhancement of $U^{235}$ concentration which is done by gaseous diffusion in the enrichment step. In the fabrication step, the enrichment product is converted to uranium dioxide which is formed into pellets that in turn are inserted into the fuel elements for the reactors.

The reprocessing step involves breaking down the fuel elements, dissolution of the spent fuel pellets, recovery of the fuel values by solvent extraction operations (Fig. 9-20), and management of the radioactive wastes.

Plutonium recovered during reprocessing can be combined with uranium to form mixed-oxide fuels for LWRs or it can be used as the primary fuel for LMFBRs. Since plutonium is highly toxic and a potential nuclear weapons material, elaborate procedures to monitor and control inventories are re-

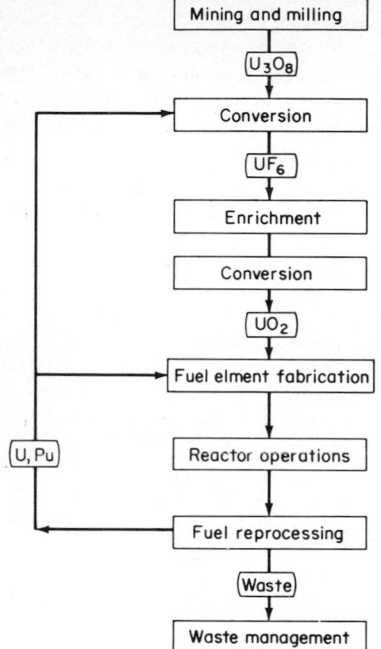

Fig. 9-19. Basic operations in the nuclear fuel cycle.

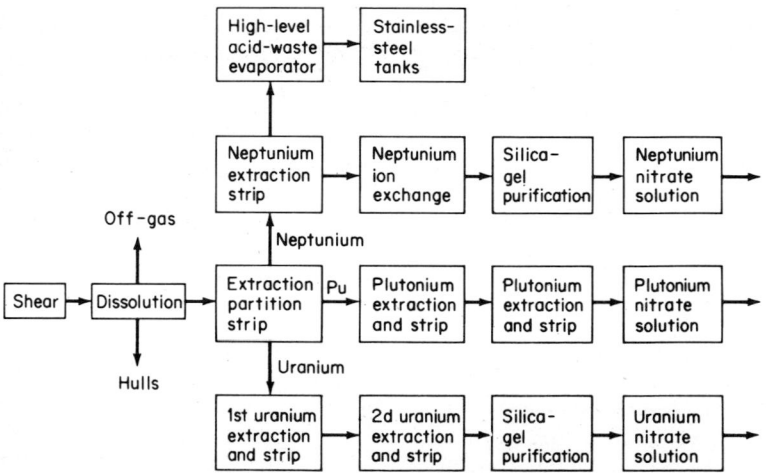

Fig. 9-20. Operations in nuclear fuel reprocessing.

quired. These "safeguard" procedures are an important feature of design and licensing of facilities that handle plutonium.

Reprocessing and other fuel-cycle operations produce a variety of radioactive wastes. The fission products are the major constituent of *high-level waste*, which contains most of the radioactivity and also has high heat-emission rates as a result of radioactive decay. Another form of waste, large in volume but low in radioactivity is the *alpha waste* which is made up of paper trash, plastics, etc., that are used in fuel-cycle operations and become lightly contaminated with uranium or plutonium. Other important wastes are the chopped cladding ("hulls") from breaking down spent fuel, and decommissioned equipment. Methods for control and disposal of these wastes are being developed.

## OTHER ENERGY CONVERSION SYSTEMS

### 9-21. Fusion

Fusion energy is released when the nuclei of light elements come together to form heavier elements. If energy conversion systems based on fusion can be developed, the supply of fuels on earth would be essentially unlimited. The most suitable fuels include the heavy isotopes of hydrogen (deuterium and tritium), the light isotope of helium (helium-3), and lithium. All are abundant in the seas.

Fusion fuels must be heated to temperatures on the order of tens to hundreds of millions of degrees in order to have fusion occur. The electrons must be stripped from the atoms and the nuclei must be brought together with force sufficient to overcome repulsive forces. These conditions must be met at densities sufficient for fusion collisions to be frequent and for times adequate to produce more energy than is consumed in the process. When all conditions are satisfied, the particles are in the plasma state, i.e., electrically charged.

Since all materials will melt at fusion temperatures, the plasma must be contained without touching the walls of the container. The basic approach for this is to confine and control the plasma with magnetic fields, i.e., in *magnetic bottles*. Many bottle configurations have been explored; one of the most promising is the TOKAMAK concept. Laser-induced fusion is also being studied.

The major problem for development of fusion systems is to confine the plasma without significant leakage from the magnetic bottle for times sufficient to have productive amounts of fusion occur. Conceptual designs for fusion systems have been developed in anticipation that this hurdle will be overcome. If fusion should become practical, environmental and health hazards are expected to be small in comparison with fission reactors. Residual radioactivity would be limited primarily to structural materials, and fusion reactors do not have the runaway potential of fission reactors. In addition, over-all plant efficiencies two or more times greater than those of fission reactors might be achieved by direct conversion to electricity.

## 9-22. Solar Power

People have used energy conversion systems based on solar power for centuries. Large-scale implementation has been impeded because solar energy is dilute and variable (thereby requiring storage systems). Cheap, abundant supplies of other fuels have been disincentives for large-scale utilization of solar-powered systems.

The simplest solar conversion systems, which are suitable for domestic use, combine adsorption of solar energy on a black surface with a storage and pumping system as illustrated by Fig. 9-21. Advanced systems based on these concepts would employ solar-tracking and focusing devices for the collector, large-scale collection areas, and large-scale storage devices.

In principle, such advanced systems could be developed by extrapolating known technology. At present, although there are technical hurdles associated with collector design, storage, and efficiency, the major impediment to development is cost. Projected capital and retail costs for electricity from a base-loaded solar conversion system are about three times current costs.

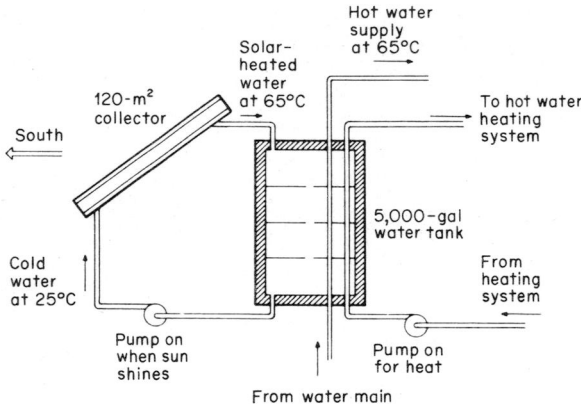

FIG. 9-21. Schematic diagram of a domestic solar-powered heating system.

An alternative for solar conversion is to use photovoltaic cells. These cells, which are widely used on spacecraft, use an $n$-$p$ junction in silicon material to convert sunlight directly into electrical current. Efficiencies are typically on the order of 10 per cent; the theoretical maximum is about 35 per cent. At 10 per cent efficiency and for current manufacturing costs, a conversion system based on photovoltaic cells would cost about $100,000 per kilowatt, i.e., more than 100 times the cost of nuclear fission systems.

Another alternative for solar conversion is to make combustible fuels after photosynthesis in trees, plants, and algae. Efficiencies would be about 3 per cent or lower, and about 3 per cent of the United States land area would be needed to supply current total energy needs. Projected costs are not unreasonable, but waste disposal may pose significant problems.

## 9-23. Magnetohydrodynamics

The basic concept for magnetohydrodynamics (MHD) is to bypass the conventional mechanical-to-electrical conversion step by making the working fluid an electricity conductor and forcing it, rather than a copper conductor, through a magnetic field. The fluid is made a conductor by burning any chemical fuel (coal, gas, or oil) at high temperature and "seeding" the combustion gases with an easily ionized material, such as potassium carbonate.

The conducting gases are expanded through a nozzle at high velocity and passed through a duct to which a strong magnetic field has been applied. Direct-current electricity is drawn off from electrodes lining the channel. The system has few moving parts and potentially high efficiency.

MHD units are expected first to be used to supply peaking and emergency power. They are expected to compete economically with gas turbines currently used for these purposes. Both are cheap to install, reliable, and capable of quick start-up.

As MHD technology advances, it is expected to be combined with conventional fossil-fired steam plants to improve efficiency as a result of higher working fluid temperatures. This application, called a *topping cycle*, would first pass the seeded combustion gases through the MHD unit and draw off electrical energy at temperatures above those that can be tolerated by turbine blades. The exhaust gases from the MHD generator would then serve as the heat source for steam production. The over-all efficiency of such units could be as high as 60 per cent; conventional fossil-fired plants have efficiencies of about 40 per cent.

MHD generators might also be used with fission or fusion systems. The technology is more difficult, however, and commercial applications are expected not to occur for several decades.

## 9-24. Fuel Cells

Fuel cells convert chemical energy directly into electrical energy with an over-all efficiency of about 60 per cent. The principle of operation is the same as that for a dry cell: a chemical reaction that produces electrons (oxidation) occurs at one electrode, and a reaction that consumes electrons (reduction) occurs at the other. This process produces a voltage between the electrodes. The fuel cell differs from the dry cell in that reactants and reaction products are supplied to and removed from the cell.

Fuel cells have been used extensively on spacecraft. They can be made to operate on a variety of fuels including natural gas and gasified coal. Cost per kilowatt is relatively independent of output; this feature makes them

attractive for individual home units.    At present, costs are not competitive.
A fuel cell can run in reverse; i.e., by supplying electricity to the cell, the
reaction products can be converted back into reactants.    An example is
electrolysis of water to produce hydrogen and oxygen.    With this mode of
operation, the fuel cell could be a device for energy storage; e.g., it could
store converted solar energy.

# ENVIRONMENTAL ENGINEERING

**Laurence J. White, B.S.;** Marketing Services Manager, Chemical Engineering; Member, American Institute of Chemical Engineers, American Chemical Society, Air Pollution Control Association

**Gary F. Bennett, Ph.D.;** Professor of Biochemical Engineering, The University of Toledo; Member, American Institute of Chemical Engineers

**Burton B. Crocker, S.M., P.E.;** Senior Engineering Fellow, Monsanto Co.; Fellow, American Institute of Chemical Engineers; Member, Air Pollution Control Association

## CONTENTS

# INTRODUCTION

The concept of environmental engineering is relatively new. Historically, governments have concerned themselves with only the most flagrant forms of pollution such as a public water supply contaminated with sewage. In general, the world's atmosphere and bodies of water were considered a limitless garbage dump, and natural purification processes were sufficient to handle human wastes for thousands of years. But now the combined impacts of continued industrial development and population growth make environmental engineering a necessity.

The field spans an enormous range of activities and embraces all of the traditional engineering disciplines. This section will concentrate on the legal and technological aspects of wastewater treatment and control of air pollution from stationary sources.

The world's industrialized nations did not give serious attention to the environment until after World War II. In the United States, federal water laws were passed in 1948, 1956, 1965, and 1970. The first federal air pollution act became law in 1955, with amendments in 1967.

Federal legislation was generally ineffective because enforcement remained in the hands of state and local governments. Few communities were willing to spend time and money on pollution control, which had not been recognized as a serious problem. (Exceptions were cities such as Pittsburgh and Los Angeles that began fighting pollution in the 1950s.)

By 1970 it was obvious that pollution would not go away by itself, and the environmental movement became popular in Congress. Among the results were passage of stringent federal laws and consolidation of all federal pollution control activities under a single agency, the Environmental Protection Agency (EPA).

## 10-1. Clean Air Act of 1970

The Clean Air Act of 1970 was the first law to require national, ambient air quality standards based on geographic regions. Air quality is regulated by two sets of standards determined by EPA (Table 10-1). The primary standards are intended to be the minimum level of air quality that is necessary to preserve human health, while the secondary standards are aimed at preventing damage to animals, plant life, and property.

State governments retain the authority to determine how ambient air quality standards are to be met within their borders. However, the state implementation plans are subject to approval by EPA, which can revise any state plan it considers unsatisfactory.

EPA also has the power to set performance standards for new stationary sources of air pollution. It has published standards for a number of major sources including fossil-fuel-fired steam generators, incinerators, cement plants, sulfuric and nitric acid manufacturing, petroleum refineries, asphalt plants, steel mills, secondary lead smelters, and various other nonferrous metal operations.

Table 10-1. National Ambient Air Quality Standards*

| | Primary standard | | Secondary standard | |
|---|---|---|---|---|
| | $\mu g/m^3$ | ppm | $\mu g/m^3$ | ppm |
| Sulfur oxides: | | | | |
|   Annual arithmetic mean............... | 80 | 0.03 | | |
|   24-hr concentration.................. | 365† | 0.14† | | |
|   3-hr concentration.................... | ..... | ..... | 1.300† | 0.5† |
| Suspended particulate matter: | | | | |
|   Annual geometric mean............... | 75 | | 60‡ | |
|   24-hr concentration.................. | 260† | ..... | 150† | |
| Carbon monoxide: | | | | |
|   8-hr concentration.................... | ..... | 9.0† | Same as primary | |
|   1-hr concentration.................... | ..... | 35.0† | Same as primary | |
| Photochemical oxidants: | | | | |
|   1-hr concentration................... | 160† | 0.08† | Same as primary | |
| Hydrocarbons (corrected for methane): | | | | |
|   3-hr concentration (6–9 am)........... | 160† | 0.24† | Same as primary | |
| Nitrogen oxides: | | | | |
|   Annual arithmetic mean.............. | 100 | 0.05 | Same as primary | |

* ppm = parts per million, $\mu g/m^3$ = micrograms per cubic meter.
† Not to be exceeded more than once a year.
‡ A guide for assessing achievement of the 24-hr standard.

Air pollutants that are particularly hazardous to human health are controlled on an individual basis under emission standards set by EPA. Asbestos, beryllium, and mercury were among the first materials to be regulated.

An early problem that developed under the 1970 act was how to treat regions where air quality was better than the ambient standards. Environmentalists argued that EPA had a duty to protect pristine areas, while others said that a certain amount of development (and pollution) was necessary for the good of the people. After several conflicting court decisions, EPA has set guidelines that allow some degradation of clean air up to the limits of the secondary air quality standards. However, the matter is still subject to further review by the courts and Congress.

As written in 1970, the Clean Air Act required compliance with the national primary standards by 1975, although the deadline could be extended for 2 years in those cases where compliance was "technologically impossible." At the beginning of 1975 there were still two major gaps in the clean air program: automobile emissions and emissions of sulfur oxides from power plants. Both problems involved not only questions of control technology, but also whether the clean air timetable should be revised because of energy considerations.

## 10-2. Water Act of 1972

The U.S. Congress tackled water pollution with the Water Pollution Control Act Amendments of 1972. This bill is considerably more stringent than any previous water legislation, and provides penalties up to $25,000 per day for willful or negligent violations. All discharges to navigable waters must be treated by best practicable technology by July 1, 1977, and best available technology by July 1, 1983. The EPA holds responsibility for determining exactly what is required under the two levels of treatment technology.

The principal control mechanism of the 1972 Amendments is the National Pollutant Discharge Elimination System (NPDES). This means that every wastewater discharger must have a permit. The permit lists the quantities of pollutants that the source may discharge in terms of the usual parameters such as biochemical oxygen demand, dissolved solids, etc. EPA has the basic authority for issuing permits, although it is required to turn over this authority to any state that has established an acceptable permit program.

During 1973 and 1974, EPA set effluent guidelines for 28 industry categories. These serve as the basis for pollutant limitations in the discharge permits. The guidelines sparked much controversy (and a number of legal suits). Further action by the Congress may be required. The objections to the guidelines involve complex legal questions, but in general they are

## Table 10-2. Regional Offices of the Environmental Protection Agency

| | |
|---|---|
| Region I: | John F. Kennedy Federal Office Bldg., Boston, MA 02203 |
| | Connecticut, Maine, Massachusetts, New Hampshire, Rhode Island, Vermont |
| Region II: | 26 Federal Plaza, New York, NY 10007 |
| | New Jersey, New York, Puerto Rico, Virgin Islands |
| Region III: | Curtis Bldg., Philadelphia, PA 19106 |
| | Delaware, Maryland, Virginia, West Virginia, Pennsylvania, District of Columbia |
| Region IV: | 1421 Peachtree St. N.E., Atlanta, GA 30309 |
| | Alabama, Florida, Georgia, Kentucky, Mississippi, North Carolina, South Carolina, Tennessee |
| Region V: | One North Wacker Dr., Chicago, IL 60606 |
| | Illinois, Indiana, Michigan, Minnesota, Ohio, Wisconsin |
| Region VI: | 1600 Patterson St., Dallas, TX 75202 |
| | Arkansas, Louisiana, New Mexico, Texas, Oklahoma |
| Region VII: | 1735 Baltimore Ave., Kansas City, MO 64108 |
| | Iowa, Kansas, Missouri, Nebraska |
| Region VIII: | 1860 Lincoln St., Denver, CO 80203 |
| | Colorado, Montana, North Dakota, South Dakota, Utah, Wyoming |
| Region IX: | 100 California St., San Francisco, CA 94111 |
| | Arizona, California, Hawaii, Nevada, American Samoa, Guam, Wake Island |
| Region X: | 1200 Sixth Ave., Seattle, WA 98101 |
| | Alaska, Idaho, Oregon, Washington |

being challenged on the grounds that EPA exceeded its statutory authority under the 1972 law.

## 10-3. Information Sources

The engineer who needs to determine the specific pollution control requirements for a particular operation in a certain location has many sources of information available—too numerous to be listed here. In general, the starting point for legal information should be at the lowest applicable level of government. Usually this is a state body, although in metropolitan areas a county or city agency may have jurisdiction.

When in doubt, the EPA has vast information resources. Established in 1970 to consolidate all federal environmental efforts under a single agency, EPA maintains 10 regional offices (Table 10-2) that are familiar with local problems and requirements.

## WASTEWATER TREATMENT

## 10-4. Preliminary Treatment

**Equalization.** Wastewater flow rates are erratic. Municipal flow rates peak during the day and reach a minimum during the night; within those periods, the flow rises and falls with the time of day and with the weather (rainfall). Industrial wastewater flows are equally variable, being a function of production rate, frequency of batch operations, etc. Thus many industries, although surprisingly few municipalities, have installed equalization tanks (holding basins) to smooth out process flows and allow the wastewater treatment plant to operate at a steady rate.

**Neutralization.** Biological wastewater treatment proceeds optimally at pH's near 7 (neutrality). Slight deviations from this value will reduce efficiency, while large differences may result in total inactivation of the bacteria.

While municipal sewage normally needs no pH adjustment, industrial waste quite often does. Using pH recorder-controllers, some incorporating analog computers, acids and bases can be added in optimum fashion. Less control equipment is needed if pH is adjusted manually in the equalization tank. Sodium hydroxide or ammonia raises the pH; sulfuric or phosphoric acid lowers the pH. Ammonia and phosphoric acid may serve the dual purpose of pH control and a source of nitrogen or phosphorus, which are important nutrients.

**Oil Removal.** Generally, oil concentrations greater than 50 mg/l inhibit biological action. Thus, gravity settling basins are usually installed to separate oil and water when they are combined in a nonemulsified form.

In general, oil-water separators are rectangular, multichannel structures that produce low flow velocity and a minimum of turbulence, while allowing time for the oil globules to float to the surface and be removed. Gravity treatment, of course, only separates free oil. When emulsified oil is present in industrial wastes, aeration followed by gravity separation is usually sufficient to reduce oil concentrations for subsequent biological treatment. Discharge to a receiving water directly from oil separators generally is not allowed by effluent standards. Secondary oil removal must follow the oil separation stage. Typical processes include dissolved air flotation, granular-media filtration, and activated carbon adsorption.

Dissolved air flotation serves the dual purpose of oil removal and clarification (removal of suspended solids). Air is dissolved in the wastewater under pressure (30 to 70 psig) and then nucleated as fine bubbles by a rapid pressure reduction when the pressurized water stream enters a flotation chamber at atmospheric pressure. The micron-size bubbles attach themselves to suspended matter and carry it to the surface for removal by skimmers. Though more expensive than separators, flotation devices produce a higher quality effluent.

Oil emulsions can be broken by chemical pretreatment with acids or aluminum and iron salts. Subsequent addition of polyelectrolytes aids in floc formation and leads to more efficient operation.

**Metallic-ion Removal.** Ions of the heavy metals (Cu, Zn, Al, Fe, Cr, etc.) in concentrations greater than 1 to 10 mg/l, depending on their nature, must be removed because they inhibit subsequent treatment processes. Adding lime, raising the pH to about 10.5, normally results in a reasonable reduction of metallic-ion concentration, because many of the metals form insoluble hydroxides at this pH level. It may even be possible to go below the level predicted by the solubility product because of adsorption of the metallic ions on the chemical floc. For cadmium, lead, and mercury, precipitation with lime may be incomplete and addition of soda ash (for lead) and sodium sulfide (for cadmium and mercury) may be needed.

## 10-5. Primary Treatment

Screens, comminutors, and/or grit chambers usually precede primary sedimentation in municipal plants. Screens, normally composed of iron bars or grates with ¾- to 3-in. openings, remove the largest suspended particles, while comminutors grind large particles into smaller ones. Grit chambers are designed to remove the heavier solids (i.e., inorganic matter such as sand whose specific gravity is greater than 2) through a slight reduction in flow velocity.

Considerable advantage is gained by removing suspended particulate matter in the primary basin rather than in the secondary. Normally, the initial sedimentation process removes 50 per cent of the suspended solids carrying with it 25 to 40 per cent of the biochemical oxygen demand (BOD); reduction of the BOD load to the secondary system is an obvious benefit in reducing air supply requirements and production of microbial solids.

Simplistic design of sedimentation systems is accomplished by choosing either a residence detention time (generally 90 to 150 min) or an overflow rate [the normal range is 600 to 1,200 gal/(day)(ft²)]. From these data, the tank depth can easily be selected.

**Chemical Addition.** Chemicals are being used increasingly in the initial stages of wastewater treatment for: (1) phosphorus removal, (2) increased efficiency of BOD removal in the primaries, and (3) proper conditioning of the wastewater for filtration, carbon adsorption, and reverse osmosis. Hence, the proper selection of chemicals that assist in the aggregation of colloidal particles (0.001 to 10 microns) is crucial to the process operation.

A number of suitable coagulating agents can be used: lime, alum, iron salts, and polyelectrolytes (Table 10-3). The first step is to test each in the laboratory. Besides determining the best additive, the laboratory study should investigate the sequence of chemical addition, as well as blending, flocculation, and detention times. Additional data sought should be estimates of effluent quality, volumetric sludge production rate, and a measure of the sludge dewatering characteristics.

Lime, ferric chloride, ferric sulfate, sodium aluminate, and alum are the most widely used inorganic coagulants. Polyelectrolytes are high-molecular-weight (15,000 to 10,000,000) polymers with multiple ionic charges along the chain; nonionic polymers have also been termed polyelectrolytes. These polymers can be linear, branched, or cross-linked, and are normally more effective as molecular weight increases.

The performance of chemical additives in wastewater treatment is affected by a number of factors: dosage and molecular weight of the coagulant, temperature, agitation rate and duration, order of addition of chemicals, nature of suspended matter, pH or amount of alkali present, and time between chemical additions. Coagulant dosage is important since the optimum concentration lies in a narrow range. Low dosages fail to destabilize while excess dosages restabilize colloids.

## 10-6. Bioxidation

Simple sedimentation, as described in the previous section, removes a major fraction of the suspended solids but none of the dissolved contaminants. Microorganisms can utilize dissolved organic material as nutrients for their growth and metabolism. The process, shown mechanistically in Fig. 10-1, could be termed flameless oxidation in the liquid state because bacteria convert carbon to carbon dioxide via metabolic pathways, therein consuming oxygen and yielding energy.

In nature, the rate of supply of oxygen may be low. Engineers circumvent this limitation by designing waste treatment systems that optimize the contact between the microorganisms (catalysts), food (dissolved organic matter), and air (oxygen source). In essence, the system is balanced so that the rates of oxygen supply and consumption are equal.

There are two different biological oxidation, or bioxidation, processes, with the distinction between them based on microbial mobility:

1. Suspended—the microorganisms are suspended in the liquid; the system operates as a continuous biological reactor, called an activated sludge plant or oxidation pond.

2. Fixed bed—the microorganisms are embedded in a gelatinous mass

## Table 10-3. Chemical Coagulant Selection*

| Chemical process | Dosage range, mg/l | Resulting pH | Sludge dewatering | Conditions favoring use |
|---|---|---|---|---|
| High lime............ | 150–600 | 11.5–12.0 | Easy | 1. For colloid coagulation and P removal<br>2. High and variable influent P<br>3. Large plants > 10 mgd, where recalcination is feasible<br>4. Wastewater with high calcium content and low noncalcium dissolved solids content |
| Low lime............ | 75–250 | 9.5–10.5 | Easy | 1. For colloid coagulation and P removal<br>2. Wastewater with low alkalinity and high and variable P<br>3. Medium-size plants, 1–20 mgd, no recalcination |
| Very low lime........ | 50–100 | 8 0–8.5 | Fair | 1. For P removal<br>2. Use in combination with the activated sludge process where a 6–8-hr detention time is available for the precipitation reaction<br>3. Wastewater with low alkalinity and high and variable P |
| Solid alum........... | 75–250 | 4.5–7.0 | Difficult | 1. For colloid coagulation and P removal<br>2. Wastewater with high alkalinity and low and stable P<br>3. Where the difficult alum sludge can be handled |
| Liquid alum......... | 75–250 | 4.5–7.0 | Difficult | 1. Same as for solid alum with the following differences:<br>  a. It is cheaper than solid alum when you are located within 100–150 mi of an alum manufacturing plant<br>  b. It is easy to feed and handle and is, therefore, well suited to small (<1 0 mgd) plants<br>  c. These differences between solid and liquid forms also apply to $FeCl_3$ |
| $FeCl_3$, $FeCl_2$.........<br>$FeSO_4$, $7H_2O$........ | 35–150<br>70–200 | 4.0–7.0<br>4.0–7.0 | Fair<br>Fair | 1. For colloid coagulation and P removal<br>2. Wastewater with high alkalinity, and low and stable P<br>3. Where leaching of iron in the effluent is allowable or can be controlled<br>4. Where economical source of waste iron is available (steel mills, etc.) |
| Cationic polymers.... | 2–5 | No change | May improve | 1. For colloid coagulation or to aid coagulation with a metal<br>2. Where the buildup of an inert chemical is to be avoided |
| Anionic and some nonionic polymers | 0.25–1.0 | No change | May improve | 1. Used as a flocculating aid to speed flocculation and settling and to toughen floc for filtration |
| Weighting aids and clays | 3–20 | No change | May improve | 1. Used for very dillute colloidal suspensions for weighting |

* mg/l = milligrams per liter, mgd = millions of gallons per day, P = phosphorus.
SOURCE: T. J. Tofflemire and L. J. Hetling, "Chemicals and Clarifiers, Which Are Best," *Water and Wastes Eng.*, **10**, F24-F26. November (1972).

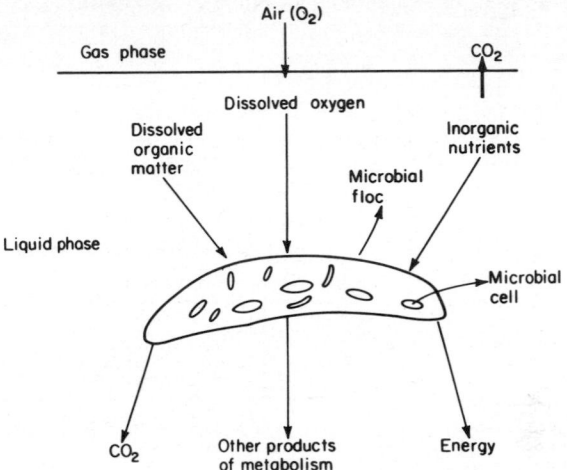

FIG. 10-1. Schematic for bioxidation of dissolved organic matter.

attached to a solid support medium.    This is referred to as a trickling filter or bioxidation tower.

The role of the bioxidation system is to remove the dissolved organic matter from the wastewater.    Because excess cells are produced, they are usually removed by sedimentation in a secondary clarifier, which is similar to the primary clarifier.

Three types of reactions take place in the bioxidation process: (1) assimilative respiration, (2) synthesis of cells, and (3) endogenous respiration. Assimilative respiration provides the necessary energy for the life processes of cells and can be represented by the following equation:

$$C_xH_yO_z + (x + \tfrac{1}{4}y - \tfrac{1}{2}z)O_2 \rightarrow CO_2 + \tfrac{1}{2}yH_2O + \text{energy} \quad (10\text{-}1)$$

Besides the energy released, the cells derive energy for the oxidation process to use in building new cells.

$$n(C_xH_yO_z) + nNH_3 + n(x - \tfrac{1}{4}y - \tfrac{1}{2}z - 5)O_2 \rightarrow (C_5H_7NO_2)n$$
$$+ n(x - 5)CO_2 + \tfrac{1}{2}n(y - 4)H_2O \quad (10\text{-}2)$$

This process is termed endogenous respiration; it is actually autoxidation or self-destruction of cellular material and can be represented by

$$(C_5H_7NO_2)n + 5nO_2 \rightarrow 5nCO_2 + 2nH_2O + nNH_3 + \text{energy} \quad (10\text{-}3)$$

The formula representative of bacterial cells is $C_5H_7NO_2$.    The need for nitrogen in order to grow cells is obvious by its inclusion in the formula. Not shown, however, is another important nutrient, phosphorus, which is needed in lesser amounts.    Phosphorus, nitrogen, and BOD should be present in a ratio of about $1:5:100$ to support growth and oxidation. Municipal sewage normally contains sufficient amounts of nitrogen and phosphorus, but many industrial wastes are deficient in one or the other and must be supplemented chemically.

**Activated Sludge System.**  The conventional activated sludge process is analogous to a continuous biological fermentor because nutrients in the untreated wastewater are continually added, oxygen is supplied through an aeration system, and cells are produced and removed in the effluent stream (Fig. 10-2).  The product is a combination of treated wastewater and

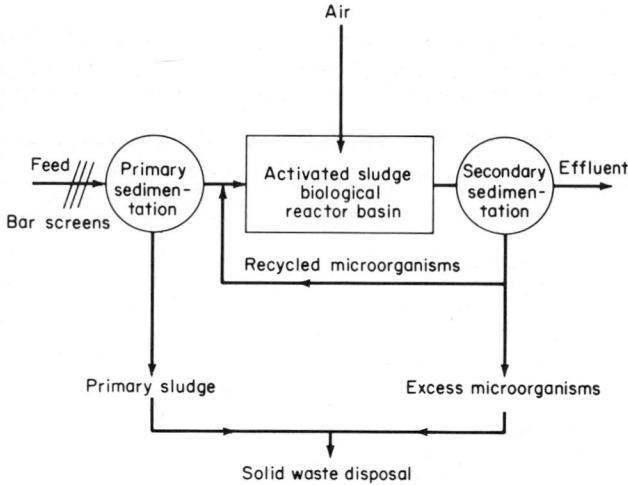

Fig. 10-2. Simplified flowsheet for activated sludge plant.

suspended cells, which are removed by sedimentation and partially recycled to the reactor.  The treated effluent overflows to discharge or reuse systems.

The activated sludge process provides a smooth running, highly efficient operation, with excellent removal of organic matter; BOD removals of 90 per cent are easily achieved.  However, many problems can develop to upset the process.  Shock loading, through changes in concentration and volumetric flow rate, abrupt variations in pH, temperature, or other environmental factors, or the presence of toxic materials are among the major causes of trouble.

The keys to the process are the active bacteria embedded in gelatinous clumps in the reactor (activated sludge basin) that oxidize biodegradable organic material.  Also present are a host of other microorganisms, fungi, protozoa, rotifers, and sometimes nematods that are necessary for a healthy, balanced population.  Important design and operating variables for the process are the food-to-microorganism ratio (the $F/M$ ratio in pounds of food added per pound of cells per day), the amount of BOD supplied per unit volume of reactor (lb BOD/1,000 ft³ of reactor volume), and oxygen consumption rate (lb $O_2$/lb BOD treated).

The conventional activated sludge process can be divided into a series of steps:

1. Mixing of the activated sludge and the sewage
2. Aeration and agitation of the mixed liquor for the prescribed period of time
3. Separation of the sludge cells from the water
4. Recycle of the proper amount of cells (to step 1)
5. Disposal of the excess sludge

Modification of these basic steps has shown the versatility of this process and has led to many variations: tapered aeration, step aeration, contact stabilization, and extended aeration to name just a few (Fig. 10-3).

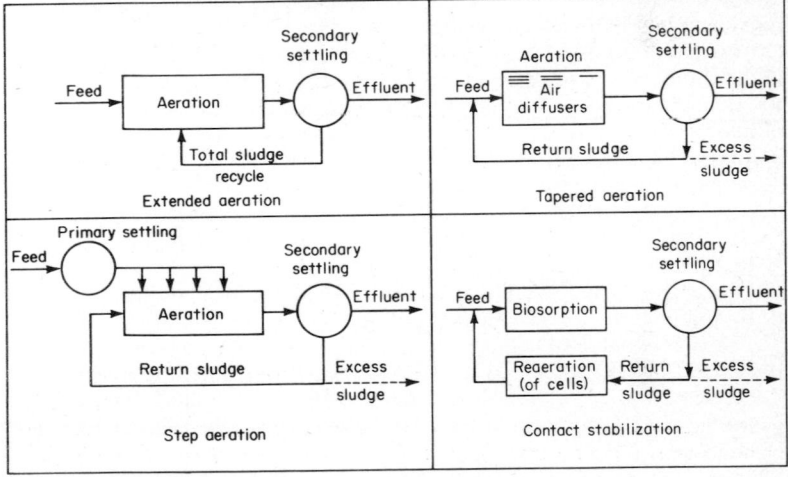

FIG. 10-3. Four variations of the activated sludge process.

**Tapered Aeration.**   As the organic contaminants are consumed by the bacteria in a long reactor, the $F/M$ ratio and oxygen demand decrease with length.   The process modification from conventional activated sludge is a variation of the airflow rate according to oxygen needs achieved by putting in decreasing numbers of diffuser tubes with increasing distance from the inlet. The obvious advantage is lower power consumption than uniform (and excess) aeration would require.

**Step Aeration.**   Another way to balance air requirements is to add the wastewater at different points along the aeration tank.   This modification keeps the $F/M$ ratio constant as a function of tank length and leads to uniform oxygen demand by the microorganisms. With these changes, BOD loading may be increased and detention time and consumption rate lowered.

**Contact Stabilization.** The adsorptive properties of activated sludge particles are advantageous when a large part of the BOD is present in suspended form. Sludge that has been conditioned by reaeration is brought into short (15 to 30 min) but unaerated contact with the wastewater. During this stabilization period, the activated sludge adsorbs and absorbs a high percentage of the colloidal, suspended, and even dissolved organic matter. The mixture is then discharged to a settling tank where solids are removed and directed to a regenerator (reactor) tank for active aeration for approximately 3 hr before reentering the cycle.

**Extended Aeration.** Extended aeration systems operate essentially without any sludge production since they include neither a primary sedimentation tank nor any allowance for continuous sludge withdrawal from the bioxidation system. However, a penalty for eliminating these units must be paid: a high oxygen supply rate is needed to oxidize all organic matter to carbon dioxide at low BOD loading values.

Aeration tanks usually are sized to provide 24- to 30-hr retention with air added mechanically or by submerged diffusers. A final settling basin where the detention is approximately 4 hr removes most solids for total recycle—although normally provision is made for intermittent sludge withdrawal to holding tanks for later off-site disposal. In general, the process functions best at BOD loadings of less than 15 lb/(day)(1,000 ft³ of aeration tank) and less than 0.1 lb BOD/lb of cells. [C. N. Sawyer, "Milestones in the Development of the Activated Sludge Process," *J. Water Pollut. Control Fed.*, **37**, 151–162 (1965).]

Advantages include simple operation, availability of prefabricated units, adaptability to a variety of topographies and climate conditions, and generally substantial BOD reduct on (an average of 87 per cent) has been reported. [R. Porges and G. L. Morris, "Small Extended Aeration Sewage Treatment Plants," *J. Environ. Health*, **25**(6), (May–June, 1967).] Disadvantages include high power cost, long retention times, and periodic loss of fine floc in the effluent.

**Oxygen Demand and Supply.** Generally, 2 mg/l of dissolved oxygen should be maintained at all points of the activated sludge basin to keep the cells working at their maximum rate. Hence, the mixed liquor in the reactor is continuously aerated by either a diffused air system or by mechanical surface aerators. The aeration process serves three functions: (1) it mixes the sludge and the sewage; (2) it keeps the sludge in suspension; and (3) it supplies the oxygen needed for bioxidation.

In diffused air systems, air is supplied by blowers and forced through diverse types of porous materials mounted in plates or tubes; the air is released as fine bubbles several feet below the surface of the liquid. In mechanical systems, surface aerators cause liquid turbulence, at which point the oxygen is dissolved and dispersed downward into the fluid.

Oxygen must be supplied at a rate sufficient to meet the demand of the cells. The amount needed varies with the waste being treated, the activity of the microbial cells, and length of aeration period, but generally it is in the range of 1 to 2 lb $O_2$/lb BOD removed.

Oxygenation economy is extremely important in sewage plant design and operation. A critical factor is the amount of oxygen dissolved in the water per unit of power, expressed in lb $O_2$ dissolved/hp-hr. Surface aerators yield an average of 3.5 lb $O_2$/hp-hr and submerged diffusers about 2.5.

**Lagoons.** The term lagoon or stabilization pond generally is applied to all bodies of water artificially created with the intention of retaining sewage or organically contaminated wastewaters until the wastes are rendered stable or unobjectionable through biological decomposition, and the waters are suitable for disposition, either by discharge into receiving waters or by way of ground seepage and evaporation.

Lagooning is a common and cheap means of eliminating dissolved or suspended organics. The feed enters at one end of the lagoon. Sedimentation occurs first, with settlable solids being deposited near the entrance. Additional sedimentation occurs throughout the pond area, with the action of soluble salts aiding precipitation of colloid matter. Added to the bottom (benthic) deposits through sedimentation are the bacteria and algae produced by biological activity.

Algae are a significant part of the purification cycle in lagoons. They use carbon dioxide, sulfates, nitrates, phosphates, water, and sunlight to synthesize their own cellular material, with free oxygen as a waste product. This process reverses the bacterial degradation sequence to generate an organic material that, if discharged, is itself a pollutant.

Treatment efficiencies vary, with indication that well-designed and well-operated ponds produce effluents equivalent to or better than conventional secondary plants. [R. Porges and K. N. Mackenthun, "Waste Stabilization Ponds: Use, Function and Biota," *Biotechnol. Bioeng.*, **5**, 255–273 (1963).]

In design of a pond, several physical factors must be considered:

1. Depth—experience indicates that a depth of 3 to 4 ft with flexibility to permit 5 to 6 ft is desirable.

2. Sidewall—sloped at a ratio of 3:1, freeboard of 2 to 5 ft.

3. Bottom—level and clear of vegetation and debris.

4. Inlet—feed should be mixed quickly and distributed throughout the pond; short-circuiting is to be eliminated.

5. Retention time—usually 2 to 30 days.

**Trickling Filters.** Trickling, or biological, filter is the name given to the process whereby microorganisms attached to stones (or plastic media) strip wastewater of its organic components as it flows (trickles) down through a packed tower. The attached growths adsorb, absorb, and oxidize suspended and colloidal matter in the wastewater. Crushed stone ($1\frac{1}{2}$ to 4 in.) normally is used as the packing medium for the bed, although recently plastic packing has found substantial application. The wastewater is applied to the filter through a central rotating distributor, which intermittently applies water to the various segments of the bed.

Approximately 60 per cent of the dissolved organic matter removed from the waste stream by the bioxidation process is utilized for growth of new microorganisms, with the remaining 40 per cent of the material being oxidized to terminal metabolic products, carbon dioxide, and water, while providing energy. Growth is most active at the liquid-slime surface where the concentration of organic material is highest. Oxygen diffuses from the liquid into the slime, but progressively lower oxygen concentrations are found at greater depths because oxygen is consumed by the cells. At some limiting slime depth, the oxygen is gone and anaerobiosis occurs.

The growing slime layer is constantly being scoured by the downward flow of liquid. After the microbial layer reaches a critical thickness, outer portions slough off (break off) and are discharged in the effluent leaving at the bottom of the tower. Sloughing will occur either intermittently or more or less continuously, depending on the hydraulic load.

Various biological processes are compared in Table 10-4.

### Table 10-4. Comparison of Biological Processes

| Process | Area, acres | Biological loading lb BOD/1,000 ft³ | BOD removal, % |
|---|---|---|---|
| Stabilization pond.......... | 57* | 0.09–0.23 | 70–90 |
| Aerated lagoon............. | 5.75† | 1.15–1.60 | 80–90 |
| Activated sludge: | | | |
|    Extended............... | 0.23 | 11.0–30.0 | 95+ |
|    Conventional........... | 0.08 | 33.0–400 | 90 |
|    High rate............... | 0.046 | 57.0–150 | 70 |
| Trickling filter: | | | |
|    Rock.................... | 0.2–0.5 | 0.7–50 | 40–70 |
|    Plastic media........... | 0.02–0.08 | 20–200 | 50–70 |

\* 5-ft deep.
† 10-ft deep.
SOURCE: *Chem. Eng.*, **76**, 63–70, April 27 (1970).

**Design Equations.** The rate of removal of BOD in a filter is a function of the BOD concentration of the wastewater and the adsorptive capacity of the biological growth. The rate of stabilization controls the adsorption capacity of the biological growth.

Eckenfelder ["Trickling Filtration Design and Performance," *J. Sanit. Eng. Div., Am. Soc. Civ. Eng.*, **87**(SA4), 33–45 (1961)] has proposed equations for design of rock filters with and without recirculation.

No recirculation:
$$\frac{L_e}{L_0} = \frac{100}{1 + 2.5(D^{0.67}/Q^{0.50})} \tag{10-4}$$

Recirculation:
$$\frac{L_e}{L_0} = \frac{1}{(1 + N)[1 + 2.5(D^{0.67}/Q^{0.5})] - N} \tag{10-5}$$

when $L_0 = \dfrac{L_a + NL_e}{N + 1}$

where $L_e$ = BOD of effluent, mg/l

$\quad L_a$ = influent BOD, mg/l

$\quad L_0$ = BOD of mixed influent plus recirculated streams, mg/l

$\quad N$ = recirculation ratio

$\quad D$ = filter depth, ft

$\quad Q$ = hydraulic application rate, million gal/(acre)(day)

Recirculation has several benefits:

1. Smooths out diurnal flows
2. Stops long retention times in settling basins at low flows
3. Provides a microbial seed
4. Brings back organic matter that escapes removal in the first pass

There is, however, a limit to the amount of organic matter that can be removed through successive treatment by increasing the recirculation rate. Each succeeding pass removes a smaller amount of the unoxidized BOD, because the remaining BOD is more refractory (i.e., resists biodegradation). As the slime is forced to use more dilute waste, it dissolves itself as endogenous respiration occurs. Additionally, as the recirculation ratio increases, the hydraulic load on the filter increases.

**Temperature Effect.** The effect of temperature on treatment efficiency is given by the formula

$$E = E_{20}1.035^{(T-20)} \tag{10-6}$$

when $E$ is the efficiency of BOD removal at any temperature $T$, °C, and $E_{20}$ is the efficiency of removal at 20°C.

**Plastic Packing.** Plastic media have found wide acceptance in municipal and industrial systems. The medium consists of corrugated plastic sheets running parallel to each other and bonded together to form a bundle or a pack. The uniform but open pattern of the plastic grid largely eliminates clogging problems that exist with rock, and allows the grid to operate at higher organic loadings.

The basic design formula for plastic-media trickling filters was developed by Germain [J. E. Germain, "Economic Treatment of Domestic Waste by Plastic-medium Trickling Filters," *J. Water Pollut. Control Fed.*, **38**, 192–203 (1966)].

$$\frac{L_e}{L_0} = e^{-kD/Q^{1/2}} \tag{10-7}$$

where $D$ = the depth of the filter, ft

$\quad Q$ = the hydraulic dosage rate of primary effluent, gal/(min)(ft²)

$\quad k$ = the reaction rate coefficient (treatability factor) in consistent units to render the exponent dimensionless

The treatability factor is important because its magnitude determines rate of BOD removal. For municipal sewage, Germain determined a value of 0.088.

Because plastic media can handle high BOD loads, it is an ideal packing for a roughing filter, i.e., to serve as a first and important step in the BOD reduction process. In a plant that has become overloaded, installation of a device of this type prior to the activated sludge tanks has proved economical and efficient.

## 10-7. Sludge Treatment

Sludge, a suspension of solids in water, must be processed further to reduce its high water content and to reduce the concentration of organic matter. If untreated and used as landfill, it will decay and produce an offensive odor. Concentration and stabilization are the primary functions of the sludge processing system that can constitute as much as 25 to 50 per cent of the operating and capital cost of a wastewater treatment plant.

According to Burd, whose FWPCA report on sludge handling and disposal has become the standard handbook in the field (R. S. Burd, "A Study of Sludge Disposal and Handling," Publication WP-20-4, U.S. Department of the Interior, Federal Water Pollution Control Administration, Washington, D.C., May, 1968), the following general observations can be made about sludge handling and disposal practices: anaerobic digestion followed by sand bed dewatering is the most common method presently employed because of its simplicity and low cost; lagooning is the most common method employed by industry; for coastal and near-coastal cities, anaerobic digestion followed by pipeline or barge transportation to an open dumping site is cheapest; marketing of dried waste sludge has been a failure; almost all the common methods now employed were used in the 1930s.

Burd (*loc cit.*) has been able to note trends in sludge treatment, among which are: (1) increasing adoption of mechanical dewatering with the centrifuge of increasing importance over vacuum filters; (2) sludge incineration is the process with the brightest future; (3) raw sludge incineration is replacing anaerobic digestion at medium and large plants; (4) the popularity of composting sludge in foreign countries is declining; (5) sludge volumes are rising and becoming more difficult to dewater.

In municipal plants, the two main sources of sludge are the primary clarifiers that produce a sludge having 1 to 4 per cent solids, and the secondary clarifiers that yield biological sludge with a solids concentration of less than 1 per cent. Of the two, primary sludge is simpler to process, dewater, stabilize, and dispose of. Sludge production characteristics of typical municipal wastewater treatment plants are shown in Table 10-5.

**Conditioning.** Sludge is most often conditioned chemically. Coalescence can be enhanced, in theory, by chemicals that neutralize charges on suspended particles, causing them to agglomerate and simultaneously lessening the particle's tendency to bind water.

Synthetic polymers have taken on a role of increasing importance in flocculation, but inorganic compounds such as alum and ferric chloride are still widely used. Other conditioning agents include fly ash and diatomaceous earth.

Elutriation has been defined by Burd (*loc. cit.*) as "a washing operation which removes sludge constituents that interfere with thickening and de-

## Table 10-5. Normal Quantities of Sludge Produced by Different Treatment Processes*

| Treatment process | Normal quantity of sludge | | | Mois-ture, % | Specific gravity of sludge solids | Specific gravity of sludge | Dry solids | |
|---|---|---|---|---|---|---|---|---|
| | Gal/ million gal of sewage | Tons/ million gal of sewage | Ft³/ 1,000 persons daily | | | | Lb/ million gal of sewage | Lb/ 1,000 persons daily |
| Primary sedimentation: | | | | | | | | |
| Undigested.............. | 2,950 | 12.5 | 39.0 | 95 | 1.40 | 1.02 | 1.250 | 125 |
| Digested in separate tanks.. | 1,450 | 6.25 | 19.0 | 94 | .... | 1.03 | 750 | 75 |
| Digested and dewatered on sand beds.............. | .... | 0.94 | 5.7 | 60 | .... | ..... | 750 | 75 |
| Digested and dewatered on vacuum filters.......... | ...... | 1.36 | 4.3 | 72.5 | .... | 1.00 | 750 | 75 |
| Trickling filter.............. | 745 | 3.17 | 9.9 | 92.5 | 1.33 | 1.025 | 476 | 48 |
| Chemical precipitation....... | 5,120 | 22.0 | 68.5 | 92.5 | 1.93 | 1.03 | 3,300 | 330 |
| Dewatered on vacuum filters.................. | ...... | 6.0 | 19.3 | 72.5 | .... | ..... | 3,300 | 330 |
| Primary sedimentation and activated sludge: | | | | | | | | |
| Undigested.............. | 6.900 | 29.25 | 92.0 | 96 | .... | 1.02 | 2.340 | 234 |
| Undigested and dewatered on vacuum filters........ | 1,480 | 5.85 | 20.0 | 80 | .... | 0.95 | 2,340 | 234 |
| Digested in separate tanks.. | 2,700 | 11.67 | 36.0 | 94 | .... | 1.03 | 1.400 | 140 |
| Digested and dewatered on sand beds.............. | ...... | 1.75 | 18.0 | 60 | .... | ..... | 1.400 | 140 |
| Digested and dewatered on vacuum filters.......... | ...... | 3.5 | 11.7 | 80 | .... | 0.95 | 1.400 | 140 |
| Activated sludge: | | | | | | | | |
| Wet sludge.............. | 19,400 | 75.0 | 258.0 | 98.5 | 1.25 | 1.005 | 2,250 | 225 |
| Dewatered on vacuum filters................. | .... | 5.62 | 19.0 | 80 | .... | 0.95 | 2,250 | 225 |
| Dried by heat dryers....... | ...... | 1.17 | 3.0 | 4 | .... | 1.25 | 2,250 | 225 |
| Septic tanks, digested........ | 900 | ..... | 12.0 | 90 | 1.40 | 1.04 | 810 | 81 |
| Imhoff tanks, digested........ | 500 | ..... | 6.7 | 85 | 1.27 | 1.04 | 690 | 69 |

* Based on a sewage flow of 100 gallon per capita per day and 300 mg/l, or 0.25 lb per capita daily, of suspended solids in sewage.

source: Metcalf and Eddy, "Wastewater Engineering: Collection, Treatment, Disposal," McGraw-Hill Book Company, New York, 1972.

watering processes." The simplest method of elutriation occurs in a single-stage batch process by a fill and draw procedure, with sedimentation and decantation performed in a single step.

**Gravity Thickening.** Gravity thickening basins are usually circular and are provided with a raking mechanism to convey the solids to the point of discharge at the center of the tank. For municipal sewage sludges, the following loading rates, given in lb/(ft²)(day), are recommended for thickeners: (1) 22 for primary sludge; (2) 15 for primary plus trickling filter biofloc; (3) 8 to 12 for a blend of primary and waste-activated sludge; (4) 4 for waste-activated sludge alone.

It is obvious that activated sludge alone releases its water very slowly, while blends of primary and secondary sludge respond better. Thus, blending is recommended when both sludges are available.

**Flotation.** Flotation is attractive for sludge thickening because of a faster dewatering rate that is not affected adversely by decomposing solids, which evolve gases and cause sludge bulking.  In flotation units, air is dissolved under pressure in the liquid being treated.  When the pressurized air/wastewater stream is discharged into an aeration tank, maintained at atmospheric pressure, the supersaturated air nucleates as small bubbles in the 10- to 100-micron range.

The air bubbles, normally carrying a charge, collide with the solid particles and form agglomerates whose specific gravity is less than the water.  The "lightened" solids rise to the surface, forming a blanket that is skimmed off by mechanical scrapers.

Processes to prepare sludge for dewatering and disposal are as diverse as any unit operation in waste treatment.  They range from those employed for many years (anaerobic digestion) to those having only research interest at present (ultrasound), from aerobic to anaerobic digestion by bacteria, and from heating to freezing.

**Aerobic Digestion.**  In aerobic digestion, waste sludge is simultaneously the food and the oxidation system, i.e., in autoxidation or endogenous systems the living cells utilize nutrients released when other cells die and dissolve.  Thus the microbial population and the amount of biodegradable organic matter are continually decreasing.  The process is carried out until the volatile suspended solids are reduced to about 50 per cent, a level at which the sludge product will not cause a nuisance.  The residence time should be about 15 days to produce a good sludge that forms a filter cake of 20 to 25 per cent solids.

For medium- to small-sized plants, aerobic digestion has real promise, and offers several advantages over anaerobic digestion including lower BOD of the supernatant fluid, production of a better dewatering sludge, and lower capital cost.  Offsetting these advantages is the major added cost of power to supply air to oxidize the cellular matter.

**Anaerobic Digestion.**  Anaerobic digestion is the most popular method of sludge stabilization.  The term applies to the process in which organic material is decomposed biologically in an environment devoid of oxygen. This decomposition results from the action of two major groups of bacteria: acid formers consisting of facultative bacteria that convert carbohydrates, fats, and proteins to organic acids and alcohols; and methane bacteria, which are strict anaerobes that convert the organic acids and alcohols produced by the acid formers into carbon dioxide and methane.  Small amounts of hydrogen and hydrogen sulfide are also formed.

The gaseous mixture produced by these symbiotic (mutually benefiting) reactions contains approximately 60 per cent methane and 35 per cent carbon dioxide, with the remainder being water vapor, hydrogen sulfide, hydrogen, ammonia, and nitrogen.  Its heat value is in the range of 600 to 650 Btu/ft³ as contrasted to 960 Btu/ft³ for pure methane.  This gas, which can be burned to produce energy, is one of the major advantages of anaerobic digestion.  Many investigators are examining ways of optimizing methane production by anaerobic processes.  Other advantages of this process include a high degree of waste stabilization, combined with production of a low amount of sludge that is inoffensive, has low hydrophilic properties and is easy to dewater.

The standard-rate digestor, shown schematically in Fig. 10-4, has wide application in sewage treatment. To obtain significant stabilization of the sludge, retention times are long—on the order of 30 to 60 days with loadings of 30 to 100 lb volatile solids/1,000 ft³. This results in large vessels. Further, only about one-third of the volume is used for actual digestion. The remainder of the space is occupied by stabilized solids, supernatant fluid, scum, and gas.

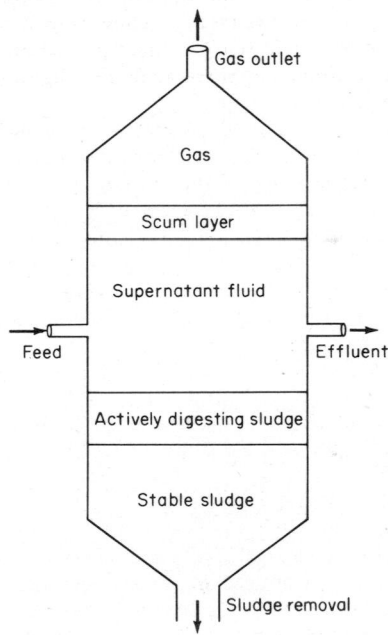

Fig. 10-4. Active regions in conventional anaerobic digestor.

The feed to the digestor and withdrawal of byproducts are intermittent. Upon entering the digestor, the new sludge rises to the scum zone and undergoes initial decomposition, thereby producing much of the gas. As decomposition proceeds, the partially decomposed solids fall to the bottom of the tank and build up a layer of digested and digesting solids. The intermediate layer is a supernatant liquid, which must be returned to the liquid treatment system because it contains a high concentration of dissolved and suspended solids.

The size of the tank is based on population. For a facility treating primary sludge, design values are 2 to 3 ft³ of volume per capita, 4 to 5 ft³ if the plant is a trickling filter, and 4 to 6 ft³ if activated sludge. The higher values should be used in each case for installations handling the waste of populations of 5,000 or less.

Temperature exerts a significant effect on anaerobic digestion. As the temperature increases, the time required to attain a specific degree of stabilization decreases. The optimum temperature for digestor operation in the mesophilic range is 98°F. Thermophilic temperatures, it has been shown, cause digestion to occur more efficiently, but temperature control is difficult and more thermal energy is required.

Reduction of residence time through continuous mixing is the basis of the high-rate digestion process. Continuous feed and withdrawal, and heating of the entire tank contents, result in stable operation with a detention time of 10 to 15 days at loadings of 100 to 500 lb volatile solids/1,000 ft³. Since there is no scum layer and no stratification, the entire tank is actively used for sludge digestion.

Although anaerobic digestion is mainly used for the stabilization of sludges, it is employed to a limited extent for the treatment of high-strength wastes. Anaerobic digestion becomes economical at effluent concentrations in excess of 4,000 mg/l and has increased economic advantages as the strength of the organic waste rises. At the upper limit (above 50,000 mg/l), evaporation and solid recovery begin to look attractive.

**Thermal Conditioning.** Heat treatment of sludge under pressure for about 30 min produces a sludge that is easily dewatered without the use of chemicals. Essentially, the process consists of grinding the sludge to eliminate large particles, and sparging the sludge with live steam in a reactor for 30 to 45 min at 250 psig.

Sludge is withdrawn to a decanter where the solid material settles rapidly to about one-third of its original volume. Vacuum filters handling heat-treated sludge average 12 lb/(ft²)(day). The filter cake, which prior to heat treatment averaged 20 to 25 per cent solids, is increased to 40 to 45 per cent.

**Dewatering.** Used for many years, applicable to a wide variety of waste, and relatively cheap, vacuum filtration is a popular method of mechanically dewatering sludge. A typical vacuum filter consists of a rotating cylindrical drum, partially submerged in sludge. A vacuum of 10 to 20 in. Hg is applied to the center of the drum, drawing the water through the filter medium but leaving the solids deposited on the periphery. Continual application of vacuum as the drum rotates out of the liquid removes more water. Just before the drum re-enters the sludge, the cake is removed.

Several factors affect the design of the vacuum filter and dictate its performance. The major ones are:

1. Sludge type, age, temperature, and compressibility
2. Solids concentration in the feed
3. Flocculant type, dosage, and procedure

These considerations affect vacuum level and vacuum pump capacity, drum size and degree of submergence, and filter medium. The vacuum level is important in determining the final solids level. Doubling the vacuum from

10 to 20 in. will increase by several percentage points the solids content of the cake. In addition, the filtration rate varies approximately as the 0.3 power of the vacuum. To size the vacuum pump, one generally designs for a flow rate of 2 $ft^3/(min)(ft^2)$ at 20 in. Hg. Typical performance of a vacuum filter is shown in Table 10-6.

**Table 10-6. Expected Performance of Vacuum Filters Handling Properly Conditioned Sludge**

| Type of Sludge | Yield, $lb/(ft^2)(hr)$ |
|---|---|
| Fresh solids: | |
| Primary | 4–12 |
| Primary + trickling filter | 4–8 |
| Primary + activated | 4–5 |
| Activated (alone) | 2.5–3.5 |
| Digested solids (with or without elutriation): | |
| Primary | 4–8 |
| Primary + trickling filter | 4–5 |
| Primary + activated | 4–5 |

SOURCE: Sewage Treatment Plant Design, "Manual of Practice," *Water Pollution Control Federation.* Vol. 8, 1959.

Another parameter affecting the filtration rate is the solids concentration. There is a direct relation between the two with the cake solids concentration increasing as the feed solids increase. Practically, however, there is a limit of 8 to 10 per cent to the feed solids concentration; if this concentration level is exceeded, chemical conditioning and sludge distribution become difficult.

**Pressure Filtration.** This essentially batch operation is relatively labor intensive, but the product is so dry (50 to 60 per cent solids) that in many cases the advantages in disposal of the resultant solid waste outweigh the cost of labor. Besides a well-dewatered product, the filtrate generally contains a low concentration (20 mg/l or less) of suspended solids.

The filter press is an alternating series of plates and frames. The feed is pumped into the interior of a frame and forced through the filter medium by the pressure applied by the pump. Filtrate leaves through channels in the plates, while the solids are retained in the frame. When full, continuous hydraulic pressure applied by the feed pump forces (presses) more liquid out. Then the press is opened and the sludge removed. Normal cycle times are in the range of 2 to 4 hr.

**Centrifugation.** In recent years, centrifuges have gained increasing acceptance in the handling of municipal and industrial sludges. Among the reasons are the development of centrifuges requiring less maintenance than in the past, while operating at higher feed rates and greater sludge recovery. Disadvantages still include the maintenance problems and noise that are inherent with high-speed equipment.

There are basically three types of centrifuges: disk, solid-bowl conveyor, and solid-bowl baskets. Each has its own operating characteristics and special advantages, requiring a study of the potential application to choose the proper design.

Construction and application details of the three types of units have been discussed by Keith and Little. [F. W. Keith and T. H. Little, "Centrifuges in Water and Waste Treatment," *Chem. Eng. Prog.*, **65**, 77–80 (1969).] In the disk centrifuge, the feed stream is divided among many narrow channels in the bowl and discharges from it through peripheral orifices. A solid-bowl centrifuge essentially consists of a cylindrical bowl in which solids are carried by an internal helical conveyor to one end of the bowl, while the clarified effluent is discharged at the other end. In the basket centrifuge, solids are retained in a perforated basket while the liquid passes through. Performance data for centrifuges on a wide variety of sludges are found in Table 10-7.

## Table 10-7. Performance of Centrifuges on Various Sludges

| Sludge source | Sludge type | Cake, % solids | Recovery of solids, % | Chemical needs, lb/ton |
|---|---|---|---|---|
| Primary | Raw | 30–40 | 70–80 | None |
| Secondary activated | Undigested | 15–20 | 85–100 | 10–15 |
| Secondary—extended aeration | Activated | 5–15 | 90–100 | 5–10 |
| Biological | Heated | 40–45 | 90–95 | None |
| Chemical/antibiotic | .......... | 20–30 | High | |
| Refinery | Blended | 15–35 | 82–86 | None |

**Sand Drying Beds.** Although mechanical methods of sludge dewatering have received the greatest research attention recently, operators, especially those of small plants, still use sand beds effectively. Drying beds are constructed by placing 4 to 6 in. of sand over graded layers of gravel or crushed stone underlaid by tile drains. The sides of the beds are made of concrete or wood; internally the beds are subdivided into cells, usually less than 20 ft wide and 100 ft long.

The digested sludge (raw sludge cannot be dried this way because it will cause an odor problem) flows into the bed to a depth of 6 to 12 in. and is allowed to dewater. The process occurs by both drainage and evaporation. Although drainage is the predominant mechanism, evaporation is also important.

Burd (*loc. cit.*) has recommended as design criteria the following sizes for open beds in 40 to 45°N latitude.

$$\text{Primary sludge} \dots\dots\dots\dots\dots\dots\dots\dots\dots 1.00 \text{ ft}^2/\text{capita}$$
$$\text{High-rate trickling filter sludge} \dots\dots\dots\dots 1.50 \text{ ft}^2/\text{capita}$$
$$\text{Activated sludge} \dots\dots\dots\dots\dots\dots\dots\dots 1.35 \text{ ft}^2/\text{capita}$$

After drying, the sludge normally is removed by hand, although mechanical handling equipment has been developed. The requirement for hand labor is one of the major disadvantages of sand beds. Another is the large land area required. But for small plants, sand beds following anaerobic digestion are the most economical treatment sequence.

## 10-8. Sludge Disposal

It should be evident that waste treatment does not destroy pollution, but merely converts the pollutant to another form. Hence, a large fraction of wastewater contaminants, both suspended and dissolved, are converted to a solid waste whose ultimate disposal location is on the land, in the atmosphere, or in the ocean. Economics, location, and environmental impact are the three factors normally controlling process selection.

**Incineration.** With suitable sanitary landfill areas rapidly disappearing and ocean dumping being closely scrutinized, incineration is becoming an increasingly popular method of sludge disposal. Burning accomplishes a significant reduction in sludge volume and produces a sterile ash. The fundamental concern is the supplementary heat required (if any) to sustain combustion. Dried solids contain considerable heat value. Based on the ultimate analyses, the heat value is:

$$Q = 14,600 \; C + 62,000 \; (H - O/8) \qquad (10\text{-}8)$$

where $Q$ = Btu/lb dried sludge
$C$ = per cent carbon
$H$ = per cent hydrogen
$O$ = per cent oxygen

A typical sewage sludge will have a heat value of 7000 Btu/lb dry solids. If inorganic coagulants are used in the thickening process, this value is reduced.

Factors affecting the amount of supplementary heat required for combustion include:

1. Heat value of the dried sludge.

2. Per cent moisture of the feed. This is an extremely important variable since approximately 1000 Btu are needed to evaporate each pound of water accompanying the solids.

3. Concentration of inert chemicals combined with the sludge as a result of their use in conditioning or phosphate removal.

4. Temperature of combustion, which determines the heat losses from the furnace by radiation and convection in addition to the heat lost in the ash and gases.

5. Excess air required and degree of preheating.

Multiple-hearth incinerators are quite popular. The furnace consists of circular hearths, one above the other, enclosed in a refractory-lined steel shell. Entering at the top of the furnace, the sludge is raked by blades that are attached to a central vertical rotating shaft. Solids follow a spiral path across the hearth to a drop hole where they fall to the next hearth. The sludge passes through three successive zones: drying, for removal of water to the point at which combustion can occur; combustion; and subsequent cooling to recover heat from the solids. The ash, freed of all organic matter, is a sterile mixture of the oxides of silicon, aluminum, magnesium, sulfur, sodium, calcium, and iron.

The prime requisite for ultimate disposal of sewage sludge, processed or unprocessed, is that it not cause a secondary pollution problem. For example, minimization of air pollution from incinerators is accomplished with wet scrubbers. Comparable steps must be taken to ensure the environmental soundness of any disposal technique for solid matter on land or into water.

**Land Spreading.** There are two principal methods of disposing of sludge on land, a process that has the advantage of recovering the fertilizer value of the sludge. Drying and sale of the material for fertilizer have had limited success. The city of Milwaukee for many years has produced Milorganite, a fertilizer that gained prominence largely due to marketing techniques. However, synthetic fertilizers have provided severe competition, because they deliver more nitrogen and phosphorus at lower cost.

The sludge also can be spread wet, thus avoiding the cost of drying. Many farmers seek this material if available locally, although there may be problems with odor and possible accumulation of heavy metals in the soil.

**Ocean Disposal.** For cities near the ocean, barging or pipeline transport of sludge to deep water has been popular because ocean disposal is cheap and simple. Although all traces of sludge disappear within $\frac{1}{2}$ hr of the discharge, and studies by large cities using the method claim no detectable degradation of the sea, there are serious concerns over the environmental impact of this method. Thus ocean dumping of sludge will probably not last.

## 10-9. Tertiary Treatment

Tertiary treatment is defined as any process that follows secondary biological systems. For the purpose of this section, it has been divided into four areas: removal of nutrients, fine solids, organic material, and dissolved solids. None of the steps to remove these pollutants is cheap in view of the incremental cost. Microscreening, for example, can reduce BOD by a further 5 per cent at a cost of 1.5 cents/1,000 gal—compared to a 90 per cent reduction in secondary treatment for 11 cents. For dissolved solids removal, the cost increases dramatically.

**Phosphorus.** Although phosphorus can be removed biologically and chemically (ion exchange included), physical processes appear to be better understood and more controllable. It is generally agreed that primary sedimentation will remove, on the average, 10 per cent of the influent phosphorus. Conventional activated sludge plants, including primary sedimentation, achieve about 46 per cent removal. There are, however, cases where significantly improved phosphorus removal has been attained biologically. Even so, chemical precipitation is the favored method.

The two most common types of chemical treatment are alkaline removal with lime and adsorption or precipitation with a metallic hydroxide such as alum. Costs vary widely, but are usually less than 10 cents/1,000 gal.

Lime, if added in proper concentration, not only precipitates phosphorus but, through increased flocculation and enhanced sedimentation, removes

BOD and suspended solids better than iron and aluminum salts.  Moreover, lime systems are less susceptible to shock loading and corrosion in the sludge handling system, and present less trouble in removing oil, grease, and scum. They also put fewer dissolved solids into the effluent (sulfate and chloride concentrations are increased by the other two chemicals mentioned). Finally, lime systems offer cost savings since the lime can be recycled by recalcination.

Lime is cheap but must be added in large doses (approximately 200 to 400 mg/l) to raise the pH to 9.5 or above.  At this level of alkalinity, a good settling floc is obtained and the phosphorus reduced to 0.5 mg/l or less.

Iron and aluminum salts have advantages, especially in the lesser amount of sludge generated and its treatability.  Aluminum is generally supplied by alum (aluminum sulfate) or sodium aluminate; removal of phosphorus by these salts is optimum at pH = 6.  Unlike lime, which must be used at high bactericidal pHs, aluminum salts can be added directly to the biological system.  Although the cost for this additive is higher, its lower dosage is a balancing factor.

Ferric salts are also employed in phosphorus removal, especially with a suitable polymer.  The required dosage is generally low and the sludge is readily settleable.  Waste pickle liquor, if available locally, should be considered as an iron source.

**Nitrogen.**  Although conventional sewage plants remove some nitrogen, there are still varying amounts released in diverse forms, such as ammonia, organic nitrogen, and nitrates.  A process for nitrogen removal receiving much attention at the present time is nitrification-denitrification.  Nitrogen is first oxidized to its highest oxidation state; the nitrates are then reduced to nitrogen, which is air-stripped.  Nitrification occurs in two steps.  The bacteria capable of oxidizing ammonia to nitrate are believed to be primarily autotrophic nitrifiers, among which *Nitrosomonas* are most common:

$$NH_4^+ + 1.50_2 \rightarrow 2H^+ + H_2O + NO_2^- \qquad (10\text{-}9)$$

The second phase of the reaction, nitrite to nitrate, is carried out by *Nitrobacter*

$$NO_2^- + 0.50_2 \rightarrow NO_3^- \qquad (10\text{-}10)$$

One factor of great importance in nitrification is the large amount of oxygen required per unit of ammonia—theoretically 4.56 mg $O_2$/mg $NH_3$.

The first step can be carried out in activated sludge basins if proper conditions are maintained: optimum pH of 8 to 8.5 and temperature 30 to 35°C (the reactor will work in the pH range 7 to 9 and temperature 15 to 35°, but at lower rates and efficiency), dissolved oxygen concentration 1 to 2 mg/l or higher, and low loading equivalent to extended aeration systems [0.2 lb BOD/(day)(lb MLSS*)].  [S. Balakrishnan and W. W. Eckenfelder, Jr., "Nitrogen Removal by Activated Sludge Process," *J. Sanit. Eng. Div., Am. Soc. Civ. Eng.*, **96**(SA2), 501–512 (1970).]

* MLSS = mixed liquor suspended solids.

Dentrification is also a biological process in which bacteria, in the absence of free molecular oxygen, reduce the nitrates to nitrogen. In order to achieve a reasonable efficiency, a supplementary source of carbon, commonly methanol, is added to accomplish the following:

$$5CH_3OH + 6NO_3^- \rightarrow 5CO_2 + 7H_2O + 3N_2 + 6OH^- \qquad (10\text{-}11)$$

The air-stripping process for ammonia removal consists in raising the pH to 10.5 to 11.5 and degassing the ammonia through air/water contact. Cooling towers have been used for this step. Important design variables for the process are pH control, the rate of interphase ammonia transfer, and the air/liquid flow rate. The process can be controlled to attain selected ammonia removals but there are several disadvantages including low efficiency at low temperatures, deposits of calcium in the tower, air pollution problems, and deterioration of wood packing of the tower at high pH.

Almost all soluble nitrogen compounds in wastewater are in ionic form and theoretically can be removed by deionizing or desalting processes such as reverse osmosis, electrodialysis, or distillation. However, none of these processes has a favorable selectivity for either the ammonium or nitrate ion, and all would require 85 to 90 per cent removal of all salts to remove 90 per cent of the nitrogen. Distillation is not useful since ammonia would be in the distillate unless the liquid was acidic, in which case nitrous acid would form.

A potentially promising process is ion exchange using a naturally occurring material called clinoptilolite, a natural zeolite that has a strong affinity for the ammonium ion. Removals of 93 to 99 per cent of the ammonia nitrogen can be obtained.

Breakpoint chlorination has been used for many years in disinfection of drinking water and wastewater. Applied to the latter in sufficient concentration, it has been shown that essentially all the ammonia can be oxidized to nitrogen gas. Side products such as nitrates and nitrogen trichloride can be produced if the pH is uncontrolled, but effective mixing and pH management can limit their concentration to satisfactory levels.

Ammonia removal proceeds according to the reaction

$$3Cl_2 + 2NH_3 \rightarrow N_2 + 6HCl \qquad (10\text{-}12)$$

The breakpoint is defined as the point where the nitrogen in the form of ammonia is eliminated and free available chlorine is detected.

**Removal of Trace Suspended Solids.**   The characterization of trace suspended solids is difficult. Particle size, concentration, and physical and chemical properties (specific gravity, toxicity, stickiness, etc.) are all highly variable and dependent on the pollution source. Sewage plant effluents, especially those from biological plants, contain small amounts of suspended solids that can have a serious impact on the receiving body of water.

A prime variable that must be considered in selecting tertiary solids removal equipment, aside from chemical composition, is the fineness of the material. The spectrum ranges from coarse particles that rapidly sink or float to colloids, macroscopic, microscopic, and submicroscopic organisms dispersed in the discharge. Of concern in this section is the removal of small amounts (generally less than 30 mg/l of organic solids) in effluents emanating from biological waste treatment.

**Microstraining.** Essentially, a microstrainer consists of a rotating drum with a fine screen around its periphery. Feedwater enters the drum through an open end and passes radially through the screen while solids are deposited on the inner surface of the screen. At the top of the drum, pressure jets of effluent water, amounting to 1 to 5 per cent of the treated flow, are directed onto the screen to remove the mat of deposited solids. The portion of backwash water that permeates the screen and dislodged solids are captured in a waste hopper and are removed through the hollow axle of the unit.

The effectiveness of removal depends on the screen pore size. The screens used in microstraining are made of a variety of plastics and stainless steel, and have extremely small openings. They have high porosity to effectuate high flow rate at low pressure drops (6 in. due to the fabric and 12 to 15 in. over-all).

**Sand Filtration.** The most common filter medium is a graded bed of silica sand. Developed in Great Britain and used for water clarification, these filters were operated at rates of 0.04 to 0.12 gal/(min)(ft²). In the United States, precoagulation has increased flow rates to 1 to 4 gal/(min)(ft²). The two processes have thus been called slow and rapid sand filtration.

A slow sand filter consists of a watertight basin containing a layer of sand 3 to 5 ft thick, supported on a layer of gravel 6 to 12 in. thick. The gravel is overlaid by a system of open-joint underdrains placed 10 to 20 ft apart, which leads the filtered water to a single outlet where the rate of flow through the filter is controlled. The most common effective size of the sand is 0.35 mm.

The filter is operated with a water depth of 3 to 5 ft above the sand surface. The solids removed from the water in a properly functioning slow sand filter are found mainly in the mat of previously trapped solids in the upper inch of sand. When the pressure drop equals the head of fluid, the filter is removed from service, drained, and cleaned.

Cleaning is traditionally done by scraping an inch of sand from the surface before returning the filter to service. The dirty sand is cleaned hydraulically and stored for future replacement in the filter. Slow sand filters remove most suspended solids except for fine clays and other colloidal solids.

With the exception of gravity sedimentation, deep-bed filtration is the most widely used unit operation for solid-liquid separation. At the present time, virtually all deep-bed filters utilized for waste treatment are the rapid sand type. This type consists of a layer of sand or other granular medium 18 to 30 in. thick, supported on an underdrain system. It can be an open-air gravity system or an enclosed pressure filter.

Rapid sand filters differ from slow sand filters in three major respects: thinner layers of granular media are used, higher loading rates are possible, and cleaning is accomplished by backwashing. Preceded by coagulation, rapid filters are more effective than slow filters and are better suited to cope with fine particles. Slow sand filters, on the other hand, can remove 99 per cent of the bacteria even without pretreatment as opposed to 80 per cent for rapid sand filters. The feedwater is filtered by forcing it through the sand layer, by either pressure or gravity. When the quantity of suspended solids in the effluent reaches an acceptable level, or the pressure drop exceeds the set point, the filter medium is washed. Washing is normally done by reversing the flow of the water through the filter at a rate adequate to lift the grains of filter medium into suspension. The deposited material, dislodged from the grains by the hydraulic shearing action of the rising water and by the abrasion of grains of filter medium rubbing against each other, is thus flushed up through the expanded bed.

**Mixed-media Filter.** One major disadvantage of a single-medium filter is that the particles tend to pack themselves so that the largest, having the fastest settling velocities, fall to the bottom of the filter after backwashing. Thus, the particle size of the medium decreases with distance from the bottom of the bed—just the opposite of what is desired.

The mixed-media filter solves this problem. Combinations of anthracite coal, sand, garnet, and ilmenite, with average specific gravities of 1.5, 2.6, 4.2, and 4.8, respectively, have been used. Properly formed after washing, the bed has the largest particles on top and the smallest on the bottom (if a downflow filter). The feed stream first meets and deposits its heaviest load of solids in the pores between the largest particles. The smaller pores, near the bottom, remove the last of the solids.

**Organic Removal by Activated Carbon.** Activated carbon selectively adsorbs dissolved organics (phenol, xylenes, etc.) and has a high capacity because of its large surface area, typically 500 to 1,400 $m^2/g$.

Activated carbon is used to produce high-quality effluents. Two processes involve tertiary treatment: passing biologically treated wastewater through carbon columns or adding powdered activated carbon to the activated sludge basin. Another method of using activated carbon is part of physical-chemical treatment that eliminates biological oxidation entirely; this process is described in the next subsection.

The engineer has a great variety of configurations to choose from to attain contact between the wastewater and the carbon: upflow or downflow, pressurized gravity flow, packed or expanded beds, series or parallel column configuration, and flat or conical bottom shapes. Key variables are contact time between the waste and the carbon, flow rate, and pressure drop. See Table 10-8 for typical design criteria.

As pollutants are adsorbed, the pores of the activated carbon become filled and finally breakthrough occurs, i.e., the point where no more contaminants can be removed from the solution. Then, the exhausted carbon must be replaced with fresh material. Normally, thermal regeneration is used to destroy adsorbed organics so the carbon can be recycled. The thermal process has three steps: drying, baking to pyrolize the adsorbates, and reactivation by oxidation at the surface to increase the number of active

## Table 10-8. Design Criteria for Activated Carbon Adsorption Columns

| Design Parameter | Design Range |
|---|---|
| Flow rate | 4–8 gpm/ft² |
| Bed depth | 8–12 ft |
| Contact time | 20–50 min (longer for some industrial wastes) |
| Media size | 8 × 30 mesh |
| Backwash frequency | Daily |
| Backwash rate: | |
| Air scour | 3–5 ft³/(min)(ft²) |
| Water wash | 14–18 ft³/(min)(ft²) |
| Number of stages: | |
| Low-quality effluent | 1 |
| High-quality effluent | 2 |
| Carbon loss | 2–10 % per cycle |
| Cost (tertiary sewage treatment) | 10–30¢/1,000 gal |

SOURCE: Eimco Processing Machinery Div., Envirotech Corp., Salt Lake City, Utah.

sites. Regeneration, like initial activation, is accomplished by heating at 910 to 940°C in an atmosphere of steam.

**Reverse Osmosis.** Filtration through semipermeable membranes under pressure removes biological and colloidal matter, as well as most dissolved organics that affect color, odor, and taste. Additionally, iron and hardness are removed as is done in conventional water treatment processes; however, reverse osmosis also reduces the concentration of all dissolved solids (chlorides, sulfates, nitrates, fluorides, trace metals, etc.) and can even extract pesticides and radioactive material.

Presently the municipal application for reverse osmosis is the desalting of brackish water having total dissolved solids between 1,000 and 15,000 mg/l. Industrially, there are a wide variety of uses. The important design parameter is the rate of transport of water across the membrane, called the *flux J*, which is

$$J = C(\Delta P_g - \Delta P_o) \tag{10-13}$$

where $\Delta P_g$ = physical (gage) pressure (feedside-productside)

$\Delta P_o$ = osmotic pressure (feed-product)

Osmosis is a phenomenon that occurs naturally. It results whenever a dilute liquid (such as freshwater) and a concentrated liquid (such as saltwater or sugar) are separated by a semipermeable, selective material that permits one kind of molecule to pass through, but not the other. The osmotic pressure is a function of the concentration of the dissolved ions. For every 100 mg/l of dissolved inorganic solids, an osmotic pressure of about 1 psi is generated.

If pressure is applied to impure water in excess of the osmotic pressure, the reverse process takes place: water molecules from the saline solution are forced through the membrane into the freshwater with the selective, semipermeable membrane acting as a barrier to passage of the contaminant.

The most popular membranes developed so far are made of cellulose acetate. They are solvent-cast like photographic films and are characterized by extremely thin, dense skins on one side of highly porous substructures. They are designed for flow from the dense side only of about 20 gal/(day)(ft²).

These membranes do, however, have problems: temperatures greater than

140°F render them impermeable; they are vulnerable to attack by enzymes and microorganisms; and if not kept wet, they crack.

Du Pont and Dow are developing hollow fiber technology to take the place of flat or rolled membranes. Dow has used cellulose acetate; Du Pont has developed a nylon fiber they say is more resistant to bacterial attack and degradation by acidic and alkaline solutions. Flow rates are a modest 0.10 gal/(day)(ft²) over an estimated lifetime of 5 years, but the packing density is high; $30 \times 10^6$ fibers with an area of 85,000 ft² is practical. These fibers typically have an outside diameter of 0.002 in. and a thickness of 0.0005 in.

The primary factors that determine the performance of these systems are: applied pressure, concentration of the dissolved solids in the feed, feed temperature, and porosity or permeability of the membrane. To assure optimum performance over long periods of time, it is important in most cases to pretreat the feed by filtration and chemical addition.

**Ion Exchange.** Ion exchange is a process in which ions, held by electrostatic force to charged functional groups on the surface of a solid, are exchanged for ions of a similar charge in a solution in which the solid is immersed. Because the charged functional groups at which exchange occurs are on the surface of the solid, and because the exchanging ions must undergo a phase transfer from the liquid phase to residence on a solid phase, ion exchange is a sorption process.

Although synthetic resins are used for most ion-exchange applications, the phenomenon of exchange is known to occur with a number of natural solids including soil, humus, cellulose, wood, active carbon, coal, lignin, metallic oxides, and living cells such as algae and bacteria.

Two important terms are used to define resin capacity. The first is the total amount of exchangeable ions per unit weight of dry resin. More important, however, is the breakthrough capacity. This is an operational term that is a function of flow rate, concentration of ions, grain size, temperature, etc. Breakthrough occurs when the ion being removed from the solution appears in the column effluent.

Ion-exchange resins adsorb different ions selectively, based on the relative strength of the ion's charge and its radius of hydration. Multicharged ions are more readily adsorbed than single-charged ions, and smaller ions more readily than larger ones. As with activated carbon, ion exchange takes place in the capillaries of the solid particles. The size and number of these porous structures are important. Most silicate-based ion exchangers can be used over only a very small pH range, whereas synthetic resins are generally quite stable. The selectivity of a resin decreases with increasing temperatures but attains equilibrium faster.

A typical application of ion exchange is in preparing high-purity waters for boilers. Recent advances have shown potential for control and recovery of metallic ions in the plating industry; nickel, gold, and silver wastes can be treated. Radioactive wastes, too, are amenable to ion-exchange treatment.

Until recently, ion exchange was thought to be impractical for municipal wastewater treatment because extensive pretreatment would be needed to remove organic matter to prevent fouling of the resins. The development of macroporous resins that can adsorb organic matter without fouling and require less-expensive regenerates makes the process worth considering.

## 10-10. Physical-chemical Processes

Physical-chemical treatment (involving coagulation, carbon adsorption, and filtration) is an alternative to biological wastewater treatment methods. It replaces microbial stabilization of the dissolved organic matter with a combination of physical treatment processes that do not rely on bacterial action. A typical flow diagram for physical-chemical treatment is shown in Fig. 10-5.

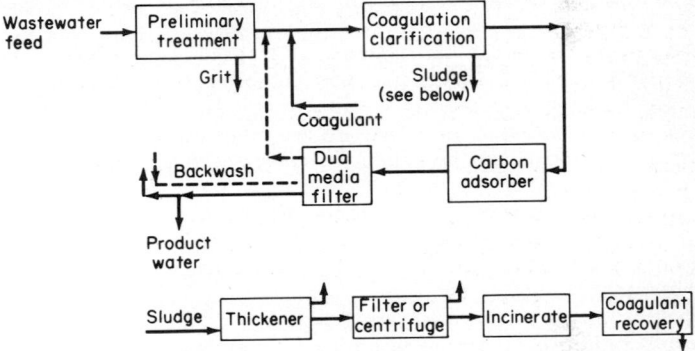

FIG. 10-5. Principal operations in physical-chemical treatment.

The primary goal of pretreatment is to remove most of the suspended solids from the raw waste. In the case of physical-chemical treatment of municipal sewage, lime and high molecular weight, water-soluble polymers, and other flocculants are combined with primary sedimentation and filtration through mixed-media filters to produce as clear an effluent as possible. An additional benefit of chemical treatment is that it takes out almost all of the phosphorus in the wastewater.

After removing the major portion of pollutants, the clarified wastewater is given a final polishing by activated carbon. Over-all, physical-chemical treatment produces an effluent essentially free of suspended solids, and with TOC (total organic carbon) in the range of 5 to 10 mg/l.

## 10-11. Disinfection

Wastewater treatment processes remove some but not nearly all of the pathogens (Table 10-9). Hence, the discharge from biological treatment

## Table 10-9. Removal of Bacteria During Wastewater Treatment

| Operation | Bacterial Removal, % |
|---|---|
| Coarse screen | 0–5 |
| Fine screen | 10–20 |
| Grit chamber | 10–25 |
| Sedimentation | 25–75 |
| Chemical precipitation | 40–80 |
| Trickling filtration | 90–95 |
| Activated sludge | 90–98 |

SOURCE: Metcalf and Eddy, "Wastewater Engineering: Collection, Treatment, Disposal." McGraw-Hill Book Company, New York, 1972.

plants, as well as natural waters contaminated by human sources, can carry diseases caused by bacteria and viruses. Thus, disinfection of sewage plant effluents is widely practiced.

Disinfection, it must be emphasized, is not the same as sterilization. The latter is total inactivation of all microbial life, while the former is selective destruction of most of the disease-causing pathogens.

Many chemical agents kill bacteria. The bactericidal effect of heavy metals (such as silver and mercury) has long been recognized, and phenols and alcohols have been used in hospitals for many years. However, the most popular chemical disinfectants are oxidizing agents such as ozone, the halogens (chlorine, iodine, bromine), hydrogen peroxide, and dyes.

**Chlorination.** Of all the chemical disinfectants, chlorine is perhaps most commonly used. It is reasonably economical, toxic to microorganisms that carry waterborne diseases, is tasteless and nonpoisonous to humans at low concentrations, and can be detected in residual amounts some time after application.

Chlorine is quite soluble in water—7,160 mg/l at 20°C and at atmospheric pressure, forming hypochlorite according to the following equation:

$$Cl_2 + H_2O \rightleftarrows HOCl + H^+ + Cl^-$$
$$\updownarrow \qquad\qquad (10\text{-}14)$$
$$H^+ + OCl^-$$

When ammonia is present it reacts to form chloramines, which are slow-acting disinfectants:

$$NH_3 + HOCl \rightarrow NH_2Cl + H_2O \qquad (10\text{-}15)$$
$$NH_3 + 2HOCl \rightarrow NHCl_2 + 2H_2O \qquad (10\text{-}16)$$
$$NH_3 + 3HOCl \rightarrow NCl_3 + 3H_2O \qquad (10\text{-}17)$$

Chlorine, present as $NH_2Cl$ or $NHCl_2$, is called combined available chlorine. When all reducing agents have been satisfied, then free chlorine is detected. This is called the breakpoint, which occurs only after the chloramines have been oxidized to $N_2O$ and $N_2$.

Chemical oxidation by chlorine depends on the usual factors affecting a chemical reaction: temperature, time of contact, concentration of chlorine, pH, and surface tension. Because it is a microbial system, other items of importance are the species of microorganisms, the nature of the suspending fluid, the number of microorganisms, and the way in which they are suspended in the fluid, i.e., as single, discrete cells, in flocs, or attached to particulates. Table 10-10 lists dosages for several types of sewage.

### Table 10-10. Chlorine Dosages Required to Yield 0.2 ppm Residual after 10–15 min Contact Time

| Sewage Type | Dosage, mg/l |
|---|---|
| Raw: | |
| Fresh to stale | 6–12 |
| Septic | 12–25 |
| Settled: | |
| Fresh to stale | 5–10 |
| Septic | 12–40 |
| Effluent chemical precipitation | 3–6 |
| Trickling filter: | |
| Normal | 3–5 |
| Poor | 5–10 |
| Activated sludge: | |
| Normal | 2–4 |
| Poor | 3–8 |
| Intermittent sand filter: | |
| Normal | 1–3 |
| Poor | 3–5 |

SOURCE: W. W. Eckenfelder, Jr., P. A. Krenkel, and C. A. Adams, "Advanced Wastewater Treatment," AIChE Today Series, AIChE, New York, 1972.

**Ozonization.** A relatively new disinfection method utilizes ozone, a molecule that consists of three atoms of elemental oxygen ($O_3$). In water, these atoms break down rapidly, so they are free to oxidize organic matter and inactivate bacteria and viruses. Additionally, ozonization adds oxygen to the system but not dissolved solids. Further, ozone will deodorize gases in a few seconds of contact time at low levels of concentration.

Ozone has a half-life of about 20 min in water. It is approximately 13 times more soluble than oxygen. The amount of ozone needed for disinfection of water or wastewater depends on the temperature, physical properties, and contaminants in the water being treated. For control purposes, one should be able to detect some residual ozone at the end of 5 min.

## AIR POLLUTION CONTROL

### 10-12. Definitions

Air pollution is defined as the "presence in the atmosphere of one or more contaminants of such quantity and duration as may be injurious to human, plant, or animal life, or property, or which may unreasonably interfere with comfortable enjoyment of life, property, or conduct of business." Air pollution results from a two-part phenomenon, a time-concentration relationship. This relationship is recognized in the National Ambient Air Quality Standards (Table 10-1).

**Classes of Pollutants.** Atmospheric pollutants may exist as gases or particulates. Frequently a third category, odors, is recognized. Chemically, odors fall into one of the previous categories. They are frequently separated because of their objectionable olfactory sensation at concentrations where they are neither toxic nor hazardous. Table 10-11 lists typical gaseous pollutants emitted to the atmosphere.

## Table 10-11. Significant Gaseous Pollutants

| Major class | Subclass | Typical pollutants |
|---|---|---|
| Organic gases | Hydrocarbons | Hexane, benzene, methane, butane, ethylene, butadiene |
|  | Aldehydes and ketones | Formaldehyde, acetone |
|  | Other organics | Alcohols, chlorinated hydrocarbons |
| Inorganic gases | Oxides of $N_2$ | NO, $NO_2$ |
|  | Oxides of S | $SO_2$, $SO_3$ |
|  | Oxides of C | CO, $CO_2$ |
|  | Other inorganics | $H_2S$, HF, HCl, $NH_3$ |

The term *particulates* denotes the release of both solid and liquid particles. The term should apply only to materials that are particles at their point of release and that could be captured by particulate control equipment. However, regulatory definitions frequently include as particulates, substances that may be vapors at the point of release, but that can produce particulates by subsequent cooling and condensation to a standard temperature such as 70°F. Figure 10-6 compares sizes of typical particulates in the atmosphere and effective ranges for particle sizing and particle collection devices.

Several terms are used to describe particulates, which causes some confusion. *Aerosol* applies to a mixture of suspended particulates, either liquid or solid, or a combination of both. It usually applies to particles which settle slowly, if at all, due to gravity, and includes particles from submicron to 10 to 20 microns in size. *Dusts* are coarser solid particles generated by entrainment from handling, crushing, grinding, and similar operations. They do not flocculate spontaneously, and generally settle due to gravity. Sizes range from 1.0 to 2,000 microns. *Smoke* generally applies to carbon and soot particles resulting from incomplete combustion, though occasionally the term may be applied to other finely divided solid particles such as ammonium chloride smoke. Smoke particles are 0.01 to 1.0 microns in size and do not settle by gravity. *Mists* are suspended liquid droplets usually resulting from condensation. *Fumes* are fine solid suspensions generated by condensation from the gaseous state, often in smelting and metallurgical operations.

Table 10-12 lists the sources of pollutants by general classes. Source emissions are best evaluated by sampling and reporting the contaminants released per unit of production. These results are called *emission factors*.

### Table 10-12. Estimated Sources of Air Pollutants, 1970

(Million tons per year)

| Source | CO | Particulates | SO$_x$ | HC* | NO$_x$ | Total |
|---|---|---|---|---|---|---|
| Transportation | 111.0 | 0.7 | 1.0 | 19.5 | 11.7 | 143.9 |
| Stationary source fuel combustion | 0.8 | 6.8 | 26.5 | 0.6 | 10.0 | 44.7 |
| Industrial processes | 11.4 | 13.1 | 6.0 | 5.5 | 0.2 | 36.2 |
| Solid-waste disposal | 7.2 | 1.4 | 0.1 | 2.0 | 0.4 | 11.1 |
| Miscellaneous | 16.8 | 3.4 | 0.3 | 7.1 | 0.4 | 28.0 |
| Total | 147.2 | 25.4 | 33.9 | 34.7 | 22.7 | 263.9 |

\* Hydrocarbons.
SOURCE: U.S. Environmental Protection Agency.

Typical factors for many uncontrolled operations are given in *U.S. Public Health Serv. Publ.* 999-AP-42.

**Effects.**  SO$_2$ and other acidic pollutants corrode architecture and sculpture (both metals and masonry, especially limestone and marble).  Deterioration of historic works has increased significantly in the last 50 years.  Carbonaceous deposits discolor buildings.  Particulates and condensation nuclei reduce atmospheric visibility, hindering aircraft and ground transportation, block off distant scenic views, and decrease the quantity of solar radiation reaching the earth.  Pollutants damage vegetation and reduce crop yields.  Jacobson and Hill ("Recognition of Air Pollution Injury to Vegetation: a Pictorial Atlas," Air Pollution Control Assn., Pittsburgh, 1970) illustrate many of these effects.

Pollutants also affect the health of animals and humans.  High concentrations of pollutants retard recovery from many diseases.  They contribute to respiratory diseases, and sufficiently high concentrations can lead to death.  Effects of various contaminants on humans have been discussed in the Air Quality Criteria Documents.  (*U.S. Public Health Serv. Publs.* AP-49, Particulates, 1969; AP-50, SO$_x$, 1969; AP-62, CO, 1970; *U.S. Environ. Protect. Agency Publs.* AP-63, Oxidants, 1970; AP-64, Hydrocarbons, 1970; AP-84, NO$_x$, 1971.)

Studies on animals have shown a synergistic effect between particulate matter and SO$_2$.  When SO$_2$ is combined with low concentrations of certain particulates, the SO$_2$ needed to produce harmful effects need be only 25 per cent of the amount required in pure air to produce the same result.

### 10-13. Transport and Meteorological Effects

Pollutants are transported through the atmosphere by wind currents from their point of release to downwind receptors.  They are dispersed and diluted so that an emission, toxic at its release point, may be harmless at ground level downwind.  The higher the release point above the surroundings, and the more buoyant the plume, the greater the dilution.

The major meteorological parameters controlling atmospheric dispersion

Particle diameter, μ

**Equivalent sizes**

| (1 mμ) 0.0001 | 0.001 | 0.01 | 0.1 | 1 | 10 | 100 | (1 mm) 1,000 | (1 cm) 10,000 |

Ångström units, Å.

10 | 100 | 1,000 | 10,000 | 100,000 | 1,000,000 | 10,000,000

Theoretical mesh (used very infrequently)
5,000 2,500 625 | 250

Tyler screen mesh
U.S. screen mesh

**Electromagnetic waves**

X-rays — Ultraviolet — Visible — Near infrared — Solar radiation — Far infrared — Microwaves (radar, etc.)

**Technical definitions**

Gas dispersoids:
Solid: Fume — Dust
Liquid: Mist — Spray

Soil: Atterberg or International Std. Classification System adopted by Internat. Soc. Soil Sci. since 1934
Clay — Silt — Fine sand — Coarse sand — Gravel

**Common atmospheric dispersoids**

Smog — Clouds and fog — Mist — Drizzle — Rain

**Typical particles and gas dispersoids**

Gas Molecules*
O₂ CO₂ C₆H₆
H₂ F₂ Cl₂
N₂ CH₄ SO₂
CO H₂O HCl C₄H₁₀

Rosin smoke
Oil smoke
Tobacco smoke
Metallurgical dusts and fumes
Ammonium chloride fumes
Carbon black
Zinc oxide fumes
Colloidal silica
Aitken nuclei
Combustion nuclei
Sea salt nuclei
Atmospheric dust
Viruses
Bacteria

Fertilizer, ground limestone
Fly ash
Coal dust
Cement dust
Sulfuric concentrator mist
Contact sulfuric mist
Paint pigments
Insecticide dusts
Spray dried milk
Ground talc
Plant spores
Pollens
Milled flour
Alkali fume
Beach sand
Pulverized coal
Flotation ores
Human hair
Plant
Lung-damaging dust
Nebulizer drops
Pneumatic nozzle drops
Hydraulic nozzle drops
Red-blood-cell diameter (adults): 7.5μ ± 0.3μ
Sulfuric mist
Paint pigments

10–36

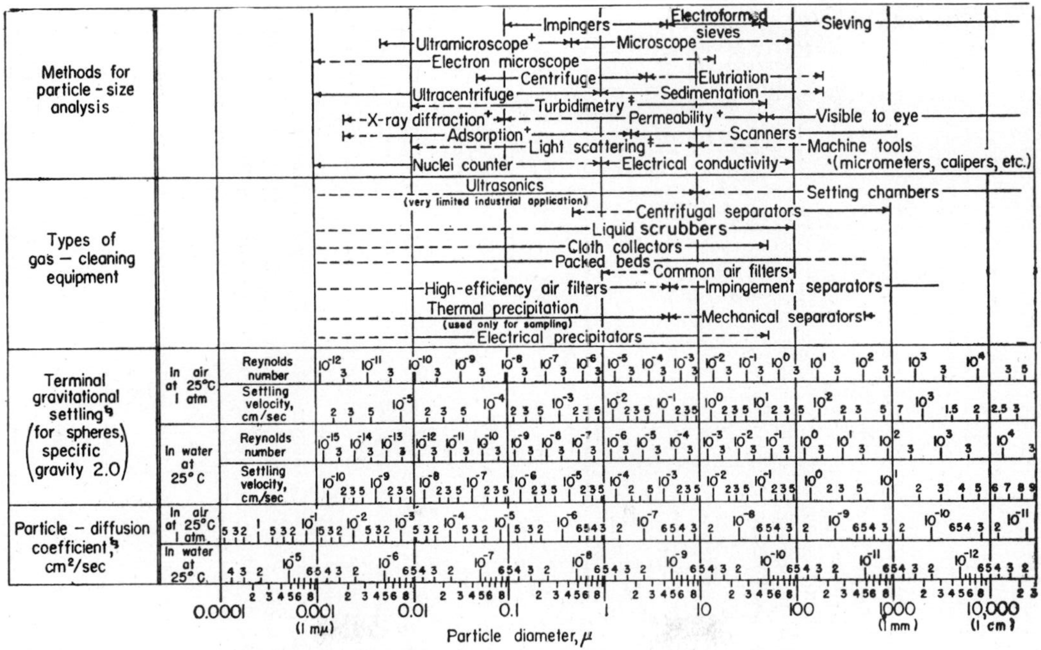

FIG. 10-6. Characteristics of particles and particle dispersoids.

* Molecular diameters calculated from viscosity data at 0°C.  
⊕ Stokes–Cunningham factor included in values given for air but not included for water.  
+ Furnishes average particle diameter but no size distribution.  
‡ Size distribution may be obtained by special calibration.

are atmospheric stability and wind velocity. Atmospheric stability is affected by solar radiation and the vertical temperature gradient of the atmosphere called *lapse rate*. Stability can be characterized by comparing the actual lapse rate to the dry adiabatic lapse rate (DALR). An atmosphere at the DALR decreases in temperature 5.4°F for each 1,000-ft increase in altitude ($-0.98$°C/100 m). An atmosphere that decreases in temperature faster than the DALR is *superadiabatic* (unstable atmosphere). One that decreases at the DALR rate is neutral. An atmosphere that decreases less rapidly than the DALR is *subadiabatic* and is more stable. An atmosphere at constant temperature is *isothermal* and is more stable than subadiabatic. When temperature increases with elevation, an *inversion* condition exists—an extremely stable atmosphere.

Best conditions for dispersion are high wind speeds and a highly unstable temperature gradient (superadiabatic). As the atmosphere becomes progressively more stable, wind speeds tend to decrease, and plume dispersion becomes progressively poorer. An indication of atmospheric stability can be gained from smoke plume behavior (Fig. 10-7).

Under superadiabatic conditions, high ground concentrations may occur close to the stack as the looping plume actually strikes the ground at times, but concentrations considerably downwind will be reduced. Under neutral conditions, dispersion is still good but less rapid. Ground concentrations close to the stack will be less than under unstable conditions, but they will be higher further downwind. Under inversion conditions, dilution occurs very slowly, and the plume meanders and spreads out horizontally. Ground concentrations will often be low, but the pollution accumulates aloft waiting to sink to ground level in large undiluted puffs at the moment of inversion breakup or fumigation. At these times, often of 0.5- to 1-hr duration, ground concentrations up to 20 times normal neutral atmosphere ground concentrations may be experienced.

Increasing wind speed aids dispersion. The plume is blown over more rapidly and rises less in the atmosphere. Because of this, higher ground concentrations may be experienced closer to the stack, but as dilution occurs concentrations farther downwind will be reduced.

Dispersion frequently follows a daily cycle. In late evening with clear sky, the earth radiates heat to space, cooling more rapidly than the air above. This creates an inversion that lasts throughout the night. Winds die down and pollutants accumulate at the base of the inversion layer. When the sun rises, the earth is heated and a decreasing temperature lapse rate builds up from the earth's surface. When this reaches the inversion ceiling, the accumulated pollutants begin to sink to ground level, producing a fumigation. As further warming occurs, the DALR is restored, wind speeds increase, and the pollutants are blown away and dispersed. Conditions for good dispersal remain until cooling sets in again near sunset.

**Dispersion Equations.** Standard gaussian distribution can predict downwind concentrations with fair accuracy. Models currently most in use are described by Turner ["Workbook of Atmospheric Dispersion Estimates,"

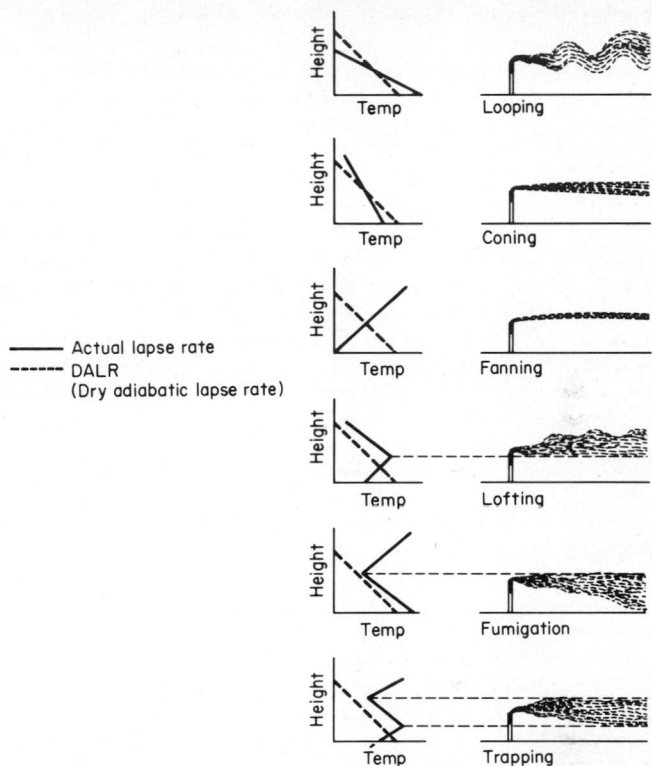

— Actual lapse rate
----- DALR
(Dry adiabatic lapse rate)

FIG. 10-7. Smoke plume behavior as a function of temperature. (*Source:* H. L. Perkins, *"Air Pollution," Fig.* 7.19, p. 169, *McGraw-Hill Book Company, New York*, 1974.)

USEPA, AP-26(1970)]. Figure 10-8 illustrates the model. The plume is discharged from a stack of height $h_s$. A buoyant plume continues to rise some distance above the stack, $\Delta h$. The over-all effective stack height $H$ is the sum of $h_s$ and $\Delta h$. The plume is assumed to be released from a point source upwind from the stack such that it spreads out in the shape of a cone as it travels downwind. The plume is free to spread horizontally about its axis and upward. However, at the point where the plume hits the ground, the assumption is made that the pollutants are reflected upward. This is handled mathematically by assuming a virtual image source beneath the ground that adds to the concentration downwind beyond the point $x_s$. Coordinates are: $x$, the downwind direction with wind speed of $\bar{U}$; $y$, the crosswind direction measured from the plume centerline; and $z$, the vertical height above the ground.

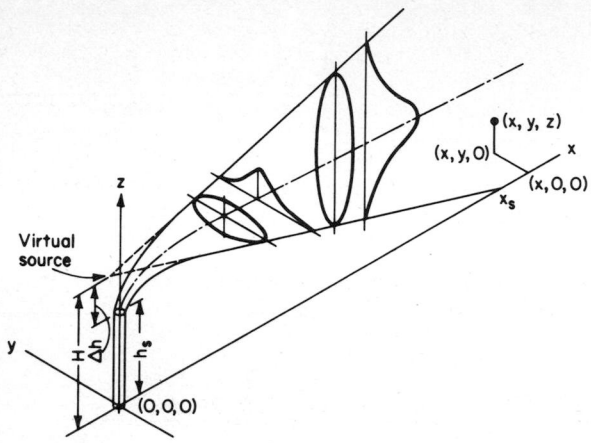

FIG. 10-8. Gaussian model for dispersion equations.

The downwind concentration $C$ at any point $(x,y,z)$ can be computed from Eqs. (10-18) to (10-22). For downwind distances up to $x_s$,

$$\frac{C_{(x,y,z)}\bar{U}}{Q} = \frac{1}{2\pi\sigma_y\sigma_z} \exp\left\{ -\frac{1}{2}\left[ \left(\frac{y}{\sigma_y}\right)^2 + \left(\frac{z-H}{\sigma_z}\right)^2 \right] \right\} \qquad (10\text{-}18)$$

Beyond $x_s$:

$$\frac{C_{(x,y,z)}\bar{U}}{Q} = \frac{1}{2\pi\sigma_y\sigma_z} \exp\left[ -\frac{1}{2}\left(\frac{y}{\sigma_y}\right)^2 \right] \left\{ \exp\left[ -\frac{1}{2}\left(\frac{z-H}{\sigma_z}\right)^2 \right] \right.$$
$$\left. + \exp\left[ -\frac{1}{2}\left(\frac{z+H}{\sigma_z}\right)^2 \right] \right\} \qquad (10\text{-}19)$$

Beyond $x_s$, where ground-level concentrations are desired $(z = 0)$,

$$\frac{C_{(x,y,0)}\bar{U}}{Q} = \frac{1}{\pi\sigma_y\sigma_z} \exp\left\{ -\frac{1}{2}\left[ \left(\frac{y}{\sigma_y}\right)^2 + \left(\frac{H}{\sigma_z}\right)^2 \right] \right\} \qquad (10\text{-}20)$$

If plume centerline concentrations only are desired $(y = 0)$,

$$\frac{C_{(x,0,0)}\bar{U}}{Q} = \frac{1}{\pi\sigma_y\sigma_z} \exp\left[ -\frac{1}{2}\left(\frac{H}{\sigma_z}\right)^2 \right] \qquad (10\text{-}21)$$

If the effluent is released at ground level $(H = 0)$, the downwind plume centerline concentration is obtained by

$$\frac{C_{(x,0,0)}\bar{U}}{Q} = \frac{1}{\pi\sigma_y\sigma_z} \qquad (10\text{-}22)$$

Any consistent set of units may be used such as $C$, concentration in $g/m^3$; $\bar{U}$, wind speed, $m/sec$; $Q$, the pollutant release, $g/sec$; and distances $H$, $x$, $y$, $z$, and diffusion coefficients $\sigma_y$ and $\sigma_z$ in meters.

For very accurate predictions of ground concentration, the values of $\sigma_y$ and $\sigma_z$ should be determined experimentally for the meteorological conditions and terrain under consideration. The model represents the real situation best in a neutral atmosphere (lapse rate close to the DALR).

There are six stability categories as defined in Table 10-13 with experi-

## Table 10-13. Stability Categories

| Surface wind speed, m/sec | Day | | | Night | |
|---|---|---|---|---|---|
| | Incoming solar radiation | | | Thin overcast or $\geq \frac{4}{8}$ cloudiness | $\leq \frac{3}{8}$ cloudiness |
| | Strong | Moderate | Slight | | |
| <2 | A | A–B | B | | |
| 2 | A–B | B | C | E | F |
| 4 | B | B–C | C | D | E |
| 6 | C | C–D | D | D | D |
| >6 | C | D | D | D | D |

The neutral class, D, should be assumed for overcast conditions during day or night. A—extremely unstable conditions, B—moderately unstable conditions, C—slightly unstable conditions, D—neutral conditions, E—slightly stable conditions, F—moderately stable conditions.

SOURCE: D. B. Turner, U.S. Environmental Protection Agency, AP-26, 1970.

mental values for $\sigma_y$ and $\sigma_z$ as shown in Figs. 10-9 and 10-10. These values of $\sigma_y$ and $\sigma_z$ are most accurate ($\pm 100$ per cent) for level farm land. Ground structures, wooded areas, and rolling or mountainous terrain will result in larger actual departures from the calculated results. Figure 10-11 may be used with the stability categories of Table 10-13 to determine downwind plume centerline distance at which the maximum ground concentration will occur and its concentration.

The calculated concentrations are typical of those which would be obtained while sampling for 10 min. A sample averaged over a longer time period will give lower values due to normal fluctuations in wind direction. Equation (10-23) may be used to predict the average ground concentrations to be expected up to periods of several hours.

$$C_t = C_0 \left(\frac{t_0}{t_t}\right)^p \qquad (10\text{-}23)$$

where $C_t$ = the concentration for a longer time period
$C_0$ = the concentration estimated for a 10-min period
$t_0$ and $t_t$ = the respective time periods
$p$ = value between 0.17 and 0.20

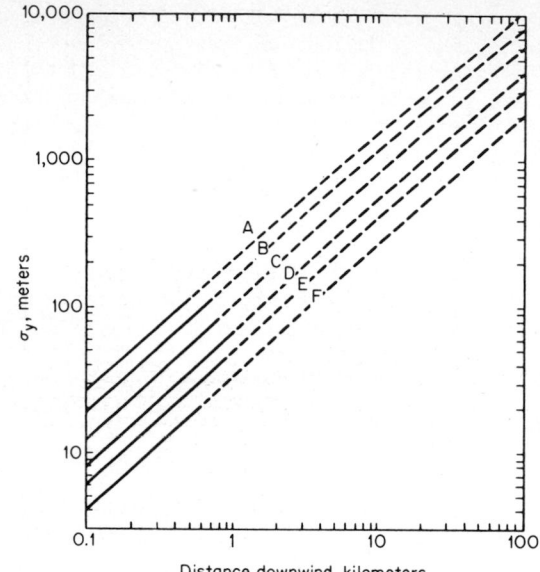

FIG. 10-9. Horizontal dispersion coefficient as a function of downwind distance from the source. Letters refer to stability category; see Table 10-13. [*Source: Turner, U.S. Environmental Protection Agency AP*-26 (1970).]

The dispersion equations apply specifically to gases and suspended particulates. Due to gravity fallout close to the stack, they should not be applied to particulates larger than 20 microns. [D. H. Slade, "Meteorology and Atomic Energy, 1968," U.S. Atomic Energy Commission (T1D-24190), 1968 and "Recommended Guide for the Prediction of the Dispersion of Airborne Effluents," ASME, New York, 1968.]

**Plume Rise.** The preceding dispersion equations require an estimate of the distance the plume continues to rise above the stack.

$$\Delta h = \frac{V_s d}{\bar{U}} \left[ 1.5 + 2.68 \times 10^{-3} pd \left( \frac{T_s - T_a}{T_a} \right) \right] \tag{10-24}$$

where $\Delta h$ = the plume rise, m
  $V_s$ = stack discharge velocity, m/sec
  $d$ = stack diameter, m
  $\bar{U}$ = wind speed, m/sec
  $p$ = atmospheric pressure, mbars
  $T_s$ = stack discharge temperature, K
  $T_a$ = ambient temperature, K
To correct plume rise for atmospheric stability, it is recommended that the

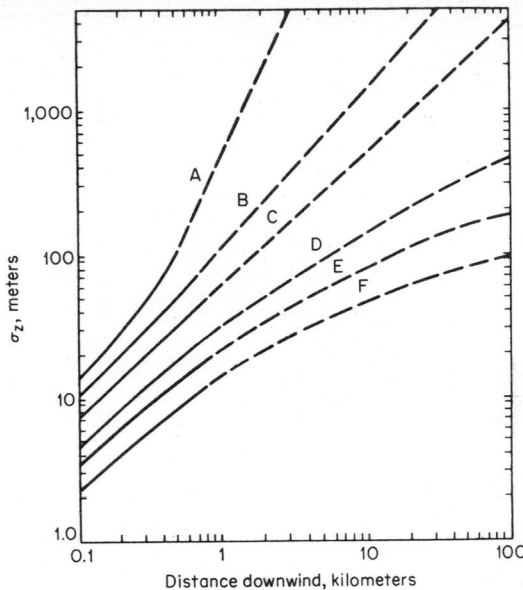

FIG. 10-10. Vertical dispersion coefficient as a function of downwind distance from the source. Letters refer to stability category; see Table 10-13. [*Source: Turner, U.S. Environmental Protection Agency AP*-26 (1970).]

value of Eq. (10-24) be multiplied by 1.15 for unstable to 0.85 for very stable conditions (1.0 for neutral).

Carson and Moses [*J. Air Pollut. Control Assoc.*, **19**, 862–866 (1969); **22**, 621–630 (1972)] evaluated plume rise from many equations and recommend several as the most accurate. Briggs ["Plume Rise," U.S. Atomic Energy Commission (TID-25075) (1969)] also evaluated plume rise and proposed a considerable number of equations which best fit specific situations. His equations for stable atmospheres require a detailed knowledge of temperature gradients with altitude.

**Pollution Potential and Climatology.** When considering a location for an operation with a major air pollution potential, locations well suited to atmospheric dispersion should be considered. Flat, open spaces are preferred. Narrow mountain valleys should be avoided. Locations adjacent to large lakes and oceans can present special problems. Studies by Hosler ["Low Level Inversion Frequency in the Contiguous US," *Mon. Weather Rev.*, **89**, 319–339 (1961)] and by Holzworth ["Estimates of Mean Mixing Depths in the Contiguous US," *Mon. Weather Rev.*, **92**, 235–242 (1964)] give indications of geographical areas with high and low potential for air pollution problems. It is desirable to pick an area with a mean mixing depth as great as possible.

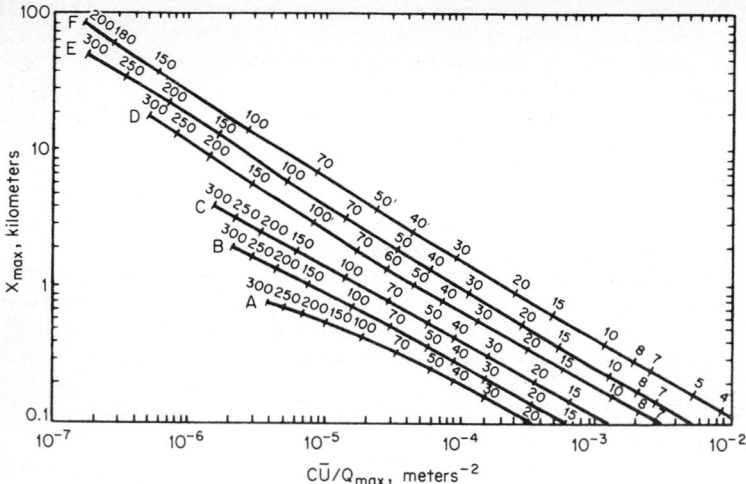

Fig. 10-11. Distance of maximum concentration as a function of stability and effective height of emission. Letters indicate stability class; numbers, the effective stack height; see Table 10-13. [*Source: Turner, U.S. Environmental Protection Agency AP-26 (1970).*]

## 10-14. Air Pollution Measurements

Sampling may be occasional, intermittent, and infrequent to check on specific problems, or it may be conducted at regular intervals or even continuously to provide frequent monitoring. Positioning of samplers must be carefully considered. For a representative area sample, the intake should not be at a busy intersection, along a dusty road, or directly downwind of a major pollution source. Sampling for gases (and some particulates) is practiced by drawing a measured sample through a series of midget impingers. Water or a suitable solution absorbs the gaseous component.

For routine sampling, there are a number of devices of varying complexity and cost. Primitive methods still in use consist of the "dust-fall" bucket and the lead peroxide planchette. These measure, respectively, dust that settles by gravity and the $SO_2$ or sulfation rate, usually on a monthly basis.

The "hi-vol" sampler is the specified method for checking ambient-air compliance with national standards for particulates. It pulls about 50 $ft^3/min$ of ambient air through a glass-fiber filter for a 24-hr period. Tape samplers that filter ambient air through a 1-in.-diameter spot on a cellulose tape for a 2-hr period are also used. After the sampling period, the tape is advanced automatically and the soil measured photometrically by reduction in light transmittance. Tape sampler results are usually reported as *coefficient of haze* (coh/1,000 ft). Unfortunately, coh is not comparable with "hi-vol" gravimetric measurements ($\mu g/m^3$). Tape samplers specific for certain gases ($H_2S$ and fluorides) have been developed in which two chemically treated tapes are brought together. Their contact activates the tapes for absorption of the gaseous component. Absorption causes discoloration of the tape which can be measured photometrically.

Atmospheric monitoring for $SO_2$ and other gases is frequently performed by taking successive 2-hr composite samples with midget impingers in a sequential sampler. The absorbing liquid is analyzed using the West-Gaeke technique. (Ozone and oxidants are determined by their oxidizing effect on KI solution.) Nitrogen oxide analysis is usually based on a chemiluminescence reaction using a continuous instrumental analyzer.

Many communities use continuous monitors for $SO_2$ measurement. The coulometric instruments are probably the preferred choice, but colorimetric, conductivity, and flame photometry instruments have also been used. The standard method for measuring CO is an instrument using nondispersive infrared.

Methods for analysis of air pollutants are discussed in *U.S. Public Health Serv. Publ.* AP-11 and by Ruch ("Chemical Detection of Gaseous Pollutants," Ann Arbor Science Publishers, Ann Arbor, 1967). Continuous monitors are discussed by Crocker [*Chem. Eng. Prog.*, **70**, 41–49 (January 1974)] and by Hollowell and McLaughlin [*Environ. Sci. Tech.*, **7**, 1011–1017, (November 1973)].

**Source Sampling.** Source sampling devices are well described (Bull. WP-50, Western Precipitation Div., Los Angeles, 1968; H. G. H. Cooper, Jr., and A. T. Rossono, Jr., "Source Testing for Air Pollution Control," Environmental Research and Applications, Inc., Wilton, Conn., 1971; Byers and Crocker, "Stack Sampling and Monitoring," AIChE, New York, 1972). There is a current tendency to use the U.S. EPA Method V sampler [*Federal Register*, **36**, 24893 (Dec. 23, 1971)], with modifications if needed, for many types of source sampling.

Process location for sample withdrawal must be carefully selected to obtain a representative sample. The EPA test Method V specifies, "a composite sample shall be taken from at least 12 different locations in the same plane in a duct with each point the centroid of an equal area section." Under this circumstance, there should be at least 8 duct diameters of straight pipe upstream of the sampling point and 2 duct diameters downstream. When a location having a greater amount of flow disturbance is the only suitable location available, the effects of poor flow distribution may be compensated for by increasing the number of traverse points (Fig. 10-12).

**Velocity Measurement.** A measure of the desirability of a particular sampling location can be obtained from the velocity distribution across the duct. The more uniform the velocity distribution, the better the chance that the pollutants will be somewhat uniformly distributed. Also, a knowledge of the velocity distribution is necessary in selection of the sample probe orifice, sampling flow rate, and total sampling time. The velocity distribution is usually determined in a round duct on two different diameters at right angles to each other using an S-type pitot tube. The velocity at each

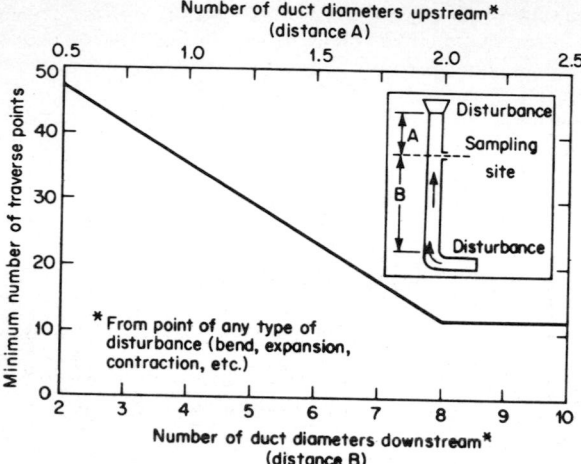

Fig. 10-12. Number of sampling points when there are duct disturbances. [*Source: Federal Register* **36**, 24882, Dec. 23 (1971).]

point is calculated with Eq. (10-25):

$$V_0 = 2.90(F) \sqrt{\left(\frac{29.92}{P}\right)\left(\frac{1.00}{G_d}\right)(\Delta h_p T_R)} \qquad (10\text{-}25)$$

where $V_0$ = local velocity, ft/sec
$P$ = absolute pressure in duct, in. Hg
$G_d$ = carrier gas specific gravity (air = 1)
$\Delta h_p$ = pitot differential pressure, in. $H_2O$
$T_R$ = absolute temperature of carrier gas, °R
The value $F$ in Eq. (10-25) is the discharge coefficient for the S-type pitot tube. This should be obtained for the velocity range being measured by calibration in a duct with clean air against a standard pitot tube having a discharge coefficient of unity. Recalibration of the S-type pitot should be performed at intervals as the value of the discharge coefficient can change with abrasion and wear.

In a circular duct, it is usual to divide the area into one-fourth the number of concentric circles as there are sample traverse points or velocity determination points. The circles are proportioned so that each has equal area. A measurement point is then located at the intersection of the traverse diameter and the centroid of area for each annulus.

Table 10-14 locates measurement points for circular ducts. This method of locating velocity measurement points weights the individual velocities so that the average velocity for the entire duct is equal to the arithmetic average of all velocities. (If the pitot readings are used, the arithmetic average of the square roots of each pitot tube reading must be used.)

## Table 10-14. Location of Traverse Points in Circular Stacks

(Per cent of stack diameter from inside wall to traverse point)

| Traverse point number on a diameter | Number of traverse points on a diameter | | | | | | | | | | | |
|---|---|---|---|---|---|---|---|---|---|---|---|---|
| | 2 | 4 | 6 | 8 | 10 | 12 | 14 | 16 | 18 | 20 | 22 | 24 |
| 1 | 14.6 | 6.7 | 4.4 | 3.3 | 2.5 | 2.1 | 1.8 | 1.6 | 1.4 | 1.3 | 1.1 | 1.1 |
| 2 | 85.4 | 25.0 | 14.7 | 10.5 | 8.2 | 6.7 | 5.7 | 4.9 | 4.4 | 3.9 | 3.5 | 3.2 |
| 3 | .... | 75.0 | 29.5 | 19.4 | 14.6 | 11.8 | 9.9 | 8.5 | 7.5 | 6.7 | 6.0 | 5.5 |
| 4 | .... | 93.3 | 70.5 | 32.3 | 22.6 | 17.7 | 14.6 | 12.5 | 10.9 | 9.7 | 8.7 | 7.9 |
| 5 | .... | .... | 85.3 | 67.7 | 34.2 | 25.0 | 20.1 | 16.9 | 14.6 | 12.9 | 11.6 | 10.5 |
| 6 | .... | .... | 95.6 | 80.6 | 65.8 | 35.5 | 26.9 | 22.0 | 18.8 | 16.5 | 14.6 | 13.2 |
| 7 | .... | .... | .... | 89.5 | 77.4 | 64.5 | 36.6 | 28.3 | 23.6 | 20.4 | 18.0 | 16.1 |
| 8 | .... | .... | .... | 96.7 | 85.4 | 65.0 | 63.4 | 37.5 | 29.6 | 25.0 | 21.8 | 19.4 |
| 9 | .... | .... | .... | .... | 91.8 | 82.3 | 73.1 | 62.5 | 38.2 | 30.6 | 26.1 | 23.0 |
| 10 | .... | .... | .... | .... | 97.5 | 88.2 | 79.9 | 71.7 | 61.8 | 38.8 | 31.5 | 27.2 |
| 11 | .... | .... | .... | .... | .... | 93.3 | 85.4 | 78.0 | 70.4 | 61.2 | 39.3 | 32.3 |
| 12 | .... | .... | .... | .... | .... | 97.9 | 90.1 | 83.1 | 76.4 | 69.4 | 60.7 | 39.8 |
| 13 | .... | .... | .... | .... | .... | .... | 94.3 | 87.5 | 81.2 | 75.0 | 68.5 | 60.2 |
| 14 | .... | .... | .... | .... | .... | .... | 98.2 | 91.5 | 85.4 | 79.6 | 73.9 | 67.7 |
| 15 | .... | .... | .... | .... | .... | .... | .... | 95.1 | 89.1 | 83.5 | 78.2 | 72.8 |
| 16 | .... | .... | .... | .... | .... | .... | .... | 98.4 | 92.5 | 87.1 | 82.0 | 77.0 |
| 17 | .... | .... | .... | .... | .... | .... | .... | .... | 95.6 | 90.3 | 85.4 | 80.6 |
| 18 | .... | .... | .... | .... | .... | .... | .... | .... | 98.6 | 93.3 | 88.4 | 83.9 |
| 19 | .... | .... | .... | .... | .... | .... | .... | .... | .... | 96.1 | 91.3 | 86.8 |
| 20 | .... | .... | .... | .... | .... | .... | .... | .... | .... | 98.7 | 94.0 | 89.5 |
| 21 | .... | .... | .... | .... | .... | .... | .... | .... | .... | .... | 96.5 | 92.1 |
| 22 | .... | .... | .... | .... | .... | .... | .... | .... | .... | .... | 98.9 | 94.5 |
| 23 | .... | .... | .... | .... | .... | .... | .... | .... | .... | .... | .... | 96.8 |
| 24 | .... | .... | .... | .... | .... | .... | .... | .... | .... | .... | .... | 98.9 |

SOURCE: *Federal Register*, **36**, 24883, Dec. 23, 1971.

**Sampling Technique.** After determining from the velocity traverse that the sampling location is suitable, the diameter of the sampling probe tip and the desired sampling rate must be selected. *Isokinetic sampling* means sampling with the same inlet velocity at the probe tip as the flowing gas stream has at the same site. Sampling isokinetically is necessary to obtain a representative sample of large particles. (In sampling gaseous pollutants, isokinetic sampling is unnecessary.) The error from anisokinetic sampling is zero for particles smaller than 3 microns and less than 5 per cent for 5-micron particles. For larger particles, the error can be very substantial.

Isokinetic sampling requirements set the velocity at the probe inlet. Efficient collection of pollutants in the sampling train may require a given flow rate through the train. The availability of a number of different size probe inlet tips is usually needed to make these two requirements compatible.

Sampling-train characteristics will usually make the pressure at which the train flow rate is measured different from that in the stack. Often the temperature is also different and the sample may be dried. Relating train flow rate to tip velocity for isokinetic sampling requires advance planning for quick computation.

The pollutant concentration in the sample must be related to its concentration and total mass in the stack. This means comparing measured sample volume to its original stack volume. With dry filtering equipment this correction is made using the ideal gas law. Where wet collection occurs, water condensed or evaporated from the sample train must also be accounted for. This is usually handled by running a material balance on the train covering the sampling period.

**Particle-size Measurement.** Particle-size measurements are usually made either by laboratory analysis or by classification into various size ranges in a flowing gas sample. Where an unclassified sample is collected, particle size is usually measured by microscopic measurement or resuspension and classification in either an air stream or a liquid. With microscopic methods, a thin deposit of solids may be collected on a dry membrane filter or a glass slide, using an electrostatic or thermal precipitator sampler. The sizes of particles are measured visually and the number in each size range counted. The particles are usually measured by equivalent area or by a predominant dimension. To obtain particle-size distribution on a mass basis, it is necessary to make assumptions about particle volume and density.

For determining particle size for cyclone performance, the American Society of Mechanical Engineers (ASME) has approved the Bahco dry classifier. A 10-g dust sample is required. The Bahco classifier will size the sample into a number of fractions between 44 and 1 microns. A number of questions arise about this technique. In collection on the filter, particles may agglomerate and the Bahco classifier may not resuspend them in the same size range they had initially. There also can be a tendency for finer particles to be lost from the sample by becoming embedded in the pores of the filter.

Particles collected wet or dry may be sized in a liquid-classifying device by measuring settling rate due to gravity or centrifugal force (Whitby centrifuge). The Coulter counter is also used to measure particles in a liquid. It measures the effect of a particle on an electric field, and the size distribution resulting is based on the particle volume.

Airborne particles often flocculate, producing chains and agglomerates held together by electrostatic charges. When the particles are dispersed in a liquid, these charges can be markedly different than in the original airstream. Because of this dispersion problem and the fact that different measurement techniques really measure different properties of the particle, it is seldom found that a single sample will give the same particle-size distribution when measured by different techniques. For comparative tests, it is essential that similar measuring techniques be used.

Probably the best way to measure particle sizes is to classify the particles by size while they are still suspended in the original gas. This is done using a classifying sampler such as a cascade impactor or an Andersen sampler. A very short sampling period should be used when dust loadings are high; otherwise coarser particles collected in the first stages can be reentrained into the finer particle-size stages as the first stages become overloaded.

**Monitoring.** For monitoring pollutant gases, a number of instruments have been developed using infrared and ultraviolet absorption. These are adaptable to $SO_2$, $NO$, $CO$, $CO_2$, hydrocarbons, and other specific gases. Membrane-type cells involving oxidation-reduction reactions with the pollutant have also been developed.

For monitoring particulates from source emissions, tape samplers have recently been developed that filter out the particulates and determine their mass by $\beta$-ray attenuation.

In many situations, regulations on plume visibility are more restrictive than mass emission regulations. In-stack instruments that measure light attenuation across the stack can be calibrated in terms of emission mass when the particle-size distribution does not change with plant operation.

There is a need to predict plume opacity for compliance with regulations when certain particulate collection efficiencies are anticipated or specified. Ensor and Pilat [*J. Air Pollut. Control Assoc.*, **21**, 496–501 (Aug. 1971)] present a theoretical equation for relating plume opacity and particle concentration and properties:

$$W = -\frac{K\rho \ln (I/I_0)}{L} \tag{10-26}$$

where $W$ = the particle mass concentration, $g/m^3$

$\rho$ = true particle density, $g/cm^3$

$I/I_0$ = ratio of light transmitted through plume to quantity transmitted if there were no emission [opacity = $1 - (I/I_0)$]

$L$ = length of light path, m

$K$ = a proportionality constant, $cm^3/m^2$, and dependent on particle properties and light wavelength

This equation is most accurate for a process in which particle-size distribution is constant and when the value of $K$ has been determined experimentally by measuring opacity and particle concentration simultaneously. However, predictions of plume opacity can be made with fair accuracy from fundamental data on the mass-median particle size, the standard particle-size deviation, the density of the particles, and their refractive index. Theoretical values of $K$ are given in the reference.

## 10-15. Pollution Control Techniques

**Minimizing Need for Collection Devices.** Table 10-15 lists means of reducing or eliminating pollutants without specific removal devices. Examples are substitution of low-sulfur fuels, nonvolatile solvents, operation at lower temperatures to reduce $NO_x$ formation or volatilization of a processed material; use of indirect rather than direct-contact heat transfer; heat transfer from radiant panels; pretreatment of natural raw materials to remove easily airborne fines or volatiles; and wetting or agglomeration of solids to reduce airborne releases during handling.

**Application of Control Devices.** Table 10-16 lists equipment types useful for reducing emission of gases, odors, and particulates. Some odors can be cancelled by other antagonistic odor molecules. This provides a means of eliminating an odor without actually preventing its release.

Equipment for control of gaseous pollutants is often not suitable for collection of solid particulates, especially when the loading is heavy. A moderate loading of solid particulates can be handled in gas absorption

**Table 10-15. Fundamental Means of Reducing or Eliminating Pollutant Emissions to the Atmosphere**

I. Eliminate the source of the pollutant
- Seal the system to prevent interchanges between system and atmosphere
  - Use pressure vessels
  - Interconnect vents on receiving and discharging containers
  - Provide seals on rotating shafts and other necessary openings
- Change raw materials, fuels, etc., to eliminate the pollutant from the process
- Change the manner of process operation to prevent or reduce formation of, or air entrainment of, a pollutant
- Change the type of process step to eliminate the pollutant
- Use a recycle gas or recycle the pollutants rather than using fresh air or venting

II. Reduce the quantity of pollutant released or the quantity of carrier gas to be treated
- Minimize entrainment of pollutants into a gas stream
- Reduce number of points in system in which materials can become airborne
- Recycle a portion of process gas
- Design hoods to exhaust the minimum quantity of air necessary to ensure pollutant capture

III. Use equipment for dual purposes, such as a fuel combustion furnace to serve as a pollutant incinerator

**Table 10-16. Types of Equipment Applicable to the Control of Various Classes of Air Pollutants**

| Equipment type | Pollutant classification | | | |
|---|---|---|---|---|
| | Gas | Odor | Particulate | |
| | | | Liquid | Solid |
| Absorption.............. | X | X | | |
| Adsorption.............. | X | X | | |
| Air dispersion (stacks)..... | X | X | X | X |
| Condensation............ | X | X | | |
| Centrifugal (dry)......... | | | X | X |
| Filtration, bags........... | | | | X |
| beds........... | | | X | X |
| fine fibers...... | | | X | X |
| Gravitational settling...... | | | | X |
| Impingement (dry)........ | | | X | X |
| Incineration.............. | X | X | X | X |
| Precipitation, electric...... | | | X | X |
| thermal..... | | | X | X |
| Wet collection............ | | | X | X |

equipment if the particulates are readily soluble in the absorbing liquid and all impacting surfaces are well flushed. A typical upper limit for handling solids in packed gas absorption devices is 5 grains/ft³.

Even though it is difficult to handle gas and solid pollutant mixtures, many situations occur where this must be done. Three alternatives are generally available:

1. Use absorption equipment such as cross-flow packed towers when the particulate solid loading is light.

2. Use wet scrubbing equipment such as venturi scrubbers, impingement tray towers, and fluidized-bed impaction spheres.

3. Use dry particulate collectors such as cyclones, bag filters, and electrostatic precipitators, followed by equipment for efficient gas collection.

In equipment selection, consider the physical form of the collected material most amenable to reuse or disposal. Collection in dry form may permit recycling to the process or blending into product. It might be the preferred form for landfill disposal, but care must be taken in handling to prevent redispersal into the atmosphere. Wetting and pugging of the dry dust may be necessary. Wet collection of insoluble particulates followed by settling and disposal of the sludge to an impoundment area by pipeline can be an inexpensive disposal technique. However, if the solids are water-soluble, disposal without causing groundwater pollution can be a major problem.

### 10-16. Collection and Removal of Gases

**Absorption.** This is one of the most frequently used methods for removal of water-soluble gases. Acidic gases (such as HCl, HF, and $SiF_4$) can readily be absorbed in water efficiently, especially if contact is with water having an alkaline pH. Less-soluble acidic gases (such as $SO_2$, $Cl_2$, and $H_2S$) can be absorbed more readily in a dilute caustic solution such as 5 to 10 per cent NaOH. Scrubbing with an ammonium salt solution is also employed; the gas is often contacted with the more alkaline solution first and a neutral or slightly acid solution last to prevent loss of $NH_3$ to the atmosphere. Lime is an inexpensive alkali but often leads to plugging problems in absorption equipment if the calcium salts have only limited solubility. A better technique is to absorb with an NaOH solution, which is then limed external to the absorption tower. The calcium salts are settled out and the regenerated NaOH returned to the absorption system.

When flue gases containing $CO_2$ are being scrubbed with an alkaline solution, $CO_2$ can also be absorbed, resulting in use of an inordinate amount of caustic. However, if the pH of the scrubbing liquid is kept below 9, the amount of $CO_2$ absorbed can be reduced to a very small value.

Alkaline gases such as $NH_3$ can be removed by scrubbing with acidic solutions such as dilute $H_2SO_4$, $H_3PO_4$, or $HNO_3$. The resulting mixtures can often be disposed of as fertilizer ingredients.

Absorption equipment is frequently favored when the pollution concentration is quite low (chemical costs can become sizeable for large-volume gas streams or high concentrations). The most common devices are packed towers and spray towers. Packed towers are smaller because of the better gas-liquid contact provided by the packing, but open spray chambers are more resistant to plugging. Bubble-cap columns, baffle towers, and sieve-plate towers are also used occasionally.

Design principles of absorption equipment are discussed in Sec. 5.

While there is theoretically no limit to the number of transfer units (see Sec. 5 for definition) that can be built into a packed tower if it is made tall enough, there is a limit to spray towers due to spray entrainment, which results in loss of countercurrentcy. Vertical spray towers with upward gas flow parallel to the tower walls have been demonstrated to be capable of at least 5.8 transfer units. A spray tower with cyclonic gas flow can have up to 7 transfer units. Horizontal spray chambers have been reported with up to 3.5 transfer units.

Venturi scrubbers are sometimes used for gas absorption. Their capability appears to be limited to about 3 transfer units.

While water is the most common scrubbing liquid, organics such as dimethylaniline and various amines (mono-, di-, and triethanol amines, methyl diethanolamine) have been used with acidic gases. In these cases, the absorbing liquid is regenerated by stripping solute off in a concentrated form at a higher temperature or by treatment with lime to produce a precipitate. The volatility of the organic solvent and its possible loss to the atmosphere (or its oxidation) are possible problems that must be considered.

**Adsorption.** Adsorption processes consist of contacting a gas with a solid. The solids are essentially porous with an affinity for certain substances. Typical materials are activated carbon, activated alumina, silica gels, and molecular sieves. An adsorbent can hold from 8 to 25 per cent of its own weight in adsorbed vapors. Generally, the adsorbate is held in liquid phase, even though physical principles would predict that its physical state should be a vapor. The adsorbate may be held in the pore structure by direct physical attraction or by the formation of chemical bonds. Since the adsorbate is held in the liquid phase, heat of condensation is released in the bed. To keep the bed from heating up, it may be necessary to precool the inlet gas. Capacity and efficiency of adsorption decrease with an increase in temperature.

Some adsorbents exhibit a selectivity for particular vapors; for instance, aluminas and silica gels have a special affinity for water vapor. Therefore, it may be necessary to dry the gas first to prevent saturating the bed with condensed water. Carbon does not selectively adsorb water vapor, which makes it useful for treating moist gases and for bed regeneration with steam.

The adsorption process is practically complete regardless of inlet gas concentration as long as the bed is not saturated. Once the bed becomes saturated, the exit concentration of the adsorbate increases exponentially. This fact makes adsorption processes especially attractive for control in those situations where extremely low exit gas concentrations must be reached.

The bed is regenerated by raising its temperature above the boiling point of the adsorbate and stripping the adsorbate from the bed with a hot gas which is recirculated to the bed after condensation of a portion of the adsorbate. The bed is then cooled and returned to service. Occasionally, the bed may be regenerated by heating and evacuating without use of a stripping gas.

**Condensation.** A number of vapors, especially hydrocarbons with low volatility, can be recovered by condensation. A tubular, water-cooled heat exchanger is adequate for many high-molecular-weight organic vapors. Where the volatility is greater, a refrigerated condenser following the water-cooled condenser may be necessary. In condensing many vapors, where heat transfer is more rapid than mass transfer, fog particles of the condensate (0.5- to 1.5-micron diameter) are apt to form in the bulk gas stream. These particles can result in plume opacity violations as well as a recovery from cooling that is not as great as predicted from vapor pressure-temperature data. Therefore, it is often necessary to follow the condenser with a mist eliminator having high efficiency on fine particles. A small-diameter in-depth fiber-bed filter or an electrostatic precipitator are good for fog control. However, electrostatic precipitators should not be used where combustible mixtures are present.

When the vapor is too volatile for efficient removal by condensation alone, compressing the gas before condensation will result in adequate recovery at an economical cost in situations where the total gas volume is small and the quantity of vapor to be recovered is an appreciable portion of the total. Condensation is attractive where the gas stream is already at an elevated pressure and can be cooled prior to pressure release.

## 10-17. Special Gaseous Pollutant Control Methods

Two widely released gaseous pollutants are $SO_2$ and $NO_x$. Major sources of these pollutants are in flue gases from combustion operations. Control of these two gases has received wide study, and a number of specialized techniques are available.

**Sulfur Dioxide.** One control technique is to substitute a low-sulfur or desulfurized fuel. Desulfurization has been commercialized for petroleum fuels. Several processes have been demonstrated for coal, but it may be several years before they are commercial.

A number of processes have been developed for removal of $SO_2$ from flue gases. Among the most economical are those that react $SO_2$ with an inexpensive alkali such as limestone or quicklime. Dry injection of ground limestone through the boiler burner also has received extensive study. The limestone is calcined to quicklime, which reacts with $SO_2$. $CaSO_4$ is recovered in the unit's electrostatic precipitator. Removal of $SO_2$ is low at stoichiometric proportions of limestone, but as excess limestone is used, recovery efficiencies increase. However, only 50 per cent recovery has been achieved with 100 per cent excess limestone.

Reaction of $SO_2$ with lime or limestone slurry in a wet scrubbing system has been demonstrated to give 70 to 90 per cent $SO_2$ removal. Difficulties with pluggage from $CaSO_4$ deposits have been experienced in some units. Scrubbing with a sodium alkali or an organic absorbent is a more pluggage-resistant method, but considerably more expensive in initial cost.

In smaller boilers, scrubbing with NaOH or $NaCO_3$ has been utilized to give efficient $SO_2$ removal, but disposal of a $Na_2SO_3$-$Na_2SO_4$ solution or solid may be troublesome unless the unit is located near a sulfite pulp mill.

The Cominco $NH_3$ scrubbing system has also been used. This is a two-stage scrubber in which the gas is contacted first with a $(NH_4)_2SO_4$ solution. The second stage, which prevents loss of $NH_3$, contains $NH_4HSO_4$ solution. The make-up $NH_3$ is added to the second stage, and a portion of the $NH_4HSO_4$ solution is fed to the first stage. The $(NH_4)_2SO_4$ produced is used in fertilizer manufacture.

Two other processes that recover $SO_2$ in the form of $H_2SO_4$ for sale are the Monsanto Cat-OX and the Chemico MgO scrubbing systems. In the Cat-OX process, particulates are removed with an electrostatic precipitator. The cleaned flue gas containing $SO_2$ is mixed with heated air and passed through a vanadium catalyst where the $SO_2$ is oxidized to $SO_3$ and absorbed in $H_2SO_4$. A fiber mist eliminator provides final clean-up, and the gas is reheated by interchange with heat from the gases leaving the catalyst bed.

In the Chemico process, MgO, suspended in a saturated solution of $MgSO_4$, contacts the flue gas in a venturi scrubber. The $SO_2$ is absorbed, reacting with MgO to produce $MgSO_3$ and further $MgSO_4$ through air oxidation. Precipitated solids are removed in a centrifuge and charged to a kiln where $MgSO_3$ is decomposed. Carbon is added to decompose $MgSO_4$. The regenerated MgO is recycled to the process and the kiln off-gas, containing 15 to 16 per cent $SO_2$, is sent to a sulfuric acid plant.

Both processes are capable of recovering 80 to 90 per cent of the $SO_2$. The sales price of the $H_2SO_4$ produced is insufficient to carry the cost of operation when using flue gases from 3 per cent sulfur coal. The cost of these plants is essentially a function of the flue-gas quantity rather than the $H_2SO_4$ produced. There is definite opportunity to improve the economics by increasing the $H_2SO_4$ production by using a high-sulfur (5 to 6 per cent) coal.

**Nitrogen Oxides.** Combustion operations are a major source of $NO_x$ pollutants. $O_2$ and $N_2$ from air react at high temperatures in the flame to produce NO. NO reacts more slowly at lower temperatures with $O_2$ to produce $NO_2$. In most furnaces, reaction rates to form NO are too slow to produce equilibrium amounts of NO corresponding to flame temperature, but it is not unusual for the flue gases from oil and coal combustion to contain 1,000 to 2,000 ppm by volume of NO when using 5 to 10 per cent excess air.

The usual control techniques consist of modifying the combustion process to minimize the formation of NO. Common methods are: (1) use of low excess air, (2) two-stage combustion, (3) flue-gas recirculation, and (4) combustion chamber modification.

Reducing the $O_2$ in the combustion products to 0.3 to 0.5 per cent can produce a two- to fivefold reduction in the NO produced. Reducing the flame temperature lowers the quantity of NO produced by making the equilibrium less favorable and decreasing the rate of reaction.   Recirculation of flue gas can be used as a means to reduce flame temperature.   Formation of NO becomes extremely slow at temperatures below 1600°F.

Two-stage combustion involves burning the majority of the fuel at a high temperature for good heat transfer with a deficiency of $O_2$ (50 to 60 per cent of theoretical) to discourage NO formation.   Combustion of CO and other products is completed downstream at a lower temperature where the remaining combustion air is added.   Flue-gas recirculation may also be used to provide flame cooling.   A combination of two-stage combustion and flue-gas recirculation can result in a 90 per cent reduction in NO formed.

Combustion chamber modifications speed up combustion rate and reduce flame temperature more quickly.   One approach is to arrange burners so they burn tangentially along radiating refractory walls.   The refractory surface speeds the reaction catalytically and absorbs heat from the flame providing rapid quenching.

Other techniques can be used where NO formation cannot be reduced adequately.   For example, the gases may be passed through a combustion catalyst maintained at 1000 to 1400°F where NO decomposes to $O_2$ and $N_2$. If the gases are too cold to keep the catalyst hot, some additional fuel is added.

NO can be scrubbed from gases with an alkaline scrubbing liquid as long as the mole ratio of $NO_2$ to NO is above unity.   Unfortunately, the reaction rate of oxidizing NO to $NO_2$ is fairly slow and equilibrium becomes more favorable as room temperature is approached.   Thus scrubbing is usually practical only if the flue gases are cooled, mixed with additional air, and held in a large reaction chamber for several seconds before scrubbing.

**Control of Pollutants by Incineration.**   Incineration is used to destroy combustible vapors such as hydrocarbons (especially unsaturated and aromatic compounds that are photochemically reactive), CO, $H_2$, $H_2S$, and mercaptans.

Consideration should be given to collecting valuable hydrocarbons and organic solvents by other means, such as condensation, rather than destroying them by combustion.   If the quantities are appreciable, recovery for fuel value may be worthwhile.   Gases containing sufficient combustibles to support combustion are burned in flares, waste-heat recovery boilers, or used for process heat.

Incineration is used to destroy odors in those cases where the odor substance can be oxidized.   It is also possible to use an incinerator, when properly designed, to burn combustible airborne liquid and solid particles using a burner much like one designed to burn pulverized coal.   In such an incinerator, it is necessary to provide rapid ignition of the particles by heating them above the kindling temperature, and providing adequate residence time and flame space for complete combustion.   The presence of noncombustible residue that can produce a solid or molten ash must also be considered.

Two types of gas incinerators are in use: direct flame and catalytic.   In the direct-flame type, gases are heated in a fuel-fired refractory chamber to their autoignition temperature, where oxidation occurs with or without a visible flame. Autoignition temperatures vary with chemical structure but are generally in the range of 1000 to 1400°F. Gases can be incinerated by indirect heating, but a higher temperature is generally required than when a direct flame is present.

For continuous use, a direct-flame incinerator is usually equipped with heat interchangers to preheat the incoming gas with the exhaust gases to save fuel. The required residence time in the combustion chamber is that needed to provide 100 per cent oxidation of the combustible materials. This varies with the substances to be oxidized and the temperature. It is specified that incinerators shall provide 0.3-sec residence time at 1300°F or above in many installations.

Catalytic incinerators oxidize substances at temperatures below which they would burn in air, usually around 500°F. Catalysts are from the platinum family of metals, but certain other metallic oxides are occasionally used. In catalytic oxidation, fuel must be provided initially to start the reaction. Once the catalyst bed is heated, frequently no further fuel is needed. Hence, the advantages of catalytic oxidation are less fuel, no NO formation, and less bulk. However, the catalyst may be poisoned by heavy metals, phosphates, and arsenic. Its activity may be decreased temporarily by halogens and sulfur compounds. Further, it may be rendered inactive by surface coatings of soot and inorganic dust. There is also danger of destroying a catalytic incinerator by overheating if the gas composition is highly variable.

Flares for burning concentrated combustible gases are generally located at high elevations away from other structures to provide protection from the radiant heat of the flame. The design must provide against flashback; a water seal or gas purge is frequently used. To prevent air pollution, the flare should burn smokelessly. Clean burning can be achieved with steam injection (0.05 to 0.3 lb steam/lb combustibles) or with multijet flares having a radiant refractory. Combustible particulates must not be fed to flares since falling burning particles can create a fire hazard. Centrifugal knock-out drums are often used to protect against entrained particulates.

## 10-18. Collection and Removal of Particulates

Six basic principles are used alone or in combination in particulate collectors:

1. Gravity settling
2. Flowline interception
3. Inertial deposition
4. Diffusional deposition
5. Electrostatic deposition
6. Thermal precipitation

Table 10-17 lists these mechanisms and their basic parameters. In addition, sonic agglomeration has been considered but has seldom become commercially practical.

## Table 10-17. Summary of Mechanisms and Parameters in Aerosol Deposition*

| Deposition mechanism‡ | Origin of force field | Deposition mechanism measurable in terms of | | System parameters |
|---|---|---|---|---|
| | | **Basic parameter** | **Specific modifying parameters** | |
| Flow-line interception‡ | Physical gradient‡ | $N_{sI} = \left(\dfrac{D_p}{D_b}\right)$ | | Geometry: $(D_{bI}/D_b)$, $(D_{bI}/D_b)$, etc. $e_b$ $\alpha$ |
| Inertial deposition | Velocity gradient | $N_{si} = \left(\dfrac{K_m \rho_p D_p^2 V_o}{18\mu D_b}\right)$ | $N_{sc} = \left(\dfrac{N^2{}_{sI}}{N_{si}N_{sd}}\right)$ $= \left(\dfrac{18\mu}{K_m \rho_p D_s}\right)$ ¶ | |
| Diffusional deposition | Concentration gradient | $N_{sd} = \left(\dfrac{D_v}{V_o D_b}\right)$ | | |
| Gravity settling | Elevation gradient | $N_{sg} = \left(\dfrac{u_t}{V_o}\right)$ | | Flow pattern: $N_{Re}$ ‖ $N_{Ma}$ $N_{Kn}$ |
| Electrostatic precipitation§ $\quad$a. Attraction $\quad$b. Induction | Electric-field gradient§ | $N_{seo} = \left(\dfrac{K_m Q_p e b}{\mu D_p V_o}\right)$ ; $N_{sei} = \left(\dfrac{\delta_p - 1}{\delta_p + 2}\right)\left(\dfrac{K_m D_p^2 \delta_o e b^2}{\mu D_b V_o}\right)$ | $\delta_p, \delta_b$† | Surface accommodation |
| Thermal precipitation | Temperature gradient | $N_{st} = \left(\dfrac{T - T_b}{T}\right)\left(\dfrac{\mu}{K_m \rho D_b V_o}\right)$ $\left(\dfrac{k_t}{2k_t + k_{tp}}\right)$ | $(T_b/T)$, $(T_p/T)$,† $(N_{Pr})$, $(k_{tp}/k_t)$, $(k_{tb}/k_t)$,† $(c_{hp}/c_h)$, $(c_{hb}/c_h)$† | |

*SOURCE: Perry and Chilton (eds.), "Chemical Engineers' Handbook," 5th ed., p. 20–80, McGraw-Hill Book Company, New York, 1973.
† Not likely to be significant contributors.
‡ This has also commonly been termed "direct interception" and in conventional analysis would constitute a physical boundary condition imposed upon particle path induced by action of other forces. By itself it reflects deposition that might result with a hypothetical particle having finite size but no mass or elasticity.
§ In cases where the body charge distribution is fixed and known, $e_b$ may be replaced with $Q_{bs}/\delta_o$.
¶ This parameter is an alternate to $N_{sI}$, $N_{si}$, or $N_{sd}$ and is useful as a measure of the interactive effect of one of these on the other two. It is comparable with the Schmidt number.
‖ When applied to the inertial deposition mechanism, a convenient alternate is $(K_{m}\rho_p/18\rho) = N_{sI}/(N_{si}^2 N_{Re})$.

**Particle Dynamics.** Larger particles encountered in air pollution obey Stokes' law [see Perry and Chilton (eds.), "Chemical Engineers Handbook," 5th ed., sec. 5, McGraw-Hill Book Company, New York, 1973]. Terminal settling velocities of particles as calculated from Stokes' law are shown in Fig. 10-13. Below 16 microns in size, the Stokes-Cunningham correction

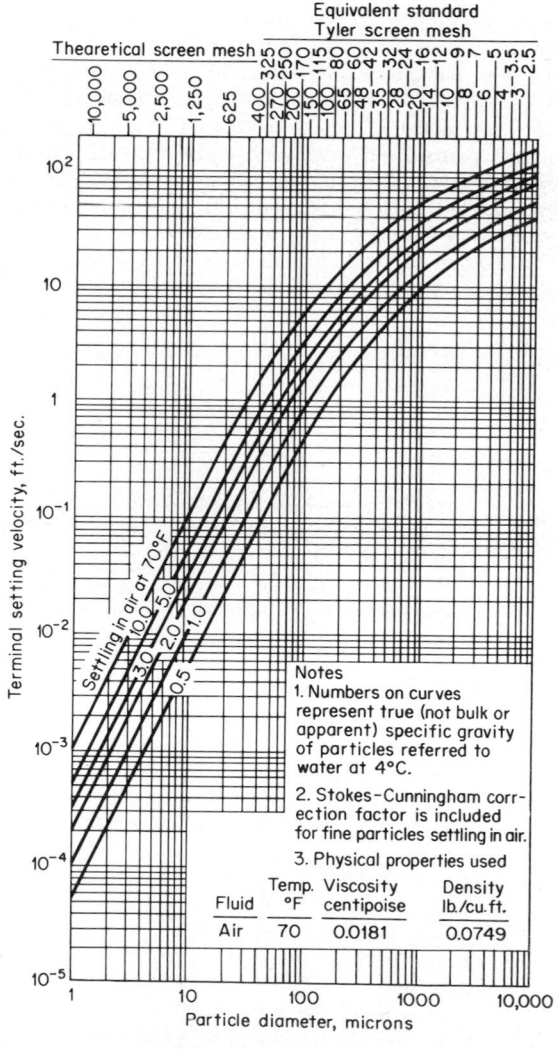

Fig. 10-13. Terminal settling velocity of spherical particles.

factor becomes important, and this correction has been applied. Below about 5 microns, particles tend to become suspended by Brownian motion because of the impact of gas molecules on the particle, and gravity settling ceases as an important influence.

**Gravity Settling Chambers.** Due to low collection efficiency on smaller particles, gravity settling is seldom useful today as a sole collection device. However, such chambers may be used as preclassifiers to remove large abrasive particles ahead of another type of collector. Settling chambers are seldom efficient on particles smaller than 50 microns.

Basically, a settling chamber is a large horizontal enlargement in the duct where the gas slows down. Turbulence should be low to prevent reentrainment. The collection efficiency for any size particle can be calculated from the terminal settling velocity of the particle as determined in Fig. 10-13, the distance the particle has to fall to settle out, and the residence time in the chamber. Equation (10-27) can be applied to each individual size particle to calculate collection efficiency for that size. Equation (10-28) gives the smallest size particle that can be collected with 100 per cent efficiency.

$$\eta = \frac{u_t L_s}{H_s V_s} \tag{10-27}$$

$$D_{p,\min} = \sqrt{\frac{18\mu H_s V_s}{g_L L_s (\rho_s - \rho_g)}} \tag{10-28}$$

where
$\eta$ = fractional collection efficiency
$u_t$ = particle terminal settling velocity, ft/sec
$V_s$ = bulk gas velocity, ft/sec
$L_s$ and $H_s$ = the length and height of the settling path, respectively, ft
$\mu$ = gas viscosity, lb/(sec)(ft)
$g_L$ = local acceleration due to gravity, ft/sec²
$(\rho_s - \rho_g)$ = difference in particle and gas density, lb/ft³

**Cyclonic Collectors.** In cyclonic collectors, centrifugal force separates particles from the gas stream. Since centrifugal force can equal many times that of gravity, much smaller particles can be collected. Figure 10-14 shows a typical dust collection cyclone in which the gas enters tangentially, spirals downward, reverses direction in the cone, and exits through the top in smaller spirals. The dust particles spiral downward along the wall and discharge at the bottom.

Cyclones are reasonably effective for collecting solid and liquid particles down to 5 to 10 microns. The smaller the diameter of the cyclone, the higher its efficiency on small particles. When cyclones are used to collect liquids, special modifications must be made to prevent reentrainment of droplets at the outlet tube.

The principles of centrifugal force are applied in many wet scrubbers in which the particles are centrifuged into a water film on a wall. Wet scrubbers have been built using a cyclone like that shown in Fig. 10-14 in which

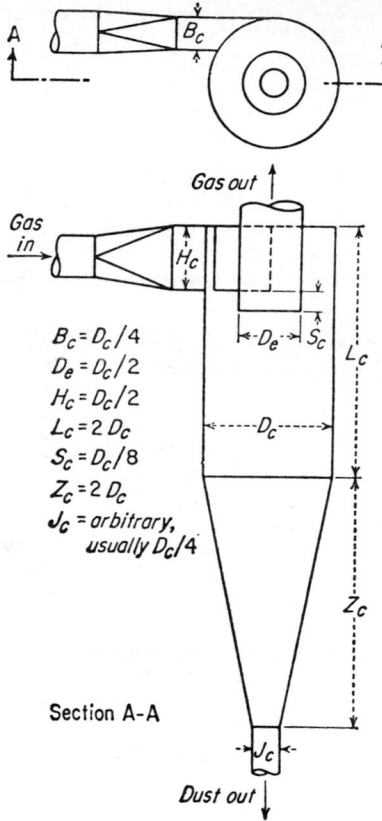

$B_c = D_c/4$
$D_e = D_c/2$
$H_c = D_c/2$
$L_c = 2 D_c$
$S_c = D_c/8$
$Z_c = 2 D_c$
$J_c$ = arbitrary, usually $D_c/4$

Section A-A

FIG. 10-14. Cyclone separator proportions.

spray nozzles are placed in the annular space at the top or a film of water is allowed to flow down the walls.

The efficiency of a cyclone is computed by integration of the collection efficiency for each individual size particle as obtained from the manufacturer's efficiency curve. It is frequent practice to calculate the particle cut size $D_{pc}$ (the diameter particle that is collected with 50 per cent efficiency). Equation (10-29) gives the cut size for a cyclone with dimensions as given in Fig. 10-14.

$$D_{pc} = \sqrt{\frac{9\mu B_c}{2\pi N_e V_c(\rho_s - \rho_g)}} \tag{10-29}$$

where $N_e$ = the number of turns the gas makes in the cyclone, often 5 to 10
$B_c$ = the width of the gas inlet, ft
$V_c$ = inlet gas velocity, ft/sec

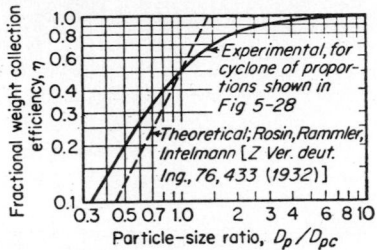

Fig. 10-15. Separation efficiency of cyclones.

All other symbols are as defined for Eq. (10-28). Figure 10-15 relates the collection efficiency for other size particles to the cut size for the Fig. 10-14 cyclone.

Pressure drop through a cyclone depends on cyclone geometry. The manufacturer's calibration curve should be used. The pressure drop of the Fig. 10-14 cyclone in terms of the number of inlet velocity heads, $F_{cv}$, is given by Eq. (10-30). The value of $K$ for this cyclone is 16. Other symbols are defined by Fig. 10-14.

$$F_{cv} = KB_cH_c/D_c^2 \qquad (10\text{-}30)$$

**Impingment Collectors.** Figure 10-16 illustrates several collection principles involved in impingement collection. If $D_b$ represents a target, all particles contained upstream within the streamlines $A$-$B$ will be collected on the target, unless the particles have sufficient mobility so that they can flow around the target with the gas. These collected particles are caught by flowline interception. Large particles just outside these streamlines, due to

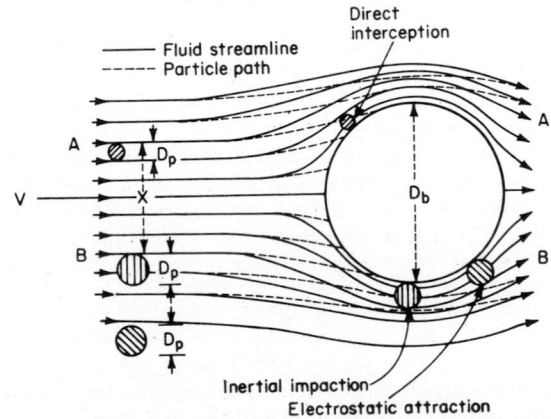

Fig. 10-16. Mechanisms involved in impingement separators.

their diameter, will have to move farther away from the target to prevent impact. However, the larger particles may have too much momentum to be deflected and will also be collected. This is known as inertial deposition. Other particles still farther away, but carrying an electric charge, may be attracted to the target, especially if the target develops an opposite charge.

Some dry impingement separators have been built in which the particulates are directed first at one target and then at another. Figure 10-17

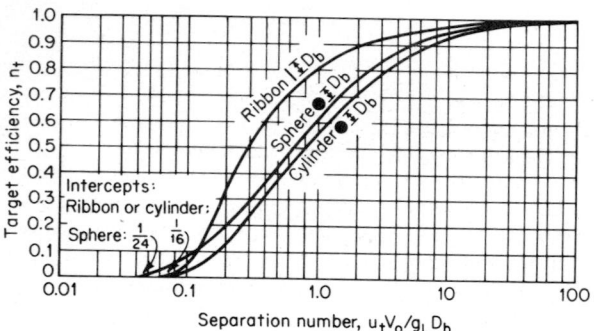

Fig. 10-17. Target efficiency for conditions where Stoke's law applies to the motion of the particle. [*Source: Langmuir and Blodgett, U.S. Army Air Force Tech. Rept.* 541S, *Feb.* 19, 1946 (*U.S. Dept. of Commerce, OTS,* PB 27565).]

gives the collection efficiency for a single row of targets of various shapes for collection by both direct and inertial impaction but does not include electrostatic attraction. Many wet scrubbers use the impaction principle to collect a particle in a liquid film on a target. The higher the upstream velocity and the larger the particles, the greater the collection efficiency of a single stage. Over-all efficiency is improved by using a number of stages such as in a fiber-bed filter.

**Granular-bed Filters.** In these particulate-removal devices, a gas is forced through a granular bed. Reported efficiencies are as high as 97 to 98 per cent on particles of 2 to 5 microns. Often the primary collection technique is direct and inertial impaction. The high efficiency on small particles results from the many successive targets.

In some beds, plastic granules that develop an electrostatic charge have been used to increase the collection efficiency. Some problems with granular-bed filters are adequate removal of the collected particles from the bed and particle reentrainment. [See Perry and Chilton (eds.), "Chemical Engineers' Handbook," 5th ed., sec. 20, McGraw-Hill Book Company, New York, 1973, for further discussion on granular-bed filters.]

**Bag Filters.** Bag filters for the collection of particulates use a woven fabric or a nonwoven felt bag. Either can be specified for collection efficiencies of 99 per cent or better. The collection process is not merely a screening or filtration of the dust since the openings in the cloth are many times the size of the dust particles collected. Efficiency of a new bag may be fairly low for a few moments until it develops a precoat of dust that serves as the filtering layer. Once precoated, the bag usually retains sufficient solids in its pores so that it does not return to its original efficiency.

Pressure drop through the fabric is usually negligible compared to that through the layer of collected dust. Pressure drop through the dust layer can be expressed by

$$\Delta p_i = K_d \mu C_d V_f^2 \theta \tag{10-31}$$

where $C_d$ = the concentration of dust in the gas stream
 $V_f$ = the gas velocity through the bag
 $\mu$ = the gas viscosity
 $\theta$ = the time since the last bag cleaning cycle

The resistance factor $K_d$ is a function of the dust size, shape, packing density, etc., and is best determined experimentally. Usual practice is to specify a maximum desired pressure drop across the bag and pick a cleaning cycle such that this pressure drop is not exceeded. Typical operating pressure drops are 2 to 6 in. $H_2O$.

With woven fabrics, cleaning must be done by mechanically shaking the bag. The cycle may be controlled manually or automatically. More frequent shaking permits a smaller bag area, but bag life is shortened by the mechanical strains of shaking. Filters are usually sized on superficial velocity through the bag and will be in the range of 1 to 8 ft³/(min)(ft²) of bag area. Longer bag life results from more generous sizing. For fine dusts, velocity should not exceed 3 ft³/(min)(ft²). Airflow through a compartment is shut off while the bags are shaken. On a continuous process this means providing one or two spare compartments for use while others are being cleaned.

With felt cloth, somewhat higher flow rates can be used, up to 15 ft³/ (min)(ft²). Cleaning can be accomplished in several ways, but a reverse-flow air blast is one of the best. Table 10-18 lists the maximum desirable operating temperatures for a number of filter fabrics, while Table 10-19 lists the chemical compatability of different fabrics.

### Table 10-18. Maximum Desirable Operating Temperatures for Filter Bags

| Fiber | Temperature, °C |
|---|---|
| Cotton | 80 |
| Wool | 100 |
| Nylon 66 and 6 | 105 |
| Dacron | 140 |
| Orlon | 120 |
| Dynel | 85 |
| Saran | 70 |
| Polyethylene | 70 |
| Glass | 290 |
| Teflon | 260 |
| Nomex nylon | 230 |

Table 10-19. Chemical Compatibility of Fibers for Dust Collector Bags

| Resistance | Acid media | Alkaline media |
|---|---|---|
| Excellent......... | Polyethylene<br>Saran<br>Teflon<br>Nomex nylon | Dynel<br>Nylon 66<br>Polyethylene<br>Teflon |
| Good............ | Dacron<br>Dynel<br>Glass<br>Wool | Cotton<br>Nylon 6<br>Saran<br>Saran |
| Unsuitable........ | Cotton<br>Nylon 66<br>Nylon 6 | Wool |

Care must be taken to operate bag filters well above the dew point to prevent caking on the bags. To avoid condensation, baghouses are often insulated and occasionally steam traced. A few steam-jacketed baghouse designs are available.

When handling very fine dusts, problems with dust leakage through the bag are sometimes encountered. Leakage can be reduced by: (1) a bag made of staple rather than monofilament fibers, (2) a napped fiber bag, (3) a finer weave, (4) less frequent and less vigorous cleaning, (5) a bag fiber better suited to attract and retain the dust particles, (6) a precoat material following cleaning.

Textile fibers tend to develop static charges as gases pass through the bags. If a leaking bag does not develop a strong charge, a bag developing a stronger charge or the opposite charge might be tested.

Occasionally, a bag retains its cake too tenaciously during cleaning and the bag is said to be subject to blinding. This problem may be improved by taking the reverse action to that used for leakage.

**Electrostatic Precipitators.** An electrostatic precipitator collects solid or liquid particulates with high efficiency and low energy utilization. The pressure drop is composed almost entirely of inlet and exit losses. Initial cost of precipitators is high, however, and their on-stream reliability may be less than that of other collection devices.

Collection is based on imparting a charge to particles in an electric field, which then causes them to migrate and deposit on a collection plate of the opposite charge. Liquids agglomerate on the collection plate and drain off, but solids must be removed by rapping. Reentrainment may occur during rapping.

When power is applied to a precipitator, no current flows until a sufficient voltage is achieved to create a corona discharge (gas ionization) at the discharge electrodes. Then particle charging and deposition begin. As voltage is increased further, the field strength becomes more intense and precipitator efficiency increases until the point where sparkover or arcing from the discharge electrode to the collecting plate occurs.

For maximum collection efficiency, the voltage should be maintained just short of sparkover. Unfortunately, sparkover potential is an operating variable affected by changes in temperature, dust resistivity, moisture content, and frequency and efficiency of removal of collected dust. For maximum operability, the difference in potential between start of corona and sparkover should be as large as possible. This difference decreases as gas temperature increases or system absolute pressure decreases. The difference is greater for negative polarity current than for positive. Thus, most industrial precipitators are operated with negative polarity.

Because of sparkover, precipitators should not be used to collect combustible or explosive mixtures. A two-stage precipitator in which the particle charging is done in one step and the deposition in a second step (developed primarily for air conditioning) is somewhat safer because lower voltages are employed. Such units have been used with hydrocarbon mists in air. However, the safety of this application is still questionable.

Equation (10-32) gives the theoretical efficiency for a precipitator:

$$\eta = 1 - e^{-[(u_e A_e)/q]} = 1 - e^{-K_e u_e} \qquad (10\text{-}32)$$

where $\eta$ = the weight fractional efficiency

$u_e$ = the velocity of migration of particle toward collecting electrode

$A_e$ = the area of collecting electrodes

$q$ = the total gas flow rate

Any consistent set of units that makes the exponent dimensionless can be used. The efficiency factor $K_e$ is basically a geometric design factor for the precipitator. For a plate-type precipitator, $K_e = L/B_e V_e$. For a tube type, $K_e = 4L/D_t V_e$, where $L$ is the length of a flow passage in the electric field, $B_e$ is the distance between adjacent plates, $D_t$ is the tube diameter, and $V_e$ is the gas velocity through an individual flow passage. The migrational velocity, $u_e$, is a function of the particle size, shape, composition, and average electric field strength. Each different size particle has a different migration velocity. For a dust of a given particle-size distribution, an average migrational velocity is often used.

With a given precipitator, it is often desired to know the effect of changes in operating conditions. Using Eq. (10-32) with geometric values for $K_e$, the effects of changes in gas velocity or residence time can be determined without any knowledge of the value of $u_e$. However, changing gas temperature or particle size will change the value of $u_e$. For a constant field strength, the relative changes in $u_e$ can be estimated by Eqs. (10-33) and (10-34).

For a change in temperature,

$$(u_e)_{t_2} = \frac{(u_e)_{t_1} (\mu_g)_{t_1}}{(\mu_g)_{t_2}} \qquad (10\text{-}33)$$

where $\mu_g$ is the gas viscosity at temperatures $t_1$ and $t_2$.

For a change in median particle size at constant particle standard diaviation,

$$u_{e_2} = \frac{u_{e_1} r_1}{r_2} \tag{10-34}$$

where $r_1$ and $r_2$ are the radius of the median particle sizes. For a detailed discussion of electric precipitators see H. J. White, "Industrial Electrostatic Precipitation," Addison-Wesley Publishing Co., Inc., Reading, Mass., 1963; S. Oglesby, "A Manual of Electrostatic Precipitation Technology," NTIS-PB-196-380, Southern Research Institute, Birmingham, Ala., 1970.

**Mist Filters.** Mists are created by condensation. They consist of small particles, from submicron size up to 10 microns. For coarser mists, 5 microns and up, knitted-wire-mesh separator pads, 4 to 6 in. thick, have collection efficiencies of 98 per cent and above with 1 to 2 in. of water-pressure drop. These separators are most efficient when installed horizontally with upflow of gas, and gravity drainage of liquid. Collection is by impingement.

Impingement collectors consisting of thin mats, 1 to 2 in. thick, of compressed glass or plastic fibers give 92 to 94 per cent collection efficiency on 1-micron particles with 4 to 8 in. of water-pressure drop. Similar fiber-bed filters, 4 to 6 in. thick, and employing Brownian diffusion as the collection principle, are used to collect submicron mists with collection efficiencies up to 99 per cent and pressure drops from 15 to 30 in. of water. These filters, when irrigated, are also used to collect submicron particles that are soluble in the irrigating liquid.

**Wet Scrubbers.** A great variety of liquid scrubbing devices are available for collection of particulates. The major collection mechanism is inertial impaction with direct interception and Brownian diffusion of lesser importance. Some units contact the gas with atomized liquids using either gravity or centrifugal force. Others impinge the gas at high velocity against

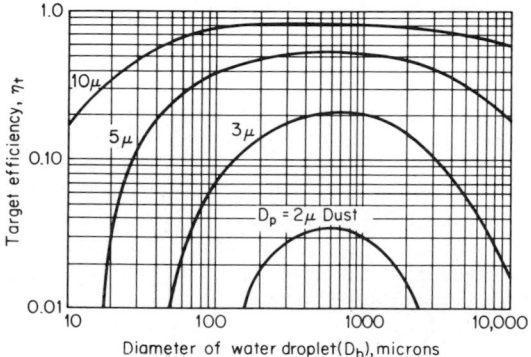

FIG. 10-18.  Target efficiency in a spray tower.  [*Source: Stairmand, J. Inst. Fuel,* **29,** 58(1956).]

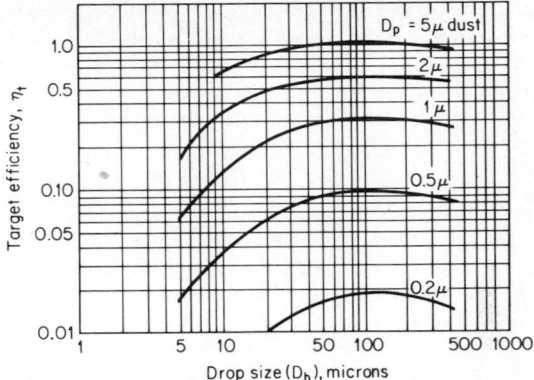

Fig. 10-19. Target efficiency in 100 times gravitational field. [*Source: Johnstone and Roberts, Ind. Eng. Chem.*, **46**, 1601 (1954).]

sheets and films of water, causing the water to shatter into droplets. In either case, the major portion of the collection occurs by collision between small liquid droplets and the particulate.

Figure 10-18 shows data on target efficiency for collection in a gravity spray tower between dust particles of specific gravity 2.0 and spray droplet size. It can be seen that the most efficient droplet sizes are from 500 to 1000 microns. The data also indicate that it would be difficult to obtain high efficiency in a gravity spray tower for particulates smaller than 3 microns.

Collection efficiency can be improved greatly by the addition of centrifugal force. Figure 10-19 shows target efficiencies for water droplets in a centrifugal field of 100g. (A centrifugal force of 100g is equivalent to a tangential velocity of 57 ft/sec at a radius of 1 ft.) It can be seen that the optimum scrubbing droplet size has been reduced to 40 to 200 microns and that sizeable collection efficiencies on particles down to 1 micron can now be obtained.

## Table 10-20. Minimum Particle Size for Various Types of Scrubbers

|  | Pressure drop, in. water | Min. particle size, microns |
|---|---|---|
| Spray towers.......................... | 0.5–1.5 | 10 |
| Cyclone spray scrubbers................ | 2–10 | 2–10 |
| Impingement scrubbers................. | 2–50 | 1–5 |
| Packed- and fluidized-bed scrubbers...... | 2–50 | 1–10 |
| Orifice scrubbers...................... | 5–100 | 1 |
| Venturi scrubbers..................... | 5–100 | 0.8 |
| Fibrous-bed scrubbers................. | 5–110 | 0.5 |

SOURCE: Perry and Chilton (eds.), "Chemical Engineer's Handbook," 5th ed., p. 20-98. McGraw-Hill Book Company, New York, 1973.

High collection efficiencies on submicron particles in wet scrubbers can only be obtained with venturi scrubbers and wetted fibrous-bed scrubbers. The latter are subject to plugging if the dust loading is high or insoluble. The former requires high energy expenditure, in the order of 30 to 50 in. of water pressure drop for high efficiency on 0.5- to 1.0-micron particles.

Table 10-20 indicates typical pressure drop and minimum particle size that can generally be collected at 80 per cent collection efficiency in various types of wet scrubbers.

# INDEX

1

| To convert from | to | multiply by |
|---|---|---|
| kilogram force (kgf) | newton | +00 9.806 65 |
| kilopond force | newton | +00 9.806 65 |
| kip | newton | +03 4.448 221 |
| knot (international) | meter/second | −01 5.144 444 |
| lambert | candela/meter² | +04 1/π |
| lambert | candela/meter² | +03 3.183 098 |
| langley | joule/meter² | +04 4.184 |
| lbf (pound force, avoirdupois) | newton | +00 4.448 221 |
| lbm (pound mass, avoirdupois) | kilogram | −01 4.535 923 |
| league (British nautical) | meter | +03 5.559 552 |
| league (international nautical) | meter | +03 5.556 |
| league (statute) | meter | +03 4.828 032 |
| light year | meter | +15 9.460 55 |
| link (engineer or ramden) | meter | −01 3.048 |
| link (surveyor or gunter) | meter | −01 2.011 68 |
| liter | meter³ | −03 1.00 |
| lux | lumen/meter² | +00 1.00 |
| maxwell | weber | −08 1.00 |
| meter | wavelengths Kr 86 | +06 1.650 763 |
| micron | meter | −06 1.00 |
| mil | meter | −05 2.54 |
| mile (U.S. statute) | meter | +03 1.609 344 |
| mile (U.K. nautical) | meter | +03 1.853 184 |
| mile (international nautical) | meter | +03 1.852 |
| mile (U.S. nautical) | meter | +03 1.852 |
| millibar | newton/meter² | +02 1.00 |
| millimeter of mercury (0° C) | newton/meter² | +02 1.333 224 |
| minute (angle) | radian | −04 2.908 882 |

| To convert from | to | multiply by |
|---|---|---|
| pound mass (troy or apothecary) | kilogram | −01 3.732 417 |
| poundal | newton | −01 1.382 549 |
| quart (U.S. dry) | meter³ | −03 1.101 220 |
| quart (U.S. liquid) | meter³ | −04 9.463 529 |
| rad (radiation dose absorbed) | joule/kilogram | −02 1.00 |
| Rankine (temperature) | kelvin | $t_K = (5/9)t_R$ |
| rayleigh (rate of photon emission) | 1/second meter² | +10 1.00 |
| rhe | meter²/newton | +01 1.00 |
| rod | meter | +00 5.0292 |
| roentgen | coulomb/kilogram | −04 2.579 76 |
| rutherford | disintegration/second | +06 1.00 |
| second (angle) | radian | −06 4.848 136 |
| second (ephemeris) | second | +00 1.000 000 |
| second (mean solar) | second (ephemeris) | Consult American Ephemeris and Nautical Almanac |
| second (sidereal) | second (mean solar) | −01 9.972 695 |
| section | meter² | +06 2.589 988 |
| scruple (apothecary) | kilogram | −03 1.295 978 |
| shake | second | −08 1.00 |
| skein | meter | +02 1.097 28 |
| slug | kilogram | +01 1.459 390 |
| span | meter | −01 2.286 |
| statampere | ampere | −10 3.335 640 |
| statcoulomb | coulomb | −10 3.335 640 |
| statfarad | farad | −12 1.112 650 |
| stathenry | henry | +11 8.987 554 |
| statmho | mho | −12 1.112 650 |
| statohm | ohm | +11 8.987 554 |